Lecture Notes in Computer Science   16503

Founding Editors

Gerhard Goos
Juris Hartmanis

Editorial Board Members

Elisa Bertino, USA
Wen Gao, China

Bernhard Steffen, Germany
Moti Yung, USA

Advanced Research in Computing and Software Science

Subline of Lecture Notes in Computer Science

Subline Series Editors

Giorgio Ausiello, *University of Rome 'La Sapienza', Italy*
Vladimiro Sassone, *University of Southampton, UK*

Subline Advisory Board

Susanne Albers, *TU Munich, Germany*
Benjamin C. Pierce, *University of Pennsylvania, USA*
Bernhard Steffen, *University of Dortmund, Germany*
Deng Xiaotie, *Peking University, Beijing, China*
Jeannette M. Wing, *Microsoft Research, Redmond, WA, USA*

Nathalie Bertrand · Stefan Milius

Editors

Foundations of Software Science and Computation Structures

29th International Conference, FoSSaCS 2026
Held as Part of the International Joint Conferences
on Theory and Practice of Software, ETAPS 2026
Turin, Italy, April 11–16, 2026
Proceedings

 Springer

Editors
Nathalie Bertrand
Inria
Rennes, France

Stefan Milius
Friedrich-Alexander-Universität
Erlangen-Nürnberg
Erlangen, Germany

ISSN 0302-9743 ISSN 1611-3349 (electronic)
Lecture Notes in Computer Science
ISBN 978-3-032-22729-4 ISBN 978-3-032-22730-0 (eBook)
https://doi.org/10.1007/978-3-032-22730-0

ETAPS Foreword

Welcome to the 29th edition of ETAPS, which took place as an on-site event in Turin, Italy during April 11–16, 2026!

ETAPS 2026 was the 29th instance of the International Joint Conferences on Theory and Practice of Software (ETAPS). ETAPS is an annual federated conference established in 1998, and consists of four main conferences: ESOP, FASE, FoSSaCS, and TACAS. Each conference has its own Program Committee (PC) and its own Steering Committee (SC). The ETAPS main conferences cover various aspects of software systems, ranging from theoretical computer science to foundations of programming languages, tools and algorithms for system analysis, and formal approaches to software engineering. Organizing these conferences in a coherent, highly synchronized conference programme enables researchers to participate in an exciting event, having the possibility to meet many colleagues working in different directions in the field, and to easily attend talks of different conferences. In addition to its four main conferences, ETAPS 2026 also hosted fifteen satellite workshops and two colocated events, which together further attracted many researchers from all over the globe.

ETAPS 2026 received 456 submissions in total, 138 of which were accepted, yielding an overall acceptance rate of 30%. Out of the 138 accepted papers, 16 papers were selected as ETAPS distinguished papers. I thank all the authors of submitted papers for their interest in ETAPS, all the reviewers for their reviewing efforts, the PC members for their contributions, and in particular the PC (co-)chairs for their hard work in running this entire intensive process in a constructive, objective and timely manner. I congratulate all authors of the ETAPS 2026 accepted papers!

ETAPS 2026 featured the unifying invited keynotes by

- Monika Henzinger (Institute of Science and Technology Austria, Austria), delivering a talk about "Guarding Privacy Over Time: Challenges and Solutions in Continuous Data Observation",
- Einar Broch Johnsen (University of Oslo, Norway), discussing "Formal Methods Meet Digital Twins: Challenges and Opportunities".

ETAPS 2026 hosted the invited keynote speakers

- Christel Baier (Technische Universität Dresden, Germany) for FoSSaCS, presenting "Verification of Infinite-horizon Properties of Dynamic Bayesian Networks",
- Guy Van den Broeck (University of California, Los Angeles, USA) for TACAS, introducing "Symbolic Reasoning in the Age of Large Language Models".

The ETAPS 2026 invited tutorials were provided by

- Mieke Massink (CNR-ISTI Pisa, Italy) on "Model Checking in Space with Applications to Medical Image Analysis",
- Leonardo de Moura (Amazon Web Services, USA) surveying "The Lean Programming Language and Theorem Prover".

The ETAPS 2026 programme also featured a lively Ask-Me-Anything session, interactive tool demos, a Diversity, Equity, and Inclusion session, SV-Comp and Test-Comp community building events, and the ETAPS industry day. The goal of the ETAPS industry day is to bring industrial practitioners into the heart of the research community and to catalyze the interaction between industry and academia. The ETAPS 2026 industry day was organized by Giorgio Audrito (University of Turin, Italy), Sean Kauffman (Queen's University, Kingston, Canada), and Nikolai Kosmatov (Thales Research and Technology, Palaiseau, France).

ETAPS 2026 was organized by the Department of Computer Science of the University of Turin, which is the center for coordinating research, teaching, dissemination and technological transfer in computer science in Turin, Italy. The department covers both methodological and application oriented aspects of computer science, and performs research in several interdisciplinary areas. This is reflected in the collaborations with other research centers and companies in many scientific areas and in its participation in national, European and international projects.

ETAPS 2026 was further supported by the following associations and societies: ETAPS e. V. (the ETAPS Association), EATCS (European Association for Theoretical Computer Science), EAPLS (European Association for Programming Languages and Systems), and EASST (European Association of Software Science and Technology).

The ETAPS Steering Committee consists of an Executive Board, and representatives of the individual ETAPS conferences, as well as representatives of EATCS, EAPLS, and EASST. The Executive Board consists of Laura Kovács (TU Wien, chair), Andrzej Wąsowski (IT University of Copenhagen, vice-chair), Thomas Noll (RWTH Aachen, treasurer), Arnd Hartmanns (University of Twente, artifact evaluation coordinator), Barbara König (University of Duisburg-Essen, proceedings coordination), Caterina Urban (Inria, PhD activities), Elizabeth Polgreen (University of Edinburgh, social media), Jan Kofroň (Charles University Prague, organisational support, website), Jan Křetínský (Masaryk University Brno and TU Munich, diversity & inclusion), and Marieke Huisman (University of Twente, blog, awards). Further members of the ETAPS Steering Committee committee are: Robbert Krebbers (Radboud University Nijmegen), Azalea Raad (Imperial College London), Luís Caires (Tecnico ULisboa), Elvira Albert (Universidad Complutense de Madrid), Corina Păsăreanu (Carnegie Mellon University), Erika Ábrahám (RWTH Aachen), Marsha Chechik (University of Toronto), Marie-Christine Jakobs (LMU Munich), Nathalie Bertrand (Inria Rennes), Stefan Milius (Friedrich-Alexander Universität Erlangen-Nürnberg), Alexandra Silva (Cornell University), Joël Ouaknine (MPI-SWS Saarbrücken), Andrzej Murawski (University of Oxford), Sebastian Junges (Radboud University Nijmegen), Guy Katz (The Hebrew University of Jerusalem), Christian Schilling (Aalborg University), Naijun Zhan (Peking University), Joost-Pieter Katoen (RWTH Aachen and University of Twente), Dirk Beyer (LMU Munich), Fabrice Kordon (Sorbonne University Paris), Laure Petrucci (Université Paris 13), Peter Y.A. Ryan (University of Luxembourg), Claudio Menghi (University of Bergamo and McMaster University Hamilton), Mark Lawford (McMaster University Hamilton), Maurice ter Beek (CNR-ISTI Pisa), Ferruccio Damiani (University of Turin), Kim Guldstrand Larsen (Aalborg University), Bernhard Beckert (KIT Karlsruhe), Mattias Ulbrich (KIT Karlsruhe), Reiko Heckel (University of Leicester), Vladimiro Sassone

(University of Southampton), Anton Wijs (Eindhoven University of Technology), and Nikolai Kosmatov (Thales Research and Technology, Palaiseau).

The ETAPS 2026 local organization team consisted of Maurice ter Beek (CNR-ISTI Pisa, general co-chair), Ferruccio Damiani (University of Turin, general co-chair), Barbara Boni (Synesthesia Turin, local organization chair), Vincenzo Ciancia (CNR-ISTI Pisa, satellite events co-chair), Luca Paolini (University of Turin, satellite events co-chair), Maria Tacconi (Synesthesia Turin, satellite events co-chair and publicity co-chair), Francesco Brocero (Synesthesia Turin, web co-chair and volunteers co-chair), José Proença (University of Porto, web co-chair), Gianluca Torta (University of Turin, publicity co-chair and local proceedings co-chair), Lucy James (Synesthesia Turin, sponsor chair), Giovanna Broccia (CNR-ISTI Pisa, local proceedings co-chair), Giorgio Audrito (University of Turin, volunteers co-chair), Riccardo Sieve (UiO Oslo, volunteers co-chair), and Reiner Hähnle (TU Darmstadt, wine chair).

I would like to take this opportunity to thank all authors, keynote speakers, invited tutorial speakers, and attendees. Special thanks goes to the organizers of the ETAPS 2026 satellite workshops and colocated events. ETAPS 2026 is grateful for the generous support of Amazon Web Services, AccessiWay, Camera di Commercio Industria Artigianato e Agricoltura di Torino, the Department of Computer Science of the University of Turin, Springer Nature, and Turismo Torino e provincia Convention Bureau. I thank our general co-chairs Maurice ter Beek (CNR-ISTI Pisa) and Ferruccio Damiani (University of Turin), who made it all happen in Turin, and their local organization team for their enormous efforts to make ETAPS 2026 a fantastic event. I am especially grateful to Barbara Boni, Maria Tacconi, and Lucy James (Synesthesia Turin) for handling the organizational process in a smooth and reliable way. Last but not least, a big thanks to Jan Kofroň for all his help as an ETAPS Fellow and providing online presence support for the ETAPS conferences and the ETAPS Association.

I hope you all enjoyed ETAPS 2026!

April 2026   Laura Kovács
ETAPS SC Chair, President of the ETAPS
Association

Preface

This volume contains the papers presented at FoSSaCS 2026, the 28th International Conference on Foundations of Software Science and Computation Structures, which was held April 13–15, 2026 in Turin, Italy.

The conference is dedicated to foundational research that is clearly significant for software science and brings together research on theories and methods to support the analysis, integration, synthesis, transformation, and verification of programs and software systems.

In addition to an invited talk by Christel Baier (Technische Universität Dresden) on "Verification of Infinite-Horizon Properties of Dynamic Bayesian Networks", the program consisted of 30 talks on contributed papers, selected from 104 submissions. Each submission was assessed by three or more Program Committee members, with the help of external reviewers. The conference management system EasyChair was used to handle the submissions, to conduct the electronic Program Committee discussions, and to assist with the assembly of the proceedings.

We wish to thank all the authors who submitted papers for consideration, the members of the Program Committee for their conscientious work, and all additional reviewers who assisted the Program Committee in the evaluation process. We would also like to thank Andrzej Murawski, the FoSSaCS Steering Committee Chair for various pieces of advice, and the members of the ESOP/FASE/FoSSaCS joint Artifact Evaluation Committee for the artifact evaluation. Finally, we would like to thank the ETAPS organization for providing an excellent environment for FoSSaCS, the other conferences and the workshops.

April 2026

Stefan Milius
Nathalie Bertrand

Organization

Program Committee Chairs

Nathalie Bertrand   Inria Rennes, France
Stefan Milius   FAU Erlangen-Nürnberg, Germany

Program Committee

Franz Baader	TU Dresden, Germany
Tegan Brennan	Stevens Institute of Technology, USA
Yijia Chen	Shanghai Jiao Tong University, China
Pierre Clairambault	CNRS, France
Pedro D'Argenio	Universidad Nacional de Córdoba, Argentina
Ugo Dal Lago	University of Bologna, Italy
Laure Daviaud	University of East Anglia, UK
Stéphane Demri	CNRS, France
Josee Desharnais	Université Laval, Canada
Jérémy Dubut	École Polytechnique, France
Maribel Fernandez	King's College London, UK
Bernd Finkbeiner	CISPA Helmholtz Center for Information Security, Germany
Wan Fokkink	Vrije Universiteit Amsterdam, The Netherlands
Jean Goubault-Larrecq	ENS Paris Saclay, France
Bart Jacobs	Radboud University Nijmegen, The Netherlands
Shin-ya Katsumata	Kyoto Sangyo University, Japan
Sławek Lasota	University of Warsaw, Poland
Alessio Mansutti	IMDEA Software Institute, Spain
Anca Muscholl	Université de Bordeaux, France
Elaine Pimentel	University College London, UK
Alexandra Silva	Cornell University, USA
Rui Soares Barbosa	International Iberian Nanotechnology Laboratory, Portugal
Pawel Sobocinski	Tallinn University of Technology, Estonia
B Srivathsan	Chennai Mathematical Institute, India
Christine Tasson	ISAE Supaero, France
Rob van Glabbeek	University of Edinburgh, UK
Frits Vaandrager	Radboud University, The Netherlands
Martin Zimmermann	Aalborg University, Denmark

Joint ESOP/FASE/FoSSaCS Artifact Evaluation Committee Chairs

Yannic Noller	Ruhr Universität Bochum, Germany (FASE)
Guillermo Alberto Perez	University of Antwerp, The Netherlands (FoSSaCS)
Michael Sammler	Institute of Science and Technology, Austria (ESOP)

Joint ESOP/FASE/FoSSaCS Artifact Evaluation Committee

Mohammad Afzal	TCS Research and IIT Bombay, India
Flavio Ascari	University of Konstanz, Germany
Aren A. Babikian	University of Toronto, Canada
Alexander Bai	New York University, USA
David Boetius	University of Konstanz, Germany
Michaël Cadilhac	DePaul University, USA
Ronaldo Canizales	Colorado State University, USA
William Eiers	Stevens Institute of Technology, USA
Kasper Engelen	University of Antwerp, Belgium
Máté Földiák	Linköping University, Sweden
Alvin George	IISc Bangalore, India
Holly Hendry	University of York, UK
Martin Kristjansen	Aalborg University, Denmark
Andrea Laretto	Tallinn University of Technology, Estonia
Megan Maton	University of Sheffield, UK
Logan Murphy	University of Toronto, Canada
Olek Osikowicz	University of Sheffield, UK
Jan-Paul Ramos-Dávila	Boston University, USA
Wojciech Rozowski	Lean FRO, USA
Chia Sabah	University of Leicester, UK
William Scarbro	Colorado State University, USA
Hitarth Singh	Hong Kong University of Science and Technology, China
Steffan Sølvsten	Aarhus University, Denmark
Stephan Spengler	Uppsala University, Sweden
Gaëtan Staquet	École Centrale Nantes, France
Abhishek U	IISc, India
Alexandra van der Spuy	Stellenbosch University, South Africa
Szumi Xie	Eötvös Loránd University, Hungary
Ekaterina Zhuchko	Tallinn University of Technology, Estonia

FoSSaCS Steering Committee

Patricia Bouyer-Decitre	Paris-Saclay, France
Javier Esparza	Munich, Germany
Naoki Kobayashi	Tokyo, Japan
Barbara König	Duisburg-Essen, Germany
Orna Kupferman	Jerusalem, Israel
Andrzej Murawski (Chair)	Oxford, UK
Jurriaan Rot	Nijmegen, The Netherlands
Alex Simpson	Ljubljana, Slovenia

Additional Reviewers

Beniamino Accattoli
Augustin Albert
Nicola Assolini
Sacha-Elie Ayoun
A. R. Balasubramanian
Thibaut Benjamin
Filippo Bonchi
James Brotherston
Roberto Bruni
Mishel Carelli
Antonio Casares
Leonardo Ceragioli
Yorgo Chamoun
Jules Chouquet
Robin Cockett
Thomas Colcombet
Alex Coleman
Pierre-Louis Curien
Fredrik Dahlqvist
Marc de Visme
Elena Di Lavore
Kyveli Doveri
Laurent Doyen
Sergei Drobyshevich
Joerg Endrullis
Uli Fahrenberg
Marie Fortin
Florian Frank
Tobias Fritz
Paul Gallot

Richard Garner
Francesco Gavazzo
Luca Geatti
Komi Golov
Daniel Gratzer
Stefano Guerrini
Julian Gutierrez
Amar Hadzihasanovic
Eric Hehner
Stefan Hetzl
Chris Heunen
Son Ho
Naohiko Hoshino
Peter Höfner
Radu Iosif
Jacques-Henri Jourdan
Bartek Klin
Roman Kniazev
Florian Kohn
Mayuko Kori
Vasileios Koutavas
Jan Kretinsky
Alexander Kurz
Adrienne Lancelot
Andrea Laretto
Francois Laroussinie
Jérémy Ledent
Engel Lefaucheux
Jean-Simon Lemay
Bruno Lopes

Fosco Loregian
Jiaheng Lu
Bas Luttik
Ian Mackie
Radu Mardare
Filip Mazowiecki
Tobias Meggendorfer
Éléanore Meyer
Łukasz Mikulski
Vincent Moreau
Ike Mulder
Michele Pagani
Hugo Paquet
Julie Parreaux
Fabio Pasquali
Nicolas Peltier
Paolo Pistone
Andrei Popescu
David Purser
Eigil Rischel
Mario Román
Maryam Rostamigiv
Jurriaan Rot
Aleksi Saarela
Suman Sadhukhan

Katsuhiko Sano
Ralph Sarkis
Tetsuya Sato
Sylvain Schmitz
Lutz Schröder
Pawel Sobocinski
Priyaa Varshinee Srinivasan
Sam Staton
Dario Stein
Márk Széles
Taichi Uemura
Henning Urbat
Femke van Raamsdonk
Gabriele Vanoni
Anton Varonka
Isa Vialard
Renaud Vilmart
Aymeric Walch
Di Wang
Sarah Winter
Thorsten Wißmann
Vladimir Zamdzhiev
Zhicheng Zhang
Yangluo Zheng
Krzysztof Ziemiański

Contents

Varieties of Quantitative Algebras Presented by 1-Basic Monads

Jiří Adámek

[1] Czech Technical University in Prague
[2] Technical University of Braunschweig `j.adamek@tu-bs.de`

Abstract. Varieties of quantitative algebras, studied by Mardare, Panangaden and Plotkin, can be presented as categories of algebras for enriched monads on the categories **Met** (of metric spaces) or **CMet** (of complete metric spaces). We solve the open problem of whether the resulting monads are precisely the strongly finitary ones. The solution is negative: the monad corresponding to two ε-close binary operations is not strongly finitary. We characterize the monads representing varieties of quantitative algebras as precisely the *1-basic monads* which are the weighted colimits of strongly finitary monads.
We conclude that strongly finitary endofunctors on **Met** are not closed under composition.

Keywords: quantitative algebra $\cdot$ 1-basic monad $\cdot$ colimits of monads

1 Introduction

Quantitative algebras were introduced by Mardare, Panangaden and Plotkin [17], [18] as a foundation for formal semantics of probabilistic and stochastic programs. A quantitative algebra of a signature Σ is a metric space A endowed, for every n-ary symbol $\sigma \in \Sigma$, with a nonexpanding operation $\sigma_A : A^n \to A$. The main tool for presenting classes of quantitative algebras are c-basic quantitative equations for a cardinal number c. The case $c = 1$ means an expression $t =_\varepsilon t'$, where t and t' are terms, and $\varepsilon \geq 0$ is a real number. These are the equations that we consider in our paper.

By a *variety* (aka 1-basic variety) of quantitative algebras we mean a class presented by quantitative equations. (One of the central examples in [17] are quantitative semilattices: upper semilattices with 0 working on complete metric spaces with nonexpanding join.) Every variety $\mathcal{V}$ is proved in loc.cit. to have free algebras. This yields a corresponding monad $T_\mathcal{V}$ on the category **Met** of metric spaces (or **CMet** of complete metric spaces). Moreover, every variety $\mathcal{V}$ is isomorphic to the Eilenberg-Moore category $\mathbf{Met}^{T_\mathcal{V}}$ (or $\mathbf{CMet}^{T_\mathcal{V}}$). (For example, the monad $T_\mathcal{V}$ on **CMet** corresponding to quantitative semilattices is the Hausdorff monad $\mathcal{H}$, where $\mathcal{H}X$ is the semilattice of compact subsets of the space X with the Hausdorff metric, see Ex. 13.) Thus the study of varieties can be viewed as the study of the corresponding monads. Our goal is a complete description of monads on **Met** or **CMet** of the form $T_\mathcal{V}$.

N. Bertrand and S. Milius (Eds.): FoSSaCS 2026, LNCS 16503, pp. 1–20, 2026.
https://doi.org/10.1007/978-3-032-22730-0_1

Every monad $T_\mathcal{V}$ is finitary (preserves filtered colimits) and enriched (locally nonexpanding). In the opposite direction, every enriched, *strongly finitary* monad (one determined by its values on finite discrete spaces) has the form $T_\mathcal{V}$ for a variety $\mathcal{V}$ (Proposition 38). Several authors asked in the recent years whether the monads $T_\mathcal{V}$ are *precisely* the strongly finitary ones, see e.g. [14], [16], [21]. The aim of our paper is a complete clarification: In Section 5 we prove the following

Theorem. *The monads on* **Met** *or* **CMet** *corresponding to varieties of quantitative algebras are precisely the 1-basic finitary monads: the weighted colimits of strongly finitary monads (in the category of finitary monads).*

In Section 6 we then exhibit a variety whose monad $T_\mathcal{V}$ is not strongly finitary: the variety of two ε-close binary operations σ_1 and σ_2. It is presented by the quantitative equation $\sigma_1(x,y) =_\varepsilon \sigma_2(x,y)$.

This is actually surprising, since in many categories, varieties bijectively correspond to strongly finitary monads: this is true e.g. for sets [15], posets [13], and even for ultrametric spaces [4].

Corollary 1. *Strongly finitary endofunctors on* **Met** *or* **CMet** *are not closed under composition.*

Indeed, Bourke and Garner introduced in [8] saturated classes of arities and proved in Theorem 43 a result that implies that, in case strongly finitary functors compose (in their terminology: all finite discrete spaces form a saturated class of arities in **Met** or **CMet**) the free-algebra monads of varieties of quantitative algebras are precisely the strongly finitary ones.

Full proofs are presented in [1].

Related Work. We have announced our example of the variety of ε-close binary operations in July 2025 at the conference CT25 in Brno [1]. Independently, a similar example has been announced by Mardare, Ghan, and Rischel [16], Example 8.3, in September 2025 at the conference GandALF 2025 in Valletta, Malta.

More general quantitative equations were used in the pioneering paper of Mardare, Panangaden and Plotkin [17]: the *c-basic* equations for a cardinal number c. They have the form $t =_\varepsilon t'$, where t and t' are elements of the free algebra $T_\Sigma V$ on a space V (of variables) of power less than c. The corresponding monads are called *c-basic monads* in [2], Corollary 4.15. For $c = \aleph_0$ their restriction to ultrametric spaces was characterized as precisely the *finitely basic monads*: monads preserving surjective morphisms and directed colimits of split monomorphisms. It is an open problem whether this characterization also holds for monads corresponding to $\aleph_0$-basic varieties in **Met** or **CMet**.

The unpublished paper [4] with the extended abstract [5] contains some incomplete arguments. The co-authors unfortunately do not intend publishing a revised version. This leads us to providing a new (and, as it happens, simpler) proofs in Section 4 and 5 below. We also repeat some of the introductory material of [4] in Section 3.

An alternative approach to varieties of quantitative algebras is presented by J. Rosický [21], who uses algebraic theories. A characterization of the corresponding monads is formulated in loc.cit. as an open problem.

Acknowledgements. The author is grateful to M. Dostál, J. Rosický, H. Urbat and J. Velebil for fruitful discussions that helped to improve the presentation of our paper. One of the anonymous referees has provided a very detailed and very helpful report for which the author is particularly grateful.

2   Varieties of Quantitative Algebras

We recall varieties of quantitative algebra from [17]. Every variety $\mathcal{V}$ is known to be isomorphic to the category $\mathbf{Met}^{T_{\mathcal{V}}}$ of Eilenberg-Moore algebras, where $T_{\mathcal{V}}$ is the free-algebra monad. We prove that $T_{\mathcal{V}}$ is an enriched, finitary monad.

Throughout this paper $\Sigma = (\Sigma_n)_{n \in \mathbb{N}}$ denotes a finitary signature: Σ_n is the set of n-ary operation symbols.

Notation 2. **Met** denotes the category of (extended) metric spaces. Objects are metric spaces extended in the sense that the distance ∞ is allowed. Morphisms are nonexpansive functions $f \colon X \to Y$: for all x, $x' \in X$ we have $d(x, x') \geq d\big(f(x), f(x')\big)$.

CMet is the full subcategory on complete spaces: every Cauchy sequence converges.

Every space X has the underlying set denoted by $|X|$. Conversely, every set is considered as the discrete space: all distinct pairs have distance ∞.

Every natural number n is considered to denote the discrete space $\{0, \dots, n-1\}$.

Remark 3. **Met** is a symmetric monoidal closed category, where the tensor product

$$X \otimes Y$$

is the cartesian product with the *sum metric*:

$$d\big((x, y), (x', y')\big) = d(x, x') + d(y, y')\,.$$

In contrast, the categorical product $X \times Y$ is the cartesian product with the *maximum metric*: the maximum of $d(x, x')$ and $d(y, y')$.) When we write X^n, we always mean the categorical product.

The monoidal unit I is a singleton space.

The hom-space $[X, Y]$ is the space of all morphisms $f \colon X \to Y$ with the *supremum metric*

$$d(f, f') = \sup_{x \in X} d\big(f(x), f'(x)\big) \quad \text{for} \quad f, f' \colon X \to Y\,.$$

A coproduct of spaces in **Met** or **CMet** is their disjoint union with distance ∞ between every pair of elements in distinct summands.

A **Met**-*enriched category* (shortly: enriched) is a category with a metric on every hom-set making composition nonexpanding (with respect to the sum metric). A **Met**-*enriched functor* F between **Met**-enriched categories is a functor which is locally nonexpanding: for all parallel pairs f, g in the domain category we have $d(Ff, Fg) \leq d(f, g)$. Enriched natural transformations are the ordinary ones (among enriched functors).

Given enriched categories $\mathcal{A}$ and $\mathcal{B}$, the category $[\mathcal{A}, \mathcal{B}]$ of all enriched functors $F: \mathcal{A} \to \mathcal{B}$ and natural transformations is enriched: the distance of natural transformations $\varphi, \psi: F \to F'$ is $d(\varphi, \psi) = \sup\limits_{X \in \mathcal{A}} d(\varphi_X, \psi_X)$.

Definition 4 ([17]). A *quantitative algebra* is a metric space A endowed with nonexpanding operations $\sigma_A: A^n \to A$ (for all $\sigma \in \Sigma_n$) with respect to the maximum metric on A^n. It is *complete* if A is a complete metric space.

Notation 5. The category of quantitative algebras and nonexpanding maps preserving the operations (shortly homomorphisms) is denoted by $\Sigma\text{-}\mathbf{Met}$. Analogously for its full subcategory $\Sigma\text{-}\mathbf{CMet}$ of complete quantitative algebras.

Examples 6 (Term algebras). In universal algebra the free Σ-algebra on a set V of variables is the algebra $T_\Sigma V$ of terms. These are either variables, or composite terms $\sigma(t_i)_{i<n}$ for $\sigma \in \Sigma_n$ and an n-tuple $(t_i)_{i<n}$ of terms.

Analogously, the free quantitative algebra $T_\Sigma X$ on a space X is described: its underlying algebra is the algebra $T_\Sigma|X|$ of terms. Let us call terms t and t' *similar* if we can obtain t' from t by changing some variables. Thus, all pairs of variables are similar. And terms similar to $\sigma(t_i)$ are precisely the terms $\sigma(t_i')$ with t_i and t_i' similar for each i. The metric d_X^* of the free algebra $T_\Sigma X$ is defined recursively as follows:

$$d_X^*(t, t') = \begin{cases} d_X(t, t') & \text{if } t, t' \in X \\ \max\limits_{i<n} d_X^*(t_i, t_i') & \text{if } t = \sigma(t_i) \text{ and } t' = \sigma(t_i') \\ \infty & \text{if } t \text{ is not similar to } t' \end{cases}$$

It follows that the endofunctor T_Σ is a coproduct of functors $(-)^n$, one summand per every similarity class of terms on n variables ($n \in \mathbb{N}$).

Lemma 7 ([2], Remark 2.4). *Given a space X, the quantitative algebra $(T_\Sigma X, d_X^*)$ is free: for every quantitative algebra A and every nonexpanding function $f: X \to A$ there is a unique nonexpanding homomorphism $f^\#: (T_\Sigma X, d_X^*) \to A$ extending f, i.e., $f = f^\# \cdot \eta_X$.*

Notation 8. Let A be a quantitative algebra. Every term $t \in T_\Sigma n$ defines an n-ary operation $t_A: A^n \to A$: to an n-tuple $f: n \to A$ it assigns $t_A(f) = f^\#(t)$.

Definition 9 ([17]). A (1-basic) *quantitative equation* is an expression $t =_\varepsilon t'$, where t and t' are terms in $T_\Sigma V$ for a finite set V of variables, and $\varepsilon \geq 0$ is a real number.

A quantitative algebra A *satisfies* this equation provided that every interpretation $f \colon V \to A$ of the variables fulfils

$$d_A\big(f^\#(t), f^\#(t')\big) \leq \varepsilon \, .$$

A *variety* (aka 1-basic variety) of quantitative algebras is a full subcategory of Σ-**Met** or Σ-**CMet** specified by a set of quantitative equations.

We write $t = t'$ in place of $t =_0 t'$, and call such equations *ordinary*. In [18] the number ε was assumed to be rational. But this makes no difference: if $\varepsilon > 0$ is irrational, choose any dicreasing sequence $\varepsilon(n)$, $n \in \mathbb{N}$, of rationals converging to ε. Then the equation $t =_\varepsilon t'$ is equivalent to the set of equations $t =_{\varepsilon(n)} t'$ for $n \in \mathbb{N}$.

Examples 10. (1) *Quantitative monoids.* This is the variety presented by the usual signature (of a binary operation and constant e), and the usual ordinary equations: $(xy)z = x(yz)$, $xe = x$, and $ex = x$.

(2) *Actions of quantitative monoids.* Let us recall that an action of a (classical) monoid M on a set X is a mapping from $M \times X$ to X (notation $(m, x) \mapsto mx$) whose curryfication is a monoid homomorphism from M to the composition monoid $\mathbf{Set}(X, X)$. Analogously, given a quantitative monoid M, its (quantitative) *action* on a metric space X is a nonexpanding homomorphism from M to $[X, X]$ in **Met**. That is a monoid action such that $d_X(mx, my) \leq d_X(x, y)$ for all $(x, y) \in X^2$, and $d_M(mx, m'x) \leq d_M(m, m')$ for all $(m, m') \in M \times M$.

We obtain a variety of quantitative Σ-algebras, where Σ consists of unary operations $m(-)$ for $m \in M$. It is presented by the usual ordinary equations: $m(m'x) = (mm')x$, and $ex = x$, together with the following quantitative equations

$$mx =_\varepsilon m'x \quad \text{where} \quad \varepsilon = d_M(m, m') \, .$$

(3) *Quantitative semilattices.* By a semilattice we mean a join-semilattice with a bottom. Equivalently, a commutative and idempotent monoid.

Quantitative semilattices are semilattices acting on a metric space with nonexpanding binary joins. In other words, commutative and idempotent quantitative monoids.

Theorem 11 ([17], Sections 6 and 7). *In every variety $\mathcal{V}$ of (complete) quantitative algebras each (complete) space X generates a free algebra $F_\mathcal{V} X$. That is, the forgetful functor $U_\mathcal{V} \colon \mathcal{V} \to \mathbf{Met}$, or $U_\mathcal{V} \colon \mathcal{V} \to \mathbf{CMet}$, has a left adjoint $F_\mathcal{V}$.*

Notation 12. We denote by $T_\mathcal{V} = U_\mathcal{V} F_\mathcal{V}$ the free-algebra monad on **Met** or **CMet**, respectively.

Examples 13. (1) A free quantitative monoid on a space X in **Met** or **CMet** is the coproduct (disjoint union) of finite powers with the maximum metric. That is, the word monoid $T_{\mathcal{V}}X = X^*$ with the following metric: $d(x, y) = \infty$ for words x, y of distinct lengths, else $d(x, y) = \max d(x_i, y_i)$.

(2) Let M be a quantitative monoid. The free action of M on a space X is the product of these spaces $T_{\mathcal{V}}X = M \times X$ with the obvious action: $m(m', x) = (mm', x)$.

(3) Free quantitative semilattices on **CMet** are given by the Haudorff functor $T_{\mathcal{V}} = \mathcal{H}$. That is, $T_{\mathcal{V}}X$ is the space of all compact sets in X with the Hausdorff metric $d_{\mathcal{H}}$: for all A, B compact we put

$$d_{\mathcal{H}}(A, B) = \max \left\{ \sup_{a \in A} d(a, B), \sup_{b \in B} d(A, b) \right\}.$$

The join operation is $A \cup B$. See [17], Section 9.

(4) Let E be a metric space (of exceptions). Moggi's exception monad [20] is defined as the coproduct $TX = X + E$. This is the free-algebra monad for the variety of nullary operations indexed by E. It is presented by the quantitative equations $e =_\varepsilon e'$ $(e, e' \in E)$ where $\varepsilon = d(e, e')$.

By a *concrete category* over **Met** we mean an enriched category $\mathcal{K}$ together with an enriched (forgetful) functor $U \colon \mathcal{K} \to$ **Met** which is faithful in the enriched sense: $d(f, g) = d(Uf, Ug)$ for all parallel pairs f, g. Example: every variety $\mathcal{V}$ is a concrete category with the obvious forgetful functor $U_{\mathcal{V}} \colon \mathcal{V} \to$ **Met**.

A *concrete functor* between concrete categories is a (necessarily enriched) functor commuting with the forgetful functors.

Analogously for **CMet**.

Theorem 14. *Every variety $\mathcal{V}$ is concretely isomorphic to the Eilenberg-Moore category* **Met**$^{T_{\mathcal{V}}}$.

Indeed, the comparison functor $K_{\mathcal{V}} \colon \mathcal{V} \to$ **Met**$^{T_{\mathcal{V}}}$ is an isomorphism commuting with the forgetful functors. This is analogous to Theorem VI.8.1 in [15].

Theorem 15 ([7], Theorem 3.6.8). *For arbitrary monads T and T' on* **Met** *there is a bijective correspondence between monad morphisms $\varphi \colon T \to T'$ and concrete functors $F \colon$* **Met**$^{T'} \to$ **Met**^{T}. *The functor F assigns to an algebra $a \colon T'A \to A$ in* **Met**$^{T'}$ *the algebra $a \cdot \varphi_A \colon TA \to A$.*

Example 16. (1) For every variety $\mathcal{V}$ the (concrete) embedding into Σ-**Met** corresponds to the *canonical monad morphism* $k_{\mathcal{V}} \colon T_\Sigma \to T_{\mathcal{V}}$. Its component $(k_{\mathcal{V}})_X$ is the unique homomorphism extending $\eta_X \colon X \to T_{\mathcal{V}}X$.

In case of **Met**, these components are surjective. Indeed, the image of $(k_{\mathcal{V}})_X$ (as a subspace of $T_{\mathcal{V}}X$) is a subalgebra of $T_{\mathcal{V}}X$ (closed under all operations) containing the image of η_X. It is clear that no proper subalgebra $m \colon A \to T_{\mathcal{V}}X$ has this property. (Indeed, let $e \colon T_{\mathcal{V}}X \to A$ be the unique homomorphism extending the restriction of η_X to $\eta'_X \colon X \to A$. Then $m \cdot e$ is an endomorphism of $T_{\mathcal{V}}X$ with $\eta_X = m \cdot e \cdot \eta_X$. Hence $m \cdot e = \mathrm{id}$, whence m is invertible.)

In case of **CMet**, the components of $k_{\mathcal{V}}$ are dense: the closure of the image of $(k_{\mathcal{V}})_X$ is a closed subalgebra containing the image of $k_{\mathcal{V}}$, this again is necessarily all of $T_{\mathcal{V}}X$.

(2) Let t be a term on n variables, say, $t \in T_\Sigma n$, and let Γ be the signature of a single n-ary opertation γ. We denote by $F_t \colon \Sigma\text{-}\mathbf{Met} \to \Gamma\text{-}\mathbf{Met}$ the correpsonding concrete funcotr. It assigns to a Σ-algebra A the operation computing t:

$$\gamma_A(f) = f^{\#}(t),$$

for all n-tuples $f \colon n \to A$.

Remark 17. For every variety $\mathcal{V}$ the monad $T_\mathcal{V}$ is enriched: Let morphisms f, $g \colon X \to Y$ have distance $d(f, g) = \delta$. We prove that $d\big(T_\mathcal{V} f(s), T_\mathcal{V} g(s)\big) \leq \delta$ holds for all $s \in T_\mathcal{V} X$; thus $d(T_\mathcal{V} f, T_\mathcal{V} g) \leq d(f, g)$. Indeed, denote by $S \subseteq |T_\mathcal{V} X|$ the set of all s satisfying the desired inequality. Then S contains the image of η_X, and it is closed under all operation. Thus $S = |T_\mathcal{V} X|$, as claimed.

Remark 18. (1) Colimits of diagrams on directed posets D (every finite subset has an upper bound in D) are called *directed*. In **Met** the colimit cocones $c_i \colon D_i \to C$ are collectively surjective: $C = \bigcup_{i \in I} c_i[D_i]$. (Indeed, if $m \colon C' \hookrightarrow C$ is the subspace of C on that union, then the codomain restrictions $c_i' \colon D_i \to C'$ of the colimit cocone form a cocone od D. The factorizing morphism $f \colon C \to C'$ with $c_i' = f \cdot c_i$, $i \in I$, fulfils $m \cdot f \cdot c_i = c_i$, thus $m \cdot f = \mathrm{id}$. Therefore m is invertible.) In **CMet** they are collectively dense: C is the closure of $\bigcup_{i \in I} c_i[D_i]$.

(2) Every space in **Met** or **CMet** is the directed colimit of the diagram of all of its finite subspaces.

Definition 19. A functor is *finitary* if it preserves directed colimits. A finitary monad is one with a finitary underlying functor.

Examples 20. (1) The endofunctor $(-)^n$ of the n-th categorical power is finitary on **Met** as well as **CMet** (for every $n \in \mathbb{N}$).

(2) A coproduct of finitary functors is finitary: coproducts commute with colimits.

(3) For every (complete) metric space M the endofunctor $M \times -$ on **Met** (or **CMet**) is finitary. This is verified analogously to finite products commuting in **Set** with directed colimits.

(4) The *Hausdorff endofunctor* $\mathcal{H} \colon \mathbf{CMet} \to \mathbf{CMet}$ of Ex. 13(3) is finitary, as proved in [6], Ex. 3.13.

Proposition 21. *Every variety of quantitative algebras is closed in Σ-**Met** or Σ-**CMet** under directed colimits.*

Proof (sketch). Using that the forgetful functor $U \colon \Sigma\text{-}\mathbf{Met} \to \mathbf{Met}$ preserves directed colimits, one can prove that for every equation $t =_\varepsilon t'$ the class of quantitative algebras satisfying it is closed under directed colimits.

Corollary 22. *For every variety $\mathcal{V}$ the monad $T_\mathcal{V}$ is finitary.*

Indeed, the monad T_Σ is finitary (Example 20(1) and (2)). Since $\mathcal{V}$ is closed under directed colimits, it follows that its forgetful functor $U_\mathcal{V}$ is finitary. Hence $T_\mathcal{V}$ is finitary: it is the composite $U_\mathcal{V} F_\mathcal{V}$, where $F_\mathcal{V}$ preserves colimits.

3   Strongly Finitary Monads

We recall here the characterization of strongly finitary functors from [4], slightly impoved in [1], and present some examples.

Let us start with the concept of a weighted colimit in an enriched category $\mathcal{K}$. Let enriched functors $D\colon \mathcal{D} \to \mathcal{K}$ (a diagram) and $W\colon \mathcal{D}^{\mathrm{op}} \to \mathbf{Met}$ (a weight) be given. The weighted colimit $C = \mathrm{colim}_W D$ is an object endowed with an isomorphism in $\mathbf{Met}$

$$\psi_S\colon \mathcal{K}(C, S) \simeq [\mathcal{D}^{\mathrm{op}}, \mathcal{K}](W, \mathcal{K}(D-, S))$$

natural in $S \in \mathcal{K}$.

Dually, for a diagram $D\colon \mathcal{D} \to \mathcal{K}$ and a weight $W\colon \mathcal{D} \to \mathbf{Met}$ the *weighted limit* $L = \mathrm{lim}_W D$ is defined.

Examples 23. Let $f, f'\colon K \to L$ be a parallel pair in an enriched category.

(1) For $\varepsilon \geq 0$ an *ε-coequalizer* is a morphism $q\colon L \to L'$ universal with respect to $d(q \cdot f, q \cdot f') \leq \varepsilon$. That is, a. every morphism $r\colon L \to R$ with $d(r \cdot f, r \cdot f') \leq \varepsilon$ factorizes through q, and b. given $u_1, u_2\colon L' \to U$, we have $d(u_1, u_2) = d(u_1 \cdot q, u_2 \cdot q)$.

This is the weighted colimit of the diagram, where $\mathcal{D}$ consists of a parallel pair, and D and W are given as follows:

$$K \underset{f}{\overset{f'}{\rightrightarrows}} L \quad \text{and} \quad 2_\varepsilon \rightleftarrows 1$$

respectively. Here 1 is a singleton space, and 2_ε is a space of two elements of distance ε.

(2) Dually, an *ε-equalizer* is a morphism $e\colon K' \to K$ universal with respect to $d(f \cdot e, f' \cdot e) \leq \varepsilon$.

Remark 24. Let $\mathcal{A}$ be an enriched category with a full enriched subcategory $K\colon \mathcal{A}_0 \hookrightarrow \mathcal{A}$. We recall the concept of (enriched) *Kan extension*. Suppose that the restriction functor $(-) \cdot K\colon [\mathcal{A}, \mathcal{A}] \to [\mathcal{A}_0, \mathcal{A}]$ has an enrihed left adjoint. Then this adjoint is denoted by

$$F \mapsto \mathbf{Lan}_K F\colon \mathcal{A} \to \mathcal{A} \quad (\text{for } F\colon \mathcal{A}_0 \to \mathcal{A})$$

and is called the left Kan extension along K. Thus $T = \mathbf{Lan}_K F$ is an enriched endofunctor equipped with a natural transformation $\tau\colon F \to TK$ with the expected universal property.

Notation 25. The full embedding of the category $\mathbf{Set}_f$ of finite sets (considered as finite discrete spaces) is denoted by

$$J\colon \mathbf{Set}_f \hookrightarrow \mathbf{Met} \quad \text{or} \quad J\colon \mathbf{Set}_f \hookrightarrow \mathbf{CMet} .$$

Definition 26 ([12]). An enriched endofunctor T of **Met** or **CMet** is *strongly finitary* if it is obtained from its restriction TJ to finite discrete spaces via the left Kan extension: $T = \mathbf{Lan}_J\, TJ$.

A monad is strongly finitary if its underlying endofunctor is.

The following condition characterizing strong finitary was proved in [4] (Propositions 2.20, 2.22 and Theorem 3.6). We first need additional notation:

Notation 27. (1) For every metric space X and every $\varepsilon > 0$ we denote by $\Delta_\varepsilon X \subseteq |X|^2$ the set of all pairs of distance at most ε. The left and right projections to $|X|$ are denoted by $l_\varepsilon, r_\varepsilon \colon \Delta_\varepsilon X \to |X|$.

(2) In [4] the *foliation* of X is defined as the weighted diagram

$$D_X \colon \mathcal{B} \to \mathbf{Met} \quad \text{with weight} \quad B \colon \mathcal{B}^{\mathrm{op}} \to \mathbf{Met}$$

as follows. The category $\mathcal{B}$ is obtained from the linearly ordered real interval $(0, +\infty)$ by adding the object 0 and two cocones, denoted by $\lambda_\varepsilon, \varrho_\varepsilon \colon \varepsilon \to 0$ ($\varepsilon > 0$), where $d(\lambda_\varepsilon, \varrho_\varepsilon) = \infty$. We define the diagram D_X as follows:

$$D_X 0 = |X|, \quad D_X \varepsilon = \Delta_\varepsilon X, \quad D\lambda_\varepsilon = l_\varepsilon \quad \text{and} \quad D\varrho_\varepsilon = r_\varepsilon.$$

The weight is
$$B0 = 1 \quad \text{and} \quad B\varepsilon = 2_\varepsilon,$$

where $B\lambda_\varepsilon, B\varrho_\varepsilon \colon 1 \to 2_\varepsilon$ are the two injections.

(3) We denote by $i_X \colon |X| \to X$ the identity-carried map of a space X.

Proposition 28 ([4] Propositions 2.20 and 2.22, and Theorem 3.6). *Every space X is a colimit of its foliation:*

$$X = \operatorname{colim}_B D_X.$$

For an enriched, finitary endofunctor T on **Met** *or* **CMet** *the following conditions are equivalent:*

(1) T is strongly finitary.

(2) T preserves colimits of foliations of all spaces.

(3) Given spaces X and Y, then

(3a) The morphism Ti_X has dense image.

(3b) Every nonexpanding map $f \colon T|X| \to Y$ satisfying

$$d(f \cdot Tl_\varepsilon, f \cdot Tr_\varepsilon) \le \varepsilon \quad \text{for all} \quad \varepsilon > 0 \tag{3.1}$$

has a nonexpanding factorization f' through Ti_X:

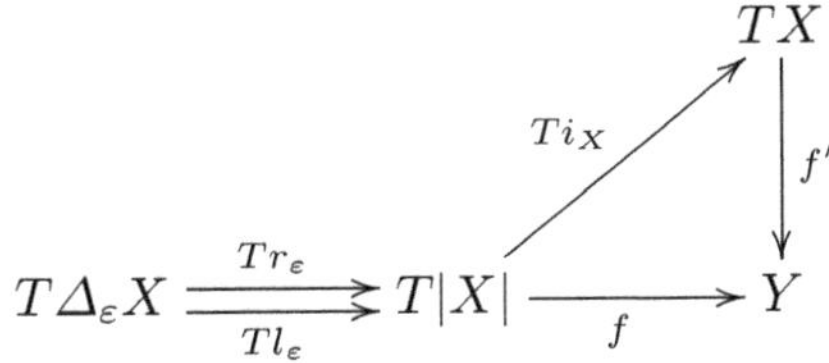

Remark 29. For endofunctors of **CMet** the same result holds except that (3a) states that Ti_X is dense, see [1], Proposition 2.10.

Examples 30. (1) The endofunctor $TX = X^n$ is strongly finitary for every $n \in \mathbb{N}$.

In contrast, if $M = \{0, 1\}$ is the space with $d(0, 1) > 0$, then the functor $T = \mathbf{Met}(M, -)$, assigning to X the subspace of $X \times X$ on $\Delta_\varepsilon M$, is not strongly finitary. Indeed, $\mathbf{Met}(M, -)$ does not take $i_M \colon |M| \to M$ to a surjective morphism: consider $\mathrm{id}_M \in \mathbf{Met}(M, M)$.

(2) A coproduct of strongly finitary functors is strongly finitary. (Colimits commute with weighted colimits.)

(3) The functor T_Σ is strongly finitary. Indeed, following Example 6, this functor is a coproduct of functors that are finite powers. Thus, we can apply Items (1) and (2).

(4) The Hausdorff endofunctor $\mathcal{H} \colon \mathbf{CMet} \to \mathbf{CMet}$ is strongly finitary ([1]).

4   1-Basic Monads

We introduce 1-basic monads, and prove that for every variety the corresponding monad has that property. We use the category of finitary monads on **Met**, where the distance of parallel monad morphisms $\varphi, \psi \colon S \to T$ is the supremum of $d(\varphi_n, \psi_n)$ over all natural numbers n. This distance forms a pseudometric: even if φ and ψ are distinct, their distance can be 0 (since the monads are finitary, but not necessarily strongly finitary). We thus need to work with categories enriched over pseudometric spaces:

Notation 31. The category of (extended) pseudometric spaces and nonexpanding maps is denoted by **PMet** . It differs from **Met** only by allowing distances of distinct elements to be 0. As **Met**, this category is symmetric monoidal closed, and the concepts of enriched category and functor over **PMet** are also completely analogous to **Met**.

Definition 32. We denote by $\mathbf{Mnd}_f(\mathbf{Met})$ the category of **Met**-enriched finitary monads on **Met** and monad morphisms. We consider it as a **PMet**-enriched category: the distance of parallel monad morphisms is

$$d(\varphi, \psi) = sup_{n \in \mathbb{N}} d(\varphi_n, \psi_n).$$

We further denote by $\mathcal{V}ar(\mathbf{Met})$ the **PMet**-enriched category of all varieties of quantitative algebras (for arbitrary signatures). Morphisms from $\mathcal{V}$ to $\mathcal{W}$ are the enriched concrete functors $H \colon \mathcal{V} \to \mathcal{W}$. The distance of concrete functors $H, H' \colon \mathcal{V} \to \mathcal{W}$ is

$$d(H, H') = \sup d(t_{HA}, t_{H'A}),$$

where the supremum ranges over algebras A of $\mathcal{V}$ and terms t in the signature of $\mathcal{W}$.

The above notation transfers to complete spaces smoothly. The **PMet**-enriched categories $\mathbf{Mnd}_f(\mathbf{CMet})$ of **CMet**-enriched finitary monads, and $Var(\mathbf{CMet})$ of varieties of complete quantitative algebras are analogous to those above.

Definition 33. A monad on **Met** or **CMet** is *1-basic* if it is a weighted colimit of strongly finitary monads in the category $\mathbf{Mnd}_f(\mathbf{Met})$ or $\mathbf{Mnd}_f(\mathbf{CMet})$, respectively.

We thus have the implication below for enriched monads:

$$\text{strongly finitary} \Rightarrow \text{1-basic} \Rightarrow \text{finitary}.$$

The latter one follows from Corollary 22 and Theorem 40. Those implications cannot be reversed: in Section 5 a 1-basic monad (on **Met** or **CMet**) is presented which is not strongly finitary. An example of a finitary monad that is not 1-basic is given by the full subcategory $\mathcal{V}$ of **Met** of all spaces where non-zero distances are all larger or equal to 1. (Here we work with the empty signature.) This subcategory is reflective because it is closed under products and (regular) subspaces. It is also closed under directed colimits, hence, the corresponding monad T is finitary. But using Theorem 38 below, it is clear that T is not 1-basic .

We recall that a **PMet**-enriched functor F is *faithful* if for every parallel pair u, u' of morphisms we have $d(u, u') = d(Fu, Fu'.)$

Proposition 34. *The following defines a* **PMet***-enriched, fully faithful functor*

$$E\colon Var(\mathbf{Met})^{\mathrm{op}} \to \mathbf{Mnd}_f(\mathbf{Met}).$$

It assigns to every variety $\mathcal{V}$ the monad $T_{\mathcal{V}}$. Given a concrete functor $H\colon \mathcal{V} \to \mathcal{W}$, we form the following (concrete) composite

$$\mathbf{Met}^{T_{\mathcal{V}}} \xrightarrow{K_{\mathcal{V}}^{-1}} \mathcal{V} \xrightarrow{H} \mathcal{W} \xrightarrow{K_{\mathcal{W}}} \mathbf{Met}^{T_{\mathcal{W}}} \ .$$

Then $E(H)\colon T_{\mathcal{W}} \to T_{\mathcal{V}}$ is the corresponding monad morphism.

Proof (sketch). The monad $E\mathcal{V}$ is enriched and finitary (Remark 17 and Corollary 22). The functor E is well defined: it clearly preserves identity morphisms and composition. Due to Theorem 15, E is full. It remains to prove, for concrete functors $H, H'\colon \mathcal{V} \to \mathcal{W}$, that $d(H, H') = d(EH, EH')$. Since $K_{\mathcal{V}}$ and $K_{\mathcal{W}}$ are concrete isomorphisms (Theorem 14), it is sufficient to prove, for finitary monads T and S, that given concrete functors $H, H'\colon \mathbf{Met}^T \to \mathbf{Met}^S$, the corresponding monad morphism $\varphi, \varphi'\colon S \to T$ have the distance $d(H, H')$. This is a lengthy computation using the above definition of the distance of monad morphisms.

Recall that a **PMet**-enriched category $\mathcal{K}$ is *complete* if it has weighted limits. Equivalently: it has ordinary limits and *cotensors* $M \pitchfork K$ (for spaces $M \in \mathbf{PMet}$ and objects $K \in \mathcal{K}$) ([10], Theorem 6.6.14). The latter are defined by a bijection

$$\frac{X \longrightarrow M \pitchfork K}{M \longrightarrow \mathcal{K}(X, K)}$$

natural in $X \in \mathcal{K}^{\mathrm{op}}$.

Dually, a **PMet**-enriched category is *cocomplete* iff it has ordinary colimits and tensors. An enriched functor preserves weighted colimits iff it preserves ordinary colimits and tensors ([10], Corollary 6.6.15).

We next prove that $\mathcal{V}ar(\mathbf{Met})$ has weighted limits. First we describe ε-equalizers since we use them later:

Proposition 35. *The ε-equalizer of concrete functors*

$$P, P' : \mathcal{V} \to \mathcal{W} \quad in \quad \mathcal{V}ar(\mathbf{Met}) \quad or \quad \mathcal{V}ar(\mathbf{CMet})$$

is the full embedding $\mathcal{V}^0 \hookrightarrow \mathcal{V}$ of the class $\mathcal{V}^0$ of all algebras $A \in \mathcal{V}$ satisfying, for the signature Γ of $\mathcal{W}$, that

$$d\big(\gamma_{PA}, \gamma_{P'A}\big) \leq \varepsilon \quad for\ every \quad \gamma \in \Gamma.$$

Proof (sketch). We verify that $\mathcal{V}^0$ is a variety of quantitative algebras, the rest is easy. Without loss of generality assume $\mathcal{W} = \Gamma\text{-}\mathbf{Met}$: indeed, the ε-equalizer of P, P' is equal to that of IP and IP' for the embedding $I : \mathcal{W} \to \Gamma\text{-}\mathbf{Met}$.

Recall the canonical monad morphism $k_{\mathcal{V}} : T_\Sigma \to T_{\mathcal{V}}$ (Ex. 16). It has the property that for every algebra $A \in \mathcal{V}$ and every interpretation $f : n \to A$, the unique extension to a homomorphism $\bar{f} : T_{\mathcal{V}} n \to A$ fulfils $f^{\#} = \bar{f} \cdot (k_{\mathcal{V}})_A$.

With every symbol $\gamma \in \Gamma_n$ we associate the following set $\mathcal{E}_\Gamma$ of equations (using φ_P and $\varphi_{P'}$ of Example **??**):

$$t =_\varepsilon t' \quad where \quad k_{\mathcal{V}}(t) = \varphi_P(\gamma) \quad and \quad k_{\mathcal{V}}(t') = \varphi_{P'}(\gamma),$$

for terms t, $t' \in T_\Sigma n$. If $\mathcal{V}$ is presented by a set $\mathcal{E}$ of equations, then we verify that $\mathcal{V}^0$ is presented by the set

$$\mathcal{E}^0 = \mathcal{E} \cup \bigcup_{\gamma \in \Gamma} \mathcal{E}_\gamma.$$

Proposition 36. *The category of varieties is complete, and the functor $E : \mathcal{V}ar(\mathbf{Met})^{\mathrm{op}} \to \mathbf{Mnd}_f(\mathbf{Met})$ preserves weighted colimits. Analogously for the category $\mathcal{V}ar(\mathbf{CMet})$.*

Proof (sketch). We prove that $\mathcal{V}ar(\mathbf{Met})$ has products and equalizers, as well as cotensors; and E^{op} preserves all these. We denote by $\mathcal{V} = (\Sigma, \mathcal{E})\text{-}\mathbf{Met}$ the variety presented by a set $\mathcal{E}$ of quantitative equations.

(1) Products of varieties $\mathcal{V}^i = (\Sigma^i, \mathcal{E}^i)\text{-}\mathbf{Met}$ for $i \in I$. Let Σ be the signature which is a disjoint union of Σ^i for $i \in I$. Thus, every term t for Σ^i is also a term for Σ. Put $\mathcal{E} = \bigcup_{i \in I} \mathcal{E}^i$. Moreover, for every term t of Σ_i and every Σ-algebra A the value t_A (Notation 8) is independent of the choice Σ or Σ^i of our signature. This follows by an easy induction in the depth of t. Then $(\Sigma, \mathcal{E})\text{-}\mathbf{Met}$ is the product of $\mathcal{V}^i$ in $\mathcal{V}ar(\mathbf{Met})$.

(2) Equalizers follow from Proposition 35 with $\varepsilon = 0$.

(3) Cotensors. Given a variety $\mathcal{V}$ and a space M, we describe the variety $M \pitchfork \mathcal{V}$. Let $\mathcal{V} = (\Sigma, \mathcal{E})$-**Met**, then the signature $\widetilde{\Sigma}$ of $M \pitchfork \mathcal{V}$ has as n-ary symbols all pairs (m, σ) where $m \in M$ and $\sigma \in \Sigma_n$. Every term $s \in T_\Sigma X$ define terms $s^m \in T_{\widetilde{\Sigma}} X$ $(m \in M)$ by the expected recursion: if s is a variable, then $s^m = s$. Else, we have $s = \sigma(s_i)_{i<k}$ for a k-ary symbol σ, and we put $s^m = (m, \sigma)(s_i^m)_{i<k}$. The variety $M \pitchfork \mathcal{V}$ of $\widetilde{\Sigma}$-algebras is presented by the following equations (a) $s^m =_\varepsilon t^m$ for $s =_\varepsilon t$ in $\mathcal{E}$ and $m \in |M|$, and (b) $(m, \sigma)(x_i)_{i<n} =_\delta (m', \sigma)(x_i)_{i<n}$ for $\sigma \in \Sigma_n$ and $d(m, m') = \delta$ in M.

Theorem 37. *For every variety $\mathcal{V}$ of (complete) quantitative algebras the monad $T_\mathcal{V}$ on* **Met** *or* **CMet** *is 1-basic.*

Proof. We present a proof for **Met**, that for **CMet** is analogous.

(1) Let t be a term with variables in the $n = \{0, \ldots, n-1\}$. Denote by Γ the signarure of a single n-ary operation γ. We obtain a concrete functor $F_t \colon \Sigma$-**Met** $\to \Gamma$-**Met** assigning to every algebra A the n-ary operation of computation of t in A (Notation 8). The corresponding monad morphism $t^* \colon T_\Gamma \to T_\Sigma$ takes a term s in $T_\Gamma X$ and substitutes every occurence of γ by t.

(2) If the variety $\mathcal{V}$ is given by a single equation $s =_\varepsilon t$, then Proposition 35 implies that the embedding I into of $\mathcal{V}$ into Σ-**Met** or Σ-**CMet** is the ε-equalizer of F_s and F_t. Thus, by Proposition 36 the ε-coequalizer of the correesponding monad morphisms $T_\Gamma \to T_\Sigma$ is the monad morphism $c \colon T_\Sigma \to T_\mathcal{V}$ corresponding to I, see Ex. 16. Since both T_Γ and T_Σ are strongly finitary (Ex. 30), we conclude that $T_\mathcal{V}$ is1-basic.

(3) An arbitrary variety $I \colon \hookrightarrow \mathcal{V}$ is given by a set $\mathcal{E}$ of equations. For every equation e in $\mathcal{E}$ let $I_e \colon \mathcal{V}(e) \hookrightarrow \Sigma$-**Met** be the embedding of the variety $\mathcal{V}(e)$ presented by e alone. Then I is the intersection (wide pullback) of these concrete functors I_e. Thus $T_\mathcal{V}$ is the wide pushout of the 1-basic monads $T_{\mathcal{V}(e)}$, which proves that it is 1-basic.

5   The Main Theorem

Here we prove that varieties of quantitative algebras correspond bijectively to 1-basic monads.

Notation 38. With every strongly finitary monad T we associate a variety $\mathcal{V}_T$ of quantitative Σ-algebras, where $\Sigma_n = |Tn|$ for $n \in \mathbb{N}$. Thus, an n-ary symbol σ is an element of Tn. We identify σ with the term $\sigma(x_0, \ldots, x_{n-1})$ in $T_\Sigma V$ where $V = \{x_i\}_{i \in \mathbb{N}}$ is a chosen set of variables.

The variety $\mathcal{V}_T$ is presented by three types of quantitative equations, where n and k range over $\mathbb{N}$. First, the ordinary equations describing the monad structure μ and η:

$$\eta_n(x_i) = x_i \qquad (i < n) \tag{i}$$

and

$$\mu_n \cdot Tf(\sigma) = \sigma\big(f(0), \ldots f(k-1)\big) \quad (f \colon k \to Tn \text{ and } \sigma \in \Sigma_k). \tag{ii}$$

Second, the quantitative equations describing the metric d_{Tn} of Tn:

$$\sigma =_\varepsilon \sigma' \quad \text{for all } \sigma, \sigma' \in \Sigma_n \text{ with } d_{Tn}(\sigma, \sigma') = \varepsilon. \tag{iii}$$

Proposition 39. *Every strongly finitary monad T on **Met** or **CMet** is the free-algebra monad for the variety $\mathcal{V}_T$.*

Proof (sketch). We present a sketch of proof for **Met**, the argument for **CMet** is analogous.

(1) With every algebra $\alpha\colon TA \to A$ in $\mathbf{Met}^T$ we associate a Σ-algebra RA on A as follows: given $\sigma \in \Sigma_n$ and $f\colon n \to |A|$, put $\sigma_{RA}(f) = \alpha \cdot Tf(\sigma)$. Then RA lies in $\mathcal{V}_T$. Moreover, the homomorphisms h from (A, α) to (B, β) in $\mathbf{Met}^T$ are precisely the homomorphisms from RA to RB in $\mathcal{V}_T$.

(2) We thus get a concrete full embedding $R\colon \mathbf{Met}^T \to \mathcal{V}_T$. To prove that T is the free-algebra monad, it is sufficient, due to Theorem 14, to verify for all metric spaces X that $R(TX, \mu_X)$ is the free algebra of $\mathcal{V}_T$ on X with respect to $\eta_X\colon X \to TX$.

(2a) Let X be finite and discrete. Say, $X = n$ ($n \in \mathbb{N}$). Given an algebra $A \in \mathcal{V}_T$ and an interpretation $v\colon n \to |A|$, the morphism $v^\#\colon Tn \to A$ assigning to $\sigma \in |Tn| = \Sigma_n$ the value $v^\#(\sigma) = \sigma_A(v)$ is a Σ-homomorphism: use that A satisfies $\mu_n \cdot Tf(\sigma) = \sigma\big(f(x_i)\big)$ in (ii) above. Due to the equations(i), it fulfils $v = v^\# \cdot \eta_n$. And it is nonexpanding due to the equations (iii).

(2b) For an arbitrary finite space X, recall that D_X is a diagram of finite discrete spaces. Thus, we can apply (2a), using that $X = \operatorname{colim}_B D_X$ and $TX = \operatorname{colim}_B TD_X$ (Proposition 28).

(2c) For an arbitrary space X, use the directed colimit $X = \operatorname{colim}_{i \in I} X_i$ where X_i ranges over finite subspaces (Remark 18). Since T is finitary, $TX = \operatorname{colim} TX_i$, and from Item (2) we conclude that $R(TX, \mu_X)$ is free on X in $\mathcal{V}_T$.

Theorem 40. *A monad on **Met** or **CMet** is the free-algebra monad of a variety of quantitative algebras iff it is 1-basic.*

Proof. We again present a proof for **Met**, that for **CMet** is analogous.

Necessity follows from Theorem 37. To prove the sufficiency, let T be 1-basic . We thus have a diagram $D\colon \mathcal{D} \to \mathbf{Mnd}_f(\mathbf{Met})$ weighted by $W\colon \mathcal{D}^{op} \to \mathbf{Met}$, whose objects Dd are strongly finitary monads, such that T is the weighted colimit of D. For every d we can, by Theorem 39 , choose a variety $\mathcal{V}(d)$ whose free-algebra monad is Dd. By Theorem 15 this yields a diagram $\bar{D}$ in the dual of $\mathcal{V}ar(\mathbf{Met})$ with D naturally isomorphic to $E \cdot \bar{D}$. We know that a colimit

$$\mathcal{V} = \operatorname{colim}_W \bar{D}$$

exists in $(\mathcal{V}ar(\mathbf{Met}))^{op}$, and is preserved by E (Theorem 36). We conclude that T and $T_\mathcal{V}$ are isomorphic monads: both are $\operatorname{colim}_W D$. Therefore, T is the free-algebra monad of $\mathcal{V}$.

Corollary 41. *The following ordinary categories are dually equivalent:*

(1) *Varieties of quantitative algebras, $Var(\mathbf{Met})$ or $Var(\mathbf{CMet})$, and concrete functors.*
(2) *Semi-strongly finitary monads on $\mathbf{Met}$ or $\mathbf{CMet}$, and monad morphisms.*

This follows from Theorem 40: the functor $\Phi\colon Var(\mathbf{Met})^{\mathrm{op}} \to \mathbf{Mnd}_f(\mathbf{Met})$ has the codomain restriction ψ to the full subcategory of $\mathbf{Mnd}_f(\mathbf{Met})$ on all 1-basic monads. Since Φ is fully faithful, so is ψ. By Theorem 40, ψ is an equivalence functor.

Remark 42. We do not claim that $\mathbf{Mnd}_f(\mathbf{Met})$ is cocomplete. But every diagram of strongly finitary monads has, for each weight, a weighted colimit. 1-Basic monads are precisely the resulting weighted colimits.

6 A Counter-Example

In the present section we prove that the free-algebra monad $T_\mathcal{V}$ for the variety of two ε-close binary operations is not strongly finitary.

Notation 43. Throughout this section $\mathcal{V}$ denotes the variety presented by the binary signature $\Sigma = \{\sigma_1, \sigma_2\}$ and the quantitative equation

$$\sigma_1(x,y) =_\varepsilon \sigma_2(x,y) \quad \text{for a fixed number} \quad 0 < \varepsilon < 1.$$

Remark 44. In universal algebra the free algebra $T_\Sigma V$ on a set V of variables can be represented as follows. The elements are all finite, ordered binary trees with leaves labelled in V and inner nodes labelled by σ_1 or σ_2. Here, two trees with a label-preserving isomorphism between them are identified. The operation σ_i is the tree-tupling with the root labelled by σ_i. Indeed, the variable $x \in V$ is represented by the root-only tree labelled by x. The composite term $\sigma_i(t_l, t_r)$ is represented by the tree below

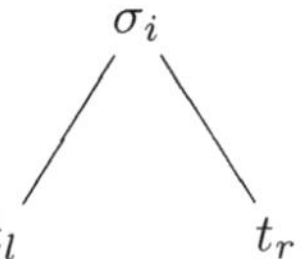

Notation 45. For every metric space X we define the following metric $\widehat{d}_X$ on the set $T_\Sigma|X|$ of all terms. For all variables x and y we use their distance in X: $\widehat{d}_X(x,y) = d_X(x,y)$, and put $\widehat{d}_X(x,t) = \infty$ if $t \notin |X|$. All other distances $\widehat{d}_X(t,t')$ are defined by recursion: Represent t and t' as the following trees

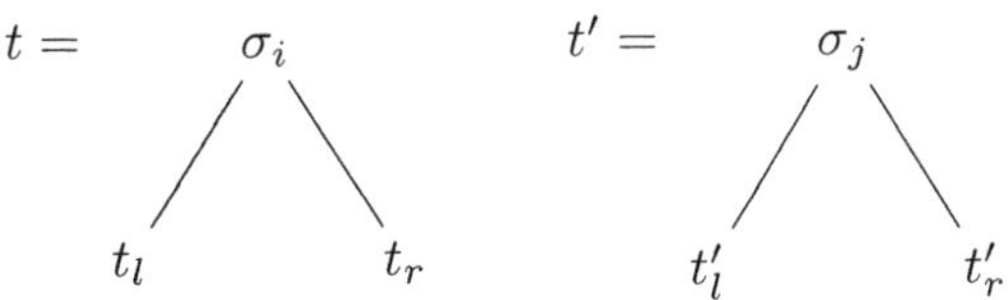

Let m denote the maximum of the distances $\widehat{d}_X(t_l, t_l')$ and $\widehat{d}_X(t_r, t_r')$. Put

$$\widehat{d}_X(t, t') = \begin{cases} m & \text{if } i = j \\ \varepsilon + m & \text{else.} \end{cases}$$

Recall that the *depth* $\delta(t)$ of a tree is defined by recursion: it is 0 for variables or constants, and $\delta\big(\sigma_i(t_l, t_r)\big) = \max\big\{\delta(t_l), \delta(t_r)\big\} + 1$.

Lemma 46. *The free algebra $T_{\mathcal{V}}X$ on a metric space X is the space $\big(T_\Sigma|X|, \widehat{d}_X\big)$ with operations given by tree-tupling. The universal map is the inclusion morphism $X \hookrightarrow T_\Sigma|X|$.*

Proof (sketch). Let A be an algebra in $\mathcal{V}$. Given a nonexpanding morphism $f\colon X \to A$, there is a unique homomorphism of the underlying Σ-algebras $f^\#\colon T_\Sigma|X| \to |A|$ extending f. It is our task to verify that $f^\#$ is nonexpanding:

$$\widehat{d}_X(t, t') \geq d\big(f^\#(t), f^\#(t')\big) \quad \text{for } t, t' \in T_\Sigma|X|.$$

This is done by induction on the maximum of the depths of t and t'.

Corollary 47. *The monad $T_{\mathcal{V}}$ is given on objects by $X \mapsto (T_\Sigma|X|, \widehat{d}_X)$. It takes a morphism $f\colon X \to Y$ to the morphism $T_{\mathcal{V}}f$ assigning to a tree $t \in T_\Sigma|X|$ the tree in $T_\Sigma|Y|$ obtained by relabelling all the leaves from x to $f(x)$.*

All metrics on a given set are partially ordered (pointwise). Given metrics d_1, d_2, we denote by $d_1 \wedge d_2$ their meet, provided that it exists.

Proposition 48. *The functor $T_{\mathcal{V}}$ is not strongly finitary.*

Proof. We are going to present spaces X and Y and a nonexpanding map $f\colon T_{\mathcal{V}}|X| \to Y$ satisfying (3.1) of Proposition 28, which does not factorize through $T_{\mathcal{V}}i_X$. This proves our proposition.

Let X be the space $\{a, b\}$ with $d(a, b) = 1$. Thus $T_{\mathcal{V}}|X|$ is the space of all terms on $\{a, b\}$ with the metric $\widehat{d}_{|X|}$. The space Y is defined as the same set of terms with the metric d which is the meet of the metrics d_X^* (Example 6) and $\widehat{d}_{|X|}$:

$$Y = \big(T_\Sigma|X|, d\big), \quad \text{where } d = d_X^* \wedge \widehat{d}_{|X|}.$$

More detailed, for terms t and t' we have

$$d(t, t') = \inf\Big(\sum_{i=0}^{n-1} d_X^*(s_{2i}, s_{2i+1}) + \sum_{i=1}^{n-1} \widehat{d}_{|X|}(s_{2i-1}, s_{2i})\Big), \qquad (6.1)$$

where the infimum is ranging over all sequences of terms $t = s_0, s_1, \ldots, s_{2n} = t'$ ($n \in \mathbb{N}$).

The function d above is easily seen to be a metric. In fact, $d(t, t') = 0$ by choosing $n = 0$. If $t \neq t'$ then $d(t, t') \neq 0$: the minimum distance of distinct trees is ε. Symmetry and triangle inequality are clear.

We have $d_X^* \geq d$ (choose $n = 1$) and $\widehat{d}_{|X|} \geq d$ (choose $n = 2$ and $s_0 = s_1$). And d is the infimum since every metric $\widetilde{d}$ with $d_X^* \geq \widetilde{d}$ and $\widehat{d}_{|X|} \geq \widetilde{d}$ fulfills $\widetilde{d} \leq d$ due to the triangle inequality.

The identity-carried map $f \colon T_V|X| \to Y$ is clearly nonexpanding. To verify (3.1), consider a number $\delta > 0$ and an arbitrary tree $u \in T_V(\Delta_\delta X)$. This is a binary tree with k leaves labelled by pairs (x_i, x_i'), $i < k$, such that $d_X(x_i, x_i') \leq \delta$. The tree $t = T_V l_\delta(u)$ is the same one except that the i-th leaf label is x_i; analogously $t' = T_V r_\delta(u)$. The definition of d_X^* yields

$$d_X^*\big(T_V l_\delta(u), T_V r_\delta(u)\big) = d_X^*(t, t') = \max_{i < k} d_X(x_i, x_i') \leq \delta \,.$$

The condition (3.1) states precisely this inequality, since f is identity-carried.

We now prove that f does not factorize through $T_V i_X$. Since both f and $T_V i_X$ are identity-carried, this means that $\widehat{d}_X(t, t') < d(t, t')$ holds for some trees. Indeed, we prove for the following trees

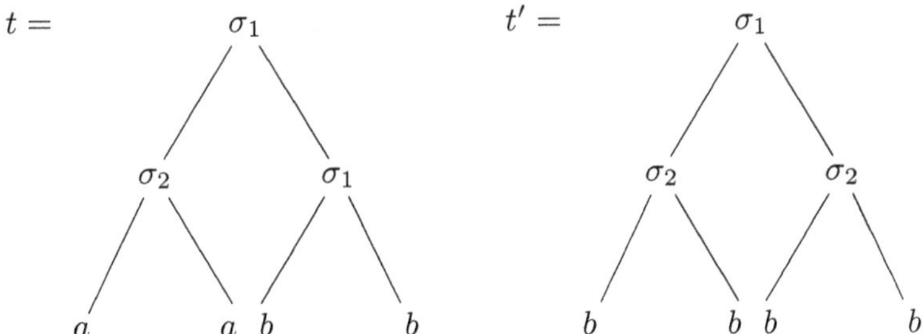

that $\widehat{d}_X(t, t') = 1$ and $d(t, t') = \varepsilon + 1$.

The first equality follows from $\widehat{d}_X(t_l, t_l') = 1$ and $\widehat{d}_X(t_r, t_r') = \varepsilon < 1$. To prove $d(t, t') = 1 + \varepsilon$, observe first that $\widehat{d}_{|X|}(t, t') = \infty = d_X^*(t, t')$. (Since $d_{|X|}(a, b) = \infty$, we get $\widehat{d}_{|X|}(t, t') = \infty$. Since t is not similar to t', we have $d_X^*(t, t') = \infty$.)

The infimum defining $d(t, t')$ thus uses a sequence $t = s_0, s_1, \ldots, s_{2n+1} = t'$ with $n > 1$ in (6.1) above. One such sequence is s_0, s_1, s_2 where $s_1 = \sigma_1\big(\sigma_2(a, a), \sigma_2(b, b)\big)$:

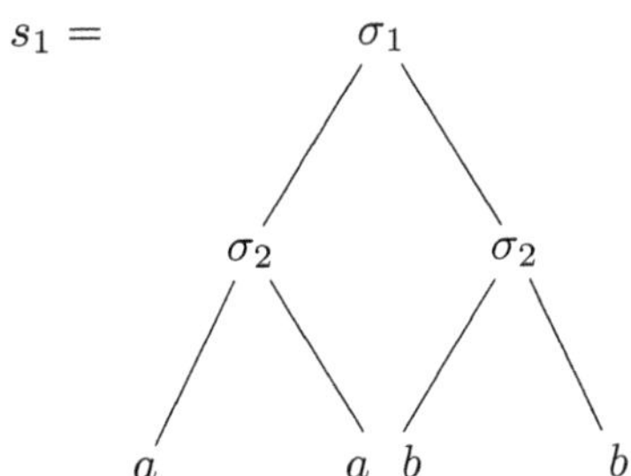

The corresponding sum of distances is $\varepsilon + 1$ due to

$$\widehat{d}_{|X|}(t, s_1) = \varepsilon$$

and
$$d_X^*(s_1, t') = 1\,.$$

The corresponding sum of distances is $\varepsilon + 1$, due to $\widehat{d}_{|X|}(t, s_1) = \varepsilon$ and $d_X^*(s_1, t') = 1$. For every other sequence the sum is at least $\varepsilon + 1$: let i be the largest index such that s_i has label a at the left-most leaf. Then s_{i+1} has label h at that leaf, consequently $d(s_i, s_{i+1}) \geq d(a, b) = 1$. Since $n \geq 2$, and ε is the smallest distance between distinct trees, $\sum\limits_{j<n} d(s_j, s_{j+1}) \geq \varepsilon + 1$. This proves $d(t, t') = \varepsilon + 1$.

In other words, f does not factorize through $T_{\mathcal{V}} i_X$, thus $T_{\mathcal{V}}$ is not a strongly finitary functor.

7   Conclusions and Open Problems

We have characterized varieties of quantitative algebras as precisely the Eilenberg-Moore categories $\mathbf{Met}^T$, where T is a 1-basic monad. This means that T is a weighted colimit of strongly finitary monads in the **PMet**-enriched category $\mathbf{Mnd}_f(\mathbf{Met})$ of finitary **Met**-enriched monads. An analogous result holds for varieties of complete quantitative algebras.

We have further presented a variety $\mathcal{V}$ for which $T_{\mathcal{V}}$ is not a strongly finitary: the variety of algebras on two binary operations of distance $\varepsilon < 1$. Independently, a similar example has been presented by Mardare et al. [16].

Open problem 49. *Which varieties $\mathcal{V}$ of quantitative algebras yield strongly finitary monads $T_{\mathcal{V}}$?*

The affirmative answer for varieties usincccg just unary operations has been proved in [1].

In the original work of Mardare, Panangaden and Plotkin, more general quantitative equations are considered: the c-basic equations for a cardinal c, using variables from metric spaces of power smaller than c. A monadic characterization of varieties presented by $\aleph_0$-basic equations is a highly interesting open problem.

For $c = \aleph_1$ these monads were characterized in [2]: they are precisely the enriched monads preserving countably directed colimits and surjective morphisms. This is generalized in [21]. For varieties with non-discrete arities see [11]. Several authors further investigate quantitative algebras based on categories of generalized metric spaces, e.g. [11] and [19].

Open problem 50. *How are monads corresponding to varieties of generalized quantitative algebras characterized?*

References

1. Adámek, J.: Strongly finitary metric monads are too strong. `arxive:2601.03180`. A short abstract was presented at the International Category Theory Conference (CT 2025), Brno, Czech Republic, July 2025

2. Adámek, J.: Varieties of quantitative algebras and their monads. Proc. Logic in Computer Science, pp. 1–12. (LICS 2022)
3. Adámek, J., Dostál, M., Velebil, J.: A categorical view of varieties of ordered algebras. Math. Structures Comput. Sci. **32**, 1–25 (2022)
4. Adámek, J., Dostál, M., Velebil, J.: Quantitative algebras and a classification of metric monads. `arXiv:2210.01565`
5. Adámek, J., Dostál, M., Velebil, J.: Strongly finitary monads for varieties of quantitative algebras. Conf. Algebra and Coalgebra Methods in Computer Science (CALCO 2023), LIPICs, vol. 270
6. Adámek, J., Milius, S., Moss, L., Urbat, H.: On finitary functors and their presentations. J. Comput. System Sci. **81**, 813–833 (2015)
7. Barr, M., Wells, C.: Toposes, triples and theories. Springer, New York (1985)
8. Bourke, J., Garner, R.: Monads and Theories. Adv. Math. **351**, 1024–1071 (2019)
9. van Breugel, F., Hermida, C., Makkai, M., Worrell, J.: Recursively defined metric spaces without contraction. Theoret. Comput. Sci. **380**, 143–163 (2007)
10. Borceux, F.: Handbook of categorical algebra 2. Cambridge Univ. Press (1994)
11. Ford, C., Milius, S., Schroeder, L.: Monads on categories of relational structures. Conf. Algebra and Coalgebra in Computer Science (CALCO 2021), LIPICs, vol. 14
12. Kelly, G.M., Lack, S.: Finite-product-preserving-functor, Kan extensions and strongly finitary 2-monads. Appl. Categ. Structures **1**, 85–94 (1993)
13. Kurz, A., Velebil, J.: Quasivarieties and varieties of ordered algebras. Math. Struct. Comput. Sci. **27**, 1–42 (2016)
14. Lucyshyn-Wright, R.B.B., Parker, J.: Enriched structure-semantics adjunctions and monad theory equivalences for subcategories of arities. Theory Appl. Categories **41**, 1873–1918 (2024)
15. Mac Lane, S.: Categories for the working mathematician. 2nd edition. Springer (1968)
16. Mardare, R., Ghani, N., Rischel, E.: Metric equational theories. G. Bacci, A. Francalanza (Eds.): Sixteenth International Symposium on Games, Automata, Logics, and Formal Verification (GandALF 2025). EPTCS 428, 144–160 (2025)
17. Mardare, R., Panaganden, P., Plotkin, G.D.: Quantitative algebraic reasoning Proc. Logic in Computer Science (LICS 2016), pp. 700–709. IEEE Computer Science (2016)
18. Mardare, R., Panaganden, P., Plotkin, G.D.: On the axiomatizability of quantitative algebras. Proc. Logic in Computer Science (LICS 2017), 12 pp. IEEE Computer Science (2017)
19. Mio, M., Sarkis, R., Vignudelli, V.: Universal quantitative algebra for fuzzy relations and generalised metric spaces. Log. Methods Comput. Sci. **20** no. 4, Paper No. 19 (2024), 56 pp.
20. Moggi, E.: Notions of computation and monads. Inform. and Comput. **93**, 55–92 (1991)
21. Rosický, J.: Metric monads, Math. Struct. Comp. Sci. **31** , 535–552 (2021)
22. Trnková, V., Adámek, J., Koubek, V., Reiterman, J.: Free algebras, input processes and free monads. Comment. Math. Univ. Carolinae **16**, 339–351 (1975)

Learning bottom-up tree automata valued in monoidal categories

Quentin Aristote and Daniela Petrişan

Université Paris Cité, CNRS, Inria, IRIF, F-75013, Paris, France
{aristote,petrisan}@irif.fr

Abstract. This paper provides a unifying framework for minimization and learning algorithms for bottom-up tree automata valued in monoidal categories. Our aim is two-fold: encompass existing algorithms for various forms of tree automata (with side-effects) – deterministic bottom-up tree automata, residual finite tree automata, tree automata weighted over a field – and instantiate the abstract framework in order to obtain new results – for tree automata weighted over principal ideal domains (PIDs).

1 Introduction

A centerpiece of computational learning theory is the active learning algorithm of deterministic finite automata (DFA) proposed by Angluin [2]. This algorithm, called L*, is based on the *minimally adequate teacher* model: it outputs the minimal deterministic finite automaton for an a priori unknown target language, by interacting with an oracle who can provide answers to *equivalence queries* and *membership queries*. To a membership query – which asks whether or not a word belongs to the target language – the oracle provides a binary answer. To an equivalence query – which asks whether a hypothesis automaton recognizes the target language – the oracle either answers "yes" and the algorithm terminates, or provides a counterexample word proving that this is not the case.

The versatility and robustness of Angluin's algorithm are witnessed by its numerous extensions to learning other forms of automata, including [19,11,26,40,30]. While learning word or tree weighted automata over fields is well established, see [6,22], extending these algorithms to automata weighted over arbitrary commutative rings remains challenging and is an active field of research. For example, [23] provides a learning algorithm for weighted word automata over principal ideal domains, but its complexity is unbounded. In the special case of $\mathbb{Z}$-weighted word automata, [13] provides a polynomial time learning algorithm, by employing a reduction procedure to learning $\mathbb{Q}$-weighted automata.

In this work, we focus on generic learning algorithms for *tree* automata interpreted in monoidal categories, and obtain as an instance a new learning algorithm for *tree* automata weighted over principal ideal domains.

The interplay between category theory and automata has a long history: tree automata were interpreted in categories already in [35,20], while early contributions to categorical automata minimization, as well as work on the duality between reachability and observability emerged in the early seventies [25,3].

© The Author(s) 2026
N. Bertrand and S. Milius (Eds.): FoSSaCS 2026, LNCS 16503, pp. 21–42, 2026.
https://doi.org/10.1007/978-3-032-22730-0_2

Recent years have seen renewed interest in applying categorical methods to automata and language theory. Notable examples include the monadic approach to language recognition developed by Bojańczyk [8,10,9] and by Blumensath [7]. The theory of coalgebras provides another approach to automata and transition systems. It proved successful in improving existing algorithms, see [12], and was extensively used for unifying learning algorithms, see for example [37,39,36] and the references therein.

The present submission is closer in spirit to the functorial approach to automata theory developed by Colcombet and Petrişan [16], which can be summarized as follows: automata are seen as machines that process some *input* – taking into account its structure (in particular, words over some finite unary alphabet in *ibid.*) – and produce as a side effect some quantity in a universe of *output values* (Boolean values, probabilities, scalars in a field or semiring, words over an output alphabet, etc.) As such, automata can be modelled as functors from an *input category* – that captures the structure of the input – to an output category. By suitably changing the output category to encompass a specific "effect", one can model deterministic automata, non-deterministic ones, weighted automata over a field or semiring, sequential transducers or probabilistic automata. This category-theoretic perspective was further refined to encompass active learning in [17]. For example, the generic learning algorithm FunL^* of [17] instantiates to the well known L^*-algorithm of Angluin [2], but also to learning algorithms for weighted automata [6], or for sequential transducers [40]. This line of research was further advanced to streaming transducers in [32,33], whereas [4] obtains new algorithms for minimizing and learning deterministic transducers with outputs in arbitrary monoids.

A limitation of the functorial framework [16] is that it is mainly concerned with automata whose inputs are words over a unary alphabet. The sole exception was the development in [16, Section 6], concerning the syntactic monoid of a language. In this paper we thus extend the scope of [16] to encompass various families of tree automata. Concretely, the plan and contributions of this work are as follows:

- To help the reader build intuitions, we start in Section 2 with a concrete instance of the later sections: we outline a new learning algorithm for tree automata weighted over principal ideal domains.
- In Section 3 we model tree automata within a monoidal category, that is, a category equipped with a notion of tensor product. We reduce the problem of minimizing such tree automata in a monoidal setting to that of minimizing an associated context automaton, obtained by restricting the transitions of the automaton to those corresponding to at most unary terms. Context automata can be minimized under mild assumptions. We identify further reasonable conditions on the ambient monoidal category so that the minimal tree automaton can be retrieved from the minimal context automaton, see Theorem 26. The main challenge there is to find assumptions strong enough on the output category for minimal tree automata to exist and be

computable, while keeping them weak enough to encompass as wide an array of concrete families of tree automata as possible.

– Once the minimal automaton is proven to exist in this framework (Theorem 26), a second challenge is to express minimization and learning algorithms in this generic language, and to prove their correctness and termination. The learning algorithm we provide in Section 4 – Algorithm 1 – instantiates in particular to known (or hinted at) learning algorithms for: bottom-up deterministic tree automata [19], residual finite tree automata [26], respectively tree automata weighted over a field [22], but also to the *new* algorithm for PID-weighted automata of Section 2.

Notations. Throughout this work, we fix a *signature* Σ, that is a finite set Σ of *symbols* with *arities*: the set of symbols with arity $n \in \mathbb{N}$ is written Σ_n. The elements of Σ_0 are called *constants*. The set $\mathcal{T}_0(\Sigma)$ of *trees over* Σ is defined inductively as the smallest set of formal expressions such that $\sigma(t_1, \ldots, t_n) \in \mathcal{T}_0(\Sigma)$ for every $n \in \mathbb{N}$, $\sigma \in \Sigma_n$ and $t_1, \ldots, t_n \in \mathcal{T}_0(\Sigma)$. In particular, $\Sigma_0 \subseteq \mathcal{T}_0(\Sigma)$. More generally, let $\square$ be a special symbol of arity 0 called a *hole* and such that $\square \notin \Sigma$. Then for $n \in \mathbb{N}$ we let $\mathcal{T}_n(\Sigma)$, the set of *n-holed trees*, be the subset of trees over $\Sigma \cup \{\square\}$ where $\square$ occurs exactly n times, and $\mathcal{T}(\Sigma) = \bigcup_{n \in \mathbb{N}} \mathcal{T}_n(\Sigma)$. A term $t \in \mathcal{T}_n(\Sigma)$ is said to have *arity n*, and we write $\mathrm{ar}(t) = n$. The elements of $\mathcal{T}_1(\Sigma)$ are in particular also called *contexts over* Σ. For any $n \in \mathbb{N}$, given an n-holed tree c and trees $t_1, \ldots, t_n \in \mathcal{T}(\Sigma)$ we write $c \circ (t_1, \ldots, t_n)$ (or just $c \circ t_1$ when $n = 1$) for the tree obtained by substituting the leftmost hole of c with t_1, the second leftmost one with t_2, etc. $t_1, \ldots, t_n$ are then called *children* of the resulting tree, while c is one of its *parent trees*. If $n \in \mathbb{N}$ and $\sigma \in \Sigma_n$ the n-holed tree $\sigma(\square, \ldots, \square)$ is also simply written σ. For instance we will consider the signature $\Sigma = \Sigma_0 \cup \Sigma_2$ given by $\Sigma_0 = \{l\}$ and $\Sigma_2 = \{n\}$ as a recurring example: $n \circ (\square, l)$ will denote the context $n(\square, l)$. Given $n \in \mathbb{N}$ and $t_1, \ldots, t_n \in \mathcal{T}(\Sigma)$, we will write the tuple $(t_1, \ldots, t_n)$ as a tensor product $t_1 \otimes \ldots \otimes t_n$. Abbreviating $\mathsf{t} = t_1 \otimes \ldots \otimes t_n$, we write $\mathsf{t}[i:=\square] = t_1 \otimes \ldots \otimes t_{i-1} \otimes \square \otimes t_{i+1} \otimes \ldots \otimes t_n$.

2 Learning tree automata weighted over PIDs

In this section we describe a learning algorithm for computing minimal tree automata weighted over principal ideal domains. This new result generalizes those for tree automata weighted over fields [22] and word automata weighted over principal ideal domains [13]. It is proved by instantiating the categorical framework developed in Sections 3 and 4 to the category $\mathbf{Mod}_R$. We present this concrete instance first so that the reader may build up intuitions about tree automata and their algorithms.

2.1 Tree automata weighted over PIDs

Recall that a *commutative ring* is a set R equipped with the structure of a commutative group, $(R, +, 0)$, and that of a commutative monoid, $(R, \times, 1)$, that are compatible: $\times$ should distribute over $+$. An R-module M is a commutative group M equipped with an action of the monoid $(R, \times, 1)$ compatible with M's group structure. Examples of R-modules include its *ideals*: an ideal of a ring R is a subgroup of $(R, +, 0)$ that is stable under multiplication.

Definition 1 (instance of Definition 11). *Let R be a ring and Σ be a signature. An R-weighted Σ-tree automaton is a family (Q, μ, λ) where Q is a set of states, μ is a Σ-indexed family such that for $\sigma \in \Sigma_n$, $\mu_\sigma \in R^{Q^n \times Q}$ is the σ-transition matrix with $|Q|^n$ rows and $|Q|$ columns, and $\lambda \in R^{Q \times 1}$ is the termination column vector.*

μ extends to a $\mathcal{T}_0(\Sigma)$-indexed family given recursively by the matrix product $\mu_{\sigma \text{ot}} = (\mu_{t_1} \otimes \cdots \otimes \mu_{t_n}) \mu_\sigma$ for $\sigma \in \Sigma_n$ and $\mathsf{t} = (t_1, \ldots, t_n) \in \mathcal{T}_0(\Sigma)^n$, and where $\otimes$ denotes the Kronecker product of matrices[1]. An automaton (Q, μ, λ) then recognizes the function $L \colon \mathcal{T}_0(\Sigma) \to R$ given by $L(t) = \mu_t \lambda$.

The R-module R^Q is called the *module of configurations* of (Q, μ, λ). The matrix μ_σ corresponds to a linear function $R^{(Q^n)} = \left(R^Q\right)^{\otimes n} \to R^Q$, or alternatively to an n-linear function $R^Q \times \cdots \times R^Q \to R^Q$ from the module of configurations to itself. Similarly, λ gives a linear function $R^Q \to R$ from the module of configurations to R.

Example 2. Let Σ be the signature consisting of a constant $l \in \Sigma_0$ and a binary symbol $n \in \Sigma_2$. Let A be the $\mathbb{Z}$-weighted Σ-tree automaton with states $\{\bullet, \circ\}$ and matrices

$$
\mu_l = \begin{array}{c} \bullet \quad \circ \\ \left(\begin{array}{cc} 1 & 0 \end{array} \right) \end{array}
\qquad
\mu_n = \begin{array}{c} \qquad \bullet \quad \circ \\ \begin{array}{c} (\bullet,\bullet) \\ (\bullet,\circ) \\ (\circ,\bullet) \\ (\circ,\circ) \end{array} \left(\begin{array}{cc} 0 & 2 \\ 1 & 0 \\ 1 & 0 \\ 0 & 1 \end{array} \right) \end{array}
\qquad
\lambda = \begin{array}{c} \bullet \\ \circ \end{array} \left(\begin{array}{c} 1 \\ 1 \end{array} \right)
$$

By induction, $\mu_t = \begin{pmatrix} 0 & 2^q \end{pmatrix}$ if the term t has $2q$ leaves, and $\mu_t = \begin{pmatrix} 2^q & 0 \end{pmatrix}$ if t has $2q + 1$ leaves. The automaton A thus recognizes the function L that maps a binary tree with $2q + r$ leaves (with $q \in \mathbb{N}$ and $r \in \{0, 1\}$) to the value 2^q. On the other hand the automaton B with same state-set but with $\mu_l = \begin{pmatrix} 0 & 1 \end{pmatrix}$ would recognize the constant function $L(t) = 1$.

Note that the module of configurations of an R-weighted tree automata is *free*, i.e. it has a basis. When attempting to minimize an automaton, we may need to compute a quotient of this module, which, in general, will not be free – and thus, will not correspond stricto sensu to an automaton. If R is a field, this problem disappears as R-modules are just R-vector spaces and are very well behaved: a vector space always has a *basis*. This is a key ingredient in proving that word and tree automata weighted over a field can be minimized. We show that R-weighted tree automata minimization works under even milder assumption on R, namely when R is a *principal ideal domain*.

---

[1] if $A \in R^{a \times b}$ and $B \in R^{c \times d}$ then $A \otimes B \in R^{ac \times bd}$ has entries $(A \otimes B)_{c(q-1)+r, d(p-1)+s} = A_{q,p} B_{r,s}$ for $1 \leq q \leq a$, $1 \leq p \leq b$, $1 \leq r \leq c$ and $1 \leq s \leq d$

Definition 3. *A* principal ideal domain (PID) *R* *is a commutative ring whose ideals are all* principal, *i.e. of the form* $(a) = aR = \{ar \mid r \in R\}$ *for some* $a \in R$.

Example 4. The ring $\mathbb{Z}$ of integers is a PID. Every field, e.g. $\mathbb{R}$ or $\mathbb{Q}$, is also a PID. More generally, the ring of polynomials over a field, e.g. $\mathbb{R}[X]$ or $\mathbb{Q}[X]$, is still a PID.

A fundamental theorem of PIDs is that their *finitely generated torsion-free modules* are free [15, Theorem 2.4.1]: in practice, given a PID R and $v_1, \ldots, v_n \in R^n$, the module $\langle v_1, \ldots, v_n \rangle$ – whose elements are linear combinations of the v's – has a basis. This fact is key in proving the following proposition, which is an instance of our generic results.

Proposition 5 (existence of the minimal R-weighted tree automata). *Let R be a PID and let $L \colon \mathcal{T}_0(\Sigma) \to R$ be a function recognized by a Σ-tree R-weighted automaton with finite state-set. Then there is an automaton* $\mathsf{Min}\, L = (Q^{min}, \mu^{min}, \lambda^{min})$ *recognizing L that has the minimum number of states among all others automata recognizing L. Moreover, up to change of basis,* $\mathsf{Min}\, L$ *is uniquely determined.*

Example 6. The weighted automaton A of Example 2 is minimal: there is no automaton with a single state that recognizes the same function. Another less obvious minimal automaton recognizing the same language will be computed in Example 9. On the other hand, the automaton B is not minimal, as a constant function is recognized by a single state automaton.

2.2   Learning $\mathbb{Z}$-weighted tree automata

For simplicity we now restrict to the case $R = \mathbb{Z}$, but the following also holds for any PID that is sufficiently effective.

Fix a function $L \colon \mathcal{T}_0(\Sigma) \to \mathbb{Z}$ recognized by some finite-state $\mathbb{Z}$-weighted tree automaton. Proposition 5 gives us the existence of a minimal weighted tree automaton recognizing L, but does not tell us how to compute it. We therefore now focus on describing an algorithm that computes $\mathsf{Min}\, L$. While the results of Section 4 provide us with two such algorithms, namely a minimization and an active learning algorithm, we only describe the latter here.

In active learning, the goal is to compute $\mathsf{Min}\, L$ with the help of an oracle able to answer two kinds of queries – *evaluation* and *equivalence queries*:

- when queried with $\mathrm{EVAL}_L(t)$ for a tree $t \in \mathcal{T}_0(\Sigma)$, the oracle outputs the value $L(t)$;
- when queried with $\mathrm{EQUIV}_L(Q, \mu, \lambda)$ for a hypothesis Σ-tree automaton (Q, μ, λ), the oracle answers whether (Q, μ, λ) recognizes L, and if this is not the case also outputs a *counterexample* tree, that is, some $t \in \mathcal{T}_0(\Sigma)$ such that $\mu_t \lambda \neq L(t)$.

The idea behind an active learning algorithm for weighted tree automata is to incrementally build two finite sets $S \subseteq \mathcal{T}_0(\Sigma)$ and $C \subseteq \mathcal{T}_1(\Sigma)$, and, once some *closure* and *consistency* conditions are satisfied, to use the partial knowledge of

L on $C \circ S = \{c \circ s \mid c \in C, s \in S\} \subseteq \mathcal{T}_0(\Sigma)$, acquired through $\textsc{Eval}_L$ queries, to build a hypothesis weighted tree automaton $(Q^{hyp}, \mu^{hyp}, \lambda^{hyp})$. This automaton is then fed to the oracle through an $\textsc{Equiv}_L$ query: if it accepts L then it is the minimal such automaton, and otherwise the counterexample tree t and all its children are added to S, starting over.

In practice the partial knowledge of L on $C \circ S$ is represented as a matrix, called the *Hankel matrix* in the weighted automata literature, see, for example, [34]:

Definition 7. *Given* $S \subseteq \mathcal{T}_0(\Sigma)$ *and* $C \subseteq \mathcal{T}_1(\Sigma)$, *the* partial Hankel matrix $H^{S,C} \in \mathbb{Z}^{S \times C}$ *with* $|S|$ *rows and* $|C|$ *columns has for* (s,c)-*entry (where* $s \in S$ *and* $c \in C$) *the value* $H^{S,C}_{s,c} = L(c \circ s)$. *The* full Hankel matrix *is the (infinite) matrix* $H = H^{\mathcal{T}_0(\Sigma), \mathcal{T}_1(\Sigma)}$.

Given a pair (S, C) with $S \subseteq \mathcal{T}_0(\Sigma)$ and $C \subseteq \mathcal{T}_1(\Sigma)$ both finite, let $r^{S,C}$ be the rank of the matrix $H^{S,C}$. Section 3.2 then requires the existence of a so-called category-theoretical factorization system: here, this is witnessed by the fact that $H^{S,C}$ can be effectively (for instance through the Smith normal form [15, Section 2.4]) factored into

$$H^{S,C} = U^{S,C} \left(\begin{array}{c|c} I_{r^{S,c}} & 0 \\ \hline 0 & 0 \end{array} \right) V^{S,C}$$

where $I_{r^{s,c}}$ denotes the identity matrix with $r^{S,C}$ rows and columns, and where $U^{S,C} \in \mathbb{Z}^{S \times S}$ is unimodular (that is, it has an inverse $\left(U^{S,C}\right)^{-1} \in \mathbb{Z}^{S \times S}$) and $V^{S,C} \in \mathbb{Z}^{C \times C}$ has non-zero determinant (and thus has a $\mathbb{Q}$-valued inverse $\left(V^{S,C}\right)^{-1} \in \mathbb{Q}^{C \times C}$).

In particular, the first $r^{S,C}$ rows $v^{S,C}_{1,\cdot}, \ldots, v^{S,C}_{r^{S,C},\cdot}$ of $V^{S,C}$ form a basis of the $\mathbb{Z}$-module generated by the rows of $H^{S,C}$. We say that the pair (S, C) is *closed* if, for every $\sigma \in \Sigma_n$ and $\mathbf{t} \in S^n$, the row vector $(L(c \circ \sigma \circ \mathbf{t}))_{c \in C}$ can also be written as a linear combination (with coefficients in $\mathbb{Z}$) of the $v^{S,C}_{i,\cdot}, 1 \leq i \leq r^{S,C}$. Informarlly, when building a hypothesis automaton $(Q^{hyp}, \mu^{hyp}, \lambda^{hyp})$, we would like its module of configurations $\mathbb{Z}^{Q^{hyp}}$ to be the module generated by the rows of $H^{S,C}$: closure ensures that σ-transitions can be constructed on this module so that they loop back into it.

Dually, the first $r^{S,C}$ columns $u^{S,C}_{\cdot,1}, \ldots, u^{S,C}_{\cdot,r^{S,C}}$ of $U^{S,C}$ form a basis of the $\mathbb{Q}$-vector space generated by the columns of $H^{S,C}$. We say that the pair (S, C) is *consistent* if, for every $c \in C$, $\sigma \in \Sigma_n$ with $n \geq 1$ and for every $\mathbf{t} \in S^n$ and $1 \leq i \leq n$, the column vector $(L(c \circ \sigma \circ \mathbf{t}[i := s]))_{s \in S}$ can also be written as a linear combination with coefficients in $\mathbb{Q}$ of the $u^{S,C}_{\cdot,i}$. Informally, when building a hypothesis automaton, consistency ensures that following a σ-transition does not distinguish any more states, i.e. that the choice of the σ-transition itself will be consistent.

The learning algorithm then formally proceeds as follows: it starts with the pair $S = \Sigma_0$ and $C = \{\square\}$. While there exist $\sigma \in \Sigma_n$ and $\mathbf{s} \in S^n$ witnessing that (S, C) is not closed, $\sigma \circ \mathbf{s}$ is added to S. Similarly, while there exist $c \in C$, $\sigma \in \Sigma_n$ with $n \geq 1$, $\mathbf{t} \in S^n$ and $1 \leq i \leq n$ witnessing that (S, C) is not consistent, $c \circ \sigma \circ \mathbf{t}[i := \square]$ is added to C. Once the pair (S, C) is both closed and consistent, a hypothesis $\mathbb{Z}$-weighted tree automaton is built (we do not give the exact formulæ so here for the sake of readability). After an EQUIV_L query, if the oracle replies that this automaton recognizes L then the algorithm stops, otherwise the counterexample t and all its children are added to S and the algorithm starts over.

Theorem 8 (instance of Theorem 32). *The algorithm described above computes* Min L. *With respect to the number of states of* Min L, *the number of equivalence queries is linear.*

Example 9. Let us showcase a run of the algorithm when learning the minimal automaton A of Example 2. We start with $S = \Sigma_0 = \{l\}$ and $C = \{\Box\}$. The partial Hankel matrix is $H^{S,C} = (1)$. Every element of $\mathbb{Z}$ is a multiple of 1, hence the pair (S, C) is both closed and consistent: the algorithms submits the automaton with a single state and matrices $\mu_l = \mu_f = \lambda = (1)$ to the oracle for an equivalence query, and the oracle replies with the counterexample tree $n(l, l)$: the corresponding value should be 2, not 1.

Now $S = \{l, n(l, l)\}$, the partial Hankel matrices $H^{S,C}$ and $H^{S,C\cup\{n(\Box,l)\}}$ are the matrices

$$
\begin{array}{cc}
 & \Box \\
\begin{array}{c} l \\ n(l,l) \end{array} & \begin{pmatrix} 1 \\ 2 \end{pmatrix}
\end{array}
\qquad
\begin{array}{cc}
 & \Box \quad n(\Box,l) \\
\begin{array}{c} l \\ n(l,l) \end{array} & \begin{pmatrix} 1 & 2 \\ 2 & 2 \end{pmatrix}
\end{array}
= \begin{pmatrix} 2 & 1 \\ 1 & 1 \end{pmatrix} \begin{pmatrix} 1 & 0 \\ 0 & 1 \end{pmatrix} \begin{pmatrix} -1 & 0 \\ 3 & 2 \end{pmatrix}
$$

The second column of $H^{S,C\cup\{n(\Box,l)\}}$ is not a multiple of its first column, so the pair (S, C) is not consistent and $n(\Box, l)$ is added to C. It is now possible to check that (S, C) is closed and consistent again, and we build the hypothesis automaton with two states and matrices

$$
\mu_l = (2\ 1) \qquad \mu_n = \left(\begin{pmatrix} 2 & 1 \\ 1 & 1 \end{pmatrix}^{-1} \right)^{\otimes 2} \begin{pmatrix} 2 & 2 \\ 2 & 4 \\ 2 & 4 \\ 4 & 4 \end{pmatrix} \begin{pmatrix} -1 & 0 \\ 3 & 2 \end{pmatrix}^{-1} = \begin{pmatrix} -5 & -1 \\ 7 & 1 \\ 7 & 1 \\ -7 & -1 \end{pmatrix} \qquad \lambda = \begin{pmatrix} -1 \\ 3 \end{pmatrix}
$$

This new hypothesis automaton recognizes the target language, and is thus accepted by the oracle and returned by the algorithm. Note that it is not A, but another minimal automaton from which A can be retrieved by change of basis.

The algorithm above could be made more efficient by carefully choosing the order in which to add trees to S, as done in [13] for PID-weighted word automata. But we believe that the proof techniques used there cannot give anything better than an exponential complexity bound (for the whole algorithm, not just the number of EQUIV_L queries) for tree automata, the main difference being that the number of nodes in a tree is exponential in its height.

3   A monoidal categorical framework for tree automata minimization

In Section 2 we described tree automata weighted over PIDs, the corresponding minimal tree automata and a learning algorithm to construct these. Proving the correctness of this algorithm could be achieved by adapting the existing proof for tree automata weighted over fields, but, while certainly doable, this would be very tedious. Instead, we now describe a categorical framework that encompasses various families of tree automata – notably, tree automata over PIDs – and define minimal automata within this framework. We will then use this to describe a generic learning algorithm in Section 4.

3.1   Tree automata in monoidal categories

We model tree automata using monoidal categories. A *monoidal category* is a category $\mathcal{C}$ equipped with a notion of tensor $\otimes_\mathcal{C}$ and unit object $1_\mathcal{C}$, a classical example of which is the category of vector spaces over some field $\mathbb{K}$ with the usual tensor product and with $\mathbb{K}$ as unit. Formally, $\otimes_\mathcal{C}$ is a *monoidal product* given as a functor $\otimes_\mathcal{C} : \mathcal{C} \times \mathcal{C} \to \mathcal{C}$ assumed to be associative up to some natural isomorphism called the *associator*. The unit object $1_\mathcal{C} \in \mathsf{Ob}(\mathcal{C})$ is acting as a unit for $\otimes_\mathcal{C}$ up to natural isomorphisms called left and right *unitors*. These natural isomorphisms are subject to certain coherence axioms, see for instance [28, Chapter VII].
In this work we focus on four concrete examples of monoidal categories.

Example 10.   – (**Set**, ×, 1): The category **Set** of sets and functions with the cartesian product and any singleton set 1 as unit is a monoidal category.
  – (**Rel**, ×, 1): The category **Rel** of sets and relations is also equipped with a monoidal product given by the (set-wise) cartesian product and having a singleton set 1 as unit.
  – (**Mod**$_R$, ⊗, R): Given a ring R, the category **Mod**$_R$ of R-modules has a monoidal structure when equipped with the usual tensor product of modules, for which R – seen as a module over itself – is the unit object. Recall that, for R-modules X, Y and Z, linear maps $X \otimes Y \to Z$ are in one-to-one correspondence with bi-linear maps $X \times Y \to Z$.
  – (**JSL**, ⊗, 2): The category **JSL** has as objects complete join-semilattices (JSLs) – posets with arbitrary joins – and as morphisms join-preserving functions. Given JSLs X, Y and Z, a bi-morphism $f : X \times Y \to Z$ is a function that preserves joins in each variable separately. For every JSLs X and Y, there exists a JSL $X \otimes Y$ and a universal bi-morphism $u : X \times Y \to X \otimes Y$, such that any bi-morphism $f : X \times Y \to Z$ factors uniquely through u. $X \otimes Y$ is called the tensor product of X and Y, and can be concretely described as a quotient of the free JSL on $X \times Y$, see [5]. The free JSL on one element, $2 \cong \mathcal{P}(1)$, is a unit for this tensor, see for example [24, Chapter I.5, Prop. 2]

Given an object Q of a monoidal category $(\mathcal{C}, \otimes_\mathcal{C}, 1_\mathcal{C})$, let $Q^{\otimes_\mathcal{C} n}$ denote the n-fold tensor $Q \otimes_\mathcal{C} \cdots \otimes_\mathcal{C} Q$, with the convention that $Q^{\otimes_\mathcal{C} 0} = 1_\mathcal{C}$. We are now ready to introduce tree automata over a monoidal category.

Definition 11. *Let Σ be a signature, $(\mathcal{C}, \otimes_\mathcal{C}, 1_\mathcal{C})$ a monoidal category and Ω an object of $\mathcal{C}$. A Σ-tree $(\mathcal{C}, \Omega)$-automaton $\mathcal{A}$ is a tuple $(\mathcal{A}(\mathtt{st}), (\mathcal{A}(\sigma))_{\sigma \in \Sigma}, \mathcal{A}(\triangleleft))$:*
  – $\mathcal{A}(\mathtt{st})$ is an object of $\mathcal{C}$, called the state object,

- *for each $\sigma \in \Sigma$ of arity n, $\mathcal{A}(\sigma)\colon \mathcal{A}(\mathtt{st})^{\otimes_{\mathcal{C}} n} \to \mathcal{A}(\mathtt{st})$ is an n-ary* transition morphism,
- $\mathcal{A}(\lhd)\colon \mathcal{A}(\mathtt{st}) \to \Omega$ *is called the* termination morphism.

Given two Σ-tree $(\mathcal{C}, \Omega)$-automata $\mathcal{A}$ and $\mathcal{B}$, a morphism of automata $\varphi\colon \mathcal{A} \to \mathcal{B}$ *is a $\mathcal{C}$-morphism $\varphi\colon \mathcal{A}(\mathtt{st}) \to \mathcal{B}(\mathtt{st})$ which preserves transition and termination morphisms: $\mathcal{B}(\sigma) \circ \varphi^{\otimes_{\mathcal{C}} n} = \varphi \circ \mathcal{A}(\sigma)$ for all $\sigma \in \Sigma_n$ and $\mathcal{B}(\lhd) \circ \varphi = \mathcal{A}(\lhd)$.*

Example 12.   – Let $2 = \{\bot, \top\}$. A Σ-tree $(\mathbf{Set}, 2)$-automaton is given by a (possibly infinite) set $Q = \mathcal{A}(\mathtt{st})$ of *states*, a subset $\mathcal{A}(\lhd)^{-1}(\top)$ of *accepting states*, and *transition functions* $\mathcal{A}(\sigma)\colon Q^n \to Q$ for each $\sigma \in \Sigma_n$: Σ-tree $(\mathbf{Set}, 2)$-automata are deterministic bottom-up Σ-tree automata.
- Let $1 = \{*\}$. A Σ-tree $(\mathbf{Rel}, 1)$-automaton consists of a set $Q = \mathcal{A}(\mathtt{st})$ of *states*, a subset $\{q \in Q \mid (q, *) \in \mathcal{A}(\lhd)\}$ of *accepting states*, and *transition relations* $\mathcal{A}(\sigma) \subseteq Q^n \times Q$ for each $\sigma \in \Sigma$ of arity n: Σ-tree $(\mathbf{Rel}, 1)$-automata are non-deterministic bottom-up Σ-tree automata.
- Let $2 = \{\bot, \top\}$ be the standard JSL with $\bot \leq \top$. A Σ-tree $(\mathbf{JSL}, 2)$-automaton consists of a JSL $Q = \mathcal{A}(\mathtt{st})$ of states, an up-closed subset $\mathcal{A}(\lhd)^{-1}(\top)$ of *accepting states*, and for each $\sigma \in \Sigma$ of arity n, a *transition function* $\mathcal{A}(\sigma)\colon Q^n \to Q$ which is join-preserving in each variable. Notice that we have a monoidal functor $\mathbf{Rel} \to \mathbf{JSL}$ mapping a set X to the JSL $\mathcal{P}(X)$ (it is indeed monoidal by [24, Chapter I.5, Prop. 2]). Such a functor can be used to transform a Σ-tree $(\mathbf{Rel}, 1)$-automaton into a Σ-tree $(\mathbf{JSL}, 2)$-automaton. Contrary to non-deterministic automata, the latter can be minimized. Considering only the join-irreducible states of a minimal $(\mathbf{JSL}, 2)$-automaton yields the minimal *residual* automaton [14,31].
- A Σ-tree $(\mathbf{Vec}_{\mathbb{K}}, \mathbb{K})$-automaton has a $\mathbb{K}$-vector-space $\mathbb{K}\langle S \rangle = \mathcal{A}(\mathtt{st})$ of *configurations*, which are given as linear combinations of some *states* forming a basis S of $\mathcal{A}(\mathtt{st})$; the linear map $\mathcal{A}(\lhd)\colon \mathbb{K}\langle S \rangle \to \mathbb{K}$ is entirely determined by its restriction to S and thus specifies an output weight $\mathcal{A}(\lhd)(s)$ for each state $s \in S$; similarly for every $\sigma \in \Sigma$ of some arity n the transition n-linear map $\mathcal{A}(\sigma)\colon \mathbb{K}\langle S \rangle^{\otimes n} \to \mathbb{K}\langle S \rangle$ is entirely determined by its restriction to S^n. As such, a Σ-tree $(\mathbf{Vec}_{\mathbb{K}}, \mathbb{K})$-automaton determines a $\mathbb{K}$-weighted Σ-tree automaton. Similarly, Σ-tree $(\mathbf{Mod}_R, R)$-automata $\mathcal{A}$ such that $\mathcal{A}(\mathtt{st})$ is a free module are R-weighted Σ-tree automata in the sense of Definition 1: for $\Sigma = \{n, l\}$, the tree automaton A of Example 2 is given by the following data: $\mathcal{A}(\mathtt{st}) = R\langle \bullet, \bigcirc \rangle$, $\mathcal{A}(l) = \mu_l$, $\mathcal{A}(n) = \mu_n$, and $\mathcal{A}(\lhd) = \lambda$.

Construction 13 (runs). *Let $\mathcal{A}$ be a Σ-tree $(\mathcal{C}, \Omega)$-automaton. We can inductively define a $\mathcal{C}$-morphism $\mathcal{A}(t)\colon \mathcal{A}(\mathtt{st})^{\otimes_{\mathcal{C}} \mathrm{ar}(t)} \to \mathcal{A}(\mathtt{st})$ for any tree $t \in \mathcal{T}(\Sigma)$. Indeed, if $\sigma \in \Sigma_n$, we have by definition $\mathcal{A}(\sigma)\colon \mathcal{A}(\mathtt{st})^{\otimes_{\mathcal{C}} n} \to \mathcal{A}(\mathtt{st})$. For $\sigma \in \Sigma_n$ and trees $t_1, \ldots, t_n \in \mathcal{T}(\Sigma)$, we obtain $\mathcal{A}(\sigma(t_1, \ldots, t_n))$ as the composite*

$$\mathcal{A}(\mathtt{st})^{\otimes_{\mathcal{C}} \mathrm{ar}(t_1)} \otimes_{\mathcal{C}} \cdots \otimes_{\mathcal{C}} \mathcal{A}(\mathtt{st})^{\otimes_{\mathcal{C}} \mathrm{ar}(t_n)} \xrightarrow{\mathcal{A}(t_1) \otimes_{\mathcal{C}} \cdots \otimes_{\mathcal{C}} \mathcal{A}(t_n)} \mathcal{A}(\mathtt{st})^{\otimes_{\mathcal{C}} n} \xrightarrow{\mathcal{A}(\sigma)} \mathcal{A}(\mathtt{st}).$$

This construction leads to the definition of the language accepted by a tree automaton. Recall that $\mathcal{C}(1_{\mathcal{C}}, \Omega)$ denotes the set of $\mathcal{C}$-morphisms from $1_{\mathcal{C}}$ to Ω.

Definition 14. *A Σ-tree $(\mathcal{C}, \Omega)$-language is a function $\mathcal{L}\colon \mathcal{T}_0(\Sigma) \to \mathcal{C}(1_\mathcal{C}, \Omega)$. The* language recognized by an automaton $\mathcal{A}$ is the function $[\![\mathcal{A}]\!]\colon \mathcal{T}_0(\Sigma) \to \mathcal{C}(1_\mathcal{C}, \Omega)$ defined by $[\![\mathcal{A}]\!](t) = \mathcal{A}(\lhd) \circ \mathcal{A}(t)$ for all $t \in \mathcal{T}_0(\Sigma)$.
We write **Auto**$(\mathcal{L})$ *for the category of Σ-tree $(\mathcal{C}, \Omega)$-automata recognizing $\mathcal{L}$.*

Example 15. The language recognized by a Σ-tree $(\mathbf{Set}, 2)$-automaton $\mathcal{A}$ is a function $[\![\mathcal{A}]\!]\colon \mathcal{T}_0(\Sigma) \to \mathbf{Set}(1, 2) \cong 2$. The trees $t \in \mathcal{T}_0(\Sigma)$ such that $\mathcal{A}(\lhd) \circ \mathcal{A}(t) = \top$ form the tree language recognized by $\mathcal{A}$ in the usual sense. Σ-tree $(\mathbf{Rel}, 1)$- and $(\mathbf{JSL}, 2)$-automata also recognize tree languages, since a subset of $\mathcal{T}_0(\Sigma)$ can respectively be encoded as a function $\mathcal{T}_0(\Sigma) \to \mathbf{Rel}(1, 1) \cong 2$ or $\mathcal{T}_0(\Sigma) \to \mathbf{JSL}(2, 2) \cong 2$. Similarly, the language recognized by a Σ-tree $(\mathbf{Mod}_R, R)$-automaton $\mathcal{A}$ is a function $[\![\mathcal{A}]\!]\colon \mathcal{T}_0(\Sigma) \to \mathbf{Mod}_R(R, R) \cong R$, and thus coincides with the R-weighted Σ-tree language recognized by the corresponding R-weighted Σ-tree automaton.

3.2   Minimal automata

Given a Σ-tree $(\mathcal{C}, \Omega)$-language, we now seek to define a generic notion of *minimal Σ-tree $(\mathcal{C}, \Omega)$-automaton* recognizing $\mathcal{L}$, which we will denote $\mathcal{M}in(\mathcal{L})$. We begin by describing the appropriate state object for this minimal automaton, obtained by generalizing the MyhillNerode equivalence relation. Constructing a tree automaton structure on this object is not straightforward. As a workaround, we therefore introduce the notion of a context automaton by restricting attention to trees with at most one hole – that is, contexts and closed trees. For context automata, we can apply the category-theoretic minimization results of [16]. Finally, Theorem 26 shows that, under mild assumptions on the monoidal category, one can construct a minimal tree automaton whose restriction to contexts coincides exactly with the minimal context automaton.

Let us start by describing what $\mathcal{M}in(\mathcal{L})(\mathtt{st})$ should be. For deterministic tree automata, the state-set of the minimal automaton recognizing a language L is obtained as the set of Myhill-Nerode equivalence classes, i.e., the image of the function $\mathcal{T}_0(\Sigma) \to 2^{\mathcal{T}_1(\Sigma)}$ that sends a closed tree t to the Brzozowski derivative $\partial_t L = \{c \in \mathcal{T}_1(\Sigma) \mid c \circ t \in L\} \subseteq \mathcal{T}_1(\Sigma)$. The standard Hankel matrix used to define minimal weighted automata can be seen as a weighted version of the above map. These constructions are generalized as follows.

Assume (in the rest of this work) that $\mathcal{C}$ has countable coproducts[2] of $1_\mathcal{C}$ and countable products[3] of Ω. Given a set X, denote by abuse of notation $\coprod_X 1_\mathcal{C}$ by X and $\prod_X \Omega$ by Ω^X. With these notations, there is a $\mathcal{C}$-morphism

---

[2] meaning for any countable set X, there is an object $\coprod_X 1_\mathcal{C}$ and injections $(\kappa_x \colon 1_\mathcal{C} \to \coprod_X 1_\mathcal{C})_{x \in X}$ such that a morphism $f \colon \coprod_X 1_\mathcal{C} \to Y$ is determined by the familiy $(f \circ \kappa_x)_{x \in X}$ and every family $(f_x \colon 1_\mathcal{C} \to Y)_{x \in X}$ comes from a morphism written $[f_x]_{x \in X} \colon \coprod_X 1_\mathcal{C} \to Y$

[3] meaning for any countable set X, there is an object $\prod_X \Omega$ and projections $(\pi_x \colon \prod_X \Omega \to \Omega)_{x \in X}$ such that a morphism $f \colon Y \to \prod_X \Omega$ is determined by the familiy $(\pi_x \circ f)_{x \in X}$ and every family $(f_x \colon Y \to \Omega)_{x \in X}$ comes from a morphism written $\langle f_x \rangle_{x \in X} \colon Y \to \prod_X \Omega$

$\partial\mathcal{L}\colon \mathcal{T}_0(\Sigma) \to \Omega^{\mathcal{T}_1(\Sigma)}$, formally $\partial\mathcal{L}\colon \coprod_{\mathcal{T}_0(\Sigma)} 1_{\mathcal{C}} \to \prod_{\mathcal{T}_1(\Sigma)} \Omega$, uniquely defined by having its t-indexed component (for $t \in \mathcal{T}_0(\Sigma)$) be the pairing of the $\mathcal{C}$-morphisms $\mathcal{L}(c \circ t)\colon 1_{\mathcal{C}} \to \Omega$ with c ranging over $\mathcal{T}_1(\Sigma)$. We now recall how to generalize the notion of image; the image of $\partial\mathcal{L}$ shall then be $\mathcal{M}in(\mathcal{L})(\mathtt{st})$.

A *factorization system* in a category $\mathcal{C}$ is the data $(\mathcal{E}, \mathcal{M})$ of two classes $\mathcal{E}$ and $\mathcal{M}$ of $\mathcal{C}$-morphisms (respectively represented with $\twoheadrightarrow$ and $\hookrightarrow$), both closed under composition and containing all isomorphisms, and such that each $f\colon X \to Y$ in $\mathcal{C}$ has, up to a unique isomorphism, a unique factorization $f = m \circ e$ with $e\colon X \twoheadrightarrow \operatorname{im}(f)$ in $\mathcal{E}$ and $m\colon \operatorname{im}(f) \rightarrowtail Y$ in $\mathcal{M}$. We call $\operatorname{im}(f)$ the $(\mathcal{E}, \mathcal{M})$-*image* of f.

Example 16. In the category **Set** there is a factorization system $(\mathrm{Surj}, \mathrm{Inj})$ consisting of surjections and injections: a function $f\colon X \to Y$ factors through its image $f(X) \subseteq Y$. Similarly, R-linear surjections and injections form a factorization system in the category $\mathbf{Mod}_R$, and join-preserving surjections and injections form a factorization system in the category **JSL**: we also write these factorization systems $(\mathrm{Surj}, \mathrm{Inj})$.

We now fix a factorization system $(\mathcal{E}, \mathcal{M})$ on $\mathcal{C}$. A natural candidate for the state-object of the minimal automaton recognizing $\mathcal{L}$ is the $(\mathcal{E}, \mathcal{M})$-image of the morphism $\partial\mathcal{L}\colon \mathcal{T}_0(\Sigma) \to \Omega^{\mathcal{T}_1(\Sigma)}$: we call it $\mathcal{M}in(\mathcal{L})(\mathtt{st})$. The next step is then to actually equip $\mathcal{M}in(\mathcal{L})(\mathtt{st})$ with the structure of a Σ-tree $(\mathcal{C}, \Omega)$-automaton recognizing $\mathcal{L}$, i.e., to define $\mathcal{M}in(\mathcal{L})(\triangleleft)$ as well as $\mathcal{M}in(\mathcal{L})(\sigma)$ for every $\sigma \in \Sigma$ so that $[\![\mathcal{M}in(\mathcal{L})]\!] = \mathcal{L}$.

Defining such an action of all symbols $\sigma \in \Sigma$ – and consequently, by Construction 13, of all trees $t \in \mathcal{T}(\Sigma)$ with any number of holes –, and doing so at this level of generality, will require some additional assumptions on $\mathcal{C}$. As an intermediate step toward this goal, we can first restrict ourselves to only handle trees with at most one hole – contexts and closed trees. This is easier because there is no need to handle the monoidal structure of $\mathcal{C}$ anymore, and so we may re-use of-the-shelf results on minimization in a category [16]. This intermediate step will be instrumental in designing the learning algorithm in Section 4.

Formally, this action of trees with at most one hole is defined as follows:

Definition 17. *Let Σ be a signature, $(\mathcal{C}, \otimes_{\mathcal{C}}, 1_{\mathcal{C}})$ a monoidal category and Ω an object of $\mathcal{C}$. A Σ-context $(\mathcal{C}, \Omega)$-automaton $\mathfrak{A}$ is a tuple $(\mathfrak{A}(\mathtt{st}), (\mathfrak{A}(c))_{c \in \mathcal{T}_1(\Sigma)}, (\mathfrak{A}(\sigma))_{\sigma \in \Sigma_0}, \mathfrak{A}(\triangleleft))$ where*

- *$\mathfrak{A}(\mathtt{st})$ is an object of $\mathcal{C}$,*   – *$\mathfrak{A}(c)\colon \mathfrak{A}(\mathtt{st}) \to \mathfrak{A}(\mathtt{st})$ for every context c,*
- *$\mathfrak{A}(\triangleleft)\colon \mathfrak{A}(\mathtt{st}) \to \Omega$,*   – *$\mathfrak{A}(\sigma)\colon 1_{\mathcal{C}} \to \mathfrak{A}(\mathtt{st})$ for every constant σ,*

and so that $\mathfrak{A}(\square) = \operatorname{id}_{\mathfrak{A}(\mathtt{st})}$, $\mathfrak{A}(c \circ c') = \mathfrak{A}(c) \circ \mathfrak{A}(c')$ and, if $c \circ \sigma = c' \circ \sigma'$, $\mathfrak{A}(c) \circ \mathfrak{A}(\sigma) = \mathfrak{A}(c') \circ \mathfrak{A}(\sigma')$.

A morphism of Σ-context $(\mathcal{C}, \Omega)$-automata $\varphi\colon \mathfrak{A} \to \mathfrak{B}$ is a $\mathcal{C}$-morphism $\varphi\colon \mathfrak{A}(\mathtt{st}) \to \mathfrak{B}(\mathtt{st})$ compatible with the actions of contexts, constants and the termination morphism: $\varphi \circ \mathfrak{A}(\sigma) = \mathfrak{B}(\sigma)$ for all $\sigma \in \Sigma_0$, $\varphi \circ \mathfrak{A}(c) = \mathfrak{B}(c) \circ \varphi$ for all $c \in \mathcal{T}_1(\Sigma)$ and $\mathfrak{A}(\triangleleft) = \mathfrak{B}(\triangleleft) \circ \varphi$.

Every tree $t \in \mathcal{T}_0(\Sigma)$ can be written as $t = c \circ \sigma$ for some $c \in \mathcal{T}_1(\Sigma)$ and $\sigma \in \Sigma_0$. For a context automaton $\mathfrak{A}$, we can thus define $\mathfrak{A}(t)\colon 1_{\mathcal{C}} \to \mathfrak{A}(\mathtt{st})$ as the composite $\mathfrak{A}(c) \circ \mathfrak{A}(\sigma)$, and by assumption this does not depend on the choice of c and σ. This allows us to define the language accepted by $\mathfrak{A}$.

Definition 18. *A Σ-context $(\mathcal{C}, \Omega)$-automaton accepts the Σ-tree $(\mathcal{C}, \Omega)$-language $[\![\mathfrak{A}]\!] \colon \mathcal{T}_0(\Sigma) \to \mathcal{C}(1_{\mathcal{C}}, \Omega)$ defined by $[\![\mathfrak{A}]\!](t) = \mathfrak{A}(\lhd) \circ \mathfrak{A}(t)$.*
Given a Σ-tree $(\mathcal{C}, \Omega)$-language $\mathcal{L}$, $\mathbf{CtxAut}(\mathcal{L})$ denotes the category of those Σ-context $(\mathcal{C}, \Omega)$-automata recognizing $\mathcal{L}$.

Of course, restricting a Σ-tree $(\mathcal{C}, \Omega)$-automaton $\mathcal{A}$ to its action of contexts yields a Σ-context $(\mathcal{C}, \Omega)$-automaton $\mathfrak{Ctr}(\mathcal{A})$ recognizing the same language. Any example of tree automaton thus yields an example of context automaton.

To equip $\mathcal{M}in(\mathcal{L})(\mathbf{st})$ with the structure of a context automaton recognizing $\mathcal{L}$, we use the classical result that factorization systems lift to categories of functors (here $\mathbf{CtxAut}(\mathcal{L})$) [16, Lemma 2.8]: if we identify $\partial\mathcal{L}$ as a morphism of context automata, then its image, $\mathcal{M}in(\mathcal{L})(\mathbf{st})$, itself canonically inherits the structure of a context automaton recognizing $\mathcal{L}$.

To make $\partial\mathcal{L} \colon \mathcal{T}_0(\Sigma) \to \Omega^{\mathcal{T}_1(\Sigma)}$ into a morphism of context automata, we equip both $\mathcal{T}_0(\Sigma) = \coprod_{\mathcal{T}_0(\Sigma)} 1_{\mathcal{C}}$ and $\Omega^{\mathcal{T}_1(\Sigma)} = \prod_{\mathcal{T}_1(\Sigma)} \Omega$ with the structure of a context automaton. Intuitively, these two context automata generalize how contexts act by composition respectively on the set of closed trees and on the set of contexts. We call them $\mathfrak{A}^{init}(\mathcal{L})$ and $\mathfrak{A}^{final}(\mathcal{L})$. That $\partial\mathcal{L} \colon \mathfrak{A}^{init}(\mathcal{L}) \to \mathfrak{A}^{final}(\mathcal{L})$ is then indeed a morphism of context automata generalizes how the Myhill-Nerode equivalence relation is a *congruence*. We call $\mathfrak{Min}(\mathcal{L})$ the image context automaton, which satisfies $\mathfrak{Min}(\mathcal{L})(\mathbf{st}) = \mathcal{M}in(\mathcal{L})(\mathbf{st})$. $\mathfrak{Min}(\mathcal{L})$ is minimal as a context automaton recognizing $\mathcal{L}$ in the following sense:

Lemma 19. *For all Σ-context $(\mathcal{C}, \Omega)$-automaton $\mathfrak{A}$, there are spans and cospans of morphisms of context automata $\mathfrak{A} \leftarrowtail \cdot \twoheadrightarrow \mathfrak{Min}([\![\mathfrak{A}]\!])$ and $\mathfrak{A} \twoheadrightarrow \cdot \leftarrowtail \mathfrak{Min}([\![\mathfrak{A}]\!])$.*

Example 20. In **Set**, if there is a span $X \leftarrowtail \cdot \twoheadrightarrow Y$ or a cospan $X \twoheadrightarrow \cdot \leftarrowtail Y$, then $|Y| \le |X|$ since surjections decrease cardinals and injections increase them. The same situation in $\mathbf{Vec}_{\mathbb{K}}$ means that $\dim Y \le \dim X$, since surjections decrease dimensions and injections increase them. In **JSL** the situation is more complicated to describe, but it should be noted that while injections increase the number of elements of a JSL, surjections decrease them but also decrease the number of join-irreducible elements.

Ideally, to now extend the context automaton structure on $\mathfrak{Min}(\mathcal{L})(\mathbf{st})$ to a tree automaton structure, we would first extend the context automata structures on $\mathfrak{A}^{init}(\mathcal{L})$ and $\mathfrak{A}^{final}(\mathcal{L})$ to tree automata structures and show that $\partial\mathcal{L}$ is a morphism of tree automata. This unfortunately cannot work as there is no obvious tree automaton structure to equip $\mathfrak{A}^{final}(\mathcal{L})(\mathbf{st}) = \Omega^{\mathcal{T}_1(\Sigma)}$ with. There is one on $\mathfrak{A}^{init}(\mathcal{L})(\mathbf{st}) = \mathcal{T}_0(\Sigma)$ given further mild assumptions on $\mathcal{C}$:

Definition 21. *A monoidal category $(\mathcal{C}, \otimes_{\mathcal{C}}, 1_{\mathcal{C}})$ with all countable coproducts is called* distributive *when for all countable families $(X_i)_{i \in I}$ and $(Y_j)_{j \in J}$ there is a natural isomorphism $\coprod_{i \in I} X_i \otimes_{\mathcal{C}} \coprod_{j \in J} Y_j \cong \coprod_{i,j \in I \times J} X_i \otimes_{\mathcal{C}} Y_j$.*

Proposition 22. *Fix a countable signature Σ, a monoidal category $(\mathcal{C}, \otimes_\mathcal{C}, 1_\mathcal{C})$ and a Σ-tree $(\mathcal{C}, \Omega)$-language $\mathcal{L}$. Assume that $\mathcal{C}$ has all countable coproducts, so that the initial Σ-context $(\mathcal{C}, \Omega)$-automaton $\mathfrak{A}^{init}(\mathcal{L})$ recognizing $\mathcal{L}$ exists.*

If $\mathcal{C}$ is moreover distributive, there is a unique Σ-tree $(\mathcal{C}, \Omega)$-automaton $\mathcal{A}^{init}(\mathcal{L})$ recognizing $\mathcal{L}$ such that $\mathfrak{Ctx}(\mathcal{A}^{init}(\mathcal{L})) = \mathfrak{A}^{init}(\mathcal{L})$.

Proof (sketch). We write $\mathcal{A}^{init}$ and $\mathfrak{A}^{init}$ for $\mathcal{A}^{init}(\mathcal{L})$ and $\mathfrak{A}^{init}(\mathcal{L})$, respectively. $\mathcal{A}^{init}(\mathtt{st})$, $\mathcal{A}^{init}(\sigma)$ for $\sigma \in \Sigma_0$, $\mathcal{A}^{init}(c)$ for $c \in \mathcal{T}_1(\Sigma)$ and $\mathcal{A}^{init}(\triangleleft)$ are forced to be those inherited from $\mathfrak{A}^{init}$. For $\mathcal{A}^{init}(\sigma)$ where σ has arity $n \geq 2$, say 2, we need a morphism $\coprod_{\mathcal{T}_0(\Sigma)} 1_\mathcal{C} \otimes_\mathcal{C} \coprod_{t \in \mathcal{T}_0(\Sigma)} 1_\mathcal{C} \to \coprod_{\mathcal{T}_0(\Sigma)} 1_\mathcal{C}$. The distributivity of the monoidal category makes this choice very natural: since $\coprod_{\mathcal{T}_0(\Sigma)^2} 1_\mathcal{C} \cong \coprod_{\mathcal{T}_0(\Sigma)} 1_\mathcal{C} \otimes_\mathcal{C} \coprod_{\mathcal{T}_0(\Sigma)} 1_\mathcal{C}$, we set $\mathcal{A}^{init}(\sigma)\colon \coprod_{\mathcal{T}_0(\Sigma)^2} 1_\mathcal{C} \to \coprod_{\mathcal{T}_0(\Sigma)} 1_\mathcal{C}$ to have the (t_1, t_2)-indexed component be sent to the $(\sigma \circ (t_1, t_2))$-indexed one. $\qquad\square$

Remark 23. In the functorial approach to automata [16], initial automata are computed as left Kan extensions. Observing that Σ-tree $(\mathcal{C}, \Omega)$-automata can be seen as strongly monoidal functors, the initial tree automaton is a left Kan extension of strongly monoidal functors, which exists because the left Kan extension of the underlying functors happens to be strongly monoidal. It is always true that the left Kan extension of strong monoidal functors from a PRO exists [29]. This left Kan extension is also known as the initial operad algebra [21].

It turns out that the tree automaton structure on $\mathfrak{A}^{init}(\mathcal{L})$ is enough to define one on $\mathfrak{Min}(\mathcal{L})$ if we assume an additional condition on the factorization system $(\mathcal{E}, \mathcal{M})$. This same condition appears in [1], where it is used to show the existence of syntactic monoids in categories of algebras.

Definition 24. *Let $(\mathcal{C}, \otimes_\mathcal{C}, 1_\mathcal{C})$ be a monoidal category with a factorization system $(\mathcal{E}, \mathcal{M})$. $\mathcal{E}$ is said to be $\otimes_\mathcal{C}$-compatible when for any two $\mathcal{E}$-morphisms $e_1\colon X_1 \twoheadrightarrow Y_1$ and $e_2\colon X_2 \twoheadrightarrow Y_2$, $e_1 \otimes_\mathcal{C} e_2 \in \mathcal{E}$ and $(e_1 \otimes_\mathcal{C} \mathrm{id}) \circ (\mathrm{id} \otimes_\mathcal{C} e_2) = (\mathrm{id} \otimes_\mathcal{C} e_2) \circ (e_1 \otimes_\mathcal{C} \mathrm{id})$ is a pushout square[4].*

Remark 25. The $\otimes_\mathcal{C}$-compatibility of $\mathcal{E}$ is rather mild assumption, and is satisfied for example by the categories of Eilenberg-Moore algebras for a commutative monad on **Set**. In particular, **Set**, $\mathbf{Vec}_\mathbb{K}$, $\mathbf{Mod}_R$ and **JSL** satisfy these assumptions.

Theorem 26. *Let $(\mathcal{C}, \otimes_\mathcal{C}, 1_\mathcal{C})$ be a distributive monoidal category with countable products and coproducts and equipped with a factorization system $(\mathcal{E}, \mathcal{M})$ so that $\mathcal{E}$ is $\otimes_\mathcal{C}$-compatible. Let $\mathcal{L}$ be a Σ-tree $(\mathcal{C}, \Omega)$-language. Then, there is a unique Σ-tree $\mathcal{C}$-automaton $\mathcal{Min}(\mathcal{L})$ recognizing $\mathcal{L}$ and such that $\mathfrak{Ctx}(\mathcal{Min}(\mathcal{L})) = \mathfrak{Min}(\mathcal{L})$.*

Proof (sketch). For simplicity, write $\mathcal{Min}$, $\mathfrak{Min}$ and $\mathcal{A}^{init}$ for $\mathcal{Min}(\mathcal{L})$, $\mathfrak{Min}(\mathcal{L})$ and $\mathcal{A}^{init}(\mathcal{L})$, respectively. First, $\mathcal{Min}(\mathtt{st})$ is forced to be $\mathfrak{Min}(\mathtt{st})$, and we have

---

[4] meaning, for any $f_1\colon Y_1 \otimes_\mathcal{C} X_2 \to Z$ and $f_2\colon X_1 \otimes_\mathcal{C} Y_2 \to Z$ such that $f_1 \circ (e_1 \otimes_\mathcal{C} \mathrm{id}) = f_2 \circ (\mathrm{id} \otimes_\mathcal{C} e_2)$, there is a unique $h\colon Y_1 \otimes Y_2 \to Z$ such that $f_1 = h \circ (\mathrm{id} \otimes_\mathcal{C} e_2)$ and $f_2 = h \circ (e_1 \otimes_\mathcal{C} \mathrm{id})$

an $\mathcal{E}$-morphism $\mathcal{A}^{init}(\mathtt{st}) \twoheadrightarrow \mathcal{M}\mathrm{in}(\mathtt{st})$. Given, for instance, a symbol $\sigma \in \Sigma_2$ of arity 2, we now want to build $\mathcal{M}\mathrm{in}(\sigma)\colon \mathcal{M}\mathrm{in}(\mathtt{st}) \otimes_{\mathcal{C}} \mathcal{M}\mathrm{in}(\mathtt{st}) \to \mathcal{M}\mathrm{in}(\mathtt{st})$.

A priori, there is no obvious way to do that. However, $\mathcal{M}\mathrm{in}$ should agree with $\mathfrak{Min}$ on unary contexts, and this determines its values on contexts of the form $\sigma \circ (t, \square)$ and $\sigma \circ (\square, t)$ for $t \in \mathcal{T}_0(\Sigma)$. The corresponding two morphisms $\mathcal{M}\mathrm{in}(\mathtt{st}) \otimes_{\mathcal{C}} \mathcal{A}^{init}(\mathtt{st}) \to \mathcal{M}\mathrm{in}(\mathtt{st})$ and $\mathcal{A}^{init}(\mathtt{st}) \otimes_{\mathcal{C}} \mathcal{M}\mathrm{in}(\mathtt{st}) \to \mathcal{M}\mathrm{in}(\mathtt{st})$ sit on the right of the diagram

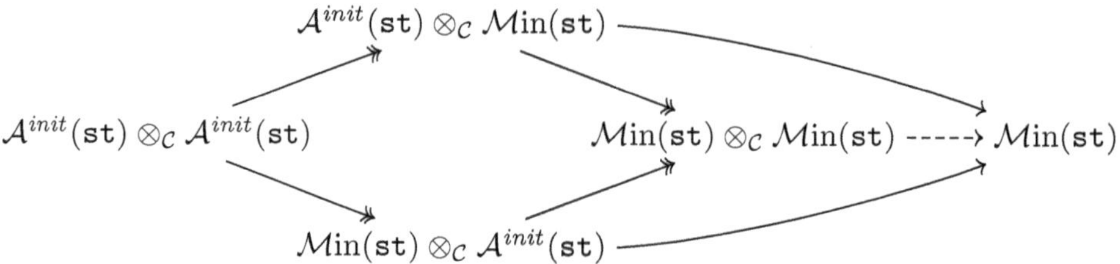

and can be thus be aggregated, using $\otimes_{\mathcal{C}}$-compatibility, to obtain the desired morphism $\mathcal{M}\mathrm{in}(\mathtt{st}) \otimes_{\mathcal{C}} \mathcal{M}\mathrm{in}(\mathtt{st}) \to \mathcal{M}\mathrm{in}(\mathtt{st})$. $\qquad\square$

When it exists, $\mathcal{M}\mathrm{in}(\mathcal{L})$ is minimal amongst all automata recognizing $\mathcal{L}$ in the sense that its associated context automaton satisfies Lemma 19 – which as we saw in Example 20 usually entails minimality of the size of its carrier object. Since this carrier object is the same as the state-object of $\mathcal{M}\mathrm{in}(\mathcal{L})$, $\mathcal{M}\mathrm{in}(\mathcal{L})$ has the smallest state-object amongst all automata recognizing $\mathcal{L}$.

Example 27. The minimal Σ-tree $(\mathbf{Set}, 2)$-automaton is the minimal deterministic bottom-up tree automaton [18], the minimal $(\mathbf{JSL}, 2)$-automaton corresponds to the minimal residual tree automaton [14] and the minimal $(\mathbf{Vec}_{\mathbb{K}}, \mathbb{K})$-automaton is the minimal multiplicity tree automaton [27]. A new example is the minimal $(\mathbf{Mod}_R, R)$-automaton described in Proposition 5.

4   A learning algorithm for tree automata

In Section 3 we showed that the minimal Σ-tree $(\mathcal{C}, \Omega)$-automaton recognizing a Σ-tree $(\mathcal{C}, \Omega)$-language exists under reasonable conditions. Its state space was constructed as the factorization of a morphism between two infinite Σ-context $(\mathcal{C}, \Omega)$-automata, i.e., in a non-computable way. In this section we describe a generic learning algorithm – parametric in the category $\mathcal{C}$ – that outputs the minimal tree automaton for a given language. In this setting we do not have access to a prior Σ-tree $(\mathcal{C}, \Omega)$-automaton recognizing $\mathcal{L}$ anymore, but instead to an oracle able to answer two kinds of queries:

- when queried with $\mathrm{Eval}_{\mathcal{L}}(t)$ for a tree $t \in \mathcal{T}_0(\Sigma)$, the oracle outputs the value $\mathcal{L}(t)$;
- when queried with $\mathrm{Equiv}_{\mathcal{L}}(\mathcal{H})$ for a hypothesis Σ-tree $\mathcal{C}$-automaton $\mathcal{H}$, the oracle answers whether $\mathcal{H}$ recognizes $\mathcal{L}$, and if this is not the case also outputs a counter-example tree $t \in \mathcal{T}_0(\Sigma)$ such that $[\![\mathcal{H}]\!](t) \neq \mathcal{L}(t)$.

The goal of active learning algorithms is to build $\mathcal{Min}(\mathcal{L})$ by querying this oracle.

From now on, fix a Σ-tree $(\mathcal{C}, \Omega)$-language $\mathcal{L}$. Recall that Theorem 26 ensures the existence of the minimal Σ-tree $(\mathcal{C}, \Omega)$-automaton recognizing $\mathcal{L}$, whose state-object $\mathcal{Min}(\mathcal{L})(\mathtt{st})$ is obtained as the $(\mathcal{E}, \mathcal{M})$-image of the unique morphism $\partial\mathcal{L}\colon \coprod_{\mathcal{T}_0(\Sigma)} 1_{\mathcal{C}} \to \prod_{\mathcal{T}_1(\Sigma)} \Omega$, in which the product and coproduct are indexed by *infinite* families of trees and contexts. Formally, $\partial\mathcal{L}$ is given as $\left[\langle \mathcal{L}(c \circ s)\rangle_{c \in \mathcal{T}_1(\Sigma)}\right]_{s \in \mathcal{T}_0(\Sigma)}$.

Instead, the algorithm incrementally builds two *finite* sets $S \subseteq \mathcal{T}_0(\Sigma)$ and $C \subseteq \mathcal{T}_1(\Sigma)$, and keeps track of the restriction of $\mathcal{L}$ to $C \circ S$, formally given as the morphism $\partial_S^C \mathcal{L} = \left[\langle \mathcal{L}(c \circ s)\rangle_{c \in C}\right]_{s \in S}\colon \coprod_S 1_{\mathcal{C}} \to \prod_C \Omega$. In concrete instances of the algorithm, this morphism is encoded as a table of size $S \times C$, whose (s, c)-indexed entry is the value $\mathcal{L}(c \circ s)$ – obtained via an evaluation query.

In keeping track of the table $\partial_S^C \mathcal{L}$, which approximates the morphism $\partial\mathcal{L}$, the algorithm therefore also keeps track of successive *finite* approximations of the Myhill-Nerode congruence: while the latter is encoded as the factorization of $\partial\mathcal{L}$, the former are encoded as the factorizations of the morphisms $\partial_S^C \mathcal{L}$. To improve readability we allow ourselves the following abuse of notation:

Notation 28. *Given a signature Σ, $S \subseteq \mathcal{T}_0(\Sigma)$, and $C \subseteq \mathcal{T}_1(\Sigma)$,*
  - *recall that S also denotes the coproduct $\coprod_S 1_{\mathcal{C}}$, and Ω^C the product $\prod_C \Omega$;*
  - *write S/Ω^C for the factorization of the map $\partial_S^C \mathcal{L}\colon S \to \Omega^C$.*

With these notations, $\mathcal{Min}(\mathcal{L})(\mathtt{st})$ is the factorization $\mathcal{T}_0(\Sigma)/\Omega^{\mathcal{T}_1(\Sigma)}$ and the approximations S/Ω^C fit in the following diagram:

$$
\begin{array}{ccccc}
S & \longrightarrow\!\!\!\!\! & S/\Omega^C & \rightarrowtail & \Omega^C \\
\downarrow & & & & \uparrow \\
\mathcal{T}_0(\Sigma) & \longrightarrow\!\!\!\!\! & \mathcal{Min}(\mathcal{L})(\mathtt{st}) & \rightarrowtail & \Omega^{\mathcal{T}_1(\Sigma)}
\end{array}
$$

When the pair (S, C) satisfies some *closure* and *consistency* conditions, the partial knowledge of $\mathcal{L}$ over $C \circ S$ suffices to build a hypothesis tree automaton $\mathcal{H}$ such that $[\![\mathcal{H}]\!](c \circ s) = \mathcal{L}(c \circ s)$ for all $c \in C$ and $s \in S$. An equivalence query is then submitted to the oracle: either $\mathcal{H}$ is the target minimal tree automaton, or the oracle gives a counter-example which is used to expand S and to guide the algorithm towards another hypothesis tree automaton.

Construction of the hypothesis automaton. We now delve into the technical details of how to build the hypothesis automaton for a pair (S, C). Compared to the categorical learning algorithm for word automata of [17], this is where the main novelty lies: although the proofs adhere to the same general schema, they are more difficult because they involve trees instead of words.

Just like in Section 3 the minimal tree automaton had for state-object the $(\mathcal{E}, \mathcal{M})$-image of $\partial\mathcal{L}$, in the learning algorithm the hypothesis tree automaton is constructed with state-object the $(\mathcal{E}, \mathcal{M})$-image S/Ω^C of $\partial_S^C \mathcal{L}$. But unlike in

Section 3, the domain and codomain of $\partial_S^C \mathcal{L}$, S and Ω^C, do not carry structures of Σ-context $(\mathcal{C}, \Omega)$-automata (let alone that of Σ-tree $(\mathcal{C}, \Omega)$-automata) that could be inherited by S/Ω^C. However, these objects do carry another structure, that of *biautomata*. Intuitively, these are automata with two state-objects and no loops that accept a language over $C \circ S = \{c \circ s \mid c \in C, s \in S\}$ – that is, a function $C \circ S \to \mathcal{C}(1_{\mathcal{C}}, \Omega)$ – and therefore capture the information contained in a $S \times C$-table.

Definition 29 ([17, Def. 17] adapted to trees). *Given a pair (S, C) obtained during the execution of the learning algorithm, an (S, C)-biautomaton is the data of two objects $\mathfrak{B}(\mathsf{st}_1)$ and $\mathfrak{B}(\mathsf{st}_2)$ of $\mathcal{C}$ and of $\mathcal{C}$-morphisms as in the diagram:*

$$1_{\mathcal{C}} \xrightarrow{\;\mathfrak{B}(s)\;} \mathfrak{B}(\mathsf{st}_1) \underset{\mathfrak{B}(\sigma \circ s[i:=\square])}{\overset{\mathfrak{B}(\square)}{\rightrightarrows}} \mathfrak{B}(\mathsf{st}_2) \xrightarrow{\;\mathfrak{B}(\triangleleft \circ c)\;} \Omega$$

where, on the left s ranges in S, in the middle σ ranges in Σ, s in $S^{\mathrm{ar}(\sigma)}$ and $1 \leq i \leq \mathrm{ar}(\sigma)$, while $\square$ is the trivial context, and on the right c ranges in C. The associated morphisms satisfy some coherence conditions deferred to a long version of this paper. The (S, C)-language recognized by a biautomaton $\mathfrak{B}$ is the function $[\![\mathfrak{B}]\!]: C \circ S \to \mathcal{C}(1_{\mathcal{C}}, \Omega)$ that maps $c \circ s \in C \circ S$ to $\mathfrak{B}(\triangleleft \circ c) \circ \mathfrak{B}(\square) \circ \mathfrak{B}(s)$.

The coherence conditions on a biautomaton ensure in particular that the language it recognizes is well-defined: if $c_1 \circ s_1 = c_2 \circ s_2$, $\mathfrak{B}(\triangleleft \circ c_1) \circ \mathfrak{B}(\square) \circ \mathfrak{B}(s_1) = \mathfrak{B}(\triangleleft \circ c_2) \circ \mathfrak{B}(\square) \circ \mathfrak{B}(s_2)$.

S and Ω^C are part of two biautomata $\mathfrak{B}_{S,C}^{init}(\mathcal{L})$ and $\mathfrak{B}_{S,C}^{final}(\mathcal{L})$ both recognizing $\mathcal{L}_{S,C}$, the restriction of $\mathcal{L}$ to $C \circ S$. To describe these biautomata concisely, we introduce the following notations.

- For $S \subseteq \mathcal{T}_0(\Sigma)$, write $\Sigma(S) = S \cup \{\sigma \circ \mathsf{s} \mid \sigma \in \Sigma_n, \mathsf{s} \in S^n, n \geq 0\}$;
- for $S \subseteq \mathcal{T}_0(\Sigma)$ and $C \subseteq \mathcal{T}_1(\Sigma)$, write $C(\Sigma, S) = C \cup \{c \circ \sigma \circ \mathsf{s}[i:=\square] \mid c \in C, \sigma \in \Sigma, \mathsf{s} \in S^{\mathrm{ar}(\sigma)}, 1 \leq i \leq \mathrm{ar}(\sigma)\}$.

$\mathfrak{B}_{S,C}^{init}(\mathcal{L})$ and $\mathfrak{B}_{S,C}^{final}(\mathcal{L})$ are then the biautomata sitting respectively at the top and bottom of the following diagram:

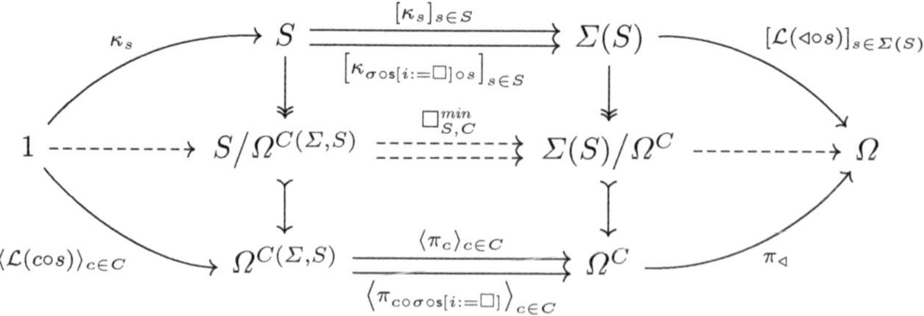

It moreover turns out that, together, the tables $\partial_S^{C(\Sigma,S)} \mathcal{L}: S \to \Omega^{C(\Sigma,S)}$ and $\partial_{\Sigma(S)}^C \mathcal{L}: \Sigma(S) \to \Omega^C$ form a *morphism of biautomata* $\mathfrak{B}_{S,C}^{init}(\mathcal{L}) \to \mathfrak{B}_{S,C}^{final}(\mathcal{L})$, meaning all the triangles and squares in the diagram above commute. Again (as in Section 3.2), because factorization systems lift to categories of functors (here the category of biautomata recognizing $\mathcal{L}_{S,C}$ and their morphisms), the images $S/\Omega^{C(\Sigma,S)}$ and $\Sigma(S)/\Omega^C$ are canonically equipped with the structure of an

(S, C)-biautomaton recognizing $\mathcal{L}_{S,C}$. This biautomaton that we call $\mathfrak{Min}_{S,C}(\mathcal{L})$ is depicted in the center of the diagram above. Let us emphasize in particular the role of the morphism $\mathfrak{Min}_{S,C}(\mathcal{L})(\square)$, that we denote $\square_{S,C}^{min}$: this morphism encodes the canonical way to move from $S/\Omega^{C(\Sigma,S)}$ to $\Sigma(S)/\Omega^{C}$. Intuitively, it maps the congruence class of a closed tree in S, with respect to the extended contexts in $C(\Sigma, S)$, to the congruence class of the same tree with respect only to C. As we will see next, $\square_{S,C}^{min}$ plays a crucial role in the categorical encoding of the notions of *closure* and *consistency*.

Since at this point we know what $\mathcal{L}$ is like on $C \circ S$, we would like the hypothesis automaton that the algorithm builds to at least agree with $\mathcal{L}$ on $C \circ S$. Because $\mathfrak{Min}_{S,C}(\mathcal{L})$ recognizes $\mathcal{L}_{S,C}$, a reasonable strategy is to merge its state-objects to obtain a Σ-context $(\mathcal{C}, \Omega)$-automaton and then additionally infer the structure of a Σ-tree $(\mathcal{C}, \Omega)$-automaton on top of it. Merging these two state-objects is in particular possible when they are isomorphic, i.e., when $\square_{S,C}^{min}$ is an isomorphism. In that case it is even possible to show that $S/\Omega^{C(\Sigma,S)} \cong S/\Omega^{C} \cong \Sigma(S)/\Omega^{C}$, and we set $\mathcal{H}(\mathrm{st}) = S/\Omega^{C}$, $\mathcal{H}(l) \cong \mathfrak{Min}_{S,C}(\mathcal{L})(l)$ for $l \in \Sigma_0$ and $\mathcal{H}(\triangleleft) \cong \mathfrak{Min}_{S,C}(\triangleleft)$.

To construct the transition morphism $\mathcal{H}(\sigma) \colon \left(S/\Omega^{C(\Sigma,S)} \right)^{\otimes n} \to \Sigma(S)/\Omega^{C}$ for $\sigma \in \Sigma$, we crucially use the $\otimes_C$-compatibility of $\mathcal{E}$ again, similarly to how it was used in the proof of Theorem 26. For every $\mathrm{s} \in S^{\mathrm{ar}(\sigma)}$ and $1 \leq i \leq \mathrm{ar}(\sigma)$, we already know what $\mathcal{H}(\sigma \circ \mathrm{s}[i:=\square]) \colon S/\Omega^{C(\Sigma,S)} \to \Sigma(S)/\Omega^{C}$ should be: the morphism $\mathfrak{Min}_{S,C}(\mathcal{L})(\sigma \circ \mathrm{s}[i:=\square])$. $\otimes_C$-compatibility then allows us to aggregate these unary morphisms into the n-ary transition morphism $\mathcal{H}(\sigma)$.

Conversely, when $\square_{S,C}^{min}$ is not an isomorphism, then either it is not an $\mathcal{E}$-morphism or it is not an $\mathcal{M}$-morphism. In the former case we say the table $\partial_S^C \mathcal{L}$ is not *closed*, while in the latter case we say it is not *consistent*. As already the case for word automata [17], these notions of closure and match precisely the ones given in the concrete learning algorithms that are instances of our categorical algorithm. Intuitively, a closure defect means there are not enough states in $\mathfrak{Min}_{S,C}(\mathrm{st}_1)$ to build the transition morphisms, in which case we thus increase the set S and set it to $\Sigma(S)$. Dually, a consistency defect means the Myhill-Nerode approximation S/Ω^{C} is not refined enough, and building the transition morphisms as above would lead to inconsistencies: in this case we thus refine the approximation more by increasing the set C and setting it to $C(\Sigma, S)$.

The learning algorithm. A basic version of the learning algorithm is given in Algorithm 1. Note that this algorithm can be improved so that the sets S and C it keeps track of are smaller.

For the algorithm to terminate, the minimal tree automaton needs to be finitely representable. Categorically, this property is expressed through the notion of $(\mathcal{E}, \mathcal{M})$-finiteness, that we now recall.

Definition 30 ([36, §4]). *In a category $\mathcal{C}$ equipped with a factorization system*

$(\mathcal{E}, \mathcal{M})$, *an object X is said to be $\mathcal{M}$-noetherian when every strict chain of its is finite: for any $(x_n \colon X_n \rightarrowtail X)_{n \in \mathbb{N}}$ and $(m_n \colon X_n \rightarrowtail X_{n+1})_{n \in \mathbb{N}}$ such that the diagram on the right commutes, then only finitely many of the m_n's are not isomorphisms.*

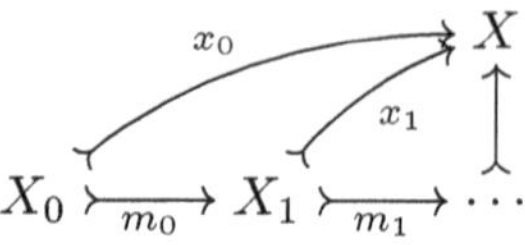

Dually, an object X of $\mathcal{C}$ is said to be $\mathcal{E}^{op}$-noetherian when every strict co-chain of its $\mathcal{E}$-quotients, as in the diagram on the right, is finite.

Finally, X is said to be $(\mathcal{E}, \mathcal{M})$-finite if it is both $\mathcal{M}$- and $\mathcal{E}^{op}$-noetherian.

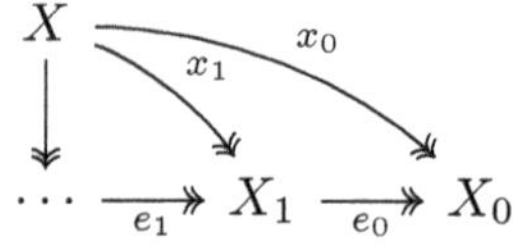

Example 31. In **Set**, the $(\mathrm{Surj}, \mathrm{Inj})$-finite objects are the finite sets. In $\mathbf{Vec}_{\mathbb{K}}$, the $(\mathrm{Surj}, \mathrm{Inj})$-finite objects are the finite-dimensional vector spaces.

Theorem 32. *Within the setting of Theorem 26, assume that $\mathrm{Min}(\mathcal{L})(\mathtt{st})$ is $(\mathcal{E}, \mathcal{M})$-finite. Then Algorithm 1 is correct and terminates.*

Instances of Theorem 32 include active learning of minimal deterministic tree automata [19], residual finite-state tree automata (which was hinted at in [26] but not described nor formally proved), multiplicity tree automata [22], and of course PID-weighted tree automata, as described in Section 2.2. Note that these instances may come with additional optimizations that rely on the specific setting they are in and cannot be accounted for by a more general framework such as ours.

5   Conclusion, related work and future work

In this work we introduced a categorical framework for the minimization and learning of various families of bottom-up tree automata in monoidal categories. One of our main contrbutions lies in proving the correctness and termination of such algorithms under suitable assumptions on the monoidal categories considered. Our aim is to strike the right balance between the variety of examples

---

Algorithm 1 The basic MonFunL* algorithm

Input: an oracle answering $\textsc{Eval}_{\mathcal{L}}$ and $\textsc{Equiv}_{\mathcal{L}}$ queries
Output: $\mathrm{Min}(\mathcal{L})$
  1: $S \leftarrow \Sigma_0$, $C \leftarrow \{\lhd\}$
  2: **loop**
  3:    **while** $\square_{S,C}^{min}$ is not an isomorphism **do**
  4:      **if** $\square_{S,C}^{min}$ is not an $\mathcal{E}$-morphism **then** $S \leftarrow \Sigma(S)$
  5:      **if** $\square_{S,C}^{min}$ is not an $\mathcal{M}$-morphism **then** $C \leftarrow C(\Sigma, S)$
  6:    **end while**
  7:    compute the hypothesis tree automaton $\mathcal{H}$ with state space $\mathcal{H}(\mathtt{st}) = S/\Omega^C$
  8:    **if** $\textsc{Equiv}_{\mathcal{L}}(\mathcal{H})$ returns some counter-example $t \in \mathcal{T}_0(\Sigma)$ **then**
  9:      $S \leftarrow S \cup \{t\} \cup \mathrm{children}(t)$ {children(t) denotes the set of children of t}
10:    **else return** $\mathcal{H}$
11: **end loop**

---

satisfying them – e.g., Eilenberg-Moore algebras for commutative **Set**- monads – and the strentgh of the theorems we can prove: existence of the minimal automaton, correctness and termination of the minimization and learning algorithms.

Complementary categorical approaches exist, notably the CALF framework [39], which encompasses minimization of various forms of tree automata with effects: they focus on classical tree automata and multiplicity tree automata [37]. This framework also covers learning of tree automata, but only based in the category **Set**, without monoidal tensor or effects [38]. Monoidal categories, including **JSL**, appear in the coalgebraic framework of [36], for which the learning algorithm does not apply to tree automata. The compatibility condition of Definition 24 also plays the crucial role in [1] – which deals with syntactic monoids for languages interpreted in commutative varieties of algebras. Our strategy of using context automata is conceptually similar to the use of automata presentations and linearizations appearing in the work on learning syntactic algebras for monad-recognizable languages [36, Section 5].

There is a plethora of families of automata and their algorithms for which we could adapt the category theoretic framework, e.g., symbolic automata, nominal automata or alternating automata. The aim for future work is to deploy this conceptual approach to obtain new algorithms by investigating suitable output categories which have the required properties in order to ensure correctness and termination for the ensuing algorithms.

References

1. Adamek, J., Milius, S., Urbat, H.: A Categorical Approach to Syntactic Monoids. Logical Methods in Computer Science **Volume 14, Issue 2** (May 2018). https://doi.org/10.23638/LMCS-14(2:9)2018, https://lmcs.episciences.org/4505

2. Angluin, D.: Learning regular sets from queries and counterexamples. Information and Computation **75**(2), 87–106 (Nov 1987). https://doi.org/10.1016/0890-5401(87)90052-6, https://www.sciencedirect.com/science/article/pii/0890540187900526

3. Arbib, M.A., Manes, E.G.: A Categorist's view of automata and systems. In: Manes, E.G. (ed.) Category Theory Applied to Computation and Control. pp. 51–64. Springer, Berlin, Heidelberg (1975). https://doi.org/10.1007/3-540-07142-3_61

4. Aristote, Q.: Active Learning of Deterministic Transducers with Outputs in Arbitrary Monoids. In: DROPS-IDN/v2/Document/10.4230/LIPIcs.CSL.2024.11. Schloss-Dagstuhl - Leibniz Zentrum für Informatik (2024). https://doi.org/10.4230/LIPIcs.CSL.2024.11, https://drops.dagstuhl.de/entities/document/10.4230/LIPIcs.CSL.2024.11

5. Banaschewski, B., Nelson, E.: Tensor products and bimorphisms. Canadian Mathematical Bulletin **19**(4), 385402 (1976). https://doi.org/10.4153/CMB-1976-060-2

6. Bergadano, F., Varricchio, S.: Learning Behaviors of Automata from Multiplicity and Equivalence Queries. SIAM Journal on Computing **25**(6), 1268–1280 (Dec 1996). https://doi.org/10.1137/S009753979326091X, https://epubs.siam.org/doi/10.1137/S009753979326091X

7. Blumensath, A.: Regular Tree Algebras. Logical Methods in Computer Science **Volume 16, Issue 1** (Feb 2020). https://doi.org/10.23638/LMCS-16(1:16)2020, https://lmcs.episciences.org/6101

8. Bojańczyk, M.: Recognisable languages over monads (Feb 2015). https://doi.org/10.48550/arXiv.1502.04898, http://arxiv.org/abs/1502.04898

9. Bojańczyk, M.: Languages recognised by finite semigroups, and their generalisations to objects such as trees and graphs, with an emphasis on definability in monadic second-order logic (Aug 2020). https://doi.org/10.48550/arXiv.2008.11635, http://arxiv.org/abs/2008.11635

10. Bojańczyk, M., Klin, B.: A non-regular language of infinite trees that is recognizable by a sort-wise finite algebra. Logical Methods in Computer Science **Volume 15, Issue 4** (Nov 2019). https://doi.org/10.23638/LMCS-15(4:11)2019, https://lmcs.episciences.org/5927

11. Bollig, B., Habermehl, P., Kern, C., Leucker, M.: Angluin-style learning of nfa. In: Proceedings of the 21st International Joint Conference on Artificial Intelligence. p. 10041009. IJCAI'09, Morgan Kaufmann Publishers Inc., San Francisco, CA, USA (2009)

12. Bonchi, F., Pous, D.: Checking NFA equivalence with bisimulations up to congruence. In: Proceedings of the 40th Annual ACM SIGPLAN-SIGACT Symposium on Principles of Programming Languages. pp. 457–468. POPL '13, Association for Computing Machinery, New York, NY, USA (Jan 2013). https://doi.org/10.1145/2429069.2429124, https://dl.acm.org/doi/10.1145/2429069.2429124

13. Buna-Marginean, A., Cheval, V., Shirmohammadi, M., Worrell, J.: On Learning Polynomial Recursive Programs. Proceedings of the ACM on Programming Languages 8(POPL), 34:1001–34:1027 (Jan 2024). https://doi.org/10.1145/3632876, https://dl.acm.org/doi/10.1145/3632876

14. Carme, J., Gilleron, R., Lemay, A., Terlutte, A., Tommasi, M.: Residual Finite Tree Automata. In: Ésik, Z., Fülöp, Z. (eds.) Developments in Language Theory. pp. 171–182. Lecture Notes in Computer Science, Springer, Berlin, Heidelberg (2003). https://doi.org/10.1007/3-540-45007-6_13

15. Cohen, H.: Algorithms for Linear Algebra and Lattices. In: Cohen, H. (ed.) A Course in Computational Algebraic Number Theory, pp. 45–107. Graduate Texts in Mathematics, Springer, Berlin, Heidelberg (1993). https://doi.org/10.1007/978-3-662-02945-9_2, https://doi.org/10.1007/978-3-662-02945-9_2

16. Colcombet, T., Petrişan, D.: Automata Minimization: A Functorial Approach. Logical Methods in Computer Science **Volume 16, Issue 1** (Mar 2020). https://doi.org/10.23638/LMCS-16(1:32)2020, https://lmcs.episciences.org/6213/pdf

17. Colcombet, T., Petrişan, D., Stabile, R.: Learning Automata and Transducers: A Categorical Approach. In: Baier, C., Goubault-Larrecq, J. (eds.) 29th EACSL Annual Conference on Computer Science Logic (CSL 2021). Leibniz International Proceedings in Informatics (LIPIcs), vol. 183, pp. 15:1–15:17. Schloss Dagstuhl–Leibniz-Zentrum für Informatik, Dagstuhl, Germany (2021). https://doi.org/10.4230/LIPIcs.CSL.2021.15, https://drops.dagstuhl.de/opus/volltexte/2021/13449

18. Comon, H., Dauchet, M., Gilleron, R., Jacquemard, F., Lugiez, D., Löding, C., Tison, S., Tommasi, M.: Tree Automata Techniques and Applications (2008), https://inria.hal.science/hal-03367725

19. Drewes, F., Högberg, J.: Learning a Regular Tree Language from a Teacher. In: Ésik, Z., Fülöp, Z. (eds.) Developments in Language Theory. pp. 279–291. Lecture Notes in Computer Science, Springer, Berlin, Heidelberg (2003). https://doi.org/10.1007/3-540-45007-6_22

20. Eilenberg, S., Wright, J.B.: Automata in general algebras. Information and Control **11**(4), 452–470 (Oct 1967). https://doi.org/10.1016/S0019-9958(67)90670-5, https://www.sciencedirect.com/science/article/pii/S0019995867906705

21. Fresse Benoit: Modules over Operads and Functors. Lecture Notes in Mathematics, Springer Berlin Heidelberg, Berlin, Heidelberg, 1st ed. 2009. edn.

22. Habrard, A., Oncina, J.: Learning Multiplicity Tree Automata. In: Sakakibara, Y., Kobayashi, S., Sato, K., Nishino, T., Tomita, E. (eds.) Grammatical Inference: Algorithms and Applications. pp. 268–280. Lecture Notes in Computer Science, Springer, Berlin, Heidelberg (2006). https://doi.org/10.1007/11872436_22

23. Heerdt, G.v., Kupke, C., Rot, J., Silva, A.: Learning Weighted Automata over Principal Ideal Domains. In: Goubault-Larrecq, J., König, B. (eds.) Foundations of Software Science and Computation Structures. pp. 602–621. Springer International Publishing, Cham (2020). https://doi.org/10.1007/978-3-030-45231-5_31

24. Joyal, A., Tierney, M.: An Extension of the Galois Theory of Grothendieck, Memoirs of the American Mathematical Society, vol. 51. American Mathematical Society (1984). https://doi.org/10.1090/memo/0309, https://www.ams.org/memo/0309

25. Kaplan, W.: Topics in Mathematical System Theory (Rudolf E. Kalman, Peter L. Falb and Michael A. Arbib). SIAM Review **12**(1), 157–158 (Jan 1970). https://doi.org/10.1137/1012030, https://epubs.siam.org/doi/10.1137/1012030

26. Kasprzik, A.: Inference of Residual Finite-State Tree Automata from Membership Queries and Finite Positive Data. In: Mauri, G., Leporati, A. (eds.) Developments in Language Theory. pp. 476–477. Lecture Notes in Computer Science, Springer, Berlin, Heidelberg (2011). https://doi.org/10.1007/978-3-642-22321-1_45

27. Kiefer, S., Marusic, I., Worrell, J.: Minimisation of Multiplicity Tree Automata. Logical Methods in Computer Science **Volume 13, Issue 1** (Mar 2017). https://doi.org/10.23638/LMCS-13(1:16)2017, https://lmcs.episciences.org/3224

28. Mac Lane, S.: 07. Monoids. In: Mac Lane, S. (ed.) Categories for the Working Mathematician, pp. 161–190. Graduate Texts in Mathematics, Springer, New York, NY (1978). https://doi.org/10.1007/978-1-4757-4721-8_8, https://doi.org/10.1007/978-1-4757-4721-8_8

29. Melliès, P.A., Tabareau, N.: Free models of T-algebraic theories computed as Kan extensions (2008), https://hal.science/hal-00339331

30. Moerman, J., Sammartino, M., Silva, A., Klin, B., Szynwelski, M.: Learning nominal automata. In: Proceedings of the 44th ACM SIGPLAN Symposium on Principles of Programming Languages. p. 613625. POPL '17, Association for Computing Machinery, New York, NY, USA (2017). https://doi.org/10.1145/3009837.3009879, https://doi.org/10.1145/3009837.3009879

31. Myers, R.S.R., Adámek, J., Milius, S., Urbat, H.: Canonical Nondeterministic Automata. In: Bonsangue, M.M. (ed.) Coalgebraic Methods in Computer Science. pp. 189–210. Lecture Notes in Computer Science, Springer, Berlin, Heidelberg (2014). https://doi.org/10.1007/978-3-662-44124-4_11

32. Nguyen, L.T.D.: Automates Implicites En Logique Linéaire et Théorie Catégorique Des Transducteurs. Ph.D. thesis, Université Paris-Nord - Paris XIII (Dec 2021), https://theses.hal.science/tel-04132636

33. Nguyên, L.T.D., Noûs, C., Pradic, C.: Implicit automata in typed λ-calculi II: Streaming transducers vs categorical semantics (Aug 2021). https://doi.org/10.48550/arXiv.2008.01050, http://arxiv.org/abs/2008.01050

34. Sakarovitch, J.: Elements of Automata Theory. Cambridge University Press (2009)

35. Thatcher, J.W., Wright, J.B.: Generalized finite automata theory with an application to a decision problem of second-order logic. The Journal of Symbolic Logic **37**(3), 619–620 (Sep 1972). https://doi.org/10.2307/2272790
36. Urbat, H., Schröder, L.: Automata Learning: An Algebraic Approach. In: Proceedings of the 35th Annual ACM/IEEE Symposium on Logic in Computer Science. pp. 900–914. LICS '20, Association for Computing Machinery, New York, NY, USA (Jul 2020). https://doi.org/10.1145/3373718.3394775, https://dl.acm.org/doi/10.1145/3373718.3394775
37. van Heerdt, G., Kappé, T., Rot, J., Sammartino, M., Silva, A.: Tree Automata as Algebras: Minimisation and Determinisation. In: Roggenbach, M., Sokolova, A. (eds.) 8th Conference on Algebra and Coalgebra in Computer Science (CALCO 2019). Leibniz International Proceedings in Informatics (LIPIcs), vol. 139, pp. 6:1–6:22. Schloss Dagstuhl–Leibniz-Zentrum fuer Informatik, Dagstuhl, Germany (2019). https://doi.org/10.4230/LIPIcs.CALCO.2019.6, http://drops.dagstuhl.de/opus/volltexte/2019/11434
38. van Heerdt, G., Kappé, T., Rot, J., Sammartino, M., Silva, A.: A Categorical Framework for Learning Generalised Tree Automata. In: Hansen, H.H., Zanasi, F. (eds.) Coalgebraic Methods in Computer Science. pp. 67–87. Lecture Notes in Computer Science, Springer International Publishing, Cham (2022). https://doi.org/10.1007/978-3-031-10736-8_4
39. van Heerdt, G., Sammartino, M., Silva, A.: CALF: Categorical Automata Learning Framework. In: DROPS-IDN/v2/Document/10.4230/LIPIcs.CSL.2017.29. Schloss Dagstuhl – Leibniz-Zentrum für Informatik (2017). https://doi.org/10.4230/LIPIcs.CSL.2017.29, https://drops.dagstuhl.de/entities/document/10.4230/LIPIcs.CSL.2017.29
40. Vilar, J.M.: Query learning of subsequential transducers. In: Miclet, L., de la Higuera, C. (eds.) Grammatical Interference: Learning Syntax from Sentences. pp. 72–83. Lecture Notes in Computer Science, Springer, Berlin, Heidelberg (1996). https://doi.org/10.1007/BFb0033343

The Complexity of Games with Randomised Control

Sarvin Bahmani, Rasmus Ibsen-Jensen, Soumyajit Paul, Sven Schewe, Friedrich Slivovsky, Qiyi Tang, Dominik Wojtczak, and Shufang Zhu

University of Liverpool, Liverpool, UK
{r.bahmani, r.ibsen-jensen, soumyajit.paul, sven.schewe, f.slivovsky,
qiyi.tang, dkw, shufang.zhu}@liverpool.ac.uk

Abstract. We study the complexity of solving two-player infinite duration games played on a fixed finite graph, where the control of a node is not predetermined but rather assigned randomly. In classic *random-turn games*, control of each node is assigned randomly every time the node is visited during a play. In this work, we study two natural variants of this where control of each node is assigned only once: (i) control is assigned randomly during a play when a node is visited for the first time and does not change for the rest of the play and (ii) control is assigned a priori before the game starts for every node by independent coin tosses and then the game is played.

We investigate the complexity of computing the winning probability with three kinds of objectives—reachability, parity, and energy. We show that the qualitative questions on all variants and all objectives are **NL**-complete. For the quantitative questions, we show that deciding whether the maximiser can win with probability at least a given threshold for every objective is **PSPACE**-complete under the first mechanism, and that computing the exact winning probability for every objective is #**P**-complete under the second. To complement our hardness results for the second mechanism, we propose randomised approximation schemes that efficiently estimate the winning probability for all three objectives, assuming a bounded number of parity colours and unary-encoded weights for energy objectives, and we empirically demonstrate their fast convergence.

Keywords: turn-based games, two-player games, random-turn games, stochastic games

1 Introduction

Graph games are fundamental framework for modelling and studying decision making scenarios in formal verification [19,25,1,2] and reactive synthesis [33,10]. In the most classic setting, these are used to capture a system interacting with an adversarial environment. This can be naturally perceived as a two-player game between the system and the environment where the system's desired requirements are formulated as a winning condition of the game. Classic winning

N. Bertrand and S. Milius (Eds.): FoSSaCS 2026, LNCS 16503, pp. 43–64, 2026.
https://doi.org/10.1007/978-3-032-22730-0_3

conditions on infinite traces include reachability, parity and energy [20]. In two-player turn-based games on graphs, a token is placed on an initial node of the graph, and the players take turns moving it along the edges, thereby generating an infinite trace. Each node is typically owned by one of the players: the system selects the successor node on its own nodes, while the environment does so on its respective ones. A standard assumption in these two-player games on graphs is that the ownership of nodes is fixed before the game begins. Each node belongs permanently to one of the players, and control never changes hands throughout the game. But this assumption disregards situations where there is some uncertainty in who controls a decision. Randomness being a natural simulator of uncertainty, the controls of a node can be decided with a coin toss, giving rise to what are known as *random-turn games*.

Random-turn games have been studied for combinatorial games [ ] as well as for infinite duration graph games [ ]. These games have also found applications in bargaining networks [ ]. In random-turn games, every time a node is visited, the ownership of this node is decided randomly and independently. While this mechanism is simple, it also grants the players an unbounded amount of leeway to make non-optimal choices in the initial stages of the game. Moreover, this does not capture scenarios where the control of a decision point depends only on some single uncertain event. In this work, we study random-turn games where ownership of a particular node is decided randomly but only once, i.e., the ownership of a node is fixed once it is assigned. This is already the case in random-turn games for some board games with acyclic state graph as studied in [ ].

We study two natural variants of this: (i) the ownership of nodes are decided gradually on the fly as the game goes on and fixed for the rest of the game, termed as *Unfolding Random-Turn Game*–these games can be perceived as the unfolding of a larger stochastic game; (ii) the ownership of all nodes are decided a priori randomly and independently before the start of the game, termed as *Random Arena Game*–this is equivalent to randomly selecting a *game arena* for the game. With randomness in the picture, the fundamental question that we are concerned with is computing the probability that the system player wins. Random-turn games, unfolding random-turn game and random arena game, all of them can be reduced to equivalent stochastic games [ ]. For random-turn games, this translation is actually in deterministic logarithmic space, resulting in a stochastic game of polynomial size. However, it remains unknown whether any stochastic game can be reduced to a structure-preserving (i.e., a simple translation in deterministic logarithmic space), probability-preserving random-turn game. On the other hand, for unfolding random-turn game and random arena game the equivalent stochastic games can suffer from an exponential blow-up in size. In this paper, we consider three classic objectives: reachability, parity and energy. Example 1 illustrates these different variants of games with randomised control on a simple graph in Fig. 1(a) with a reachability objective.

Example 1. In the game graph in Fig. 1(a), the token is initially on state s_0. The players are Max and Min, and Max's objective is to maximise the probability of

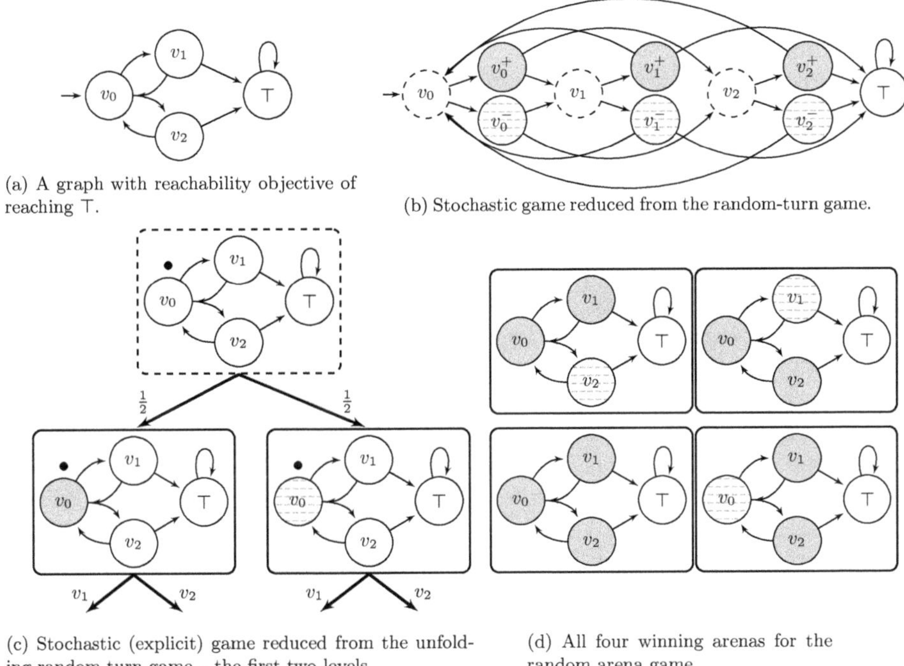

(a) A graph with reachability objective of reaching ⊤.

(b) Stochastic game reduced from the random-turn game.

(c) Stochastic (explicit) game reduced from the unfolding random-turn game – the first two levels.

(d) All four winning arenas for the random arena game.

Fig. 1: Blue nodes are controlled by Max, pink nodes with dashed patterns by Min, white nodes with dashed contours are random, and white nodes with solid contours are unassigned. In (c) and (d), boxes denote game states. Black dots above some nodes indicate token positions. Transition probabilities from each random node are uniformly distributed.

reaching the target node ⊤. Let us consider the games with randomised control on this graph where ownership is always determined uniformly randomly. First, in the random arena game, each node from $\{v_0, v_1, v_2\}$ is assigned to Max or Min with probability $\frac{1}{2}$. Hence, each of the $2^3 = 8$ possible *game arenas* is obtained with probability $\frac{1}{8}$. Among these, Max has a winning strategy in 4 of them, with ownership among one of $\{v_0, v_1\}, \{v_0, v_2\}, \{v_0, v_1, v_2\}, \{v_1, v_2\}$; see Fig. 1(d). Hence, Max wins random arena game with probability $\frac{1}{2}$ on this graph.

On the other hand, in the unfolding random-turn game on this graph, the initial node v_0 is assigned to Max or Min with probability $\frac{1}{2}$. After the assignment, the respective player chooses a node among v_1 or v_2. By symmetry, since both options are equally good, suppose w.l.o.g. they choose v_1. Next, v_1 is assigned randomly. If v_0 was assigned to Min, then Max loses when Min gets v_1, as Min can play the cycle v_0, v_1, v_0. Max wins in the other case, when Max gets v_1. If v_0 was assigned to Max, then Max wins when Max gets v_1, whereas if Min gets v_1, they would move the token back to v_0. Since Max already owns v_0, they will then move to the still unassigned v_2. In the random assignment of v_2, Max wins if v_2 is assigned to Max and loses otherwise. Consequently, the probability of Max winning is $\frac{1}{2}(\frac{1}{2} + \frac{3}{4}) = \frac{5}{8}$.

Moreover, for the random-turn game on this graph, Max wins almost surely, that is, with probability one, since Max wins whenever they obtain v_1 or v_2, which occurs with probability one. $\qquad\square$

Our Contributions. We begin by considering the problem of whether there exists a deterministic logarithmic-space translation from a stochastic game to a random-turn game that preserves the winning probability. We show that this is unlikely: first, we prove that the qualitative question for the random-turn game with a reachability objective is **NL**-complete; then, we infer that a simple structure- and probability-preserving translation would imply **NL** = **P**.

The main focus of this paper is on two new and distinct ways of assigning control: (i) ownership is decided dynamically when a node is visited for the first time, and (ii) ownership is determined a priori by independent coin tosses for each node. These models differ from standard stochastic games, and we show that they exhibit markedly different computational complexities: the quantitative problems are **PSPACE**-complete for (i) and #**P**-complete for (ii), while both retain **NL**-completeness for the qualitative questions. These complexity results hold for safety, reachability, parity, and energy objectives. For (ii), we complement the #**P**-completeness result with a randomised approximation scheme that can efficiently estimate the probability of winning for all three objectives, assuming a bounded number of parity colours and unary-encoded weights for energy objectives.

Missing proofs can be found in the full version [ ].

Related Work. Previous work has investigated *random-turn games* [ ], where ownership of states is (re-)assigned randomly each time a state is visited. Such games can be transformed, in deterministic logarithmic space, into *stochastic games* [ ] with the same winning probabilities, and are therefore often regarded as a subclass of stochastic games. Since stochastic games with the objectives we consider can be solved in **NP** ∩ co-**NP** [, , ,], the same complexity bound applies to random-turn games. It has been shown that random-turn games with out-degree 2 can be solved in polynomial time [ ], implying that there is no simple degree-preserving reduction in the opposite direction, as *simple stochastic games*—that is, stochastic games where all random vertices have out-degree 2— are already as hard as general stochastic games [ ].

A rich body of work also explores *graph games* in which control is decided through alternative mechanisms. In *bidding games*, introduced by [,], each turn involves an "auction" between the two players to determine who moves the token. Different bidding rules have been studied, including *Richman bidding* [ ] (named after David Richman), where the higher bidder pays the lower bidder, and *poorman bidding* [ ], where the higher bidder pays the bank and the payment is lost. It was observed in [ ] that *reachability* Richman-bidding games (and only those) are equivalent to random-turn games.

Another related model is that of *pawn games* [ ], where control of vertices also evolves dynamically during play. Here, control is determined by "pawns": each pawn owns a subset of vertices in the game graph, and pawns are distributed among players, with ownership possibly changing throughout the game. As noted in the seminal work on pawn games [ ], these games differ fundamentally from bidding games: pawn games support richer and more flexible mechanisms for transferring control, whereas bidding games rely solely on strict auction-based mechanisms.

2   Preliminaries

Game Graph and Arena. In this work, we study the turn-based zero-sum games that are played between two players —Max and Min—on finite directed graphs, which are called *game arenas*. A finite directed graph is $G = (V, E)$, where V is a non-empty set of nodes and $E \subseteq V \times V$ is a set of non-empty edges. Since our games are potentially infinite duration games, we only consider graphs with no dead ends, i.e., for each $u \in V$, for some v in V, $(u, v) \in E$.

A game *arena* on a graph $G = (V, E)$ is a tuple $\mathcal{A} = (G, V_0, V_1, v_0)$ where V is the set of nodes, $E \subseteq V \times V$ is the set of edges, and v_0 is the start node. The set V is partitioned into a set V_0 of nodes controlled by player Max and a set V_1 of nodes controlled by player Min. A play of such a game starts by placing a token on the start node v_0. The player controlling this node then chooses a successor node v_1 such that $(v_0, v_1) \in E$ and the token is moved to this successor node. In the next turn the player controlling the node v_1 chooses the successor node v_2 with $(v_1, v_2) \in E$ and the token is moved accordingly. Both players move the token over the arena in this manner and thus form a play of the game. Note that, on a graph G with node set V, the total number of possible arenas is $2^{|V|}$.

Play and Strategies. Formally, a *play* on a game arena $\mathcal{A}$ is a finite (resp. an infinite) sequence of nodes $\langle v_0, v_1, \dots \rangle \in V^*$ (resp. $\langle v_0, v_1, \dots \rangle \in V^\omega$) such that, for all $i \geq 0$, we have that $(v_i, v_{i+1}) \in E$. We write $\mathsf{Plays}_{\mathcal{A}}(v)$ for the set of plays on arena $\mathcal{A}$ that start from a node $v \in V$ and $\mathsf{Plays}_{\mathcal{A}}$ for the set of plays on the arena. We omit the subscript when the arena is clear from the context.

A *strategy* for player Max on a given game arena, is a function $\sigma : V^* V_0 \to V$ such that $\big(v, \sigma(\rho, v)\big) \in E$ for all $\rho \in V^*$ and $v \in V_0$. A strategy σ is called memoryless if σ only depends on the last state ($\sigma(\rho, v) = \sigma(\rho', v)$ for all $\rho, \rho' \in V^*$ and $v \in V_0$). A play $\langle v_0, v_1, \dots \rangle$ is consistent with σ if, for every initial sequence $\rho_n = v_0, v_1, \dots, v_n$ of the play that ends in a state of player Max ($v_n \in V_0$), $\sigma(\rho_n) = v_{n+1}$ holds. Strategies for player Min are defined analogously.

Game Objectives. We consider three objectives: reachability, parity and energy. For *reachability* games, we have a target node $\top \in V$. If $\top$ is visited during a play, player Max is declared the winner and the game immediately terminates. Otherwise, if $\top$ is never visited, player Min is the winner.

For *parity* objectives, we equip the game with a priority function $\mathrm{pri} : V \to \mathbb{N}$. A play $\langle v_0, v_1, \dots \rangle$ is won by player Max if the maximal priority seen infinitely often is even, that is, $\limsup_{i \to \infty} \mathrm{pri}(v_i)$ is even, and by player Min otherwise.

For *energy* objectives, the game has an initial credit $c \in \mathbb{N}$ and a weight function $w : V \to \mathbb{Z}$. The goal of player Max is to construct an infinite play $\langle v_0, v_1, \ldots \rangle$ such that

$$c + \sum_{i=0}^{j} w(v_i) \geq 0 \text{ for all } j \geq 0. \tag{1}$$

The quantity $c + \sum_{i=0}^{j} w(v_i)$ is called the *energy level* of the play prefix $\langle v_0, v_1, \ldots, v_j \rangle$. A play $\langle v_0, v_1, \ldots \rangle$ is won by player Max if it satisfies (1), otherwise it is won by player Min.

A node $v \in V$ is winning for player Max (resp. Min) if there exists a winning strategy for player Max (resp. Min) from v. All three games are memoryless determined [18,22,29,19,11]; that is, for all $v \in V$, v is winning for either player Max or player Min, and memoryless strategies are sufficient.

Games with Randomised Control. In a game with randomised control, we play on a fixed game graph G with a designated starting node v_0, where the objective may be reachability, parity, or energy. Although the graph G is fixed, the control of its nodes is not predetermined; initially there is no partition of V into V_0 and V_1. Instead, the control of each node $v \in V$ is determined randomly and independently by a coin toss: given a function $t : V \to (0, 1)$, the node v is assigned to player Max with probability $t(v)$ (i.e., $v \in V_0$) and to player Min with probability $1 - t(v)$ (i.e., $v \in V_1$). Thus, the partition (V_0, V_1) is determined entirely by these coin tosses.

Let own $: V \rightharpoonup \{\mathsf{Max}, \mathsf{Min}\}$ be a partial function that, when defined, assigns a player to a node of the graph. The game starts with an empty ownership function, own $= \emptyset$. The ownership of the nodes is determined randomly during gameplay. In a *random-turn game* [30] the ownership of a node is decided every time it is visited by independent coin tosses. We consider two new variants of the classic random-turn games. In an *unfolding random-turn game*, the ownership of a node is decided upon its first visit by t and remains fixed thereafter. In a *random arena game*, the ownership of all nodes is determined at the start of the game using t.

Stochastic Game. A stochastic game is also a turn-based game played between players Max and Min on a (stochastic) game arena. In contrast to non-stochastic games, stochastic games include random nodes. Formally, the set of nodes V is partitioned into three disjoint subsets: V_0, controlled by Max; V_1, controlled by Min; and V_r, representing random nodes. Transitions from random nodes are probabilistic, specified by a transition function $\delta : V_r \to \mathrm{Dist}(V)$, where $\mathrm{Dist}(V)$ denotes the set of probability distributions over V. Strategies are defined similarly to deterministic games. A pair of strategies σ, τ of Max and Min respectively induces a probability measure on any measurable subset S of infinite plays denoted by $\mathcal{P}_{\sigma,\tau}(S)$. For objective $\mathsf{obj} \in \{reachability, parity, energy\}$, let $\mathsf{Plays}^{\mathsf{obj}}$ be the set of all plays satisfying objective obj. Then the probability that objective obj is satisfied under σ, τ is given by $\mathcal{P}_{\sigma,\tau}(\mathsf{Plays}^{\mathsf{obj}})$. The value of the game is defined as $\sup_\sigma \inf_\tau \mathcal{P}_{\sigma,\tau}(\mathsf{Plays}^{\mathsf{obj}})$. Intuitively, this value captures how likely Max is to achieve the objective under optimal play.

A random-turn game can be viewed as a special case of a stochastic game. An unfolding random-turn game, as we will show later, can be transformed into an exponentially large stochastic game, called the explicit game, while random arena game can be transformed into a trivial stochastic game with an initial random node that transitions to game arenas with different control assignments, where the transition probability corresponds to the likelihood of generating each random assignment. The values of these games are defined in respective equivalent stochastic games.

3   Qualitative Analysis and Random-Turn Games

In this section, we discuss the qualitative analysis of games with randomised control, focusing on almost-sure and sure winning for Max in random-turn games, unfolding random-turn games, and random arena games. A game is almost-surely won by player Max if it is won with probability one, and surely won if Max wins regardless of the outcomes of the coin tosses. We show that it is **NL**-complete to decide whether Max wins surely or almost-surely for random-turn games with a reachability objective, and for unfolding random-turn games and random arena games with reachability, parity, or energy objectives. We also discuss reductions between random-turn games and stochastic games.

We first show deciding whether Max almost surely wins a random-turn game with a reachability objective is **NL**-complete. The reduction is from st-connectivity, a well-known **NL**-complete problem.

Theorem 2. *It is **NL**-complete to decide whether* Max *almost surely wins a random-turn game with a reachability objective.*

For random-turn games, almost-sure winning and sure winning differ in general — for instance, Max wins the random-turn game played on Fig. 1(a) almost surely but not surely, see the equivalent stochastic game in Fig. 1(b): if it happens that every time v_1 and v_2 are assigned to Min when visited, then Max loses; however, this event occurs with probability zero. Nevertheless, deciding whether Max surely wins is also **NL**-complete.

This result extends to parity and energy objectives: for parity, Max does not win surely iff there is a "bad" lasso-path whose cycle has an odd maximum priority; for energy, Max does not win surely iff there is a "bad" lasso-path where the energy level drops below zero anywhere on this path or the cycle is negative. In both cases, checking the existence of such a lasso-path can be done in **NL**.

Theorem 3. *It is **NL**-complete to decide whether* Max *surely wins a random-turn game with a reachability, parity, or energy objective.*

As a consequence of the **NL**-completeness of the qualitative solutions to random-turn games, there is no simple structure[1] and probability-preserving

---

[1] With structure preserving we mean any simple translation, such as gadgets, that can be calculated in deterministic logarithmic space.

translation from stochastic games to random-turn games, even when the stochastic games are restricted to non-stochastic two-player games.

Theorem 4. *If* $\mathbf{NL} \neq \mathbf{P}$, *then there is no structure- and probability-preserving translation from two-player reachability games to random-turn games.*

Proof. Solving two-player reachability games is $\mathbf{P}$-complete [ ]. A deterministic logarithmic space (logspace) reduction that guarantees probability preservation to an $\mathbf{NL}$-complete problem would therefore imply that we can solve this reachability problem in $\mathbf{NL}$, and thus $\mathbf{NL} = \mathbf{P}$.

While deterministic logspace reductions sound unusual for more general games, we note that the reductions from parity to mean-payoff and from mean-payoff to discounted-payoff in [ ] are deterministic logspace reductions; as $\mathbf{P}$-hardness is not an issue for these games, the inclusion in $\mathbf{L}$ is usually not emphasised. The argument of non-reducibility extends to parity and energy games, as they can encode reachability with 2 priorities and weights 0 and -1, respectively.

Corollary 5. *If* $\mathbf{NL} \neq \mathbf{P}$, *then there is no structure- and probability-preserving translation from two-player parity or energy games to random-turn games.*

We now consider almost-sure and sure winning for unfolding random-turn games and random arena games. We first observe that sure winning for unfolding random-turn game and random arena game is the same as sure winning for random-turn games. The next thing we observe is that almost-sure winning coincides with sure winning for unfolding random-turn games and random arena games. By Theorem 3, we immediately have:

Theorem 6. *It is* $\mathbf{NL}$-*complete to decide whether* Max *wins an unfolding random-turn game or a random arena game almost surely or surely, for reachability, parity, or energy objectives.*

4   Unfolding Random-Turn Game

In this section, we show that deciding whether the winning probability of Max for an unfolding random-turn game is at least a given threshold θ, with any of the objectives—reachability, parity, or energy—is $\mathbf{PSPACE}$-complete.

We state the main complexity results below; the upper bound is proved in Section 4.1 and the lower bound in Section 4.2, respectively.

Theorem 7. *Checking whether player* Max *can win the unfolding random-turn game with a given reachability, parity, or energy objective with a probability at least a given rational threshold θ, encoded in binary, is* $\mathbf{PSPACE}$-*complete.*

4.1   Membership in PSPACE

To determine the winning probability of player Max in unfolding random-turn games, one can transform the game into an equivalent stochastic reachability, parity, or energy game, called the *explicit game*, and then solve this explicit game directly. We formally define the explicit game as follows.

Explicit Game Let $G = (V, E)$ be the game graph of an unfolding random-turn game, and let t be the associated coin-toss function. We assume that t assigns rational probabilities to vertices, each given by a finite binary encoding. The *explicit game* $\mathcal{G}$ is a stochastic game whose states are of the form (G, own, u), where own is an ownership function and $u \in V$ is the current token position. The initial state is $(G, \emptyset, v_0)$, where no ownership is assigned to any node and $v_0 \in V$ is the initial node in the original unfolding random-turn game.

The transitions in the explicit game are defined as follows. At a game state (G, own, u),

- if $\mathrm{own}(u)$ is undefined, then the state is random. It transitions with probability $\mathsf{t}(u)$ to (G, own', u) where $\mathrm{own}' = \mathrm{own} \cup \{u \mapsto \mathsf{Max}\}$, and with probability $1 - \mathsf{t}(u)$ to (G, own'', u) where $\mathrm{own}'' = \mathrm{own} \cup \{u \mapsto \mathsf{Min}\}$;
- otherwise, $\mathrm{own}(u)$ is defined, (G, own', u) is owned by player $\mathrm{own}(u)$ and has successors (G, own, v) for all $(u, v) \in E$.

The first two levels of the explicit game constructed from the unfolding random-turn game on the game graph in Fig. 1(a) is shown in Fig. 1(c). A play $\langle (G, \emptyset, v_0), (G, \mathrm{own}_1, v_1), (G, \mathrm{own}_2, v_2), \ldots \rangle$ is won by player Max if the projected play on V, $\langle v_0, v_1, v_2, \ldots \rangle$, is won by Max. For example, for the *reachability* objective, the game states $(G, \mathrm{own}, \top)$ are the target states. The probability that player Max wins an unfolding random-turn game on G is thus equal to the probability that Max wins the corresponding stochastic game on $\mathcal{G}$.

Directly solving the explicit game—namely, a stochastic reachability, parity, or energy game—yields decidability. However, this approach is inefficient, as the explicit game contains $3^{|V|} \cdot |V|$ states: there are $3^{|V|}$ possible ownership functions and $|V|$ possible token positions. Nevertheless, we show that the winning probability for player Max can be decided in alternating polynomial time (**APTIME**), which is equivalent to **PSPACE** [ ], matching the **PSPACE**-hard lower bound established in Section 4.2.

We first observe that although the explicit game is large, it has a simple structure and directly inherits memoryless optimal strategies from the deterministic versions of the reachability, parity, and energy games. Furthermore, when both players follow their respective memoryless optimal strategies, any path leading to a winning cycle has polynomial length. Moreover, the winning probabilities for Max and Min at each game state can be represented in polynomial size. These properties together allow the winning probability at the initial state of the explicit game to be decided in **APTIME**.

Lemma 8. *Our unfolding random-turn games have (a) optimal memoryless strategies. Moreover, (b) the probability of winning can be written as a fraction with a denominator, which is the product of the denominators of the coin tosses of all unassigned states. If the* Max *(resp.* Min*) player plays optimally they can (c) enforce that the first cycle[2] reached and closed is winning with at least*

---

[2] Note that, for reachability, we assume that target states have only target states as successors.

(resp. at most) their optimal winning probability, (d) has a length of at most $|V|$, and (e) the path to closing the cycles is of length at most $(|V|+1) \cdot (|V|+2)/2$.

Proof. To analyse the unfolding random-turn games, we observe that it has a natural DAG structure, as the set of unassigned states can only fall. We analyse the game inductively from the bottom SCC upwards, starting with the $2^{|V|}$ copies where the ownership of all nodes are assigned (induction basis). For the length of the path until a cycle is closed, we show the stronger claim this to be (f) $\sum_{i=0}^{m}(|V|-i+1)$ if there are m unassigned nodes. Note that (f) entails (e).

For this **induction basis**, the game is an ordinary non-stochastic two player game; they are memoryless determined with memoryless optimal strategies [18,22,29,19,14] (a,c), which guarantee to reach and close a winning cycle with length at most $|V|$ (d) in $|V|+1$ steps (f) for the winning player. Obviously, such strategies are also optimal. As this only allows for the winning probabilities 0 and 1, they are fractions with a denominator being the empty product, showing (b). For the **induction step**, we assume that the properties have been shown for up to m unassigned states and move to $m+1$ unassigned states. From an unassigned node s, the next step is a coin toss that assigns ownership. The probability of winning is $\mathsf{t}(s)$ times the probability of winning when starting from s in the explicit game state that only differs from the current state in that s is assigned to Max (Max winning the coin toss), plus $1 - \mathsf{t}(s)$ times the probability of winning when starting from s in the explicit game state that only differs from the current state in that s is assigned to Min (Min winning the coin toss). By induction hypothesis, the probability can be written as a fraction with a denominator, which is the product of the denominators of the coin tosses of all unassigned states (b).

We also inherit that optimal memoryless policies of Max (resp. Min) guarantee that the first cycle reached is winning with at least (at most) this probability (a). The length of a path until closing this cycle is just one longer than the path from either successor state (c,d,f).

For states that are assigned an owner, we focus on the $|V|$ game states with a fixed partial assignment. For them, we design a game we call a strong (resp. weak) p-game for any $p \in [0, 1]$ by turning all unassigned states with probability of winning $\geq p$ (resp. $> p$) into winning[3] sinks, and those with probability of winning $< p$ (resp. $\leq p$) into losing sinks.

Max (resp. Min) wins with probability at least p (resp. $1 - p$) if they win the strong (resp. weak) p-game by following their memoryless strategy in the strong (resp. weak) p-game on the game with the fixed partial assignment, and the optimal strategy that exists by induction hypothesis after leaving it. This is because, when following this strategy, either a winning cycle is closed within this area (which is then of length $\leq |V| - m$ is reached and closed in $\leq |V| - m + 1$ steps, or a state from which there is an assignment is made by coin toss is reached

---

[3] For reachability games a winning (losing) sink is a final (non-final) state with only a self-loop, for parity games these sink states get an even (odd) priority, for energy games a positive (negative) weight.

within $|V| - m$ steps. From this state, Max (resp. Min) wins with a probability of at least p (resp. $1 - p$) (see above), providing (d,f).

By their construction, for every state there is clear monotonicity in p-games, Max wins the strong 1-game, Min wins the weak 0-game, or there is a p such that Max wins the strong and Min the weak p-game; in each of the three cases, this probability is the one each player can guarantee (a,c). In the latter case, this p is a probability that arises for one of the unassigned states, providing (b) as shown above.

This completes the inductive argument and the proof. $\square$

With Lemma 8 at hand, we obtain an algorithm for deciding whether the winning probability meets a given threshold in **APTIME = PSPACE**.

Theorem 9. *Checking whether player* Max *can win the unfolding random-turn game with a given reachability, parity, or energy objective with a probability at least a given rational threshold θ, encoded in binary, can be done in alternating polynomial time.*

Proof. We can simply play the explicit game forwards with an alternating machine, which first guesses the probability p of winning, given by guessing the numerator and implicitly using the implicit denominator as the product of all denominators of the probability of all unassigned states (initially all states) given by t. We then check that $p \geq \theta$ holds and set the current node v to v_0.

We then repeat the following loop:

1. Write the current node on a string.
2. For states assigned to Max (resp. Min), we existentially (resp. universally) choose a successor, update the current node v to this successor.
3. For unassigned nodes, existentially guess the numerator[4] of the probability of winning for the case that Max wins the coin toss (p_+) and the chance of winning when Min wins the coin toss (p_-). We then universally choose whether to check that the weighted sum of these probabilities is at least the current probability of winning, or to universally make one of these assignments and update the probability accordingly to p_+ resp. p_-.
4. We then offer both players in turn the option to claim that they have closed a winning cycle. A player claiming this wins if they can and loses if they cannot show this. (Note that checking for having closed a winning cycle is an easy deterministic check.)

If the actual probability is $p \geq \theta$, Max can always guess the correct probability and follow their optimal strategy. If the actual probability is $p < \theta$, player Min can always follow their optimal strategy and either challenge in case there if there is a local inconsistency and otherwise challenge p_+ if p_+ is higher than the actual probability, and challenge p_- otherwise. $\square$

---

[4] When representing fractions by the numerator only, using the product of all denominators of unassigned nodes, then we can check $p \geq p_+ \cdot n + p_- \cdot (d - n)$, where d/n is the probability of Max winning the coin toss.

4.2 PSPACE-hardness

To prove **PSPACE**-hardness of computing the winning probability in the unfolding random-turn game, we reduce from the satisfiability problem for quantified Boolean formulas (QBF) in prenex conjunctive normal form (PCNF).

For simplicity, we assume the quantifier prefix of the QBF is strictly alternating, so that the input has the form

$$\Phi \;=\; \forall x_1 \exists y_1 \; \forall x_2 \exists y_2 \; \ldots \; \forall x_n \exists y_n \; \varphi,$$

where the matrix $\varphi = C_1 \wedge C_2 \wedge \cdots \wedge C_m$ is in CNF, and only contains variables occurring in the quantifier prefix. It is well known that deciding satisfiability of such QBFs is **PSPACE**-complete [ ]. Moreover, their semantics can be understood in terms of the following (evaluation) game played by a universal and an existential player. The players assign variables following the order in the quantifier prefix, with the universal player assigning universally quantified variables x_i, and the existential player assigning existentially quantified variables y_i. The existential player wins if the resulting truth assignment satisfies the matrix, and the universal player wins if the assignment falsifies the matrix. A QBF is satisfiable (true) if the existential player has a winning strategy for the evaluation game, and unsatisfiable (false) if the universal player has a winning strategy.

To reduce from QBF satisfiability, we will construct a game graph G_Φ and show that the probability of the Max player winning is at least θ if Φ is satisfiable, and strictly less than θ if Φ is unsatisfiable, for a threshold value θ only depending on n.

Game graph construction. Given Φ, we construct a directed game graph G_Φ with nodes as follows: for each *universally quantified* variable x_i, nodes $\forall x_i$, x_i, $\neg x_i$; for each *existentially quantified* variable y_i, nodes $\exists y_i$, y_i, y_i', $\neg y_i$, $\neg y_i'$, y_i'', $\neg y_i''$; for each *clause* C_j of φ, a node C_j; one conjunction node $\wedge$; and two *sink* nodes $\top$ and $\bot$. The initial node is $\forall x_1$, and the target node is $\top$. Edges are added as follows:

Universal choice. From $\forall x_i$, add edges to x_i and $\neg x_i$.

Existential choice. From $\exists y_i$, add edges to y_i'' and $\neg y_i''$. From y_i'' add an edge to y_i', and from $\neg y_i''$ an edge to $\neg y_i'$. Further, add edges from y_i' to y_i and from $\neg y_i'$ to $\neg y_i$.

Quantifier progression. From x_i and $\neg x_i$, add edges to $\exists y_i$. For $1 \le i < n$, add edges from y_i and $\neg y_i$ to $\forall x_{i+1}$. For $i = n$, add edges from y_n and $\neg y_n$ to $\wedge$.

Clause choice. From $\wedge$, add an edge to each clause C_j.

Literal choice. A literal is a variable or its negation. For each clause C_j and each literal $\ell \in \{x_i, \neg x_i, y_i, \neg y_i\}$, if $\overline{\ell} \in C_j$ add an edge from C_j to ℓ (so each clause connects to the *negations* of its literals).

Sink connections. Add edges to $\top$ from literal nodes x_i, $\neg x_i$, y_i, $\neg y_i$, from every $\forall x_i$, and from $\wedge$. Add edges to $\bot$ from each $\exists y_i$, the nodes y_i', $\neg y_i'$, y_i'', $\neg y_i''$, and each C_j. Finally, add self-loops on $\top$ and $\bot$.

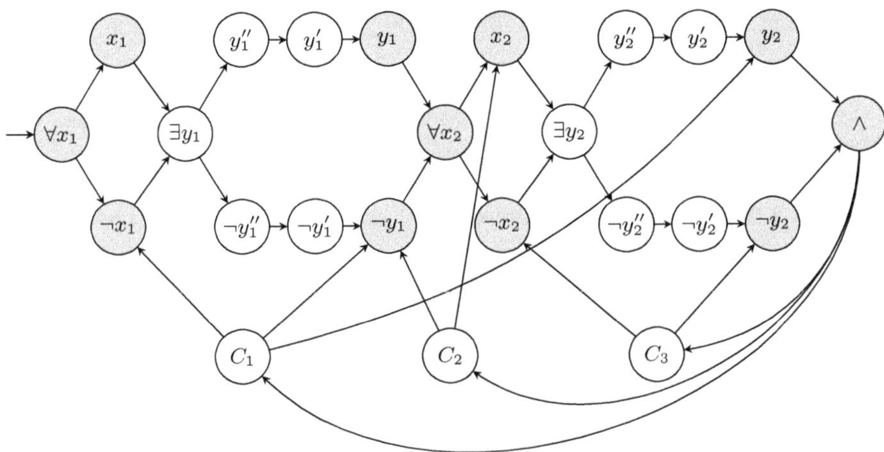

Fig. 2: Simplified game graph for the QBF in Example 10. Shaded nodes have transitions to the target node $\top$, while all other nodes have transitions to $\bot$.

Example 10. Consider the QBF

$$\Phi = \forall x_1 \exists y_1 \forall x_2 \exists y_2 \, \varphi$$

with matrix

$$\varphi = \underbrace{(x_1 \vee y_1 \vee \neg y_2)}_{C_1} \wedge \underbrace{(\neg x_2 \vee y_1)}_{C_2} \wedge \underbrace{(x_2 \vee y_2)}_{C_3}.$$

Fig. 2 shows a simplified drawing of its game graph G_Φ where the sinks and their incident edges are omitted. $\qquad\square$

Consider the unfolding random-turn game with a reachability objective played on G_Φ where each coin toss is of probability $\frac{1}{2}$, that is, each node is assigned to player Max or Min with equal chance. If a node with an edge to $\top$ is assigned to the Max player, Max can win immediately by following that edge. Similarly, we can assume that the Min player always transitions to $\bot$ when given the chance. Conversely, it is clearly never in the interest of the Max player to go to $\bot$, or for the Min player to go to $\top$, so we can assume they never choose such a transition if any other transition is available.

The game only gets interesting once the token arrives at a clause node that is assigned to the Max player. That only happens if, up to that point, any visited node with a $\top$-transition is assigned to Min, and any visited node with a $\bot$-transition is assigned to Max. We now argue that the probability of this event is independent of players' choices up to that point (assuming they play rationally) and only depends on the number of variables in the quantifier prefix.

- Let p_i denote the probability of the Max player winning before the token reaches quantifier node $\forall x_i$ for $1 \leq i \leq n$, and let p_{n+1} be the probability of them winning before the token gets to the $\wedge$-node.

- Further, let q_i be the probability of the token arriving at node $\forall x_i$ for $1 \leq i \leq n$, and let q_{n+1} be the probability of it reaching the $\wedge$-node.
- Finally, let q be the probability of the token reaching a clause node C_j and the Max player gaining control, and let p be the probability of the Max player winning before the token gets to a clause node C_j.

Lemma 11. *Given n, the probabilities defined above can be computed as follows:*

$$q_1 = 1, \quad q_{i+1} = 2^{-6} q_i, \qquad\qquad q = 2^{-2} q_{n+1},$$
$$p_1 = 0, \quad p_{i+1} = p_i + q_i(2^{-1} + 2^{-2} + 2^{-6}), \quad p = p_{n+1} + 2^{-1} q_{n+1}.$$

Proof. Since $\forall x_1$ is the initial state, $q_1 = 1$ and $p_1 = 0$. Reaching $\forall x_{i+1}$ for $1 \leq i < n$ requires reaching $\forall x_i$ with probability q_i, assigning $\forall x_i$, $(\neg)x_i$ and $(\neg)y_i$ to Min, as well as assigning $\exists y_i$, $(\neg)y_i''$ and $(\neg)y_i'$ to Max, so the probability is $q_{i+1} = 2^{-6} q_i$. By the same argument, $q_{n+1} = 2^{-6} q_n$. From node $\forall x_i$ for $1 \leq i \leq n$, the Max player can win before the token arrives at $\forall x_{i+1}$ (for $1 \leq i < n$) or $\wedge$ (for $i = n$) by having one of the nodes with edges to $\top$ assigned to them, which happens with probability $2^{-1} + 2^{-2} + 2^{-6}$. Thus, $p_{i+1} = p_i + q_i(2^{-1} + 2^{-2} + 2^{-6})$. From the $\wedge$-node, the Max player wins if that node is assigned to them, so $p = p_{n+1} + 2^{-1} q_{n+1}$, and the token is passed on to a clause node with Max in control with probability $\frac{1}{4}$, so $q = 2^{-2} q_{n+1}$. $\qquad\square$

The above lemma allows us to compute the probability p of the Max player winning the game before gaining control of a clause node. Next, we will show that they can win with probability $\frac{1}{2}$ from a clause node if they follow an existential winning strategy in the evaluation game for the QBF Φ.

Lemma 12. *If the existential player has a winning strategy in the evaluation game for Φ, then the Max player can win the unfolding random-turn game with a reachability objective on G_Φ with probability at least $p + \frac{1}{2}q$.*

Proof. Interpreting a transition from $\forall v$ to v as setting variable v to true, and a transition to $\neg v$ as setting v to false, before reaching a clause node, the Max player mimics the existential winning strategy for Φ. Assuming rational moves by the Min player, the Max player wins with probability p before making it to a clause node C_j. With probability q, they end up in a clause node C_j that they control. Since they mimicked an existential winning strategy, there is a literal $\ell \in C_j$ that is satisfied by the assignment induced by the players' moves. And so by construction of G_Φ, there is an edge from C_j to the negated literal node $\bar{\ell}$ that has not been visited before. The Max player moves to $\bar{\ell}$, and wins with probability $\frac{1}{2}$ by gaining control of this node and moving to $\top$. Overall, the probability of the Max player winning is at least $p + \frac{1}{2}q$. $\qquad\square$

On the other hand, if the universal player has a winning strategy in the evaluation game, the Min player can mimic this strategy and force the Max player to visit a previously seen node, lowering their chances of reaching $\top$.

Lemma 13. *If the universal player has a winning strategy in the evaluation game for Φ, then the* Max *player can win the unfolding random-turn game with a reachability objective on G_Φ with probability at most $p + \frac{1}{4}q$.*

Proof. In this case, the strategy of the Min player is to mimic the universal winning strategy. As a consequence, upon gaining control of the $\wedge$-node, the Min player can transition to a clause C_j that is falsified by the assignment induced by the players' choices. Assuming rational moves, the Max player wins with probability p before visiting a clause C_j, and arrives at C_j and gains control with probability q. Since C_j is falsified, and by construction of G_Φ, there are only edges to literal-nodes ℓ that have been visited before and that are controlled by the Min player. Assume the Min player simply repeats their previous moves. The only chance for the Max player to win against this strategy is to eventually gain control of a literal node y_i or $\neg y_i$ that has not been visited before and go to the target node $\top$ from there. To do that, they must first go through literal nodes y_i'' and y_i' or $\neg y_i''$ and $\neg y_i'$, and gain control of both, which happens with probability $\frac{1}{4}$. Overall, the Max player can either win before they gain control of a clause node C_j with probability p, or gain control of clause node C_j with probability q, reach a previously unvisited literal node y_i or $\neg y_i$ with probability at most $\frac{1}{4}$, and win from there. Thus the Max player can win with probability at most $p + \frac{1}{4}q$. $\qquad\square$

Theorem 14. *Deciding whether player* Max *can win an unfolding random-turn game with a reachability, parity, or energy objective with probability at least a given threshold θ is* **PSPACE**-*hard.*

Proof. Given a QBF Φ, we construct the game graph G_Φ and compute the probabilities p and q as defined above. Clearly, this can be done in polynomial time. Let $\theta = p + \frac{1}{2}q$.

It follows from Lemma 13 and Lemma 12 that player Max wins the unfolding random-turn game with a reachability objective on the game graph G_Φ with probability at least θ if, and only if, the QBF Φ is satisfiable. This establishes **PSPACE**-hardness for deciding whether the winning probability for player Max in an unfolding random-turn game with a reachability objective is at least a given threshold θ.

This result can be easily extended to unfolding random-turn games with parity or energy objectives. In particular, an unfolding random-turn game with a reachability objective can be transformed into one with a parity or energy objective while preserving Max's winning probability. For parity objective, one can add a self-loop to each target node, assign them an even priority of 2, and assign all other nodes an odd priority of 1. For energy objectives, a self-loop can similarly be added to each target node with a weight of 1, while all other nodes are assigned a weight of -1; the initial energy level can then be set to $(|V| + 1) \cdot (|V| + 2)/2$, which is the upper bound of a path to close a winning cycle according to Lemma 8. $\qquad\square$

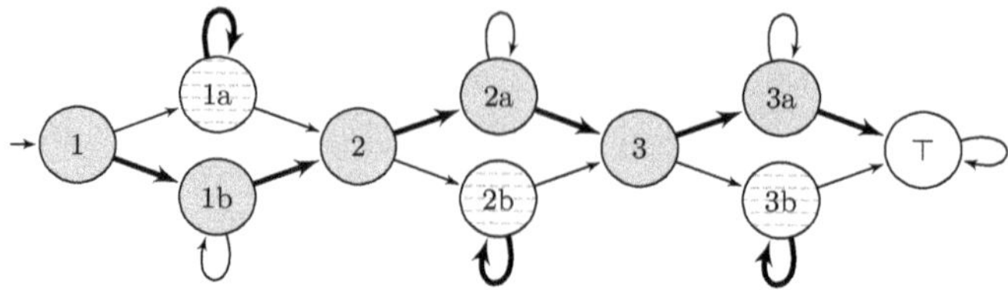

Fig. 3: A random arena game with a reachability objective in which player Max requires distinct strategies for exponentially many different arenas. The figure illustrates one assignment for $\mathcal{G}_3$. The optimal strategies for both players are highlighted using thick edges.

5   Random Arena Game

We show that computing the exact winning probability of Max for random arena game, with any of the objectives—reachability, parity, or energy—is computationally hard. In fact, the problem is #**P**-complete. We then propose approximation algorithms that efficiently estimate this probability, and demonstrate empirically that they achieve fast convergence.

We begin with an example that sheds light on why computing the winning probability for player Max in random arena game is hard. A strategy in random arena game of Max, provides a strategy for each arena on G. Our objective is to compute the probability that player Max has a winning strategy in the randomly generated arena. Note that it is possible that two arenas $\mathcal{A}$ and $\mathcal{A}'$ have the same strategy prescribed by Max. However, for maximising winning probability, Max might require exponentially many different strategies. This is demonstrated by a simple class of games with reachability objectives, $\mathcal{G}_n$ which is a chain as shown in Fig. 3 for $\mathcal{G}_3$. In this game, Max needs a specific path strategy which is following her own nodes, to win this particular game. And there are at least 2^n ways to assign the top and bottom nodes, such the path is unique, requiring a unique distinct strategy. This hints towards the fact that computing optimal strategy can be costly.

5.1   #**P**-Completeness

The hardness follows from a reduction from a variant of the two-terminal reliability problem, which is known to be #**P**-complete [ ]. Given a directed graph $G = (V, E)$ with designated terminals $s, t \in V$, the two-terminal reliability problem asks for the probability that s can reach t, assuming each edge $(u, v) \in E$ is independently present with probability p. This problem remains #**P**-complete even for undirected or acyclic source-sink planar graphs of maximum degree three. We consider an *adjusted* variant, also #**P**-complete, where the start node s has an incoming edge that exists with probability p; the probability of a path from s to t is thus $p \cdot \alpha$, where α is the reachability probability from s to t given that the incoming edge exists. To establish #**P**-hardness for random arena games, we give a logspace reduction from this adjusted problem to a random arena game with a reachability objective, and extend the argument to parity and energy objectives. The intuition behind this reduction is that we construct a random arena game on a graph where the ownership of each node mimics the presence or absence of a corresponding edge in G.

For membership in $\#\mathbf{P}$, the probability that Max wins a random arena game can be computed in $\#\mathbf{P}$ by enumerating all possible ownership assignments, checking for each whether Max has a winning strategy, and counting the fraction of assignments where Max wins. The key difference between reachability objectives and parity or energy objectives is that, while a reachability game induced by a fixed ownership assignment can be solved in polynomial time, no such polynomial-time algorithm is known for parity or energy games. To obtain the $\#\mathbf{P}$ upper bound, we rely on the classic result that solving parity or energy games lies in $\mathbf{UP} \cap \text{co-}\mathbf{UP}$ [,], which allows to non-deterministically guess a unique short certificate for each game. Note that if the subproblem of deciding the winner in a parity or energy game were merely in $\mathbf{NP}$, this approach would fail: multiple short certificates could exist for a positive instance, so counting certificates would not correctly count winning assignments.

Theorem 15. *Computing the winning probability of* Max *in random arena game with a reachability, parity, or energy objective is $\#\mathbf{P}$-complete.*

5.2 FPRAAS

Although computing the exact probability of winning in random arena game is $\#\mathbf{P}$-complete, this does not rule out the possibility of efficiently approximating it. However, it was shown in [ ] that approximation of the two-terminal reliability problem within a given $\varepsilon > 0$ additive error (specified as a rational number in the binary notation) is $\#\mathbf{P}$-hard, so the same holds for approximating the value of our game. This still does not rule out a possibility of approximating this value with a high probability. Indeed, we show that such a scheme exists by leveraging a Monte Carlo method (see, e.g. [ ]) to randomly sample two-player games from the given distribution and checking how often in them player Max wins the game. When this is done for "long enough", the frequency of winning the game in randomly chosen instances is a good approximation of the actual probability winning, but only with probability some less than 1. In order to formally define "long enough", we will make use of a special case of the Hoeffding's inequality when applied to the case where all the random variables are identically distributed and whose value can only be 0 or 1.

Theorem 16 (Hoeffding's inequality [ ]). *Let $X_1, \dots, X_n$ be independent and identically distributed random variables with $X_i \in \{0, 1\}$ almost surely. Then, for any $\varepsilon > 0$, $\Pr\left(\left|\frac{1}{n}\sum_{i=1}^{n} X_i - \mathbb{E}[X_1]\right| \geq \varepsilon\right) \leq 2\,e^{-2\varepsilon^2 n}.$*

Definition 17 (Fully Polynomial Randomised Additive Approximation Scheme (FPRAAS)). *Let $\mathcal{V}$ be a function mapping problem instances x to a*

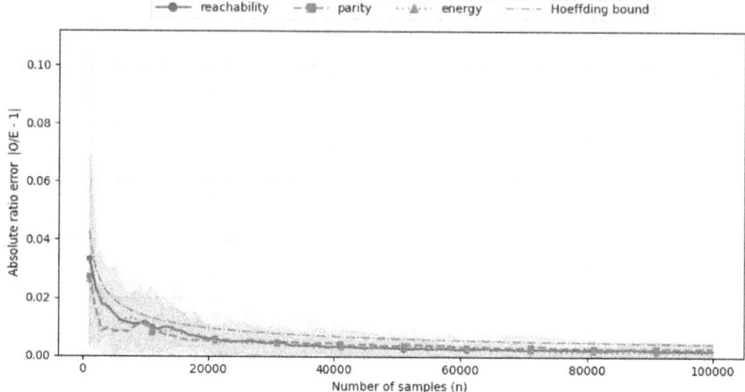

Fig. 4: Error convergence of the FPRAAS for random arena game with reachability, parity, and energy objectives. Each line shows the mean absolute ratio error across 10 random game graphs, with shaded regions indicating variance.

nonnegative real value $\mathcal{V}(x)$. *A randomised algorithm* A *is an additive* (ε, δ)-approximation scheme *for* $\mathcal{V}$ *if, for every instance* x *and every* $\varepsilon, \delta \in (0, 1)$, *the output* $\hat{\mathcal{V}} = A(x, \varepsilon, \delta)$ *satisfies* $\Pr\left[|\hat{\mathcal{V}} - \mathcal{V}(x)| \leq \varepsilon\right] \geq 1 - \delta$.

The algorithm A *is called a* Fully Polynomial Randomised Additive Approximation Scheme (FPRAAS) *if, for any input* (x, ε, δ), A *is an additive* (ε, δ)-*approximation scheme and its running time is polynomial in* $|x|$, $\frac{1}{\varepsilon}$, *and* $\log \frac{1}{\delta}$.

By Hoeffding's inequality (cf. Theorem 16), choosing $n \geq \frac{1}{2\varepsilon^2} \ln \frac{2}{\delta}$ ensures that A is indeed an additive (ε, δ)-approximation scheme, and the running time is polynomial in $1/\varepsilon$ and $\ln(1/\delta)$.

Now, to get an FPRAAS each instance of the game on G has to be solvable in polynomial time. Reachability games can be solved in linear time. Energy games can be solved in polynomial time when the input weights are given in unary. Finally, parity games can be solved in polynomial time if the number of priorities is $\mathcal{O}(\log |G|)$ [28].

Theorem 18. *Random arena games with a reachability objective, a parity objective with* $\mathcal{O}(\log |G|)$ *priorities or an energy objective where weights are given in unary all admit an FPRAAS.*

Implementation and Experiments. We implemented the FPRAAS for random arena game with reachability, parity, and energy objectives [ ]. The associated artifact includes the complete source code, experiment scripts, and datasets necessary to reproduce the results in this paper. It builds on two external tools: ggg [ ] for reachability and parity games, whose solvers are based on [,], and egsolver [ ] for energy games, whose solver is based on [ ]. All experiments were performed on a machine with an Apple M3 8-core CPU and 16 GB of RAM running macOS. For each objective, we generated 10 random game graphs with

20 nodes and maximum outdegree of 20; parity objectives used up to 6 priorities, and energy objectives used an initial credit of 0 with weights in $\{-10, \ldots, 10\}$. We computed the exact winning probability for Max by exhaustively solving all 2^{20} ownership assignments and compared it to the approximate probability from 100,000 random samples. The mean absolute ratio error, $|O/E - 1|$, was computed across all 10 graphs. Fig. 4 shows the error (y-axis) against accumulated samples (x-axis), with blue solid, orange dashed, and green dotted lines for reachability, parity, and energy, respectively. Empirical errors converge much faster than Hoeffding's bound ($\varepsilon = 0.005$, $\delta = 0.05$, dash-dotted grey line), and the variance decreases with increasing samples, as reflected by the narrowing shaded regions. While exact computation takes over an hour for reachability and parity and over 24 hours for energy, our FPRAAS achieves comparable accuracy in about 6 minutes for reachability and parity, and 2 hours for energy, demonstrating that the FPRAAS is both efficient and effective.

6   Discussion

We studied the computational complexity of two-player infinite-duration games in which control of nodes is assigned randomly—either dynamically upon first visit or statically before play begins. These settings differ from the classical random-turn games, where control may be reassigned each time a node is visited. Our results establish a clear complexity landscape across these variants and objectives. While most qualitative questions are **NL**-complete, quantitative problems exhibit higher complexity: **PSPACE**-complete for unfolding random-turn game and #**P**-complete for random arena game. To complement these hardness results, we developed efficient approximation schemes for the latter case and demonstrated their practical effectiveness.

We also explored the possibility of efficiently reducing stochastic games to random-turn games; although our findings provide partial insights, the existence of such a reduction remains an open question. A promising direction for future work is to investigate intermediate mechanisms between random-turn games and unfolding random-turn game, where ownership is fixed only on the k-th visit. Studying how increasing k influences the winning probabilities could reveal a more nuanced understanding of the relationship between randomness in control and strategic outcomes.

Beyond theoretical interest, random arena games have potential applications in game design, particularly in ensuring fairness in competition. Our methods for computing exact winning probabilities can be used to assess whether a given game design between two competitive players is fair—i.e., whether both players have nearly equal chances of winning. Moreover, by adjusting node control probabilities or fixing control on selected nodes, one could systematically balance strategic advantages and improve fairness in game mechanics.

Acknowledgments. This work was supported by the EPSRC through grants EP/X03688X/1 and EP/X042596/1.

References

1. de Alfaro, L., Henzinger, T.A., Majumdar, R.: From verification to control: Dynamic programs for omega-regular objectives. In: 16th Annual IEEE Symposium on Logic in Computer Science, Boston, Massachusetts, USA, June 16-19, 2001, Proceedings. pp. 279–290. IEEE Computer Society (2001). https://doi.org/10.1109/LICS.2001.932504
2. Alur, R., Henzinger, T.A., Kupferman, O.: Alternating-time temporal logic. Journal of the ACM (JACM) **49**(5), 672–713 (2002)
3. Austin, P., Dell'Erba, D., Totzke, P.: Game graph gym. https://github.com/gamegraphgym/ggg, commit: 5d31ac68e36cd781bfa00b9672ea0861872c45cf, Accessed: 2025-10-02
4. Avni, G., Ghorpade, P., Guha, S.: A game of pawns. Log. Methods Comput. Sci. **21**(2) (2025). https://doi.org/10.46298/LMCS-21(2:3)2025
5. Avni, G., Henzinger, T.A., Chonev, V.: Infinite-duration bidding games. Journal of the ACM (JACM) **66**(4), 1–29 (2019)
6. Avni, G., Henzinger, T.A., Ibsen-Jensen, R.: Infinite-duration poorman-bidding games. In: Web and Internet Economics - 14th International Conference, WINE 2018, Oxford, UK, December 15-17, 2018, Proceedings. pp. 21–36 (2018)
7. Bahmani, S., Ibsen-Jensen, R., Paul, S., Schewe, S., Slivovsky, F., Tang, Q., Wojtczak, D., Zhu, S.: The complexity of games with randomised control (2026), https://arxiv.org/abs/2601.07775
8. Bahmani, S., Ibsen-Jensen, R., Paul, S., Schewe, S., Slivovsky, F., Tang, Q., Wojtczak, D., Zhu, S.: Experiments for random arena games (Jan 2026). https://doi.org/10.5281/zenodo.18166538
9. Benerecetti, M., Dell'Erba, D., Mogavero, F.: Solving parity games via priority promotion. Formal Methods Syst. Des. **52**(2), 193–226 (2018). https://doi.org/10.1007/S10703-018-0315-1
10. Bloem, R., Chatterjee, K., Jobstmann, B.: Graph Games and Reactive Synthesis, pp. 921–962. Springer (2018). https://doi.org/10.1007/978-3-319-10575-8_27
11. Bouyer, P., Fahrenberg, U., Larsen, K.G., Markey, N., Srba, J.: Infinite runs in weighted timed automata with energy constraints. In: Cassez, F., Jard, C. (eds.) Formal Modeling and Analysis of Timed Systems, 6th International Conference, FORMATS 2008, Saint Malo, France, September 15-17, 2008. Proceedings. Lecture Notes in Computer Science, vol. 5215, pp. 33–47. Springer (2008). https://doi.org/10.1007/978-3-540-85778-5_4
12. Brim, L., Chaloupka, J., Doyen, L., Gentilini, R., Raskin, J.: Faster algorithms for mean-payoff games. Formal Methods Syst. Des. **38**(2), 97–118 (2011). https://doi.org/10.1007/S10703-010-0105-X
13. Celis, L.E., Devanur, N.R., Peres, Y.: Local dynamics in bargaining networks via random-turn games. In: Proceedings of the 6th International Conference on Internet and Network Economics. p. 133–144. WINE'10, Springer-Verlag, Berlin, Heidelberg (2010)
14. Chandra, A.K., Kozen, D.C., Stockmeyer, L.J.: Alternation. J. ACM **28**(1), 114–133 (Jan 1981). https://doi.org/10.1145/322234.322243
15. Chatterjee, K., Henzinger, T.A.: Reduction of stochastic parity to stochastic mean-payoff games. Inf. Process. Lett. **106**(1), 1–7 (2008). https://doi.org/10.1016/J.IPL.2007.08.035
16. Chatterjee, K., Jurdzinski, M., Henzinger, T.A.: Quantitative stochastic parity games. In: Munro, J.I. (ed.) Proceedings of the Fifteenth Annual ACM-SIAM

Symposium on Discrete Algorithms, SODA 2004, New Orleans, Louisiana, USA, January 11-14, 2004. pp. 121–130. SIAM (2004), http://dl.acm.org/citation.cfm?id=982792.982808

17. Condon, A.: The complexity of stochastic games. Information and Computation pp. 203–224 (1992)
18. Ehrenfeucht, A., Mycielski, J.: Positional strategies for mean payoff games. International Journal of Game Theory **8**, 109–113 (1979). https://doi.org/10.1007/BF01768705
19. Emerson, E.A., Jutla, C.S.: Tree automata, mu-calculus and determinacy (extended abstract). In: FOCS. pp. 368–377. IEEE Computer Society (1991), http://dblp.uni-trier.de/db/conf/focs/focs91.html#EmersonJ91
20. Fijalkow, N., Aiswarya, C., Avni, G., Bertrand, N., Bouyer, P., Brenguier, R., Carayol, A., Casares, A., Fearnley, J., Gastin, P., Gimbert, H., Henzinger, T.A., Horn, F., Ibsen-Jensen, R., Markey, N., Monmege, B., Novotný, P., Ohlmann, P., Randour, M., Sankur, O., Schmitz, S., Serre, O., Skomra, M., Sznajder, N., Vandenhove, P.: Games on graphs: From logic and automata to algorithms (2025). https://doi.org/10.48550/arXiv.2305.10546
21. Goldschlager, L.M.: The monotone and planar circuit value problems are log space complete for P. SIGACT News **9**(2), 25–29 (Summer 1977)
22. Gurvich, V., Karzanov, A., Khachivan, L.: Cyclic games and an algorithm to find minimax cycle means in directed graphs. USSR Computational Mathematics and Mathematical Physics **28**(5), 85–91 (1988). https://doi.org/https://doi.org/10.1016/0041-5553(88)90012-2
23. Hoeffding, W.: Probability inequalities for sums of bounded random variables. Journal of the American statistical association **58**(301), 13–30 (1963)
24. Jurdzinski, M.: Deciding the winner in parity games is in UP ∩ co-UP. Inf. Process. Lett. **68**(3), 119–124 (1998). https://doi.org/10.1016/S0020-0190(98)00150-1
25. Kupferman, O., Vardi, M.Y.: Module checking revisited. In: International Conference on Computer Aided Verification. pp. 36–47. Springer (1997)
26. Lazarus, A.J., Loeb, D.E., Propp, J.G., Stromquist, W.R., Ullman, D.H.: Combinatorial games under auction play. Games and Economic Behavior **27**(2), 229–264 (1999)
27. Lazarus, A.J., Loeb, D.E., Propp, J.G., Ullman, D.: Richman games. Games of no chance **29**, 439–449 (1996)
28. Lehtinen, K., Parys, P., Schewe, S., Wojtczak, D.: A recursive approach to solving parity games in quasipolynomial time. Logical Methods in Computer Science **18** (2022)
29. Mostowski, A.: Games with Forbidden Positions. Preprint - Uniwersytet Gdański. Instytut Matematyki, UG (1991), https://books.google.co.uk/books?id=clvwtgAACAAJ
30. Papadimitriou, C.H.: Computational Complexity. Addison-Wesley, Reading, MA (1994)
31. Peres, Y., Schramm, O., Sheffield, S., Wilson, D.B.: Tug-of-war and the infinity laplacian. Journal of the American Mathematical Society **22**(1), 167–210 (2009). https://doi.org/10.1090/S0894-0347-08-00606-1
32. Peres, Y., Schramm, O., Sheffield, S., Wilson, D.B.: Random-turn hex and other selection games. The American Mathematical Monthly **114**, 373 – 387 (2005), https://api.semanticscholar.org/CorpusID:15583858
33. Pnueli, A., Rosner, R.: On the synthesis of a reactive module. In: Symposium on Principles of Programming Languages (POPL 1989). p. 179–190. Association for Computing Machinery (1989). https://doi.org/10.1145/75277.75293

34. Provan, J.S.: The complexity of reliability computations in planar and acyclic graphs. SIAM Journal on Computing **15**(3), 694–702 (1986). https://doi.org/10.1137/0215050
35. Provan, J.S., Ball, M.O.: The complexity of counting cuts and of computing the probability that a graph is connected. SIAM Journal on Computing **12**(4), 777–788 (1983)
36. Rubinstein, R.Y., Kroese, D.P.: Simulation and the Monte Carlo Method. Wiley Series in Probability and Statistics, Wiley, 2 edn. (2007). https://doi.org/10.1002/9780470230381
37. Totzke, P.: Egsolver. https://github.com/pazz/egsolver, commit:4f8c500f892c10073a8e3b52cf6700ddbf2deeda, Accessed: 2025-10-10
38. Zwick, U., Paterson, M.: The complexity of mean payoff games on graphs. Theor. Comput. Sci. **158**(1&2), 343–359 (1996). https://doi.org/10.1016/0304-3975(95)00188-3

Bridging the Gap Between Plain VASS and Branching VASS

Clotilde Bizière ◉, Jérôme Leroux ◉, and Grégoire Sutre ◉

Univ. Bordeaux, CNRS, Bordeaux INP, LaBRI, UMR 5800, F-33400 Talence, France
`{clotilde.biziere, jerome.leroux, gregoire.sutre}@labri.fr`

Abstract. Vectors addition systems with states (VASS), a model equivalent to Petri nets, are finite-state machines with finitely many counters ranging over the natural numbers. The decidable reachability problem for VASS has many applications in logic, automata, and verification. In this paper we study the reachability problem for BVASS, a branching generalization of VASS. We show that BVASS reachability sets are very similar to VASS reachability sets, namely that they are sections of VASS. Our proof relies on a new well-quasi-order (wqo) on BVASS runs that generalizes the well-known wqo on VASS runs. By leveraging an amalgamation property, we prove that every BVASS run can be transformed into an equivalent one of bounded branching complexity. This allows us to derive several results on the geometry of BVASS reachability sets. As an application we obtain that reachability sets of 5-dimensional BVAS are effectively semilinear, as is the case for 5-dimensional VAS.

Keywords: Branching VASS · Reachability problem · Strahler number · Wqo · Semilinear set.

1 Introduction

Vectors addition systems with states (VASS), a model equivalent to Petri nets, are finite-state m achines w ith fi nitely ma ny co unters ra nging ov er th e natural numbers. Operations on counters are limited to increment and guarded decrement (a counter can be decremented only if it remains non-negative). A central decision problem for VASS is reachability: whether there exists a run from an initial configuration to a final one. This problem was shown decidable more than forty years ago [28] and was recently revisited [26,8,25]. The decidability of the reachability problem for VASS is the cornerstone of many decidability results in logic, automata, and verification [33].

Several VASS extensions have been proposed to increase the expressive power of the model, most notably VASS with nested zero tests [31,14], unordered data nets [20], pushdown VASS [2,19], and branching VASS [13,35]. While the reachability problem is known to be decidable for VASS with nested zero tests, the problem is still open for the last three models.

In this paper, we investigate the reachability problem for branching VASS (shortly called BVASS in the sequel), a branching generalization of VASS. More

N. Bertrand and S. Milius (Eds.): FoSSaCS 2026, LNCS 16503, pp. 65–87, 2026.
https://doi.org/10.1007/978-3-032-22730-0_4

precisely, BVASS extend VASS with special branching transitions that merge configurations (by summing their vectors). Thus runs for BVASS are trees of configurations (whereas they are sequences of configurations for plain VASS). The BVASS model has gained a lot of interest recently due to strong links with several fields in computer science such as cryptographic protocols [35], linear logic [13,21], recursively parallel programs [5], timed pushdown systems [6], computational linguistic [30,32], game semantics [7], equational tree automata [29,27] and data logics [17,4]. As mentioned before, the reachability problem is still open in arbitrary dimension for BVASS. In dimensions one and two, the reachability problem is decidable [3] and the exact complexity is known in dimension one [12,11].

Contributions. We introduce a well-quasi-order (wqo) on the set of runs of a BVASS that generalizes the well-known wqo on VASS runs. As in the case of VASS, this wqo satisfies the *amalgamation property*. Using this property, we prove that reachable configurations are reachable by runs of bounded branching complexity. The idea is to transfer complex computations from one node to a descendant of a sibling node. From this simple form of runs, we deduce several results.

- Reachability sets of BVASS are VASS sections, i.e., projections on a subset of counters of VASS reachable sets intersected with semilinear sets. From this characterization, we derive that BVASS reachability sets are almost semilinear, as for VASS. Moreover, we prove that when the reachability set of a BVASS is semilinear then it is effectively computable, again as for VASS.
- The so-called Strahler-bounded reachability problem is decidable. This problem asks whether a given configuration is reachable by a run whose Strahler number is at most a given bound. We solve that problem by reduction to the reachability problem of VASS with nested zero tests [31]. The Strahler-bounded reachability problem can be seen as a precise under-approximation of the BVASS reachability problem since we prove that for every BVASS, there exists a uniform bound such that every reachable configuration is reached by a run whose Strahler number is at most that bound.
- We then focus on small dimensions and provide an iterative fix-point algorithm based on semilinear sets. The termination of this algorithm guaranteed by the previously shown uniform bound. This iterative algorithm shows that reachability sets are effectively semilinear for 2-dimensional BVASS and 5-dimensional BVAS (BVAS are BVASS with a single state). For 2-BVASS, another algorithm computing the semilinear reachability set was given in [3], but the proof of termination was significantly longer and more involved.

Related Work. In general dimension, the coverability problem (a weak version of the reachability problem) and the boundedness problem are decidable for BVASS [35], and their precise complexity is known [9,21]. The *Strahler number* was introduced by Horton and Strahler in the 1950s in the context of hydrology, and reinvented many times since then [10]. In [2], it is shown that the reachability problem for VASS along finite-index context-free languages is decidable by reduction to the reachability problem for VASS with nested zero tests. As observed in [10], for a context-free grammar in Chomsky normal form, the index of a derivation tree is equal to its Strahler number minus one.

2   Preliminaries

2.1   Generalities

We let $\mathbb{Z}$, $\mathbb{N}$ and $\mathbb{Q}$ denote the usual sets of integers, natural numbers and rational numbers, respectively. We write $\mathbb{Q}_{>0}$ (resp. $\mathbb{Q}_{\geq 0}$) the set of positive (resp. nonnegative) rational numbers. For every $a, b \in \mathbb{N}$, we denote by $[a, b] := \{n \in \mathbb{N} \mid a \leq n \leq b\}$ the integer interval between a and b. Throughout the paper we use the notation $\sqcup$ instead of $\cup$ whenever we want to emphasize that the sets in the union are pairwise disjoint.

Vectors. Given a finite set I of indices and a set $\mathbb{S} \in \{\mathbb{N}, \mathbb{Z}, \mathbb{Q}, \mathbb{Q}_{>0}, \mathbb{Q}_{\geq 0}\}$, an *I-vector over* $\mathbb{S}$ is a function $I \to \mathbb{S}$. Most often, we take $I = [1, d]$, yielding the usual set $\mathbb{S}^d$ of *d-dimensional vectors*. Vectors are written in boldface and, for $i \in I$, the *component* i of the vector $\mathbf{v}$ is denoted $\mathbf{v}(i)$. The restriction of an I-vector $\mathbf{v}$ to a subset $J \subseteq I$ is denoted $\mathbf{v}_{|J}$. The zero I-vector is denoted $\mathbf{0}_I$, $\mathbf{0}_d$ when $I = [1, d]$ or $\mathbf{0}$ when I is clear from context. The product ordering of the usual ordering on $\mathbb{S}$ is denoted $\leq$, that is $\mathbf{u} \leq \mathbf{v}$ if and only if for all $i \in I$, $\mathbf{u}(i) \leq \mathbf{v}(i)$.

Semilinear Sets. A *linear set* of $\mathbb{N}^d$ is a set of the form $\mathbf{L} = \mathbf{b} + \mathbf{P}$ where $\mathbf{b} \in \mathbb{N}^d$ is called the *basis* and $\mathbf{P} = \{n_1 \mathbf{v}_1 + \cdots n_k \mathbf{v}_k \mid n_1, \ldots, n_k \in \mathbb{N}\}$ is called the *periodic set spanned* by the vectors $\mathbf{v}_1, \ldots, \mathbf{v}_k \in \mathbb{N}^d$ called the *periods*. The pair $\gamma = (\mathbf{b}, \mathbf{V})$ is called a *presentation* of the linear set $\mathbf{L}$, and we denote by $[\![\gamma]\!]$ the linear set $\mathbf{L}$ presented by γ. A *semilinear set* of $\mathbb{N}^d$ is a set of the form $\mathbf{S} = \mathbf{L}_1 \cup \ldots \cup \mathbf{L}_k$ where $\mathbf{L}_j$ is a linear set for every $1 \leq j \leq k$. A finite set $\Gamma = \{\gamma_1, \ldots, \gamma_k\}$ where γ_j is a presentation of the linear set $\mathbf{L}_j$ for every $1 \leq j \leq k$ is called a *presentation* of $\mathbf{S}$. In that case, we denote by $[\![\Gamma]\!]$ the semilinear set $\mathbf{S}$ presented by Γ. A semilinear set is said to be *effectively computable* if a presentation of the semilinear set is *computable*.

Multisets. For a finite set Q, we denote by $\mathrm{Mult}(Q)$ the set of all (finite) multisets over Q. A *multiset* M on Q is a function $M : Q \to \mathbb{N}$, where $M(q)$ is the *multiplicity* of $q \in Q$. The empty multiset (multiplicity 0 for every element) is denoted $\emptyset$, like the empty set. Given a list $q_1, \ldots, q_k$ of elements of Q, we denote $\{\!\!\{q_1, \ldots, q_k\}\!\!\}$ the multiset containing every element with the same multiplicity as in the list. The *cardinality* of M, written $|M|$, is defined as $|M| := \sum_{q \in Q} M(q)$. The *union* of two multisets $M \uplus N$ is defined by $(M \uplus N)(q) := M(q) + N(q)$ for all $q \in Q$.

Rooted labelled trees. A *(finite rooted directed labelled) tree* α is a finite labelled directed graph $(\mathcal{N}_\alpha, \to_\alpha, \lambda_\alpha)$, with set of *nodes* $\mathcal{N}_\alpha$, binary *child* relation $\to_\alpha$ and labelling function λ_α over the domain $\mathcal{N}_\alpha$, that contains a special node, called the *root* and denoted by root_α, that has no predecessor and that has a unique directed path to every other node. We write $\alpha \approx \beta$ when the trees α and β are (graph) isomorphic. We use the standard terminology of leaves, internal nodes, depth, arity, children, parents, descendants, and ancestors. We call a node of a tree *branching* if it has at least two children and *unary* if it has exactly one child (otherwise, it has no child and it is a leaf).

For all $n, m \in \mathcal{N}_\alpha$, we write $n \xrightarrow{*}_\alpha m$ and $n \xrightarrow{+}_\alpha m$ if m is respectively a descendant or a strict descendant of n. Given a node $n \in \mathcal{N}_\alpha$, we denote by $\alpha_{|n}$ the subtree of α rooted at n, that is, the tree whose set of nodes is $\{m \in \mathcal{N}_\alpha \mid n \xrightarrow{*}_\alpha m\}$ and whose child relation and labelling functions are those of α restricted to this set. For proofs by induction on the structure of trees, we will write that a tree is of the form $n\{\alpha_1, \dots, \alpha_k\}$, where $n \in \mathcal{N}_\alpha$ is the root of the tree and $\alpha_1, \dots, \alpha_k$, possibly empty, is the set of subtrees rooted at the children of n.

Branching depth and Strahler number. We use two measures of the structural complexity of a tree, namely the *branching depth* and the *Strahler number*.

The *branching depth* of a tree α is the depth of the tree obtained from α by merging every unary node with its child. It is defined recursively by $\mathrm{bd}(n\{\}) = 0$, $\mathrm{bd}(n\{\alpha_1\}) = \mathrm{bd}(\alpha_1)$ for trees whose root has arity 1 and $\mathrm{bd}(n\{\alpha_1, \dots, \alpha_k\}) = 1 + \max_{i=1,\dots,k} \mathrm{bd}(\alpha_i)$ if $k \geq 2$.

The second measure of tree complexity that we use is the *Strahler number*, a less well-known concept with an interesting history. It was introduced by Horton and Strahler in the 1950s in hydrology as a tool to quantify the branching complexity of river networks. Since then, Strahler numbers have been reinvented many times under different names and applied in a wide range of fields within computer science. We refer the reader to the survey [10] for a detailed account of these developments.

The Strahler number of a tree α is defined recursively as follows. Leaves have Strahler number 0. If the root has children with Strahler numbers $s_1, \dots, s_k$, let $s = \max\{s_1, \dots, s_k\}$. Then the Strahler number of the root is $s+1$ if s occurs at least twice among the s_i, and it is s otherwise. An example of a tree with nodes labelled by their Strahler number is given on the left of Figure 1.

2.2 Wqo

This article relies on results about well quasi-orders, but requires no prior familiarity with the topic. We therefore keep the presentation minimal and refer the reader to [34] for more background and for complete proofs.

A *quasi-order* is a pair $(X, \leq)$ where $\leq$ is a reflexive and transitive relation on X. Two elements $x, y \in X$ are said to be $\leq$-*equivalent* if $x \leq y$ and $y \leq x$. The $\leq$-*upward-closure* $\uparrow_\leq(Y)$ of a subset $Y \subseteq X$ is the set $\{x \in X \mid \exists y \in Y, y \leq x\}$. We

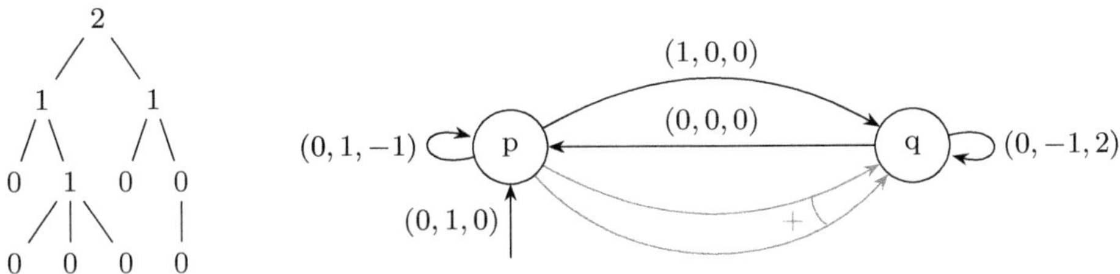

Fig. 1. On the left: a tree where nodes are labelled by their Strahler number (node identifiers omitted). On the right: a 3-BVASS with states $Q := \{p, q\}$ and transitions $\Delta :=$ $\{(\emptyset, (0, 1, 0), p), (\{\!| p |\!\}, (0, 1, -1), p), (\{\!| p |\!\}, (1, 0, 0), q), (\{\!| q |\!\}, (0, -1, 2), q), (\{\!| q |\!\}, \mathbf{0}, p),$ $(\{\!| p, p |\!\}, \mathbf{0}, q)\}$. Without the branching transition depicted by two red arrows, we would have the VASS of states Q, initial configurations $\mathbf{S} := \{p(0, 1, 0)\}$ and transitions $\Delta' := \{(p, (0, 1, -1), p), (p, (1, 0, 0), q), (q, (0, -1, 2), q), (q, \mathbf{0}, p)\}$

say that Y is $\leq$-*upward-closed* when $Y = \uparrow_{\leq}(Y)$. A $\leq$-*basis* of a $\leq$-upward-closed subset $Y \subseteq X$ is a set $B \subseteq Y$ such that $Y = \uparrow_{\leq}(B)$.

A quasi-order $(X, \leq)$ is a *well quasi-order* (wqo) if for every infinite sequence $(x_i)_{i \in \mathbb{N}}$ with $x_i \in X$, there exist indices $i < j$ such that $x_i \leq x_j$. Equivalently, $(X, \leq)$ is a wqo if every $\leq$-upward-closed subset $Y \subseteq X$ admits a finite $\leq$-basis.

Example 1. $(\mathbb{N}, \leq)$ is a wqo. Every finite quasi-order is a wqo (in particular, $(Q, =)$ where Q is a finite set).

Many natural constructions preserve wqos. We recall below the ones used in this paper.

Order-reflecting functions. If $f : (X, \leq_X) \to (Y, \leq_Y)$ satisfies $f(x) \leq_Y f(x') \implies x \leq_X x'$, and $(Y, \leq_Y)$ is a wqo then $(X, \leq_X)$ is also a wqo.

Monotonic functions. If a surjective function $f : (X, \leq_X) \to (Y, \leq_Y)$ satisfies $x \leq_X x' \implies f(x) \leq_Y f(x')$, and $(X, \leq_X)$ is a wqo then $(Y, \leq_Y)$ is also a wqo.

Cartesian products. If $(X, \leq_X)$ and $(Y, \leq_Y)$ are wqo, then so is their product $(X \times Y, \leq_{X \times Y})$, where $(x, y) \leq_{X \times Y} (x', y')$ iff $x \leq_X x'$ and $y \leq_Y y'$.

Words. If $(X, \leq)$ is a wqo, then so is the set X^* of finite words is ordered by the *subsequence embedding*: $x_1 \cdots x_m \leq_{X^*} y_1 \cdots y_n$ if there exist indices $1 \leq i_1 < \cdots < i_m \leq n$ such that $x_j \leq y_{i_j}$ for all j.

Trees. If $(X, \leq)$ is a wqo, then so is the set $\mathcal{T}(X)$ of finite rooted trees labelled by X, ordered by *homeomorphic embedding*: a tree t is embedded into t', written $t \leq_{\mathcal{T}(X)} t'$, if there is an injective map $f : \mathcal{N}_t \to \mathcal{N}_{t'}$ such that (i) for every $n \in \mathcal{N}_t, \lambda_t(n) \leq \lambda_{t'}(f(n))$ and (ii) for every distinct children m and m' of a node $n \in \mathcal{N}_t$, $f(m)$ and $f(m')$ belong to distinct subtrees of $f(n)$.

Finally, given a wqo $(X, \leq_X)$, we define a wqo $\leq_{\mathrm{Mult}(X)}$ on $\mathrm{Mult}(X)$ in the following way. Consider the function $f : x_1 x_2 ... x_\ell \in X^* \mapsto \{\!| x_1, ..., x_\ell |\!\} \in \mathrm{Mult}(X)$. For every $M, M' \in \mathrm{Mult}(X)$, we say that $M \leq_{\mathrm{Mult}(X)} M'$ if there exist $w, w' \in X^*$ such that $f(w) = M, f(w') = M'$ and $w \leq_{X^*} w'$. The function f is a monotonic surjective function from the wqo $(X^*, \leq_{X^*})$ to the quasi-order $(\mathrm{Mult}(X), \leq_{\mathrm{Mult}(X)})$, so $(\mathrm{Mult}(X), \leq_{\mathrm{Mult}(X)})$ is a wqo.

2.3   VASS and extensions

VASS. A *d-dimensional VASS* (*d*-VASS) is a triple $\mathcal{V} := (Q, \Delta, \mathbf{S})$ where Q is a nonempty finite set of *states*, $\Delta \subseteq Q \times \mathbb{Z}^d \times Q$ is a finite set of *transitions*, and $\mathbf{S} \subseteq Q \times \mathbb{N}^d$ is a finite set of initial configurations. The *configurations* of $\mathcal{V}$ are pairs of the form $\mathbf{c} := (q, \mathbf{x})$, usually written $q(\mathbf{x})$, with $q \in Q$ and $\mathbf{x} \in \mathbb{N}^d$. Intuitively, a VASS is an automaton equipped with counters which can be incremented and decremented. Counters must always remain nonnegative, and any transition that would make a counter negative is disabled.

A *run* of $\mathcal{V}$ is a finite sequence $\rho := q_1(\mathbf{x}_1)q_2(\mathbf{x}_2)\ldots q_k(\mathbf{x}_k)$ of configurations such that the first configuration $q_1(\mathbf{x}_1)$, called *source* and denoted $\mathbf{src}(\rho)$, belongs to $\mathbf{S}$, and every other configuration is obtained from the previous one by applying a transition, *i.e.* for each $i \in [1, k-1]$, $(q_i, \mathbf{x}_{i+1} - \mathbf{x}_i, q_{i+1}) \in \Delta$. The last configuration $q_k(\mathbf{x}_k)$ is called the *target* and denoted $\mathbf{tgt}(\rho)$. The set of runs of $\mathcal{V}$ is denoted $\mathbf{Runs}(\mathcal{V})$. A finite sequence that satisfies the two conditions above but whose elements are in $Q \times \mathbb{Z}^d$ instead of $Q \times \mathbb{N}^d$ is called a $\mathbb{Z}$-*run*. The *reachability set* $\mathbf{Reach}(\mathcal{V})$ is the set of targets of runs of $\mathcal{V}$, *i.e.* $\mathbf{Reach}(\mathcal{V}) := \{\mathbf{tgt}(\rho) \mid \rho \in \mathbf{Runs}(\mathcal{V})\}$. The *sections* of $\mathcal{V}$ are the sets obtained from $\mathbf{Reach}(\mathcal{V})$ by intersecting with a semilinear set and projecting. In this article, we only consider sections of the form $\mathbf{Sect}_I^{Q'} := \{q(\mathbf{x}_{|I}) \mid q \in Q', \mathbf{x}_{|[1,d]\setminus I} = \mathbf{0}, q(\mathbf{x}) \in \mathbf{Reach}(\mathcal{V})\}$ where $Q' \subseteq Q$ and $I \subseteq [1, d]$. The *reachability problem* asks, given a VASS $\mathcal{V}$ and a configuration $\mathbf{c}$, whether $\mathbf{c} \in \mathbf{Reach}(\mathcal{V})$.

BVASS. A *d-dimensional branching VASS* (*d*-BVASS) is a pair $\mathcal{B} := (Q, \Delta)$ where Q is a finite set of *states* and $\Delta \subseteq \mathrm{Mult}(Q) \times \mathbb{Z}^d \times Q$ is a finite set of *transitions (rules)*. The *arity* $\mathrm{ar}(\delta)$ of a transition rule $\delta := (P, \mathbf{a}, q)$ is $|P|$. We say that δ is *branching* if $\mathrm{ar}(\delta) \geq 2$. As for VASS, we call *configurations* of $\mathcal{V}$ the pairs $(q, \mathbf{x}) \in Q \times \mathbb{N}^d$, and we write them $\mathbf{c} := q(\mathbf{x})$. Configurations $q(\mathbf{a})$ where $(\emptyset, \mathbf{a}, q) \in \Delta$ are called *initial configurations*. An example of a BVASS is given on the right of Figure 1.

A *run* of $\mathcal{B}$ is a tree $\alpha \in \mathcal{T}(Q \times \mathbb{N}^d)$ such that for every node $n \in \mathcal{N}_\alpha$ with label $q(\mathbf{x})$ and children labelled $p_1(\mathbf{x}_1), \ldots, p_r(\mathbf{x}_r)$, $(\{\!| p_1, \ldots, p_r |\!\}, \mathbf{x} - \sum_{i=1}^r \mathbf{x}_i, q) \in \Delta$, *i.e.* leaves are labelled by initial configurations and interior nodes are obtained from their children by applying a transition rule. The label of a node $n \in \mathcal{N}_\alpha$ is denoted by $\mathbf{config}_\alpha(n)$, its state by $\mathrm{state}_\alpha(n)$ and its counters by $\mathbf{cnt}_\alpha(n)$. The subscript α is dropped when clear from context. The configuration labelling the root of α is called the *target* of α, written $\mathbf{tgt}(\alpha)$. A tree that satisfies the two conditions above but has labels in $Q \times \mathbb{Z}^d$ instead of $Q \times \mathbb{N}^d$ is called a $\mathbb{Z}$-*run*. Reachability sets and sections are defined and denoted exactly as in the case of VASS. The BVASS reachability problem is also defined similarly.

A set of configurations $\mathbf{I} \subseteq Q \times \mathbb{N}^d$ is called an *inductive invariant* for a BVASS $\mathcal{B} = (Q, \Delta)$ if for every sequence $p_1(\mathbf{x}_1), \ldots, p_r(\mathbf{x}_r) \in \mathbf{I}$ and every configuration $q(\mathbf{x})$ such that $(\{\!|\, p_1, \ldots, p_r \,|\!\}, \mathbf{x} - \sum_{i=1}^{r} \mathbf{x}_i, q) \in \Delta$, we have $q(\mathbf{x}) \in \mathbf{I}$. Observe that the reachability set of $\mathcal{B}$ is an inductive invariant and is contained in every inductive invariant. In particular a configuration is not reachable for a BVASS, if and only if, there exists an inductive invariant that does not contain that configuration.

3   A wqo on BVASS runs

3.1   Background: the wqo on VASS runs

We start by informally recalling Jančar's wqo on VASS runs [18,24]—of which our wqo on BVASS runs is a generalisation—and by giving intuitions common to both.[1]

Let $\mathcal{V}$ be a VASS. A *cycle* of $\mathcal{V}$ is defined like a $\mathbb{Z}$-run, except that it starts and ends in the same state and its source is not required to be an initial configuration. The *effect* of a cycle π is the difference $\mathbf{effect}(\pi) := \mathbf{y} - \mathbf{x}$ where $q(\mathbf{x})$ and $q(\mathbf{y})$ are the source and target of π, respectively. Given a run $\rho := q_1(\mathbf{x}_1) \ldots q_k(\mathbf{x}_k)$ and a family $\boldsymbol{\pi} := (\pi_i)_{i \in [1,k]}$ where π_i is a cycle starting and ending in q_i for each $i \in [1, k]$, we denote $\rho\langle\boldsymbol{\pi}\rangle$ the $\mathbb{Z}$-run obtained by inserting in ρ each cycle π_i (translated by an appropriate vector) after the configuration $q_i(\mathbf{x}_i)$. In other words, $\rho\langle\boldsymbol{\pi}\rangle$ first executes the cycle π_1 (possibly translated by some vector in order to start from $q_1(\mathbf{x}_1)$), then the transition $(q_1, \mathbf{x}_2 - \mathbf{x}_1, q_2)$, then the cycle π_2 (translated by the appropriate vector), etc. We say that $\boldsymbol{\pi}$ is *prefix-positive* if for every $j \in [1, k]$, $\sum_{i=1}^{j} \mathbf{effect}(\pi_i) \geq \mathbf{0}$. The wqo is defined by $\rho \trianglelefteq \rho'$ if $\rho' = \rho\langle\boldsymbol{\pi}\rangle$ for some prefix-positive family $\boldsymbol{\pi}$.

To prove that this ordering is a wqo, we view it as a subword embedding over a suitably chosen wqo alphabet and then use the wqo on words (see Section 2.2). The prefix-positive condition in the definition is essential. It implies the following *amalgamation property*: given a run ρ, we can combine two runs $\rho_1 := \rho\langle\boldsymbol{\pi_1}\rangle$ and $\rho_2 := \rho\langle\boldsymbol{\pi_2}\rangle$ such that $\boldsymbol{\pi_1}$ and $\boldsymbol{\pi_1}$ are *prefix-positive* into the run $\rho\langle\boldsymbol{\pi_1\pi_2}\rangle$, which satisfies both $\rho_1 \trianglelefteq \rho\langle\boldsymbol{\pi_1\pi_2}\rangle$ and $\rho_2 \trianglelefteq \rho\langle\boldsymbol{\pi_1\pi_2}\rangle$. Here, we let $\boldsymbol{\pi_1\pi_2}$ stand for the prefix-positive family defined by $(\pi_1\pi_2)_i := \pi_{1,i}\pi_{2,i}$.

This wqo has in particular been used as a kind of pumping lemma to prove that some functions are not weakly computable by VASS [23], nor by their extension with a pushdown [22], and to establish geometric properties of VASS reachability sets [24]. In this paper, we exploit the wqo in a different, novel way. Given a BVASS run α, we do not look for a pattern that can be pumped to produce runs with a larger target. Instead, we aim to rearrange α into a new run α' with the same target but lower structural complexity (see Section 4).

---

[1] Our presentation slightly differs from [18,24] as they allow comparable runs to have distinct (but comparable) source configurations. In our setting, runs always start from an initial configuration, of which there are only finitely many, so we require the source configurations to be the same for simplicity.

3.2 Contexts

We fix a d-BVASS $\mathcal{B} := (Q, \Delta)$ until the end of this section.

A *context* of $\mathcal{B}$ is defined like a run, except for a distinguished *hole* leaf, which is not required to be an initial configuration and whose ancestors may carry negative counters. In addition the hole and the root must have the same state. Formally, a context is a pair (γ, n) with $\gamma \in \mathcal{T}(Q \times \mathbb{Z}^d)$ and $n \in \mathcal{N}_\gamma$ such that: (i) every node that is not an ancestor of n has its label in $Q \times \mathbb{N}^d$; (ii) for every node $m \in \mathcal{N}_\gamma \setminus \{n\}$ with label $q(\mathbf{x})$ and children labelled $p_1(\mathbf{x}_1), \ldots, p_r(\mathbf{x}_r)$, we have $(\{\!\vert p_1, \ldots, p_r \vert\!\}, \mathbf{x} - \sum_{i=1}^r \mathbf{x}_i, q) \in \Delta$; and (iii) it holds that $\mathrm{state}(n) = \mathrm{state}(\mathrm{root}_\gamma)$. For conciseness, we will slightly abuse notation and write "let γ be a context" while denoting its hole by hole_γ. We also define $\mathcal{N}'_\gamma := \mathcal{N}_\gamma \setminus \{\mathrm{hole}_\gamma\}$. We define the *effect* of a context γ by $\mathbf{effect}(\gamma) := \mathbf{cnt}(\mathrm{root}_\gamma) - \mathbf{cnt}(\mathrm{hole}_\gamma)$. The sum of a label $q(\mathbf{x})$ in $Q \times \mathbb{Z}^d$ and a vector $\mathbf{y} \in \mathbb{Z}^d$ is defined by $q(\mathbf{x}) + \mathbf{y} := q(\mathbf{x} + \mathbf{y})$.

We say that two contexts γ_1 and γ_2 are *equivalent*, written $\gamma_1 \equiv \gamma_2$, if they have same nodes, same child relation, same hole, and if there is $\mathbf{a} \in \mathbb{Z}^d$ such that, for each node n, $\mathbf{config}_{\gamma_1}(n) = \mathbf{config}_{\gamma_2}(n) + \mathbf{a}$ if n belongs to the branch from the root to the hole and $\mathbf{config}_{\gamma_1}(n) = \mathbf{config}_{\gamma_2}(n)$ otherwise. Clearly, for every context γ and $\mathbf{z} \in \mathbb{Z}^d$, there is a unique context $\gamma' \equiv \gamma$ whose hole has counters $\mathbf{z}$. Note that two equivalent contexts have the same effect.

We say that a run α is *pluggable* in a context γ if the state of hole_γ is the same as that of root_α and $\mathcal{N}_\alpha \cap \mathcal{N}_\gamma = \emptyset$. In that case, we denote $\gamma[\alpha]$ the $\mathbb{Z}$-run obtained by taking the context $\gamma' \equiv \gamma$ with $\mathbf{config}(\mathrm{hole}_\gamma) = \mathbf{tgt}(\alpha)$ and merging hole_γ with root_α. Formally, $\mathcal{N}_{\gamma[\alpha]} := \mathcal{N}'_{\gamma'} \sqcup \mathcal{N}_\alpha$, every node has the same label in $\gamma[\alpha]$ as in its original tree, and the child relation of $\gamma[\alpha]$ is the union of that of γ and that of α, except that root_α replaces hole_γ. Observe that $\mathbf{tgt}(\alpha)$ and $\mathbf{tgt}(\gamma[\alpha])$ have the same state as hole_γ and root_γ have the same state.

Similarly, we say that a context γ_2 is *concatenable* with another context γ_1 if $\mathrm{state}(\mathrm{root}_{\gamma_2}) = \mathrm{state}(\mathrm{hole}_{\gamma_1})$ and $\mathcal{N}_\alpha \cap \mathcal{N}_\gamma = \emptyset$. In that case, the *concatenation* $\gamma_1 \gamma_2$ of γ_2 with γ_1 is obtained by taking the context $\gamma'_1 \equiv \gamma_1$ with $\mathbf{config}(\mathrm{hole}_{\gamma_1}) = \mathbf{tgt}(\gamma_2)$ and merging hole_{γ_1} with root_{γ_2}.

Fact 2.   1. For context γ and run α pluggable in γ, $\mathbf{tgt}(\gamma[\alpha]) = \mathbf{tgt}(\alpha) + \mathbf{effect}(\gamma)$.

   2. For every concatenable contexts γ_1 and γ_2, $\mathbf{effect}(\gamma_1 \gamma_2) = \mathbf{effect}(\gamma_1) + \mathbf{effect}(\gamma_2)$.

Let α be a run and $\boldsymbol{\gamma} := (\gamma_n)_{n \in \mathcal{N}_\alpha}$ be a family of contexts indexed by the nodes of α. We say that $\boldsymbol{\gamma}$ is *compatible* with α if $\mathrm{state}(\mathrm{hole}_{\gamma_n}) = \mathrm{state}(n)$ for every $n \in \mathcal{N}_\alpha$, and if all the sets $\mathcal{N}_\alpha$ and $\mathcal{N}_{\gamma_n}$ for $n \in \mathcal{N}_\alpha$ are pairwise disjoint. In that case, we define recursively the *insertion* of the context family $\boldsymbol{\gamma}$ into the run α, written $\alpha\langle\boldsymbol{\gamma}\rangle$, by $\alpha\langle\boldsymbol{\gamma}\rangle := \gamma_n[n\{\alpha_1\langle(\gamma_m)_{m \in \mathcal{N}_{\alpha_1}}\rangle, \ldots, \alpha_k\langle(\gamma_m)_{m \in \mathcal{N}_{\alpha_k}}\rangle\}]$ if $\alpha = n\{\alpha_1, \ldots, \alpha_k\}$ with $k \geq 0$. An illustration is provided in Figure 2. In general, $\alpha\langle\boldsymbol{\gamma}\rangle$ is a $\mathbb{Z}$-run, but not necessarily a run.

Fact 3. For every run α, family of contexts $\boldsymbol{\gamma} := (\gamma_n)_{n \in \mathcal{N}_\alpha}$ compatible with α:

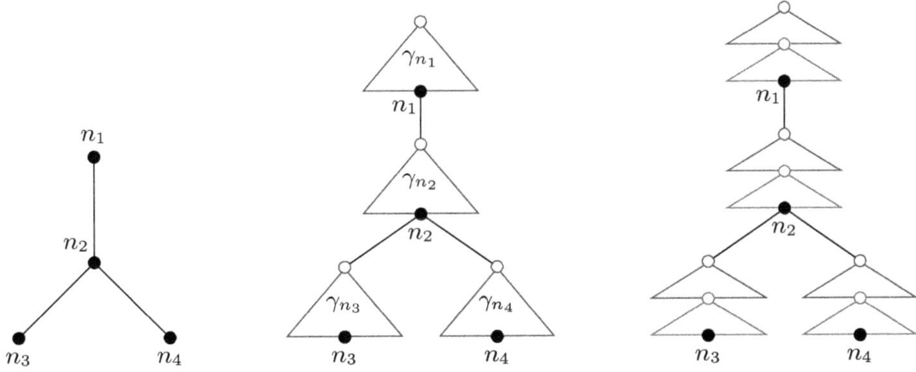

Fig. 2. Illustration of the binary relation $\trianglelefteq$ (Definition 4) and of the amalgamation property (Proposition 5). The left part depicts a run α and the middle part depicts the run $\alpha\langle\gamma\rangle$ obtained by inserting a context family $\gamma := (\gamma_n)_{n\in\mathcal{N}_\alpha}$ into α. If γ is α-hereditarily positive then $\alpha \trianglelefteq \beta$ for every run $\beta \approx \alpha\langle\gamma\rangle$. Notice that γ is α-hereditarily positive iff $\sum_{i\in I} \mathbf{effect}(\gamma_{n_i}) \geq \mathbf{0}$ for every $I \in \{\{3\}, \{4\}, \{2,3,4\}, \{1,2,3,4\}\}$. The right part depicts the amalgamation of two runs β and β' such that $\alpha \trianglelefteq \beta$ and $\alpha \trianglelefteq \beta'$.

1. for every node $n \in \mathcal{N}_\alpha$, $\mathbf{cnt}_{\alpha\langle\gamma\rangle}(n) = \mathbf{cnt}_\alpha(n) + \sum_{m\in\mathcal{N}_{\alpha_{|n}}\setminus\{n\}} \mathbf{effect}(\gamma_m)$.
2. $\mathbf{tgt}(\alpha\langle\gamma\rangle) = \mathbf{tgt}(\alpha) + \mathbf{effect}(\gamma)$, where $\mathbf{effect}(\gamma) := \sum_{n\in\mathcal{N}_\alpha} \mathbf{effect}(\gamma_n)$

We say that the family is α-*hereditarily positive* if for every $n \in \mathcal{N}_\alpha$, we have $\sum_{m\in\mathcal{N}_{\alpha_{|n}}} \mathbf{effect}(\gamma_m) \geq \mathbf{0}$.

3.3 A wqo on the BVASS runs

Definition 4 (Wqo on BVASS runs). *We define the binary relation $\trianglelefteq$ on the runs of $\mathcal{B}$ by $\alpha \trianglelefteq \beta$ if there is a hereditarily positive family of contexts $\gamma := (\gamma_n)_{n\in\mathcal{N}_\alpha}$ such that[2] $\beta \approx \alpha\langle\gamma\rangle$.*

Proposition 5 (Amalgamation). *Given a run α and two hereditarily positive families of contexts γ_1 and γ_2 compatible with α such that $\alpha\langle\gamma_1\rangle$ and $\alpha\langle\gamma_2\rangle$ are runs, $\alpha\langle\gamma_1\gamma_2\rangle$ is also a run (where $(\gamma_1\gamma_2)_n := \gamma_{1,n}\gamma_{2,n}$). Moreover, we have $\alpha\langle\gamma_1\rangle \trianglelefteq \alpha\langle\gamma_1\gamma_2\rangle$ and $\alpha\langle\gamma_2\rangle \trianglelefteq \alpha\langle\gamma_1\gamma_2\rangle$.*

We provide a visual description of Definition 4 and Proposition 5 in Figure 2. The rest of this subsection is devoted to the proof that $(\mathbf{Runs}(\mathcal{B}), \trianglelefteq)$ is a wqo.

An *instantiated rule* of $\mathcal{B}$ is a triple $(\{\!|\, q_1(\mathbf{x}_1), \ldots, q_k(\mathbf{x}_k) \,|\!\}, \delta, q(\mathbf{x}))$ such that $q_1(\mathbf{x}_1), \ldots, q_k(\mathbf{x}_k), q(\mathbf{x})$ are configurations and $\delta = (\{\!|\, q_1, \ldots, q_k \,|\!\}, \mathbf{x} - \sum_{i=1}^{k} \mathbf{x}_i, q)$ is in Δ. Let Σ be the set of instantiated rules of $\mathcal{B}$. We equip Σ with the product order $\leq_\Sigma := \leq_{\mathrm{Mult}(Q\times\mathbb{N}^d)} \times =_\Delta \times \leq_{Q\times\mathbb{N}^d}$, where $\leq_{Q\times\mathbb{N}^d}$ is the product order

---

[2] Recall that $\approx$ denotes (graph) isomorphism on trees.

$=_Q \times \leq_{\mathbb{N}^d}$. By Cartesian product (see Section 2.2), both $(Q \times \mathbb{N}^d, \leq_{Q \times \mathbb{N}^d})$ and $(\Sigma, \leq_\Sigma)$ are wqos. Using the wqo on trees (see Section 2.2), we lift the wqo $(\Sigma, \leq_\Sigma)$ to a wqo $(\mathcal{T}(\Sigma), \leq_{\mathcal{T}(\Sigma)})$.

We associate to every run α of $\mathcal{B}$ a tree of instantiated rules decorate(α). The tree decorate(α) has same nodes and same child relation as α, but its labels are enriched: if a node $n \in \mathcal{N}_\alpha$ is labelled by $q(\mathbf{x})$ in α, and its children are labelled by $q_1(\mathbf{x}_1), \ldots, q_k(\mathbf{x}_k)$ in α, then n is labelled in decorate(α) by $(\{\!\mid q_1(\mathbf{x}_1), \ldots, q_k(\mathbf{x}_k) \mid\!\}, \delta, q(\mathbf{x}))$ where $\delta = (\{\!\mid q_1, \ldots, q_k \mid\!\}, \mathbf{x} - \sum_{i=1}^{k} \mathbf{x}_i, q)$. Note that δ is guaranteed to be in Δ by definition of a BVASS run, so the labels of decorate(α) are indeed instantiated rules.

Lemma 6. *Let α and β be two runs. It holds that $\alpha \trianglelefteq \beta$ if, and only if, $\mathbf{tgt}(\alpha) \leq_{Q \times \mathbb{N}^d} \mathbf{tgt}(\beta)$ and decorate$(\alpha) \leq_{\mathcal{T}(\Sigma)}$ decorate(β).*

Corollary 7. *The relation $\trianglelefteq$ is a wqo.*

Proof. By the previous lemma, the function that maps each run α of $\mathcal{B}$ to the pair $(\mathbf{tgt}(\alpha), \text{decorate}(\alpha))$ is an order-reflecting function from $(\mathbf{Runs}(\mathcal{B}), \trianglelefteq)$ to the Cartesian product wqo $((Q \times \mathbb{N}^d) \times \mathcal{T}(\Sigma), \leq_{Q \times \mathbb{N}^d} \times \leq_{\mathcal{T}(\Sigma)})$. $\qquad\square$

4   Putting BVASS runs on a diet

In this section, we establish the main technical tool that will allow us to derive our results in the next sections. Roughly speaking, we prove that every reachable configuration of a BVASS can be obtained by a run of bounded branching complexity. The main ingredient to obtain this result is the transformation performed in the following lemma. In short, given two disjoint subruns of the form $\sigma\langle\gamma_1\rangle$ and $\sigma\langle\gamma_2\rangle$, we move the context family γ_1 into the subrun $\sigma\langle\gamma_2\rangle$, obtaining a small σ and a big $\sigma\langle\gamma_1\gamma_2\rangle$. An illustration is provided in Figure 3. By repeating this transformation, we progressively make the run less balanced and more path-like. This transformation preserves the size of BVASS runs and is based on the amalgamation property (see Proposition 5). The *size* of a run α is the number of its nodes and is written $|\alpha|$.

Lemma 8. *Let α, β and σ be runs of a BVASS $\mathcal{B}$, and let $q_\alpha(\mathbf{x}_\alpha)$, $q_\beta(\mathbf{x}_\beta)$ and $q_\sigma(\mathbf{x}_\sigma)$ denote their respective targets. Assume that $\sigma \trianglelefteq \alpha$ and that $\sigma \trianglelefteq \beta_{|n}$ for some node $n \in \mathcal{N}_\beta$. Then there exists a run β' of $\mathcal{B}$, with target $q_{\beta'}(\mathbf{x}_{\beta'})$, such that $q_\alpha = q_\sigma$, $q_\beta = q_{\beta'}$, $\mathbf{x}_\alpha + \mathbf{x}_\beta = \mathbf{x}_\sigma + \mathbf{x}_{\beta'}$ and $|\alpha| + |\beta| = |\sigma| + |\beta'|$.*

Proof. Let $\nu := \beta_{|n}$ and let $q_\nu(\mathbf{x}_\nu)$ denote its target. Assume that $\sigma \trianglelefteq \alpha$ and $\sigma \trianglelefteq \nu$. By definition, there exist two hereditarily positive families of contexts γ_1 and γ_2, both compatible with σ, such that $\alpha \approx \sigma\langle\gamma_1\rangle$ and $\nu \approx \sigma\langle\gamma_2\rangle$. Note that $q_\alpha = q_\sigma = q_\nu$, and, by the second point of Fact 3, $\mathbf{x}_\alpha = \mathbf{x}_\sigma + \mathbf{effect}(\gamma_1)$ and $\mathbf{x}_\nu = \mathbf{x}_\sigma + \mathbf{effect}(\gamma_2)$. By Proposition 5, $\nu' := \sigma\langle\gamma_1\gamma_2\rangle$ is a run. Its target $q_{\nu'}(\mathbf{x}_{\nu'})$ verifies $q_\nu = q_{\nu'}$ and $\mathbf{x}_{\nu'} = \mathbf{x}_\nu + \mathbf{effect}(\gamma_1)$. Note that $\mathbf{effect}(\gamma_1) \geq \mathbf{0}$. We construct the run β' from β by, firstly, replacing in β the subrun $\beta_{|n}$ by ν',

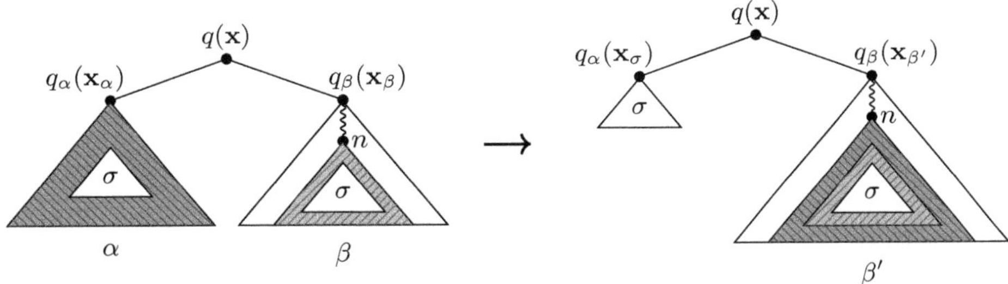

Fig. 3. Illustration and application of Lemma 8. The premises of the lemma are depicted on the left-hand side and the run β' provided by the lemma is depicted on the right-hand side. Intuitively, the context family witnessing that $\sigma \trianglelefteq \alpha$ is "extracted" from α (so only σ remains) and merged together, by amalgamation, with the context family witnessing that $\sigma \trianglelefteq \beta_{|n}$. To apply the lemma, assume that, in addition to the lemma premises, α and β are subruns rooted at two distinct children n_α and n_β of a node m with label $q(\mathbf{x})$ of a BVASS run. Since $\mathbf{x}_\alpha + \mathbf{x}_\beta = \mathbf{x}_\sigma + \mathbf{x}_{\beta'}$, we may replace α by σ and β by β', and the resulting tree is still a BVASS run.

and secondly, replacing the vector $\mathbf{cnt}_\beta(m)$ by $\mathbf{cnt}_\beta(m) + \mathbf{effect}(\gamma_1)$ for each strict ancestor m of n. It is readily seen that the resulting tree is a run of $\mathcal{B}$. Moreover, the target $q_{\beta'}(\mathbf{x}_{\beta'})$ of β' verifies $q_{\beta'} = q_\beta$ and $\mathbf{x}_{\beta'} = \mathbf{x}_\beta + \mathbf{effect}(\gamma_1)$. We derive that $q_\alpha = q_\sigma$, $q_\beta = q_{\beta'}$ and $\mathbf{x}_\alpha + \mathbf{x}_\beta = \mathbf{x}_\sigma + \mathbf{x}_{\beta'}$. It remains to prove that $|\alpha| + |\beta| = |\sigma| + |\beta'|$.

Let us define the *size* $|\gamma|$ of a context γ as the number of its nodes minus one (we ignore the hole in the size of a context). Given a run ζ of $\mathcal{B}$, the size $|\gamma|$ of a family of contexts $\gamma := (\gamma_n)_{n \in \mathcal{N}_\zeta}$ is $\sum_{n \in \mathcal{N}_\zeta} |\gamma_n|$. If $\zeta\langle\gamma\rangle$ is defined then its size verifies $|\zeta\langle\gamma\rangle| = |\zeta| + |\gamma|$. So we have $|\alpha| = |\sigma| + |\gamma_1|$, $|\nu| = |\sigma| + |\gamma_2|$ and $|\nu'| = |\sigma| + |\gamma_1| + |\gamma_2|$. It also holds by construction that $|\beta'| - |\beta| = |\nu'| - |\nu|$. The equality $|\alpha| + |\beta| = |\sigma| + |\beta'|$ follows. $\square$

Our bounded branching complexity statement can be formulated more conveniently by introducing an extension of the BVASS under consideration. Given a BVASS $\mathcal{B} := (Q, \Delta)$ and a finite set $\mathbf{F} \subseteq \mathbf{Reach}(\mathcal{B})$, we define the *extension of $\mathcal{B}$ with the configurations in $\mathbf{F}$* as $\mathrm{Extend}(\mathcal{B}, \mathbf{F}) := (Q, \Delta')$ where Δ' is the (finite) set of triples $(P', \mathbf{a}', q)$ in $\mathrm{Mult}(Q) \times \mathbb{Z}^d \times Q$ such that there exists a transition rule $(P, \mathbf{a}, q) \in \Delta$ and a family $\{q_i(\mathbf{x}_i)\}_{i \in I}$ of configurations in $\mathbf{F}$ satisfying $P = P' \uplus \{\!\{ q_i \mid i \in I \}\!\}$ and $\mathbf{a}' = \mathbf{a} + \sum_{i \in I} \mathbf{x}_i$. Note that $\Delta \subseteq \Delta'$. Intuitively, $\mathrm{Extend}(\mathcal{B}, \mathbf{F})$ has the same runs as $\mathcal{B}$, except that some subtrees rooted at configurations in $\mathbf{F}$ can be removed. More precisely, a transition rule $\delta \in \Delta$ applied with k input configurations in $\mathbf{F}$ can be replaced by a transition rule $\delta' \in \Delta'$ of arity $\mathrm{ar}(\delta') = \mathrm{ar}(\delta) - k$.

Lemma 9. *Let $\mathcal{B}$ be BVASS, let $\mathcal{S}$ be a finite $\trianglelefteq$-basis of the set of runs of $\mathcal{B}$, and let $\mathbf{F} := \{\mathbf{tgt}(\sigma) \mid \sigma \in \mathcal{S}\}$. For every run α of $\mathcal{B}$, there is a run η of $\mathrm{Extend}(\mathcal{B}, \mathbf{F})$ with same target as α and whose branching depth is at most $|\mathcal{S}| - 1$.*

Proof. Let $\mathcal{B}$, $\mathcal{S}$, and $\mathbf{F}$ be as in the lemma. We introduce the equivalence relation $\simeq$ on runs of $\mathcal{B}$ defined by $\alpha \simeq \beta$ if $\mathbf{tgt}(\alpha) = \mathbf{tgt}(\beta)$ and $|\alpha| = |\beta|$. For every run α of $\mathcal{B}$, we introduce the set

$$\mathcal{R}(\alpha) := \{\sigma \in \mathcal{S} \mid \exists \beta \in \mathbf{Runs}(\mathcal{B}), \exists n \in \mathcal{N}_\beta : \alpha \simeq \beta \wedge \sigma \trianglelefteq \beta_{|n}\}$$

We start with three easy observations. Firstly, $\mathcal{R}(\alpha) = \mathcal{R}(\beta)$ for all runs α, β of $\mathcal{B}$ with $\alpha \simeq \beta$. This observation directly follows from the definition of $\mathcal{R}(\alpha)$. Secondly, $\mathcal{R}(\alpha) \neq \emptyset$ for every run α of $\mathcal{B}$. This observation follows from the assumption that $\mathcal{S}$ is a $\trianglelefteq$-basis of the set of runs of $\mathcal{B}$. Thirdly, for every run α of $\mathcal{B}$ and every node $n \in \mathcal{N}_\alpha$, we have $\mathcal{R}(\alpha_{|n}) \subseteq \mathcal{R}(\alpha)$. Indeed, if $\sigma \in \mathcal{R}(\alpha_{|n})$ then there exist a run β' of $\mathcal{B}$ and a node $m \in \mathcal{N}_{\beta'}$ such that $\alpha_{|n} \simeq \beta'$ and $\sigma \trianglelefteq \beta'_{|m}$. We may replace in α the subrun $\alpha_{|n}$ by β' as they have the same target. The resulting run β has same size and same target as α, hence, $\alpha \simeq \beta$. Moreover, $m \in \mathcal{N}_\beta$ and $\sigma \trianglelefteq \beta'_{|m} = \beta_{|m}$. We get that $\sigma \in \mathcal{R}(\alpha)$. The lemma is an immediate consequence of the following claim.

Claim. For every run α of $\mathcal{B}$, there is a run η of $\mathrm{Extend}(\mathcal{B}, \mathbf{F})$ with same target as α and whose branching depth satisfies $\mathrm{bd}(\eta) < |\mathcal{R}(\alpha)|$.

We prove the claim by induction on $|\alpha|$. If $|\alpha| = 1$ then $\mathrm{bd}(\alpha) = 0$, hence, $\mathrm{bd}(\alpha) < |\mathcal{R}(\alpha)|$ since $\mathcal{R}(\alpha)$ is not empty according to the second observation shown before the claim. So the run $\eta := \alpha$ satisfies the claim. Now let $K \geq 1$ and assume that the claim holds for every run α of size $|\alpha| \leq K$. Pick a run α of size $|\alpha| = K + 1$. As $|\alpha| \geq 2$, the root of α has at least one child. We distinguish the children of $r := \mathrm{root}_\alpha$ depending on whether or not their label is in $\mathbf{F}$. Let $\{n_i\}_{i \in I}$ denote the children of r whose label is not in $\mathbf{F}$. According to the first observation shown before the claim, we may assume w.l.o.g. that we picked a run α that minimizes $|I|$ among the runs with same target and same size as α. We consider two cases depending on $|I|$.

The first case is when $|I| \leq 1$. So there exists a child n of r such that the label of every other child of r is in $\mathbf{F}$. We have $\mathcal{R}(\alpha_{|n}) \subseteq \mathcal{R}(\alpha)$ according to the third observation shown before the claim. As $|\alpha_{|n}| \leq |\alpha| - 1 = K$, we get from the induction hypothesis that there is a run η' of $\mathrm{Extend}(\mathcal{B}, \mathbf{F})$ with same target as $\alpha_{|n}$ and whose branching depth satisfies $\mathrm{bd}(\eta') < |\mathcal{R}(\alpha_{|n})|$. We transform α into a run η of $\mathrm{Extend}(\mathcal{B}, \mathbf{F})$ in two steps. First, we replace in α the subrun $\alpha_{|n}$ by η'. Second, we remove the subtree $\alpha_{|m}$ for every child m of r with $m \neq n$. It is readily seen that the resulting tree η is a run of $\mathrm{Extend}(\mathcal{B}, \mathbf{F})$. Moreover, the root of η is unary, so $\mathrm{bd}(\eta) = \mathrm{bd}(\eta') < |\mathcal{R}(\alpha_{|n})| \leq |\mathcal{R}(\alpha)|$.

The second case is when $|I| \geq 2$. We have $\mathcal{R}(\alpha_{|n_i}) \subseteq \mathcal{R}(\alpha)$ for each $i \in I$, according to the third observation shown before the claim. Let us show that this inclusion is strict. Let $i \in I$. There exists $j \in I$ with $i \neq j$, as $|I| \geq 2$. Since $\mathcal{S}$ is a $\trianglelefteq$-basis of $\mathbf{Runs}(\mathcal{B})$, there exists a run $\sigma \in \mathcal{S}$ such that $\sigma \trianglelefteq \alpha_{|n_j}$. Observe that $\sigma \in \mathcal{R}(\alpha_{|n_j}) \subseteq \mathcal{R}(\alpha)$. Let us prove that $\sigma \notin \mathcal{R}(\alpha_{|n_i})$. By contradiction, suppose that there exist a run β'' of $\mathcal{B}$ and a node $m \in \mathcal{N}_{\beta''}$ such that $\alpha_{|n_i} \simeq \beta''$ and $\sigma \trianglelefteq \beta''_{|m}$. Let β' denote the run obtained by applying Lemma 8 on $\alpha_{|n_j}, \beta''$ and

σ. We transform α by replacing the subrun $\alpha_{|n_j}$ by σ and the subrun $\alpha_{|n_i}$ by β'. Let α' denote the resulting tree. The properties of β' ensure that α' is a run of $\mathcal{B}$ with same size and same target as α. However, the number of children of the root whose label is not in $\mathbf{F}$ is strictly less in α' than in α. This contradicts the assumption that our choice of α minimizes $|I|$. We have shown that $\mathcal{R}(\alpha_{|n_i})$ is strictly contained in $\mathcal{R}(\alpha)$, for every $i \in I$. Now, as $|\alpha_{|n_i}| \leq |\alpha| - 1 = K$, we get from the induction hypothesis that, for each $i \in I$, there is a run η_i of $\mathrm{Extend}(\mathcal{B}, \mathbf{F})$ with same target as $\alpha_{|n_i}$ and whose branching depth satisfies $\mathrm{bd}(\eta_i) < |\mathcal{R}(\alpha_{|n_i})|$. It follows that $\mathrm{bd}(\eta_i) \leq |\mathcal{R}(\alpha)| - 2$. As before, we transform α into a run η of $\mathrm{Extend}(\mathcal{B}, \mathbf{F})$ in two steps. First, we replace in α the subrun $\alpha_{|n_i}$ by η_i, for each $i \in I$. Second, we remove the subtree $\alpha_{|m}$ for every child m of r whose label is in $\mathbf{F}$. It is readily seen that the resulting tree is a run of $\mathrm{Extend}(\mathcal{B}, \mathbf{F})$. The root of η is branching, so $\mathrm{bd}(\eta) = 1 + \max_{i \in I} \mathrm{bd}(\eta_i) < |\mathcal{R}(\alpha)|$. This concludes the proof of the claim, and the proof of the lemma. $\qquad\square$

Corollary 10. *For every BVASS $\mathcal{B}$, there exists $s \in \mathbb{N}$ such that every configuration reached by $\mathcal{B}$ is reached by a run whose Strahler number is at most s.*

5 BVASS reachability sets are VASS sections

We now use the technical results of Section 4 to show that BVASS reachability sets are VASS sections. Consider a d-BVASS $\mathcal{B} := (Q, \Delta)$ and a bound $b \in \mathbb{N}$. We introduce the set $\mathbf{Runs}_{\mathrm{bd} \leq b}(\mathcal{B})$ of runs of $\mathcal{B}$ whose branching depth is at most b, and the set $\mathbf{Reach}_{\mathrm{bd} \leq b}(\mathcal{B}) := \{\mathbf{tgt}(\rho) \mid \rho \in \mathbf{Runs}_{\mathrm{bd} \leq b}(\mathcal{B})\}$. In this section, we describe an algorithm computing a VASS $\mathcal{V}_b$ that precisely captures the set $\mathbf{Reach}_{\mathrm{bd} \leq b}(\mathcal{B})$, i.e., such that $\mathbf{Sect}^{Q}_{[1,d]}(\mathcal{V}_b) = \mathbf{Reach}_{\mathrm{bd} \leq b}(\mathcal{B})$.

We first observe that VASS sections are effectively closed under finite unions. In fact, we can define an algorithm that takes as input two VASS $\mathcal{V}$ and $\mathcal{V}'$ and returns a VASS denoted by $\mathcal{V} \sqcup \mathcal{V}'$ such that $\mathbf{Sect}^{Q}_{[1,d]}(\mathcal{V} \sqcup \mathcal{V}') = \mathbf{Sect}^{Q}_{[1,d]}(\mathcal{V}) \cup \mathbf{Sect}^{Q}_{[1,d]}(\mathcal{V}')$. The VASS $\mathcal{V} \sqcup \mathcal{V}'$ is obtained as follows. First we consider the union of $\mathcal{V}$, $\mathcal{V}'$ and the set of states Q. In this union the states of $\mathcal{V}$ and $\mathcal{V}'$ are renamed in such a way states are pairwise disjoint. By adding to this VASS zero-effect transitions from the two copies of each state $q \in Q$ to q, we get $\mathcal{V} \sqcup \mathcal{V}'$ satisfying the required property.

Next, we provide an algorithm that produces a VASS capturing the effect of a unique transition rule $\delta = (P, \mathbf{a}, q)$. Given a set $\mathbf{A}$ of configurations, we introduce the set $\delta[\mathbf{A}]$ of configurations $q(\mathbf{x})$ such that there exists $q_1(\mathbf{x}_1), \ldots, q_r(\mathbf{x}_r) \in \mathbf{A}$ satisfying $\delta = (\{\!| q_1, \ldots, q_r |\!\}, \mathbf{x} - \sum_{i=1}^{r} \mathbf{x}_i, q)$. Starting from a VASS $\mathcal{V}$ such that $\mathbf{A} = \mathbf{Sect}^{Q}_{[1,d]}(\mathcal{V})$, we construct a VASS $\delta[\mathcal{V}]$ such that $\delta[\mathbf{A}] = \mathbf{Sect}^{Q}_{[1,d]}(\delta[\mathcal{V}])$, as follows. Consider the Cartesian product of r copies of $\mathcal{V}$ with disjoint sets of counters and disjoint from $[1, d]$. As expected, the states of this Cartesian product are tuples $(q_1, \ldots, q_r) \in Q^r$. We add to this product the state q (the target of δ) and a fresh state $q_!$ corresponding to an intermediate state between the Cartesian product and q. We also add zero-effect transitions from states $(q_1, \ldots, q_r) \in Q^r$ such that $P = \{\!| q_1, \ldots, q_r |\!\}$ to $q_!$, and a transition from $q_!$ to q that adds $\mathbf{a}$ to the counters $[1, d]$. In order to transfer the counters from each copy of $\mathcal{V}$ to $[1, d]$, we add dr loops on $q_!$ that decrement a counter of some copy of $\mathcal{V}$ while increasing the corresponding counter in $[1, d]$. It is routinely checked that the VASS $\delta[\mathcal{V}]$ defined in this way satisfies $\delta[\mathbf{A}] = \mathbf{Sect}^Q_{[1,d]}(\delta[\mathcal{V}])$.

Now, let us provide the algorithm computing inductively on b the VASS $\mathcal{V}_b$. The VASS $\mathcal{V}_0$ is simply the VASS obtained from $\mathcal{B}$ by removing all branching transitions. The VASS $\mathcal{V}_{b+1}$ is defined by first defining the VASS $\mathcal{V}'_b = \sqcup_{\delta \in \Delta} \delta[\mathcal{V}_b]$. By adding to this VASS the unary transition rules of $\mathcal{B}$, we get the VASS $\mathcal{V}_{b+1}$.

Lemma 11. $\mathbf{Sect}^Q_{[1,d]}(\mathcal{V}_b) = \mathbf{Reach}_{\mathrm{bd} \leq b}(\mathcal{B})$.

We can now state and prove our main theorem:

Theorem 12. *Every reachability set of a BVASS is a section of VASS.*

Proof. Consider a d-BVASS $\mathcal{B}$. The set of runs of $\mathcal{B}$ admits a finite $\trianglelefteq$-basis $\mathcal{S}$ since $\trianglelefteq$ is a wqo. Let $b := |\mathcal{S}| - 1$, let $\mathbf{F} := \{\mathbf{tgt}(\sigma) \mid \sigma \in \mathcal{S}\}$, and let $\mathcal{E} := \mathrm{Extend}(\mathcal{B}, \mathbf{F})$. By Lemma 9, we have $\mathbf{Reach}(\mathcal{B}) \subseteq \mathbf{Reach}_{\mathrm{bd} \leq b}(\mathcal{E})$. It follows that $\mathbf{Reach}(\mathcal{B}) = \mathbf{Reach}_{\mathrm{bd} \leq b}(\mathcal{E})$ since $\mathbf{Reach}_{\mathrm{bd} \leq b}(\mathcal{E}) \subseteq \mathbf{Reach}(\mathcal{E}) = \mathbf{Reach}(\mathcal{B})$ by definition. By applying Lemma 11 to $\mathcal{E}$, we conclude that $\mathbf{Reach}(\mathcal{B}) = \mathbf{Reach}_{\mathrm{bd} \leq b}(\mathcal{E})$ is a section of VASS. $\square$

The proof of Theorem 12 is based on a $\trianglelefteq$-basis of the set of runs of a BVASS $\mathcal{B}$. Such a basis is not computable even for plain VASS of dimension two [1, Appendix A]. Still, Lemmas 9 and 11 provide a way to compute a sequence of VASS $(\mathcal{V}_n)_{n \in \mathbb{N}}$ such that the sequence of sections $(\mathbf{Sect}^Q_{[1,d]}(\mathcal{V}_n))_{n \in \mathbb{N}}$ eventually stabilizes to the reachability set of $\mathcal{B}$. Intuitively, it is sufficient to consider any fair enumeration of the reachable configurations $(\mathbf{c}_n)_{n \in \mathbb{N}}$ of $\mathcal{B}$ and define the VASS $\mathcal{V}_n$ such that $\mathbf{Sect}^Q_{[1,d]}(\mathcal{V}_n)$ is the set of targets of runs of $\mathrm{Extend}(\mathcal{B}, \mathbf{F}_n)$ of branching depth at most n, where $\mathbf{F}_n = \{\mathbf{c}_0, \ldots, \mathbf{c}_n\}$. This sequence of VASS is not sufficient for deciding the BVASS reachability problem since we cannot decide the equality of VASS reachability sets [15]. However, when the reachability set of $\mathcal{B}$ is semilinear, there exists $n \in \mathbb{N}$ such that $\mathbf{Sect}^Q_{[1,d]}(\mathcal{V}_n)$ is a semilinear inductive invariant. Since we can decide the semilinearity of a VASS section [14, Corollary VII.7, Theorem VII.11] and we can decide if a semilinear set is inductive, we deduce the following result.

Proposition 13. *When the reachability set of a BVASS is semilinear, it is effectively computable.*

The reachability set of a BVASS $\mathcal{B}$ is clearly an inductive invariant. Even if such a set is not semilinear in general, we prove in that section that it is almost semilinear; a geometrical characterization used in [24] to prove that for every VASS and for every configuration not reachable, there exists a *semilinear inductive invariant* that does not contain the configuration.

Formally, an *almost semilinear set* of $Q \times \mathbb{N}^d$ is a finite union of sets of the form $q(\mathbf{b} + \mathbf{P})$ where $q \in Q$ is a state, $\mathbf{b}$ is a vector in $\mathbb{N}^d$ called the *basis*, and $\mathbf{P}$ if a subset of $\mathbb{N}^d$ called the *periodic set* and satisfying $\mathbf{0} \in \mathbf{P}, \mathbf{P} + \mathbf{P} \subseteq \mathbf{P}$ and such that the *cone* $\mathbb{Q}_{\geq 0}\mathbf{P} = \{\lambda\mathbf{p} \mid \lambda \in \mathbb{Q}_{\geq 0} \wedge \mathbf{p} \in \mathbf{P}\}$ spanned by $\mathbf{P}$ is definable in $FO(\mathbb{Q}_{\geq 0}, +, =)$. In [24], intersections of VASS reachability sets with semilinear sets were proved to be almost semilinear. Since VASS sections are intersections of VASS reachability sets with semilinear sets and semilinear sets are stable by finite intersections, we deduce that intersections of VASS sections with semilinear sets are also almost semilinear. From the previous Theorem 12, we deduce the following corollary.

Corollary 14. *Intersections of BVASS reachability sets with semilinear sets are almost semilinear.*

6 Strahler-Bounded Reachability

We prove that the following bounded-variant of the reachability problem for BVASS is decidable.

STRAHLER-BOUNDED REACHABILITY
Instance: A BVASS $\mathcal{B}$, a configuration $q(\mathbf{x})$, and $s \in \mathbb{N}$
Question: Is the configuration $q(\mathbf{x})$ reachable by a run of Strahler number at most s ?

This problem is motivated by Corollary 10 since for every BVASS $\mathcal{B}$ there exists a minimal $s \in \mathbb{N}$, called the *Strahler number* of $\mathcal{B}$, such that every reachable configuration is reachable by a run of Strahler number at most s.

Given a d-BVASS $\mathcal{B} := (Q, \Delta)$, we define a VASS with nested zero tests $\mathcal{Z}_s$ such that $\mathbf{Sect}^Q_{[1,d]}(\mathcal{Z}_s)$ coincides with the set of configurations of $\mathcal{B}$ reachable by a run whose Strahler number is at most s. *VASS with nested zero tests* (abbreviated *VASSnz*) extend classical VASS by allowing restricted forms of zero tests: transitions may simultaneously test for zero all counters belonging to some counter subset I_k in an increasing sequence $I_1 \subset I_2 \subset \cdots \subset I_K$. Membership in a VASSnz section is decidable [14, corollary VII.7, theorem VII.11], hence this construction is enough to prove decidability of the above problem.

Instead of describing $\mathcal{Z}_s$ by explicitly listing its states and transitions, we specify it by means of Algorithms 1 and 2. These algorithms can easily be compiled into a VASSnz, as they only involve state comparisons, constant additive counter updates, zero tests, finite branching, nondeterministic choices, and function calls with bounded nesting depth. The $\star$ symbol at lines 4 and 10 denotes a non-deterministic choice condition (this condition is non-deterministically evaluated to false or true at each iteration of the **while**-loop).

Given a run of $\mathcal{B}$ (or any tree), we call *main branch* any branch that can be constructed top-down by repeatedly selecting a child of maximum Strahler number. Intuitively, a main branch is one that "carries" every increase of the

---

Algorithm 1: $\text{REACH}_t^{(s)}$ for $t \in Q$ and $s \in \mathbb{N}$

---

1 **pick** $(P, \mathbf{a}, q)$ in Δ such that $P = \emptyset$ and $\mathbf{a} \geq \mathbf{0}$
2 $\texttt{state}^{(s)} := q$
3 $(\texttt{c}_1^{(s)}, \ldots, \texttt{c}_d^{(s)}) \mathrel{+}= (\mathbf{a}(1), \ldots, \mathbf{a}(d))$
4 **while** $\star$ **do**
5   **pick** $(P, \mathbf{a}, q)$ in Δ such that $\texttt{state}^{(s)} \in P$
6      $\texttt{M}^{(s)} := P - \left\{\!\left|\, \texttt{state}^{(s)} \,\right|\!\right\}$
7   **while** $\texttt{M}^{(s)} \neq \emptyset$ **do**
8   **pick** p in $\texttt{M}^{(s)}$
9   **call** $\text{REACH}_p^{(s-1)}$
10   **while** $\star$ **do**
11   **pick** i in $[1, d]$
12             $(\texttt{c}_i^{(s-1)}, \texttt{c}_i^{(s)}) \mathrel{+}= (-1, 1)$
13   **assert** $\texttt{c}_1^{(1)} = \cdots = \texttt{c}_d^{(1)} = \cdots = \texttt{c}_1^{(s-1)} = \cdots = \texttt{c}_d^{(s-1)} = 0$
14         $\texttt{M}^{(s)} := \texttt{M}^{(s)} - \left\{\!\left|\, p \,\right|\!\right\}$
15     $\texttt{state}^{(s)} := q$
16     $(\texttt{c}_1^{(s)}, \ldots, \texttt{c}_d^{(s)}) \mathrel{+}= (\mathbf{a}(1), \ldots, \mathbf{a}(d))$
17 **if** $\texttt{state}^{(s)} = t$ **then**
18   **return**
19 **else**
20   **abort**

Algorithm 2: $\text{REACH}_t^{(-1)}$ for $t \in Q$

---

1 **abort**

Strahler number: when traversing it bottom-up, each time the Strahler number increases, this branch contributes to that increase by providing one of the siblings of maximal Strahler numbers. For example, on Figure 1, all branches are main branches, except the leftmost branch $2 - 1 - 0$, because its leaf has Strahler number 0 and a sibling of Strahler number 1. It follows immediately that, in a run of Strahler number s, all nodes outside any main branch have Strahler number at most $s - 1$.

The algorithm $\text{REACH}_t^{(s)}$ simulates a bottom-up traversal of a main branch in a run of $\mathcal{B}$ whose target state is t and whose Strahler number is at most s. It first nondeterministically selects an initial configuration (lines 1–3), then performs an arbitrary number of transitions of $\mathcal{B}$ (lines 4–17), and finally returns the final configuration if the current control state is t (lines 17–20).

To simulate a branching transition of $\mathcal{B}$, $\text{REACH}_t^{(s)}$ makes the required number of calls to $\text{REACH}_p^{(s-1)}$ (namely, the arity of the branching rule minus one) on auxiliary counters. It then transfers the content of these auxiliary counters back to the main ones (line 12), and finally uses a zero test (line 13) to ensure that all auxiliary counters have been completely emptied. Zero tests play a crucial role here: since auxiliary counters may be reused for several calls to $\mathrm{REACH}_p^{(s-1)}$, any residual value left in them could incorrectly enable transitions during subsequent calls.

We denote by $\mathcal{Z}_s$ the VASS obtained by compiling the algorithms $\mathrm{REACH}_t^{(s)}$ for all $t \in Q$ and taking their disjoint union—that is, the disjoint union of their sets of states, transitions, and initial configurations—while sharing the counters $(\mathsf{c}_i^{(j)}, i \in [1,d], j \in [0,s])$. We index the counters so that $(\mathsf{c}_1^{(s)}, \ldots, \mathsf{c}_d^{(s)})$ correspond to indices $1, \ldots, d$, and we assign the control states so that t is the state associated with line 18 (the **return** instruction) of $\mathrm{REACH}_t^{(s)}$. In particular, the set of counter values $\mathbf{x}$ such that $t(\mathbf{x}) \in \mathbf{Reach}(\mathcal{Z}_s)$ coincides exactly with the set of values that $(\mathsf{c}_i^{(j)}, i \in [1,d], j \in [0,s])$ may hold when executing line 18 in $\mathrm{REACH}_t^{(s)}$.

Lemma 15. $\mathbf{Sect}_{[1,d]}^Q(\mathcal{Z}_s)$ *is the set of configuration reachable by a run of $\mathcal{B}$ of Strahler number at most s.*

Observe that the existence of a computable upper bound on the Strahler number of a BVASS would imply decidability of the reachability problem for BVASS. It is still open whether such a bound exists. However, an easy reduction from inclusion of VASS reachability sets already shows that exact computation of the Strahler number is impossible.

Proposition 16. *For every $s \in \mathbb{N}$, it is undecidable whether, given a BVASS $\mathcal{B}$, the Strahler number of $\mathcal{B}$ is inferior or equal to s.*

7 Application to 5-BVAS and 2-BVASS

A d-BVAS (resp. a d-VAS) is a d-BVASS (resp. a d-VASS) with one state. By discarding this unique state, formally, a d-BVAS is given as a finite set of transitions $\Delta \subseteq \mathbb{N} \times \mathbb{Z}^d$ where a transition $\delta = (r, \mathbf{a})$ is defined by a natural number r denoting the arity of δ and a vector $\mathbf{a}$ denoting the effect of δ. Symmetrically, a d-VAS is a pair $(\Delta, \mathbf{S})$ where Δ is a finite subset of vectors in $\mathbb{Z}^d$ called *actions* and $\mathbf{S}$ is a finite set of initial configurations in $\mathbb{N}^d$. Based on this observation, notions defined for d-BVASS and d-VASS are naturally propagated over d-BVAS and d-VAS.

Hopcroft and Pansiot proved in [16] that reachability sets of 5-VAS are effectively semilinear while there exists a 3-VASS with two states having a non-semilinear reachability set. Since reachability sets of d-VASS are sections of $d+3$-VAS reachability sets (see for instance [16, lemma 2.1]), there also exists a 6-VAS with a non-semilinear reachability set. In this section, we push those results on branching VAS and VASS by proving that reachability sets of 5-BVAS and 2-BVASS are effectively semilinear. The computation is performed by a fix-point algorithm. We only present the computation for 5-BVAS since for 2-BVASS the computation is similar (it is based on the same fix-point algorithm but instantiated with 2-BVASS). Our computation for 5-BVAS is based on an extended model of 5-VAS, called 5-LVAS that allow linear actions.

Formally, a *d-dimensional linear VAS* (*d-LVAS* for short) is a pair $\mathcal{V} = (\Delta, \Gamma_{init})$ where Δ is a finite set of pairs $(\mathbf{z}, \mathbf{V})$ where $\mathbf{z} \in \mathbb{Z}^d$ and $\mathbf{V}$ is a finite subset of $\mathbb{N}^d$, and Γ_{init} is a presentation of a semilinear set of $\mathbb{N}^d$. We associate with a pair $\delta = (\mathbf{z}, \mathbf{V})$ in Δ the set $[\![\delta]\!] = \mathbf{z} + \mathbf{P}$ where $\mathbf{P}$ is the periodic set spanned by $\mathbf{V}$. We also introduce $[\![\Delta]\!] = \bigcup_{\delta \in \Delta}[\![\delta]\!]$. A *configuration* is a vector in $\mathbb{N}^d$ and a *run* is a sequence $\rho = \mathbf{x}_0\mathbf{x}_1 \ldots \mathbf{x}_k$ of configurations satisfying $\mathbf{x}_0 \in [\![\Gamma_{init}]\!]$, and $\mathbf{x}_j - \mathbf{x}_{j-1} \in [\![\Delta]\!]$ for every $1 \leq j \leq k$. Its last configuration is called *target* and is denoted $\mathbf{tgt}(\rho)$. The set of runs of $\mathcal{V}$ is denoted $\mathbf{Runs}(\mathcal{V})$. The *reachability set* $\mathbf{Reach}(\mathcal{V})$ is the set of targets of runs of $\mathcal{V}$, *i.e.* $\mathbf{Reach}(\mathcal{V}) = \{\mathbf{tgt}(\rho) \mid \rho \in \mathbf{Runs}(\mathcal{V})\}$.

Reachability sets of d-VAS are clearly reachability sets of d-LVAS. Conversely, reachability sets of d-LVAS are sections of reachability sets of d-VASS. In fact, any transition $\delta = (\mathbf{z}, \mathbf{V})$ of a d-LVAS can be encoded in a d-VASS by two transitions $(q_0, \mathbf{z}, r)$ and $(r, \mathbf{0}, q_0)$ and loops $(r, \mathbf{v}, r)$. Such a construction cannot be used for proving that reachability sets of 5-LVAS are effectively semilinear since it introduces several control states. Nevertheless, in the following lemma, by decomposing runs of a 5-LVAS with respect to the first time a transition $(\mathbf{z}, \mathbf{V})$ with $\mathbf{V} \neq \emptyset$ is used, we provide a way to prove that reachability sets of 5-LVAS are effectively semilinear.

Lemma 17. *Reachability sets of 5-LVAS are effectively semilinear.*

We associate with a d-BVAS Δ and a presentation Γ of a semilinear set $\mathbf{S} \subseteq \mathbb{N}^d$, the d-LVAS $\Delta\langle\Gamma\rangle = (T, \{\mathbf{a} \in \mathbb{N}^d \mid (0, \mathbf{a}) \in \Delta\})$ where T is the following set of transitions:

$$T = \bigcup_{(r,\mathbf{a})\in\Delta \mid r \geq 1} \{(\mathbf{a} + \sum_{\ell=1}^{r-1}\mathbf{b}_\ell, \bigcup_{\ell=1}^{r-1}\mathbf{V}_\ell) \mid (\mathbf{b}_1, \mathbf{V}_1), \ldots, (\mathbf{b}_{r-1}, \mathbf{V}_{r-1}) \in \Gamma\}$$

Clearly, if $\mathbf{S} \subseteq \mathbf{Reach}(\Delta)$ then $\mathbf{Reach}(\Delta\langle\Gamma\rangle) \subseteq \mathbf{Reach}(\Delta)$. In fact, from a run ρ of $\Delta\langle\Gamma\rangle$ we obtain a run α of Δ with $\mathbf{tgt}(\rho) = \mathbf{tgt}(\alpha)$ just by considering ρ as a (single branch) tree and by inserting as new children of nodes of that branch some runs $\alpha_\mathbf{s}$ of Δ such that $\mathbf{tgt}(\rho_\mathbf{s}) = \mathbf{s}$.

Computation of 5-BVAS reachability sets is based on the previous construction. We introduce a computable function POSTSTAR that maps a 5-LVAS $\mathcal{V}$ on a presentation POSTSTAR($\mathcal{V}$) of the semilinear set $\mathbf{Reach}(\mathcal{V})$. Lemma 17 shows that such a function exists. We introduce the sequence $(\Gamma_i)_{i\in\mathbb{N}}$ of presentations of semilinear sets defined by $\Gamma_0 = \emptyset$ and by induction for every $i \in \mathbb{N}$ as follows:

$$\Gamma_{i+1} = \text{POSTSTAR}(\Delta\langle\Gamma_i\rangle)$$

Let $\mathbf{S}_i$ be the semilinear presented by Γ_i. By induction on i we deduce that $\mathbf{S}_i \subseteq \mathbf{S}_{i+1}$ and $\mathbf{S}_i \subseteq \mathbf{Reach}(\Delta)$. Notice that if $\mathbf{S}_{i+1} = \mathbf{S}_i$ then $\mathbf{S}_i$ is an inductive invariant for Δ. It follows that $\mathbf{S}_i = \mathbf{Reach}(\Delta)$ for such an index i. As we can decide the inclusion of two semilinear sets given by presentations, we deduce that the reachability set of 5-BVAS are effectively semilinear if such an index i exists. We first prove the following lemma.

Lemma 18. $\mathbf{S}_i$ *is the set of configurations reachable by a run of Δ of Strahler number at most i.*

Corollary 10 shows that there exists $s \in \mathbb{N}$ such that every configuration reached by Δ is reached by a run whose Strahler number is at most s. From the previous lemma, we deduce that for every $i \geq s$ we have $\mathbf{S}_{i+1} = \mathbf{S}_i$. We have proved the following theorem.

Theorem 19. *Reachability sets of 5-BVAS and 2-BVASS are effectively semilinear.*

8 Conclusion

In this paper, we showed how to bridge the gap between VASS and branching VASS. In small dimensions, we generalized the effective semilinearity of reachability sets of 2-dimensional VASS and 5-dimensional VAS to the branching case, via a simple iterative fix-point algorithm. We obtained those results by applying techniques that are valid in arbitrary dimension.

We introduced a well-quasi-order (wqo) on the set of runs of BVASS that generalizes the well-known wqo on VASS runs and that satisfies, as in the case of VASS, an *amalgamation property*. Using this property, we have proved that every BVASS can be extended with an appropriate finite set $\mathbf{F}$ of reachable configurations to get an equivalent BVASS with bounded branching depth complexity. This result has three consequences. First, reachability sets of BVASS are VASS sections, and hence, are almost semilinear. Second, if the reachability set of a BVASS is semilinear then it is effectively computable. Third, the Strahler number of a BVASS is finite, but not computable. Whether such an appropriate set $\mathbf{F}$ is computable is an open problem. Similarly, we do not know if an upper bound on the Strahler number is computable. Finally, we proved that the Strahler-bounded reachability problem is decidable, hence the computability of such an upper-bound would entail the decidability of the reachability problem for BVASS. The latter problem is still open, but hopefully not for a long time!

Acknowledgments. The authors wish to thank the anonymous reviewers for their constructive and valuable comments. This work was supported by the grant ANR-25-CE48-6933 of the French National Research Agency (project CoqoPetri).

References

1. Anand, A., Schmitz, S., Schütze, L., Zetzsche, G.: Verifying unboundedness via amalgamation. In: Proceedings of the 39th Annual ACM/IEEE Symposium on

Logic in Computer Science. LICS '24, Association for Computing Machinery, New York, NY, USA (2024). https://doi.org/10.1145/3661814.3662133, https://doi.org/10.1145/3661814.3662133

2. Atig, M.F., Ganty, P.: Approximating Petri net reachability along context-free traces. In: Chakraborty, S., Kumar, A. (eds.) IARCS Annual Conference on Foundations of Software Technology and Theoretical Computer Science, FSTTCS 2011, December 12-14, 2011, Mumbai, India. LIPIcs, vol. 13, pp. 152–163. Schloss Dagstuhl - Leibniz-Zentrum für Informatik (2011). https://doi.org/10.4230/LIPICS.FSTTCS.2011.152

3. Bizière, C., Hilaire, T., Leroux, J., Sutre, G.: On the reachability problem for two-dimensional branching VASS. In: Gawrychowski, P., Mazowiecki, F., Skrzypczak, M. (eds.) 50th International Symposium on Mathematical Foundations of Computer Science, MFCS 2025, August 25-29, 2025, Warsaw, Poland. LIPIcs, vol. 345, pp. 22:1–22:19. Schloss Dagstuhl - Leibniz-Zentrum für Informatik (2025). https://doi.org/10.4230/LIPICS.MFCS.2025.22, https://doi.org/10.4230/LIPIcs.MFCS.2025.22

4. Bojanczyk, M., David, C., Muscholl, A., Schwentick, T., Segoufin, L.: Two-variable logic on data trees and XML reasoning. In: Vansummeren, S. (ed.) Proceedings of the Twenty-Fifth ACM SIGACT-SIGMOD-SIGART Symposium on Principles of Database Systems, June 26-28, 2006, Chicago, Illinois, USA. pp. 10–19. ACM (2006). https://doi.org/10.1145/1142351.1142354, https://doi.org/10.1145/1142351.1142354

5. Bouajjani, A., Emmi, M.: Analysis of recursively parallel programs. ACM Trans. Program. Lang. Syst. 35(3), 10:1–10:49 (2013). https://doi.org/10.1145/2518188, https://doi.org/10.1145/2518188

6. Clemente, L., Lasota, S., Lazic, R., Mazowiecki, F.: Timed pushdown automata and branching vector addition systems. In: 32nd Annual ACM/IEEE Symposium on Logic in Computer Science, LICS 2017, Reykjavik, Iceland, June 20-23, 2017. pp. 1–12. IEEE Computer Society (2017). https://doi.org/10.1109/LICS.2017.8005083, https://doi.org/10.1109/LICS.2017.8005083

7. Cotton-Barratt, C., Murawski, A.S., Ong, C.L.: ML and extended branching VASS. In: Yang, H. (ed.) Programming Languages and Systems - 26th European Symposium on Programming, ESOP 2017, Held as Part of the European Joint Conferences on Theory and Practice of Software, ETAPS 2017, Uppsala, Sweden, April 22-29, 2017, Proceedings. Lecture Notes in Computer Science, vol. 10201, pp. 314–340. Springer (2017). https://doi.org/10.1007/978-3-662-54434-1_12, https://doi.org/10.1007/978-3-662-54434-1_12

8. Czerwinski, W., Lasota, S., Lazic, R., Leroux, J., Mazowiecki, F.: The reachability problem for Petri nets is not elementary. J. ACM 68(1), 7:1–7:28 (2021). https://doi.org/10.1145/3422822, https://doi.org/10.1145/3422822

9. Demri, S., Jurdzinski, M., Lachish, O., Lazic, R.: The covering and boundedness problems for branching vector addition systems. J. Comput. Syst. Sci. 79(1), 23–38 (2013). https://doi.org/10.1016/J.JCSS.2012.04.002

10. Esparza, J., Luttenberger, M., Schlund, M.: A brief history of Strahler numbers. In: Language and Automata Theory and Applications - 8th International Conference, LATA 2014, Madrid, Spain, March 10-14, 2014. Proceedings. Lecture Notes in Computer Science, vol. 8370, pp. 1–13. Springer (03 2014). https://doi.org/10.1007/978-3-319-04921-2_1

11. Figueira, D., Lazic, R., Leroux, J., Mazowiecki, F., Sutre, G.: Polynomial-space completeness of reachability for succinct branching VASS in dimension one.

In: Chatzigiannakis, I., Indyk, P., Kuhn, F., Muscholl, A. (eds.) 44th International Colloquium on Automata, Languages, and Programming, ICALP 2017, July 10-14, 2017, Warsaw, Poland. LIPIcs, vol. 80, pp. 119:1–119:14. Schloss Dagstuhl - Leibniz-Zentrum für Informatik (2017). https://doi.org/10.4230/LIPICS.ICALP.2017.119

12. Göller, S., Haase, C., Lazić, R., Totzke, P.: A polynomial-time algorithm for reachability in branching VASS in dimension one. In: ICALP. LIPIcs, vol. 55, pp. 105:1–105:13. Schloss Dagstuhl (2016). https://doi.org/10.4230/LIPIcs.ICALP.2016.105

13. de Groote, P., Guillaume, B., Salvati, S.: Vector addition tree automata. In: 19th IEEE Symposium on Logic in Computer Science (LICS 2004), 14-17 July 2004, Turku, Finland, Proceedings. pp. 64–73. IEEE Computer Society (2004). https://doi.org/10.1109/LICS.2004.1319601

14. Guttenberg, R., Czerwinski, W., Lasota, S.: Reachability and related problems in vector addition systems with nested zero tests. CoRR abs/2502.07660 (2025). https://doi.org/10.48550/ARXIV.2502.07660, https://doi.org/10.48550/arXiv.2502.07660

15. Hack, M.: The equality problem for vector addition systems is undecidable. Theoretical Computer Science 2(1), 77–95 (1976). https://doi.org/https://doi.org/10.1016/0304-3975(76)90008-6, https://www.sciencedirect.com/science/article/pii/0304397576900086

16. Hopcroft, J.E., Pansiot, J.: On the reachability problem for 5-dimensional vector addition systems. Theor. Comput. Sci. 8, 135–159 (1979)

17. Jacquemard, F., Segoufin, L., Dimino, J.: Fo2(<, +1, ~) on data trees, data tree automata and branching vector addition systems. Log. Methods Comput. Sci. 12(2) (2016). https://doi.org/10.2168/LMCS-12(2:3)2016

18. Jančar, P.: Decidability of a temporal logic problem for Petri nets. Theoretical Computer Science 74(1), 71–93 (1990). https://doi.org/https://doi.org/10.1016/0304-3975(90)90006-4, https://www.sciencedirect.com/science/article/pii/0304397590900064

19. Lazic, R.: The reachability problem for vector addition systems with a stack is not elementary. CoRR abs/1310.1767 (2013). https://doi.org/10.48550/arXiv.1310.1767

20. Lazic, R., Newcomb, T.C., Ouaknine, J., Roscoe, A.W., Worrell, J.: Nets with tokens which carry data. Fundam. Informaticae 88(3), 251–274 (2008), http://content.iospress.com/articles/fundamenta-informaticae/fi88-3-03

21. Lazić, R., Schmitz, S.: Nonelementary complexities for branching VASS, MELL, and extensions. ACM Trans. Comput. Log. 16(3), 20:1–20:30 (2015). https://doi.org/10.1145/2733375

22. Leroux, J., Praveen, M., Schnoebelen, P., Sutre, G.: On functions weakly computable by pushdown Petri nets and related systems. Logical Methods in Computer Science Volume 15, Issue 4, 15 (Dec 2019). https://doi.org/10.23638/LMCS-15(4:15)2019, https://lmcs.episciences.org/5362

23. Leroux, J., Schnoebelen, P.: On functions weakly computable by Petri nets and vector addition systems. In: Ouaknine, J., Potapov, I., Worrell, J. (eds.) Reachability Problems. pp. 190–202. Springer International Publishing, Cham (2014)

24. Leroux, J.: Vector addition systems reachability problem (a simpler solution). In: Voronkov, A. (ed.) Turing-100. The Alan Turing Centenary. EPiC Series in Computing, vol. 10, pp. 214–228. EasyChair (2012). https://doi.org/10.29007/bnx2, /publications/paper/Blr

25. Leroux, J.: The reachability problem for Petri nets is not primitive recursive. In: 62nd IEEE Annual Symposium on Foundations of Computer Science, FOCS 2021, Denver, CO, USA, February 7-10, 2022. pp. 1241–1252. IEEE (2021). https://doi.org/10.1109/FOCS52979.2021.00121, https://doi.org/10.1109/FOCS52979.2021.00121
26. Leroux, J., Schmitz, S.: Reachability in vector addition systems is primitive-recursive in fixed dimension. In: 34th Annual ACM/IEEE Symposium on Logic in Computer Science, LICS 2019, Vancouver, BC, Canada, June 24-27, 2019. pp. 1–13. IEEE (2019). https://doi.org/10.1109/LICS.2019.8785796, https://doi.org/10.1109/LICS.2019.8785796
27. Lugiez, D.: Counting and equality constraints for multitree automata. In: Gordon, A.D. (ed.) Foundations of Software Science and Computational Structures, 6th International Conference, FOSSACS 2003 Held as Part of the Joint European Conference on Theory and Practice of Software, ETAPS 2003, Warsaw, Poland, April 7-11, 2003, Proceedings. Lecture Notes in Computer Science, vol. 2620, pp. 328–342. Springer (2003). https://doi.org/10.1007/3-540-36576-1_21, https://doi.org/10.1007/3-540-36576-1_21
28. Mayr, E.W.: An algorithm for the general Petri net reachability problem. SIAM J. Comput. **13**(3), 441–460 (1984). https://doi.org/10.1137/0213029, https://doi.org/10.1137/0213029
29. Ohsaki, H.: Beyond regularity: Equational tree automata for associative and commutative theories. In: Fribourg, L. (ed.) Computer Science Logic, 15th International Workshop, CSL 2001. 10th Annual Conference of the EACSL, Paris, France, September 10-13, 2001, Proceedings. Lecture Notes in Computer Science, vol. 2142, pp. 539–553. Springer (2001). https://doi.org/10.1007/3-540-44802-0_38, https://doi.org/10.1007/3-540-44802-0_38
30. Rambow, O.: Multiset-valued linear index grammars: Imposing dominance constraints on derivations. In: Pustejovsky, J. (ed.) 32nd Annual Meeting of the Association for Computational Linguistics, 27-30 June 1994, New Mexico State University, Las Cruces, New Mexico, USA, Proceedings. pp. 263–270. Morgan Kaufmann Publishers / ACL (1994). https://doi.org/10.3115/981732.981768
31. Reinhardt, K.: Reachability in Petri nets with inhibitor arcs. Electronic Notes in Theoretical Computer Science **223**, 239–264 (2008). https://doi.org/https://doi.org/10.1016/j.entcs.2008.12.042, https://www.sciencedirect.com/science/article/pii/S1571066108005057, proceedings of the Second Workshop on Reachability Problems in Computational Models (RP 2008)
32. Schmitz, S.: On the computational complexity of dominance links in grammatical formalisms. In: Hajic, J., Carberry, S., Clark, S. (eds.) ACL 2010, Proceedings of the 48th Annual Meeting of the Association for Computational Linguistics, July 11-16, 2010, Uppsala, Sweden. pp. 514–524. The Association for Computer Linguistics (2010), https://aclanthology.org/P10-1053/
33. Schmitz, S.: The complexity of reachability in vector addition systems. ACM SIGLOG News **3**(1), 4–21 (2016). https://doi.org/10.1145/2893582.2893585, https://doi.org/10.1145/2893582.2893585
34. Schmitz, S., Schnoebelen, P.: Algorithmic Aspects of WQO Theory (Aug 2012), https://cel.hal.science/cel-00727025, lecture
35. Verma, K.N., Goubault-Larrecq, J.: Karp-Miller trees for a branching extension of VASS. Discret. Math. Theor. Comput. Sci. **7**(1), 217–230 (2005). https://doi.org/10.46298/DMTCS.350

Tapes as Stochastic Matrices of String Diagrams

Filippo Bonchi[1] and Cipriano Junior Cioffo[1]

University of Pisa, Pisa, Italy
`filippo.bonchi@unipi.it`, `ciprianojunior.cioffo@di.unipi.it`

Abstract. Tape diagrams provide a graphical notation for categories equipped with two monoidal products, $\otimes$ and $\oplus$, where $\oplus$ is a biproduct. Recently, they have been generalised to handle Kleisli categories of arbitrary monoidal monads. In this work, we show that for the subdistribution monad, tapes are isomorphic to stochastic matrices of subdistributions of string diagrams. We then exploit this result to provide a complete axiomatisation of probabilistic Boolean circuits.

Keywords: string diagrams, probabilistic systems, rig categories

1 Introduction

Driven by the growing interest in compositional semantics for probabilistic systems, an increasing body of work [11,27,39,6,16,15,28,36,21,20,23,42,38,22,19] employs *string diagrams* [30,41], which formally represent morphisms in a strict symmetric monoidal category freely generated by a monoidal signature. These diagrams are typically interpreted in $\mathcal{KL}(\mathcal{D}_\leq)$, the Kleisli category of the subdistribution monad $\mathcal{D}_\leq$ (or suitable variants), with the monoidal structure induced by the cartesian product $\otimes$. Much research has focused on the role of the *copier* $-\!\!\!\prec$ and the *discharger* $-\!\!\bullet$, and their correspondence to marginals and joint distributions.

In contrast, significantly less attention [18] has been devoted to the second monoidal tensor $\oplus$ in $\mathcal{KL}(\mathcal{D}_\leq)$, which corresponds to disjoint union of sets, even though its interaction with $\otimes$ plays a pivotal role in many probabilistic frameworks [42,12,33]. One possible reason is that standard string diagrams are inherently limited to representing a single monoidal structure at a time. To overcome this limitation, we focus on *tape diagrams* [5], which can be informally understood as string diagrams of string diagrams. This more expressive graphical language supports simultaneous reasoning about both $\otimes$ and $\oplus$, in particular in categories where $\oplus$ forms a *biproduct*, i.e., serves as both categorical product and coproduct. Moreover, as recently shown [3], tape diagrams extend naturally to Kleisli categories of arbitrary monoidal monads and thus, in particular, to $\mathcal{KL}(\mathcal{D}_\leq)$, which is the main focus of this work.

Our starting observation is that the coproduct $\oplus$ in $\mathcal{KL}(\mathcal{D}_\leq)$ enjoys an additional universal property, akin to that of a product, which makes it resemble a biproduct. We refer to categories where $\oplus$ satisfies this property as *convex biproduct categories* (Definition 2), and we study their associated monoidal algebras. While categories with finite biproducts are equipped with natural monoids and *comonoids*, convex biproduct categories carry natural monoids and *co-pointed convex algebras*, which are the dual

N. Bertrand and S. Milius (Eds.): FoSSaCS 2026, LNCS 16503, pp. 88–109, 2026.
https://doi.org/10.1007/978-3-032-22730-0_5

of pointed convex algebras (pca)–the Eilenberg-Moore algebras for the subdistribution monad [43,7,8,35].

There is a further striking analogy with the theory of finite biproduct categories: given any category $\mathbf{C}$ enriched over monoids, one can construct the category $\mathbf{Mat(C)}$ of matrices over $\mathbf{C}$ [34], which possesses finite biproducts. Analogously, starting from a category $\mathbf{C}$ enriched over pcas, one can define the convex biproduct category $\mathbf{StMat(C)}$ of *stochastic matrices* over $\mathbf{C}$ (Proposition 3). Composition is defined via matrix multiplication: sums are given by the convex sums of the pca-enrichment and products by composition of arrows in $\mathbf{C}$. We show that this construction yields a functor $\mathbf{StMat}(-)$ from the category $\mathbf{PCACat}$ of pca-enriched categories to the category $\mathbf{CBCat}$ of convex biproduct categories, which is left adjoint to the forgetful functor U (Theorem 2):

The leftmost adjunction is standard (see, e.g., [9,44]): the functor $(-)^+$ freely enriches a locally small category $\mathbf{C}$ over pcas. The resulting category $\mathbf{C}^+$ has arrows given by subdistributions of arrows in $\mathbf{C}$.

Our key result is that the composite of the two adjunctions admits a syntactic presentation in terms of generators and equations. Specifically, we define a functor $\mathbf{T}(-)\colon \mathbf{Cat} \to \mathbf{CBCat}$ that freely adds to each category $\mathbf{C}$ a monoidal structure with natural monoids, and co-pointed convex algebras. We show that $\mathbf{T}(-)$ is left adjoint to the forgetful functor $U\colon \mathbf{CBCat} \to \mathbf{Cat}$ (Theorem 4) and we thus conclude that $\mathbf{T(C)}$ is isomorphic to $\mathbf{StMat(C^+)}$ (Corollary 1).

Importantly, when $\mathbf{C}$ is a category of string diagrams $\mathbf{Diag}_\Sigma$, $\mathbf{T}(\mathbf{Diag}_\Sigma)$ coincides with the category of tape diagrams for the monad $\mathcal{D}_\leq$ introduced in [3]. A result from [3] shows that $\mathbf{T}(\mathbf{Diag}_\Sigma)$ admits two monoidal products, $\oplus$ and $\otimes$, forming a *rig category* (also known as a bimonoidal category) [32,29]. The isomorphism $\mathbf{T}(\mathbf{Diag}_\Sigma) \cong \mathbf{StMat}(\mathbf{Diag}_\Sigma^+)$ then provides a concrete interpretation of tapes as stochastic matrices whose entries are probability distributions of string diagrams. Interestingly, $\oplus$ is the direct sum of matrices and $\otimes$ is (an extended) Kronecker product.

We argue that this isomorphism can be fruitfully applied to give more principled axiomatisations of probabilistic languages based on string diagrams, and to streamline their completeness proofs. As a case study, we revisit the complete axiomatisation of *probabilistic Boolean circuits* from [39]. These can be encoded into tape diagrams of Boolean circuits resulting in a more effective realisation of probabilistic control (see [3, Example 30]). We show that (Corollary 3) adding a single extra axiom–originally appearing in [14]– to the axioms of Boolean circuits and the tape structure (i.e., natural monoids and co-pcas) suffices to derive an alternative complete axiomatisation. The proof of completeness amounts to several categorical observations, two almost trivial lemmas and the universal properties of convex biproducts.

Synopsis. We begin in Section 2 by recalling the notions of pointed convex algebras, pca-enriched categories, and the leftmost adjunction in the diagram above. Section 3 introduces convex biproduct categories and their associated monoidal algebraic structures. In Section 4, we define stochastic matrices and establish the rightmost adjunction. The syntactic construction $\mathbf{T}(-)$, corresponding to the composite adjunction, is presented in Section 5. This construction is then applied to categories of string diagrams in Section 6 to obtain tape diagrams. Finally, Section 7 revisits probabilistic Boolean circuits from [39], explains their encoding as probabilistic tapes from [3], and presents the alternative axiomatisation along with the completeness proof.

An extended version of this article, including full proofs, is available in [2].

2   Pointed Convex Algebras

We commence our exposition by recalling several well-known algebraic structures.

A *pointed convex algebra* (pca) [43] consists of a set X, a designated element $\star \in X$ and, for all p in the open real interval $(0, 1)$, a function $+_p \colon X \times X \to X$ such that, by fixing $\tilde{p} \stackrel{\text{def}}{=} pq$ and $\tilde{q} \stackrel{\text{def}}{=} \frac{p(1-q)}{1-pq}$, the following laws hold for all $x_1, x_2, x_3 \in X$.

$$(x_1 +_q x_2) +_p x_3 = x_1 +_{\tilde{p}} (x_2 +_{\tilde{q}} x_3) \qquad x_1 +_p x_2 = x_2 +_{1-p} x_1 \qquad x_1 +_p x_1 = x_1 \quad (1)$$

A morphism of pcas is a function preserving $\star$ and $+_p$. Pointed convex algebras and their morphisms form a category, hereafter denoted by $\mathbf{PCA}$.

In any pca, $+_p$ can be defined for all $p \in [0, 1]$ by setting $x +_1 y \stackrel{\text{def}}{=} x$ and $x +_0 y \stackrel{\text{def}}{=} y$. For all $n \in \mathbb{N}$, $p_1, \ldots, p_n \in [0, 1]$ such that $\sum_i p_i \le 1$, one defines $\sum_{i=1}^n p_i \cdot (-)_i \colon X^n \to X$ inductively as

$$\sum_{i=1}^0 p_i \cdot (-)_i \stackrel{\text{def}}{=} \star \qquad \sum_{i=1}^{n+1} p_i \cdot (-)_i \stackrel{\text{def}}{=} (-)_1 +_{p_1} \sum_{j=1}^n q_j \cdot (-)_j. \quad (2)$$

where $(-)_j = (-)_{i+1}$ and q_j is $\frac{p_{i+1}}{1-p_1}$ if $p_1 \ne 1$ and 0 otherwise. Note that when $n = 1$, $\sum_{i=1}^1 p_i \cdot (-)_i$ is, by definition, $(-)_1 +_{p_1} \star$. Since this operation will play a crucial role we name it *multiplication by a scalar $p \in [0, 1]$* and we fix the following notation.

$$p \cdot x \stackrel{\text{def}}{=} x +_p \star$$

The standard example of a pca is provided by the set $\mathcal{D}_{\le}(X)$ of finitely supported probability subdistributions over some set X. Recall that a finitely supported subdistribution over X is a function $d \colon X \to [0, 1]$ such that $\sum_{x \in X} d(x) \le 1$ and $d(x) \ne 0$ for finitely many x. The designed element $\star$ in $\mathcal{D}_{\le}(X)$ is the null subdistribution, i.e., $\star(x) \stackrel{\text{def}}{=} 0$ for all $x \in X$; given $d_1, d_2 \in \mathcal{D}_{\le}(X)$, for all $p \in (0, 1)$, $(d_1 +_p d_2)(x) \stackrel{\text{def}}{=} p \cdot d_1(x) + (1 - p) \cdot d_2(x)$. At this point, it is convenient to fix some extra notation: $\mathcal{D}(X)$ is the set of all finitely supported distributions (i.e., $\sum_{x \in X} d(x) = 1$) and, for all $x \in X$, δ_x is the Dirac distribution at x (i.e., $\delta_x(x') = 1$ if $x = x'$ and 0 otherwise).

A category $\mathbf{C}$ is $\mathbf{PCA}$-*enriched* if every homset $\mathbf{C}[X, Y]$ carries a pca and composition of arrows is a pca morphism, i.e., for all $p \in (0, 1)$ and properly typed arrows e, f, g, h, the followings hold.

$$e; (f +_p g) = (e; f) +_p (e; g) \qquad (f +_p g); h = (f; h +_p g; h) \qquad f; \star = \star = \star; f \quad (3)$$

A functor F is **PCA**-enriched if it preserves the structure of pca over each homset. **PCA**-enriched categories and functors form a category, denoted by **PCACat**. All categories considered hereafter are tacitly assumed to be locally small: **Cat** stands for the category of locally small categories. Note that there is a forgetful functor $U\colon$ **PCACat** $\to$ **Cat**.

Any category $\mathbf{C}$ gives rise to the **PCA**-enriched category $\mathbf{C}^+$: objects of $\mathbf{C}^+$ are those of $\mathbf{C}$; for all objects X, Y, the homset is defined as $\mathbf{C}^+[X,Y] \stackrel{\text{def}}{=} \mathcal{D}_\leq(\mathbf{C}[X,Y])$. For $d_1\colon X \to Y$ and $d_2\colon Y \to Z$, their composition $d_1;d_2\colon X \to Z$ is defined for all $h \in \mathbf{C}[X,Z]$ as $d_1;d_2(h) \stackrel{\text{def}}{=} \sum_{\{(f,g)\mid f;g=h\}} d_1(f)\cdot d_2(g)$; The identity $id_X\colon X \to X$ is given by δ_{id_X}. One can easily see that $\mathbf{C}^+$ is a **PCA**-enriched category. Moreover, the assignment $\mathbf{C} \mapsto \mathbf{C}^+$ gives rise to a functor $(-)^+\colon$ **Cat** $\to$ **PCACat** which is left adjoint to the forgetful functor U (see e.g. [9, Prop. 6.4.7] or [44, Cor. 1]).

Theorem 1 ([9,44]). $(-)^+\colon$ **Cat** $\to$ **PCACat** *is left adjoint to* $U\colon$ **PCACat** $\to$ **Cat**.

Example 1. Let $\mathbf{1}$ be a category with a single object and a single arrow. The **PCA**-enriched category $\mathbf{1}^+$ has a single object, arrows are $p \in [0,1]$ and composition is given by multiplication.

For a set A, the set of words over A (hereafter denoted by A^*) carries the structure of a category with a single object. The **PCA**-enriched category $(A^*)^+$ has a single object and arrows are subdistributions over A^*. Composition of $d_1, d_2 \in \mathcal{D}_\leq(A^*)$ is defined for all words $w \in A^*$ as $d_1;d_2(w) \stackrel{\text{def}}{=} \sum_{u;v=w} d_1(u) \cdot d_2(v)$.

Our main example of **PCA**-enriched category is $\mathcal{KL}(\mathcal{D}_\leq)$, the Kleisli category of the monad $\mathcal{D}_\leq\colon$ **Set** $\to$ **Set** (see e.g., [24]). Objects are sets; morphisms $f\colon X \to Y$ are functions $X \to \mathcal{D}_\leq(Y)$. We often write $f(y \mid x)$ for $f(x)(y)$ as this number represents the probability that f returns y given the input x. Identities $id_X\colon X \to \mathcal{D}_\leq(X)$ map each element $x \in X$ to δ_x. For two functions $f\colon X \to \mathcal{D}_\leq(Y)$ and $g\colon Y \to \mathcal{D}_\leq(Z)$, their composition in $\mathcal{KL}(\mathcal{D}_\leq)$ is defined as $f;g(z \mid x) \stackrel{\text{def}}{=} \sum_{y\in Y} f(y \mid x) \cdot g(z \mid y)$.

In our work, it is fundamental that $\mathcal{KL}(\mathcal{D}_\leq)$ carries two symmetric monoidal structures: $(\mathcal{KL}(\mathcal{D}_\leq), \otimes, 1)$ and $(\mathcal{KL}(\mathcal{D}_\leq), \oplus, 0)$. The monoidal product $\otimes$ is defined on objects as the cartesian product of sets, often denoted by $\times$, with unit the singleton $1 \stackrel{\text{def}}{=} \{\bullet\}$; $\oplus$ is the disjoint union of sets with unit the empty set $0 \stackrel{\text{def}}{=} \{\}$. Hereafter we denote the disjoint union of two sets X and Y by $X \oplus Y \stackrel{\text{def}}{=} \{(x,0) \mid x \in X\} \cup \{(y,1) \mid y \in Y\}$ where 0 and 1 are tags used to distinguish the set of provenance. For arrows $f\colon X \to Y$ and $g\colon X' \to Y'$, $f \otimes g\colon X \otimes X' \to Y \otimes Y'$ and $f \oplus g\colon X \oplus X' \to Y \oplus Y'$ are defined, for all $x \in X, x' \in X', y \in Y, y' \in Y', u \in X \oplus X'$ and $v \in Y \oplus Y'$, as follows.

$$f \otimes g(y, y' \mid x, x') \stackrel{\text{def}}{=} f(y \mid x) \cdot g(y' \mid x')$$

$$f \oplus g(v \mid u) \stackrel{\text{def}}{=} \begin{cases} f(y \mid x) & \text{if } u = (x,0) \text{ and } v = (y,0) \\ g(y' \mid x') & \text{if } u = (x',1) \text{ and } v = (y',1) \\ 0 & \text{otherwise} \end{cases} \tag{4}$$

Symmetries $\sigma^\otimes_{X,Y}\colon X \otimes Y \to Y \otimes X$ and $\sigma^\oplus_{X,Y}\colon X \oplus Y \to Y \oplus X$ are defined as in **Set**. Actually, $(\mathcal{KL}(\mathcal{D}_\leq), \oplus, \otimes, 0, 1)$ forms a rig (aka bimonoidal) category in the sense of [32]. We will come back to rig categories in Section 6. Until then, we will focus only on the monoidal structure provided by $\oplus$.

3   Convex Biproduct Categories

As is the case for all Kleisli categories of monads on **Set**, the category $\mathcal{KL}(\mathcal{D}_\leq)$ inherits coproducts from **Set**: the operation $\oplus$ in (4) serves as a coproduct in $\mathcal{KL}(\mathcal{D}_\leq)$. Our initial observation is that $\oplus$ satisfies an additional universal property, which we refer to as the *convex product*, described below.

Definition 1. *Let X_1, X_2 be two objects of a* **PCA**-*enriched category* **C**. *The* convex product *of X_1 and X_2 is an object Z with two arrows $\pi_1 \colon Z \to X_1$ and $\pi_2 \colon Z \to X_2$ satisfying the following property: for all $p_1, p_2 \in [0,1]$ such that $p_1 + p_2 \leq 1$ and all arrows $f \colon A \to X_1$, $g \colon A \to X_2$, there exists a unique arrow $h \colon A \to Z$ such that $h; \pi_1 = p_1 \cdot f$ and $h; \pi_2 = p_2 \cdot g$.*

Similarly, the convex product of n objects $X_1, \ldots, X_n$ is an object Z with arrows $\pi_i \colon Z \to X_i$ for $i = 1, \ldots, n$ satisfying the following property: for all $p_1, \ldots, p_n \in [0,1]$ where $\sum_{i=1}^{n} p_i \leq 1$ and arrows $f_i \colon A \to X_i$, there exists a unique arrow $h \colon A \to Z$ such that $h; \pi_i = p_i \cdot f_i$ for all $i = 1, \ldots, n$. Observe that, by definition, the 0-ary convex product is a final object. Hereafter, we will denote the unique arrow h by $\langle f_1, \ldots, f_n \rangle_p$ where p is a compact notation for $p_1, \ldots, p_n$.

Definition 2. *A convex biproduct category is a* **PCA**-*enriched category* **C** *with an object 0 which is both initial and final and, for every pair of objects X_1, X_2, an object $X_1 \oplus X_2$ and morphisms $\pi_i \colon X_1 \oplus X_2 \to X_i$ and $\iota_i \colon X_i \to X_1 \oplus X_2$ such that $(X_1 \oplus X_2, \iota_1, \iota_2)$ is a coproduct, $(X_1 \oplus X_2, \pi_1, \pi_2)$ is a convex product and*

$$\iota_i; \pi_j = \delta_{i,j} \tag{5}$$

where $\delta_{i,j} \colon X_i \to X_j$ is defined as $\delta_{i,j} \overset{def}{=} id_{X_i}$ if $i = j$ and $\delta_{i,j} \overset{def}{=} \star_{X_i, X_j}$ otherwise.

A morphism of convex biproduct categories is a **PCA**-*enriched functor $F \colon$* **C** $\to$ **D** *preserving finite coproducts. We write* **CBCat** *for the category of convex biproduct categories and their morphisms.*

As for categories with finite biproducts, morphisms of convex biproduct categories preserve (convex) products. Moreover, the following two elementary properties hold.

Lemma 1. *In a convex biproduct category* **C**, *the enrichment is compatible with respect to the (convex-co)products, namely, for all $f, g \colon X \to Y$ and $p \in [0,1]$,*

$$f +_p g = \langle f, g \rangle_{p,1-p}; [id_Y, id_Y]. \tag{6}$$

Lemma 2. *Let* **C** *be a convex biproduct category. Then* **C** *has n-ary convex products.*

Since any convex biproduct category **C** has finite coproducts, by Fox's theorem [17], it carries a symmetric monoidal category structure $(\mathbf{C}, \oplus, 0)$ where every object X is equipped with a commutative monoid that is natural and coherent, i.e., morphisms $\triangleright_X \colon X \oplus X \to X$ and $\mathord{\text{\textsc{i}}}_X \colon 0 \to X$ satisfying the laws in Table 1 where $\alpha_{X,Y,Z}$, λ_X, ρ_Y denote associators, left and right unitors. Recall that $\triangleright_X$ is given as the copairing of the identities $[id_X, id_X]$ while $\mathord{\text{\textsc{i}}}_X$ is the unique arrow from the initial object.

$$\alpha_{P,P,P};(id_P \oplus \triangleright_P);\triangleright_P = (\triangleright_P \oplus id_P);\triangleright_P \qquad (\triangleright\text{-as}) \qquad\qquad \sigma_{P,P};\triangleright_P = \triangleright_P \qquad (\triangleright\text{-sym})$$

$$(\mathord{\circ}_P \oplus id_P);\triangleright_P = \lambda_P \qquad (id_P \oplus \mathord{\circ}_P);\triangleright_P = \rho_P \qquad (\triangleright\text{-un})$$

$$\triangleright_{P\oplus Q} = \alpha_{P,Q,P\oplus Q};(id_P \oplus \alpha^-_{Q,P,Q});(id_P \oplus (\sigma_{Q,P} \oplus id_Q));(id_P \oplus \alpha_{P,Q,Q});\alpha^-_{P,P,Q\oplus Q};(\triangleright_P \oplus \triangleright_Q) \qquad (\triangleright\text{-coh})$$

$$\mathord{\circ}_{P\oplus Q} = \lambda^-_0;(\mathord{\circ}_P \oplus \mathord{\circ}_Q) \qquad (\mathord{\circ}\text{-coh}) \qquad \mathord{\circ}_0 = id_0 \qquad \triangleright_0 = \lambda_0 \qquad (\mathord{\circ}_0,\triangleright_0\text{-coh})$$

$$\triangleright_P;f = (f \oplus f);\triangleright_Q \qquad (\triangleright\text{-nat}) \qquad\qquad \mathord{\circ}_P;f = \mathord{\circ}_Q \qquad (\mathord{\circ}\text{-nat})$$

Table 1: Axioms for natural and coherent monoid objects.

$$\mathord{^p\triangleleft}_P;(\mathord{^q\triangleleft}_P \oplus id_P) = \mathord{^{\tilde p}\triangleleft}_P;(id_P \oplus \mathord{^{\tilde q}\triangleleft});\alpha^-_{P,P,P} \qquad (\triangleleft\text{-as}) \qquad\qquad \tilde p = pq \qquad \tilde q = \frac{p(1-q)}{1-pq}$$

$$\mathord{^p\triangleleft}_P;\triangleright_P = id_P \qquad (\triangleleft\text{-idem}) \qquad\qquad \mathord{^p\triangleleft}_P\, \sigma_{P,P} = \mathord{^{1-p}\triangleleft}_P \qquad (\triangleleft\text{-sym})$$

$$\mathord{^p\triangleleft}_{P\oplus Q} = (\mathord{^p\triangleleft}_P \oplus \mathord{^p\triangleleft}_Q);\alpha_{P,P,Q\oplus Q};(id_P \oplus \alpha^-_{P,Q,Q});(id_P \oplus (\sigma_{P,Q} \oplus id_Q));(id_P \oplus \alpha_{Q,P,Q});\alpha^-_{P,Q,P\oplus Q} \qquad (\triangleleft\text{-coh})$$

$$\mathord{\downarrow}_{P\oplus Q} = (\mathord{\downarrow}_P \oplus \mathord{\downarrow}_Q);\lambda_0 \qquad (\mathord{\downarrow}\text{-coh}) \qquad \mathord{\downarrow}_0 = id_0 \qquad \mathord{^p\triangleleft}_0 = \lambda^-_0 \qquad (\mathord{\downarrow}_0,\mathord{^p\triangleleft}_0\text{-coh})$$

$$f;\mathord{^p\triangleleft}_Q = \mathord{^p\triangleleft}_P;(f \oplus f) \qquad (\triangleleft\text{-nat}) \qquad\qquad f;\mathord{\downarrow}_Q = \mathord{\downarrow}_P \qquad (\mathord{\downarrow}\text{-nat})$$

Table 2: Axioms for natural and coherent co-pca objects.

Similarly, the convex product structure equips any object X with a natural and coherent *co-pca*: for all $p \in (0,1)$, there exist an arrow $\mathord{^p\triangleleft}_X \colon X \to X \oplus X$, defined as $\langle id_X, id_X \rangle_{p,1-p}$, and an arrow $\mathord{\downarrow}_X \colon X \to 0$, defined as the unique map to the final object 0, satisfying the laws in Table 2.

Proposition 1. *Let $\mathbf{C}$ be a convex biproduct category. Then for every object X, the triple $(X, \mathord{^p\triangleleft}_X, \mathord{\downarrow}_X)$ is a natural and coherent co-pca, i.e., the axioms in Table 2 hold.*

Examples of convex biproduct categories are $\mathcal{KL}(\mathcal{D}_\le)$ and its continuous analogue (see [2] for a proof). The most relevant for our work is $\mathcal{KL}(\mathcal{D}_\le)$. For all sets X, co-pcas and monoids in $\mathcal{KL}(\mathcal{D}_\le)$ are illustrated below.

$$
\begin{array}{llll}
\mathord{^p\triangleleft}_X \colon X \to X \oplus X & \mathord{\downarrow}_X \colon X \to 0 & \mathord{\circ}_X \colon 0 \to X & \triangleright_X \colon X \oplus X \to X \\
\quad x \mapsto \delta_{(x,0)} +_p \delta_{(x,1)} & \quad x \mapsto \star & & \quad (x,i) \mapsto \delta_x
\end{array}
\tag{7}
$$

Monoids and co-pcas will be useful later to freely generate convex biproduct categories. In particular, morphisms of convex biproduct categories can be characterised as follows.

Proposition 2. *A functor $F \colon \mathbf{C} \to \mathbf{D}$ is a morphism of convex biproduct categories if and only if it is a strong monoidal functor preserving monoids and co-pcas.*

From now on, we assume monoidal categories to be strict, meaning that the structural isomorphisms $\alpha_{X,Y,Z}$, λ_X, ρ_X are identities. This assumption is made without loss of generality [34], it greatly simplifies calculations and allows for diagrammatic notations: the last three rows of Figure 1 provide a diagrammatic representation of the axioms of natural monoids and co-pcas in a strict monoidal category illustrated in Tables 3 and 4.

4 Stochastic Matrices over PCA-enriched categories

It is well known (see, e.g., [34, Exercises VIII.2.5-6]) that from a category enriched over commutative monoids, one can freely generate the finite biproduct category of its

$(id_P \oplus \rhd_P); \rhd_P = (\rhd_P \oplus id_P); \rhd_P$	($\rhd$-as)	$(!_P \oplus id_P); \rhd_P = id_P$	($\rhd$-un)
$!_0 = id_0 \qquad \rhd_0 = id_0$	($!_0, \rhd_0$-coh)	$\sigma_{P,P}; \rhd_P = \rhd_P$	($\rhd$-sym)
$!_{P\oplus Q} = !_P \oplus !_Q$	($!$-coh)	$\rhd_{P\oplus Q} = (id_P \oplus \sigma_{P,Q} \oplus id_Q); (\rhd_P \oplus \rhd_Q)$	($\rhd$-coh)
$!_P; f = !_Q$	($!$-nat)	$\rhd_P; f = (f \oplus f); \rhd_Q$	($\rhd$-nat)

Table 3: Axioms for natural and coherent monoids in a strict monoidal category.

$\lhd_P; (\lhd_P \oplus id_P) = \tilde{\lhd}_P; (id_P \oplus \tilde{\lhd})$	($\lhd$-as)	$\tilde{p} = pq \qquad \tilde{q} = \frac{p(1-q)}{1-pq}$	
$\lhd_P; \rhd_P = id_P$	($\lhd$-idem)	$^{p}\!\lhd_P\, \sigma_{P,P} = {}^{1-p}\!\lhd_P$	($\lhd$-sym)
$\lhd_0 = id_0$	($\lhd_0$-coh)	$\flat_0 = id_0$	($\flat_0$-coh)
$\flat_{P\oplus Q} = \flat_P \oplus \flat_Q$	($\flat$-coh)	$\lhd_{P\oplus Q} = (\lhd_P \oplus \lhd_Q); (id_P \oplus \sigma_{P,Q} \oplus id_Q)$	($\lhd$-coh)
$f; \flat_Q = \flat_P$	($\flat$-nat)	$f; {}^{p}\!\lhd_Q = {}^{p}\!\lhd_P; (f \oplus f)$	($\lhd$-nat)

Table 4: Axioms for natural and coherent co-pcas in a strict monoidal category.

matrices. In this section, we show that a similar construction holds for **PCA**-enriched categories: given such a category **C**, one can construct the convex biproduct category **StMat(C)** of *stochastic matrices* over **C**.

Objects of **StMat(C)** are words in $Ob(\mathbf{C})^*$. We will write $\bigoplus_{k=1}^{m} U_k$ for the word $U_1 \dots U_m$ and 0 for the empty word. We will denote objects of **C** by U, V and those of **StMat(C)** by P, Q. In **StMat(C)**, an arrow $M: \bigoplus_{k=1}^{n} U_k \to \bigoplus_{k=1}^{m} V_k$ is an equivalence class of $m \times n$ matrices with (j, i)-entries given by pairs (p_{ji}, f_{ji}) where $f_{ji} \in \mathbf{C}[U_i, V_j]$ and $p_{ji} \in [0, 1]$ satisfy $\sum_{j=1}^{m} p_{ji} \leq 1$. Two matrices M and M' are equivalent, in symbols $M \equiv M'$, if $p_{ji} \cdot f_{ji} = p'_{ji} \cdot f'_{ji}$ in $\mathbf{C}[U_i, V_j]$ for all i, j.

Remark 1. The use of pairs (p_{ji}, f_{ji}) as entries of matrices is necessary to specify the constraints $\sum_{j=1}^{m} p_{ji} \leq 1$. However, $\equiv$ ensures that one can safely write $M_{ji} = p_{ji} \cdot f_{ji}$.

The composition of two morphisms $M: \bigoplus_{k=1}^{n} U_k \to \bigoplus_{k=1}^{m} V_k$ and $M': \bigoplus_{k=1}^{m} V_k \to \bigoplus_{k=1}^{l} W_k$ is obtained by matrix multiplication $M'M$: for all $u \in \{1, \dots, l\}$ and $i \in \{1, \dots, n\}$, the entry at (u, i) is

$$(M'M)_{ui} \stackrel{\mathrm{def}}{=} r_{ui} \cdot \left(\sum_{j=1}^{m} \frac{p'_{uj} p_{ji}}{r_{ui}} \cdot (f_{ji}; f'_{uj}) \right) = \sum_{j=1}^{m} p'_{uj} p_{ji} \cdot (f_{ji}; f'_{uj}) \tag{8}$$

where the convex sums $\sum_k p_k \cdot (-)_k$ are those provided by the **PCA** enrichment of **C**, the composition ; is in **C** and $r_{ui} = (\sum_{j=1}^{m} p'_{uj} p_{ji})$. Simple computations confirm that $\sum_{u=1}^{l} r_{ui} \leq 1$ and $\frac{p'_{uj} p_{ji}}{r_{ui}} \in [0, 1]$. For all objects $P = \bigoplus_{k=1}^{n} U_k$, id_P is the $n \times n$ matrix with entries $(id_P)_{jj} = 1 \cdot id_{U_j}$ and, for $i \neq j$, $(id_P)_{ji} = 0 \cdot \star$.

Example 2. Recall $\mathbf{1}^+$ and $(A^*)^+$ from Example 1. In **StMat($\mathbf{1}^+$)** objects are natural numbers and arrows $n \to m$ are the usual $m \times n$ substochastic matrices (i.e., $\sum_{j=1}^{m} p_{ji} \leq 1$).

In **StMat**$((A^*)^+)$ objects are natural numbers and arrows $n \to m$ are $m \times n$ matrices with entries in $\mathcal{D}_{\leq}(A^*)$. Consider for instance the matrices $N\colon 2 \to 3$ and $M\colon 2 \to 2$ on the left below. The composition $M; N\colon 2 \to 3$ is the matrix on the right.

$$
\begin{pmatrix}
\frac{1}{2} \cdot a & \frac{1}{2} \cdot c \\
\frac{1}{3} \cdot ab & 0 \cdot \star \\
0 \cdot \star & \frac{1}{3} \cdot id
\end{pmatrix}
\begin{pmatrix}
\frac{1}{2} \cdot a & 1 \cdot c \\
\frac{1}{2} \cdot ab & 0 \cdot \star
\end{pmatrix}
=
\begin{pmatrix}
\frac{1}{2} \cdot (\frac{1}{2} \cdot aa + \frac{1}{2} abc) & \frac{1}{2} \cdot ca \\
\frac{1}{6} \cdot aab & \frac{1}{3} \cdot cab \\
\frac{1}{6} \cdot ab & 0 \cdot \star
\end{pmatrix}
$$

For arbitrary matrices M and N, $M \oplus N$ is defined as below on the left

$$
M \oplus N \overset{\text{def}}{=} \begin{pmatrix} M & \varnothing \\ \varnothing & N \end{pmatrix}
\qquad
\sigma_{U,Q} \overset{\text{def}}{=} \begin{pmatrix} \varnothing & id_Q \\ id_P & \varnothing \end{pmatrix}
\tag{9}
$$

where $\varnothing$ is the matrix with all entries $0 \cdot \star$. The symmetry $\sigma^{\oplus}_{P,Q}\colon P \oplus Q \to Q \oplus P$ is defined as on the right above. Every object P is equipped with co-pca and monoid structures. These are defined for all objects $U \in \mathbf{C}$ as follows, where $!_U$ (respectively $?_U$) is the unique matrix with 0 rows (columns).

$$
{}^{p}\!\vartriangleleft_U \overset{\text{def}}{=} \begin{pmatrix} p \cdot id_U \\ (1-p) \cdot id_U \end{pmatrix}
\qquad
\flat_U \overset{\text{def}}{=} !_U
\qquad
\bar{\flat}_U \overset{\text{def}}{=} ?_U
\qquad
\vartriangleright_U \overset{\text{def}}{=} \begin{pmatrix} 1 \cdot id_U & 1 \cdot id_U \end{pmatrix}
\tag{10}
$$

For arbitrary objects P of **StMat**$(\mathbf{C})$, co-pca and monoids are defined inductively as:

$$
\begin{array}{ll}
{}^{p}\!\vartriangleleft_0 \overset{\text{def}}{=} id_0 \quad \flat_0 \overset{\text{def}}{=} id_0 \quad \flat_{U \oplus P} \overset{\text{def}}{=} \flat_U \oplus \flat_P & \vartriangleright_0 \overset{\text{def}}{=} id_0 \quad \bar{\flat}_0 \overset{\text{def}}{=} id_0 \quad \bar{\flat}_{U \oplus P} \overset{\text{def}}{=} \bar{\flat}_U \oplus \bar{\flat}_P \\[4pt]
{}^{p}\!\vartriangleleft_{U \oplus P} \overset{\text{def}}{=} ({}^{p}\!\vartriangleleft_U \oplus {}^{p}\!\vartriangleleft_P); (id_U \oplus \sigma_{U,P} \oplus id_P) & \vartriangleright_{U \oplus P} \overset{\text{def}}{=} (id_U \oplus \sigma_{P,U} \oplus id_P); (\vartriangleright_U \oplus \vartriangleright_P)
\end{array}
\tag{11}
$$

Proposition 3. **StMat**$(\mathbf{C})$ *is a convex biproduct category.*

Proof. We have seen in (9), (10) and (11) monoid and co-pca structure on each object of **StMat**$(\mathbf{C})$. The enrichment over **PCA** is given by defining, for all $M, N\colon P \to Q$ and $p \in [0,1]$, the convex sum $M +_p N$ through the composition ${}^{p}\!\vartriangleleft_P; (M \oplus N); \vartriangleright_Q$. While $\star_{P,Q}\colon P \to Q$ is the matrix with all entries equal to $\varnothing$. For convex binary products, given the $m_i \times n$ matrices $M_i\colon P \to Q_i$ for $i = 1, 2$, and $p, q \in [0, 1]$ with $p + q \leq 1$, the mediating arrow $\langle M_1, M_2 \rangle_{p,q}\colon P \to Q_1 \oplus Q_2$ is given by ${}^{p,q}\!\vartriangleleft_P; (M_1 \oplus M_2)$, where ${}^{p,q}\!\vartriangleleft_P$ is the $2n \times n$ matrix whose upper block is given by $p \cdot id_P$ and the bottom one by $q \cdot id_P$. Projections $\pi_i\colon Q_1 \oplus Q_2 \to Q_i$, for $i = 1, 2$, are defined respectively as $\pi_1 \overset{\text{def}}{=} id_{Q_1} \oplus \flat_{Q_2}$ and $\pi_2 \overset{\text{def}}{=} \flat_{Q_1} \oplus id_{Q_2}$. Simple matrix multiplications show that this is the unique arrow with the required properties and that all the axioms of convex biproduct categories are satisfied. See [2] for further details.

Just as categories of matrices are freely generated finite biproduct categories [34], similarly **StMat**$(\mathbf{C})$ is the convex biproduct category freely generated by $\mathbf{C}$. We are going to illustrate this below.

For all **PCA**-enriched functors $F\colon \mathbf{C} \to \mathbf{D}$, one can define **StMat**$(F)\colon$ **StMat**$(\mathbf{C}) \to$ **StMat**$(\mathbf{D})$ as the functor mapping an object $\bigoplus_{k=1}^{n} U_k$ to $\bigoplus_{k=1}^{n} F(U_k)$ and a matrix $M\colon \bigoplus_{k=1}^{n} U_k \to \bigoplus_{k=1}^{m} V_k$ with entries $M_{ji} = p_{ji} \cdot f_{ji}$ into the matrix with entries **StMat**$(F)(M)_{ji} \overset{\text{def}}{=} p_{ji} \cdot F(f_{ji})$.

$$\begin{array}{l}
\mathbin{R}_U : U \to U \oplus U \quad \mathbin{\delta}_U : U \to 0 \quad \sigma^{\oplus}_{U,V} : U \oplus V \to V \oplus U \quad \triangleright_U : U \oplus U \to U \quad \mathbin{?}_U : 0 \to U \\[4pt]
id_0 : 0 \to 0 \quad id_U : U \to U \quad \dfrac{c : U \to V}{\overline{c} : U \to V} \quad \dfrac{t : P \to Q \quad s : Q \to R}{t; s : P \to R} \quad \dfrac{t : P_1 \to Q_1 \quad s : P_2 \to Q_2}{t \oplus s : P_1 \oplus P_2 \to Q_1 \oplus Q_2}
\end{array}$$

$$\begin{array}{l}
(f; g); h = f; (g; h) \qquad id_P; f = f = f; id_Q \qquad id_0 \oplus f = f = f \oplus id_0 \qquad (f \oplus g) \oplus h = f \oplus (g \oplus h) \\[4pt]
(f_1 \oplus f_2); (g_1 \oplus g_2) = (f_1; g_1) \oplus (f_2; g_2) \qquad \sigma_{P,Q}; \sigma_{Q,P} = id_{P \oplus Q} \qquad (s \oplus id_R); \sigma_{Q,R} = \sigma_{P,R}; (id_R \oplus s)
\end{array}$$

Table 5: Typing rules (top) and axioms (bottom) for freely generated strict symmetric monoidal categories.

Proposition 4. *Let* $F : \mathbf{C} \to \mathbf{D}$ *be a* **PCA***-enriched functor.* $\mathbf{StMat}(F) : \mathbf{StMat}(\mathbf{C}) \to \mathbf{StMat}(\mathbf{D})$ *is a morphism of convex biproduct categories.*

Then, it is easy to check that the assignment $\mathbf{C} \mapsto \mathbf{StMat}(\mathbf{C})$ and $F \mapsto \mathbf{StMat}(F)$ provides a functor $\mathbf{StMat}(-) : \mathbf{PCACat} \to \mathbf{CBCat}$ which is left adjoint to the forgetful $U : \mathbf{CBCat} \to \mathbf{PCACat}$.

Theorem 2. $\mathbf{StMat}(-) : \mathbf{PCACat} \to \mathbf{CBCat}$ *is left adjoint to* $U : \mathbf{CBCat} \to \mathbf{PCACat}$

We conclude this section with a simple observation that will be useful later in Section 7.

Proposition 5. *For any convex biproduct category* $\mathbf{C}$*, the counit of the adjunction above* $\epsilon_{\mathbf{C}} : \mathbf{StMat}(U(\mathbf{C})) \to \mathbf{C}$ *is a full and faithful morphism of convex biproduct categories.*

5   An equational presentation of stochastic matrices

So far, we have seen that given a category, one can freely enrich it over **PCA**, and then obtain a convex biproduct category by taking its category of stochastic matrices. Now, we give a syntactic description, in terms of generators and equations, of such category.

For a category $\mathbf{C}$, we consider terms generated by the following grammar

$$t ::= \mathbin{R}_U \mid \mathbin{\delta}_U \mid \overline{c} \mid \mathbin{?}_U \mid \triangleright_U \mid id_U \mid id_0 \mid \sigma^{\oplus}_{U,V} \mid t; t \mid t \oplus t \tag{12}$$

where $p \in (0, 1)$, $U, V \in Ob(\mathbf{C})$, and c is an arrow in $\mathbf{C}$. Terms are typed according to the rules at the top of Table 5: each type is an arrow $P \to Q$ where $P, Q \in Ob(\mathbf{C})^*$ are regarded as sums of objects of $\mathbf{C}$. As expected, we consider only those terms that are typable. For arbitrary $P \in Ob(\mathbf{C})^*$, we define $\mathbin{R}_P, \mathbin{\delta}_P, \mathbin{?}_P, \triangleright_P$ as in (11). Analogous inductive definitions give us id_P and $\sigma^{\oplus}_{P,Q}$.

The category $\mathbf{T}(\mathbf{C})$ has as set of objects $Ob(\mathbf{C})^*$. Arrows in $\mathbf{T}(\mathbf{C})[P, Q]$ are terms modulo the axioms of natural and coherent monoids, co-pcas, strict symmetric monoidal categories (respectively in Tables 3, 4 and 5 bottom) and the following two axioms.

$$\overline{id_P} = id_P \qquad \overline{c; d} = \overline{c}; \overline{d} \tag{Tape}$$

Identities, composition, symmetries and monoidal product are defined as on terms. It is thus immediate to see that $(\mathbf{T}(\mathbf{C}), \oplus, 0)$ forms a symmetric monoidal category. Moreover such category is enriched over **PCA**: for all objects P, Q and arrows $f, g \colon P \to Q$

$$f +_p g \stackrel{\text{def}}{=} {}^p\!\triangleleft_P ; (f \oplus g) ; \triangleright_Q \qquad \star_{P,Q} \stackrel{\text{def}}{=} \downarrow_P ; \uparrow_Q \tag{13}$$

By naturality of $\uparrow$ and $\downarrow$, 0 is both initial and final object. By the Fox theorem [17], $\oplus$ is a coproduct with injections $id_{P_1} \oplus \uparrow_{P_2} \colon P_1 \to P_1 \oplus P_2$ and $\uparrow_{P_1} \oplus id_{P_2} \colon P_2 \to P_1 \oplus P_2$.

Our main technical effort consists of proving that $\oplus$ is a convex product with projections $id_{P_1} \oplus \downarrow_{P_2} \colon P_1 \oplus P_2 \to P_1$ and $\downarrow_{P_1} \oplus id_{P_2} \colon P_1 \oplus P_2 \to P_2$. By the coproduct property, any $\mathsf{t} \colon \bigoplus_{j=1}^{m} V_j \to \bigoplus_{i=1}^{n} U_i$ is the copairing $[\mathsf{t}_1, \ldots, \mathsf{t}_m]$ for $\mathsf{t}_j = \iota_j; \mathsf{t}$. One can thus restrict to consider the case of an arrow of type $V_j \to \bigoplus_{i=1}^{n} U_i$. First, we extend $\triangleleft$ to arbitrary $p \in [0, 1]$: $^0\!\triangleleft_U \stackrel{\text{def}}{=} \uparrow_U \oplus id_U$ and $^1\!\triangleleft_U \stackrel{\text{def}}{=} id_U \oplus \uparrow_U$. Then, for all $n \in \mathbb{N}$ and $\boldsymbol{p} = p_1, \ldots, p_n$ with $p_i \in [0, 1]$ such that $\sum_{i=1}^{n} p_i \leq 1$, we inductively define $^{\boldsymbol{p}}\!\triangleleft_U^{\,n} \colon U \to \bigoplus_{i=1}^{n} U$ as follows

$$^{\boldsymbol{p}}\!\triangleleft_U^{\,0} \stackrel{\text{def}}{=} \downarrow_U \qquad ^{\boldsymbol{p}}\!\triangleleft_U^{\,n+1} \stackrel{\text{def}}{=} {}^{p}\!\triangleleft_U ; (id_U \oplus {}^{\boldsymbol{q}}\!\triangleleft_U^{\,n}) \tag{14}$$

where $\boldsymbol{q} = q_1, \ldots q_n$ for $q_i = 0$ if $p_1 = 1$ and $q_i = \frac{p_{i+1}}{1-p_1}$ otherwise. Given n arrows $\mathsf{t}_i \colon U \to U_i$, simple computations confirm that $(^{\boldsymbol{p}}\!\triangleleft_U^{\,n} ; \bigoplus_{i=1}^{n} \mathsf{t}_i); \pi_i = p_i \cdot \mathsf{t}_i$. Thus, $^{\boldsymbol{p}}\!\triangleleft_U^{\,n} ; \bigoplus_{i=1}^{n} \mathsf{t}_i \colon U \to \bigoplus_{i=1}^{n} U_i$ is a mediating arrow satisfying the constraints of n-ary convex products. The hard part consists of proving its uniqueness.

Theorem 3. $\mathbf{T}(\mathbf{C})$ *is a convex biproduct category. In particular, for all* $\boldsymbol{p} = p_1, \ldots, p_n$ *and* $\mathsf{t}_i \colon U \to U_i$, $\langle \mathsf{t}_1, \ldots \mathsf{t}_n \rangle_{\boldsymbol{p}} = {}^{\boldsymbol{p}}\!\triangleleft_U^{\,n} ; \bigoplus_{i=1}^{n} \mathsf{t}_i.$

Proof. To prove the statement, we rely on two key properties of $\mathbf{T}(\mathbf{C})$. The first one is that every arrow $\mathsf{t} \colon U \to \bigoplus_{i=1}^{n} U_i$ can be decomposed as $\mathsf{t} = {}^{\boldsymbol{p}}\!\triangleleft_U^{\,n} ; \bigoplus_{i=1}^{n} \mathsf{t}_i$ for some $\boldsymbol{p} = p_1, \ldots, p_n$ and $\mathsf{t}_i \colon U \to U_i$. The second one is cancellativity: for all $r \in (0, 1)$, for all s, t, if $r \cdot \mathsf{s} = r \cdot \mathsf{t}$ then $\mathsf{s} = \mathsf{t}$. These to properties provide existence and uniqueness of mediating arrows for convex products. See [2] for further details.

The assignment $\mathbf{C} \mapsto \mathbf{T}(\mathbf{C})$ gives rise to a functor $\mathbf{T}(-) \colon \mathbf{Cat} \to \mathbf{CBCat}$ which is left adjoint to the forgetfull functor $U \colon \mathbf{CBCat} \to \mathbf{Cat}$.

Theorem 4. $\mathbf{T}(-) \colon \mathbf{Cat} \to \mathbf{CBCat}$ *is left adjoint to* $U \colon \mathbf{CBCat} \to \mathbf{Cat}$.

Proof. For a functor $F \colon \mathbf{C} \to \mathbf{D}$, $\mathbf{T}(F) \colon \mathbf{T}(\mathbf{C}) \to \mathbf{T}(\mathbf{D})$ is defined on objects as $\mathbf{T}(F)(\bigoplus_{i=1}^{n} U_i) = \bigoplus_{i=1}^{n} F(U_i)$. On arrows, it is defined inductively as follows.

$$\mathbf{T}(F)(\mathsf{t}_1 ; \mathsf{t}_2) \stackrel{\text{def}}{=} \mathbf{T}(F)(\mathsf{t}_1); \mathbf{T}(F)(\mathsf{t}_2) \qquad \mathbf{T}(F)(\mathsf{t}_1 \oplus \mathsf{t}_2) \stackrel{\text{def}}{=} \mathbf{T}(F)(\mathsf{t}_1) \oplus \mathbf{T}(F)(\mathsf{t}_2)$$

$$\mathbf{T}(F)(^p\!\triangleleft_P) \stackrel{\text{def}}{=} {}^p\!\triangleleft_{\mathbf{T}(F)(P)} \qquad \mathbf{T}(F)(\triangleright_P) \stackrel{\text{def}}{=} \triangleright_{\mathbf{T}(F)(P)}$$

$$\mathbf{T}(F)(\downarrow_P) \stackrel{\text{def}}{=} \downarrow_{\mathbf{T}(F)(P)} \qquad \mathbf{T}(F)(\uparrow_P) \stackrel{\text{def}}{=} \uparrow_{\mathbf{T}(F)(P)}$$

$$\mathbf{T}(F)(id_U) \stackrel{\text{def}}{=} id_{\mathbf{T}(F)(U)} \qquad \mathbf{T}(F)(id_0) \stackrel{\text{def}}{=} id_{\mathbf{T}(F)(0)}$$

$$\mathbf{T}(F)(\sigma^{\oplus}_{P,Q}) \stackrel{\text{def}}{=} \sigma^{\oplus}_{\mathbf{T}(F)(P),\mathbf{T}(F)(Q)} \qquad \mathbf{T}(F)(\underline{c}) \stackrel{\text{def}}{=} \underline{F(c)}$$

Since by definition $\mathbf{T}(F)$ is a monoidal functor preserving co-pcas and monoids, by Proposition 2, it is a morphism of convex biproduct categories. Thus, we have a functor $\mathbf{T}(-)\colon \mathbf{Cat} \to \mathbf{CBCat}$. Below, we prove the adjunction.

For all categories $\mathbf{C}$, the unit of the adjunction $\eta_{\mathbf{C}}\colon \mathbf{C} \to \mathbf{T}(\mathbf{C})$ is the identity-on-objects functor mapping each arrow $c\colon U \to V$ in $\mathbf{C}$ into $\overline{c}$. The axioms in (Tape) force $\eta_{\mathbf{C}}$ to be a functor. Naturality of the unit is straightforward.

Now, take a convex biproduct category $\mathbf{D}$ and consider a functor $F\colon \mathbf{C} \to U(\mathbf{D})$. By Proposition 1, $\mathbf{D}$ is a symmetric monoidal category where every object X carries a co-pca $({}^{\wp}\!\!\triangleleft_X, \delta_X)$ and a natural and coherent monoid $(\triangleright_X, \mathring{\imath}_X)$. One can use these structures to define $F^{\sharp}\colon \mathbf{T}(\mathbf{C}) \to \mathbf{D}$ inductively in the same way as $\mathbf{T}(F)$: e.g., $F^{\sharp}({}^{\wp}\!\!\triangleleft_P) \overset{\text{def}}{=} {}^{\wp}\!\!\triangleleft_{F^{\sharp}(P)}$. The base case $F^{\sharp}(\overline{c}) = F(c)$ ensures that $\eta_{\mathbf{C}}; F^{\sharp} = F$. Since by definition $F^{\sharp}$ is a strict symmetric monoidal functor preserving co-pcas and monoids, by Proposition 2, it is a morphism of convex biproduct categories.

To prove uniqueness, take a morphism of convex biproduct categories $H\colon \mathbf{T}(\mathbf{C}) \to \mathbf{D}$ such that $\eta_{\mathbf{C}}; H = F$, i.e., $H(\overline{c}) = F(c)$. By Proposition 2, H has to preserve co-pcas and monoids. Thus $H = F^{\sharp}$.

Corollary 1. *For all categories* $\mathbf{C}$, $\mathbf{T}(\mathbf{C})$ *is isomorphic to* $\mathbf{StMat}(\mathbf{C}^{+})$ *in* $\mathbf{CBCat}$.

Proof. By composition of adjoints, their uniqueness and Theorems 1, 2 and 4.

The isomorphism $\mathbf{T}(\mathbf{C}) \to \mathbf{StMat}(\mathbf{C}^{+})$ maps $\overline{c}$ into the 1×1 matrix $\left(1 \cdot c\right)$ and ${}^{\wp}\!\!\triangleleft_U$, $\delta_U, \mathring{\imath}_U, \triangleright_U$ into the matrices defined in (10). Compositions and sums of arrows in $\mathbf{T}(\mathbf{C})$ are mapped into multiplications and direct sums of matrices as defined in (8) and (9). The same holds for identities and symmetries. For instance, $(\tfrac{1}{2}\triangleleft_U \oplus id_U); (\overline{a} \oplus \underline{ab} \oplus \overline{c}); (id_U \oplus \sigma^{\oplus}_{U,U}); (\triangleright_U \oplus id_U)$ is mapped into the matrix M of Example 2.

The characterisation of $\mathbf{StMat}(\mathbf{C}^{+})$ by means of generators and equations provided by the above corollary, paves the way to study further equational theories (such as the one in Section 7). The following result guarantees that the category obtained by quotienting $\mathbf{T}(\mathbf{C})$ by additional axioms is still a convex biproduct category.

Proposition 6. *Let* $\mathbf{C}$ *be a category and* $\sim$ *be a congruence relation (w.r.t ; and* $\oplus$*) on* $\mathbf{T}(\mathbf{C})$. *Let* $\mathbf{T}(\mathbf{C})_{\sim}$ *be the category obtained as the quotient of* $\mathbf{T}(\mathbf{C})$ *by* $\sim$ *and* $Q_{\sim}\colon \mathbf{T}(\mathbf{C}) \to \mathbf{T}(\mathbf{C})_{\sim}$ *be the functor mapping each arrow to its* $\sim$*-equivalence class. Then* $\mathbf{T}(\mathbf{C})_{\sim}$ *is a convex biproduct category and* $Q_{\sim}$ *is a morphism of convex biproduct categories.*

6 Probabilistic Tape Diagrams

Recall that a *monoidal signature* is a tuple $(S, \Sigma, ar, coar)$ where S is a set of basic sorts, hereafter denoted by $A, B, \ldots$, Σ is a set of generators, denoted by $s, t \ldots$, and $ar, coar\colon \Sigma \to S^{*}$ assign to each symbol its arity and coarity (words over S). From a monoidal signature Σ, one can freely generate the strict symmetric monoidal category $\mathbf{Diag}_{\Sigma}$: objects are words in S^{*}; arrows are *string diagrams* [30,41]. These can be regarded as the terms generated by the following grammar (where $A, B \in S$ and $s \in \Sigma$)

$$c ::= id_A \mid id_1 \mid s \mid \sigma^{\otimes}_{A,B} \mid c;c \mid c \otimes c \tag{15}$$

modulo the axioms of strict symmetric monoidal categories. A *monoidal theory* $\mathbb{T} = (\Sigma, E)$ consists of a monoidal signature equipped with a set E of pairs of arrows of $\mathbf{Diag}_\Sigma$ with same source and target. Let $=_\mathbb{T}$ be the equivalence relation on arrows of $\mathbf{Diag}_\Sigma$ obtained as the congruence closure (w.r.t. ; and $\otimes$) of E. We write $\mathbf{Diag}_\mathbb{T}$ for the category obtained as the quotient of $\mathbf{Diag}_\Sigma$ by $=_\mathbb{T}$.

The category $\mathbf{T}(\mathbf{Diag}_\mathbb{T})$, obtained by applying the construction from Section 5 to $\mathbf{Diag}_\mathbb{T}$, coincides by definition with the category of *tape diagrams* from [3]. Objects of $\mathbf{T}(\mathbf{Diag}_\mathbb{T})$ are elements of $(S^*)^*$ which we often write as polynomials $P = \bigoplus_{i=1}^n \bigotimes_{j=1}^{m_i} A_{i,j}$. We call *monomials* of P the n words $\bigotimes_{j=1}^{m_i} A_{i,j}$. For instance, the monomials of $(A \otimes B) \oplus 1$ are $A \otimes B$ and 1. We denote monomials by $U, V, \dots$ Arrows of $\mathbf{T}(\mathbf{Diag}_\mathbb{T})$ enjoy an intuitive diagrammatic representation specified by the following two-layered grammar.

The first layer corresponds to (15) while the second to (12); diagrams from the first layer are string diagrams, those from the second are called tapes. Note that (a) string diagrams can occur inside tapes, (b) string diagrams have type $U \to V$, for $U, V \in S^*$ and vertical composition corresponds to $\otimes$, (c) tapes have type $P \to Q$ for $P, Q \in (S^*)^*$ and vertical composition corresponds to $\oplus$.

The identity id_0 is rendered as the empty tape, while id_1 is: a tape filled with the empty string diagram. For a monomial $U = A_1 \dots A_n$, id_U is depicted as a tape containing n wires labelled by A_i. For instance, id_{AB} is rendered as ${}^A_B \!=\!=\! {}^A_B$. When clear from the context, we will simply represent it as a single wire $U \!=\!=\! U$ with the appropriate label. Similarly, for a polynomial $P = \bigoplus_{i=1}^n U_i$, id_P is obtained as a vertical composition of tapes, as illustrated below on the left of (16).

$$
id_{AB \oplus 1 \oplus C} = \qquad \triangleright_{A \oplus B \oplus C} = \qquad \Upsilon_{AB \oplus B \oplus C} = \tag{16}
$$

The codiagonal $\triangleright_U : U \oplus U \to U$ is represented as a merging of tapes, $\triangleleft_U : U \oplus U \to U$ as a splitting of tapes labeled with $p \in (0, 1)$, the cobang $\Upsilon_U : 0 \to U$ is a tape closed on its left boundary, while $\curlywedge_U : U \to 0$ is closed on the right. Exploiting the definitions in (11), we can construct $\triangleright_P, \triangleleft_P, \Upsilon_P, \curlywedge_P$ for arbitrary polynomials. For example, $\triangleright_{A \oplus B \oplus C}$ and $\Upsilon_{AB \oplus B \oplus C}$ are depicted as the second and third diagrams above. The last diagram is the tape for $\triangleleft_A ; (\triangleleft_A \oplus id_A) ; (\curlywedge_A \oplus \triangleright_A)$. For an arbitrary $t : P \to Q$ we write ${}^P \boxed{t} {}^Q$.

The graphical representation embodies several axioms such as (Tape) and those of monoidal categories. The remaining axioms are shown in Figure 1.

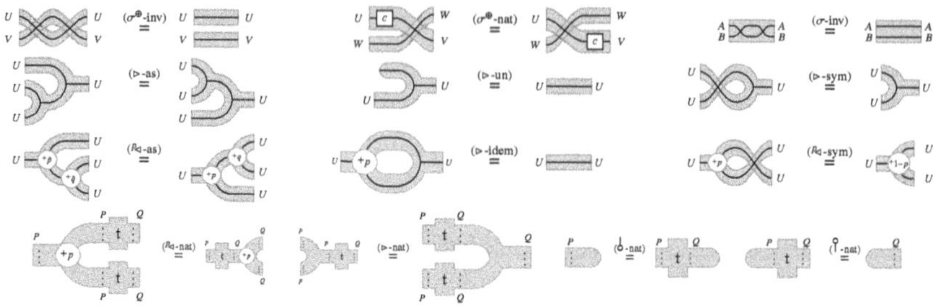

Fig. 1: Axioms for tape diagrams.

By Corollary 1, we know that $\mathbf{T}(\mathbf{Diag}_\mathbb{T})$ is isomorphic to $\mathbf{StMat}(\mathbf{Diag}_\mathbb{T}^+)$: tapes represent exactly stochastic matrices of subprobability distributions of string diagrams in $\mathbf{Diag}_\mathbb{T}$. For instance, the rightmost diagram in (16) corresponds to the 1×1 matrix $\left((1-p)(1-q) \cdot \; {}_A \!\!-\!\! {}_A \right)$ while the following tape diagram on the left represents the 2×2 stochastic matrix on the right.

It is now crucial to recall from [3] that $\otimes$ can be defined not only on string diagrams but also on tapes: for $t_1 : P \to Q$, $t_2 : R \to S$, $t_1 \otimes t_2 \stackrel{\text{def}}{=} \mathsf{L}_P(t_2); \mathsf{R}_S(t_1)$ where $\mathsf{L}_P(\cdot)$, $\mathsf{R}_S(\cdot)$ are the left and right whiskerings, defined as in Table 4 of [3]. For objects $P = \bigoplus_i U_i$ and $Q = \bigoplus_j V_j$, $P \otimes Q \stackrel{\text{def}}{=} \bigoplus_i \bigoplus_j U_i V_j$.

Theorem 27 in [3] guarantees that $(\mathbf{T}(\mathbf{Diag}_\mathbb{T}), \otimes, 1)$ is a symmetric monoidal category. Such category is monoidally enriched over $\mathbf{PCA}$:

$$t \otimes (s_1 +_p s_2) = (t \otimes s_1) +_p (t \otimes s_2) \qquad (s_1 +_p s_2) \otimes t = (s_1 \otimes t) +_p (s_2 \otimes t) \qquad \star \otimes t = t = t \otimes \star$$

Most importantly, the two monoidal categories $(\mathbf{T}(\mathbf{Diag}_\mathbb{T}), \oplus, 0)$ and $(\mathbf{T}(\mathbf{Diag}_\mathbb{T}), \otimes, 1)$ interact through the laws of (right strict) rig category (see [32,29]).

Theorem 5 (From [3]). $(\mathbf{T}(\mathbf{Diag}_\mathbb{T}), \oplus, \otimes, 0, 1)$ *is a right strict rig category.*

By means of the isomorphism in Corollary 1, we can equip $\mathbf{StMat}(\mathbf{Diag}_\mathbb{T}^+)$ with an additional product $\otimes$. This is described as follows. On objects $P \otimes Q$ is defined as above for tapes. On arrows, first define, for all $d_1 \in \mathbf{Diag}_\mathbb{T}^+[U_1, V_1]$ and $d_2 \in \mathbf{Diag}_\mathbb{T}^+[U_2, V_2]$, $d_1 \otimes^+ d_2 \in \mathbf{Diag}_\mathbb{T}^+[U_1 U_2, V_1 V_2]$ as

$$d_1 \otimes^+ d_2(c) \stackrel{\text{def}}{=} \sum_{\{(c_1, c_2) \mid c_1 \otimes c_2 = c\}} d_1(c_1) \cdot d_2(c_2) \qquad \text{for } c \in \mathbf{Diag}_\mathbb{T}[U_1 U_2, V_1 V_2].$$

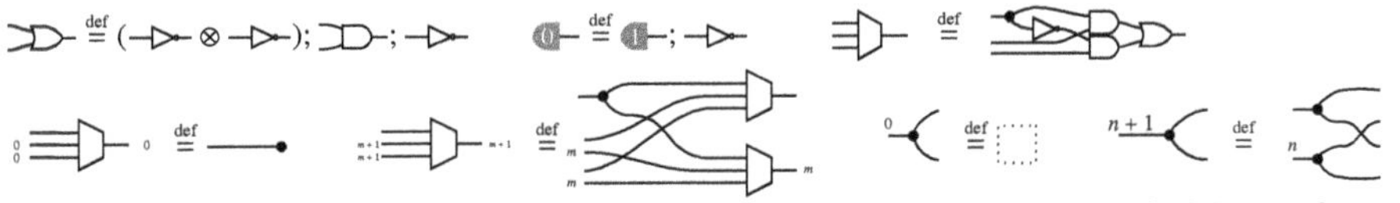

Table 6: Definitions of $\sqsupset\!\!D\text{-}$, $\textcircled{0}\text{-}$ and multiplexer (top); inductive definitions of m-ary multiplexer and n-ary copier (bottom).

(recall from Section 2 that $\mathbf{Diag}_\mathbb{T}^+[U, V] = \mathcal{D}_\leq(\mathbf{Diag}_\mathbb{T}[U, V])$). Then for matrices, it is defined as the Kronecker product using $\otimes^+$ as multiplication. More precisely, let $M\colon \bigoplus_{i=1}^n U_i \to \bigoplus_{i'=1}^{n'} U_{i'}$ and $N\colon \bigoplus_{j=1}^m V_j \to \bigoplus_{j'=1}^{m'} V_{j'}$, $M \otimes N$ is a matrix of size $n'm' \times nm$ whose rows are indexed by pairs $(i', j') \in \{1, \ldots, n'\} \times \{1, \ldots m'\}$ and columns by pairs $(i, j) \in \{1, \ldots, n\} \times \{1, \ldots m\}$. Assuming that $M_{i',i} = p_{i',i} \cdot d_{i',i}$ and $N_{j',j} = q_{j',j} \cdot e_{j',j}$, $M \otimes N$ is defined for all indexes (i', j'), (i, j) as $(M \otimes N)_{(i',j'),(i,j)} \overset{\text{def}}{=} (p_{i',i} \cdot q_{j',j}) \cdot (d_{i',i} \otimes^+ e_{j',j})$.

Corollary 2. $\mathbf{T}(\mathbf{Diag}_\mathbb{T})$ *and* $\mathbf{StMat}(\mathbf{Diag}_\mathbb{T}^+)$ *are isomorphic as rig categories.*

7 Probabilistic Boolean Circuits

Now, we quickly recall causal probabilistic Boolean circuits from [39]. Consider the monoidal signature with a single sort, $\mathcal{S} = \{A\}$, and generators

$$PB \overset{\text{def}}{=} \{\sqsupset\!\!D\text{-}, \text{-}\!\!\triangleright\text{-}, \textcircled{1}\text{-}, \text{-}\!\!\prec, \text{-}\!\!\bullet\} \cup \{\textcircled{p}\text{-} \mid p \in (0, 1)\}.$$

Arities and coarities are determined by the number of ports on the left and on the right: for instance $\sqsupset\!\!D\text{-}$ has arity $A^2 = AA$ and coarity A, while $\text{-}\!\!\bullet$ has arity A and coarity $A^0 = 1$. The first three generators represent operations and constants of boolean algebras $(\wedge, \neg, 1)$, $\text{-}\!\!\prec$ receives a boolean signal on the left and emits two copies on the right, $\text{-}\!\!\bullet$ receives one signal on the left and discards it, while $\textcircled{p}\text{-}$ denotes the distribution on the set $2 \overset{\text{def}}{=} \{0, 1\}$ mapping $1 \mapsto p$ and $0 \mapsto 1 - p$.

Formally, the semantics of diagrams is the monoidal functor $\langle\!\langle -\rangle\!\rangle \colon \mathbf{Diag}_{PB} \to \mathcal{KL}(\mathcal{D}_\leq)$ defined on objects as $A^n \mapsto 2^n$; for arrows it is defined by first interpreting the generators as arrows in $\mathcal{KL}(\mathcal{D}_\leq)$

$$\langle\!\langle \sqsupset\!\!D\text{-}\rangle\!\rangle\colon 2 \times 2 \to 2 \qquad \langle\!\langle \text{-}\!\!\triangleright\text{-}\rangle\!\rangle\colon 2 \to 2 \qquad \langle\!\langle \textcircled{1}\text{-}\rangle\!\rangle\colon 1 \to 2$$
$$(x, y) \mapsto \delta_{x \wedge y} \qquad\qquad x \mapsto \delta_{\neg x} \qquad\qquad \bullet \mapsto \delta_1$$

$$\langle\!\langle \text{-}\!\!\prec\rangle\!\rangle\colon 2 \to 2 \times 2 \qquad \langle\!\langle \text{-}\!\!\bullet\rangle\!\rangle\colon 2 \to 1 \qquad \langle\!\langle \textcircled{p}\text{-}\rangle\!\rangle\colon 1 \to 2$$
$$x \mapsto \delta_{(x,x)} \qquad\qquad x \mapsto \delta_\bullet \qquad\qquad \bullet \mapsto \delta_1 +_p \delta_0$$

and then inductively by means of the structure of the monoidal category $(\mathcal{KL}(\mathcal{D}_\leq), \otimes, 1)$: for instance $\langle\!\langle c_1 \otimes c_2 \rangle\!\rangle = \langle\!\langle c_1 \rangle\!\rangle \otimes \langle\!\langle c_2 \rangle\!\rangle$ where the second $\otimes$ is the one in (4).

Figure 4 in [39] illustrates a sound and complete equational axiomatisation for probabilistic Boolean circuits. In a nutshell, these axioms are those of Boolean algebras (reported in Table 7) together with 4 extra sophisticated axioms to deal with 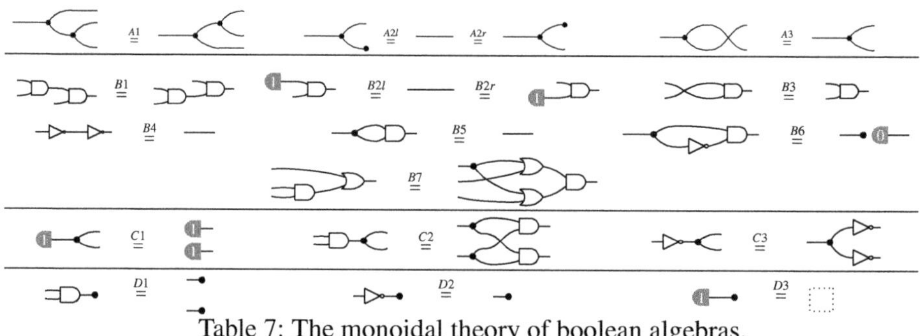. It is worth mentioning that such axioms crucially rely on *multiplexer*: the $A^3 \to A$ diagram in Table 6 denoting the boolean function mapping (x, y, z) into y if $x = 1$ and into z if $x = 0$. Table 6 also illustrates the definition of a m-ary multiplexer $A^{2m+1} \to A^m$. This has been used in [14] to provide an alternative axiomatisation which is based on the following.[1]

$$ n \underset{}{\overset{}{\longrightarrow}} \boxed{f}\, m \;=\; {}_n \boxed{f}\, m \;. \tag{17} $$

Table 7: The monoidal theory of boolean algebras.

In [3, Example 30], an encoding of probabilistic Boolean circuits into tape diagrams is presented. Let B be the signature obtained from PB by dropping and let $\mathbb{B}$ be the monoidal theory in Table 7 consisting of the usual axioms of Boolean algebras. For later use, it is worth recalling (see e.g. [39, Theorem 3.2]) that $\mathbf{Diag}_\mathbb{B}$ is isomorphic to $\mathbf{Set}_2$, the full subcategory of $\mathbf{Set}$ having as objects the sets 2^n for all $n \in \mathbb{N}$. We call Boolean circuits the arrows of $\mathbf{Diag}_B$ and probabilistic tapes of Boolean circuits those of $\mathbf{T}(\mathbf{Diag}_B)$. The semantics of such tapes is given in [3, Example 30] by the morphism of rig categories $[\![-]\!] : \mathbf{T}(\mathbf{Diag}_B) \to \mathcal{KL}(\mathcal{D}_\leq)$ reported in Figure 2. The encoding of probabilistic Boolean circuits into tapes is the unique monoidal functor $\mathcal{E}(-): (\mathbf{Diag}_{PB}, \otimes, 1) \to (\mathbf{T}(\mathbf{Diag}_B), \otimes, 1)$ mapping

where

As expected, the encoding preserves the semantics, in the sense that $[\![\mathcal{E}(c)]\!] = \langle\!\langle c \rangle\!\rangle$ for all c in $\mathbf{Diag}_{PB}$.

In this section, we show that the axioms of Boolean algebras and (17) are enough to characterise the equivalence induced by $[\![-]\!]$. More precisely, we take the encoding

---

[1] More precisely, in [14], the law (17) appears in the equivalent form of naturality of multiplexer.

$$\llbracket id_A \rrbracket \overset{\text{def}}{=} id_2 \qquad \llbracket id_1 \rrbracket \overset{\text{def}}{=} id_1 \qquad \llbracket \sigma^\otimes_{A,A} \rrbracket \overset{\text{def}}{=} \sigma^\otimes_{2,2} \qquad \llbracket c; d \rrbracket \overset{\text{def}}{=} \llbracket c \rrbracket ; \llbracket d \rrbracket \qquad \llbracket c \otimes d \rrbracket \overset{\text{def}}{=} \llbracket c \rrbracket \otimes \llbracket d \rrbracket$$

$$\llbracket s \rrbracket \overset{\text{def}}{=} \langle\!\langle s \rangle\!\rangle \qquad \llbracket \overline{c} \rrbracket \overset{\text{def}}{=} \llbracket c \rrbracket \qquad \llbracket {}^{R}\!\triangleleft_{A^n} \rrbracket \overset{\text{def}}{=} {}^{R}\!\triangleleft_{2^n} \qquad \llbracket {\downarrow}_{A^n} \rrbracket \overset{\text{def}}{=} {\downarrow}_{2^n} \qquad \llbracket {\uparrow}_{A^n} \rrbracket \overset{\text{def}}{=} {\uparrow}_{2^n} \qquad \llbracket {\triangleright}_{A^n} \rrbracket \overset{\text{def}}{=} {\triangleright}_{2^n}$$

$$\llbracket id_{A^n} \rrbracket \overset{\text{def}}{=} id_{2^n} \qquad \llbracket id_0 \rrbracket \overset{\text{def}}{=} id_0 \qquad \llbracket \sigma^\oplus_{A^n,A^m} \rrbracket \overset{\text{def}}{=} \sigma^\oplus_{2^n,2^m} \qquad \llbracket s; t \rrbracket \overset{\text{def}}{=} \llbracket s \rrbracket ; \llbracket t \rrbracket \qquad \llbracket s \oplus t \rrbracket \overset{\text{def}}{=} \llbracket s \rrbracket \oplus \llbracket t \rrbracket$$

Fig. 2: The semantics $\llbracket - \rrbracket : \mathbf{T}(\mathbf{Diag}_B) \to \mathcal{KL}(\mathcal{D}_\leq)$. Here s is a generator in B.

of (17) in tapes: for all $t \colon A^{n+1} \to A^m$, $t = (id_A \otimes (\,\rule[0.3ex]{1.2em}{0.1ex}\!\!\!\bullet\!\!<\;; t_1 \otimes t_0)\,); \boxed{}\!\!\!-^m$ where $t_0 = (\,\boxed{0}\!\!-\, \otimes id_{A^n}); t$ and $t_1 = (\,\boxed{1}\!\!-\, \otimes id_{A^n}); t$. Simple computations confirm that the axiom is sound: $\llbracket - \rrbracket$ maps the left and the right hand side of the equation in the same arrow in $\mathcal{KL}(\mathcal{D}_\leq)$.

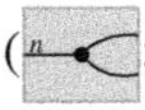

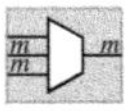

We write $\sim$ for the congruence (w.r.t. $\oplus$, $\otimes$ and ;) on $\mathbf{T}(\mathbf{Diag}_\mathbb{B})$ generated by the above axiom and $\mathbf{T}(\mathbf{Diag}_\mathbb{B})_\sim$ for the category obtained as the quotient of $\mathbf{T}(\mathbf{Diag}_\mathbb{B})$ by $\sim$. We denote by $Q_\sim \colon \mathbf{T}(\mathbf{Diag}_\mathbb{B}) \to \mathbf{T}(\mathbf{Diag}_\mathbb{B})_\sim$ the functor mapping each tape into its $\sim$ equivalence class and $Q_\mathbb{B} \colon \mathbf{Diag}_B \to \mathbf{Diag}_\mathbb{B}$ the functor mapping each diagram into its $=_\mathbb{B}$ equivalence class.

Since the axioms are sound $\llbracket - \rrbracket : \mathbf{T}(\mathbf{Diag}_B) \to \mathcal{KL}(\mathcal{D}_\leq)$ can be factored as

$$\mathbf{T}(\mathbf{Diag}_B) \xrightarrow{\;\mathbf{T}(Q_\mathbb{B})\;} \mathbf{T}(\mathbf{Diag}_\mathbb{B}) \xrightarrow{\;Q_\sim\;} \mathbf{T}(\mathbf{Diag}_\mathbb{B})_\sim \overset{I}{\dashrightarrow} \mathcal{KL}(\mathcal{D}_\leq)$$

for some functor I. Proving completeness amounts to proving that I is faithful.

By Corollary 1, we know that $\mathbf{T}(\mathbf{Diag}_\mathbb{B}) \cong \mathbf{StMat}(\mathbf{Diag}_\mathbb{B}^+) \cong \mathbf{StMat}(\mathbf{Set}_2^+)$. Now, let $\mathcal{KL}(\mathcal{D}_\leq)_2$ be the full subcategory of $\mathcal{KL}(\mathcal{D}_\leq)$ having as objects the sets 2^n for all $n \in \mathbb{N}$, and let $J \colon \mathbf{Set}_2 \to \mathcal{KL}(\mathcal{D}_\leq)_2$ be the obvious embedding. Since $\mathcal{KL}(\mathcal{D}_\leq)_2$ is $\mathbf{PCA}$-enriched, by Theorem 1, there exists a $\mathbf{PCA}$-enriched functor $J^\sharp \colon \mathbf{Set}_2^+ \to \mathcal{KL}(\mathcal{D}_\leq)_2$. This enjoys the following key property.

Lemma 3. *For all $d_1, d_2 \in \mathbf{Set}_2^+[1, 2^m]$, $J^\sharp(d_1) = J^\sharp(d_2)$ iff $d_1 = d_2$.*

Proof. It is convenient to give the explicit definition of $J^\sharp$. Recall that an arrow $d \colon 2^n \to 2^m$ in $\mathbf{Set}_2^+$ is a subprobability distribution on $\mathbf{Set}_2[2^n, 2^m]$, i.e., an element of $\mathcal{D}_\leq(\mathbf{Set}_2[2^n, 2^m])$. $J^\sharp$ is the identity on objects functor mapping $d \colon 2^n \to 2^m$ into the Kleisli arrow defined for all $b \in 2^n, b' \in 2^m$ as $J^\sharp(d)(b'|b) = \sum_{\{f \mid f(b)=b'\}} d(f)$. It is trivial to see that for $n = 0$, $J^\sharp$ provides the obvious bijection

$$\mathcal{D}_\leq(\mathbf{Set}_2[1, 2^m]) \cong \mathcal{D}_\leq(2^m) \cong \mathcal{KL}(\mathcal{D}_\leq)_2[1, 2^m].$$

We call $F \colon \mathbf{T}(\mathbf{Diag}_\mathbb{B}) \to \mathbf{StMat}(\mathcal{KL}(\mathcal{D}_\leq)_2)$ the morphism of convex biproduct categories obtained by composing the isomorphism $\mathbf{T}(\mathbf{Diag}_\mathbb{B}) \to \mathbf{StMat}(\mathbf{Set}_2^+)$ with $\mathbf{StMat}(J^\sharp)$.

We call $G \colon \mathbf{StMat}(\mathcal{KL}(\mathcal{D}_\leq)_2) \to \mathcal{KL}(\mathcal{D}_\leq)$ the morphism of convex biproduct categories obtained by composing the embedding $\mathbf{StMat}(\mathcal{KL}(\mathcal{D}_\leq)_2) \to \mathbf{StMat}(\mathcal{KL}(\mathcal{D}_\leq))$ with the counit of the adjunction in Theorem 2, $\epsilon_{\mathcal{KL}(\mathcal{D}_\leq)} \colon \mathbf{StMat}(\mathcal{KL}(\mathcal{D}_\leq)) \to \mathcal{KL}(\mathcal{D}_\leq)$.

The latter is, by Proposition 5, full and faithful and, therefore, so is G.

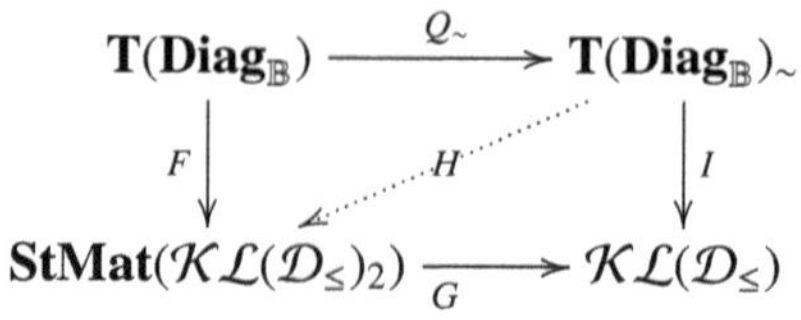

$$\begin{array}{ccc} \mathbf{T}(\mathbf{Diag}_\mathbb{B}) & \xrightarrow{\ Q_\sim\ } & \mathbf{T}(\mathbf{Diag}_\mathbb{B})_\sim \\ {\scriptstyle F}\big\downarrow & \ {}^{H} & \big\downarrow{\scriptstyle I} \\ \mathbf{StMat}(\mathcal{KL}(\mathcal{D}_\leq)_2) & \xrightarrow[\ G\ ]{} & \mathcal{KL}(\mathcal{D}_\leq) \end{array}$$

Since $Q_\sim$ is the identity on objects and G is full and faithful, by the property of orthogonal factorisation systems (see e.g. [37]), there exists a unique $H\colon \mathbf{T}(\mathbf{Diag}_\mathbb{B})_\sim \to \mathbf{StMat}(\mathcal{KL}(\mathcal{D}_\leq)_2)$ making the above diagram commute. Below we prove that H is faithful to conclude that also I is faithful.

Hereafter, we write $\mathbf{B}[1, A^n]$ for the set n-ary Boolean vectors, i.e., all those tapes in $\mathbf{T}(\mathbf{Diag}_\mathbb{B})[1, A^n]$ of the form $\bigotimes_{i=1}^{n} b_i$ for $b_i \in \{\ \boxed{0-},\ \boxed{1-}\ \}$. The following is proved by a simple inductive argument using (17) in the induction step.

Lemma 4. *Let* $s, t \in \mathbf{T}(\mathbf{Diag}_\mathbb{B})[A^n, A^m]$. *If, for all* $b \in \mathbf{B}[1, A^n]$, $b; s \sim b; t$, *then* $s \sim t$.

Theorem 6. $H\colon \mathbf{T}(\mathbf{Diag}_\mathbb{B})_\sim \to \mathbf{StMat}(\mathcal{KL}(\mathcal{D}_\leq)_2)$ *is faithful.*

Proof. Let $s, t\colon A^n \to A^m$ be arrows in $\mathbf{T}(\mathbf{Diag}_\mathbb{B})$. Denote by $[s]_\sim$ and $[t]_\sim$ their $\sim$ equivalence classes.

$$\begin{aligned}
H([s]_\sim) = H([t]_\sim) &\Rightarrow F(s) = F(t) & (F = Q_\sim; H) \\
&\Rightarrow \text{for all } b \in \mathbf{B}[1, A^n],\ F(b); F(s) = F(b); F(t) & \\
&\Rightarrow \text{for all } b \in \mathbf{B}[1, A^n],\ F(b; s) = F(b; t) & (\text{Functoriality}) \\
&\Rightarrow \text{for all } b \in \mathbf{B}[1, A^n],\ b; s = b; t & (\text{Lemma 3}) \\
&\Rightarrow s \sim t & (\text{Lemma 4})
\end{aligned}$$

The above derivation proves that H is faithful for arrows of type $A^n \to A^m$. This fact easily entails that H is faithful for arrows $s, t\colon A^k \to \bigoplus_{i=1}^{n} A^{m_i}$. Indeed, by Theorem 3,

$$s = \langle s_1, \ldots, s_n \rangle_q \quad t = \langle t_1, \ldots, t_n \rangle_p \tag{18}$$

for some $q = q_1, \ldots, q_n$, $p = p_1, \ldots, p_n$, $s_i\colon A^k \to A^{m_i}$ and $t_i\colon A^k \to A^{m_i}$. Thus

$$\begin{aligned}
H([s]_\sim) = H([t]_\sim) &\Rightarrow F(s) = F(t) & (F = Q_\sim; H) \\
&\Rightarrow \text{for all } i,\ F(s); F(\pi_i) = F(t); F(\pi_i) & \\
&\Rightarrow \text{for all } i,\ F(s; \pi_i) = F(t; \pi_i) & (\text{Functoriality}) \\
&\Rightarrow \text{for all } i,\ s; \pi_i \sim t; \pi_i & (\text{Previous implication}) \\
&\Rightarrow \text{for all } i,\ q_i \cdot s_i \sim p_i \cdot t_i & (18) \\
&\Rightarrow \text{for all } i,\ [q_i \cdot s_i]_\sim = [p_i \cdot t_i]_i &
\end{aligned}$$

By Proposition 6, $[s]_\sim = \langle [s_1]_\sim, \ldots, [s_n]_\sim \rangle_q$ and $[t]_\sim = \langle [t_1]_\sim, \ldots, [t_n]_\sim \rangle_p$. Since for all i, $[t]; [\pi_i]_\sim$ and $[s]; [\pi_i]_\sim$ are both equal to $p_i \cdot [t_i]_\sim = [p_i \cdot t_i]_\sim = [q_i \cdot s_i]_\sim$, the universal property of n-ary convex products in $\mathbf{T}(\mathbf{Diag}_\mathbb{B})_\sim$ implies that $[s]_\sim = [t]_\sim$.

For arrows of arbitrary type $\bigoplus_{j=1}^{o} A^{k_j} \to \bigoplus_{i=1}^{n} A^{m_i}$, one can easily rely on the universal property of coproducts and the case that we just proved.

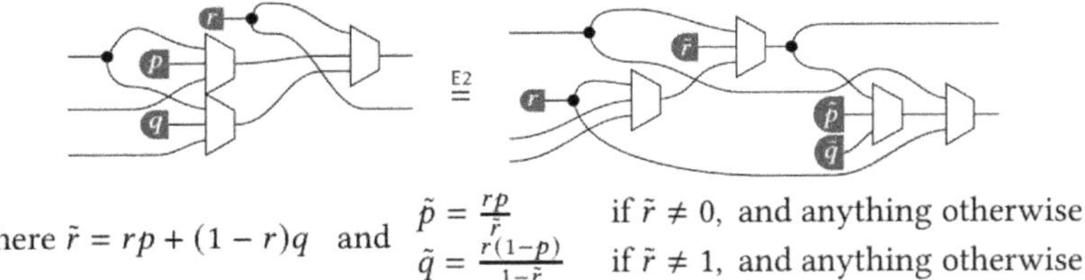

where $\tilde{r} = rp + (1-r)q$ and $\begin{aligned}\tilde{p} &= \frac{rp}{\tilde{r}} & \text{if } \tilde{r} \neq 0, \text{ and anything otherwise}\\ \tilde{q} &= \frac{r(1-p)}{1-\tilde{r}} & \text{if } \tilde{r} \neq 1, \text{ and anything otherwise}\end{aligned}$

Fig. 3: Axiom E2 from [39].

Corollary 3. $I\colon \mathbf{T}(\mathbf{Diag}_{\mathbb{B}})_{\sim} \to \mathcal{KL}(\mathcal{D}_{\leq})$ *is faithful.*

Remark 2. Our proof of completeness for probabilistic Boolean circuits via tape diagrams differs substantially from the approach in [39], which is based on a normal-form argument. Instead, we rely on the isomorphism $\mathbf{T}(\mathbf{Diag}_{\mathbb{B}}) \cong \mathbf{StMat}(\mathbf{Diag}_{\mathbb{B}}^{+})$ established in Corollary 1. The axioms for tape diagrams involve only monoid and co-pca structures, the laws of Boolean algebra, and the additional multiplexer axiom (17). By contrast, the axioms for $p\text{–}$ in [39] are more elaborate, as they act directly on the probabilistic structure. In particular, axioms E1, E3, and E4 of [39] follow readily from the monoid, co-pca, and Boolean axioms, while axiom E2 (see Figure 3) requires the multiplexer axiom (17) from [14].

8   Conclusion

We have introduced the notion of convex biproduct categories (Def. 2) and the construction $\mathbf{StMat}(\mathbf{C})$ of stochastic matrices over a $\mathbf{PCA}$-enriched category $\mathbf{C}$, showing that $\mathbf{StMat}(\mathbf{C})$ is the free convex biproduct category generated by $\mathbf{C}$ (Thm. 2).

We then presented a syntactic construction: for any category $\mathbf{C}$, the category $\mathbf{T}(\mathbf{C})$ is obtained by freely adding a monoidal structure with natural monoids and co-pointed convex algebras. We showed that $\mathbf{T}(\mathbf{C})$ is the free convex biproduct category over $\mathbf{C}$ (Thm. 4). Combining this with a known result (Thm. 1), we derived the isomorphism $\mathbf{T}(\mathbf{C}) \cong \mathbf{StMat}(\mathbf{C}^{+})$ (Cor. 1), where arrows in $\mathbf{C}^{+}$ are subdistributions of arrows in $\mathbf{C}$.

This construction is particularly relevant when $\mathbf{C} = \mathbf{Diag}_{\Sigma}$, a category of string diagrams: in this case, $\mathbf{T}(\mathbf{Diag}_{\Sigma})$ yields the category of probabilistic tape diagrams introduced in [3]. The isomorphism $\mathbf{T}(\mathbf{Diag}_{\Sigma}) \cong \mathbf{StMat}(\mathbf{Diag}_{\Sigma}^{+})$ thus provides a concrete interpretation of tape diagrams as stochastic matrices whose entries are probability subdistributions over string diagrams.

Relying on [3, Ex. 30], we applied this framework to the complete axiomatisation of probabilistic Boolean circuits in [39]. We proposed an alternative axiomatisation, combining the axioms for Boolean circuits, those for probabilistic tapes, and one additional law from [14]. Thanks to the isomorphism $\mathbf{T}(\mathbf{Diag}_{\Sigma}) \cong \mathbf{StMat}(\mathbf{Diag}_{\Sigma}^{+})$, the resulting completeness proof becomes conceptually simpler and structurally more transparent.

Related and future work. The construction of stochastic matrices over pca-enriched categories is conceptually akin to the classical construction of matrices over categories enriched in commutative monoids [34,13]. However, while the latter is straightforward enough to be left as an exercise in [34], the former involves significantly more work. In particular, finite biproduct categories enjoy a bijective correspondence between arrows $f \colon A \oplus B \to C \oplus D$ and quadruples of morphisms $(f_{A,C}, f_{B,C}, f_{A,D}, f_{B,D})$, where each $f_{X,Y} \colon X \to Y$. This correspondence breaks down in convex biproduct categories.

Our probabilistic tape diagrams appear to be closely related to effectuses [12,10]. Effectuses are based on finitely *partially additive categories* [1], that is, categories enriched over partial commutative monoids, rather than pointed convex algebras. A precise understanding of the relationship between these two frameworks constitutes an interesting direction for future research.

In [44], a general framework is introduced to enrich string diagrams over arbitrary algebraic theories. This approach fundamentally differs from the tape-based construction in [3], in that it does not account for the structure induced by the coproduct $\oplus$. As a consequence, the correspondence with matrix-based semantics does not arise in [44].

We plan to extend probabilistic tape diagrams with traces for $\oplus$, following the approach developed in [4]. That work showed that a specific property of monoidal traces–called *uniformity*–corresponds to reasoning by invariants in Hoare-style program logics [25]. Interestingly, as demonstrated in [26], uniformity holds for traces over $\oplus$ in $\mathcal{KL}(\mathcal{D}_{\leq})$. Preliminary results suggest that, in the probabilistic setting, invariants correspond to sub-martingales, thus offering a promising link between diagrammatic semantics and probabilistic program verification.

Moreover, probabilistic tape diagrams extended with traces are expressive enough to encode probabilistic regular expressions from [40], which provide a complete axiomatisation for the regular behaviours of generative probabilistic systems–that is, coalgebras in $\mathcal{KL}(\mathcal{D}_{\leq})$ [24]. We expect that tapes may again offer a more principled and modular axiomatisation, akin to Kleene algebra in [31]. Interestingly, the completeness proof in [31] relies heavily on matrix representations over languages; as we hint in Example 2, our tape formalism naturally supports stochastic matrices over probabilistic languages.

To conclude, we mention that extending our axiomatisation to probabilistic circuits with ⊃– (used in [39] for explicit conditioning) is challenging future work. Indeed, our proof crucially relies on the well-known isomorphism $\mathbf{Diag}_{\mathbb{B}} \cong \mathbf{Set}_2$, while we are not aware of a similar result for partial functions, necessary to accommodate ⊃–.

Acknowledgements This research was partly funded by the Advanced Research + Invention Agency (ARIA) Safeguarded AI Programme. Bonchi is supported by the Ministero dell'Università e della Ricerca of Italy grant PRIN 2022 PNRR No. P2022HXNSC - RAP (Resource Awareness in Programming). This study was carried out within the National Centre on HPC, Big Data and Quantum Computing - SPOKE 10 (Quantum Computing) and received funding from the European Union Next-GenerationEU - National Recovery and Resilience Plan (NRRP) MISSION 4 COMPONENT 2, INVESTMENT N. 1.4 CUP N. I53C22000690001.

References

1. Arbib, M.A., Manes, E.G.: Partially additive categories and flow-diagram semantics. Journal of Algebra **62**(1), 203–227 (1980)

2. Bonchi, F., Cioffo, C.J.: Tapes as stochastic matrices of string diagrams (2026), `https://arxiv.org/abs/2601.01472`

3. Bonchi, F., Cioffo, C.J., Di Giorgio, A., Di Lavore, E.: Tape diagrams for monoidal monads (2025), `https://arxiv.org/abs/2503.22819`

4. Bonchi, F., Di Giorgio, A., Di Lavore, E.: A diagrammatic algebra for program logics. In: FoSSaCS 2025 (to appear) (2025)

5. Bonchi, F., Di Giorgio, A., Santamaria, A.: Deconstructing the calculus of relations with tape diagrams. Proceedings of the ACM on Programming Languages **7**(POPL), 1864–1894 (2023)

6. Bonchi, F., Di Lavore, E., Román, M.: Effectful mealy machines: Bisimulation and trace. To appear in LICS 2025 **abs/2410.10627** (2024). `https://doi.org/10.48550/ARXIV.2410.10627`, `https://doi.org/10.48550/arXiv.2410.10627`

7. Bonchi, F., Silva, A., Sokolova, A.: The power of convex algebras. In: 28th International Conference on Concurrency Theory (CONCUR 2017). pp. 23–1. Schloss Dagstuhl–Leibniz-Zentrum für Informatik (2017)

8. Bonchi, F., Sokolova, A., Vignudelli, V.: The theory of traces for systems with nondeterminism, probability, and termination. Log. Methods Comput. Sci. **18**(2) (2022). `https://doi.org/10.46298/LMCS-18(2:21)2022`, `https://doi.org/10.46298/lmcs-18(2:21)2022`

9. Borceux, F.: Handbook of categorical algebra. 2, Encyclopedia of Mathematics and its Applications, vol. 51. Cambridge University Press, Cambridge (1994), categories and structures

10. Cho, K.: Effectuses in categorical quantum foundations. arXiv preprint arXiv:1910.12198 (2019)

11. Cho, K., Jacobs, B.: Disintegration and bayesian inversion via string diagrams. Mathematical Structures in Computer Science **29**(7), 938–971 (2019)

12. Cho, K., Jacobs, B., Westerbaan, B., Westerbaan, A.: An introduction to effectus theory. ArXiv **abs/1512.05813** (2015), `https://api.semanticscholar.org/CorpusID:14013233`

13. Coecke, B., Selby, J., Tull, S.: Two Roads to Classicality **266**, 104–118 (Feb 2018). `https://doi.org/10.4204/EPTCS.266.7`

14. Di Giorgio, A., Sobocinski, P., Voorneveld, N.: Parametric iteration in resource theories. In: 34th EACSL Annual Conference on Computer Science Logic (CSL 2026). Leibniz International Proceedings in Informatics (LIPIcs), Schloss Dagstuhl–Leibniz-Zentrum für Informatik (2026), to appear

15. Di Lavore, E., Román, M.: Evidential decision theory via partial markov categories. CoRR **abs/2301.12989** (2023). `https://doi.org/10.48550/ARXIV.2301.12989`, `https://doi.org/10.48550/arXiv.2301.12989`

16. Di Lavore, E., Román, M., Sobocinski, P.: Partial markov categories. CoRR **abs/2502.03477** (2025). `https://doi.org/10.48550/ARXIV.2502.03477`, `https://doi.org/10.48550/arXiv.2502.03477`

17. Fox, T.: Coalgebras and cartesian categories. Communications in Algebra **4**(7), 665–667 (1976). `https://doi.org/10.1080/00927877608822127`

18. Fritz, T.: A presentation of the category of stochastic matrices. arXiv preprint arXiv:0902.2554 (2009)

19. Fritz, T., Gadducci, F., Perrone, P., Trotta, D.: Weakly markov categories and weakly affine monads. arXiv preprint arXiv:2303.14049 (2023)

20. Fritz, T., Gonda, T., Houghton-Larsen, N.G., Lorenzin, A., Perrone, P., Stein, D.: Dilations and information flow axioms in categorical probability. Mathematical Structures in Computer Science **33**(10), 913–957 (2023)

21. Fritz, T., Gonda, T., Perrone, P.: De Finetti's theorem in categorical probability. arXiv preprint arXiv:2105.02639 (2021)

22. Fritz, T., Klingler, A.: The d-separation criterion in categorical probability. Journal of Machine Learning Research **24**(46), 1–49 (2023)
23. Fritz, T., Perrone, P.: Bimonoidal structure of probability monads. Electronic notes in theoretical computer science **341**, 121–149 (2018)
24. Hasuo, I., Jacobs, B., Sokolova, A.: Generic trace semantics via coinduction. Logical Methods in Computer Science **3** (2007)
25. Hoare, C.A.R.: An axiomatic basis for computer programming. Communications of the ACM **12**(10), 576–580 (1969)
26. Jacobs, B.: From coalgebraic to monoidal traces. Electronic Notes in Theoretical Computer Science **264**(2), 125–140 (2010)
27. Jacobs, B., Kissinger, A., Zanasi, F.: Causal inference by string diagram surgery. In: International conference on foundations of software science and computation structures. pp. 313–329. Springer (2019)
28. Jacobs, B., Zanasi, F., et al.: The logical essentials of Bayesian reasoning. Cambridge University Press (2021)
29. Johnson, N., Yau, D.: Bimonoidal categories, e_n-monoidal categories, and algebraic k-theory. arXiv preprint arXiv:2107.10526 (2021)
30. Joyal, A., Street, R.: The geometry of tensor calculus, I. Advances in Mathematics **88**(1), 55–112 (Jul 1991). `https://doi.org/10.1016/0001-8708(91)90003-P`
31. Kozen, D.: A completeness theorem for kleene algebras and the algebra of regular events. Information and Computation **110**, 366–390 (1994)
32. Laplaza, M.L.: Coherence for distributivity. In: Kelly, G.M., Laplaza, M., Lewis, G., Mac Lane, S. (eds.) Coherence in Categories. pp. 29–65. Lecture Notes in Mathematics, Springer, Berlin, Heidelberg (1972). `https://doi.org/10.1007/BFb0059555`
33. Liell-Cock, J., Staton, S.: Compositional imprecise probability: A solution from graded monads and markov categories. Proc. ACM Program. Lang. **9**(POPL), 1596–1626 (2025). `https://doi.org/10.1145/3704890`, `https://doi.org/10.1145/3704890`
34. Mac Lane, S.: Categories for the Working Mathematician, Graduate Texts in Mathematics, vol. 5. Springer-Verlag, New York, second edn. (1978), `//www.springer.com/gb/book/9780387984032`
35. Mio, M., Sarkis, R., Vignudelli, V.: Combining nondeterminism, probability, and termination: Equational and metric reasoning. In: 36th Annual ACM/IEEE Symposium on Logic in Computer Science, LICS 2021, Rome, Italy, June 29 - July 2, 2021. pp. 1–14. IEEE (2021). `https://doi.org/10.1109/LICS52264.2021.9470717`, `https://doi.org/10.1109/LICS52264.2021.9470717`
36. Moss, S., Perrone, P.: A category-theoretic proof of the ergodic decomposition theorem. Ergodic Theory and Dynamical Systems **43**(12), 4166–4192 (2023)
37. nLab authors: (bo, ff) factorization system. `https://ncatlab.org/nlab/show/%28bo%2C+ff%29+factorization+system` (Jul 2025), Revision 12
38. Perrone, P.: Markov categories and entropy. IEEE Transactions on Information Theory **70**(3), 1671–1692 (2023)
39. Piedeleu, R., Torres-Ruiz, M., Silva, A., Zanasi, F.: A complete axiomatisation of equivalence for discrete probabilistic programming. In: ESOP 2025 (to appear) (2025)
40. Rozowski, W., Silva, A.: A completeness theorem for probabilistic regular expressions. In: Sobocinski, P., Lago, U.D., Esparza, J. (eds.) Proceedings of the 39th Annual ACM/IEEE Symposium on Logic in Computer Science, LICS 2024, Tallinn, Estonia, July 8-11, 2024. pp. 66:1–66:14. ACM (2024). `https://doi.org/10.1145/3661814.3662084`, `https://doi.org/10.1145/3661814.3662084`
41. Selinger, P.: A survey of graphical languages for monoidal categories. In: New structures for physics, pp. 289–355. Springer (2010)

42. Stein, D., Staton, S.: Probabilistic programming with exact conditions. Journal of the ACM **71**(1), 1–53 (2024)
43. Stone, M.H.: Postulates for the barycentric calculus. Annali di Matematica Pura ed Applicata **29**(1), 25–30 (1949)
44. Villoria, A., Basold, H., Laarman, A.: Enriching diagrams with algebraic operations. In: International Conference on Foundations of Software Science and Computation Structures. pp. 121–143. Springer (2024)

Inquisitive Team Semantics of LTL

Laura Bozzelli[1], Tadeusz Litak[2],
Munyque Mittelmann[3], and
Aniello Murano[2]

[1] University of Pegaso, Naples, Italy
[2] University of Naples Federico II, Italy
[3] CNRS, LIPN, Université Sorbonne Paris Nord, France

Abstract. In recent years, hyperproperties (i.e., properties of sets of traces) have been identified as a crucial concept for capturing information-flow security policies and knowledge of agents in distributed systems. In the synchronous linear-time setting, a well-known approach for the specification of hyperproperties is based on set semantics of standard LTL. The resulting logic (TeamLTL) inherits the powerful split interpretation of disjunction from dependency logic. In this paper, we introduce a novel team semantics of LTL inspired by inquisitive logic, and whose main features are the intuitionistic interpretation of implication and the inquisitive semantics of disjunction. We investigate expressiveness, decidability, and complexity issues of the novel logic that we call InqLTL and its extension with strong negation. We show that InqLTL with strong negation is highly undecidable and strictly less expressive than TeamLTL with strong negation. On the positive side, we identify a meaningful fragment of InqLTL with a decidable model-checking problem which can express relevant classes of hyperproperties. To the best of our knowledge, this fragment represents the unique hyper logic with a decidable model-checking problem which allows unrestricted use of temporal modalities and universal second-order quantification over traces.

1 Introduction

Hyperproperties [22] are a specification paradigm that generalizes trace properties to properties of sets of traces by allowing the comparison of distinct execution traces of a system. They play a crucial role in capturing information-flow security requirements such as noninterference [30,44] and observational determinism [56], which relate observations of an external low-security agent along distinct traces resulting from different values of not directly observable inputs. These requirements are not regular properties and so cannot be expressed in traditional temporal logics like LTL, CTL, and CTL* [46,25]. Other relevant examples of hyperproperties include epistemic properties specifying the knowledge of agents in distributed systems [36,35], bounded termination of programs, the symmetrical access to critical resources in distributed protocols [27], and diagnosability of critical systems [50,4].

N. Bertrand and S. Milius (Eds.): FoSSaCS 2026, LNCS 16503, pp. 110–132, 2026.
https://doi.org/10.1007/978-3-032-22730-0_6

Two main approaches have been proposed for the formal specification, analysis, and automatic verification (model-checking) of hyperproperties in a synchronous setting. The first extends standard temporal logics like Linear Temporal Logic (LTL), CTL*, QPTL [52], and PDL [28] with explicit first-order quantification over traces (and trace variables to refer to multiple traces at the same time) yielding logics like HyperLTL [21], HyperCTL* [21], HyperQPTL [47,23], and HyperPDL$-\Delta$ [34]. These logics enjoy a decidable, although nonelementary, model checking problem and have a synchronous semantics: temporal modalities advance by a lockstepwise traversal of all the quantified traces. The second approach adopts a set semantics of temporal logics, in particular LTL, resulting in the logic TeamLTL [38], where the semantical entities are sets of traces (*teams*) instead of single traces, and temporal operators advance in lockstep on all the traces of the current team. Moreover, TeamLTL inherits the powerful split interpretation of disjunction from dependency logic, which also allows expressing *existential* quantification over subteams of the current team. The team approach preserves the modal nature of temporal logics and enables a more readable and compact formulation of hyperproperties. A distinctive expressivity feature of team temporal logics is the ability to relate an unbounded number of traces, which makes global bounded-time requirements expressible. An example is bounded termination, where one has to check whether a system terminates in bounded time, i.e. for the (possibly infinite) set of system traces, there is a time t such that each trace has terminated at time t. On the other hand, very few positive decidability results are known for TeamLTL and its extensions [38,55]. For example, model checking the extension with strong/Boolean negation $\sim$ is highly undecidable (allowing encoding of *third-order* arithmetic) [42], while for TeamLTL itself and its extensions with dependence and/or inclusion atoms of dependence logic, the corresponding decidability problem is an intriguing open question. The only known positive results have been achieved by imposing drastic syntactical or semantical restrictions [38,55]. Asynchronous variants of HyperLTL and TeamLTL have been recently investigated [3,7,33]. These logics have an undecidable model checking problem, so the research focuses on better behaved fragments [3,7,33] or more tractable *lax* semantics [37].

Inquisitive logic. Propositional inquisitive logic [12,14,16] allows for a unified treatment of declarative and interrogative statements using *inquisitive disjunction*. One can see it as a conservative extension of *classical* logic replacing classical objects of evaluation (e.g., first-order relational structures) with collections (*teams* or *information sets*) of such objects. Intuitively, a team encodes the informational uncertainty about the current state of affairs. If we use de Morgan laws to remove classical disjunction from the primitives, the formalism becomes a specialization of *intuitionistic* logic over teams, with the reverse inclusion being the intuitionistic partial order. As inquisitively atoms (but not arbitrary formulas) satisfy the double negation law[4], disjunction-free inquisi-

---

[4] This is just one example of how our InqLTL has more theorems than intuitionistic LTL of Balbiani et al [2]. Note that our semantics can be encoded into a strict subclass of the birelational models of *op.cit.*

tive propositional formulas are indeed classical.[5] The intuitionistic character of team semantics [1] and inquisitive aspects of dependencies [13] have been noted before, and inquisitive first-order [11,16,17,19,31,32,41] and modal formalisms [11,15,16,18,20,43,45] (without *non-declarative* modalities) have been studied. However, inquisitive extensions of temporal logics have not been considered yet.

Our contribution. In this paper, we advance the research on temporal logics with set semantics by introducing an inquisitive synchronous team semantics of LTL. The resulting team logic InqLTL is interpreted on sets of traces (or *teams*) and enjoys both downward closure and related meta-properties from the inquisitive tradition. Its distinguishing feature is that it replaces the split disjunction of TeamLTL with inquisitive disjunction and intuitionistic implication, thereby capturing the dynamics of information-seeking behaviour. While TeamLTL without split disjunction was already investigated in [38] and with both split and inquisitive disjunctions and strong negation in [42,55], none of these works consider *intuitionistic* implication combined with inquisitive disjunction.

We investigate expressiveness, decidability, and complexity issues of InqLTL and its extension with Boolean negation, denoted InqLTL($\sim$). We show that the inquisitive team semantics can be expressed in TeamLTL($\sim$), and that the logic TeamLTL($\sim$) turns out to be strictly more expressive than InqLTL($\sim$). In particular, while it is known that there are satisfiable TeamLTL($\sim$) formulas whose models are uncountable teams [42], we establish that InqLTL($\sim$) has the countable model property. Moreover, we prove that satisfiability and model checking of InqLTL($\sim$) are highly undecidable by a reduction from *second-order* arithmetic.

As a main contribution, we identify a meaningful fragment of InqLTL, which we call *left-positive* InqLTL, where the nested use of implication in the left side of an implication formula is disallowed. Left-positive InqLTL can formalize relevant information-flow security requirements and, unlike TeamLTL, is able to express dependency atoms and *universal subteam* quantification. We show that model checking left-positive InqLTL is decidable, although with a nonelementary complexity in the nesting depth of implication. For the upper bounds, we introduce an abstract semantics of InqLTL where teams of paths are abstracted away by paths of sets of states (*macro-paths*). We then prove that this abstraction is sound and complete for left-positive InqLTL, and provide an automata-theoretic approach for solving model checking under the macro-path semantics.

2   Preliminaries

Let $\mathbb{N}$ be the set of natural numbers. For all $n, h \in \mathbb{N}$ and integer constants $c > 1$, $\mathsf{Tower}_c(h, n)$ denotes a tower of exponentials of base c, height h, and argument n: $\mathsf{Tower}_c(0, n) = n$ and $\mathsf{Tower}_c(h + 1, n) = c^{\mathsf{Tower}_c(h,n)}$. For each $h \in \mathbb{N}$, h-EXPSPACE is the class of languages decided by deterministic Turing machines

---

[5] In this context, recall the Gödel-Gentzen translation [10, Ch. 2] [26,40] [53, Ch. 6–7] [54, Ch. 2.3].

bounded in space by functions of n in $O(\mathsf{Tower}_c(h, n^d))$ for some integer constants $c > 1$ and $d \geq 1$. Note that 0-EXPSPACE coincides with PSPACE.

Given a (finite or infinite) word w over some alphabet, $|w|$ is the length of w ($|w| = \infty$ if w is infinite). For each $0 \leq i < |w|$, $w(i)$ is the $(i+1)^{th}$ symbol of w and $w_{\geq i}$ is the suffix of w from position i, that is, the word $w(i)w(i+1)\ldots$. For a set $\mathcal{L}$ of infinite words over some alphabet and $i \geq 0$, $\mathcal{L}_{\geq i}$ is the set of suffixes of the words in $\mathcal{L}$ from position i: $\mathcal{L}_{\geq i} := \{w_{\geq i} \mid w \in \mathcal{L}\}$.
We fix a finite set AP of atomic propositions. A *trace* is an infinite word on 2^{AP}.

Kripke Structures. We describe the dynamic behaviour of systems by Kripke structures over AP which are tuples $K = \langle S, S_0, R, Lab \rangle$ where S is a nonempty set of states, $S_0 \subseteq S$ is a set of initial states, $R \subseteq S \times S$ is a *left-total* transition relation (i.e., for each $s \in S$, there is $s' \in S$ such that $(s, s') \in R$), and $Lab : S \to 2^{AP}$ assigns to each state the propositions in AP which hold at s. For a state s, we write $R[s]$ to mean the set of successors of state s, i.e., the nonempty set of states s' such that $(s, s') \in R$. A *path* π of K is an infinite word $\pi = s_1 s_2 \ldots$ over S such that $(s_i, s_{i+1}) \in R$ for each $i \geq 1$. The path π is *initial* if it starts at some initial state, that is, $s_1 \in S_0$. The path π induces the trace $Lab(s_0)Lab(s_1)\ldots$. We denote by $\mathcal{L}(K)$ the set of traces induced by the *initial* paths of K.

In this paper, we consider logics whose interpretations (models) are sets $\mathcal{L}$ of traces. For such logics, we consider the following decision problems.

- *Satisfiability*: checking, given a formula φ, whether there is a *nonempty* set of traces satisfying φ (the notion of satisfaction depends on the specific logic).
- *Model checking*: checking, given a finite Kripke structure K and a formula φ, whether $\mathcal{L}(K)$ satisfies φ.

As usual for two formulas φ and φ', we write $\varphi \equiv \varphi'$ to mean that φ and φ' are equivalent, i.e., they are fulfilled by the same interpretations.

TeamLTL. We recall TeamLTL [38] whose syntax is the same as that of LTL [46] in negation normal form. Formally, TeamLTL formulas φ (on AP) are defined as:

$$\varphi ::= p \mid \neg p \mid \varphi \vee \varphi \mid \varphi \wedge \varphi \mid \mathsf{X}\varphi \mid \varphi\mathsf{U}\varphi \mid \varphi\mathsf{R}\varphi$$

where $p \in AP$ and X, U and R are the *next*, *until*, and *release* temporal modalities, respectively. The logical constants $\top$ and $\bot$ are defined as usual (e.g., $\bot :=p \wedge \neg p$). We also use the following abbreviations: $\mathsf{F}\varphi := \top\mathsf{U}\varphi$ (*eventually*) and $\mathsf{G}\varphi := \bot\mathsf{R}\varphi$ (*always*). TeamLTL formulas are interpreted over *arbitrary* (possibly infinite) sets $\mathcal{L}$ of traces (also called *teams* in the terminology of TeamLTL). The satisfaction relation $\mathcal{L} \models \varphi$ is defined as follows:

$$
\begin{aligned}
\mathcal{L} \models p \quad &\Leftrightarrow \text{ for each } w \in \mathcal{L}, p \in w(0) \\
\mathcal{L} \models \neg p \quad &\Leftrightarrow \text{ for each } w \in \mathcal{L}, p \notin w(0) \\
\mathcal{L} \models \varphi_1 \vee \varphi_2 \quad &\Leftrightarrow \text{ for some teams } \mathcal{L}_1, \mathcal{L}_2 \text{ with } \mathcal{L} = \mathcal{L}_1 \cup \mathcal{L}_2: \\
&\qquad\quad \mathcal{L}_1 \models \varphi_1 \text{ and } \mathcal{L}_2 \models \varphi_2 \\
\mathcal{L} \models \varphi_1 \wedge \varphi_2 \quad &\Leftrightarrow \mathcal{L} \models \varphi_1 \text{ and } \mathcal{L} \models \varphi_2 \\
\mathcal{L} \models \mathsf{X}\varphi \quad &\Leftrightarrow \mathcal{L}_{\geq 1} \models \varphi \\
\mathcal{L} \models \varphi_1\mathsf{U}\varphi_2 \quad &\Leftrightarrow \text{ for some } i \geq 0 : \mathcal{L}_{\geq i} \models \varphi_2 \text{ and } \mathcal{L}_{\geq k} \models \varphi_1 \text{ for all } 0 \leq k < i \\
\mathcal{L} \models \varphi_1\mathsf{R}\varphi_2 \quad &\Leftrightarrow \text{ for each } i \geq 0 : \mathcal{L}_{\geq i} \models \varphi_2 \text{ or } \mathcal{L}_{\geq k} \models \varphi_1 \text{ for some } 0 \leq k < i
\end{aligned}
$$

It is worth noting that while in LTL the logical constant $\bot$ has no model, in TeamLTL, $\bot$ has as its unique model the empty team. The *model checking problem for* TeamLTL is checking given a finite Kripke structure K and a TeamLTL formula φ, whether $\mathcal{L}(K) \models \varphi$.

We consider some semantical properties of formulas φ from the team and inquisitive semantics literature:

- *Downward closed*: if $\mathcal{L} \models \varphi$ and $\mathcal{L}' \subseteq \mathcal{L}$ then $\mathcal{L}' \models \varphi$.
- *Empty property*: $\emptyset \models \varphi$.
- *Flatness*: $\mathcal{L} \models \varphi$ iff $w \models_{\mathsf{LTL}} \varphi$ for each $w \in \mathcal{L}$.
- *Singleton equivalence*: $w \models_{\mathsf{LTL}} \varphi$ iff $\{w\} \models \varphi$ for each trace w.

where $w \models_{\mathsf{LTL}} \varphi$ means that the trace w satisfies φ under the standard LTL semantics [46]. One can easily check that TeamLTL formulas satisfy downward closure, singleton equivalence, and empty properties [38,55]. However, TeamLTL formulas do not satisfy flatness in general. A standard example [55] is the formula $\mathsf{F}p$ asserting that there is a timestamp i such that p uniformly holds at position i along each trace of the given team $\mathcal{L}$. This formula is not flat (even if the teams are assumed to be finite) and cannot be expressed in HyperLTL [6]. Due to the downward closure property and singleton equivalence, TeamLTL satisfiability corresponds to standard LTL satisfiability (which is PSPACE-complete). On the other hand, the decidability status of model checking TeamLTL is an intriguing open question.

We also consider the known extension $\mathsf{TeamLTL}(\sim)$ of TeamLTL with the *contradictory negation*, or *strong* negation, which is denoted by $\sim$ to distinguish it from $\neg$ [42]. The semantics of $\sim$ is as follows: $\mathcal{L} \models \sim\varphi \Leftrightarrow \mathcal{L} \not\models \varphi$.

Note that for each atomic proposition p, $\sim p$ and $\neg p$ are not equivalent. In particular, $\sim p$ does not satisfy downward closure. Moreover, $\sim\top$ characterizes the nonempty teams.

3   Inquisitive LTL

In this section, we introduce *Inquisitive* LTL (InqLTL) which is the natural LTL counterpart of inquisitive first-order logic [19,11,16,17,31,32,41]. Like TeamLTL, InqLTL provides an alternative semantics of LTL where the interpretations are sets of traces (teams). The main differences between TeamLTL and InqLTL are the intuitionistic semantics of implication in InqLTL and the fact that the split disjunction connective $\vee$ of TeamLTL is replaced with inquisitive disjunction in InqLTL. In the literature on team temporal logics, inquisitive disjunction is denoted by $\varovee$. Formally, formulas φ of InqLTL (over AP) are generated by the following grammar, where $p \in AP$:

$$\varphi ::= \bot \mid p \mid \varphi \varovee \varphi \mid \varphi \wedge \varphi \mid \varphi \to \varphi \mid \mathsf{X}\varphi \mid \varphi\mathsf{U}\varphi \mid \varphi\mathsf{R}\varphi$$

Negation of φ is defined as $\neg\varphi := \varphi \to \bot$. For a set $\mathcal{L}$ of traces, the satisfaction relation $\mathcal{L} \models \varphi$, is inductively defined as follows (the semantics of atomic propositions, conjunction, and temporal operators is the same as TeamLTL):

$$\mathcal{L} \models \bot \qquad \Leftrightarrow \mathcal{L} = \emptyset$$
$$\mathcal{L} \models \varphi_1 \lozenge \varphi_2 \Leftrightarrow \mathcal{L} \models \varphi_1 \text{ or } \mathcal{L} \models \varphi_2$$
$$\mathcal{L} \models \varphi_1 \to \varphi_2 \Leftrightarrow \text{for all } \mathcal{L}' \subseteq \mathcal{L},\ \mathcal{L}' \models \varphi_1 \text{ implies } \mathcal{L}' \models \varphi_2$$

Note that $\varphi_1 \to \varphi_2$ is checked at all the subsets (subteams) of the given team $\mathcal{L}$. Moreover, $\mathcal{L} \models \neg\varphi$ iff for each $\mathcal{L}' \subseteq \mathcal{L}$ with $\mathcal{L}' \neq \emptyset$, $\mathcal{L}' \not\models \varphi$. We also consider the extension InqLTL($\sim$) of InqLTL with strong negation $\sim$. *Classical implication* with antecedent φ_1 and consequent φ_2 is expressible in InqLTL($\sim$) as $\sim\varphi_1 \lozenge \varphi_2$. Moreover, note that in InqLTL($\sim$), the temporal modalities U and R are interdefinable. The following can be easily checked.

Proposition 1. InqLTL *formulas satisfy downward closure, empty, and single-ton equivalence properties. Hence,* InqLTL *satisfiability reduces to* LTL *satisfiability. Moreover, for all* InqLTL($\sim$) *formulas φ and teams $\mathcal{L}$, it holds that $\mathcal{L} \models \neg\neg\varphi$ iff for each $w \in \mathcal{L}$, $\{w\} \models \varphi$.*

It is worth noting that since an InqLTL formula φ is downward closed, it holds that for all teams $\mathcal{L}$, $\mathcal{L} \models \neg\varphi$ iff for each $w \in \mathcal{L}$, $\{w\} \not\models \varphi$. Moreover, $\emptyset \models \varphi$. Hence, the InqLTL($\sim$) formula $\sim\bot$ cannot be expressed in InqLTL.

Example 1. Let us consider the InqLTL formula φ given by $\varphi := (\neg\neg Fp) \to Fp$. Evidently, for each trace w, $\{w\} \models \varphi$. However, there are teams that are not models of φ: an example is the team $\mathcal{L} = \{w_n\}_{n \geq 0}$ where for each n, p holds exactly at position n along the trace w_n (i.e, $p \in w_n(i)$ iff $i = n$ for all $i, n \geq 0$).

Investigated fragments of InqLTL. We consider the so called *positive fragment* and the *left-positive fragment* of InqLTL. The positive fragment is defined by the following grammar:

$$\varphi ::= \bot \mid p \mid \neg p \mid \varphi \lozenge \varphi \mid \varphi \wedge \varphi \mid X\varphi \mid \varphi U\varphi \mid \varphi R\varphi$$

Intuitively, the positive fragment of InqLTL corresponds to the logic obtained from TeamLTL by replacing split disjunction $\vee$ with inquisitive disjunction $\lozenge$. The left-positive fragment of InqLTL subsumes the positive fragment and is defined as follows:

$$\varphi ::= \bot \mid p \mid \neg\xi \mid \varphi \lozenge \varphi \mid \varphi \wedge \varphi \mid \psi \to \varphi \mid X\varphi \mid \varphi U\varphi \mid \varphi R\varphi$$

where ξ is an arbitrary InqLTL formula and the antecedent ψ in the implication $\psi \to \varphi$ is a positive InqLTL formula. Thus, in left-positive InqLTL, we allow an unrestricted use of negation $\neg$ and a restricted use of intuitionistic implication where the left operand has to be a positive InqLTL formula whenever the right operand is distinct from $\bot$. For each $k \geq 0$, InqLTL$_k$ denotes the fragment of InqLTL where the nesting depth of the implication connective—occurrences of negation $\neg$ are *not counted*—is at most k. Note that InqLTL$_0$ and left-positive InqLTL$_0$ coincide and allow an unrestricted use of intuitionistic negation.

Derived operators in InqLTL and InqLTL($\sim$). We now show that some well-known operators from the team logic literature can be expressed in InqLTL($\sim$), and some even in the left-positive fragment of InqLTL. In particular, we recall the universal subteam quantifier A and the universal singleton quantifier A_1 whose semantics are as follows: (i) $\mathcal{L} \models A\varphi \Leftrightarrow \mathcal{L}' \models \varphi$ for all $\mathcal{L}' \subseteq \mathcal{L}$, and (ii) $\mathcal{L} \models A_1\varphi \Leftrightarrow \{w\} \models \varphi$ for all $w \in \mathcal{L}$. The quantifiers A and A_1 can be expressed in left-positive InqLTL as: $A\varphi \equiv \top \rightarrow \varphi$ and $A_1\varphi \equiv \neg\neg\varphi$. Thus, by using strong negation, we can formalize in InqLTL($\sim$) the dual E of A (existential subteam quantifier) and the dual E_1 of A_1 (existential singleton quantifier): $E\varphi \equiv \sim A\sim\varphi$ and $E_1\varphi \equiv \sim A_1\sim\varphi$.

Throughout the paper, we will use the notation $\mathsf{card}_{\leq 1}$ as a shorthand for the left-positive InqLTL formula $\bigwedge_{p \in AP} G(p \oslash \neg p)$ which characterizes the teams of cardinality at most one: $\mathcal{L} \models \mathsf{card}_{\leq 1}$ iff for all $i \geq 0$ and $w, w' \in \mathcal{L}$, $w(i) = w'(i)$.

3.1 Expressiveness results on InqLTL($\sim$)

We show that each satisfiable InqLTL($\sim$) formula φ has a countable model, that is, a countable team satisfying φ. In fact, we prove a stronger result by using a normal form of InqLTL($\sim$).

Proposition 2. *Let φ be an* InqLTL($\sim$) *formula. Then, for each uncountable model $\mathcal{L}_u$ of φ, there is a countable model $\mathcal{L}_c$ of φ such that $\mathcal{L}_c \subseteq \mathcal{L}_u$ and for each team $\mathcal{L}$ such that $\mathcal{L}_c \subseteq \mathcal{L} \subseteq \mathcal{L}_u$, $\mathcal{L}$ is still a model of φ.*

Proof. We exploit a normal form of InqLTL($\sim$) which is defined as follows:

$$\varphi ::= \bot \mid {\sim}\bot \mid p \mid {\sim}p \mid \varphi \oslash \varphi \mid \varphi \wedge \varphi \mid A\varphi \mid E\varphi \mid X\varphi \mid \varphi U\varphi \mid \varphi R\varphi$$

We observe that this normal form is expressively complete for InqLTL($\sim$). Indeed, $\varphi_1 \rightarrow \varphi_2 \equiv A(\sim\varphi_1 \oslash \varphi_2)$. Hence, since modalities A and E are duals and the temporal modalities U and R are duals, by pushing strong negation $\sim$ inward, one can convert an InqLTL($\sim$) formula φ into an equivalent InqLTL($\sim$) formula in normal form. Thus, we can assume that the given InqLTL($\sim$) formula φ is in normal form. Let $\mathcal{L}_u$ be an uncountable model of φ. We prove Proposition 2 by induction on the structure of φ. We focus on the case where φ is of the form $\varphi_1 R\varphi_2$. The other cases are similar or simple. Since $\mathcal{L}_u$ is a model of φ, either (i) $(\mathcal{L}_u)_{\geq i} \models \varphi_2$ for each $i \geq 0$, or (ii) there is $k \geq 0$ such that $(\mathcal{L}_u)_{\geq k} \models \varphi_1$ and $(\mathcal{L}_u)_{\geq i} \models \varphi_2$ for each $0 \leq i \leq k$. We focus on the first case (the other case being similar). By the induction hypothesis, for each $i \geq 0$, there is a countable subteam $\mathcal{L}_{c,i}$ of $(\mathcal{L}_u)_{\geq i}$ such that each team $\mathcal{L}$ satisfying $\mathcal{L}_{c,i} \subseteq \mathcal{L} \subseteq (\mathcal{L}_u)_{\geq i}$ is a model of φ_2. For each $i \geq 0$, let $\mathcal{L}'_{c,i}$ be any subset of $\mathcal{L}_u$ such that $(\mathcal{L}'_{c,i})_{\geq i} = \mathcal{L}_{c,i}$. Since the number of words of length i over the finite alphabet 2^{AP} is finite, $\mathcal{L}'_{c,i}$ is countable. We set $\mathcal{L}_c := \bigcup_{i \geq 0} \mathcal{L}'_{c,i}$. Since the countable union of countable sets is still countable, by construction the result follows. $\square$

By using Proposition 2 and known results on TeamLTL($\sim$) [42], we now establish the following expressiveness results.

Proposition 3. InqLTL($\sim$) *is more expressive than* InqLTL, *and* InqLTL($\sim$) *is less expressive than* TeamLTL($\sim$).

Proof. We recall that the InqLTL($\sim$) formula $\sim\!\perp$ is not expressible in InqLTL which proves the first part of Proposition 3. For the second part, we first show that InqLTL($\sim$) is subsumed by TeamLTL($\sim$). We observe that $\varphi_1 \to \varphi_2 \equiv \mathsf{A}(\sim\!\varphi_1 \varoslash \varphi_2)$, $\varphi_1 \varoslash \varphi_2 \equiv \sim(\sim\!\varphi_1 \wedge \sim\!\varphi_2)$, $\mathsf{A}\varphi \equiv \sim\!\mathsf{E}\!\sim\!\varphi$, and $\mathsf{E}\varphi \equiv \top \vee \varphi$ (recall that $\vee$ is split disjunction in TeamLTL). Hence, each InqLTL($\sim$) formula can be converted in linear time into an equivalent TeamLTL($\sim$) formula.

It remains to prove that there are TeamLTL($\sim$) formulas which cannot be expressed in InqLTL($\sim$). It is known that satisfiability of TeamLTL($\sim$) is hard for truth in third-order arithmetics [42]. The proof in [42] entails the existence of satisfiable formulas whose unique models are uncountable. On the other hand, by Proposition 2, each satisfiable InqLTL($\sim$) has a countable model, and we are done. $\square$

3.2 Examples of specifications

InqLTL and its left-positive fragment can express relevant information-flow security properties. An example is *noninterference* [30] which requires that all the traces which globally agree on the low-security inputs also globally agree on the low-security outputs, independently of the values of high-security inputs. Noninterference can be expressed in left-positive InqLTL as follows, where LI (resp., LO) is the set of propositions describing low-security inputs (resp., low-security outputs):

$$[\bigwedge_{p \in LI} \mathsf{G}(p \varoslash \neg p)] \to [\bigwedge_{p \in LO} \mathsf{G}(p \varoslash \neg p)]$$

Another example is *observational determinism* [56], which states that traces which have the same initial low inputs are indistinguishable to a low user. This can be expressed by the left-positive InqLTL formula:

$$[\bigwedge_{p \in LI} (p \varoslash \neg p)] \to [\bigwedge_{p \in LO} \mathsf{G}(p \varoslash \neg p)]$$

More flexible noninterference policies allow controlled releases of secret information (information *declassification* [49]). For example, a password checker must reveal whether the entered password is correct or not. Let φ be an InqLTL formula describing facts about high-security inputs which may be released. Noninterference with declassification policy φ is expressible as:

$$[\varphi \wedge \bigwedge_{p \in LI} \mathsf{G}(p \varoslash \neg p)] \to [\bigwedge_{p \in LO} \mathsf{G}(p \varoslash \neg p)]$$

Note that unlike left-positive InqLTL, unrestricted InqLTL allows expressing a hierarchy of preconditions on the current subset of traces. *Refinement verification.* Unlike TeamLTL and HyperLTL [21], InqLTL allows to enforce properties on all the refinements (subsets of traces) of the given Kripke structure that satisfy certain conditions. As an example, we consider the team version of the classical response property $\mathsf{G}(q \to \mathsf{F}p)$. Under the inquisitive team semantics, this left-positive InqLTL formula asserts that for each refinement $\mathcal{L}_r$, whenever the request q occurs uniformly (i.e., q holds at the current time i on all the traces of $\mathcal{L}_r$), then a response p will be given uniformly too (i.e., for some $j \geq i$, p holds on all the traces of $\mathcal{L}_r$). We conjecture that this property can be expressed neither in TeamLTL nor in known extensions of HyperLTL such as HyperQPTL [47,23]. Intuitively, the motivation is that TeamLTL and HyperQPTL do not allow universal subteam quantification.

Expressing dependence atoms. TeamLTL is usually enriched with novel atomic statements describing properties of teams. The most studied ones are *dependence atoms* $dep(\varphi_1, ..., \varphi_n, \psi)$, where $\varphi_1, \ldots, \varphi_n, \psi$ are LTL formulas, stating that for each trace w the truth value $\|\psi\|_w$ of ψ is functionally determined by the truth values $\|\varphi_1\|_w, \ldots, \|\varphi_n\|_w$ of $\varphi_1, \ldots, \varphi_n$. Formally, $\mathcal{L} \models dep(\varphi_1, ..., \varphi_n, \psi)$ *iff*:

$$\text{for all } w, w' \in \mathcal{L}, \left(\bigwedge_{i=1}^{i=n} \|\varphi_i\|_w = \|\varphi_i\|_{w'}\right) \Rightarrow \|\psi\|_w = \|\psi\|_{w'}$$

The atom $dep(\varphi_1, ..., \varphi_n, \psi)$ is expressible in InqLTL as:

$$\left[\bigwedge_{i=1}^{i=n} (\neg\varphi_i \otimes \neg\neg\varphi_i)\right] \rightarrow \left[(\neg\psi \otimes \neg\neg\psi)\right]$$

Note that if φ_i is propositional, then the formula above is in left-positive InqLTL.

4   Undecidability of InqLTL($\sim$)

In this section, we show that model checking and satisfiability of InqLTL($\sim$) are highly undecidable since they can encode truth in second-order arithmetic.

Recall that second-order arithmetic (see e.g. [48]) is second-order predicate logic with equality over the signature $(<, +, *, \in)$ evaluated over the the set $\mathbb{N}$ of natural numbers, where $<$ is interpreted as the standard ordering over $\mathbb{N}$, $+$ and $*$ are interpreted as standard addition and multiplication in $\mathbb{N}$, respectively, and $\in$ is the set membership operator. Note that first-order variables range over natural numbers, while second-order variables range over sets of natural numbers. W.l.o.g. we assume that arithmetical formulas are in prenex normal form, i.e., consisting of a prefix of existential ($\exists$) or universal ($\forall$) quantifiers, applied to first-order or second-order variables, followed by a Boolean combination of atomic formulas of the form $x < y$ or $x = y + z$ or $x = y * z$ or $x \in X$ for first-order variables x, y, z and second-order variables X. Truth in second-order arithmetic is the decision problem consisting of checking whether an arithmetical sentence (i.e., a formula with no free variables) is true over $\mathbb{N}$.

Fix an arithmetic sentence $\Phi = Q_1\nu_1 \ldots Q_k\nu_k.\Psi$ where Ψ is quantifier-free and for each $1 \leq i \leq k$, $Q_i \in \{\exists, \forall\}$ and ν_i is a first-order or second-order variable. We build an InqLTL($\sim$) formula $enc(\Phi)$ which is satisfiable iff Φ is true over $\mathbb{N}$. Moreover, at the end of the section, we show that InqLTL($\sim$) model checking is at least as hard as InqLTL($\sim$) satisfiability.

Encoding of natural numbers and arithmetic operations. Given an atomic proposition p, each natural number $n \in \mathbb{N}$ can be encoded by the trace over $2^{\{p\}}$ where proposition p holds exactly at position n. Subsets of natural numbers can then be encoded by teams consisting of traces of the previous form. However, in the valuation of the quantifier prefix $Q_1\nu_1 \ldots Q_k\nu_k$ of the given arithmetic sentence Φ, we need to distinguish the natural numbers (resp., the sets of natural numbers) which are assigned to distinct first-order variables (resp., distinct second-order variables). This justifies the following definition. Let $AP_{num} := \{\nu_1, \ldots, \nu_k, \#\}$. For each $1 \leq i \leq k$, a ν_i-*trace* is a trace of the form $\{\nu_i\}^{n-1}\{\#, \nu_i\}\{\nu_i\}^\omega$ for some $n \in \mathbb{N}$ (i.e., ν_i holds at each position and $\#$ holds exactly at position n). The previous trace encodes the natural number n.

For the encoding of addition and multiplication in $\mathsf{InqLTL}(\sim)$, we use a coloured variant of the encoding considered in [29] for second-order $\mathsf{HyperLTL}$. In this latter logic, there is explicit first-order quantification over traces and propositions are parameterized by first-order variables so that one can distinguish the single traces that compose a team of traces. Since this is not directly possible in $\mathsf{InqLTL}(\sim)$, we use different colors to distinguish the traces in the encoding.

Let $AP_{arith} := \{arg_1, arg_2, res, +, *, 0, 1\}$. For all $c \in \{0, 1\}$ and $op \in \{+, *\}$, an *op-trace with colour c* is a trace w over $2^{\{c, op, arg_1, arg_2, res\}}$ satisfying:

- $c \in w(i)$ and $op \in w(i)$ for all $i \geq 0$;
- there are unique $n_1, n_2, n_3 \in \mathbb{N}$ with $arg_1 \in w(n_1)$, $arg_2 \in w(n_2)$, and $res \in w(n_3)$. We write $arg_1(w)$ (resp., $arg_2(w)$) to mean n_1 (resp., n_2), and $res(w)$ to mean n_3.

An *op-trace* is an *op*-trace with colour c for some $c \in \{0, 1\}$. An *op*-trace w is *well-formed* if $res(w) = arg_1(w) + arg_1(w)$ when op is $+$, and $res(w) = arg_1(w) * arg_1(w)$ otherwise.
The used set AP of propositions is then defined as $AP := AP_{num} \cup AP_{arith}$.

A trace w is *consistent* if either w is a ν_i-trace for some $1 \leq i \leq k$, or w is an *op*-trace for some $op \in \{+, *\}$ (note that we do *not* require that the *op*-trace is well-formed). A team is *consistent* if it contains only consistent traces. One can trivially construct an LTL formula ψ_{con} characterizing the consistent traces. Hence, $\varphi_{con} := \neg\neg\psi_{con}$ characterizes the consistent teams under the InqLTL semantics. Moreover, for each $1 \leq i \leq k$, let $\mathcal{L}_{all}^{\nu_i}$ be the team consisting of all ν_i-traces. By Proposition 4, $\mathcal{L}_{all}^{\nu_i}$ is the unique model of the formula $\varphi_{all}^{\nu_i} := \varphi_{con} \wedge \mathsf{G}(\nu_i \wedge \mathsf{E}_1\#)$. Thus, we obtain the following result.

Proposition 4. *One can build two InqLTL formulas φ_{con} and $\varphi_{all}^{\nu_i}$ such that φ_{con} captures the consistent teams and $\varphi_{all}^{\nu_i}$ has a unique model which is $\mathcal{L}_{all}^{\nu_i}$.*

Let $\mathcal{L}_{arith}$ be the team consisting of the well-formed $+$-traces and the well-formed $*$-traces. The following result will allow us to implement addition and multiplication in $\mathsf{InqLTL}(\sim)$.

Proposition 5. *One can construct an $\mathsf{InqLTL}(\sim)$ formula φ_{arith} such that $\mathcal{L} \models \varphi_{arith}$ iff $\mathcal{L}$ is a consistent team whose set of $+$-traces and $*$-traces is $\mathcal{L}_{arith}$.*

Proof (Sketched proof.). Formula φ_{arith} is defined as follows:

$$\varphi_{arith} := \varphi_{con} \wedge \bigwedge_{op \in \{+,*\}} (\varphi_{all}^{op} \wedge \varphi_{wf}^{op})$$

where φ_{con} is the formula of Proposition 4 capturing the consistent teams. Conjunct φ_{all}^{op} ensures that for all natural numbers n_1 and n_2 and for each colour $c \in \{0,1\}$, there is an op-trace w with colour c whose arguments $arg_1(w)$ and $arg_2(w)$ are n_1 and n_2, respectively:

$$\varphi_{all}^{op} := \bigwedge_{c \in \{0,1\}} \bigwedge_{\ell \in \{1,2\}} \mathsf{GE}(op \wedge c \wedge arg_\ell \wedge \mathsf{GE}_1 arg_{3-\ell}).$$

Conjuncts φ_{wf}^{+} and φ_{wf}^{*} activate recursion by encoding the inductive definition of addition and multiplication. Here, we focus on φ_{wf}^{+}. We use the formula $\theta_{0,1}^{+}$ requiring that for each consistent team $\mathcal{L}$, $\mathcal{L}$ consists of one $+$-trace with colour 0 and one $+$-trace with colour 1:

$$\theta_{0,1}^{+} := {+} \wedge \bigwedge_{c \in \{0,1\}} (\mathsf{E}_1 c \wedge (c \rightarrow \mathsf{card}_{\leq 1})).$$

Then, the formula φ_{wf}^{+} enforces the following requirements for each colour c.
- For each $+$-trace w with colour c such that $arg_1(w) = arg_2(w) = 0$, it holds that $res(w) = 0$. This is trivially expressible.
- For all $\ell \in \{1,2\}$, $+$-traces w with colour c and $+$-traces w' with colour $1-c$ such that $arg_\ell(w) = arg_\ell(w')$ and $arg_{3-\ell}(w') = arg_{3-\ell}(w) + 1$, it holds that $res(w') = res(w) + 1$. This can be expressed as:

$$\bigwedge_{c \in \{0,1\}} \bigwedge_{\ell \in \{1,2\}} ([\mathsf{F}arg_\ell \wedge \theta_{0,1}^{+} \wedge \psi(c, arg_{3-\ell})] \rightarrow \psi(c, res))$$

$$\psi(c,p) := \mathsf{F}(\mathsf{E}_1(c \wedge p) \wedge \mathsf{XE}_1((1-c) \wedge p))$$

We use two distinct colours for ensuring that for the two compared $+$-traces, the one having the greatest argument $arg_{3-\ell}$ has also the greatest result res. $\square$

Let $\mathcal{L}_{all}$ be the team defined as $\mathcal{L}_{all} := \mathcal{L}_{arith} \cup \bigcup_{i=1}^{i=k} \mathcal{L}_{all}^{\nu_i}$. Note that for each variable ν_i, $\mathcal{L}_{all}$ contains all the ν_i-traces. By Propositions 4–5, $\mathcal{L}_{all}$ is the unique model of the $\mathsf{InqLTL}(\sim)$ formula $\varphi_{con} \wedge \varphi_{arith} \wedge \bigwedge_{i=1}^{i=k} \mathsf{E}\varphi_{all}^{\nu_i}$. Hence:

Proposition 6. *One can construct an* $\mathsf{InqLTL}(\sim)$ *formula* φ_{all} *whose unique model is* $\mathcal{L}_{all}$.

Encoding of variable valuations. Let g be a variable valuation over the set $\{\nu_1, \ldots, \nu_k\}$, i.e., a mapping assigning to each variable ν_i a natural number if ν_i is a first-order variable, and a subset of natural numbers otherwise. We encode g by the consistent team $\mathcal{L}_g := \mathcal{L}_{arith} \cup \mathcal{L}'_g$, where $\mathcal{L}'_g$ does not contain $+$-traces and $*$-traces and for each variable ν_i, we have:
- if ν_i is a first-order variable, then $\mathcal{L}_g$ contains exactly one ν_i-trace. Moreover, this trace encodes the natural number $g(\nu_i)$. If instead ν_i is a second-order variable, then the set of natural numbers encoded by the ν_i-traces which belong to $\mathcal{L}_g$ is exactly $g(\nu_i)$.

By exploiting the previous encoding, we now show how to express the evaluation of quantifier-free arithmetic formulas over $\{\nu_1, \ldots, \nu_k\}$ in $\mathsf{InqLTL}(\sim)$.

Proposition 7. *Given a quantifier-free arithmetic formula Ψ with variables in $\{\nu_1, \ldots, \nu_k\}$, one can construct an $\mathsf{InqLTL}(\sim)$ formula $enc(\Psi)$ such that for each variable valuation g: g satisfies Ψ iff $\mathcal{L}_g \models enc(\Psi)$.*

Proof. For each nonempty set $P \subseteq \{\nu_1, \ldots, \nu_k, +, *\}$, we first build an $\mathsf{InqLTL}(\sim)$ formula χ_P such that a consistent team $\mathcal{L}$ is a model of χ_P *iff* $\mathcal{L}$ has cardinality $|P|$ and for each $t \in P$, there is exactly one t-trace in $\mathcal{L}$:

$$\chi_P := (\mathsf{A}_1 \bigvee_{t \in P} t) \wedge \bigwedge_{t \in P} (\mathsf{E}_1 t \wedge (t \to \mathsf{card}_{\leq 1})).$$

Fix a quantifier-free arithmetic formula Ψ with variables in $\{\nu_1, \ldots, \nu_k\}$. Since Boolean connectives can be expressed in $\mathsf{InqLTL}(\sim)$, w.l.o.g. we can assume that Ψ is an atomic formula. There are the following cases:

- Ψ is of the form $x < y$: $enc(\Psi) := \mathsf{E}[\chi_{\{x,y\}} \wedge \mathsf{F}(\mathsf{E}_1(x \wedge \#) \wedge \mathsf{XFE}_1(y \wedge \#))]$.
- Ψ is of the form $x = y + z$: $enc(\Psi)$ is given by

$$\mathsf{E}\big(\chi_{\{x,y,z,+\}} \wedge \mathsf{F}[\mathsf{E}_1(x \wedge \#) \wedge \mathsf{E}_1 res] \wedge \mathsf{F}[\mathsf{E}_1(y \wedge \#) \wedge \mathsf{E}_1 arg_1] \wedge \mathsf{F}[\mathsf{E}_1(z \wedge \#) \wedge \mathsf{E}_1 arg_2]\big).$$

- Ψ is of the form $x = y * z$: this case is similar to the previous one.
- Ψ is of the form $x \in X$: $enc(\Psi) := \mathsf{E}[\chi_{\{x,X\}} \wedge \mathsf{F}\#]$.

Correctness of the construction easily follows. $\square$

For the given arithmetic sentence $\Phi = Q_1\nu_1 \ldots Q_k\nu_k. \Psi$, where Ψ is quantifier-free, the arithmetical quantifiers $Q_i\nu_i$ are emulated in $\mathsf{InqLTL}(\sim)$ as follows. We start with the consistent team $\mathcal{L}_{all}$ which is the unique model of the $\mathsf{InqLTL}(\sim)$ formula φ_{all} of Proposition 6. Recall that $\mathcal{L}_{all} = \mathcal{L}_{arith} \cup \bigcup_{i=1}^{i=k} \mathcal{L}_{all}^{\nu_i}$. Then by exploiting the $\mathsf{InqLTL}(\sim)$ formulas $\varphi_{all}^{\nu_1}, \ldots, \varphi_{all}^{\nu_k}, \varphi_{arith}$ of Propositions 4 and 5, we can select, existentially or universally (depending on the polarity of $Q_1 \in \{\exists, \forall\}$), a subteam $\mathcal{L}_1 \subseteq \mathcal{L}_{all}$ of the form $\mathcal{L}_1 = (\mathcal{L}_{all} \setminus \mathcal{L}_{all}^{\nu_1}) \cup T_1$ where $T_1 \subseteq \mathcal{L}_{all}^{\nu_1}$ and T_1 is a singleton if ν_1 is a first-order variable. Then, we proceed with the team $\mathcal{L}_1$ and apply the previous procedure by selecting a subteam of the form $(\mathcal{L}_1 \setminus \mathcal{L}_{all}^{\nu_2}) \cup T_2$ where $T_2 \subseteq \mathcal{L}_{all}^{\nu_2}$ and T_2 is a singleton if ν_2 is a first-order variable, and so on. At the end of this process, we obtain a subteam $\mathcal{L}_g$ of $\mathcal{L}_{all}$ encoding a variable valuation g on $\{\nu_1, \ldots, \nu_k\}$.

Let $enc(\Psi)$ be the $\mathsf{InqLTL}(\sim)$ formula of Proposition 7 for the quantifier-free arithmetic formula Ψ. Moreover, let $\theta_{k+1}, \ldots, \theta_1$ be the $\mathsf{InqLTL}(\sim)$ formulas defined as follows: $\theta_{k+1} := enc(\Psi)$ and for each $i = k, \ldots, 1$,

$$\theta_i := \begin{cases} \mathsf{E}\,\xi_i & \text{if } Q_i \text{ is } \exists \\ \mathsf{A}\,\xi_i & \text{otherwise} \end{cases} \qquad \xi_i := \theta_{i+1} \wedge \varphi_{arith} \wedge select(\nu_i) \wedge \bigwedge_{\ell=i+1}^{\ell=k} \mathsf{E}\,\varphi_{all}^{\nu_\ell}$$

$$select(\nu_i) := \begin{cases} (\nu_i \to \mathsf{card}_{\leq 1}) \wedge \mathsf{E}_1\,\nu_i & \text{if } \nu_i \text{ is first-order} \\ \top & \text{otherwise} \end{cases}$$

Finally, define $enc(\Phi) := \varphi_{all} \wedge \theta_1$. By Propositions 4–7, we obtain that $enc(\Phi)$ is satisfiable iff $\mathcal{L}_{all}$ is a model of $enc(\Phi)$ iff Φ is true over $\mathbb{N}$. Now, let us consider the Kripke structure $K_{AP} = \langle 2^{AP}, 2^{AP}, 2^{AP} \times 2^{AP}, Lab \rangle$, where Lab is the identity mapping. Evidently, an $\mathsf{InqLTL}(\sim)$ formula θ is satisfiable if $K_{AP} \models \mathsf{E}\theta$. Hence, $\mathsf{InqLTL}(\sim)$ satisfiability is reducible to $\mathsf{InqLTL}(\sim)$ model checking. Thus, we obtain the following result.

Theorem 1. *Model checking and satisfiability of* $\mathsf{InqLTL}(\sim)$ *are undecidable. In particular, the truth in second-order arithmetic is reducible to* $\mathsf{InqLTL}(\sim)$ *model checking and to* $\mathsf{InqLTL}(\sim)$ *satisfiability.*

Remark. For ease of presentation, the proposed reduction uses a number of propositions which depends on the length k of the quantifier prefix. A slightly more involved reduction allows us to use a fixed and small set of propositions. The idea is to exploit just one proposition, say p, for encoding the different types of traces: in particular, for each $1 \leq i \leq k$, a ν_i-trace is replaced with a trace over $2^{\{p,\#\}}$ whose projection over $\{p\}$ is the word $(\{p\}^i \emptyset)^\omega$.

5   Decidability results

In this section, we show that for left-positive InqLTL, model checking is decidable. Moreover, we prove that for each $k \geq 0$, model checking left-positive InqLTL_k formulas is exactly k-EXPSPACE-complete. The upper bounds are obtained in two steps. In the first step, we define an abstract semantics of InqLTL on Kripke structures, which we call *macro-path semantics*. In this setting, for a given Kripke structure K, InqLTL formulas are interpreted over infinite sequences of subsets of K-states (*macro-paths*), which provide a word-encoding of sets (teams) of K-paths. Not all the teams of K-paths can be represented by macro-paths. However, we show that for left-positive InqLTL, the macro-path semantics captures the teams semantics over Kripke structures. Then, in the second step, we provide an automata-theoretic approach for solving the InqLTL model-checking problem under the macro-path semantics.

5.1   Macro-path semantics for InqLTL

Fix a Kripke structure $K = \langle S, S_0, R, Lab \rangle$. For a set Π of paths of K, we denote by $\mathcal{L}_K(\Pi)$ the set of traces induced by the paths in Π. For an InqLTL formula φ, we write $\Pi \models_K \varphi$ to mean that $\mathcal{L}_K(\Pi) \models \varphi$.

A *macro-state* of K is a (possibly empty) set S' of states of K, that is $S' \subseteq S$. Given two macro-states S' and S'', we say that S'' *is a successor of* S' if:
- for each $s' \in S'$, there is $s'' \in R[s'] \cap S''$,
- for each $s'' \in S''$, there is $s' \in S'$ such that $s'' \in R[s']$.

A *macro-path* ρ of K is an infinite sequence of macro-states $\rho = S_1 S_2 \ldots$ such that S_{i+1} is a successor of S_i for each $i \geq 1$. A *macro-path* ρ encodes a set $Paths_K(\rho)$ of paths defined as the set of K-paths π such that $\pi(i) \in \rho(i)$ for each $i \geq 0$. Given two macro-paths ρ and ρ', we write $\rho \sqsubseteq \rho'$ to mean that for each $i \geq 0$, $\rho(i) \subseteq \rho'(i)$. Evidently, if $\rho \sqsubseteq \rho'$, then $Paths_K(\rho) \subseteq Paths_K(\rho')$. A *singleton* macro-path is a macro-path ρ such that $Paths_K(\rho)$ is a singleton: note that $\rho(i)$ is a singleton for each $i \geq 0$.

Not all the sets of paths can be encoded by macro-paths. Intuitively, the macro-paths can only represent *memoryless* sets Π of paths where for all $i \geq 0$ and $\pi, \pi' \in \Pi$ such that $\pi(i) = \pi'(i)$, the following holds for all $\nu \in S^\omega$: $\pi(0) \ldots \pi(i) \cdot \nu \in \Pi$ iff $\pi'(0) \ldots \pi'(i) \cdot \nu \in \Pi$. As an example, assume that $S = \{s_0, s_1\}$ and $(s_i, s_j) \in R$ for all $i, j \in \{0, 1\}$. Then, there is no macro-path ρ such that $Paths_K(\rho) = \{s_0^\omega, s_1^\omega\}$. Indeed, the unique macro-path encoding a set of paths which contains both the paths s_0^ω and s_1^ω is S^ω, but $Paths_K(S^\omega)$ is the set of all the paths.

However, each set Π of paths can be abstracted away by the macro-path, denoted by $\mathsf{mp}(\Pi)$, whose i^{th} macro state is the collection of states associated with the ith position of the paths in Π. Formally, for each $i \geq 0$:

$$\mathsf{mp}(\Pi)(i) := \{\pi(i) \mid \pi \in \Pi\}.$$

Note that $Paths_K(\mathsf{mp}(\Pi)) \supseteq \Pi$ and, in general, $\Pi \neq Paths_K(\mathsf{mp}(\Pi))$. With reference to the previous example, let $\Pi = \{s_0^\omega, s_1^\omega\}$. We have that $\mathsf{mp}(\Pi) = \{s_0, s_1\}^\omega$ and $Paths_K(\mathsf{mp}(\Pi))$ is the set of all the paths of K. Hence, $\Pi \subset Paths_K(\mathsf{mp}(\Pi))$.

Macro-path semantics. We now provide a semantics of InqLTL interpreted over macro-paths of the given Kripke structure K. For a macro-path ρ and an InqLTL formula φ, the satisfaction relation $\rho \models_K \varphi$ is inductively defined as follows (we omit the semantics of temporal modalities which is defined as for LTL but we replace traces w with macro-paths ρ):

$$
\begin{aligned}
\rho &\models_K \bot & &\Leftrightarrow Paths_K(\rho) = \emptyset \\
\rho &\models_K p & &\Leftrightarrow \text{for each } s \in \rho(0), p \in Lab(s) \\
\rho &\models_K \varphi_1 \varovee \varphi_2 & &\Leftrightarrow \rho \models_K \varphi_1 \text{ or } \rho \models_K \varphi_2 \\
\rho &\models_K \varphi_1 \wedge \varphi_2 & &\Leftrightarrow \rho \models_K \varphi_1 \text{ and } \rho \models_K \varphi_2 \\
\rho &\models_K \varphi_1 \rightarrow \varphi_2 & &\Leftrightarrow \text{for each macro-path } \rho' \sqsubseteq \rho, \rho' \models_K \varphi_1 \text{ implies } \rho' \models_K \varphi_2
\end{aligned}
$$

Note that $Paths_K(\rho) = \emptyset$ iff $\rho(i) = \emptyset$ for each $i \geq 0$. The macro-path semantics is downward closed, that is for all macro-paths ρ, ρ' such that $\rho \sqsubseteq \rho'$, $\rho' \models_K \varphi$ implies $\rho \models_K \varphi$.

Recall that the positive fragment of InqLTL is defined as follows.

$$\varphi ::= \bot \mid p \mid \neg p \mid \varphi \varovee \varphi \mid \varphi \wedge \varphi \mid \mathsf{X}\varphi \mid \varphi\mathsf{U}\varphi \mid \varphi\mathsf{R}\varphi$$

By construction for each set of paths Π of the given Kripke structure K, it holds that $\Pi \models_K p$ iff $\mathsf{mp}(\Pi) \models_K p$. Moreover, $\Pi \models_K \neg p$ iff $\mathsf{mp}(\Pi) \models_K \neg p$. Additionally, for each $i \geq 0$, $\mathsf{mp}(\Pi_{\geq i}) = (\mathsf{mp}(\Pi))_{\geq i}$. Thus, by a straightforward induction on the structure of the given formula, we obtain the following result.

Proposition 8. *Let K be a Kripke structure and φ be a positive* InqLTL *formula φ. Then, for each set Π of K-paths, $\Pi \models_K \varphi$ iff $\mathsf{mp}(\Pi) \models_K \varphi$.*

We now show that for the left-positive fragment of InqLTL, the team semantics over Kripke structures and the macro-path semantics are equivalent.

Proposition 9. *Let K be a Kripke structure and φ be a left-positive* InqLTL *formula φ. Then, for each macro-path ρ of K, $\rho \models_K \varphi$ iff $Paths_K(\rho) \models_K \varphi$.*

Proof. The proof is by induction on the structure of φ. The cases where $\varphi = p$ and $\varphi = \bot$ easily follow from the macro-path semantics. The cases where the root modality is a temporal modality or a connective in $\{\lozenge, \wedge\}$ directly follow from the induction hypothesis and the fact that $(Paths_K(\rho))_{\geq i} = Paths_K(\rho_{\geq i})$. For the remaining cases, where the root modality of φ is $\neg$ or $\rightarrow$, we proceed as follows:

- $\varphi = \neg\psi$, where ψ is an arbitrary InqLTL formula: by downward closure of InqLTL formulas, we have that $Paths_K(\rho) \models_K \neg\psi$ iff for each path $\pi \in Paths_K(\rho)$, $\{\pi\} \not\models_K \psi$. Moreover, since the macro-path semantics is downward closed, we have that $\rho \models_K \neg\psi$ iff for each *singleton* macro-path ρ' with $\rho' \sqsubseteq \rho$, $\rho' \not\models_K \psi$ iff (by the macro-path semantics) for each $\pi \in Paths_K(\rho)$, $\{\pi\} \not\models_K \psi$. Hence, the result follows.
- $\varphi = \psi_1 \rightarrow \psi_2$, where ψ_1 is a positive InqLTL formula: first assume that $Paths_K(\rho) \models_K \psi_1 \rightarrow \psi_2$. Let ρ' be a macro-path such that $\rho' \sqsubseteq \rho$ and $\rho' \models_K \psi_1$. We need to show that $\rho' \models_K \psi_2$. By the induction hypothesis, $Paths_K(\rho') \models_K \psi_1$, Since $Paths_K(\rho') \subseteq Paths_K(\rho)$ and $Paths_K(\rho) \models_K \psi_1 \rightarrow \psi_2$, it follows that $Paths_K(\rho') \models_K \psi_2$. Thus, by the induction hypothesis, we obtain that $\rho' \models_K \psi_2$.

  For the converse direction, let $\rho \models_K \psi_1 \rightarrow \psi_2$ and Π be a set of K-paths such that $\Pi \subseteq Paths_K(\rho)$ and $\Pi \models_K \psi_1$. We need to show that $\Pi \models_K \psi_2$. We note that $\mathsf{mp}(\Pi) \sqsubseteq \rho$. Moreover, since ψ_1 is a positive InqLTL formula, by Proposition 8, it follows that $\mathsf{mp}(\Pi) \models_K \psi_1$. Thus, being $\mathsf{mp}(\Pi) \sqsubseteq \rho$ and $\rho \models_K \psi_1 \rightarrow \psi_2$, we have that $\mathsf{mp}(\Pi) \models_K \psi_2$ and by the induction hypothesis, $Paths_K(\mathsf{mp}(\Pi)) \models_K \psi_2$. Since $\Pi \subseteq Paths_K(\mathsf{mp}(\Pi))$ and ψ_2 is downward closed, we conclude that $\Pi \models_K \psi_2$, and we are done. $\square$

Given a Kripke structure $K = \langle S, S_0, R, Lab \rangle$, the *initial macro-path* of K is the macro-path ρ_0 starting at S_0 of the form $\rho_0 = S_0, S_1, \dots$ where $S_{i+1} := \{s' \in S \mid s' \in R[s] \text{ for some } s \in S_i\}$ for each $i \geq 0$. We crucially observe that for the initial macro-path ρ_0 of K, $Paths_K(\rho_0)$ is the set of initial paths of K. Hence, by Proposition 9, we obtain the following result.

Corollary 1. *Given a Kripke structure K with initial macro-path ρ_0 and a left-positive InqLTL formula φ, $\rho_0 \models_K \varphi$ iff $\mathcal{L}(K) \models \varphi$.*

5.2   Model checking of left-positive InqLTL

In this section, we provide an asymptotically optimal automata-theoretic approach for checking whether the initial macro-path of a finite Kripke structure K satisfies an InqLTL formula φ under the macro-path semantics. In particular, we show how to construct an *hesitant alternating word automaton* (HAA) [39] $\mathcal{A}_{K,\varphi}$ accepting the set of macro-paths of K satisfying φ. As a consequence, the considered problem is reduced to the membership problem $\rho_0 \in \mathcal{L}(\mathcal{A}_{K,\varphi})$, where ρ_0 is the initial macro-path of K. The latter problem can be reduced to nonemptiness of one-letter HAA which in turn can be efficiently solved in logarithmic space in the number of states [39]. Thus, by Corollary 1, for each $k \geq 0$, we obtain that model checking the left-positive fragment of InqLTL_k is decidable and precisely in k-EXPSPACE. A matching lower bound is proved in [5].

Syntax and semantics of HAA [39]. An HAA is a tuple $\mathcal{A} = \langle \Sigma, Q, q_0, \delta, \mathcal{F} \rangle$, where Σ is a finite input alphabet, Q is a finite set of states, $q_0 \in Q$ is the initial state, $\delta : Q \times \Sigma \to \mathbb{B}^+(Q)$ is the transition function, with $\mathbb{B}^+(Q)$ being the set of positive Boolean formulas over Q (we also allow the formulas true and false), and the acceptance condition $\mathcal{F}$ is encoded as an ordered set $\mathcal{F} = \{(Q_1, F_1, t_1), \ldots, (Q_h, F_h, t_h)\}$ of *strata*, where $Q_i \subseteq Q$, $F_i \subseteq Q_i$, and $t_i \in \{\mathsf{b}, \mathsf{c}, \mathsf{t}\}$. Each stratum (Q_i, F_i, t_i) is classified either as transient ($t_i = \mathsf{t}$) or *Büchi* ($t_i = \mathsf{b}$) or *coBüchi* ($t_i = \mathsf{c}$). Moreover, we require that the components $Q_1, \ldots, Q_k$ form a partition of Q and moves from states in Q_i lead to states in components Q_j so that $j \geq i$ (*partial-order requirement*): formally, for each $(q, \sigma) \in Q_i \times \Sigma$, $\delta(q, \sigma)$ contains only states in components Q_j with $j \geq i$. Additionally, for each component Q_i and $(q, \sigma) \in Q_i \times \Sigma$, the following holds (*hesitant requirement*):

- if Q_i is transient, $\delta(q, \sigma)$ has no states in Q_i;
- if Q_i is Büchi, each conjunct in the disjunctive normal form of $\delta(q, \sigma)$ contains at most one state in Q_i;
- if Q_i is coBüchi, each disjunct in the conjunctive normal form of $\delta(q, \sigma)$ contains at most one state in Q_i.

Intuitively, when $\mathcal{A}$ is in state q, reading the symbol $\sigma \in \Sigma$, then $\mathcal{A}$ chooses a set of states $\{q_1, \ldots, q_k\}$ satisfying $\delta(q, \sigma)$ and splits in k copies such that the ith copy moves to the next input symbol in state q_i. Formally, a run over an infinite word $w \in \Sigma^\omega$ is a $Q \times \mathbb{N}$-labeled tree T_r such that the root is labeled by $(q_0, 0)$ and for each T_r-node x with label $(q, i) \in Q \times \mathbb{N}$ (describing a copy of $\mathcal{A}$ in state q which reads $w(i)$), there is a (possibly empty) set $H = \{q_1, \ldots, q_k\} \subseteq Q$ satisfying $\delta(q, w(i))$ such that x has k children $x_1, \ldots, x_k$, and for $\ell \in [1, k]$, x_ℓ has label $(q_\ell, i + 1)$.

The hesitant and partial-order requirements ensure that every infinite path π of the run gets trapped in some Büchi or coBüchi component. Then, the run T_r is accepting if for every infinite path π, denoting with Q_i the Büchi/coBüchi component in which π gets trapped, π satisfies the Büchi/coBüchi acceptance condition F_i associated with Q_i: formally, π visits infinitely (resp., finitely) many times nodes labeled by states in F_i if $t_i = \mathsf{b}$ (resp., $t_i = \mathsf{c}$). We denote by $\mathcal{L}(\mathcal{A})$ the set of inputs $w \in \Sigma^\omega$ such that there is an accepting run over w. The dual $\widetilde{\mathcal{A}}$ of $\mathcal{A}$ is the HAA obtained from $\mathcal{A}$ by dualizing the transition function and by converting each Büchi (resp., coBüchi) stratum into a coBüchi (resp., Büchi) stratum. The *depth* of $\mathcal{A}$ is the number of $\mathcal{A}$-components. A 1-letter HAA is an HAA over a singleton alphabet. It is known [39] that nonemptiness of one-letter can be solved efficiently. In particular, we exploit the following known results.

Proposition 10. [39] *Given an HAA $\mathcal{A}$, the dual $\widetilde{\mathcal{A}}$ of $\mathcal{A}$ is an HAA accepting the complement of $\mathcal{L}(\mathcal{A})$. Moreover, nonemptiness of 1-letter HAA with n states and depth k can be solved in space $O(k \cdot \log^2 n)$.*

Translation into HAA. For a finite Kripke structure K and an InqLTL formula φ, we denote by $\mathsf{mp}(K)$ the set of macro-paths of K and by $\mathsf{mp}(K, \varphi)$ the set of macro-paths ρ of K such that $\rho \models_K \varphi$. In the following, for the given finite Kripke structure K, we consider HAA over the alphabet 2^S, where S is the set of K-states. The following result is straightforward.

Proposition 11. *Let K be a finite Kripke structure with set of states S and $\mathcal{A}$ be an HAA over 2^S with n states and depth k. Then, one can construct in time $O(n + 2^{|S|})$ an HAA $\mathcal{A}'$ with depth $k + 2$ such that $\mathcal{L}(\mathcal{A}') = \mathcal{L}(\mathcal{A}) \cap \mathsf{mp}(K)$.*

By [24,51], given an HAA with n states, one can construct an equivalent Büchi nondeterministic word automaton (NWA) in time $2^{O(n \cdot \log n)}$. Moreover, a Büchi NWA corresponds to an HAA with just one Büchi stratum. Thus, since Büchi NWA are closed under projection and intersection, by Proposition 10, we easily obtain the following result which allows to handle intuitionistic implication under the macro-path semantics.

Proposition 12. *Let K be a finite Kripke structure with set of states S and for each $i = 1, 2$, let φ_i be an InqLTL formula and $\mathcal{A}_i$ be an HAA with n_i states accepting $\mathsf{mp}(K, \varphi_i)$. Then, one can construct in time $2^{O(n)}$, where $n = n_1 \log n_1 + n_2 \log n_2 + |S|$, an HAA $\mathcal{A}$ with depth $O(1)$ such that $\mathcal{L}(\mathcal{A}) = \mathsf{mp}(K, \varphi_1 \to \varphi_2)$.*

Intuitionistic negation can be managed by a generalization of the standard automata-theoretic approach for LTL.

Proposition 13. *Given a finite Kripke structure K with set of states S and an InqLTL formula φ, one can build in time $2^{O(|\varphi|+|S|)}$ an HAA $\mathcal{A}$ with depth 1 s.t. $\mathcal{L}(\mathcal{A}) = \mathsf{mp}(K, \neg\varphi)$.*

By exploiting Propositions 11–13, we deduce the following result.

Proposition 14. *Let $k \geq 0$, K be a finite Kripke structure with set of states S, and φ be an InqLTL_k formula. Then, one can construct in time $\mathsf{Tower}_2(k + 1, |S| + |\varphi|)$ an HAA with depth $O(|\varphi|)$ accepting $\mathsf{mp}(K, \varphi)$.*

Proof. The proof is by induction on $k \geq 0$. Let $FS(\varphi)$ be the set of subformulas ψ of φ such that some occurrence of ψ is not preceded by the connectives in $\{\neg, \to\}$ in the syntax tree of φ. Moreover, let $H_\neg$ be the set of formulas in $FS(\varphi)$ of the form $\neg\theta$, and $H_\to$ the set of formulas in $FS(\varphi)$ of the form $\theta_1 \to \theta_2$. Note that $H_\to = \emptyset$ if $k = 0$. By Proposition 13, for each $\psi \in H_\neg$, one can construct in time $2^{O(|S|+|\psi|)}$ an HAA $\mathcal{A}_\psi$ accepting $\mathsf{mp}(K, \psi)$. Moreover, if $k > 0$, then by the induction hypothesis and Proposition 12, for each $\psi \in H_\to$, one can construct in time $\mathsf{Tower}_2(k + 1, |S| + |\psi|)$ an HAA $\mathcal{A}_\psi$ accepting $\mathsf{mp}(K, \psi)$. Then, by an easy generalization of the standard linear-time translation of LTL formulas into Büchi alternating word automata and by using the HAA $\mathcal{A}_\psi$ with $\psi \in H_\neg \cup H_\to$, one can construct in time $\mathsf{Tower}_2(k + 1, O(|S| + |\varphi|))$ an HAA $\mathcal{A}_\varphi$ such that $\mathcal{L}(\mathcal{A}_\varphi) \cap \mathsf{mp}(K) = \mathsf{mp}(K, \varphi)$. Hence, by Proposition 11, the result follows. Intuitively, given an input macro-path of K, each copy of $\mathcal{A}_\varphi$ keeps track of the current subformula in $FS(\varphi)$ which needs to be evaluated. The evaluation simulates the macro-path semantics of InqLTL, but when the current subformula ψ is in $H_\neg \cup H_\rightarrow$, then the current copy of $\mathcal{A}_\varphi$ activates a copy of $\mathcal{A}_\psi$ in the initial state. Formally, for each $\psi \in H_\neg \cup H_\rightarrow$, let $\mathcal{A}_\psi = \langle 2^S, Q_\psi, q_\psi, \delta_\psi, \mathcal{F}_\psi \rangle$. W.l.o.g., we assume that the state sets of the HAA $\mathcal{A}_\psi$ are pairwise distinct. Then, $\mathcal{A}_\varphi = \langle 2^S, Q, q_0, \delta, \mathcal{F} \rangle$, where:

- $Q := FS(\varphi) \cup \bigcup_{\psi \in H_\neg \cup H_\rightarrow} Q_\psi$ and $q_0 = \varphi$;
- The transition function δ is defined as follows: $\delta(q, \sigma) = \delta_\psi(q, \sigma)$ if $q \in Q_\psi$ for some $\psi \in H_\neg \cup H_\rightarrow$. If instead $q \in FS(\varphi)$, then $\delta(q, \sigma)$ is inductively defined as follows:
  - $\delta(p, \sigma) = \mathsf{true}$ if $p \in Lab(s)$ for each $s \in \sigma$, and $\delta(p, \sigma) = \mathsf{false}$ otherwise;
  - $\delta(\phi_1 \otimes \phi_2, \sigma) = \delta(\phi_1, \sigma) \vee \delta(\phi_2, \sigma)$;
  - $\delta(\phi_1 \wedge \phi_2, \sigma) = \delta(\phi_1, \sigma) \wedge \delta(\phi_2, \sigma)$;
  - $\delta(\mathsf{X}\phi, \sigma) = \phi$;
  - $\delta(\phi_1 \mathsf{U}\phi_2, \sigma) = \delta(\phi_2, \sigma) \vee (\delta(\phi_1, \sigma) \wedge \phi_1 \mathsf{U}\phi_2)$;
  - $\delta(\phi_1 \mathsf{R}\phi_2, \sigma) = \delta(\phi_2, \sigma) \wedge (\delta(\phi_1, \sigma) \vee \phi_1 \mathsf{R}\phi_2)$;
  - for each $\psi \in H_\neg \cup H_\rightarrow$, $\delta(\psi, \sigma) = \delta(q_\psi, \sigma)$.
- $\mathcal{F} = \bigcup\limits_{\psi \in H_\neg \cup H_\rightarrow} \mathcal{F}_\psi \cup \bigcup\limits_{\phi \in FS(\varphi)} \{\mathcal{S}_\phi\}$, where for each $\phi \in FS(\varphi)$, the stratum $\mathcal{S}_\phi$ is defined as:
  - if ϕ has the form $\psi_1 \mathsf{U}\psi_2$, then $\mathcal{S}_\phi$ is the Büchi stratum $(\{\phi\}, \emptyset, \mathsf{b})$;
  - if ϕ has the form $\psi_1 \mathsf{R}\psi_2$, then $\mathcal{S}_\phi$ is the coBüchi stratum $(\{\phi\}, \emptyset, \mathsf{c})$;
  - otherwise, $\mathcal{S}_\phi$ is the transient stratum $(\{\phi\}, \emptyset, \mathsf{t})$. $\square$

Let $k \geq 0$, $K = \langle S, S_0, R, Lab \rangle$ be a finite Kripke structure with initial macro-path ρ_0, and φ be a left-positive InqLTL_k formula. By Corollary 1 and Proposition 14, $\mathcal{L}(K) \models \varphi$ iff $\rho_0 \in \mathcal{L}(\mathcal{A}_\varphi)$, where $\mathcal{A}_\varphi = \langle 2^S, Q, q_0, \delta, \mathcal{F} \rangle$ is the HAA of Proposition 14 accepting $\mathsf{mp}(K, \varphi)$. We construct a 1-letter HAA $\mathcal{A}'_\varphi$ which simulates the behaviour of $\mathcal{A}_\varphi$ over ρ_0 and accepts iff $\mathcal{A}_\varphi$ accepts ρ_0. Formally, $\mathcal{A}'_\varphi = \langle \{1\}, Q \times 2^S, (q_0, S_0), \delta', \mathcal{F}' \rangle$, where:
- for all $(q, T) \in Q \times 2^S$, $\delta'((q, T), 1)$ is obtained from $\delta(q, T)$ by replacing each state q' occurring in $\delta(q, T)$ with (q', T'), where $T' := \{s' \in S \mid s' \in R[s] \text{ for some } s \in T\}$;
- $\mathcal{F}'$ is obtained from $\mathcal{F}$ by replacing each stratum $(Q', F', t) \in \mathcal{F}$ with $(Q' \times 2^S, F' \times 2^S, t)$.

By Proposition 14, $\mathcal{A}'_\varphi$ has depth $O(|\varphi|)$ and size $\mathsf{Tower}_2(k+1, |S|+|\varphi|)$. Hence, by Proposition 10, we obtain the following result, where for the lower-bounds, we provide a detailed proof in [5]. Note that InqLTL_0 allows an unrestricted use of intuitionistic negation and subsumes the split-disjunction-free fragment of $\mathsf{TeamLTL}$ for which model checking is known to be in PSPACE [38].

Theorem 2. *For each $k \geq 0$, model checking of left-positive InqLTL_k is k-EXPSPACE-complete. In particular, model checking of InqLTL_0 is PSPACE-complete.*

6   Conclusion

We have introduced InqLTL, a team semantics for LTL inspired by inquisitive logic. The logic replaces the split disjunction of TeamLTL with inquisitive disjunction and intuitionistic implication. We have shown that, when enhanced with strong negation $\sim$, the logic has the countable model property, is highly undecidable, and is strictly less expressive than TeamLTL($\sim$). We then have identified a fragment of InqLTL, called left-positive InqLTL, with a decidable model checking problem, which does not allow for nesting of implication in the left side of an implication. We have illustrated how left-positive InqLTL can capture meaningful classes of hyperproperties such as information-flow security properties. It is worth noting that the decidability status of model checking TeamLTL and fragments of TeamLTL($\sim$) is a challenging open question. The known positive results [38,55] have been achieved by imposing drastic syntactical or semantical restrictions. This paper shows, for the first time, a hyper logic, namely left-positive InqLTL, with unrestricted use of temporal modalities and universal second-order quantification over traces with a decidable model-checking problem. Our results were obtained with non-trivial proof techniques. The abstraction technique introduced for left-positive InqLTL is a notable contribution in its own right, with promising potential for generalization to other logics such as TeamLTL and unrestricted InqLTL.

A possible direction for future research involves analysing the complexity of model-checking within more constrained fragments, such as those limited to unary temporal operators. Moreover, we plan to investigate extensions or variants of InqLTL for the specification of asynchronous hyperproperties where traces of a team progress with different speed. Finally, a natural direction is to develop branching-time and strategic counterparts of InqLTL. Existing intuitionistic branching-time and strategic logics, such as ICTL and IATL (the intuitionistic version of the logic CTL and ATL, respectively), provide relevant semantic and algorithmic starting points for combining temporal branching/strategic quantification with constructive notions of information growth [8,9].

Acknowledgments. We would like to thank the reviewers of several subsequent versions of the paper for their comments, and Katsuhiko Sano for some inspiring early discussions. Laura Bozzelli, Tadeusz Litak and Aniello Murano would like to acknowledge the support of the PNRR MUR projects PE0000013-FAIR and ECS00000037-MUSA.

Disclosure of Interests. The authors have no competing interests to declare that are relevant to the content of this article.

References

1. Samson Abramsky and Jouko A. Väänänen. From IF to BI. a tale of dependence and separation. *Synth.*, 167(2):207–230, 2009. URL: https://doi.org/10.1007/s11229-008-9415-6, doi:10.1007/S11229-008-9415-6.

2. Philippe Balbiani, Joseph Boudou, Martín Diéguez, and David Fernández-Duque. Intuitionistic linear temporal logics. *ACM Trans. Comput. Log.*, 21(2):14:1–14:32, 2020. `doi:10.1145/3365833`.

3. J. Baumeister, N. Coenen, B. Bonakdarpour, B. Finkbeiner, and C. Sánchez. A Temporal Logic for Asynchronous Hyperproperties. In *Proc. 33rd CAV*, volume 12759 of *LNCS 12759*, pages 694–717. Springer, 2021. `doi:10.1007/978-3-030-81685-8_33`.

4. B. Bittner, M. Bozzano, A. Cimatti, M. Gario, S. Tonetta, and V. Vozárová. Diagnosability of fair transition systems. *Artif. Intell.*, 309:103725, 2022. `doi:10.1016/J.ARTINT.2022.103725`.

5. L. Bozzelli, T. Litak, M. Mittelmann, and A. Murano. Inquisitive team semantics of LTL. *CoRR*, abs/2505.10700, 2025. `doi:10.48550/ARXIV.2505.10700`.

6. L. Bozzelli, B. Maubert, and S. Pinchinat. Unifying Hyper and Epistemic Temporal Logics. In *Proc. 18th FoSSaCS*, LNCS 9034, pages 167–182. Springer, 2015. `doi:10.1007/978-3-662-46678-0_11`.

7. L. Bozzelli, A. Peron, and C. Sánchez. Asynchronous Extensions of HyperLTL. In *Proc. 36th LICS*, pages 1–13. IEEE, 2021. `doi:10.1109/LICS52264.2021.9470583`.

8. Laura Bozzelli, Andrea Capone, Davide Catta, and Aniello Murano. An Intuitionistic Version of Alternating-Time Temporal Logic. In *Proceedings of the 22nd International Conference on Principles of Knowledge Representation and Reasoning*, pages 185–195, 10 2025. `doi:10.24963/kr.2025/19`.

9. Andrea Capone, Laura Bozzelli, Davide Catta, Vadim Malvone, and Aniello Murano. An intuitionistic version of computation tree logic. In *EUMAS 2025*, To appear, 2025.

10. Alexander Chagrov and Michael Zakharyaschev. *Modal Logic*. Number 35 in Oxford Logic Guides. Clarendon Press, 1997.

11. I. Ciardelli. *Questions in Logic*. PhD thesis, University of Amsterdam, 2016.

12. I. Ciardelli and F. Roelofsen. Inquisitive logic. *J. Philos. Log.*, 40(1):55–94, 2011. `doi:10.1007/S10992-010-9142-6`.

13. Ivano Ciardelli. Dependency as question entailment. In Samson Abramsky, Juha Kontinen, Jouko Väänänen, and Heribert Vollmer, editors, *Dependence Logic, Theory and Applications*, pages 129–181. Springer, 2016. `doi:10.1007/978-3-319-31803-5_8`.

14. Ivano Ciardelli. Propositional inquisitive logic: a survey. *Comput. Sci. J. Moldova*, 24(3):295–311, 2016.

15. Ivano Ciardelli. Describing neighborhoods in inquisitive modal logic. In David Fernández-Duque, Alessandra Palmigiano, and Sophie Pinchinat, editors, *Advances in Modal Logic, AiML 2022, Rennes, France, August 22-25, 2022*, pages 217–236. College Publications, 2022. URL: `http://www.aiml.net/volumes/volume14/16-Ciardelli.pdf`.

16. Ivano Ciardelli. *Inquisitive Logic: Consequence and Inference in the Realm of Questions*. Springer International Publishing, 2022. `doi:10.1007/978-3-031-09706-5`.

17. Ivano Ciardelli and Gianluca Grilletti. Coherence in inquisitive first-order logic. *Annals of Pure and Applied Logic*, 173(9):103–155, October 2022. `doi:10.1016/j.apal.2022.103155`.

18. Ivano Ciardelli and Martin Otto. Bisimulation in inquisitive modal logic. In *TARK*, volume 251 of *EPTCS*, pages 151–166, 2017.

19. Ivano A Ciardelli. Inquisitive semantics and intermediate logics. Master's thesis, Universiteit van Amsterdam, 2009.

20. Ivano A. Ciardelli and Floris Roelofsen. Inquisitive dynamic epistemic logic. *Synthese*, 192(6):1643–1687, 2015. `doi:10.1007/s11229-014-0404-7`.

21. M.R. Clarkson, B. Finkbeiner, M. Koleini, K.K. Micinski, M.N. Rabe, and C. Sánchez. Temporal Logics for Hyperproperties. In *Proc. 3rd POST*, LNCS 8414, pages 265–284. Springer, 2014. `doi:10.1007/978-3-642-54792-8_15`.

22. M.R. Clarkson and F.B. Schneider. Hyperproperties. *Journal of Computer Security*, 18(6):1157–1210, 2010. `doi:10.3233/JCS-2009-0393`.

23. N. Coenen, B. Finkbeiner, C. Hahn, and J. Hofmann. The hierarchy of hyperlogics. In *Proc. 34th LICS*, pages 1–13. IEEE, 2019. `doi:10.1109/LICS.2019.8785713`.

24. C. Dax and F. Klaedtke. Alternation elimination by complementation (extended abstract). In *Proc. 15th LPAR*, LNCS 5330, pages 214–229. Springer, 2008. `doi:10.1007/978-3-540-89439-1_16`.

25. E.A. Emerson and J.Y. Halpern. "Sometimes" and "Not Never" revisited: on branching versus linear time temporal logic. *J. ACM*, 33(1):151–178, 1986. `doi:10.1145/4904.4999`.

26. Gilda Ferreira and Paulo Oliva. On the relation between various negative translations. In Ulrich Berger and Helmut Schwichtenberg, editors, *Logic, Construction, Computation*, volume 3 of *Mathematical Logic Series*, pages 227–258. Ontos-Verlag, 2012.

27. B. Finkbeiner, M.N. Rabe, and C. Sánchez. Algorithms for Model Checking HyperLTL and HyperCTL*. In *Proc. 27th CAV Part I*, volume 9206 of *LNCS 9206*, pages 30–48. Springer, 2015. `doi:10.1007/978-3-319-21690-4_3`.

28. M.J. Fischer and R.E. Ladner. Propositional Dynamic Logic of Regular Programs. *J. Comput. Syst. Sci.*, 18(2):194–211, 1979. `doi:10.1016/0022-0000(79)90046-1`.

29. H. Frenkel and M. Zimmermann. The complexity of second-order HyperLTL. In *Proc. 33rd CSL*, volume 326 of *LIPIcs*, pages 10:1–10:23. Schloss Dagstuhl - Leibniz-Zentrum für Informatik, 2025. `doi:10.4230/LIPICS.CSL.2025.10`.

30. J.A. Goguen and J. Meseguer. Security Policies and Security Models. In *IEEE Symposium on Security and Privacy*, pages 11–20. IEEE Computer Society, 1982. `doi:10.1109/SP.1982.10014`.

31. Gianluca Grilletti. Disjunction and existence properties in inquisitive first-order logic. *Studia Logica*, 107(6):1199–1234, 2019.

32. Gianluca Grilletti. Completeness for the classical antecedent fragment of inquisitive first-order logic. *Journal of Logic, Language and Information*, 30(4):725–751, October 2021. URL: `http://doi.org/10.1007/s10849-021-09341-y`.

33. J.Oliver. Gutsfeld, A. Meier, C. Ohrem, and J. Virtema. Temporal Team Semantics Revisited. In *Proc. 37th LICS*, pages 44:1–44:13. ACM, 2022. `doi:10.1145/3531130.3533360`.

34. J.Oliver. Gutsfeld, M. Müller-Olm, and C. Ohrem. Propositional dynamic logic for hyperproperties. In *Proc. 31st CONCUR*, LIPIcs 171, pages 50:1–50:22. Schloss Dagstuhl - Leibniz-Zentrum für Informatik, 2020. `doi:10.4230/LIPIcs.CONCUR.2020.50`.

35. J.Y. Halpern and K.R. O'Neill. Secrecy in multiagent systems. *ACM Trans. Inf. Syst. Secur.*, 12(1), 2008.

36. J.Y. Halpern and M.Y. Vardi. The Complexity of Reasoning about Knowledge and Time: Extended Abstract. In *Proc. 18th STOC*, pages 304–315. ACM, 1986. `doi:10.1145/12130.12161`.

37. J. Kontinen, M. Sandström, and J. Virtema. Set semantics for asynchronous teamltl: Expressivity and complexity. *Inf. Comput.*, 304:105299, 2025. `doi:10.1016/J.IC.2025.105299`.

38. A. Krebs, A. Meier, J. Virtema, and M. Zimmermann. Team Semantics for the Specification and Verification of Hyperproperties. In *Proc. 43rd MFCS*, LIPIcs

117, pages 10:1–10:16. Schloss Dagstuhl - Leibniz-Zentrum für Informatik, 2018. `doi:10.4230/LIPIcs.MFCS.2018.10`.

39. O. Kupferman, M.Y. Vardi, and P. Wolper. An Automata-Theoretic Approach to Branching-Time Model Checking. *J. ACM*, 47(2):312–360, 2000. `doi:10.1145/333979.333987`.

40. Tadeusz Litak, Miriam Polzer, and Ulrich Rabenstein. Negative Translations and Normal Modality. In Dale Miller, editor, *Proceedings of FSCD 2017*, volume 84 of *LIPIcs*, pages 27:1–27:18, Dagstuhl, Germany, 2017. Schloss Dagstuhl–Leibniz-Zentrum fuer Informatik. URL: `http://drops.dagstuhl.de/opus/volltexte/2017/7741`, `doi:10.4230/LIPIcs.FSCD.2017.27`.

41. Tadeusz Litak and Katsuhiko Sano. Bounded Inquisitive Logics: Sequent Calculi and Schematic Validity, 2025. A modified and expanded version of a paper accepted for TABLEAUX 2025. URL: `https://arxiv.org/abs/2507.13946`, `arXiv:2507.13946`.

42. M. Lück. On the complexity of linear temporal logic with team semantics. *Theor. Comput. Sci.*, 837:1–25, 2020. `doi:10.1016/j.tcs.2020.04.019`.

43. Stipe Marić and Tin Perkov. Decidability of inquisitive modal logic via filtrations. *Studia Logica*, 2024. `doi:10.1007/s11225-024-10134-0`.

44. J. McLean. A General Theory of Composition for a Class of "Possibilistic" Properties. *IEEE Trans. Software Eng.*, 22(1):53–67, 1996. `doi:10.1109/32.481534`.

45. Karl Nygren. Free choice in modal inquisitive logic. *J. Philos. Log.*, 52(2):347–391, 2023.

46. A. Pnueli. The Temporal Logic of Programs. In *Proc. 18th FOCS*, pages 46–57. IEEE Computer Society, 1977. `doi:10.1109/SFCS.1977.32`.

47. M.N. Rabe. *A temporal logic approach to information-flow control*. PhD thesis, Saarland University, 2016.

48. H. Rogers. *Theory of Recursive Functions and Effective Computability*. MIT Press, Cambridge,MA, US, 1987.

49. A. Sabelfeld and D. Sands. Dimensions and principles of declassification. In *Proc. 18th CSFW*, pages 255–269. IEEE Computer Society, 2005. `doi:10.1109/CSFW.2005.15`.

50. M. Sampath, R. Sengupta, S. Lafortune, K. Sinnamohideen, and D. Teneketzis. Diagnosability of discrete-event systems. *IEEE Trans. Autom. Control.*, 40(9):1555–1575, 1995. `doi:10.1109/9.412626`.

51. C. Sánchez and J. Samborski-Forlese. Efficient regular linear temporal logic using dualization and stratification. In *Proc. 19th TIME*, pages 13–20. IEEE Computer Society, 2012. `doi:10.1109/TIME.2012.25`.

52. A.P. Sistla, M.Y. Vardi, and P. Wolper. The Complementation Problem for Büchi Automata with Applications to Temporal Logic. *Theoretical Computer Science*, 49:217–237, 1987. `doi:10.1016/0304-3975(87)90008-9`.

53. Morten Heine Sørensen and Pawel Urzyczyn. *Lectures on the Curry-Howard Isomorphism*, volume 149 of *Stud. Logic Found. Math.* Elsevier Science Inc., 2006.

54. Anne S. Troelstra and Dirk van Dalen. *Constructivism in Mathematics: An Introduction*, volume 121 of *Studies in Logic and the Foundations of Mathematics*. Elsevier, 1988.

55. J. Virtema, J. Hofmann, B. Finkbeiner, J. Kontinen, and F. Yang. Linear-Time Temporal Logic with Team Semantics: Expressivity and Complexity. In *Proc. 41st IARCS FSTTCS*, LIPIcs 213, pages 52:1–52:17. Schloss Dagstuhl - Leibniz-Zentrum für Informatik, 2021. `doi:10.4230/LIPIcs.FSTTCS.2021.52`.

56. S. Zdancewic and A.C. Myers. Observational Determinism for Concurrent Program Security. In *Proc. 16th IEEE CSFW-16*, pages 29–43. IEEE Computer Society, 2003. `doi:10.1109/CSFW.2003.1212703`.

Synthesising Asynchronous Automata from Fair Specifications

Béatrice Bérard[1], Benjamin Monmege[2], B Srivathsan[3,4], and Arnab Sur[3]

[1] Sorbonne Université, CNRS, LIP6, F-75005 Paris, France
`beatrice.berard@lip6.fr`
[2] Aix-Marseille Univ, CNRS, LIS, Marseille, France
`benjamin.monmege@univ-amu.fr`
[3] Chennai Mathematical Institute, Chennai, India
`{sri, arnabs}@cmi.ac.in`
[4] CNRS IRL 2000, ReLaX, Chennai, India

Abstract. Asynchronous automata are a model of distributed finite state processes synchronising on shared actions. A celebrated result by Zielonka shows how a deterministic asynchronous automaton (AA) can be synthesised, starting from two inputs: a global specification given as a deterministic finite-state automaton (DFA) and a distribution of the alphabet into local alphabets for each process. The DFA to AA translation is particularly complex and has been revisited several times, with no complete prototype tool provided for the full construction. In this work, we revisit this construction on a restricted class of "fair" specifications: a DFA describes a fair specification if in every loop, all processes participate in at least one action — so, no process is starved. For fair specifications, we present a new construction to synthesise an AA. Our construction results in an AA where every process has a number of local states that is linear in the number of states of the DFA, and where the only exponential explosion is related to a fairness parameter: the length of the longest word that can be read in the DFA in which not every process participates. We have implemented a prototype tool showing how it can be applied to some examples, in particular a concrete one: the dining philosophers problem. Finally, we show how this construction can be combined with an existing construction for hierarchical process architectures, in order to relax the fairness assumption.

1 Introduction

Asynchronous automata (AA) are a foundational model for distributed systems, where independent processes synchronise on shared actions. These models enable the design and analysis of systems with decentralised control, making them critical in theoretical computer science and applications like parallel computing and distributed algorithms. Figure 1 gives an example (from [17]) of an AA. It contains two processes p_1, p_2 which control letters $\{a, c\}$ and $\{b, c\}$ respectively: so a is local to p_1, b is local to p_2, and c is shared by p_1 and p_2. The key point is the synchronisation mechanism on shared actions. Here, the AA is defined in

N. Bertrand and S. Milius (Eds.): FoSSaCS 2026, LNCS 16503, pp. 133–152, 2026.
https://doi.org/10.1007/978-3-032-22730-0_7

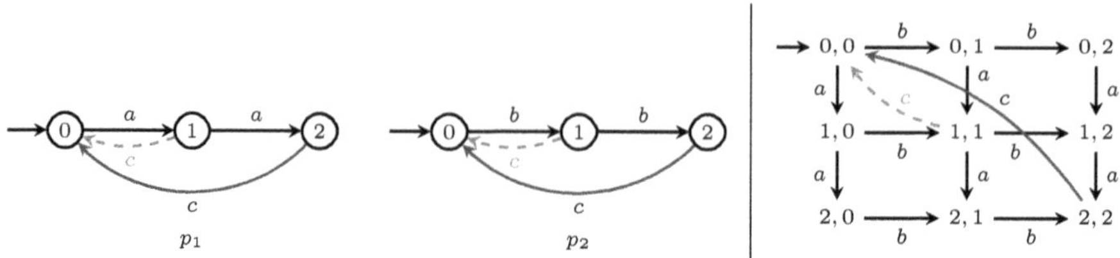

Fig. 1. Left: An AA. Right: its semantics seen as a DFA

a way that the two red-dashed c transitions can synchronise, as well as the two blue c transitions. However, the red-dashed c in one cannot be done together with the blue c of the other. The behaviour of the AA is described as a deterministic finite-state automaton (DFA) on the right. The language of this AA is $((([ab] + [aabb]).c)^*$ where $[ab]$ stands for either ab or ba, and $[aabb]$ denotes the set of words obtained by shuffling two a's and two b's.

From an AA, it is straightforward to construct a language-equivalent DFA, as shown in the example. The reverse process — that is, given a DFA (global specification), construct a language-equivalent AA (distributed implementation) — is highly non-trivial. A cornerstone result is Zielonka's theorem, which provides a method for synthesising a (deterministic) AA from a given DFA, and an alphabet distribution among processes [21]. However, Zielonka's construction is known to be extremely complex and has been the subject of numerous refinements and optimisations (e.g. [8,18,12]), leading to an optimal Zielonka-type construction [11]. In all these constructions, each process keeps track of what it believes to be the latest "information" available to every other process. When there is a shared action, the processes share their local information, reconcile them and update their local states. This underlying idea is implemented using a *gossip automaton* in [18,17], and so-called *zones* in [12,11]. This idea yields a number of local states that is exponential in the number of processes (whereas the construction can be kept polynomial in the number of states of the DFA).

Several solutions have been considered in order to avoid the resolution of this *gossip problem*. For instance, compositional methods have been used in order to construct non-deterministic AA [4,19,5], still requiring a number of states exponential in the number of processes. Another solution, leading to a quadratic number of states, is to restrict the topology of communications to be acyclic, and with any letter synchronising at most two processes [14]. This special case has recently been extended to remove the restriction on the number of processes synchronising on each letter, and to allow for a reconfiguration of the topology along the executions [13]. Another recent work [1] has proposed a logic-based route for Zielonka's theorem by going through a local, past-oriented fragment of propositional dynamic logic. The AA is obtained by a cascade product of localised AA, which essentially operate on a single process.

Our goal in this work is to find a "meaningful" restriction on the DFA specifications which can result in a conceptually simpler and more efficient AA synthesis, that can be implemented in a prototype tool. Indeed, there are very few attempts of implementations of the synthesis algorithm based on Zielonka's theorem, partly due to its intricacy. The work of [20] proposes a notion of safe asynchronous automata and states that the problem of synthesizing a safe AA whose language is contained in a DFA specification is undecidable. They go on to study syntactic conditions on DFAs which make them equivalent to safe AA and then propose a method to find sub-automata inside the DFA specification that satisfy these conditions. From such sub-automata, safe AA are synthesized using a modified Zielonka's procedure, which is implemented and experimented. Another implementation has been proposed in [2] for a fragment of asynchronous automata that are called *realistic*, that is both deadend-free and locally accepting. Both studies define restrictions on the DFA specification which are orthogonal to those we consider.

In this article, we revisit AA synthesis with a focus on *fair* specifications. As an example, consider the dining philosophers problem, introduced in [10]. The problem is modeled for n philosophers (the processes), sitting around a table with a single chopstick lying between two consecutive philosophers. A philosopher's actions (picking up/putting down the left/right chopstick) are shared actions, involving the philosopher and its two neighbors. The description of the authorised sequence of actions, as well as the objective for all philosophers to eat (i.e. have both chopsticks at some point in the future, before releasing them), can be given as a DFA where each state stores the current status of each philosopher. One of the objectives is to design a distributed algorithm to avoid deadlock situations (where nothing else can be done), and the starvation of any of the philosophers. In our setting, we want to design an AA where every philosopher eats infinitely often, which is traditionally called *fairness* in the realm of infinite words (every process that wants to play infinitely often should be scheduled infinitely often). We strengthen the requirement to *bounded-fairness* (though we will drop the *bounded-* in the rest of the article) where we not only want that every process plays infinitely often, but also frequently enough. Scheduling policies like round-robin scheduling impose bounded-fairness naturally (see Section 2 of [3], for instance). Similar bounds are also frequently used in timed systems to require, for instance, bounded response times.

From a DFA perspective, this bound on fairness requires that in every loop of the DFA, every process participates (and thus in the long run, no process is starved). The DFA in Figure 1 is fair, since every loop is closed on c, which is a global action where both processes participate. We introduce a natural number k to measure the quality of fairness: a DFA is k-fair whenever every word of length k that can be read in the DFA makes every process participate at least once. The DFA in Figure 1 is 3-fair. Notice that this parameter k is a priori independent of the number of processes and the size of the DFA: there are examples (see Example 10 for one) of DFA with arbitrary size and with a distribution of actions among arbitrarily many processes that are k-fair for a fixed parameter k.

After studying the fairness of trace languages in Section 3, we introduce a novel construction that simplifies the synthesis process for fair specifications in

Section 4. A crucial technical ingredient of our construction is the notion of Foata Normal Form (FNF) [7]. Our construction does not require the use of gossip automata, or other sophisticated tools used in other constructions. As a result, our approach ensures that the resulting AA is linear in the size of the DFA, polynomial in the size of the alphabet, and the only exponential explosion is related to the parameter k. The complexity is independent of the number of processes. We can even extend our synthesis procedure to synthesise an AA recognising all k-fair traces of a (potentially non-fair) DFA specification: in the dining philosophers problem, this allows us to ensure that every philosopher is active from time to time. We present a tool implementing our synthesis procedures in Section 6.

We finally weaken the fairness restriction in Section 5, by showing that our new construction can be combined with results known for hierarchical process architectures (as studied in [14,13]). In these architectures, the set of processes is organised as a tree where each process can communicate only with its parent or its children in the tree. We show that each node in this tree can be enlarged into a *bag* of processes. If the DFA restricted to the alphabet of each of these bags is fair, then our fairness construction can be coupled with the construction known for hierarchical architectures. As a possible application, we can imagine a pool of servers that are communicating in a hierarchical fashion, each linked to some clients where a server and its clients should interplay in a fair fashion (ignoring what happens outside). We once again obtain a construction of polynomial complexity for a fixed value of fairness parameter k.

Detailed proofs of all results are presented in the extended version [6].

2   Preliminaries

Let Σ be a finite alphabet. A *concurrent alphabet* is a pair (Σ, I) with $I \subseteq \Sigma \times \Sigma$ the *independence relation*, being a symmetric and irreflexive relation. The corresponding dependence relation is the set $D = (\Sigma \times \Sigma) \setminus I$. A concrete independence relation can be obtained by distributing the letters into a finite set P of processes, formally defined by a function $\mathrm{loc} \colon \Sigma \to 2^P$ where $\mathrm{loc}(a)$ is the set of processes that can read the letter $a \in \Sigma$. We further define $\Sigma_p = \{a \in \Sigma \mid p \in \mathrm{loc}(a)\}$ as the alphabet of process $p \in P$. We call (Σ, loc) a *distributed alphabet*. Such a distribution naturally leads to an independence relation $I_{\mathrm{loc}} = \{(a, b) \in \Sigma \times \Sigma \mid \mathrm{loc}(a) \cap \mathrm{loc}(b) = \emptyset\}$. We extend the function loc to words by letting $\mathrm{loc}(\varepsilon) = \emptyset$, and $\mathrm{loc}(ua) = \mathrm{loc}(u) \cup \mathrm{loc}(a)$ for all $u \in \Sigma^*$ and $a \in \Sigma$.

Example 1. Let $\Sigma = \{a, b, c, d\}$ and $P = \{p_1, p_2, p_3\}$ be the processes. Consider the distribution $\mathrm{loc}(a) = \{p_1, p_2\}$, $\mathrm{loc}(b) = \{p_1, p_3\}$, $\mathrm{loc}(c) = \{p_2\}$, and $\mathrm{loc}(d) = \{p_3\}$. Then $\Sigma_{p_1} = \{a, b\}$, $\Sigma_{p_2} = \{a, c\}$ and $\Sigma_{p_3} = \{b, d\}$. The independence relation is $I_{\mathrm{loc}} = \{(a, d), (d, a), (d, c), (c, d), (c, b), (b, c)\}$.

Traces. A *trace* over a concurrent alphabet (Σ, I) is a labelled partial order $t = (\mathcal{E}, \leq, \lambda)$ where $\mathcal{E}$ is a set of *events*, $\lambda \colon \mathcal{E} \to \Sigma$ labels each event by a letter, and $\leq$ is a partial order of $\mathcal{E}$ satisfying the following conditions:

- $(\lambda(e), \lambda(f)) \notin I$ implies $e \leq f$ or $f \leq e$;
- $e \lessdot f$ implies $(\lambda(e), \lambda(f)) \notin I$ where $\lessdot\; = \;<\; \setminus\; <^2$ is the immediate successor relation induced by the partial order: $\{(e, f) \mid e < f \text{ and } \neg\exists g.\ e < g < f\}$.

The number of events in the trace t is denoted as $|t|$, and is called its *length*.

A word $w = a_0 \cdots a_{n-1} \in \Sigma^*$ gives rise to a unique trace by putting one event per position in the word, and defining the successor relation $<$ as all the pairs (i, j) of positions such that $i < j$, $(a_i, a_j) \notin I$ and there are no positions k such that $i < k < j$ and $(a_i, a_k), (a_k, a_j) \notin I$. We call this word a *linearisation* of the trace. Two words w and w' mapped to the same trace are said to be *equivalent*, denoted by $w \sim w'$. We write $[w]$ for the equivalence class of w with respect to $\sim$. In the following, we use both representations (partial orders and equivalence classes) of traces interchangeably.

Minimal (respectively, maximal) elements of a trace t are all the events that have no smaller (respectively, larger) events. The set of minimal (respectively, maximal) events of t is denoted by $\min(t)$ (respectively, $\max(t)$). Given two traces $t_1 = (\mathcal{E}_1, \leq_1, \lambda_1)$ and $t_2 = (\mathcal{E}_2, \leq_2, \lambda_2)$, the concatenation $t_1 t_2$ is the trace $(\mathcal{E}', \leq', \lambda')$ where $\mathcal{E}' = \mathcal{E}_1 \cup \mathcal{E}_2$, $\lambda'(e) = \lambda_1(e)$ if $e \in \mathcal{E}_1$, and $\lambda'(e) = \lambda_2(e)$ otherwise, and $\leq'$ is $\leq_1 \cup \leq_2 \cup \{(x, y) \mid x \text{ is maximal in } t_1, y \text{ is minimal in } t_2, \text{ and } (x, y) \notin I\}$.

Example 2. A trace over the distributed alphabet of Example 1 can be depicted as follows, where the arrows between the events denote the relation $\lessdot$:

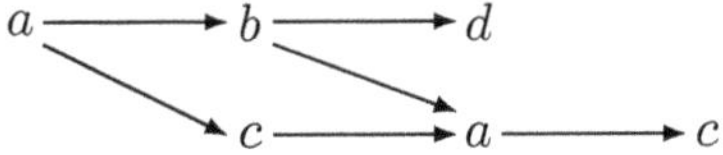

It is the trace associated with the equivalence class $[abcacd]$, that is equal, for instance, to the equivalence class $[acbdac]$. The minimal event is the one labelled by a on the left. The maximal events are those labelled by d and c on the right.

Views. For a trace $t = (\mathcal{E}, \leq, \lambda)$, a subset $J \subseteq \mathcal{E}$ is called an *ideal* of t if for all $e \in J$, and $f \in \mathcal{E}$ such that $f \leq e$, we have $f \in J$. An ideal can also be seen as a trace by keeping the same partial order and labelling as in the original trace. For a subset $X \subseteq \mathcal{E}$ of events, we let $X{\downarrow}$ be the ideal $\{f \in \mathcal{E} \mid \exists x \in X\ f \leq x\}$. From a linearisation perspective, an ideal s of a trace t is related to a prefix of one of the linearisations: there must exist a linearisation w of t that can be written as $w = uv$ where s is the trace $[u]$.

We identify special ideals called *views*: the *view* of a process p in a trace t is the ideal of t consisting of all events currently known by the process p. Formally, we let $\max_p(t)$ be the largest event of t that is labelled in Σ_p. Then, the view of process p in a trace t, denoted as $\mathrm{view}_p(t)$, is the ideal $\max_p(t){\downarrow}$. The view of a set of processes $X \subseteq P$ in a trace t is the ideal $\{\max_p(t) \mid p \in X\}{\downarrow}$, obtained as the union of the views of the processes in X.

Example 3. Consider again the trace $t = [abcacd]$ presented in Example 2. Then, $\mathrm{view}_{p_1}(t) = [abca]$, $\mathrm{view}_{p_2}(t) = [abcac]$, $\mathrm{view}_{p_3}(t) = [abd]$, and $\mathrm{view}_{\{p_1, p_3\}}(t) = [abcad]$.

Foata normal form. The Foata normal form (FNF) of a trace encodes a maximal parallel execution of the trace [7]. To define this notion formally, we use the concept of *steps* from [9]. A *step* is a non-empty subset $S \subseteq \Sigma$ of pairwise independent letters: it is sometimes called a *clique*, from the point of view of the graph of the independence relation. In a step, all letters can be executed in parallel. Observe that the set of labels of the minimal elements of a trace t provides a maximal step for t. Then, the FNF $\mathsf{F}(t)$ of a trace t is a sequence of steps $\varphi = S_1 S_2 \cdots S_m$ where S_1 is the set of minimal elements of t, and $S_2 \cdots S_m = \mathsf{F}(t')$ where t' is the trace obtained by removing all minimal elements from t. The FNF is a unique decomposition of the trace into maximal steps.

Example 4. The FNF of the trace $[abcacd]$ from Example 2 is $\{a\}\{b,c\}\{a,d\}\{c\}$.

We denote by φ_i the i-th step of the FNF φ. We shall also denote by $|t|_\mathsf{F}$, the number of steps in the FNF decomposition of the trace t, and we call it the *Foata length of t*. In order to simplify further explanations, we suppose that a step φ_i exists as the empty set, for all i greater than the length of φ: we do not depict this infinite sequence of empty sets when we give examples of FNF.

Interestingly, FNF are increasing with respect to ideals: if a trace s is an ideal of t, there is an injective correspondence from steps of s to the $|s|_\mathsf{F}$ first steps of t. Essentially, the FNF $\mathsf{F}(s)$ is obtained by deleting from $\mathsf{F}(t)$ the events not in s.

Lemma 1. *Let t be a trace and s an ideal of t. Then for all $i \leq |s|_\mathsf{F}$, $\mathsf{F}(s)_i \subseteq \mathsf{F}(t)_i$.*

Example 5. Consider again the trace $t = [abcacd]$ of Example 2. The FNF of the ideal $s = [abd]$ is $\{a\}\{b\}\{d\}$. We can extend the ideal s by the remainder of the the trace t, $[cac]$ letter by letter. The first letter c commutes with b and d but not with a and thus must belong to the second step. Arguing similarly we see that the next letter a must belong to the third step and the final letter c must be appended in a new step. We thus end up with the FNF of t as expected.

Even more interestingly, if we know the FNF of the views of two (or more) processes p_1 and p_2, it is possible to deduce the FNF of the view of $\{p_1, p_2\}$: they can accumulate their knowledge simply by taking pairwise unions of the steps. This gives us an algorithm to compute $\mathrm{view}_X(t)$ for some $X \subseteq P$, given $\mathrm{view}_p(t)$ for all $p \in X$. In the following, given two FNF φ and φ', we write $\varphi \cup \varphi'$ for the sequence of steps $(\varphi_1 \cup \varphi'_1) \cdots (\varphi_m \cup \varphi'_m)$ where m is the maximal length of φ and φ' (remember that we have added empty steps at the end of the FNF so that φ_k has a meaning, even for k greater than the length of φ).

Lemma 2. *Let t be a trace and $X \subseteq P$. Then, $\mathsf{F}(\mathrm{view}_X(t)) = \bigcup_{p \in X} \mathsf{F}(\mathrm{view}_p(t))$.*

Example 6. Continuing Example 3, we have $\mathsf{F}(\mathrm{view}_{p_1}(t)) = \{a\}\{b,c\}\{a\}$ and $\mathsf{F}(\mathrm{view}_{p_3}(t)) = \{a\}\{b\}\{d\}$, from which we can indeed deduce $\mathsf{F}(\mathrm{view}_{\{p_1,p_3\}}(t)) = \{a\}\{b,c\}\{a,d\}$.

Regular Trace-Closed Languages and Asynchronous Automata. A language $L \subseteq \Sigma^*$ is said to be *trace-closed* (for the independence relation I) if for all $w, w' \in \Sigma^*$ such that $w \in L$ and $w' \sim w$, we have $w' \in L$. A trace-closed language is *regular* if it is accepted by a finite state automaton $\mathcal{A} = (Q, \Sigma, Q_0, \Delta, Q_f)$, where Q is the set of states, Q_0 are the initial states, Q_f the final ones and $\Delta \subseteq Q \times \Sigma \times Q$ are the transitions. We denote by $q \xrightarrow{a} q'$ the transition $(q, a, q') \in \Delta$. In the following, we let $\Delta(q, w)$ be the set of states q' such that there is a sequence of transitions $q = q_0 \xrightarrow{a_0} q_1 \xrightarrow{a_1} \cdots \xrightarrow{a_{n-1}} q_n = q'$, if $w = a_0 a_1 \dots a_{n-1}$. As usual, the language recognised by $\mathcal{A}$ is the set of words w such that there exists $q \in Q_0$ with $\Delta(q, w) \cap Q_f \neq \emptyset$. Trace-closed regular languages are recognised by automata having a special syntactical property, the *diamond property*.

Definition 1. *For a concurrent alphabet (Σ, I), a finite state automaton $\mathcal{A} = (Q, \Sigma, Q_0, \Delta, Q_f)$ satisfies the* diamond property *if for all $q, q', q'' \in Q$ and $(a, b) \in I$ such that $q \xrightarrow{a} q' \xrightarrow{b} q''$, there exists $q''' \in Q$ such that $q \xrightarrow{b} q''' \xrightarrow{a} q''$.*

The diamond property is a sufficient condition for an automaton to recognise a trace-closed language. Indeed, if $\mathcal{A}$ is a finite state automaton satisfying the diamond property then for all $q \in Q$ and $w, w' \in \Sigma^*$ such that $w \sim w'$, we have $\Delta(q, w) = \Delta(q, w')$.

In particular, $\Delta(q, t)$ is well-defined even for traces t, since every linearisation of t goes to the same state. In the following, we let $L_{\mathsf{tr}}(\mathcal{A})$ be the set of traces accepted by $\mathcal{A}$.

A finite state automaton $\mathcal{A} = (Q, \Sigma, Q_0, \Delta, Q_f)$ is said to be deterministic if Q_0 is a singleton, and for all $q \in Q$, and $a \in \Sigma$, $\Delta(q, a)$ is of cardinality at most 1. We generally write $\mathcal{A} = (Q, \Sigma, q_0, \delta, Q_f)$ with $Q_0 = \{q_0\}$, and $\delta(q, a) = q'$ for all $(q, a, q') \in \Delta$. Every finite state automaton satisfying the diamond property can be transformed into an equivalent deterministic finite state automaton (DFA) satisfying the diamond property.

Zielonka's theorem aims at distributing a trace-closed regular language to the individual processes. This can be formally stated with (deterministic) asynchronous automata.

Definition 2. *An asynchronous automaton (AA) over the distributed alphabet (Σ, loc) is a tuple $\mathcal{B} = ((Q_p)_{p \in P}, \Sigma, q^0, (\delta_a)_{a \in \Sigma}, F)$ where*

- *Q_p is the set of local states of a process $p \in P$,*
- *$\delta_a \colon \Pi_{p \in \mathrm{loc}(a)} Q_p \to \Pi_{p \in \mathrm{loc}(a)} Q_p$ is the transition function associated with $a \in \Sigma$,*
- *$q^0 \in \Pi_{p \in P} Q_p$ is the global initial state, and*
- *$F \subseteq \Pi_{p \in P} Q_p$ is the set of global final states.*

The AA is said to be finite *if each process has a finite number of states.*

The semantics of an AA is given by a DFA whose set of states is the product $\Pi_{p \in P} Q_p$. We call these states the *global states* of $\mathcal{B}$. Its initial global state is q^0, and its final global states are given by F. Moreover, for each letter $a \in \Sigma$, we let $(q_p)_{p\in P} \xrightarrow{a} (q'_p)_{p\in P}$ if for all $p' \notin \mathrm{loc}(a)$, $q'_p = q_p$, and $\delta_a((q_p)_{p\in\mathrm{loc}(a)}) = (q'_p)_{p\in\mathrm{loc}(a)}$. We can show that this DFA satisfies the diamond property, and thus, we let $L_{\mathsf{tr}}(\mathcal{B})$ be its trace language, and say that the AA $\mathcal{B}$ recognises $L_{\mathsf{tr}}(\mathcal{B})$.

Theorem 1 ([21]). *For every DFA $\mathcal{A}$ over the concurrent alphabet (Σ, I) satisfying the diamond property, and every distributed alphabet (Σ, loc) such that $I_{\mathrm{loc}} = I$, there exists a finite AA $\mathcal{B}$ over (Σ, loc) such that $L_{\mathsf{tr}}(\mathcal{A}) = L_{\mathsf{tr}}(\mathcal{B})$.*

3   Fair specifications

Fairness is an important condition in the verification of distributed systems, that imposes conditions on distributed runs such that no process starves in the long run. In our situation where finite-state automata and finite traces are considered, we strengthen the fairness condition to obtain a notion of *bounded-fairness*: we require that no process lags behind the other ones more than a bounded number of steps. Formally, for a positive integer k, a trace t is k-*fair* if for every factor u of any linearisation w of t, that has length at least k, we have $\mathrm{loc}(u) = P$ (where P is the set of all processes).

Example 7. Consider the trace $[abcacd]$ over the distributed alphabet of Example 2. Notice that in every factor of the linearisation $abcacd$ with length at least 4, all processes participate. This property can be verified to be true in every linearisation. Hence $[abcacd]$ is 4-fair. However, $[abcacd]$ is not 3-fair since the factor cac in the linearisation $abcacd$ does not involve process p_3.

As intended, the k-fairness of a trace ensures that views of different processes do not differ too much.

Lemma 3. *Let t be a k-fair trace. For all pairs of processes $p, p' \in P$, we have $||\mathrm{view}_p(t)| - |\mathrm{view}_{p'}(t)|| \leq k - 1$.*

The situation is indeed even better. The view of a process, when written in FNF, coincides with the whole trace, except in the last steps that contain at least $k - 1$ letters in total. To describe such a property more easily, we define a measure on traces: for a natural number ℓ and a trace t of length at least ℓ, we let $f(t, \ell)$ be the largest natural number i such that the steps $\mathsf{F}(t)_i \cdots \mathsf{F}(t)_{|t|_\mathsf{F}}$ contain at least ℓ letters in total. Notice in particular that, by maximality of i, the steps $\mathsf{F}(t)_{i+1} \cdots \mathsf{F}(t)_{|t|_\mathsf{F}}$ contain in total at most $\ell - 1$ letters.

Lemma 4. *Let t be a k-fair trace and $p \in P$ be a process with a view of length at least $k - 1$. For all $i < f(\mathrm{view}_p(t), k - 1)$, $\mathsf{F}(t)_i = \mathsf{F}(\mathrm{view}_p(t))_i$.*

Example 8. The bound given in Lemma 4 is optimal in the following sense. Consider the alphabet $\{a, b\}$ with two processes p_1 and p_2 such that $\mathrm{loc}(a) = \{p_1\}$, $\mathrm{loc}(b) = \{p_2\}$. The trace t whose linearisation is ba^{k-1} is k-fair, and the view of process p_1 has linearisation a^{k-1} of length $k - 1$, with FNF where all steps are singletons. The FNF of t is $\{a, b\}\{a\}^{k-2}$, and thus p_1 is not fully aware of the first step $\{a, b\}$. In particular, $f(\mathrm{view}_{p_1}(t), k - 1) = 1$, and indeed the first step is such that $\mathsf{F}(t)_1 \neq \mathsf{F}(\mathrm{view}_{p_1}(t))_1$.

We need a slightly more refined result in the following, since the full trace is known by no process, and thus we need to understand for a process p, what guarantee it has on the view of another process p', to enable a successful synchronisation between them.

Lemma 5. *Let t be a k-fair trace and $p \in P$ be a process with a view of length at least $2k - 2$. Then for all $i < f(\text{view}_p(t), 2k - 2)$ and for all $p' \in P$, $\mathsf{F}(t)_i = \mathsf{F}(\text{view}_{p'}(t))_i$.*

Thus, for a k-fair trace t, the views of two different processes $\text{view}_p(t)$ and $\text{view}_{p'}(t)$, when considered in their FNF, have an identical prefix, and differ only in the last few steps. The identical prefix in fact corresponds to the given full trace t. In our construction of an AA, we use this fact crucially to maintain only a finite suffix of the trace in the local states.

Example 9. Once again, the result of Lemma 5 is optimal. Consider an extension of the previous example where $\Sigma = \{a, b, c, d\}$, $\text{loc}(a) = \{p_1\}$, $\text{loc}(b) = \{p_2\}$, $\text{loc}(c) = \{p_1, p_2\}$, and $\text{loc}(d) = \{p_1, p_3\}$. Consider the trace $t = ba^{k-2}dca^{k-2}$, whose FNF is $\{a, b\}\{a\}^{k-3}\{d\}\{c\}\{a\}^{k-2}$. The views of the 3 processes are: $\text{view}_{p_1}(t) = t$, $\text{view}_{p_2}(t) = ba^{k-2}dc$, and $\text{view}_{p_3}(t) = a^{k-2}d$. The trace t is k-fair: indeed all factors of any linearisation of length k either pick both the letters d and b, or the letters d and c. Notice that it is not $(k-1)$-fair, since it contains the factor $a^{k-2}d$ in which process p_2 does not participate. Finally, p_1 has a view of length $2k - 1$, and $f(\text{view}_{p_1}(t), 2k - 2) = 1$. Indeed, p_3 is not fully aware of the first step of the FNF of t.

A DFA satisfying the diamond-property is said to be *fair* if in every loop, all processes participate in at least one action. Since this must be true already on loops without repetition of states, we can refine this definition to introduce a parameter k of fairness, as before. We first do it on the languages: a trace language X is k-fair if for all $t \in X$, t is k-fair. We can also refine the definition of fairness of a DFA to take the parameter k into account.

Definition 3. *A DFA $\mathcal{A} = (Q, \Sigma, q_0, \delta, Q_f)$ satisfying the diamond property is said to be k-fair if for each state $q \in Q$ reachable from an initial state, and for every word $u \in \Sigma^*$ such that $|u| \geq k$ and $\delta(q, u)$ is co-reachable (i.e. there is a path from $\delta(q, u)$ to a final state), we have $\text{loc}(u) = P$.*

This definition is indeed a characterisation of k-fair trace languages:

Proposition 1. *Let $\mathcal{A}$ be a DFA satisfying the diamond property. Then, $\mathcal{A}$ is k-fair if and only if the language $L_{\text{tr}}(\mathcal{A})$ of traces is k-fair.*

As a corollary, we see that for a language L of traces, there exists k such that it is k-fair if and only if in every loop of any DFA recognising L (satisfying the diamond-property), all processes participate. This is thus easy (polynomial-time) to test if a given DFA recognises a fair language (i.e. k-fair for some k). It is also possible to test if a DFA is k-fair, for a given k, since it is enough to check the property of Definition 3 for every word u of length k. This naive algorithm requires an exponential-time complexity with respect to k. In [6], we propose a dynamic-programming algorithm that solves this problem in polynomial-time.

We end this section with an example of a family of languages where the alphabet size, the required number of states in a DFA, and the number of processes increase arbitrarily, while k remains constant. This shows that the fairness parameter can be helpful to describe the complexity of a trace language, in a different way than other parameters. This is in particular motivating for the contribution we present in the next section, where the only exponential component in the complexity depends on the parameter k.

Example 10. Let $\Sigma = \{c, a_1, a_2, \ldots, a_n\}$, $P = \{p_1, \ldots, p_n\}$ and $\Sigma_{p_i} = \{a_i, c\}$ for all i. Consider the trace language described by $L_n = ((\bigcup_{1 \leq i < j \leq n}[a_i a_j]).[c])^*$: it is the alternation of two among the n letters $\{a_1, \ldots, a_n\}$ (where two distinct processes participate concurrently), and the global synchronisation letter c (where all processes participate). Hence, L_n is 3-fair for all n, though it requires a DFA with $\Omega(n)$ states.

4   A Zielonka's theorem for fair trace languages

Our first contribution is a new, and more efficient, proof of Zielonka's theorem in the specific case where the specification trace language is fair:

Theorem 2. *For every k-fair DFA $\mathcal{A}$ over the distributed alphabet (Σ, loc) satisfying the diamond property, there exists a finite AA $\mathcal{B}$ over (Σ, loc) such that $L_{\mathrm{tr}}(\mathcal{A}) = L_{\mathrm{tr}}(\mathcal{B})$, where every set of local states (for each process) is of size bounded by $O(n \times k \times |\Sigma|^{3k-3})$, where n is the number of states of $\mathcal{A}$.*

Notice that the complexity is linear in the size of the DFA, and polynomial in the size of the alphabet — with the degree of the polynomial only depending linearly in the fairness parameter. We now describe an overview of the construction over an example. A complete proof of Theorem 2 is provided in [6].

This construction starts from $\mathcal{A}$ and proceeds in three steps, each building an AA accepting $L_{\mathrm{tr}}(\mathcal{A})$. In the first step, we build an infinite AA whose states are tuples of the FNF of process views. In the second step, we explain how to cut these views, only keeping suitable suffixes but adding local states and unbounded counters. The third and final AA is then obtained by bounding the counter values using modulo counting.

Consider the DFA specification $\mathcal{A}$ in Figure 2 over the distributed alphabet $(\{a, b, c, d\}, \mathrm{loc})$ with $\mathrm{loc}(a) = \{p_1, p_2\}$, $\mathrm{loc}(b) = \{p_1, p_3\}$, $\mathrm{loc}(c) = \{p_2, p_3\}$, and $\mathrm{loc}(d) = \{p_1\}$ (note that the distributed alphabet is different from the one previously used in our examples). The associated independence relation is $I_{\mathrm{loc}} = \{(c, d), (d, c)\}$. The word $abcdbdcb$ can be read from state 0, reaching back to state 0. It is a linearisation of the trace depicted on the right of the figure, which also gives its FNF.

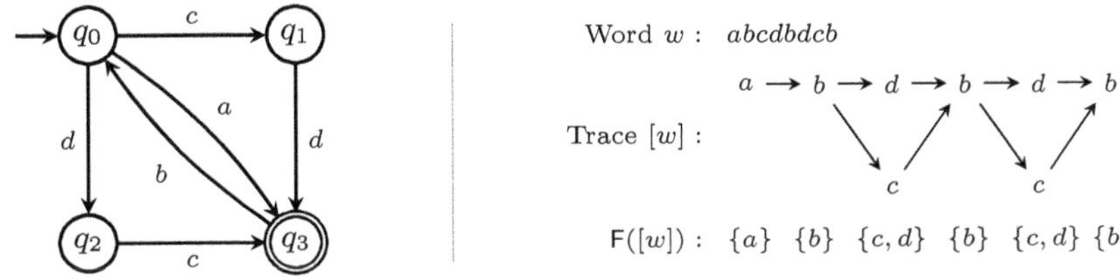

Fig. 2. (Left) A DFA satisfying the diamond property; (Right) a word w, the trace $[w]$ and its Foata normal form $\mathsf{F}([w])$.

	a	b	c	d	b	d	c	b
p_1	$\{a\}$	$\{a\}\{b\}$	$\{a\}\{b\}$	$\{a\}\{b\}\{d\}$	$\{a\}\{b\}\{c,d\}\{b\}$	$\{a\}\{b\}\{c,d\}\{b\}\{d\}$	$\{a\}\{b\}\{c,d\}\{b\}\{d\}$	$\{a\}\{b\}\{c,d\}\{b\}\{c,d\}\{b\}$
p_2	$\{a\}$	$\{a\}$	$\{a\}\{b\}\{c\}$	$\{a\}\{b\}\{c\}$	$\{a\}\{b\}\{c\}$	$\{a\}\{b\}\{c\}$	$\{a\}\{b\}\{c,d\}\{b\}\{c\}$	$\{a\}\{b\}\{c,d\}\{b\}\{c\}$
p_3	$\{\}$	$\{a\}\{b\}$	$\{a\}\{b\}\{c\}$	$\{a\}\{b\}\{c\}$	$\{a\}\{b\}\{c,d\}\{b\}$	$\{a\}\{b\}\{c,d\}\{b\}$	$\{a\}\{b\}\{c,d\}\{b\}\{c\}$	$\{a\}\{b\}\{c,d\}\{b\}\{c,d\}\{b\}$

Fig. 3. Run of the infinite asynchronous automaton corresponding to the finite state automaton of Figure 2, on the word $abcdbdcb$

Step 1. Even without the fairness assumption, it is always possible to define an *infinite* AA recognising $L_{\mathsf{tr}}(\mathcal{A})$, in which each process keeps its *view of the current trace* as its local state. Furthermore, maintaining the view in the FNF allows to compute the synchronised views easily during shared actions. We illustrate this idea in Figure 3. The beginning of the run of this infinite AA is depicted in Figure 3. Initially, the views of all the processes are empty. When the letter a is read, processes p_1 and p_2 update their view, while process p_3 is not aware of it. When the letter b is read next, processes p_1 and p_3 synchronise. This implies that p_3 catches up by learning about the letter a read before. Since a and b are dependent, we add a new step in the FNF of both views. We then read letters c and d in the same manner (note that we would obtain the same result by reading first d and then c). When letter b is read next, processes p_1 and p_3 learn from each other about the previous letters c and d. They merge their views by making stepwise unions of their FNF (as justified by Lemma 2), before adding a new step with letter b. The run goes on like this, increasing the number of steps in the FNF. At the end, the acceptance status of the current run is obtained by once again merging the FNF of all processes (through a stepwise union), getting the FNF of the actual full trace, over which we can simply run the DFA to know whether the last state is final or not. Our objective now is to make this construction finite using the fairness assumption.

Step 2. The DFA of Figure 2 is 4-fair, since every sequence of 4 transitions makes every process participate: notice that it is not 3-fair because of the word dbd that can be read from state q_1, where process p_2 does not participate. The crux of our construction is Lemma 5. As an example, let $t = abcdbdcb$, and consider $\mathrm{view}_{p_1}(t)$ as shown in the last column of Figure 3. We have $k = 4$, and hence $2k - 2 = 6$, and $j = f(\mathrm{view}_{p_1}(t), 2k - 2) = 3$. Observe that the prefix up to $j - 1$ (given by $\{a\}\{b\}$ in this case) is known to every other process. Hence the

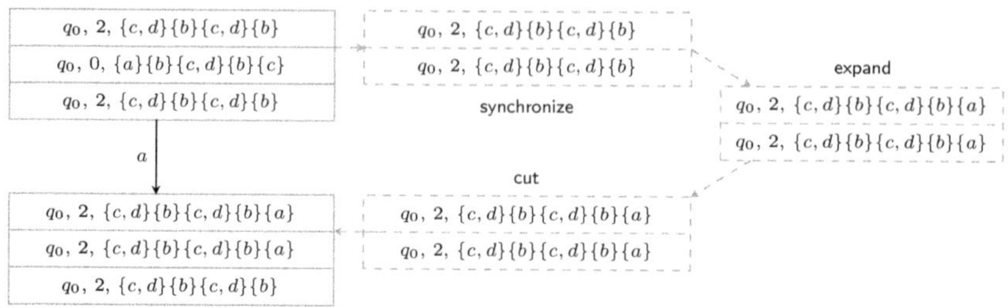

Fig. 4. Illustrating one transition in the AA that maintains an unbounded counter

process does not need to maintain this prefix in the local state. We can instead cut out these first steps of the FNF, and update the state of the DFA.

A process p thus only stores the steps of the FNF of its view t_p starting with the one of index $f(t_p, 2k-2)$. However, for later synchronisations, processes need to keep a way to align their views in order to do the stepwise union. We first choose to do it by using an *unbounded counter*, remembering how many letters have been forgotten so far. The beginning of computation in Figure 3 remains identical, simply adding the initial state q_0, and a counter value 0, before the view. As seen before, in the last configuration though, processes p_1 and p_3 can cut two letters still keeping the last steps having $2k-2 = 6$ letters. They thus update the state component (though it comes back to q_0 again), and increment the counter component by 2, and we reach the configuration:

$$
\begin{array}{l|l}
p_1 & q_0, 2, \{c,d\}\{b\}\{c,d\}\{b\} \\
p_2 & q_0, 0, \{a\}\{b\}\{c,d\}\{b\}\{c\} \\
p_3 & q_0, 2, \{c,d\}\{b\}\{c,d\}\{b\}
\end{array}
$$

While reading an additional letter a, Figure 4 presents the update performed by processes p_1 and p_2. They first synchronise their local views: process p_2 that had counter value 0 first removes the first two letters of its view to obtain the same counter value 2 as p_1, and then it aligns the rest of its view with p_1 and learns about the events d and b. Then they expand their views by adding the step $\{a\}$. Even if the steps contain $7 > 2k - 2$ letters in total, the first step, that contains two letters, cannot be cut, since the four last steps contain only $5 < 2k - 2$ letters in total. If we read the letter b afterwards, processes p_1 and p_3 synchronise to the view of p_1 that had the most recent information, and the first step can be cut while incrementing the counter value twice (since the step that is removed contains 2 letters), to obtain the new configuration

$$
\begin{array}{l|l}
p_1 & q_3, 4, \{b\}\{c,d\}\{b\}\{a\}\{b\} \\
p_2 & q_0, 2, \{c,d\}\{b\}\{c,d\}\{b\}\{a\} \\
p_3 & q_3, 4, \{b\}\{c,d\}\{b\}\{a\}\{b\}
\end{array}
$$

For each letter read, the corresponding processes first synchronise, then expand, and finally make the necessary cuts.

Step 3. So far, only the counter values make the set of local states infinite. We now propose to only remember these values modulo $2k$. The crux is Lemma 3:

processes cannot lag behind one another by more than $k-1$ letters. Because of this observation, the range of possible counter values of all processes, when added to the number of letters not yet forgotten, is included in an interval of values of the form $\{c, c+1, c+2, \ldots, c+(k-1)\}$, i.e., k values. When considered modulo $2k$, this becomes a cyclic interval of k values. In particular, there is a full range of the other k values (modulo $2k$) that cannot be taken by any process. We can thus unambiguously find which of the processes is ahead of the other when synchronising (even if its counter value seems lower than others, due to the modulo counting). For instance, let us continue reading letters dcb after which we reach the following configuration still with counter values at most $2k-1$:

$$
\begin{array}{l|l}
p_1 & q_3, 7, \{b\}\{a\}\{b\}\{c,d\}\{b\} \\
p_2 & q_0, 5, \{c,d\}\{b\}\{a\}\{b\}\{c\} \\
p_3 & q_3, 7, \{b\}\{a\}\{b\}\{c,d\}\{b\}
\end{array}
$$

When reading the next letter a, the cutting of the first step would increase the counter value to 8, which is rounded modulo $2k = 8$ to 0. We thus get:

$$
\begin{array}{l|l}
p_1 & q_0, 0, \{a\}\{b\}\{c,d\}\{b\}\{a\} \\
p_2 & q_0, 0, \{a\}\{b\}\{c,d\}\{b\}\{a\} \\
p_3 & q_3, 7, \{b\}\{a\}\{b\}\{c,d\}\{b\}
\end{array}
$$

We arrive in the situation depicted on the top left of Figure 5. When reading the next letter b, processes p_1 and p_3 have respective counter values 0 and 7. They add 6 (the length of the suffix of the view which they explicitly remember) to get values 6 and 5 modulo 8. As mentioned before, the list of counter values added to the length of the remaining suffix, falls under a continuous range of at most $k = 4$ values. These values should include 6 and 5 — hence p_1 is indeed ahead of p_3 (despite having a smaller counter value originally). If it was the other way around, that is, p_1 is behind p_3, the range of values would include $6, 7, 0, 1, 2, 3, 4, 5$, which has more than $k = 4$ values: this is not possible. Thus, p_3 can discard its first step to synchronise. They expand the configuration by adding b in a new step of the FNF. They cut the first element, and update their state to q_3 and counter value to 1. This way, using modulo counting, the processes are able to determine who is ahead and then perform the synchronisation, expanding and cutting to get to the new local states that maintain a suffix of their views.

Computing the state in the DFA. The final task is to be able to check whether a given global state of the built AA $\mathcal{B}$ is accepting or not. To do so, we map each global state of $\mathcal{B}$ to a unique state of the DFA $\mathcal{A}$ as follows. First we synchronise all the processes: first p_1 and p_2 synchronise, then the joint state of $\{p_1, p_2\}$ is reconciled with the local state of p_3 and so on. During the synchronisation, the same care is taken on the modulo counter values. Once this is done, all the local states are the same. From the state q stored in a local state, we compute the state q' reached in $\mathcal{A}$ by reading all the letters of the common suffix, step by step, in any order. We declare the global state accepting in the AA $\mathcal{B}$ if and only if the state q' of $\mathcal{A}$ is accepting.

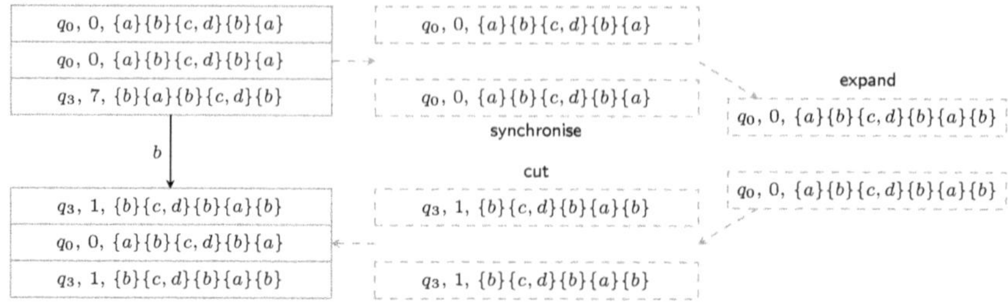

Fig. 5. Illustrating one transition in the finite AA $\mathcal{B}$ that counts modulo $2k$

The entire construction is carefully presented in [6] for the general case. The crucial point in the proof is to maintain that the state q' computed above to decide the acceptance condition in $\mathcal{B}$ is indeed the state in $\mathcal{A}$ that would be reached if we had read the full trace.

The number of local states for each process is bounded by $n \times 2k \times |\Sigma|^M$ where n is the number of states of $\mathcal{A}$, and M is the maximum number of letters that should be maintained in the suffix of views. This value M is bounded by $2k - 2 + (k - 1) = 3k - 3$ since we may have to add up to $k - 1$ letters to the first step in order to obtain a full step: indeed in a k-fair trace, every step of its FNF contains at most k letters (otherwise there would be a letter a independent of k other letters in a factor u of the trace, meaning that all processes reading a would be starved during the factor u).

Lower bound. For the general case considering any trace-closed regular language as a specification, [11] presents a lower bound on the number of local states in any *locally-rejecting* AA: a (deterministic) AA $\mathcal{B}$ is said to be locally-rejecting if for every process p, there is a set of reject states R_p s.t. for every trace t, $\text{view}_p(t) \notin \text{pref}(L_{\text{tr}}(\mathcal{B}))$ iff the local state reached by p on t is in R_p, where $\text{pref}(L_{\text{tr}}(\mathcal{B}))$ denotes the set of all ideals of all traces accepted by $\mathcal{B}$. If an AA synthesis construction maintains enough information so that the state reached by the global specification DFA $\mathcal{A}$ on $\text{view}_p(t)$ can be deduced, then the synthesised AA is locally rejecting by a suitable choice of states R_p. It is remarked in [11] that every known Zielonka-type AA synthesis construction until then had induced a locally-rejecting automaton. After [11], the only AA synthesis procedure that we are aware of is [1], but it is a different setting since the specification is written as a logical formula and the implementation is a cascade product of AAs. Notice that our algorithm also synthesises a locally-rejecting automaton since we can indeed derive the state reached by $\mathcal{A}$ on $\text{view}_p(t)$. We are then able to show the following lower-bound, whose proof is presented in [6].

Theorem 3. *There is a family of n-fair languages L_n for which every locally-rejecting AA has size at least $2^{\frac{n}{4}}$.*

5   Combining fairness with acyclic communication

In a fair DFA, every process participates in every cycle. In a DFA specification that is not fair, some set of processes can perform an unbounded number of actions, while the other processes wait. This could lead to processes having diverging views with an unbounded number of events in the difference. However, Krishna and Muscholl [14] have proposed an efficient way to synthesise an asynchronous automaton when the communication between processes happens in a hierarchical manner. Given a set P of processes, and a distributed alphabet (Σ, loc), an undirected graph called the *communication graph* $G_{(\Sigma,\mathrm{loc})}$ is constructed as follows: vertices of $G_{(\Sigma,\mathrm{loc})}$ are the processes in P; for $p_1, p_2 \in P$, there is an edge (p_1, p_2) if p_1 and p_2 share a common action, i.e. $\Sigma_{p_1} \cap \Sigma_{p_2} \neq \emptyset$. [14] provides an efficient AA synthesis when the communication graph is a tree. In particular, this entails that every process synchronises only with its parent or its child in the communication graph. In their construction, the local states of a process p are pairs of states $(\overleftarrow{q_p}, q_p)$ of the specification DFA $\mathcal{A}$ so that: $\overleftarrow{q_p}$ is the state reached by $\mathcal{A}$ on reading $\overleftarrow{\mathrm{view}}_p(t)$, the smallest ideal of $\mathrm{view}_p(t)$ containing all actions where p synchronises with its parent in $G_{(\Sigma,\mathrm{loc})}$; and q_p is the state reached by the DFA $\mathcal{A}$ on reading $\mathrm{view}_p(t)$. Their key technique is an elegant method to construct the state of $\mathcal{A}$ reached on a trace t by looking at the local states $(\overleftarrow{q_p}, q_p)$ as above. We recall the details of the construction in [6].

In this section, we propose a *tree-of-bags* architecture where essentially each node in the communication graph is replaced with a set (bag) of processes. When we impose a fairness condition on the DFA restricted to each of these bags, we are able to incorporate the construction of Section 4 into the method of [14].

Tree-of-bags architecture. A distributed alphabet (Σ, loc) is said to form a *tree-of-bags* architecture if the following conditions are all satisfied:

- the set of processes P can be partitioned into bags $\{B_1, B_2, \ldots, B_\ell\}$,
- each bag B_j has a special process o_j called its *outer process* — the set of all outer processes is denoted as O, and the rest of the processes are called *inner processes*;
- the communication graph restricted to O forms a tree — hence, one outer process can be designated as the root, and for every other outer process o_j, there is a unique outer process which is the parent in this tree, which we denote as $\mathsf{parent}(o_j)$;
- every inner process communicates within its bag: for a process $\iota \in B_j \setminus \{o_j\}$, there is no edge outside B_j in the communication graph.

On the right, we depict an example where the red dashed bubbles represent the bags, each bubble has an outer process shown as a black circle, and the blue squares are the inner processes. We write $(\Sigma, \mathrm{loc}, \mathfrak{B})$ to denote a tree-of-bags architecture as above, with $\mathfrak{B} = \{B_1, B_2, \ldots, B_\ell\}$. For a bag $B \in \mathfrak{B}$, we write $o(B)$ for the outer process of B. Also, we let $\Sigma^{\mathrm{in}}(B) = \{a \in \Sigma \mid$

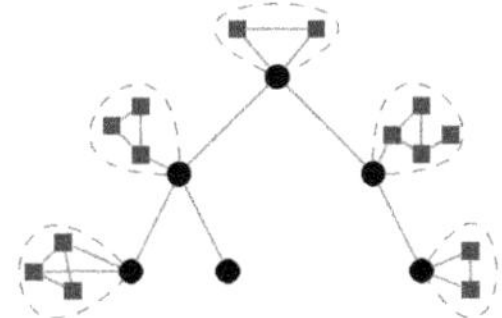

$\text{loc}(a) \subseteq B\}$ be the set of actions for which only the processes in B participate. We will sometimes refer to $\Sigma^{\text{in}}(B)$ as the bag alphabet of B.

Definition 4. *Let (Σ, loc) be a distributed alphabet forming a tree-of-bags architecture $(\Sigma, \text{loc}, \mathfrak{B})$. A DFA specification $\mathcal{A}$ over (Σ, loc) is said to be fair for the tree-of-bags architecture $(\Sigma, \text{loc}, \mathfrak{B})$ if in every loop of $\mathcal{A}$, either all processes in a bag participate, or none of them participates.*

As for the notion of fairness studied in Section 3, one can test if the DFA $\mathcal{A}$ is fair for a tree-of-bags architecture $(\Sigma, \text{loc}, \mathfrak{B})$. For each bag $B \in \mathfrak{B}$ replace the transitions labelled with $\Sigma \setminus \Sigma^{\text{in}}(B)$ by ε-transitions, and find if there exists a natural number k_B such that the resulting automaton is k_B-fair. Such constants must exist for each bag $B \in \mathfrak{B}$ for $\mathcal{A}$ to be fair for $(\Sigma, \text{loc}, \mathfrak{B})$. In other words, for every trace in $L(\mathcal{A})$, the trace obtained by only keeping events labelled by $\Sigma^{\text{in}}(B)$ is k_B-fair. It is also possible to compute in polynomial time such constants k_B for all bags (by using the same algorithm described in [6] for computing the fairness parameter for a DFA). In our later explanations, we suppose that such constants k_B are known, and we use them to give a complexity bound of our construction.

The construction. Here is the main challenge: on reading a trace t, we need for each bag B, the state reached by $\mathcal{A}$ on $\text{view}_B(t)$. Let us write $\delta(q_0, \text{view}_B(t))$ for this state. Once we know it, we can use the STATE function of [14] to compute the state reached by $\mathcal{A}$ on the whole trace t. So let us now focus on how we can compute $\delta(q_0, \text{view}_B(t))$. Firstly, recall that $\text{view}_B(t) = \bigcup_{\iota \in B} \text{view}_\iota(t)$. If we maintain the individual views $\text{view}_\iota(t)$ for each $\iota \in B$, in FNF, we could take the stepwise union to get $\text{view}_B(t)$. Since we assumed $\mathcal{A}$ is fair when restricted to each bag, we are tempted to apply the construction of Section 4 and maintain only a finite suffix of these views. However, there are new challenges to overcome. Since the outer process $o(B)$ communicates with processes outside bag B, $\text{view}_o(t)$ includes letters outside the bag alphabet. Further, as other inner processes $\iota \in B$ can potentially synchronise with o, $\text{view}_\iota(t)$ may have also letters from outside the bag alphabet Σ_B. Hence we cannot directly cut the prefixes of the views based on the parameter k_B.

Here are the key ideas. Firstly, we consider $t{\downarrow}_{\Sigma^{\text{in}}(B)}$, the trace t restricted to the bag alphabet and maintain $\text{view}_\iota(t{\downarrow}_{\Sigma^{\text{in}}(B)})$ in FNF. Each process $\iota \in B$ maintains $\text{view}_\iota(t{\downarrow}_{\Sigma^{\text{in}}(B)})$. With this information, we can compute $\text{view}_B(t{\downarrow}_{\Sigma^{\text{in}}(B)})$. Secondly, we observe that $\text{view}_B(t)$ can be written with $\text{view}_o(t)$, $\text{view}_B(t{\downarrow}_{\Sigma^{\text{in}}(B)})$ and $\text{view}_o(t{\downarrow}_{\Sigma^{\text{in}}(B)})$.

Lemma 6. *If t is a trace, $B \in \mathfrak{B}$, and $o = o(B)$, we have $\text{view}_B(t) = \text{view}_o(t) \cdot t'$ where t' is the trace obtained by removing $\text{view}_o(t{\downarrow}_{\Sigma^{\text{in}}(B)})$ from $\text{view}_B(t{\downarrow}_{\Sigma^{\text{in}}(B)})$.*

Algorithm 1 shows how to compute $\delta(q_0, \text{view}_B(t))$ using $\delta(q_0, \text{view}_o(t))$ and $\text{view}_\iota(t{\downarrow}_{\Sigma^{\text{in}}(B)})$ for all $\iota \in B$. It identifies the trace t' in Line 3, and then does the required computation in Line 4. The last task is to be able to implement Algorithm 1 using finite information as states. This is where the fairness assumption on each bag helps, allowing us to maintain only a bounded suffix

---

Algorithm 1 Computing the state $\delta(q_0, \mathrm{view}_B(t))$

Require: A bag B, a trace $\mathrm{view}_\iota(t\!\downarrow_{\Sigma^{\mathrm{in}}(B)})$ for all $\iota \in B$, $q_o = \delta(q_0, \mathrm{view}_o(t))$ where
    $o = o(B)$
Ensure: $q = \delta(q_0, \mathrm{view}_B(t))$
  1: $\hat{t} \leftarrow$ pointwise union of $\mathsf{F}(\mathrm{view}_\iota(t\!\downarrow_{\Sigma^{\mathrm{in}}(B)}))$ over all $\iota \in B$
  2: $e \leftarrow$ maximal element of o in $\hat{t}$
  3: $t' \leftarrow$ trace obtained by removing $e\!\downarrow$ from $\hat{t}$
  4: $q \leftarrow \delta(q_o, t')$

---

of $\mathrm{view}_\iota(t)$ for each ι. For outer processes we also maintain a pair $(\overleftarrow{q_o}, q_o)$ where $q_o = \delta(q_0, \mathrm{view}_o(t))$ is the state reached by $\mathcal{A}$ on reading $\mathrm{view}_o(t)$ and $\overleftarrow{q_o} = \delta(q_0, \overleftarrow{\mathrm{view}_o}(t))$, to mimick the construction of [14] for the outer processes. All the details of the construction are presented in [6].

Theorem 4. *For a DFA $\mathcal{A}$ that is fair for a tree-of-bags architecture $(\Sigma, \mathrm{loc}, \mathfrak{B})$, there exists a finite AA $\mathcal{B}$ over (Σ, loc) such that $L_{\mathsf{tr}}(\mathcal{A}) = L_{\mathsf{tr}}(\mathcal{B})$, where the set of local states for outer processes is of size bounded by $O(n^3 \times k \times |\Sigma|^{3k-3})$, and that of inner processes is bounded by $O(k \times |\Sigma|^{3k-3})$, where n is the number of states of $\mathcal{A}$, and k is the maximum of k_B over all bags B.*

6   Implementation and experimentation

We have implemented the synthesis algorithm of Section 4. It is available both as a docker container [16], and as a git repository https://gitlab.lis-lab.fr/benjamin. monmege/faast. The tool is implemented in Python. Each Python class contains tests that clarify the usage of the tool, and a Jupyter notebook is provided that shows how to use the code, and some of the results we are able to obtain from it. There are no dependencies other than Python 3.

Thanks to the tool, our algorithm was applied to the classic *Dining Philosophers* problem [10], a fundamental case study for concurrency and fairness issues that we have recalled in the introduction. The DFA modelling every possible sequence of actions of the $n \geq 2$ philosophers is trace closed but not fair (unless there are $n = 2$ philosophers). One possible goal is to synthesise a distributed implementation of the dining philosophers that guarantees liveness (absence of starvation of philosophers, here meant in the sense that every philosopher should be able to do an action often enough).

However the DFA is not fair. We can still adapt the algorithm of Section 4 to allow for an unfair DFA in input, as well as a fairness parameter $k \in \mathbb{N}$. The modified algorithm produces an AA that accepts all k-fair traces in the language of the trace-closed DFA (possibly none). The construction is modified to only generate a state if its FNF part is k-fair (i.e. all infixes of length k of all possible linearisations of the FNF involve all processes). We are then able to obtain an AA for the dining philosophers by using as fairness parameter $2n$. The value $2n$

is chosen to ensure that at least one complete set of actions for any philosopher occurs over any sufficiently long execution, thereby ensuring fairness.[5]

Notice that the tool proposes to apply a small optimisation, modifying the *cut* step: instead of keeping at least $2k - 2$ letters in the FNF, we cut as many steps as possible in the FNF, by removing all steps containing letters that are all known by every process (i.e. that are fully in the views of all processes). This is a correct optimisation, and we depict the various results we obtain with the tool on a small example in [6].

As an alternative solution for the dining philosophers, we could apply directly the construction of Section 4 to the intersection of the DFA with the set of all $2n$-fair traces. Notice that the set of all $2n$-fair traces is indeed a regular trace language that is fair, but the DFA that recognises it is huge (at least exponential in n). Thus, the intersection results in a huge DFA that is long to produce, and thus impracticable for the synthesis algorithm afterwards.

On-the-Fly Generation Optimisation. Our implementation also introduces a crucial optimisation to manage the potential state-space explosion of the AA, whose size is exponential with respect to the fairness parameter k. It consists of an on-the-fly generation of the AA, that computes and stores the states and transitions of the AA only when they are effectively needed to compute an actual run from the initial global state. Instead of computing the transition functions δ_a for all letters a, we only write the recipe to generate all possible transitions. Afterwards, the collections of states and transitions are populated dynamically only: when we run the AA (to check whether a given word is accepted, or via a random exploration of the paths in the global semantics), it is only at this moment that the new local states, global transitions, and global accepting states are computed and stored. This approach also ensures that the constructed AA is limited to its reachable portion from the initial state, which is essential in practice for analyzing complex distributed systems where the non-reachable global state-space would be too large to compute.

7   Conclusion

We have studied Zielonka's theorem to synthesise AA from DFA specifications, in the context of fair specifications. We have strengthened this result to restrict fairness to a subset of processes with an outer process, where outer processes have no cyclic dependencies. In these special cases, we obtain AA whose sets of local states do not depend on the number of processes, contrary to all previous methods where an exponential dependency in this number of processes is unavoidable. Furthermore, our construction is, arguably, simpler to understand conceptually. It would be interesting to study how the fairness restriction could simplify the construction of the gossip automaton, and thus the other proofs of Zielonka's theorem. As future works, we could also consider infinite traces, where

---

[5] If, e.g., the DFA forbids a philosopher to put back a chopstick as long as they did not eat, the built AA ensures that every philosopher eats often.

the fairness condition makes even more sense. Similarly, we could investigate the fairness condition in the context of a specification that is stated as a logical formula, for instance LTL or PDL, hoping that our techniques can lead to beneficial results in the context of translating logical formulae into AA. We would also like to study extensions of our techniques to the setting of k-connectedly communicating processes [15], where the fairness restriction can be relaxed for processes that are no longer communicating with each other in the future.

References

1. Adsul, B., Gastin, P., Kulkarni, S., Weil, P.: An expressively complete local past propositional dynamic logic over Mazurkiewicz traces and its applications. In: LICS (2024). https://doi.org/10.1145/3661814.3662110
2. Akshay, S., Dinca, I., Genest, B., Stefanescu, A.: Implementing realistic asynchronous automata. In: 33rd International Conference on Foundations of Software Technology and Theoretical Computer Science (FSTTCS 2013). Lipics (2013)
3. Alur, R., Henzinger, T.A.: Finitary fairness. ACM Trans. Program. Lang. Syst. **20**(6), 1171–1194 (Nov 1998). https://doi.org/10.1145/295656.295659, https://doi.org/10.1145/295656.295659
4. Baudru, N.: Compositional synthesis of asynchronous automata. Theoretical Computer Science **412** (2011)
5. Baudru, N., Morin, R.: Unfolding synthesis of asynchronous automata. In: Computer Science–Theory and Applications: First International Computer Science Symposium in Russia, CSR 2006, St. Petersburg, Russia, June 8-12. 2006. Proceedings 1. pp. 46–57. Springer (2006)
6. Bérard, B., Monmege, B., Srivathsan, B., Sur, A.: Synthesising asynchronous automata from fair specifications (2025), https://arxiv.org/abs/2504.14623
7. Cartier, P., Foata, D.: Problèmes combinatoires de commutation et réarrangements. Springer (1969)
8. Cori, R., Métivier, Y., Zielonka, W.: Asynchronous mappings and asynchronous cellular automata. Information and Computation **106**, 159–202 (1993)
9. Diekert, V., Muscholl, A.: Trace theory. In: Padua, D.A. (ed.) Encyclopedia of Parallel Computing, pp. 2071–2079. Springer (2011). https://doi.org/10.1007/978-0-387-09766-4_491
10. Dijkstra, E.W.: Hierarchical ordering of sequential processes. Acta Informatica **1**, 115–138 (1971)
11. Genest, B., Gimbert, H., Muscholl, A., Walukiewicz, I.: Optimal Zielonka-type construction of deterministic asynchronous automata. In: ICALP (2010)
12. Genest, B., Muscholl, A.: Constructing exponential-size deterministic Zielonka automata. In: ICALP'06. LNCS, vol. 4052. Springer (2006)
13. Hausmann, D., Lehaut, M., Piterman, N.: Distribution of reconfiguration languages maintaining tree-like communication topology. In: ATVA (2024)
14. Krishna, S., Muscholl, A.: A quadratic construction for Zielonka automata with acyclic communication structure. Theoretical Computer Science **503** (2013). https://doi.org/10.1016/j.tcs.2013.07.015
15. Madhusudan, P., Thiagarajan, P.S., Yang, S.: The MSO theory of connectedly communicating processes. In: FSTTCS'05. LNCS, vol. 3821. Springer (2005)

16. Monmege, B., Bérard, B., Sur, A., Srivathsan, B., Mohandi, A.: Faast: Fair asynchronous automata synthesis tool (Jan 2026). https://doi.org/10.5281/zenodo.18165965
17. Mukund, M.: Automata on distributed alphabets. In: Modern Applications of Automata Theory, IISc Research Monographs Series, vol. 2, pp. 257–288. World Scientific (2012)
18. Mukund, M., Sohoni Milind, A.: Keeping track of the latest gossip in a distributed system. Distributed Computing **10**(3), 137–148 (1997)
19. Pighizzini, G.: Synthesis of nondeterministic asynchronous automata. In: Semantics of programming languages and model theory. vol. 5, pp. 109–126. Gordon & Breach New York (1993)
20. Stefanescu, A., Esparza, J., Muscholl, A.: Synthesis of distributed algorithms using asynchronous automata. In: CONCUR (2003)
21. Zielonka, W.: Notes on finite asynchronous automata. Theoretical Informatics and Applications **21**(2), 99–135 (1987), http://www.numdam.org/item?id=ITA_1987__21_2_99_0

Abstract Lipschitz Continuity
Combining Semantic and Quantitative Approximations

Marco Campion[1,2], Isabella Mastroeni[3], Michele Pasqua[3], and Caterina Urban[2]

[1] LIP6, Sorbonne Université, Paris, France
marco.campion@lip6.fr
[2] Inria & ENS | PSL, Paris, France
caterina.urban@inria.fr
[3] University of Verona, Department of Computer Science, Verona, Italy
{isabella.mastroeni,michele.pasqua}@univr.it

Abstract. We introduce *Abstract Lipschitz Continuity* (ALC), an extensional (i.e., input/output) property that ensures proportionally bounded differences in the *semantic approximations* of the output of a function (e.g., a program semantics) when the *semantic approximations* of the input differ slightly. ALC explicitly discerns between two complementary notions of approximation: *quantitative* differences, expressed via pre-metrics, and *qualitative* (or *semantic*) differences, captured through upper closure operators. This explicit separation of approximations has two main advantages. First, it enables ALC to be related to other important extensional *program properties*, including partial abstract non-interference in language-based security, partial completeness in abstract interpretation, and abstract robustness in machine learning. Second, ALC enables reasoning about its validity for programs through inductive reasoning on their syntax and on the chosen semantic abstractions. To this end, we propose a sound deductive system, parameterized by the quantitative and semantic approximations of interest, for proving ALC of programs. This proof system makes explicit the assumptions required for ALC, thereby ensuring a compositional proof approach.

Keywords: Abstract Lipschitz Continuity, Abstract Interpretation, Partial Abstract Non-Interference, Partial Completeness, Abstract Robustness

1 Introduction

Lipschitz continuity (LC) is a fundamental property in calculus and computational analysis, providing a strong guarantee: A Lipschitz continuous function is one whose output changes at most linearly with respect to changes in its input. Its utility is wide-ranging, serving as a cornerstone for convergence guarantees in optimization [46] and a key tool for analyzing robustness and stability in machine learning and program analysis, particularly in adversarial settings [21,30,31,56].

© The Author(s) 2026
N. Bertrand and S. Milius (Eds.): FoSSaCS 2026, LNCS 16503, pp. 153–177, 2026.
https://doi.org/10.1007/978-3-032-22730-0_8

When reasoning about software programs, LC provides an essential measure of *program robustness* for code operating on uncertain input [10,11,12].

The standard definition of LC elegantly captures variations through metrics or other weaker forms of distances on the raw input and output spaces. In this work, we argue that making the *semantic dimension* of data explicit provides a natural and complementary perspective. Indeed, in complex systems such as machine learning models or programs, small changes in the *syntactic representation* of data (e.g., quantitative differences such as raw bit flips) may correspond to negligible or irrelevant *semantic differences*. By reasoning directly in terms of semantic properties, we can capture robustness and stability in a way that more closely aligns with the intended behavior of such systems. This enriched view of LC opens new possibilities for applications in areas ranging from program verification [16,25,52,53] to machine learning robustness [1,26,32].

For instance, in the context of security, there are several areas of application, such as code/SQL injection or file integrity, that motivate such explicit discrimination between quantitative and semantic properties of data. More concretely, SQL injection vulnerabilities arise from an unchecked interaction between untrusted input and the execution of a SQL query, allowing an attacker to execute arbitrary queries on the database. A trivial example is given on the left of Fig. 1 where the attacker is able to dump the whole `users` table by injecting the value `"3 OR 1 = 1"` in the (unsanitized) input `argv[0]`, forcing the query `q` condition into a tautology. In such a scenario, we could require that strings with a *similar* number of boolean operators do not yield a *difference* in the number of selected tuples. In other words, we aim at checking the distance between *properties* of inputs and outputs, rather than between concrete values. As another example, consider the scenario of malware detection, where detection tools typically aim to locate malicious patterns (e.g., signatures or byte sequences) in the files under inspection, or identify pattern presence to pinpoint potential malware infections in the files. A simple example is given by the code on the right of Fig. 1, which counts the occurrences of *"a"*, e.g., representing a malicious pattern, in a file. In this context, we could be interested in an integrity check: small changes in file dimension yield proportionally small changes in *"a"* (the pattern) count range, consistent with controlled, non-malicious file modifications. Dually, a small change in file dimension that yields a large change in the number of patterns, potentially out of the allowed range, may indicate a malicious infection in the file. Thus, we again need to compare distances between properties of data, rather than distances of data. Ordinary LC does not suit these scenarios, as distance metrics (e.g., the Levenshtein string distance) can only model quantitative data variations. One could define a tailored distance function that operates directly on semantic properties; however, this approach would embed the semantic abstraction within the distance calculation itself, making the resulting property less comparable with other established notions in the literature.

We introduce *Abstract Lipschitz Continuity* (ALC for short), a novel property that generalizes the notion of LC to the semantic domain (Sec. 3). ALC formally ensures that *differences in the semantic approximations* of inputs lead

```php
<?php

$id = $argv[0];
$q = "SELECT * FROM users WHERE id = $id;";
$result = pg_query($conn, $q);

?>
```

```c
int count-a(FILE *F) {
  int c, nc = 0;
  while ((c=fgetc(F)) != EOF) {
    if (c == 'a') nc++;
  }
  return nc;
}
```

Fig. 1: Example of SQL injection (left) and counting pattern program (right).

to *proportionally bounded differences in the semantic approximations* of outputs. In other words, ALC represents a generalized continuity-like requirement that explicitly incorporates semantic approximations of input and output spaces.

Our formalization is grounded in the theory of *abstract interpretation* [14,15]. We thus model *semantic* (or *qualitative*) approximations using upper closure operators, which effectively abstract concrete values (such as a string) to sets of values sharing a common property (e.g., all strings having the same length). ALC is defined by combining these semantic approximations with *quantitative* approximations via a distance function over a pre-metric space. Unlike the definition of LC, the definition of ALC is parametrized by these two forms of approximations.

Thanks to this explicit and parametrized distinction of approximations, we establish a formal correlation between ALC in the context of program semantics, and other foundational program properties (Sec. 4). We show how ALC relates to *(partial) completeness* [5,9,27] in abstract interpretation, a property essential for bounding imprecision in static analysis. By connecting quantitative stability to qualitative precision, we provide new insights into the interplay between these domains. For instance, we prove that, by fixing a program semantics of interest, any ALC program admits a *complete* best correct approximation over the abstract domain corresponding to the semantic approximation, when the chosen distance satisfies a structural property. Similarly, we demonstrate the utility of ALC in machine learning by relating it to *(abstract) robustness* [26,36], where ALC provides a stronger, quantified measure of semantic stability, generalizing simple qualitative robustness guarantees, but also in language-based security by relating ALC to *(partial) abstract non-interference* [8,25]. Overall, these results demonstrate that ALC is a stronger quantitative program property than the notions considered above. Properties such as abstract robustness are weaker as they can capture quantitative behaviors beyond those enforced by continuity-like properties; conversely, ALC excludes functions that satisfy quantitative constraints only in a weaker, non-continuous sense.

Finally, we propose a novel sound *deductive system* for verifying ALC for programs through inductive reasoning on their syntax (Sec. 5). Our system is parametric with respect to the chosen input and output semantic approximations and distance functions. It is designed to make explicit the assumptions required for ALC, thereby ensuring a compositional proof approach.

Related Work. We formally discuss the relationship between ALC and (partial) abstract non-interference [8,28,25] or approximate non-interference [48], (par-

$$\mathsf{Stm} \ni \mathsf{c} ::= \mathbf{skip} \mid x := \mathsf{a} \mid \mathsf{b}? \qquad\qquad [\![\mathsf{P}_1 \,;\, \mathsf{P}_2]\!]c \stackrel{\text{def}}{=} [\![\mathsf{P}_2]\!] \circ [\![\mathsf{P}_1]\!]c$$

$$\mathsf{Prog} \ni \mathsf{P} ::= \mathsf{c} \mid \mathsf{P}\,;\,\mathsf{P} \mid \mathsf{P} \oplus \mathsf{P} \mid \mathsf{P}^* \qquad [\![\mathsf{P}_1 \oplus \mathsf{P}_2]\!]c \stackrel{\text{def}}{=} [\![\mathsf{P}_1]\!]c \vee [\![\mathsf{P}_2]\!]c$$

$$\mathsf{a} \in \mathsf{AExp}, \ x \in \mathbb{X}, \ \mathsf{b} \in \mathsf{BExp} \qquad\qquad [\![\mathsf{P}^*]\!]c \stackrel{\text{def}}{=} \bigvee\{[\![\mathsf{P}]\!]^n c \mid n \in \mathbb{N}\}$$

Fig. 2: Syntax (left) and semantics (right) of Prog.

tial) completeness [5,9,27], and (abstract) robustness [26,36] in Sec. 4. The idea of software doping, technically introduced in [17], looks for a similar effect as ALC with the aim to detect intentionally developed ill-programs. The authors propose a Lipschitz-style condition for defining software doping which is tailored to the Hausdorff lifting. This is then further generalized via a functional modulus of continuity. Their work, like ours, discerns a property of interest, expressed through the Hausdorff lifting, while, in our approach, we are discerning a semantic property that can be formalized as a closure operator (or Galois connection). Other potential related works are cited in Sec. 6, since a better understanding of the relation with ALC deserves further research.

Our deductive system for verifying ALC in Sec. 5 generalizes and extends the one proposed by Chaudhuri et al. [11,12] for proving program robustness [26]. Specifically, when the semantic approximations are the identity function (namely, no semantic approximation is involved), our system recovers (an extension, due to the corresponding (**star**) rule, of) the deductive system in [11].

2   Background

In the following, we introduce the relevant background on programs, as well as approximations for simplifying reasoning about properties of their semantics.

2.1   Program Syntax and Semantics

We consider programs written in the language **Prog** of regular commands [4,47], which encompasses deterministic as well as non-deterministic and probabilistic computations. The syntax of **Prog** is on the left of Fig. 2, where $\oplus$ denotes non-deterministic choice and $*$ is the Kleene closure. For our purposes, we instantiate the basic commands $\mathsf{c} \in \mathsf{Stm}$ with **skip**, variable assignments, and Boolean tests. We assume a standard grammar for arithmetic expressions in **AExp** and Boolean expressions in **BExp**. Variables range from a denumerable set $\mathbb{X}$ while values range from a denumerable set $\mathbb{V}$ (e.g., integer or natural numbers).

Suppose we have a semantics $[\![\mathsf{c}]\!] : C \to C$ for basic commands $\mathsf{c} \in \mathsf{Stm}$ on a complete lattice $\langle C, \preceq, \vee, \wedge, \top, \bot \rangle$, where $\preceq$ is the partial order, $\vee$ is the least upper bound, $\wedge$ the greatest lower bound, $\top$ is the supremum of C and $\bot$ is the infimum of C. Then, the *semantics for programs* $[\![\cdot]\!] : \mathsf{Prog} \to C \to C$ is inductively defined on program syntax as on the right of Fig 2.

A notable instance of such concrete semantics is the *collecting semantics*, central in abstract interpretation [14,15], expressing the set of all possible program states that could occur at each program point. Formally, consider a complete lattice $\langle \wp(\mathsf{M}), \subseteq, \cup, \cap, \mathsf{M}, \varnothing \rangle$ of program memories, where $\mathsf{m} \in \mathsf{M}$ maps variables to values, namely $\mathsf{m} \colon \mathbb{X} \to \mathbb{V}$. We can define a collecting big-step semantics $[\![\mathsf{P}]\!] \colon \wp(\mathsf{M}) \to \wp(\mathsf{M})$ for a program $\mathsf{P} \in \mathsf{Prog}$ as the standard predicate transformer semantics on sets of program memories $\wp(\mathsf{M})$, collecting all possible program memories that reach the end of the program. Assume a big-step evaluation semantics $\Downarrow_{\mathsf{a}}$ for arithmetic expressions and $\Downarrow_{\mathsf{b}}$ for Boolean expressions. Given $S \in \wp(\mathsf{M})$, the semantics of basic commands is defined as:

$$[\![\mathsf{skip}]\!]S \stackrel{\text{def}}{=} S \quad [\![x := \mathsf{a}]\!]S \stackrel{\text{def}}{=} \{\mathsf{m}[x \hookleftarrow v] \mid \mathsf{m} \in S \wedge \mathsf{m} \Downarrow_{\mathsf{a}} v\} \quad [\![\mathsf{b}?]\!]S \stackrel{\text{def}}{=} \{\mathsf{m} \in S \mid \mathsf{m} \Downarrow_{\mathsf{b}} \mathsf{tt}\}$$

The collecting semantics for basic commands is monotone by construction on the powerset lattice of program memories, and so $[\![\mathsf{P}]\!]$ is also monotone for any program in Prog. We will extensively use this semantics in the examples of Sec. 4, 5.

2.2 Abstractions and Distances

In program analysis approximations are fundamental for simplifying reasoning while preserving essential properties. Following [8], we consider *qualitative* or *semantic* approximations, formalized by *upper closure operators*, and *quantitative* approximations, formalized by *pre-metrics*. Their combination yields a general approximation parametrized by both an abstraction and a distance.

Semantic Approximations via Upper Closure Operators. Qualitative or semantic approximations preserve certain *semantic properties* of the approximated data. Semantics approximations are at the hearth of abstract interpretation for approximating computations by evaluating functions (e.g., program semantics) over an abstract domain—a partially ordered set (poset, for short)—$\langle A, \preceq_A \rangle$ instead of the concrete domain $\langle C, \preceq_C \rangle$. A (monotone) concretization function $\gamma \colon A \to C$ relates abstract elements to their concrete counterparts, preserving the (partial) ordering of information. When paired with a (monotone) abstraction function $\alpha \colon C \to A$ such that $\alpha(c) \preceq_A a \Leftrightarrow c \preceq_C \gamma(a)$, $\forall a \in A$ and $\forall c \in C$, the pair forms a *Galois Connection* (GC) between the two domains. A GC is a *Galois Insertion* (GI) when $\alpha \circ \gamma = id$, where $id \stackrel{\text{def}}{=} \lambda x. \, x$.

Given a concrete domain $\langle C, \preceq_C \rangle$, GIs can be equivalently formulated in terms of *upper closure operators* [15] (ucos or closures, for short), namely a function $\rho \colon C \to C$ with the following properties $\forall c, c' \in C$: monotonicity ($c \preceq_C c' \Rightarrow \rho(c) \preceq_C \rho(c')$), extensivity ($c \preceq_C \rho(c)$), idempotence ($\rho(\rho(c)) = \rho(c)$). Ucos are uniquely determined by the set of their fixpoints: $\rho(C) = \{c \in C \mid \rho(c) = c\}$. For instance, the composition $\gamma \circ \alpha$ is an uco of C. In the following, we will often write ρf for the composition of any two functions $\rho \circ f$. The set of all ucos on a poset C is denoted by $uco(C)$. As an example, the closure $\mathsf{Sign} \in uco(\wp(\mathbb{Z}))$ abstracts a set of integers by discarding all information except the sign of its

	pre-	quasipseudo-	quasisemi-	semi-	quasi-	pseudo-	metric
(if-identity)	✓	✓	✓	✓	✓	✓	✓
(iff-identity)	✗	✗	✓	✓	✓	✗	✓
(symmetry)	✗	✗	✗	✓	✗	✓	✓
(triangle-inequality)	✗	✓	✗	✗	✓	✓	✓
Example		$\delta_\subseteq^{\mathsf{Int}}$			$\delta_\subseteq$	$\delta_{siz}, \delta_{\mathrm{DIM}}, \delta_-$	δ_2
Reference		Ex. 5			Ex. 3	Ex. 2, 6, 11	

Fig. 3: Metrics and their weakenings.

values, unless the set contains only the value 0. The closure is defined by the set of fixpoints $\mathsf{Sign}(\wp(\mathbb{Z})) \stackrel{\text{def}}{=} \{\varnothing, \{0\}, \{z \in \mathbb{Z} \mid z \leq 0\}, \{z \in \mathbb{Z} \mid z \geq 0\}, \mathbb{Z}\}$.

Definition 1 (Semantic Approximation [8]). *Given a poset $\langle C, \preceq_c \rangle$ and an uco $\rho \in uco(C)$, an element $x \in C$ is* semantically approximated *by $\rho(x)$.*

Example 1. Let $\mathsf{Int} \in uco(\wp(\mathbb{Z}))$ be the interval abstraction [13], mapping a set of integers $S \in \wp(\mathbb{Z})$ to the smallest interval $[l, u] \stackrel{\text{def}}{=} \{i \in \mathbb{Z} \mid l \leq i \leq u\}$ such that $S \subseteq [l, u]$, where $l \in \mathbb{Z} \cup \{-\infty\}$, $u \in \mathbb{Z} \cup \{+\infty\}$ and $l \leq u$. The set of integers $\{0, 1, 4\}$ can be semantically approximated by the interval $[0, 4]$ through Int. Moreover, the set $\{\{0,4\}, \{0,1,4\}, \{0,2,4\}, \{0,3,4\}, \{0,1,2,4\}, \{0,2,3,4\}, \{0,1,3,4\}, \{0,1,2,3,4\}\}$ contains all sets of integers S such that $\mathsf{Int}(S) = [0, 4]$. $\blacksquare$

Quantitative Approximations via Pre-Metrics. Quantitative approximations preserve *closeness* of the approximated data, here measured through a distance function between elements of a poset. We model distance functions using (pre-)metrics. Let $\mathbb{R}^\infty \stackrel{\text{def}}{=} \mathbb{R} \cup \{\infty\}$ such that for all $r \in \mathbb{R}$, $r < \infty$, and let $\mathbb{R}_{\geq 0}$ be the restriction of $\mathbb{R}$ to values greater or equal to 0.

Definition 2 (Pre-Metric [9]). *Given a non-empty set L, a* pre-metric *is a binary function $\delta \colon L \times L \to \mathbb{R}^\infty_{\geq 0}$ satisfying the only axiom:*

$$(\textit{if-identity}) \quad \forall x, y \in L.\ x = y \implies \delta(x, y) = 0.$$

The pair $\langle L, \delta \rangle$ is called a pre-metric space.

We will occasionally use the subscript δ_{L} in cases where the set L may not be immediately clear from the context. The same convention will be adopted for orderings $\preceq$. Pre-metrics are among the less restrictive formal notion of distances. On the other hand, *metrics* could be considered as very restrictive formal notion of distance, since they further satisfy the following axioms:

$$(\textit{iff-identity}) \quad x = y \iff \delta(x, y) = 0;$$
$$(\textit{symmetry}) \quad \delta(x, y) = \delta(y, x);$$
$$(\textit{triangle-inequality}) \quad \delta(x, y) \leq \delta(x, z) + \delta(z, y).$$

A classic example of metric is the Euclidean distance $\delta_2(x, y) \overset{\text{def}}{=} |x - y|$. Fig. 3 summarizes other distance notions lying between pre-metrics and metrics in terms of the properties that they satisfy. The last two rows display the distance symbol and the example in which the distance is defined and used for the first time. Understanding the type of distance function we are manipulating is essential for proving some implications between properties of programs (cf. Sec. 4). From this point forward, whenever we say that a function δ is a distance, we assume that it satisfies, at least, the axiom of a pre-metric.

Def. 2 is general enough to be instantiated with several practical notions of distance proposed in the abstract interpretation literature (e.g., [5,33,34,49,54]). The following examples illustrate two of these notions, used extensively throughout the rest of the paper. For additional examples of pre-metrics and their applications in domains used within the context of program analysis, we refer to [9].

Example 2 (Size Distance). Consider the powerset $\wp(L)$ of a set L. We write $size(S)$ for the number of elements in $S \in \wp(L)$. We define the size distance $\delta_{siz} : \wp(L) \times \wp(L) \to \mathbb{R}^{\infty}_{\geq 0}$ between two sets $S_1, S_2 \in \wp(L)$ as follows: $\delta_{siz}(S_1, S_2) \overset{\text{def}}{=} 0$ if $S_1 = S_2$, $\delta_{siz}(S_1, S_2) \overset{\text{def}}{=} |size(S_2) - size(S_1)|$ otherwise where $|\infty - \infty| = \infty$. In other words, δ_{siz} calculates the absolute value of the difference in their size. Note that δ_{siz} is a pseudo-metric: two sets may have the same size yet being different. In program analysis, δ_{siz} could be used to count, for instance, the number of spurious elements added by an abstract computation with respect to the abstraction of a concrete computation. For instance, if $[0, 0]$ is the (interval abstraction of the) strongest numerical invariant of a program variable x at certain program point, while $[0, 10]$ is the abstract invariant generated by an abstract interpretation over Int, then $\delta_{siz}(\{0\}, \{0, 1, \ldots, 10\}) = 10$ indicates that the abstract interpretation added 10 more spurious values with respect to the (interval abstraction of the) concrete execution. ∎

Example 3 (Inclusion Distance). Given a poset $\langle \wp(\mathbb{Z}), \subseteq \rangle$, we define the inclusion distance $\delta_{\subseteq} : \wp(\mathbb{Z}) \times \wp(\mathbb{Z}) \to \mathbb{N}^{\infty}$ such that $\delta_{\subseteq}(S_1, S_2) \overset{\text{def}}{=} k$ with $k \in \mathbb{N}$ if $S_1 \subseteq S_2$ and S_2 has k more elements than S_1. For all other cases, the distance is ∞. For instance, $\delta_{\subseteq}(\{0, 1, 4\}, \{0, 1, 4, 10\}) = 1$ while $\delta_{\subseteq}(\{0, 1, 4\}, \{1, 4, 10\}) = \infty$ because $\{0, 1, 4\} \not\subseteq \{1, 4, 10\}$. This is another distance, like δ_{siz}, that could be used in program analysis to count the number of spurious elements added by an abstract computation with respect to the abstraction of the concrete execution. Note that $\delta_{\subseteq}$ may differ from δ_{siz} even between comparable sets: $\delta_{\subseteq}(\mathbb{Z}_{>0}, \mathbb{Z}_{\geq 0}) = 1 \neq \infty = \delta_{siz}(\mathbb{Z}_{>0}, \mathbb{Z}_{\geq 0})$. The pair $\langle \wp(\mathbb{Z}), \delta_{\subseteq} \rangle$ forms a quasi-metric space. It is not a metric space because $\delta_{\subseteq}$ does not satisfy (*symmetry*). ∎

Definition 3 (Quantitative Approximation [8]). *Given a pre-metric space* $\langle C, \delta \rangle$ *and a fixed constant* $\varepsilon \in \mathbb{R}^{\infty}_{\geq 0}$, *an element* $x \in C$ *is* quantitatively approximated *by any element* $y \in C$ *such that* $\delta(x, y) \leq \varepsilon$.

Example 4. Continuing Ex. 1, we may approximate sets of integer numbers by the size distance δ_{siz} defined in Ex. 2. For instance, $\{0, 1, 4\}$ can be quantitatively approximated by any set of integers whose maximum distance from it is at most $\varepsilon = 1$. Examples of such approximations include sets $\{0, 1\}$ and $\{5, 6, 8, 10\}$. ∎

Here, the admitted noise concerns elements that are "close", according to the chosen δ, to the original one but that may share no semantic property.

Combining Semantic and Quantitative Approximations. By *combining* the two forms of approximation, we obtain a general approximation that incorporates a quantitative error within a qualitative abstraction. Given $\rho \in uco(C)$ and a pre-metric space $\langle C, \delta \rangle$, we define the distance $\delta^{\rho} \colon C \times C \to \mathbb{R}^{\infty}_{\geq 0}$ as

$$\delta^{\rho}(x, y) \stackrel{\text{def}}{=} \delta(\rho(x), \rho(y))$$

which calculates the distance between the semantic approximations of x and y.

Definition 4 (General Approximation [8]). *Let $\langle C, \preceq \rangle$ be a poset and $\langle C, \delta \rangle$ be a pre-metric space, and let $\rho \in \mathrm{uco}(C)$. An element $x \in C$ is semantically approximated with ρ and quantitatively approximated up to $\varepsilon \in \mathbb{R}^{\infty}_{\geq 0}$ by δ, by any element $y \in C$ such that $\delta^{\rho}(x, y) \leq \varepsilon$.*

Example 5. By considering the interval abstraction $\mathsf{Int} \in uco(\wp(\mathbb{Z}))$, we can combine the two forms of approximation, namely $\delta_{\subseteq}$ and Int, into $\delta_{\subseteq}^{\mathsf{Int}}$: this new distance calculates the number of more elements between two comparable interval abstractions rather than considering the original input sets. Note that $\delta_{\subseteq}^{\mathsf{Int}}$ loses the (*iff-identity*) axiom as one interval might represent more than one set in $\wp(\mathbb{Z})$, thus $\langle \wp(\mathbb{Z}), \delta_{\subseteq}^{\mathsf{Int}} \rangle$ forms a quasipseudo-metric space.

Continuing Ex. 3, the set $\{0, 1, 4\}$ can be semantically and quantitatively approximated by $\delta_{\subseteq}^{\mathsf{Int}}$ and $\varepsilon = 1$ in any set in $\{S \in \wp(\mathbb{Z}) \mid \delta_{\subseteq}^{\mathsf{Int}}(\{0, 1, 4\}, S) \leq 1\}$.

∎

3   Abstract Lipschitz Continuity

In mathematical analysis and calculus, Lipschitz continuity is a strong form of uniform continuity of functions that establishes a quantitative relationship between changes to the input of a function and the resulting changes in its output. Specifically, it imposes that perturbations to the input of a function lead to at most linear changes to its output. Lipschitz continuity is usually defined assuming that the input and output domains coincide and that both are metric spaces. In our setting, we allow for *distinct pre-metric input and output spaces*.

Definition 5 (Lipschitz Continuity). *Let $\langle C, \delta_C \rangle$ and $\langle D, \delta_D \rangle$ be pre-metric spaces. Let $k \in \mathbb{R}_{\geq 0}$. A function $f : C \to D$ satisfies k-Lipschitz continuity (k-LC for short) w.r.t. $\langle \delta_C, \delta_D \rangle$ if and only if:*

$$\forall x, y \in C. \ \delta_D(f(x), f(y)) \ \leq \ k\delta_C(x, y).$$

A function f satisfies Lipschitz continuity (LC) w.r.t. $\langle \delta_C, \delta_D \rangle$ if and only if there exists $k \in \mathbb{R}_{\geq 0}$ such that f satisfies k-Lipschitz continuity w.r.t. $\langle \delta_C, \delta_D \rangle$.

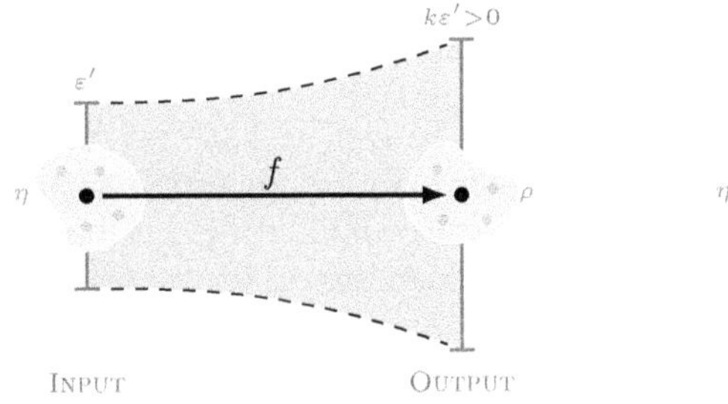
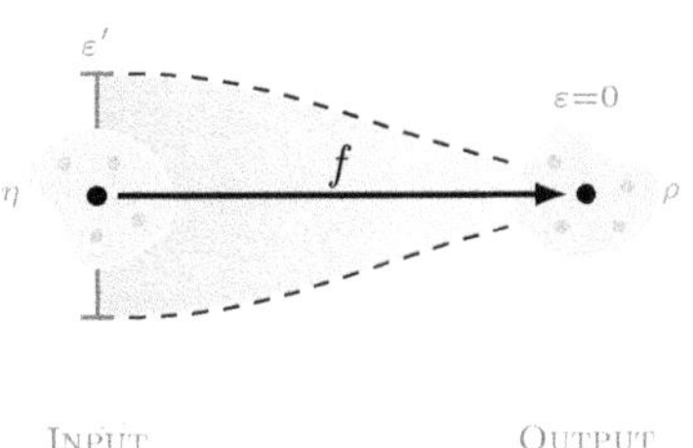

(a) Controlled input/output approximations.   (b) Suppression of input approximation.

Fig. 4: Abstract Lipschitz Continuity.

The Lipschitz constant k provides an upper bound on the rate of change for the output of the function f, i.e., $\delta_\mathrm{D}(f(x),f(y))/\delta_\mathrm{C}(x,y) \leq k$. Note that, k-Lipschitz continuity can be equivalently formulated as follows:

$$\forall x, y \in C.\ \forall \varepsilon' \geq 0.\ \delta_\mathrm{C}(x, y) \leq \varepsilon' \Rightarrow \delta_\mathrm{D}(f(x), f(y)) \leq k\varepsilon'$$

When approximations are introduced to the input of a function, they propagate through its computations, affecting the output. Understanding how approximations evolve during computations provides insight into the behavior of the function and, consequently, into program executions when the function represents program semantics [10,11,57,35,7].

In this work, we generalize the definition of Lipschitz continuity (Def. 5), which relies on quantitative approximations (Def. 3), to work with general approximations (Def. 4), *explicitly discerning semantic approximations (i.e., ucos) from quantitative approximations (i.e., distances)*. This yields the novel notion of *abstract Lipschitz continuity*, which enforces a *controlled (linear) error propagation* from a *general approximation of the input* to a *general approximation of the output* of a function computation (Fig. 4a).

Definition 6 (Abstract Lipschitz Continuity). *Let $\langle C, \preceq_\mathrm{C} \rangle$ and $\langle D, \preceq_\mathrm{D} \rangle$ be posets, $\langle C, \delta_\mathrm{C} \rangle$ and $\langle D, \delta_\mathrm{D} \rangle$ be pre-metric spaces, and let $\eta \in \mathrm{uco}(C)$, $\rho \in \mathrm{uco}(D)$, and $k \in \mathbb{R}_{\geq 0}$. A function $f \colon C \to D$ satisfies* Abstract k-Lipschitz Continuity *(k-ALC, for short) w.r.t. $\langle \delta_\mathrm{C}^\eta, \delta_\mathrm{D}^\rho \rangle$ when:*

$$\forall x, y \in C.\ \delta_\mathrm{D}^\rho(f(x), f(y)) \leq k\delta_\mathrm{C}^\eta(x, y)$$

A function f satisfies Abstract Lipschitz Continuity *(ALC) if and only if there exists $k \in \mathbb{R}_{\geq 0}$ such that f satisfies abstract k-Lipschitz continuity.*

In other words, f satisfies ALC when it satisfies LC w.r.t. $\langle \delta_\mathrm{C}^\eta, \delta_\mathrm{D}^\rho \rangle$. When k-ALC holds, k will be called the *abstract Lipschitz constant*. 0-ALC represents a special case in which the function computation completely suppresses the input semantic approximation (Fig. 4b).

It is worth remarking that Def. 5 can be instantiated to model Def. 6 by incorporating the distance calculus between semantic approximations within δ_C and δ_D. However, explicitly discerning semantic approximations (i.e., ucos) from quantitative approximations (i.e., distances) as in Def. 6, has two main advantages. First it allows us to relate ALC to other extensional program properties (i.e., relative to the input/output behavior of programs) studied in the program analysis literature (Sec. 4). Second, it allows reasoning on its validity over programs by an inductive reasoning on their syntax and on the chosen ucos (Sec. 5).

Similarly to k-LC, k-ALC can be equivalently reformulated as follows:

Proposition 1. *Consider the premises of Def. 6. A function $f \colon C \to D$ satisfies k-ALC w.r.t. $\langle \delta_C^\eta, \delta_D^\rho \rangle$ if and only if:*

$$\forall x, y \in C.\ \forall \varepsilon' \geq 0.\ \delta_C^\eta(x, y) \leq \varepsilon' \ \Rightarrow\ \delta_D^\rho(f(x), f(y)) \leq k\varepsilon'.$$

Example 6. Consider the fragment of C code in Fig. 1 (on the right), which counts the occurrences of "a" in a file $F \in \textsc{Files}$. Let $[\![\texttt{count-a}]\!]$ denote its semantics. As described in the introduction, "a" could stand for a specific malicious pattern (e.g., signature or byte sequence) that a malware detection tool may look for. In this context, ALC could serve as an integrity check: small changes in file dimension yield proportionally small changes in the "a" (pattern) count range, consistent with controlled, non-malicious modifications. On the other hand, it is suspicious if a small change in file dimension causes a big change in the number of patterns. More formally, let $\textsc{Dim}(F) \colon \textsc{Files} \to \mathbb{N}$ be the function extracting the dimension of a file $F \in \textsc{Files}$. We define an (input) abstraction η by additive lift to sets of files: $\eta \stackrel{\text{def}}{=} \lambda F.\ \{F' \mid \textsc{Dim}(F) = \textsc{Dim}(F')\}$. Moreover, given sets of files $\mathbf{F}, \mathbf{F'}$, we define the pseudo-metric $\delta_{\textsc{Dim}}(\mathbf{F}, \mathbf{F'}) = |\max\{\textsc{Dim}(F) \mid F \in \mathbf{F}\} - \max\{\textsc{Dim}(F) \mid F \in \mathbf{F'}\}|$, measuring the difference between file dimensions. Finally, we define an (output) abstraction ρ that only keeps track of reasonable output values, say, between -1000 and 1000 pattern occurrences: $\rho \stackrel{\text{def}}{=} \lambda n.\ n$ if $-1000 \leq n \leq 1000$, and $\rho \stackrel{\text{def}}{=} \lambda n.\top$ otherwise. We formalize our integrity property of interest using ALC:

$$\forall F_1, F_2.\ \delta_2(\rho[\![\texttt{count-a}]\!](F_1), \rho[\![\texttt{count-a}]\!](F_2)) \leq k\delta_{\textsc{Dim}}(\eta(F_1), \eta(F_2))$$

where δ_2 is the Euclidean distance. Now, if a slight change in the file dimension caused a non-linear increase of "a", i.e., if $\exists F_1, F_2 \in \textsc{Files}$, and $\varepsilon \in \mathbb{R}_{\geq 0}$ such that $\delta_{\textsc{Dim}}(\eta(F_1), \eta(F_2)) \leq \varepsilon$ but $\delta_2(\rho[\![\texttt{count-a}]\!](F_1), \rho[\![\texttt{count-a}]\!](F_2)) \not\leq k\varepsilon$, then something suspicious has happened, e.g., a malware attack transforming file F_1 into file F_2 by replicating the malicious pattern "a". $\blacksquare$

4   ALC and Other Extensional Program Properties

If we consider the set of all programs whose semantics (to be specified) satisfy Def. 6, then, ALC represents an extensional property of programs, i.e., relative to their input/output behavior. By explicitly discerning the role of qualitative and quantitative approximations in its definition, we can relate ALC to other important extensional properties in the literature that capture input/output relations in comparable yet distinct ways: *partial abstract non-interference* (Eq. 1), *partial completeness* (Eq. 2), and *abstract robustness* (Eq. 3).

Partial Abstract Non-Interference. Partial Abstract Non-Interference [8] (PANI for short) is a relaxation of (abstract) non-interference [28,25] that combines both semantic and quantitative approximations on the output of a function computation. This notion can be seen as a generalization of approximate non-interference [48], originally introduced in a probabilistic process algebra, requiring the *observable* behaviors of two agents to be under a similarity threshold ε. Specifically, PANI observes indistinguishable (distance-wise) *properties* of input data, while allowing a *bounded distance* between the observed output *properties*. This enables a more nuanced treatment of security policies, especially when small, bounded differences in outputs are tolerable. In the same premises of Def. 6, ε-PANI$[\delta_C^\eta, \delta_D^\rho]$ has been formalized as [8]:

$$\forall x, y \in C.\ \delta_C^\eta(x, y) = 0 \implies \delta_D^\rho(f(x), f(y)) \leq \varepsilon \tag{1}$$

Given inputs with zero property distance under δ_C^η, ε-PANI$[\delta_C^\eta, \delta_D^\rho]$ allows the function to produce different outputs, potentially with different properties under δ_D^ρ, as long as the variation remains bounded, i.e., not exceeding the threshold ε.

It turns out that ALC is a strictly stronger property than PANI, as stated by the following theorem.

Theorem 1. *Let $\langle C, \delta_C \rangle$ and $\langle D, \delta_D \rangle$ be pre-metric spaces. If $f \colon C \to D$ satisfies ALC w.r.t. $\langle \delta_C^\eta, \delta_D^\rho \rangle$, then, for any $\varepsilon \in \mathbb{R}_{\geq 0}$, f also satisfies ε-PANI$[\delta_C^\eta, \delta_D^\rho]$.*

In the context of programs where f represents the semantics $[\![P]\!]$ of a program P, requiring $[\![P]\!]$ to satisfy ALC is a stronger condition than requiring ε-PANI.

Example 7. Let $R = (x > 0?\,;\ x := x - 1) \oplus (x \leq 0?\,;\ x := x + 1)$ be a program which increments all non-negative values by 1 and decrements all non-positive values by 1. Let us consider the pseudo-metric space $\langle \wp(\mathbb{Z}), \delta_{siz} \rangle$ as the input and output domain, and the interval closure $\mathsf{Int} \in uco(\wp(\mathbb{Z}))$. The collecting semantics $[\![R^*]\!] : \wp(\mathbb{Z}) \to \wp(\mathbb{Z})$ of the Kleene closure of R, R^*, satisfies 1-ALC w.r.t. $\langle \delta_{siz}^{\mathsf{Int}}, \delta_{siz}^{\mathsf{Int}} \rangle$. Indeed, $[\![R^*]\!]$ is monotone by definition, thus preserving the inclusion relation, and either reduces the distance of input intervals or leaves them unchanged. For instance:

$\delta_{siz}^{\mathsf{Int}}([\![R^*]\!]([2,6]), [\![R^*]\!]([0,7])) = \delta_{siz}^{\mathsf{Int}}([0,6], [0,7]) = 1 \leq 3 = \delta_{siz}^{\mathsf{Int}}([2,6], [0,7])$

$\delta_{siz}^{\mathsf{Int}}([\![R^*]\!]([-5,-2]), [\![R^*]\!]([-7,0])) = \delta_{siz}^{\mathsf{Int}}([-5,1], [-7,1]) = 2 \leq 5 = \delta_{siz}^{\mathsf{Int}}([-5,-2], [-7,0])$

$\delta_{siz}^{\mathsf{Int}}([\![R^*]\!]([-2,3]), [\![R^*]\!]([-5,3])) = \delta_{siz}^{\mathsf{Int}}([-2,3], [-5,3]) = 3 \leq 3 = \delta_{siz}^{\mathsf{Int}}([-2,3], [-5,3])$

By Thm. 1, the program semantics $[\![R^*]\!]$ of R^* also satisfies 0-PANI$[\delta_{siz}^{\mathsf{Int}}, \delta_{siz}^{\mathsf{Int}}]$. $\blacksquare$

As a corollary result, when δ_D additionally satisfies the (*iff-identity*), ALC implies the Abstract Non-Interference (ANI) property, where distances are converted as equality requirements [25]: $\forall x, y \in C.\ \eta(x) = \eta(y) \implies \rho f(x) = \rho f(y)$.

Corollary 1. *Let $\langle C, \delta_\mathsf{C} \rangle$ and $\langle D, \delta_\mathsf{D} \rangle$ be quasisemi-metric spaces. If $f \colon C \to D$ satisfies ALC w.r.t. $\langle \delta_\mathsf{C}^\eta, \delta_\mathsf{D}^\rho \rangle$, then f satisfies also ANI.*

Partial Completeness. Partial completeness [5,9] is a weakening of the standard notion of completeness in abstract interpretation. In the classical abstract interpretation framework [14], the computation of a concrete (possibly uncomputable) monotone function $f \colon C \to C$ over a poset $\langle C, \preceq_\mathsf{C} \rangle$ is replaced by an abstract (possibly computable) sound computation $f^\natural \colon \rho(C) \to \rho(C)$ over an abstract domain $\rho(C)$, where $\rho \in uco(C)$. For example, if P is a program with a single variable then f could be the collecting big-step semantics $[\![\mathsf{P}]\!] \colon \wp(\mathbb{Z}) \to \wp(\mathbb{Z})$ defined in Sec. 2.1 over $\langle \wp(\mathbb{Z}), \subseteq \rangle$, $\rho = \mathsf{Int}$ the interval abstract domain (Ex. 1) and $[\![\mathsf{P}]\!]^\natural \colon \mathsf{Int}(\wp(\mathbb{M})) \to \mathsf{Int}(\wp(\mathbb{M}))$ is the abstract interval semantics soundly computed on intervals. An abstract function $f^\natural \colon \rho(C) \to \rho(C)$ is *sound* when $\rho f \preceq_\mathsf{C} f^\natural \rho$, and is said to be *complete* when $\rho f = f^\natural \rho$, i.e., $f^\natural$ coincides with the *best correct approximation* (bca) [15] $\bar{f} \stackrel{\text{def}}{=} \rho f \rho$ of f. Completeness is the best possible scenario where no imprecision is introduced by the abstract computation of $f^\natural$. In practice, however, completeness is rarely achieved. For this reason, Campion et al. [5,9] introduced a weaker notion of completeness, called ε-*partial completeness* which allows some degree of imprecision limited by ε. This imprecision is measured by pre-metrics $\delta_{\rho(\mathsf{C})} \colon \rho(C) \times \rho(C) \to \mathbb{R}^\infty_{\geq 0}$ that are $\preceq_\mathsf{C}$-*compatible* with the abstract domain, i.e., pre-metrics additionally satisfying [9]:

$$x \preceq_\mathsf{C} y \preceq_\mathsf{C} z \;\Rightarrow\; \delta_{\rho(\mathsf{C})}(x, y) \leq \delta_{\rho(\mathsf{C})}(x, z) \wedge \delta_{\rho(\mathsf{C})}(y, z) \leq \delta_{\rho(\mathsf{C})}(x, z)$$

Partial completeness was originally defined as a local property (namely, defined on a strict subset of the input domain) [5]. Here, for our purposes, we consider a *global* version of ε-partial completeness:

$$\forall x \in C. \; \delta_{\rho(\mathsf{C})}(\rho f(x), f^\natural \rho(x)) \leq \varepsilon \tag{2}$$

To relate ALC w.r.t. $\langle \delta_\mathsf{C}^\rho, \delta_\mathsf{C}^\rho \rangle$ for a monotone function $f \colon C \to C$, to partial completeness, we observe that, given a pre-metric $\preceq_\mathsf{C}$-compatible δ_C defined over C, since $\rho(C) \subseteq C$ we can always restrict δ_C to elements of $\rho(C)$, i.e., $\forall x, y \in \rho(C) \colon \delta_{\rho(\mathsf{C})}(x, y) \stackrel{\text{def}}{=} \delta_\mathsf{C}(x, y)$ without losing the $\preceq_\mathsf{C}$-compatible property.

Theorem 2. *Let $\langle C, \delta_\mathsf{C} \rangle$ be a $\preceq_\mathsf{C}$-compatible pre-metric space. If $f \colon C \to C$ satisfies ALC w.r.t. $\langle \delta_\mathsf{C}^\rho, \delta_\mathsf{C}^\rho \rangle$, then the bca $\bar{f}$ of f satisfies ε-partial completeness w.r.t. f for any $\varepsilon \in \mathbb{R}_{\geq 0}$.*

Corollary 2. *Let $\delta_{\rho(\mathsf{C})}$ be a quasisemi-metric. If $f \colon C \to C$ satisfies ALC w.r.t. $\langle \delta_\mathsf{C}^\rho, \delta_\mathsf{C}^\rho \rangle$, then the bca $\bar{f}$ of f satisfies completeness w.r.t. f.*

Thm. 2 and Cor. 2 reveal a novel relationship between the ALC of a function f and the partial completeness of its bca, particularly in the context of program analysis where f represents a (monotone) semantics $[\![\mathsf{P}]\!]$ of a given program P.

Notably, a proof of ALC of $[\![P]\!]$ w.r.t. the distance $\delta_{\mathsf{C}}^{\rho}$ implies that the imprecision introduced by approximating $[\![P]\!]$ with its bca $\rho[\![P]\!]\rho$ is bounded by any constant factor. Consequently, $\rho[\![P]\!]\rho$ also satisfies 0-partial completeness, namely *no imprecision is generated by the bca according to δ_{C}.* Thus, ALC of a program $[\![P]\!]$ *ensures the existence of an abstract interpreter capable of approximating the computation of $[\![P]\!]$ over $\rho(C)$ with no imprecision* according to the imprecision measured by δ_{C}. Furthermore, when the distance $\delta_{\rho(\mathsf{C})}$ is a quasisemi-metric, then the bca $\rho[\![P]\!]\rho$ is guaranteed to satisfy completeness, namely, *no imprecision is generated by the bca.* Another way to interpret Cor. 2 (and, analogously, Thm. 2) is as follows: if a program $[\![P]\!]$ does not admit a complete bca $\rho[\![P]\!]\rho$, then $[\![P]\!]$ *cannot* satisfy ALC for any quasisemi-metric.

Example 8. Consider again the collecting semantics $[\![R^*]\!]$ of Ex. 7 which has been proved to satisfy 1-ALC w.r.t. $\langle \delta_{siz}^{\mathsf{Int}}, \delta_{siz}^{\mathsf{Int}} \rangle$. It is easy to note that $[\![R^*]\!]$ also satisfies 1-ALC w.r.t. $\langle \delta_{\sqsubseteq}^{\mathsf{Int}}, \delta_{\sqsubseteq}^{\mathsf{Int}} \rangle$, where $\delta_{\sqsubseteq}$ is the quasi-metric defined in Ex. 3. Then by Thm. 2, the bca $\mathsf{Int}[\![R^*]\!]\mathsf{Int}$ also satisfies 0-partial completeness w.r.t. $\delta_{\sqsubseteq}^{\mathsf{Int}}$, i.e., $\delta_{\sqsubseteq}^{\mathsf{Int}}([\![R^*]\!]S, [\![R^*]\!]\mathsf{Int}(S)) \le 0$, for any $S \in \wp(\mathbb{Z})$. Moreover, since $\delta_{\sqsubseteq}$ is also a quasisemi-metric, we can conclude that $\mathsf{Int}[\![R^*]\!]\mathsf{Int}$ is complete, namely it does not add any imprecision when approximating $[\![R^*]\!]$.

Consider instead the pseudo-metric space $\langle \wp(\mathbb{Z}), \delta_{siz} \rangle$ and the semantics $[\![R]\!]$ of program R. The bca $\mathsf{Int}[\![R]\!]\mathsf{Int}$ does not satisfy 0-partial completeness w.r.t. $\delta_{siz}^{\mathsf{Int}}$: given $X = \{-1, 1\}$, we have that $[\![R]\!]$ cannot satisfy ALC for $\langle \delta_{siz}^{\mathsf{Int}}, \delta_{siz}^{\mathsf{Int}} \rangle$ since $\delta_{siz}^{\mathsf{Int}}([\![R]\!]X, [\![R]\!]\mathsf{Int}(X)) = \delta_{siz}^{\mathsf{Int}}([0,0],[0,1]) = 1 \ne 0$. In fact, it is easy to note that $[\![R]\!]$ satisfies 1-partial completeness for all inputs.   ∎

Abstract Robustness. The idea of abstract robustness is to formally characterize the amount of error that an analyst may tolerate in a machine learning classification result [26]. This error, generated by an input perturbation, either a semantic or a quantitative approximation [26], is modeled as a semantic approximation of the classification outcome, expressed through a uco ρ. Given an admissible margin of error $\varepsilon \in \mathbb{R}_{\ge 0}$, and assuming f to be a classifier, abstract robustness with a quantitative input perturbation is defined as [26]:

$$\forall x, y \in C. \; \delta_{\mathsf{C}}(x, y) \le \varepsilon \;\Rightarrow\; \rho f(x) = \rho f(y) \tag{3}$$

If we look at this definition in terms of ALC, then we observe that this notion requires that the error introduced by the perturbation of the input must be neutralized, at least for the ρ observation of the output classification. In fact, if we consider in output a quasi-metric the equality means distance smaller than 0, and this corresponds precisely to 0-ALC.

Theorem 3. *Let $\langle D, \delta_{\mathsf{D}} \rangle$ be a quasisemi-metric space. 0-ALC w.r.t. $\langle \delta_{\mathsf{C}}^{id}, \delta_{\mathsf{D}}^{\rho} \rangle$ holds if and only if abstract robustness holds.*

Setting the abstract Lipschitz constant to zero is necessary to stabilize the output of the classifier under the same semantic abstraction ρ. The distance δ_{D}

must also be a quasisemi-metric to ensure that equal elements can be distinguished when the distance evaluates to zero. What is interesting in this result is that abstract robustness, and therefore standard robustness, is a specific instance of ALC, the one where the potential error in input does not affect in any way the classification of the output (at least in the observed output property). In this sense, ALC becomes a generalization of robustness by *admitting* an effect in the output, but this effect must be linearly bounded to the input error. From this perspective, ALC serves as a model for further weakening robustness in ML, enabling us to better adapt it to specific contexts or fields of application.

5 Proving ALC for Programs

Deductive systems for the verification of completeness [23], partial completeness [7] and Lipschitz continuity [11,12] properties of programs have already been formalized in the literature. In this section we introduce a novel deductive system, inductively defined on the program's syntax, that is able to soundly prove the new ALC notion of a program semantics w.r.t. the input and output abstractions $\langle \eta, \rho \rangle$ and a given pre-metric δ. Our objective in designing this deductive system is to identify and track the assumptions necessary to establish a proof of ALC. In a Hoare-style approach, the proof system also allows to weaken or strengthen both the input and output abstractions when attempting to complete a proof of ALC for a program. Soundness here means that when the semantics $[\![\mathsf{P}]\!] \colon C \to C$ of a program P is typed as k-ALC w.r.t. $\langle \delta^\eta, \delta^\rho \rangle$ by the deductive system, then $[\![\mathsf{P}]\!]$ is certainly k-ALC for $\langle \delta^\eta, \delta^\rho \rangle$.

Given $k \in \mathbb{R}_{\geq 0}^\infty$, $\eta, \rho \in uco(C)$, $\delta \colon C^2 \to \mathbb{R}_{\geq 0}^\infty$, we introduce the following set

$$k\text{-}ALip\langle \delta^\eta, \delta^\rho \rangle \overset{\text{def}}{=} \{ f \in C \to C \mid f \text{ is } k\text{-ALC w.r.t. } \langle \delta^\eta, \delta^\rho \rangle \}$$

of all abstract k-Lipschitz continuous functions on a complete lattice $\langle C, \preceq \rangle$ for $\langle \delta^\eta, \delta^\rho \rangle$. In order to preserve the soundness of derivations in the proof system, we allow k to take the value ∞, since this is an admissible sound (albeit possibly imprecise) result. Clearly, any function is ∞-ALC. The following lemma outlines some basic properties of $k\text{-}ALip\langle \delta^\eta, \delta^\rho \rangle$ that will be later exploited.

Lemma 1. *The following hold for all functions* $f \in C \to C$, *closures* $\eta, \rho \in uco(C)$, *pre-metric* δ *and* $k \in \mathbb{R}_{\geq 0}^\infty$:

 (i) $k \geq 1 \;\Rightarrow\; \rho \in k\text{-}ALip\langle \delta^\rho, \delta^\rho \rangle$

 (ii) f *is* k-LP *w.r.t.* $\langle \delta, \delta \rangle \;\Leftrightarrow\; f \in k\text{-}ALip\langle \delta^{id}, \delta^{id} \rangle$

 (iii) $\rho \in k\text{-}ALip\langle \delta^{id}, \delta^{id} \rangle \;\Leftrightarrow\; \rho \in k\text{-}ALip\langle \delta^{id}, \delta^\rho \rangle$

(i) states that, when considering the same input-output abstractions (i.e. $\eta = \rho$), then the abstraction function is k-ALC for any $k \geq 1$. Moreover, for the statement (ii), when both input-output abstractions are the identity function id, then the class $k\text{-}ALip\langle \delta^{id}, \delta^{id} \rangle$ corresponds precisely to the set of all k-LP functions. Finally, (iii) shows that, when a closure ρ satisfies ALC w.r.t.

$$\frac{[\![\mathsf{c}]\!] \in k\text{-}ALip\langle \delta^\eta, \delta^\rho \rangle}{k \vdash [\delta^\eta]\,\mathsf{c}\,[\delta^\rho]}\;(\textbf{base})$$

$$\frac{k' \vdash [\delta^{\eta'}]\,\mathsf{P}\,[\delta^{\rho'}] \quad k' \leq k \quad \eta' \in t\text{-}ALip\langle \delta^\eta, \delta^{\eta'} \rangle \quad \rho \in s\text{-}ALip\langle \delta^{\rho'}, \delta^\rho \rangle}{stk \vdash [\delta^\eta]\,\mathsf{P}\,[\delta^\rho]}\;(\textbf{weaken})$$

$$\frac{k_1 \vdash [\delta^\eta]\,\mathsf{P}_1\,[\delta^\rho] \quad k_2 \vdash [\delta^\eta]\,\mathsf{P}_2\,[\delta^\rho] \quad \eta \in t\text{-}ALip\langle \delta^\rho, \delta^\eta \rangle}{k_1 tk_2 \vdash [\delta^\eta]\,\mathsf{P}_1;\mathsf{P}_2\,[\delta^\rho]}\;(\textbf{seq})$$

$$\frac{k_1 \vdash [\delta^\eta]\,\mathsf{P}_1\,[\delta^\rho] \quad k_2 \vdash [\delta^\eta]\,\mathsf{P}_2\,[\delta^\rho]}{k_1 \uplus k_2 \vdash [\delta^\eta]\,\mathsf{P}_1 \oplus \mathsf{P}_2\,[\delta^\rho]}\;(\textbf{join})$$

$$\frac{\eta \in t\text{-}ALip\langle \delta^\rho, \delta^\eta \rangle \quad k \vdash [\delta^\eta]\,\mathsf{P}\,[\delta^\rho] \quad \rho \in v\text{-}ALip\langle \delta^\eta, \delta^\rho \rangle \quad *\text{-Bound}(\mathsf{P}^*, m)}{K_m \vdash [\delta^\eta]\,\mathsf{P}^*\,[\delta^\rho]}\;(\textbf{bound-star})$$

$$\frac{k \vdash [\delta^\eta]\,\mathsf{P}\,[\delta^\rho] \quad \eta \in t\text{-}ALip\langle \delta^\rho, \delta^\eta \rangle \quad \rho \in v\text{-}ALip\langle \delta^\eta, \delta^\rho \rangle}{K_\infty \vdash [\delta^\eta]\,\mathsf{P}^*\,[\delta^\rho]}\;(\textbf{star})$$

Fig. 5: A deductive system for proving ALC for Prog.

$\langle \delta^{id}, \delta^{id} \rangle$, then ρ also satisfies k-ALC for $\langle \delta^{id}, \delta^\rho \rangle$, and vice-versa. This is due to the idempotence property of closure operators.

From now on, we fix a program semantics of interest $[\![\cdot]\!]: \mathsf{Prog} \to C \to C$ as well as a complete lattice $\langle C, \preceq, \vee, \wedge, \top, \bot \rangle$, and we will also use the statement "P is k-ALC w.r.t. $\langle \delta^\eta, \delta^\rho \rangle$" to indicate that the semantics $[\![\mathsf{P}]\!]$ is abstract k-Lipschitz continuous w.r.t. $\langle \delta^\eta, \delta^\rho \rangle$, i.e. $[\![\mathsf{P}]\!] \in k\text{-}ALip\langle \delta^\eta, \delta^\rho \rangle$.

The deductive rules are provided in Fig. 5. The judgments take the form:

$$k \vdash [\delta^\eta]\,\mathsf{P}\,[\delta^\rho]$$

We will later show that deriving a judgment $k \vdash [\delta^\eta]\,\mathsf{P}\,[\delta^\rho]$ through the deductive rules in Fig. 5, implies that $[\![\mathsf{P}]\!] \in k\text{-}ALip\langle \delta^\eta, \delta^\rho \rangle$. Let us examine each rule and provide an intuitive, informal explanation.

The (**base**) rule allows deriving the triple $k \vdash [\delta^\eta]\,\mathsf{c}\,[\delta^\rho]$ for all basic commands $\mathsf{c} \in \mathsf{Stm}$ (i.e., for **skip**, assignments and Boolean guards) assuming that we have a proof of k-ALC of them, encoded by the predicate $[\![\mathsf{c}]\!] \in k\text{-}ALip\langle \delta^\eta, \delta^\rho \rangle$.

The (**weaken**) rule allows weakening both the abstract Lipschitz constant and the abstractions considered. In particular, when we are able to derive the k'-ALC for program P w.r.t. $\langle \delta^{\eta'}, \delta^{\rho'} \rangle$, then we can always deduce a higher abstract Lipschitz constant $k \geq k'$ without changing the validity of the triple. For the input abstraction η', we can consider a new input abstraction η whenever η' is proved to be t-ALC w.r.t. $\langle \delta^\eta, \delta^{\eta'} \rangle$ with η as input abstraction. This weakening comes at the cost of multiplying the already deduced constant k' with the new constant t. This could happen, for instance, when η is in fact widening the distance $\delta^{\eta'}(c_1, c_2)$ between any two elements $c_1, c_2 \in C$, by a constant factor of t, namely by $t\delta^\eta(c_1, c_2)$. On the other hand, we can weaken the output abstraction ρ' by a new abstraction ρ whenever ρ is proved to be ALC for $\langle \delta^{\rho'}, \delta^\rho \rangle$ namely with ρ' as input abstraction. Here ρ could represent a narrow output abstraction in terms of distance δ between elements in C with respect to ρ', namely having distance $\delta^\rho(c_1, c_2) \le s\delta^{\rho'}(c_1, c_2)$ and thus introducing a new abstract Lipschitz constant s. Note that (**weaken**) allows also for selecting which weakening we want to apply. For instance, if we want to weaken the abstract Lipschitz constant k' only, then we can set $\eta' \in 1\text{-}ALip\langle \delta^{\eta'}, \delta^{\eta'} \rangle$ and $\rho' \in 1\text{-}ALip\langle \delta^{\rho'}, \delta^{\rho'} \rangle$ in the premises as they always hold (by Lem. 1(i)) without modifying any abstraction.

Composition of programs is treated by the (**seq**) rule. Although it is well known that composing two functions f_1 and f_2 which are k_1-LP and k_2-LP, respectively, gives as result a new $k_1 k_2$-LP function $f_2 \circ f_1$, this in general does not always hold for ALC as abstractions come into play. However, when we have a derivation for P_1 and P_2 with abstract Lipschitz constants k_1 and k_2, respectively, and we are able to prove that the input abstraction η is t-ALC w.r.t. $\langle \delta^\rho, \delta^\eta \rangle$, then this is a sufficient condition for deriving the $k_2 t k_1$-ALC of the composition $\mathsf{P}_1; \mathsf{P}_2$. Requiring $\eta \in t\text{-}ALip\langle \delta^\rho, \delta^\eta \rangle$ corresponds to require $\delta^\eta(c_1, c_2) \le t\delta^\rho(c_1, c_2)$, namely that we have a linear relation between their distances: when $t \ge 1$ then ρ is widening the distance, while when $0 < t < 1$ then ρ is narrowing their distances, both cases with a constant factor of t. Note that, when the input and output abstractions coincide, i.e. $\eta = \rho$, then $\rho \in 1\text{-}ALip\langle \delta^\rho, \delta^\rho \rangle$ holds trivially (by Lem. 1(i)). As a consequence, for the case $\eta = \rho$, the ALC property is closed under composition, analogously to the LP property.

The rule (**join**) involves the join operator. Similarly for the composition, the join of two ALC functions is not necessarily ALC. The problem here stems in the fact that the resulting abstract Lipschitz constant bound could not be determined by knowing only the abstract Lipschitz constants of both P_1 and P_2. This is because the distance between the execution of $\mathsf{P}_1 \oplus \mathsf{P}_2$ and the join of the two post-conditions, relies on the underlying structure of the input and output abstractions considered. Our solution, inspired by [5,7], consists in parametrizing the proof system with a binary operation $\uplus : \mathbb{R}_{\ge 0}^\infty \times \mathbb{R}_{\ge 0}^\infty \to \mathbb{R}_{\ge 0}^\infty$, called an $\oplus$-bound, which yields a new abstract Lipschitz constant.

Definition 7 ($\oplus$-Bound). *Let $\uplus : \mathbb{R}_{\ge 0}^\infty \times \mathbb{R}_{\ge 0}^\infty \to \mathbb{R}_{\ge 0}^\infty$ be a binary operator such that $\langle \mathbb{R}_{\ge 0}^\infty, \uplus, \cdot \rangle$ forms a complete semiring, where $\cdot$ is the standard multiplication. $\uplus$ is a $\oplus$-bound when the following holds for all $\mathsf{P}_1, \mathsf{P}_2 \in \mathsf{Prog}$ and $k_1, k_2 \in \mathbb{R}_{\ge 0}^\infty$:*

$$\begin{aligned} [\![\mathsf{P}_1]\!] \in k_1\text{-}ALip\langle \delta^\eta, \delta^\rho \rangle \\ \text{and}\ \ [\![\mathsf{P}_2]\!] \in k_2\text{-}ALip\langle \delta^\eta, \delta^\rho \rangle \end{aligned} \ \Rightarrow\ \rho[\![\mathsf{P}_1]\!] \vee \rho[\![\mathsf{P}_2]\!] \in k_1 \uplus k_2\text{-}ALip\langle \delta^\eta, \delta^\rho \rangle$$

Example 9. Consider the pseudo-metric space $\langle \wp(\mathbb{Z}), \delta_{siz} \rangle$ and the collecting semantics $[\![\cdot]\!]$. Let the input and output abstractions be $\rho = \eta = \mathsf{Int}$. Then the standard addition $+$ is a $\oplus$-bound. In other words, having an ALC proof for both P_1 and P_2, with abstract Lipschitz constants k_1, k_2, respectively, gives:

$$\delta_{siz}((\mathsf{Int}[\![\mathsf{P}_1]\!] \vee \mathsf{Int}[\![\mathsf{P}_2]\!])c_1, (\mathsf{Int}[\![\mathsf{P}_1]\!] \vee \mathsf{Int}[\![\mathsf{P}_2]\!])c_2) \le$$
$$k_1 \delta_{siz}(\mathsf{Int}(c_1), \mathsf{Int}(c_2)) + k_2 \delta_{siz}(\mathsf{Int}(c_1), \mathsf{Int}(c_2)) = (k_1 + k_2)\delta_{siz}(\mathsf{Int}(c_1), \mathsf{Int}(c_2))$$

This is because, when considering δ_{siz} as distance and Int as input and output abstractions, the size of the join of two intervals can be over-approximated by the sum of the number of the elements inside the two intervals. A similar reasoning holds for the quasisemi-metric space $\langle \wp(\mathbb{Z}), \delta_\subseteq \rangle$ defined in Ex. 3. $\blacksquare$

The premise of the (**join**) rule asks for the validity of the following predicates: assume we have an ALC derivation $k_1 \vdash [\delta^\eta] \, \mathsf{P}_1 \, [\delta^\rho]$ for P_1 and $k_2 \vdash [\delta^\eta] \, \mathsf{P}_2 \, [\delta^\rho]$ for P_2, then we can soundly conclude that the join $\mathsf{P}_1 \oplus \mathsf{P}_2$ is $k_1 \uplus k_2$-ALC.

For the Kleene star operator, we introduce two rules: (**bound-star**) and (**star**). Both rules require that the program P satisfies k-ALC w.r.t. $\langle \delta^\eta, \delta^\rho \rangle$. Moreover, in order to compose programs, the input abstraction η is required to satisfy t-ALC w.r.t. $\langle \delta^\rho, \delta^\eta \rangle$ (as required by (**seq**)). The additional requirement $\rho \in v\text{-}ALip\langle \delta^\eta, \delta^\rho \rangle$ enforces a linear relationship between η and ρ which is necessary to handle the case of zero iterations. Note that these requirements are trivially satisfied when $\eta = \rho$ (by Lem. 1(i)). The (**bound-star**) rule further requires the validity of the assertion $*$-$\mathrm{Bound}(\mathsf{P}^*, m)$, with $m \in \mathbb{N}$, which holds when $[\![\mathsf{P}^*]\!]c = \bigvee \{ [\![\mathsf{P}]\!]^n c \mid n \in \mathbb{N}_{\leq m} \}$ for any $c \in C$. In other words, given any input $c \in C$, the result of $[\![\mathsf{P}^*]\!]c$ can be obtained by executing P at most m times. (**star**) rule, instead, has no constraints on the number of iterations of P. Rules (**bound-star**) and (**star**) rely on the following definitions:

$$K_0 \overset{\mathrm{def}}{=} v \,, \qquad K_n \overset{\mathrm{def}}{=} v \uplus \biguplus_{i=1}^{n} k^i t^{i-1} \quad (n \geq 1) \,, \qquad K_\infty \overset{\mathrm{def}}{=} \sup_{n \geq 0} K_n$$

where the supremum is with respect to the standard order induced by $\uplus$ (i.e., $x \preceq_\uplus y$ iff $x \uplus y = y$). For rule (**star**), the completeness of the semiring guarantees that K_∞ always exists and, depending on the definition of $\uplus$, as well as the values of k and t (see, e.g., Ex. 11), it is either a finite value in $\mathbb{R}_{\geq 0}$ or ∞.

The following theorem shows that our proposed deductive system is sound, namely, if $k \vdash [\delta^\eta] \, \mathsf{P} \, [\delta^\rho]$ can be derived by applying the rules of Fig. 5, then $[\![\mathsf{P}]\!]$ satisfies k-ALC w.r.t. $\langle \delta^\eta, \delta^\rho \rangle$.

Theorem 4 (Soundness). $k \vdash [\delta^\eta] \, \mathsf{P} \, [\delta^\rho] \ \Rightarrow \ [\![\mathsf{P}]\!] \in k\text{-}ALip\langle \delta^\eta, \delta^\rho \rangle$

Example 10. Consider the program $(x < 0? \,; x := 0) \oplus (x \geq 0? \,; \mathbf{skip})$ implementing the ReLU function used in artificial neural networks [45], that filters the input below 0. Consider the quasi-metric space $\langle \wp(\mathbb{Z}), \delta_\subseteq \rangle$ and the input and output abstraction $\eta = \rho = \mathsf{Int}$. We want to prove that the collecting semantics $[\![\mathsf{ReLU}]\!] : \wp(\mathbb{Z}) \to \wp(\mathbb{Z})$ satisfies 1-ALC for $\langle \delta_\subseteq^{\mathsf{Int}}, \delta_\subseteq^{\mathsf{Int}} \rangle$. Let us start by analyzing the base commands on the left of $\oplus$. Because the Boolean guard $x < 0?$ is either preserving or removing values from the input, by the (**base**) rule, we can derive $1 \vdash [\delta_\subseteq^{\mathsf{Int}}] \, x < 0? \, [\delta_\subseteq^{\mathsf{Int}}]$. The command $x := 0$ is neutralizing any distance between input sets since $\delta_\subseteq^{\mathsf{Int}}([\![x := 0]\!]S_1, [\![x := 0]\!]S_2) = 0$ for any $S_1, S_2 \in \wp(\mathbb{Z})$. We can derive $0 \vdash [\delta_\subseteq^{\mathsf{Int}}] \, x := 0 \, [\delta_\subseteq^{\mathsf{Int}}]$ by the (**base**) rule. Since $\mathsf{Int} \in 1\text{-}ALip\langle \delta_\subseteq^{\mathsf{Int}}, \delta_\subseteq^{\mathsf{Int}} \rangle$ follows from Lem. 1, we can infer $0 \vdash [\delta_\subseteq^{\mathsf{Int}}] \, x < 0? \,; x := 0 \, [\delta_\subseteq^{\mathsf{Int}}]$ by the (**seq**) rule. For the base commands on the right of $\oplus$, we get $1 \vdash [\delta_\subseteq^{\mathsf{Int}}] \, x \geq 0? \, [\delta_\subseteq^{\mathsf{Int}}]$ with rule (**base**). The **skip** command does not modify the distance of the input sets, so $1 \vdash [\delta_\subseteq^{Int}]$ **skip** $[\delta_\subseteq^{Int}]$ can be derived by (**base**). Since $Int \in 1\text{-}ALip\langle \delta_\subseteq^{Int}, \delta_\subseteq^{Int}\rangle$ holds, the rule (**seq**) derives $1 \vdash [\delta_\subseteq^{Int}]\, x \geq 0?\,;\, $ **skip** $[\delta_\subseteq^{Int}]$. Now for the $\oplus$ operation, we consider $\uplus = +$ (as in Ex. 9), thus guaranteeing a sound upper bound for the abstract Lipschitz constants on the program join. By the two derivations on the left and right parts of $\oplus$, we can conclude by the (**join**) rule: $1 \vdash [\delta_\subseteq^{Int}]$ ReLU $[\delta_\subseteq^{Int}]$. By Thm. 4, $[\![$ReLU$]\!]$ satisfies 1-ALC for $\langle \delta_\subseteq^{Int}, \delta_\subseteq^{Int}\rangle$. ∎

Example 11. Consider the program $\mathsf{F}\colon (x > 0?\ x := x/4) \oplus (x \leq 0?\,;\, x := 0)$ corresponding to ReLU with a scaling factor of $\frac{1}{4}$, and the psueudo-metric space $\langle \wp(\mathbb{R}), \delta_-\rangle$, where $\delta_-(S_1, S_2) \overset{\text{def}}{=} 0$ if $S_1 = S_2$, $\delta_-(S_1, S_2) \overset{\text{def}}{=} |(\max(S_2) - \min(S_2)) - (\max(S_1) - \min(S_1))|$ otherwise. Let $\eta = \rho = \mathsf{Int}$. Intuitively, when calculating the distance δ_- over two sets in $\mathsf{Int}(\wp(\mathbb{R}))$, the distance returns the absolute value of the difference between the length of the two intervals. We derive the abstract Lipschitz constant of the collecting semantics $[\![\mathsf{F}^*]\!]\colon \wp(\mathbb{R}) \to \wp(\mathbb{R})$. By following derivations similar to those in Ex. 10, we derive $\frac{1}{4} \vdash [\delta_-^{Int}]\, \mathsf{F}\, [\delta_-^{Int}]$ for F. To apply rule (**star**), we consider the valid premises $\mathsf{Int} \in 1\text{-}ALip\langle \delta^{Int}, \delta^{Int}\rangle$ and $\mathsf{Int} \in 1\text{-}ALip\langle \delta^{Int}, \delta^{Int}\rangle$, and we choose $\uplus = +$ as a valid $\oplus$-bound. Then, K_∞ turns into the infinite series $1 + \sum_{i=0}^{\infty}(\frac{1}{4})^i$ which is a geometric series converging to $\frac{7}{3}$. Thus, by rule (**star**), we derive the judgment $\frac{7}{3} \vdash [\delta_-^{Int}]\, \mathsf{F}^*\, [\delta_-^{Int}]$ and, by Thm. 4, we can conclude $[\![\mathsf{F}^*]\!] \in \frac{7}{3}\text{-}ALip\langle \delta_-^{Int}, \delta_-^{Int}\rangle$. Note that (**bound-star**) cannot by applied since there is no m such that $*\text{-Bound}(\mathsf{F}^*, m)$ holds. ∎

As a direct consequence of Thm. 4, if we instantiate the abstractions with $\eta = \rho = id$, then the deductive rules of Fig. 5 derive judgments for the LP of programs (Def. 5). This is because all the predicates on abstractions, such as $\eta \in t\text{-}ALip\langle \delta^\rho, \delta^\eta\rangle$, becomes trivially true by Lem. 1.

Corollary 3. $k \vdash [\delta^{id}]\, \mathsf{P}\, [\delta^{id}] \;\Rightarrow\; [\![\mathsf{P}]\!]$ *is* k-*LP w.r.t.* $\langle \delta, \delta\rangle$

6 Conclusion

Our work is not the first to adapt the LC notion to specific user-needs. For instance, ϵ-*sensitivity* [51] can be seen as the probabilistic counterpart of LC, requiring an ϵ-bounded maximum divergence of output distributions for probabilistic functions. Sensitivity allows to model and reason about security-relevant properties like ϵ-differential privacy [20], while it does not suite more complex scenarios, like (ϵ, δ)-differential privacy [19], requiring distance metrics not satisfying the triangle inequality. Thus, de Amorin et al. [18] proposed a categorical framework to encode relational properties, like (ϵ, δ)-differential privacy, into equivalent sensitivity-like properties over a suitable metric space. In other words, the framework lifts sensitivity from the standard Max Divergence metric to more complex, non-metric divergences like Skew Divergence. These generalizations account for probabilistic computations but not for qualitative aspects of (deterministic) computations like ALC. Qualitative aspects are considered by Bonchi et al. [2], who proved the soundness of *up-to techniques* [50] from the qualitative bisimilarity setting to the quantitative metric setting. Specifically, the paper gives sufficient conditions for the soundness of the quantitative version of up-to contexts in systems that have a (co-)algebraic structure (e.g., a bialgebra). That is, contexts are *non-expansive* with respect to the behavioral metric, which is the quantitative analogue of behavioral equivalence being a congruence. This mixed qualitative and quantitative formalization is obtained by adopting quantale-based metric spaces, thus generalizing the ordinary LC setting. Still, non-expansiveness requires a non-increasing distance change, thus potentially generalizing LC only when $k = 1$. We plan to formally investigate the relation between ALC and the aforementioned approaches as a future work.

As stated in Sec. 2.1, basic commands could be instantiated differently, for instance with probabilistic or nondeterministic assignments. In this work we considered only deterministic assignments and $\oplus$ is defined simply as the semantic join. Extending ALC to a richer probabilistic language is an interesting direction, and suitable adjustments to the join rule should make this possible. In this context we would expect standard metrics for probabilistic programs (e.g., the total variation or Kantorovich metrics) to work as well.

The proposed ALC notion is a *global* property, in the sense that it is universally quantified over all inputs. As a future work, we plan to formalize its *local* version, namely requiring ALC over a strict subset of inputs, and study its relation with other local properties in the context of abstract interpretation [3,4,6]. Dropping the universal quantification may invalidate the correlation already established between the global counterparts. Also, reasoning about local properties may be more challenging, as the proposed deductive system requires nontrivial modifications to be used for proving ALC on a subset of executions.

Another interesting future direction consists in considering weaker abstraction notions able to formalize properties that do not necessarily admit a best abstraction function, such as the domain of convex polyhedra [29], and more generally, as the weak closures defined in [41]. Finally, in [41] it is proved that, under certain assumptions, there is a correspondence between the completeness property in abstract interpretation and ANI in language-based security [24,25,37,22]. Such a connection suggests a potential relation between ALC and ANI, as well as to other quantitative program properties [38,39,40,42,43,44,55], which deserves further investigation in several directions, both theoretical and practical.

Finally, in Thm. 4, we proved that the proof system presented in Fig. 5 is sound. However, in its current form, the proof system is incomplete. As a consequence, the tightest Lipschitz constant derivable may be strictly greater than the actual one. The main source of this limitation lies in the treatment of the Kleene star when the body of a program P is not contracting (i.e., when $k \geq 1$). In such cases, even the (**bound-star**) rule may be ineffective, since the $*$-Bound predicate is not guaranteed to hold (see, e.g., Ex. 11). A more thorough investigation is required and we leave it for a future extension of this work.

Acknowledgements We wish to thank the anonymous reviewers of FoSSaCS 2026 for their detailed comments. This work was partially supported by the project SERICS (PE00000014) under the MUR National Recovery and Resilience Plan funded by the European Union - NextGenerationEU; by the project PRIN2022PNRR "RAP-ARA" (PE6) - codice MUR: P2022HXNSC; by the SAIF project, funded by the "France 2030" government investment plan managed by the French National Research Agency, under the reference ANR-23-PEIA-0006.

References

1. Kumail Alhamoud, Hasan Abed Al Kader Hammoud, Motasem Alfarra, and Bernard Ghanem. Generalizability of adversarial robustness under distribution shifts. *Trans. Mach. Learn. Res.*, 2023, 2023. URL: https://openreview.net/forum?id=XNFo3dQiCJ.

2. Filippo Bonchi, Barbara König, and Daniela Petrisan. Up-to techniques for behavioural metrics via fibrations. *Math. Struct. Comput. Sci.*, 33(4-5):182–221, 2023. doi:10.1017/S0960129523000166.

3. Roberto Bruni, Roberto Giacobazzi, Roberta Gori, and Francesco Ranzato. A logic for locally complete abstract interpretations. In *36th Annual ACM/IEEE Symposium on Logic in Computer Science, LICS 2021, Rome, Italy, June 29 - July 2, 2021*, pages 1–13. IEEE, 2021. doi:10.1109/LICS52264.2021.9470608.

4. Roberto Bruni, Roberto Giacobazzi, Roberta Gori, and Francesco Ranzato. A correctness and incorrectness program logic. *J. ACM*, 70(2):15:1–15:45, 2023. doi:10.1145/3582267.

5. Marco Campion, Mila Dalla Preda, and Roberto Giacobazzi. Partial (in)completeness in abstract interpretation: limiting the imprecision in program analysis. *Proc. ACM Program. Lang.*, 6(POPL):1–31, 2022. doi:10.1145/3498721.

6. Marco Campion, Mila Dalla Preda, Roberto Giacobazzi, and Caterina Urban. Monotonicity and the precision of program analysis. *Proc. ACM Program. Lang.*, 8(POPL):1629–1662, 2024. doi:10.1145/3632897.

7. Marco Campion, Mila Dalla Preda, Roberto Giacobazzi, and Caterina Urban. A logic for the imprecision of abstract interpretations. *Proc. ACM Program. Lang.*, 10(POPL), 2026. doi:10.1145/3776707.

8. Marco Campion, Isabella Mastroeni, and Caterina Urban. Relating distances and abstractions - an abstract interpretation perspective. In Hakjoo Oh and Yulei Sui, editors, *Static Analysis - 32nd International Symposium, SAS 2025, Singapore, October 13-14, 2025, Proceedings*, volume 16100 of *Lecture Notes in Computer Science*, pages 249–277. Springer, 2025. doi:10.1007/978-3-032-07106-4_11.

9. Marco Campion, Caterina Urban, Mila Dalla Preda, and Roberto Giacobazzi. A formal framework to measure the incompleteness of abstract interpretations. In Manuel V. Hermenegildo and José F. Morales, editors, *Static Analysis - 30th International Symposium, SAS 2023, Cascais, Portugal, October 22-24, 2023, Proceedings*, volume 14284 of *Lecture Notes in Computer Science*, pages 114–138. Springer, 2023. doi:10.1007/978-3-031-44245-2_7.

10. Swarat Chaudhuri, Sumit Gulwani, and Roberto Lublinerman. Continuity Analysis of Programs. In *Proceedings of the 37th Annual ACM SIGPLAN-SIGACT Symposium on Principles of Programming Languages*, POPL '10, page 5770, New York, NY, USA, 2010. Association for Computing Machinery. doi:10.1145/1706299.1706308.

11. Swarat Chaudhuri, Sumit Gulwani, and Roberto Lublinerman. Continuity and Robustness of Programs. 55(8):107115, 2012. doi:10.1145/2240236.2240262.

12. Swarat Chaudhuri, Sumit Gulwani, Roberto Lublinerman, and Sara NavidPour. Proving Programs Robust. In Tibor Gyimóthy and Andreas Zeller, editors, *SIGSOFT/FSE'11 19th ACM SIGSOFT Symposium on the Foundations of Software Engineering (FSE-19) and ESEC'11: 13th European Software Engineering Conference (ESEC-13), Szeged, Hungary, September 5-9, 2011*, pages 102–112. ACM, 2011. `doi:10.1145/2025113.2025131`.

13. Patrick Cousot. *Principles of Abstract Interpretation*. The MIT Press, Cambridge, Mass., 2021.

14. Patrick Cousot and Radhia Cousot. Abstract interpretation: A unified lattice model for static analysis of programs by construction or approximation of fixpoints. In Robert M. Graham, Michael A. Harrison, and Ravi Sethi, editors, *Conference Record of the Fourth ACM Symposium on Principles of Programming Languages, Los Angeles, California, USA, January 1977*, pages 238–252. ACM, 1977. `doi:10.1145/512950.512973`.

15. Patrick Cousot and Radhia Cousot. Systematic design of program analysis frameworks. In Alfred V. Aho, Stephen N. Zilles, and Barry K. Rosen, editors, *Conference Record of the Sixth Annual ACM Symposium on Principles of Programming Languages, San Antonio, Texas, USA, January 1979*, pages 269–282. ACM Press, 1979. `doi:10.1145/567752.567778`.

16. Patrick Cousot and Radhia Cousot. An abstract interpretation-based framework for software watermarking. In *Proceedings of the 31st ACM SIGPLAN-SIGACT Symposium on Principles of Programming Languages*, POPL '04, pages 173–185, New York, NY, USA, 2004. Association for Computing Machinery. `doi:10.1145/964001.964016`.

17. Pedro R. D'Argenio, Gilles Barthe, Sebastian Biewer, Bernd Finkbeiner, and Holger Hermanns. Is your software on dope? - formal analysis of surreptitiously "enhanced" programs. In Hongseok Yang, editor, *Programming Languages and Systems - 26th European Symposium on Programming, ESOP 2017, Held as Part of the European Joint Conferences on Theory and Practice of Software, ETAPS 2017, Uppsala, Sweden, April 22-29, 2017, Proceedings*, volume 10201 of *Lecture Notes in Computer Science*, pages 83–110. Springer, 2017. `doi:10.1007/978-3-662-54434-1_4`.

18. Arthur Azevedo de Amorim, Marco Gaboardi, Justin Hsu, and Shin-ya Katsumata. Probabilistic relational reasoning via metrics. In *34th Annual ACM/IEEE Symposium on Logic in Computer Science, LICS 2019, Vancouver, BC, Canada, June 24-27, 2019*, pages 1–19. IEEE, 2019. `doi:10.1109/LICS.2019.8785715`.

19. Cynthia Dwork, Krishnaram Kenthapadi, Frank McSherry, Ilya Mironov, and Moni Naor. Our data, ourselves: privacy via distributed noise generation. In *Proceedings of the 24th Annual International Conference on The Theory and Applications of Cryptographic Techniques*, EUROCRYPT'06, pages 486–503, Berlin, Heidelberg, 2006. Springer-Verlag. `doi:10.1007/11761679_29`.

20. Cynthia Dwork, Frank McSherry, Kobbi Nissim, and Adam Smith. Calibrating noise to sensitivity in private data analysis. In Shai Halevi and Tal Rabin, editors, *Theory of Cryptography*, pages 265–284, Berlin, Heidelberg, 2006. Springer Berlin Heidelberg.

21. Mahyar Fazlyab, Alexander Robey, Hamed Hassani, Manfred Morari, and George J. Pappas. Efficient and accurate estimation of lipschitz constants for deep neural networks. In Hanna M. Wallach, Hugo Larochelle, Alina Beygelzimer, Florence d'Alché-Buc, Emily B. Fox, and Roman Garnett, editors, *Advances in Neural Information Processing Systems 32: Annual Conference on Neural Infor-*

mation Processing Systems 2019, NeurIPS 2019, December 8-14, 2019, Vancouver, BC, Canada, pages 11423–11434, 2019. URL: https://proceedings.neurips.cc/paper/2019/hash/95e1533eb1b20a97777749fb94fdb944-Abstract.html.

22. R. Giacobazzi and I. Mastroeni. Adjoining classified and unclassified information by abstract interpretation. *Journal of Computer Security*, 18(5):751 – 797, 2010.

23. Roberto Giacobazzi, Francesco Logozzo, and Francesco Ranzato. Analyzing program analyses. In Sriram K. Rajamani and David Walker, editors, *Proceedings of the 42nd Annual ACM SIGPLAN-SIGACT Symposium on Principles of Programming Languages, POPL 2015, Mumbai, India, January 15-17, 2015*, pages 261–273. ACM, 2015. doi:10.1145/2676726.2676987.

24. Roberto Giacobazzi and Isabella Mastroeni. Abstract non-interference: parameterizing non-interference by abstract interpretation. In Neil D. Jones and Xavier Leroy, editors, *Proceedings of the 31st ACM SIGPLAN-SIGACT Symposium on Principles of Programming Languages, POPL 2004, Venice, Italy, January 14-16, 2004*, pages 186–197. ACM, 2004. doi:10.1145/964001.964017.

25. Roberto Giacobazzi and Isabella Mastroeni. Abstract non-interference: A unifying framework for weakening information-flow. *ACM Trans. Priv. Secur.*, 21(2):9:1–9:31, 2018. doi:10.1145/3175660.

26. Roberto Giacobazzi, Isabella Mastroeni, and Elia Perantoni. Adversities in abstract interpretation - accommodating robustness by abstract interpretation. *ACM Trans. Program. Lang. Syst.*, 46(2):1–31, 2024. doi:10.1145/3649309.

27. Roberto Giacobazzi, Francesco Ranzato, and Francesca Scozzari. Making abstract interpretations complete. *J. ACM*, 47(2):361–416, 2000. doi:10.1145/333979.333989.

28. Joseph A. Goguen and José Meseguer. Security policies and security models. In *1982 IEEE Symposium on Security and Privacy, Oakland, CA, USA, April 26-28, 1982*, pages 11–20. IEEE Computer Society, 1982. doi:10.1109/SP.1982.10014.

29. Branko Grünbaum, Victor Klee, Micha A Perles, and Geoffrey Colin Shephard. *Convex polytopes*, volume 16. Springer, 1967.

30. Yujia Huang, Huan Zhang, Yuanyuan Shi, J. Zico Kolter, and Anima Anandkumar. Training certifiably robust neural networks with efficient local lipschitz bounds. In Marc'Aurelio Ranzato, Alina Beygelzimer, Yann N. Dauphin, Percy Liang, and Jennifer Wortman Vaughan, editors, *Advances in Neural Information Processing Systems 34: Annual Conference on Neural Information Processing Systems 2021, NeurIPS 2021, December 6-14, 2021, virtual*, pages 22745–22757, 2021. URL: https://proceedings.neurips.cc/paper/2021/hash/c055dcc749c2632fd4dd806301f05ba6-Abstract.html.

31. Ziwei Ji and Matus Telgarsky. Directional convergence and alignment in deep learning. In Hugo Larochelle, Marc'Aurelio Ranzato, Raia Hadsell, Maria-Florina Balcan, and Hsuan-Tien Lin, editors, *Advances in Neural Information Processing Systems 33: Annual Conference on Neural Information Processing Systems 2020, NeurIPS 2020, December 6-12, 2020, virtual*, 2020. URL: https://proceedings.neurips.cc/paper/2020/hash/c76e4b2fa54f8506719a5c0dc14c2eb9-Abstract.html.

32. Lin Li, Yifei Wang, Chawin Sitawarin, and Michael W. Spratling. Oodrobustbench: a benchmark and large-scale analysis of adversarial robustness under distribution shift. In *Forty-first International Conference on Machine Learning, ICML 2024, Vienna, Austria, July 21-27, 2024*. OpenReview.net, 2024. URL: https://openreview.net/forum?id=kAFevjEYsz.

33. Dennis Liew, Tiago Cogumbreiro, and Julien Lange. Sound and partially-complete static analysis of data-races in GPU programs. *Proc. ACM Program. Lang.*, 8(OOPSLA2):2434–2461, 2024. `doi:10.1145/3689797`.

34. Francesco Logozzo. Towards a quantitative estimation of abstract interpretations. In *Workshop on Quantitative Analysis of Software*. Microsoft, June 2009. URL: `https://www.microsoft.com/en-us/research/publication/towards-a-quantitative-estimation-of-abstract-interpretations/`.

35. Ulrike von Luxburg and Olivier Bousquet. Distance-based classification with lipschitz functions. *Journal of Machine Learning Research*, 5(Jun):669–695, 2004.

36. Luca Marzari, Isabella Mastroeni, and Alessandro Farinelli. Advancing neural network verification through hierarchical safety abstract interpretation. *CoRR*, abs/2505.05235, 2025. URL: `https://doi.org/10.48550/arXiv.2505.05235`, `arXiv:2505.05235`, `doi:10.48550/ARXIV.2505.05235`.

37. I. Mastroeni. On the rôle of abstract non-interference in language-based security. In K. Yi, editor, *Third Asian Symp. on Programming Languages and Systems (APLAS '05)*, volume 3780 of *Lecture Notes in Computer Science*, pages 418–433. Springer-Verlag, 2005.

38. Isabella Mastroeni. Abstract domain adequacy. *Int. J. Softw. Tools Technol. Transf.*, 26(6):747–765, 2024. URL: `https://doi.org/10.1007/s10009-024-00774-x`, `doi:10.1007/S10009-024-00774-X`.

39. Isabella Mastroeni. Abstract local completeness. In Krishna Shankaranarayanan, Sriram Sankaranarayanan, and Ashutosh Trivedi, editors, *Verification, Model Checking, and Abstract Interpretation*, pages 3–25, Cham, 2025. Springer Nature Switzerland.

40. Isabella Mastroeni and Michele Pasqua. Statically analyzing information flows: an abstract interpretation-based hyperanalysis for non-interference. In Chih-Cheng Hung and George A. Papadopoulos, editors, *Proceedings of the 34th ACM/SIGAPP Symposium on Applied Computing, SAC 2019*, pages 2215–2223. ACM, 2019. `doi:10.1145/3297280.3297498`.

41. Isabella Mastroeni and Michele Pasqua. Domain precision in galois connection-less abstract interpretation. In Manuel V. Hermenegildo and José F. Morales, editors, *Static Analysis - 30th International Symposium, SAS 2023, Cascais, Portugal, October 22-24, 2023, Proceedings*, volume 14284 of *Lecture Notes in Computer Science*, pages 434–459. Springer, 2023. `doi:10.1007/978-3-031-44245-2_19`.

42. Isabella Mastroeni and Michele Pasqua. Abstract interpretation-based verification for confidentiality: Information hiding and code protection by abstract interpretation. *ACM Trans. on Privacy and Security*, pages 1–28, 2025. `doi:10.1145/3786347`.

43. Denis Mazzucato, Marco Campion, and Caterina Urban. Quantitative input usage static analysis. In Nathaniel Benz, Divya Gopinath, and Nija Shi, editors, *NASA Formal Methods - 16th International Symposium, NFM 2024, Moffett Field, CA, USA, June 4-6, 2024, Proceedings*, volume 14627 of *Lecture Notes in Computer Science*, pages 79–98. Springer, 2024. `doi:10.1007/978-3-031-60698-4_5`.

44. Denis Mazzucato, Marco Campion, and Caterina Urban. Quantitative static timing analysis. In Roberto Giacobazzi and Alessandra Gorla, editors, *Static Analysis - 31st International Symposium, SAS 2024, Pasadena, CA, USA, October 20-22, 2024, Proceedings*, volume 14995 of *Lecture Notes in Computer Science*, pages 268–299. Springer, 2024. `doi:10.1007/978-3-031-74776-2_11`.

45. Vinod Nair and Geoffrey E. Hinton. Rectified linear units improve restricted boltzmann machines. In Johannes Fürnkranz and Thorsten Joachims, editors,

Proceedings of the 27th International Conference on Machine Learning (ICML-10), June 21-24, 2010, Haifa, Israel, pages 807–814. Omnipress, 2010. URL: `https://icml.cc/Conferences/2010/papers/432.pdf`.

46. Yurii Nesterov. *Lectures on Convex Optimization.* Springer Publishing Company, Incorporated, 2nd edition, 2018. `doi:10.1007/978-3-319-91578-4`.

47. Peter W. O'Hearn. Incorrectness logic. *Proc. ACM Program. Lang.,* 4(POPL):10:1–10:32, 2020. `doi:10.1145/3371078`.

48. Alessandra Di Pierro, Chris Hankin, and Herbert Wiklicky. Approximate non-interference. *J. Comput. Secur.,* 12(1):37–82, 2004. `doi:10.3233/JCS-2004-12103`.

49. Alessandra Di Pierro and Herbert Wiklicky. Measuring the precision of abstract interpretations. In Kung-Kiu Lau, editor, *Logic Based Program Synthesis and Transformation, 10th International Workshop, LOPSTR 2000 London, UK, July 24-28, 2000, Selected Papers,* volume 2042 of *Lecture Notes in Computer Science,* pages 147–164. Springer, 2000. `doi:10.1007/3-540-45142-0_9`.

50. Damien Pous. Complete lattices and up-to techniques. In *Proceedings of the 5th Asian Conference on Programming Languages and Systems,* APLAS'07, pages 351–366, Berlin, Heidelberg, 2007. Springer-Verlag.

51. Jason Reed and Benjamin C. Pierce. Distance makes the types grow stronger: a calculus for differential privacy. *SIGPLAN Not.,* 45(9):157–168, September 2010. `doi:10.1145/1932681.1863568`.

52. Xavier Rival and Kwangkeun Yi. *Introduction to static analysis: an abstract interpretation perspective.* Mit Press, 2020.

53. Daniel Schoepe and Andrei Sabelfeld. Understanding and enforcing opacity. In *2015 IEEE 28th Computer Security Foundations Symposium,* pages 539–553, 2015. `doi:10.1109/CSF.2015.41`.

54. Pascal Sotin. Quantifying the Precision of Numerical Abstract Domains. Research report, February 2010. URL: `https://inria.hal.science/inria-00457324`.

55. Caterina Urban and Peter Müller. An abstract interpretation framework for input data usage. In Amal Ahmed, editor, *Programming Languages and Systems - 27th European Symposium on Programming, ESOP 2018, Held as Part of the European Joint Conferences on Theory and Practice of Software, ETAPS 2018, Thessaloniki, Greece, April 14-20, 2018, Proceedings,* volume 10801 of *Lecture Notes in Computer Science,* pages 683–710. Springer, 2018. `doi:10.1007/978-3-319-89884-1_24`.

56. Bohang Zhang, Du Jiang, Di He, and Liwei Wang. Rethinking lipschitz neural networks and certified robustness: A boolean function perspective. In Sanmi Koyejo, S. Mohamed, A. Agarwal, Danielle Belgrave, K. Cho, and A. Oh, editors, *Advances in Neural Information Processing Systems 35: Annual Conference on Neural Information Processing Systems 2022, NeurIPS 2022, New Orleans, LA, USA, November 28 - December 9, 2022,* 2022. URL: `http://papers.nips.cc/paper_files/paper/2022/hash/7b04ec5f2b89d7f601382c422dfe07af-Abstract-Conference.html`.

57. Bohang Zhang, Du Jiang, Di He, and Liwei Wang. Rethinking lipschitz neural networks and certified robustness: A boolean function perspective. In S. Koyejo, S. Mohamed, A. Agarwal, D. Belgrave, K. Cho, and A. Oh, editors, *Advances in Neural Information Processing Systems,* volume 35, pages 19398–19413. Curran Associates, Inc., 2022. URL: `https://proceedings.neurips.cc/paper_files/paper/2022/file/7b04ec5f2b89d7f601382c422dfe07af-Paper-Conference.pdf`.

Realization of relational presheaves

Yorgo Chamoun and Samuel Mimram

LIX, CNRS, École polytechnique, Institut Polytechnique de Paris, 91120 Palaiseau, France `{yorgo.chamoun,samuel.mimram}@polytechnique.edu`

Abstract. Relational presheaves generalize traditional presheaves by going to the category of sets and relations (as opposed to sets and functions) and by allowing functors which are lax. This added generality is useful because it intuitively allows one to encode situations where we have representables without boundaries or with multiple boundaries at once. In particular, the relational generalization of precubical sets has natural application to modeling concurrency. In this article, we study categories of relational presheaves, and construct realization functors for those. We begin by observing that they form the category of set-based models of a cartesian theory, which implies in particular that they are locally finitely presentable categories. By using general results from categorical logic, we then show that the realization of such presheaves in a cocomplete category is a model of the theory in the opposite category, which allows characterizing situations in which we have a realization functor. Finally, we explain that our work has applications in the semantics of concurrency theory. The realization namely allows one to compare syntactic constructions on relational presheaves and geometric ones. Thanks to it, we are able to provide a syntactic counterpart of the blowup operation, which was recently introduced by Haucourt on directed geometric semantics, as way of turning a directed space into a manifold.

Keywords: Presheaves · Relations · Realization · Categorical logic · Precubical sets · Higher dimensional automata.

1 Introduction

Presheaves. The notion of *presheaf* is omnipresent in modern mathematics and theoretical computer science. For instance, simplicial sets are at the heart of modern algebraic topology [27] and higher category theory [24], and cubical sets are central in modeling truly concurrent processes through the notion of higher dimensional automaton [13,30] or achieving constructive approaches to univalent type theory [5] to name a few of the myriad of occurrences of those. Here, we will be mostly interested in their ability to model transition systems, such as those arising from computational processes, concurrent ones in particular.

Formally, a presheaf is a functor $P : \mathcal{C}^{op} \to \mathbf{Set}$, which can be understood as encoding algebraically a geometric object. Namely, an object c of $\mathcal{C}$ abstractly describes a shape, the maps of $\mathcal{C}$ describe the face operations, and the sets $P(c)$

© The Author(s) 2026
N. Bertrand and S. Milius (Eds.): FoSSaCS 2026, LNCS 16503, pp. 178–197, 2026.
https://doi.org/10.1007/978-3-032-22730-0_9

encode the elements of shape c. Moreover, if we have a functor $\mathcal{C} \to \mathbf{Top}$ which describes how to associate an actual topological space to each abstract shape, we get an induced canonical functor $\mathbf{Psh}(\mathcal{C}) \to \mathbf{Top}$, called the *geometric realization*, which associates a topological space to any presheaf, and this construction actually generalizes to any cocomplete category in place of $\mathbf{Top}$ [26]. Above, the notation $\mathbf{Psh}(\mathcal{C})$ denotes the category of presheaves over $\mathcal{C}$ and natural transformations between those. Such a category is always a Grothendieck topos (with trivial topology), and thus has very nice properties: it is always cocomplete (it is in fact the free cocompletion of $\mathcal{C}$), complete, cartesian closed, has a subobject classifier, and so on.

As a simple example, consider the following category $\mathcal{G}$ with two objects and two non-trivial morphisms:

$$\mathcal{G} \quad = \quad 0 \underset{t}{\overset{s}{\rightrightarrows}} 1$$

A presheaf $P : \mathcal{G}^{op} \to \mathbf{Set}$ precisely corresponds to a (directed) *graph* with $P(0)$ as set of vertices, $P(1)$ as set of edges, the maps $P(s), P(t) : P(1) \to P(0)$ respectively associating to each edge its source and target. Moreover, if one considers the functor $\mathcal{G} \to \mathbf{Top}$ sending 0 to a point and 1 to an interval, the image of s and t respectively being the inclusion of the point into the endpoints of the interval, the geometric realization functor $\mathbf{Psh}(\mathcal{G}) \to \mathbf{Top}$ associates to a graph the corresponding topological graph.

Presheaves to relations. In order to be able to take into account some more situations, it is natural to generalize presheaves by replacing the category $\mathbf{Set}$ by some other category $\mathcal{V}$ [22]. We will refrain from adopting such a general point of view here, and will be mostly interested in the case where $\mathcal{V} = \mathbf{Rel}$, which allows accounting for situations where some elements of given shape have no boundary, or multiple boundaries. For instance, consider the two "graphs" below:

On the left, we have drawn a graph $G : \mathcal{G}^{op} \to \mathbf{Set}$ with $G(0) = \{x, y, z_1, z_2\}$, $G_1(1) = \{a, b_1, b_2\}$, $G(s)(a) = x$, $G(t)(a) = y$, and so on. By post-composition with the canonical functor $\mathbf{Set} \to \mathbf{Rel}$ (sending a set to itself and a function to its graph), this graph can also be seen as functor $\mathcal{G}^{op} \to \mathbf{Rel}$. By opposition, the picture on the right represents a variant of this graph where the edge a has no vertex as source and two vertices as target. It can be represented as the presheaf $H : \mathcal{G}^{op} \to \mathbf{Rel}$ with

$$
\begin{array}{lll}
H(0) = \{y_1, y_2, z_1, z_2\} & H(s)(a) = \emptyset & H(s)(b_i) = \{y_i\} \\
H(1) = \{a, b_1, b_2\} & H(t)(a) = \{y_1, y_2\} & H(t)(b_i) = \{z_i\}
\end{array}
$$

for $i \in \{0,1\}$ (as customary, we write a relation between $H(1)$ and $H(0)$ as a Kleisli map, i.e. a function from $H(1)$ to the powerset of $H(0)$). We can think of the edge a as representing a transition in some system which has no starting point, and makes a choice during the execution of the transition, about whether it wants to end on y_1 or y_2: depending on this choice, only b_1 or b_2 can be executed afterward. As illustrated in the previous example, going to relational presheaves allows for representing both shapes which are "partial" in the sense that they are lacking some boundary (a has no source for instance) and "multiple" in the sense that they have multiple boundaries (a has both y_1 and y_2 as target).

Higher dimensional automata. In this article, we will not only be interested in presheaves over the category $\mathcal{G}$, but rather on the *cube category* $\square$ which generalizes the previous situation, in the sense that $\mathcal{G}$ can be recovered as a full subcategory. The objects of $\square$ are natural numbers, where $n \in \mathbb{N}$ encodes the shape of an n-dimensional cube. Morphisms are generated by $d_{n,i}^{\varepsilon} : n \to n+1$ with $\varepsilon \in \{-,+\}$, $n, i \in \mathbb{N}$ with $0 \le i \le n$, subject to the relations

$$d_{n+1,j}^{\varepsilon'} \circ d_{n,i}^{\varepsilon} = d_{n+1,i}^{\varepsilon} \circ d_{n,j-1}^{\varepsilon'}$$

for $0 \le i < j \le n+1$. A morphism $d_{n,i}^{\varepsilon}$ encodes the canonical inclusion of the n-cube into the $(n+1)$-cube as source or target (depending on ε) in direction i. For instance, we have the following inclusions of the 1-cube into the 2-cube:

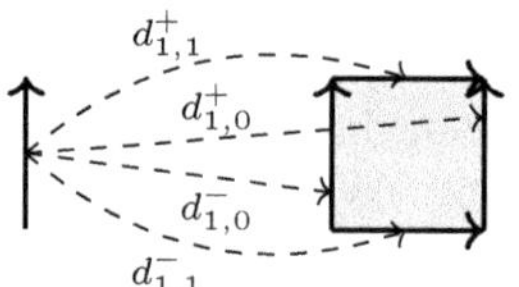

As expected, we can recover $\mathcal{G}$ as the full subcategory of $\square$ on the objects 0 and 1.

Presheaves over $\square$ are called *precubical sets* and are widely used in models of true concurrency. Namely, the presence of a square can be thought of as encoding a commutation between the transitions corresponding to its edges, and n-cubes similarly encode commutations between n transitions at once. This explains why they are at the basis of the definition of *higher-dimensional automata* [13,30] (HDA) which generalize traditional automata, and constitute a very expressive model of true concurrency [14]: an HDA consists of a precubical set, with a labeling of edges (1-cubes), and identified initial and final states (0-cubes). For instance, the HDA on the right encodes a system executing a transition a, followed by two transitions b and c executed in parallel.

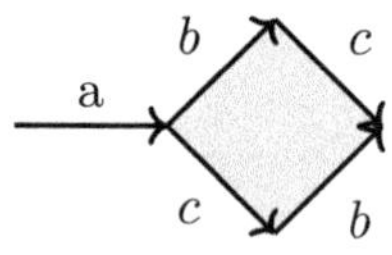

Relational presheaves. As indicated above, a *relational presheaf* on $\mathcal{C}$ is a functor $P : \mathcal{C}^{op} \to \mathbf{Rel}$. Our definition is actually slightly more liberal than this: it allows functors to be *lax*. By this, we mean that for every composable functions f and g, we have $P(g) \circ P(f) \subseteq P(g \circ f)$ and $\mathrm{id} \subseteq P(\mathrm{id})$, i.e. we allow an inclusion where an equality would have been required for traditional functors. Let us illustrate what this extra level of generality brings by illustrating it on HDA, i.e. presheaves on $\square$. As explained above, a square as (A) below encodes a situation where we have two transitions a and b running in parallel:

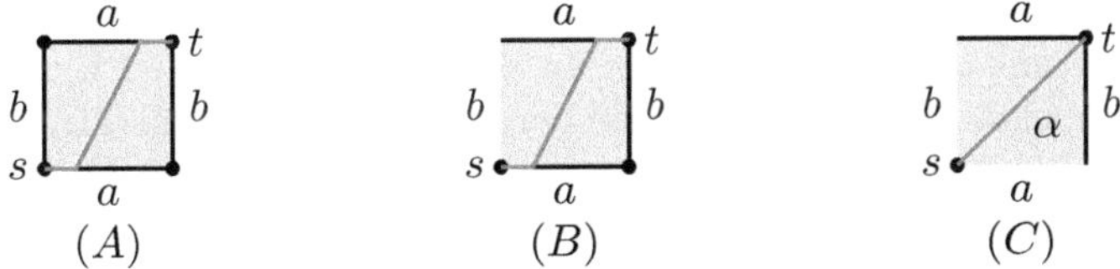

Moreover, an actual interleaving of the two transitions can be represented by a path such as the one in red in (A), which corresponds to an execution where a starts, then both a and b run in parallel, b stops and then a stops. In (B), we have represented a variant of this situation where we require that b cannot start before a does, which amounts to removing the vertical boundary edge on the left. In (C), we have represented a third variant where both a and b have to start at the same time (like in the red diagonal path). This last situation can be encoded as a precubical relational presheaf $H : \square \to_{\mathrm{lax}} \mathbf{Rel}$ with $H(0) = \{s,t\}$, $H(1) = \{a,b\}$ and $H(2) = \{\alpha\}$. The face relations are such that $(s,\alpha) \in H(d_{0,0}^- \circ d_{1,1}^-)$ but $H(d_{1,1}^-(\alpha)) = \emptyset$ so that $H(d_{0,0}^- \circ d_{1,1}^-) \subsetneq H(d_{0,0}^-) \circ H(d_{1,1}^-)$, which illustrates the need for lax functors: the point s is in the iterated source of the square α, but it is not a source of a 1-cell in the source of α.

Realization. The figures $(A) - (C)$ above should be handled formally: it is useful to associate, to each HDA, a topological space. More precisely, the topological space we construct should be "directed", i.e. equipped with a notion of "time direction", in such a way that paths which are increasing in time correspond to actual executions (as for the red paths above), see [9] for a presentation of this approach based on directed algebraic topology, which has important applications to concurrency theory. This process can be encoded by a functor called *geometric realization* constructed as follows. We have a canonical functor $M : \square \to \mathbf{Top}$, sending an object n to the canonical n-cube $[0,1]^n$. By left Kan extension along the Yoneda embedding $\sharp : \square \to \mathbf{Psh}(\square)$, this functor induces a continuous functor $R : \mathbf{Psh}(\square) \to \mathbf{Top}$ called the *realization*, which to a presheaf P associates the space obtained by gluing standard cubes as described by the presheaf. This process works more generally with any cocomplete category $\mathcal{D}$ instead of $\mathbf{Top}$ (such as the category of directed topological spaces).

Here, we are interested in realizing relational presheaves, i.e. defining functors $\mathbf{RelPsh}(\mathcal{C}) \to \mathcal{D}$ to some cocomplete category $\mathcal{D}$. It turns out that the situation is more complicated than above: intuitively, instead of simply specifying how to realize abstract shapes, i.e. objects of $\mathcal{C}$ by the functor M, we also need to specify the realization of every pair of objects related by a morphism. For example, to realize relational graphs in $\mathbf{RelPsh}(\mathcal{G})$, we still have to choose a realization of 0 and 1, which we take to be the terminal set and the open interval $]0,1[$ respectively, but we also need to specify how to realize an egde with a source or with a target endpoint, the natural choices being $[0, 1[$ and $]0, 1]$ respectively:

$$M(0) = \bullet \qquad M(1) = \text{——} \qquad M(0 \xrightarrow{s} 1) = \bullet\!\text{——} \qquad M(0 \xrightarrow{t} 1) = \text{——}\!\!\bullet$$

We can then realize every relational graph as before, by gluing these building blocks together. In general, we should not expect all assignments of this kind to work (for instance, the realization of an edge and an edge with a target should be somehow related), and we provide here conditions for this. In particular, we will see that the natural first try does not satisfy those conditions and has to be modified so that it is the case.

Blowup. As an application of relational presheaves, we explain here how they can be used to provide a nice description of the operation of *blowup* on geometric realizations of precubical sets introduced in [4,17]. This operation canonically turns such a space into a manifold, enabling the use of differential geometry to study the corresponding concurrent system. We show here that the blowup can be obtained as the geometric realization of a combinatorial blowup construction, turning a precubical set into a relational one. In this way, we obtain a description of the blowup operation in purely combinatorial terms, which is simpler to define and study than the topological one.

This construction can be illustrated on graphs. The blowup of a topological directed graph is a 1-dimensional manifold (i.e. a space in which every point has a neighborhood which looks like a line), which behaves like the original graph. The way this is constructed informally consists in replacing every singular point by as many points as there are ways to traverse the singularity, equipped with a suitable topology. For instance, in the lower graph on the right the point in the middle is replaced by four points because there are four ways to go from the left to the right, thus obtaining the space above (and a projection map to the original graph). This construction can be described as an operation transforming a precubical set into a relational one. Indeed, the upper graph is the realization of the relational graph $H : \mathcal{G}^{op} \to \mathbf{Rel}$ with elements and faces

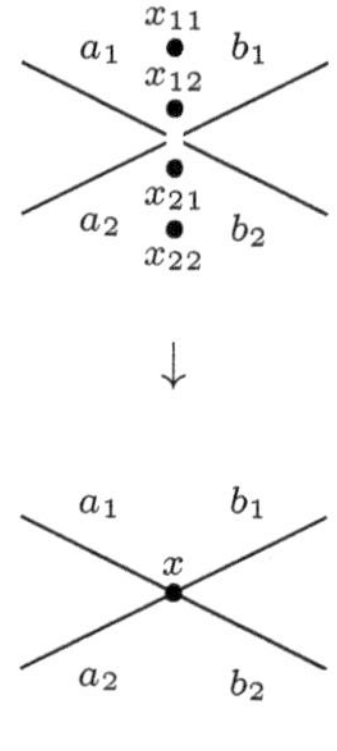

$$H(0) = \{x_{11}, x_{12}, x_{21}, x_{22}\} \quad H(t)(a_1) = \{x_{11}, x_{12}\} \quad H(s)(b_1) = \{x_{11}, x_{21}\}$$
$$H(1) = \{a_1, a_2, b_1, b_2\} \qquad H(t)(a_2) = \{x_{21}, x_{22}\} \quad H(s)(b_2) = \{x_{12}, x_{22}\}$$

and no other boundaries (the point x_{ij} corresponds to passing from a_i to b_j).

Previous work. The idea of using functors to $\mathbf{Rel}$ as semantics of non-deterministic programs is quite old [2]. The formal theory of relational presheaves was developed by Rosenthal in connection to nondeterministic automata theory [31,32] and independently by Ghilardi and Meloni in connection to modal logic [10,11]. Relational presheaves have found several applications in computer science, notably for labelled transition systems (see for example [34] for recent developments). The category of relational presheaves over $\mathcal{C}$ has been shown to be equivalent to several well-studied categories. In [31], Rosenthal showed that this category is equivalent to the category of categories enriched in the free quantaloid over $\mathcal{C}$. Then, in [28,29], using an analogue of the Grothendieck construction, Niefield established a correspondence between relational presheaves over $\mathcal{C}$ and faithful functors over $\mathcal{C}$, which led to a better understanding of exponentiability in the category of relational presheaves.

Higher dimensional automata (HDA) have been introduced as models for true concurrency [13,30] and bear close relationship with geometric models for concurrency [9]. The variant of *partial* HDA was studied in [6,8] to model priorities, and constitutes a subclass of relational presheaves, corresponding to allowing partial functions for face maps. A way of realizing partial precubical sets (in topological spaces) was suggested in [6], based on the idea of completing a partial precubical set into an ordinary one. However, this procedure is not satisfying because it is not cocontinuous (consider for example the pushout square of [6, §4.1]; in fact we will see that it does not even preserve all colimits of partial presheaves which are colimits in the category of relational presheaves).

Connections between relational presheaves and categorical logic have existed since the introduction of the former. For example, models of algebraic theories in **Rel** where already studied in [32], and the internal logic of relational presheaves was studied in [12]. However, to the best of our knowledge, the use of categorical logic in order to study relational presheaves, as well as the definition of their realization, and the combinatorial definition of the blowup are original contributions of this paper.

Plan of the paper. Based on some classical results in categorical logic (Section 2), we define relational presheaves and show that they are models of a given theory (Section 3). We then define a notion of realization, show that it can also be formulated in terms of models of a theory, and that it generalizes the usual notion of realization (Section 4). Some concrete instances of realizations, such as in topological spaces, are then provided (Section 5). We briefly introduce variants of relational presheaves and explain their use (Section 6). Finally, we show that the blowup construction can be expressed as the realization of one on relational presheaves (Section 7).

2   Preliminaries: some categorical logic

We begin by recalling some classical definitions and results on categorical logic that will be used in the following. We refer to [3] for details. Let Σ be a *multisorted signature*, consisting of a set Σ_0 of sorts, Σ_1 of function symbols with sorted arities and a set Σ_R of relation symbols with sorted arities. Recall that a *regular formula* is a first order formula built from atomic formulas using the connectives $\wedge$ and $\exists$. A *regular formula in context* is a finite tuple of free variables $\mathbf{x}$ together with a regular formula φ with free variables in $\mathbf{x}$, noted $\{\mathbf{x}.\varphi\}$, and considered up to α-conversion. A *regular sequent* is a pair of formulas in context φ and ψ, sharing the same context $\mathbf{x}$, noted $\varphi \vdash_{\mathbf{x}} \psi$. Its interpretation in full first order logic is $\forall \mathbf{x}.(\varphi \Rightarrow \psi)$. A *regular theory* is a set of regular sequents, called the axioms of the theory.

Definition 1. *A cartesian formula* with respect to a regular theory $\mathbb{T}$ *(or* $\mathbb{T}$*-cartesian formula) is a regular formula such that every existential quantification is provably unique with respect to* $\mathbb{T}$. *A cartesian theory is a regular theory* $\mathbb{T}$ *such that its axioms can be well-ordered in such a way that every axiom is cartesian with respect to the theory consisting of the preceding axioms.*

In particular, a theory whose axioms are sequents of formulas built from atomic ones only using $\wedge$ is cartesian. Now, for every cartesian (i.e. finitely complete) category $\mathcal{C}$ and cartesian theory $\mathbb{T}$, we can define a category $\mathbb{T}\text{-}\mathbf{Mod}(\mathcal{C})$ of *models* of $\mathbb{T}$ in $\mathcal{C}$. Informally, a model consists in an interpretation of sorts as objects of $\mathcal{C}$, of function symbols as morphisms, and of relation symbols as subobjects, in a way which respects the axioms. Writing Lex for the category of cartesian categories and functors which are *left-exact* (i.e. product-preserving), we have that $\mathbb{T}\text{-}\mathbf{Mod}(-)$ defines a functor on Lex and $\mathbb{T}\text{-}\mathbf{Mod}(\mathbf{Set})$ is the usual category of set-based models of $\mathbb{T}$ and homomorphisms between them.

Definition 2. *A set-based model M of a cartesian theory T is* finitely presented *when there is a* $\mathbb{T}$*-cartesian formula in context $\{\mathbf{x}.\varphi\}$ and a tuple $\mathbf{a} \in M$ such that $M \models \varphi(\mathbf{a})$ and for every model N of $\mathbb{T}$, for every tuple $\mathbf{b} \in N$ such that $N \models \varphi(\mathbf{b})$, there is a unique homomorphism $M \to N$ sending $\mathbf{a}$ to $\mathbf{b}$.*

Proposition 3. *Let $\mathbb{T}$ be a cartesian theory. There is a cartesian category $\mathcal{C}_{\mathbb{T}}$, called the* cartesian syntactic category *of $\mathbb{T}$, such that there is an equivalence*

$$\mathbb{T}\text{-}\mathbf{Mod}(\mathcal{D}) \cong \mathrm{Lex}(\mathcal{C}_{\mathbb{T}}, \mathcal{D})$$

for every cartesian category $\mathcal{D}$, natural in $\mathcal{D}$. Moreover, $\mathcal{C}_{\mathbb{T}}^{op}$ is equivalent to the category of finitely presented set-based models of $\mathbb{T}$.

For a detailed study of syntactic categories, see [19, §D1.4]. We mention that $\mathcal{C}_{\mathbb{T}}$ is built in the following way: its objects are formulas in context which are cartesian with respect to $\mathbb{T}$, and its morphisms are $\mathbb{T}$-provably functional cartesian formulas between them (more precisely, equivalence classes of such with respect to $\mathbb{T}$-provable equivalence). Cartesian theories correspond to essentially algebraic theories, for which similar theorems are established in [1]. Finally, we mention that every cartesian category is the syntactic category of some cartesian theory [3, Theorem 1.4.13], so that cartesian theories can actually be defined as cartesian categories.

3   Relational presheaves as models of a cartesian theory

We now introduce relational presheaves and show that they form the category of models of a particular cartesian theory. We write **Rel** for the category whose objects are sets and whose morphisms are relations. We suppose fixed a small category $\mathcal{C}$.

Definition 4. *A relational presheaf P over $\mathcal{C}$ is a lax functor $\mathcal{C}^{op} \to$* **Rel***. A morphism $\alpha : P \Rightarrow Q$ of relational presheaves is an oplax natural transformation. We write* **RelPsh**$(\mathcal{C})$ *for the corresponding category.*

A relational presheaf P amounts to the data of a family of sets $P(c)$ indexed by objects $c \in \mathcal{C}$ and a family of relations $P(f)$ indexed by morphisms f, with $P(f) \subseteq P(c) \times P(d)$ for $f : d \to c$, such that

$$\mathrm{id}_{P(c)} \subseteq P(\mathrm{id}_c) \qquad\qquad P(g) \circ P(f) \subseteq P(f \circ g) \tag{1}$$

for every $c \in \mathcal{C}$ and every composable morphisms f and g. We write $a \to_f b$ for $(a, b) \in P(f)$. The first condition states that $P(\mathrm{id}_c)$ is reflexive for every $c \in \mathcal{C}$ and the second condition that if $a \to_f b$ and $b \to_g c$ then $a \to_{g \circ f} c$. Similarly, a morphism $\alpha : P \Rightarrow Q$ amounts to a family of functions $(\alpha_c : P(c) \to Q(c))_{c \in \mathcal{C}}$ such that $a \to_f b$ implies $\alpha_c(a) \to_f \alpha_d(b)$ for every $f : d \to c$.

Definition 5. *We write $\Sigma_{\mathcal{C}}^{\mathbf{Rel}}$ for the signature whose sorts are the objects of $\mathcal{C}$, with a relation symbol $R_f \subseteq c \times d$ for every morphism $f : d \to c$, and no function symbol. We write $\mathbb{T}_{\mathcal{C}}^{\mathbf{Rel}}$ for the theory with*

1. for every object $c \in \mathcal{C}$, an axiom

$$x = y \vdash_{x:c,y:c} R_{\mathrm{id}_c}(x, y)$$

2. for all morphisms $f : d \to c$, $g : e \to d$ of $\mathcal{C}$, an axiom

$$R_f(x, y) \wedge R_g(y, z) \vdash_{x:c,y:d,z:e} R_{g \circ f}(x, z)$$

Proposition 6. **RelPsh**$(\mathcal{C})$ *is the category of set-based models of $\mathbb{T}_{\mathcal{C}}^{\mathbf{Rel}}$.*

Proof. A model P amounts to the data of a set $P(c)$ for each sort c and a relation $f \subseteq P(c) \times P(d)$ for every relation R_f with $f : c \to d$, and the two families of axioms ensure that the inclusions of (1) are satisfied. This is thus precisely a relational presheaf in the sense of Definition 4. Similarly, model homomorphisms correspond to oplax natural transformations. $\qquad\square$

Recall that locally finitely presentable categories are exactly the categories of models of cartesian theories [21, Definition 9.2 and Theorem 9.8].

Corollary 7. *For any category $\mathcal{C}$,* **RelPsh**$(\mathcal{C})$ *is locally finitely presentable. Hence, it is complete and cocomplete.*

In the following, we write $\tilde{\mathcal{C}}$ for the cartesian syntactic category of $\mathbb{T}_{\mathcal{C}}^{\mathbf{Rel}}$. By Propositions 3 and 6, the category of relational presheaves is precisely $\mathrm{Lex}(\tilde{\mathcal{C}}, \mathbf{Set})$. Every object x of $\tilde{\mathcal{C}}$ can be seen as a relational presheaf $\,\sharp_{\tilde{\mathcal{C}}^{op}}(x)$.

Example 8. Consider the category $\mathcal{G}$ of the introduction, whose (relational) presheaves are (relational) directed graphs. The signature $\Sigma_{\mathcal{G}}^{\mathbf{Rel}}$ contains two sorts 0 and 1 and two relations $R_s, R_t \subseteq 1 \times 0$. The category $\tilde{\mathcal{G}}$ contains the following morphisms:

$$\{(x:1,y:0).R_s(x,y)\}$$
$$\downarrow$$
$$\{(y:0).\top\} \longleftarrow \{(x:1,y:0).\top\} \longrightarrow \{(x:1).\top\}$$
$$\uparrow$$
$$\{(x:1,y:0).R_t(x,y)\}$$

In a model G, i.e. a left exact functor $\tilde{\mathcal{G}} \to \mathbf{Set}$, the object on the left and on the right are respectively mapped to sets G_0 and G_1, the object in the middle is mapped to the product $G_0 \times G_1$, and the top and bottom object are mapped to subsets $G_s, G_t \subseteq G_0 \times G_1$. We see that the object $\{(x:1,y:0).R_s(x,y)\}$ represents the "proofs of relations" between 0 and 1 via s.

Remark 9. If we apply the same procedure to the presheaf category $\mathbf{Psh}(\mathcal{C})$, seeing it as the category of models of a theory $\mathbb{T}_\mathcal{C}$ similar to the relational case but with function symbols instead of relation symbols, the cartesian syntactic category will just be the free finite limit completion of $\mathcal{C}^{op}$. In fact, $\tilde{\mathcal{C}}$ is also some kind of completion of (a modified version of) $\mathcal{C}$, see Corollary 15.

4   Realizing relational presheaves

A *realization* of a category $\mathcal{C}$ in a cocomplete category $\mathcal{D}$ is a functor $\mathcal{C} \to \mathcal{D}$ which is cocontinuous, i.e. colimit-preserving. The fact that such a functor preserves colimits can be understood as the fact that the image of a "complex" element can be obtained by gluing the images of "simpler" elements. We write $\mathrm{coCont}(\mathcal{C},\mathcal{D})$ for the corresponding category. Typically, for a presheaf category $\mathcal{C} = \mathbf{Psh}(\mathcal{B})$, one can define a realization $\mathbf{Psh}(\mathcal{B}) \to \mathcal{D}$ by starting from a functor $F : \mathcal{B} \to \mathcal{D}$ and taking its left Kan extension $F_!$ along the Yoneda embedding $\sharp : \mathcal{B} \to \mathbf{Psh}(\mathcal{B})$:

$$
\begin{array}{ccc}
 & \mathcal{D} & \\
F \nearrow & \Rightarrow & \nwarrow F_! \\
\mathcal{B} & \xrightarrow{\quad \sharp \quad} & \mathbf{Psh}(\mathcal{B})
\end{array}
$$

More explicitly, the realization can be computed on a presheaf $P \in \mathbf{Psh}(\mathcal{B})$ as the colimit $F_!(P) = \mathrm{colim}_{(b,p)\in\int P} F(b)$, where $\int P$ denotes the category of elements of P [25]. The original notion of geometric realization was essentially defined in this way with $\mathcal{B}$ being the simplicial category and $\mathcal{D}$ the category of topological spaces. Given a small category $\mathcal{C}$ and a cocomplete category $\mathcal{D}$, we explain how to construct a realization functor for the category $\mathbf{RelPsh}(\mathcal{C})$ of relational presheaves and show that the situation is essentially analogous to the case of presheaves. We begin by observing that constructing a realization amounts to constructing a model:

Proposition 10. *The category of realizations of* $\mathbf{RelPsh}(\mathcal{C})$ *in* $\mathcal{D}$ *is isomorphic to the category* $\mathrm{Lex}(\tilde{\mathcal{C}}, \mathcal{D}^{op})$ *of models of* $\mathbb{T}_\mathcal{C}^{\mathbf{Rel}}$ *in* $\mathcal{D}^{op}$.

Proof. We have that relational presheaves form the Ind-completion (whose definition is recalled below) of $\tilde{\mathcal{C}}^{op}$:

$$\mathbf{RelPsh}(\mathcal{C}) \cong \mathbb{T}_{\mathcal{C}}^{\mathbf{Rel}}\text{-}\mathbf{Mod}(\mathbf{Set}) \qquad \text{by Proposition 3}$$
$$\cong \mathrm{Lex}(\tilde{\mathcal{C}}, \mathbf{Set}) \qquad \text{by Proposition 6}$$
$$\cong \mathrm{Ind}(\tilde{\mathcal{C}}^{op})$$

The last step can be justified as follows. Recall that the *ind-completion* $\mathrm{Ind}(\mathcal{C})$ of a category $\mathcal{C}$ is its free cocompletion by filtered colimits [20]. It can be obtained by first forming the free cocompletion $\mathbf{Psh}(\mathcal{C})$ of $\mathcal{C}$ and then taking the full subcategory on presheaves which are filtered colimits of representables, or equivalently on presheaves whose category of elements is filtered, or equivalently on functors $\mathcal{C}^{op} \to \mathbf{Set}$ which are *flat* [26, Theorem VII.6.3]. It is a standard fact that if $\mathcal{C}^{op}$ has all small limits, then flat functors can equivalently be defined as left-exact functors [26, Corollary VII.6.4], i.e. $\mathrm{Lex}(\mathcal{C}^{op}, \mathbf{Set}) \cong \mathrm{Ind}(\mathcal{C})$. Recall that $\tilde{\mathcal{C}}$ indeed has all finite limits, by Proposition 3.

Finally, from the above sequence of equivalences, we deduce

$$\mathrm{coCont}(\mathbf{RelPsh}(\mathcal{C}), \mathcal{D}) \cong \mathrm{coCont}(\mathrm{Ind}(\tilde{\mathcal{C}}^{op}), \mathcal{D}) \qquad \text{by the above}$$
$$\cong \mathrm{Rex}(\tilde{\mathcal{C}}^{op}, \mathcal{D}) \qquad \text{by [1, Proposition 1.45 (ii)]}$$
$$\cong \mathrm{Lex}(\tilde{\mathcal{C}}, \mathcal{D}^{op}) \qquad \text{by duality}$$
$$\cong \mathbb{T}_{\mathcal{C}}^{\mathbf{Rel}}\text{-}\mathbf{Mod}(\mathcal{D}^{op}) \qquad \text{by Proposition 3}$$

This concludes the proof. $\qquad\qquad\qquad\qquad\qquad\qquad\qquad\qquad\qquad\qquad$ $\square$

As a consequence, in order to define a realization of $\mathbf{RelPsh}(\mathcal{C})$ in $\mathcal{D}$, we need to construct a model M of $\mathbb{T}_{\mathcal{C}}^{\mathbf{Rel}}$ in $\mathcal{D}^{op}$. This amounts to the following data:

Proposition 11. *A model in $\mathbb{T}_{\mathcal{C}}^{\mathbf{Rel}}\text{-}\mathbf{Mod}(\mathcal{D}^{op})$ is uniquely determined by the following data:*

A. for every object $c \in \mathcal{C}$, an object $M(c)$ of $\mathcal{D}$,
B. for every morphism $f : d \to c$ of $\mathcal{C}$, an object $M(R_f)$ of $\mathcal{D}$ and morphisms $\iota_f^0 : M(c) \to M(R_f)$ and $\iota_f^1 : M(d) \to M(R_f)$,

such that for every $f : d \to c$ and $g : e \to d$:

0. $\iota_f := (\iota_f^0, \iota_f^1) : M(c) \sqcup M(d) \to M(R_f)$ is an epimorphism,
1. the codiagonal $M(c) \sqcup M(c) \to M(c)$ factors through ι_{id_c},
2. the canonical morphism $M(e) \sqcup M(c) \sqcup M(d) \to M(R_f) \sqcup_{M(d)} M(R_g)$ factors through $\iota_{f \circ g} \sqcup \mathrm{id}_{M(d)}$.

Proof. The data of $M(c)$ and $M(R_f)$ is precisely the interpretation of the sorts and relation symbols of the signature (there is no function symbol). The conditions can be explained as follows, see [3, §1.3.3] for details. Condition 0 ensures that M is a $\Sigma_{\mathcal{C}}^{\mathbf{Rel}}$-structure. Indeed, $M(c)$ is the object of elements of sort c, and $M(R_f)$ represents $\{(x,y) \in M(c) \times M(d) \mid x \to_f y\}$, so it should be a subobject of $M(c) \times M(d)$ in $\mathcal{D}^{op}$, which becomes an epimorphism out of the coproduct in $\mathcal{D}$, which by universal property of the coproduct has to be characterized by a pair of morphisms ι_f^0 and ι_f^1. The other two conditions ensure that M satisfies the axioms of $\mathbb{T}_{\mathcal{C}}^{\mathbf{Rel}}$. Every formula in context $\{\mathbf{x}.\varphi\}$ gives rise to a subobject $M(\{\mathbf{x}.\varphi\})$. An axiom $\varphi \vdash_{\mathbf{x}} \psi$ is satisfied in a structure if and only if the subobject corresponding to φ factors through the one corresponding to ψ, and we must again write this in the opposite category,. This gives conditions 1 and 2 respectively associated to the two families of axioms of Definition 5 (which, in turn, correspond to the two conditions of (1)). $\qquad\square$

We define $\mathcal{C}_{\mathbf{Rel}}$ as the full subcategory of $\tilde{\mathcal{C}}$ on the objects of the form $\{(x : c).\top\}$ and $\{(x : c, y : d).R_f(x, y)\}$ for every $f : d \to c$. This category is interesting because functors out of it encode precisely the data needed to describe a model:

Lemma 12. *The data of A and B in Proposition 11 amounts precisely to defining a functor $M : \mathcal{C}_{\mathbf{Rel}} \to \mathcal{D}^{op}$.*

Proof. Since $\tilde{\mathcal{C}}^{op}$ is the category of finitely presented models of $\mathbb{T}_{\mathcal{C}}^{\mathbf{Rel}}$, this category is just the category $\mathcal{C}$ where we have replaced every morphism by a span. More precisely, $\mathrm{Ob}(\mathcal{C}_{\mathbf{Rel}}) = \mathrm{Ob}(\mathcal{C}) \cup \mathrm{Mor}(\mathcal{C})$, and there is exactly one morphism $\pi_f^1 : f \to \mathrm{dom}(f)$ and one morphism $\pi_f^0 : f \to \mathrm{cod}(f)$ for every $f \in \mathrm{Mor}(\mathcal{C})$. In particular, we do not have to close by composition, because no two such morphisms are composable. $\qquad\square$

By precomposition with the inclusion functor $I : \mathcal{C}_{\mathbf{Rel}} \hookrightarrow \tilde{\mathcal{C}}$, we obtain a functor $F_{\mathcal{D}} : \mathrm{Lex}(\tilde{\mathcal{C}}, \mathcal{D}^{op}) \to \mathbf{Fun}(\mathcal{C}_{\mathbf{Rel}}, \mathcal{D}^{op})$ sending a model M (i.e. a realization, by Proposition 10) to the functor $M \circ I$. In particular, we simply write $F : \mathrm{Lex}(\tilde{\mathcal{C}}, \mathbf{Set}) \to \mathbf{Psh}(\mathcal{C}_{\mathbf{Rel}}^{op})$ for $F_{\mathbf{Set}}$. We write $\mathbf{Fun}_{\mathbf{Mod}}(\mathcal{C}_{\mathbf{Rel}}, \mathcal{D}^{op})$ for the image of the functor $F_{\mathcal{D}}$: its objects consist of functors $M : \mathcal{C}_{\mathbf{Rel}} \to \mathcal{D}^{op}$ (i.e. interpretations of the signature by Lemma 12) satisfying conditions 1, 2 and 3 of Proposition 11.

Lemma 13. *The functor F preserves filtered colimits.*

Proof. The functor F is the composite $\mathrm{Ind}(\tilde{\mathcal{C}}^{op}) \to \mathbf{Psh}(\tilde{\mathcal{C}}^{op}) \to \mathbf{Psh}(\mathcal{C}_{\mathbf{Rel}}^{op})$ where the first functor is the inclusion (see the proof of Proposition 10) and preserves filtered colimits by [20, Theorem 6.1.8], and the second is precomposition and preserves small colimits since it admits a right adjoint [16, §5]. $\qquad\square$

We can now give an explicit way of going from the category of models $\mathrm{Lex}(\tilde{\mathcal{C}}, \mathcal{D}^{op})$ to the category of realizations $\mathrm{coCont}(\mathbf{RelPsh}(\mathcal{C}), \mathcal{D})$. In fact, we can show that this realization is essentially a left Kan extension, in the following sense. Consider a model $M \in \mathrm{Lex}(\tilde{\mathcal{C}}, \mathcal{D}^{op})$. As we have seen in Proposition 10, M corresponds to a realization $\hat{M} : \mathrm{Lex}(\tilde{\mathcal{C}}, \mathbf{Set}) \to \mathcal{D}$. Recall that we have a functor $F_{\mathcal{D}}(M) : \mathcal{C}_{\mathbf{Rel}} \to \mathcal{D}^{op}$, which encodes the data of the model M. By left Kan extension of $F_{\mathcal{D}}(M)$ along the Yoneda embedding $\sharp : \mathcal{C}_{\mathbf{Rel}}^{op} \to \mathbf{Psh}(\mathcal{C}_{\mathbf{Rel}}^{op})$, we can extend this functor to $\mathbf{Psh}(\mathcal{C}_{\mathbf{Rel}}^{op})$, obtaining a functor $F_{\mathcal{D}}(M)_!$ which is cocontinuous. Note that the category $\mathbf{Psh}(\mathcal{C}_{\mathbf{Rel}}^{op})$ contains the category of relational presheaves on $\mathcal{C}$: intuitively, it is a category of *spans* over $\mathcal{C}$, because $\mathcal{C}_{\mathbf{Rel}}$ is a "span version" of $\mathcal{C}$, and a relation is a special kind of span (for which the induced map to the product is injective). Now this inclusion is exactly given by F, so $\hat{M}$ can be compared to $F_{\mathcal{D}}(M)_! \circ F$: we show that in fact both coincide.

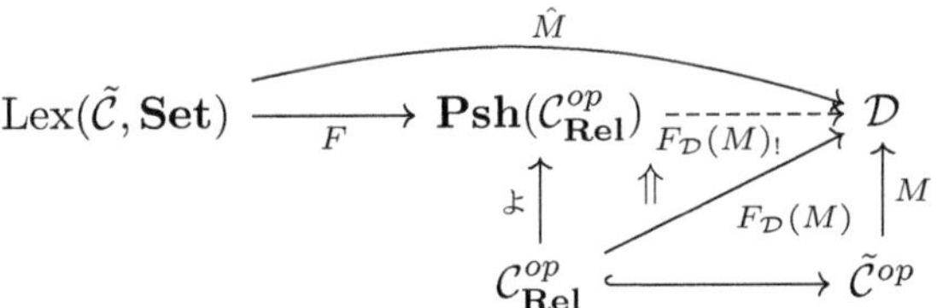

Proposition 14. *For every $M \in \mathrm{Lex}(\tilde{\mathcal{C}}, \mathcal{D}^{op})$, we have $F_{\mathcal{D}}(M)_! \circ F = \hat{M}$, i.e.*

$$\hat{M}(P) = \mathrm{colim}_{(x,a)\in \int F(P)} M(x)$$

Proof. By [19, Lemma D1.4.4(ii)], any object of $\tilde{\mathcal{C}}$ is isomorphic to a formula in context which is a conjunction of atomic formulas. We first see that in fact any object of $\tilde{\mathcal{C}}$ is isomorphic to a conjunction of formulas of the form R_f, and no equalities. Indeed, we have

$$\{(x_1 : c, x_2 : c, \mathbf{x}).(x_1 = x_2) \wedge \varphi\} \cong \{(x : c, \mathbf{x}).\varphi[x_1 \mapsto x, x_2 \mapsto x]\}$$

(where $\varphi[x_1 = x, x_2 = x]$ is the formula φ where we have substituted the variables x_1 and x_2 by the variable x), because the "function" from left to right sending x_1 and x_2 to x is in fact functional in both directions (see [19, Lemma D1.4.4(i)]). Repeating this procedure a finite number of times, we get a formula with no equalities: we say that such a formula in context is *normalized*. Since $\tilde{\mathcal{C}}^{op}$ is the category of finitely presented models (Definition 2) of $\mathbb{T}_{\mathcal{C}}^{\mathbf{Rel}}$, the morphisms between two such formulas are just maps of contexts which preserve the relations. First note that F sends objects of $\mathcal{C}_{\mathbf{Rel}}$ to representable presheaves, since $\mathcal{C}_{\mathbf{Rel}}$ is a full subcategory of $\tilde{\mathcal{C}}$. Now we show that the functors coincide on $\tilde{\mathcal{C}}$, by explicit comparison. This is enough to conclude, by Lemma 13 and the fact that $\mathrm{Ind}(\tilde{\mathcal{C}}^{op})$ is the free completion of $\tilde{\mathcal{C}}^{op}$ by filtered colimits. Let $\{\mathbf{x}.\varphi\}$ be an normalized object of $\tilde{\mathcal{C}}$. Then $\{\mathbf{x}.\varphi\}$ is a pullback over $\{\mathbf{x}.\top\}$ of all the $\{\mathbf{x}.R_f(x_i, x_j)\}$ such that $\varphi \vdash_{\mathbf{x}} R_f(x_i, x_j)$ is provable in $\mathbb{T}_{\mathcal{C}}^{\mathbf{Rel}}$, by definition of the pullback in syntactic categories. Now, $\{\mathbf{x}.\top\}$ is the product of the $\{x_i.\top\}$, so we have a colimit in $(\tilde{\mathcal{C}})^{op}$:

$$\{\mathbf{x}.\varphi\} = \mathrm{colim}_{(x,a)\in \int F(\{\mathbf{x}.\varphi\})} x$$

Indeed, let $D : \int F(\{\mathbf{x}.\varphi\}) \to \tilde{\mathcal{C}}^{op}$ be the corresponding diagram. Since the model $\{\mathbf{x}.\varphi\}$ is presented by the formula φ, for a model N of $\mathbb{T}_{\mathcal{C}}^{\mathbf{Rel}}$, a morphism $\{\mathbf{x}.\varphi\} \to N$ is equivalent to a tuple $\mathbf{y}$ such that $N \models \varphi(\mathbf{y})$, which in turn is equivalent to a family of elements $(y_i)_{1 \leq i \leq n}$ with n the length of $\mathbf{x}$, such that $\varphi \vdash_{\mathbf{x}} R_f(x_i, x_j)$ implies $N \models R_f(y_i, y_j)$. Now $F(\{\mathbf{x}.\varphi\})(\{(y : c).\top\})$ is a finite set counting the number of $(x_i : c)$ in $\mathbf{x}$, and $F(\{\mathbf{x}.\varphi\})(\{(y : c, z : d).R_f(y, z)\})$ is isomorphic the set of pairs (i, j) such that $(x_i : c)$ and $(x_j : d)$ in $\mathbf{x}$ and $\varphi \vdash_{\mathbf{x}} R(x_i, x_j)$ in $\mathbb{T}_{\mathcal{C}}^{\mathbf{Rel}}$. So the family $(y_i)_i$ exactly defines a morphism of diagrams from D to the constant diagram at N. This colimit is sent by $\hat{M}$ to a colimit, so we exactly get $\hat{M}(\{\mathbf{x}.\varphi\}) = F_{\mathcal{D}}(M)_!(F(\{\mathbf{x}.\varphi\}))$. The proof extends to morphisms of $\tilde{\mathcal{C}}$, so we are done. $\qquad\qquad\square$

This is the formalization of the intuition we started with: we realize a relational presheaf by gluing the building blocks chosen for objects and for pairs of objects related by a morphism. We can sum up the previous propositions as:

Corollary 15. *A functor $M : \mathcal{C}_{\mathbf{Rel}} \to \mathcal{D}^{op}$ with $\mathcal{D}$ cocomplete induces a cocontinuous functor $\mathbf{RelPsh}(\mathcal{C}) \to \mathcal{D}$ if and only if it satisfies conditions 0, 1 and 2 of Proposition 11.*

Now it is easy to see that the geometric realization of relational graphs suggested in the introduction is indeed a realization. We will give a more general and detailed construction, for precubical sets, in section 5.2 below.

Remark 16. One could also consider, instead of relational presheaves, the category $\mathrm{Pseudo}(\mathcal{C}^{op}, \mathrm{Span}(\mathbf{Set}))$ of pseudofunctors from $\mathcal{C}^{op}$ to the category whose objects are sets and whose morphisms are spans of functions: namely spans of sets can be seen as a "quantitative" variant of relations, where two elements have a set of relations between them (instead of simply being in relation or not). It is proven in [18,29] that, under some condition (called CFI in [18] and IG in [29]), this is a Grothendieck topos, whose site can be obtained by putting a (non-trivial) Grothendieck topology on the twisted arrows category $\mathcal{C}_{tw}$ of $\mathcal{C}$. The idea is similar to what we just presented: going from $\mathcal{C}$ to $\mathcal{C}_{tw}$ essentially replaces every morphism of $\mathcal{C}$ by a "formal" span. But $\mathcal{C}_{tw}$ additionaly contains coherence data for composition of morphisms in $\mathcal{C}$. The twisted arrow category approach can also be used to realize relational (more precisely, span-valued) presheaves by Kan extension. However, in the cases that interest us, mainly those related to precubical sets, the condition IG is not satisfied, and in fact taking pseudofunctors instead of lax functors would significantly reduce the situations which we would be able to capture, as argued in the introduction.

We end this section by explaining the relationship between presheaves and relational ones on a fixed category $\mathcal{C}$, and the relationship between their realizations. There is a comparison functor $U : \mathbf{Psh}(\mathcal{C}) \to \mathbf{RelPsh}(\mathcal{C})$, induced by post-composition with the canonical functor $\mathbf{Set} \to \mathbf{Rel}$ sending a function to the corresponding functional relation. Since both categories are locally presentable (Corollary 7), we can use the special adjoint functor theorem to construct adjoints.

Theorem 17. *The comparison functor $U : \mathbf{Psh}(\mathcal{C}) \to \mathbf{RelPsh}(\mathcal{C})$ is full and faithful and admits right and left adjoint.*

We deduce that $\mathbf{Psh}(\mathcal{C})$ is a full subcategory of $\mathbf{RelPsh}(\mathcal{C})$ which is reflective and coreflective, hence closed under limits and colimits. In particular:

Theorem 18. *Every realization* **RelPsh**$(\mathcal{C}) \to \mathcal{D}$ *induces, by precomposition with* U, *a realization* **Psh**$(\mathcal{C}) \to \mathcal{D}$.

For instance, it is not difficult to see that the geometric realization of relational graphs induces the usual geometric realization of graphs. Again, the more general case of precubical sets is treated below.

5   Realizations of relational precubical sets

In this section we extend some standard operations on precubical sets to relational precubical sets, and in particular define a first notion of realization.

5.1   Barycentric subdivision

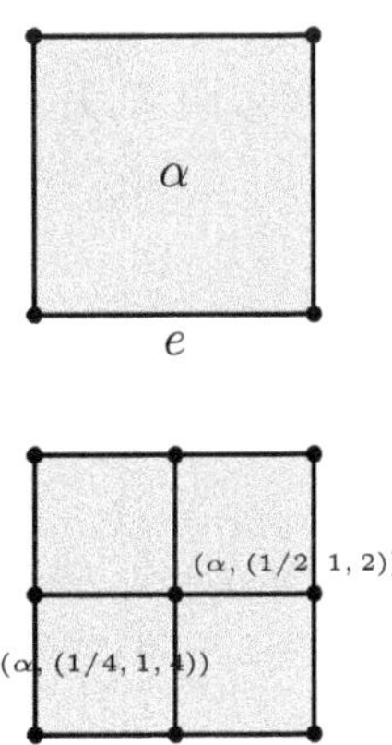

As a first application of previous work, let us define the barycentric subdivision of relational precubical sets. Let I be the graph $\cdot \to \cdot$ with one edge, and J be $\cdot \to \cdot \to \cdot$, its barycentric subdivision (the graph with two consecutive edges). Recall that the category of precubical sets can be equipped with a tensor product [9, §3.4.1], denoted $\otimes$. The classical barycentric subdivision functor $\frac{1}{2}$ can be defined as a realization functor on **Psh**$(\square)$, by $n \mapsto J^{\otimes n}$. In particular, $J^{\otimes n}$ is the barycentric subdivision of $I^{\otimes n}$. For a precubical set P, we can see the elements of $\frac{1}{2}P$ as pairs $(x, (t_1, \ldots, t_m))$ with $x \in P(m)$ and $t_j \in \{1/4, 1/2, 3/4\}$. The dimension of $(x, (t_1, \ldots, t_m))$ is given by the number of indices j such that $t_j \neq 1/2$. See [4, §5.1] for details. We say that $(x, (t_1, \ldots, t_m))$ *belongs to* x.

By Corollary 15, defining a subdivision realization functor amounts to defining a model M of $\mathbb{T}_{\square}^{\mathbf{Rel}}$ in **RelPsh**$(\square)$, which we now do. To an object $n \in \square$, we associate the relational precubical subset of $J^{\otimes n}$ consisting of the cubes belonging to the unique n-cube of $I^{\otimes n}$, so that this is the subdivision of the open n-cube $\text{\textasciidieresis}_{\tilde{\mathcal{C}}^{op}}(\{(x : n).\top\})$. We also set $M(R_{\mathrm{id}_n}) := M(n)$. Any other morphism $f : n \to n+k$ in $\square$ can be seen as an inclusion of an n-cube into an $(n+k)$-cube, and we define $M(R_f)$ to be the subdivision of the open $(n+k)$-cube together with the corresponding subdivided open n-cube in its boundary. More precisely, $\text{\textasciidieresis}_{\tilde{\mathcal{C}}^{op}}(\{(x : n+k, y : n).R_f(x, y)\})$ is canonically a subobject of the representable $(n+k)$-cube $I^{\otimes n}$ seen as a relational precubical set. We then define $M(R_f)$ to be the relational precubical subset of $\frac{1}{2}I^{\otimes n}$ consisting of the cubes wich belong to $\text{\textasciidieresis}_{\tilde{\mathcal{C}}^{op}}(\{(x : c, y : d).R_f(x, y)\})$. For example:

$$M(2) = \quad M(R_{d_0^-}) = \quad M(R_{d_1^+}) = \quad M(R_{d_1^+ d_0^-}) =$$

We can check without difficulty that this indeed defines a model, thus a realization. In addition, the induced realization on **Psh**$(\square)$ is exactly the usual barycentric subdivision.

5.2 Sequential geometric realization

Now, let us define a geometric realization functor for relational precubical sets: we will see that the situation there is quite subtle. Again, we do it by defining a model M of $\mathbb{T}_\square^{\mathbf{Rel}}$ in $\mathbf{Top}^{op}$. Following the idea of [6] for realizing partial precubical sets, it is natural to start by trying the following construction. To an object $n \in \square$, we associate the topological open n-cube $M(n) :=]0,1[^n$. A morphism $f : n \to n+k$ in $\square$ can be seen as an inclusion of an n-cube into an $(n+k)$-cube and $M(R_f)$ is defined as an $(n+k)$-cube together with the corresponding n-cube in its boundary. More precisely, such a morphism decomposes uniquely as

$$f = d_{n+k-1,i_k}^{\varepsilon_k} \circ \ldots \circ d_{n+1,i_2}^{\varepsilon_2} \circ d_{n,i_1}^{\varepsilon_1}$$

with $0 \leq i_1 < i_2 < \ldots < i_k < n+k$, see [15]. We define $M(R_f)$ to be the subspace of $[0,1]^n$ consisting of points $(x_1,\ldots,x_n)$ such that either all the x_i belong to $]0,1[$, or for every index i we have $x_i = 0$ (resp. $x_i = 1$) if $i = i_j$ for some j with $\varepsilon_j = -$ (resp. $\varepsilon_j = +$) and $x_i \in]0,1[$ otherwise. In particular, $M(R_{\mathrm{id}_n}) = M(n)$. For instance,

$$M(2) = \qquad M(R_{d_0^-}) = \qquad M(R_{d_1^+}) = \qquad M(R_{d_1^+ d_0^-}) =$$

However, this construction does not work, because M is not a model of $\mathbb{T}_\square^{\mathbf{Rel}}$. For instance, the condition 2 of Proposition 11 is not satisfied. Indeed, consider the obvious inclusion $f : M(R_{d_1^- d_0^-}) \hookrightarrow M(R_{d_0^-}) \sqcup_{M(1)} M(R_{d_0^-})$:

This map is not continuous, since the space in the target has more open sets than the one in the source. For instance, on the space in the source we have a diagonal path (drawn in red) starting from the lower-left corner and going inside the square, but its image in the space in the target is not continuous. In fact, the image of f is topologically equivalent to the space $M(0) \sqcup M(1)$. There are essentially two solutions to this problem, we provide a first one here, and another one in next section.

A first solution is to slightly modify the above definition in the following way. We consider a model M of $\mathbb{T}_\square^{\mathbf{Rel}}$ in $\mathbf{Top}^{op}$, which is defined as above on objects and on morphisms of the form R_{id_n} and $R_{d_{n,i}^\varepsilon}$. However, for any other morphism $f : m \to n$, we define $M(R_f) := M(m) \sqcup M(n)$, as suggested by the above observation. This indeed satisfies condition 2 of Proposition 11, because the topology on $M(R_{f \circ g})$ for $f, g \neq \mathrm{id}$ will now always be the coproduct topology, i.e. the topology with the most possible open sets, so that no continuity problems can occur. This solution is interesting, because we get a space where continuous paths can only go from (the realization of) a cube to an adjacent cube of dimension $+1$ or -1. This is coherent with the combinatorial definition of paths of partial HDAs in [6,8], where one is only allowed to go from a cube to an adjacent cube of dimension $+1$ or -1, which in turn is crucial for the interpretation of partial HDAs in formalizing priorities. Indeed, going directly from a vertex to a square means starting two transitions at the same time, so it is impossible to force one of them to start *strictly* before the other if this is allowed. Since in a path, the creation of multiple processes has do be done in a sequential way, we introduce the following terminology:

Definition 19. *The* sequential geometric realization *is the functor induced by the above model.*

6 Variants of relational presheaves

Many variants of the notion of precubical relation can be thought of. For instance, one could take colax functors instead of lax ones, i.e. we reverse the inclusions of (1) in the definition. In this section, we will be mostly interested in another variant that we call *relational families*, which is the variant of Definition 4 where we simply drop both conditions (1): a relational family on a category $\mathcal{C}$ consists of a family of sets indexed by objects of $\mathcal{C}$ and a family of relations indexed by morphisms of $\mathcal{C}$. We still keep oplax natural transformations as morphisms and write **RelFam**$(\mathcal{C})$ for the resulting category. All the results of the paper hold for the variants of relational presheaves if we modify appropriately the conditions. In particular, for relational families, one should remove both families of axioms in Definition 5 and both conditions 1 and 2 in Proposition 11 in order to have the above theorems holding for relational families instead of relational presheaves.

The category of relational families is interesting for the semantics of concurrency. For example, consider the square on the right. We can now allow a path in an HDA to jump from a cube to some adjacent cube of arbitrary dimension, which is more standard. Namely, since there are no conditions on composites, we can 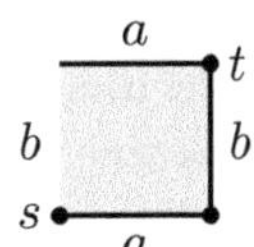 choose whether s should be or not the 0-face of the square, which amounts to determining whether the transitions a and b can or cannot start exactly at the same time. The $\Sigma_\square^{\mathbf{Rel}}$-structure M defined at the beginning of Section 5.2 (before modifications) is a model in this new setting, and therefore defines a realization **RelFam**$(\square) \to$ **Top**.

Definition 20. *The* geometric realization $|-|_{\mathbf{Top}} :$ **RelFam**$(\square) \to$ **Top** *is the functor induced by the above model.*

We use the same notation and terminology for the functor **RelPsh**$(\square) \to$ **Top** obtained by precomposing $|-|_{\mathbf{Top}}$ with **RelPsh**$(\square) \hookrightarrow$ **RelFam**$(\square)$ the canonical inclusion. This realization induces a (cocontinuous) realization of precubical sets **Psh**$(\square) \to$ **Top**. Unlike for the sequential geometric realization, we get:

Proposition 21. $|-|_{\mathbf{Top}} \circ U$ *is the geometric realization of precubical sets.*

7 Blowup commutes with realization

Suppose fixed $n \in \mathbb{N}$. The blowup of a (locally ordered) space is a best approximation of it by an n-euclidean space:

Definition 22. *Given a locally ordered space X, a blowup is an n-euclidean locally ordered space $\tilde{X}$ equipped with a local embedding $\beta_X : \tilde{X} \to X$ such that for every n-euclidean local order E and every local embedding $f : E \to X$, there is a unique continuous lift $\tilde{f}$ such that $\beta_X \circ \tilde{f} = f$.*

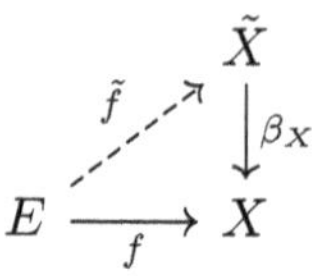

The blowup of a locally ordered space always exists and is unique up to isomorphism [4]. A combinatorial description of the blowup is given in [4, section 5], in the case where X is the realization of a precubical set P. In this section, we show how this description can be expressed as a relational precubical set $\tilde{P}$ over P, and then we prove that the realization of $\tilde{P}$ as defined here is exactly the underlying topological space of $\tilde{X}$..

Definition 23. *Let P be a relational precubical set, c a cube of P. The* neighborhood $N(c)$ *of c is the relational precubical subset of P consisting of the cubes c' such that there exists f with $c' \to_f c$.*

Recall that I is the graph $\cdot \to \cdot$ with one edge, and J is $\cdot \to \cdot \to \cdot$ the graph with two consecutive edges, and write $I_{k,m} := I^{\otimes k} \otimes J^{\otimes m}$. Recall that a *symmetric* precubical set is a precubical set P such that each $P(k)$ is endowed with an action of the symmetric group $\mathfrak{S}_k$, such that the face maps are compatible with these actions, see [15, §6]. Every precubical set can canonically be seen as a symmetric precubical set, by freely adding the images of these actions. Intuitively, a k-cube of a symmetric precubical set P is represented by the orbit of an element of $P(k)$, formalizing the idea that the order on the dimensions does not matter. For example, $I_{1,1}$ is not isomorphic to the precubical set in the upper right corner. However, they are isomorphic as symmetric precubical sets.

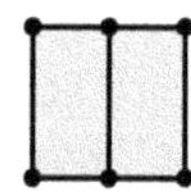

Definition 24. *A (n,k)-euclidean brick is the neighborhood of the minimal cube of some precubical set which is isomorphic to $I_{k,n-k}$ as symmetric precubical sets.*

For example:

$$N(\min(I_{2,0})) = \qquad N(\min(I_{1,1})) = \text{———} \qquad N(\min(I_{0,2})) = +$$

Definition 25. *A morphism of relational precubical sets $\alpha : P \to Q$ is a local embedding if $a \to_f c$ and $b \to_f c$ implies $\alpha(a) \neq \alpha(b)$ for cubes a, b and c of P.*

Euclidean bricks are typically precubical sets whose geometric realization is isomorphic to $\mathbb{R}^n$, and the realization of a local embedding is a local embedding. Note that $I_{k,m}$ has a unique cube of dimension k, and so does any euclidean brick B: we write $\min(B)$ for this cube, since all other cubes of B have dimension $> k$.

Definition 26. *Let P be a precubical set of maximum dimension n. The* blowup *of P is the relational precubical set $\tilde{P}$ given by*

$$\tilde{P}(k) = \{S \subseteq P \mid \exists \alpha : B \to S \text{ surjective local embedding,}$$
$$\text{for some } (n,k)\text{-euclidean brick } B\}$$
$$\tilde{P}(f) = \{(S, S') \mid \exists \alpha : B \to S, \, \exists \alpha' : B' \to S' \text{as above, } \exists \iota : B \to B' \text{ monic}$$
$$\text{such that } \alpha(\min(B)) \to_f \alpha'(\min(B')) \text{ and } \alpha' \circ \iota = \alpha\}$$

where '$\subseteq$' means 'subobject of'. The blowup map *$\beta_P : \tilde{P} \to P$ is defined by $\beta_P(S) := \alpha(\min(B))$ for $\alpha : B \to S$ as above.*

This corresponds to the combinatorial blowup of [4]. More precisely, the combinatorial blowup of a precubical set P is expressed there as a pointwise subset of a presheaf $\mathsf{Comb}_P : \int P \to \mathbf{Set}$ on the opposite of the category of elements $\int P$ of P. And this corresponds to a relational presheaf, because we have an equivalence between this category of presheaves and a subcategory of relational presheaves over P, which generalizes the classical equivalence between presheaves and discrete fibrations [23, Theorem 2.1.2]. Namely, define the category $\mathbf{DFib}(P)$ of *discrete fibrations* over P as the full subcategory of the slice category $\mathbf{RelPsh}(\mathcal{C})/P$ whose objects have the unique right lifting property with respect to the inclusions $\curlywedge_{\tilde{\mathcal{C}}^{op}}(\{(y : d).\top\}) \hookrightarrow \curlywedge_{\tilde{\mathcal{C}}^{op}}(\{(x : c, y : d).R_f(x, y)\})$. We then have,

Theorem 27. *Let P be a relational presheaf over a category $\mathcal{C}$. There is an equivalence of categories $\varphi : \mathbf{Psh}((\int P)^{op}) \cong \mathbf{DFib}(P)$.*

Transforming Comb_P into a relational presheaf over P by φ and taking the relational precubical subset corresponding to the combinatorial blowup, we get $\tilde{P}$. Thanks to this, we are able to compare the realization of $\tilde{P}$ and the "realization" of Comb_P as defined in [4], which leads to the desired theorem:

Theorem 28. *Let P be a precubical set of maximum dimension n. The geometric realization of the blowup $\tilde{P}$ of P is the underlying space of the blowup $\tilde{X}$ of the locally ordered realization X of P.*

This suggests that our realization procedure is indeed the "good" one, and serves as a proof of concept.

8   Conclusion

This paper is a first exploration of relational presheaves in concurrency theory from the point of view of their realizations. Since the category of relational presheaves on a fixed category is well-behaved (finitely locally presentable), it is a good setting to categorically and homotopically study some objects of interest, especially (higher dimensional) automata, which will be the subject of a future work. In particular, relational automata seem to be a good alternative to automata with ε-transitions [33, §1.1.4]. This could be for instance exploited in proving Kleene-like theorems (for example [7]). The simple combinatorial expression of the blowup also suggests to study directly this construction in relational precubical sets, and a homotopical (in the sense of Quillen) definition of euclidean relational precubical sets seems plausible. Finally, on an abstract level, it would be interesting to develop furthur the theory of relational presheaves, for example by adding a Grothendieck topology on the base category and trying to define relational *sheaves*.

References

1. Adamek, J., Rosicky, J.: Locally presentable and accessible categories. Cambridge University Press (1994)
2. Burstall, R.M.: An algebraic description of programs with assertions, verification and simulation. In: Proceedings of ACM conference on Proving assertions about programs. pp. 7–14 (1972)
3. Caramello, O.: Theories, Sites, Toposes. Oxford University Press (2018)
4. Chamoun, Y., Haucourt, E.: Non-hausdorff manifolds over locally ordered spaces via sheaf theory. arXiv preprint arXiv:2505.12087 (2025)
5. Cohen, C., Coquand, T., Huber, S., Mörtberg, A.: Cubical type theory: A constructive interpretation of the univalence axiom. In: 21st International Conference on Types for Proofs and Programs (2018)
6. Dubut, J.: Trees in partial higher dimensional automata. In: FoSSaCS. vol. 11425, pp. 224–241 (2019)
7. Fahrenberg, U., Johansen, C., Struth, G., Ziemiański, K.: Kleene theorem for higher-dimensional automata. Logical Methods in Computer Science **20** (2024)
8. Fahrenberg, U., Legay, A.: Partial higher-dimensional automata. In: 6th Conference on Algebra and Coalgebra in Computer Science. vol. 35 (2015)
9. Fajstrup, L., Goubault, E., Haucourt, E., Mimram, S., Raussen, M.: Directed algebraic topology and concurrency, vol. 138. Springer (2016)
10. Ghilardi, S., Meloni, G.C.: Modal logics with n-ary connectives. Mathematical Logic Quarterly **36**(3) (1990)
11. Ghilardi, S., Meloni, G.: Relational and topological semantics for temporal and modal predicative logic. In: Atti del congresso "Nuovi problemi della logica e della scienza". vol. 2, pp. 59–77. Viareggio (1990)
12. Ghilardi, S., Meloni, G.: Relational and partial variable sets and basic predicate logic. The Journal of Symbolic Logic **61**(3), 843–872 (1996)
13. van Glabbeek, R.J.: Bisimulations for Higher Dimensional Automata. Manuscript available electronically at `http://theory.stanford.edu/~rvg/hda` (1991)
14. van Glabbeek, R.J.: On the Expressiveness of Higher Dimensional Automata. Theoretical Computer Science **356**(3), 265–290 (2006)
15. Grandis, M., Mauri, L.: Cubical sets and their site. Theory and Applications of Categories **11**(8), 185–211 (2003)
16. Grothendieck, A., Verdier, J.: SGA 4 Exposé I. Théorie des topos et cohomologie étale des schemas. Séminaire de géometrie algébrique du Bois-Marie 1963-1964 (SGA 4): Tome 1 **269** (2006)
17. Haucourt, E.: Non-Hausdorff parallelized manifolds over geometric models of conservative programs. Mathematical Structures in Computer Science **35** (2025)
18. Johnstone, P.: A note on discrete Conduché fibrations. Theory and Applications of Categories **5**(1), 1–11 (1999)

19. Johnstone, P.T.: Sketches of an elephant, vol. 2. Oxford University Press (2002)
20. Kashiwara, M., Schapira, P.: Categories and sheaves. Springer (2006)
21. Kelly, G.M.: Structures defined by finite limits in the enriched context, i. Cahiers de Topologie et Géométrie Différentielle **23**(1), 3–42 (1982)
22. Kelly, G.M.: Basic concepts of enriched category theory, Lecture Notes in Mathematics, vol. 64. Cambridge University Press (1982)
23. Loregian, F., Riehl, E.: Categorical notions of fibration. Expositiones Mathematicae **38**(4), 496–514 (2020)
24. Lurie, J.: Higher topos theory. Princeton University Press (2009)
25. Mac Lane, S.: Categories for the working mathematician, vol. 5. Springer Science & Business Media (1998)
26. MacLane, S., Moerdijk, I.: Sheaves in geometry and logic: A first introduction to topos theory. Springer Science & Business Media (2012)
27. May, J.P.: Simplicial objects in algebraic topology, vol. 11. University of Chicago Press (1992)
28. Niefield, S.: Change of base for relational variable sets. Theory and Applications of Categories **12**(7), 248–261 (2004)
29. Niefield, S.: Lax presheaves and exponentiability. Theory and Applications of Categories **24**(12), 288–301 (2010)
30. Pratt, V.: Modeling concurrency with geometry. In: Wise, D.S. (ed.) Proceedings of the 18th annual ACM symposium on Principles of Programming Languages. pp. 311–322 (1991)
31. Rosenthal, K.I.: Free quantaloids. Journal of Pure and Applied Algebra **72**(1), 67–82 (1991)
32. Rosenthal, K.I.: Quantaloids, enriched categories and automata theory. Applied Categorical Structures **3**, 279–301 (1995)
33. Sakarovitch, J.: Elements of automata theory. Cambridge university press (2009)
34. Sobociński, P.: Relational presheaves, change of base and weak simulation. Journal of Computer and System Sciences **81**(5), 901–910 (2015)

𝕂 Definitions as Matching Logic Theories, Formally

Xiaohong Chen[1], Horaţiu Cheval[2,3],
Dorel Lucanu[1,4], and Grigore Roşu[1,5]

[1] Pi Squared Inc. `xiaohong.chen@pi2.network`
[2] LOS Center, University of Bucharest, Romania `horatiu.cheval@unibuc.ro`
[3] Institute for Logic Data Science, Romania
[4] Alexandru Ioan Cuza University of Iaşi, Romania `dorel.lucanu@info.uaic.ro`
[5] University of Illinois at Urbana-Champaign, USA `grosu@illinois.edu`

Abstract. The 𝕂 Framework is a state-of-the-art tool for designing and analyzing programming languages, relying on a frontend tool, kompile, to translate high-level 𝕂 definitions into a low-level Kore representation based on Matching Logic (𝕄𝕃). Currently, this compilation process lacks a formal specification, making it a trusted but unverified component in the toolchain. This paper addresses this gap by proposing a formal mechanism for obtaining the denotational semantics for 𝕂 definitions directly as 𝕄𝕃 theories. A strong feature of the proposed approach is that it respects the abstraction and modularity principles of 𝕂.

Keywords: 𝕂 Framework · Matching Logic · Context Theory.

1 Introduction

The 𝕂 Framework (https://kframework.org) allows for the design and modeling of programming languages and software/hardware systems. It is based on the programming, modeling, and specification language 𝕂 [ ]. The 𝕂 Framework comes with tools for building interpreters, model checkers, verifiers, related documentation, and more. These tools are implemented generically once and for all and then instantiated by a language definition specified using 𝕂.

The 𝕂 process can be broadly separated into two phases: the *frontend phase* and the *backend phase* (see Figure 1, the white part). During the frontend stage, kompile compiles a 𝕂 language definition into an intermediate representation known as the Kore format, used to specify matching logic (𝕄𝕃) representations of the 𝕂 definitions. It also infers the omitted parts of configurations in semantic rules and the types of all the variables. The tool kompile outputs a source file `definition.kore` that includes the entire matching logic theory encoding the formal semantics of the language. This Kore file is then passed to 𝕂s backend to instantiate the corresponding language tools.

How kompile generates the 𝕄𝕃 encoding (in Kore format) of the formal language semantics is magic for 𝕂 users. For example, the below 𝕂 syntax declaration (taken from Figure 2)

N. Bertrand and S. Milius (Eds.): FoSSaCS 2026, LNCS 16503, pp. 198–219, 2026.
https://doi.org/10.1007/978-3-032-22730-0_10

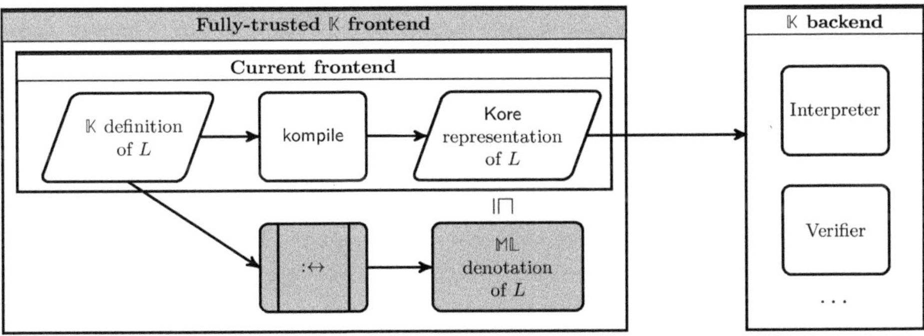

Fig. 1. 𝕂 process: the white boxes represent the current approach, the blue boxes the contribution of this paper. The relation :↔ represents a formal specification of kompile.

```
syntax Stmt ::= "inc" "(" Counter "," Label ")"
```

is compiled into a Kore[6] statement of the form

```
symbol Lblinc{}(SortCounter{}, SortLabel{}) : SortStmt{}
    [constructor{}(), functional{}(), injective{}]
```

together with the axioms corresponding to the attributes `constructor` (indicating that `Lblinc` is a constructor of `SortStmt`), `functional`, and `injective` (indicating that `Lblinc` is an injective function).

The kompile tool is fully developed and tested on many real-life programming languages (see Section 5). However, in order to completely trust that its output `definition.kore` represents indeed the matching logic denotation of the 𝕂 definition, we need a transparent formalization of this process.

The main contribution of this paper is to propose such a formalization, where the 𝕂 frontend definition can be seen as a *domain specific notation* of the 𝕄𝕃 theory it represents (blue part in Figure 1, where the notation relation is :↔). This formalization can be used further to show that kompile is correct by proving that the Kore representation it produces is a refinement (notation ⊑ in Figure 1) of the matching logic theory[7].

There are several significant challenges that must be addressed when formalizing 𝕂 definitions as matching logic theories, briefly explained below.

Axiomatization of abstract rules. 𝕂 allows to specify the rewrite rules describing the semantics in a very abstract way. A main advantage of this abstraction mechanism is modularity: if several languages share some functionality, then this functionality can be specified once and for all and imported by the corresponding language definitions. The tool kompile generates specific axioms for each language. The challenge is to have just one axiomatization for such rules to be shared by all language theories requiring it. The solution we propose is to use the 𝕄𝕃 *theory of contexts* (Sections 3 and 4.5).

---

[6] Kore uses a many-sorted version of matching logic [,].

[7] The correctness aspect is outside the scope of this paper.

Axiomatization of configurations. Configuration specification in $\mathbb{K}$ uses an XML-like notation, describing a tree structure of cells. The challenge is to find an appropriate axiomatic representation for such structures. Our solution consists in a specific abstract datatype theory for each kind of cells (Section 4.4).

Axiomatization of dynamic attributes. Abstraction is also obtained in $\mathbb{K}$ using dynamic attributes. For instance, arguments' order is specified by **strict**-like attributes. Since these attributes are themselves abstractions of certain semantic rules, their axiomatization is based on contexts as well (Sections 3 and 4.2).

Axiomatization of equality on sorts whose constructors satisfy some axioms. The $\mathbb{K}$ definition of a programming language can include datatypes whose constructors satisfy some axioms, such as associativity or commutativity. Hence, there are two kinds of equalities for these data structures: syntactic and semantic (modulo axioms). Our solution is based on using *quotient sorts* (Sections 2.2 and 4.3).

Structure of the paper Section 2 provides background on the two key formalisms. It first describes the components of a $\mathbb{K}$ definition, then introduces the functional variant of matching logic ($\mathbb{ML}$), covering relevant information on the theories used in the paper. Section 3 includes a central contribution: a formal theory of contexts within $\mathbb{ML}$. This reified specification is used to abstract away parts of a configuration, which is crucial for defining the denotations of $\mathbb{K}$'s dynamic features. Section 4 is the core technical contribution, explaining how to translate a $\mathbb{K}$ definition into an $\mathbb{ML}$ theory. We conclude with a discussion on related work (Section 5) and future directions (Section 6).

2 Preliminaries

2.1 $\mathbb{K}$ Frontend Definitions

In $\mathbb{K}$, a definition of a programming language consists of three main components:

Concrete syntax A conventional BNF grammar defining the structure of valid programs in the language. We distinguish three basic kinds of BNF productions:
- *sort declaration*: `syntax` $\langle nonterminal \rangle$;
- *subsorting*: `syntax` $\langle nonterminal \rangle ::= \langle nonterminal \rangle$, and
- *symbol declaration*:
    `syntax` $\langle nonterminal \rangle ::= \langle list\text{-}nonterminals\text{-}and\text{-}terminals \rangle \, [\langle attributes \rangle]$.

A *function symbol* is a symbol declared with the attribute **function**. A non-function symbol is considered as being a *constructor* of the sort given by the left-hand side non-terminal.

Execution state (configuration) A representation of the state maintained by a program during execution. Its declaration is of the form

$$\texttt{configuration} \ \langle cfg \rangle$$

where $\langle cfg \rangle$ is a tree-like structure of cells, declared using an XML-like syntax.

Operational semantics A set of rewrite rules defining how the program's state changes over time. A rewrite rule

$$\texttt{rule}\ \langle rl \rangle\ [\ \texttt{requires}\ \langle precond \rangle\ ]\ [\ \texttt{ensures}\ \langle postcond \rangle\ ]$$

can be a *simplification rule*, where $\langle rl \rangle$ is of the form $\langle lhs \rangle$ => $\langle rhs \rangle$ and used to compute function symbols, or a *one-step transition* specification, where $\langle rl \rangle$ is of the form $C_1[\langle lhs_1 \rangle$ => $\langle rhs_1 \rangle] \ldots C_k[\langle lhs_k \rangle$ => $\langle rhs_k \rangle]$ and C_i specifies a sub-configuration where the local change $\langle lhs_i \rangle$ => $\langle rhs_i \rangle$ holds.

𝕂 uses rewrite rules to define execution in a precise way. A rewrite rule takes the current configuration and transforms it into a new configuration based on a predefined mechanism, ensuring that the execution follows the formal semantics.

Running Example. A minimal computation model, Turing complete, and including all ingredients of a programming language, is the Minsky Machine [20]. It consists of two counters and the following statements: *inc(C, L)*—increments counter C and jumps to statement L; *decjz(C, L_1, L_2)*—if $C > 0$ decrements C and jumps to L_1, otherwise jumps to L_2; and *halt*—stops the execution.

A 𝕂 definition of the Minsky Machine is given in Figure 2. Module **MM-SYNTAX** includes the syntax for statements and counter names. Attribute **strict(1)** refers to the *strict* evaluation of the first argument (**Exp**), a.k.a. eager evaluation or call-by-value. This kind of evaluation is obtained through heating/cooling rules, which are automatically generated. Module **CONFIG** describes the configuration (execution state) consisting of three cells: <k>—for the sequential list of tasks to be executed, <counters>—for the two counters, and <stmts>—for the set of statements represented as pairs *label* $\mapsto$ *statement*. Module **MM-SEMANTICS** includes the rewrite rules describing the execution steps. The rule in line 22 is of the form $C_1[\langle lhs_1 \rangle$ => $\langle rhs_1 \rangle]\ C_2[\langle lhs_2 \rangle$ => $\langle rhs_2 \rangle]$, where C_1 is the sub-configuration given by cell <k>, and C_2 is the sub-configuration given by cell <stmts>. The rule in line 27 is syntactic sugar for <k>**decjz(0, L1, _) => L1** ...</k>. Each such rule represents an execution step *current-config* $\rightarrow$ *next-config* in an abstract way, using a configuration-abstraction/concretization mechanism. The rules in lines 30 and 31 are *simplification* rules because they are used to compute the function symbol **max**; they do not represent semantic execution steps, but function evaluations.

In Figure 3 we extend the 𝕂 definition of the Minsky Machine with multi-threading. Note that this extension is very modular: the syntax and semantics are augmented with new specific rules, and only the configuration is changed significantly. The multiplicity attribute in <thread multiplicity = "*" ...> says that cell <thread> can have zero, one, or more occurrences in a concrete configuration. *The configuration-abstraction/concretization mechanism allows to use the rewrite rules of the sequential machine in a multi-threading configuration.*

2.2   Matching Logic (𝕄𝕃)

In this paper, we use the functional variant of Matching Logic (𝕄𝕃) [8].

An 𝕄𝕃 signature (Σ, EV, SV) consists of a set of *constant symbols* Σ, a set of *element variables* EV, and a set of *set variables* SV. The formulas, called

```
module MM-SYNTAX
  imports INT
  syntax Counter ::= "c1" | "c2"
  syntax Label ::= Int
  syntax Exp ::= Counter | Int
  syntax Stmt ::= "inc" "(" Counter "," Label ")"
    | "decjz" "(" Exp "," Label "," Label ")" [strict(1)]
    | "halt"
  syntax Int ::= max(Int, Int) [function]
endmodule
module CONFIG
  imports MM-SYNTAX
  imports MAP
  configuration <T>
      <k> 0 </k>
      <counters> c1 |-> 0 c2 |-> 0 </counters>
      <stmts> $PGM:Map </stmts>
    </T>
endmodule
module MM-SEMANTICS
  imports CONFIG
  rule <k>L:Label => S</k> <stmts>...L |-> S:Stmt...</stmts>
  rule <k> inc(C:Counter, L:Label) => L ...</k>
      <counters>... C |-> (V => V +Int 1)  </counters>
  rule <k> C:Counter => max(0, V) ...</k>
      <counters>... C |-> (V => max(0,V -Int 1)) </counters>
  rule decjz(0, L1, _) => L1
  rule decjz(V, _, L2) => L2 requires V >Int 0
  rule halt => .K
  rule max(I1:Int, I2:Int) =>  I1 requires (I1 >Int I2)
  rule max(I1:Int, I2:Int) =>  I2 requires (I1 <=Int I2)
endmodule
```

Fig. 2. $\mathbb{K}$ definition of a Minsky Machine

patterns, are defined as follows:

$$\varphi ::= x \mid X \mid \sigma \mid \varphi_1\, \varphi_2 \mid \bot \mid \varphi_1 \to \varphi_2 \mid \exists x.\varphi \mid \mu X.\varphi \text{ if } \varphi \text{ is positive in } X$$

A pattern φ is positive in X if every free occurrence of X is under an even number of negations in φ, where a negation $\neg\psi$ is the notation $\psi \to \bot$ and sub-formulas $\psi_1 \to \psi_2$ are seen as $\neg\psi_1 \vee \psi_2$. Let PATTERN denote the set of patterns.

Definition 1 (Models). *Given* $\Sigma = (EV, SV, \Sigma)$, *a* Σ-*model (or simply model) is a tuple* $(M, \_\cdot\_, \{M_\sigma\}_{\sigma\in\Sigma})$, *where*

1. M *is a carrier set, required to be nonempty;*
2. $\_\cdot\_: M \times M \to \mathcal{P}(M)$ *is a function, called the* interpretation of application; *here,* $\mathcal{P}(M)$ *is the powerset of* M;
3. $M_\sigma \subseteq M$ *is a subset of* M, *called the* interpretation of σ in M, $\sigma \in \Sigma$.

```
1   module MT-SYNTAX
2     imports MM-SYNTAX
3     syntax Stmt ::= "fork" "(" Label "," Label ")"
4                   | "join" "(" Counter "," Label ")"
5                   | "exit"
6   endmodule
7   module CONFIG
8     imports MT-SYNTAX
9     imports MAP
10    configuration <T>
11        <threads>
12          <thread multiplicity = "*" type = "List">
13            <k> 0 </k>
14          </thread>
15        </threads>
16        <counters> c1 |-> 0 c2 |-> 0 </counters>
17        <stmts> $PGM:Map </stmts>
18      </T>
19  endmodule
20  module MT-SEMANTICS
21    imports MM-SEMANTICS
22    rule <threads>.Bag => <thread> <k>0</k> </thread></threads>
23    rule <k> fork(L1, L2) => L2 ...</k>
24        ( .Bag => <thread> <k> L1 </k> </thread> )
25    rule <k> join(C, L) => L ...</k>
26        <counters>... C |-> 0 </counters>
27    rule <thread> <k> exit ...</k> </thread> => .Bag
28  endmodule
```

Fig. 3. 𝕂 definition of a Minsky Machine extended with multi-threading

By abuse of notation, we write M to denote the above model.

We further extend $\_\bullet\_$ *pointwise* from elements to sets as follows:

$$\_\bullet\_: \mathcal{P}(M) \times \mathcal{P}(M) \to \mathcal{P}(M) \qquad A \bullet B = \bigcup_{a \in A, b \in B} a \bullet b \text{ for } A, B \subseteq M$$

Note that $\emptyset \bullet A = A \bullet \emptyset = \emptyset$ for any $A \subseteq M$.

Definition 2 (Pattern interpretation). *Given $\Sigma = (EV, SV, \Sigma)$ and a model M, an (M-)valuation is a function $\rho: (EV \cup SV) \to (M \cup \mathcal{P}(M))$ that maps element variables to elements in M and set variables to subsets of M; i.e., $\rho(x) \in M$ for all $x \in EV$ and $\rho(X) \subseteq M$ for all $X \in SV$. We define a* pattern *interpretation $|\_|_\rho: \text{PATTERN} \to \mathcal{P}(M)$ inductively as follows:*

$$|x|_\rho = \{\rho(x)\} \quad |X|_\rho = \rho(X) \quad |\sigma|_\rho = M_\sigma \quad |\bot|_\rho = \emptyset \quad |\varphi_1 \, \varphi_2|_\rho = |\varphi_1|_\rho \bullet |\varphi_2|_\rho$$

$$|\varphi_1 \to \varphi_2|_\rho = M \setminus (|\varphi_1|_\rho \setminus |\varphi_2|_\rho) \qquad |\exists x.\, \varphi|_\rho = \bigcup_{a \in M} |\varphi|_{\rho[a/x]} \qquad |\mu X.\, \varphi|_\rho = \mu \mathcal{F}^\rho_{X,\varphi}$$

where $\rho[a/x]$ is the valuation ρ' such that $\rho'(x) = a$, $\rho'(y) = \rho(y)$ for any $y \in EV$ distinct from x, and $\rho'(X) = \rho(X)$ for any $X \in SV$. Here, $\mathcal{F}^\rho_{X,\varphi}: \mathcal{P}(M) \to$

$\mathcal{P}(M)$ *is the function defined as* $\mathcal{F}^{\rho}_{X,\varphi}(A) = |\varphi|_{\rho[A/X]}$ *for every* $A \subseteq M$, *where* $\rho[A/X]$ *is the valuation* ρ' *such that* $\rho'(X) = A$, $\rho'(Y) = \rho(Y)$ *for any* $Y \in SV$ *distinct from* X, *and* $\rho'(x) = \rho(x)$ *for any* $x \in EV$.

Definition 3.   *1.* φ *is* valid *in* M *(equivalently,* φ *holds in* M*), written* $M \models \varphi$, *if* $|\varphi|_{\rho} = M$, *for any* M-*valuation* ρ.
   2. If Γ *is a set of patterns, then* $M \models \Gamma$ *if* $M \models \varphi$ *for any* $\varphi \in \Gamma$.
   3. $\Gamma \models \varphi$ *if* $M \models \varphi$ *for any model* M *with* $M \models \Gamma$. *We say that* φ *is a* semantic consequence *of* Γ. *If* Γ *is the empty set, we simply write* $\models \varphi$.

Notations [ ] is a mechanism that allows flexible use of $\mathbb{ML}$. In particular, notations can be used to define a domain-specific logic inside $\mathbb{ML}$. It uses a less conventional notion of theory given as a triple $(\Sigma, \Phi, \vdash)$, where Σ is the alphabet of constant symbols, Φ a set of patterns, and $\vdash$ an entailment relation.

A *notation-based specification* is given as a conservative inclusion theory morphism $(\Sigma, \Phi, \vdash) \hookrightarrow (\Sigma, \Phi', \vdash')$, such that each *new formula* $\varphi' \in \Phi' \setminus \Phi$ is a *notation* of a formula $\varphi \in \Phi$, written as $\varphi' :\leftrightarrow \varphi$ and expressed by two new axioms: $\vdash' \varphi' \to \varphi$ and $\vdash' \varphi \to \varphi'$. Φ' might also contain additional axioms that limit the use of notations. By abuse of notation, we often write $(\Sigma, \Phi, \vdash) :\leftrightarrow (\Sigma, \Phi', \vdash')$ to express that $(\Sigma, \Phi', \vdash')$ is an extension by notation of $(\Sigma, \Phi, \vdash)$.

The simplest example is the definition of the derived operators as notations: $\neg\varphi :\leftrightarrow \varphi \to \bot$, $\varphi_1 \vee \varphi_2 :\leftrightarrow \neg\varphi_1 \to \varphi_2$, $\top :\leftrightarrow \neg\bot$, etc.

A Hilbert-style proof system exists for $\mathbb{ML}$ [,], establishing the provability relation $\Gamma \vdash \varphi$, where Γ is a matching logic theory and φ is a pattern. This proof system is very expressive as a variety of common logics can be derived from it. Therefore, we believe that it is suitable for deriving specific proof rules for $\mathbb{K}$ definitions. However, familiarity with this proof system is not required to understand the technical results presented in this paper.

2.3   Basic $\mathbb{ML}$ Theories

The $\mathbb{ML}$ denotation of a $\mathbb{K}$ definition is built on top of several theories.

Theory of equality   It includes a symbol def axiomatically defined by the axiom $\forall x. \mathsf{def}\, x$ and having the following meaning: $|\mathsf{def}\, \varphi|_{M,\rho} = $ **if** $|\varphi|_{M,\rho} \neq \emptyset$ **then** M **else** $\emptyset$. Totality, equality, inclusion, and membership are specified as notations: $\lceil \varphi \rceil :\leftrightarrow \mathsf{def}\, \varphi$ (definedness), $\lfloor \varphi \rfloor :\leftrightarrow \neg\lceil \neg\varphi \rceil$ (totality), $\varphi_1 = \varphi_2 :\leftrightarrow \lfloor \varphi_1 \leftrightarrow \varphi_2 \rfloor$ (equality), $\varphi_1 \subseteq \varphi_2 :\leftrightarrow \lfloor \varphi_1 \to \varphi_2 \rfloor$ (set inclusion), and $x \in \varphi :\leftrightarrow x \subseteq \varphi$ (membership). Using equality, e.g., we may define that a pattern φ is *functional* if axiom $\exists x.\varphi = x$ holds, i.e., it is always evaluated to singleton. An example of functional patterns is given by the (algebraic) terms.

Theory of sorts   A symbol inh is used such that the pattern $(\mathsf{inh}\, s)$ represents the set of all inhabitants of the sort with name s. Notation $\top_s :\leftrightarrow (\mathsf{inh}\, s)$ is often used for the pattern representing the set of inhabitants. Symbol Sort is used to represent the sort of all sort names, i.e., its inhabitants are the sort names; in particular, $\mathsf{Sort} \in \top_{\mathsf{Sort}}$.

$\mathbb{ML}$ allows to specify operations over sorts [,]. Next, we present only some of them.

Product sorts. Given two sorts s_1 and s_2 (i.e., $s_1, s_2 \in \top_{\mathsf{Sort}}$), their *product sort* is specified by a sort name $s_1 \otimes s_2$, $s_1 \otimes s_2 \in \top_{\mathsf{Sort}}$, a constant symbol $\langle\_,\_\rangle$ in Σ, together with a notation $\langle x, y \rangle :\leftrightarrow \langle\_,\_\rangle \, x \, y$ and the following axioms:

$$\forall x{:}s_1.\forall y{:}s_2.\exists z{:}s_1 \otimes s_2.\langle x, y \rangle = z \qquad \langle x_1, y_1 \rangle = \langle x_2, y_2 \rangle \rightarrow x_1 = x_2 \wedge y_1 = y_2$$

$$\top s_1 \otimes s_2 = \exists x{:}s_1.\exists y{:}s_2.\langle x, y \rangle \qquad \forall x{:}s_1.\forall y{:}s_2.\forall z{:}s_3.\langle\langle x, y \rangle, z \rangle = \langle x, \langle y, z \rangle\rangle$$

Due to associativity, $\langle x, y, z \rangle$ equally denotes either $\langle\langle x, y \rangle, z \rangle$ or $\langle x, \langle y, z \rangle\rangle$. This naturally extends to tuples of arbitrary lengths: $\langle x_1, \ldots, x_k \rangle$, $k \geq 2$.

Remark 1. If $\varphi{:}s_1$ and $\psi{:}s_2$ are two (nonfunctional) patterns, then $\langle \varphi, \psi \rangle$ denotes the cartesian product of φ and ψ, as

$$|\langle \varphi, \psi \rangle|_\rho = \bigcup_{a \in |\varphi|_\rho, b \in |\psi|_\rho} (a, b).$$

Power sorts. Given a sort s ($s \in \top_{\mathsf{Sort}}$), its *power sort* is specified by a sort name 2^s, $2^s \in \top_{\mathsf{Sort}}$, two constant symbols, extension and intension in Σ, together with the following axioms:

$$\forall \alpha{:}2^s.\mathsf{extension}\ \alpha \subseteq \top_s \qquad \forall \alpha{:}2^s.\forall \beta{:}2^s.\mathsf{extension}\ \alpha = \mathsf{extension}\ \beta \rightarrow \alpha{=}\beta$$

$$X{\subseteq}\top_s \rightarrow \exists \alpha{:}2^s.\mathsf{extension}\ \alpha{=}X \qquad \mathsf{intension}\ \varphi :\leftrightarrow \exists \alpha{:}2^s.\alpha \wedge (\mathsf{extension}\ \alpha = \varphi)$$

A *relation* $R \subseteq \top_{s_1} \times \top_{s_2}$ is presented by an element $r : 2^{s_1 \otimes s_2}$ such that extension $r = R$. The *inverse* R^{-1} of a relation $R \subseteq \top_s \times \top_s$ is specified by a constant symbol $\_^{-1}$, a notation, and two axioms:

$$r^{-1} :\leftrightarrow \_^{-1}\, r \qquad\qquad \forall r : 2^{s \otimes s}.\exists r' : 2^{s \otimes s}.r^{-1} = r'$$

$$\forall r : 2^{s \otimes s}.\mathsf{extension}\ r^{-1} = \exists x, y{:}s.\langle y, x \rangle \wedge \langle x, y \rangle \in \mathsf{extension}\ r$$

The *composition* $R_1 \circ R_2$ of two relations $R_1 \subseteq \top_{s_1} \times \top_{s_2}$ and $R_2 \subseteq \top_{s_2} \times \top_{s_3}$ is specified by a constant symbol $\_\circ\_$, a notation, and two axioms:

$$r_1 \circ r_2 :\leftrightarrow \_\circ\_\, r_1\, r_2 \qquad \forall r_1 : 2^{s_1 \otimes s_2}.\forall r_2 : 2^{s_2 \otimes s_3}.\exists r : 2^{s_1 \otimes s_3}.r = r_1 \circ r_2$$

$$\forall r_1 : 2^{s_1 \otimes s_2}.\forall r_2 : 2^{s_2 \otimes s_3}.\mathsf{extension}\ r_1 \circ r_2 = \exists x{:}s_1.\exists y{:}s_2.\exists z{:}s_3.$$
$$\langle x, z \rangle \wedge \langle x, y \rangle \in \mathsf{extension}\ r_1 \wedge \langle y, z \rangle \in \mathsf{extension}\ r_2$$

Function Sorts. Given sorts s_1, s_2 ($s_1, s_2 \in \top_{\mathsf{Sort}}$), the functions $\top_{s_1} \rightarrow \top_{s_2}$ are specified by a *function sort* $[s_1 \rightarrow s_2]$, where the functions are represented by their graphs, i.e., $\top_{[s_1 \rightarrow s_2]} \subseteq \top_{2^{s_1 \otimes s_2}}$, and the following domain-specific axioms:

$$\forall f{:}[s_1 \rightarrow s_2].\forall x{:}s_1.\exists y{:}s_2.\langle x, y \rangle \in \mathsf{extension}\ f$$

$$\forall f{:}[s_1 \rightarrow s_2].\forall x{:}s_1.\forall y_1, y_2{:}s_2.$$
$$(\langle x, y_1 \rangle \in \mathsf{extension}\ f \wedge \langle x, y_2 \rangle \in \mathsf{extension}\ f) \rightarrow y_1 = y_2$$

$$\forall f{:}[s_1 \rightarrow s_2].\forall x{:}s_1.f\ x = \exists y.y \wedge \langle x, y \rangle \in \mathsf{extension}\ f$$

Quotient Sorts. Given a sort s ($s \in \top_{\mathsf{Sort}}$) and an equivalence relation $\equiv$ on $\top_s \times \top_s$, the *quotient sort* is specified by a sort name $s/{\equiv}$ ($s/{\equiv} \in \top_{\mathsf{Sort}}$), a constant symbol $[\_]_\equiv$ together with a notation $[x]_\equiv :\leftrightarrow [\_]_\equiv\, x$, and the axioms:

$$\forall x{:}s.[x]_\equiv \in \top_{2^s} \qquad \top_{s/\equiv} = \exists x{:}s.[x]_\equiv \qquad \forall x{:}s.\mathsf{extension}\ [x]_\equiv = \exists y{:}s.y \wedge x \equiv y$$

The pattern $[x]_\equiv$ uniquely identifies the equivalence class of x, represented by extension $[x]_\equiv$: if $x \equiv y$ then $[x]_\equiv = [y]_\equiv$. The following hold:

$$\forall f : [s \rightarrow s'].(\forall x, y{:}s.x \equiv y \rightarrow (f\ x = f\ y)) \rightarrow \exists \bar{f}{:}[(s/_\equiv) \rightarrow s'].(\forall x{:}s.\bar{f}\ [x]_\equiv = f\ x)$$
$$\forall \bar{f} : [(s/_\equiv) \rightarrow s'].\exists f{:}[s \rightarrow s'].(\forall x{:}s.\bar{f}\ [x]_\equiv = f\ x)$$

Specification of rewriting The one-step rewriting (transition) relation is specified using a *one-path next* symbol $\bullet$ having the following interpretation: for each $x \in \bullet\varphi'$ there is a $y \in \varphi'$ such that $x \Rightarrow y$ is a one-step rewriting relation. Patterns φ' and $\bullet\varphi'$ must have the same sort: $\varphi' \subseteq \top_s \rightarrow \bullet\varphi' \subseteq \top_s$. Furthermore, we assume that $\bullet\varphi'$ is defined pointwise: $\bullet\varphi' = \exists y{:}s. \bullet y \wedge y \in \varphi'$. A rewrite rule, specifying a particular one-step rewriting relation, is a pattern of the form $\varphi \rightarrow \bullet\varphi'$. In order to emphasize that $\varphi \rightarrow \bullet\varphi'$ indeed represents a relation, we introduce the following notation:

$$\mathsf{CH}(\varphi \rightarrow \bullet\varphi') :\leftrightarrow \exists x{:}s.\exists y{:}s.\langle x, y \rangle \wedge x \in (\varphi \wedge \bullet y) \wedge y \in \varphi'$$

where a rewriting step $x \Rightarrow y$ can be seen as a notation for the pair $\langle x, y \rangle$ in the above pattern. This can be interpreted as a "particular case" of the CurryHoward correspondence, where the implication corresponds to a relation (instead of a function). Note that $\mathsf{CH}(\varphi \rightarrow \bullet\varphi')$ is a pattern of sort $s \otimes s$. In particular, when φ is $\top_s$, we have $\mathsf{CH}(\bullet\varphi') :\leftrightarrow \exists x{:}s.\exists y{:}s.\langle x, y \rangle \wedge x \in \bullet y \wedge y \in \varphi'$.

Remark 2. If a rewrite pattern $\varphi \Rightarrow \varphi' :\leftrightarrow \varphi \rightarrow \bullet\varphi'$ is an axiom of a theory Γ, then we have $M \models \varphi \rightarrow \bullet\varphi'$ iff $M \models \varphi \subseteq \bullet\varphi'$ for any Γ-model M.

3   A Theory of Contexts

A *context* is intended to represent a pattern φ with a hole, usually denoted by $\Box$, which can be filled by some other pattern ψ. Although it would be straightforward to define contexts as meta-theoretical constructs, e.g., operations on the syntax, such an approach would not be aligned with our goal of formalizing the semantics of $\mathbb{K}$ as an $\mathbb{ML}$ theory. We need a reified specification, given through a signature and some axioms, together with notations on top of it. For this, we apply the general mechanism for defining binders introduced in [ ], which starts with the following notation

$$[\Box{:}s_1]\varphi :\leftrightarrow \mathsf{intension}\ (\exists\Box{:}s_1.\langle \Box, \varphi \rangle)$$

where $\Box$ is an element variable of sort s_1 and φ is a functional pattern of sort s_2. The $[\Box{:}s_1]\varphi$ notation, of sort $2^{s_1 \otimes s_2}$, is used to indicate that variable $\Box$ is bound in φ, and is the building block for encoding the binders in $\mathbb{ML}$.

The extension of a binding is a functional relation:

$$\mathsf{extension}\ [\Box{:}s_1]\varphi :\leftrightarrow \mathsf{extension}\ (\mathsf{intension}\ (\exists\Box{:}s_1.\langle \Box, \varphi \rangle)) \quad \text{(by notation)}$$
$$= \exists\Box{:}s_1.\langle \Box, \varphi \rangle \quad \text{(by power sort def.)}$$

If ρ is a variable valuation, then $|\exists\Box{:}s_1.\langle \Box, \varphi \rangle|_\rho = \bigcup_{a \in M_{s_1}} |\langle \Box, \varphi \rangle|_{\rho[a/\Box]} = \bigcup_{a \in M_{s_1}} \langle a, |\varphi|_{\rho[a/\Box]} \rangle$, where $M_{s_1} = |\top_{s_1}|_\rho$. Since φ is functional, it follows that the relation is the graph of a function.

Starting with the theory of sorts from Section 2.2, we add the symbols

$$\text{plug, gamma, and Context}^{s_2}_{s_1} \text{ for any sorts } s_1, s_2.$$

$\text{Context}^{s_2}_{s_1}$ stands for the sort of contexts of sort s_2 with a hole of sort s_1. Constant symbol gamma defines a binder used to specify that a certain element variable $\Box$ is treated as a hole in a pattern, thus forming a context, while plug is used to substitute the bound hole for another pattern. We use the following notations:

$$\gamma\Box{:}s_1.\varphi :\leftrightarrow \text{gamma}([\Box{:}s_1]\varphi) \qquad\qquad C[x] :\leftrightarrow \text{plug}(C, x)$$

The specification Γ_{context} of the contexts is given by the following axioms:

$$\forall s_1, s_2{:}\text{Sort}.\text{Context}^{s_2}_{s_1} \in \top_{\text{Sort}} \tag{Ax.1}$$

$$\forall s_1, s_2{:}\text{Sort}.\forall \alpha_1, \alpha_2{:}[s_1 \to s_2].(\text{gamma } \alpha_1 = \text{gamma } \alpha_2) \to \alpha_1 = \alpha_2 \tag{Ax.2}$$

$$\forall s_1, s_2{:}\text{Sort}.\top_{\text{Context}^{s_2}_{s_1}} = \exists \alpha{:}[s_1 \to s_2].\text{gamma } \alpha \tag{Ax.3}$$

$$\forall s_1, s_2{:}\text{Sort}.\forall C_1, C_2{:}\text{Context}^{s_2}_{s_1}.(\forall x{:}s_1.C_1[x] = C_2[x]) \to C_1 = C_2 \tag{Ax.4}$$

$$\forall s_1, s_2{:}\text{Sort}.\forall C{:}\text{Context}^{s_2}_{s_1}.\forall x{:}s_1.$$
$$C[x] = \exists \alpha{:}[s_1 \to s_2].\exists y{:}s_2.y \wedge (C = \text{gamma } \alpha \wedge \langle x, y\rangle \in \text{extension } \alpha) \tag{Ax.5}$$

Most of the axioms are self-explanatory, considering the meaning of gamma and plug. The only rather technical axiom is Ax.5, which, for $C :\leftrightarrow \gamma\Box{:}s_1.\varphi$, allows us to compute the plugging $C[z]$ in terms of φ:

$$C[z] = \exists y{:}s_2.y \wedge (\langle z, y\rangle \in \exists\Box{:}s_1.\langle\Box, \varphi\rangle).$$

By Axiom Ax.3, each context C is of the form gamma α. We may obtain $C :\leftrightarrow \gamma\Box{:}s_1.\varphi$, where $\varphi :\leftrightarrow \alpha(\Box)$ and $\alpha(\Box) :\leftrightarrow \exists y.y \wedge \langle\Box, y\rangle \in \text{extension } \alpha$. Since α is a function, it follows that y is uniquely determined. Therefore, we always consider the contexts to be of the form $\gamma\Box{:}s_1.\varphi$, with φ functional.

The term 'context' is often used for constructs supporting a plugging operation that can be performed without constraints on capture avoidance. By contrast, in our formulation, plugging behaves like a capture avoiding substitution. While this is not strictly necessary, it is a consequence of our choice to internalize their definition as a theory, rather than presenting them as syntactical objects. This restriction helps avoid soundness issues related to accidental variable capture, and it does not affect the use of contexts for $\mathbb{K}$ definitions.

Crucially, from these axioms we are able to prove that the plugging operation corresponds to substitution in the following sense, which is equivalent to saying that β-reduction relative to plug holds.

Theorem 1. *Let s_1, s_2 be two sorts, and φ be a pattern of sort s_2. Then,*

$$\Gamma_{\text{context}} \models (\gamma\Box{:}s_1.\varphi)[\psi] = \varphi[\psi/\Box]$$

for any functional pattern ψ of sort s_1 not having $\Box$ as a free element variable.

Given $C_1{:}\text{Context}^{s_2}_{s_1}$ and $C_2{:}\text{Context}^{s_3}_{s_2}$, the composition of C_1 and C_2 is defined as the following notation of sort $\text{Context}^{s_3}_{s_1}$, and behaves similarly to function composition:

$$C_2 \circ C_1 :\leftrightarrow \gamma\Box{:}s_1.C_2[C_1[\Box]] \qquad\qquad (C_2 \circ C_1)[\psi] = C_2[C_1[\psi]]$$

Because $\mathbb{K}$ allows multiple rewrites to appear in the same rule (as long as they are not nested), we will need a notion of *multihole context* to formalize this feature, which is straightforward to define via currying.

An n-hole context on sorts $s_1, \ldots, s_n$ is a pattern of the form

$$C :\leftrightarrow \gamma\square_1{:}s_1.\ldots.\gamma\square_n{:}s_n.\varphi$$

where φ is of sort s. We will abbreviate the sort of such patterns by $\mathsf{Context}^s_{s_1,\ldots,s_n}$. As in currying, plugging into multihole contexts is simply iterated single-hole plugging: $C[\psi_1, \ldots, \psi_n] :\leftrightarrow C[\psi_1] \ldots [\psi_n]$.

The composition is naturally extended to contexts with multiple holes.

Definition 4 (Multiple context composition). *Let $C{:}\mathsf{Context}^s_{s_1,\ldots,s_n}$ be an n-hole context, and $C_i{:}\mathsf{Context}^{s_i}_{s'_i}$ with $i = 1, \ldots, n$ be contexts. We define the multiple composition of C and $C_1, \ldots, C_n$ as*

$$C \circ \langle C_1, \ldots, C_n \rangle :\leftrightarrow \gamma\square_1{:}s'_1 \ldots \gamma\square_n{:}s'_n.C[C_1[\square_1], \ldots, C_n[\square_n]],$$

where $\square_1, \ldots, \square_n$ are fresh element variables.

For $n = 1$, the multiple composition becomes *functional composition*: $C \circ C_1 \leftrightarrow C \circ C_1$. Note that $C \circ \langle C_1, \ldots, C_n \rangle$ will be of sort $\mathsf{Context}^s_{s'_1,\ldots,s'_n}$. The following result shows how to plug a multiple context composition.

Proposition 1 (Plugging multiple context composition). *Consider the contexts $C{:}\mathsf{Context}^s_{s_1,\ldots,s_n}$ and $C_i{:}\mathsf{Context}^{s_i}_{s'_i}$, $i = 1, \ldots, n$. For any functional patterns $\psi_i{:}s'_i$ with $i = 1, \ldots, n$, we have that*

$$\Gamma_{context} \models (C \circ \langle C_1, \ldots, C_n \rangle)[\psi_1, \ldots, \psi_n] = C[C_1[\psi_1], \ldots, C_n[\psi_n]].$$

In this section, we focused on the aspects of the theory of contexts that are strictly required for the $\mathbb{ML}$ denotations in Section 4. A more comprehensive presentation can be found in [ ].

4 $\mathbb{K}$ Definitions as $\mathbb{ML}$ Theories

In this section, we formally specify a notation $:\leftrightarrow$ such that

$$\mathbb{K} \text{ definition of } L :\leftrightarrow L^{\mathbb{ML}}$$

where $L^{\mathbb{ML}}$ is the matching logic theory represented by L.

4.1   The Main Idea

The main idea to generate the theory $L^{\mathbb{ML}}$ is as follows:

1. Let Sort be the constant symbol whose inhabitants are given by the set of sorts used in $L^{\mathbb{ML}}$.

2. Each non-terminal N denotes a sort N subsorted to KItem (see Section 4.3):

$$\texttt{syntax } N \quad :\leftrightarrow \quad \frac{\exists y{:}\mathsf{Sort}.N = y}{\top_N \subseteq \top_{\mathsf{KItem}}}$$

3. Each subsorting BNF production denotes a subsort axiom:

$$\texttt{syntax } A ::= B \; :\leftrightarrow \top_B \subseteq \top_A$$

We also use notation $B <: S$ for the subsort relation.

4. Each non-subsorting BNF production denotes a distinct symbol $\sigma \in \Sigma(L^{\mathsf{ML}})$, which is functional:

$$\texttt{syntax } N ::= T_1 \; N_1 \dots T_k \; N_k \; T_{k+1} \; :\leftrightarrow \; \begin{array}{c} \forall x_1{:}N_1.\dots.\forall x_k{:}N_k.\exists y{:}N. \\ \sigma(x_1,\dots,x_k) = y \end{array}$$

In this way, each K term t denotes, via the notation, an ML pattern t^{ML}, i.e.,

$$t \; :\leftrightarrow t^{\mathsf{ML}}$$

5. Each attribute of a BNF production, excepting **symbol** and **function**, denotes a set of axioms (see Section 4.2). So, the syntax $BNF(L)$ of L denotes an ML theory $BNF(L)^{\mathsf{ML}}$ [ ]:

$$BNF(L) \; :\leftrightarrow BNF(L)^{\mathsf{ML}}$$

6. The cells of the configuration denote a theory including specific sets of sorts, symbols, and axioms (see Section 4.4). So, the configuration declaration of L denotes an ML theory $CONFIG(L)^{\mathsf{ML}}$:

$$CONFIG(L) \; :\leftrightarrow CONFIG(L)^{\mathsf{ML}}$$

7. The K simplification rewrite rules, which compute the value of a function symbol, denote equalities of the corresponding ML patterns.

8. The K semantic rewrite rules denote one-step transition ML patterns, built using contexts (see Section 4.5). In this way, the semantic rewrite rules of L, $SEM(L)$, denote an ML theory $SEM(L)^{\mathsf{ML}}$:

$$SEM(L) \; :\leftrightarrow SEM(L)^{\mathsf{ML}}$$

9. The builtin modules are specified as abstract datatypes. An example is given in Section 4.3. Let $BLTN(L)^{\mathsf{ML}}$ denote the theory including the specifications of the builtin datatypes.

10. The ML theory denoted by L will be $L^{\mathsf{ML}} = BNF(L)^{\mathsf{ML}} \cup CONFIG(L)^{\mathsf{ML}} \cup SEM(L)^{\mathsf{ML}} \cup BLTN(L)^{\mathsf{ML}}$ and satisfies $L \; :\leftrightarrow L^{\mathsf{ML}}$.

4.2 Attributes of Symbols

Each BNF production defining a symbol may have, implicitly or explicitly, one or more attributes, grouped as follows:

- *definitional attributes*: **symbol**, **function**;
- *static semantic attributes*: **constructor**, **functional**, **total**, **injective**, **assoc**, **comm**, **unit**, **idem**;
- *dynamic semantic attributes*: **strict**, **seqstrict**, **strict**$(i_1,\dots,i_k)$.

We use $\pi_\sigma(N_1,\dots,N_k,N)$ to denote a non-subsorting BNF production for the non-terminal N with the attribute **symbol**(σ), where $N_1,\dots,N_k$ are the non-terminals occurring in the BNF expression (in this order).

Definitional attributes The attribute **symbol**, designates the symbol name associated with that rule. It is mentioned implicitly or explicitly. The attribute **function** specifies that the symbol associated to that rule denotes a function.

Such symbols should have associated (simplification) rules that compute the value of the function, denoting equalities of the corresponding $\mathbb{ML}$ patterns:

$$\texttt{rule } t \Rightarrow t' \texttt{ [simplification]} \;:\leftrightarrow\; t^{\mathbb{ML}} = t'^{\mathbb{ML}}$$

Example 1. Since the BNF production for `max` in Figure 2 has the attribute `function`, the rules computing it are simplification rules and their $\mathbb{ML}$ denotations are equalities. For example,

$$\texttt{rule max(I1:Int, I2:Int) => I1 requires (I1 >Int I2)}$$

$$:\leftrightarrow$$

$$\forall i_1, i_2 \mathsf{:Int.} >_{\mathsf{Int}} (i_1, i_2) \to \mathsf{max}(i_1, i_2) = i_1$$

Static Semantic Attributes We present only the axioms denoted by the attribute `constructor`. The $\mathbb{ML}$ denotations of the other static semantic attributes are intuitive and therefore they are omitted.

1. For each $\pi_\sigma(N_1, \ldots, N_k, N)$ having the attribute `constructor`, an axiom saying that σ is injective is added, which is equivalent to the slogan "no confusion, the same constructor":

$$\forall x_1, x_1' \mathord{:} N_1 \ldots \forall x_k, x_k' \mathord{:} N_k. (\sigma(x_1, \ldots, x_k) = \sigma(x_1', \ldots, x_k')) \to$$
$$(x_1 = x_1' \wedge \cdots \wedge x_k = x_k')$$

2. For each pair $\pi_\sigma(N_1, \ldots, N_k, N)$ and $\pi_{\sigma'}(N_1', \ldots, N_{k'}', N)$ with $\sigma \neq \sigma'$, having the attribute `constructor`, an axiom formalizing the slogan "no confusion, different constructors" is added:

$$\forall x_1 \mathord{:} N_1 \ldots \forall x_k \mathord{:} N_k. \forall x_1' \mathord{:} N_1' \ldots \forall x_{k'}' \mathord{:} N_{k'}'. \sigma(x_1, \ldots, x_k) \neq \sigma'(x_1', \ldots, x_{k'}'))$$

3. If $\pi_{\sigma^i}(N_1^i, \ldots, N_{k_i}^i, N)$, $i \in I$, are all the productions having (explicitly or implicitly) the `constructor` attribute for the sort N, then the following axiom specifies the carrier set of N, formalizing the slogan "no junk":

$$\top_N = \mu X.(\bigvee_{s <: N} \top_s) \vee (\bigvee_{i \in I} \sigma^i(Y_1, \ldots, Y_{k_i})) \qquad \text{(No-Junk)}$$

where $Y_j = X$ if $N_j^i = N$, and $Y_j = \top_{N_j^i}$ if $N_j^i \neq N$.

Remark 3. 1. We assume that all sorts N are non-void: $\lceil \top_N \rceil$.

2. If the BNF productions are mutually recursive, then computing the carrier set using (No-Junk) is a bit tricky. For such cases, the technique described by Chen et al. [ ] can be applied.

If some constructors of sort N have attributes `assoc`, `comm`, `idem`, and/or `unit`, then axioms defining the quotient sort $N/{\cong_N}$ are added, where $\cong_N$ is the congruence generated by these axioms. An example is given in Section 4.3.

Dynamic Semantic Attributes We consider only the case of the attribute `strict`$(i_1, \ldots, i_k)$ for $\pi_\sigma(N_1, \ldots, N_n, N)$, the other ones being particular cases (syntactic sugar) of this. In 𝕂, sort KResult is used to specify the "values" of the programming language. We use notation $KResult(x) :\leftrightarrow x \in \top_{KResult}$ to say that x is a value already evaluated.

The expressions to be evaluated always lie in the `<k>` cells, therefore we have to use the contexts in order to abstract the configurations over which the computations are executed. The content of a cell `<k>` is a sequential list $t_1 \curvearrowright t_2 \curvearrowright \ldots$, meaning that t_1 is evaluated/computed first, then t_2, and so on.

For each $i \in \{i_1, \ldots, i_k\}$ of `strict` we have a *heating* rewriting axiom:

$$\forall C : \mathsf{Context}^{\mathsf{Cell}\langle\mathsf{T}\rangle}_{\mathsf{Cell}\langle\mathsf{k}\rangle}.\forall\kappa{:}\mathsf{K}.\forall x_1{:}N_1, \ldots, x_n{:}N_n.$$

$$(C \circ C_{\mathbf{k}})[\sigma(x_1, \ldots, x_n)] \wedge \neg\mathsf{KResult}(x_i) \rightarrow (\bullet C \circ C_{\mathbf{k}})[x_i \curvearrowright C_{\sigma,i}]$$

and a *cooling* rewriting axiom:

$$\forall C : \mathsf{Context}^{\mathsf{Cell}\langle\mathsf{T}\rangle}_{\mathsf{Cell}\langle\mathsf{k}\rangle}.\forall\kappa{:}\mathsf{K}.\forall x_1{:}N_1, \ldots, x_n{:}N_n.$$

$$(C \circ C_{\mathbf{k}})[x_i \curvearrowright C_{\sigma,i}] \wedge \mathsf{KResult}(x_i) \rightarrow \bullet(C \circ C_{\mathbf{k}})[C_{\sigma,i}[x_i]]$$

where $C_{\mathbf{k}} :\leftrightarrow \gamma\Box_{\mathbf{k}}{:}\mathsf{KItem}.\langle\!|\mathbf{k}|\!\rangle(\Box_{\mathbf{k}} \curvearrowright \kappa)$ and $C_{\sigma,i} :\leftrightarrow \gamma\Box_i{:}N_i.\sigma(x_1, \ldots, \Box_i, \ldots, x_n)$. Sorts K and KItem are described in Section 4.3 and Cell$\langle_\rangle$ in Section 4.4.

The heating rules correspond to the "split" operation, where a term is split into a context and a subterm (to be evaluated), while the cooling rules correspond to the "plug" operation, where the value obtained by the evaluation of the subterm is plugged into the context.

Example 2. Here are the axioms corresponding to `strict(1)` on `decjz` in the syntax of the Minsky Machine (see Figure 2):

$$\forall C{:}\mathsf{Context}^{\mathsf{Cell}\langle\mathsf{T}\rangle}_{\mathsf{Cell}\langle\mathsf{k}\rangle}.\forall\kappa{:}\mathsf{K}.\forall x_1{:}\mathsf{Exp}.\forall x_2, x_3{:}\mathsf{Label}.$$

$$(C \circ C_{\mathbf{k}})[decjz(x_1, x_2, x_3)] \wedge \neg\mathsf{KResult}(x_1) \rightarrow \bullet(C \circ C_\kappa)[x_1 \curvearrowright C_{\mathsf{decjz},1}]$$

$$\forall C{:}\mathsf{Context}^{\mathsf{Cell}\langle\mathsf{T}\rangle}_{\mathsf{Cell}\langle\mathsf{k}\rangle}.\forall\kappa{:}\mathsf{K}.\forall x_1{:}\mathsf{Exp}.\forall x_2, x_3{:}\mathsf{Label}.$$

$$(C \circ C_{\mathbf{k}})[x_1 \curvearrowright C_{\mathsf{decjz},1}] \wedge \mathsf{KResult}(x_1) \rightarrow \bullet(C \circ C_{\mathbf{k}})[C_{\mathsf{decjz},1}[x_1]]$$

where $C_{\mathsf{decjz},1}$ is the context $\gamma\Box{:}\mathsf{Exp}.decjz(\Box, x_2, x_3) \in \mathsf{Context}^{\mathsf{Stmt}}_{\mathsf{Exp}}$. These axioms can be instantiated in both configurations defined in Figures 2 and 3.

4.3   Builtin Theories

𝕂 includes a set of builtin datatypes [ ], whose operations are implemented using a *hooking* mechanism. The datatypes are also specified in 𝕂 [ ], so we can use these specifications to define their 𝕄𝕃 denotations. Here we consider the K datatype, which is an important component used in almost any 𝕂 definition.

K Datatype   The K datatype is an associative list $t_1 \curvearrowright t_2 \curvearrowright \ldots$ [ ] and is used to populate the cells `<k>`. Its 𝕂 specification and 𝕄𝕃 denotation are given in Figure 4, where *equality modulo axioms* $\cong_\mathsf{K}$ is specified by a constant symbol Eq_K, a notation $x \cong_\mathsf{K} y :\leftrightarrow \langle x, y \rangle \in \mathsf{Eq}_\mathsf{K}$, and the axiom in Figure 5.

The specification of the quotient sort $\mathsf{K}/\!\cong_\mathsf{K}$ corresponds to the attributes `assoc` and `unit` of $\curvearrowright$.

$$
\begin{array}{lll}
\texttt{syntax K [hook(K.K)]} & :\leftrightarrow & \exists y{:}\mathsf{Sort.K} = y \\[4pt]
\texttt{syntax KItem [hook(K.KItem)]} & :\leftrightarrow & \exists y{:}\mathsf{Sort.KItem} = y \\[4pt]
\texttt{syntax K ::= KItem} & :\leftrightarrow & \top_{\mathsf{KItem}} \subseteq \top_{\mathsf{K}} \\[4pt]
\texttt{syntax K ::= ".K"} & & \\
\texttt{[symbol(dotK)]} & :\leftrightarrow & \exists z{:}\mathsf{K.dotK} = z \\[4pt]
\end{array}
$$

```
syntax K ::= K "~>" K
  [symbol(⤳),
   assoc, unit(dotK)]
```

$$
:\leftrightarrow
\begin{array}{c}
x \curvearrowright y :\leftrightarrow (\curvearrowright\ x\ y) \\
\forall x, y{:}\mathsf{K}. \exists z{:}\mathsf{K}. x \curvearrowright y = z \\
\forall x, y{:}\mathsf{K}. x \curvearrowright y \neq \mathsf{dotK} \\
\forall x, x', y, y'{:}\mathsf{K}. x \curvearrowright y = x' \curvearrowright y' \to \\
x = x' \wedge y = y' \\
\text{the axioms of the quotient sort } \mathsf{K}/{\cong_{\mathsf{K}}}
\end{array}
$$

Fig. 4. Axioms for the K datatype

$$
\begin{aligned}
\mathsf{Eq}_{\mathsf{K}} = \mu R{:}\mathsf{K}\otimes\mathsf{K}.\, & \exists x{:}\mathsf{K}.\langle x, x\rangle \vee && \text{/* reflexive */} \\
& \exists x, y{:}\mathsf{KItem}.\langle x, y\rangle \wedge x \cong_{\mathsf{KItem}} y \vee && \text{/* } \cong_{\mathsf{KItem}} \subseteq \cong_{\mathsf{K}} \text{ */} \\
& \exists x, y, z{:}\mathsf{K}.(x \curvearrowright y) \curvearrowright z \cong_{\mathsf{K}} x \curvearrowright (y \curvearrowright z) \vee && \text{/* associative */} \\
& \exists x{:}\mathsf{K}.x \curvearrowright \mathsf{dotK} \cong_{\mathsf{K}} x \vee && \text{/* right unit */} \\
& \exists y{:}\mathsf{K}.\mathsf{dotK} \curvearrowright y \cong_{\mathsf{K}} y \vee && \text{/* left unit */} \\
& (\text{intension } R)^{-1} \vee && \text{/* symmetric */} \\
& (\text{intension } R) \circ (\text{intension } R) \vee && \text{/* transitive */} \\
& \exists x_1, x_2, y_1, y_2{:}\mathsf{K}.\langle x_1 \curvearrowright y_1, x_2 \curvearrowright y_2\rangle && \text{/* congruence */} \\
& \quad \wedge \langle x_1, x_2\rangle \in R \wedge \langle y_1, y_2\rangle \in R &&
\end{aligned}
$$

Fig. 5. Definition of Eq_{K}

4.4 Configurations

The formal syntax for configuration declarations is quite complex, as it allows one to specify various structures needed to define the semantics of programming languages or software systems.

The main component of a configuration is that of a cell, which can be singleton, optional, multiple, and may include other cells. Here we include only the ML denotation of single cells that include other cells. The specification is given in Figure 6, where S_i is the sort of the cell C_i, i.e., $S_i = \mathsf{Cell}\langle name(C_i)\rangle$. The specification includes a return sort $\mathsf{Cell}\langle cn\rangle$ and a constructor $\langle\!| cn |\!\rangle$ for the content of the cell.

The denotations of cells having the attributes `<cn multiplicity = *, type = ...` are more complex because they have to include specifications of specific data structures mentioned by the **type** attribute, which can be Set, List, or Map. For example, the following cell declaration from Figure 3

```
1  <threads>
2    <thread multiplicity = "*" type = "List">  ... </thread>
3  </threads>
```

$$\exists y{:}\mathsf{Sort}.\mathsf{Cell}\langle cn\rangle = y$$
$$\exists z.(\!|cn|\!) = z$$

`<cn>` $C_1\dots C_k$ `</cn>`
an instance of
`<cell> ::=`
  `<single-cell>`

$$\forall x_1{:}S_1,\dots,x_k{:}S_k.\exists z{:}\mathsf{Cell}\langle cn\rangle.(\!|cn|\!)(x_1,\dots,x_k) = z$$
$${:}\!\leftrightarrow \forall x_1{:}S_1,\dots,x_k{:}S_k.\forall y_1{:}S_1,\dots,y_k{:}S_k.$$
$$(\!|cn|\!)(x_1,\dots,x_k) = (\!|cn|\!)(y_1,\dots,y_k) \rightarrow$$
$$x_1 = y_1 \wedge \cdots \wedge x_k = y_k$$
$$\top_{\mathsf{Cell}\langle cn\rangle} = \exists x_1{:}S_1,\dots,x_k{:}S_k.(\!|cn|\!)(x_1,\dots,x_k)$$

Fig. 6. Axioms for single cells

requires that the contents of the cell `<threads>` be of sort $\mathsf{ListCell}\langle\mathsf{thread}\rangle$.

Example 3. Here are the axioms for the `<T>` cell from the configuration of the Minsky Machine (Figure 2):

`<T> <k> ... </k> <counters> ... </counters> <stmts> ... </stmts>`
$${:}\!\leftrightarrow$$

$$\mathsf{Cell}\langle T\rangle \in \top_{\mathsf{Sort}} \qquad \exists z.(\!|T|\!) = z$$
$$\forall x_1{:}\mathsf{Cell}\langle k\rangle.\forall x_2{:}\mathsf{Cell}\langle counters\rangle.\forall x_3{:}\mathsf{Cell}\langle stmts\rangle.\exists z{:}\mathsf{Cell}\langle T\rangle.(\!|T|\!)(x_1,x_2,x_3) = z$$

$$\forall x_1{:}\mathsf{Cell}\langle k\rangle.\forall x_2{:}\mathsf{Cell}\langle counters\rangle.\forall x_3{:}\mathsf{Cell}\langle stmts\rangle.(\!|T|\!)(x_1,x_2,x_3) = (\!|T|\!)(y_1,y_2,y_3) \rightarrow$$
$$x_1 = y_1 \wedge x_2 = y_2 \wedge x_3 = y_3$$

$$\top_{\mathsf{Cell}\langle T\rangle} = (\!|T|\!)(\top_{\mathsf{Cell}\langle k\rangle}, \top_{\mathsf{Cell}\langle counters\rangle}, \top_{\mathsf{Cell}\langle stmts\rangle})$$

The specification of the `<k>` cell, e.g., is similar but its content is of sort K:

`<k> ... </k>`
$${:}\!\leftrightarrow$$

$$\mathsf{Cell}\langle k\rangle \in \top_{\mathsf{Sort}} \qquad \exists z.(\!|k|\!) = z \qquad \forall x{:}\mathsf{K}.\exists z{:}\mathsf{Cell}\langle k\rangle.(\!|k|\!)(x) = z$$
$$\top_{\mathsf{Cell}\langle k\rangle} = \exists x{:}\mathsf{K}.(\!|k|\!)(x) \qquad \forall x,y{:}\mathsf{K}.(\!|k|\!)(x) = (\!|k|\!)(y) \rightarrow x = y$$

4.5 Semantic Rewrite Rules

These rules define the dynamic semantics of a language/system: each such a rule specifies a one-step execution. We first explain the most general form of a rule. Assuming that we already have

$$cfg\ \texttt{=>}\ cfg'\ \text{an instance of}\ \texttt{<rule-body>}\ {:}\!\leftrightarrow \forall\bar{x}.cfg^{\mathsf{ML}} \rightarrow \bullet cfg'^{\mathsf{ML}}$$

then

$$\texttt{rule}\ cfg\ \texttt{=>}\ cfg'\ \texttt{requires}\ precond\ \texttt{ensures}\ postcond\ \text{ an instance of}$$

`<rule-declaration> ::=`
    `"rule" <rule-body> "requires" <condition>`
                    `"ensures" <condition>`
$${:}\!\leftrightarrow$$
$$\forall\bar{x}.(cfg^{\mathsf{ML}} \wedge precond^{\mathsf{ML}}) \rightarrow \bullet(cfg'^{\mathsf{ML}} \wedge postcond^{\mathsf{ML}})$$

If either the precondition or the postcondition (or both) is omitted from the rule declaration, it can be assumed to be equal to $\top$.

We explain in the following how $\forall \bar{x}.cfg^{\mathbb{ML}} \to \bullet cfg'^{\mathbb{ML}}$ is obtained for various kinds of rewrites $cfg \Rightarrow cfg'$. Here is a place where the axiomatic definition of contexts plays a major role.

K-Contextual Rewrite These rules describe changes in `<k>`-cells without mentioning any cell/context.

$$t \Rightarrow t' \text{ an instance of } \texttt{<rule-body> ::= <k-term> "=>" <k-term>}$$

$$:\leftrightarrow$$

$$\forall C : \mathsf{Context}^{\mathsf{Cell}\langle\mathsf{T}\rangle}_{\mathsf{Cell}\langle\mathsf{k}\rangle}.\forall\kappa{:}\mathsf{K}.C[\langle\!\!\langle\mathsf{k}\rangle\!\!\rangle(t^{\mathbb{ML}} \curvearrowright \kappa)] \to \bullet C[\langle\!\!\langle\mathsf{k}\rangle\!\!\rangle(t'^{\mathbb{ML}} \curvearrowright \kappa)]$$

Example 4. For the $\mathbb{K}$ definition in Figure 2 we have

$$\texttt{rule decjz(0, L1, \_) => L1}$$

$$:\leftrightarrow$$

$$\forall C{:}\mathsf{Context}^{\mathsf{Cell}\langle\mathsf{T}\rangle}_{\mathsf{Cell}\langle\mathsf{k}\rangle}.\forall l_1, l_2{:}\mathsf{Label}.(C \circ C_\kappa)[\mathsf{decjz}(0, l_1, l_2)] \to \bullet(C \circ C_\kappa)[l_1]$$

Local Rewrites These rules specify one-step executions by mentioning only those configuration components that are changed (using abstraction mechanism).

We start by specifying the components of the the $\mathbb{ML}$ denotation given by the local rewrites in a single cell:

$$\texttt{<cn>}\, c[t_1 \Rightarrow t'_1] \cdots [t_k \Rightarrow t'_k]\texttt{</cn>} \text{ an instance of}$$

$$\texttt{<local-changes-under-local-multi-context>}$$

$$:\leftrightarrow$$

$$\gamma\square_1{:}s_1 \ldots \gamma\square_k{:}s_k.c^{\mathbb{ML}}{:}\mathsf{Context}^{\mathsf{Cell}\langle cn\rangle}_{s_1,\ldots,s_k}$$
$$t_1^{\mathbb{ML}},\ldots,t_k^{\mathbb{ML}}$$
$$t'_1{}^{\mathbb{ML}},\ldots,t'_k{}^{\mathbb{ML}}$$

where s_i is the sort of t_i and t'_i and local rewrites $t_i \Rightarrow t'_i$ in c are replaced by the corresponding holes $\square_i$ in $c^{\mathbb{ML}}$.

Now we extend the specification to all local rewrites declared in a rule body:

$$\texttt{<cn}^1\texttt{>}c^1[t_1^1 \Rightarrow t'^1_1] \ldots [t^1_{k_1} \Rightarrow t'^1_{k_1}]\texttt{</cn}^1\texttt{>} \ldots \texttt{<cn}^\ell\texttt{>}c^\ell[t_1^\ell \Rightarrow t'^\ell_1] \ldots [t^\ell_{k_\ell} \Rightarrow t'^\ell_{k_\ell}]\texttt{</cn}^\ell\texttt{>}$$
$$\text{an instance of}$$

$$\texttt{<rule-body> ::= <local-changes-under-local-multi-contexts>}$$

$$:\leftrightarrow$$

$$\forall C{:}\mathsf{Context}^{\mathsf{Cell}\langle\mathsf{T}\rangle}_{\mathsf{Cell}\langle cn_1\rangle,\ldots,\mathsf{Cell}\langle cn_\ell\rangle}.$$
$$(C \odot \langle c_1^{\mathbb{ML}},\ldots,c_\ell^{\mathbb{ML}}\rangle)[t_1^1{}^{\mathbb{ML}},\ldots,t^1_{k_1}{}^{\mathbb{ML}},\ldots,t_1^\ell{}^{\mathbb{ML}},\ldots,t^\ell_{k_\ell}{}^{\mathbb{ML}}] \to$$
$$\bullet(C \odot \langle c_1^{\mathbb{ML}},\ldots,c_\ell^{\mathbb{ML}}\rangle)[t'^1_1{}^{\mathbb{ML}},\ldots,t'^1_{k_1}{}^{\mathbb{ML}},\ldots,t'^\ell_1{}^{\mathbb{ML}},\ldots,t'^\ell_{k_\ell}{}^{\mathbb{ML}}]$$

where $c_i^{\mathbb{ML}}$ are defined as above.

Example 5. For the $\mathbb{K}$ definition in Figure 2 we have

```
rule <k> inc(C:Counter, L:Label) => L ... </k>
     <counters> ... C | -> ( V => V +Int 1) </counters>
```

$$: \leftrightarrow$$

$$\forall C{:}\mathsf{Context}^{\mathsf{Cell}\langle\mathsf{T}\rangle}_{\mathsf{Cell}\langle\mathsf{k}\rangle,\mathsf{Cell}\langle\mathsf{counters}\rangle}.\forall c{:}\mathsf{Counter}.\forall l{:}\mathsf{Label}.\forall m{:}\mathsf{Map}.\forall v{:}\mathsf{Int}.$$

$$(C \circ \langle C_1, C_2 \rangle)[\mathsf{inc}(c, l), v] \to \bullet(C \circ \langle C_1, C_2 \rangle)[l, v + 1]$$

where $C_1 := \gamma\square{:}\mathsf{K}.\langle\!\mathsf{k}\!\rangle(\square \curvearrowright \kappa)$, $C_2 := \gamma\square{:}\mathsf{Int}.\langle\!\mathsf{counters}\!\rangle(m\ \mathsf{MAP}\ c \mapsto \square)$. Here, MAP is the concatenation operation on the sort Map and $c \mapsto \square$ represents the singleton map with key c and value $\square$, which are part of the theory of builtins.

5   Related Work

The $\mathbb{K}$ Framework's rule-based approach is conceptually similar to Structural Operational Semantics (SOS) [ ]. However, $\mathbb{K}$ introduces powerful abstraction mechanisms, such as implicit configuration contexts, inspired from evaluation contexts [,]. The practical importance of providing a solid formal foundation for $\mathbb{K}$ is emphasized by its widespread use in developing complete formal definitions for numerous real-world languages, including C [ ], Java [ ], JavaScript [ ], x86 [ ], LLVM [ ], as well as emerging blockchain languages such as EVM [ ] and IELE [ ]. Recently, $\mathbb{K}$ was used to generate proof certificates for a language-agnostic interpreter, where each (concrete) execution of a program is certified by a machine-checkable mathematical proof [,].

The first version of the matching logic was used to specify static properties of programs in reachability logic [,]. A sound and complete Hilbert-style proof system of (structural) matching logic is given in [ ]. The matching logic was then extended with a least fixpoint-binder μ [ ], which subsumes not only reachability logic, but also a variety of common logics/calculi that are used to reason about fixpoints and induction. These versions follow a many-sorted approach and are used in the design of the intermediate representation Kore. In this paper, we use the functional variant of matching logic [,] that adopts a minimalist design and defines only the simplest building blocks, from which more complex and notationally heavier concepts can be axiomatized as theories, and intended to be used as a basic reference.

A central theme of our work is the formal treatment of contexts. While the notion of a context as a term-with-a-hole is fundamental to term rewriting, it is typically treated as a metatheoretical concept. Our work reifies contexts as first-class objects within the logic, an approach that shares conceptual similarities with other formalisms such as higher-order abstract syntax (HOAS) in frameworks such as Twelf [ ], refinement calculus [,], and program transformation languages such as Stratego [ ]. Our formalization in the paper opens the door for providing a common matching logic foundation for these related approaches.

While in our approach we chose to represent binders using the intension technique [ ], another possibility would have been to use nominal logic, which was axiomatized as an $\mathbb{ML}$ theory [,]. Regardless of this choice, the use of context for denoting $\mathbb{K}$ definitions would work in the same way, after establishing the main properties of the plugging operation.

There have been other theoretical attempts to formalize $\mathbb{K}$, e.g., using double-pushout graph transformations [ ], Isabelle [ ], Rocq [ ]; however, these approaches introduce a significant representational distance from the original definitions. In contrast, our approach provides a direct denotational semantics, mapping a $\mathbb{K}$ definition to a formal theory in $\mathbb{ML}$. This is distinct from the classic denotational semantics [ ], which maps programs to mathematical domains.

The ultimate motivation for this work is to enable the formal verification of the kompile tool, placing it in the broader context of verified software toolchains. The most notable project in this area is CompCert [ ], a formally verified C compiler that guarantees semantic preservation from source code to assembly. While CompCert focuses on verifying the translation of programs, our work provides the basis for addressing the unique challenge of verifying the translation of language definitions.

6   Conclusion

In this paper, we have presented a formal denotational semantics for the $\mathbb{K}$ frontend, establishing a direct and transparent link between high-level $\mathbb{K}$ language definitions and their underlying foundation in matching logic ($\mathbb{ML}$). We have systematically demonstrated how each component of a $\mathbb{K}$ definition can be formally translated into a corresponding $\mathbb{ML}$ theory.

Our approach is centered on a novel, reified theory of contexts within $\mathbb{ML}$, which proved essential for elegantly capturing $\mathbb{K}$'s complex context-dependent features, including the configuration abstraction mechanism and the heating/cooling rules for dynamic strictness attributes. By formalizing these concepts, we replace the opaque "magic" of kompile with a clear and verifiable specification.

This work provides the formal foundation necessary to verify the correctness of the $\mathbb{K}$ toolchain. An immediate direction for future work is to leverage this specification to formally prove that the Kore generated by the kompile tool is a correct refinement of the $\mathbb{ML}$ theory we have defined. To this end, we would first need a definition of the Kore language as an $\mathbb{ML}$ theory, which is well-known and was for example formalized in the Metamath proof system [ ]. Then, we would need to prove that the Kore theory generated by kompile is indeed a refinement of the $\mathbb{ML}$ theory defined using our approach.

Acknowledgments. The authors would like to thank the reviewers for their insightful and constructive comments. Many thanks to Iulia Bastys for helping in the improvement of the presentation of this paper. This study was funded by Pi Squared.

References

1. Back, R.J., von Wright, J.: Refinement Calculus: A Systematic Introduction. Graduate Texts in Computer Science, Springer-Verlag (1998). https://doi.org/10.1007/978-1-4612-1672-7

2. Bogdănaş, D., Roşu, G.: K-Java: A complete semantics of Java. In: Proceedings of the 42nd Symposium on Principles of Programming Languages (POPL'15). pp. 445–456. ACM (2015)

3. Chen, X., Cheval, H., Lucanu, D., Rou, G.: A Matching Logic theory of multihole contexts (2026), submitted

4. Chen, X., Lin, Z., Trinh, M.T., Roşu, G.: Towards a trustworthy semantics-based language framework via proof generation. In: Proceedings of the 33rd International Conference on Computer-Aided Verification. ACM (July 2021)

5. Chen, X., Lucanu, D., Roşu, G.: Initial algebra semantics in matching logic. Tech. rep., University of Illinois at Urbana-Champaign (Jul 2020), http://hdl.handle.net/2142/107781

6. Chen, X., Lucanu, D., Roşu, G.: Matching logic explained. J. Log. Algebraic Methods Program. **120**, 100638 (2021). https://doi.org/10.1016/J.JLAMP.2021.100638

7. Chen, X., Roşu, G.: Applicative matching logic: Semantics of K. Tech. rep., University of Illinois at Urbana-Champaign (July 2019), http://hdl.handle.net/2142/104616

8. Chen, X., Roşu, G.: A general approach to define binders using matching logic. In: Proceedings of the 25th ACM SIGPLAN International Conference on Functional Programming (ICFP'20). vol. 4, pp. 1–32. ACM/IEEE (Aug 2020), https://doi.org/10.1145/3408970

9. Chen, X., Rou, G.: Matching μ-logic. In: 2019 34th Annual ACM/IEEE Symposium on Logic in Computer Science (LICS). pp. 1–13 (2019). https://doi.org/10.1109/LICS.2019.8785675

10. Cheney, J., Fernández, M.: Nominal matching logic. In: PPDP 2022: 24th International Symposium on Principles and Practice of Declarative Programming, Tbilisi, Georgia, September 20 - 22, 2022. pp. 5:1–5:15. ACM (2022). https://doi.org/10.1145/3551357.3551375

11. Ştefănescu, A., Ciobâcă, Ş., Mereuţă, R., Moore, B.M., Şerbănuţă, T.F., Roşu, G.: All-path reachability logic. In: Proceedings of the Joint 25th International Conference on Rewriting Techniques and Applications and 12th International Conference on Typed Lambda Calculi and Applications (RTA-TLCA'14). vol. 8560, pp. 425–440. Springer (2014)

12. Dasgupta, S., Park, D., Kasampalis, T., Adve, V.S., Roşu, G.: A complete formal semantics of x86-64 user-level instruction set architecture. In: Proceedings of the 40th ACM SIGPLAN Conference on Programming Language Design and Implementation (PLDI'19). ACM (2019)

13. Felleisen, M., Findler, R.B., Flatt, M.: Semantics engineering with PLT Redex. Mit Press (2009)

14. Felleisen, M., Hieb, R.: The revised report on the syntactic theories of sequential control and state. Theoretical Computer Science **103**(2), 235–271 (1992). https://doi.org/https://doi.org/10.1016/0304-3975(92)90014-7

15. Hathhorn, C., Ellison, C., Roşu, G.: Defining the undefinedness of C. In: Proceedings of the 36th annual ACM SIGPLAN Conference on Programming Language Design and Implementation (PLDI'15). pp. 336–345. ACM (2015)

16. Hildenbrandt, E., Saxena, M., Zhu, X., Rodrigues, N., Daian, P., Guth, D., Moore, B., Zhang, Y., Park, D., Ştefănescu, A., Roşu, G.: KEVM: A complete semantics of the Ethereum virtual machine. In: Proceedings of the 2018 IEEE Computer Security Foundations Symposium (CSF'18). IEEE (2018), http://jellopaper.org

17. Kasampalis, T., Guth, D., Moore, B., Şerbănuţă, T.F., Zhang, Y., Filaretti, D., Şerbănuţă, V., Johnson, R., Roşu, G.: IELE: A rigorously designed language and tool ecosystem for the blockchain. In: Proceeding of the 23rd International Symposium on Formal Methods (FM'19) (2019)

18. Leroy, X.: A formally verified compiler back-end. Journal of Automated Reasoning **43**(4), 363–446 (2009). https://doi.org/10.1007/s10817-009-9155-4

19. Li, L., Gunter, E.: IsaK-static A complete static semantics of K. In: Formal Aspects of Component Software. pp. 196–215. Springer (2018)

20. Li, L., Gunter, E.L.: K-LLVM: A Relatively Complete Semantics of LLVM IR. In: Hirschfeld, R., Pape, T. (eds.) 34th European Conference on Object-Oriented Programming (ECOOP 2020). Leibniz International Proceedings in Informatics (LIPIcs), vol. 166, pp. 7:1–7:29. Schloss Dagstuhl – Leibniz-Zentrum für Informatik, Dagstuhl, Germany (2020). https://doi.org/10.4230/LIPIcs.ECOOP.2020.7

21. Lin, Z., Chen, X., Trinh, M.T., Wang, J., Roşu, G.: Generating proof certificates for a language-agnostic deductive program verifier. In: Proceedings of OOPSLA 2023. ACM (July 2023)

22. Minsky, M.L.: Recursive unsolvability of post's problem of "tag" and other topics in theory of turing machines. Annals of Mathematics **74**(3), 437–455 (1961), http://www.jstor.org/stable/1970290

23. Moore, B., Peña, L., Roşu, G.: Program verification by coinduction. In: Proceedings of the 27th European Symposium on Programming (ESOP'18). pp. 589–618. Springer (2018)

24. Morgan, C.: Programming from Specifications. International Series in Computer Science, Prentice Hall International, 2nd edn. (1998)

25. Park, D., Ştefănescu, A., Roşu, G.: KJS: A complete formal semantics of JavaScript. In: Proceedings of the 36th annual ACM SIGPLAN Conference on Programming Language Design and Implementation (PLDI'15). pp. 346–356. ACM (2015)

26. Pfenning, F., Schürmann, C.: System description: Twelf — a meta-logical framework for deductive systems. In: Proceedings of the 16th International Conference on Automated Deduction (CADE-16). Lecture Notes in Computer Science, vol. 1632, pp. 202–206. Springer (1999)

27. Pi Squared: Proof of Proof: Verifiable computing for all languages. https://pi2.network/papers/proof-of-proof-whitepaper (2025)

28. Plotkin, G.D.: A structural approach to operational semantics. Tech. Rep. FN-19, DAIMI, Aarhus University (1981)

29. Roşu, G.: Matching logic. Logical Methods in Computer Science **13**(4), 1–61 (December 2017). https://doi.org/https://doi.org/10.23638/LMCS-13(4:28)2017

30. Roşu, G., Ştefănescu, A., Ciobâcă, Ş., Moore, B.M.: One-path reachability logic. In: Proceedings of the 28th Symposium on Logic in Computer Science (LICS'13). pp. 358–367. IEEE (2013)

31. Runtimeverification: K Language Features. https://github.com/runtimeverification/k/blob/master/k-distribution/include/kframework/builtin/kast.md

32. Runtimeverification: K User Manual. https://github.com/runtimeverification/k/blob/master/docs/user_manual.md

33. Runtimeverification: Basic builtin types in K. https://github.com/runtimeverification/k/blob/master/k-distribution/include/kframework/builtin/domains.md (2024)

34. Scott, D., Strachey, C.: Toward a mathematical semantics for computer languages. In: Proceedings of the Symposium on Computers and Automata. pp. 19–46. Polytechnic Institute of Brooklyn Press, New York, NY, USA (1971)
35. Sebe, M., Fernández, M., Cheney, J.: Nominal matching logic with fixpoints. In: Stark, K., Timany, A., Blazy, S., Tabareau, N. (eds.) Proceedings of the 14th ACM SIGPLAN International Conference on Certified Programs and Proofs, CPP 2025, Denver, CO, USA, January 20-21, 2025. pp. 17–33. ACM (2025). https://doi.org/10.1145/3703595.3705872, https://doi.org/10.1145/3703595.3705872
36. Şerbănuţă, T.F., Roşu, G.: A truly concurrent semantics for the K framework based on graph transformations. In: Proceedings of the 6th International Conference on Graph Transformation (ICGT'12). pp. 294–310. Springer (2012)
37. Visser, E.: Stratego: A language for program transformation based on rewriting strategies. In: International Conference on Rewriting Techniques and Applications (RTA'01). Lecture Notes in Computer Science, vol. 2051, pp. 357–361. Springer (2001)

Partial Reductions for Kleene Algebra with Linear Hypotheses

Liam Chung and Tobias Kappé

LIACS, Leiden University, The Netherlands
{l.w.chung,t.w.j.kappe}@liacs.leidenuniv.nl

Abstract. Kleene algebra (KA) is an important tool for reasoning about general program equivalences, with a decidable and complete equational theory. However, KA cannot always prove equivalences between specific programs. For this purpose, one adds hypotheses to KA that encode program-specific knowledge. Traditionally, a map on regular expressions called a *reduction* then lets us lift decidability and completeness to these more expressive systems. Explicitly constructing such a reduction requires significant labour. Moreover, due to regularity constraints, a reduction may not exist for all combinations of expression and hypothesis.

We describe an automaton-based construction to mechanically derive reductions for a wide class of hypotheses. These reductions can be partial, in which case they yield *partial completeness*: completeness for expressions in their domain. This allows us to automatically establish the provability of more equivalences than what is covered in existing work.

1 Introduction

Program equivalence is a high-value but generally undecidable problem. In spite of this, there are alternatives for the enterprising computer scientist, by modifying the problem or restricting its domain. One possibility is to model program behaviour using *regular expressions*, and then use the laws of *Kleene algebra* (KA) to prove equivalence, typically with respect to a standard, language-based semantics. Kozen [12] proved that KA is sound and complete with respect to this semantics. Due to their connection with finite automata, equivalence of regular expressions is decidable, and so we can decide provable equivalence, too.

Unfortunately, many true equivalences are unprovable by pure KA, as the standard language semantics is concerned with *propositional equivalences*, i.e., those that hold for *any* interpretation of the programs' primitive statements. In particular, this excludes the reordering of independent operations. KA thus trades expressive power for decidability: one can prove equivalence of propositionally similar programs, while complex equivalences remain unprovable by KA alone.

At the same time, often there is more program-specific information available, in the form of (in)equations of the form $e \leq f$ for e, f regular expressions. By introducing these into the proof system as *hypotheses*, new equivalences become provable. Examples include the properties of Boolean assertions [13] as well as the behaviour of network packets [2] and concurrent programs [10].

© The Author(s) 2026
N. Bertrand and S. Milius (Eds.): FoSSaCS 2026, LNCS 16503, pp. 220–239, 2026.
https://doi.org/10.1007/978-3-032-22730-0_11

By altering KA to make new equivalences provable, however, the system is no longer sound with respect to regular language semantics. For example, ab and ba have different languages, though we might add a hypothesis to say that ab = ba. A new semantics is therefore necessary; for example in the case of *Kleene algebra with tests* [13] the *guarded string semantics* [15] was proposed. In general, significant work is required to develop such a semantics and subsequently recover soundness, completeness, and decidability — cf. [15,9,10,2].

To make matters worse, not every extension of KA admits such results. For instance, *commutativity hypotheses* like the aforementioned ab = ba can be used to encode Post's correspondence problem as an equivalence in KA with hypotheses, dashing any hope of decidability [13,16,1]. This is unfortunate, as commutativity hypotheses are especially useful in program equivalence reasoning [13].

In recent literature a meta-theory has developed around *Kleene algebra with hypotheses*. The cornerstone of this approach is the *hypothesis closure* operation on languages [8], which gives rise to a sound-by-construction semantics of KA with additional equations. The goal then becomes to recover completeness and decidability, by constructing a *reduction*: a map on regular expressions that syntactically implements hypothesis closure, such that the regular expression semantics of a reduced expression coincides with its hypothesis-closed semantics. Existing work discusses how to construct reductions for some kinds of hypotheses [14,8], or combine reductions for different hypotheses [18].

The Achilles heel of this approach is that a reduction must be defined on *all* expressions. Sometimes, however, hypothesis closure does not preserve regularity, preventing a syntactic realization of its effects; this includes the aforementioned combination of expressions and commutativity hypotheses that encode Post's correspondence problem [13]. In such cases, the method falters, even though regularity might be preserved for *some* expressions. Our key idea is to define *partial reductions*, which in turn give completeness over their domain of definition. Using this relaxed notion of reduction, we can recover a limited form of completeness for KA with hypotheses in a wider variety of scenarios.

The following comprise the core technical contributions of this work:

- a translation of the hypothesis closure on languages into a construction on automata, for hypotheses of the form $e \leq w$ (where w is a word);
- a proof that, when this operation produces a finite automaton, it gives rise to a reduction to KA; and
- a new technique to automatically check and prove equivalence of expressions in KA with hypotheses in this format, limited to expressions on whose automaton our construction has finite output.

If our construction always outputs a finite automaton for a given hypothesis, our method yields a full completeness proof for KA extended with that hypothesis.

Overview. The remainder of this paper is organised as follows. Section 2 gives context and necessary definitions. Section 3 discusses and verifies an initial version of our construction, which is improved in Section 4. We conclude in Section 5. For brevity's sake, proofs are sketched; details appear in the full version [5].

2   Background

Kleene algebra (KA) models a program as a set of instructions that, depending on various factors (the input, ambient environment) may execute in different ways. The instructions of the program (e.g., `print("hi")` or `x := 5`) are represented as *letters* in a finite *alphabet* $\Sigma = \{a, b, \dots\}$. Each possible program execution is a finite sequence of these letters, which creates a *word*, or element of Σ^*. The semantics of a program is then a set of words, known as a *language*.

By focusing on the sequences of events, KA abstracts from their specific meaning — equivalence in KA is only concerned with *which actions happen, and in what order*. We refer to this as *propositional equivalence*.

Example 2.1. The following programs are propositionally equivalent — they behave the same regardless of the effects of T_1, P_A, P_B, and P_C.

<table>
<tr><td>

Program 1

```
1  if T₁
2       Pₐ
3       P_B
4  else
5       P_C
6       P_B
```

</td><td>

Program 2

```
1  if T₁
2       Pₐ
3  else
4       P_C
5  P_B
```

</td></tr>
</table>

To represent the programs themselves, we use *regular expressions*. These are formed over the alphabet Σ, and composed inductively using sequencing ($e \cdot f$), non-deterministic choice ($e + f$) and iteration (e^*). Together, these operations can be used to model traditional program composition — often in conjunction with so called "tests" to represent boolean control flow [13]. For instance, the latter program can be modelled as $(T_1 \cdot P_A + \overline{T_1} \cdot P_C) \cdot P_B$, where T_1 (resp. $\overline{T_1}$) should be read an instruction asserting that T_1 does (resp. does not) hold, which aborts the non-deterministic branch on failure.

We formalise these operations, and their connection to languages, here. From here on, we fix a finite alphabet Σ.

Definition 2.2 (Regular expressions, regular language). *The set of regular expressions* $\mathbb{E}$ *is formed by the following grammar:*

$$e, f ::= 0 \mid 1 \mid a \in \Sigma \mid e \cdot f \mid e + f \mid e^*$$

Every regular expression $e \in \mathbb{E}$ *can be assigned a regular language* $[\![e]\!] \subseteq \Sigma^*$:

$$[\![0]\!] = \emptyset \qquad [\![1]\!] = \{\varepsilon\} \qquad [\![a]\!] = \{a\} \qquad [\![e + f]\!] = [\![e]\!] \cup [\![f]\!]$$

$$[\![e \cdot f]\!] = \{wx : w \in [\![e]\!], x \in [\![f]\!]\} \qquad [\![e^*]\!] = \{w_0 w_1 \cdots w_n : w_0, \dots, w_n \in [\![e]\!]\}$$

Here, ww' denotes the concatenation of w and w', and ε denotes the empty word. A language $L \subseteq \Sigma^*$ such that $L = [\![e]\!]$ for some $e \in \mathbb{E}$ is called regular.

We can also represent potential orderings of events using *(finite) automata.*

Definition 2.3 (Automaton). *An* automaton $\mathcal{X}$ *is a tuple* $(X, \to, x_\oplus, x_0)$, *in which X is a set of* states, $\to\, \subseteq X \times (\Sigma \cup \{\varepsilon\}) \times X$ *is the* transition relation, *and $x_\oplus, x_0 \in X$ are the* final *and* initial state *respectively. $\mathcal{X}$ is* finite *when X is; the collection of automata (resp. finite automata) is written* NA^∞ *(resp.* NA*).*

When $(x, a, x') \in\, \to$ for some $a \in \Sigma \cup \{\varepsilon\}$, we write $x \xrightarrow{a} x'$. Using similar notation, we can then define $\to^*$ as the smallest subset of $X \times \Sigma^* \times X$ satisfying:

$$\frac{x \xrightarrow{a} x'}{x \xrightarrow{a}{}^* x'} \qquad\qquad \frac{x \xrightarrow{\varepsilon} x'}{x \xrightarrow{\varepsilon}{}^* x'} \qquad\qquad \frac{x \xrightarrow{u}{}^* x' \quad x' \xrightarrow{v}{}^* x''}{x \xrightarrow{uv}{}^* x'}$$

An element $(x, w, x') \in\, \to^*$ is known as an $\mathcal{X}$-trace for w; it is called accepting *when $x' = x_\oplus$, in which case we say x* accepts w. *For a state $x \in X$, the* language accepted *by x is the set of words it accepts:*

$$l_{\mathcal{X}}(x) := \{w \in \Sigma^* : x \xrightarrow{w}{}^* x_\oplus\}$$

The language accepted by $\mathcal{X}$, denoted $L_{\mathcal{X}}$, is the language of its initial state x_0.

The intuitive link between automata and regular expressions as program specifications is legitimised by *Kleene's theorem* [11], which states that any automaton can be converted into a regular expression with the corresponding language, and vice versa — thus, they describe the same set of languages.

Theorem 2.4 (Kleene's theorem). *A language $L \in \mathcal{P}(\Sigma^*)$ is* regular *if and only if it is the language of a finite automaton. That is, there is an $e \in \mathbb{E}$ such that $[\![e]\!] = L$ if and only if there is a finite automaton $\mathcal{X}$ such that $L_{\mathcal{X}} = L$.*

Kleene's theorem allows us to move seamlessly between regular expressions and finite automata. It becomes straightforward to decide equivalence of regular expressions, by simply checking whether their automata are equivalent.

One can also reason algebraically about equivalence of regular expressions, using the rules that make up the proof system KA [12]. These include familiar laws such as associativity and commutativity of $+$, in addition to the *fixpoint rules* that dictate the behaviour of the Kleene star.

Definition 2.5 (Kleene algebra). *The proof system* KA *has these axioms:*

$$e + 0 = e \qquad e + e = e \qquad e + f = f + e \qquad e + (f + g) = (e + f) + g$$

$$e \cdot 0 = 0 \cdot e = 0 \qquad e \cdot 1 = 1 \cdot e = e \qquad e \cdot (f \cdot g) = (e \cdot f) \cdot g$$

$$e \cdot (f + g) = e \cdot f + e \cdot g \qquad (e + f) \cdot g = e \cdot g + f \cdot g \qquad 1 + e \cdot e^* = 1 + e^* \cdot e = e^*$$

$$e + f \cdot g \leq g \implies f^* \cdot e \leq g \qquad\qquad e + f \cdot g \leq f \implies e \cdot g^* \leq f$$

where in the last two rules, we use $e \leq f$ as a shorthand for $e + f = f$.

If $\mathsf{KA} \vdash e = f$, *we say that $e \equiv f$; equivalently, $\equiv$ is the least congruence on $\mathbb{E}$ satisfying the rules of* KA. *We write $e \leq f$ as a shorthand for $e + f \equiv f$.*

It is not too hard to show that these rules are *sound* for language equivalence. A landmark result by Kozen [12] shows that KA is *complete* for the standard language semantics — i.e., every true $[\![-]\!]$ equivalence is provable in KA.

Theorem 2.6 (Kozen's theorem). KA *is sound and complete w.r.t.* $[\![-]\!]$*: for all expressions* $e, f \in \mathbb{E}$*, we have that* $[\![e]\!] = [\![f]\!]$ *if and only if* $e \equiv f$*.*

Combining decidability of semantic equivalence for regular expressions with completeness of KA, it follows that equivalence in KA is decidable.

The key insight necessary to prove Theorem 2.6 is that every automaton can be thought of as a system of equations in KA, and that solving this system yields an expression representing the language of a state [11,7,12]. We will leverage this idea heavily in this paper, and so we take a moment to discuss it in depth.

Definition 2.7 (Solution to automaton). *Let* $\mathcal{X} = (X, \rightarrow, x_\oplus, x_0)$ *be an automaton. A solution to* $\mathcal{X}$ *is a map* $s : X \rightarrow \mathbb{E}$ *satisfying the following rules:*

$$\frac{}{1 \leqq s(x_\oplus)} \qquad \frac{x \xrightarrow{\mathsf{a}} x'}{\mathsf{a} \cdot s(x') \leqq s(x)} \qquad \frac{x \xrightarrow{\varepsilon} x'}{s(x') \leqq s(x)}$$

A least *solution* s *is a solution where* $s(x) \leqq s'(x)$ *for all solutions* s' *and* $x \in X$*.*

Since least solutions are unique up to $\equiv$, we often speak of *the* least solution to an automaton. Solutions are connected to the languages in the following way.

Lemma 2.8. *For any solution* $s : X \rightarrow \mathbb{E}$ *to an automaton* $\mathcal{X}$*, we have that for every* $x \in X$*,* $l_{\mathcal{X}}(x) \subseteq [\![s(x)]\!]$*. This inclusion is an equality when* s *is least.*

Crucially, for any finite automaton a least solution exists, and we can compute it. This is typically done with a matrix representation of the transitions of the automaton that we will not cover here; we refer to [12,7] for more information.

Lemma 2.9. *For any finite automaton, we can construct a least solution* $s_{\mathcal{X}}$*.*

2.1 Hypotheses for Kleene Algebra

Because of KA's focus on propositional equivalence, it concedes significant expressive abilities, leaving many program equivalences on the table.

Example 2.10. Suppose P_B and P_A are interchangeable in order with no effect on program behaviour, perhaps because they operate on distinct parts of memory; and P_B has no effect on whether T_1 holds. Then these programs are equivalent:

<table>
<tr><td>

Program 3

```
1  P_B
2  while T_1
3      P_A
```

</td><td>

Program 4

```
1  while T_1
2      P_A
3  P_B
```

</td></tr>
</table>

Still, the corresponding regular expressions have different languages:

$$\llbracket P_B \cdot (T_1 \cdot P_A)^* \cdot \overline{T_1} \rrbracket \neq \llbracket (T_1 \cdot P_A)^* \cdot \overline{T_1} \cdot P_B \rrbracket$$

It *is* however possible to prove equivalence of these expressions using KA, under assumptions about the primitives such as $P_A \cdot P_B = P_B \cdot P_A$, $P_B \cdot T_1 = T_1 \cdot P_B$, $P_B \cdot \overline{T_1} = \overline{T_1} \cdot P_B$. For a proof, see [5, Appendix A].

This example illustrates that when we represent a program as a sequence of instructions, we forget that some programs have special relationships with others. In spite of this, KA can still be used to prove relevant equivalences, provided we incorporate assumptions into our reasoning in the form of (in)equations.

To a significant extent, the situation can be improved by identifying general subclasses of programs that satisfy additional equations, and incorporating those into the theory. A prime example is *Kleene algebra with tests* (KAT) [13] which distinguishes a subset of the atomic programs as Boolean "tests", and adds Boolean reasoning principles into the proof system. KAT can be given a semantics in terms of *guarded languages*, w.r.t. which it is complete and decidable [15].

While this approach is feasible, it is an enormous amount of work to develop such a theory for an individual use-case where we have information that is not indicative of some deeper lack of expressiveness (e.g., Boolean assertions in the case of KAT), but is rather just useful case-specific information that we want to employ in our reasoning. Recent literature has developed a meta-theory called *Kleene algebra with hypotheses* [14,8,10,18], which parameterises over the specific hypotheses used. Through this abstraction, we can ask: for which hypotheses are classical results about completeness and decidability recoverable?

Definition 2.11. *A Kleene algebra* hypothesis *is an (in)equation of two regular expressions, that is, $e \leq f$ for some $e, f \in \mathbb{E}$. A hypothesis is called a* linear hypothesis *if it is of the form $e \leq w$ for some $w \in \Sigma^*$.*

Given some set of hypotheses H, the proof system KA augmented by the equations in H as additional axioms is called KA_H. We define $\equiv_H$ as the smallest congruence that satisfies both the axioms of KA and those asserted by H, with the $-_H$ subscript extended to define $\leq_H$ in a similar manner to $\leq$.

Remark 2.12. Letters in a hypothesis H are *not* universally quantified! That is, if $H = \{aa \leq a\}$, the a is a specific letter; bb $\leq_H$ b is only true when a = b.

We would like to show completeness and decidability of KA_H for as many sets of hypotheses H as possible. In doing so, we could be optimistic and hope for a "silver bullet" theorem that would work for any set of hypotheses, but that would contradict undecidability of program equivalence. Indeed, even with commutativity hypotheses, equivalence between certain expressions is undecidable [13,16,1]. This underscores that we cannot carelessly add axioms into KA.

Setting decidability and completeness aside for a moment, it is important to be clear about what semantics we are working with for KA_H. Indeed, adding any non-trivial hypotheses to the system will render it unsound with respect to the standard regular language semantics. For example, if $H = \{\mathsf{ab} \le \mathsf{ba}\}$, then clearly $\mathsf{ab} \le_H \mathsf{ba}$, even though $[\![\mathsf{ab}]\!] \not\subseteq [\![\mathsf{ba}]\!]$. So, to show completeness of KA_H (or indeed for that to mean much of anything) we first need a new semantics. Doumane et al. introduced the *hypothesis closure semantics* [8] for this purpose.

Definition 2.13. *We define the* one-step hypothesis closure *for hypotheses H:*

$$H : \mathcal{P}(\Sigma^*) \to \mathcal{P}(\Sigma^*) \quad \text{given by} \quad L \mapsto L \cup \bigcup \{u[\![e]\!]v : u[\![f]\!]v \subseteq L, e \le f \in H\}$$

Where $u[\![f]\!]v = \{uwv : w \in [\![f]\!]\}$, and similarly for $[\![e]\!]$.

The hypothesis closure *of L, written $H^*(L)$, is the smallest superset of L such that $H(H^*(L)) = H^*(L)$. We can now define the* hypothesis closure semantics*:*

$$[\![-]\!]_H : \mathbb{E} \to \mathcal{P}(\Sigma^*) \qquad \text{given by} \qquad e \mapsto H^*([\![e]\!])$$

For example, the H-closure of $\{\mathsf{aaaa}\}$ under $\mathsf{a} \le \mathsf{aa}$ is $\{\mathsf{a}, \mathsf{aa}, \mathsf{aaa}, \mathsf{aaaa}\}$. KA_H is sound w.r.t. $[\![-]\!]_H$ [8], but completeness is generally much harder. Methods for some hypotheses were proposed in [8], which were later developed to recover completeness of KAT and NetKAT [18]. Notably, the hypothesis closure semantics is isomorphic to the established semantics for these systems.

2.2   Reductions to Kleene Algebra

The approach to completeness in existing work [8,18,10,15,6,14], has been to leverage Theorem 2.6 by realising the semantics of the expanded system in syntax. Formally, this takes the form of a *reduction* [18]. We call such reductions "total", allowing for a "partial" version with the totality requirement relaxed.

Definition 2.14 (Reduction). *A total reduction for a set of hypotheses H is a total map $r : \mathbb{E} \to \mathbb{E}$ such that for every e in $\mathbb{E}$, the following hold:*

$$[\![e]\!]_H \subseteq [\![r(e)]\!] \qquad\qquad r(e) \le_H e$$

A partial reduction for H is a partial map $r' : \mathbb{E} \rightharpoonup \mathbb{E}$ that satisfies the above two conditions for every e in its domain.

Partial reductions give us completeness and decidability over their domain; we record this restatement of a well-known fact [8,10] below.

Lemma 2.15. *Let H be a set of hypotheses, and r a partial reduction for H. Then KA_H is complete (w.r.t. $[\![-]\!]_H$) and decidable over the domain of r: if r is defined on e and f, then $[\![e]\!]_H = [\![f]\!]_H$ implies $e \equiv_H f$, and $e \equiv_H f$ is decidable. In particular, if r happens to be total, the above is true for all $e, f \in \mathbb{E}$.*

Proof. Note that by definition, for any $f \in \mathbb{E}$, $[\![f]\!] \subseteq [\![f]\!]_H$. Since r is a reduction, $r(e) \le_H e$; then by soundness of KA_H, $[\![r(e)]\!]_H \subseteq [\![e]\!]_H$, so $[\![r(e)]\!] \subseteq [\![e]\!]_H$. Again since r is a reduction, $[\![e]\!]_H \subseteq [\![r(e)]\!]$, so $[\![e]\!] \subseteq [\![r(e)]\!]$. Therefore by Theorem 2.6, $e \le r(e)$. So if r is a reduction, $[\![e]\!]_H = [\![r(e)]\!]$ and $e \equiv_H r(e)$.

Now let $e, f \in \mathbb{E}$ in the domain of r be such that $[\![e]\!]_H = [\![f]\!]_H$. Then:

$$[\![r(e)]\!] = [\![e]\!]_H = [\![f]\!]_H = [\![r(f)]\!].$$

We can then apply Theorem 2.6 to find that $r(e) \equiv r(f)$ — these expressions are equivalent by KA alone. Since KA_H has all of the rules of KA, it follows that $r(e) \equiv_H r(f)$. Lastly, we can then combine this with the other requirement for partial reductions to conclude that $e \equiv_H r(e) \equiv_H r(f) \equiv_H f$.

As for decidability, note that the above (in combination with soundness) tells us that to decide $e \equiv_H f$ is to decide whether $[\![r(e)]\!] = [\![r(f)]\!]$, and equivalence of regular expressions is decidable. $\square$

Various works have defined reductions tailored to specific hypotheses. For example, in [9] (see also [14]) the "contraction" hypothesis $\mathsf{a} \le \mathsf{aa}$ is realised by applying "transitive closure" to automata, as detailed below.

Example 2.16. Let $H = \{\mathsf{a} \le \mathsf{aa}\}$. This hypothesis can quite naturally be seen as transitive closure at the automaton level. Any two states connected by a sequence of two a-transitions should also be connected by just one a-transition:

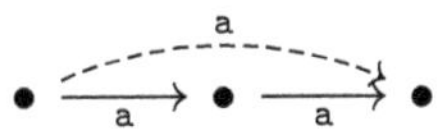

For any expression e, converting e to an automaton, applying this "transitive closure", and converting back to regular expression constitutes a total reduction.

Indeed many reductions are defined through automata in some fashion: we can think of a hypothesis $e \le f$ as "if we can move between two states with any word from f, we should be able to do the same while reading any word from e". In contrast, working directly on expressions to define a reduction is difficult, because their inductive structure is of little help.

We therefore approach the general reduction problem by converting a regular expression g into an automaton, applying a closure construction there, and then converting the result back into an expression $r(g)$. We can then use algebraic representations of the automata to prove that $r(g) \le_H g$.

Of course, for a reduction of g to be feasible at all, $[\![g]\!]_H$ needs to be a regular language; if it is not, then there is no hope of finding a sensible expression for $r(g)$. This can happen for commutativity hypotheses.

Example 2.17. Let $H := \{\mathsf{ab} \le \mathsf{ba}\}$. Then $[\![(\mathsf{ab})^*]\!]_H$ cannot be regular, because otherwise $[\![(\mathsf{ab})^*]\!]_H \cap [\![\mathsf{a}^*\mathsf{b}^*]\!]$ would also be regular, and the latter is precisely $\{\mathsf{a}^n\mathsf{b}^n : n \in \mathbb{N}\}$, which is a well-known non-regular (but context free) language.

While a total reduction, which gives a decision procedure and completeness all in one, is undoubtedly the best case scenario, it fails for certain combinations of hypothesis and expression, as in Example 2.17. However, this also means giving up on many useful equivalences, even if all of the (hypothesis-closed) languages involved are regular. Our main innovation is a technique to construct reductions for *some* expressions w.r.t. hypotheses H, even if it might fail for others. We thereby widen the perspective beyond systems we can show to be decidable and complete via a total reduction.

3   Partial Reduction: Patching

We now define a notion of hypothesis closure on automata. When paired with standard algorithms converting between regular expressions and automata (e.g. [19]), this operation yields a partial reduction for the given hypothesis.

Assumption 3.1. *From here on, unless specified, we use a hypothesis set $H = \{e \leq w\}$, and an automaton $\mathcal{X} := (X, \rightarrow, x_\oplus, x_0)$.*

We first state the desired effect on one state. Recall that the hypothesis $e \leq w$ enforces that if there is a word in L with w as a subword — i.e., $uwv \in L$ — then we can replace w with any word from $[\![e]\!]$, giving $uw'v \in H(L)$ for each $w' \in [\![e]\!]$. To mimic this effect on automata, we will seek out states where we can read w to reach some other state. We then add an alternate path: instead of following w, step to a new automaton recognising $[\![e]\!]$, and from the accepting state of that automaton, go back to any of the states we could have gone to with w.

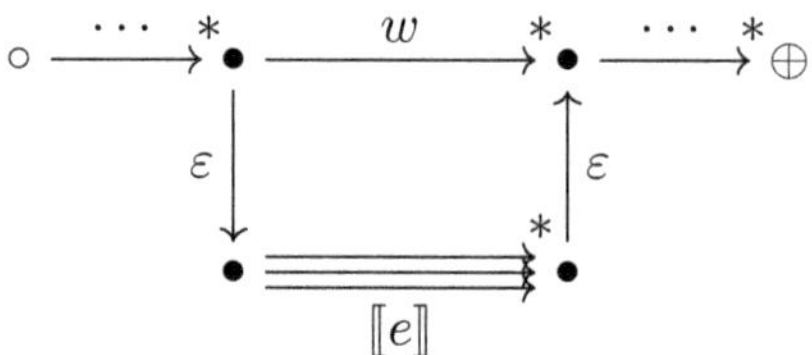

Definition 3.2 (Patching).   *Let $x \in X$, let $\mathcal{Z} := (Z, \rightarrow_\mathcal{Z}, z_\oplus, z_0)$ be an automaton. We define the automaton $\mathcal{X}[\mathcal{Z}/w\,@\,x]$ (read: $\mathcal{X}$ with $\mathcal{Z}$ w-patched at x) as $(X \cup Z, \rightarrow_P, x_\oplus, x_0)$,[1] where $\rightarrow_P$ is the smallest subset of $(X \cup Z) \times (\Sigma \cup \{\varepsilon\}) \times (X \cup Z)$ satisfying the following rules, where $a \in \Sigma \cup \{\varepsilon\}$:*

$$\frac{y \xrightarrow{a}_\mathcal{X} y'}{y \xrightarrow{a}_P y'} \qquad \frac{y \xrightarrow{a}_\mathcal{Z} y'}{y \xrightarrow{a}_P y'} \qquad \frac{}{x \xrightarrow{\varepsilon}_P z_0} \qquad \frac{x \xrightarrow{w}{}^*_\mathcal{X} x'}{z_\oplus \xrightarrow{\varepsilon}_P x'}$$

Patching mimics hypothesis closure w.r.t. H on automata, local to one state.

Example 3.3. As seen in Example 2.16, an automaton construction was used in [9] to realise the hypothesis $\mathsf{a} \leq \mathsf{aa}$. The hypothesis replaces any instance of the subword aa with a, and in automaton terms, any instance of two a-transitions can also be traversed by a single a-transition: a kind of "transitive closure". As can be seen below, patching replicates this notion, up to ε-removal:

Just like the one-step language closure operation is iterated to define language closure, multiple patching operations may be necessary to achieve the desired

---

[1] Without loss of generality, the state sets X and Z are assumed to be disjoint.

closure — either to the newly added states, or to states already in $\mathcal{X}$. The latter happens when, after patching one state, opportunities to patch other states become apparent, as shown in the example below.

Example 3.4. Let $H = \{\mathsf{a} \leq \mathsf{ba}\}$, and consider the expression bba. Then, applying our patching construction one time, we obtain an automaton whose language is not yet closed: the output automaton needs to accept a, as $\mathsf{a} \in [\![\mathsf{bba}]\!]_{\mathsf{a} \leq \mathsf{ba}}$. So we patch again, obtaining the desired output:

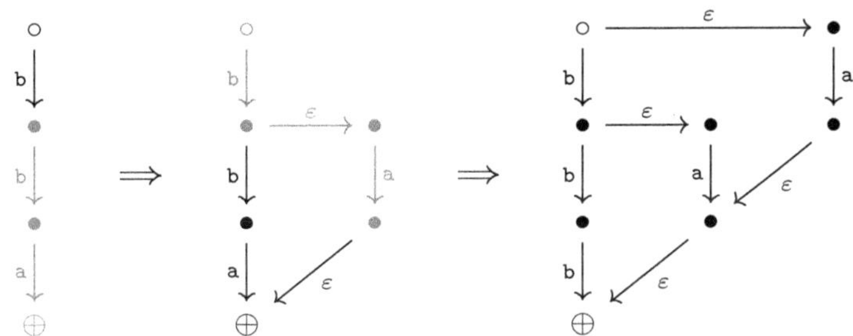

For each step, the part to be patched next is highlighted. Clearly, we can force patching to be necessary up to n times if we like, by using $H = \{\mathsf{a} \leq \mathsf{ba}\}$ on an automaton for the expression $\mathsf{b}^n \mathsf{a}$.

Of course, repeated patching opens up the possibility that we never stop patching, despite the fact that the constructed automaton must be finite to extract an expression. Indeed, no correct automaton closure construction (in the sense that it outputs an automaton for the closed language) always preserves finite automata, since hypothesis closure does not preserve regularity (cf. Example 2.17).

While the construction we seek to define may need to iterate patching to reach a correct output, it should terminate where possible. Therefore, it needs a criterion to judge if a state should be patched, and indeed if its work on the automaton as a whole is done. To this end, we recall the *Brzozowski derivative*, as well as its generalisation, which operates on languages.

Definition 3.5 (Brzozowski derivative). *Let L be a language, and $w \in \Sigma^*$ a word. The* Brzozowski derivative *of L with respect to w is defined as follows:*

$$w^{-1}L := \{u \in \Sigma^* : wu \in L\}.$$

Let L, M be languages. The generalised Brzozowski derivative *(also known as the* left residual*) of L with respect to M is defined as follows:*

$$M^{-1}L := \bigcap_{w \in M} w^{-1}L = \{v \in \Sigma^* : wv \in L, \text{ for all } w \in M\}.$$

The following characterisation can then be used to check whether an automaton requires any more patching — that is, if the languages of each of the automaton's states are closed with respect to the hypothesis.

Lemma 3.6. *For any automaton* $\mathcal{X}$,

$$\forall x \in X, \ w^{-1} l_{\mathcal{X}}(x) \subseteq [\![e]\!]^{-1} l_{\mathcal{X}}(x) \ \text{if and only if} \ \forall x \in X, \ l_{\mathcal{X}}(x) = H(l_{\mathcal{X}}(x)).$$

Note the location of the quantifiers: the language containment holds at *all* states if and only if *all* states are hypothesis closed. This equivalence relates hypothesis-closure of the language of each state to a containment of one Brzozowski derivative in another. Since regular languages are closed under (generalised) Brzozowski derivatives, the latter is computable.

We now define the first version of our construction on automata. First we patch all states in the automaton as needed according to the criterion from Lemma 3.6. This will imitate the action of H at every state in $\mathcal{X}$ at once.

Definition 3.7 (T_0). *Write* $X = \{x_0, \ldots, x_{n-1}\}$; *we define* $T_0(\mathcal{X}) = \mathcal{X}_n$, *where*

$$\mathcal{X}_0 = \mathcal{X} \qquad \mathcal{X}_{i+1} = \begin{cases} \mathcal{X}_i & w^{-1} l_{\mathcal{X}}(x_i) \subseteq [\![e]\!]^{-1} l_{\mathcal{X}}(x_i) \\ \mathcal{X}_i[\mathcal{Z}/w @ x_i] & otherwise \end{cases}$$

Remark 3.8. The languages used to check whether T_0 pastes onto a state x_i are from $\mathcal{X}$, not the intermediate automata $\mathcal{X}_i$. Thus all patches occur simultaneously, and the output of T_0 is independent of the order states are considered.

One application of T_0 to $\mathcal{X}$ introduces new states and grows the language of existing ones, potentially necessitating further patching. Consequently, T_0 must be iterated (potentially infinitely often) until a hypothesis-closed automaton arises. To rigorously define this process and formalise the notion of "building" the output of the construction, we introduce an order on automata.

Definition 3.9 (Automaton order). *Let* $\sqsubseteq$ *be the order on automata where*

$$(X, \to, x_\oplus, x_0) \sqsubseteq (X', \to', x'_\oplus, x'_0) \ \text{iff} \ X \subseteq X', \to \ \subseteq \ \to', x_0 = x'_0, x_\oplus = x'_\oplus.$$

Clearly, if $\mathcal{X} \sqsubseteq \mathcal{X}'$, then $l_{\mathcal{X}}(x) \subseteq l_{\mathcal{X}'}(x)$ for every state x in $\mathcal{X}$.

When equipped with $\sqsubseteq$, automata form an ω-complete partial order, meaning any chain $\mathcal{X}_0 \sqsubseteq \mathcal{X}_1 \sqsubseteq \cdots$ must have a *limit* $\mathcal{X}_*$, which is the $\sqsubseteq$-least automaton such that $\mathcal{X}_i \sqsubseteq \mathcal{X}_*$ for each i. This allows us to define the output of the construction as the limit of a chain — with possibly infinitely many states.

Definition 3.10 (T_0^*). $T_0^*(\mathcal{X})$ *is the limit of* $\mathcal{X} \sqsubseteq T_0(\mathcal{X}) \sqsubseteq T_0^2(\mathcal{X}) \sqsubseteq \cdots$.

With standard constructions, T_0^* lifts to a partial function on regular expressions.

Definition 3.11 (Candidate partial reduction: r_0**).** *Given a regular expression* g, *let* $\mathcal{X}$ *be its finite automaton. If* $T_0^*(\mathcal{X})$ *is finite, then* $r_0(g)$ *is the regular expression for the initial state of* $T_0^*(\mathcal{X})$. *Otherwise, it is undefined.*

For r_0 to be a partial reduction, Definition 2.14 tells us we need to show two things when it is defined on a regular expression g. First, $[\![g]\!]_H$ should be contained in $[\![r_0(g)]\!]$; this boils down to showing that the H-closure of the language of an automaton $\mathcal{X}$ is contained in the (plain) language of $T_0^*(\mathcal{X})$.

Lemma 3.12. *For every $x \in X$, it holds that $H^*(l_{\mathcal{X}}(x)) \subseteq l_{T_0^*(\mathcal{X})}(x)$.*

Proof sketch. One can more easily show the inclusion at one step: for every $x \in X$, $H(l_{\mathcal{X}(x)}) \subseteq l_{T_0}(\mathcal{X})(x)$. This requires taking a word of the form $uwv \in l_{\mathcal{X}}(x)$ and verifying that $uw'v \in l_{T_0(\mathcal{X})}(x)$ for any $w' \in [\![e]\!]$, which is simply a matter of applying the criterion in Lemma 3.6 and tracing the patch as it is constructed. $\square$

To validate the second part of Definition 2.14, we must show that $r_0(g) \leq_H g$. We will achieve this using the systems of equations for the automata corresponding to these expressions, $\mathcal{X}$ and $T_0^*(\mathcal{X})$. To this end, we first extend the machinery around least solutions to KA_H.

Definition 3.13. *An H-solution to $\mathcal{X}$ is a map $s : X \to \mathbb{E}$ satisfying the rules from Definition 2.7, but with $\leq_H$ instead of $\leq$.*

Lemma 3.14. *The (plain) least solution $s_{\mathcal{X}}$ to $\mathcal{X}$ from Lemma 2.9 is also an H-solution to $\mathcal{X}$, and in fact it is* least *among all H-solutions to $\mathcal{X}$.*

Proof sketch. The first claim follows because $\leq_H$ is weaker than $\leq$. One can then use the same construction used for Lemma 2.9 to obtain a least H-solution, but the expression constructed is the same as that obtained for the least solution. $\square$

Of course, (least) H-solutions may not be (least) solutions in the sense of Definition 2.7. Lemma 3.14 provides leverage, relating H-solutions of an automaton to its (plain) least solution. To use this, we first show that T_0^* does not (up to H) perturb a solution of $\mathcal{X}$ — we only need to include the new states.

Lemma 3.15. *Suppose $T_0^*(\mathcal{X})$ is finite, and s is a solution to $\mathcal{X}$. We can construct an H-solution s^* to $T_0^*(\mathcal{X})$, which moreover agrees with s on all $x \in X$.*

Proof sketch. Since $T_0^*(\mathcal{X})$ is finite, it is T_0 applied to $\mathcal{X}$ finitely many times. We prove the claim just for $T_0(\mathcal{X})$; the main claim then follows by induction. Let $s_{\mathcal{Z}}$ be the least solution to $\mathcal{Z}$. $T_0(\mathcal{X})$ has $\mathcal{Z}_1, \ldots \mathcal{Z}_n$ (all copies of $\mathcal{Z}$) with state sets $Z_1, \ldots, Z_n$ w-patched on at states $x_1, \ldots x_n$. One defines s^* as follows.

$$s^*(y) = \begin{cases} s(y) & y \in X \\ s_{\mathcal{Z}}(y) \cdot r_i & y \in Z_i \end{cases} \qquad \text{where} \qquad r_i := \sum_{x_i \xrightarrow{w}_* x'} s(x')$$

For a state in $\mathcal{X}$, s^* just looks at s; for a state in $\mathcal{Z}_i$, it uses $s_{\mathcal{Z}}$ while accounting for the fact that we need to re-enter $\mathcal{X}$, by composing with r_i.

The majority of the equations required for s^* to be an H-solution to $T_0^*(\mathcal{X})$ (i.e., the ones resulting from the first two rules in Definition 3.2) follow by definition. The only ones that remain correspond to ε-transitions out to, and back from, the patched on automata (the last two rules in Definition 3.2). We first look at equations resulting from the fourth rule, where an ε-transition from some $\mathcal{Z}_i$ goes back into $\mathcal{X}$. Here, we must prove that $s^*(x') \leq s^*(z_\oplus^i)$ where x' is one of the states where $\mathcal{Z}_i$ transitions back to $\mathcal{X}$. This is proved as follows:

$$s^*(x') \equiv 1 \cdot s^*(x') \leq s_{\mathcal{Z}}(z_\oplus^i) \cdot s^*(x') \leq s_{\mathcal{Z}}(z_\oplus^i) \cdot r_i = s^*(z_\oplus^i)$$

For the second-to-last step, we use the definition of r_i to observe that $s^*(x') = s(x') \leqq r_i$ because $x_i \xrightarrow{w}_* x'$. All other steps follow by the axioms of KA, the fact that $z_\oplus^i$ is the accepting state of $\mathcal{Z}_i$ with $s_\mathcal{Z}$ as solution, and the definition of s^*.

Lastly, each equation corresponding to an ε-transition from $\mathcal{X}$ out to some $\mathcal{Z}_i$ is of the form $s^*(z_0^i) \leqq s^*(x_i)$, which is proved as follows:

$$s^*(z_0^i) \equiv s_\mathcal{Z}(z_0^i) \cdot r_i \equiv e \cdot r_i \leqq_H w \cdot r_i \leqq s(x_i) = s^*(x_i). \tag{1}$$

For the second-to-last step, i.e., $w \cdot r_i \leqq s(x_i)$, it suffices to show that for all x' such that $x_i \xrightarrow{w}_* x'$, $w \cdot s(x') \leqq s(x_i)$. This follows directly from the definition of solution, applied in an induction on the construction of $x_i \xrightarrow{w}_* x'$.

All other steps in (1) follow by the definition of s^*, the definition of $\mathcal{Z}_i$ as representing e, and the (only) hypothesis in H. The use of H in the middle is the reason that s^* is an H-solution of $T_0(\mathcal{X})$, not necessarily a plain solution. $\quad\square$

Since the least solution of $\mathcal{X}$ is part of an H-solution for $T_0^*(\mathcal{X})$, it is above the least solution for $T_0^*(\mathcal{X})$, up to H. Because $r_0(g)$ is created from the least solution (and thus the least H-solution) to $T_0^*(\mathcal{X})$, we can relate g to $r_0(g)$.

Theorem 3.16. r_0 *is a partial reduction.*

Proof. We show that for every $g \in \mathbb{E}$ on which r_0 is defined, we have $[\![g]\!]_H \subseteq [\![r_0(g)]\!]$ and $r_0(g) \leqq_H g$. The former follows from Lemma 3.12:

$$[\![g]\!]_H = H^*(L_\mathcal{X}) \subseteq L_{T_0^*(\mathcal{X})} = [\![r_0(g)]\!]$$

For the latter, let s be the least solution to $\mathcal{X}$, the automaton representing g; by Theorem 2.6, $s(x_0) \equiv g$. Let s' be the least solution to $T_0^*(\mathcal{X})$; by definition, $r_0(g) = s'(x_0)$. Lemma 3.15 extends s to an H-solution s^* for $T_0^*(\mathcal{X})$. Since s' is the *least* solution to $T_0^*(\mathcal{X})$, it is a least H-solution, allowing us to derive:

$$r_0(g) = s'(x_0) \leqq_H s^*(x_0) = s(x_0) \equiv g \qquad\square$$

With our partial reduction in hand, we can now conclude decidability and completeness for the cases where it is defined, by way of Lemma 2.15.

Corollary 3.17. *Let g, h be expressions for which r_0 is defined. If $[\![g]\!]_H = [\![h]\!]_H$, then $g \equiv_H h$; moreover, the latter is decidable. Thus, if r_0 is known to output finite automata for all expressions in $\mathbb{E}$, then KA_H is complete and decidable.*

In the next section we will improve on T_0, so that its output is finite in more situations. This will allow for wider domains of reduction for some hypotheses. However, we can already apply T_0^* to recover some known results.

For contraction hypotheses $\mathsf{a} \leq \mathsf{aa}$ (so-named in [10]), T_0^* always outputs finitely. As such, the reduction given in op. cit., and the completeness achieved as a result, are recoverable using this method. In [10, Lemma 4.35], linear hypotheses (called "grounded" there) are also shown to lift to a concurrent setting. Finite output generalises to other "contraction hypotheses" $\mathsf{a} \leq w$, as T_0 emulates the "descendent construction" described in the proof of [3, Theorem 4.1.2].

While T_0^* only operates on one hypothesis at a time, we can still make use of it for "independent" sets of hypotheses, where closure with respect to all of them at once is the same as closure w.r.t. each hypothesis individually, in some order [10,18]. Then we can use T_0^* to construct a reduction from KA_H down to KA, removing one hypothesis at a time. This is easily shown to be the case for a set of contraction hypotheses, for example $H := \{\mathsf{a} \leq \mathsf{aa},\ \mathsf{b} \leq \mathsf{bb},\ \ldots\}$.

4   Improved Partial Reduction: Saturation

Corollary 3.17 is predicated on the domain of r_0, and by extension on whether T_0^* outputs a finite automaton when fed the automata for g and h, but does not tell us when this might be the case. While we cannot expect T_0^* to *always* produce a finite automaton (per Example 2.17), we wish for it to do so *when possible* — i.e., when the hypothesis closure of the language in question is regular. Unfortunately, there are cases where T_0^* unnecessarily outputs an infinite automaton.

Example 4.1. Let $H := \{\mathsf{ba} \leq \mathsf{a}\}$, and consider the expression a. We can easily calculate that the desired output expression is $\mathsf{b}^*\mathsf{a}$, but T_0^* proceeds infinitely:

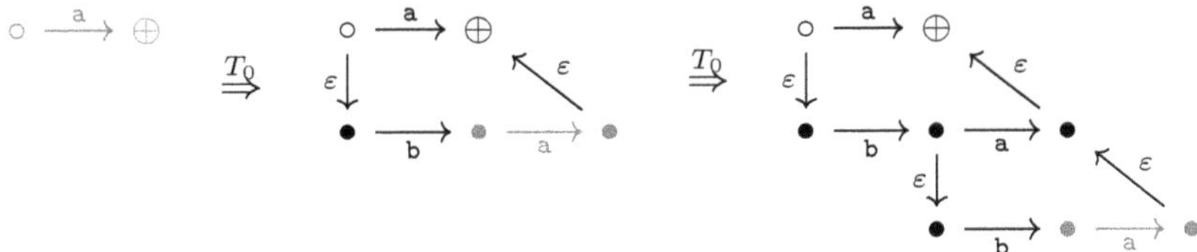

Since each step adds an a-transition that needs patching, T_0^* will be infinite.

There is also a symmetric counterexample: with $H := \{\mathsf{ab} \leq \mathsf{a}\}$ applied to $\{\mathsf{a}\}$, the output should be ab^* but T_0^* will again produce an infinite automaton.

To deal with these cases we will improve the construction.[2] Examining the target expression $\mathsf{b}^*\mathsf{a}$, our output should have a b-loop. Indeed, the (sub)word to be expanded, a, appears in in the expansion ba; on the automaton level, there is a state with a in its language. So, rather than starting a new patch, we reset the current one to its initial state. We call this an *initial state reset*.

Using this new automaton, the automaton in Example 4.1 needs only one patch.

---

[2] The reader may think to fix this by first "saturating" the hypothesis itself, to skip redundant steps; in the above case, obtaining $\mathsf{b}^*\mathsf{a} \leq \mathsf{a}$. While this works in principle, it is circular: to calculate a hypothesis closed expression, we construct the automaton; to do that, we calculate a hypothesis closed expression. Instead, we seek to create a similar effect on the automaton level, where we can be certain it will terminate.

For the symmetric hypothesis $\mathsf{ab} \leq \mathsf{a}$, we apply a similar tactic — in this case we search for states that have an a path *from the initial state*. For such states introduce a *final state reset*, back to that state from the final state.

Both initial and final state resets prevent the need to patch a patched on automaton $\mathcal{Z}$, when that work could have been done before even attaching $\mathcal{Z}$ to $\mathcal{X}$. We now define the general case of adding these transitions, then use it in place of the automaton for e when patching.

Definition 4.2 (Saturated automaton). *Let $\mathcal{Z} = (Z, \to, z_\oplus, z_0)$ be an automaton, and w a word. The w-saturation of $\mathcal{Z}$, written $\mathcal{Z}^w$, is $(Z, \to_S, z_\oplus, z_0)$, where $\to_S$ is the smallest subset of $Z \times (\Sigma \times \{\varepsilon\}) \times Z$ satisfying the rules:*

$$\frac{z \xrightarrow{a} z'}{z \xrightarrow{a}_S z'} \qquad\qquad \frac{z \xrightarrow{w}{}^*_S z_\oplus}{z \xrightarrow{\varepsilon}_S z_0} \qquad\qquad \frac{z_0 \xrightarrow{w}{}^*_S z}{z_\oplus \xrightarrow{\varepsilon}_S z}$$

The latter rules correspond to initial and final state resets respectively.

Saturation adds only transitions, so it will preserve finiteness of automata. Note that the transition relation of $\mathcal{Z}^w$ is well-defined, as adding a new ε-transition does not prevent any other ε-transition from being added.

Example 4.3. We compute the a-saturation of the automaton for $\mathsf{ab} + \mathsf{ba}$.

The left $\xrightarrow{\varepsilon}$ is added because $\bullet_1 \xrightarrow{\mathsf{a}} \oplus$ in the original automaton; it is an initial state reset. The right $\xrightarrow{\varepsilon}$ is added because $\circ \xrightarrow{\mathsf{a}} \bullet_2$; it is a final state reset.

Incorporating the saturated automaton into the existing construction gives a revised version of T_0, which patches using $\mathcal{Z}^w$ in place of $\mathcal{Z}$.

Definition 4.4 (T_H). *Write $X = \{x_0, \ldots, x_{n-1}\}$; we define $T_H(\mathcal{X}) = \mathcal{X}_n$, with*

$$\mathcal{X}_0 = \mathcal{X} \qquad\qquad \mathcal{X}_{i+1} = \begin{cases} \mathcal{X}_i & w^{-1}l_\mathcal{X}(x_i) \subseteq [\![e]\!]^{-1}l_\mathcal{X}(x_i) \\ \mathcal{X}_i[\mathcal{Z}^w/w @ x_i] & otherwise \end{cases}$$

Finally, $T_H^*(\mathcal{X})$ is the limit of the chain $\mathcal{X} \sqsubseteq T_H(\mathcal{X}) \sqsubseteq T_H^2(\mathcal{X}) \sqsubseteq \ldots$.

Definition 4.5 (Candidate partial reduction: r_H). *Given a regular expression g, let $\mathcal{X}$ be its finite automaton. If $T_H^*(\mathcal{X})$ is finite, then $r_H(g)$ is the regular expression for the initial state of $T_H^*(\mathcal{X})$. Otherwise, it is undefined.*

To show that r_H is a partial reduction, we first show $[\![g]\!]_H \subseteq [\![r_H(g)]\!]$:

Lemma 4.6. *For every $x \in X$, it holds that $H^*(l_{\mathcal{X}}(x)) \subseteq l_{T_H^*(\mathcal{X})}(x)$.*

Proof. The proof follows fairly directly from the corresponding proof for T_0 (Lemma 3.12). There, we proved that $H(l_{\mathcal{X}}(x)) \subseteq l_{T_0(\mathcal{X})}(x)$. Since $\mathcal{Z} \sqsubseteq \mathcal{Z}^w$, we have $T_0(\mathcal{X}) \sqsubseteq T_H(\mathcal{X})$; therefore, for any $x \in X$, it follows that $l_{T_0(\mathcal{X})}(x) \subseteq l_{T_H(\mathcal{X})}(x)$. Thus, the desired inclusion follows by transitivity. $\square$

Now we verify that $r_H(g) \leq_H g$. Similarly to T_0, we leverage completeness of KA (see Theorem 2.6). However, the proof for T_0 required that the least solution to the patched on $\mathcal{Z}$ at the initial state z_0 is equivalent to e. In contrast, for T_H, the patched on automaton is w-saturated, so its least solution is generally above e. This makes sense: the automaton was w-saturated so that it would accept more words, so its least solution *should* grow w.r.t. $\leq$. However, it should *not* change w.r.t. $\leq_H$, which is exactly what we will show: though the least solution at the initial state is not equivalent to e, it is equivalent to e *up to H*.

The proof requires that we briefly introduce the *reverse* operation on words, languages, etc. We write w^r for the reverse of a word w; this lifts to languages: $L^r = \{w^r : w \in L\}$. The reverse g^r of an expression g sends $g_1 \cdot g_2$ to $g_2^r \cdot g_1^r$, and acts homomorphically on the other operators, so $[\![g^r]\!] = [\![g]\!]^r$. The laws of KA are unperturbed under reversal, so $\mathsf{KA} \vdash f = g$ iff $\mathsf{KA} \vdash f^r = g^r$.

The *reverse automaton* $\mathcal{X}^r$ refers to $\mathcal{X}$ with all transitions reversed, and initial and final states swapped; clearly, $L_{\mathcal{X}}^r = L_{\mathcal{X}^r}$. We can then speak of *reverse solutions*, i.e., solutions to the reversed automaton $\mathcal{X}^r$. There is a tight relationship between solutions and reverse solutions at the initial and final states.

Lemma 4.7. *If s and s^r are respectively the least and least reverse solution to some automaton $\mathcal{Z}$, then $s^r(z_\oplus) \equiv s(z_0)^r$*

Proof. Since s is the least solution to $\mathcal{Z}$, we know by Lemma 2.8 that $[\![s(z_0)]\!] = L_{\mathcal{Z}}$; similarly, $[\![s^r(z_\oplus)]\!] = L_{\mathcal{Z}^r}$. With the observations above, we then derive:

$$[\![s^r(z_\oplus)]\!] = L_{\mathcal{Z}^r} = L_{\mathcal{Z}}^r = [\![s(z_0)]\!]^r = [\![s(z_0)^r]\!]$$

The claim follows from the above by Theorem 2.6. $\square$

Lemma 4.8. *Suppose that $\mathcal{Z}$ is an automaton representing the expression e. Let s^+ be the least solution to the w-saturated automaton $\mathcal{Z}^w$. Then $e \equiv_H s^+(z_0)$.*

Proof Sketch. $\mathcal{Z}^w$ can be constructed by adding one ε-transition to $\mathcal{Z}$ at a time. One can then show that when an initial state reset is added, solutions are preserved, but only up to H. Similarly, when a final state reset is added, *reverse* solutions are preserved, up to H^r, the reveresed hypothesis.

Lemma 4.7 is then applied to relate the least reverse solution at the final state to the reverse of the least solution at the initial state, with its result expanded to H and H^r-least solutions. Therefore, at each step the least solution at the initial state does not change, up to H. $\square$

Now we can replay the proof strategy for Lemma 3.15 to obtain a corresponding lemma that T_H preserves H-solutions.

Lemma 4.9. *Suppose $T_H^*(\mathcal{X})$ is finite, and s is a solution to $\mathcal{X}$. We can construct an H-solution s^* to $T_H^*(\mathcal{X})$, which moreover agrees with s on all $x \in X$.*

Proof Sketch. Define s^* as in the proof of Lemma 3.15, but for states in $\mathcal{Z}_i^w$ use s^w, the least solution to $\mathcal{Z}^w$, rather than $s_\mathcal{Z}$. This map satisfies most conditions to be a solution of $T_H(\mathcal{X})$, in particular those in $\mathcal{Z}_i^w$, even for reset transitions.

It remains to validate the conditions originating from ε-transitions into and out of the patched on automata, and these can be solved in the same way as in Lemma 3.15 — with one alteration. To show that $s^*(z_0^i) \leq s^*(x_i)$ for each x_i, instead of using $s_\mathcal{Z}(z_0^i) \equiv e$, we now use that $s_{\mathcal{Z}^w}(z_0^i) \equiv_H e$ (per Lemma 4.7). $\square$

The proof that r_H is a partial reduction is effectively the same as that of Theorem 3.16, using analogous lemmas about T_H, so we do not spell it out here.

Theorem 4.10. *r_H is a partial reduction.*

Using Lemma 2.15 we then conclude decidability and completeness for expressions in the domain of r_H, i.e., expressions where T_H^* outputs a finite automaton.

Corollary 4.11. *Let g, h be expressions for which r_H is defined. If $[\![g]\!]_H = [\![h]\!]_H$, then $g \equiv_H h$; moreover, the latter is decidable. Thus, if r_H is known to output finite automata for all expressions in $\mathbb{E}$, then KA_H is complete and decidable.*

T_0^* never outputs a finite automaton when T_H^* does not, and T_H^* outputs finitely for the hypothesis/expression pairs from Example 4.1. But does T_0^* output a finite automaton *when it can?* Unfortunately, this is not the case.

Example 4.12. Let $H := \{ab \leq ba\}$, and consider ba^*. The hypothesis closure of this expression is a^*ba^*. T_H^* proceeds infinitely, as sketched below:

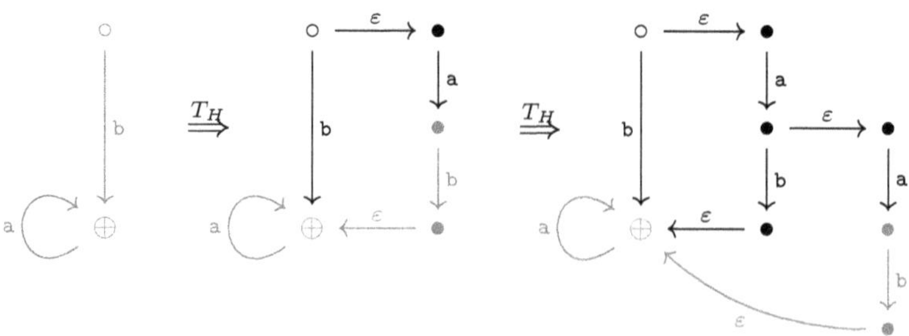

Since ba does not appear as a path in the patched-on automaton, there is nothing to saturate. Indeed, it is only *in the context of $\mathcal{X}$* that the need to introduce a loop becomes clear. This separates this example from saturation as an operation independent from patching. A further alteration to T_H to deal with this sort of scenario is an exciting endeavour, but is left for now to future work.

Similar to T_0^*, T_H^* can be applied to "independent" sets of hypotheses. There is already significant work towards creating "modular" completeness proofs by composing reductions [18]. However, this work is orthogonal to ours — composing reductions benefits from a formula for creating new ones, which T_H^* offers.

These techniques can also be used to construct reductions to *any system that reduces to* KA, via an existing reduction. For example, as per [18], both KAT and NetKAT can be regarded as KA with infinitely many hypotheses. Reductions introduced there can be combined with one produced by T_H^* to show completeness of KAT with some hypotheses added, or decide equivalence of two NetKAT-expressions with hypotheses.

5   Conclusion & Further Work

We have presented a general strategy to construct reductions for Kleene algebra with a linear hypothesis, through a construction on automata. Decidability and completeness of the relevant system follow for the cases where the construction terminates. This strategy can then be used for sets of independent linear hypotheses as well, and recover known reductions in the literature [10,18].

Hypothesis closure provides a useful canonical language semantics, but they are no substitute for the insight that a "hand-crafted" semantics provides. It remains an important task to develop these for sufficiently ubiquitous systems.

There are also sets of hypotheses that the tools given here do not cover, namely those where hypotheses directly interact with one another. Another difficulty is with general hypotheses of the form $e \leq f$, where f is some general regular expression. It is a natural direction for future work to expand the automaton construction to deal with these uncovered cases. This would allow for creating reductions to KA, or other known-complete systems like KAT, in a wider array of Kleene algebras with hypotheses.

Another direct line of future work is studying when the construction "succeeds", i.e., outputs a finite automaton; or indeed, when that is even possible. Characterising when the construction can succeed would allow for more precise decidability and completeness results. Commutativity hypotheses in particular have already seen investigation, in [17], where they allow for modelling "parallel actions" in the style of Mazurkiewicz traces.

Finally, a word on a possible application of this work. In [4], Braibant and Pous develop a tactic in Rocq to automatically prove equivalence of regular expressions using completeness and decidability of KA. The techniques introduced in this paper could be used to expand this tactic, where the user designates relevant facts in the context as hypotheses.

Acknowledgments. The authors want to thank Yde Venema for valuable comments early on, and Bas Laarakker and Sean Prendi for proofreading. This work was supported by the Dutch research council (NWO) under grant no. VI.Veni.232.286 (ChEOpS).

References

1. de Amorim, A.A., Zhang, C., Gaboardi, M.: Kleene algebra with commutativity conditions is undecidable. In: CSL (2025). `https://doi.org/10.4230/LIPICS.CSL.2025.36`
2. Anderson, C.J., Foster, N., Guha, A., Jeannin, J., Kozen, D., Schlesinger, C., Walker, D.: NetKAT: semantic foundations for networks. In: POPL (2014). `https://doi.org/10.1145/2535838.2535862`
3. Book, R.V., Otto, F.: String-Rewriting Systems. Springer (1993)
4. Braibant, T., Pous, D.: Deciding Kleene algebras in Coq. Log. Methods Comput. Sci. **8**(1) (2012). `https://doi.org/10.2168/LMCS-8(1:16)2012`
5. Chung, L., Kappé, T.: Partial reductions for Kleene algebra with linear hypotheses (2026), `https://arxiv.org/abs/2601.14114`
6. Cohen, E.: Hypotheses in Kleene algebra. Tech. rep., Bellcore (1994)
7. Conway, J.H.: Regular Algebra and Finite Machines. Chapman and Hall, London (1971)
8. Doumane, A., Kuperberg, D., Pous, D., Pradic, C.: Kleene algebra with hypotheses. In: FoSSaCS (2019). `https://doi.org/10.1007/978-3-030-17127-8_12`
9. Kappé, T., Brunet, P., Rot, J., Silva, A., Wagemaker, J., Zanasi, F.: Kleene algebra with observations. In: CONCUR (2019). `https://doi.org/10.4230/LIPIcs.CONCUR.2019.41`
10. Kappé, T., Brunet, P., Silva, A., Wagemaker, J., Zanasi, F.: Concurrent Kleene algebra with observations: from hypotheses to completeness. FoSSaCS (2020). `https://doi.org/10.1007/978-3-030-45231-5_20`
11. Kleene, S.C.: Representation of events in nerve nets and finite automata. Automata Studies pp. 3–41 (1956)
12. Kozen, D.: A completeness theorem for Kleene algebras and the algebra of regular events. Inf. Comput. **110**(2), 366–390 (1994)
13. Kozen, D.: Kleene algebra with tests and commutativity conditions. In: TACAS (1996). `https://doi.org/10.1007/3-540-61042-1_35`
14. Kozen, D., Mamouras, K.: Kleene algebra with equations. In: ICALP (2014). `https://doi.org/10.1007/978-3-662-43951-7_24`
15. Kozen, D., Smith, F.: Kleene algebra with tests: Completeness and decidability. In: CSL (1996). `https://doi.org/10.1007/3-540-63172-0_43`
16. Kuznetsov, S.L.: On the complexity of reasoning in Kleene algebra with commutativity conditions. In: ICTAC (2023). `https://doi.org/10.1007/978-3-031-47963-2_7`
17. Maarand, H., Uustalu, T.: Reordering derivatives of trace closures of regular languages. In: CONCUR (2019). `https://doi.org/10.4230/LIPICS.CONCUR.2019.40`
18. Pous, D., Rot, J., Wagemaker, J.: On tools for completeness of Kleene algebra with hypotheses. Log. Methods Comput. Sci. **20**(2) (2024). `https://doi.org/10.46298/LMCS-20(2:8)2024`
19. Thompson, K.: Regular expression search algorithm. Commun. ACM **11**(6), 419–422 (1968). `https://doi.org/10.1145/363347.363387`

Diagrammatic Reasoning with Control as a Constructor, Applications to Quantum Circuits

Noé Delorme and Simon Perdrix

Université de Lorraine, CNRS, INRIA, LORIA, 54000 Nancy, France

Abstract. Control is a fundamental concept in quantum and reversible computational models. It enables the conditional application of a transformation to a system, depending on the state of another system. We introduce a general framework for diagrammatic reasoning featuring control as a constructor. To this end, we provide an elementary axiomatisation of control functors, extending the standard formalism of props to controlled props. As an application, we show that controlled props facilitate diagrammatic reasoning for quantum circuits by introducing a simple and complete set of relations involving at most three qubits, whereas in the standard prop setting any complete axiomatisation necessarily requires relations acting on arbitrarily many qubits.

1 Introduction

Diagrammatic reasoning has been very successful in recent years across various domains. These include quantum computing [14,24,20,4,1,12], linear algebra [31,7], control theory [6,5], natural language processing [13,32] and symbolic dynamics [23] to cite a few. The natural mathematical framework for diagrammatic reasoning is that of symmetric monoidal categories, and more precisely, props. This formalism allows representing processes and how they interact in space-time via sequential and parallel compositions.

Control is a central concept in quantum and reversible computational models. Roughly speaking, a transformation can be conditioned on an additional system such that, depending on its state, the transformation is either applied or not. For instance, controlling a NOT gate yields the well-known CNOT gate, and adding another control leads to the Toffoli gate. Notice that various quantum programming languages [19,18,28,29,9] provide primitives that allow controlling gates or even circuits.

In this paper, we introduce a formalisation of control as a constructor, i.e. as a *control functor* in a prop: every diagram f with n inputs and n outputs yields a new diagram $\mathsf{C}(f)$ with $1+n$ inputs and $1+n$ outputs, which represents its controlled version. A *controlled prop* is then defined as a prop equipped with a control functor. Similarly to props, controlled props can be used to define graphical languages by generators and relations.

The controlled prop formalism is a versatile language that captures the fundamental properties of control and facilitates reasoning on processes. For instance, if an equation $f = g$ is derivable, so is its controlled version $\mathsf{C}(f) = \mathsf{C}(g)$,

© The Author(s) 2026
N. Bertrand and S. Milius (Eds.): FoSSaCS 2026, LNCS 16503, pp. 240–261, 2026.
https://doi.org/10.1007/978-3-032-22730-0_12

$$\textcircled{2\pi} \;=\; \boxed{} \quad (1) \qquad \textcircled{\alpha_1}\,\textcircled{\alpha_2} \;=\; \textcircled{\alpha_1 + \alpha_2} \quad (2) \qquad -\!\boxed{H}\!-\!\boxed{H}\!- \;=\; \underline{\qquad} \quad (3)$$

$$\times \;=\; \text{(CNOT circuit)} \quad (4)$$

$$-\!\boxed{H}\!-\!\boxed{Z(\alpha_1)}\!-\!\boxed{H}\!-\!\boxed{Z(\alpha_2)}\!-\!\boxed{H}\!- \;=\; \textcircled{\beta_0}\;-\!\boxed{Z(\beta_1)}\!-\!\boxed{H}\!-\!\boxed{Z(\beta_2)}\!-\!\boxed{H}\!-\!\boxed{Z(\beta_3)}\!- \quad (5)$$

Fig. 1: Complete set of relations for quantum circuits defined as a conjugated prop generated by the Hadamard gate $-\!\boxed{H}\!-$ together with the global phases gates $\textcircled{\alpha}$ for any $\alpha \in \mathbb{R}$, where $-\!\boxed{Z(\alpha)}\!-$ is a notation for $\mathsf{C}(\textcircled{\alpha})$ and $-\!\oplus\!-$ is a notation for $-\!\boxed{H}\!-\!\boxed{Z(\pi)}\!-\!\boxed{H}\!-$. See Corollary 1 in Section 5 for details.

whereas the latter can be cumbersome to derive when using the standard prop formalism. Controlled props rely on three constructors – sequential composition, parallel composition, and control – and specify how they interact. In particular, control does not preserve parallel composition, i.e., $\mathsf{C}(f \otimes g) \neq \mathsf{C}(f) \otimes \mathsf{C}(g)$ since the latter acts on one more wire. However, it distributes in a natural way, as captured by the axioms of control functors.

We point out a fundamental property – the *conjugation law* – which is central when dealing with reversing and control. Given f and an invertible g, one can conjugate f by g with the *compute-uncompute* pattern $g^{-1} \circ f \circ g$. This pattern is widely used in algorithm design, e.g. for a basis change, for ancilla preparation and release, for oracle implementation, and also for circuit optimisation [18]. Controlling such a pattern can be reduced to controlling only f, leading to the notion of *conjugated prop*, namely a controlled prop satisfying the following so-called conjugation law.

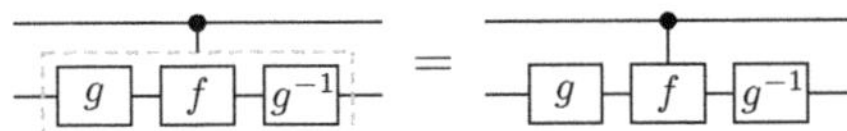

An important motivation for our framework comes from quantum computing, where control is extensively used. Complete sets of relations for quantum circuits have been established in earlier works using the formalism of props [12,11,10]. These complete equational theories do not rely on control as a constructor, but however point out the pivotal role of controlled operations in diagrammatic reasoning, and motivate the introduction of controlled props. As our main application in this present paper, we define the *controllable quantum circuits* and give a complete set of relations for it. In its simplest form (see Figure 1), where controllable quantum circuits are seen as a conjugated prop, this complete set of relations is made of four basic equations (Equations (1) to (4)), together with the standard Euler decomposition (Equation (5)). Notice that this equational theory is provably minimal. We also provide a complete set of equations for controllable circuits seen as a controlled prop, with only a few additional relations corresponding to particular instances of the conjugation law.

Remarkably, in the complete set of relations we introduce for controllable quantum circuits, all relations involve circuits acting on at most three qubits, whereas it has been shown in [10] that any complete set of relations for (unitary) quantum circuits, defined as a prop, requires at least one relation involving circuits acting on n qubits for any $n \in \mathbb{N}$. Hence, controlled props circumvent the unboundedness issue of props, making them particularly well suited to diagrammatic reasoning.

As another application, we consider controllable quantum circuits with auxiliary qubits, a.k.a. *ancillae*, for which we also introduce a simple complete set of relations. It illustrates that controlled props are not necessarily made up only of endomorphisms (even if the control functor only applies to endomorphisms). Throughout the paper, we also give several examples illustrating the versatility of the formalism, including classical reversible circuits and qudit systems.

1.1 Related work

In graphical languages, control is a central notion that has been used in various contexts, but not as a constructor. For quantum circuits, the completeness procedure is based on a back and forth translation with optical circuits [12] that extensively relies on controlled operations, and their basic algebra. In this context, controlled operations are considered as shorthand notations, as they can be decomposed into generators. In the ZX-calculus, controlled diagrams have also been extensively used, in particular for defining normal forms [25,2,26]. Notice that the expressivity of the language yields a slightly different definition of control, allowing the control of arbitrary diagrams (not necessarily with the same number of inputs and outputs). Again in this context, controlled operations are considered as shorthand notations. Another interesting approach is to extend the quantum circuit model with a new kind of composition that graphically consists in vertically connecting two gates, where the behaviour of the resulting circuit depends on the hamiltonians of the original ones. This work in progress allows, for instance, to represent a CNOT by connecting a Z gate and a X gate [30].

Additionally, various quantum programming languages (especially circuit description languages) allow for control of quantum circuits. In particular, Quipper supports control and a *withcompute* primitive, which consists in conjugating a circuit with another. These features have been semantically formalised using controllable categories [18] which involves control functions – not necessarily functors – and *withcompute* as a constructor, to provide a categorical model for the high level quantum programming language Quipper. Another quantum programming language introduced recently in [22] allows any arbitrary unitaries to be generated from some simple global phases combined with a quantum analogue of *if let* primitive, which can be embedded in our formalism (see Section 7). $\sqrt{\varPi}$ [8] and Hadamard-$\varPi$ [16] are two powerful quantum programming languages for approximatively universal fragments of quantum computing, that are equipped with sound and complete equational theories. These languages differ from our approach in several ways: first, they consider fragments of unitary transformations, second, they are not graphical, and third, they rely on the direct sum as a primitive rather than control functor. Indeed, this promising approach based on rig categories relies on two constructors $\otimes$ and $\oplus$ corresponding to the tensor product and the direct sum respectively. These two products provide a means of expressing control and enable powerful decompositions of unitary transformations, which, however, cannot always be interpreted as graphical circuit-like transformations. In contrast, control functors categorify the intuitive representation of control in circuits, in particular in the reversible and quantum cases. Finally, very recently, [21] introduces a construction to turn props with an identified involution into (poly)controlled props with two control functors. These control functors are commutative and exhaustive[1]. This recent pre-print relies on our completeness result on controlled quantum circuits (Theorem 2) to provide a variant of this result in their framework, among other completeness results for various fragments of quantum circuits. Notice that the rig-based approach is currently specific to the binary case, whereas controlled props are well-suited for more general settings (see Examples 2 and 8).

1.2   Outline of the paper

Due to space constraints, we only provide proof intuitions in this version; full and detailed proofs are available in the extended version [15]. In Section 2, we fix the notations and recall the usual notions used to define graphical languages in the prop formalism. In Section 3, we introduce the controlled prop formalism and its underlying notion of conjugation. In Section 4, we review how vanilla quantum circuits can be defined as a prop and recall their already-known completeness result. In Section 5, we define controllable quantum circuits as a controlled prop, prove that they satisfy the conjugation law, and give a complete set of relations for them. In Section 6, we add ancillae to the controllable quantum circuits and give a complete set of relations in these settings. In Section 7, we define some notable properties that multiple coexisting control functors may satisfy, and give examples. In Section 8, we provide our concluding remarks.

2   Background on props and graphical languages

A *prop* is, in category-theoretic terms, a strict symmetric monoidal category whose objects are generated by a single object, or equivalently, with $(\mathbb{N}, +)$ as a monoid of objects. Concretely, a prop $\mathbf{P}$ consists of collections of morphisms $\mathbf{P}(n, m)$ for any objects $n, m \in \mathbb{N}$. We write $f : n \to m$ when f is a morphism in $\mathbf{P}(n, m)$. Morphisms can be composed sequentially ($g \circ f : n \to m$, for any $f : n \to k$, $g : k \to m$) and in parallel ($f \otimes g : n + k \to m + \ell$, for any $f : n \to m$, $g : k \to \ell$). A prop has specific morphisms: identities $\mathrm{id}_0 : 0 \to 0$, $\mathrm{id}_1 : 1 \to 1$ and a symmetry $\sigma_{1,1} : 2 \to 2$. Identity on any object $n \in \mathbb{N}$ is inductively defined as $\mathrm{id}_n := \mathrm{id}_{n-1} \otimes \mathrm{id}_1$ when $n > 1$. Symmetry on any objects $n, m \in \mathbb{N}$ is inductively defined as $\sigma_{n,m} := (\mathrm{id}_1 \otimes \sigma_{n-1,m-1} \otimes \mathrm{id}_1) \circ (\sigma_{n-1,1} \otimes \sigma_{1,m-1}) \circ (\mathrm{id}_{n-1} \otimes \sigma_{1,1} \otimes \mathrm{id}_{m-1})$ when $n \neq 0 \neq m$ and $\sigma_{n,0} := \sigma_{0,n} := \mathrm{id}_n$ otherwise.

---

[1] $\mathsf{C}_2(g) \circ \mathsf{C}_1(f) = \mathsf{C}_1(f) \circ \mathsf{C}_2(g)$ and $\mathsf{C}_2(f) \circ \mathsf{C}_1(f) = \mathrm{id}_1 \otimes f$ (see Definitions 12 and 13).

Props have a nice graphical representation as string diagrams where objects are depicted as wires and morphisms as boxes. Here are some examples.

$$\mathrm{id}_0 \qquad \mathrm{id}_1 \qquad \sigma_{1,1} \qquad f \qquad g \circ f \qquad f \otimes g$$

Strings diagrams take advantage of a second dimension by drawing the sequential composition horizontally and the parallel compositions vertically. Finally, the morphisms have to satisfy the *coherence laws* of Figure 2. Depicted graphically, many coherence laws are trivialised or correspond to diagram deformations. We denote $\mathbf{P}_{\mathrm{endo}}$ the sub-prop of endomorphisms of $\mathbf{P}$, defined as the collections of morphisms $\mathbf{P}(n, n)$ for any object $n \in \mathbb{N}$.

Example 1. Given $d \in \mathbb{N}$, let $\mathbf{FdHilb}_d$ be the prop of qudit linear maps, meaning that $\mathbf{FdHilb}_d(n, m) = \mathcal{L}(\mathbb{C}^{d^n}, \mathbb{C}^{d^m})$ is the collection of linear maps from n qudits to m qudits, and where the parallel composition is the usual tensor product. Moreover, let $\mathbf{Iso}$ be the sub-prop of $\mathbf{FdHilb}_2$ restricted to isometries[2], and $\mathbf{Qubit} := \mathbf{Iso}_{\mathrm{endo}}$ its restriction to unitary evolutions.

Given two props $\mathbf{P}$ and $\mathbf{Q}$, a *functor* $F : \mathbf{P} \to \mathbf{Q}$ is a map that assigns to each object n of $\mathbf{P}$ an object $F(n)$ of $\mathbf{Q}$, and to each morphism $f \in \mathbf{P}(n, m)$ a morphism $F(f) \in \mathbf{Q}(F(n), F(m))$ while preserving the compositional structure, meaning $F(\mathrm{id}_0) = \mathrm{id}_{F(0)}$, $F(\mathrm{id}_1) = \mathrm{id}_{F(1)}$, and $F(g \circ f) = F(g) \circ F(f)$ whenever $g \circ f$ is defined. We say that F is *monoidal* when it also preserves the parallel composition, meaning $F(0) = 0$, $F(n + m) = F(n) + F(m)$ and $F(f \otimes g) = F(f) \otimes F(g)$. A *prop functor* is a monoidal functor that also preserves the symmetry, meaning $F(\sigma_{1,1}) = \sigma_{F(1),F(1)}$. A *dagger functor* $(\cdot)^\dagger : \mathbf{P} \to \mathbf{P}$ is an identity-on-object involutive contravariant prop functor, meaning that every morphism $f : n \to m$ has a dagger $f^\dagger : m \to n$ satisfying $f^{\dagger\dagger} = f$, $(g \circ f)^\dagger = f^\dagger \circ g^\dagger$, $(f \otimes g)^\dagger = f^\dagger \otimes g^\dagger$ and additionally $\sigma_{1,1}^\dagger = \sigma_{1,1}$.

Graphical languages, such as boolean or quantum circuits, can be defined as props by generators and relations. That is, the diagrams (or morphisms) are generated inductively by sequential and parallel compositions of generators, quotiented by the smallest congruence[3] that satisfies the set of relations as well as the coherence laws of props. Given a graphical language $\mathbf{P}$, the corresponding congruence is denoted $\mathbf{P} \vdash D_1 = D_2$ where D_1 and D_2 are two diagrams that can be transformed one into the other using the relations and the coherence laws.

A graphical language $\mathbf{P}$ often comes with an interpretation (or semantics) functor $[\![\cdot]\!] : \mathbf{P} \to \mathbf{S}$. The graphical language $\mathbf{P}$ is universal when $[\![\cdot]\!]$ is full, i.e. for any morphism f in $\mathbf{S}$, there exists a diagram D in $\mathbf{P}$ such that $[\![D]\!] = f$; and $\mathbf{P}$ is complete when $[\![\cdot]\!]$ is faithful, i.e. $[\![D_1]\!] = [\![D_2]\!]$ implies $\mathbf{P} \vdash D_1 = D_2$ for any $D_1, D_2 \in \mathbf{P}(n, m)$. We say that $\mathbf{P}$ is universally complete when $[\![\cdot]\!]$ is full and faithful. Completeness guarantees that all fundamental properties of the semantic domain $\mathbf{S}$ are graphically captured by $\mathbf{P}$.

---

[2] $f : n \to m$ is an isometry if its adjoint $f^\dagger : m \to n$ is its left inverse, i.e. $f^\dagger \circ f = \mathrm{id}_n$.

[3] A congruence is an equivalence relation $\mathcal{R}$ on the set of morphisms such that if $f_1 \mathcal{R} f_2$ and $g_1 \mathcal{R} g_2$, then $(g_1 \circ f_1)\mathcal{R}(g_2 \circ f_2)$ and $(f_1 \otimes g_1)\mathcal{R}(f_2 \otimes g_2)$.

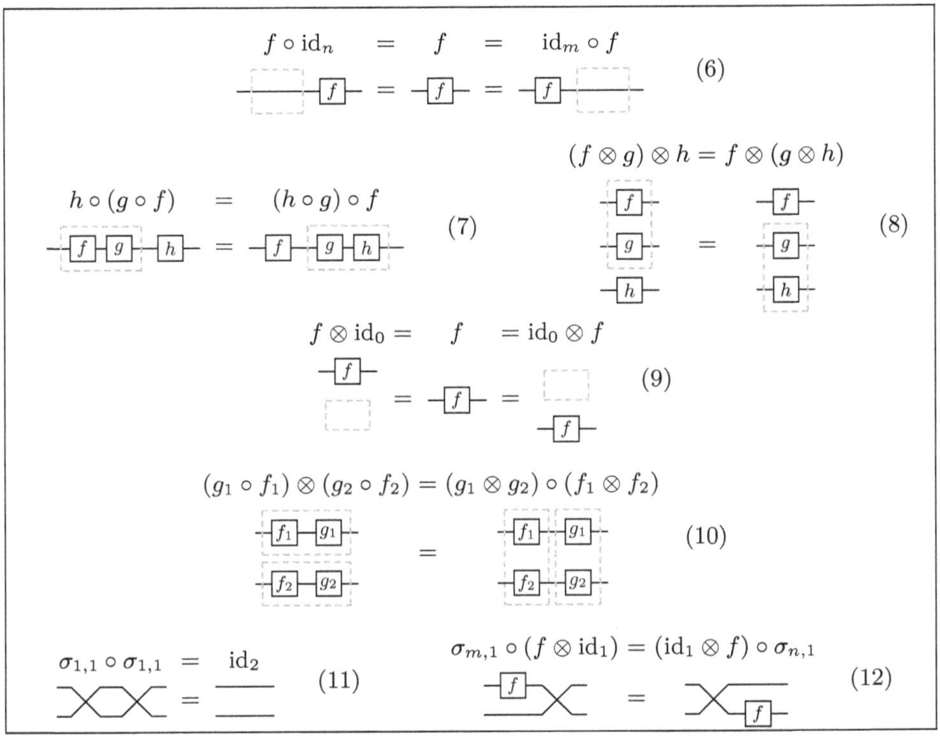

$$f \circ \mathrm{id}_n \;=\; f \;=\; \mathrm{id}_m \circ f \tag{6}$$

$$h \circ (g \circ f) \;=\; (h \circ g) \circ f \tag{7}$$

$$(f \otimes g) \otimes h = f \otimes (g \otimes h) \tag{8}$$

$$f \otimes \mathrm{id}_0 = \; f \; = \mathrm{id}_0 \otimes f \tag{9}$$

$$(g_1 \circ f_1) \otimes (g_2 \circ f_2) = (g_1 \otimes g_2) \circ (f_1 \otimes f_2) \tag{10}$$

$$\sigma_{1,1} \circ \sigma_{1,1} \;=\; \mathrm{id}_2 \tag{11}$$

$$\sigma_{m,1} \circ (f \otimes \mathrm{id}_1) = (\mathrm{id}_1 \otimes f) \circ \sigma_{n,1} \tag{12}$$

Fig. 2: Coherence laws of props. Notice that the dotted boxes are just here to highlight the correspondence with the syntax.

3 Control and conjugation

In this section, we formalise the notion of control as a constructor in a prop. Formally, this is achieved by equipping a prop with a control functor that satisfies some coherence laws.

Definition 1 (control functor). *Given a prop* **P**, *a control functor is a functor* $\mathsf{C} : \mathbf{P}_{\mathrm{endo}} \to \mathbf{P}_{\mathrm{endo}}$ *mapping any object* $n \in \mathbb{N}$ *to the object* $1 + n$ *and mapping any morphism* $f : n \to n$ *to a morphism* $\mathsf{C}(f) : (1 + n) \to (1 + n)$ *while satisfying the coherence laws depicted in Figure 3.*

Graphically, functors can be depicted using boxes [27]. Thus, $\mathsf{C}(f)$ is depicted as a bullet point on the additional wire connected to the diagram f, enclosed within a dotted box. This dotted box is omitted when the diagram is made of a single morphism or when there are nested controls (thanks to Equation (14)).

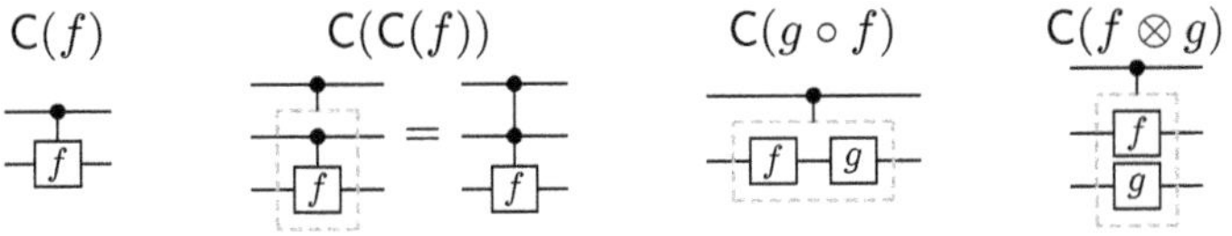

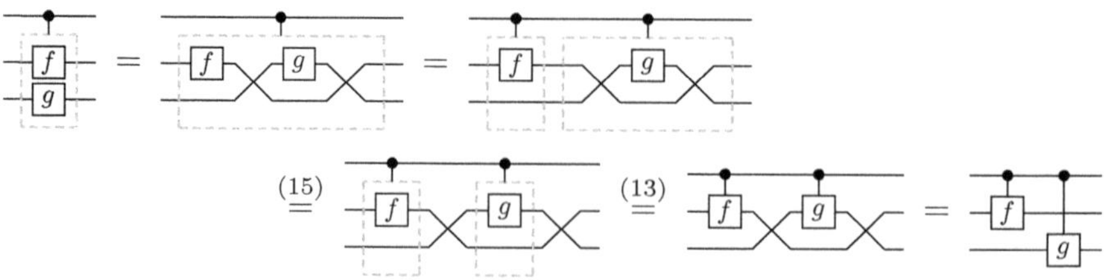

$$C(f \otimes \mathrm{id}_1) = C(f) \otimes \mathrm{id}_1 \qquad\qquad C(C(f)) \circ (\sigma_{1,1} \otimes \mathrm{id}_n) = (\sigma_{1,1} \otimes \mathrm{id}_n) \circ C(C(f))$$

$$(13) \qquad\qquad (14)$$

$$C((\mathrm{id}_k \otimes \sigma_{1,1} \otimes \mathrm{id}_\ell) \circ f \circ (\mathrm{id}_k \otimes \sigma_{1,1} \otimes \mathrm{id}_\ell)) = (\mathrm{id}_{k+1} \otimes \sigma_{1,1} \otimes \mathrm{id}_\ell) \circ C(f) \circ (\mathrm{id}_{k+1} \otimes \sigma_{1,1} \otimes \mathrm{id}_\ell)$$

$$(15)$$

Fig. 3: Coherence laws of control functors. Equations (13) and (14) is defined for any $n \in \mathbb{N}$ and any $f \in \mathbf{P}(n,n)$. Equation (15) is defined for any $k,\ell \in \mathbb{N}$ and any $f \in \mathbf{P}(k+2+\ell, k+2+\ell)$. Several wires are sometimes depicted as a single wire for simplicity.

We may write $C^n(f)$ for n nested application of the control functor C from f. Formally, $C^0(f) := f$ and $C^{n+1}(f) := C(C^n(f))$. Also, the coherence laws of props (Figure 2) allow us to simplify the drawings by stretching the line to depict morphisms that are controlled by several non-consecutive wires. The following diagrams are the same modulo the coherence laws.

The coherence laws of control functors capture the basic behavior of a control: identities are not affected by control (Equation (13)), nested controls commute (Equation (14)), and finally, the symmetry, which is left-invertible by definition in a prop, satisfies the conjugation law (Equation (15)). Notice that the functoriality of C additionally implies $C(g \circ f) = C(g) \circ C(f)$ (whenever $g \circ f$ is defined), and $C(\mathrm{id}_n) = \mathrm{id}_{1+n}$ (in particular, $C(\ ) = $ —). However, C is not monoidal because $C(f \otimes g) \neq C(f) \otimes C(g)$ in general, but the expected behavior of the control functor on parallel compositions is provided by the coherence laws of control functor, as illustrated in the following derivation.

Example 2. In **FdHilb$_2$**, the standard notion of control is to apply a linear map $U \in \mathcal{L}(\mathbb{C}^{2^n}, \mathbb{C}^{2^n})$ on the targeted system when the control qubit is in the state $|1\rangle$, and to apply the identity I_{2^n} when the control qubit is in the state $|0\rangle$. This can be embedded as the following control functor.

$$\mathsf{C}_{|1\rangle} : U \mapsto |0\rangle\langle 0| \otimes I_{2^n} + |1\rangle\langle 1| \otimes U$$

However, this is not the only control functor. For instance,

$$\mathsf{C}_{|0\rangle} : U \mapsto |0\rangle\langle 0| \otimes U + |1\rangle\langle 1| \otimes I_{2^n} \quad \text{and} \quad \mathsf{C}_{|-\rangle} : U \mapsto |+\rangle\langle +| \otimes I_{2^n} + |-\rangle\langle -| \otimes U$$

are also control functors in **FdHilb$_2$** (where $|+\rangle := {}^1\!/\!\sqrt{2}(|0\rangle + |1\rangle)$ and $|-\rangle := {}^1\!/\!\sqrt{2}(|0\rangle - |1\rangle)$). In higher dimensions, **FdHilb$_d$** admits a control functor

$$\mathsf{C}_{|k\rangle} : U \mapsto |k\rangle\langle k| \otimes U + \sum_{\ell \neq k} |\ell\rangle\langle \ell| \otimes I_{d^n}$$

for any $0 \leq k < d$. Notice however that the map $U \mapsto \sum_k |k\rangle\langle k| \otimes U^k$ is not a control functor as it fails to be functorial in general when $d > 2$.

We now focus on props equipped with a distinguished control functor, that we call *controlled props*. The general case of props with several control functors is described in Section 7. Notice that a single control functor already offers a rich structure as other control functors can be obtained by conjugating the control wire with some invertible morphism. For instance in **FdHilb$_2$** one can recover the various examples described in Example 2 by conjugating the control qubit with NOT or Hadamard gates.

Definition 2 (controlled prop). *A controlled prop is a prop equipped with a control functor.*

Throughout the paper, we denote by C the control functor of a controlled prop, unless a different notation is specified.

Similarly to props, a controlled prop **P** can be defined by generators and relations. In this setting, the diagrams are inductively defined as follows: the generators are morphisms, and for any morphisms $f_1 : n \to k$, $g_1 : k \to m$, $f_2 : n \to m$, $g_2 : k \to \ell$ and $f : n \to n$, we have that $g_1 \circ f_1$, $f_2 \otimes g_2$ and $\mathsf{C}(f)$ are morphisms. The morphisms are then quotiented by the smallest congruence $\mathbf{P} \vdash \cdot = \cdot$ that satisfies the given relations as well as the coherence laws in Figure 2 and Figure 3, where the congruence does not only preserve the parallel and sequential compositions but also the control, i.e. if $\mathbf{P} \vdash f = g$ then $\mathbf{P} \vdash \mathsf{C}(f) = \mathsf{C}(g)$. In other words, if an equation $f = g$ can be derived, its controlled version $\mathsf{C}(f) = \mathsf{C}(g)$ follows directly by construction.

Example 3. We consider the controlled prop **CNOT** generated by $\oplus : 1 \to 1$ with relations $\oplus\oplus = {-}\!\!-$, $\oplus\!\!\times\!\!\oplus = \times\!\oplus\!\times$ and $\oplus\!{\bullet}\!\oplus = {\bullet\bullet}$. Notice that $\oplus$ is not a generator but a diagram obtained by applying the control functor C to $\oplus$, moreover the equation $\oplus\oplus = {-}\!\!-$ can be derived.

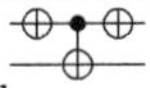

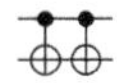

Definition 3 (conjugated prop). *A* conjugated prop **P** *is a controlled prop such that Equation* (16) *is satisfied for any* $f \in \mathbf{P}(m, m), g \in \mathbf{P}(n, m), h \in \mathbf{P}(m, n)$ *such that* $\mathbf{P} \vdash h \circ g = \mathrm{id}_n$.

$$\mathsf{C}(h \circ f \circ g) = (\mathrm{id}_1 \otimes h) \circ \mathsf{C}(f) \circ (\mathrm{id}_1 \otimes g) \tag{16}$$

Equation (16) is known as the *conjugation law*. This law is particularly important in algorithm design where the *compute-uncompute* pattern $g^{-1} \circ f \circ g$ is often used to prepare and release ancillary bits (or qubits) and implement oracles. This law can be used as much as possible in some compilation processes to minimize the scope of control as explained in [18].

Example 4. **FdHilb**$_2$ with $\mathsf{C}_{|1\rangle}$ as control functor is a conjugated prop.

Similarly to controlled props, conjugated props can be defined with generators and relations. The morphisms are freely generated from the generators using sequential compositions, parallel compositions and control, and then quotiented by the given relations together with the coherence laws of Figure 2 and Figure 3 as well as the conjugation law (Equation (16)). Notice that Equation (16) is somehow circular as it applies to morphisms with left-inverse, as a consequence, one can define a sequence of congruences $(\mathbf{P} \vdash_k \cdot = \cdot)_{k \in \mathbb{N}}$ as follows: $\mathbf{P} \vdash_0 \cdot = \cdot$ is the congruence of the corresponding controlled prop (i.e. using all equations but the conjugated law), and $\mathbf{P} \vdash_{k+1} \cdot = \cdot$ is the minimal congruence of the controlled prop augmented with the equations $\mathsf{C}(h \circ f \circ g) = (\mathrm{id}_1 \otimes h) \circ \mathsf{C}(f) \circ (\mathrm{id}_1 \otimes g)$ for any $f : m \to m, g : n \to m, h : m \to n$ s.t. $\mathbf{P} \vdash_k h \circ g = \mathrm{id}_n$. Tarski fix point theorem guarantees the existence of a fix point congruence $\mathbf{P} \vdash \cdot = \cdot$.

Notice that any prop admits two trivial control functors $f \mapsto \mathrm{id}_1 \otimes f$ and $f \mapsto \mathrm{id}_{1+n}$ (when $f : n \to n$) which correspond to the two degenerate cases of a control, namely when f is respectively always or never applied. The notion of *points*, we introduce below, can be used to rule out these two degenerate cases. Intuitively, the points of a control functor are two specific morphisms *false* ($\rhd\!\!- : 0 \to 1$) and *true* ($\blacktriangleright\!\!- : 0 \to 1$) that respectively fires and annihilates the controlled operation when they are plugged to the control wire.

Definition 4 (points). *A control functor* $\mathsf{C} : \mathbf{P}_{\mathrm{endo}} \to \mathbf{P}_{\mathrm{endo}}$ *has points if there exist two morphisms* $\rhd\!\!- : 0 \to 1$ *and* $\blacktriangleright\!\!- : 0 \to 1$ *such that the following equations are satisfied for any* $f \in \mathbf{P}(n, n)$.

$$\mathsf{C}(f) \circ (\rhd\!\!- \otimes \mathrm{id}_n) = \rhd\!\!- \otimes \mathrm{id}_n \qquad \mathsf{C}(f) \circ (\blacktriangleright\!\!- \otimes \mathrm{id}_n) = \blacktriangleright\!\!- \otimes f$$

Example 5. In **FdHilb**$_2$, $|0\rangle$ and $|1\rangle$ are points of the control functor $\mathsf{C}_{|1\rangle}$. In **FdHilb**$_d$, $|\ell\rangle$ and $|k\rangle$ are points of the control functor $\mathsf{C}_{|k\rangle}$ whenever $\ell \neq k$.

Notice however that points do not necessarily exist, in particular when all morphisms of the prop are endomorphisms, like in **Qubit**. In Section 5, we consider the controlled prop of quantum circuits without ancillae, where all generators are endomorphisms, and which has thus no points, whereas Section 6 is dedicated to quantum circuits with ancillae, leading to controlled props with points. We first review the properties of quantum circuits defined as a prop in the following section.

4 Vanilla quantum circuits

In this section, following [12,10], we consider vanilla quantum circuits defined as a prop (without control functor) by generators and relations and interpreted as unitary maps in **Qubit** (see Example 1). We also recall the already-known completeness result.

Definition 5 (vanilla quantum circuits). *Let* **QC** *be the prop of* vanilla quantum circuits *generated by the following generators (where $\alpha \in \mathbb{R}$) and the relations $\mathcal{R}_v$ depicted in Figure 4.*

$$\text{@} : 0 \to 0 \qquad \boxed{H} : 1 \to 1 \qquad \boxed{Z(\alpha)} : 1 \to 1 \qquad : 2 \to 2$$

Definition 6 (interpretation). *Let $\llbracket \cdot \rrbracket_v : \mathbf{QC} \to \mathbf{Qubit}$ be the interpretation of vanilla quantum circuits inductively defined as the following identity-on-object prop functor.*

$$\llbracket \boxed{H} \rrbracket_v := |+\rangle\langle 0| + |-\rangle\langle 1| \qquad\qquad \llbracket \boxed{Z(\alpha)} \rrbracket_v := |0\rangle\langle 0| + e^{i\alpha}|1\rangle\langle 1|$$

$$\llbracket \text{@} \rrbracket_v := e^{i\alpha} \qquad\qquad \left\llbracket \ \ \right\rrbracket_v := \sum_{x,y\in\{0,1\}} |x, x\oplus y\rangle\langle x,y|$$

The functor $\llbracket \cdot \rrbracket_v : \mathbf{QC} \to \mathbf{Qubit}$ assigns a unitary to all vanilla quantum circuits. Conversely, it is known that any unitary can be expressed as a vanilla quantum circuit. Thus, **QC** is universal for **Qubit**, meaning that $\llbracket \cdot \rrbracket_v$ is full.

In particular, we can implement any controlled operation by composing basic gates. For instance, the multi-controlled Z-rotations can be implemented for any $\alpha \in \mathbb{R}$ and $n \in \mathbb{N}$ by the circuit $\lambda^n(\alpha) \in \mathbf{QC}(n,n)$ inductively defined as follows.

$$\lambda^0(\alpha) := \text{@} \qquad\qquad \lambda^1(\alpha) := \boxed{Z(\alpha)}$$

$$\lambda^{n+2}(\alpha) := \lambda^{n+1}(\alpha/2) \quad \lambda^{n+1}(\alpha/2) \quad \lambda^{n+1}(-\alpha/2) \tag{27}$$

We can check that its interpretation (see Definition 6) is the desired Z-rotation with parameter α and controlled by the first $n-1$ qubits.

$$\llbracket \lambda^n(\alpha) \rrbracket_v = e^{i\alpha}|1\ldots 1\rangle\langle 1\ldots 1| + \sum_{x\neq 1\ldots 1} |x\rangle\langle x|$$

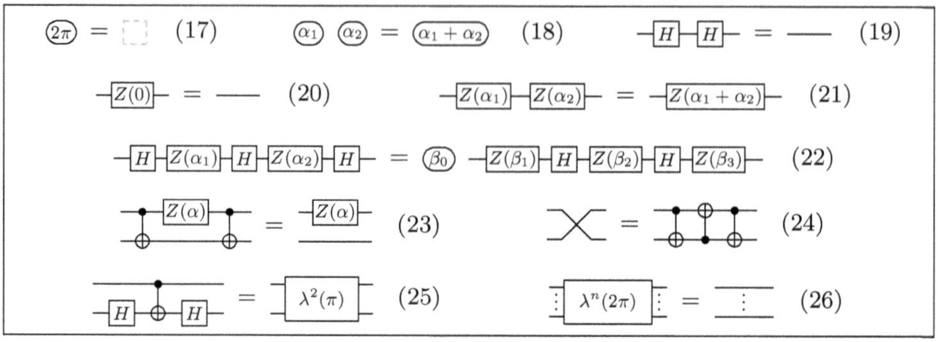

$$(2\pi) = \quad (17) \qquad (\alpha_1)\,(\alpha_2) = (\alpha_1 + \alpha_2) \quad (18) \qquad -H-H- = \; \underline{\quad} \quad (19)$$

$$-Z(0)- = \; \underline{\quad} \quad (20) \qquad -Z(\alpha_1)-Z(\alpha_2)- = -Z(\alpha_1 + \alpha_2)- \quad (21)$$

$$-H-Z(\alpha_1)-H-Z(\alpha_2)-H- = (\beta_0)\,-Z(\beta_1)-H-Z(\beta_2)-H-Z(\beta_3)- \quad (22)$$

$$-Z(\alpha)- = -Z(\alpha)- \quad (23) \qquad \times = \quad (24)$$

$$-H-\oplus-H- = \lambda^2(\pi) \quad (25) \qquad \lambda^n(2\pi) = \; \underline{\quad} \quad (26)$$

Fig. 4: Relations $\mathcal{R}_v$ of **QC**, containing an instance of Equation (26) for any number of qubits $n \geq 3$, and where $\lambda^n(\alpha)$ is defined by Equation (27). The relations between α_1, α_2 and $\beta_0, \beta_1, \beta_2, \beta_3$ is explained in Section 4.

As a consequence, each instance of Equation (26) is semantically trivial ($[\![\lambda^n(2\pi)]\!]_v = I_{2^n}$) but syntactically involved as the circuit $\lambda^n(\pi)$ contains a number of $-Z(\pi/2^{n-1})-$ gates which is exponential in n.

It has been shown in [10] that **QC** is complete.

Theorem 1 ([10]). *The prop* **QC** *is universally complete for* **Qubit**, *meaning that the functor* $[\![\cdot]\!]_v : \mathbf{QC} \to \mathbf{Qubit}$ *is full and faithful.*

All the relations in $\mathcal{R}_v$, except Equation (26), are fairly simple and commonly used in the literature. Surely the most powerful relation is Equation (22). This relation follows from the well-known Euler decomposition which states that any unitary can be decomposed, up to a global phase, into basic X- and Z-rotations. The angles α_1, α_2 are arbitrary in $\mathbb{R}$ whereas the angles $\beta_0, \beta_1, \beta_2, \beta_3$ are restricted to $[0, 2\pi)$ and are computed by explicit functions as follows using the intermediate complex numbers $u, v \in \mathbb{C}$.

$$u := -\sin\left(\alpha_1 + \alpha_2/2\right) + i\cos\left(\alpha_1 - \alpha_2/2\right)$$
$$v := +\cos\left(\alpha_1 + \alpha_2/2\right) - i\sin\left(\alpha_1 - \alpha_2/2\right) \quad \text{and} \quad \beta_0 := \frac{(\pi + \alpha_1 + \alpha_2 - \beta_1 - \beta_2 - \beta_3)}{2}$$

	β_1	β_2	β_3
if $v = 0$	$2\arg(u)$	0	0
if $u = 0$	$2\arg(v)$	π	0
otherwise	$\arg(u) + \arg(v)$	$2\arg\left(i + |u/v|\right)$	$\arg(u) - \arg(v)$

It has been shown in [10] that every relation of $\mathcal{R}_v$ is necessary. This means that none can be removed without losing completeness. Notice that there is an instance of Equation (26) for any number of qubits $n \geq 3$. All such instances are necessary. More generally, any set of relations yielding a universally complete graphical language (defined as a prop) for unitary maps requires at least one relation acting on n qubits for all $n \in \mathbb{N}$. As we will see in the following, the use of controlled prop simplifies the diagrammatic reasoning on quantum circuits, and is, in particular, a way to go around the necessity of relations acting on an unbounded number of qubits.

5   Controllable quantum circuits

In this section we define the *controllable quantum circuits* as a controlled prop, prove that the conjugation law holds in these settings, and give a completeness result for it. Using the controlled prop formalism, we can use much simpler gate sets to define quantum circuits. This is because a Z-rotation gate can be seen as a controlled global phase gate, and a CNOT gate can be seen as a controlled NOT gate. In fact, we can just take the global phases together with the Hadamard gate as gate set and still capture all unitaries in **Qubit**.

Definition 7 (controllable quantum circuits). *Let* **CQC** *be the controlled prop of* controllable quantum circuits *generated by the following generators (where $\alpha \in \mathbb{R}$) and the relations $\mathcal{R}_c$ depicted in Figure 5.*

$$\text{\textcircled{α}} : 0 \to 0 \qquad \text{\ -\!\boxed{H}\!-\ } : 1 \to 1$$

There are two kinds of relations in $\mathcal{R}_c$: On the one hand, Equations (28) to (32) directly correspond to some relations in $\mathcal{R}_v$. In particular the parameters of Equation (32) are computed with same functions as in Equation (22). Notice that some relations $\mathcal{R}_v$ do not appear in $\mathcal{R}_c$ as they are trivialised within the controlled prop formalism. On the other hand, Equations (33) to (36) are new and are all instances of the conjugation law.

We extend the interpretation of vanilla quantum circuits (Definition 6) to embed the control. This is achieved with the following prop functor.

Definition 8 (interpretation). *Let $[\![\cdot]\!]_c : $ **CQC** $\to$ **Qubit** *be the interpretation of controllable quantum circuits inductively defined as the following identity-on-object prop functor.*

$$[\![\,\text{\textcircled{α}}\,]\!]_c := e^{i\alpha} \qquad [\![\,\text{\ -\!\boxed{H}\!-\ }\,]\!]_c := |+\rangle\langle 0| + |-\rangle\langle 1|$$

$$[\![\mathsf{C}(C)]\!]_c := |0\rangle\langle 0| \otimes I_{2^n} + |1\rangle\langle 1| \otimes [\![C]\!]_c$$

We show in the following that the conjugation law can actually be derived for any **CQC** circuits. To do so, we need two ingredients. The first one being to equip **CQC** with the dagger functor $(\cdot)^\dagger$ defined as follows, and which provides a natural way to associate an inverse to any quantum circuit.

$$(\,\text{\ -\!\boxed{H}\!-\ }\,)^\dagger = \text{\ -\!\boxed{H}\!-\ } \qquad (\,\text{\textcircled{α}}\,)^\dagger = \text{\textcircled{$-\alpha$}} \qquad (\mathsf{C}(C))^\dagger = \mathsf{C}(C^\dagger)$$

Proposition 1. **CQC** $\vdash C^\dagger \circ C = \mathrm{id}_n = C \circ C^\dagger$ *for any $C \in$ **CQC**(n, n).*

Proof. By induction on C where the base case $C = \text{\ -\!\boxed{H}\!-\ }$ is directly proved by Equation (30) and the base case $C = \text{\textcircled{$\alpha$}}$ is proved by $\text{\textcircled{$\alpha$}}\ \text{\textcircled{$-\alpha$}} = \text{\textcircled{$0$}} = \text{\textcircled{$0$}}\ \text{\textcircled{$2\pi$}} = \text{\textcircled{$2\pi$}} = \mathrm{id}_0$ using Equations (28) and (29). $\qquad\square$

The second ingredient to prove the conjugation law is the ability to provably reduce the number of controls of a gate. To do so, we derive the following equations that intuitively reduce the number of controls of a gate at the cost

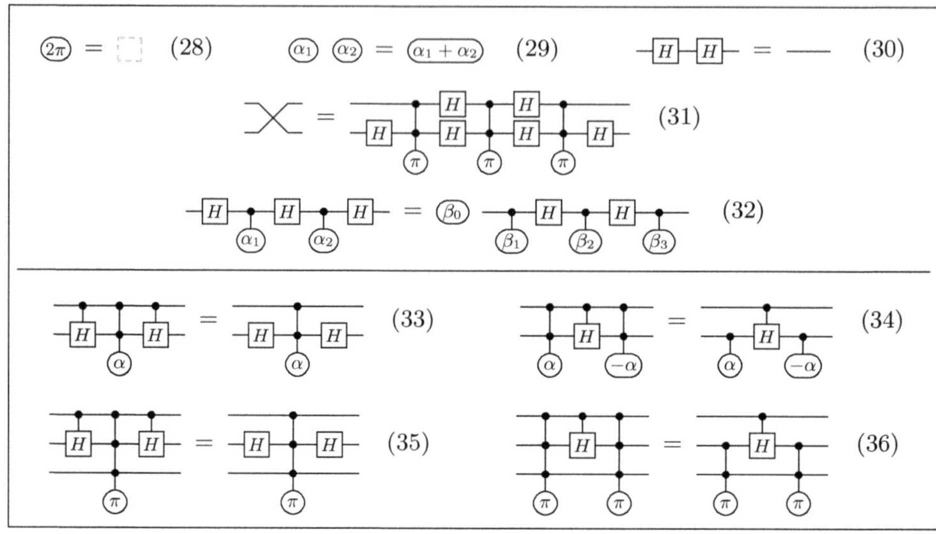

Fig. 5: Relations $\mathcal{R}_c$ of **CQC**.

of introducing other gates with fewer controls. This is particularly useful for establishing the correspondence with the vanilla quantum circuits. Indeed, we can transform any controllable quantum circuit into one that contains only gates controlled by up to two qubits, and all such gates correspond to a certain vanilla quantum circuit.

$$(37)$$

$$(38)$$

$$(39)$$

Proposition 2. CQC *is a conjugated prop.*

Proof. We prove a sufficient condition for the conjugation law to hold in the case of a unitary controlled prop, such as **CQC** (see [15]). The idea is to transform any controllable quantum circuit C into an equivalent one C' that contains only the gates ⓐ, $\mathsf{C}($ⓐ$)$, $\mathsf{C}(\mathsf{C}(\pi))$ and $-\boxed{H}-$. This is achieved using Equations (37) to (39) and their controlled versions. Then the sufficient condition states that it is enough to prove the conjugation law for individual gates.   □

The conjugation law is very powerful and can be used to prove many equations. In particular, we can generalize Equations (38) and (39) by defining the circuit $\mu^n(\alpha) \in \mathbf{CQC}(n,n)$ for any $\alpha \in \mathbb{R}$ and $n \in \mathbb{N}$ as follows.

$$\mu^0(\alpha) := \text{@} \qquad \mu^1(\alpha) := \mathsf{C}(\text{@})$$

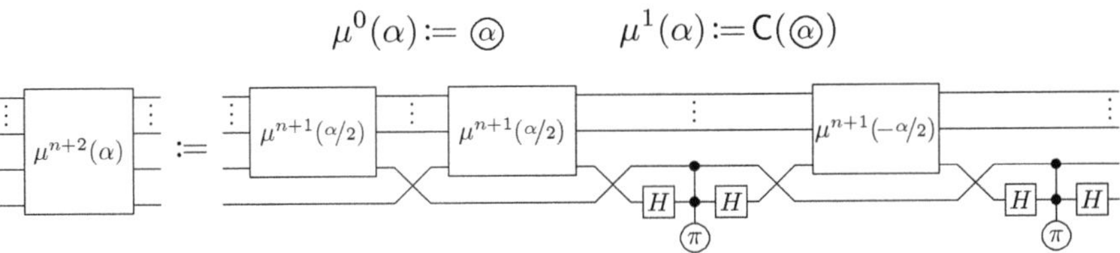

Intuitively, $\mu^n(\alpha)$ is similar to $\lambda^n(\alpha)$ and implements a Z-rotation gate with parameter α controlled by $n-1$ first qubits. It is provably equivalent to directly controlling an @ gate.

Proposition 3. $\mathbf{CQC} \vdash \mathsf{C}^n(\text{@}) = \mu^n(\alpha)$ *for any* $n \in \mathbb{N}$.

Proof. By induction on n. The base cases $n = 0$ and $n = 1$ are trivial and the case $n = 2$ is proved as Equation (38). The induction case $n+2$ with $n > 0$ is proved using the conjugation law (Proposition 2). $\qquad\square$

We are now ready to prove completeness for controllable quantum circuits by reducing it to the completeness of vanilla quantum circuits.

Theorem 2. *The controlled prop* $\mathbf{CQC}$ *is universally complete for* $\mathbf{Qubit}$, *meaning that the functor* $[\![\cdot]\!]_{\mathrm{c}} : \mathbf{CQC} \to \mathbf{Qubit}$ *is full and faithful.*

Proof. Universality is already known. The completeness of $\mathbf{CQC}$ is reduced to the completeness of $\mathbf{QC}$ (see Theorem 1). To do so, we define an encoding map $E : \mathbf{CQC} \to \mathbf{QC}$ and a decoding map $D : \mathbf{QC} \to \mathbf{CQC}$ to transform a proof in $\mathbf{QC}$ into a proof in $\mathbf{CQC}$. $\qquad\square$

It is remarkable that $\mathcal{R}_{\mathrm{c}}$ only contains relations acting on at most three qubits, whereas this was not possible in the prop formalism. This is mainly possible thanks to the context rule saying that if we have $\mathbf{CQC} \vdash C_1 = C_2$ then $\mathbf{CQC} \vdash \mathsf{C}(C_1) = \mathsf{C}(C_2)$ comes at free cost by definition of the control as a constructor. In some sense, the controlled prop formalism extracts the structural equations that are specific to control.

As Equations (33) to (36) are instances of the conjugated law, one can get rid of these equations by directly considering controllable quantum circuits as a conjugated prop.

Corollary 1. *The conjugated prop* $\mathbf{CCQC}$ *generated by* @ *and* $-\boxed{H}-$ *and the relations (28) to (32) is universally complete.*

This means that (28) to (32) are sufficient to have a complete graphical language of quantum circuits defined as a conjugated prop. Moreover, we show that all these relations are necessary, meaning that none can be removed without loosing completeness.

6 Controllable quantum circuits with ancillae

In this section, we introduce the controllable quantum circuits with auxiliary qubits, a.k.a. ancillae. This provides an example of a controlled prop that does not consist solely of endomorphisms and demonstrates that, similarly to the vanilla quantum circuit case [11], ancillae can be embedded into controllable quantum circuits.

Definition 9 (controllable quantum circuits with ancillae). *Let* **AQC** *be the controlled prop of* controllable quantum circuits with ancillae *generated by the following generators (where $\alpha \in \mathbb{R}$) and the relations $\mathcal{R}_\mathrm{c} \cup \mathcal{R}_\mathrm{a}$ depicted in Figure 5 and Figure 6.*

$$\text{ⓐ} : 0 \to 0 \qquad -\boxed{H}- \, : 1 \to 1 \qquad \rhd\!- \, : 0 \to 1 \qquad -\!\lhd \, : 1 \to 0$$

The additional equations of Figure 6 can be interpreted as follows: $\rhd\!-$ is a point that acts as an annihilator of the control in two particular instances (Equations (41) and (42)), moreover this point has a left inverse (Equation (40)), and, as a consequence, is subject to the conjugation law (Equation (43)).

Definition 10 (interpretation). *Let* $\llbracket \cdot \rrbracket_\mathrm{a} : \mathbf{AQC} \to \mathbf{FdHilb}_2$ *be the interpretation of controllable quantum circuits with ancillae inductively defined as the following identity-on-object prop functor.*

$$\llbracket\, \text{ⓐ} \,\rrbracket_\mathrm{a} := e^{i\alpha} \qquad \llbracket\, \rhd\!- \,\rrbracket_\mathrm{a} := |0\rangle \qquad \llbracket\, -\!\lhd \,\rrbracket_\mathrm{a} := \langle 0|$$

$$\llbracket\, -\boxed{H}- \,\rrbracket_\mathrm{a} := |+\rangle\langle 0| + |-\rangle\langle 1| \qquad \llbracket \mathsf{C}(C) \rrbracket_\mathrm{a} := |0\rangle\langle 0| \otimes I_{2^n} + |1\rangle\langle 1| \otimes \llbracket C \rrbracket_\mathrm{a}$$

The two new generators $\rhd\!-$ and $-\!\lhd$ correspond respectively to a qubit initialisation and a post-selected measurement, hence the semantics of an arbitrary **AQC** circuit is a contraction. It is standard to consider quantum circuits with *clean* ancillae, i.e. a circuit where all $-\!\lhd$ are performed on qubits which states have been returned to $|0\rangle$, to be safely removed without disturbing the state of the other qubits. Such clean releases are achieved if and only if the semantics of the overall circuit is an isometry[4]. Since there is an inclusion $\mathbf{Iso} \hookrightarrow \mathbf{FdHilb}_2$, we can define $\mathbf{AQC}_\mathrm{clean}$ and its interpretation $\llbracket \cdot \rrbracket_\mathrm{i} : \mathbf{AQC}_\mathrm{clean} \to \mathbf{Iso}$ as the following pullback, which states that $\mathbf{AQC}_\mathrm{clean}$ is the restriction of $\mathbf{AQC}$ to circuits having isometries as interpretation.

$$\begin{array}{ccc} \mathbf{CQC} \hookrightarrow & \mathbf{AQC}_\mathrm{clean} \hookrightarrow & \mathbf{AQC} \\ & \downarrow \llbracket\cdot\rrbracket_\mathrm{i} & \downarrow \llbracket\cdot\rrbracket_\mathrm{a} \\ & \mathbf{Iso} \hookrightarrow & \mathbf{FdHilb}_2 \end{array}$$

We also add to the above commutative diagram the inclusion of **CQC** in $\mathbf{AQC}_\mathrm{clean}$, and we may write $C \in \mathbf{CQC}(n,n)$ when $C \in \mathbf{AQC}(n,n)$ contains no $\rhd\!-$ or $-\!\lhd$ gates. Moreover, we define some useful notations.

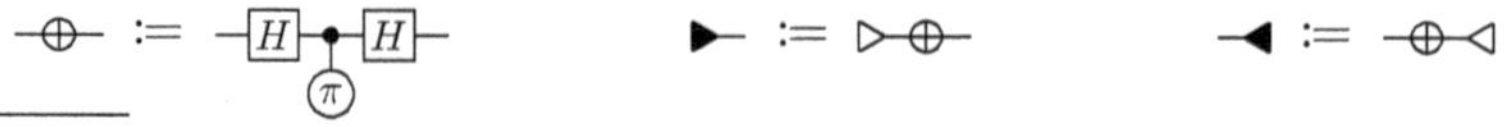

$$-\!\oplus\!- := -\boxed{H}\!-\!\bullet\!-\boxed{H}\!- \atop \;\;\text{ⓟ} \qquad\qquad \blacktriangleright\!- := \rhd\!-\!\oplus\!- \qquad\qquad -\!\blacktriangleleft := -\!\oplus\!-\!\lhd$$

---

[4] Indeed, a postselected evolution being an isometry means that all the performed measurements were actually determinstic, hence whenever $-\!\lhd$ has been applied, the corresponding qubit was already in the state $|0\rangle$.

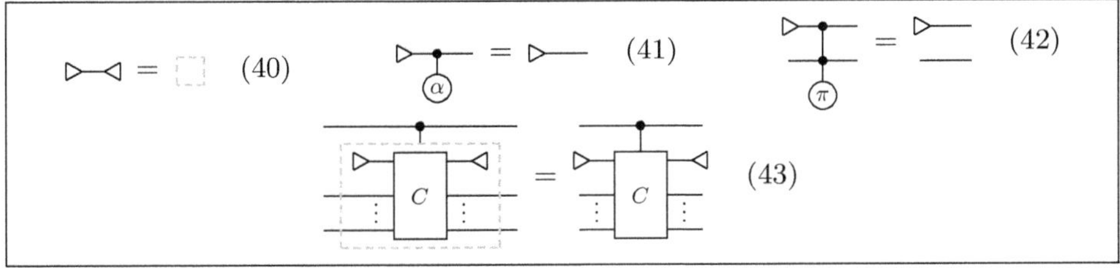

Fig. 6: Relations $\mathcal{R}_a$ of $\mathbf{AQC}$.

Proposition 4. *For any $C \in \mathbf{AQC}_{\mathrm{clean}}(k, k + n)$ there exists $C' \in \mathbf{CQC}(m + n + k, m + n + k)$ such that the following equation is derivable.*

$$\mathbf{AQC} \vdash k\Big\{ \boxed{\ :\ C\ :\ } \Big\}k+n \quad = \quad C'$$

Proof. We can always bend the wires to put all ▷— and —◁ gates on top of the circuit. Moreover, every ▷— and —◁ that are inside a control can be taken out using Equation (43) (only endomorphisms can be controlled, which implies that there are always as many ▷— as there are —◁ inside a control). ◻

Proposition 5. *▷— and ▶— are points in $\mathbf{AQC}_{\mathrm{clean}}$, meaning that the following equations are derivable for any $C \in \mathbf{AQC}_{\mathrm{clean}}(n, n)$.*

$$\mathbf{AQC} \vdash \quad = \quad \qquad\qquad \mathbf{AQC} \vdash \quad = \quad$$

Proof. By induction on C and using Proposition 4. ◻

Proposition 6. *Given $C \in \mathbf{CQC}(n + k, n + k)$, the following equation is derivable whenever it is sound with respect to $[\![\cdot]\!]_i$.*

$$\mathbf{AQC} \vdash \quad n\Big\{\ \boxed{C}\ \Big\}^n_k \quad = \quad n\Big\{\ \Big\}^n_k$$

Proof. The proof uses the variant of the cosine-sine decomposition of [11]. ◻

Theorem 3. *The controlled prop $\mathbf{AQC}_{\mathrm{clean}}$ is universally complete for $\mathbf{Iso}$, meaning that the functor $[\![\cdot]\!]_i : \mathbf{AQC}_{\mathrm{clean}} \to \mathbf{Iso}$ is full and faithful.*

Proof. Universality is already known, we only need to prove completeness. To do so, let $C_1, C_2 \in \mathbf{AQC}_{\mathrm{clean}}(k, n + k)$ be such that $[\![C_1]\!]_i = [\![C_2]\!]_i$. Applying Proposition 4 to C_i gives the circuit $C'_i \in \mathbf{CQC}(m_i + n + k, m_i + n + k)$. Assume w.l.o.g. that $m_1 \geq m_2$. Then, the following derivation proves completeness, where the four wires depict in reality $m_1 - m_2$, m_2, n and k wires respectively.

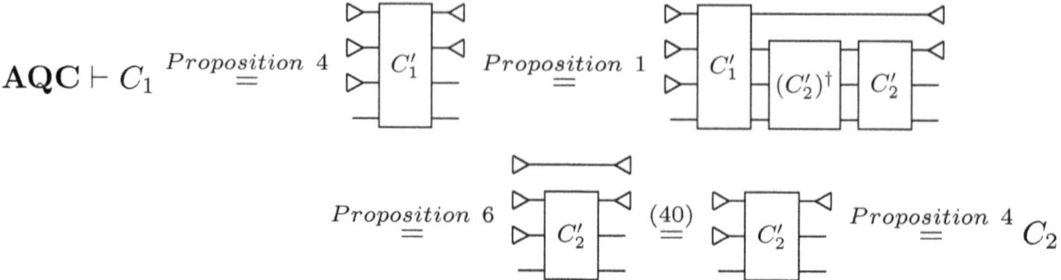

where Proposition 6 is applicable because the equality

$$[\![C_1]\!]_i = \left[\!\!\left[\begin{array}{c} m_1+n\{\quad C_1' \quad \}m_1 \\ k\{ \qquad \}n+k \end{array} \right]\!\!\right]_i = \left[\!\!\left[\begin{array}{c} m_2+n\{\quad C_2' \quad \}m_2 \\ k\{ \qquad \}n+k \end{array} \right]\!\!\right]_i = [\![C_2]\!]_i$$

implies the following equality.

$$\left[\!\!\left[\begin{array}{c} m_1+n\{\quad C_1' \quad \}m_1-m_2 \\ k\{ \quad (C_2')^\dagger \quad \}m_2+n+k \end{array} \right]\!\!\right]_i = \left[\!\!\left[\begin{array}{c} m_1+n\{\qquad \}m_1+n \\ k\{ \qquad \}k \end{array} \right]\!\!\right]_i$$

$\square$

Similarly to the ancilla-free case, one can consider quantum circuit with ancillae as a conjugated prop and get rid of the equations that are instances of the conjugated law.

Corollary 2. *Let* **CAQC** *be the conjugated prop generated by* $@$, $-\boxed{H}-$, $\triangleright-$ *and* $-\triangleleft$ *and the relations (28) to (32) together with (40) to (42). Then, its restriction to isometric circuit* **CAQC**$_{\text{clean}}$ *is universally complete.*

7 Polycontrolled prop

Throughout the paper, we focused on props that are equipped with a single control functor. This is natural to consider *polycontrolled props*, i.e. props that have multiple control functors.

Considering polycontrolled props instead of controlled prop can sometimes simplify diagrammatic reasoning. For instance, while controllable quantum circuits are defined using two generators and one control functor (see Section 5), we can also define them using one generator and two control functors. This idea as first been explained in [22], and can be expressed in the controlled prop formalism as follows: if we only have global phases $@$ for any $\alpha \in \mathbb{R}$, and two control functors C_Z and C_X, we can still get a universal graphical language for **Qubit** with the following interpretation.

$$[\![@]\!] := e^{i\alpha} \qquad [\![\mathsf{C}_Z(C)]\!] := |0\rangle\langle 0| \otimes I_{2^n} + |1\rangle\langle 1| \otimes [\![C]\!]$$

$$[\![\mathsf{C}_X(C)]\!] := |+\rangle\langle +| \otimes I_{2^n} + |-\rangle\langle -| \otimes [\![C]\!]$$

Intuitively C_Z allow to implement Z-rotations while C_X allow to implement X-rotations. Moreover, Z- and X-rotations together with their multi-controlled version can express any unitary. For instance, the Hadamard gate can be implemented by $\boxed{-\pi/4} \otimes \mathsf{C}_Z(\boxed{\pi/2}) \circ \mathsf{C}_X(\boxed{\pi/2}) \circ \mathsf{C}_Z(\boxed{\pi/2})$.

In the following, we define some notable properties that multiple coexisting control functors may satisfy.

Definition 11 (compatibility). *Given a polycontrolled prop* **P**, *we say that its control functors* C_1 *and* C_2 *are* compatible *if the following equation is satisfied for any* $f \in \mathbf{P}(n, n)$.

$$\mathsf{C}_1(\mathsf{C}_2(f)) \circ (\sigma_{1,1} \otimes \mathrm{id}_n) = (\sigma_{1,1} \otimes \mathrm{id}_n) \circ \mathsf{C}_2(\mathsf{C}_1(f)) \tag{44}$$

Example 6. In $\mathbf{FdHilb}_2$, $\mathsf{C}_{|1\rangle}$ is compatible with $\mathsf{C}_{|0\rangle}$ because each control can be obtained by conjugating the control wire of the other control with a NOT gate that slides through the swaps. However, $\mathsf{C}_{|1\rangle}$ is not compatible with the control functor $\mathsf{C}_\sharp : U \mapsto I_2 \otimes X_n U X_n$ where U acts on n qubits and where X_n applies NOT gates to n qubits in parallel. Intuitively, $\mathsf{C}_\sharp$ is independent of the state of the control qubit and always conjugates the target by a series of NOT gates in parallel.

Definition 12 (commutativity). *Given a polycontrolled prop* **P**, *we say that its control functors* C_1 *and* C_2 commute *if the following equation is satisfied for any* $f, g \in \mathbf{P}(n, n)$.

$$\mathsf{C}_2(g) \circ \mathsf{C}_1(f) = \mathsf{C}_1(f) \circ \mathsf{C}_2(g) \tag{45}$$

Example 7. In $\mathbf{FdHilb}_2$, $\mathsf{C}_{|1\rangle}$ commutes with $\mathsf{C}_{|0\rangle}$ whereas it does not with $\mathsf{C}_{|-\rangle}$.

Definition 13 (exhaustivity). *Given a polycontrolled prop* **P** *with a family of* ℓ *commuting control functors* $(\mathsf{C}_k)_{k \in [\ell]}$, *we say that the family is* exhaustive *if the following equation is satisfied for any* $f \in \mathbf{P}(n, n)$.

$$\mathsf{C}_\ell(f) \circ \ldots \circ \mathsf{C}_1(f) = \mathrm{id}_1 \otimes f \tag{46}$$

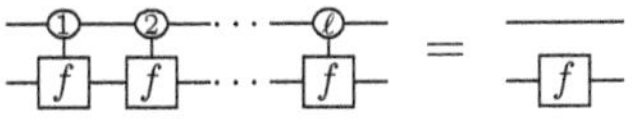

Example 8. In $\mathbf{FdHilb}_d$, the family of control functors $(\mathsf{C}_{|k\rangle})_{0 \leq k < d}$ is exhaustive. Intuitively, each controlled operation $\mathsf{C}_{|k\rangle}(f)$ fires f if the control qubit is in the state $|k\rangle$. Thus, if we apply all these controlled operations sequentially, then f is always fired exactly once.

8   Conclusion

In this paper, we introduced controlled props as an extension of the standard prop formalism enabling diagrammatic reasoning with control as a constructor. Beyond quantum computing, we expect controlled props to be useful in other domains where conditional behavior plays a central role, such as classical reversible computing. Note that the notion of control functor can be straightforwardly generalised to arbitrary symmetric monoidal categories and also to weaker structures like braided monoidal categories.

We considered, in this paper, the control as a constructor, independently of its actual physical implementation. We leave a full development of this question for future investigation, however we mention a no-go theorem concerning the possibility of controlling unknown operations using standard quantum circuits formalism [3]. As pointed out by the authors of this result, the no-go theorem does not apply in practice, as any physical implementation of an unknown unitary provides additional information that makes control possible. Moreover, there is evidence that such control can be implemented in practice [17,33]. The authors of the no-go theorem even argue that the quantum circuit formalism should be extended to capture the possibility of controlling unknown unitaries, which is a feature of the controlled prop of controllable quantum circuits we introduced. Furthermore, in the present paper, this no-go theorem does not apply, as we work in a white-box setting, i.e. we assume that the implementation of the unitary as a circuit is known.

Finally, another direction for future research is to consider controllable quantum circuits for quantum circuit optimization tasks, since they provide simpler rewriting rules than those of vanilla quantum circuit.

Acknowledgements

The authors want to thank Alexandre Clément, Emmanuel Jeandel, Louis Lemonnier, William Schober, and Scott Wesley for fruitful discussions. This work is supported by the the Plan France 2030 through the PEPR integrated project EPiQ ANR-22-PETQ-0007 and the HQI platform ANR-22-PNCQ-0002; and by the European Union through the MSCA Staff Exchange project Qcomical HORIZON-MSCA-2023-SE-01. The project is also supported by the Maison du Quantique MaQuEst.

References

1. Abramsky, S., Coecke, B.: A categorical semantics of quantum protocols. In: Proceedings of the 19th Annual IEEE Symposium on Logic in Computer Science (LICS) (2004). https://doi.org/10.1109/LICS.2004.1319636, https://arxiv.org/abs/quant-ph/0402130
2. Agnew, E., Yeh, L., Yeung, R.: Algebraic structure of controlled states and operators in the zxw calculus. Poster at QPL 2025 (2025)

3. Araújo, M., Feix, A., Costa, F., Brukner, v.: Quantum circuits cannot control unknown operations. New Journal of Physics (2014). https://doi.org/10.1088/1367-2630/16/9/093026, https://arxiv.org/abs/1309.7976

4. Backens, M., Kissinger, A.: Zh: A complete graphical calculus for quantum computations involving classical non-linearity. In: Proceedings of the 15th International Conference on Quantum Physics and Logic (QPL) (2018). https://doi.org/10.4204/EPTCS.287.2, https://arxiv.org/abs/1805.02175

5. Baez, J.C., Erbele, J.: Categories in control. arXiv preprint (2014), https://arxiv.org/abs/1405.6881

6. Bonchi, F., Sobociński, P., Zanasi, F.: A categorical semantics of signal flow graphs. In: Proceedings of the 25th International Conference on Concurrency Theory (CONCUR) (2014). https://doi.org/10.1007/978-3-662-44584-6_30, https://hal.science/hal-02134182

7. Bonchi, F., Sobociński, P., Zanasi, F.: Interacting bialgebras are frobenius. In: Proceedings of the 17th International Conference on Foundations of Software Science and Computation Structures (FoSSaCS) (2014). https://doi.org/10.1007/978-3-642-54830-7_23, https://hal.science/hal-00989174

8. Carette, J., Heunen, C., Kaarsgaard, R., Sabry, A.: With a few square roots, quantum computing is as easy as pi. In: Proceedings of the 51st ACM SIGPLAN Symposium on Principles of Programming Languages (POPL) (2024). https://doi.org/10.1145/3632861, https://arxiv.org/abs/2310.14056

9. Cirq: A python framework for creating, editing, and invoking noisy intermediate scale quantum (nisq) circuits. Zenodo (2021). https://doi.org/10.5281/zenodo.16867504, https://cirq.readthedocs.io

10. Clément, A., Delorme, N., Perdrix, S.: Minimal equational theories for quantum circuits. In: Proceedings of the 39th Annual ACM/IEEE Symposium on Logic in Computer Science (LICS) (2024). https://doi.org/10.1145/3661814.3662088, https://arxiv.org/abs/2311.07476

11. Clément, A., Delorme, N., Perdrix, S., Vilmart, R.: Quantum circuit completeness: Extensions and simplifications. In: Proceedings of the 32nd EACSL Annual Conference on Computer Science Logic (CSL) (2024). https://doi.org/10.4230/LIPIcs.CSL.2024.20, https://arxiv.org/abs/2303.03117

12. Clément, A., Heurtel, N., Mansfield, S., Perdrix, S., Valiron, B.: A complete equational theory for quantum circuits. In: Proceedings of the 38th Annual ACM/IEEE Symposium on Logic in Computer Science (LICS) (2023). https://doi.org/10.1109/LICS56636.2023.10175801

13. Coecke, B.: The mathematics of text structure. Joachim Lambek: The Interplay of Mathematics, Logic, and Linguistics (2021). https://doi.org/10.1007/978-3-030-66545-6_6, https://arxiv.org/abs/1904.03478

14. Coecke, B., Duncan, R.: Interacting quantum observables. In: Proceedings of the the 35th International Colloquium on Automata, Languages and Programming (ICALP) (2008). https://doi.org/10.1007/978-3-540-70583-3_25, https://arxiv.org/abs/0906.4725

15. Delorme, N., Perdrix, S.: Diagrammatic reasoning with control as a constructor, applications to quantum circuits. arXiv preprint (2025), https://arxiv.org/abs/2508.21756

16. Fang, W., Heunen, C., Kaarsgaard, R.: Hadamard-π: Equational quantum programming. arXiv preprint (2025), https://arxiv.org/abs/2506.06835

17. Friis, N., Dunjko, V., Dür, W., Briegel, H.J.: Implementing quantum control for unknown subroutines. Phys. Rev. A (2014). https://doi.org/10.1103/PhysRevA.89.030303, https://arxiv.org/abs/1401.8128

18. Fu, P., Kishida, K., Ross, N.J., Selinger, P.: Proto-quipper with reversing and control. In: Proceedings of the 22nd International Conference on Quantum Physics and Logic (QPL) (2025). https://doi.org/10.4204/EPTCS.426.1, https://arxiv.org/abs/2410.22261

19. Green, A.S., Lumsdaine, P.L., Ross, N.J., Selinger, P., Valiron, B.: Quipper: A scalable quantum programming language. In: Proceedings of the 34th ACM SIGPLAN Conference on Programming Language Design and Implementation (PLDI) (2013). https://doi.org/10.1145/2499370.2462177, https://arxiv.org/abs/1304.3390

20. Hadzihasanovic, A.: A diagrammatic axiomatisation for qubit entanglement. In: Proceedings of the 30th Annual ACM/IEEE Symposium on Logic in Computer Science (LICS) (2015). https://doi.org/10.1109/LICS.2015.59, https://arxiv.org/abs/1501.07082

21. Heunen, C., Kaarsgaard, R., Lemonnier, L.: One rig to control them all. arXiv preprint (2025), https://arxiv.org/abs/2510.05032

22. Heunen, C., Lemonnier, L., McNally, C., Rice, A.: Quantum circuits are just a phase. arXiv preprint (2025), https://arxiv.org/abs/2507.11676

23. Jeandel, E.: Strong shift equivalence as a category notion. arXiv preprint (2021), https://arxiv.org/abs/2107.10734

24. Jeandel, E., Perdrix, S., Vilmart, R.: A complete axiomatisation of the ZX-calculus for Clifford+T quantum mechanics. In: Proceedings of the 33rd Annual ACM/IEEE Symposium on Logic in Computer Science LICS (2018). https://doi.org/10.1145/3209108.3209131, https://arxiv.org/abs/1705.11151

25. Jeandel, E., Perdrix, S., Vilmart, R.: A generic normal form for zx-diagrams and application to the rational angle completeness. In: Proceedings of the 34th Annual ACM/IEEE Symposium on Logic in Computer Science (LICS) (2019). https://doi.org/10.5555/3470152.3470196, https://arxiv.org/abs/1805.05296

26. McDowall-Rose, H., Shaikh, R.A., Yeh, L.: From fermions to qubits: A zx-calculus perspective. arXiv preprint (2025), https://arxiv.org/abs/2505.06212

27. Melliès, P.A.: Functorial boxes in string diagrams. In: Proceedings of the 20th International Workshop on Computer Science Logic (CSL) (2006). https://doi.org/10.1007/11874683, https://hal.science/hal-00154243v1

28. Microsoft: Q# programming guide (2025), https://learn.microsoft.com/en-us/azure/quantum/user-guide/

29. Qiskit: An open-source framework for quantum computing. Zenodo (2019). https://doi.org/10.5281/zenodo.2562111, https://qiskit.org

30. Schober, W.: Extended quantum circuit diagrams. arXiv preprint (2024), https://arxiv.org/abs/2410.02946

31. Sobocinski, P.: Graphical linear algebra. Mathematical blog. (2015), https://graphicallinearalgebra.net

32. Zeng, W., Coecke, B.: Quantum algorithms for compositional natural language processing. In: Proceedings of the Workshop on Semantic Spaces at the Intersection of NLP, Physics and Cognitive Science (SLPCS) (2016). https://doi.org/10.4204/EPTCS.221.8, https://arxiv.org/abs/1608.01406

33. Zhou, X.Q., Ralph, T.C., Kalasuwan, P., Zhang, M., Peruzzo, A., Lanyon, B.P., O'Brien, J.L.: Adding control to arbitrary unknown quantum operations. Nature Communications (2011). https://doi.org/10.1038/ncomms1392, https://arxiv.org/abs/1006.2670

Lambda Galore

Mariangiola Dezani-Ciancaglini[1] ✉dezani@di.unito.it,
Besik Dundua[2] Besik.Dundua@kiu.edu.ge, and
Furio Honsell[3] furio.honsell@uniud.it

[1] Dipartimento di Informatica, Università di Torino, Italy
[2] Kutaisi International University, Kutaisi, Georgia
VIAM, Tbilisi State University, Tbilisi, Georgia
[3] DSMIF, Università di Udine, Italy

Abstract. In the present paper we solve two open problems in the theory of λ-calculus and intersection type theories. In particular we prove that there exist models which equate all unsolvable terms, but nonetheless separate fixed point combinators, *i.e.* terms which have the same Böhm tree. Moreover we show how the results concerning recursive types in second-order λ-calculus for strong normalisation change significantly when head normalisation in untyped λ-calculus endowed with type assignment systems is considered. We achieve this by generalising intersection type theory to an algebraic framework of meet-semilattices, thereby assigning an algebraic notion to each applicative structure. In the style of algebraic topology we establish transfer results between the theories of the models capitalising on the morphisms between the corresponding meet-semilattices. This machinery permits also to yield a thorough analysis of the potential and the limitations of computability arguments.

Keywords: λ-calculus, Intersection Types, Filter Models.

1 Introduction

Since the 1960s λ-calculus has been extensively studied as the stylised paradigm of functional programming [33,43,50]. One of its distinguishing features is the ease with which it admits a logical interpretation through the notion of *type*. Type systems for λ-calculus, both in the style *à la* Curry [20], *i.e.* as type-assignment systems, and *à la* Church [15], *i.e.* as typed calculi, stand today at the foundations, or have been the template, of most program logics following the slogan *typed programs cannot go wrong* [38], such as session types in communication based programming [25], and of proof development environments informed by the principle *proposition as types, λ-terms as programs*, such as Rocq [19,44].

We think, however, that the dynamic notion of computation as captured by λ-calculus and its relation to static or invariant logical notions as captured by types are still far from being completely understood. This is apparent in the semantics of untyped λ-calculus and its finitary accounts in terms of intersection type disciplines. The literature on the subject is huge [1,16,46], [11, Part III] and

N. Bertrand and S. Milius (Eds.): FoSSaCS 2026, LNCS 16503, pp. 262–284, 2026.
https://doi.org/10.1007/978-3-032-22730-0_13

spreads over more than half a century. Yet, in our view, the number of open problems continues to outweigh by far the corpus of settled results.

Addressing these open questions is important, since even the most innovative and emerging results in quantitative and relational reasoning [6,7,23] ultimately rely on the *finitary* descriptions, which type assignment systems permit to give to the interpretations of extensions or variants of λ-calculus in the appropriate mathematical structures. These in turn establish tight connections between proof rules and computation rules, thereby permitting to use proof-theoretic techniques to study denotational structures. Therefore any advancement in the conceptual understanding of type assignments in λ-calculus might spill over useful techniques and conceptual frameworks in many of the modern specialised topics. Just to give two examples: the seminal example of a higher order relational reasoning is the computability argument of Tait-Girard-Krivine-Mendler [26,32,37,49]; and Böhm trees [9, Chapter 10] are the archetype of *fully abstract models*.

This paper addresses the issue of transferring computational properties between models by chasing the morphisms between their algebraic accounts as type assignment systems. This is technically rather simple but it was never brought to the limelight. And precisely its simplicity makes it very flexible and easily extendable to more sophisticated mathematical structures. We apply it to the issue of when recursive types trigger *vicious circles* and generate non-sensible models and when, on the contrary, they produce *virtuous circles* and yield solid fixed points. Thereby we solve the open problem of generalising to type assignment systems for untyped λ-calculus the results by Mendler [37], as well as separating different fixed points, still keeping all unsolvable terms at bay. Building on old and very recent results on the subject [17,21,29] we carry out a refined analysis on the technique based on Tait-Girard-Krivine-Mendler computability [26,32,37,49] for type assignment systems for untyped λ-calculus [18,28,32], thereby achieving a λ-*results galore* on λ-theories and intersection type theories, which we hope might foster further research on this intriguing foundational topic.

More specifically, we generalise the notion of *intersection type theory* to the general notion of *meet-semilattice*, thus associating algebraic structures to λ-*models* or more generally to applicative structures, in the style of algebraic geometry. We go beyond the finitary denotational semantics developed in [17] and later fully expanded into a duality in [1]. The present algebraic setting permits to transfer computational behaviours between models leveraging on morphisms between the associated algebraic structures. Namely we define *transfer theorems* which permit to relate the λ-theories induced by the models. This greatly simplifies the task of studying the theory of a model from scratch, using perhaps difficult approximation theorems, and it permits to factor out arguments which can then be done once and for all. This setting also allows for a deep anatomy and generalisation of the "computability argument" and its *dangerous connections* to *union* types. Thus we achieve a rich and flexible kit of methods for transferring properties of λ-theories, such as sensibility and separability, which yields rather unexpected results. In particular we prove that there exist *sensible* filter models [11, Chapter 16] where all continuous functions are representable and whose theory does not extend the Böhm tree theory, a result which escaped for decades. Moreover we show how to extend Mendler's result [37] to type assignment systems for untyped λ-calculus w.r.t. head normalisation, thus breaking the myth that negative occurrences in recursive types yield non-sensible theories.

Synopsis In Section 2 we give basic definitions and present the generalisation of intersection type theories to meet-semilattices. Models are discussed in Section 3. Embeddings and the Transfer Theorem are in Section 4. The algebraic analysis of the computability argument is discussed in Section 5. The type theories which provide solutions to the open questions appear in Section 6, together with the extension of Mendler's result to type assignment systems. Related work and conclusions appear in Section 7.

2   Type Theories

We try to be self-contained, and provide throughout this section basic definitions on λ-calculus and intersection type assignment systems. The less familiar reader may refer to Barendregt's books [9,11] for more details.

We start out with the definition of the basic *algebraic structure* that we shall assign to applicative structures, which we call Type Theory.

Definition 1 (Type Theory). *A* type theory (TT) *is a non trivial meet-semilattice* $\Theta = \langle \mathscr{D}_\Theta, \sqsubseteq_\Theta \rangle$ *with maximal element* $\top_\Theta$, *closed under a binary arrow type constructor* $\Rightarrow_\Theta$. *We denote by* $\sqcap_\Theta$ *the meet, by* $\equiv_\Theta$ *the equivalence induced by* $\sqsubseteq_\Theta$, *and we use* α, β, ... *to range over the elements of* $\mathscr{D}_\Theta$, *which we call types.*

Notational Conventions. We omit the subscript Θ when clear from the context. As particular cases of TTs we consider domains which are sets of sets (see Examples 1 and 2) and domains which are syntactic intersection types (see Definition 3). For sets we denote by $\subseteq$ the partial order, by $\Rightarrow$ the arrow constructor, by $\cap$ the meet, and by $=$ the equivalence induced by $\subseteq$, and we use $X, Y,\ldots$ to range over elements. For syntactic intersection type theories (ranged over by $\mathcal{T}$) we denote by $\leq$ the partial order, by $\rightarrow$ the arrow constructor, by $\wedge$ the meet, by $\sim$ the equivalence induced by $\leq$, and we use $A, B, \ldots$ to range over elements. The following table summarises correspondences among these notations.

TT	structure	element	order	meet	arrow	top element	equivalence
algebraic	Θ	α, β	$\sqsubseteq$	$\sqcap$	$\Rightarrow$	$\top$	$\equiv$
set theoretic	Θ	X, Y	$\subseteq$	$\cap$	$\Rightarrow$	$\top$	$=$
syntactic	$\mathcal{T}$	A, B	$\leq$	$\wedge$	$\rightarrow$	U	$\sim$

The following list of heterogeneous denotational and syntactic examples should illuminate on the general perspective with which we deal with types. First, we recall the seminal definition of solvable term.

Definition 2 (Solvable terms [9, Definition 2.2.10]). *A λ-term M is solvable if there are n λ-terms $N_1,\ldots,N_n$ such that $(\lambda x_1 \ldots x_m.M)N_1 \ldots N_n \to_\beta^* \mathbf{I}$, where $x_1,\ldots,x_m$ are the variables which occur free in M and $\mathbf{I} = \lambda x.x$ is the identity combinator.*

Example 1. 1. Let $\langle \mathscr{D},\cdot \rangle$ be an applicative structure and let $\mathscr{P}(\mathscr{D})$ be the set of subsets of the domain. Take the standard *set theoretic* interpretation of $\Rightarrow$, namely $X \Rightarrow Y = \{d \mid \forall a \in X \ d \cdot a \in Y\}$. The lattice $\langle \mathscr{P}(\mathscr{D}), \subseteq \rangle$ is a TT.
2. Let $\langle \mathscr{D},\cdot, [\![\ ]\!]^{\mathscr{D}} \rangle$ be a CPO λ-model and let $\mathscr{C}(\mathscr{D})$ be the set of all compact Scott-open subsets of the domain. Taking the arrow defined in Item 1, we have that the meet semilattice $(\mathscr{C}(\mathscr{D}), \subseteq)$ is a TT [17].
3. Let $\langle \mathscr{D},\cdot, [\![\ ]\!]^{\mathscr{D}} \rangle$ be a λ-model and let $\mathscr{S}(\mathscr{D})$ be the set of all subsets of the domain (all Scott-open subsets, all subsets of the interior of $\mathscr{D}$). Clearly $(\mathscr{S}(\mathscr{D}), \subseteq)$ is a TT, with the arrow defined in Item 1.

In the following we call β-*nice* a set of λ-terms closed under β-conversion. The first example is crucial to this paper.

Example 2. 1. Let Λ denote the set of λ-terms, $\mathscr{S}$ the set of solvable terms, and $\mathscr{B} = \{M \in \Lambda \mid M \to_\beta^* x\overrightarrow{M}\}$. A set X is *saturated* if $\mathscr{B} \subseteq X \subseteq \mathscr{S}$ and it is β-nice. Define $X \Rightarrow Y = \{M \in \Lambda \mid \forall N \in X \ MN \in Y\}$, where X, Y range over β-nice sets. It is easy to verify that if Y is saturated, then $X \Rightarrow Y$ is saturated and that $X \Rightarrow \Lambda = \Lambda$. Moreover if X and Y are saturated, then $X \cap Y$ is saturated and $X \cap \Lambda = X$ for all X. Let $\mathcal{SAT} = \{X \mid X \text{ is saturated}\}$, then the lattice $\mathscr{SA} = \langle \mathcal{SAT} \cup \{\Lambda\}, \subseteq \rangle$ with top Λ, meet $\cap$ and arrow $\Rightarrow$ is a TT.
2. Let $\mathbb{M}$ be the set of subsets of Λ which are β-nice. The lattice $\langle \mathbb{M}, \subseteq \rangle$ with top Λ, meet $\cap$ and arrow $\Rightarrow$ defined in Item 1 is a TT.
3. Let $\mathscr{P} = \{M \mid M \to_\beta^* \lambda x_1,\ldots x_n.yM_1 \ldots M_m\}$, where the variable y is free, be the set of *persistently head normalising terms*, and let $\mathscr{N}$ be the set of β-nice subsets of Λ including $\mathscr{P}$. The lattice $\langle \mathscr{N}, \subseteq \rangle$ with top Λ, meet $\cap$ and arrow $\Rightarrow$ defined in Item 1 is a TT.
4. Let $\mathscr{C}$ be the set of β-nice subsets of Λ whose elements are *closable*, namely reduce to closed terms, and Λ. The lattice $\langle \mathscr{C}, \subseteq \rangle$ with top Λ, meet $\cap$ and arrow $\Rightarrow$ defined in Item 1 is a TT.
5. Let $\mathscr{I}$ be the set of β-nice subsets of Λ whose elements are $\mathbf{I}$-*terms*, *i.e.* reduce to terms of the $\lambda \mathbf{I}$-calculus, namely terms in which all abstracted variables occur in the abstraction bodies, and Λ. The lattice $\langle \mathscr{I}, \subseteq \rangle$ with top Λ, meet $\cap$ and arrow $\Rightarrow$ defined in Item 1 is a TT.

Notice that Λ is the top of all lattices defined in previous example.

Clearly the familiar notion of intersection type theory, together with the standard axioms and rules for subtyping [21] can be viewed as a TT.

Definition 3. *A syntactic type theory (shortly STT) $\mathcal{T} = \langle \mathbb{T}_\mathbb{A}, \leq_\mathcal{T} \rangle$ is a TT, where $\mathbb{T}_\mathbb{A}$ is the set of types, for $\mathsf{c} \in \mathbb{A}$ (the set of constants), given by*
$$A, B ::= \mathsf{c} \mid \mathsf{U} \mid A \to A \mid A \wedge A$$
and $\leq_\mathcal{T}$ is a subtyping relation which turns $\sim$ into an equivalence, U into top and $\wedge$ into meet.

The syntactic notion of type assignment system can be naturally extended to the algebraic notion of TT as follows.

Definition 4 (Type Assignment System). *The* intersection type assignment system *over a TT* $\Theta = \langle \mathscr{D}_\Theta, \sqsubseteq_\Theta \rangle$ *is a formal system deriving judgements of the shape* $\Upsilon \vdash_\Theta M : \alpha$, *where* $\alpha \in \Theta$ *and a basis* Υ *is a finite mapping from term variables to types in* Θ:

$$\Upsilon ::= \emptyset \mid \Upsilon, x : \alpha.$$

The axioms and rules of the type system are the following

$$\frac{}{\Upsilon, x : \alpha \vdash x : \alpha} \text{ (Ax)} \qquad\qquad \frac{}{\Upsilon \vdash M : \top_\Theta} \text{ (}\top\text{)}$$

$$\frac{\Upsilon, x : \beta \vdash M : \alpha}{\Upsilon \vdash \lambda x.M : \beta \Rightarrow_\Theta \alpha} \text{ (}\Rightarrow\text{I)} \qquad \frac{\Upsilon \vdash M : \beta \Rightarrow_\Theta \alpha \quad \Upsilon \vdash N : \beta}{\Upsilon \vdash MN : \alpha} \text{ (}\Rightarrow\text{E)}$$

$$\frac{\Upsilon \vdash M : \beta \quad \Upsilon \vdash M : \alpha}{\Upsilon \vdash M : \beta \sqcap_\Theta \alpha} \text{ (}\sqcap\text{I)} \qquad \frac{\Upsilon \vdash M : \beta \quad \beta \sqsubseteq_\Theta \alpha}{\Upsilon \vdash M : \alpha} \text{ (}\sqsubseteq\text{)}$$

The significance of the above type assignment system lies in the following crucial theorem, which is a substantial part of the "computability" arguments.

Theorem 1. *Let* $\mathfrak{R} = (\mathscr{D}_\mathfrak{R}, \subseteq)$ *be a set of β-nice sets of λ-terms closed under intersection and arrow (defined in Example 2(1)). Then $\mathfrak{R}$ is a TT. Moreover* $\{x_i : Y_i \mid 1 \leq i \leq n\} \vdash_\mathfrak{R} M : X$ *implies* $M[x_i := N_i \mid 1 \leq i \leq n] \in X$ *if* $N_i \in Y_i$ *for all i ($1 \leq i \leq n$).*

Proof. The proof is by induction on type derivations. The most interesting case is when the last applied rule is Rule $(\Rightarrow\text{I})$. In this case $M = \lambda x.M'$ and $X = Y \Rightarrow X'$ and $\{x_i : Y_i \mid 1 \leq i \leq n\} \cup \{x : Y\} \vdash_\mathfrak{R} M' : X'$. Let $N \in Y$, then by induction hypothesis we have that

$$M'[x_i := N_i \mid 1 \leq i \leq n][x := N] \in X',$$

which implies $M[x_i := N_i \mid 1 \leq i \leq n]N \in X'$, since β-nice sets are closed under β-conversion. Since $N \in Y$ is arbitrary by definition we conclude

$$M[x_i := N_i \mid 1 \leq i \leq n] \in Y \Rightarrow X' = X.$$

The applicability of Theorem 1 lies in that the inverse images under $[\![\ ]\!]^\mathscr{D}$ of the elements in the TTs of Example 1 are β-nice sets. The converse, however, does not hold in general, for example for the identity $\mathbf{I}$ we get $\mathbf{I} \in \{\mathbf{I}\}$, but $\nvdash \mathbf{I} : \{\mathbf{I}\}$ in the type assignment systems on the TTs defined in Example 2.

The *structural results* for intersection type theories, as *e.g.* in [11, Theorem 14.1.1], hold also for TTs, suitably rephrasing the proofs. We list them without further comments.

Lemma 1 (Inversion Lemma). *1. If* $\Upsilon \vdash_\Theta x : \alpha$ *and* $\alpha \not\equiv_\Theta \top_\Theta$, *then* $\Upsilon(x) \sqsubseteq_\Theta \alpha$;
2. If $\Upsilon \vdash_\Theta MN : \alpha$ *and* $\alpha \not\equiv_\Theta \top_\Theta$, *then there are I and β_i, γ_i for $i \in I$ such that* $\sqcap_{i \in I} \gamma_i \sqsubseteq_\Theta \alpha$ *and* $\Upsilon \vdash_\Theta M : \beta_i \Rightarrow_\Theta \gamma_i$ *and* $\Upsilon \vdash_\Theta N : \beta_i$ *for all $i \in I$;*

$$\top \equiv \alpha \Rrightarrow \top \quad (\Rrightarrow \top) \qquad\qquad (\beta \Rrightarrow \alpha) \sqcap (\beta \Rrightarrow \alpha') \equiv \beta \Rrightarrow \alpha \sqcap \alpha' \quad (\Rrightarrow \sqcap)$$

$$\frac{\beta' \sqsubseteq \beta \quad \alpha \sqsubseteq \alpha'}{\beta \Rrightarrow \alpha \sqsubseteq \beta' \Rrightarrow \alpha'} \ (\Rrightarrow) \qquad\qquad \frac{\top \sqsubseteq \beta \Rrightarrow \alpha}{\top \sqsubseteq \alpha} \ (\top \sqsubseteq)$$

Fig. 1: Subtyping Axioms and Rules.

3. If $\Upsilon \vdash_\Theta \lambda x.M : \alpha$, then there are I and β_i, γ_i for $i \in I$ such that $\bigsqcap_{i \in I}(\beta_i \Rrightarrow_\Theta \gamma_i) \sqsubseteq_\Theta \alpha$ and $\Upsilon, x : \beta_i \vdash_\Theta M : \gamma_i$ for all $i \in I$.

Theorem 2 (Subject Expansion). *If $M \to_\beta M'$ and $\Upsilon \vdash_\Theta M' : \alpha$, then $\Upsilon \vdash_\Theta M : \alpha$.*

Theorem 3 (Subject Reduction).
$$M \to_\beta M' \text{ and } \Upsilon \vdash_\Theta M : \alpha \text{ imply } \Upsilon \vdash_\Theta M' : \alpha$$
if and only if
$$\Upsilon \vdash_\Theta \lambda x.N : \beta \Rrightarrow_\Theta \gamma \text{ implies } \Upsilon, x : \beta \vdash_\Theta N : \gamma.$$

The following condition holds in all TTs arising from λ-models, where all continuous functions are representable [17,47], and it is the easiest way to check that filter structures (see Definition 95) built from TTs are λ-models, as we will see many times.

Definition 5 (β-soundness). *A TT Θ is β-sound if $\alpha \not\equiv_\Theta \mathsf{U}$ and*
$$\bigsqcap_{i \in I}(\beta_i \Rrightarrow_\Theta \alpha_i) \sqsubseteq_\Theta \beta \Rrightarrow_\Theta \alpha$$
imply that there is $J \subseteq I$ such that $\beta \sqsubseteq_\Theta \bigsqcap_{j \in J} \beta_j$ and $\bigsqcap_{j \in J} \alpha_j \sqsubseteq_\Theta \alpha$.

Repeatedly using the construction in Example 1(2), we can adapt proofs and counterexamples in [3,17], thus establishing the following result.

Theorem 4. *β-soundness is sufficient but not necessary for Subject Reduction.*

The following condition is natural, but it is not satisfied by TTs arising from the *graph*-models of λ-calculus, once we turn the basic elements into a TT [21].

Definition 6 ($\Rrightarrow$-soundness). *A TT Θ is $\Rrightarrow$-sound if $\sqsubseteq_\Theta$ satisfies the axioms and rules in Figure 1.*

Example 3. 1. The TT $\mathscr{A}$ defined in Example 2(1) is $\Rrightarrow$-sound.
2. The STT $\mathcal{T}_{Scott}$ with one constant c_{Scott} and the axiom $\mathsf{c}_{Scott} \sim \mathsf{U} \to \mathsf{c}_{Scott}$, defined in [17], is $\to$-sound.
3. The STT $\mathcal{T}_{CDZ}$ with two constants c_{CDZ} and d_{CDZ} and the three axioms $\mathsf{c}_{CDZ} \sim \mathsf{d}_{CDZ} \to \mathsf{c}_{CDZ}$, $\mathsf{d}_{CDZ} \sim \mathsf{c}_{CDZ} \to \mathsf{d}_{CDZ}$, and $\mathsf{c}_{CDZ} \leq \mathsf{d}_{CDZ}$, defined in [18], is $\to$-sound.
4. The STT $\mathcal{T}_{Park}$ with one constant c_{Park} and the axiom $\mathsf{c}_{Park} \sim \mathsf{c}_{Park} \to \mathsf{c}_{Park}$, defined in [17], is $\to$-sound.
5. The STT $\mathcal{T}_{HR}$ with two constants $\mathsf{c}_{HR}, \mathsf{d}_{HR}$ and axioms $\mathsf{d}_{HR} \sim \mathsf{c}_{HR} \to \mathsf{d}_{HR}$,

$d_{HR} \leq c_{HR}$, $c_{HR} \sim (c_{HR} \to c_{HR}) \wedge (d_{HR} \to d_{HR})$, defined in [29], is $\to$-sound.
6. The STT $\mathcal{T}_{AO}$ without constants and with only Axiom ($\to \wedge$) and Rules ($\to$), ($\sqcup \leq$), defined in [4], is not $\to$-sound.

We end this section recalling one definition and two theorems which will be used in Sections 5 and 6.

Definition 7 ([9, Definition 6.5.4(ii)]). *A Barendregt's fixed point combinator is a λ-term $\lambda f.\mathbf{H}\mathbf{H}M$ where $\mathbf{H} = \lambda xy.f(xxy)$ and M is arbitrary.*

Theorem 5 ([11, Theorem 17.4.3]). *A λ-term M has type c_{Park} from the empty basis in the type system over the STT $\mathcal{T}_{Park}$ defined in Example 3(4) if and only if M is closable, in the sense of Example 2(4).*

Theorem 6 ([11, Theorem 17.4.9]). *A λ-term M has type c_{HR} from the basis assigning type c_{HR} to all free variables of M, in the type system over the STT $\mathcal{T}_{HR}$ defined in Example 3(5) if and only if M is an $\mathbf{I}$-term, in the sense of Example 2(5).*

3 Filter Models

In this section we relate the basic definitions of λ-model [27] and filter model [10] and generalise the latter to TTs.

Definition 8 ([27]). *A λ-model is a triple: domain, application on the domain, interpretation of λ-terms in the domain $\langle \mathscr{D}, \cdot, [\![\ ]\!]^{\mathscr{D}} \rangle$ which satisfies*
1. $[\![x]\!]_{\rho}^{\mathscr{D}} = \rho(x)$;
2. $[\![MN]\!]_{\rho}^{\mathscr{D}} = [\![M]\!]_{\rho}^{\mathscr{D}} \cdot [\![N]\!]_{\rho}^{\mathscr{D}}$;
3. $[\![\lambda x.M]\!]_{\rho}^{\mathscr{D}} = [\![\lambda y.M[y/x]]\!]_{\rho}^{\mathscr{D}}$;
4. $\forall d \in \mathscr{D}.[\![M]\!]_{\rho[x:=d]}^{\mathscr{D}} = [\![N]\!]_{\rho[x:=d]}^{\mathscr{D}}$ *implies* $[\![\lambda x.M]\!]_{\rho}^{\mathscr{D}} = [\![\lambda x.N]\!]_{\rho}^{\mathscr{D}}$;
5. $\rho(x) = \rho'(x)$ *for all variables x which occur free in M implies* $[\![M]\!]_{\rho}^{\mathscr{D}} = [\![M]\!]_{\rho'}^{\mathscr{D}}$;
6. $[\![\lambda x.M]\!]_{\rho}^{\mathscr{D}} \cdot d = [\![M]\!]_{\rho[x:=d]}^{\mathscr{D}}$.

Definition 9. *1. Let $\Theta = \langle \mathscr{D}_{\Theta}, \sqsubseteq_{\Theta} \rangle$ be a TT. A set of elements $F \subseteq \mathscr{D}_{\Theta}$ is a Θ-filter if:*
$- \top_{\Theta} \in F$
$- \alpha, \beta \in F$ *imply* $\alpha \sqcap_{\Theta} \beta \in F$
$- \alpha \in F$ *and* $\alpha \sqsubseteq_{\Theta} \beta$ *imply* $\beta \in F$.
2. We use $\mathcal{F}_{\Theta}$ to denote the set of Θ-filters.
3. Application between filters is defined by $F \cdot G = \{\alpha \mid \exists \beta \in G.\ \beta \Rrightarrow_{\Theta} \alpha \in F\}$.
4. The term interpretation is defined by
$$[\![M]\!]_{\rho}^{\mathcal{F}_{\Theta}} = \{\alpha \in \mathscr{D}_{\Theta} \mid \exists \Upsilon \models \rho.\ \Upsilon \vdash_{\Theta} M : \alpha\}$$
where $\Upsilon \models \rho$ if $x : \alpha \in \Upsilon$ implies $\alpha \in \rho(x)$.
5. We call the triple $\langle \mathcal{F}_{\Theta}, \cdot, [\![\ ]\!]^{\mathcal{F}_{\Theta}} \rangle$ a filter structure.

The following theorem is the natural, and easy, extension to TTs of the corresponding momentous results for filter models, see *e.g.* [17,21].

Theorem 7. *1. Let $\Upsilon \models \rho$, then $\Upsilon \vdash_\Theta M : \alpha$ if and only if $\alpha \in [\![M]\!]_\rho^{\mathcal{F}_\Theta}$.*
2. All representable functions in $\mathcal{F}_\Theta$ are monotone w.r.t. filter inclusion.
3. A filter structure $\langle \mathcal{F}_\Theta, \cdot, [\![\ ]\!]^{\mathcal{F}_\Theta} \rangle$ is a λ-model iff $[\![\lambda x.M]\!]_\rho^{\mathcal{F}_\Theta} \cdot F \subseteq [\![M]\!]_{\rho[x:=F]}^{\mathcal{F}_\Theta}$.
Hence Subject Reduction holds in the TT Θ.
4. If Θ is β-sound, then all Scott-continuous functions are representable in $\mathcal{F}_\Theta$.
5. If Θ is β-sound, then $\langle \mathcal{F}_\Theta, \cdot, [\![\ ]\!]^{\mathcal{F}_\Theta} \rangle$ is a λ-model.

Example 4. 1. The STT $\mathcal{T}_{Scott}$ in Example 3(2) generates a filter model isomorphic to Scott model [45].
2. The STT $\mathcal{T}_{CDZ}$ in Example 3(3) generates a filter model analysed in [18].
3. The STT $\mathcal{T}_{Park}$ in Example 3(4) generates a filter model isomorphic to Park model [42].
4. The STT $\mathcal{T}_{HR}$ in Example 3(5) generates a filter model analysed in [29].
5. The STT $\mathcal{T}_{AO}$ in Example 3(6) generates a filter model isomorphic to Abramsky-Ong model [2].
6. Let Θ be the TT in Example 1(1). Define the filter-structure $\mathcal{F}_\Theta$ over Θ. Clearly $\mathscr{D}$ can be embedded in $\mathcal{F}_\Theta$ by $\iota(d) =\uparrow \{\alpha \mid d \in \alpha\}$. We get $\iota(d)\cdot\iota(e) \supseteq \iota(d\cdot e)$.
7. Let Θ be the TT in Example 1(2). We have that $\mathscr{D}$ is isomorphic to the filter structure over Θ. Moreover we have that if $\Upsilon \models \rho$, then
$$\Upsilon \vdash_\Theta M : \alpha \text{ if and only if } [\![M]\!]_\rho^{\mathscr{D}} \in \alpha.$$
8. Let Θ be the TT in Example 1(3). We have that if $\Upsilon \models \rho$, then
$$\Upsilon \vdash_\Theta M : \alpha \text{ implies } [\![M]\!]_\rho^{\mathscr{D}} \in \alpha.$$

4 Embeddings

In this section we introduce the main tools for transferring results between λ-models leveraging on the algebraic TT structures of meet-semilattice which can be associated to them.

The main property of λ-models which we will transfer is the fundamental property of *sensibility*, namely that unsolvable terms, *i.e.* terms which are not solvable (Definition 2), are all equated in the bottom element. This notion can be expressed also in terms of TTs as follows:

Definition 10 (Sensibility). *A TT is* sensible *if in the generated type assignment system any unsolvable term can be typed only by elements (types) α such that $\alpha \equiv_\Theta \top_\Theta$.*

The definition of sensible model in the literature is weaker, namely that all closed unsolvable terms are equated [9, Definition 4.1.7(ii)]. Our notion amounts to the stronger definition [12, Definition 12.1(ii)(3)] (where it is called order-sensibility). We do not know if there exist filter models which equate all unsolvable terms to an element which is not the bottom.

We immediately have that:

Proposition 1. *A filter model on a TT is sensible iff the TT is sensible.*

Example 5. 1. The TT $\mathscr{S\!A}$, defined in Example 2(1), is sensible, as proved in Theorem 9.

2. The STT $\mathcal{T}_{Scott}$, in Example 3(2), is sensible, see [17].

3. The STT $\mathcal{T}_{CDZ}$, in Example 3(3), is sensible, see [18].

4. The STT $\mathcal{T}_{Park}$, in Example 3(4), is non-sensible, see [17].

5. The STT $\mathcal{T}_{HR}$, in Example 3(5), is non-sensible, see [29].

6. The STT $\mathcal{T}_{AO}$, in Example 3(6), is non-sensible, see [4].

The key notion for relating two TTs is the natural notion of *embedding*.

Definition 11 (Embedding). *Let* $\Theta = \langle \mathscr{D}_\Theta, \sqsubseteq_\Theta \rangle$ *and* $\Theta' = \langle \mathscr{D}_{\Theta'}, \sqsubseteq_{\Theta'} \rangle$ *be* TTs, *then* Θ *is* embeddable *in* Θ' *if there is a morphism* $\kappa : \mathscr{D}_\Theta \to \mathscr{D}_{\Theta'}$ *such that:*

1. $\kappa(\alpha) = \top_{\Theta'}$ *if and only if* $\alpha \equiv_\Theta \top_\Theta$;

2. $\kappa(\alpha \Rightarrow_\Theta \beta) = \kappa(\alpha) \Rightarrow_{\Theta'} \kappa(\beta)$;

3. $\kappa(\alpha \sqcap_\Theta \beta) = \kappa(\alpha) \sqcap_{\Theta'} \kappa(\beta)$;

4. $\alpha \sqsubseteq_\Theta \beta$ *implies* $\kappa(\alpha) \sqsubseteq_{\Theta'} \kappa(\beta)$.

Example 6. 1. The STT $\mathcal{T}_{Scott}$, defined in Example 3(2), can be embedded in the TT $\mathscr{S\!A}$, defined in Example 2(1), by putting $\kappa(\mathsf{U}) = \Lambda$ and $\kappa(\mathsf{c}_{Scott}) = \mathscr{P}$ ($\mathscr{P}$ is defined in Example 2(3)). This embedding agrees with the axiom $\mathsf{c}_{Scott} \sim \mathsf{U} \to \mathsf{c}_{Scott}$, since $\mathscr{P} = \Lambda \Rightarrow \mathscr{P}$ as shown in Proposition 2(3).

2. The STT $\mathcal{T}_{CDZ}$, defined in Example 3(3), can be embedded in the TT $\mathscr{S\!A}$ by putting $\kappa(\mathsf{U}) = \Lambda$ and $\kappa(\mathsf{c}_{CDZ}) = \mathscr{S}$ ($\mathscr{S}$ is defined in Example 2(1)) and $\kappa(\mathsf{d}_{CDZ}) = \mathscr{P}$. This embedding agrees with the axioms, since $\mathscr{S} = \mathscr{P} \Rightarrow \mathscr{S}$ and $\mathscr{P} = \mathscr{S} \Rightarrow \mathscr{P}$ as shown in Proposition 2(2) and (3).

3. The TT $\mathscr{S\!A}$ can be embedded in $\mathcal{T}_{Park}$, defined in Example 3(4), by putting $\kappa(\Lambda) = \mathsf{U}$ and $\kappa(X) = \mathsf{c}_{Park}$ for all saturated sets X.

4. The TT $\mathscr{C}$, defined in Example 2(4), whose elements are β-nice sets of closable terms, can be embedded in $\mathcal{T}_{Park}$ by putting $\kappa(\Lambda) = \mathsf{U}$ and $\kappa(X) = \mathsf{c}_{Park}$.

5. The TT $\mathscr{I}$, defined in Example 2(5), whose elements are β-nice sets of **I**-terms, can be embedded in the STT $\mathcal{T}_{HR}$, defined in Example 3(5), by $\kappa(\Lambda) = \mathsf{U}$ and

$$\kappa(X) = \begin{cases} \mathsf{d}_{HR} & \text{if } X \text{ contains only non-closable terms,} \\ \mathsf{c}_{HR} & \text{otherwise.} \end{cases}$$

Embeddings can transfer both sensibility and non-sensibility properties from one TT to another, as shown in the theorem below.

Theorem 8 (Transfer). *Let* Θ *be* embeddable *in* Θ'. *We get:*

1. if Θ' *is sensible, then* Θ *is sensible.*

2. if Θ *is non-sensible, then* Θ' *is non-sensible.*

Proof. Let $\Upsilon \vdash_\Theta M : \alpha$, where M is unsolvable and $\alpha \not\equiv_\Theta \top_\Theta$. Then $\kappa(\Upsilon) \vdash_{\Theta'} M : \kappa(\alpha)$, where $\kappa(\Upsilon) = \{x : \kappa(\beta) \mid x : \beta \in \Upsilon\}$ by the conditions in Definition 11. Condition (1) of Definition 11 and $\alpha \not\equiv_\Theta \top_\Theta$ imply $\kappa(\alpha) \not\equiv_{\Theta'} \top_{\Theta'}$. Therefore $\langle \Theta', \sqsubseteq_{\Theta'} \rangle$ sensible gives $\langle \Theta, \sqsubseteq_\Theta \rangle$ sensible and $\langle \Theta, \sqsubseteq_\Theta \rangle$ non-sensible gives $\langle \Theta', \sqsubseteq_{\Theta'} \rangle$ non-sensible.

Theorem 9 in Section 5 proves that $\mathscr{S\!\Lambda}$ is sensible. Hence applying the Transfer Theorem to the embeddings of Example 6(1) and (2) we conclude that the STTs $\mathcal{T}_{Scott}$ and $\mathcal{T}_{CDZ}$ are sensible. In the literature the sensibility of Scott models was usually shown using the Approximation Theorem [9, Section 19.2], apart from [39]. Embeddings give an alternative tool, indeed, for proving sensibility and non-sensibility of λ-models.

Sometimes it is useful to require only a weaker condition on the embedding function.

Definition 12 (α-embedding). *Let $\Theta = \langle \mathscr{D}_\Theta, \sqsubseteq_\Theta \rangle$ and $\Theta' = \langle \mathscr{D}_{\Theta'}, \sqsubseteq_{\Theta'} \rangle$ be TTs and $\alpha \in \mathscr{D}_\Theta$ with $\alpha \not\approx_\Theta \top_\Theta$. Then Θ is α-embeddable in Θ' if there is a morphism $\kappa_\alpha : \mathscr{D}_\Theta \to \mathscr{D}_{\Theta'}$ such that:*
1. $\kappa_\alpha(\top_\Theta) = \top_{\Theta'}$;
2. $\kappa_\alpha(\alpha) \not\approx_{\Theta'} \top_{\Theta'}$;
3. $\kappa_\alpha(\beta \Rrightarrow_\Theta \gamma) = \kappa_\alpha(\beta) \Rrightarrow_{\Theta'} \kappa_\alpha(\gamma)$;
4. $\kappa_\alpha(\beta \sqcap_\Theta \gamma) = \kappa_\alpha(\beta) \sqcap_{\Theta'} \kappa\alpha(\gamma)$;
5. $\beta \sqsubseteq_\Theta \gamma$ *implies* $\kappa_\alpha(\beta) \sqsubseteq_{\Theta'} \kappa_\alpha(\gamma)$.

Example 7. The STT $\mathcal{T}_\star$ with the set of constants $\{c, d_1, d_2, d_3\}$, the only axioms $(\to \mathsf{U})$ of Figure 1 and $c \sim ((c \to d_1) \to d_2) \to d_3$ can be d_i-embedded for $i = 2, 3$ in the STT $\mathcal{T}'_\star$ with the same constants, the axioms $c \sim (\mathsf{U} \to d_2) \to d_3$ and $(\to \mathsf{U})$ via the function $\kappa(d_1) = \mathsf{U}$ and the identity otherwise.

We remark that if a term M has type $\alpha \not\approx_\Theta \top_\Theta$ in $\vdash_\Theta$ and Θ is α-*embeddable* in Θ_α, then M has type $\kappa_\alpha(\alpha) \not\approx_{\Theta_\alpha} \top_{\Theta_\alpha}$ in $\vdash_{\Theta_\alpha}$. This fact will be used in the proof of Theorem 19 and in Example 10 which extends Example 7.

5   The Set Theoretic TT $\mathscr{S\!\Lambda}$

The Transfer Theorem 8 of the previous section allows for numberless *relative sensibility* results, but we need some absolute result to start from. We achieve this by means of the fundamental TT $\mathscr{S\!\Lambda}$ defined in Example 2(1), which we study in this section together with the filter structure $\langle \mathcal{F}_{\mathscr{S\!\Lambda}}, \cdot, [\![\ ]\!]^{\mathcal{F}_{\mathscr{S\!\Lambda}}} \rangle$ which it generates. The TT $\mathscr{S\!\Lambda}$, together with Theorem 1, captures the essence of all the techniques which are at the heart of the various *computability arguments*, which have been used for establishing strong normalisation of various typed calculi and untyped calculi equipped with type assignment systems, see [18,26,28,32,37,49]. Namely we have immediately the following crucial result which follows from Theorem 1.

Theorem 9. *The TT $\mathscr{S\!\Lambda} = \langle \mathcal{SAT} \cup \{\Lambda\}, \subseteq \rangle$ is sensible.*

Proof. By Theorem 1 $\{x_i : Y_i \mid 1 \leq i \leq n\} \vdash_{\mathscr{S\!\Lambda}} M : X$ implies
$$M[x_i := N_i \mid 1 \leq i \leq n] \in X$$
if $N_i \in Y_i$ for all i ($1 \leq i \leq n$). Then $X \subsetneq \Lambda$ gives X saturated, *i.e.* M solvable.

Sensibility of many filter models can now be *reduced* to devising appropriate embeddings in $\mathscr{SA}$, as pointed out after Theorem 8. The structure of the lattice TT $\mathscr{SA}$ is very rich indeed and, being a complete lattice, embeddings can be defined also using fixed points as in [22,28,37] and Theorem 18.

The following proposition gives interesting inclusions between sets belonging to $\mathcal{SAT} \cup \Lambda$. We recall that $\mathscr{S}$, $\mathscr{B}$ are defined in Example 2(1) and $\mathscr{P}$ is defined in Example 2(3). We put

Definition 13. *For each solvable term M and saturated set X we introduce the saturated set $X_M \equiv X \cup \{N \mid N =_\beta M\}$.*

Proposition 2. *1. $\Lambda = \Lambda \Rightarrow \Lambda$.*
2. $\mathscr{S} = \mathscr{P} \Rightarrow \mathscr{S}$.
3. $\mathscr{P} = \Lambda \Rightarrow \mathscr{P} = \mathscr{S} \Rightarrow \mathscr{P} \subsetneq \mathscr{B} \Rightarrow \mathscr{P}$.
4. $\mathscr{B} \subsetneq \Lambda \Rightarrow \mathscr{B}$.
5. $\Lambda \Rightarrow \mathscr{B} = \mathscr{S} \Rightarrow \mathscr{B} = \mathscr{P} \Rightarrow \mathscr{B}$.
6. $\mathscr{P} \Rightarrow \mathscr{B} \subseteq \mathscr{P} \subseteq \mathscr{P} \Rightarrow \mathscr{S}$.
7. Let $Y = \{N \mid \forall M \in X.MN \in \mathscr{S}\}$ and $Z = \{P \mid P =_\beta MN \ \& \ M \in X \ \& \ N \in Y\}$, then $X \subseteq Y \Rightarrow Z$.
8. The inclusion in the item (7) may indeed be strict.
9. $\Lambda \Rightarrow \mathscr{B} \subseteq X \Rightarrow Y$.
10. $(X \Rightarrow Y) \cap (X \Rightarrow Z) = X \Rightarrow Y \cap Z$.
11. $(X \Rightarrow Z) \cap (Y \Rightarrow Z) = X \cup Y \Rightarrow Z$.
12. $(X \Rightarrow Y) \cup (X \Rightarrow Z) = X \Rightarrow Y \cup Z$.

Proof. The first two items and Item (7) are straightforward.
(3). Inclusions are straightforward, strictness follows from $\mathbf{I} \in \mathscr{B} \Rightarrow \mathscr{P}$.
(4). Since $\lambda x.y \in \Lambda \Rightarrow \mathscr{B}$.
(5). Any term with more than one abstraction or with a bound head variable cannot be in $\mathscr{P} \Rightarrow \mathscr{B}$. But any term in $\mathscr{B}$ or with just one abstraction and a free head variable belongs to $\Lambda \Rightarrow \mathscr{B}$. Hence
$\Lambda \Rightarrow \mathscr{B} = \mathscr{S} \Rightarrow \mathscr{B} = \mathscr{P} \Rightarrow \mathscr{B} = \{M \mid M \twoheadrightarrow^*_\beta \lambda x.yM_1 \ldots M_n\} \cup \mathscr{B}$.
(6). Item (3) gives $\mathscr{P} = \Lambda \Rightarrow \mathscr{P}$. Then $\mathscr{B} \subseteq \mathscr{P}$, $\mathscr{P} \subseteq \Lambda$ and $\mathscr{P} \subseteq \mathscr{S}$ imply $\Lambda \Rightarrow \mathscr{B} \subseteq \mathscr{P} \subseteq \mathscr{P} \Rightarrow \mathscr{S}$ by the contravariance and covariance of $\Rightarrow$.
(8). Consider $\mathscr{B}_\mathbf{I}$. We have that $\mathscr{S} = \{N \mid \forall M \in \mathscr{B}_\mathbf{I}.MN \in \mathscr{S}\}$ and $\mathscr{S} = \{MN \mid M \in \mathscr{B}_\mathbf{I} \ \& \ N \in \mathscr{S}\}$, but $\mathscr{B}_\mathbf{I} \subsetneq \mathscr{S} \Rightarrow \mathscr{S}$, since $\lambda x.y \in \mathscr{S} \Rightarrow \mathscr{S}$.
(10). $\{M \mid \forall N \in X.MN \in Y\} \cap \{M \mid \forall N \in X.MN \in Z\} =$
$$\{M \mid \forall N \in X.MN \in Y \cap Z\}.$$
(11). $\{M \mid \forall N_1 \in X.MN_1 \in Z\} \cap \{M \mid \forall N_2 \in Y.MN_2 \in Z\} =$
$$\{M \mid \forall N \in X \cup Y.MN \in Z\}.$$
(12). $\{M \mid \forall N \in X.MN \in Y\} \cup \{M \mid \forall N \in X.MN \in Z\} =$
$$\{M \mid \forall N \in X.MN \in Y \cup Z\}.$$

The filter structure generated by $\mathscr{SA}$, namely $\langle \mathcal{F}_{\mathscr{SA}}, \cdot, [\![\ ]\!]^{\mathcal{F}_{\mathscr{SA}}} \rangle$, is intriguing. It sets some of the stepping stones for solving a long standing open problem concerning the theories which can be modelled by sensible λ-models in which all continuous functions are representable, *e.g.* filter models. To the best of our knowledge, it was not known whether sensible filter structures necessarily equate λ-terms with the same Böhm tree, and hence all fixed point combinators. Of course, *e.g.* Park model [42] separates Curry's and Barendregt's fixed point combinators, for a proof see [29], but this model is hopelessly non-sensible, equating every term to an unsolvable term. The following result is therefore remarkable.

Theorem 10. $\langle \mathcal{F}_{\mathscr{SA}}, \cdot, [\![\ ]\!]^{\mathcal{F}_{\mathscr{SA}}} \rangle$ *separates Curry's and Barendregt's fixed point combinators.*

Proof. It is easy to derive $(\mathscr{S} \Rightarrow \mathscr{B}) \Rightarrow \mathscr{B}$ for Curry's fixed point combinator using the subset inclusions proved in Proposition 2(6). Let $\lambda f.\mathbf{HH}M$ be a Barendregt's fixed point combinator, as in Definition 7, such that M is non-closable. Assume by contradiction that $\vdash_{\mathscr{SA}} \lambda f.\mathbf{HH}M : (\mathscr{S} \Rightarrow \mathscr{B}) \Rightarrow \mathscr{B}$. The TT $\mathscr{SA}$ can be embedded in $\mathcal{T}_{Park}$ (defined in Example 3(4)) by putting $\kappa(\mathscr{S}) = \kappa(\mathscr{P}) = \kappa(\mathscr{B}) = \mathsf{c}_{Park}$. We could then derive type c_{Park} for $\lambda f.\mathbf{HH}M$ in the type system over $\mathcal{T}_{Park}$. But this is impossible, since $\lambda f.\mathbf{HH}M$ is a non-closable term which cannot have type c_{Park} from the empty basis, by Theorem 5.

We would be done if $\langle \mathcal{F}_{\mathscr{SA}}, \cdot, [\![\ ]\!]^{\mathcal{F}_{\mathscr{SA}}} \rangle$, which looks similar to many filter structures which are λ-models, were a λ-model. But defying Leibniz, ours *is not the best of all possible worlds*, and for much more dramatic reasons. In effect, also in this case there are "dangerous connections" with union types which imply that the condition of Theorem 3 *fails* in the TT $\mathscr{SA}$.

Theorem 11. $\langle \mathcal{F}_{\mathscr{SA}}, \cdot, [\![\ ]\!]^{\mathcal{F}_{\mathscr{SA}}} \rangle$ *is not a λ-model.*

Proof. Subject Reduction does not hold, since the condition in Theorem 3 fails. Let $\mathbf{0} = \lambda xy.y$, $\mathbf{C}_\star = \lambda xy.yx$ and $\mathbf{D} = \lambda xy.x(y\mathbf{II})(y\mathbf{0})$, then $\mathbf{D}MN \in \mathscr{B}_\mathbf{0}$ when either $M, N \in \mathscr{B}_\mathbf{I}$ or $M, N \in \mathscr{B}_{\mathbf{C}_\star}$. This implies
$$\mathscr{B}_\mathbf{D} \subseteq (\mathscr{B}_\mathbf{I} \Rightarrow \mathscr{B}_\mathbf{I} \Rightarrow \mathscr{B}_\mathbf{0}) \cap (\mathscr{B}_{\mathbf{C}_\star} \Rightarrow \mathscr{B}_{\mathbf{C}_\star} \Rightarrow \mathscr{B}_\mathbf{0}).$$
Using this inclusion we can derive
$$x : \mathscr{B}_\mathbf{D} \vdash_{\mathscr{SA}} \lambda y.xyy : (\mathscr{B}_\mathbf{I} \Rightarrow \mathscr{B}_\mathbf{0}) \cap (\mathscr{B}_{\mathbf{C}_\star} \Rightarrow \mathscr{B}_\mathbf{0}),$$
then by Proposition 2(11) we get $x : \mathscr{B}_\mathbf{D} \vdash_{\mathscr{SA}} \lambda y.xyy : \mathscr{B}_\mathbf{I} \cup \mathscr{B}_{\mathbf{C}_\star} \Rightarrow \mathscr{B}_\mathbf{0}$.
We prove that to derive $x : \mathscr{B}_\mathbf{D}, y : \mathscr{B}_\mathbf{I} \cup \mathscr{B}_{\mathbf{C}_\star} \vdash_{\mathscr{SA}} xyy : \mathscr{B}_\mathbf{0}$ we need
$$\mathscr{B}_\mathbf{D} \subseteq \mathscr{B}_\mathbf{I} \cup \mathscr{B}_{\mathbf{C}_\star} \Rightarrow \mathscr{B}_\mathbf{I} \cup \mathscr{B}_{\mathbf{C}_\star} \Rightarrow \mathscr{B}_\mathbf{0}.$$
By Lemma 1(2) $x : \mathscr{B}_\mathbf{D}, y : \mathscr{B}_\mathbf{I} \cup \mathscr{B}_{\mathbf{C}_\star} \vdash_{\mathscr{SA}} xyy : \mathscr{B}_\mathbf{0}$ implies that for some I, X_i, Y_i, with $i \in I$ we get $x : \mathscr{B}_\mathbf{D}, y : \mathscr{B}_\mathbf{I} \cup \mathscr{B}_{\mathbf{C}_\star} \vdash_{\mathscr{SA}} xy : X_i \Rightarrow Y_i$ and $x : \mathscr{B}_\mathbf{D}, y : \mathscr{B}_\mathbf{I} \cup \mathscr{B}_{\mathbf{C}_\star} \vdash_{\mathscr{SA}} y : X_i$ and $\bigcap_{i \in I} Y_i \subseteq \mathscr{B}_\mathbf{0}$. By Lemma 1(1) $x : \mathscr{B}_\mathbf{D}, y : \mathscr{B}_\mathbf{I} \cup \mathscr{B}_{\mathbf{C}_\star} \vdash_{\mathscr{SA}} y : X_i$ implies $\mathscr{B}_\mathbf{I} \cup \mathscr{B}_{\mathbf{C}_\star} \subseteq X_i$ for all $i \in I$. Then $x : \mathscr{B}_\mathbf{D}, y : \mathscr{B}_\mathbf{I} \cup \mathscr{B}_{\mathbf{C}_\star} \vdash_{\mathscr{SA}} xy : X_i \Rightarrow Y_i$ implies
$$x : \mathscr{B}_\mathbf{D}, y : \mathscr{B}_\mathbf{I} \cup \mathscr{B}_{\mathbf{C}_\star} \vdash_{\mathscr{SA}} xy : \mathscr{B}_\mathbf{I} \cup \mathscr{B}_{\mathbf{C}_\star} \Rightarrow Y_i$$
for all $i \in I$ using Rule ($\subseteq$) together with the contravariance of the arrow, which holds since $\mathscr{SA}$ is $\Rightarrow$-sound (see Example 3(1)). Using Rule ($\cap$) we derive $x : \mathscr{B}_\mathbf{D}, y : \mathscr{B}_\mathbf{I} \cup \mathscr{B}_{\mathbf{C}_\star} \vdash_{\mathscr{SA}} xy : \bigcap_{i \in I}(\mathscr{B}_\mathbf{I} \cup \mathscr{B}_{\mathbf{C}_\star} \Rightarrow Y_i)$, which by Proposition 2(10) implies $x : \mathscr{B}_\mathbf{D}, y : \mathscr{B}_\mathbf{I} \cup \mathscr{B}_{\mathbf{C}_\star} \vdash_{\mathscr{SA}} xy : \mathscr{B}_\mathbf{I} \cup \mathscr{B}_{\mathbf{C}_\star} \Rightarrow \bigcap_{i \in I} Y_i$. Applying Rule ($\subseteq$) with the covariance of the arrow we derive

$x : \mathscr{B}_{\mathbf{D}}, y : \mathscr{B}_{\mathbf{I}} \cup \mathscr{B}_{\mathbf{C}_*} \vdash_{\mathscr{SA}} xy : \mathscr{B}_{\mathbf{I}} \cup \mathscr{B}_{\mathbf{C}_*} \Rightarrow \mathscr{B}_{\mathbf{0}}$. Similarly we can show that $x : \mathscr{B}_{\mathbf{D}}, y : \mathscr{B}_{\mathbf{I}} \cup \mathscr{B}_{\mathbf{C}_*} \vdash_{\mathscr{SA}} xy : \mathscr{B}_{\mathbf{I}} \cup \mathscr{B}_{\mathbf{C}_*} \Rightarrow \mathscr{B}_{\mathbf{0}}$ requires

$$x : \mathscr{B}_{\mathbf{D}}, y : \mathscr{B}_{\mathbf{I}} \cup \mathscr{B}_{\mathbf{C}_*} \vdash_{\mathscr{SA}} x : \mathscr{B}_{\mathbf{I}} \cup \mathscr{B}_{\mathbf{C}_*} \Rightarrow \mathscr{B}_{\mathbf{I}} \cup \mathscr{B}_{\mathbf{C}_*} \Rightarrow \mathscr{B}_{\mathbf{0}}.$$

By Lemma 1(1) we need $\mathscr{B}_{\mathbf{D}} \subseteq \mathscr{B}_{\mathbf{I}} \cup \mathscr{B}_{\mathbf{C}_*} \Rightarrow \mathscr{B}_{\mathbf{I}} \cup \mathscr{B}_{\mathbf{C}_*} \Rightarrow \mathscr{B}_{\mathbf{0}}$, but this inclusion does not hold, since $\mathbf{D} \in \mathscr{B}_{\mathbf{D}}$, $\{\mathbf{I}, \mathbf{C}_*\} \subseteq \mathscr{B}_{\mathbf{I}} \cup \mathscr{B}_{\mathbf{C}_*}$ and $\mathbf{DC_*I} = \mathbf{I} \notin \mathscr{B}_{\mathbf{0}}$. We conclude, since we have established that $x : \mathscr{B}_{\mathbf{D}}, y : \mathscr{B}_{\mathbf{I}} \cup \mathscr{B}_{\mathbf{C}_*} \nvdash_{\mathscr{SA}} xyy : \mathscr{B}_{\mathbf{0}}$.

Remark 1. The proof of Theorem 11 is inspired by the proof that *Subject Reduction fails in a* STT *extended with a union type constructor*, see [8]. The same argument as in Theorem 11 essentially applies to all the TTs which are closed under union, *e.g.* those in Example 1.

Since the filter structure $\langle \mathcal{F}_{\mathscr{SA}}, \cdot, [\![\ ]\!]^{\mathcal{F}_{\mathscr{SA}}} \rangle$ is so fundamental, it would be interesting to study which functions are representable in it. We give some preliminary results. The smallest $\mathscr{SA}$-filter containing X is denoted by $\uparrow^{\mathscr{SA}} X$.

Theorem 12. *1.* $\langle \mathcal{F}_{\mathscr{SA}}, \cdot, [\![\ ]\!]^{\mathcal{F}_{\mathscr{SA}}} \rangle$ *is not extensional. Moreover* $M \in X$ *and* M *closed do not imply* $\vdash_{\mathscr{SA}} M : X$.
2. All constant functions can be represented in $\langle \mathcal{F}_{\mathscr{SA}}, \cdot, [\![\ ]\!]^{\mathcal{F}_{\mathscr{SA}}} \rangle$.
3. Not all step functions can be represented in $\langle \mathcal{F}_{\mathscr{SA}}, \cdot, [\![\ ]\!]^{\mathcal{F}_{\mathscr{SA}}} \rangle$.

Proof. (1). The filters $\uparrow^{\mathscr{SA}} \{\mathscr{B} \Rightarrow \mathscr{B}\}$ and $\uparrow^{\mathscr{SA}} \{\mathscr{B} \Rightarrow \mathscr{B}\} \cup \{\mathscr{B}_{\mathbf{I}}\}$ are different, since $\mathscr{B}_{\mathbf{I}}$ is not functional (see the proof of Proposition 2(8)), yet are extensionally equal. This same argument proves that $\nvdash_{\mathscr{SA}} \mathbf{I} : \mathscr{B}_{\mathbf{I}}$.
(2). By Theorem 16.2.19(ii) of [11] it is enough to show that $\Lambda \Rightarrow X \subseteq Z \Rightarrow Y$ implies $X \subseteq Y$, where Z is arbitrary. Let $N \in X$, then $\mathbf{K}N$ (where $\mathbf{K} = \lambda xy.x$) is in $\Lambda \Rightarrow X$ and hence also in $Z \Rightarrow Y$, but then for any $P \in Z$, we have $\mathbf{K}NP =_\beta N \in Y$.
(3). By Theorem 16.2.19(iii) of [11], since the equalities in Proposition 2(3) and (5) hold. For example $\uparrow^{\mathscr{SA}} (\mathscr{P} \Rightarrow \mathscr{B}) \cdot \uparrow^{\mathscr{SA}} \mathscr{B}_{\mathbf{I}} = \uparrow^{\mathscr{SA}} \mathscr{B}$.

6 STTs Galore

In this section we capitalise on all the results which have been laid down in the previous ones. It consists of two subsections. In the first we discuss separability of fixed point operators in sensible theories, thereby solving in the positive the open question on the existence of a sensible λ-model in which all continuous functions are representable and which separates Böhm trees. In the second subsection we discuss sensibility of models which arise from recursive types where the recurring variable might occur negatively, thereby busting the long standing myth that this would not be the case. In both subsections we make abundant use of the Transfer Theorem 8 in multiple and novel ways.

6.1 Separating Fixed Points, Sensibly

We can now lay down the remaining stepping stones for proving that fixed point operators can be separated in sensible models. Namely, we will *sensibly* separate Curry's fixed point combinator from Barendregt's fixed point combinators,

$\lambda f.\mathbf{H}\mathbf{H}M$, both when M is non-closable, and when M is not an $\mathbf{I}$-term.

First of all making use of Theorem 5 we prove that

Theorem 13. *Let $\mathcal{T}$ be a STT whose only axioms/rules involving U (if any) are those in Figure 1 and which satisfies Rule $(\to)$ of Figure 1. Moreover let M be a closable term and N be a non-closable term. If we can derive $\vdash_{\mathcal{T}} M : A$ where $A \not\approx_{\mathcal{T}} \mathsf{U}$ and U does not occur negatively in A, then M and N cannot be equated in the filter structure over $\mathcal{T}$.*

Proof. Let $\{\mathsf{c}_i \mid i \in I\}$ be the set of constants of $\mathcal{T}$. If $A \not\approx_{\mathcal{T}} \mathsf{U}$ and U does not occur negatively in A, then $A \leq_{\mathcal{T}} B$ for some $B \approx_{\mathcal{T}} \mathsf{U}$ without occurrences of U using Rule $(\to)$. Assume by contradiction that $\vdash_{\mathcal{T}} N : B$. We can embed $\mathcal{T}$ in $\mathcal{T}_{Park}$ defined in Definition 3(4) by taking $\kappa(\mathsf{c}_i) = \mathsf{c}_{Park}$ for $i \in I$. But then we could derive from the empty basis type c_{Park} for N in the type system over $\mathcal{T}_{Park}$. This is impossible by Theorem 5 since N is a non-closable term. $\qquad\blacksquare$

Mimicking the proof of Theorem 10, we can finally prove that:

Theorem 14. *The $\to$-sound STT $\mathcal{T}_\flat = \langle \mathsf{c}, \mathsf{d}, \mathsf{e}, \leq_\flat \rangle$ with the only axioms $\mathsf{c} \to \mathsf{d} \leq \mathsf{c} \leq \mathsf{c} \to \mathsf{e}$ and $\mathsf{d} \leq \mathsf{e}$ is β-sound, sensible and it separates Curry's and non-closable Barendregt's fixed point combinators.*

Proof. The β-soundness of $\mathcal{T}_\flat$ can be easily proved by induction on $\leq_\flat$. The STT $\mathcal{T}_\flat$ can be embedded in $\mathscr{S}\Lambda$ by taking $\kappa(\mathsf{c}) = \mathscr{P}$, $\kappa(\mathsf{d}) = \mathscr{B}$ and $\kappa(\mathsf{e}) = \mathscr{S}$ thanks to the inclusions shown in Proposition 2(6). This implies that $\mathcal{T}_\flat$ is sensible. In the proof of Theorem 10 we derive $(\mathscr{S} \to \mathscr{B}) \to \mathscr{B}$ for Curry's fixed point combinator. This derivation using the inverse mapping κ^{-1} implies $\vdash_{\mathcal{T}_\flat} \lambda f.(\lambda x.f(xx))(\lambda x.f(xx)) : (\mathsf{e} \to \mathsf{d}) \to \mathsf{d}$. Non-closable Barendregt's fixed point combinators do not have this type by Theorem 13. $\qquad\blacksquare$

The above proof of the sensibility of $\mathcal{T}_\flat$ is a clear example of the power of embeddings. A direct proof, possibly using some form of indexed reduction [34], could exist, but it would require a much more involved argument. Sensibility of $\mathscr{S}\Lambda$, on the other hand, is straightforward, being the very essence of the computability argument.

In order to establish the same separability result for Barendregt's fixed point combinators, which are not $\mathbf{I}$-terms, we essentially carry out a similar argument as above, making use of Theorem 6 instead of Theorem 5.

Theorem 15. *If $\{x_i : A_i \mid i \in I\} \vdash_{\mathcal{T}} M : A$ and $\mathcal{T}$ can be embedded in $\mathcal{T}_{\mathcal{HR}}$ via a function κ such that $\kappa(A_i) \sim_{\mathcal{HR}} \kappa(A) \sim_{\mathcal{HR}} \mathsf{c}_{\mathcal{HR}}$ for $i \in I$ and N is not an $\mathbf{I}$-term, then the filter structure over $\mathcal{T}$ separates M and N.*

Proof. If we assume $\{x_i : A_i \mid i \in I\} \vdash_{\mathcal{T}} N : A$ the embedding gives $\{x_i : \mathsf{c}_{\mathcal{HR}} \mid i \in I\} \vdash_{\mathcal{T}_{\mathcal{HR}}} N : \mathsf{c}_{\mathcal{HR}}$, which is impossible by Theorem 6. $\qquad\blacksquare$

Theorem 16. *The STT $\mathcal{T}_\flat$ defined in Theorem 14 separates the application to a free variable of Curry's and Barendregt's fixed point combinators which are not $\mathbf{I}$-terms.*

Proof. Let Γ be the basis which associates type c to all variables which occur free in M and $e \to d$ to the variable z which does not occur in M. As in the proof of Theorem 14 we have that we can derive the type $(e \to d) \to d$ for Curry fixed-point combinator. Then we can derive type d for Curry fixed-point combinator applied to z. Assume that we can derive the type d from Γ for Barendregt's fixed point combinators $\lambda f.\mathbf{HH}M$ applied to z, where M is not an $\mathbf{I}$-term. Embed $\mathcal{T}_\flat$ into $\mathcal{T}_{\mathcal{HR}}$ with the mapping $\kappa(c) = \kappa(e) = c_{\mathcal{HR}}$ and $\kappa(d) = d_{\mathcal{HR}}$. Then in the type system over $\mathcal{T}_{\mathcal{HR}}$ we can derive type $d_{\mathcal{HR}}$ for $\mathbf{H'H'}M$, where $\mathbf{H'} = \lambda xy.z(xxy)$, from the basis which associates type $c_{\mathcal{HR}}$ to all variables which occur free in M and $d_{\mathcal{HR}}$ to the variable z. By the Approximation Theorem [11, Theorem 17.3.17] there must be an approximant of $\mathbf{H'H'}M$ which can be typed by $d_{\mathcal{HR}}$. By definition [11, Definition 17.3.2] the approximants of $\mathbf{H'H'}M$ are the terms $z^n(\Phi\mathbf{H'H'}M)$ with $n \in \mathbb{N}$, where Φ is a constant term typed by $c_{\mathcal{HR}}$. Notice that $c_{\mathcal{HR}} \to c_{\mathcal{HR}} \to c_{\mathcal{HR}} \to c_{\mathcal{HR}}$ is the type $\geq_{\mathcal{HR}} c_{\mathcal{HR}}$ with 3 arrows and less demanding for typing the arguments. Then since $\mathcal{T}_{\mathcal{HR}}$ is β-sound, by the Inversion Lemma (Lemma 1(2)) typability of $\Phi\mathbf{H'H'}M$ requires to derive type $c_{\mathcal{HR}}$ for M from the basis which associates type $c_{\mathcal{HR}}$ to all variables which occur free in M. But this is impossible by Theorem 6.

6.2   Sensible Recursions

The following discussion on the sensibility of STTs with recursive types is a counterpoint to the result for second order λ-calculus in [37]. We consider only recursive types where the intersection constructor cannot occur in the path leading to the recurring constant. This is enough to outplay the prejudice that negative recurrences imply non-sensibility.

We introduce notation to highlight a specific occurrence of the recurring constant:

Definition 14. *1.* $\mathbf{D}^{r_j} \to E$ *is short for* $D_1^j \to \cdots \to D_{r_j}^j \to E$.

$$C_0 = \mathbf{B}^{n_0} \to c$$
$$C_1 = \mathbf{B}^{n_1} \to C_0 \to \mathbf{A}^{p_1} \to d_1$$
$$\cdots$$
$$C_m = \mathbf{B}^{n_m} \to C_{m-1} \to \mathbf{A}^{p_m} \to d_m$$

where the occurrence of c in C_m, which we focus on, is the one in C_0.
2. The characteristic pair *of C_m is* $(\langle n_0, \cdots, n_m \rangle, \langle p_1, \cdots, p_m \rangle)$.
3. The target set *of C_m is* $\{d_1, \cdots, d_m\}$.

Clearly c occurs positively in C_m if m is even and negatively in C_m if m is odd.

We start by showing that a STT in which there exists an equivalence $c \sim C_m$ with m odd and the target set of C_m is a singleton is non-sensible (Theorem 17). This is done by building a λ-term which is unsolvable, but can be typed in the STT under consideration. The constructions of such λ-terms are inspired by the not strongly normalising second order λ-terms in Mendler [37].

Definition 15. *The λ-terms M_i for $0 \leq i \leq 2k+1$ are defined in Figure 2. Notice that the free variable y in M_1 is bound in M_{2k+1}.*

$$M_1 \quad = \quad \lambda\mathbf{x}^{n_1} t\mathbf{v}^{p_1}.t\mathbf{z}^{n_0}\mathbf{z}^{n_{2k+1}} y\mathbf{u}^{p_{2k+1}}$$

$$\cdots$$

$$M_{2h+1} = \quad \lambda\mathbf{x}^{n_{2h+1}} t\mathbf{v}^{p_{2h+1}}.t\mathbf{z}^{n_{2h}} M_{2h-1}\mathbf{u}^{p_{2h}}$$

$$\cdots$$

$$M_{2k+1} = \quad \lambda\mathbf{x}^{n_{2k+1}} y\mathbf{v}^{p_{2k+1}}.y\mathbf{z}^{n_{2k}} M_{2k-1}\mathbf{u}^{p_{2k}}$$

$$M_0 \quad = \quad \lambda\mathbf{x}^{n_0}.M_{2k+1}$$

$$\cdots$$

$$M_{2h} \quad = \quad \lambda\mathbf{x}^{n_{2h}} t\mathbf{v}^{p_{2h}}.t\mathbf{z}^{n_{2h-1}} M_{2h-2}\mathbf{u}^{p_{2h-1}}$$

$$\cdots$$

$$M_{2k} \quad = \quad \lambda\mathbf{x}^{n_{2k}} t\mathbf{v}^{p_{2k}}.t\mathbf{z}^{n_{2k-1}} M_{2k-2}\mathbf{u}^{p_{2k-1}}$$

where $1 \leq h \leq k-1$ and $\mathbf{w}^{r_j}$ denotes the sequence $w_1^j \cdots w_{r_j}^j$ with the appropriate concatenation operator.

Fig. 2: Definition of M_i for $0 \leq i \leq 2k+1$.

Lemma 2. *Let $\mathcal{T}_\clubsuit$ be a STT in which $\mathsf{c} \sim C_{2k+1}$ holds, the target set of C_{2k+1} is a singleton and $(\langle n_0, \cdots, n_{2k+1}\rangle, \langle p_1, \cdots, p_{2k+1}\rangle)$ is the characteristic pair of C_{2k+1}. Let M_m for $0 \leq m \leq 2k+1$ be as in Definition 15, where the y shown in M_1 is free in M_{2h+1} for $0 \leq h \leq k-1$ and bound in M_{2k+1} and M_{2h} for $0 \leq h \leq k$. Define the basis $\Gamma_\clubsuit$ as*

$$\{z_{j_i}^i : B_{j_i}^i, u_{q_l}^l : A_{q_l}^l \mid 0 \leq i \leq 2k+1, 1 \leq j_i \leq n_i, 1 \leq l \leq 2k+1, 1 \leq q_l \leq p_l\}.$$

Then we can derive

1. $\Gamma_\clubsuit, y : C_{2k} \vdash_{\mathcal{T}_\clubsuit} M_{2h+1} : C_{2h+1}$ *for* $0 \leq h \leq k-1$ *and* $\Gamma_\clubsuit \vdash_{\mathcal{T}_\clubsuit} M_{2k+1} : C_{2k+1}$.
2. $\Gamma_\clubsuit \vdash_{\mathcal{T}_\clubsuit} M_{2h} : C_{2h}$ *for* $0 \leq h \leq k$.

Proof. (1). By induction on h.
(2). By induction on h using (1) for the case $h = 0$.

Theorem 17. *The λ-term $M_{2k+1}\mathbf{z}^{n_{2k+1}} M_{2k}\mathbf{u}^{p_{2k+1}}$ is unsolvable and has type d from the basis $\Gamma_\clubsuit$ defined in Lemma 2 in the type system over a STT in which $\mathsf{c} \sim C_{2k+1}$ holds and $\{\mathsf{d}\}$ is the target set of C_{2k+1}.*

Proof. Lemma 2 implies $\Gamma_\clubsuit \vdash_{\mathcal{T}_\clubsuit} M_{2k+1}\mathbf{z}^{n_{2k+1}} M_{2k}\mathbf{u}^{p_{2k+1}} : \mathsf{d}$.
The λ-term $M_{2k+1}\mathbf{z}^{n_{2k+1}} M_{2k}\mathbf{u}^{p_{2k+1}}$ is unsolvable, since it reduces to itself as follows:

$$M_{2k+1}\mathbf{z}^{n_{2k+1}} M_{2k}\mathbf{u}^{p_{2k+1}} \rightarrow_\beta^* M_{2k}\mathbf{z}^{n_{2k}} M_{2k-1}[y := M_{2k}]\mathbf{u}^{p_{2k}} \rightarrow_\beta^*$$

$$M_1[y := M_{2k}]\mathbf{z}^{n_1} M_0\mathbf{u}^{p_1} \rightarrow_\beta^* M_0\mathbf{z}^{n_0}\mathbf{z}^{n_{2k+1}} M_{2k}\mathbf{u}^{p_{2k+1}} \rightarrow_\beta^* M_{2k+1}\mathbf{z}^{n_{2k+1}} M_{2k}\mathbf{u}^{p_{2k+1}}.$$

Example 8. For example a STT with $\mathsf{c} \sim ((\mathsf{c} \rightarrow \mathsf{d}) \rightarrow \mathsf{e} \rightarrow \mathsf{d}) \rightarrow \mathsf{d}$ types the unsolvable term $M_3(\lambda tv.tM_3)$ where $M_3 = \lambda y.y(\lambda t.ty)u$.

In order to discuss conditions for the sensibility of STTs crucial is the restriction to safe STTs defined as follow. Clearly each STT in which the axiom $(\rightarrow \mathsf{U})$ of Figure 1 does not hold is trivially non-sensible.

Definition 16. *A STT is* safe *if it satisfies the axiom $(\rightarrow \mathsf{U})$ and only equivalences of the shape $\mathsf{c} \sim C_m$ for some C_m as defined in Definition 14(1).*

As expected safe STTs where all constants are equivalent to types in which the constants occur positively are sensible. This can be proved by means of a non-constructive embedding in $\mathscr{SA}$ built by solving recursive equations using fixed points of suitable monotone operators in the complete lattice $\mathscr{SA}$.

Theorem 18. *A safe STT is sensible if m is even in all possible equivalences* $\mathsf{c} \sim C_m$.

Proof. Let $\mathcal{T}_\diamond$ be a safe STT such that $\mathsf{c}_i \sim C_{m_i}$ only holds with m_i even for $i \in I$. We restrict to a finite subset I' of I. Let $\{\mathsf{d}_j \mid j \in J\}$ be the set of the remaining constants. That this kind of compactness result is enough for dealing even with infinite sets of equivalences was first noticed by Mendler [37], since all but a number of constants are ever used in any type derivation. Moreover, if an equivalence on a constant is not used in a derivation where that constant appears, then that constant can be safely taken to be equal just to itself. Let $\widehat{}$ be the mapping $\widehat{\mathsf{c}_i} = \phi_i$ for $i \in I'$ and $\widehat{\mathsf{d}_j} = \psi_j$ for $j \in J$. It is standard to verify that the set of equation $\phi_i = \widehat{C_{m_i}}$, $\psi_j = \psi_j$ has a solution in $\mathscr{SA}$, namely Φ_i, Ψ_j for $i \in I'$ and $j \in J$. Then the embedding $\kappa(\mathsf{c}_i) = \Phi_i$, $\kappa(\mathsf{d}_j) = \Psi_j$ for $i \in I'$ and $j \in J$ proves that $\mathcal{T}_\diamond$ is sensible.

Example 9. The need to consider finite sets of types at a time, in building the monotone operator in Theorem 18, is apparent in the STT whose axioms are:
$$\{\mathsf{c}_n \sim \mathsf{d}_n \to \mathsf{c}_{n+1} \to \mathsf{d}_{n+1} \to A, \mathsf{d}_n \sim \mathsf{c}_n \to \mathsf{c}_{n+1} \to \mathsf{d}_{n+1} \to A \mid n \in \mathbb{N}\}.$$
A monotone operator can only be defined piecemeal, if we consider at a time only the finite set of constants with indexes up to the indexes used in a given derivation. Then a "compactness argument" yields sensibility for the whole STT.

The following theorem uses the notion of α-embedding, Definition 12, to show that a safe STT is sensible if all C_m with m odd, occurring in derivable equivalences, have target sets with at least two constants, thus preventing to apply the construction of Theorem 17.

Theorem 19. *A safe STT is sensible if in all possible equivalences,* $\mathsf{c} \sim C_m$ *with m odd, the target sets of C_m are not singletons.*

Proof. Let $\mathcal{T}_\heartsuit$ be a STT satisfying the conditions of the theorem and suppose that $\mathcal{T}_\heartsuit$ is non-sensible. Then there exists M unsolvable and $A \not\approx_{\mathcal{T}_\heartsuit} \mathsf{U}$ such that $\vdash_{\mathcal{T}_\heartsuit} M : A$. Applying M to suitably typed variables to get rid of arrows and using Rule $(\wedge E)$ to get rid of intersections we can find an unsolvable N typed by a constant, let it be d. Let $\mathsf{c}_i \sim C_{m_i}$ for $i \in I$ be the equivalences with m_i odd which hold in $\mathcal{T}_\heartsuit$. For the target set of each C_{m_i} we can find an element $\mathsf{d}_i \neq \mathsf{d}$ with $i \in I$ by the hypothesis that the target set is not a singleton. Putting $\kappa_\mathsf{d}(\mathsf{d}_i) = \mathsf{U}$ for $i \in I$, and κ_d equal to the identity otherwise, we can d-embed $\mathcal{T}_\heartsuit$ in the type theory $\mathcal{T}'_\heartsuit$ obtained from $\mathcal{T}_\heartsuit$ by adding the axiom $\mathsf{d}_i \sim \mathsf{U}$ for $i \in I$. The function κ_d maps C_{m_i} into a type which does not contain occurrence of c_i for $i \in I$. So the STT $\mathcal{T}'_\heartsuit$ satisfies the conditions of Theorem 18, then $\mathcal{T}'_\heartsuit$ is sensible. Notice that $\vdash_{\mathcal{T}'_\heartsuit} N : \mathsf{d}$ by the remark after Definition 12. Contradiction.

Example 10. Theorem 19 implies that the STT $\mathcal{T}_\star$ defined in Example 7 is sensible. Moreover, since $\mathcal{T}_\star$ is β-sound, the filter model over $\mathcal{T}_\star$ is a sensible filter model in which a circular type has a negative occurrence.

The results of Theorems 17, 18 and 19 can be summarised as follows.

Theorem 20. *A safe* STT *is non-sensible if and only if there exists a derivable equivalence* $c \sim C_m$ *with* m *odd where the target set of* C_m *is a singleton.*

The restriction to safe STTs is crucial to get the sharp results of this subsection. Intersections blur the picture. They can either sterilise the negative occurrences as in the equivalence $c \sim c \wedge (c \to c)$, which yields a sensible theory, or not as in the STT $\mathcal{T}_{HR}$ defined in Example 3(5).

7   Related Work and Conclusions

Since Dana Scott and Christopher Strachey, more than half a century ago, introduced denotational semantics [45,48], the semantics of λ-calculus has been a *very hot topic* for quite a few decades. Many of the most established researchers in Theoretical Computer Science, today, have written in their "coming of age" important papers on the subject. Nowadays, the conceptual frameworks, logical formats, methodologies and techniques which were introduced and perfected in the context of pure λ-calculus stand as the foundational tools in many recent research fields of Theoretical Computer Science and Program Logics, ranging from rewriting systems, *e.g.* [24], to session types, *e.g.* [25]. In particular the finitary framework of *intersection type theories* as developed in [1,11,17,30] has become the paradigm for analysing the fine structure of all sorts of mathematical structures arising when different reduction strategies are considered, *e.g.* [41], when complexity or resource awareness type systems are introduced, *e.g.* [13], and when relational, quantitative and metric semantics are taken into account, *e.g.* [5,31,35,40].

The theory of λ-calculus *per se* still provides exciting challenges whose solutions should bear fruits also in the study of the more recent mathematical structures for computation. We think that one such example is the very simple notion of *embedding* (Definition 11), between finitary descriptions of mathematical models of λ-calculus, which we introduce in the present paper. This notion permits us to harness the apparently wild behaviour of non-terminating terms, either unsolvable or just head normalising, *e.g.* fixed point combinators. Embeddings naturally arise once we generalise the notion of STT to the concrete algebraic structures of meet semilattice TT. These can be functorially assigned to applicative structures and, as is the case in algebraic topology, permit to study the properties of the former category of objects, in terms of the properties of the latter. *Embeddings* are the algebraic morphisms between TTs, by means of which we transfer computational properties between the applicative structures. This amounts to the general logical paradigm of proving a property by "reducing" the problem relative to a structure to a problem we have already solved for another structure. Hence we can prove a Transfer Theorem 8, using which we establish a number of results of "relative" sensibility, or "relative" insensibility between models. The Archimedean point, necessary to turn *relative sensibility* into *absolute sensibility*, is the fundamental TT which we call $\mathscr{SA}$. This algebraic structure was always in the background in all *computability arguments* for strong normalisation [18,26,28,32,37,49], but it was never brought to the limelight so explicitly before. The notion of embedding, however, is a tool for transferring also other properties between applicative structure, apart from sensibility or non-sensibility. In Theorem 14 we use it to show that if a type theory $\mathcal{T}$ is embedded in $\mathcal{T}_{Park}$, then $\mathcal{T}$ satisfies the property that "no open term is equated to a closable term" in the filter model $\mathscr{F}_{\mathcal{T}}$ as is the case in $\mathscr{F}_{\mathcal{T}_{Park}}$.

The main results of this paper are the solutions of two long standing open problems in the theory of pure λ-calculus: the existence of a sensible model which separates fixed point combinators and that of a sensible model built over recursive types with negative occurrences. The latter is a fall-out of our, surprisingly diverging, extension to *sensible* theories in untyped λ-calculus of Mendler's analysis of normalising theories in 2nd order λ-calculus.

We conclude this paper addressing a few issues not considered so far and we raise some conjectures. We hope that this paper might foster future research.

Curry's and non-closable Barendregt's fixed-point combinators have the same Böhm tree, because all free variables are "pushed to infinity" in the sense of [14]. Clearly their *Ohana* trees [14] are different, since such trees remember all variables that are hidden in meaningless subtrees or pushed to infinity. Ohana trees give an interesting theory of the $\lambda\mathbf{I}$-calculus. We conjecture that there exists a filter model whose theory is the equality of Ohana trees, thus answering the question raised in [14]. We also conjecture that, along the lines of Theorem 14, Böhm's, Currys and Turings fixed-point combinators can be separated from Barendregt's, Klop's and Polonsky's fixed-point combinators [36], as well as Barendregt's, Klop's and Polonsky's fixed-point combinators among themselves. In all these models, however, the interpretation of Turings fixed-point combinator should coincide with that of Currys fixed-point combinator. Separating the two needs a new idea, but we are confident that it can be done.

In our investigation on recursive types we did not consider intersection because, as we have shown at the end of Section 6, it has a mysterious impact on the sensibility of STTs. We leave the reader with the following puzzle.

Let $A = \mathsf{d}_1 \wedge \mathsf{d}_2 \wedge (\mathsf{d}_1 \to \mathsf{d}_2 \to \mathsf{e}) \to \mathsf{e}$. Is the STT with the axioms $\mathsf{c} \sim \mathsf{c} \wedge A \to \mathsf{f}$ and $(\to \mathsf{U})$ sensible? We can derive $\vdash \Omega_2 : \mathsf{c}$ and $\vdash \Omega_3 : A$, but $\nvdash \Omega_2\Omega_3 : \mathsf{f}$, where $\Omega_2 = \lambda x.xx$ and $\Omega_3 = \lambda x.xxx$. However, if we put $A \leq \mathsf{c}$, then $\mathsf{c} \sim A \to \mathsf{f}$ and so the negative occurrence is not sterilised, since we prove $\vdash \Omega_2\Omega_3 : \mathsf{f}$. Moreover notice that if we put $\mathsf{c} \leq A$, then $\mathsf{c} \sim \mathsf{c} \to \mathsf{f}$ and $\vdash \Omega_2\Omega_2 : \mathsf{f}$.

Acknowledgments We are deeply grateful to all three referees for their valuable comments and useful suggestions. Taking them into account significantly improved the quality and clarity of our paper.

References

1. Abramsky, S.: Domain theory in logical form. Annals of Pure and Applied Logic **51**(1-2), 1–77 (1991). https://doi.org/10.1016/0168-0072(91)90065-T
2. Abramsky, S., Ong, C.L.: Full abstraction in the lazy lambda calculus. Information and Computation **105**(2), 159–267 (1993). https://doi.org/10.1006/INCO.1993.1044
3. Alessi, F.: An irregular filter model. Theoretical Computer Science **398**(1-3), 129–149 (2008). https://doi.org/10.1016/J.TCS.2008.01.047
4. Alessi, F., Barbanera, F., Dezani-Ciancaglini, M.: Intersection types and lambda models. Theoretical Computer Science **355**(2), 108–126 (2006). https://doi.org/10.1016/j.tcs.2006.01.004
5. Alves, S., Kesner, D., Ramos, M.: Extending the quantitative pattern-matching paradigm. In: Kiselyov, O. (ed.) APLAS. LNCS, vol. 15194, pp. 84–105. Springer (2024). https://doi.org/10.1007/978-981-97-8943-6$_5$
6. Azevedo de Amorim, A., Gaboardi, M., Hsu, J., Katsumata, S.: Metric semantics for probabilistic relational reasoning. CoRR **abs/1807.05091** (2018), `http://arxiv.org/abs/1807.05091`
7. Antonelli, M., Dal Lago, U., Pistone, P.: Towards logical foundations for probabilistic computation. Annals of Pure and Applied Logic **175**(9), 103341 (2024). https://doi.org/10.1016/J.APAL.2023.103341
8. Barbanera, F., Dezani-Ciancaglini, M., de' Liguoro, U.: Intersection and union types: syntax and semantics. Information and Computation **119**, 202–230 (1995). https://doi.org/10.1006/INCO.1995.1086
9. Barendregt, H.: The Lambda Calculus - its Syntax and Semantics, Studies in logic and the foundations of mathematics, vol. 103. North-Holland (1985), `https://api.pageplace.de/preview/DT0400.9780080933757_A23543814/preview-9780080933757_A23543814.pdf`
10. Barendregt, H., Coppo, M., Dezani-Ciancaglini, M.: A filter lambda model and the completeness of type assignment. Journal of Symbolic Logic **48**(4), 931–940 (1983). https://doi.org/10.2307/2273659
11. Barendregt, H., Dekkers, W., Statman, R.: Lambda Calculus with Types. Cambridge University Press (2013). https://doi.org/10.1017/CBO9781139032636
12. Barendregt, H., Manzonetto, G.: A Lambda Calculus Satellite. College Publications (2022), `https://www.collegepublications.co.uk/logic/mlf/?00035`
13. Bucciarelli, A., Kesner, D., Ventura, D.: Non-idempotent intersection types for the lambda-calculus. Logic Journal of the IGPL **25**, 433–464 (2017). https://doi.org/10.1093/jigpal/jzx018
14. Cerda, R., Manzonetto, G., Saurin, A.: Ohana trees and Taylor expansion for the λI-calculus: No variable gets left behind or forgotten! In: Fernández, M. (ed.) FSCD. LIPIcs, vol. 337, pp. 12:1–12:20. Schloss Dagstuhl (2025). https://doi.org/10.4230/LIPICS.FSCD.2025.12
15. Church, A.: A formulation of the simple theory of types. Journal of Symbolic Logic **5**, 56 – 68 (1940). https://doi.org/10.2307/2266170
16. Coppo, M., Dezani-Ciancaglini, M.: An extension of the basic functionality theory for the λ-calculus. Notre Dame Journal of Formal Logic **21**(4), 685–693 (1980). https://doi.org/10.1305/ndjfl/1093883253
17. Coppo, M., Dezani-Ciancaglini, M., Honsell, F., Longo, G.: Extended type structures and filter lambda models. In: Lolli, G., Longo, G., Marcja, A. (eds.) Logic Colloquium'82, Studies in Logic and the Foundations of Mathematics, vol. 112, pp. 241–262. Elsevier (1984). https://doi.org/10.1016/S0049-237X(08)71819-6

18. Coppo, M., Dezani-Ciancaglini, M., Zacchi, M.: Type theories, normal forms and D_∞-lambda-models. Information and Computation **72**(2), 85–116 (1987). https://doi.org/10.1016/0890-5401(87)90042-3

19. Coquand, T., Huet, G.P.: The calculus of constructions. Information and Computation **76**(2/3), 95–120 (1988). https://doi.org/10.1016/0890-5401(88)90005-3

20. Curry, H.B.: Functionality in combinatory logic. Proceedings of the National Academy of Science of the USA **20**, 584–590 (1934). https://doi.org/10.1073/pnas.20.11.584

21. Dezani-Ciancaglini, M., Giannini, P., Honsell, F.: Unsolvable terms in filter models. In: Fernández, M. (ed.) FSCD. LIPIcs, vol. 337, pp. 3:1–3:24. Schloss Dagstuhl (2025). https://doi.org/10.4230/LIPICS.FSCD.2025.3

22. Dezani-Ciancaglini, M., Honsell, F., Motohama, Y.: Compositional characterisations of lambda-terms using intersection types. Theoretical Computer Science **340**(3), 459–495 (2005). https://doi.org/10.1016/J.TCS.2005.03.011

23. Espírito Santo, J., Kesner, D., Peyrot, L.: A faithful and quantitative notion of distant reduction for the lambda-calculus with generalized applications. Logical Methods in Computer Science **20**(3) (2024). https://doi.org/10.46298/LMCS-20(3:10)2024

24. Fernández, M., Nantes-Sobrinho, D., Santaguida, D.: A completion procedure for equational rewriting systems with binders. In: Escobar, S., Titolo, L. (eds.) LOPSTR. LNCS, vol. 16117, pp. 94–112. Springer (2025). https://doi.org/10.1007/978-3-032-04848-6_6

25. Gay, S.J., Vasconcelos, V.T.: Session Types. Cambridge University Press (2025). https://doi.org/10.1017/9781009000062

26. Girard, J.Y.: Une extension de l'interpretation de Gödel a l'analyse, et son application a l'elimination des coupures dans l'analyse et la theorie des types. Studies in Logic and the Foundations of Mathematics **63**, 63–92 (1971). https://doi.org/10.1016/S0049-237X(08)70843-7

27. Hindley, R., Longo, G.: Lambda calculus models and extensionality. Mathematical Logic Quarterly **26**(19-21), 289–310 (1980). https://doi.org/10.1002/malq.19800261902

28. Honsell, F., Lenisa, M.: Semantical analysis of perpetual strategies in lambda-calculus. Theoretical Computer Science **212**(1-2), 183–209 (1999). https://doi.org/10.1016/S0304-3975(98)00140-6

29. Honsell, F., Ronchi Della Rocca, S.: An approximation theorem for topological lambda models and the topological incompleteness of lambda calculus. Journal of Computer and System Sciences **45**(1), 49–75 (1992). https://doi.org/10.1016/0022-0000(92)90040-P

30. Jacobs, B., Margaria, I., Zacchi, M.: Filter models with polymorphic types. Theoretical Computer Science **95**(1), 143–148 (1992). https://doi.org/10.1016/0304-3975(92)90070-V

31. Kerinec, A., Manzonetto, G., Ronchi Della Rocca, S.: Call-by-value, again! In: Kobayashi, N. (ed.) FSCD. LIPIcs, vol. 195, pp. 7:1–7:18. Schloss Dagstuhl (2021). https://doi.org/10.4230/LIPICS.FSCD.2021.7

32. Krivine, J.L.: Lambda-calculus, types and models. Ellis Horwood (1993), `https://www.irif.fr/~krivine/articles/Lambda.pdf`, translated from the 1990 French original by René Cori

33. Landin, P.J.: A correspondence between ALGOL 60 and Church's lambda notation: Part II. Communications of the ACM **8**, 89–101,158–165 (1965). https://doi.org/10.1145/363791.363804

34. Lévy, J.: An algebraic interpretation of the lambda beta-calculus and a labeled lambda-calculus. In: Böhm, C. (ed.) Lambda-Calculus and Computer Science Theory. LNCS, vol. 37, pp. 147–165. Springer (1975). https://doi.org/10.1007/BFB0029523
35. Manzonetto, G., Pagani, M., Ronchi Della Rocca, S.: New semantical insights into call-by-value λ-calculus. Fundamenta Informaticae **170**(1-3), 241–265 (2019). https://doi.org/10.3233/FI-2019-1862
36. Manzonetto, G., Polonsky, A., Saurin, A., Simonsen, J.G.: The fixed point property and a technique to harness double fixed point combinators. Journal of Logic and Computation **29**(5), 831–880 (2019). https://doi.org/10.1093/LOGCOM/EXZ013
37. Mendler, N.P.: Inductive types and type constraints in the second-order lambda calculus. Annals of Pure and Applied Logic **51**(1-2), 159–172 (1991). https://doi.org/10.1016/0168-0072(91)90069-X
38. Milner, R.: A theory of type polymorphism in programming. Journal of Computer and System Sciences **17**, 348–375 (1978). https://doi.org/10.1016/0022-0000(78)90014-4
39. Mosses, P.D., Plotkin, G.D.: On proving limiting completeness. SIAM Journal of Computing **16**(1), 179–194 (1987). https://doi.org/10.1137/0216015
40. Paolini, L., Piccolo, M., Ronchi Della Rocca, S.: Essential and relational models. Mathematical Structures in Computer Science **27**(5), 626–650 (2017). https://doi.org/10.1017/S0960129515000316
41. Paolini, L., Pimentel, E., Ronchi Della Rocca, S.: Strong normalization from an unusual point of view. Theoretical Computer Science **412**(20), 1903–1915 (2011). https://doi.org/10.1016/J.TCS.2010.12.018
42. Park, D.: The Y-combinator in Scott's lambda-calculus models (1976), `http://wrap.warwick.ac.uk/46310/`, Theory of Computation Report, University of Warwick. Department of Computer Science
43. Plotkin, G.D.: Call-by-name, call-by-value and the λ-calculus. Theoretical Computer Science **1**(2), 125–159 (1975). https://doi.org/10.1016/0304-3975(75)90017-1
44. Rocq (2025), `https://github.com/rocq-prover/rocq/releases/tag/V9.0.0`
45. Scott, D.S.: Continuous lattices. In: Lawvere, F.W. (ed.) Toposes, Algebraic Geometry and Logic. LNM, vol. 274, pp. 97–136. Springer (1972). https://doi.org/10.1007/BFb0073967
46. Scott, D.S.: Domains for denotational semantics. In: Nielsen, M., Schmidt, E.M. (eds.) ICALP. LNCS, vol. 140, pp. 577–613. Springer (1982). https://doi.org/10.1007/BFB0012801
47. Scott, D.S.: Lectures on a mathematical theory of computation. In: Broy, M., Schmidt, G. (eds.) Theoretical Foundations of Programming Methodology, NATO Advanced Study Institutes, vol. 91, pp. 145–292. Springer (1982). https://doi.org/10.1007/978-94-009-7893-5$_9$
48. Scott, D.S., Strachey, C.: Toward a mathematical semantics for computer languages. Tech. Rep. PRG06, OUCL (1971), `https://www.cs.cmu.edu/~crary/819-f09/Scott71.pdf`
49. Tait, W.W.: Intensional interpretations of functionals of finite type I. The Journal of Symbolic Logic **32**, 198–212 (1967). https://doi.org/10.2307/2271658
50. Wegner, P.: The Vienna definition language. ACM Comput.ing Surveys **4**(1), 563 (1972). https://doi.org/10.1145/356596.356598

Complexity of Model Checking Second-Order Hyperproperties on Finite Structures[*]

Bernd Finkbeiner[1], Hadar Frenkel[2], and Tim Rohde[3]

[1] CISPA Helmholtz Center for Information Security, Saarbrücken, Germany
[2] Bar Ilan University, Ramat Gan, Israel
[3] Saarland University, Saarbrücken, Germany
finkbeiner@cispa.de, hadar.frenkel@biu.ac.il,
tiro00001@stud.uni-saarland.de

Abstract. We study the model checking problem of Hyper^2LTL over finite structures. Hyper^2LTL is a second-order hyperlogic, that extends the well-studied logic HyperLTL by adding quantification over sets of traces, to express complex hyperproperties such as epistemic and asynchronous hyperproperties. While Hyper^2LTL is very expressive, its expressiveness comes with a price, and its general model checking problem is undecidable. This motivates us to study the model checking problem for Hyper^2LTL over finite structures – tree-shaped or acyclic graphs, which are particularly useful for monitoring purposes. We show that Hyper^2LTL model checking is decidable on finite structures: It is in PSPACE (in the size of the model) on tree-shaped models and in EX-PSPACE on acyclic models. Additionally, we show that for an expressive fragment of Hyper^2LTL, namely the Fixpoint $\text{Hyper}^2\text{LTL}_{fp}$ fragment, the model checking problem is much simpler and is P-complete on tree-shaped models and EXP-complete on acyclic models. Last, we present some preliminary results that take into account not only the size of the model, but also the formula size.

1 Introduction

Hyperproperties [12] generalize trace properties to reason about sets of sets of traces. Hyperproperties capture a wide range of properties, from information-flow and security policies [22], to complex epistemic properties such as knowledge and common knowledge [15]. HyperLTL [11] is the prominent logic to capture temporal hyperproperties, and it extends LTL (Linear Temporal Logic [35]) by adding first-order quantification over trace variables. This way, HyperLTL can express many important information-flow and security-related properties, as well as knowledge in multi-agent systems. However, first-order quantification over traces is not enough to express all properties of interest. Two main examples that HyperLTL cannot express are common knowledge [8] and asynchronous hyperproperties [23,9,2]. Many efforts have been made in recent years to capture

---

[*] This work was supported in part by the Israel Science Foundation (ISF grant No. 655/25) and by the European Union (ERC Grant HYPER, No. 101055412).

N. Bertrand and S. Milius (Eds.): FoSSaCS 2026, LNCS 16503, pp. 285–306, 2026.
https://doi.org/10.1007/978-3-032-22730-0_14

such complex hyperproperties, demonstrated by the range of logics suggested to express knowledge and common knowledge [8,24,15,32], and even more so, the wide range of logics for asynchronous hyperproperties [1,2,3,6,9,10,23,27,28,29].

Recently, Beutner et al. [4] observed that to express such complex hyperproperties, a second-order quantification, that is, quantification over *sets of traces*, is needed. To this end, [4] suggest the logic $Hyper^2LTL$, which extends Hyper-LTL with second-order quantification, and is able to express a wide range of hyperproperties such as common knowledge, asynchronous hyperproperties, and Mazurkiewicz trace theory [13]. Yet, the expressiveness of $Hyper^2LTL$ comes with a price, and its model checking problem is in general undecidable [4]. To handle $Hyper^2LTL$ model checking algorithmically, [4] identify a fragment of $Hyper^2LTL$, namely Fixpoint $Hyper^2LTL_{fp}$, for which the model checking problem is still undecidable, but due to its definition using a fixpoint construction, it allows to compute sound over- and under-approximations of the second order sets. Using these approximations, [4] provides a sound but necessarily incomplete model checking algorithm for an expressive fragment of $Hyper^2LTL$.

In this work, motivated by the inherent undecidability of the logic, but also by practical applications such as monitoring [5], we focus on the complexity of model checking second-order hyperproperties over finite structures. In particular, we analyze the complexity for two types of data structures: tree-shaped and acyclic models, both for the full $Hyper^2LTL$ and for the Fixpoint $Hyper^2LTL_{fp}$ fragment. This study completes the complexity analysis picture for $Hyper^2LTL$: The complexity of HyperLTL has been studied for general structures [19] and for finite structures [7]; for second-order hyperproperties, Frenkel and Zimmermann [20] provide a detailed complexity analysis for $Hyper^2LTL$ and its fragments over general structures, and show that the problem is highly undecidable. However, no analysis is done for finite structures. Here, we address this gap, and show that when restricting the system to finite structures, the model checking problem becomes decidable, and its complexity drops significantly.

This study is also motivated by the complexity of the monitoring problem, particularly for second-order hyperproperties. In monitoring (in contrast to model checking), we do not have access to the whole system, but only to its executions. Based on the observed executions, we wish to decide whether the property holds or is violated in the (unknown) system. Since we can never observe infinite executions, the monitoring problem is defined over finite traces. A (sequential) monitor for hyperproperties uses a finite structure to store the set of traces seen so far, and repeatedly model checks this growing set against the specification. In [5], Beutner et al. address the monitoring problem of second-order hyperproperties by defining the problem and suggesting monitoring algorithms. However, they do not address the complexity aspect.

The data structures we study here – tree-shaped and acyclic structures – are used in practical monitoring tools, such as [17,5]; the trace logs seen in the monitoring process may be stored in the form of a simple linear collection of the traces seen so far or, for space efficiency, organized by common prefixes into a tree-shaped structure, or by both prefixes and suffixes into an acyclic structure.

	Tree-shaped	Acyclic
Fixpoint Hyper^{2}LTL$_{fp}$	P-complete (Thm. 2)	EXP-complete (Thm. 4)
Hyper^{2}LTL $(\exists\forall)^k$	Σ^p_{k+1}-complete (Thm. 5)	Σ^{EXP}_{k+1}-complete (Thm. 7)
Hyper^{2}LTL $(\forall\exists)^k$	Π^p_{k+1}-complete (Thm. 5)	Π^{EXP}_{k+1}-complete (Thm. 7)
Hyper^{2}LTL	PSPACE (Cor. 4)	EXPSPACE (Cor. 6)

Table 1: Complexity of Hyper^{2}LTL model checking in the size of the structure. $(\exists\forall)^k$ and $(\forall\exists)^k$ denote formulas with k second-order quantifier alternations.

In this work, we mostly analyze the complexity in terms of the size of the model, again motivated by applications such as monitoring and bounded model checking [26], where the model grows iteratively, yet the specification remains unchanged. Therefore, all complexities given in this work are in the size of the model, except when mentioned otherwise. We summarize the results of our complexity analysis in Table 1. Note that our upper bound proofs in fact also provide concrete model checking algorithms for the respective problem. In general, our results show that the model checking problem is exponentially easier on tree-shaped models than on general acyclic models, which conforms with the fact that trees have exponentially less traces, and are a less efficient representation of trace-sets. The more interesting observation following our results, is that model checking Fixpoint Hyper^{2}LTL$_{fp}$ formulas is much easier than model checking general Hyper^{2}LTL formulas. Our results demonstrate that the advantages of the fixpoint fragment are not limited only to infinite traces, but also appear on finite models, making the logic very appealing for finite settings, such monitoring and bounded model checking. Since the reductions we provide are very technical, we use the body of the paper to present the main parts of our constructions, and refer the reader to the full version of this paper [16] for the full technical details and proofs.

Related Work For LTL, Kuhtz and Finkbeiner [30] studied the impact of structural restrictions on the complexity of LTL model checking, and Fionda and Greco [18] studied model checking complexity for different fragments of LTL over finite traces. For HyperLTL, Bonakdarpour and Finkbeiner [7] provided a complexity analysis over finite structures, and showed that the complexity of HyperLTL model checking (generally NONELEMENTARY) drops significantly when the model is restricted to be acyclic or tree-shaped.

Over general structures, Frenkel and Zimmermann [20] showed that model checking Hyper^{2}LTL is highly undecidable (equivalent to truth in 3rd order arithmetic), but when restricting to the Fixpoint Hyper^{2}LTL$_{fp}$ fragment, the complexity reduces significantly, to Σ^1_1 (yet remains undecidable). Attempts to identify easier fragments for model checking were made in the context of epistemic and asynchronous logics subsumed by Hyper^{2}LTL, either by restricting the formula, or the model structure: The logic LTL$_{K,C}$ extends LTL with epistemic operators to reason about knowledge and common knowledge. While in general its model checking problem is undecidable, it becomes decidable when removing either the *until* temporal operator, or the *common knowledge* epistemic operator [33]. Baumeister et al. [2] identified syntactic fragments of A-HLTL, an asynchronous extension of HyperLTL, for which model checking is decidable. Hsu et al. [25] showed that A-HLTL model checking is decidable over acyclic graphs.

Our work studies Hyper^{2}LTL, that subsumes all of the above, and concerns not only decidability, but completes the picture by providing a detailed complexity analysis for Hyper^{2}LTL and its fragments, over finite structures.

2   Preliminaries

Kripke Structures Let AP be a finite set of atomic propositions. A *Kripke structure* is a tuple $K = (S, s_0, \delta, L)$ such that S is a finite set of states; $s_0 \in S$ is the initial state; $\delta \subseteq S \times S$ is the transition relation; and $L : S \mapsto 2^{AP}$ is a labeling function. If $(s, s') \in \delta$, we say that s' is a *successor* of s. We require that $\forall s \in S.\exists s' \in S.(s, s') \in \delta$. That is, every state has at least one successor. Given a Kripke structure K, its underlying graph is the directed graph with the set S as vertices, and edges defined by the transition relation δ.[4]

- A Kripke structure is *acyclic* if the underlying graph (S, δ) is acyclic, except for states that have no outgoing transitions other than a self-loop. That is, if $(s, s) \in \delta$ then there is no $s' \neq s \in S$ such that $(s, s') \in \delta$.
- A Kripke structure is *tree-shaped* if: (1) for every state $s' \neq s_0$ there is exactly one state $s \neq s'$ such that $(s, s') \in \delta$; (2) there is no s such that $(s, s_0) \in \delta$; and (3) $(s, s) \in \delta$ iff there is no $s' \neq s$ such that $(s, s') \in \delta$.

A *trace* $t \in (2^{AP})^\omega$ is an infinite sequence over the alphabet 2^{AP}. We denote by $t[i]$ the i-th position of t, and by $t[i, \infty]$ the suffix of t starting at position i. Here, we only consider traces of the form $t = t[0] \cdots t[k]t[k+1] \cdots$ s.t. $\forall i \geq k.t[i] = t[i+1]$ for some $k \in \mathbb{N}$. The smallest such k is the *length* of t. We denote the set of all traces of a Kripke structure K by *Traces(K)*.

Hyper^{2}LTL Hyper^{2}LTL [4] extends the logic HyperLTL [11] by second-order quantification, that is, quantification over sets of traces. Let $\mathcal{V}$ be a finite set of trace variables, $\mathfrak{V}$ be a finite set of trace-set variables containing a special variable $\mathfrak{G}$ (which is reserved to denote the set of all traces we reason about), and let AP be a finite set of atomic propositions. Hyper^{2}LTL formulas are defined using the following grammar, for $\pi \in \mathcal{V}, X \in \mathfrak{V}, a \in AP$.

$$\psi ::= a_\pi \mid \neg\psi \mid \psi \vee \psi \mid \psi \, \mathcal{U} \, \psi \mid \bigcirc \psi$$
$$\varphi ::= \forall X.\varphi \mid \exists X.\varphi \mid \forall \pi \in X.\varphi \mid \exists \pi \in X.\varphi \mid \psi$$

where $\mathcal{U}, \bigcirc$ are the temporal operators *until* and *next* (their semantics is given below). We also allow the usual Boolean derivations *true, false*, $\wedge, \rightarrow, \leftrightarrow$; and the temporal derivations $\Diamond \psi \equiv true \, \mathcal{U} \, \psi$ (*eventually*), and $\Box \psi \equiv \neg \Diamond \neg \psi$ (*globally*).

---

[4] While the data structures we study yield finite traces, to stick with the infinite trace semantics of Hyper^{2}LTL, and with the usual definition of Kripke structures, we use infinite traces with repeating last state to represent finite traces, similar to the choice of [7], instead of defining finite trace semantics as done e.g. for LTL in [31,14].

Semantics: Given a set of traces T, we interpret Hyper²LTL formulas w.r.t. a trace assignment $\Pi : \mathcal{V} \rightharpoonup T$, and a trace-set assignment $\Delta : \mathfrak{V} \rightharpoonup 2^T$ (both are partial functions). For the trace assignment Π, we write $\Pi[i, \infty]$ for the assignment $\Pi'(\pi) = \Pi(\pi)[i, \infty]$, and $\Pi[\pi \mapsto t]$ for the assignment that maps π to t, and maps every $\pi' \neq \pi$ to $\Pi(\pi')$. Similarly, for the trace-set assignment Δ, we write $\Delta[X \mapsto A]$ for the assignment that maps X to A, and every trace-set variable $Y \neq X$ to $\Delta(Y)$. The semantics of Hyper²LTL are defined over the relation $\models_T$, w.r.t. the set T of traces, and the assignments Π and Δ, as follows.

$$
\begin{aligned}
&\Pi, \Delta \models_T a_\pi &&\text{iff } a \in \Pi(\pi)[0] \\
&\Pi, \Delta \models_T \neg\psi &&\text{iff } \Pi, \Delta \not\models_T \psi \\
&\Pi, \Delta \models_T \psi_1 \vee \psi_2 &&\text{iff } \Pi, \Delta \models_T \psi_1 \text{ or } \Pi, \Delta \models_T \psi_2 \\
&\Pi, \Delta \models_T \bigcirc\psi &&\text{iff } \Pi[1, \infty], \Delta \models_T \psi \\
&\Pi, \Delta \models_T \psi_1 \,\mathcal{U}\, \psi_2 &&\text{iff } \exists i \geq 0 : \Pi[i, \infty], \Delta \models_T \psi_2, \text{ and} \\
&&&\quad \forall 0 \leq j < i : \Pi[j, \infty], \Delta \models_T \psi_1 \\
&\Pi, \Delta \models_T \forall\pi \in X.\varphi &&\text{iff for all } t \in \Delta(X) : \Pi[\pi \mapsto t], \Delta \models_T \varphi \\
&\Pi, \Delta \models_T \exists\pi \in X.\varphi &&\text{iff there exists } t \in \Delta(X) : \Pi[\pi \mapsto t], \Delta \models_T \varphi \\
&\Pi, \Delta \models_T \forall X.\varphi &&\text{iff for all } A \subseteq T : \Pi, \Delta[X \mapsto A] \models_T \varphi \\
&\Pi, \Delta \models_T \exists X.\varphi &&\text{iff there exists } A \subseteq T : \Pi, \Delta[X \mapsto A] \models_T \varphi
\end{aligned}
$$

When the set of traces T is clear from the context, we use $\models$ instead of $\models_T$. We say that a Kripke structure K with traces $Traces(K)$ satisfies Hyper²LTL formula φ iff $\emptyset, [\mathfrak{G} \mapsto Traces(K)] \models_{Traces(K)} \varphi$, where $\emptyset$ stands for the empty trace assignment, and $[\mathfrak{G} \mapsto Traces(K)]$ is the trace-set assignment that only maps the special trace-set variable $\mathfrak{G}$ to $Traces(K)$, and non of the other trace-set variables. This is since we are only interested in satisfaction of Hyper²LTL formulas with no free variables (other than $\mathfrak{G}$). In this case, we write $K \models \varphi$.[5]

Second-Order Quantifier Alternations The second-order quantifier alternations of a Hyper²LTL formula φ, is the number of times φ alternates from universal to existential second-order quantification or vice versa. All first-order quantifications are ignored. For example, the formula $\exists A.\forall\pi \in A.\exists B.\forall C.\exists D.\psi$, where ψ is quantifier-free, has two second-order quantifier alternations: one from existential to universal set quantification, and one back to existential quantification.

We denote Hyper²LTL formulas with an outermost second-order existential quantifier by Σ-Hyper²LTL. Such formulas that have k second-order quantifier alternations, we denote by Σ_k-Hyper²LTL. Similarly, Π-Hyper²LTL is the fragment of Hyper²LTL formulas with outermost second-order universal quantifier,

---

⁵ The semantics we use slightly differs from the original Hyper²LTL semantics introduced in [4], which uses an additional special trace-set variable $\mathfrak{U}$ to represent the set of all possible traces (not restricted only to the given model). Since we analyze the model checking problem for finite models, semantics that reason over all possible traces are not suited for our approach, and so we do not allow the use of this variable. This matches the closed-world semantics defined by [20].

and Π_k-Hyper^{2}LTL are such formulas with k alternations. We say that a formula is *quantifier free* if it contains neither first-order nor second-order quantifiers.

Syntactic Sugar Before we proceed to the definition of the Fixpoint Hyper^{2}LTL$_{fp}$ fragment, we define some abbreviations we will use in the remainder of the paper.

- $\pi =_{AP} \pi'$ stands for $\bigwedge_{a \in AP} \Box(a_\pi \leftrightarrow a_{\pi'})$ denoting equality between π and π'.
- $\exists! \pi \in X. \varphi(\pi)$ stands for $\exists \pi \in X. \forall \pi' \in X. \varphi(\pi) \wedge (\varphi(\pi') \rightarrow \pi =_{AP} \pi')$, i.e., there exists exactly one trace for which φ holds.
- $\pi \triangleright X$ stands for $\exists \pi' \in X. \pi' =_{AP} \pi$, i.e., π is a member of the set X. If $\triangleright$ is not under the scope of temporal operators, we can bring the formula into a proper Hyper^{2}LTL syntax.
- A trace t *encodes a (binary) number b with proposition p* if: p holds in the i-th step of t iff b has a 1 at position i. For traces t, t' of equal length, the formula $\Box(p_t \leftrightarrow p_{t'})$ holds iff t and t' encode the same number b with proposition p. All our encodings are in binary, and we start with the least significant bit.
- A trace π is *marked* with some atomic proposition a if π satisfies $\Diamond a_\pi$.

Additionally to this syntactic sugar, we often give our formulas with quantifiers under boolean operators. It improves readability and allows us to nicely split formulas into subformulas. Syntactically, this is not permitted but such a formula can be easily transformed into valid Hyper^{2}LTL syntax by pulling all quantifiers to the front. Note that such a transformation is not possible if a quantifier appears under a temporal operator.

Fixpoint Hyper^{2}LTL Along with Hyper^{2}LTL, [4] introduces two sub-fragments: Hyper^{2}LTL$_{fp}$ and Fixpoint Hyper^{2}LTL$_{fp}$. Hyper^{2}LTL$_{fp}$ restricts second-order quantification to largest or smallest sets that satisfy a given formula. Fixpoint Hyper^{2}LTL$_{fp}$ further restricts second-order quantification so that the quantified sets are uniquely defined by least fixpoints. This makes it more applicable to model checking, as demonstrated by [4], that proposed a partial model checking algorithm for the Fixpoint Hyper^{2}LTL$_{fp}$ fragment, based on over- and under-approximations of the fixpoints.

In this work, we analyze the complexities of the full Hyper^{2}LTL logic, and of the Fixpoint Hyper^{2}LTL$_{fp}$ fragment, but not the complexity of the intermediate Hyper^{2}LTL$_{fp}$ fragment. We expect that the model checking problem for Hyper^{2}LTL$_{fp}$ is of similar complexity as for Hyper^{2}LTL. This is since Hyper^2-LTL$_{fp}$ restricts second-order quantification, but neither gives a constructive way to compute all quantified sets nor restricts the number of sets that a formula can quantify over significantly. This is also demonstrated in the results of [20], where, over general (not finite) structures, Hyper^{2}LTL and Hyper^{2}LTL$_{fp}$ are at the same level of the arithmetical hierarchy, whereas the Fixpoint Hyper^{2}LTL$_{fp}$ fragment is much less complex (see Table 1 of [20]). In contrast to Hyper^{2}LTL$_{fp}$, Fixpoint Hyper^{2}LTL$_{fp}$ restricts the number of sets that a formula can quantify over to be exactly one and gives a constructive way to compute this set. This restrictive quantification is still expressive enough to express complex properties such as common knowledge, and yields an undecidable logic for the general case.

The syntax of Fixpoint Hyper^{2}LTL$_{fp}$ is defined as follows:

$$\psi ::= a_\pi \mid \neg\psi \mid \psi \vee \psi \mid \psi\,\mathcal{U}\,\psi \mid \bigcirc\psi$$
$$\varphi ::= (X, \curlyvee, \varphi_{fp}).\varphi \mid \forall\pi \in X.\varphi \mid \exists\pi \in X.\varphi \mid \psi$$

Where φ_{fp} is a conjunction of formulas of the form: $\forall\pi_1 \in X_1.\ldots.\forall\pi_n \in X_n.\psi_{step} \rightarrow \pi_M \rhd X$, where $X_1, \ldots, X_n$ are previously quantified trace-set variables, or X itself; ψ_{step} is a quantifier free formula; and $1 \leq M \leq n$. The semantics of Fixpoint Hyper^{2}LTL$_{fp}$ are the same as the semantics of Hyper^{2}LTL, except for second-order quantification, which is defined as follows.

$$\Pi, \Delta \models_T (X, \curlyvee, \varphi_{fp}).\varphi \text{ iff there exists } A \in sol(\Pi, \Delta, (X, \curlyvee, \varphi_{fp})) \text{ s.t.}$$
$$\Pi, \Delta[X \mapsto A] \models_T \varphi$$

where sol function returns the smallest sets (w.r.t. to set inclusion) satisfying φ_{fp}:

$$sol(\Pi, \Delta, (X, \curlyvee, \varphi_{fp})) :=$$
$$\{A \subseteq T \mid \Pi, \Delta[X \mapsto A] \models_T \varphi_{fp} \wedge \forall A' \subsetneq A.\Pi, \Delta[X \mapsto A'] \not\models_T \varphi_{fp}\}$$

Due to its definition using the ψ_{step} formulas, each fixpoint iteration adds a unique set of traces to the constructed set. The resulting fixpoint is therefore unique, i.e., sol consists of only one set ([20] Sec.7). Note that this defines a *least* fixpoint, but for conciseness, and since we only analyze least fixpoints in this work, we neglect the *least* from the name of the fragment.[6] As discussed in [4], the fixpoint fragment is very appealing as, together with the fact that it is expressive enough to express properties such as common knowledge and asynchronous hyperproperties, it also allows (over general structures) to compute sound approximations of the second-order sets, thus providing a partial model checking algorithm for the fragment. In this work, we establish the efficiency of the fixpoint fragment also on finite structures, making it even more appealing.

Example 1. The Fixpoint Hyper^{2}LTL$_{fp}$ formula $\varphi = (X, \curlyvee, \forall\pi \in X.\forall\pi' \in \mathfrak{G}.(\Box a_{\pi'} \vee \Box(a_\pi \leftrightarrow a_{\pi'}) \vee \Box(b_\pi \leftrightarrow b_{\pi'})) \rightarrow \pi' \rhd X).\forall\pi \in X.\neg b_\pi$ defines a set X that initially contains all traces satisfying $\Box a$. With each fixpoint iteration, a trace π' is added to X if there is a trace $\pi \in X$ which globally agrees with π' on a or on b. See Fig. 1 for concrete traces. The formula φ can be used to reason about common knowledge in multi agent systems, as shown in [4].

Problem Statement The model checking problem is to decide, given a Kripke structure K and a Hyper^{2}LTL formula φ, whether $K \models \varphi$. We study the following model checking (MC) problems. If not stated otherwise, we study the *data complexity*, i.e., the complexities are w.r.t. the size of K and not the size of φ.

- MC[FIXPOINT, TREE]: K is tree-shaped and φ is in Fixpoint Hyper^{2}LTL$_{fp}$.
- MC[FIXPOINT, ACYCLIC]: K is acyclic and φ is in Fixpoint Hyper^{2}LTL$_{fp}$.

---

[6] In [4], it is termed the least fixpoint fragment. A largest fixpoint fragment is not defined in [4]. Still, the analysis for a largest fixpoint fragment should be dual.

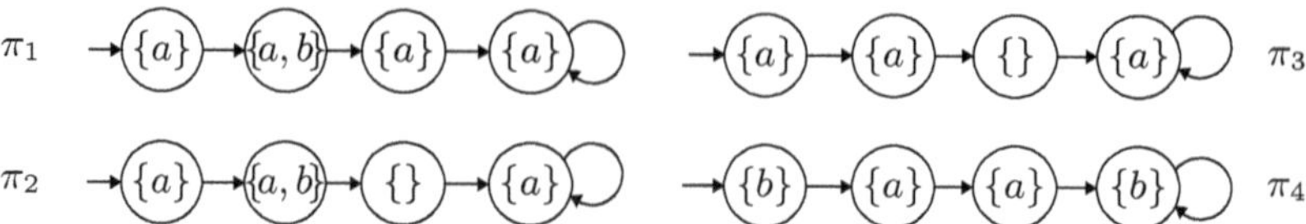

Fig. 1: The formula from Example 1 is satisfied by the above set of traces $T = \{\pi_1, \pi_2, \pi_3, \pi_4\}$. The set X initially contains π_1, π_2 is added as it agrees with π_1 on b, and π_3 is added since it agrees with π_2 on a. π_4 does not agree with any other trace on any proposition, thus $X = \{\pi_1, \pi_2, \pi_3\}$.

- MC[Hyper^{2}LTL, Tree] : K is tree-shaped and φ is not further restricted.
- MC[Hyper^{2}LTL, Acyclic] : K is acyclic and φ is not further restricted.
- For the full logic, we also define the following problems, where struct can be Tree or acyclic, and the structure K is according to struct:
  - MC[Σ_k-Hyper^{2}LTL, struct]: φ is in Σ_k-Hyper^{2}LTL.
  - MC[Π_k-Hyper^{2}LTL, struct]: φ is in Π_k-Hyper^{2}LTL.

3   P-completeness of MC[Fixpoint, Tree]

In this section, we show that MC[Fixpoint, Tree] is P-complete. We start by proving that it is in P. Then we give a reduction in logarithmic space from the Horn-satisfiability problem [34] to show that it is P-hard. All reductions we present in the paper are polynomial time.

Observation 1 *A tree-shaped Kripke structure with n states has at most n traces. In addition, each leaf of the tree corresponds to exactly one trace.*

Lemma 1 ([7]). *Let K be a tree-shaped Kripke structure. Under a fixed trace assignment, a quantifier-free HyperLTL formula ψ can be evaluated on K in polynomial time in the size of K.*

Corollary 1. *Let K be a tree-shaped Kripke structure. Under a fixed trace and trace-set assignments, a quantifier-free Hyper^{2}LTL formula ψ can be evaluated on K in polynomial time in the size of K.*

Corollary 1 follows from the fact that given a concrete trace-set assignment, verifying Hyper^{2}LTL reduces to verifying HyperLTL formulas.

Since we can represent a finite set of traces T as a tree with $|T|$ branches, we get:

Corollary 2. *Let T be a finite set of traces. Under a fixed trace assignment, a quantifier-free Hyper^{2}LTL formula ψ can be evaluated on T in polynomial time in the size of T.*

Lemma 2. MC[Fixpoint, Tree] *is in P.*

Proof. Let φ be a fixed Fixpoint Hyper^{2}LTL$_{fp}$ formula and let K be a tree-shaped Kripke structure. We recursively evaluate φ over K as follows.

- Every second-order (fixpoint) quantifier $(X, \Upsilon, \varphi_{fp}).\varphi'$ quantifies over exactly one set X, which we iteratively construct by following the fixpoint formula φ_{fp}. We can keep track of X by marking leaves in the tree that correspond to traces in X. We evaluate the inner formula φ' under the current instantiation of X.
- For every first-order quantifier $\mathbb{Q}\pi \in X.\varphi'$, we iterate over every possible instantiation of π and evaluate if φ' holds for the current instantiation.

We prove that this process is polynomial in the size of the structure, by induction over Fixpoint Hyper^{2}LTL$_{fp}$ formulas:

- If φ is quantifier-free, it can be evaluated in polynomial time under fixed trace and trace-set assignments (Cor. 1).
- For $\varphi = \forall\pi \in X.\varphi'$ or $\varphi = \exists\pi \in X.\varphi'$: By induction hypothesis, φ' can be evaluated in polynomial time under a fixed trace assignment. The number of possible assignments for π is polynomial in the size of the Kripke structure (due to Obs. 1), and therefore the satisfaction of first-order quantifiers can be evaluated in polynomial time in K.
- For $\varphi = (X, \Upsilon, \varphi_{fp}).\varphi'$: We only have restricted fixpoint sets, which are uniquely defined by the inner formula φ_{fp}. Each set can contain at most polynomially many traces (Obs. 1), so we need at most polynomially many fixpoint iterations to evaluate it. Each such iteration can be done in polynomial time, by the induction hypothesis. $\qquad\square$

3.1 P-hardness of MC[FIXPOINT, TREE]

To show P-hardness we reduce from the Horn satisfiability problem.

Definition 1 (Horn-satisfiability [34]). *Let $L = \{x_1, \ldots, x_k\}$ be a set of k variables, and let $h = c_1 \wedge c_2 \wedge \ldots \wedge c_n$ be a formula consisting of a conjunction of n Horn clauses, each of them of the form $c_i = (\neg l_1 \vee \neg l_2 \vee l_3)$, where $l_1, l_2, l_3 \in L \cup \{\top, \bot\}$. The Horn-satisfiability problem is to find a Boolean assignment for the variables $x_1, \ldots, x_k$ such that h is satisfied.*

Theorem 2. MC[FIXPOINT, TREE] *is P-complete.*

Proof. Containment in P follows from Lemma 2. We show P-hardness by a logspace and polynomial time reduction from the Horn Satisfiability problem, which is P-hard [34]. Given a Horn formula h, we construct a tree structure K and a Fixpoint Hyper^{2}LTL$_{fp}$ formula φ (which does not depend on h) such that $K \models \varphi$ iff h is true. We hereby describe our construction and refer to the full version [16] for its full correctness proof. Here, and in the following proofs throughout the paper, when we use **encode** we mean **encode in binary** (as we stated in Sec. 2, all our encodings are in binary). Intuitively, we build a tree structure K that has, for each variable x_i, one branch representing a positive assignment to x_i, and one branch representing a negative assignment to x_i. Additionally, K has one branch for every clause c_i. For simplicity, we add variables x_{k+1} and x_{k+2} to represent $\top$ and $\bot$, and restrict the valid assignments such that x_{k+1} has

294   Finkbeiner, Frenkel, and Rohde

to be assigned true and x_{k+2} has to be assigned false. On every branch of K, we encode the index of the corresponding variable(s) with atomic propositions pos, neg_1, neg_2.

The Fixpoint Hyper^{2}LTL$_{fp}$ formula constructs a set A representing an assignment to h. Intuitively, if A contains the positive trace of a variable x_i, it means that the corresponding assignment assigns x_i with true; and if A contains the negative trace of x_i, then x_i is assigned with false. We define A iteratively by adding an assignment trace to A only if there is a clause that cannot be satisfied if that variable would be assigned to the other value.

We now describe our reduction formally. Let $h = c_1 \wedge c_2 \wedge \ldots \wedge c_n$ be a Horn formula with k variables. We build a Kripke structure $K = (S, s_0, \delta, L)$ whose underlying graph is a rooted tree. We refer to Fig. 2 for an example. The state-space S consists of: the initial state s_0, $2 \cdot \lceil \log_2(k + 3) \rceil$ states for each variable (including $\top, \bot$), and $\lceil \log_2(k + 3) \rceil$ states for each clause in h. We denote the states corresponding to a variable x_i by $s_{i,1}, s_{i,2}, \cdots$ and $s'_{i,1}, s'_{i,2}, \cdots$. The states corresponding to the i-th clause of h are denoted $t_{i,1}, t_{i,2}, \cdots$. The relation δ is:

$$\delta(s_0) = \{s_{i,1} \mid i \in [1, k]\} \cup \{s'_{i,1} \mid i \in [1, k]\} \cup \{t_{j,1} \mid j \in [1, n]\}$$

$$\delta(\sigma_{i,m}) = \begin{cases} \{\sigma_{i,m+1}\} & \text{if } m < \lceil \log_2(k + 3) \rceil \\ \{\sigma_{i,m}\} & \text{otherwise} \end{cases} \quad \forall \sigma \in \{s, s', t\}$$

That is, there is a transition from s_0 to the states $s_{i,1}, s'_{i,1}$ for $i \in [1, k]$ and $t_{j,1}$ for $j \in [1, n]$, and there are branches of the type s, s', t for each i, j. Note that each trace can contain states only of one of the types s, s', or t, as this is a tree-shaped structure. The set AP is $\{pos, neg_1, neg_2, c, a\}$ where c denotes clause branches, and a denotes the branches corresponding to $\top$ and $\bot$. Traces consisting of states $s_{i,m}$ encode i with pos, and traces consisting of states $s'_{i,m}$ encode i with neg_1. A trace consisting of the states $t_{i,m}$ representing the i-th clause $(\neg x_d \vee \neg x_e \vee x_f)$ encodes: d with neg_1, e with neg_2, and f with pos. This results in a tree-shaped Kripke structure: For each variable x_i, we have two branches representing a positive and negative assignment to x_i, and for each clause we have a branch labeled with c. The branches representing correct assignment to $\top$ and $\bot$ are marked with a. The Fixpoint Hyper^{2}LTL$_{fp}$ formula φ, for which we model check K, is: $\varphi := (A, \curlyvee, \varphi_{fp} \wedge \varphi'_{fp}).\forall \pi \in A.\forall \pi' \in A.\neg\Box(pos_\pi \leftrightarrow neg_{1,\pi'})$. Where $\varphi'_{fp} = \forall \pi \in \mathfrak{G}. \bigcirc a_\pi \to \pi \triangleright A$, and φ_{fp} is defined as follows:

$$\forall \pi \in \mathfrak{G}.\forall \alpha \in A.\forall \alpha' \in A.\forall \beta \in \mathfrak{G}.\bigcirc c_\pi \wedge \neg\bigcirc c_\alpha \wedge \neg\bigcirc c_{\alpha'} \wedge \neg\bigcirc c_\beta \tag{1}$$

$$\wedge\, (\Box(pos_\alpha \leftrightarrow neg_{1,\pi}) \wedge \Box(pos_{\alpha'} \leftrightarrow neg_{2,\pi}) \wedge \Box(pos_\beta \leftrightarrow pos_\pi) \tag{2}$$

$$\vee\, \Box(pos_\alpha \leftrightarrow neg_{1,\pi}) \wedge \Box(neg_{1,\alpha'} \leftrightarrow pos_\pi) \wedge \Box(neg_{1,\beta} \leftrightarrow neg_{2,\pi}) \tag{3}$$

$$\vee\, \Box(pos_\alpha \leftrightarrow neg_{2,\pi}) \wedge \Box(neg_{1,\alpha'} \leftrightarrow pos_\pi) \wedge \Box(neg_{1,\beta} \leftrightarrow neg_{1,\pi})) \to \beta \triangleright A \tag{4}$$

φ quantifies over the smallest set A, representing the current assignment to h, and is only satisfied if there is no variable for which both the positive the negative traces are in A. φ_{fp} ensures that a trace β is only added to A (line 4, right) if there exist assignment traces α, α' which are already in A and a clause

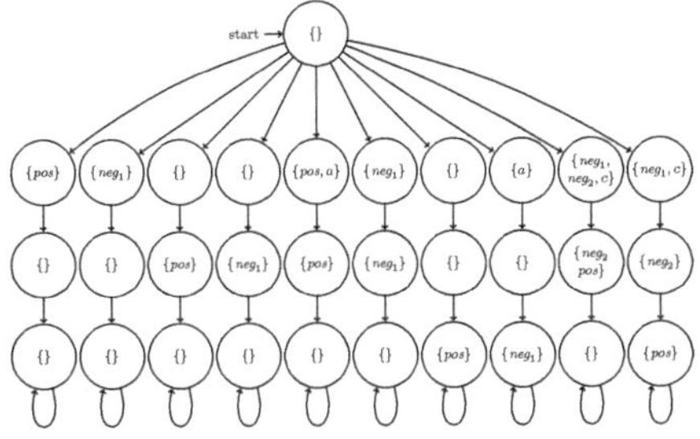

Fig. 2: The tree-structure for the Horn-formula $(\neg x_1 \vee \neg \top \vee x_2) \wedge (\neg x_1 \vee \neg x_2 \vee \bot)$. The two rightmost branches represent the two clauses. The rest of the branches (from left to right) represent positive and negative values for $x_1, x_2, \top$ (i.e., x_3), and $\bot$ (i.e., x_4).

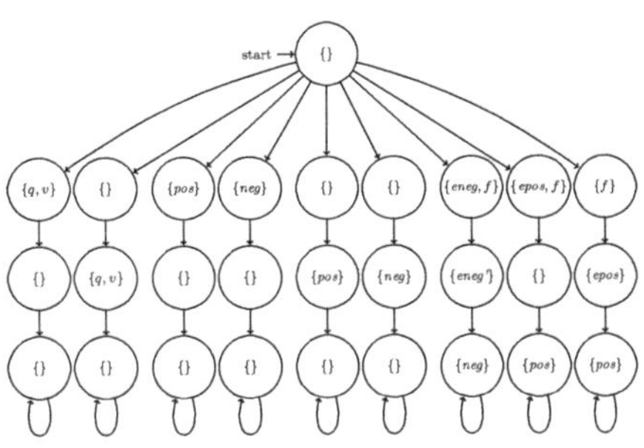

Fig. 3: Kripke structure built for the QBF formula $\forall x_1.\exists x_2.x_1 \vee x_2$

trace π (line 1), which satisfy the condition in lines 2-4. This condition (lines 2-4) holds exactly if α and α' represent an assignment to two of the variables in the clause c represented by π, such that c can only be satisfied if the third variable, represented by β, is assigned according to β. This is the case only if α and α' represent a positive assignment to both negative variables and β represents a positive assignment to the positive variable (line 2) or α represents a positive assignment to one of the negative variables and α' represents a negative assignment to the positive variable and β represents a negative assignment to the other negative variable (lines 3-4). Then h is satisfiable iff φ holds in K, and since K is polynomial in the size of h, we conclude the proof. $\square$

4   EXP-completeness of MC[FIXPOINT, ACYCLIC]

In this section, we show that MC[FIXPOINT, ACYCLIC] is EXP-complete. The full proofs for Lemma 3 and Thm. 4 can be found in the full version [16].

Observation 3 *An acyclic Kripke structure with n states has at most 2^n traces.*

Lemma 3. MC[FIXPOINT, ACYCLIC] *is in EXP.*

Proof (Sketch). Let K be an acyclic Kripke structure. We can unroll K to a tree-structure T, in which the initial state branches to the 2^n traces of K (cf. Obs. 3): the initial state of T has 2^n transitions, and each branch is a linear branch corresponding to one trace of K. We can then apply the algorithm for MC[FIXPOINT, TREE] which is polynomial in the size of T, and exponential in the size of K. $\square$

Theorem 4. MC[FIXPOINT, ACYCLIC] *is EXP-complete.*

Proof (Sketch). We provide a polynomial time reduction from the Succinct Circuit Value problem [34]. We only provide intuition here. For the full definition of the Succinct Circuit Value problem, and for the full proof, see the full version [16].

In the Succinct Circuit Value problem we are given a succinct Boolean representation C_S of a Boolean circuit C. The circuit C has no input gate and a single output gate, and the decision problem is to decide, given the succinct representation C_S, whether the output gate of C is evaluated to true. The inputs to the circuit C_S are two binary numbers i and j, and its outputs are two numbers q and r, where: i and q are indices of gates in C such that q is the j-th neighbor of gate i; r represents which kind of gate i is. This results in an encoding where every input-output pair of C_S represents exactly one edge in C.

Now, given a succinct circuit C_S, we build a Fixpoint Hyper^{2}LTL$_{fp}$ formula φ (that does not depend on C or C_S) and a Kripke structure K such that $K \models \varphi$ if and only if C evaluates to true. The reduction consists of three phases, A, B, and C, constructing subformulas that define sets X, Y, and Z. The first two phases collect the output of C_S for every input, and the last phase solves the Circuit Value problem on C. At each phase, we describe both subformulas and substructures, and explain how they correspond. However, the construction of the formulas and the construction of the structures are independent.

Phase A starts by evaluating C_S under all possible inputs. For that, our fixpoint formula defines a set X containing traces that represent the values for all edges in C_S for all possible inputs. For each edge (u, v) in C_S we add a substructure in K, in which for every possible input e to C_S there are two traces encoding e and (u, v), one is labeled *pos* and one is labeled *neg*. Both traces additionally encode all relevant information about (u, v) such as the gates that it connects and their kinds. The formula φ then defines a set X such that a trace labeled with *pos*, encoding input e and representing edge (u, v) of C_S is contained in X if (u, v) carries positive value when e is given as input to C_S. Similarly, a trace labeled with *neg*, encoding input e and representing edge (u, v) of C_S is contained in X if (u, v) carries negative value when e is given as input to C_S. Thus, X captures the complete state of C_S under each possible input.

In phase B, we build a set Y that contains, for each input, a trace encoding the input and the corresponding output of C_S. To do so, we add a substructure to K containing branches encoding arbitrary values with *inp* and *outp*, and can be labeled with *incomplete* at some point. A trace encoding e with *inp*, o with *outp* and that visits a state labeled with *incomplete* at the $k + 1$-st state, encodes that for input e, the first k output bits of C_S are the first k bits of o. The constructed Fixpoint Hyper^{2}LTL$_{fp}$ formula incrementally adds traces whose *incomplete* label appears later: Let π be a trace encoding e with *inp*, o with *outp* and visiting a state labeled with *incomplete* at step $k + 1$. We add π to Y if there exists a different trace $\pi' \in Y$ that visits a state labeled with *incomplete* at the k-th step and encoding the same o in the first $k - 1$ steps and the same e as π. Additionally, the k-th output bit on π has to be the k-th output bit of C_S on input e. Therefore, we additionally require that there is a trace in X that indicates that the k-th output gate of $\mathcal{C}_S$ for input e has the same value as the k-th bit encoded with o on π. Complete input-output pairs of $\mathcal{C}_S$ can be found on all traces in Y where the *incomplete* mark is pushed the furthest.

In phase C, which is almost analogous to phase A, we are left to evaluate the circuit $\mathcal{C}$. We add another substructure to the structure K, annotating traces from Y with *pos* and *neg*. The fixpoint formula defines a set Z representing the state of $\mathcal{C}$. It adds a trace of Y representing an edge (u, v) of $\mathcal{C}$ and labeled with *pos* if the edge carries a positive value in $\mathcal{C}$, and the trace labeled with *neg* otherwise. As $\mathcal{C}$ has no inputs, we do not encode an input to $\mathcal{C}$ on the traces.

The final, second-order quantifier free part of the constructed Fixpoint Hyper^2-LTL$_{fp}$ formula, is satisfied if and only if there exists a trace in Z indicating that the edge leaving the output gate of $\mathcal{C}$ carries a positive value. $\qquad\square$

5 The Complexity of MC[HYPER^{2}LTL, TREE]

We now turn to study the complexity of the model checking problem for the full logic Hyper^{2}LTL, starting with the analysis of tree-shaped models.

Theorem 5. *The problem of* MC[Σ_k-HYPER^{2}LTL, TREE] *is* Σ^p_{k+1}*-complete, and the problem of* MC[Π_k-HYPER^{2}LTL, TREE] Π^p_{k+1}*-complete.*

We prove Thm. 5 by proving a sequence of claims: we show the Σ^p_{k+1} (resp. Π^p_{k+1}) completeness by first showing containment in the respective class. Then we prove hardness in these classes by reducing from the QBF problem. Lemma 4 below reasons about a general trace-set T. We will use it also in section 6, to prove containment in Σ^{EXP}_{k+1} and Π^{EXP}_{k+1} .

Lemma 4. *For assignments Δ, Π, a set of traces T, and a formula φ, we have that $\Pi, \Delta \models_T \varphi$ can be decided in Σ^p_{k+1} in the size of T if φ is in Σ_k-Hyper^{2}LTL, and it can be decided in Π^p_{k+1} in the size of T if φ is in Π_k-Hyper^{2}LTL.*

Observation 6 *For each formula φ we can define a corresponding formula $\hat{\varphi}$ where each quantifier is converted into its dual and the innermost quantifier free formula is negated. The formula $\hat{\varphi}$ can be constructed from φ in polynomial time, and it has the same number of (second-order) quantifier alternations as φ. The outermost second-order quantifier in $\hat{\varphi}$ is the dual to the outermost second-order quantifier in φ: If $\Pi, \Delta \models \hat{\varphi}$ can be decided in some complexity class C, then $\Pi, \Delta \models \varphi$ can be decided in co-C and vice versa.*

Proof (of Lemma 4). By induction on the number k of quantifier alternations:
Base Case $k = 0$: φ contains only either existential or universal second-order quantifiers. We first prove, by structural induction over φ, that if φ only contains existential second-order quantifiers, then $\Pi, \Delta \models \varphi$ can be decided in Σ^p_1. Our structural induction hypothesis is that for all Π and Δ, the model checking problem $\Pi, \Delta \models \varphi'$ can be decided in Σ^p_1.
  − Base Case: For a quantifier-free formula φ, model checking can be decided in polynomial time (Cor. 2).

- If $\varphi = \forall \pi \in X.\varphi'$ or $\varphi = \exists \pi \in X.\varphi'$: By induction hypothesis, for any t we can decide in Σ_1^p whether $\Pi[\pi \mapsto t], \Delta \models \varphi'$ holds. The number of traces in $\Delta(X)$ is at most linear in $|T|$. Evaluating $\Pi[\pi \mapsto t], \Delta \models \varphi'$ for all possible values of t is therefore still in Σ_1^p.

- If $\varphi = \exists X.\varphi'$: We need to show that $\Pi, \Delta \models \varphi$ can be decided in Σ_1^p. By induction hypothesis we know that, if A is given, then $\Pi, \Delta[X \mapsto A] \models \varphi'$ can be decided in Σ_1^p. If we first non-deterministically guess A by guessing for each of the linearly many traces whether it is in A and then evaluate $\Pi, \Delta[X \mapsto A] \models \varphi'$ we can decide $\Pi, \Delta \models \varphi$ in Σ_1^p.

This concludes the analysis for formulas only containing existential second-order quantifiers. If φ only contains universal second-order quantifiers, we can use the dual formula $\hat{\varphi}$ which only contains existential second-order quantifiers, and $\Pi, \Delta \models \hat{\varphi}$ can be decided in Σ_1^p. By Obs. 6 we have that $\Pi, \Delta \models \varphi$ can be decided in Π_1^p. This concludes the base case.

Induction Step: We need to show that if φ contains $k+1$ second-order quantifier alternations, then $\Pi, \Delta \models \varphi$ can be decided in Σ_{k+2}^p (resp. Π_{k+2}^p) in the size of T. By induction hypothesis we know that for any Δ, Π and φ' with k second-order quantifier alternations $\Pi, \Delta \models \varphi'$ can be decided in Σ_{k+1}^p (resp. Π_{k+1}^p) in the size of T. Again, we first prove the claim for a formula φ whose outermost second-order quantifier is existential, by structural induction over φ. We have two induction hypotheses here: First, the model checking problem for Hyper^{2}LTL with k second-order quantifier alternations is in Σ_{k+1}^p (resp. Π_{k+1}^p). Second, for all Π and Δ the model checking problem $\Pi, \Delta \models \varphi'$ can be decided in Σ_{k+2}^p.

- φ is quantifier-free: it can be decided in polynomial time in $|T|$ (Cor. 2).

- If $\varphi = \forall \pi \in X.\varphi'$ or $\varphi = \exists \pi \in X.\varphi'$: By induction hypothesis, for any t we can decide in Σ_{k+2}^p whether $\Pi[\pi \mapsto t], \Delta \models \varphi'$ holds. The number of traces in $\Delta(X)$ is at most the size of T. Evaluating $\Pi[\pi \mapsto t], \Delta \models \varphi'$ for all possible values of t is therefore still in Σ_{k+2}^p.

- If $\varphi = \exists X.\varphi'$: We need to show that $\Pi, \Delta \models \varphi$ can be decided in Σ_{k+2}^p in the size of T. By induction hypothesis, if φ' still has $k+1$ quantifier alternations then its outermost second-order quantifier is existential and $\Pi, \Delta[X \mapsto A] \models \varphi'$ can be decided in Σ_{k+2}^p for any A. By the first induction hypothesis it follows that if φ' has k quantifier alternations then its outermost second-order quantifier is universal and $\Pi, \Delta[X \mapsto A] \models \varphi'$ can be decided in Π_{k+1}^p. For both cases holds the following: If we first non-deterministically guess A by guessing for each of the linearly many traces whether it is in A and then evaluate $\Pi, \Delta[X \mapsto A] \models \varphi'$ we can decide $\Pi, \Delta \models \varphi$ in Σ_{k+2}^p.

This concludes the analysis for formulas only containing existential second-order quantifiers. If the outermost second-order quantifier of φ is universal, then, by Obs. 6, the outermost second-order quantifier of $\hat{\varphi}$ is existential and $\Pi, \Delta \models \hat{\varphi}$ can be decided in Σ_{k+2}^p. Therefore φ can be decided in Π_{k+2}^p. $\qquad\square$

By Obs. 1, for a tree-shaped structure K, the size of $Traces(K)$ is polynomial in the size of K. Therefore, from Lemma 4 we have the following.

Corollary 3. $\mathrm{MC}[\Sigma_k\text{-HYPER}^2\text{LTL},\text{TREE}]$ *is in* Σ_{k+1}^p *in the size of K, and* $\mathrm{MC}[\Pi_k\text{-HYPER}^2\text{LTL},\text{TREE}]$ *is in* Π_{k+1}^p.

Corollary 4. $\mathrm{MC}[\mathrm{HYPER}^2\mathrm{LTL}, \mathrm{TREE}]$ *is in PSPACE.*

5.1   Lower bound

For the hardness result, We reduce from the Quantified Boolean Formula Problem (QBF), which we define as in [21].

Definition 2 (Quantified Boolean Formula [21]). *Let $k, m_1, \ldots, m_{k+1} \in \mathbb{N}$, and let y be $y = \exists x_{1,1}.\ldots.\exists x_{1,m_1}.\forall x_{2,1}.\ldots.\forall x_{2,m_2}.\ldots.\mathbb{Q}x_{k+1,1}.\ldots.\mathbb{Q}x_{k+1,m_{k+1}} E$ or $y = \forall x_{1,1}.\ldots.\forall x_{1,m_1}.\exists x_{2,1}.\ldots.\exists x_{2,m_2}.\ldots.\mathbb{Q}x_{k+1,1}.\ldots.\mathbb{Q}x_{k+1,m_{k+1}} E$ where E is an arbitrary Boolean formula over the variables $x_{1,1}, \ldots, x_{k+1,m_{k+1}}$, and where the last quantifier $\mathbb{Q}$ is the same at the first (i.e., existential for the first option and universal for the second) iff k is even. In both cases we say that y has k quantifier alternations. The QBF problem is then to decide whether y is valid.*

The QBF problem with k quantifier alternations and outermost existential quantifier is Σ_{k+1}^p-complete, and it is Π_{k+1}^p-complete for formulas with outermost universal quantification [21]. Thus, the QBF problem with k quantifier alternations, behaves similarly to $\mathrm{MC}[\mathrm{HYPER}^2\mathrm{LTL}, \mathrm{TREE}]$ with k second-order quantifier alternations (cf. Cor. 3). We use this intuition in the following where we reduce QBF with k alternations to $\mathrm{MC}[\mathrm{HYPER}^2\mathrm{LTL}, \mathrm{TREE}]$ with k second-order quantifier alternations, thus providing a lower bound and proving Thm. 5.

Proof (of Theorem 5). We provide a polynomial time reduction from the QBF problem. Given a QBF formula y with k quantifier alternations, we build a tree-shaped Kripke structure K in polynomial time, and a Hyper^{2}LTL formula φ which only depends on k and on the first quantifier of y, such that $K \models \varphi$ iff y is valid. The AP set of K is $\{q,v,\ pos,\ neg,\ eneg,\ eneg',\ epos,\ epos',\ f\}$. The structure K consists of an initial state that is connected to several branches, which do not branch further. Every branch has length l, where l is the sum of the number of variables and Boolean operators in y. Note that $k \leq l$.

We associate every variable $x_{i,j}$ and sub-formula e in y with indices $N(x_{i,j})$ and $N(e)$, respectively. We index the variables and sub-expressions subsequently, such that the index of every variable is smaller than the index of every subexpression. We then construct K as follows.

– For every variable $x_{i,j}$ K contains two branches: One is marked with *pos* on its $N(x_{i,j})$-th state, and the other is marked with *neg* on its $N(x_{i,j})$-th state.
– For sub-expressions $E_1 \otimes E_2$, where $\otimes \in \{\wedge, \vee\}$, K contains three branches:
  - If $\otimes = \wedge$: the first branch is labeled with *pos* on its $N(E_1 \otimes E_2)$-th state, with *epos* on its $N(E_1)$-th state, and with *epos'* on its $N(E_2)$-th state. The second branch is labeled with *neg* on its $N(E_1 \otimes E_2)$-th state and with *eneg* on its $N(E_1)$-th state. The third branch is labeled with *neg* on its $N(E_1 \otimes E_2)$-th state and with *eneg* on its $N(E_2)$-th state.
  - If $\otimes = \vee$: the first branch is labeled with *pos* on its $N(E_1 \otimes E_2)$-th state and with *epos* on its $N(E_1)$-th state. The second branch is labeled with *pos* on its $N(E_1 \otimes E_2)$-th state and with *epos* on its $N(E_2)$-th state. The third branch is labeled with *neg* on its $N(E_1 \otimes E_2)$-th state, with *eneg* on its $N(E_1)$-th state, and with *eneg'* on its $N(E_2)$ th state.

- For a negated sub-expression $E = \neg E_1$ in y, K has two branches: One labeled with *pos* on its $N(E)$-th state and *eneg* on its $N(E_1)$-th state. The second is labeled with *neg* on its $N(E)$-th state and *epos* on its $N(E_1)$-th state.
- K contains $k + 1$ additional branches: The i-th branch is labeled at the i-th position with atomic proposition q. The j-th state of the i-th branch is labeled with v if and only if there exists an $a \le m_i$ such that $N(x_{i,a}) = j$. Therefore, the state is labeled if and only if there exists a variable in the i-th quantifier sequence that is associated with j.

Thus, each branch for a sub-formula of y indicates its value with *pos* or *neg* if its sub-formulas have the value indicated by *epos, epos', eneg* and *eneg'*. All branches of the outermost Boolean operator of y are labeled with f in the first state. See Fig. 3 for an example. We now describe the formula φ, which has $k + 2$ second-order quantifications. We divide φ into sub-formulas, each contains exactly one second-order quantification: $\varphi = \varphi_1 \oplus_1 \varphi_2 \oplus_2 \cdots \varphi_{k+1} \oplus_{k+1} \varphi_{k+2}$ where connective $\oplus_i$ is $\wedge$ if φ_i contains an existential second-order quantifier and $\rightarrow$ otherwise. For all $1 \le i \le k + 1$ formula φ_i is of the following form:

$$\mathbb{Q}X_i.(\forall \pi \in \mathfrak{G}.\forall \pi' \in \mathfrak{G}.\exists! \pi'' \in X_i. \quad \bigcirc^i q_{\pi'} \wedge \Box(\neg epos_\pi \wedge \neg eneg_\pi \wedge \neg q_\pi)$$
$$\wedge \Diamond(v_{\pi'} \wedge (pos_\pi \vee neg_\pi)) \quad \rightarrow \Box((pos_\pi \vee neg_\pi) \leftrightarrow (pos_{\pi''} \vee neg_{\pi''})))$$

If, in y, variables $x_{i,1}$ to x_{i,m_i} are quantified universally, then $\mathbb{Q} := \forall$. If they are quantified existentially, then $\mathbb{Q} := \exists$.

An instantiation of X_i satisfies φ_i if it contains for each variable $x_{i,1}$ to x_{i,m_i} one trace assigning it positive or negative value. For each pair of traces π and π', the set X_i has to contain exactly one trace π''. The sub-formula $\bigcirc^i q_{\pi'}$ restricts instantiations for π' such that it can only be the trace that is labeled with v at all positions corresponding to the relevant variables. Instantiations for π are restricted such that π has to represent some variable $x_{i,1}$ to x_{i,m_i}. All instantiations for π'' therefore also represent the positive or negative assignment to some variable $x_{i,1}$ to x_{i,m_i}.

The formula φ_{k+2} should not introduce another quantifier alternation. Therefore, its quantifier depends on the second-order quantifier in φ_{k+1}.

$$\varphi_{k+2} = \mathbb{Q}Z.(\forall \pi_1 \in X_1.\ldots.\forall \pi_{k+1} \in X_{k+1}.\pi_1 \triangleright Z \wedge \cdots \wedge \pi_{k+1} \triangleright Z) \tag{1}$$
$$\wedge (\exists \pi \in Z.\Diamond f_\pi) \tag{2}$$
$$\wedge (\forall \pi \in Z.\exists \pi' \in Z.\exists \pi'' \in Z.\Box((epos_\pi \leftrightarrow pos_{\pi'}) \wedge (eneg_\pi \leftrightarrow neg_{\pi'}) \tag{3}$$
$$\wedge (epos'_\pi \leftrightarrow pos_{\pi''}) \wedge (eneg'_\pi \leftrightarrow neg_{\pi''}))) \tag{4}$$
$$\wedge (\forall \pi \in Z.\Box(\neg eneg_\pi \wedge \neg epos_\pi) \rightarrow \bigvee_{i=1}^{k+1} \pi \triangleright X_i) \oplus \exists \pi \in Z.\Diamond f_\pi \wedge \Diamond pos_\pi \tag{5}$$

If φ_{k+1} is universally quantified, then $\mathbb{Q} = \forall$ and $\oplus = \rightarrow$. If φ_{k+1} is existentially quantified, then $\mathbb{Q} = \exists$ and $\oplus = \wedge$. There is only one unique instantiation for Z that satisfies φ_{k+2}. The instantiation is a superset of $X_1, \ldots, X_{k+1}$ and contains a trace representing the whole quantifier free part of y (line 2). Similar to the proof of Thm. 4, the instantiation may only contain a trace if the subformulas of the represented formula have the correct value (lines 3-4). The instantiation thus evaluates E, and it remains to see whether there is a trace π indicating that E has a positive value (line 5, right). While φ is not given in formal Hyper^{2}LTL syntax it can be easily transformed into it. See the full version [16] for full proof.

$\square$

6   The Complexity of MC[HYPER^{2}LTL, ACYCLIC]

We show that the problem of MC[Σ_k-HYPER^{2}LTL, ACYCLIC] is in Σ_{k+1}^{EXP}, and MC[Π_k-HYPER^{2}LTL, ACYCLIC] is in Π_{k+1}^{EXP}. From this, it follows that Hyper^{2}LTL model checking on acyclic models is decidable and is in EXPSPACE. We then use reduction from acceptance problems for alternating Turing machines, and show that MC[HYPER^{2}LTL, ACYCLIC] is complete in these classes (Thm. 7). Containment follows directly from Lemma 4 and Obs. 3. Therefore, we have:

Corollary 5. *The problem of* MC[Σ_k-HYPER^{2}LTL, ACYCLIC] *is in* Σ_{k+1}^{EXP}, *and the problem of* MC[Π_k-HYPER^{2}LTL, ACYCLIC] *is in* Π_{k+1}^{EXP}.

Corollary 6. MC[HYPER^{2}LTL, ACYCLIC] *is in EXPSPACE.*

Theorem 7. MC[Σ_k-HYPER^{2}LTL, ACYCLIC] *is* Σ_{k+1}^{EXP}*-complete, and* MC[Π_k-HYPER^{2}LTL, ACYCLIC] *is* Π_{k+1}^{EXP}*-complete.*

Proof (Sketch). We reduce (using a polynomial time reduction) from the corresponding acceptance problem of alternating Turing machines: given an alternating Turing machine M and a $m \in \mathbb{N}$ (encoded in binary), decide whether M accepts after m steps. We assume that M alternates at most k times between universal and existential states (we call each such alternation *alternation block*). Additionally, its initial state is universal if and only if we prove completeness for Π_{k+1}^{EXP}. See the full version [16] for a formal proof.

Given an alternating Turing machine M and $m \in \mathbb{N}$ in binary, we build a structure K and a formula φ such that $K \models \varphi$ iff M accepts in at most m steps. φ only depends on k and whether the initial state of M is universal or existential.

We encode M into K using the following substructures: A substructure representing the traces for a counter, encoding all possible numbers (up to m); One substructure per transition in M, using encodings for states and tape letters; A substructure containing traces for the tape entries; A substructure containing traces representing that all tape cells are filled with blanks at time stamp 0; and A substructure containing exactly one trace encoding the number m. We use atomic propositions to encode all of the above, and to differentiate between the different substructures. All traces in K have length $l + 1$ which we define as the number of bits needed to encode $2 \cdot m$. Thus, K encodes all valid transitions of M, some initial configuration, and additional traces such that every possible state and tape content can be represented as a set of traces.

The formula φ collects valid computation fragments (i.e. a fragment that is taken within the same alternation block) into sets. It quantifies over sets $X_1, \ldots, X_{k+1}$ simulating an execution of M. The instantiations of a set X_i are restricted such that each instantiation of X_i is a valid execution branch through alternation block i of M. Each taken transition in the simulated execution is represented by a trace encoding the transition and a trace encoding the written letter to the tape. Additionally, a computation fragment must be connectable to the last step of the previous fragment and the first step of the next fragment. Further, it is only allowed to contain either transition starting in universal states or in existential but it may not alternate. We describe all of these using the formula φ as we can freely quantify over the different sets. Then, the answer to the model checking problem is equal to the question whether there is a computation step in the last fragment which visits the accepting state.

The alternations of M are encoded in the quantification over the sets collecting computation fragments; If the computation fragment is universal then φ accepts if all sets that represent a valid computation fragment lead to an accepting state. If the computation fragment is existential then φ accepts if there exists a set that represents a valid computation to an accepting state.   □

7 Complexity in the Size of the Model and Formula

In this section, we show that if the size of the formula is not assumed to be constant, then MC[HYPER²LTL, TREE] as well as MC[FIXPOINT, TREE] are PSPACE-complete in the combined size of the Kripke structure and formula.

Lemma 5. MC[HYPER²LTL, TREE] *is in PSPACE in the combined size of the structure and formula.*

Proof. Let φ be a Hyper²LTL formula of size m and with q quantifiers, and let K be a tree-shaped Kripke structure with n states. We evaluate φ recursively from the outermost quantifier to the innermost one. To evaluate a quantifier we iterate over every possible instantiation and recursively evaluate its inner formula. We track instantiations of trace or trace-set quantifiers by marking the corresponding leaf or leaves respectively. When all traces are fixed, the inner, quantifier-free formula, can be decided in polynomial space (Cor. 1). To every time-point, we track at most q quantifier instantiations. Tracking a set of traces needs $\mathcal{O}(n \cdot \log(n))$ space, as there are at most n traces in K and tracking a trace needs $\mathcal{O}(\log n)$ space. The space required to find the next set of traces or the next trace is polynomial in n and constant in m. Thus, we need at most $\mathcal{O}(q \cdot n \cdot \log(n))$ space - polynomial in the combined size of structure and formula.   □

Theorem 8. MC[FIXPOINT, TREE] *is PSPACE-complete in the combined input of the Kripke structure and formula.*

Theorem 9. MC[HYPER²LTL, TREE] *is PSPACE-complete in the combined input of the Kripke structure and formula.*

For both theorems, containment in PSPACE follows from Lemma 5. Since HyperLTL model checking on trees is PSPACE-complete in the combined input consisting of structure and formula [7], and since HyperLTL is subsumed by Fixpoint Hyper^{2}LTL$_{fp}$ and thus by Hyper^{2}LTL, Theorems 8 and 9 follow.

8 Conclusion

In this work, we analyzed the complexity of model checking Hyper^{2}LTL and Fixpoint Hyper^{2}LTL$_{fp}$ on finite structures. This problem is particularly relevant for monitoring purposes, in which tree-shaped and acyclic models are used as data structures to maintain the (repeatedly growing) set of traces seen so far. We showed that the model checking complexity for Fixpoint Hyper^{2}LTL$_{fp}$ is polynomial in the number of traces in the model. It follows from our analysis, that model checking Fixpoint Hyper^{2}LTL$_{fp}$ over finite structures is not much more complex than HyperLTL model checking, despite the fact that Fixpoint Hyper^{2}LTL$_{fp}$ is much more expressive.[7] Contrary to Fixpoint Hyper^{2}LTL$_{fp}$, unrestricted Hyper^{2}LTL reasons about all possible sets of traces in the system, significantly increasing time complexity, while preserving space complexity. Our complexity results validate the efficiency of the Fixpoint Hyper^{2}LTL$_{fp}$ fragment, motivating its use in finite-trace settings, such as monitoring.

The main contribution of this work is establishing lower bounds for the respective model checking problems. In particular, in the monitoring setting, the model (current set of traces) is changing repeatedly, and therefore we focused in complexity analysis in the size of the model. As future work, we intend to complete the analysis of Section 7 to consider also the size of the formula.

References

1. Bartocci, E., Henzinger, T.A., Nickovic, D., da Costa, A.O.: Hypernode automata. In: Pérez, G.A., Raskin, J. (eds.) CONCUR 2023. LIPIcs, vol. 279, pp. 21:1–21:16. Schloss Dagstuhl - Leibniz-Zentrum für Informatik (2023). https://doi.org/10.4230/LIPICS.CONCUR.2023.21
2. Baumeister, J., Coenen, N., Bonakdarpour, B., Finkbeiner, B., Sánchez, C.: A temporal logic for asynchronous hyperproperties. In: Silva, A., Leino, K.R.M. (eds.) CAV 2021, Part I. LNCS, vol. 12759, pp. 694–717. Springer (2021). https://doi.org/10.1007/978-3-030-81685-8_33
3. Beutner, R., Finkbeiner, B.: AutoHyper: Explicit-state model checking for HyperLTL. In: International Conference on Tools and Algorithms for the Construction and Analysis of Systems, TACAS 2023. vol. 13993. Springer (2023). https://doi.org/10.1007/978-3-031-30823-9_8
4. Beutner, R., Finkbeiner, B., Frenkel, H., Metzger, N.: Second-order hyperproperties. In: Enea, C., Lal, A. (eds.) CAV 2023, Part II. LNCS, vol. 13965, pp. 309–332. Springer (2023). https://doi.org/10.1007/978-3-031-37703-7_15

---

[7] On tree-shaped models, HyperLTL model checking is L-complete [7] and Fixpoint Hyper^{2}LTL$_{fp}$ model checking is P-complete. On acyclic models, model checking for HyperLTL is PSPACE-complete [7] and for Fixpoint Hyper^{2}LTL$_{fp}$ is EXP-complete.

5. Beutner, R., Finkbeiner, B., Frenkel, H., Metzger, N.: Monitoring second-order hyperproperties. In: Dastani, M., Sichman, J.S., Alechina, N., Dignum, V. (eds.) Proceedings of the 23rd International Conference on Autonomous Agents and Multiagent Systems, AAMAS 2024, Auckland, New Zealand, May 6-10, 2024. pp. 180–188. International Foundation for Autonomous Agents and Multiagent Systems / ACM (2024). https://doi.org/10.5555/3635637.3662865, https://dl.acm.org/doi/10.5555/3635637.3662865

6. Bombardelli, A., Bozzelli, L., Sánchez, C., Tonetta, S.: Unifying asynchronous logics for hyperproperties. In: Barman, S., Lasota, S. (eds.) 44th IARCS Annual Conference on Foundations of Software Technology and Theoretical Computer Science, FSTTCS 2024, December 16-18, 2024, Gandhinagar, Gujarat, India. LIPIcs, vol. 323, pp. 14:1–14:18. Schloss Dagstuhl - Leibniz-Zentrum für Informatik (2024). https://doi.org/10.4230/LIPICS.FSTTCS.2024.14

7. Bonakdarpour, B., Finkbeiner, B.: The complexity of monitoring hyperproperties. In: 2018 IEEE 31st Computer Security Foundations Symposium (CSF). pp. 162–174 (July 2018). https://doi.org/10.1109/CSF.2018.00019

8. Bozzelli, L., Maubert, B., Pinchinat, S.: Unifying hyper and epistemic temporal logics. In: Pitts, A.M. (ed.) FoSSaCS 2015. LNCS, vol. 9034, pp. 167–182. Springer (2015). https://doi.org/10.1007/978-3-662-46678-0_11

9. Bozzelli, L., Peron, A., Sánchez, C.: Asynchronous extensions of HyperLTL. In: LICS 2021. pp. 1–13. IEEE (2021). https://doi.org/10.1109/LICS52264.2021.9470583

10. Bozzelli, L., Peron, A., Sánchez, C.: Expressiveness and decidability of temporal logics for asynchronous hyperproperties. In: Klin, B., Lasota, S., Muscholl, A. (eds.) CONCUR 2022. LIPIcs, vol. 243, pp. 27:1–27:16. Schloss Dagstuhl - Leibniz-Zentrum für Informatik (2022). https://doi.org/10.4230/LIPIcs.CONCUR.2022.27

11. Clarkson, M.R., Finkbeiner, B., Koleini, M., Micinski, K.K., Rabe, M.N., Sánchez, C.: Temporal logics for hyperproperties. In: Abadi, M., Kremer, S. (eds.) POST 2014. LNCS, vol. 8414, pp. 265–284. Springer (2014). https://doi.org/10.1007/978-3-642-54792-8_15

12. Clarkson, M.R., Schneider, F.B.: Hyperproperties. J. Comput. Secur. **18**(6), 1157–1210 (2010). https://doi.org/10.3233/JCS-2009-0393

13. Diekert, V., Rozenberg, G. (eds.): The Book of Traces. World Scientific (1995). https://doi.org/10.1142/2563

14. Eisner, C., Fisman, D., Havlicek, J., Lustig, Y., McIsaac, A., Van Campenhout, D.: Reasoning with temporal logic on truncated paths. In: Hunt, W.A., Somenzi, F. (eds.) Computer Aided Verification. pp. 27–39. Springer Berlin Heidelberg, Berlin, Heidelberg (2003)

15. Fagin, R., Halpern, J.Y., Moses, Y., Vardi, M.Y.: Reasoning About Knowledge. MIT Press (1995). https://doi.org/10.7551/mitpress/5803.001.0001

16. Finkbeiner, B., Frenkel, H., Rohde, T.: Complexity of model checking second-order hyperproperties on finite structures (2026), https://arxiv.org/abs/2601.12361

17. Finkbeiner, B., Hahn, C., Stenger, M., Tentrup, L.: Efficient monitoring of hyperproperties using prefix trees. Int. J. Softw. Tools Technol. Transf. **22**(6), 729–740 (2020). https://doi.org/10.1007/S10009-020-00552-5

18. Fionda, V., Greco, G.: The complexity of LTL on finite traces: Hard and easy fragments. In: Schuurmans, D., Wellman, M.P. (eds.) Proceedings of the Thirtieth AAAI Conference on Artificial Intelligence, February 12-17, 2016, Phoenix, Arizona, USA. pp. 971–977. AAAI Press (2016). https://doi.org/10.1609/AAAI.V30I1.10104

19. Fortin, M., Kuijer, L.B., Totzke, P., Zimmermann, M.: Hyperltl satisfiability is highly undecidable, hyperctl* is even harder. Log. Methods Comput. Sci. **21**(1), 3 (2025). https://doi.org/10.46298/LMCS-21(1:3)2025

20. Frenkel, H., Zimmermann, M.: The complexity of second-order hyperltl. In: Endrullis, J., Schmitz, S. (eds.) 33rd EACSL Annual Conference on Computer Science Logic, CSL 2025, February 10-14, 2025, Amsterdam, Netherlands. LIPIcs, vol. 326, pp. 10:1–10:23. Schloss Dagstuhl - Leibniz-Zentrum für Informatik (2025). https://doi.org/10.4230/LIPICS.CSL.2025.10

21. Garey, M.R., Johnson, D.S.: Computers and intractability. A series of books in the mathematical sciences, Freeman, New York, NY [u.a.] (1979), `http://www.ulb.tu-darmstadt.de/tocs/5617215X.pdf`

22. Goguen, J.A., Meseguer, J.: Security policies and security models. In: 1982 IEEE Symposium on Security and Privacy, Oakland, CA, USA, April 26-28, 1982. pp. 11–20. IEEE Computer Society (1982). https://doi.org/10.1109/SP.1982.10014

23. Gutsfeld, J.O., Müller-Olm, M., Ohrem, C.: Automata and fixpoints for asynchronous hyperproperties. Proc. ACM Program. Lang. **5**(POPL), 1–29 (2021). https://doi.org/10.1145/3434319

24. Halpern, J.Y., Moses, Y.: Knowledge and common knowledge in a distributed environment. J. ACM **37**(3), 549–587 (1990). https://doi.org/10.1145/79147.79161

25. Hsu, T.H., Bonakdarpour, B., Finkbeiner, B., Sánchez, C.: Bounded model checking for asynchronous hyperproperties (2023)

26. Hsu, T., Sánchez, C., Bonakdarpour, B.: Bounded model checking for hyperproperties. In: Groote, J.F., Larsen, K.G. (eds.) Tools and Algorithms for the Construction and Analysis of Systems - 27th International Conference, TACAS 2021, Held as Part of the European Joint Conferences on Theory and Practice of Software, ETAPS 2021, Luxembourg City, Luxembourg, March 27 - April 1, 2021, Proceedings, Part I. Lecture Notes in Computer Science, vol. 12651, pp. 94–112. Springer (2021). https://doi.org/10.1007/978-3-030-72016-2_6

27. Kontinen, J., Sandström, M., Virtema, J.: Set semantics for asynchronous TeamLTL: Expressivity and complexity. In: Leroux, J., Lombardy, S., Peleg, D. (eds.) MFCS 2023. LIPIcs, vol. 272, pp. 60:1–60:14. Schloss Dagstuhl - Leibniz-Zentrum für Informatik (2023). https://doi.org/10.4230/LIPICS.MFCS.2023.60

28. Kontinen, J., Sandström, M., Virtema, J.: A remark on the expressivity of asynchronous TeamLTL and HyperLTL. In: Meier, A., Ortiz, M. (eds.) FoIKS 2024. LNCS, vol. 14589, pp. 275–286. Springer (2024). https://doi.org/10.1007/978-3-031-56940-1_15

29. Krebs, A., Meier, A., Virtema, J., Zimmermann, M.: Team semantics for the specification and verification of hyperproperties. In: Potapov, I., Spirakis, P.G., Worrell, J. (eds.) MFCS 2018. LIPIcs, vol. 117, pp. 10:1–10:16. Schloss Dagstuhl - Leibniz-Zentrum für Informatik (2018). https://doi.org/10.4230/LIPIcs.MFCS.2018.10

30. Kuhtz, L., Finkbeiner, B.: Weak kripke structures and LTL. In: Katoen, J., König, B. (eds.) CONCUR 2011 - Concurrency Theory - 22nd International Conference, CONCUR 2011, Aachen, Germany, September 6-9, 2011. Proceedings. Lecture Notes in Computer Science, vol. 6901, pp. 419–433. Springer (2011). https://doi.org/10.1007/978-3-642-23217-6_28

31. Kupferman, O., Vardi, M.Y.: Model checking of safety properties. In: Halbwachs, N., Peled, D.A. (eds.) Computer Aided Verification, 11th International Conference, CAV '99, Trento, Italy, July 6-10, 1999, Proceedings. Lecture Notes in Computer Science, vol. 1633, pp. 172–183. Springer (1999). https://doi.org/10.1007/3-540-48683-6_17

32. van der Meyden, R.: Common knowledge and update in finite environments. Inf. Comput. **140**(2) (1998). https://doi.org/10.1006/inco.1997.2679
33. van der Meyden, R., Shilov, N.V.: Model checking knowledge and time in systems with perfect recall. In: FSTTCS. vol. 1738, pp. 432–445. Springer (1999)
34. Papadimitriou, C.H.: Computational complexity. Theoretical computer science, Addison-Wesley, Reading, Massachusetts ([1994]), `http://www.gbv.de/dms/ilmenau/toc/12708035X.PDF`
35. Pnueli, A.: The temporal logic of programs. In: FOCS 1977

Karp's NP-Complete Problems over First-Order Definable Structures

Aidan Healy and Bartek Klin

University of Oxford, UK
`{aidan.healy,bartek.klin}@cs.ox.ac.uk`

Abstract. We determine the decidability of Karp's NP-complete problems on structures which are first-order definable over the theory of equality, also known as orbit-finite sets with atoms or nominal sets.

Keywords: Sets with atoms · Nominal sets · NP-complete problems

1 Introduction

We wish to take a range of classical decision problems that are usually considered for finite structures, and study them on the class of structures which are infinite but definable by first-order formulas that use equality only. Precise definitions will follow, but let us begin with two illustrative examples. Here and in the following, fix a countably infinite set $\mathbb{A}$, whose elements we call atoms.

Example 1. Consider $X = \binom{\mathbb{A}}{2}$, the set of two-element sets of atoms, and the family $\mathcal{S}$ of all three-element subsets of X of the form:

$$\{\{a,b\},\{a,c\},\{b,c\}\} \qquad \text{for all distinct } a,b,c \in \mathbb{A}.$$

This family admits an *exact cover*: there is a sub-family of $\mathcal{S}$ that forms a partition of X. However, to find such a cover we must break the pleasant symmetry of $\mathcal{S}$. For example, we may fix an enumeration of X, and proceed by induction starting with the empty family: in each step, take the first $\{a,b\}$ that has not been covered yet, choose some c such that neither $\{a,c\}$ nor $\{b,c\}$ have been covered, and add this to the family. The limit of this process is an exact cover.

Example 2. Now let X be the disjoint union of $\binom{\mathbb{A}}{2}$ and $\mathbb{A}$. The family $\mathcal{S}$ of all subsets of X of the form:

$$\{\{a,b\},\{a,c\},a\} \qquad \text{for all distinct } a,b,c \in \mathbb{A},$$

does *not* admit an exact cover of X. To see why, notice that to cover a pair $\{a,b\}$ we must include a set that contains either a or b. Any atom in $\mathbb{A}$ can be included only once in this way, so it can be used to cover only two pairs. More generally, k atoms can be used to cover only $2k$ pairs. For $k = 6$, this is not enough to cover all the $\binom{6}{2} = 15$ pairs that are built of the k atoms and need to be covered.

© The Author(s) 2026
N. Bertrand and S. Milius (Eds.): FoSSaCS 2026, LNCS 16503, pp. 307–327, 2026.
https://doi.org/10.1007/978-3-032-22730-0_15

Both these arguments are quite simple, but substantial enough to make one wonder whether or not the classical problem EXACTCOVER is decidable for structures of this kind. Indeed, one of our main results is that it is not.

By "structures of this kind" we mean structures (graphs, hypergraphs, formulas, families...) where all components and relations are defined by first-order formulas that only compare atoms for equality. Such structures are usually infinite, but they are finite up to bijective renaming of atoms. We will find it convenient to describe them within the framework of *sets with atoms*, but in the parlance of model theory (see e.g. [15]), they are simply relational structures which are first-order interpretable in $(\mathbb{A}, =)$.

The purpose of this paper is to determine the decidability, over such structures, of appropriately extended versions of the classical NP-complete problems listed by Karp in 1972 [16].

The idea of transporting computational problems from finite to infinite structures has been explored in several variants, with the main difference being the class of infinite structures considered. The broadest setting is that of *recursive structures* [13,14], where nodes are natural numbers and arbitrary decidable relations are permitted. Of course, in this setting all nontrivial questions become undecidable, and the focus is on determining where particular problems lie in the arithmetical hierarchy.

A smaller class is that of *automatic structures* [18,2,12], where nodes are represented as words over a finite alphabet, and relations are recognisable by multi-tape finite-state automata that read those words in parallel. For this class, the model checking problem for first-order logic extended with a limited form of second-order quantification, is decidable [24]. As a result, natural extensions of a few of Karp's problems become decidable [23,24], including CLIQUE, SETPACKING and a variant of SETCOVER. Several other problems remain undecidable, including HAMILTONICITY and EXACTCOVER [23,24]. Some problems that are polynomial-time decidable in the finite case, including 2-SAT, 2-COLORABILITY [23], and checking whether a graph is connected [3], are undecidable on automatic structures.

Another restricted class is that of *doubly periodic structures* [8]. These are constructed by placing infinitely many copies of a fixed finite structure on a two-dimensional grid, and imposing additional relations on nodes from neighbouring copies only, in a periodic manner. On this class, k-SAT and k-COLORABILITY are decidable for $k = 2$ and undecidable for $k > 2$ [9], and one could hope that this could be a setting where the P vs. NP gap is blown up to the gap between decidable and undecidable problems. However, all doubly periodic structures are automatic, so decidability results mentioned above hold here as well.

As we said above, our focus is on structures which are first-order interpretable in a pure set. In the following we will simply call such structures *definable*. All such structures are automatic, since the class of automatic structures is closed under first-order interpretations [2]. They are incomparable with doubly periodic ones. (For example, an infinite clique is definable but not doubly periodic, and an infinite square grid is doubly periodic but not definable.)

A few classical decision problems have been studied in the framework of definable structures. In [19], it was proved that every Constraint Satisfaction Problem for a fixed finite template is decidable in this setting. This includes k-SAT and k-COLORABILITY, for every k. On the other hand, checking whether there exists a homomorphism from one given definable structure to another is undecidable [20]. For solvability of systems of linear equations, [19] shows decidability over finite fields and where every equation contains finitely many variables. Both these assumptions are dropped in [10], at the price of searching for definable solutions only. In [11], linear programming over definable structures is shown to be decidable, and integer linear programming undecidable, again under the assumption that only definable solutions are considered.

The decidability landscape of Karp's NP-complete problems on definable structures turns out to be surprisingly varied. Out of the original 21 problems:

- Seven are undecidable, including CNF-SAT, EXACTCOVER, and HAMILTONICITY. We show this by a sequence of reductions, starting from a reduction of the Wang tiling problem to EXACTCOVER.
- Eleven are decidable. Two of these, 3-SAT and COLORABILITY, are known from [19], and we generalise the technique used there to cover a few more, including VERTEXCOVER. Decidability of CLIQUE and SETPACKING is known for automatic structures (see above), but for definable structures we provide more direct arguments.
- Three problems essentially rely on adding up unboundedly many numbers, and we see no natural way to extend them to an infinite setting.

2 Preliminaries: definable structures

We assume general familiarity with the classic paper [16] and the basic concepts discussed there such as graphs, cliques, formulas etc. We will recall along the way the 21 computational problems studied there. In this preliminary section we focus on *definable structures*, which we will use as instances for those problems. We will work in the framework of *sets with atoms* [4], also called *nominal sets* [26]. Our presentation follows [19,20]; see there and [4, Chap. 10] for more details.

Let $\mathbb{A}$ be a countably infinite set of *atoms*. We want to define sets that are somehow built of atoms and are perhaps infinite, but presented in a finite way and highly symmetric under bijective atom renaming. To this end, given some fixed infinite set of atom variables, an *expression* is either a variable or a formal finite (perhaps empty) union of set-builder expressions of the form

$$\{e \mid v_1, \ldots, v_k \in \mathbb{A}, \ \phi\}$$

where e is an expression (where the v_i as well as other variables may occur), the v_i are bound variables, and ϕ is a first-order formula over the set of atom variables, with equality as the only relation symbol. Free variables in this expression are those free variables in e and ϕ which are not among the $v_1, \ldots, v_k$. For an expression e with free variables V, any valuation $\sigma : V \to \mathbb{A}$ defines a value

$X = e[\sigma]$ in an obvious way, by induction on the structure of e. This value is either an atom or a set. We will call X a *definable set with atoms*. If the image of σ is some (necessarily finite, and perhaps empty if e has no free variables) $S \subseteq \mathbb{A}$, then we may say that X is *S-definable*. $\emptyset$-definable sets are called *equivariant*. For example, the equivariant set $\binom{\mathbb{A}}{2}$ from Example 1 is formally defined by the expression

$$\{\{u\} \cup \{v\} \mid u, v \in \mathbb{A}, u \neq v\},$$

where the subexpressions $\{u\}$ and $\{v\}$ use a little syntactic sugar: the empty list of bound variables, as well as a formula ϕ which is always true, can be elided.

The above language of expressions is rudimentary, but using standard set-theoretic machinery it is easy to extend it with pairs, tuples, integers, constants from some given finite sets, atom constants, etc. We will use such syntactic sugar without further warning. In particular, in Examples 1-2, not only the sets X and $\mathcal{S}$ but also the pair $(X, \mathcal{S})$ is definable (in fact, equivariant).

Furthermore (see [4, Chap. 10] for a detailed discussion), it is routine to prove by induction on the structure of expressions that definable sets are closed under Boolean combinations, Cartesian products, images and inverse images of definable functions, quotients under definable equivalence relations, and intersections and unions of definable families, and all these constructions are effectively computable as operations on the defining expressions. Also relations such as set equality and membership are decidable. As a result, set-builder expressions can be safely extended with more syntactic sugar by allowing bound variables to range not only over $\mathbb{A}$ but over any definable set, and allowing in ϕ set relations $\in$ and $\subseteq$ (in addition to atom equality) and quantifiers of the form $\exists x \in X$ and $\forall x \in X$, where X is a set defined by an expression. For example, for $(X, \mathcal{S})$ as in Examples 1-2, the sets

$$\{(x, Y) \mid x \in X, Y \in \mathcal{S}, x \in Y\} \qquad \text{and} \qquad \{\{Y \in \mathcal{S} \mid x \in Y\} \mid x \in X\}$$

are definable (and equivariant).

Definable structures are highly symmetric. The group $\mathrm{Aut}(\mathbb{A})$ of *atom automorphisms* (meaning: arbitrary bijections on $\mathbb{A}$) has a canonical action on the class of definable sets: for a definable set $X = e[\sigma]$ and $\pi \in \mathrm{Aut}(\mathbb{A})$, define $X \cdot \pi = e[\sigma; \pi]$, where $\sigma; \pi$ denotes function composition: $(\sigma; \pi)(v) = \pi(\sigma(v))$. This amounts to consistently renaming all the atoms throughout X according to π. For a finite $S \subseteq \mathbb{A}$, a $\pi \in \mathrm{Aut}(\mathbb{A})$ is called an *S-automorphism* if $\pi(a) = a$ for all $a \in S$. We say that S *supports* a definable set X if $X \cdot \pi = X$ for every S-automorphism π. It is easy to see that every S-definable set is supported by S, so every definable set has a finite support. Finite supports of a given set are closed under intersection (see [4, Thm. 4.13] or [26, Prop. 2.3]), so every definable set X has a *least support*, denoted $\mathrm{supp}(X)$.

We say that sets X and Y are *S-equivalent* if there is an S-automorphism π such that $X \cdot \pi = Y$. This is an equivalence relation, and its equivalence classes are called *S-orbits*, or simply *orbits* if $S = \emptyset$. Every definable set X is *orbit-finite*, i.e., it is a finite union of $\mathrm{supp}(X)$-orbits. For example:

- $\mathbb{A}$ is a single-orbit set. For any finite $S \subseteq \mathbb{A}$, the set S has $|S|$ S-orbits (every element of S is a singleton orbit), and $\mathbb{A} \setminus S$ has one S-orbit.
- For any k, the sets $\mathbb{A}^{(k)}$ (of non-repeating k-tuples) and $\binom{\mathbb{A}}{k}$ (of sets of size k) are single-orbit sets. More generally, for any finite permutation group $G \leq \mathrm{Sym}(k)$, the set $\mathbb{A}^{(k)}/G$ of non-repeating k-tuples of atoms up to permutations from G, is a single-orbit set.
- The set $\mathbb{A}^2$ (of possibly repeating pairs) consists of two orbits; more generally $\mathbb{A}^k$ has number of orbits equal to the k-th Bell number.
- The set of finite subsets of $\mathbb{A}$ is not orbit-finite, as sets of different sizes fall into different orbits; therefore this set is not definable.

There is an alternative but equivalent way to introduce definable sets, where actions of $\mathrm{Aut}(\mathbb{A})$ and finite supports are the basic concepts, with orbit-finiteness imposed as an additional condition. It then becomes a representation theorem that every S-supported orbit-finite set is (in an S-suppported bijection with) an S-definable set. This approach is taken in [4,5,6]. Other representations exist: as shown in [6], every equivariant single-orbit set is in equivariant bijection with a set of the form $\mathbb{A}^{(k)}/G$ as mentioned above. This implies that definable structures are first-order interpretable (in the sense of model theory) in the pure set $(\mathbb{A}, =)$.

So far we have focused on *equality atoms*, where $\mathbb{A}$ is a pure set, without any structure imposed on the atoms. This is our main subject of study here, and in the following we will study Karp's problems only on structures definable over equality atoms. However, in Section 4, as in [19], we will need to make a brief excursion to a richer structure of *ordered atoms*, where $\mathbb{A} = (\mathbb{Q}, \leq)$ is the total order of rationals, with $\mathrm{Aut}(\mathbb{A})$ restricted to order-preserving bijections. This extends the language of formulas ϕ in set-builder expressions with a binary order relation $\leq$. The notions of definability, support and orbit-finiteness are defined as for equality atoms, and the representation theorems mentioned above work analogously. The general framework of sets over any relational structure of atoms is studied in detail in [4,6].

3   Undecidable problems

We will list Karp's problems that become undecidable over definable structures.

3.1   Exact cover

Problem 1 (EXACTCOVER).
Input: Definable set X, definable set $\mathcal{S}$ of subsets of X
Question: Is there a pairwise-disjoint subset of $\mathcal{S}$ whose union is X?

In the classical setting, where both X and $\mathcal{S}$ are finite, Karp [16] proved NP-hardness of this problem by reduction from graph colorability. Looking for a similar reduction here would be pointless, since colorability of definable graphs is decidable. Instead, we prove undecidability by a reduction from the well-known domino tiling problem, posed by Wang [28] and shown undecidable by Berger [1].

To describe that problem, fix constants N, S, E and W. For a finite set C of colors, a *tile* is a function $t : \{\mathsf{N}, \mathsf{S}, \mathsf{E}, \mathsf{W}\} \to C$, and for a set T of tiles, a *tiling* of the plane is a function $\tau : \mathbb{Z}^2 \to T$ such that, for all $a, b \in \mathbb{Z}$:

$$\tau(a, b)(\mathsf{N}) = \tau(a, b+1)(\mathsf{S}) \qquad \text{and} \qquad \tau(a, b)(\mathsf{E}) = \tau(a+1, b)(\mathsf{W}).$$

Problem 2 (DOMINO).

Input: A finite set C of colors, a finite set T of tiles over C

Question: Is there a tiling with T?

We will reduce DOMINO to EXACTCOVER. Given an input (C, T), we will construct definable X and $\mathcal{S}$ such that an exact cover of $(X, \mathcal{S})$ exists if and only if a tiling of the plane does.

For some intuition, imagine an infinite clique with all atoms as vertices. The set X will contain 4 elements for each atom (the *vertex elements*), and a few elements for each pair of distinct atoms (the *edge elements*). The family $\mathcal{S}$ will contain two kinds of (finite) sets. An *edge set* will contain all the edge elements associated to a specific pair. A *tile set*, defined over a quadruple of distinct atoms a, b, c, d, will contain some of the vertex elements associated to a, b, c and d, as well as some edge elements associated to the pairs $(a, b), (b, c), (c, d)$ and (d, a). We can consider it associated with the quadrilateral a, b, c, d.

Only tile sets contain vertex elements, so an exact cover in $\mathcal{S}$ must use infinitely many of them. Each tile set will only partially fill its four edges (i.e. it will not include an entire edge set), so in an exact cover each tile set must match other tile sets complementing it on its edges, in a way that can be unrolled to a tiling of the infinite square grid. Edge sets are used to cover all the edge elements that do not correspond to edges in that grid.

We shall now make these ideas precise.

Theorem 1. *EXACTCOVER is undecidable.*

Proof. By reduction from DOMINO. Given as input finite sets C of colors and T of tiles, we shall construct an instance $(X, \mathcal{S})$ of EXACTCOVER. Define:

$$X = \bigcup_{a \in \mathbb{A}} \{\mathsf{NW}_a, \mathsf{NE}_a, \mathsf{SE}_a, \mathsf{SW}_a\} \ \cup \ \bigcup_{(a,b) \in \mathbb{A}^{(2)}} \{\to_{(a,b)}\}$$

$$\cup \ \bigcup_{\{a,b\} \in \binom{\mathbb{A}}{2}} \{1_{\{a,b\}}, 2_{\{a,b\}}, 3_{\{a,b\}}, 4_{\{a,b\}}\} \cup \{c_{\{a,b\}} : c \in C\}.$$

Note that X depends on C but not on T. For $\{a, b\} \in \binom{\mathbb{A}}{2}$, define the *edge set* $e_{\{a,b\}}$ by:

$$e_{\{a,b\}} = \{1_{\{a,b\}}, 2_{\{a,b\}}, 3_{\{a,b\}}, 4_{\{a,b\}}, \to_{(a,b)}, \to_{(b,a)}\} \cup \{c_{\{a,b\}} : c \in C\}.$$

For a tile $t \in T$ and atoms $(a, b, c, d) \in \mathbb{A}^{(4)}$, define the *tile set* $t_{(a,b,c,d)}$ by:

$$t_{(a,b,c,d)} = \{\mathsf{SE}_a, \mathsf{SW}_b, \mathsf{NW}_c, \mathsf{NE}_d, \rightarrow_{(a,b)}, \rightarrow_{(b,c)}, \rightarrow_{(c,d)}, \rightarrow_{(d,a)},$$
$$1_{\{a,b\}}, 2_{\{a,b\}}, 1_{\{b,c\}}, 3_{\{b,c\}}, 3_{\{c,d\}}, 4_{\{c,d\}}, 2_{\{d,a\}}, 4_{\{d,a\}}\}$$
$$\cup \{c_{\{a,b\}} : c \in C \wedge t(\mathsf{N}) = c\} \cup \{c_{\{b,c\}} : c \in C \wedge t(\mathsf{E}) = c\}$$
$$\cup \{c_{\{c,d\}} : c \in C \wedge t(\mathsf{S}) \neq c\} \cup \{c_{\{d,a\}} : c \in C \wedge t(\mathsf{W}) \neq c\}$$

(see Figure 1 for intuition). Let $\mathcal{S}$ contain all the edge and tile sets as above.

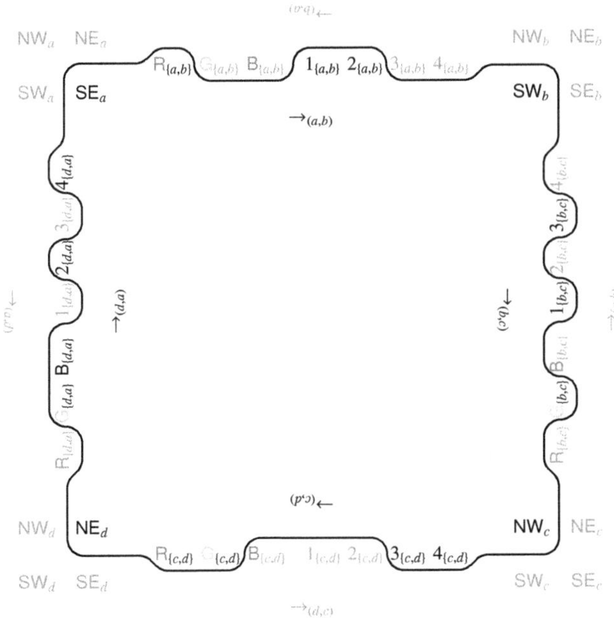

Fig. 1: The set $t_{(a,b,c,d)}$ for $C = \{\mathsf{R}, \mathsf{G}, \mathsf{B}\}$, $t = (\mathsf{N} \mapsto \mathsf{R}, \mathsf{E} \mapsto \mathsf{G}, \mathsf{S} \mapsto \mathsf{B}, \mathsf{W} \mapsto \mathsf{R})$, with the excluded neighbouring elements marked as gray

We shall show that if $(X, \mathcal{S})$ admits an exact cover then T tiles the plane.

Fix an exact cover $E \subseteq \mathcal{S}$. Let F be the set of tile sets in E, and note it is infinite (because edge sets don't contain elements associated with atoms, only pairs of atoms, and tile sets are finite). Consider a tile set $t_{(a,b,c,d)} \in F$. It covers $\rightarrow_{(a,b)}$ but not $\rightarrow_{(b,a)}$, so there is a unique tile set $n(t_{(a,b,c,d)})$ covering its complement on the elements associated to both of a and b. We can similarly define $e(t_{(a,b,c,d)})$, $s(t_{(a,b,c,d)})$ and $w(t_{(a,b,c,d)})$ by looking at elements associated respectively to b, c, to c, d, and to d, a. Doing this for all tile sets in F defines functions $n, e, s, w : F \rightarrow F$.

Moreover, for any $t_{(a,b,c,d)} \in F$ we have $n(t_{(a,b,c,d)}) = t'_{(e,f,b,a)}$ for some $t' \in T$ and $e, f \in \mathbb{A}$ (otherwise $t_{(a,b,c,d)}$ and $n(t_{(a,b,c,d)})$ intersect on at least one of $1_{\{a,b\}}, 2_{\{a,b\}}, 3_{\{a,b\}}, 4_{\{a,b\}}, \rightarrow_{(a,b)}$). Hence $s \circ n = \mathrm{id}$, and similar arguments show $n \circ s = w \circ e = e \circ w = \mathrm{id}$. Furthermore, $e(n(t_{(a,b,c,d)}))$ and $n(e(t_{(a,b,c,d)}))$ both contain NE_b, so they are equal and $n \circ e = e \circ n$. Similarly $n \circ w = w \circ n$, $s \circ e = e \circ s$ and $s \circ w = w \circ s$.

Lastly, note that if $n(t_{(a,b,c,d)}) = t'_{(e,f,b,a)}$ then $t(\mathsf{N}) = t'(\mathsf{S})$ (else the tile sets intersect on $c_{\{a,b\}}$ for some $c \in C$) and similarly in the east-west direction.

But then take any $t_{(a,b,c,d)} \in F$ and define $\phi : \mathbb{Z}^2 \to F$ by $\phi(0,0) = t_{(a,b,c,d)}$ and $\phi(x, y+1) = n(\phi(x,y))$, $\phi(x+1, y) = e(\phi(x,y))$ for all $x, y \in \mathbb{Z}$. Composing with the map taking a tile set to its associated tile, we obtain a tiling of $\mathbb{Z}^2$. Note that ϕ may not be injective, but this is not a problem. It only implies that the encoded tiling may be periodic.

Next we must show that a tiling gives rise to an exact cover. Fix a tiling $\tau : \mathbb{Z}^2 \to T$. Choose a bijection $\alpha : \mathbb{Z}^2 \to \mathbb{A}$ and (with $\|\cdot\|$ the Euclidean norm on $\mathbb{Z}^2$) define:

$$E = \left\{ (\tau(x,y))_{(\alpha(x,y+1),\alpha(x+1,y+1),\alpha(x+1,y),\alpha(x,y))} : (x,y) \in \mathbb{Z}^2 \right\}$$
$$\cup \left\{ e_{\{a,b\}} : a, b \in \mathbb{A} \wedge \|\tau^{-1}(b) - \tau^{-1}(a)\| > 1 \right\}.$$

Intuitively, α arranges the atoms into a grid. For every unit square in the grid, E contains the tile set corresponding to $\tau(x,y)$ and to the atoms associated to the four corners of the square. This covers (with pairwise disjoint sets) all elements associated to atoms or to those pairs of atoms which τ maps to adjacent points of the grid (i.e. of distance 1). The remaining elements of X are all associated with pairs of atoms not (in this sense) adjacent. We cover these with edge sets.

Finally, X and $\mathcal{S}$ are definable and the construction is effective, so the reduction is complete. $\qquad\square$

Remark 1. In our construction, all sets in $\mathcal{S}$ are finite. Hence EXACTCOVER under this restriction is undecidable as well. More can be said: a finite set of bounded size can be replaced, in a definable way, by all possible linear orderings of it. As a result, the following variant of the problem, where $\mathcal{S}$ consists of non-repeating tuples rather than subsets, remains undecidable:

Problem 3 (EXACTTUPLECOVER).
Input: Definable set X, number n, definable set $\mathcal{S} \subseteq X^{(n)}$
Question: Is there $\mathcal{S}' \subseteq \mathcal{S}$ where every $x \in X$ occurs in exactly one tuple?

We will use this variant in Theorems 5 and 7.

3.2   Hitting set, Satisfiability, 0-1 Integer Linear Programming

A few problems allow straightforward reductions from EXACTCOVER.

Problem 4 (HITTINGSET).
Input: Definable set P, definable set $\mathcal{F}$ of subsets of P
Question: Is there a subset $Q \subseteq P$ such that $|Q \cap C| = 1$ for every $C \in \mathcal{F}$?

In the finite setting [16], Karp demonstrated NP-hardness of this problem, noticing that it is essentially EXACTCOVER in disguise. The same reduction works here.

Theorem 2. *HITTINGSET is undecidable.*

Proof. Given an instance $(X, \mathcal{S})$ of EXACTCOVER, define $P = \mathcal{S}$ and

$$\mathcal{F} = \{\{Y \in \mathcal{S} \mid x \in Y\} \mid x \in X\}.$$

Then exact covers for $(X, \mathcal{S})$ are exactly hitting sets for $(P, \mathcal{F})$.   □

A CNF propositional formula over a set X of variables can be represented as a family of subsets of $X \times \{0, 1\}$, each of the subsets representing a disjunctive clause. We say that the formula is definable if the family is definable. The notion of a satisfying assignment is as in the finite case.

Problem 5 (CNFSAT).
Input: Definable CNF formula ϕ; **Question:** Is ϕ satisfiable?

Theorem 3. *CNFSAT is undecidable.*

Proof. By reduction from HITTINGSET. Given an instance $(P, \mathcal{F})$, write the following formula over variables from P:

$$\left(\bigwedge_{C \in \mathcal{F}} \bigvee_{p \in C} p \right) \wedge \left(\bigwedge_{C \in \mathcal{F}} \bigwedge_{p \neq q \in C} (\neg p \vee \neg q) \right).$$

Its satisfying assignments correspond to hitting sets for $\mathcal{F}$.   □

There is an alternative proof of undecidability for CNFSAT that avoids our reduction from Theorem 1, and proceeds directly by a straightforward reduction from the satisfiability problem for the $\forall\exists$-fragment of first order logic [7]. [1] The reduction from Theorem 3 is still worthwhile though. Later, in Theorem 9, we will show that 3-SAT, and more generally CNFSAT where all clauses are finite, is decidable. Our reduction then implies that, as a counterpoint to Remark 1, the following problems are decidable:

- HITTINGSET for instances $(P, \mathcal{F})$ where all sets in $\mathcal{F}$ are finite,
- EXACTCOVER for instances $(X, \mathcal{S})$ where every $x \in X$ belongs to finitely many sets in S.

One could try to use the reduction from $\forall\exists$-FO to dispense with our reduction from Theorem 1 altogether. Indeed, there is an easy reduction from CNFSAT to HITTINGSET (and thus to EXACTCOVER, by reversing the construction from Theorem 2): given a formula ϕ, begin by creating a HITTINGSET instance over

---

[1] We are grateful to M. Bojańczyk for pointing this out. We do not give the details, not to deprive the reader of the pleasure of solving the corresponding exercise in Section 3.3 of [4].

the variables and their formal negations, and add a set $\{v, \neg v\}$ for each v. It then suffices to add, for each clause C of ϕ, extra elements and sets which force at least one literal in C to be included in any hitting set. This is easy to do: the gadget shown in Figure 2 (where elements are shown as dots and sets are circled) ensures that at least one of the blue elements is included in every hitting set.

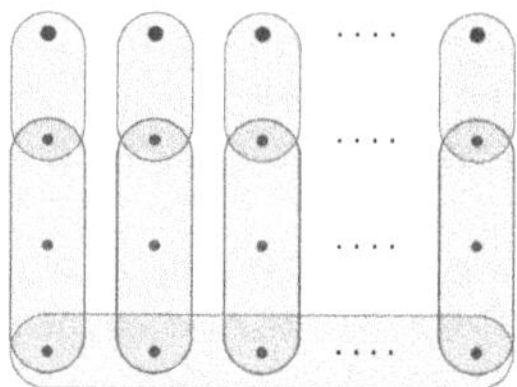

Fig. 2: A HITTINGSET instance forcing the inclusion of a blue element

This proves that HITTINGSET, and by extension EXACTCOVER, are undecidable. However, this construction would not let us show undecidability of EXACTTUPLECOVER (or EXACTCOVER on instances $(X, \mathcal{S})$ where all sets in $\mathcal{S}$ are finite). Our construction from Theorem 1 does prove that (see Remark 1), and it will be an essential starting point of further reductions in Theorems 5 and 7.

We now turn attention to 0-1 Integer Linear Programming. For a definable set X of variables, an equation is a pair (c, n), where $c : X \to \mathbb{Z}$ is a definable function and $n \in \mathbb{Z}$. This is intended to represent the equation $\sum_{x \in X} c_x x = n$. A valuation $v : X \to \{0, 1\}$ solves the equation if (i) $v(x)$ and c_x are simultaneously non-zero for only finitely many x's, and (ii) $\sum_{x \in X} c_x v(x)$ equals n. (The first condition makes the sum well defined.) The problem is then posed as:

Problem 6 (0-1-ILP).
Input: Definable set X, definable set E of equations over X
Question: Does E have a solution?

Theorem 4. *0-1-ILP is undecidable.*

Proof. Given an instance $(P, \mathcal{F})$ of HITTINGSET, put $X = P$ and for each $C \in \mathcal{F}$ add the equation $\sum_{p \in C} p = 1$ to E. Solutions to E are hitting sets for $\mathcal{F}$. $\square$

This result is not entirely new. In [11, Sec. 9], undecidability is proved for Integer Linear Programming but without the 0-1 restriction, with inequalities allowed in addition to equations, and with the additional condition that only finite solutions are sought. However, an inspection of that proof shows that only 0-1 variables are actually used and inequalities can be encoded as equations. Moreover, the proof can be adapted to deal with arbitrary solutions.[2]

---

[2] We are grateful to A. Ghosh, P. Hofman and S. Lasota for a helpful discussion about this issue.

3.3 Hamiltonicity

In the finite setting [16], Karp considered the existence of a Hamiltonian cycle, both in directed and undirected graphs. It is not clear what an infinite cycle would mean, so in the definable setting we study paths which are doubly-infinite, i.e. extending infinitely in both directions. The problem makes sense both for directed and undirected graphs:

Problem 7 ((UN)DIRECTEDHAMILTONICITY).
Input: Definable (un)directed graph G
Question: Does G have a doubly-infinite Hamiltonian path?

Theorem 5. *DIRECTEDHAMILTONICITY is undecidable.*

Proof. By a reduction from EXACTTUPLECOVER. Take as input an instance $(X, \mathcal{S})$ (with $\mathcal{S} \subseteq X^{(n)}$ a set of non-repeating tuples). We may assume that $\mathcal{S}$ is infinite. Let us construct $G = (V, E)$ which has a doubly-infinite Hamiltonian path if and only if $(X, \mathcal{S})$ admits an exact cover.

We intend G to consist of a big independent set of vertices corresponding to elements of X, together with a gadget for each $\overline{x} \in \mathcal{S}$. The gadgets will mediate how a doubly-infinite Hamiltonian path can visit vertices corresponding to elements of X. Formally, define:

$$V = X \uplus \{(\overline{x}, i) \mid \overline{x} \in \mathcal{S}, \ 1 \le i \le 3n + 3\}.$$

The vertices $(\overline{x}, i)$ will form the gadget corresponding to $\overline{x}$; the first and the last of them will be called the *ends* of the gadget.

For edges, first connect vertices in each gadget as illustrated in Figure 3; then connect all ends of all gadgets, joining them into a single infinite clique.

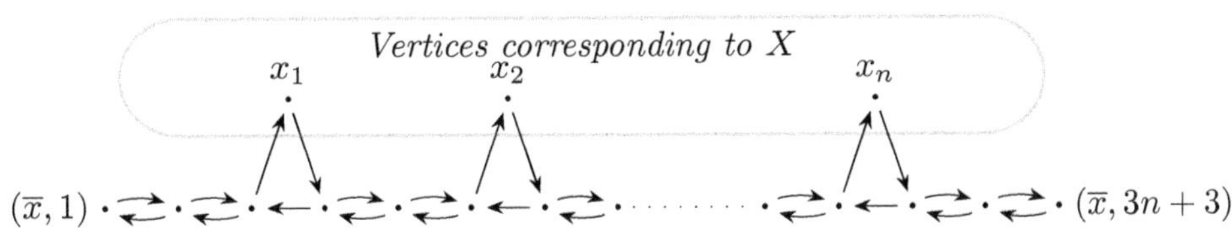

Fig. 3: The gadget corresponding to $\overline{x} = (x_1, \ldots, x_n)$

Consider a doubly-infinite Hamiltonian path in this graph. Note it can visit the gadget corresponding to $(x_1, \ldots, x_n)$ in two ways: from first vertex to last, passing through all of $x_1, \ldots, x_n$, or from last to first, visiting no other vertex along the way (see Figure 4). Hence every $x \in X$ has a unique gadget joined to it which is visited first-to-last. The corresponding tuples form an exact cover.

Conversely, fix an exact cover E. $\mathcal{S}$ is infinite, so there are infinitely many gadgets. Consider a doubly-infinite path which visits gadgets corresponding to E from first to last, and others from last to first. Correctness is clear.   □

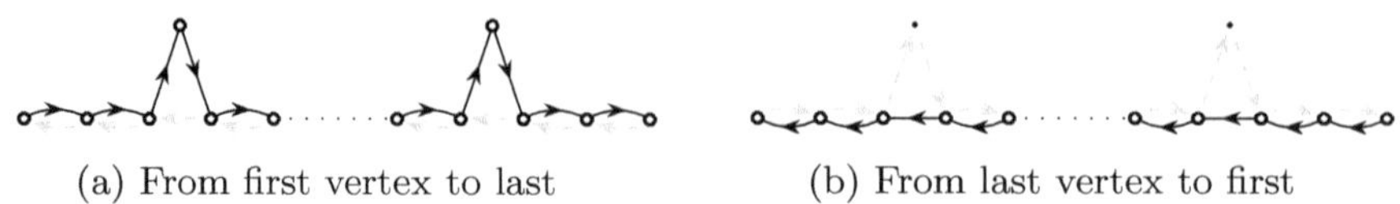

| (a) From first vertex to last | (b) From last vertex to first |

Fig. 4: The two ways to visit a gadget

There is an easy reduction from directed to undirected Hamiltonicity in the classical case [16], and it goes through in the definable setting with no change.

Theorem 6. UNDIRECTEDHAMILTONICITY *is undecidable.*

Proof. By reduction from DIRECTEDHAMILTONICITY. Given a definable digraph $G = (V, E)$, form an undirected graph G' with three vertices v_{in}, v_{mid}, v_{out} for each $v \in V$. For each $v \in V$ add edges $\{v_{in}, v_{mid}\}, \{v_{mid}, v_{out}\}$, and for each $(u, v) \in E$ add an edge $\{u_{out}, v_{in}\}$. Then a Hamiltonian path in G' must - up to reversal - always visit v_{out} after v_{mid} after v_{in}. Hence these correspond to Hamiltonian paths in G. $\qquad\square$

It is not difficult to modify our constructions to work for *singly*-infinite Hamiltonian paths, which have a starting point and extend to infinity in one direction. The corresponding problems remain undecidable.

3.4   3D matching

3D matching is a generalisation of bipartite matching to 3-hypergraphs. Some classical formulations of this problem do not adapt to an infinite setting well, but the following one does.

Problem 8 (3DMATCHING).
Input: Definable sets A, B, C, definable set $R \subseteq A \times B \times C$
Question: Is there a subset of R whose projections onto A, B, C are bijections?

Karp in [16] proved hardness by a reduction from EXACTCOVER. That reduction does not quite work in the definable setting, but we give one that works, based on the same general idea.

Theorem 7. 3DMATCHING *is undecidable.*

Proof. By reduction from EXACTTUPLECOVER. Given an instance $(X, \mathcal{S})$ (with $\mathcal{S} \subseteq X^{(n)}$ a set of non-repeating tuples), define:

$$Y = \{(\overline{x}, i) \mid \overline{x} \in \mathcal{S}, 0 \le i < n\}, \qquad A = Y \uplus \mathbb{A}, \qquad B = X \uplus \mathbb{A}.$$

Then let $R \subseteq A \times A \times B$ contain three types of triples:

(i) $((\overline{x}, i), (\overline{x}, i), x_i)$, where $(\overline{x}, i) \in Y$ for $\overline{x} = (x_0, \ldots, x_{n-1})$;
(ii) $((\overline{x}, i), (\overline{x}, i + 1), a)$, where $(\overline{x}, i) \in Y$, $a \in \mathbb{A}$, and the $+1$ is modulo n,
(iii) all triples from $\mathbb{A}^3$.

In a 3D matching for R, each $(\overline{x}, i)$ (from either of the first two components) can be covered by a triple of type (i) or of type (ii). The type (i) triple covers x_i in the third component as well, while the type (ii) triple does not. Importantly, if $(\overline{x}, i)$ is covered by a type (ii) triple, then all $(\overline{x}, j)$ in both components must be covered by type (ii) triples too. Hence for each $x \in X$ there is exactly one tuple in $\mathcal{S}$ which contains x and is associated to elements $(\overline{x}, i)$ covered by triples of type (i). These tuples form an exact cover.

Conversely, take an exact cover $E \subseteq S$. For each $\overline{x} \in E$, include all the triples of type (i) associated to it in the matching. For all other tuples, include a minimal set of triples of type (ii), each taking a fresh atom as the third component. We may assume there are infinitely many atoms left unused in B; choose any bijection σ with all of $\mathbb{A}$ and include $(\sigma a, \sigma a, a) \in \mathbb{A}^3$ for each such unused a. $\square$

4   Decidability from extreme amenability

We now turn attention to those problems from Karp's list that remain decidable in the definable setting. First we mildly generalise the technique used in [19], used there to prove that finite-template Constraint Satisfaction Problems are decidable. This generalisation will let us prove decidability for a whole range of Karp's problems in one stroke. The few remaining problems, which are not covered by the technique, will be dealt with in Sections 5-6.

4.1   Compact sets of structures

In this section we need to consider both equality and ordered atoms. The following lemmas apply to both these structures (indeed to all oligomorphic [15] atom structures).

Fix a finite relational signature σ and a definable, equivariant set X. The set Σ of all σ-structures on X comes equipped with a topology, with a basis of neighbourhoods of the form:

$$\mathcal{B}_Y(M) = \{N \in \Sigma \mid N|_Y = M|_Y\}$$

for $Y \subseteq X$ finite and $M \in \Sigma$, where $M|_Y$ is the induced substructure of M.

Lemma 1. *The space Σ is compact, and the canonical action of $\mathrm{Aut}(\mathbb{A})$ on Σ is continuous.*

Proof. The set Σ is in bijection with $\prod_{R \in \sigma} \{0, 1\}^{X^{\mathrm{arity}(R)}}$, and if σ is finite then our topology is the product topology on this set, which is compact by Tychonoff's theorem. The action is continuous since $\mathcal{B}_Y(M) \cdot \pi = \mathcal{B}_{Y \cdot \pi}(M \cdot \pi)$ for every π. $\square$

For a structure $\mathcal{M} \in \Sigma$ we write $\mathrm{Age}_{\mathbb{A}}(M)$ to denote the $\mathrm{Aut}(\mathbb{A})$-closure of the set of finite induced substructures of M, and define a preorder $\sqsubseteq$ on Σ by

$$M \sqsubseteq N \iff \mathrm{Age}_{\mathbb{A}}(M) \subseteq \mathrm{Age}_{\mathbb{A}}(N).$$

For a structure M, the down-set $M{\downarrow}$ is the set of all N such that $N \sqsubseteq M$.

Lemma 2. *For every $M \in \Sigma$, the space $M{\downarrow} \subseteq \Sigma$ is equivariant and compact.*

Proof. Equivariance is obvious: if M, N are in the same $\mathrm{Aut}(\mathbb{A})$-orbit of Σ then they are $\sqsubseteq$-equivalent. For compactness, by Lemma 1 it is enough to show that $M{\downarrow}$ is closed. Take any $N \not\sqsubseteq M$. Then for some finite Y the structure $N|_Y$ is not in $\mathrm{Age}_\mathbb{A}(M)$ so $N|_Y \neq M|_Y$ and the open $\mathcal{B}_Y(N)$ is disjoint from $M{\downarrow}$. $\square$

The following result would apply to any atoms with the Ramsey property [17], but we will only need it for ordered atoms with finitely many pairwise distinct constants, so this is how we formulate it.

Theorem 8. *Over atoms $\mathbb{A} = (\mathbb{Q}, \leq, a_1, \ldots, a_n)$, for any X and σ as above, every down-set in Σ contains an equivariant structure.*

Proof. Consider any down-set $M{\downarrow} \subseteq \Sigma$. By Lemma 2 it is equivariant, so the continuous action from Lemma 1 restricts to it. By Pestov's theorem [25], the group $\mathrm{Aut}(\mathbb{Q}, \leq)$ is *extremely amenable*: every continuous action of it on a nonempty compact space has a fixpoint. Fixpoints of the canonical action of $\mathrm{Aut}(\mathbb{A})$ are exactly equivariant elements. This proves the theorem for $n = 0$. For the general statement, note that $\mathrm{Aut}(\mathbb{A})$ is extremely amenable. This follows from the isomorphism $\mathrm{Aut}(\mathbb{A}) \cong (\mathrm{Aut}(\mathbb{Q}, \leq))^{(n+1)}$, since the class of extremely amenable groups is closed under direct products (see [17, Lem. 6.7]). $\square$

4.2 Decidability

Our decidability proofs in this section will follow the pattern used in [19]. Given a problem instance S-definable over equality atoms, we:

(i) See it as an equivariant structure over ordered atoms with constants from S;

(ii) Understand solutions to that instance as structures over a finite relational signature;

(iii) Show that if the set of legal solutions is non-empty then it contains a down-set in the sense of Sec. 4.1 (this is true e.g. if the set of legal solutions is downwards-closed with respect to $\sqsubseteq$);

(iv) Infer from Theorem 8 that if a solution exists then an equivariant one exists;

(v) Show that looking for an equivariant solution is a decidable problem.

As a first example, consider the restriction of CNFSAT to formulas with finite clauses. The decidability of this problem follows from [19, Thm. 3], but we reprove it here as an illustrative application of Thm. 8.

Problem 9 (3-SAT).
Input: Definable CNF formula ϕ with at most 3 literals per clause
Question: Is ϕ satisfiable?

Theorem 9. *3-SAT is decidable.*

Proof. Take a 3-CNF formula ϕ, S-definable over equality atoms, over an S-definable set X of variables, for $S \subseteq \mathbb{A}$. (i) If we impose on $\mathbb{A}$ any ordering isomorphic to the total order of the rationals, ϕ remains S-definable over ordered atoms. Adding the atoms from S to $\mathbb{A}$ as constants, ϕ becomes equivariant.

(ii) A valuation for ϕ (satisfying or not) can be seen as a predicate on X, i.e. a structure over a signature σ with a single predicate symbol. (iii) The set of satisfying valuations is described by an equivariant set of forbidden finite substructures, namely the partial valuations which invalidate one of the clauses of ϕ. As a result, the set of satisfying valuations for ϕ is downwards-closed, so (iv) by Theorem 8 it either is empty or contains an equivariant valuation.

To check if ϕ is satisfiable, it is therefore enough to look for a satisfying valuation which is equivariant over our extended atoms or, equivalently, S-definable over ordered atoms. (v) This is easy to do: there are only finitely many S-orbits of variables in X, and an S-definable valuation must be constant in every orbit, so there are finitely many valuations to consider, and for each of them it is easy to decide whether it satisfies every clause in ϕ. $\square$

The same argument shows that k-SAT is decidable for any k. It also means that CNFSAT restricted to formulas with finite clauses, is decidable. This is because every such formula ϕ, having an orbit-finite set of clauses, has an upper bound on the size of a clause, so it is an instance of k-SAT for some k which can be computed from ϕ.

It may be illustrative to see where the argument fails for the unrestricted CNFSAT. Consider an equivariant CNF-formula with two clauses over $X = \{x_a \mid a \in \mathbb{A}\}$:

$$\phi = \left(\bigvee_{a \in \mathbb{A}} x_a \right) \wedge \left(\bigvee_{a \in \mathbb{A}} \neg x_a \right).$$

Consider any satisfying valuation, understood as a structure M on X over a single predicate symbol. Assume that in the valuation infinitely many variables are false. (This does not lose generality: the symmetric case is that infinitely many variables are *true*.) Then $\mathrm{Age}_{\mathbb{A}}(M)$ contains (among other things) all finite substructures where no element satisfies the predicate. Let N be a structure on X where no element satisfies the predicate. Then $N \sqsubseteq M$. But the valuation corresponding to N does not satisfy ϕ, so the set of satisfying valuations does not contain $M{\downarrow}$ and Theorem 8 does not apply. Indeed, even though ϕ has plenty of satisfying valuations, none are equivariant.

The same machinery directly applies to a few more problems. In all these, the input is a definable graph G.

Problem 10 (VERTEXCOVER).
Question: Does G have a finite vertex cover?

Problem 11 (FEEDBACKVERTEXSET).
Question: Is there a finite set of vertices whose removal makes G acyclic?

Problem 12 (FEEDBACKARCSET).
Question: Is there a finite set of edges whose removal makes G acyclic?

In the finite setting of [16], the input to these problems includes a number k, and one asks whether a vertex cover (etc.) of size at most k exists. We could do the same here, but those problems would be decidable for very easy reasons: for a fixed k, there are only orbit-finitely many sets of vertices/edges of size k, and they can be effectively enumerated in the search for a solution. The statements as above are more interesting, but nevertheless:

Theorem 10. *VERTEXCOVER, FEEDBACKVERTEXSET and FEEDBACKARCSET are decidable.*

Proof. We proceed as for Thm. 9, arguing only for part (iii). Consider VERTEXCOVER. A choice of vertices in G can be seen as a predicate M on the set of all vertices. If M is a finite vertex cover (of size, say, k) then $\mathrm{Age}_{\mathbb{A}}(M)$ can be described by a set of forbidden finite structures: (a) those where more than k vertices satisfy the predicate, and (b) those where neither end of some edge satisfies the predicate. This implies that $M{\downarrow}$ contains only finite vertex covers, and Thm. 8 applies.

The arguments for the remaining two problems are similar. $\qquad\square$

Problem 13 (COLORABILITY).
Question: Does G have a vertex coloring with finitely many colors?

Problem 14 (CLIQUECOVER).
Question: Is V a disjoint union of finitely many cliques (where $G = (V, E)$)?

These two problems are equivalent (replace G with its complement to reduce one to the other), so let us focus on colorability. As for Probs. 10-12, in the finite setting of [16] the input includes a number k and a k-coloring is sought for. Unlike for Probs. 10-12, that problem remains interesting in our setting. Decidability of k-COLORABILITY follows from [19, Thm. 3], but it is easy to give an argument analogous to Thm. 9. Here, k-colorings can be seen as structures over a signature with k predicate symbols, and the set of legal k-colorings for any fixed G is downwards-closed.

Unrestricted colorability easily follows from this:

Theorem 11. *COLORABILITY (therefore also CLIQUECOVER) is decidable.*

Proof. Every finite coloring is a k-coloring for some k, so if one exists then an equivariant k-coloring exists. An equivariant coloring must be constant on every orbit of vertices, so it cannot use more colors than there are orbits. The number of orbits can be computed from G, and the set of candidate equivariant colorings can be enumerated and checked for legality. $\qquad\square$

5   Other decidable cases

The key feature of the problems from Sec. 4 was that, whenever they admit a solution to an S-definable instance, then an S-definable solution also exists. Some problems do not have this property, for example:

Problem 15 (CLIQUE).
Input: Definable graph G; **Question:** Does G contain an infinite clique?

Consider for instance a graph with ordered pairs from $\mathbb{A}^{(2)}$ as vertices, with edges between exactly those vertices that share the first component. This equivariant graph contains an infinite clique, indeed it is a disjoint union of infinite cliques, but none of these cliques are equivariant.

Nevertheless, the problem is decidable. This has already been proved for automatic graphs [24,27], but for definable graphs a more direct argument exists.

Theorem 12. *CLIQUE is decidable.*

Proof. A definable graph G is orbit-finite, so any infinite clique in it must have an infinite intersection with one of the orbits. We may therefore focus on graphs with one orbit of vertices. Further, in an infinite clique, every edge $\{v_1, v_2\}$ can be colored with its orbit. There are finitely many orbits of edges, so by Ramsey's theorem the clique contains an infinite sub-clique where every edge is in the same orbit. It is therefore enough to decide, for a given orbit of edges in G, whether G contains infinitely many vertices whose every pair belongs to that orbit.

Over equality atoms, this is easy to do: the condition holds for the orbit of $\{v_1, v_2\}$ if and only if v_1 and v_2 are in the same S-orbit, where $S = \mathrm{supp}(v_1) \cap \mathrm{supp}(v_2)$. (This condition is easy to decide by looking at v_1 and v_2.)

To see this, first assume that $v_1 \neq v_2$ are in the same S-orbit. Since $v_1 \neq v_2$, this S-orbit is infinite. Moreover, one can find infinitely many vertices $v_1, v_2, v_3 \ldots$ in this S-orbit so that $\mathrm{supp}(v_i) \cap \mathrm{supp}(v_j) = S$ for all $i \neq j$. For equality atoms, this implies that all pairs $\{v_i, v_j\}$ are in the same S-orbit.

In the other direction, assume an infinite set $v_1, v_2, \ldots$ of vertices such that the pairs $\{v_i, v_j\}$ are in the same orbit for all $i \neq j$. The size of $\mathrm{supp}(v_i)$ and $S_{ij} = \mathrm{supp}(v_i) \cap \mathrm{supp}(v_j)$ does not depend on i and j, so by basic combinatorics the intersection S_{ij} itself does not depend on i and j; call this shared intersection S. For any fixed $i \neq j$, since v_i and v_j are in the same orbit, there is an atom automorphims π_{ij} that preserves S set-wise and such that $v_i \cdot \pi_{ij} = v_j$. Keeping i fixed and choosing more than $|S|!$ j's, there are some $j \neq j'$ such that π_{ij} and $\pi_{ij'}$ agree on S. Then the composition of π_{ij}^{-1} and $\pi_{ij'}$ maps v_j to $v_{j'}$ and fixes S pointwise, hence v_j and $v_{j'}$ are in the same S-orbit. $\square$

The following was proved decidable for automatic structures in [23] with a direct argument, rather than by an immediate reduction to CLIQUE:

Problem 16 (SETPACKING).
Input: Definable set X, definable family $\mathcal{S}$ of subsets of X
Question: Does $\mathcal{S}$ contain an infinite, pairwise disjoint subfamily?

Theorem 13. *SETPACKING is decidable.*

Proof. Given X and $\mathcal{S}$, solve CLIQUE for a graph with $\mathcal{S}$ as the set of vertices, and an edge between two sets if and only if they are disjoint. $\square$

Here is another problem that does not fall into the scope of Sec. 4:

Problem 17 (SETCOVERING).
Input: Definable set X, definable family $\mathcal{S}$ of subsets of X
Question: Does $\mathcal{S}$ contain a finite subfamily C whose union is X?

In [24], a variant of this problem where C is required to be co-infinite (i.e. such that $\mathcal{S} \setminus C$ is infinite) rather than finite, was proved decidable for all automatic structures. The technique used there does not seem applicable to our formulation, but at least on definable instances the following easy argument works.

Theorem 14. *SETCOVERING is decidable.*

Proof. Clearly we can reduce to the case where X is a single infinite orbit. Assume that a finite family $C \subseteq \mathcal{S}$ covers X. Every set in C has some finite support, and the union of these supports for all sets in C is also finite, so there must be some $x \in Y \in C$ such that $\mathrm{supp}(x)$ and $\mathrm{supp}(Y)$ are disjoint modulo $\mathrm{supp}(X, \mathcal{S})$ (by which we mean their intersection lies in $\mathrm{supp}(X, \mathcal{S})$; this is empty if the instance is equivariant).

On the other hand, assume that such $x \in Y \in \mathcal{S}$ exist. Then Y contains *all* $z \in X$ such that $\mathrm{supp}(z)$ and $\mathrm{supp}(Y)$ are disjoint (again, and in the following, this is modulo $\mathrm{supp}(X, \mathcal{S})$). Take $Y_1, Y_2 \ldots, Y_{|\mathrm{supp}(x)|+1}$ in the orbit of Y with pairwise disjoint supports. Then every element of X has support disjoint from some of the Y_i, so it belongs to Y_i. Hence all the Y_i's jointly cover X.

Finally, it is easy to effectively search for $x \in Y \in \mathcal{S}$ as above. $\qquad\square$

This gives an alternative decidability proof of VERTEXCOVER (Prob. 10), via a straightforward reduction used already by Karp [16]. The argument in Thm. 10 is still worth making though, as it exhibits additional structure in the space of vertex covers that is missing in the more general case of set coverings.

6   Weighted problems

A few problems on Karp's list involve adding up sets of numbers, be it weights of graph edges, penalties in job sequencing etc. This poses an obvious difficulty in generalising these problems to infinite structures. One may try to follow the idea that we used in Secs. 4-5 and, rather than comparing various quantities to a fixed input number k, require them to be finite (or infinite). In the process, however, the essence of a weighted problem usually seems to be lost.

Consider, for example, the problem of finding a minimal Steiner tree in a weighted graph. In the finite formulation [16], given an edge-weighted graph G, a subset W of its vertices and a number k, one asks whether G has a tree of total weight at most k that spans all the vertices in W. In an infinite setting, for a "total weight" to make sense, one has to restrict attention to graphs with non-negative weights. Then one could ask whether a spanning tree with a *finite* total weight exists. (The problem with a given bound k is easily decidable for the same reason as Probs. 10-12.) This is a valid question, but the values of weights are lost in it: it is equivalent to asking whether a spanning tree exists that uses only finitely many edges with non-zero weights. The problem then simplifies to:

Problem 18 (STEINERTREE).

Input: Definable graph $G = (V, E)$, definable subsets $W \subseteq V$ and $F \subseteq E$

Question: Does G have a tree that spans all vertices from W and uses finitely many edges from F?

Theorem 15. *STEINERTREE is decidable.*

Proof. One needs to check if (G is connected and) the vertices from W fall into finitely many connected components of the graph $(V, E \setminus F)$. To this end, compute its transitive closure (such fixpoint calculations are effective; see [21] for a general study or [22] for an implementation), restrict to the connected components with vertices from W, and solve CLIQUECOVER (Prob. 14).   □

Admittedly, the essence of the original weighted problem is lost to some extent in this formulation. For other problems, it gets worse. Karp's problem of finding a maximal cut in a weighted graph becomes:

Problem 19 (MAXCUT).

Input: Definable graph $G = (V, E)$, definable subset $F \subseteq E$ of edges

Question: Is there a subset $W \subseteq V$ such that F contains infinitely many edges between W and $V \setminus W$?

Here even the set E becomes irrelevant. It is easy to see that the condition holds if and only if F is infinite, which is obviously decidable.

The remaining three problems on Karp's list [16] are: knapsack, job sequencing and number partitioning. These seem even less open to infinite generalisations than the two above, so we leave them untreated.

7   Conclusion

We chose to consider Karp's 21 problems simply because they are famous and well-studied. But, in a sense, the choice was principled: by fixing the set in advance, we forced our hand to look at a whole range of possible properties, in the definable setting, of problems which all behave similarly over finite structures. So it is interesting how different these properties turned out to be: some problems (e.g. EXACTCOVER) are undecidable for nontrivial reasons, some (e.g. 3-SAT) are decidable by easy algorithms which are correct for deep reasons, some (e.g. SET-COVERING) are decidable by algorithms which are correct for mundane reasons, and some (e.g. KNAPSACK) do not seem to generalise to the definable setting at all. Some of Karp's reductions (e.g. EXACTCOVER to HITTINGSET) remain in force, others (e.g. CNFSAT to 3-SAT) break down. This paints quite an interesting landscape, and shows that we now have a small arsenal of techniques to analyse simple computational problems in the definable setting.

References

1. Berger, R.: The undecidability of the domino problem. Mem. Amer. Math. Soc. **66**, 72 (1966). https://doi.org/10.1090/memo/0066
2. Blumensath, A., Grädel, E.: Automatic structures. In: Procs. 15th Symposium on Logic in Computer Science. pp. 51–62 (2000). https://doi.org/10.1109/LICS.2000.855755
3. Blumensath, A., Grädel, E.: Finite presentations of infinite structures: Automata and interpretations. Theory of Computing Systems **37**, 641–674 (2004). https://doi.org/10.1007/s00224-004-1133-y
4. Bojańczyk, M.: Slightly infinite sets (2025), https://www.mimuw.edu.pl/~bojan/papers/notes-July3.pdf, draft from July 3, 2025
5. Bojańczyk, M., Fijalkow, J., Klin, B., Moerman, J.: Orbit-finite-dimensional vector spaces and weighted register automata. Theoretics **13** (2024). https://doi.org/10.46298/theoretics.24
6. Bojańczyk, M., Klin, B., Lasota, S.: Automata theory in nominal sets. Logical Methods in Computer Science **10** (2014). https://doi.org/10.2168/LMCS-10(3:4)2014
7. Börger, E., Grädel, E., Gurevich, Y.: The classical decision problem. Springer (1997)
8. Burr, S.A.: Some undecidable problems involving the edge-coloring and vertex-coloring of graphs. Discrete Mathematics **50**, 171–177 (1984). https://doi.org/10.1016/0012-365X(84)90046-3
9. Freedman, M.H.: K-sat on groups and undecidability. In: Procs. 30th ACM Symposium on Theory of Computing. p. 572576 (1998). https://doi.org/10.1145/276698.276871
10. Ghosh, A., Hofman, P., Lasota, S.: Solvability of orbit-finite systems of linear equations. In: Procs. 37th Symposium on Logic in Computer Science (2022). https://doi.org/10.1145/3531130.3533333
11. Ghosh, A., Hofman, P., Lasota, S.: Orbit-finite linear programming. J. ACM **72**(1) (2025). https://doi.org/10.1145/3703909
12. Grädel, E.: Automatic structures: Twenty years later. In: Procs. 35th Symposium on Logic in Computer Science. p. 2134 (2020). https://doi.org/10.1145/3373718.3394734
13. Hirst, J.L., Lempp, S.: Infinite versions of some problems from finite complexity theory. Notre Dame J. Formal Logic **37**(4), 545–553 (1996). https://doi.org/10.1305/ndjfl/1040046141
14. Hirst, T., Harel, D.: Taking it to the limit: On infinite variants of NP-complete problems. Journal of Computer and System Sciences **53**(2), 180–193 (1996). https://doi.org/10.1006/jcss.1996.0060
15. Hodges, W.: Model Theory. Cambridge University Press (1993). https://doi.org/10.1017/CBO9780511551574
16. Karp, R.M.: Reducibility among combinatorial problems. In: Procs. Symposium on the Complexity of Computer Computations. pp. 85–103 (1972). https://doi.org/10.1007/978-1-4684-2001-2_9
17. Kechris, A.S., Pestov, V.G., Todorcevic, S.: Fraïssé limits, Ramsey theory, and topological dynamics of automorphism groups. GAFA, Geom. Funct. Anal. **15** (2005). https://doi.org/10.1007/s00039-005-0503-1
18. Khoussainov, B., Nerode, A.: Automatic presentations of structures. In: Procs. International Workshop on Logic and Computational Complexity. pp. 367–392 (1995). https://doi.org/10.1007/3-540-60178-3_93

19. Klin, B., Kopczyński, E., Ochremiak, J., Toruńczyk, S.: Locally finite constraint satisfaction problems. In: Procs. 30th Symposium on Logic in Computer Science. pp. 475–486 (2015). https://doi.org/10.1109/LICS.2015.51
20. Klin, B., Lasota, S., Ochremiak, J., Toruńczyk, S.: Homomorphism problems for first-order definable structures. In: Procs. 36th Conference on Foundations of Software Technology and Theoretical Computer Science. LIPIcs, vol. 65, pp. 14:1–14:15 (2016). https://doi.org/10.4230/LIPIcs.FSTTCS.2016.14
21. Klin, B., Łełyk, M.: Scalar and vectorial μ-calculus with atoms. Logical Methods in Computer Science 15 (2019). https://doi.org/10.23638/LMCS-15(4:5)2019
22. Klin, B., Szynwelski, M.: SMT solving for functional programming over infinite structures. In: Procs. 6th Workshop on Mathematically Structured Functional Programming. Electronic Proceedings in Theoretical Computer Science, vol. 207, pp. 57–75 (2016). https://doi.org/10.4204/EPTCS.207.3
23. Köcher, C.: Analyse der Entscheidbarkeit diverser Probleme in automatischen Graphen. Bachelor's thesis, TU Ilmenau (2014)
24. Kuske, D., Lohrey, M.: Some natural decision problems in automatic graphs. The Journal of Symbolic Logic 75(2), 678710 (2010). https://doi.org/10.2178/jsl/1268917499
25. Pestov, V.: On free actions, minimal flows, and a problem by Ellis. Trans. Amer. Math. Soc. 350 (1998). https://doi.org/10.1090/S0002-9947-98-02329-0
26. Pitts, A.M.: Nominal sets: names and symmetry in computer science. Cambridge University Press (2013). https://doi.org/10.1017/CBO9781139084673
27. Rubin, S.: Automata presenting structures: A survey of the finite string case. Bull. Symbolic Logic 14 (2008). https://doi.org/10.2178/bsl/1208442827
28. Wang, H.: Proving theorems by pattern recognition II. The Bell System Technical Journal 40, 1–41 (1961). https://doi.org/10.1002/j.1538-7305.1961.tb03975.x

Complete FSM Testing Using Strong Separability

Robert M. Hierons[1] and Mohammad Reza Mousavi[2]

[1] School of Computer Science
University of Sheffield, UK, `r.hierons@sheffield.ac.uk`
[2] Department of Informatics
King's College London, UK, `mohammad.mousavi@kcl.ac.uk`

Abstract. Apartness is a concept developed in constructive mathematics, which has resurfaced in the areas of model learning and model-based testing. We identify some fundamental shortcomings of apartness in quantitative models, such as in hybrid and stochastic systems. We propose a closely-related alternative, called strong separability and show that using it to replace apartness addresses the identified shortcomings. We adapt a well-known complete model-based testing method, the Harmonized State Identifiers (HSI) method, to adopt strong separability. We prove that the adapted HSI method is complete. As far as we are aware, this is the first work to show how complete test suites can be generated for quantitative models such as those found in the development of cyber-physical systems.

Keywords: Finite state machine testing · cyber-physical systems · apartness · strong separability

1 Introduction

1.1 Background

Testing is an important yet costly part of system quality assurance. There has been significant interest in systematic, automated test generation techniques. Model-based testing (MBT) is an important class of such techniques in which test generation is based on a formal model (or *specification*) that defines the set of behaviours of a correct *system under test (SUT)*. Typically, a behaviour is a sequence of inputs and outputs and testing is black-box: we cannot observe the state of the SUT. There are many MBT techniques, with corresponding tools and evidence of effectiveness when applied to industrial systems (see examples of applications [10, 15, 17, 22, 24, 28, 34]; and also survey papers for an overview [18, 20, 25, 30]).

To formally reason about MBT effectiveness, one assumes a common semantic domain between the specification and the SUT [16]. Most MBT work has concerned state-based semantic domains: labelled transition systems (LTSs) [28] or finite state machines (FSMs) [21]. However, the developer might use a different formalism, with a test tool mapping a model to an FSM or LTS and an adapter handling differences in abstraction between a model and the SUT.

© The Author(s) 2026
N. Bertrand and S. Milius (Eds.): FoSSaCS 2026, LNCS 16503, pp. 328–349, 2026.
https://doi.org/10.1007/978-3-032-22730-0_16

The behaviour of an FSM is defined by transitions between states, with a transition having a corresponding input and output. FSM-based test generation techniques typically aim to find two types of faults: output faults (a transition produces the wrong output); and state-transfer faults (a transition takes the SUT to the wrong state). State-transfer faults can lead to the SUT having more states than the specification FSM.

There are many FSM-based test generation techniques that aim to find state-transfer faults in addition to output faults [8]; these techniques utilise input sequences that *separate* states[3] of the specification M in order to *identify* (or *check*) states. Most FSM-based test techniques concern specification FSMs that are minimal, deterministic, and completely-specified (cf. Section 2) and we also consider such FSMs (we discuss further extensions in the Conclusions). Of particular interest are test generation techniques that produce m-complete test suites: test suites that are guaranteed to determine correctness as long as the SUT behaves like an (unknown) FSM that has at most m states. Classical examples of such m-complete methods include the W-method [10, 34], Wp-method [15], and HSI method [22].

1.2 Problem Definition

This paper aims to expand the applicability of complete FSM-based test techniques to a range of modern systems for which test generation is particularly challenging. We are interested in generating m-complete tests for systems in which outputs are drawn from a continuous domain and observations are noisy. This is characteristic of cyber-physical systems [19] and the conformance problem for such systems has been extensively researched [6, 1, 12]. For example, in testing automotive [29] and healthcare [26] systems, it is necessary to cater for error margins and noise. For example, $(\tau,\ \epsilon)$-conformance [1] incorporates error margins in time and space. However, it has not previously been possible to develop $(m\text{-})$complete testing theories for such notions of conformance. Our solution can further serve as a building block for active FSM learning of cyber-physical systems using, e.g., variants of the celebrated L^* [2] algorithm, making it possible to model-check complex implementations through their learned FSM models. We will elaborate on this in the discussion of future work.

1.3 Contributions

To achieve this, we need to reconsider what we mean by correctness (conformance) in scenarios such as those sketched above. It no longer makes sense to require that, for each input sequence $\bar{x}$, the SUT and specification produce the *same* output sequence. Instead, we require that the output sequence produced by the SUT (i.e., observed in testing) is *similar* to that produced by the specification

---

[3] An input sequence $\bar{x}$ separates two states s_1 and s_2 of M if the input of $\bar{x}$ leads to different output sequences when applied in s_1 and s_2.

[6, 1, 12], e.g., using a distance metric. This leads to a different conformance relation and so the proofs of completeness, for current FSM-based test techniques, do not apply. In this paper we also show that the W, Wp, and HSI methods need not be complete in this context.

We took as a starting point recent work by Vaandrager [31] that shows that it is possible to reason about the completeness of several FSM-based test techniques in terms of *apartness*. Apartness is a form of inequality used in constructive mathematics. Vaandrager [31] showed how the notion of apartness can be used in the generation of an m-complete test set. We found, however, that the notion of apartness is too strong for the scenario of interest in this paper (see Section 3).

In this paper, we weaken apartness in a way that allows us to reason about FSM-based testing when conformance is based on a metric μ and threshold t. We start by specifying a corresponding notion of conformance, which is reminiscent of the notions of robustness [14, 9], metric bisimulation [7, 6, 11] and conformance testing [1, 12, 4, 5]. We then define strong separability, which is strictly weaker than apartness and stronger than separability. We consider the case where the states of the specification FSM M are pairwise strongly separable. We give a sufficient condition for a test suite to be m-complete, defined in terms of strong separability. We adapt the HSI-method to the scenarios we consider and show that this returns m-complete test suites. This completeness result extends to the W-method and the Wp-method. Note that these are exactly the FSM-based test generation techniques considered by Vaandrager [31].

To summarise, the paper makes the following main contributions.

- We show that there are scenarios (classes of system, with a corresponding definition of conformance) in which the notion of apartness is too strong.
- We define strong separability and demonstrate that this is applicable in the identified scenarios.
- We give a sufficient condition for a test suite to be m-complete and show that test suites produced by the well-known Harmonized State Identifiers (HSI) method (and similar ones such as the Wp and W-methods) satisfy this condition if the states of the specification FSM are pairwise strongly separable and, for state identification, we use sequences that strongly separate states.
- We illustrate all concepts through our running example and use this to show that the W, Wp and HSI methods are not complete if we do not adapt them (i.e., if we use separability rather than strong separability).

1.4 Running Example

To illustrate the concepts and their potential applicability, we use the following hybrid systems model. The same example can be extended to cover stochastic aspects, but for the sake of readability we use a minimal running example.

Example 1. Consider a thermostat that senses the temperatures and uses it to switch a heater on/off and control the room temperature. Figure 1 gives a diagrammatic representation of the behaviour. The system starts at time zero and

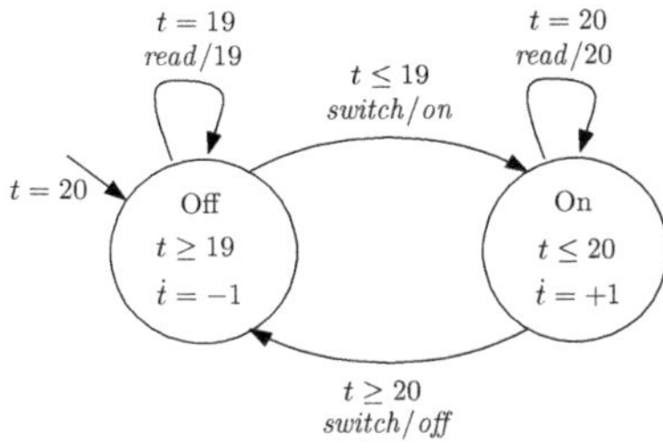

Fig. 1. Continuous Thermostat System

the room temperature at 20°C (in the discrete state Off) and the temperature decreases linearly with a first derivative -1. When the temperature reaches 19°C, the display can be updated through a "read" action (which outputs 19) and the heater is switched on through the "switch" action, which outputs "on". The behaviour in the state labelled "On" is similar, with the main difference being that the temperature increases linearly with the first derivative equal to 1.

Consider an implementation in which the initial condition and the guards differ by 0.5°C. We will compare such an implementation (as well as another faulty one) against the specification based on a notion of conformance in the remainder of the paper. The idea is to use an approximate notion of conformance that allows for an error margin and yet, produce test-cases that are complete, i.e., test cases that will fail on any non-conforming implementation with at most m states. Consider, for example, a notion of conformance that has a threshold of 0.5°C for temperature differences (and no difference in the switching outputs) and thus, it accepts the implementation that differs only 0.5°C in temperature with the specification in Figure 1, while rejecting those with temperature differences that exceed the threshold or have incorrect switching behaviour. In this paper we show how complete test suites can be produced for this specification.

1.5 Structure

The paper is structured as follows. Section 2 provides preliminary definitions and Section 3 extends FSM concepts and definitions to the scenarios of interest. Within this we define separability and apartness and show that there are classes of system, including the Thermostat example, where separability is not an apartness relation. This leads to us defining strong separability. Section 4 gives a sufficient condition for a test suite to be m-complete, with a consequence being that the well-known W, Wp, and HSI methods return m-complete test suites for the notion of conformance and class of system considered if we base the choice of state identifiers on strong separability. We also show that separability is insufficient for these test generation techniques. Finally, in Section 5, we draw conclusions and describe possible lines of future work.

2   Preliminaries

In this section, we define what we mean by a Finite State Machine and provide classical definitions. In the next section, we adapt this to the context of interest where, for example, outputs may be drawn from a metric space.

Definition 1. *A Finite State Machine (FSM) M is defined by a tuple $(S, s_0, X, Y, \delta, \lambda)$ in which: S is the finite set of states; $s_0 \in S$ is the initial state; X is the finite input alphabet; Y is the output alphabet; $\delta : S \times X \to S$ is the transition function; and $\lambda : S \times X \to Y$ is the output function.*

The transition function and the output function are inductively lifted to sequences of inputs, by defining $\delta(s_i, \varepsilon) = s_i$, $\delta(s_i, x'.\bar{x}) = \delta(\delta(s_i, x'), \bar{x})$ and $\lambda(s_i, \varepsilon) = \varepsilon$, $\lambda(s_i, x'.\bar{x}) = \lambda(s_i, x').\lambda(\delta(s_i, x'), \bar{x})$, where ε is the empty sequence, $x' \in X$, $\bar{x} \in X^*$. These FSMs are deterministic: for each state and input there is only one possible next state and output. In addition, they are completely-specified: δ and λ are total functions. We restrict attention to deterministic and completely-specified FSMs, leaving other classes of FSM to future work.

In order to reason about test effectiveness, it is normal to assume that the SUT behaves like an unknown FSM [16]. We also assume that we have an upper bound m on the number of states of the SUT.

Assumption 1 *The specification is an FSM $M = (S, s_0, X, Y, \delta, \lambda)$ with n states and the behaviour of the SUT can be represented by an FSM $M_I = (Q, q_0, X, Y, \delta_I, \lambda_I)$ with at most m states.*

Note that we do not assume that the FSM M_I is known. In addition, Definition 1 relaxes the standard definition of an FSM, which requires that the output alphabet Y is finite. This may not seem to be an important point since the set of outputs that can be produced by M is finite. However, allowing Y to be infinite has an impact on the set of potential SUTs being tested, which becomes infinite, and hence, challenges the completeness of tests.

Example 2. Consider our thermostat in Example 1 (Figure 1); assume that we would like to focus on the input/output behaviour and abstract from the internal continuous dynamics of the states. This abstraction is key to all well-known MBT techniques based on FSMs and LTSs. It is essential to note that in all such techniques, the states are not observable in testing, and only input-output behaviours are used to distinguish different behaviour.

The FSM arising from the specification in Figure 1 is depicted in Figure 2. The FSM (as well as the three implementations depicted in the same figure), have the set of input symbols $\{r, s\}$, representing reading the temperature and switching the heater, respectively. The output set is $\mathbb{R} \cup \{on, off\}$, i.e., the (uncountable) set of real numbers for temperature and the status values *on* and *off*. In our example, the relaxation to allow for infinite outputs is essential here to represent the infinite set of possible faulty implementations with deviating outputs, while a finite set of inputs would be sufficient to represent any discretisation of the passing of time.

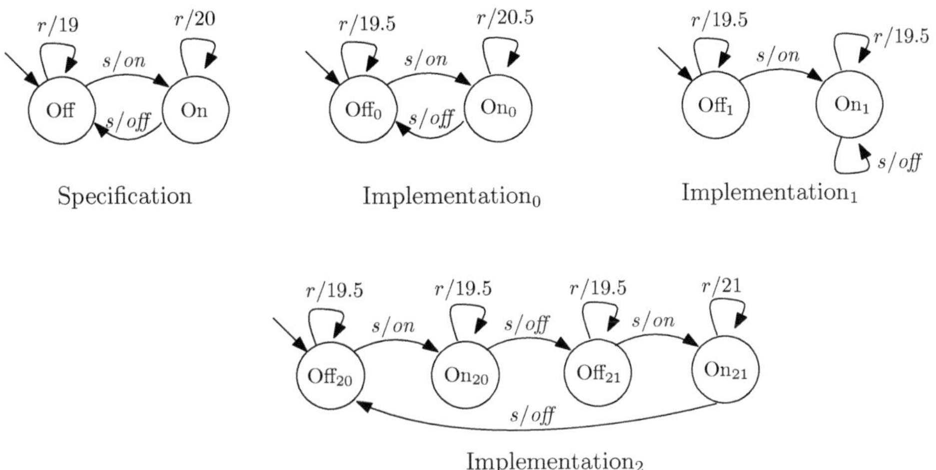

Fig. 2. Discretised Thermostat System: A Specification and Three Implementations

We use $\mathcal{F}$ to denote the set of FSMs with input alphabet X and output alphabet Y. Given integer m, we use $\mathcal{F}^m$ to denote the set of FSMs in $\mathcal{F}$ with at most m states. The following shows that if Y is finite then so is $\mathcal{F}^m$; then it is not difficult to show that there must be a finite m-complete test suite.

Proposition 1. *Given FSM $M = (S, s_0, X, Y, \delta, \lambda)$ with finite output alphabet Y, and integer m, $\mathcal{F}^m$ is finite.*

There is a correspondence between FSMs and states: if M is an FSM and s_i is a state of M then we can also see s_i as being an FSM: the FSM formed by making s_i the initial state of M. We can thus use $\mathcal{F}$ to denote the set of possible states of the unknown FSM M_I that represents the SUT.

Traditional FSM-based testing typically involves applying an input sequence to the SUT and checking that the output sequence produced is that specified. We formalise this in terms of separating states and FSMs; we use the term d-separate in order to distinguish between the classical notion and that required for the scenarios of interest (defined in Section 3).

Definition 2. *Given FSM M and states $s_1, s_2 \in S$, an input sequence $\bar{x}$ d-separates s_1 and s_2 if and only if $\lambda(s_1, \bar{x}) \neq \lambda(s_2, \bar{x})$. If $\bar{x}$ d-separates s_1 and s_2 then we can write $s_1 \not\equiv_{\bar{x}} s_2$ and say that s_1 and s_2 are d-separable. If $\bar{x}$ does not d-separate s_1 and s_2 then we write $s_1 \equiv_{\bar{x}} s_2$. Similarly, given a set W of input sequences, we write $s_1 \not\equiv_W s_2$ if an input sequence in W d-separates s_1 and s_2; otherwise $s_1 \equiv_W s_2$.*

It is straightforward to see that $\equiv_{\bar{x}}$ is an equivalence relation. In addition, it can be applied when comparing two FSMs M and M_I since one can define an FSM that is the disjoint union of M and M_I. The following is the notion of conformance used in traditional FSM-based testing, which we call d-conformance.

Definition 3. *Given FSMs $M = (S, s_0, X, Y, \delta, \lambda)$ and $M_I = (Q, q_0, X, Y, \delta_I, \lambda_I)$, M_I* d-conforms *to M if and only if for all $\bar{x} \in X^*$ we have that $M_I \equiv_{\bar{x}} M$.*

3   Conformance and Strong Separation

In this section we explain how separability and conformance can be naturally translated to the scenarios of interest. We then define strong separability.

3.1   Separability and Conformance

In the scenarios of interest an output might, for example, be the speed of a vehicle as measured by a sensor, which is an estimate of the actual speed. An output in Y could also be a probability distribution; in testing we apply a test case multiple times, observe a set of values drawn from an unknown distribution, and use statistical hypothesis testing techniques. In such situations, we need a way of comparing observations made with the expected output and we formalise this as a similarity relation $\sim$ on outputs. As previously explained, similarity will often be defined in terms of a metric μ and a threshold t; outputs y and y' are similar if $\mu(y, y') \leq t$ (see Example 3 below). Note that $\sim$ normally will not be transitive and so need not be an equivalence relation. The following lifts $\sim$ to output sequences.

Definition 4. *Given similarity relation $\sim$ on Y, $y, y' \in Y$, and $\bar{y}, \bar{y}' \in Y^*$ we have that:*

* $\varepsilon \sim \varepsilon$;
* $y\bar{y} \sim y'\bar{y}'$ *if and only if $y \sim y'$ and $\bar{y} \sim \bar{y}'$;*

We can define conformance in terms of $\sim$ as follows.

Definition 5. *Given FSMs $M = (S, s_0, X, Y, \delta, \lambda)$ and $M_I = (Q, q_0, X, Y, \delta_I, \lambda_I)$, M_I conforms to M if and only if for all $\bar{x} \in X^*$ we have that $\lambda(s_0, \bar{x}) \sim \lambda_I(q_0, \bar{x})$.*

Example 3. Consider the FSMs given in Figure 2. We instantiate the metric μ for real-valued outputs to be the absolute difference between them; the distance between *on* and *off* (as well as the distance between real numbers on one hand, and status values) is defined to be ∞. Define two outputs $o, o' \in \mathbb{R}$ to be ϵ-conforming, denoted by $t \sim_\epsilon t'$, when $|o - o'| \leq \epsilon$. This leads to a natural notion of conformance, inspired by the hybrid conformance literature [1], on FSMs.

According to this notion of conformance, *Specification $\sim_{0.5}$ Implementation_0*. This conformance relation holds, because for all input sequences, the difference between the respective outputs is bounded by 0.5.

Consider the specification *Specification* and implementation *Implementation_1* in Figure 2. In fact, *Implementation_1* does not ϵ-conform to *Specification* for any value of ϵ. This is witnessed by the output to input sequence *s.s.s*, where the difference between the last output in the two systems is ∞.

Consider again the specification *Specification* and implementation *Implementation*$_2$ in Figure 2. It does *not* hold that *Specification* $\sim_{0.5}$ *Implementation*$_2$ either. This is witnessed by the output to input sequence *s.s.s.r* (the difference between the last output in the two systems is 1).

We now define the notion of separating two states for the scenario of interest.

Definition 6. *Given FSM M and states $s_1, s_2 \in S$, an input sequence $\bar{x}$ separates s_1 and s_2 if and only if $\lambda(s_1, \bar{x}) \not\sim \lambda(s_2, \bar{x})$. If $\bar{x}$ separates s_1 and s_2 then we can write $s_1 \not\approx_{\bar{x}} s_2$. If $\bar{x}$ does not separate s_1 and s_2 then we write $s_1 \approx_{\bar{x}} s_2$. Similarly, given a set W of input sequences, we write $s_1 \not\approx_W s_2$ if there is a input sequence in W that separates s_1 and s_2 and otherwise we write $s_1 \approx_W s_2$.*

Where the input sequence of interest is clear, we often use $\not\approx$ and $\approx$, avoiding reference to the input sequences used. Conformance and separability are related.

Proposition 2. *Given FSMs $M = (S, s_0, X, Y, \delta, \lambda)$ and $M_I = (Q, q_0, X, Y, \delta_I, \lambda_I)$, M_I conforms to M if and only if no input sequence separates s_0 and q_0.*

When FSMs M and M_I do not conform to each other, we denote this by $M \not\approx M_I$; following Proposition 2, this can only happen if there exists $\bar{x} \in X^*$ that separates the initial states of M and M_I, which we denote by $M \not\approx_{\bar{x}} M_I$.

Example 4. Consider again the FSMs given in Figure 2 and the notion of similarity $\sim_{0.5}$; we have that the states *Off* in *Specification* and *Off*$_0$ in *Implementation*$_0$ are not separable by any input sequence and the two FSMs conform to each other with respect to $\sim_{0.5}$.

However, *Specification* and *Implementation*$_2$ do not conform to each other. The input sequence *s.s.s.r* separates the states *Off* and *Off*$_{20}$, because the former produces the output sequence *on.off.on.20*, while the later produces *on.off.on.21* and the two output sequences have a distance of 1. Note that *s.s.r.s* does not separate these states even though *s.s.r.s* takes *Specification* and *Implementation*$_2$ to states that do not correspond to one another; testing is black-box and we *cannot* observe the internal state reached.

3.2   Apartness and Strong Separability

Previous work has used the notion of an apartness relation and shown how one can use this to reason about test effectiveness when there is a finite set of outputs and conformance is defined in terms of equality of outputs rather than similarity.

Definition 7. *A binary relation $\#$ on a set Z is an apartness relation if and only if it satisfies the following properties*

- *It is irreflexive: for all $x \in Z$, $x \# x$ does not hold;*
- *It is symmetric: for all $x, y \in Z$, if $x \# y$ then $y \# x$; and*
- *It is co-transitive: for all $x, y, z \in Z$, if $x \# y$ then either $x \# z$ or $y \# z$.*

Consider the classical notion of separability for FSMs defined in Section 2 (d-separability). It is clear that d-separability is both irreflexive and symmetric. Now let us suppose that input sequence $\bar{x}$ d-separates states s_1 and s_2 of M and so $\bar{x}$ leads to different output sequences $\bar{y}_1$ and $\bar{y}_2$ when applied in s_1 and s_2. Further, let us suppose that we input $\bar{x}$ in some other state s_i and this leads to output sequence $\bar{y}$. Since $\bar{x}$ d-separates s_1 and s_2 we have that $\bar{y}_1 \neq \bar{y}_2$. We therefore must have that either $\bar{y} \neq \bar{y}_1$ or $\bar{y} \neq \bar{y}_2$ (or both) and so $\bar{x}$ must d-separate s_i from at least one of s_1 and s_2. As a result, d-separability is co-transitive and so is an apartness relation when testing a deterministic SUT against a completely-specified deterministic FSM M, with identity rather than similarity. The results of Vaandrager [31], regarding a test suite being m-complete, can therefore be applied.

Now consider what we mean by separability when similarity is defined in terms of a metric μ and a threshold t. Then an input sequence separates states s_1 and s_2 if the resultant observations o_1 and o_2 are such that $\mu(o_1, o_2) > t$. In such scenarios, separability is not an apartness relation since it need not be co-transitive, as shown by the following. Note that since we do not know the FSM model M_I of the SUT, in the definition of apartness (Definition 7), Z will be the set of all states of all FSMs in $\mathcal{F}$ and not just the states of M.

Example 5. Consider the FSM thermostat models in Example 2 (Figure 2); there, consider the states *Off* and *On* in the *Specification* FSM. They are separable using both inputs r and s. Let us consider what happens if we use r to separate these states and we have some state s_i of the SUT. Then s_i can have $r/19.5$, e.g., the state Off_0 in $Implementation_0$, that cannot be separated using only input r from either of the two states. This shows that separability is not co-transitive and so is not an apartness relation.

Since we cannot use apartness, we explore how this can be weakened for the scenarios of interest. We first discuss how d-separability is used in FSM-based testing. In most FSM-based test techniques that aim to find state-transfer faults, input sequences that d-separate states of a specification FSM M are used to check the current state of the SUT. To see how this works, let us suppose that input sequence $\bar{x}$ d-separates states s_1 and s_2 of M and that the (expected) output sequences produced are $\bar{y}_1$ and $\bar{y}_2$ respectively. Let us suppose that $\bar{x}$ is applied to the SUT when the SUT is in some state s_i. There are three possibilities:

1. The SUT produces $\bar{y}_2$ in response to $\bar{x}$: $\bar{x}$ d-separates s_i and s_1.
2. The SUT produces $\bar{y}_1$ in response to $\bar{x}$: $\bar{x}$ d-separates s_i and s_2.
3. The SUT produces another output sequence $\bar{y}_3$ in response to $\bar{x}$: $\bar{x}$ d-separates s_i and from both s_1 and s_2.

FSM-based test techniques utilise such information provided by applying input sequences that d-separate states of the specification. However, in Example 5 we saw that the above does not hold in the scenarios considered in this paper: if two states of the specification are separated by an input sequence $\bar{x}$ then it is possible that a state of the SUT is not separated from either of them by $\bar{x}$.

We now strengthen the notion of separability, to what we call strong separability, in a manner that ensures that strong separability satisfies the property (of d-separability in traditional FSM-based testing scenarios) described above.

Definition 8. *Given FSM M and states $s_1, s_2 \in S$, input sequence $\bar{x}$ is a witness that states s_1 and s_2, $s_1 \neq s_2$, are* strongly separable, *denoted $s_1 \#_{\bar{x}}^w s_2$, if and only if for all $s_3 \in \mathcal{F}$ we have that either $s_1 \not\approx_{\bar{x}} s_3$ or $s_2 \not\approx_{\bar{x}} s_3$. We also say that s_1 and s_2 are* strongly separable *and denote this $s_1 \#^w s_2$.*

If similarity is defined by a metric μ and threshold t then strong separability essentially requires us to double the threshold. The following is the situation for input sequences of length one and follows immediately from metrics satisfying the triangle inequality; it is straightforward to generalise the result to input sequences of arbitrary length.

Proposition 3. *If similarity is defined in terms of a metric μ and threshold t then an input x strongly separates states s_1 and s_2 of M if and only if $\mu(\lambda(s_1, x), \lambda(s_2, x)) > 2t$.*

Example 6. Consider the FSM thermostat models in Example 2 (Figure 2). Recall that states *Off* and *On* in the *Specification* FSM are separable using both inputs r and s. However, we have seen that they are not strongly separable using input r. Namely, there can be a state that has $r/19.5$, e.g., the state Off_0 in *Implementation*$_0$, that cannot be separated using only input r from either of the two states. In such cases, one can fix this issue, i.e., obtain strong separability, by either choosing different input sequences (in this case s) or decreasing the threshold of similarity, e.g., to 0.1, so that all possible states in $\mathcal{F}$ are separable (by r) from at least one of these two states. In practice, the use of a lower threshold might require changes to the equipment used to observe the SUT in testing or to additional test runs with the same input sequence (so that more precise estimates are produced).

The following states and proves basic properties of strong separability.

Proposition 4. *Given FSM M and input sequence $\bar{x}$, the relation $\#_{\bar{x}}^w$ is symmetric and irreflexive.*

Using the following, we can compare strong separability and separability.

Proposition 5. *Given FSM M, input sequence $\bar{x}$ and states s_1 and s_2 of M such that $s_1 \#_{\bar{x}}^w s_2$, we have that $\bar{x}$ separates s_1 and s_2.*

The following is immediate from the definition of strong separability and corresponds to the property (described above) of d-separability in traditional FSM-based testing.

Proposition 6. *Given states s_1 and s_2 of M with $s_1 \#_{\bar{x}}^w s_2$, if $s_3 \in \mathcal{F}$ then at least one of the following must hold.*

1. *$\bar{x}$ separates s_3 from s_1;*
2. *$\bar{x}$ separates s_3 from s_2.*

4   Test Generation

In this section, we explore the problem of test generation when, for every pair of states of specification M, we have a specific input sequence that provides a witness that these states are strongly separable.

Assumption 2 *For every pair s_1, s_2 of distinct states of M, there is an input sequence $\bar{x}$ that is a witness that s_1 and s_2 are strongly separable. We will use $w(s_1, s_2)$ to represent such a witness and require that $w(s_1, s_2) = w(s_2, s_1)$.*

Recall that M_I conforms to M if and only if for all $\bar{x} \in X^*$ we have that $\lambda(s_0, \bar{x}) \sim \lambda_I(q_0, \bar{x})$ (Definition 5). In testing, we will apply input sequences (*test sequences*) to the SUT and, for each such test sequence $\bar{x}$, check whether the response of the SUT to $\bar{x}$ is related to the specified response to $\bar{x}$ under $\sim$; namely, whether $\lambda_I(q_0, \bar{x}) \sim \lambda(s_0, \bar{x})$. If $\lambda(s_0, \bar{x}) \not\sim \lambda_I(q_0, \bar{x})$ then the SUT *fails* $\bar{x}$ and otherwise it *passes* $\bar{x}$. A test suite will thus be a set of input sequences. We are interested in generating test suites with guaranteed fault detection ability.

Definition 9. *Given an FSM M, a set T of input sequences is m-complete if for all $M_I \in \mathcal{F}^m$ we have that M_I conforms to M if and only if M_I does not fail any test sequence in T.*

We will retain the notion of a state cover used in classical FSM-based test generation techniques: a minimal prefix-closed set of input sequences that, between them, reach all states of M.

Definition 10. *Given a finite state machine M, a set V of input sequences is a state cover for M if V is prefix closed and for all $s_i \in S$ there is exactly one input sequence $v_i \in V$ such that $\delta(s_0, v_i) = s_i$.*

Example 7. Consider the *Specification* FSM in Example 2 (Figure 2); a state cover set V for *Specification* is $\{\varepsilon, s\}$.

In the following, for a state s_i of M we use $W(s_i)$ to denote the set of input sequences used to show that s_i is strongly separable from other states of M.

Definition 11. *Given state s_1 of M, we let $W(s_1)$ denote the* state identification *set*

$$W(s_1) = \{w(s_1, s_2) | s_2 \in S \setminus \{s_1\}\}$$

Note that only a single witness is used for a pair of states and typically one will use a relatively short such witness. Later (Proposition 11) we provide a polynomial upper bound on the lengths of shortest such witnesses.

Example 8. Consider FSM *Specification* in Example 2 (Figure 2). Both r and s pairwise separate the states of *Specification* but we have seen that only s strongly separates these states: $Off \#_s^w On$ and $\neg Off \#_r^w On$. Since $Off \#_s^w On$, we can choose $\{s\}$ to be the state identification set for Off (and also for On).

Recall that we do not know the FSM M_I that models the SUT but we will want to reason about the states of M_I that are met during testing. Given state s_i of M and input sequence $\bar{x}$, we use $B_{\bar{x}}(s_i)$ to denote the ball (set) of states of M_I that are not separated from s_i by $\bar{x}$.

Definition 12. *Given state s_i of M, input sequence $\bar{x}$, and FSM M_I with state set Q, $B_{\bar{x}}(s_i) = \{q \in Q | s_i \approx_{\bar{x}} q\}$.*

Example 9. Consider the *Specification* and *Implementation$_0$* FSMs in Example 2 (Figure 2), comprising the set of states $\{Off, On, Off_0, On_0\}$. We have that $Q = \{On_0, Off_0\}$. Given identification set $\{s\}$, we have that $B_s(Off) = \{Off_0\}$ and $B_s(On) = \{On_0\}$. Clearly we have that $B_s(On)$ and $B_s(Off)$ are disjoint.

The following is a consequence of strong separability and tells us that if we take two states s_1 and s_2 of M that are strongly separated by $\bar{x}$ then $B_{\bar{x}}(s_1)$ and $B_{\bar{x}}(s_2)$ are pairwise disjoint.

Proposition 7. *Given states s_1 and s_2 of M with $s_1 \#_{\bar{x}}^w s_2$, if $\bar{x}$ strongly separates s_1 and s_2, $s_1' \in B_{\bar{x}}(s_1)$ and $s_2' \in B_{\bar{x}}(s_2)$ then $s_1' \neq s_2'$.*

Observe that $B_{\bar{x}}(s_i)$ is defined in terms of separability and *not* strong separability. However, Proposition 7 concerns input sequences that strongly separate states. The following shows what can happen if we were to instead use input sequences that separate, but do not strongly separate, the states of the specification.

Example 10. Consider again the *Specification* and *Implementation$_0$* FSMs in Example 2 (Figure 2), comprising the set of states $\{Off, On, Off_0, On_0\}$, with $Q = \{On_0, Off_0\}$. Now consider the input r, which separates the states of the specification, and the sets $B_1 = \{q \in Q | q \sim_r Off\}$ and $B_2 = \{q \in Q | q \sim_r On\}$. Here $B_1 = \{Off_0\}$ and $B_2 = \{On_0, Off_0\}$, so B_1 and B_2 are not disjoint.

The test generation approach will be based on placing a lower bound on the number of states M_I must have if the SUT is faulty and it passes all the test cases in a given test suite. The approach will use Proposition 7, which tells us that if input sequences $\bar{x}_1$ and $\bar{x}_2$ reach states s_1 and s_2 of M respectively, w strongly separates s_1 and s_2 (ie $s_1 \#_w^w s_2$), and the SUT conforms to the specification on both $\bar{x}_1.w$ and $\bar{x}_2.w$ then $\bar{x}_1$ and $\bar{x}_2$ must reach *different* states of M_I.

We now consider what we know about the states of the SUT reached by the state cover V if M_I conforms to M on the set of input sequences formed by following each sequence in V by the corresponding state identification set.

Proposition 8. *Let V be a state cover for FSM M whose states are pairwise strongly separable. Further, let us suppose that for every $v \in V$ we have that the states $s_i = \delta(s_0, v)$ and $q_i = \delta_I(q_0, v)$ satisfy $s_i \approx_{W(s_i)} q_i$. Then V reaches n separate states of M_I.*

Example 11. Following up on Examples 7, 8, and 9, consider the state cover set $\{\varepsilon, s\}$ and the state identification set $\{s\}$ for which the states of the specification are strongly separable. Now consider *Implementation*$_0$; the state cover set of the specification will take *Implementation*$_0$ to *Off*$_0$ and *On*$_0$, which are clearly different states. The same holds for *Implementation*$_1$, which has the same number of states.

For *Implementation*$_2$, since it has more states, we need more inputs to reach the remaining states: using s, *Specification* and *Implementation*$_2$ will arrive in state (On, On_{20}), further inputs are needed to bring the pair of *Specification* and *Implementation*$_2$ into (On, On_{21}) and then identify the output fault on input r. This is formalised in the following test generation algorithm.

The test generation techniques that we build upon use input sequences of the form $v.\bar{x}.w$ such that v is a sequence in the state cover V, $\bar{x}$ is any input sequence of a given length ℓ (determined by the maximum number of extra states in the SUT), and w is a state identifier for the state $\delta(s_0, v.\bar{x})$ of M. There are two ways in which such an input sequence can lead to a failure being observed.

1. The SUT M_I does not conform to the specification M on the initial sequence $v.\bar{x}$. If the SUT has output faults (one or more transitions produce the wrong output) then these output faults will lead to such failures.
2. There are v, $\bar{x}$, and w such that the SUT M_I conforms M on the initial sequence $v.\bar{x}$ but does not conform to the specification on $v.\bar{x}.w$. Here, a state identifier for $\delta(s_0, v.\bar{x})$ separates the state of the SUT M_I reached by $v.\bar{x}$ from the state of the specification M reached by $v.\bar{x}$ (the test finds a state transfer fault).

The following introduces corresponding notation, for a fixed length ℓ of input sequence $\bar{x}$.

Definition 13. *Let us suppose that all states of FSM M are strongly separable and M_I is an FSM with the same input and output alphabets as M. Further, let us suppose that V is a state-cover for M and for all $s_i \in S$ we have that $W(s_i)$ is a state identification set for s_i. Given integer $\ell \geq 0$:*

$$D(M, M_I, V, \ell) \quad = \{(v, \bar{x}) | v \in V \wedge |\bar{x}| = \ell \wedge M \not\approx_{v\bar{x}} M_I\}$$
$$DW(M, M_I, V, W, \ell) = \{(v, \bar{x}) | v \in V \wedge |\bar{x}| = \ell \wedge \exists w \in W(\delta(s_0, v\bar{x})).$$
$$M \not\approx_{v\bar{x}w} M_I\}$$

We now provide two results that show what one can deduce from the value of the *smallest* integer ℓ such that $D(M, M_I, V, \ell) \cup DW(M, M_I, V, W, \ell) \neq \emptyset$. Given this ℓ, we will place a *lower* bound on the number of different states of M_I reached by V and sequences of the form $v.\bar{x}$, for $v \in V$ and input sequence $\bar{x}$ of length at most ℓ (Proposition 9). We then show that this implies an *upper* bound on the value of ℓ that we need to use in testing (Proposition 10).

Lemma 1. *Let us suppose that all states of FSM M are strongly separable, M_I does not conform to M, and $\ell > 0$ is the smallest value such that $D(M, M_I, V, \ell) \cup$*

$DW(M, M_I, V, W, \ell) \neq \emptyset$. If $(v, \bar{x}) \in D(M, M_I, V, \ell) \cup DW(M, M_I, V, W, \ell)$ and $\bar{x}_1$ is a non-empty proper prefix of $\bar{x}$ then the state of M_I reached by $v\bar{x}_1$ is not a state of M_I reached by an input sequence in V.

Lemma 2. *Let us suppose that all pairs of states of FSM M are strongly separable, M_I does not conform to M, $\ell > 0$, and ℓ is the smallest value such that $D(M, M_I, V, \ell) \cup DW(M, M_I, V, W, \ell) \neq \emptyset$. If $(v, \bar{x}) \in D(M, M_I, V, \ell) \cup DW(M, M_I, V, W, \ell)$ and $\bar{x}_1, \bar{x}_2$ are different non-empty proper prefixes of $\bar{x}$ then $v\bar{x}_1$ and $v\bar{x}_2$ reach different states of M_I.*

Example 12. For non-conforming implementation $Implementation_1$, we have that it is sufficient to take $\bar{x}$ to be s since after $s \in V$ and $\bar{x} = s$, a further input $s \in W$ can reach a non-conformance verdict between the specification and the implementation. This is because the faulty implementation does not have any additional state with respect to the specification. In terms of Definition 13 this is summarised as follows, where, and for brevity we refer to the specification as M and to $Implementation_1$ as M_I:

$$D(M, M_I, V, 1) \qquad = \emptyset$$
$$DW(M, M_I, V, W, 1) = \{(s, s)\}$$

$D(M, M_I, V, 1)$ is the empty set because the pair with the shortest sequence to distinguish M and M_I is $(s, s.s)$, of which the second component is beyond the limit of $l = 1$. However, $DW(M, M_I, V, W, 1)$ contains the pair (s, s), because after $s.s$ by performing a further $s \in W$, we can distinguish the two systems.

For non-conforming implementation $Implementation_2$, the situation is different because it contains more states than the specification. Namely, we need three further inputs, in addition to the state cover, to identify the output fault (on input r) at state On_{21}: after performing input s from the state cover, we need the additional inputs $s.s.r$ to first bring the implementation to On_{21} (using $s.s$) and then identify the faulty output by performing an additional input r. This is generalised in the following results, where we identify an upper bound on the length of intermediate sequences, between the state cover and the state identification.

In terms of Definition 13 this is summarised as follows, where, and for brevity we refer to the specification as M and to $Implementation_1$ as M_I:

$$D(M, M_I, V, 1) = D(M, M_I, V, 2) = \emptyset$$
$$D(M, M_I, V, 3) = \{(s, s.s.r)\}$$

$D(M, M_I, V, 1)$ and $D(M, M_I, V, 2)$ are both the empty set because the pair with the shortest sequence to distinguish M and M_I is $(s, s.s.r)$, of which the second component is beyond the limit of $l = 2$.

However $DW(M, M_I, V, W, 3)$ is the empty set; had we had a transition fault, instead of an output fault at the final state, then $DW(M, M_I, V, W, 3)$ would have been non-empty.

We can use the above-given Lemmas to place a lower bound on the number of states an SUT must have if the SUT is faulty and we know the smallest value ℓ such that $D(M, M_I, V, \ell) \cup DW(M, M_I, V, W, \ell) \neq \emptyset$.

Proposition 9. *Let us suppose that all pairs of states of FSM M are strongly separable, M_I does not conform to M, and for all $v \in V$ and $w \in W(\delta(s_0, v))$ we have that $M \approx_{vw} M_I$. If ℓ is the smallest value such that $D(M, M_I, V, \ell) \cup DW(M, M_I, V, W, \ell) \neq \emptyset$ and $\ell > 0$ then M_I has at least $n + \ell - 1$ states.*

Now consider the standard FSM-based testing context in which we assume that the FSM M_I that models the SUT has at most $n + k$ states and we wish to generate sufficient tests to determine whether M_I conforms to M. We can use Proposition 9 to provide an upper bound on the value of ℓ that we require to use when testing with sequences of the form $v.\bar{x}$ and $v.\bar{x}.w$, with $|\bar{x}| \leq \ell$.

Proposition 10. *Let us suppose that all pairs of states of FSM M are strongly separable and M_I does not conform to M. If M_I has at most k more states than M then there exists some $v \in V$, and $\bar{x} \in X^*$ such that $|\bar{x}| \leq k + 1$ and one of the following holds:*

- *$M_I \not\approx_{v\bar{x}} M$; or*
- *there is some $w \in W(\delta(s_0, v\bar{x}))$ such that $M_I \not\approx_{v\bar{x}w} M$.*

We now show what this implies regarding test completeness.

Theorem 1. *Let us suppose that the states of FSM M are pairwise strongly separable, M has n states, and M_I has at most $m = n + k$ states. If $D(M, M_I, V, \ell) \cup D(M, M_I, V, \ell) = \emptyset$ for all $\ell \leq k + 1$ then M_I conforms to M.*

This result shows how test generation can proceed: one simply uses the test sequences implicit in the definitions of $D(M, M_I, V, \ell)$ and $DW(M, M_I, V, W, \ell)$. Algorithm 1 gives the corresponding test generation algorithm. Note that for a state s_i of M, $W(s_i)$ is a Harmonized State Identifier for s_i as used in the HSI-method except that we require $W(s_i)$ to strongly separate s_i from other states of M; it is not sufficient for $W(s_i)$ to separate s_i from other states of M. As a result, Algorithm 1 is essentially the HSI-method, with the only difference being how we define conformance and the $W(s_i)$. Since the test set produced by the HSI-method is a subset of those produced by the W-method and the Wp-method, the above demonstrates that these test generation techniques also produce m-complete test sets when we have a notion of similarity of outputs rather than equality and we use state identifiers that strongly separate states.

Example 13. Consider the *Specification* FSM in Figure 2. We established in Example 7 that $V = \{\varepsilon, s\}$ and in Example 8 that the state-identification set that strongly separates the specification states is $W = \{s\}$. Hence, assuming that we are only interested in implementations that have the same state count as the specification, by applying Algorithm 1, we obtain:

$$T = \{v.\bar{x}.w | v \in \{\varepsilon, s\} \wedge \bar{x} \in \{\varepsilon, r, s\} \wedge w \in \{s\}\}$$
$$T = \{\varepsilon.\varepsilon.s, \varepsilon.r.s, \varepsilon.s.s, s.r.s, s.s.s\}$$
$$= \{s, r.s, s.s, s.r.s, s.s.s\}$$

Algorithm 1 Test Generation Using Strong Separability

Input: FSM $M = (S, s_0, X, Y, \delta, \lambda)$ where $S = \{s_1, \ldots, s_n\}$, $m \geq n$
Derive state cover V
Derive state identification sets $W(s_1), \ldots, W(s_n)$ based on strong separability
Derive $T = \{v.\bar{x}.w | v \in V, \bar{x} \in X^*, 0 \leq |\bar{x}| \leq m - n + 1, w \in W(\delta(s_0, v.\bar{x}))\}$
Remove from T all prefixes
return T

If we wish to find faulty implementations such as $Implementation_2$ then we need to use a larger value for k. If we use $k = 2$ then we obtain the following.

$$T = \{v.\bar{x}.w | v \in \{\varepsilon, s\} \wedge \bar{x} \in \{\varepsilon, r, s, rr, rs, sr, ss, rrr, rrs, rsr, rss, srr, srs,$$
$$ssr, sss\} \wedge w \in \{s\}\}$$

Note that the key difference between these sets of generated inputs and the traditional HSI method is that the adapted algorithm forces the state-identification set to provide strong separability. Otherwise, the traditional HSI method can use r to separate states and generate $r.r$, $s.s.r$ and $s.r.r$ as the longest sequences (for $k = 0$) and these sequences will miss the fault in $Implementation_1$. This demonstrates that the HSI method, with state identifiers that do not strongly separate the states of the specification, is not complete. Using this specification FSM, state cover, and state identifiers, the W-method and Wp-method return the same test suite and so are also incomplete.

A final observation is that, as usual, if a test suite contains input sequences $\bar{x}_1$ and $\bar{x}_2$ and $\bar{x}_2$ is a prefix of $\bar{x}_1$ then we can remove $\bar{x}_2$ from the test suite without reducing effectiveness. In the first example above, this final optimisation reduces the test suite to $\{r.s, s.r.s, s.s.s\}$.

It is now worth considering how large the test suite can be for an FSM specification M with n states. Clearly V contains n input sequences and these have length at most $n - 1$: we can generate such input sequences through a breadth-first search. Further, each $W(s_i)$ contains at most $n - 1$ sequences. The following places an upper bound on the lengths of sequences in $W(s_i)$, assuming we use shortest sequences that suffice.

Proposition 11. *Let us suppose that M is an FSM with n states, which are pairwise strongly separable. For every pair s_1, s_2 of distinct states of M, there is an input sequence of length at most $\frac{n(n-1)}{2}$ that strongly separates s_1 and s_2.*

Thus, the sizes of the state cover and state identification sets are low-order polynomial. Naturally, the size of the test suite grows exponentially as the upper bound, k, on the number of extra states grows. However, this is also the case with techniques such as the HSI-method when applied in the classical FSM context. In practice, the value of k used might depend upon domain knowledge and a cost/benefit analysis.

It is finally worth commenting on the computational complexity of deriving the state cover and state identifiers. Clearly, a state cover can be derived in low-order polynomial time since breadth-first search takes linear time. It is also possible to derive the state identifiers in polynomial time. In particular, it is possible to define a simple iterative algorithm based on the proof of Proposition 11, with this operating as follows. First, we determine which pairs of states can be strongly separated by a single input and record these (we could say that these are strongly 1-separable). We then determine which of the remaining pairs $\{s_i, s_j\}$ of states are strongly separated by input sequences of length 2 (they are strongly 2-separable). In order to check this, for each pair $\{s_i, s_j\}$ of states (that are not strongly 1-separable) we check whether there is an input x such that states $\delta(s_i, x)$ and $\delta(s_j, x)$ are strongly 1-separable. This process continues: in each iteration, for each pair $\{s_i, s_j\}$ of states not already strongly separated, we determine whether there is an input x such that states $\delta(s_i, x)$ and $\delta(s_j, x)$ have already been strongly separated in an earlier iteration. Clearly, each iteration takes polynomial time and, from Proposition 11, we know that there are at most $\frac{n(n-1)}{2}$ iterations. Bringing together the above information, we have that Algorithm 1 has polynomial time complexity as long as we bound k. Note, however, that the test suite can size grow exponentially with k but this is true also of classical FSM test generation techniques such as the W, Wp, and HSI-methods.

Finally, we identify a condition under which separability and strong separability coincide.

Definition 14 (Strongly different). *Two outputs $y_1, y_2 \in Y$ are strongly different if for all $y_3 \in Y$ we have that either $y_1 \not\sim y_3$ or $y_2 \not\sim y_3$. Further, an FSM $M = (S, s_0, X, Y, \delta, \lambda)$ has strongly different outputs if for all $s_1, s_2 \in S$ and $x \in X$ we have that either $\lambda(s_1, x) = \lambda(s_2, x)$ or $\lambda(s_1, x)$ and $\lambda(s_2, x)$ are strongly different.*

Proposition 12. *Given an FSM $M = (S, s_0, X, Y, \delta, \lambda)$, if M has strongly different outputs then an input sequence $\bar{x}$ strongly separates states s_1 and s_2 of M if and only if $\bar{x}$ separates s_1 and s_2.*

Recall that Theorem 1 tells us that we can use the HSI-method (and so the W and Wp-methods) to produce m-complete test suites as long as the state identification sequences strongly separate the states of the specification M. Proposition 12 thus tells us that if M has strongly different outputs then we can use the W-method, Wp-method and HSI-method in the usual way (and so any tools that implement these). In some scenarios, we might be able to design the testing approach (eg the number of times that test execution is repeated) in order to reduce the threshold so that the specification has strongly different outputs.

5 Conclusions

This paper proposed the notion of strong separability, inspired by apartness in constructive mathematics. This allows for separating states according to approximate notions of conformance, that allow for a margin of error, in quantitative finite-state machine models of systems. We showed the applicability of our notion by adopting it in a well-known model-based testing technique, the HSI method. We proved that using our notion, the HSI method is complete for approximate notions of conformance. To illustrate our approach, we used a simple thermostat example and three implementations. We also demonstrated that complete test suites need not be produced if we use separability rather than strong separability. We gave (low-order) complexity results for test generation and polynomial upper bound on test size.

Recall that the proposed approach operates by reasoning about the number of states of the SUT met during testing and does this by strongly separating states. We have seen that this is consistent with the W-method, the Wp-method and the HSI-method. These three approaches differ in the choice of state identification sequences used but they keep this choice fixed throughout test generation. Dorofeeva et al. [13] introduced the H-method, which allows different state identification sequences to be used for different input sequences. It may be possible to extend the results given in this paper to the H-method by allowing the input sequence $w(s_1, s_2)$ used to strongly separate two states of M to vary.

There are several lines of possible future work corresponding to the following limitations. First, we abandoned the use of the observation tree used by Vaandrager [31] since a correct SUT might produce traces that are not traces of the specification M; it may be possible to reason about observation trees that are 'similar' to those of M. Second, we considered deterministic, completely-specified FSMs and the test generation algorithm assumes that the states of M are pairwise strongly separable. Although these requirements are weaker than those imposed by the work that uses apartness, there is the question of how they might be further weakened. Finally, we will look into using strong separability to develop novel automata learning algorithms by integrating algorithms that use apartness [32, 33] with those that consider quantitative extensions of state machines [23, 3, 27]. Our threshold would allow for adjusting the level of abstraction we admit in learning and can provide an adjustable trade-off between the accuracy and the size of the learned model.

Acknowledgements Robert M. Hierons and Mohammad Reza Mousavi have been partially supported by the UKRI Trustworthy Autonomous Systems Node in Verifiability, Grant Award Reference EP/V026801/2. Mohammad Reza Mousavi has been partially supported by the EPSRC project on Verified Simulation for Large Quantum Systems (VSL-Q), grant reference EP/Y005244/1 and the EPSRC project on Robust and Reliable Quantum Computing (RoaRQ), Investigation 009 Model-based monitoring and calibration of quantum computations (ModeMCQ), grant reference EP/W032635/1 and ITEA/InnovateUK projects GENIUS and GreenCode.

The authors have no competing interests to declare that are relevant to the content of this article.

References

1. Abbas, H., Mittelmann, H.D., Fainekos, G.: Formal property verification in a conformance testing framework. In: Twelfth ACM/IEEE International Conference on Formal Methods and Models for Codesign, MEMOCODE 2014, Lausanne, Switzerland, October 19-21, 2014. pp. 155–164. IEEE (2014). https://doi.org/10.1109/MEMCOD.2014.6961854, https://doi.org/10.1109/MEMCOD.2014.6961854
2. Angluin, D.: Learning regular sets from queries and counterexamples. Inf. Comput. **75**(2), 87–106 (1987). https://doi.org/10.1016/0890-5401(87)90052-6, https://doi.org/10.1016/0890-5401(87)90052-6
3. Bacci, G., Ingólfsdóttir, A., Larsen, K.G., Reynouard, R.: Active learning of Markov Decision Processes using Baum-Welch algorithm. In: Wani, M.A., Sethi, I.K., Shi, W., Qu, G., Raicu, D.S., Jin, R. (eds.) 20th IEEE International Conference on Machine Learning and Applications, ICMLA 2021, Pasadena, CA, USA, December 13-16, 2021. pp. 1203–1208. IEEE (2021). https://doi.org/10.1109/ICMLA52953.2021.00195, https://doi.org/10.1109/ICMLA52953.2021.00195
4. Biewer, S., D'Argenio, P.R., Hermanns, H.: Doping tests for cyber-physical systems. ACM Trans. Model. Comput. Simul. **31**(3), 16:1–16:27 (2021). https://doi.org/10.1145/3449354, https://doi.org/10.1145/3449354
5. Biewer, S., Dimitrova, R., Fries, M., Gazda, M., Heinze, T., Hermanns, H., Mousavi, M.R.: Conformance relations and hyperproperties for doping detection in time and space. Log. Methods Comput. Sci. **18**(1) (2022). https://doi.org/10.46298/LMCS-18(1:14)2022, https://doi.org/10.46298/lmcs-18(1:14)2022
6. van Breugel, F., Hermida, C., Makkai, M., Worrell, J.: An accessible approach to behavioural pseudometrics. In: Caires, L., Italiano, G.F., Monteiro, L., Palamidessi, C., Yung, M. (eds.) Automata, Languages and Programming, 32nd International Colloquium, ICALP 2005, Lisbon, Portugal, July 11-15, 2005, Proceedings. Lecture Notes in Computer Science, vol. 3580, pp. 1018–1030. Springer (2005). https://doi.org/10.1007/11523468_82, https://doi.org/10.1007/11523468_82
7. van Breugel, F., Worrell, J.: Towards quantitative verification of probabilistic transition systems. In: Orejas, F., Spirakis, P.G., van Leeuwen, J. (eds.) Automata, Languages and Programming, 28th International Colloquium, ICALP 2001, Crete, Greece, July 8-12, 2001, Proceedings. Lecture Notes in Computer Science, vol. 2076, pp. 421–432. Springer (2001). https://doi.org/10.1007/3-540-48224-5_35, https://doi.org/10.1007/3-540-48224-5_35
8. Broy, M., Jonsson, B., Katoen, J.P., Leucker, M., Pretschner, A.: Model-Based Testing of Reactive Systems, Lecture Notes in Computer Science, vol. 3472. Springer (2005)
9. Chaudhuri, S., Gulwani, S., Lublinerman, R.: Continuity and robustness of programs. Commun. ACM **55**(8), 107–115 (Aug 2012). https://doi.org/10.1145/2240236.2240262, https://doi.org/10.1145/2240236.2240262
10. Chow, T.S.: Testing software design modelled by finite state machines. IEEE Transactions on Software Engineering **4**, 178–187 (1978)
11. Desharnais, J., Edalat, A., Panangaden, P.: Bisimulation for Labelled Markov Processes. Information and Computation **179**(2), 163–193 (2002). https://doi.org/https://doi.org/10.1006/inco.2001.2962, https://www.sciencedirect.com/science/article/pii/S0890540101929621

12. Deshmukh, J.V., Majumdar, R., Prabhu, V.S.: Quantifying conformance using the Skorokhod metric. Formal Methods Syst. Des. **50**(2-3), 168–206 (2017). https://doi.org/10.1007/S10703-016-0261-8, https://doi.org/10.1007/s10703-016-0261-8

13. Dorofeeva, R., El-Fakih, K., Yevtushenko, N.: An improved conformance testing method. In: 25th IFIP WG 6.1 International Conference on Formal Techniques for Networked and Distributed Systems (FORTE 2005). Lecture Notes in Computer Science, vol. 3731, pp. 204–218. Springer (2005)

14. Fainekos, G.E., Pappas, G.J.: Robustness of temporal logic specifications for continuous-time signals. Theor. Comput. Sci. **410**(42), 4262–4291 (2009). https://doi.org/10.1016/J.TCS.2009.06.021, https://doi.org/10.1016/j.tcs.2009.06.021

15. Fujiwara, S., v. Bochmann, G., Khendek, F., Amalou, M., Ghedamsi, A.: Test selection based on finite state models. IEEE Transactions on Software Engineering **17**(6), 591–603 (1991)

16. Gaudel, M.C.: Testing can be formal too. In: 6th International Joint Conference CAAP/FASE Theory and Practice of Software Development (TAPSOFT'95). Lecture Notes in Computer Science, vol. 915, pp. 82–96. Springer (1995)

17. Hennie, F.C.: Fault-detecting experiments for sequential circuits. In: Proceedings of Fifth Annual Symposium on Switching Circuit Theory and Logical Design. pp. 95–110. Princeton, New Jersey (November 1964)

18. Hierons, R.M., Bogdanov, K., Bowen, J.P., Cleaveland, R., Derrick, J., Dick, J., Gheorghe, M., Harman, M., Kapoor, K., Krause, P., Lüttgen, G., Simons, A.J.H., Vilkomir, S.A., Woodward, M.R., Zedan, H.: Using formal specifications to support testing. ACM Computing Surveys **41**(2), 9:1–9:76 (2009)

19. Khakpour, N., Mousavi, M.R.: Notions of conformance testing for cyber-physical systems: Overview and roadmap (invited paper). In: Aceto, L., de Frutos-Escrig, D. (eds.) 26th International Conference on Concurrency Theory, CONCUR 2015, Madrid, Spain, September 1.4, 2015. LIPIcs, vol. 42, pp. 18–40. Schloss Dagstuhl - Leibniz-Zentrum für Informatik (2015). https://doi.org/10.4230/LIPICS.CONCUR.2015.18, https://doi.org/10.4230/LIPIcs.CONCUR.2015.18

20. Lee, D., Yannakakis, M.: Testing finite-state machines: State identification and verification. IEEE Transactions on Computers **43**(3), 306–320 (1994)

21. Lee, D., Yannakakis, M.: Principles and methods of testing finite-state machines - a survey. Proceedings of the IEEE **84**(8), 1089–1123 (1996)

22. Luo, G., Petrenko, A., v. Bochmann, G.: Selecting test sequences for partially-specified nondeterministic finite state machines. In: The 7th IFIP Workshop on Protocol Test Systems. pp. 95–110. Chapman and Hall, Tokyo, Japan (November 8–10 1994)

23. Medhat, R., Ramesh, S., Bonakdarpour, B., Fischmeister, S.: A framework for mining hybrid automata from input/output traces. In: Girault, A., Guan, N. (eds.) 2015 International Conference on Embedded Software, EMSOFT 2015, Amsterdam, Netherlands, October 4-9, 2015. pp. 177–186. IEEE (2015). https://doi.org/10.1109/EMSOFT.2015.7318273, https://doi.org/10.1109/EMSOFT.2015.7318273

24. Miller, T., Strooper, P.A.: A case study in model-based testing of specifications and implementations. Software Testing, Verification and Reliability **22**(1), 33–63 (2012)

25. Mohd-Shafie, M.L., Kadir, W.M.N.W., Lichter, H., Khatibsyarbini, M., Isa, M.A.: Model-based test case generation and prioritization: a systematic literature review. Software and Systems Modeling **21**(2), 717–753 (2022). https://doi.org/10.1007/s10270-021-00924-8, https://doi.org/10.1007/s10270-021-00924-8

26. Sankaranarayanan, S., Kumar, S.A., Cameron, F., Bequette, B.W., Fainekos, G., Maahs, D.M.: Model-based falsification of an artificial pancreas control system. SIGBED Rev. **14**(2), 24–33 (2017). https://doi.org/10.1145/3076125.3076128, https://doi.org/10.1145/3076125.3076128

27. Tappler, M., Aichernig, B.K., Bacci, G., Eichlseder, M., Larsen, K.G.: L*-based learning of Markov decision processes (extended version). Formal Aspects Comput. **33**(4-5), 575–615 (2021). https://doi.org/10.1007/S00165-021-00536-5, https://doi.org/10.1007/s00165-021-00536-5

28. Tretmans, J.: Model based testing with labelled transition systems. In: Formal Methods and Testing. Lecture Notes in Computer Science, vol. 4949, pp. 1–38. Springer (2008)

29. Tuncali, C.E., Pavlic, T.P., Fainekos, G.: Utilizing S-Taliro as an automatic test generation framework for autonomous vehicles. In: 19th IEEE International Conference on Intelligent Transportation Systems, ITSC 2016, Rio de Janeiro, Brazil, November 1-4, 2016. pp. 1470–1475. IEEE (2016). https://doi.org/10.1109/ITSC.2016.7795751, https://doi.org/10.1109/ITSC.2016.7795751

30. Utting, M., Pretschner, A., Legeard, B.: A taxonomy of model-based testing approaches. Software Testing, Verification and Reliability **22**(5), 297–312 (2012)

31. Vaandrager, F.W.: A new perspective on conformance testing based on apartness. In: Capretta, V., Krebbers, R., Wiedijk, F. (eds.) Logics and Type Systems in Theory and Practice - Essays Dedicated to Herman Geuvers on The Occasion of His 60th Birthday. Lecture Notes in Computer Science, vol. 14560, pp. 225–240. Springer (2024). https://doi.org/10.1007/978-3-031-61716-4_15, https://doi.org/10.1007/978-3-031-61716-4_15

32. Vaandrager, F.W., Garhewal, B., Rot, J., Wißmann, T.: A new approach for active automata learning based on apartness. In: Fisman, D., Rosu, G. (eds.) Tools and Algorithms for the Construction and Analysis of Systems - 28th International Conference, TACAS 2022, Held as Part of the European Joint Conferences on Theory and Practice of Software, ETAPS 2022, Munich, Germany, April 2-7, 2022, Proceedings, Part I. Lecture Notes in Computer Science, vol. 13243, pp. 223–243. Springer (2022). https://doi.org/10.1007/978-3-030-99524-9_12, https://doi.org/10.1007/978-3-030-99524-9_12

33. Vaandrager, F.W., Sanders, M.: L$^{\#}$ for DFAs. In: Jansen, N., Junges, S., Kaminski, B.L., Matheja, C., Noll, T., Quatmann, T., Stoelinga, M., Volk, M. (eds.) Principles of Verification: Cycling the Probabilistic Landscape - Essays Dedicated to Joost-Pieter Katoen on the Occasion of His 60th Birthday, Part III. Lecture Notes in Computer Science, vol. 15262, pp. 155–172. Springer (2024). https://doi.org/10.1007/978-3-031-75778-5_8, https://doi.org/10.1007/978-3-031-75778-5_8

34. Vasilevskii, M.P.: Failure diagnosis of automata. Cybernetics **4**, 653–665 (1973)

A No-go Theorem
for Coalgebraic Product Construction[*]

Mayuko Kori[1] and Kazuki Watanabe[2,3]

[1] Research Institute for Mathematical Sciences, Kyoto University, Japan
[2] National Institute of Informatics, Japan
[3] The Graduate University for Advanced Studies (SOKENDAI), Japan

Abstract. Verifying traces of systems is a central topic in formal verification. We study model checking of Markov chains (MCs) against temporal properties represented as (finite) automata. For instance, given an MC and a deterministic finite automaton (DFA), a simple but practically useful model checking problem asks for the probability of (terminating) traces accepted by the DFA, which can be computed via a product MC of the given MC and DFA and reduced to a simple reachability problem. Recently, Watanabe, Junges, Rot, and Hasuo proposed *coalgebraic product constructions*, a categorical framework that uniformly explains such coalgebraic constructions using distributive laws. This framework covers a range of instances, including the model checking of MCs against DFAs. In this paper, on top of their framework we first present a no-go theorem for product constructions, showing a case when we *cannot* do product constructions for model checking. Specifically, we show that there are *no* coalgebraic product MCs of MCs and nondeterministic finite automata for computing the probability of the accepting traces. The proof relies on a characterisation of natural transformations between certain functors that determine the type of branching, including nondeterministic or probabilistic branching.

Second, we present a coalgebraic product construction of MCs and multiset finite automata (MFAs) as a new instance within our framework. This construction addresses a model checking problem that asks for the expected number of accepting runs on MFAs over traces of MCs. We show that this problem is solvable in polynomial time.

1 Introduction

For decades, model checking has been extensively studied as a verification technique for ensuring the correctness of systems or programs adhere to a given specification. A standard model for representing systems with uncertainties are

---

[*] We would like to thank the anonymous reviewers for their valuable comments and suggestions, which significantly improve this article. The authors were supported by the ASPIRE grant No. JPMJAP2301, JST. M. K. is supported by the JST grant No. JPMJAX25CD, and K. W. is supported by the JST grants No. JPMJAX23CU and JPMJPR25KD.

N. Bertrand and S. Milius (Eds.): FoSSaCS 2026, LNCS 16503, pp. 350–371, 2026.
https://doi.org/10.1007/978-3-032-22730-0_17

Markov chains (MCs) [6]. The behaviour of an MC is observed through a trace $w \in A^+$ until it reaches a target state, at which point it terminates; throughout the paper, we consider only terminating traces. These traces are then checked against a specification represented by a finite automaton, determining acceptance of each trace. Formally, given an MC and a finite automaton, model checking asks for the probability that traces of the MC are accepted by the automaton. This probability is mathematically expressed as $\sigma(L) = \sum_{w \in L} \sigma(w) \in [0,1]$, where σ is the subdistribution of traces on the MC, and $L \in \mathcal{P}(A^+)$ is the recognised language of the automaton (excluding the empty string). Model checking of MCs against such linear-time properties has been actively explored in the literature, including in [5, 6, 9, 38].

A well-established approach to efficient model checking algorithms is the so-called *product construction* [30, 39]. In this approach, given an MC $\mathbb{M}$ and a finite automaton $\mathcal{A}$, one first constructs an equivalent finite deterministic automaton (DFA) $\mathcal{A}_d$ by determinisation. We then constructs a product MC $\mathbb{M} \otimes \mathcal{A}_d$, where the reachability probability to a designated target state exactly coincides with the original probability $\sigma(L)$. While the determinisation construction explodes the number of states exponentially in general, it has been shown that the resulting product MC with the DFA is efficiently solvable in poly-logarithmic parallel time (NC) w.r.t. the size of its underlying graph (see e.g. [7]), which belongs to the complexity class P.

To pursue an efficient model checking algorithm, it is natural to identify conditions under which product constructions can be applied to the target model checking problem *without requiring determinisation*. Indeed, such conditions for MCs and non-deterministic (ω-regular) automata have been explored for decades. Notable examples include products for MCs with separated unambiguous automata [13], unambiguous automata [7], and limit-deterministic automata [12, 33, 38]. While these automata are not fully deterministic, they impose restrictions on their non-deterministic behaviour, placing them between deterministic and non-deterministic automata. To the best of our knowledge, no product construction is currently known for MCs and nondeterministic finite automata (NFAs) without determinisation. It is fair to say that this is unlikely to be possible, although no formal proof of impossibility has been established previously (see also [40, Remark 4.4]).

Recently, Watanabe et al. proposed a unified approach to product constructions for model checking [40], establishing a structural theory for it. In their framework, both systems and specifications are modelled as *coalgebras* [22, 31], and a generic product construction, called the *coalgebraic product construction*, provides a unified method for constructing products via *distributive laws*, natural transformations that distribute the Cartesian product over the category of sets. Within this abstract formulation, they introduced a *correctness criterion*, which is a simple yet powerful sufficient condition ensuring the correctness of the coalgebraic product construction—the solution of the product coincides with that of the original model checking problem. This framework covers a wide range of in-

Table 1: Sets that are isomorphic to $\mathrm{Nat}(F_A, F_B)$ via a reasonably simple isomorphism. These functors, including the covariant finite powerset functor $\mathcal{P}_f$ and the multiset functor $\mathcal{M}$, are formally defined in Example 6. For instance, the set $\mathrm{Nat}(\mathcal{M}, \mathcal{P}_f)$ of natural transformations $\lambda\colon \mathcal{M} \Rightarrow \mathcal{P}_f$ is isomorphic to $\mathbf{2}^{\mathbb{N}_{\geq 1}}$ with a simple isomorphism described explicitly in Example 7. The characterisations of $\lambda\colon \mathcal{P}_f \Rightarrow \mathcal{P}_f$ and $\lambda\colon \mathcal{M} \Rightarrow \mathcal{M}$ have been found by Dahlqvist and Neves in [14].

$F_A \backslash F_B$	$\mathcal{P}_f$	$\mathcal{M}$	$\mathcal{R}^+_{\geq 0}$	$\mathcal{R}^\times_{\geq 0}$
$\mathcal{P}_f$	$\mathbf{2}$	$\mathbf{1}$	$\mathbf{1}$	$\mathbf{2}$
$\mathcal{M}$	$\mathbf{2}^{\mathbb{N}_{\geq 1}}$	$\mathbb{N}^{\mathbb{N}_{\geq 1}}$	$\mathbb{R}_{>0}^{\mathbb{N}_{\geq 1}}$	$\mathbb{R}_{\geq 0}^{\mathbb{N}_{\geq 1}}$
$\mathcal{R}^+_{\geq 0}$	$\mathbf{2}^{\mathbb{R}_{>0}}$	$\mathbf{1}$	$\mathbb{R}_{\geq 0}^{\mathbb{R}_{>0}}$	?

stances, including the aforementioned problem (e.g. [5,6]) and the cost-bounded reachability probability [18, 34].

On top of their coalgebraic framework, we further explore product constructions of MCs and automata. Our main result in this paper is a *no-go theorem* showing that **no** coalgebraic product MCs for MCs and NFAs satisfy the correctness criterion for the model checking problem that asks for the probability of accepting traces (Thm. 2). To prove this, we characterise natural transformations between certain functors that govern branching types, including nondeterministic and probabilistic branching (Thm. 1).

Beyond its application to product constructions, our characterisation of natural transformations is of independent interest. A summary of our characterisation is provided in Tbl. 1. For instance, the following isomorphism for $\mathrm{Nat}(\mathcal{M}, \mathcal{D}_{\leq 1})$ exemplifies our characterisation (Prop. 3):

Corollary 1. *There is an isomorphism* $\lambda^{(-)}\colon [0,1]^{\mathbb{N}_{\geq 1}} \to \mathrm{Nat}(\mathcal{M}, \mathcal{D}_{\leq 1})$ *given by*

$$\lambda_X^b(f)(x) = \begin{cases} \dfrac{f(x)}{\sum_{x \in X} f(x)} \cdot b\big(\sum_{x \in X} f(x)\big) & \text{if } \sum_{x \in X} f(x) > 0, \\ 0 & \text{if } \sum_{x \in X} f(x) = 0. \end{cases}$$

Second, we introduce a new model checking problem for MCs and multiset finite automata (MFAs), which asks for the expected number of accepting runs on MFAs over traces on MCs. We show that a unique coalgebraic product construction exists for this setting (Props. 6 and 7). As an immediate consequence, this result provides an alternative proof of the correctness of the known product construction [7] for MCs and unambiguous finite automata, which can be seen as a special case of MFAs. Lastly, we show that model checking of products of MCs and MFAs are solvable in P (Prop. 8). Our proof is indeed very simple: we reduce the problem to the computation of expected *multiplicative rewards* on MCs (without non-trivial bottom strongly connected components), which is solvable in P due to the very recent result by Baier et al. [3].

In summary, our contributions are as follows:

- We characterise natural transformations between functors for branching, which plays an important role in the proof of our no-go theorem (Thm. 1).

- We present a no-go theorem for coalgebraic *product MCs* of MCs and NFAs (Thm. 2).
- We propose a new model checking problem of MCs and MFAs, and show that the problem is solvable in polynomial time (Prop. 8).

Structure. In §2, we recall preliminaries related to coalgebraic semantics of transition systems, including MCs and DFAs. In §3, we review the coalgebraic product constructions and the correctness criterion, which is a sufficient condition that ensures the correctness of coalgebraic products with respect to a given model checking problem [40]. In §4, we show a characterisation of natural transformations between certain functors determining the type of branching. In §5, we show a no-go theorem for coalgebraic product MCs of MCs and NFAs. In §6, we study a new model checking problem for MCs and MFAs. In §7, we discuss related work, and in §8 we conclude this paper.

Notation. We write $\mathcal{D}$ for the distribution functor (with finite supports), and $\mathcal{D}_{\leq 1}$ and $\mathcal{D}_{\leq 1,c}$ for the subdistribution functor with finite supports and countable supports, respectively. We also write $\mathcal{M}$ and $\mathcal{M}_c$ for the multiset functors with finite supports and countable supports, respectively. The constant function mapping every element to a fixed value a is written as Δ_a. We write the unit interval as $[0,1]$, and the set of booleans $\{\bot, \top\}$ as $\mathbb{B}$. We use π_1 and π_2 for the first and second projections, respectively.

2   Preliminaries

We recall coalgebraic semantics for transition systems, which has been widely used in the literature (e.g. [21, 28, 35, 40]). The semantics used in this paper is based on least fixed-point semantics, analogous to weakest precondition semantics [1, 19], actively employed in studies of formal verification [21, 26, 35]. Given a coalgebra $c \colon X \to FX$ of an endofunctor F modelling a system, the semantics of c is given by a least fixed point of a specific *predicate transformer* of c that is induced by a *modality*.

Throughout the paper, we consider coalgebras $c \colon X \to FX$ on the category of sets (**Sets**), thus F is an endofunctor on the category of sets. A *semantic structure* for F is given by a pair (Ω, τ) such that (i) Ω is an ω-complete partially ordered set $(\Omega, \preceq)$ with the least element $\bot$; and (ii) τ is a function $\tau \colon F\Omega \to \Omega$. We call Ω and τ a *semantic domain* and a *modality* (for F), respectively.

Definition 1 (predicate transformer). *Given a coalgebra $c \colon X \to FX$ and a semantic structure (Ω, τ) for F, the* predicate transformer Φ_c *of c is a function $\Phi_c \colon \Omega^X \to \Omega^X$ given by $\Phi_c(u) := \tau \circ F(u) \circ c$ for each $u \in \Omega^X$. Additionally, we assume that Φ_c is ω-continuous with respect to the pointwise order in Ω^X.*

The predicate transformer Φ_c is the composition of the coalgebra c with the predicate lifting $\tau \circ F(\_)$. The latter corresponds bijectively to the modality $\tau \colon F\Omega \to \Omega$, which is commonly used in coalgebraic modal logic (cf. [32]).

By the Kleene-fixed point theorem, the predicate transformer Φ_c has the least fixed point $\mu\Phi_c = \bigvee_{n \in \mathbb{N}} \Phi^n(\bot)$.

Definition 2 (semantics). *Given a coalgebra c and a semantic structure (Ω, τ), the* semantics *is the least fixed point $\mu\Phi_c \in \Omega^X$ of the predicate transformer Φ_c.*

We recall the coalgebraic semantics of (labelled) Markov chains (MCs) and *deterministic finite automaton (DFA).*

Example 1 (semantics of MC). We define a *(labelled) MC* as a coalgebra $c\colon X \to \mathcal{D}(X+\{\checkmark\})\times A$. The semantic structure (Ω, τ) is given by (i) $\Omega := (\mathcal{D}_{\leq 1,c}(A^+), \preceq)$, where $\preceq$ is the pointwise order; and (ii) $\tau\colon \mathcal{D}(\mathcal{D}_{\leq 1,c}(A^+) + \{\checkmark\}) \times A \to \mathcal{D}_{\leq 1,c}(A^+)$ is given by

$$\tau(\sigma, a)(w) := \begin{cases} \sigma(\checkmark) & \text{if } w = a, \\ \sum_{\mu \in \mathcal{D}_{\leq 1,c}(A^+)} \sigma(\mu) \cdot \mu(w') & \text{if } w = a \cdot w' \text{ for some } w' \in A^+, \\ 0 & \text{otherwise.} \end{cases}$$

The least fixed point of Φ_c gives the reachability probability $\mu\Phi_c(x)(w) \in [0,1]$ from each state $x \in X$ to the target $\checkmark$ with the trace $w \in A^+$.

Example 2 (semantics of DFA). We define a DFA[4] as a coalgebra $d\colon Y \to (Y \times \mathbb{B})^A$. The semantic structure (Ω, τ) is given by (i) $\Omega := (\mathcal{P}(A^+), \subseteq)$, where $\subseteq$ is the inclusion order; and (ii) $\tau\colon (\mathcal{P}(A^+) \times \mathbb{B})^A \to \mathcal{P}(A^+)$ is given by

$$\tau(\delta) := \{a \in A \mid \pi_2(\delta(a)) = \top\} \cup \{a \cdot w \in A^+ \mid w \in \pi_1(\delta(a))\}.$$

The least fixed point of the predicate transformer Φ_d yields the recognized language excluding the empty string, as $\mu\Phi_d(y) \in \mathcal{P}(A^+)$ for each state $y \in Y$.

3 Coalgebraic Product Construction

We introduce the coalgebraic product construction [40], which is the foundation for our study of product constructions. The coalgebraic product construction employs a natural transformation λ called a *distributive law* to merge behaviours of two coalgebras; see, e.g., [29] for preliminaries on category theory.

Definition 3 (distributive law, coalgebraic product). *Let F_S, F_R, and $F_{S\otimes R}$ be endofunctors. A* distributive law *λ from F_S and F_R to $F_{S\otimes R}$ is a natural transformation $\lambda\colon \times \circ (F_S \times F_R) \Rightarrow F_{S\otimes R} \circ \times$. Given two coalgebras $c\colon X \to F_S X$ and $d\colon Y \to F_R Y$, the* coalgebraic product $c \otimes_\lambda d$ *induced by the distributive law λ is defined as the coalgebra:*

$$c \otimes_\lambda d\colon X \times Y \to F_{S\otimes R}(X \times Y), \quad c \otimes_\lambda d := \lambda_{X,Y} \circ (c \times d).$$

---

[4] Following [40], the acceptance condition is given by the output, as in Mealy machines.

Example 3 (product of MC and DFA). Define a distributive law λ from $\mathcal{D}((\_)+\{\checkmark\}) \times A$ and $((\_) \times \mathbb{B})^A$ to $\mathcal{D}_{\leq 1}((\_)+\{\checkmark\})$ by

$$\lambda_{X,Y}(z)(x,y) := \begin{cases} \sigma(x) & \text{if } y = \pi_1\big(\delta(a)\big), \\ 0 & \text{otherwise,} \end{cases}$$

$$\lambda_{X,Y}(z)(\checkmark) := \begin{cases} \sigma(\checkmark) & \text{if } \top = \pi_2\big(\delta(a)\big), \\ 0 & \text{otherwise,} \end{cases}$$

where $z = (\sigma, a, \delta) \in \mathcal{D}(X+\{\checkmark\}) \times A \times (Y \times \mathbb{B})^A$. Then the coalgebraic product $c \otimes_\lambda d$ of an MC $c\colon X \to \mathcal{D}_{\leq 1}(X + \{\checkmark\})$ and a DFA $d\colon Y \to (Y \times \mathbb{B})^A$ is the standard product of the MC and the DFA, e.g. [6,40].

It is worth emphasizing that the coalgebraic product $c \otimes_\lambda d$ is itself a coalgebra. This allows its semantics to be defined in the same manner as for its components, using a semantic structure $(\mathbf{\Omega}_{S \otimes R}, \tau_{S \otimes R})$.

Example 4 (semantics of the product). Consider Example 3. We define the semantic structure $(\mathbf{\Omega}_{S \otimes R}, \tau_{S \otimes R})$ as follows: (i) $\mathbf{\Omega}_{S \otimes R} := ([0,1], \leq)$, where $\leq$ is the standard order; and (ii) $\tau_{S \otimes R}\colon \mathcal{D}_{\leq 1}([0,1] + \{\checkmark\}) \to [0,1]$ is given by $\tau_{S \otimes R}(\sigma) := \sigma(\checkmark) + \sum_{r \in [0,1]} r \cdot \sigma(r)$. Under this semantic structure, the semantics $\mu\Phi_{c \otimes_\lambda d}$ of the product gives, for each (x,y), the reachability probability from (x,y) to the target state $\checkmark$.

Remark 1 (final coalgebra semantics). In Example 4, one might wonder whether the semantics arises as a (unique) coalgebra morphism into a final coalgebra in the Kleisli category of the monad $\mathcal{D}_{\leq 1}(\_ + \{\checkmark\})$, thereby instantiating the generic trace semantics of Hasuo et al. [20]. This, however, is not the case: the semantics is defined as the least fixed point of a predicate transformer, whereas the transformer may in general admit multiple fixed points. Uniqueness of the coalgebra morphism can be ensured only under additional assumptions such as finite state spaces and almost-sure termination of coalgebras [6].

We then move on to the *correctness* of coalgebraic products. This notion is defined with respect to an *inference map* $q\colon \Omega_S \times \Omega_R \to \Omega_{S \otimes R}$, where Ω_S and Ω_R are the underlying sets of semantic domains $\mathbf{\Omega}_S$ and $\mathbf{\Omega}_R$ of c and d, respectively.

Definition 4 (inference map, correctness). *Let $c\colon X \to F_S X$ and $d\colon Y \to F_R Y$ be two coalgebras, λ be a distributive law $\lambda_{X,Y}\colon (F_S X) \times (F_R Y) \to F_{S \otimes R}(X \times Y)$, and $(\mathbf{\Omega}_S, \tau_S)$, $(\mathbf{\Omega}_R, \tau_R)$, and $(\mathbf{\Omega}_{S \otimes R}, \tau_{S \otimes R})$ be semantic structures for F_S, F_R and $F_{S \otimes R}$, respectively. An* inference map *is a function $q\colon \Omega_S \times \Omega_R \to \Omega_{S \otimes R}$ such that*

1. $q(\bot, \bot) = \bot$; and
2. q is ω-continuous, that is, $\bigvee_{l \in \mathbb{N}} q(u_l, v_l) = q\big(\bigvee_{m \in \mathbb{N}} u_m, \bigvee_{n \in \mathbb{N}} v_n\big)$ for each ω-chains $(u_m)_{m \in \mathbb{N}}, (v_n)_{n \in \mathbb{N}}$.

The coalgebraic product $c \otimes_\lambda d$ is said to be correct w.r.t. q *if $q \circ (\mu\Phi_c \times \mu\Phi_d) = \mu\Phi_{c \otimes_\lambda d}$ holds.*

Example 5. Consider Examples 3 and 4. The coalgebraic product $c \otimes_\lambda d$ is correct w.r.t. $q \colon \mathcal{D}_{\leq 1, c}(A^+) \times \mathcal{P}(A^+) \to [0, 1]$ defined by $q(\sigma, L) := \sum_{w \in L} \sigma(w)$. This means that the semantics of the products—the reachability probabilities— coincides with the probability that traces of MCs are accepted by DFAs.

A simple *correctness criterion* that ensures the correctness of the coalgebraic product is proposed in [40]. This criterion is indeed general enough to capture a wide range of known product constructions in the literature [2, 4–6], including the one illustrated in Example 5.

Proposition 1 (correctness criterion [40]). *Assume the following data:*

- *a distributive law λ from F_S and F_R to $F_{S \otimes R}$,*
- *semantic structures $(\mathbf{\Omega}_S, \tau_S)$, $(\mathbf{\Omega}_R, \tau_R)$, and $(\mathbf{\Omega}_{S \otimes R}, \tau_{S \otimes R})$ for F_S, F_R, and $F_{S \otimes R}$, respectively,*
- *an inference map $q \colon \Omega_S \times \Omega_R \to \Omega_{S \otimes R}$.*

Then for any coalgebras $c \colon X \to F_S X$ and $d \colon Y \to F_R Y$, the coalgebraic product $c \otimes_\lambda d$ is correct w.r.t. q if the following equation holds:

$$q \circ (\tau_S \times \tau_R) = \tau_{S \otimes R} \circ F_{S \otimes R}(q) \circ \lambda_{\Omega_S, \Omega_R}.$$

The correctness criterion requires that a semantic structure $(\mathbf{\Omega}_{S \otimes R}, \tau_{S \otimes R})$ for products be explicitly specified. This requirement arises from the practical need for efficient computation of the semantics of products. Model checking problems should ideally be solvable by mature techniques. For instance, the semantics described in Example 4 can be efficiently computed by solving linear equation systems, or applying value iterations (see [6]).

Remark 2 (the choice of data). The coalgebraic product construction and its correctness criterion require several pieces of data to specify both the problem and its correctness. These data are not necessarily canonical. Ideally, one would like to determine these data via universal properties—for instance, through final coalgebras. We leave this direction for future work.

4   Natural Transformations for Coalgebraic Product Constructions

As demonstrated in Prop. 1, the coalgebraic product construction requires a distributive law λ, semantic structures (Ω_i, τ_i) for $i \in \{S, R, S \otimes R\}$, and an inference map q. Before exploring these structures in concrete settings, we begin by studying natural transformations between endofunctors on **Sets** that arise from commutative monoids (see Def. 5). The results established in this section enable us to prove a no-go theorem and uniqueness of distributive laws for coalgebraic products in later sections.

We write **CMon** for the category of commutative monoids. For a commutative monoid A, the binary operator is denoted by $+_A$ and the unit element by 0_A; when the context is clear, we omit these subscripts for simplicity. For $n \in \mathbb{N}$ and $a \in A$, we define $n \cdot a$ as the sum of n copies of a, with $0 \cdot a = 0$.

Definition 5. *We define a functor* $F_{(\_)} \colon \mathbf{CMon} \to [\mathbf{Sets}, \mathbf{Sets}]$ *as follows. For each* $A \in \mathbf{CMon}$*, the functor* F_A *is defined by*

$$F_A(X) := \{h \colon X \to A \mid \operatorname{supp}(h) \text{ is finite}\}, \qquad F_A(g)(f) := \sum_{x \in g^{-1}(\_)} f(x),$$

for each $X \in \mathbf{Sets}$*,* $g \colon X \to Y$*, and* $f \in F_A(X)$*, where* $\operatorname{supp}(h) = \{x \in X \mid h(x) \neq 0\}$*. For each* $i \colon A \to B$ *in* $\mathbf{CMon}$*, the* natural transformation F_i *is defined by* $(F_i)_X(f) := i \circ f$ *for each* $X \in \mathbf{Sets}$ *and* $f \in F_A(X)$*.*

For a fixed monoid A, the functor F_A has appeared in the literature on coalgebraic transition systems labelled by monoids [17].

Example 6.   1. The functor $F_{(\mathbb{N},+,0)}$ is the multiset functor $\mathcal{M}$.
 2. The functor $F_{(\mathbb{B},\vee,\perp)}$ is the covariant finite powerset functor $\mathcal{P}_f$ where $\vee$ is the logical OR.
 3. The functors $F_{(\mathbb{R}_{\geq 0},+,0)}$ and $F_{(\mathbb{R}_{\geq 0},\times,1)}$ both map a set X to a set of non-negative real-valued functions on X with finite support, while they map functions in a different way. We write $\mathcal{R}^+_{\geq 0}$ and $\mathcal{R}^\times_{\geq 0}$ for these functors $F_{(\mathbb{R}_{\geq 0},+,0)}$ and $F_{(\mathbb{R}_{\geq 0},\times,1)}$, respectively.

For $n \in \mathbb{N}$, we use $\mathbf{n}$ to represent the set $\{1, \cdots, n\}$, with the convention that $\mathbf{0}$ refers to the empty set.

Proposition 2. *Any natural transformation* $\lambda \colon F_A \Rightarrow F_B$ *is uniquely determined by its component at* $\mathbf{2}$*, that is, for each* $\lambda, \lambda' \colon F_A \Rightarrow F_B$*,* $\lambda_{\mathbf{2}} = \lambda'_{\mathbf{2}}$ *if and only if* $\lambda = \lambda'$*.* $\square$

See [27, Appendix A.1] for the proof. Note that each collection $\operatorname{Nat}(F_A, F_B)$ is a set.

In this section, we aim to precisely characterise this set $\operatorname{Nat}(F_A, F_B)$, providing explicit constructions for such natural transformations.

Remark 3. By analogy with Yoneda's lemma, it makes sense to speculate that $\operatorname{Nat}(F_A, F_B)$ can be characterised by monoid morphisms from A to B. However, this is not the case in general: while the functor $F_{(\_)}$ is faithful (as shown in [27, Appendix A.2]), it fails to be full. To see this, consider $A = (\mathbb{N}, +, 0)$ and $B = (\mathbb{R}_{\geq 0}, +, 0)$. It will be shown in Example 7 that each natural transformation $\lambda \in \operatorname{Nat}(F_{(\mathbb{N},+,0)}, F_{(\mathbb{R}_{\geq 0},+,0)})$ is of the form $\lambda_X(f)(x) = f(x) \cdot b\left(\sum_{x \in X} f(x)\right)$ for some $b \colon \mathbb{N} \to \mathbb{R}_{\geq 0}$. Assume, for contradiction, that there exists $i \colon \mathbb{N} \to \mathbb{R}_{\geq 0}$ such that $\lambda = F_i$. Since i is uniquely determined by the component λ_1, we have $\lambda_1(\Delta_n)(1) = i(n) = n \cdot b(n)$. Since $F_i(f)(x) = i(f(x)) = f(x) \cdot b(f(x))$, $F_i(f)(x)$ is in general different from $\lambda_X(f)(x)$.

After examining some fundamental properties of these natural transformations, we proceed to explore their explicit forms in certain cases.

Lemma 1. *Let A and B be commutative monoids, and let $\lambda\colon F_A \Rightarrow F_B$ be a natural transformation. Then, for each $f \in F_A(X)$ and $x, x' \in X$, the following statements hold.*

1. *$f(x) = f(x')$ implies $\lambda_X(f)(x) = \lambda_X(f)(x')$.*
2. *For each $n \in \mathbb{N}$, $n \cdot f(x) = f(x)$ implies $n \cdot \lambda_X(f)(x) = \lambda_X(f)(x)$.*

Proof Sketch. (1) This is easy to prove by taking a function $g\colon X \to X$ that swaps x and x' and by the naturality of λ.

(2) The cases $n = 0$ and $n = 1$ are easy to prove. We sketch the proof for the case $n \geq 2$. Let κ_2 be the second coprojection of a binary coproduct. We define $f' \in F_A(X + (\mathbf{2n} - \mathbf{2}))$ as $[f, \Delta_{f(x)}]$ and show that

$$\lambda_X(f)(x) = (2n - 1) \cdot \lambda_{X+(\mathbf{2n}-\mathbf{2})}(f')(\kappa_2(n)), \text{ and}$$
$$\lambda_{X+(\mathbf{2n}-\mathbf{2})}(f')(\kappa_2(n)) = n \cdot \lambda_{X+(\mathbf{2n}-\mathbf{2})}(f')(\kappa_2(n)).$$

Then, we can see that $\lambda_X(f)(x) = \lambda_{X+(\mathbf{2n}-\mathbf{2})}(f')(\kappa_2(n))$ because

$$\begin{aligned}
\lambda_X(f)(x) &= (2n - 1) \cdot \lambda_{X+(\mathbf{2n}-\mathbf{2})}(f')(\kappa_2(n)) \\
&= n \cdot \lambda_{X+(\mathbf{2n}-\mathbf{2})}(f')(\kappa_2(n)) + (n - 1) \cdot \lambda_{X+(\mathbf{2n}-\mathbf{2})}(f')(\kappa_2(n)) \\
&= \lambda_{X+(\mathbf{2n}-\mathbf{2})}(f')(\kappa_2(n)) + (n - 1) \cdot \lambda_{X+(\mathbf{2n}-\mathbf{2})}(f')(\kappa_2(n)) \\
&= \lambda_{X+(\mathbf{2n}-\mathbf{2})}(f')(\kappa_2(n)).
\end{aligned}$$

This concludes that $\lambda_X(f)(x) = n \cdot \lambda_X(f)(x)$. See [27, Appendix A.3] for the full proof. $\qquad\square$

The second statement places a restriction on the possible values of $\lambda_X(f)(x)$ when $n \cdot f(x) = f(x)$. In particular, setting $n = 0$ implies that $\lambda_X(f)(x) = 0$ if $f(x) = 0$.

The following result tells how λ respects the n-fold operation. To formalize this, we introduce a preorder on a commutative monoid A defined by $a \leq a'$ if and only if there exists $a'' \in A$ such that $a + a'' = a'$.

Lemma 2. *Let A and B be commutative monoids, and $\lambda\colon F_A \Rightarrow F_B$ be a natural transformation. For each $f \in F_A(X)$, $x \in X$, $a \in A$, $n, m \in \mathbb{N}$, and $\rhd_1, \rhd_2 \in \{=, \geq\}$, we assume that*

$$f(x) \rhd_1 n \cdot a, \quad \text{and} \quad \sum_{x' \in X \setminus \{x\}} f(x') \rhd_2 m \cdot a.$$

Then there are $a_1, a_2 \in A$ such that $f(x) = n \cdot a + a_1$ and $\sum_{x' \in X \setminus \{x\}} f(x') = m \cdot a + a_2$ by definition of $\rhd_1$ and $\rhd_2$. For these a_1 and a_2, the following relations hold:

$$\lambda_X(f)(x) \rhd_1 n \cdot \lambda_{n+m+2}(f')(1), \quad \sum_{x' \in X \setminus \{x\}} \lambda_X(f)(x') \rhd_2 m \cdot \lambda_{n+m+2}(f')(1),$$

where $f' \in F_A(\mathbf{n} + \mathbf{m} + \mathbf{2})$ is defined by $f'(n+m+1) := a_1$, $f'(n+m+2) := a_2$, and $f'(i) := a$ for each $i \in \mathbf{n} + \mathbf{m}$.

Proof. Define $g_1\colon \mathbf{n}+\mathbf{m}+\mathbf{2} \to \mathbf{2}$ by $g_1(i) = 1$ if $i \in \mathbf{n} \cup \{n+m+1\}$ and 2 otherwise, and $g_2\colon X \to \mathbf{2}$ by $g_2(x') = 1$ if $x' = x$ and 2 otherwise. Then the following equalities hold.

$$\lambda_X(f)(x) = \lambda_{\mathbf{2}}(F_A(g_2)(f))(1) \qquad\qquad \text{by naturality for } g_2,$$
$$= \lambda_{\mathbf{2}}(F_A(g_1)(f'))(1) \qquad\qquad \text{since } F_A(g_2)(f) = F_A(g_1)(f'),$$
$$= n \cdot \lambda_{\mathbf{n+m+2}}(f')(1) + \lambda_{\mathbf{n+m+2}}(f')(n+m+1) \quad \text{by naturality and Lem. 1.1.}$$

If $\rhd_1$ is equal to $=$, then we can choose $a_1 = 0$, and then $\lambda_X(f)(x) = n \cdot \lambda_{\mathbf{n+m+2}}(f')(1)$ since $\lambda_{\mathbf{n+m+2}}(f')(n+m+1) = 0$, which follows from Lem. 1.2 with $n = 0$. If $\rhd_1$ is equal to $\geq$, then $\lambda_X(f)(x) \geq n \cdot \lambda_{\mathbf{n+m+2}}(f')(1)$ clearly holds. One can also prove $\sum_{x' \in X \setminus \{x\}} \lambda_X(f)(x') \rhd_2 m \cdot \lambda_{\mathbf{n+m+2}}(f')(1)$ by considering $\lambda_{\mathbf{2}}(F_A(g_2)(f))(2)$ instead of $\lambda_{\mathbf{2}}(F_A(g_2)(f))(1)$. $\qquad\square$

4.1 For Natural Transformations from F_A Where A is Singly Generated

Let us focus on a commutative monoid A generated by a single element $a \in A$. In this setting, each $a' \in A$ can be written by the form $l \cdot a$ for some $l \in \mathbb{N}$. We define $N\colon A \to \mathbb{N}$ that assigns to each $a' \in A$ the least natural number l such that $l \cdot a = a'$.

For a singly generated commutative monoid A, Lem. 2 provides an insight into the explicit form of λ. Informally, given a natural transformation $\lambda\colon F_A \Rightarrow F_B$ and $f \in F_A(X)$, the value $\lambda_X(f)(x)$ can be expressed as $\lambda_X(f)(x) = N(f(x)) \cdot O(f)$ for any $x \in X$, where $O\colon F_A(X) \to B$ is a certain function. Moreover, the function O depends only on the sum of values of f. This follows from the fact that the naturality of λ imposes conditions only on functions whose total sums are equal because for any $g \in X \to Y$, $F_A(g)$ preserves the sum of function values, i.e. $\sum_{x \in X} f(x) = \sum_{y \in Y}(F_A(g)(f))(y)$.

Theorem 1. *Let A be a commutative monoid generated by a single element $a \in A$, and let B be a commutative monoid. In the following two cases, the set $\mathrm{Nat}(F_A, F_B)$ of natural transformations can be explicitly characterised:*

1. *Case 1: for each $n, m \in \mathbb{N}$, $n \neq m$ implies $n \cdot a \neq m \cdot a$.*
   There exists an isomorphism $\lambda^{(-)}\colon \{b \in B^{\mathbb{N}} \mid b(0) = 0\} \to \mathrm{Nat}(F_A, F_B)$:
   $$\lambda_X^b(f)(x) := N(f(x)) \cdot b\Big(\sum_{x \in X} N(f(x))\Big).$$

2. *Case 2: $a \neq 0$ and there is $n \in \mathbb{N}_{>1}$ such that $n \cdot a = a$ and $(n-1) \cdot a \neq 0$.*
   Let n be the least natural number such that $n > 1$ and $n \cdot a = a$. There exists an isomorphism $\lambda^{(-)}\colon \{c \in B \mid n \cdot c = c\}^{\{0, \cdots, n-2\}} \to \mathrm{Nat}(F_A, F_B)$ given by
   $$\lambda_X^b(f)(x) := N(f(x)) \cdot b\Big(\big[\sum_{x \in X} N(f(x))\big]\Big),$$

   where $[l]$ is the remainder of l modulo $n - 1$.

Proof Sketch. 1) For each $n \in \mathbb{N}_{\geq 1}$, we use the function $f'_n \in F_A(\mathbf{n}+\mathbf{2})$ given by $f'_n(i) = a$ for each $i \in \mathbf{n}$ and $f'_n(n+1) = f'_n(n+2) = 0$. We define the inverse $(\lambda^{(-)})^{-1} \colon \mathrm{Nat}(F_A, F_B) \to \{b \in B^{\mathbb{N}} \mid b(0) = 0\}$ as follows: Given a natural transformation λ, the inverse $b^\lambda \colon \mathbb{N} \to B$ is given by $b(0) = 0$ and $b(n) = \lambda_{\mathbf{n}+\mathbf{2}}(f'_n)(1)$ for each $n \in \mathbb{N}_{\geq 1}$. We can show that the mapping $b^{(-)}$ is indeed the inverse by Lem. 2.

2) The construction of the inverse is more involved. We use the same function f'_n for each $n \in \mathbb{N}_{\geq 1}$. We define the inverse $(\lambda^{(-)})^{-1} \colon \mathrm{Nat}(F_A, F_B) \to \{c \in B \mid n \cdot c = c\}$ as follows: Given a natural transformation λ, the inverse $b^\lambda \colon \{0, \cdots, n-2\} \to \{c \in B \mid n \cdot c = c\}$ is given by $b^\lambda(m) := d^\lambda([m-1]+1)$, where the function $d^\lambda \colon \mathbb{N}_{\geq 1} \to \{c \in B \mid n \cdot c = c\}$ is defined by $d^\lambda(m) := \lambda_{\mathbf{m}+\mathbf{2}}(f'_m)(1)$. We can show that this mapping is the inverse by Lem. 1.2 and Lem. 2.

See [27, Appendix A.4] for the full proof. $\qquad\square$

Let us instantiate this theorem with concrete examples.

Example 7 ($\mathcal{M} \Rightarrow F_B$). Consider the multiset functor $\mathcal{M} = F_{(\mathbb{N},+,0)}$ (see Example 6). The monoid $(\mathbb{N}, +, 0)$ is generated by 1, and this falls under Case 1 of Thm. 1. Applying the theorem, we derive the following results:

1. There exists an isomorphism $\lambda \colon \{b \in \mathbb{B}^{\mathbb{N}} \mid b(0) = \bot\} \to \mathrm{Nat}(\mathcal{M}, \mathcal{P}_{\mathrm{f}})$ defined by $\lambda^b_X(f) := \{x \in X \mid f(x) > 0 \text{ and } b(\sum_{x \in X} f(x))\}$.
2. There exists an isomorphism $\lambda \colon \{b \in (\mathbb{R}_{\geq 0})^{\mathbb{N}} \mid b(0) = 0\} \to \mathrm{Nat}(\mathcal{M}, \mathcal{R}^+_{\geq 0})$ defined by $\lambda^b_X(f)(x) := f(x) \cdot b(\sum_{x \in X} f(x))$.
3. There exists an isomorphism $\lambda \colon \{b \in \mathbb{R}^{\mathbb{N}}_{\geq 0} \mid b(0) = 1\} \to \mathrm{Nat}(\mathcal{M}, \mathcal{R}^\times_{\geq 0})$ defined by $\lambda^b_X(f)(x) := \left(b\left(\sum_{x \in X} f(x)\right)\right)^{f(x)}$.

Example 8 ($\mathcal{P}_{\mathrm{f}} \Rightarrow F_B$). Consider the finite powerset functor $\mathcal{P}_{\mathrm{f}} = F_{(\mathbb{B}, \vee, \bot)}$ (see Example 6). This falls under Case 2 of Thm. 1 since $(\mathbb{B}, \vee, \bot)$ is generated by $\top$ that is idempotent. By Thm. 1, natural transformations are constrained by the idempotent property of $\top$, and the isomorphism takes the form: $\lambda^{(-)} \colon \{c \in B \mid c^2 = c\} \to \mathrm{Nat}(\mathcal{P}_{\mathrm{f}}, F_B)$. We then derive the following results:

1. There exists a unique natural transformation $\lambda \colon \mathcal{P}_{\mathrm{f}} \Rightarrow \mathcal{R}^+_{\geq 0}$ (respectively, $\lambda \colon \mathcal{P}_{\mathrm{f}} \Rightarrow \mathcal{M}$), which is given by $\lambda_X(S)(x) = 0$.
2. There exists only two natural transformations $\lambda \colon \mathcal{P}_{\mathrm{f}} \Rightarrow \mathcal{R}^\times_{\geq 0}$. One is given by $\lambda_X(S)(x) := 1$; and the other is by $\lambda_X(S)(x) := 0$ if $x \in S$, and 1 otherwise. $\qquad\square$

The following lemma allows us to extend our results about natural transformations between F_A and F_B to their respective subfunctors. See [27, Appendix A.5] for its proof.

Lemma 3. *Let F and G be subfunctors of F_A and F_B, respectively, meaning that there exist $\iota_A \colon F \Rightarrow F_A$ and $\iota_B \colon G \Rightarrow F_B$ given by inclusions. Assume that for all $g \colon X \to Y$ in **Sets** and $f \in F_A(X)$, $F_A(g)(f) \in F(Y)$ implies $f \in F(X)$. Then for any natural transformation $\lambda \colon F \Rightarrow G$, there exists a natural transformation $\lambda' \colon F_A \Rightarrow F_B$ such that $\iota_B \circ \lambda = \lambda' \circ \iota_A$.* $\qquad\square$

Proposition 3. *1. No natural transformation $\mathcal{P}_f \Rightarrow \mathcal{D}$ exists.*
2. *There exists a unique natural transformation $\lambda \colon \mathcal{P}_f \Rightarrow \mathcal{D}_{\leq 1}$.*
3. *No natural transformation $\mathcal{M} \Rightarrow \mathcal{D}$ exists.*
4. *There is an isomorphism $\lambda^{(\_)} \colon [0,1]^{\mathbb{N}_{\geq 1}} \to \mathrm{Nat}(\mathcal{M}, \mathcal{D}_{\leq 1})$ defined by*

$$\lambda_X^b(f)(x) = \begin{cases} \frac{f(x)}{\sum_{x \in X} f(x)} \cdot b\big(\sum_{x \in X} f(x)\big) & \text{if } \sum_{x \in X} f(x) > 0, \\ 0 & \text{if } \sum_{x \in X} f(x) = 0. \end{cases}$$

We provide a proof of statement 1 below. The detailed proofs of statements 2, 3, and 4 are given in [27, Appendix A.6].

Proof of 1. Suppose that a natural transformation $\lambda \colon \mathcal{P}_f \Rightarrow \mathcal{D}$ exists. Since $\mathcal{D}$ is a subfunctor of $\mathcal{R}_{\geq 0}^+$, Lem. 3 and Thm. 1.2 imply that λ should be the constant zero transformation. It may have an image that is not contained in $\mathcal{D}$. Consequently, no natural transformation $\mathcal{P}_f \Rightarrow \mathcal{D}$ exists. $\qquad\square$

4.2 For Natural Transformations from $\mathcal{R}_{\geq 0}^+$

We further investigate natural transformations λ from $\mathcal{R}_{\geq 0}^+ = F_{(\mathbb{R}_{\geq 0},+,0)}$ (see Example 6) to some F_B. Thm. 1 is not applicable here, as $(\mathbb{R}_{\geq 0}, +, 0)$ is not generated by a single element. However, we can still determine the explicit form of λ by utilizing the property that the value $\lambda_X(f)(x)$ is constrained by the sum $\sum_{x' \in X} \lambda_X(f)(x')$ and the ratio of $f(x)$ to the total sum $\sum_{x \in X} f(x)$.

Lemma 4. *Let B be a commutative monoid, $\lambda \colon \mathcal{R}_{\geq 0}^+ \Rightarrow F_B$ be a natural transformation. For any $f \in \mathcal{R}_{\geq 0}^+(X)$, $x \in X$, and $m, n \in \mathbb{N}$ with $n < m$ such that $\frac{n}{m} \cdot \sum_{x' \in X} f(x') \leq f(x) \leq \frac{n+1}{m} \cdot \sum_{x' \in X} f(x')$, the following statements hold.*

1. *There is $b \in B$ such that $(m - n) \cdot b = \sum_{x' \in X \setminus \{x\}} \lambda_X(f)(x')$ and $n \cdot b \leq \lambda_X(f)(x)$, and*
2. *There is $b \in B$ such that $(n + 1) \cdot b = \lambda_X(f)(x)$ and $(m - n - 1) \cdot b \leq \sum_{x' \in X \setminus \{x\}} \lambda_X(f)(x')$.*

Proof. Let $r := \sum_{x' \in X} f(x')$. Note that $\frac{n}{m} \cdot r \leq f(x) \leq \frac{n+1}{m} \cdot r$ is equivalent to $\frac{n}{m-n} \cdot (r - f(x)) \leq f(x)$ and $\frac{m-n-1}{n+1} \cdot f(x) \leq r - f(x)$. Lem. 2 for $(f, x, \frac{r-f(x)}{m-n}, n, m-n, \geq, =)$ implies that $n \cdot \lambda_{\mathbf{m+2}}(f')(1) \leq \lambda_X(f)(x)$ and $(m-n) \cdot \lambda_{\mathbf{m+2}}(f')(1) = \sum_{x' \in X \setminus \{x\}} \lambda_X(f)(x')$ where f' is defined in Lem. 2. Similarly, Lem. 2 for $(f, x, \frac{f(x)}{n+1}, n+1, m-n-1, =, \geq)$ implies the second statement. $\quad\square$

When $F_B = \mathcal{P}_f$, Lem. 4 allows us to conclude that for any $f \in \mathcal{R}_{\geq 0}^+(X)$ and $x \in X$, $\lambda_X(f)(x) = \bigvee_{x' \in X} \lambda_X(f)(x')$ if $f(x) > 0$. This leads to the following result. See [27, Appendix A.7] for the proof.

Proposition 4. *There is an isomorphism $\lambda^{(\_)} \colon \{b \in \mathbb{B}^{\mathbb{R}_{\geq 0}} \mid b(0) = \perp\} \to \mathrm{Nat}(\mathcal{R}_{\geq 0}^+, \mathcal{P}_f)$ defined by $\lambda_X^b(f) = \{x \in X \mid f(x) > 0 \text{ and } b(\sum_{x \in X} f(x))\}$.* $\quad\square$

Similarly, by Lem. 4 when $F_B = \mathcal{R}^+_{\geq 0}$, we can prove that the ratio of $\lambda_X(f)(x)$ to $\sum_{x' \in X} \lambda_X(f)(x')$ is equal to the ratio of $f(x)$ to $\sum_{x' \in X} f(x')$. This allows us to establish the following result.

Proposition 5. *There is an isomorphism* $\lambda^{(-)} \colon \{b \in (\mathbb{R}_{\geq 0})^{\mathbb{R}_{\geq 0}} \mid b(0) = 0\} \to$ $\mathrm{Nat}(\mathcal{R}^+_{\geq 0}, \mathcal{R}^+_{\geq 0})$ *defined by* $\lambda^b_X(f)(x) = f(x) \cdot b\big(\sum_{x \in X} f(x)\big)$.

Proof. The inverse $b^{(-)}$ of $\lambda^{(-)}$ is defined as follows: given $\lambda \colon \mathcal{R}^+_{\geq 0} \Rightarrow \mathcal{R}^+_{\geq 0}$, the function $b^\lambda \colon \mathbb{R}_{\geq 0} \to \mathbb{R}_{\geq 0}$ is defined by $b(0) := 0$ and $b^\lambda(r) := \frac{1}{r} \cdot \lambda_1(\Delta_r)(1)$. We first prove that the mapping $\lambda^{(-)}$ is the left inverse of $b^{(-)}$. Let $\lambda \colon \mathcal{R}^+_{\geq 0} \Rightarrow \mathcal{R}^+_{\geq 0}$, $f \in \mathcal{R}^+_{\geq 0}(X)$, and $x \in X$. If $f(x) = 0$, then clearly $\lambda_X(f)(x) = 0 = \lambda^{b^\lambda}_X(f)(x)$ by Lem. 1.2 with $n = 0$. Otherwise (i.e. $f(x) > 0$), define $r := \sum_{x' \in X} f(x')$. By Lem. 4, for any natural numbers m and n such that $n < m$ and $\frac{n}{m} \leq \frac{f(x)}{r} \leq \frac{n+1}{m}$, we obtain the inequalities:

$$\frac{n}{m} \cdot \left(\sum_{x' \in X} \lambda_X(f)(x') \right) \leq \lambda_X(f)(x) \leq \frac{n+1}{m} \cdot \left(\sum_{x' \in X} \lambda_X(f)(x') \right).$$

Taking limits as $m \to \infty$, these inequalities yield $\frac{f(x)}{r} \cdot \sum_{x' \in X} \lambda_X(f)(x') = \lambda_X(f)(x)$. Furthermore, by the naturality of λ for $!_X$, we have $\sum_{x' \in X} \lambda_X(f)(x') = \lambda_1(\Delta_r)(1)$, implying the equation $\lambda_X(f)(x) = \frac{f(x)}{r} \cdot \lambda_1(\Delta_r)(1) = \lambda^{b^\lambda}_X(f)(x)$. It remains to show that $\lambda^{(-)}$ is the right inverse, which is easy to check. $\square$

This result, together with Lem. 3, allows us to analyze natural transformations for important cases involving the (sub)distribution functor and the multiset functor. See [27, Appendix A.8] for the omitted proof.

Corollary 2.   *1. There is an isomorphism* $\lambda^{(-)} \colon \{b \in (\mathbb{R}_{\geq 0})^{[0,1]} \mid b(0) = 0\} \to$ $\mathrm{Nat}(\mathcal{D}_{\leq 1}, \mathcal{R}^+_{\geq 0})$ *defined by* $\lambda^b_X(f)(x) = f(x) \cdot b(\sum_{x \in X} f(x))$.
  2. There is an isomorphism $\lambda^{(-)} \colon \mathbb{R}_{\geq 0} \to \mathrm{Nat}(\mathcal{D}, \mathcal{R}^+_{\geq 0})$ *defined by* $\lambda^b_X(f)(x) = f(x) \cdot b$.
  3. There is an isomorphism $\lambda^{(-)} \colon [0,1]^{(0,1]} \to \mathrm{Nat}(\mathcal{D}_{\leq 1}, \mathcal{D}_{\leq 1})$ *defined by*

$$\lambda^b_X(f)(x) = \begin{cases} \frac{f(x)}{\sum_{x \in X} f(x)} \cdot b\big(\sum_{x \in X} f(x)\big) & \text{if } \sum_{x \in X} f(x) > 0, \\ 0 & \text{if } \sum_{x \in X} f(x) = 0. \end{cases} \qquad \square$$

Corollary 3. *There exists a unique natural transformation* $\lambda \colon \mathcal{R}^+_{\geq 0} \to \mathcal{M}$ *given by* $\lambda_X(f)(x) = 0$.

Proof. Let $\lambda \colon \mathcal{R}^+_{\geq 0} \Rightarrow \mathcal{M}$ be a natural transformation. By Lem. 3 and Prop. 5, there exists a unique $b' \colon \mathbb{R}_{\geq 0} \to \mathbb{R}_{\geq 0}$ such that $b'(0) = 0$ and $\lambda_X(f)(x) = f(x) \cdot b'\big(\sum_{x \in X} f(x)\big)$ for each $f \in \mathcal{R}^+_{\geq 0}(X)$ and $x \in X$. Now, assume for contradiction that there exists $r > 0$ such that $b'(r) > 0$. We can construct a function $f \in \mathcal{R}^+_{\geq 0}(2)$ such that $\sum_{x \in 2} f(x) = r$, $f(2) \neq 0$, and $\frac{f(1)}{f(2)}$ is an irrational number. Since $\lambda_2(f)(1)$ and $\lambda_2(f)(2)$ must be natural numbers, it follows that their ratio must be rational: $\frac{\lambda_2(f)(1)}{\lambda_2(f)(2)} = \frac{f(1) \cdot b'(r)}{f(2) \cdot b'(r)} = \frac{f(1)}{f(2)}$. This is a contradiction. Therefore, $b'(r)$ must be 0 for each $r \in \mathbb{R}_{> 0}$. $\square$

5   Markov Chains and Non-deterministic Finite Automata

To handle a specific model checking problem using the coalgebraic product construction, it is necessary to specify the structural components described in Prop. 1: a distributive law, semantic structures for both components and their composite, and an inference map. Assume that the semantic structures for the individual components and the composite system, together with the appropriate inference map, are already determined. This assumption is reasonable: the composite system is expected to be well studied and readily solvable by existing methods, and the model checking problem typically determines the relevant inference map. For the coalgebraic product construction to be valid, we must identify a distributive law that satisfies the correctness criterion with the above data. In this section, we prove that no such distributive law exists for coalgebraic product constructions involving MCs and NFAs.

Definition 6 (semantic structure of NFAs). *An NFA is a coalgebra* $d\colon Y \to \mathcal{P}_{\mathrm{f}}(Y + \{\checkmark\})^A$, *where the underlying set Y is finite. The* semantic structure $(\mathbf{\Omega}_R, \tau_R)$ *of NFAs is defined by (i)* $\mathbf{\Omega}_R := (\mathcal{P}(A^+), \subseteq)$; *and (ii)* $\tau_R\colon \mathcal{P}_{\mathrm{f}}(\mathcal{P}(A^+) + \{\checkmark\})^A \to \mathcal{P}(A^+)$ *is given by*

$$\tau_R(\delta) := \{a \in A \mid \checkmark \in \delta(a)\} \cup \{a \cdot w \mid S \in \delta(a), w \in S\}.$$

The semantics of NFAs is their recognized languages. We write $\mathbb{R}^\infty_{\geq 0} := \{r \in \mathbb{R} \mid r \geq 0\} \cup \{+\infty\}$ with the convention $\infty \cdot 0 = 0 \cdot \infty := 0$.

Definition 7 (semantic structure of the product). *The* semantic structure $(\mathbf{\Omega}_{S \otimes R}, \tau_{S \otimes R})$ *is defined by (i)* $\mathbf{\Omega}_{S \otimes R} := (\mathbb{R}^\infty_{\geq 0}, \leq)$, *where $\leq$ is the standard order; and (ii)* $\tau_{S \otimes R}\colon \mathcal{R}^+_{\geq 0}(\mathbb{R}^\infty_{\geq 0} + \{\checkmark\}) \to \mathbb{R}^\infty_{\geq 0}$ *is given by*

$$\tau_{S \otimes R}(\sigma) := \sigma(\checkmark) + \sum_{r \in \mathbb{R}^\infty_{\geq 0}} r \cdot \sigma(r).$$

We now present our first main result, a no-go theorem for coalgebraic product MCs for MCs and NFAs. The proof relies on the characterisation of natural transformations from $\mathcal{P}_{\mathrm{f}}$ to $\mathcal{R}^+_{\geq 0}$ presented in §4.1. The characterisation imposes strict constraints on possible forms of distributive laws, and we demonstrate that no such distributive law satisfies the correctness criterion.

Theorem 2 (no-go theorem under correctness criterion). *Consider semantic structures defined in Example 1, Def. 6, and Def. 7. There is no distributive law λ from $\mathcal{D}_{\leq 1}(\_) \times A$ and $\mathcal{P}_{\mathrm{f}}((\_) + \{\checkmark\})^A$ to $\mathcal{R}^+_{\geq 0}((\_) + \{\checkmark\})$ such that for each MC c and NFA d, the coalgebraic product $c \otimes_\lambda d$ satisfies the correctness criterion with the inference map $q\colon \mathcal{D}_{\leq 1,\mathrm{c}}(A^+) \times \mathcal{P}(A^+) \to \mathbb{R}^\infty_{\geq 0}$ defined by*

$$q(\sigma, L) := \sum_{w \in L} \sigma(w), \qquad \textit{for each } \sigma \in \mathcal{D}_{\leq 1,\mathrm{c}}(A^+) \textit{ and } L \in \mathcal{P}(A^+).$$

Proof. Suppose that there exists a distributive law λ from $\mathcal{D}_{\leq 1}(\_) \times A$ and $\mathcal{P}_{\mathrm{f}}((\_) + \{\checkmark\})^A$ to $\mathcal{R}^+_{\geq 0}((\_) + \{\checkmark\})$ satisfying the correctness criterion. Since there is a natural isomorphism $\mathcal{R}^+_{\geq 0}((\_) + \{\checkmark\}) \Rightarrow \mathcal{R}^+_{\geq 0}(\_) \times \mathbb{R}_{\geq 0}$, there is a bijective correspondence between $\bar{\lambda}$ and a pair of natural transformations:

- $\{\lambda'_{X,Y} : \mathcal{D}_{\leq 1}(X) \times A \times \mathcal{P}_{\mathrm{f}}(Y + \{\checkmark\})^A \to \mathbb{R}_{\geq 0}\}_{X,Y}$, and
- $\{\lambda''_{X,Y} : \mathcal{D}_{\leq 1}(X) \times A \times \mathcal{P}_{\mathrm{f}}(Y + \{\checkmark\})^A \to \mathcal{R}^+_{\geq 0}(X \times Y)\}_{X,Y}$.

We now analyze the properties of λ' and λ''.

- By the naturality of λ', for each $(\sigma, a) \in \mathcal{D}_{\leq 1}(X) \times A$ and $\delta \in \mathcal{P}_{\mathrm{f}}(Y + \{\checkmark\})^A$, we have:
$$\lambda'_{X,Y}(\sigma, a, \delta) = \lambda'_{1,1}\left(\Delta_r,\, a,\, \mathcal{P}_{\mathrm{f}}(!_Y + \{\checkmark\})^A(\delta)\right), \tag{1}$$
where $r := \sum_{x \in X} \sigma(x)$.
- For each $(\sigma, a) \in \mathcal{D}_{\leq 1}(X) \times A$, define a natural transformation $\rho \colon \mathcal{P}_{\mathrm{f}} \Rightarrow \mathcal{R}^+_{\geq 0}$ by $\{\rho_Y := \mathcal{R}^+_{\geq 0}(\pi_2) \circ \lambda''_{X,Y} \circ \langle \sigma,\, a,\, \Delta_{\mathcal{P}_{\mathrm{f}}(\kappa_1)(\_)}\rangle\}_Y$. By Example 8, it is the constant natural transformation to 0. Thus it follows that $\lambda''_{X,Y}(\sigma, a, \Delta_L)(x, y) = 0$ for each $L \in \mathcal{P}_{\mathrm{f}}(X)$, $x \in X$, and $y \in Y$ since
$$\rho_Y(L)(y) = \sum_{x \in X} \lambda''_{X,Y}(\sigma, a, \Delta_L)(x, y) = 0.$$

By Prop. 1, for each $(\sigma, a) \in \mathcal{D}_{\leq 1}(\mathbf{\Omega}_S) \times A$ and $\delta \in \mathcal{P}_{\mathrm{f}}(\mathbf{\Omega}_R + \{\checkmark\})^A$,

$$\left(1 - \sum_{\mu \in \mathbf{\Omega}_S} \sigma(\mu)\right) \cdot \delta(a)(\checkmark) + \sum_{w \in A^+} \left(\sum_{\mu \in \mathbf{\Omega}_S} \sigma(\mu) \cdot \mu(w)\right) \cdot \left(\bigvee_{L \in \delta(a)} L(w)\right)$$
$$= \lambda'_{\mathbf{\Omega}_S, \mathbf{\Omega}_R}(\sigma, a, \delta) + \sum_{\mu \in \mathbf{\Omega}_S, L \in \delta(a)} \sum_{w \in A^+} \mu(w) \cdot L(w) \cdot \lambda''_{\mathbf{\Omega}_S, \mathbf{\Omega}_R}(\sigma, a, \delta)(\mu, L). \tag{2}$$

For simplicity, we write $\delta(a)$ and L for their characteristic functions in the equation above.

For each $r \in [0, 1]$, $a \in A$, and $\delta \in \mathcal{P}_{\mathrm{f}}(1 + \{\checkmark\})^A$, define $\sigma_1 \in \mathcal{D}_{\leq 1}(\mathbf{\Omega}_S)$ and $\delta_1 \in \mathcal{P}_{\mathrm{f}}(\mathbf{\Omega}_R + \{\checkmark\})^A$ by

$$\sigma_1(x) := \begin{cases} r & \text{if } x = \Delta_0 \\ 0 & \text{otherwise} \end{cases}, \qquad \delta_1 := \mathcal{P}_{\mathrm{f}}(\Delta_\emptyset + \{\checkmark\})^A(\delta).$$

Then (2) for (σ_1, a) and δ_1 gives that $(1 - r) \cdot \delta(a)(\checkmark) = \lambda'_{\mathbf{\Omega}_S, \mathbf{\Omega}_R}(\sigma_1, a, \delta_1) = \lambda'_{1,1}(\Delta_r, a, \delta)$. Therefore, for each $(\sigma, a) \in \mathcal{D}_{\leq 1}(X) \times A$ and $\delta \in \mathcal{P}_{\mathrm{f}}(Y + \{\checkmark\})^A$, (1) induces $\lambda'_{X,Y}(\sigma, a, \delta) = \left(1 - \sum_{x \in X} \sigma(x)\right) \cdot \delta(a)(\checkmark)$.

Let $r \in (0, 1]$, $w \in A^+$, $a \in A$, and $\mu \in \mathbf{\Omega}_S$ defined by $\mu(w) = 1$ and $\mu(w') = 0$ for each $w' \in A^+ \setminus \{w\}$. Define $\sigma_2 \in \mathcal{D}_{\leq 1}(\mathbf{\Omega}_S)$ and $\delta_2 \in \mathcal{P}_{\mathrm{f}}(\mathbf{\Omega}_R + \{\checkmark\})^A$ by

$$\sigma_2(x) := \begin{cases} r & \text{if } x = \mu \\ 0 & \text{otherwise} \end{cases}, \qquad \delta_2 := \Delta_{\{w\}}.$$

Then (2) for (σ_2, a) and δ_2 gives that

$$
\begin{aligned}
r &= \sum_{w \in A^+} \left(\sum_{\mu \in \Omega_S} \sigma_2(\mu) \cdot \mu(w) \right) \cdot \left(\bigvee_{L \in \delta_2(a)} L(w) \right) \\
&= \sum_{\mu \in \Omega_S, L \in \delta_2(a)} \sum_{w \in A^+} \mu(w) \cdot L(w) \cdot \lambda''_{\Omega_S, \Omega_R}(\sigma_2, a, \delta_2)(\mu, L) = 0.
\end{aligned}
$$

This leads a contradiction, proving that no such distributive law λ can exist. $\qquad\square$

Note that this immediately implies the no-go theorem for distributive laws from $\mathcal{D}_{\leq 1}(\_) \times A$ and $\mathcal{P}_f((\_) + \{\checkmark\})^A$ to $\mathcal{D}_{\leq 1}((\_) + \{\checkmark\})$ as well.

6 Markov Chains and Multiset Finite Automata

In this section, we present a coalgebraic product construction of MCs and *multiset finite automata (MFAs)*, and show the coalgebraic product construction is correct w.r.t. the inference map that takes the expectation of the number of accepting paths. This immediately shows the correctness of the existing product construction of MCs and unambiguous finite automata [7, 8] as a special case. We further show that the coalgebraic product of MCs and MFAs is solvable in polynomial time.

Definition 8 (semantic structure of MFAs). *An MFA is a coalgebra $d \colon Y \to \mathcal{M}(Y + \{\checkmark\})^A$, where the underlying set Y is finite. The semantic structure (Ω_R, τ_R) of MFAs is defined by (i) $\Omega_R := (\mathcal{M}_c(A^+), \preceq)$, where $\preceq$ is the pointwise order; and (ii) $\tau_R \colon \mathcal{M}(\mathcal{M}_c(A^+) + \{\checkmark\})^A \to \mathcal{M}_c(A^+)$ is given by*

$$
\tau_R(\delta)(w) := \begin{cases}
\delta(a)(\checkmark) & \text{if } w = a, \\
\sum_{\mu \in \mathcal{M}_c(A^+)} \delta(a)(\mu) \cdot \mu(w') & \text{if } w = a \cdot w', \\
0 & \text{otherwise.}
\end{cases}
$$

The semantics of MFAs is the number of accepting paths for each word. MFAs are indeed a weighted automaton with the standard commutative ring $\mathbb{N}$.

Definition 9. *The distributive law $\lambda_{X,Y} \colon \mathcal{D}(X + \{\checkmark\}) \times A \times \mathcal{M}(Y + \{\checkmark\})^A \to \mathcal{R}_{\geq 0}^+(X \times Y + \{\checkmark\})$ is given by*

$$
\lambda_{X,Y}(\sigma, a, \delta)(x, y) := \delta(a)(y) \cdot \sigma(x), \qquad \lambda_{X,Y}(\sigma, a, \delta)(\checkmark) := \delta(a)(\checkmark) \cdot \sigma(\checkmark).
$$

Importantly, the coalgebraic product $c \otimes_\lambda d \colon X \times Y \to \mathcal{R}_{\geq 0}^+(X \times Y + \{\checkmark\})$ is not substochastic, that is, the sum $\sum_{(x', y')} (c \otimes_\lambda d)(x, y)(x', y') + (c \otimes_\lambda d)(x, y)(\checkmark)$ may be strictly greater than 1, for some $(x, y) \in X \times Y$.

We now establish the correctness of the coalgebraic product for MCs and MFAs w.r.t. the model checking problem that computes the expectation of the number of accepting paths.

Proposition 6 (correctness). *Consider semantic structures defined in Example 1, Def. 7, and Def. 8. Then the coalgebraic product $c \otimes_\lambda d$ is correct w.r.t. $q\colon \mathcal{D}_{\leq 1,c}(A^+) \times \mathcal{M}_c(A^+) \to \mathbb{R}^\infty_{\geq 0}$ defined by*

$$q(\sigma, \mu) := \sum_{w \in A^+} \mu(w) \cdot \sigma(w), \quad \text{for each } \sigma \in \mathcal{D}_{\leq 1,c}(A^+) \text{ and } \mu \in \mathcal{M}_c(A^+). \quad \Box$$

See [27, Appendix B.1] for the proof. As a direct consequence, the product of an MC c and an unambiguous finite automaton d [7, 8] is correct w.r.t. the inference map that computes the probability of paths that are accepting. In fact, $(\mu\Phi_d)(w) = 1$ iff l is an accepting word, and $(\mu\Phi_d)(w) = 0$ otherwise, because d is unambiguous, implying that the model checking $q \circ (\mu\Phi_c \times \mu\Phi_d)$ precisely gives the probability of paths that are accepting.

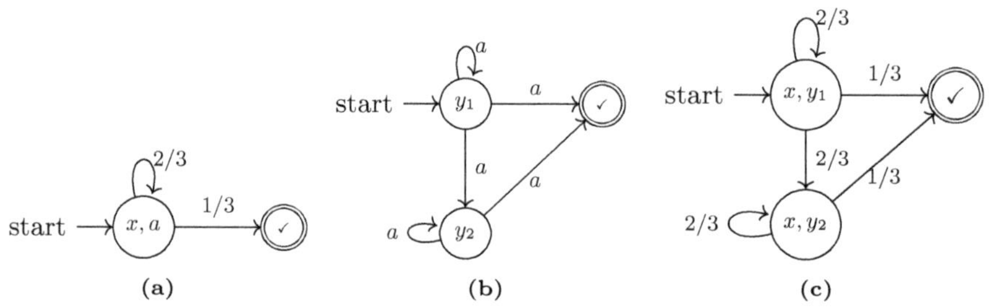

Fig. 1: Examples: a) an MC, b) an MFA, and c) their product.

Example 9. To illustrate the model checking problem with MFAs, we present a very simple example that is nevertheless rich enough to convey the core idea. Consider the MC and MFA shown in Fig. 1. The MC models a server that repeatedly requests a response from a user until it reaches the state $\checkmark$, while the user's behavior is modeled by the MFA. In the MFA, the target state $\checkmark$ represents an undesirable (bad) status for the user, which we would like to prevent the user from reaching as much as possible.

The model checking problem in this setting does not ask for the probability of such undesirable user behaviors. Instead, it concerns the expected number of these behaviors. In this way, the analysis provides a quantitative measure of the safety guarantees on the user's behavior. The model checking result is $\big(q \circ (\mu\Phi_c \times \mu\Phi_d)\big)(x, y_1) = \sum_{n=1}^{\infty} (2/3)^{n-1} \cdot (1/3) \cdot n = 3$. This indicates that we should expect around three undesirable user behaviours until the server terminates. The product of the MC and MFA is described in Fig. 1: note that this is not an MC because $2/3 + 2/3 + 1/3 = 5/3 > 1$.

We remark that the distributive law λ is the unique one:

Proposition 7 (uniqueness under correctness criterion). *The distributive law λ defined in Def. 9 is the unique one satisfying the following properties.*

1. *There is a distributive law ρ from $\mathcal{D}_{\leq 1}$ and $\mathcal{M}((\_)+1)$ to $\mathcal{R}^+_{\geq 0}((\_)+\{\checkmark\})$ s.t. $\lambda = \rho \circ (\mathrm{id}_{\mathcal{D}_{\leq 1}(\_)} \times \mathrm{ev}_{A,\mathcal{M}((\_)+1)})$, where $\mathrm{ev}_{A,(\_)} \colon A \times \mathrm{id}^A \Rightarrow \mathrm{id}$ is given by evaluation maps.*
2. *Any $c \otimes_\lambda d$ satisfies the correctness criterion in the setting of Prop. 6.* $\qquad\square$

See [27, Appendix B.2] for the proof.

6.1 Model Checking of the Product

Lastly, we study how to solve the product of MCs and MFAs. The product can be naturally regarded as an MC with *multiplicative rewards*, which is very recently proposed in [3]. This is because the number of accepting runs defined by MFAs can be seen as multiplicative rewards. Since we are only interested in traces that eventually reaches $\checkmark$, we can assume that the product MC with multiplicative rewards only has trivial bottom strongly connected components without loss of generality, and this is solvable in polynomial time as shown in [3, Theorem 3.18 and Remark 3.20]. We summarise our complexity result as follows; in the following we assume that all probabilistic transitions in MCs are rational values.

Proposition 8. *Given an MC $c \colon X \to \mathcal{D}(X + \{\checkmark\}) \times A$, an MFA $d \colon Y \to \mathcal{M}(Y + \{\checkmark\})^A$ and an initial state $(x,y) \in X \times Y$ of their product $c \otimes_\lambda d$, the semantics $\mu\Phi_{c\otimes_\lambda d}(x,y)$ is computable in polynomial time, implying that the model checking $q \circ (\mu\Phi_c \times \mu\Phi_d)(x,y)$ is solvable in polynomial time.*

Proof Sketch. The semantics of products can be seen as the expected multiplicative reward on the corresponding MC; see [27, Appendix B.3] for the reduction. By [3, Theorem 3.18 and Remark 3.20], there is a linear equation system such that (i) the semantics is a part of the unique non-negative solution of the linear equation system iff the linear equation system has a non-negative solution, and (ii) the semantics diverges ($\mu\Phi_{c\otimes_\lambda d}(x,y) = \infty$) iff there is no non-negative solution in the linear equation system. Constructing and solving such a linear equation system can be done in polynomial time, concluding the proof. $\qquad\square$

7 Related Work

Product constructions for probabilistic systems, including MCs, have been extensively studied over the decades. In particular, efficient product constructions with ω-regular automata have been a key focus of research [7, 12, 13, 33, 38]. Handling with ω-regular properties in a unified manner within the coalgebraic framework remains a challenge, which we leave for future work. An intriguing direction for future research is to build upon existing studies on coalgebraic ω-regular automata [10, 36] to extend our framework further.

Cîrstea and Kupke [11] study product constructions of quantitative systems, including probabilistic systems, and nondeterministic parity automata. They show that model checking can be reduced to solving nested fixed-point equations, while leaving the complexity analysis as future work. In this paper, we present the first no-go theorem for product constructions, and we introduce a new polynomial-time algorithm for model checking with MCs and MFAs, reducing the problem to computation of expected multiplicative rewards on MCs [3]. On the other hand, our framework does not support ω-regular properties, which are the main focus of [11].

No-go theorems for structures that enable computation by combining different semantics have been studied in the context of computational effects, particularly in relation to the distributive laws of monads. Varacca and Winskel, following a proof attributed to Plotkin, demonstrated the non-existence of distributive laws between the powerset monad and distribution monad [37]. Subsequently, Zwart and Marsden [41] developed a unified theory for no-go theorem on distributive laws, encompassing known results such as those in [25, 37]. We aim to further investigate our no-go theorem on product constructions, exploring potential connections with these no-go theorems for distributive laws of monads. Notably, distributive laws do exist between the multiset monad and the distribution monad [15, 16, 23, 24], suggesting that multisets interact well with distributions, as observed in Prop. 7.

Dahlqvist and Neves [14] show a series of characterisation of natural transformations $\alpha\colon T^n \Rightarrow T$ for a certain type of functor T, including the finite powerset functor and the multiset functor. Our characterisation covers natural transformations $\beta\colon F_A \Rightarrow F_B$, where F_A and F_B are different functors (see Tbl. 1).

8 Conclusion

We present a no-go theorem, demonstrating the incompatibility of coalgebraic product constructions with MCs and NFAs. To establish this result, we develop a novel characterisation of natural transformations for certain functors. We then introduce a new model checking problem for MCs and MFAs, based on our coalgebraic product construction.

As a future direction, one could further extend the characterisation of natural transformations form F_A to F_B given in Thm. 1, since there are cases not covered even when A is singly generated, as well as cases when A is finitely generated.

Another interesting direction is to study model checking of MCs with ω-regular multiset automata. Investigating the computational complexity of this problem——possibly by applying the results in [3]——would be an interesting future work.

References

1. Alejandro Aguirre, Shin-ya Katsumata, and Satoshi Kura. Weakest preconditions in fibrations. *Math. Struct. Comput. Sci.*, 32(4):472–510, 2022.
2. Suzana Andova, Holger Hermanns, and Joost-Pieter Katoen. Discrete-time rewards model-checked. In *FORMATS*, volume 2791 of *Lecture Notes in Computer Science*, pages 88–104. Springer, 2003.

3. Christel Baier, Krishnendu Chatterjee, Tobias Meggendorfer, and Jakob Piribauer. Multiplicative rewards in Markovian models. In LICS, pages 499–512. IEEE, 2025.

4. Christel Baier, Marcus Daum, Clemens Dubslaff, Joachim Klein, and Sascha Klüppelholz. Energy-utility quantiles. In *NASA Formal Methods*, volume 8430 of *Lecture Notes in Computer Science*, pages 285–299. Springer, 2014.

5. Christel Baier, Luca de Alfaro, Vojtech Forejt, and Marta Kwiatkowska. Model checking probabilistic systems. In *Handbook of Model Checking*, pages 963–999. Springer, 2018.

6. Christel Baier and Joost-Pieter Katoen. *Principles of model checking*. MIT Press, 2008.

7. Christel Baier, Stefan Kiefer, Joachim Klein, David Müller, and James Worrell. Markov chains and unambiguous automata. *J. Comput. Syst. Sci.*, 136:113–134, 2023.

8. Michael Benedikt, Rastislav Lenhardt, and James Worrell. Model checking Markov chains against unambiguous Buchi automata. *CoRR*, abs/1405.4560, 2014.

9. Doron Bustan, Sasha Rubin, and Moshe Y. Vardi. Verifying omega-regular properties of Markov chains. In *CAV*, volume 3114 of *Lecture Notes in Computer Science*, pages 189–201. Springer, 2004.

10. Vincenzo Ciancia and Yde Venema. Omega-automata: A coalgebraic perspective on regular omega-languages. In *CALCO*, volume 139 of *LIPIcs*, pages 5:1–5:18. Schloss Dagstuhl - Leibniz-Zentrum für Informatik, 2019.

11. Corina Cîrstea and Clemens Kupke. Measure-theoretic semantics for quantitative parity automata. In *CSL*, volume 252 of *LIPIcs*, pages 14:1–14:20. Schloss Dagstuhl - Leibniz-Zentrum für Informatik, 2023.

12. Costas Courcoubetis and Mihalis Yannakakis. The complexity of probabilistic verification. *J. ACM*, 42(4):857–907, 1995.

13. Jean-Michel Couvreur, Nasser Saheb, and Grégoire Sutre. An optimal automata approach to LTL model checking of probabilistic systems. In *LPAR*, volume 2850 of *Lecture Notes in Computer Science*, pages 361–375. Springer, 2003.

14. Fredrik Dahlqvist and Renato Neves. Compositional semantics for new paradigms: probabilistic, hybrid and beyond. *CoRR*, abs/1804.04145, 2018.

15. Fredrik Dahlqvist, Louis Parlant, and Alexandra Silva. Layer by layer - combining monads. In *ICTAC*, volume 11187 of *Lecture Notes in Computer Science*, pages 153–172. Springer, 2018.

16. Swaraj Dash and Sam Staton. A monad for probabilistic point processes. In *ACT*, volume 333 of *EPTCS*, pages 19–32, 2020.

17. H. Peter Gumm and Tobias Schröder. Monoid-labeled transition systems. In *CMCS*, volume 44 of *Electronic Notes in Theoretical Computer Science*, pages 185–204. Elsevier, 2001.

18. Ernst Moritz Hahn and Arnd Hartmanns. A comparison of time- and reward-bounded probabilistic model checking techniques. In *SETTA*, volume 9984 of *Lecture Notes in Computer Science*, pages 85–100, 2016.

19. Ichiro Hasuo. Generic weakest precondition semantics from monads enriched with order. *Theor. Comput. Sci.*, 604:2–29, 2015.

20. Ichiro Hasuo, Bart Jacobs, and Ana Sokolova. Generic trace semantics via coinduction. *Log. Methods Comput. Sci.*, 3(4), 2007.

21. Wataru Hino, Hiroki Kobayashi, Ichiro Hasuo, and Bart Jacobs. Healthiness from duality. In *LICS*, pages 682–691. ACM, 2016.

22. Bart Jacobs. *Introduction to Coalgebra: Towards Mathematics of States and Observation*, volume 59 of *Cambridge Tracts in Theoretical Computer Science*. Cambridge University Press, 2016.
23. Bart Jacobs. From multisets over distributions to distributions over multisets. In *LICS*, pages 1–13. IEEE, 2021.
24. Klaus Keimel and Gordon D. Plotkin. Mixed powerdomains for probability and nondeterminism. *Log. Methods Comput. Sci.*, 13(1), 2017.
25. Bartek Klin and Julian Salamanca. Iterated covariant powerset is not a monad. In *MFPS*, volume 341 of *Electronic Notes in Theoretical Computer Science*, pages 261–276. Elsevier, 2018.
26. Mayuko Kori, Natsuki Urabe, Shin-ya Katsumata, Kohei Suenaga, and Ichiro Hasuo. The lattice-theoretic essence of property directed reachability analysis. In *CAV (1)*, volume 13371 of *Lecture Notes in Computer Science*, pages 235–256. Springer, 2022.
27. Mayuko Kori and Kazuki Watanabe. A no-go theorem for coalgebraic product construction. *CoRR*, abs/2504.06592v3, 2025. Full version.
28. Mayuko Kori, Kazuki Watanabe, Jurriaan Rot, and Shin-ya Katsumata. Composing codensity bisimulations. In *LICS*, pages 52:1–52:13. ACM, 2024.
29. Saunders Mac Lane. *Categories for the working mathematician*, volume 5. Springer Science & Business Media, 1998.
30. Amir Pnueli. The temporal logic of programs. In *FOCS*, pages 46–57. IEEE Computer Society, 1977.
31. Jan J. M. M. Rutten. Universal coalgebra: a theory of systems. *Theor. Comput. Sci.*, 249(1):3–80, 2000.
32. Lutz Schröder. Expressivity of coalgebraic modal logic: The limits and beyond. *Theor. Comput. Sci.*, 390(2-3):230–247, 2008.
33. Salomon Sickert, Javier Esparza, Stefan Jaax, and Jan Kretínský. Limit-deterministic Büchi automata for linear temporal logic. In *CAV (2)*, volume 9780 of *Lecture Notes in Computer Science*, pages 312–332. Springer, 2016.
34. Marcel Steinmetz, Jörg Hoffmann, and Olivier Buffet. Goal probability analysis in probabilistic planning: Exploring and enhancing the state of the art. *J. Artif. Intell. Res.*, 57:229–271, 2016.
35. Natsuki Urabe, Masaki Hara, and Ichiro Hasuo. Categorical liveness checking by corecursive algebras. In *LICS*, pages 1–12. IEEE Computer Society, 2017.
36. Natsuki Urabe, Shunsuke Shimizu, and Ichiro Hasuo. Coalgebraic trace semantics for Büchi and parity automata. In *CONCUR*, volume 59 of *LIPIcs*, pages 24:1–24:15. Schloss Dagstuhl - Leibniz-Zentrum für Informatik, 2016.
37. Daniele Varacca and Glynn Winskel. Distributing probability over nondeterminism. *Math. Struct. Comput. Sci.*, 16(1):87–113, 2006.
38. Moshe Y. Vardi. Automatic verification of probabilistic concurrent finite-state programs. In *FOCS*, pages 327–338. IEEE Computer Society, 1985.
39. Moshe Y. Vardi and Pierre Wolper. Automata-theoretic techniques for modal logics of programs. *J. Comput. Syst. Sci.*, 32(2):183–221, 1986.
40. Kazuki Watanabe, Sebastian Junges, Jurriaan Rot, and Ichiro Hasuo. A unifying approach to product constructions for quantitative temporal inference. *Proc. ACM Program. Lang.*, 9(OOPSLA1):1575–1603, 2025.
41. Maaike Zwart and Dan Marsden. No-go theorems for distributive laws. *Log. Methods Comput. Sci.*, 18(1), 2022.

Interaction Improvement

Adrienne Lancelot[1,2,3] ⋆,

Giulio Manzonetto[3], Guy McCusker[4],
and Gabriele Vanoni[3]

[1] Inria, France and LIX, Ecole Polytechnique, Palaiseau
[2] Università di Bologna, Italy `adrienne.lancelot@unibo.it`
[3] Université Paris Cité, CNRS, IRIF, F-75013, Paris, France
`{gmanzone,gabriele.vanoni}@irif.fr`
[4] Department of Computer Science, University of Bath, United Kingdom
`G.A.McCusker@bath.ac.uk`

Abstract. The relational semantics of linear logic is a powerful framework for defining *resource-aware* models of the λ-calculus. However, its *quantitative* aspects are not reflected in the preorders and equational theories induced by these models. Indeed, they can be characterized in terms of (in)equalities between Böhm trees up to extensionality, which are *qualitative* in nature. We employ the recently introduced checkers calculus to define a quantitative contextual preorder on λ-terms, and demonstrate that it coincides with the preorder associated to the relational semantics.

Keywords: λ-calculus · denotational semantics · program equivalence

1 Introduction

Much of programming language theory is based upon the powerful notion of *observational equivalence* [ ]. A program phrase may be placed in an execution context, yielding a whole program which can be executed. By specifying what *observations* one may make of a whole program, we obtain a theory of observational equivalence: two program phrases are said to be observationally equivalent if, in any execution context, they give rise to the same observations. That is to say, one cannot distinguish them, no matter what context they are placed in. Typical observations might be whether the programs terminate, return a particular integer or string, and so on. This notion of equivalence is both intuitively appealing and naturally *compositional*: a program occurring as a fragment of a larger piece of code can always be replaced by an equivalent one, yielding equivalent code.

It is natural to refine such a notion of equivalence to an observational *preorder*, typically defined by asking that the observations made of one program are a subset of those made of another. For example, if one observes only termination, then a program *simulates another one* if, whenever they are both placed in the

---

⋆ The omitted proofs can be found in the technical report [ ].

N. Bertrand and S. Milius (Eds.): FoSSaCS 2026, LNCS 16503, pp. 372–395, 2026.
https://doi.org/10.1007/978-3-032-22730-0_18

same execution context, if the simulated program terminates, then the simulator does as well. When the simulated program does not terminate, the simulator is free to display any behavior. The observational preorder allows semanticists to talk about how to refine a program to handle more cases of execution contexts and which *updates* of code are sound. Formally, taking as our notion of observation the termination predicate $\Downarrow$, we define the observational preorder as:

$$t \sqsubseteq^{\mathrm{ctx}} u \ \text{ if, for all contexts } C,\ C\langle t\rangle\Downarrow \ \Rightarrow\ C\langle u\rangle\Downarrow$$

Observational equivalences and preorders provide a robust methodology that may be applied to a wide variety of programming languages, from λ-calculus [ ] to sophisticated higher-order languages with computational effects [, , , ,], with relatively little change to the basic definitions. A large body of theory supports reasoning about these theories: they may be captured by denotational semantics [,], simplified by means of *context lemmas* that characterize a subset of contexts sufficient to make all possible distinctions [,2,], or analyzed by coinductive bisimulation-style techniques [ ]. However, in most formulations, the available observations are *extensional*: they report *what* programs compute but not *how*. Intensional information such as how many computation steps are used, and what resources are consumed, are ignored. There is a challenging need for a robust way to express the fact that one program actually *optimizes* another.

Improvements. Sands proposed *improvement theory* [, ,] as an execution-time-sensitive refinement of the observational preorder. He formulated a notion of observational improvement, where the observation is given not by mere termination $\Downarrow$ but by a quantitative notion $\Downarrow^k$ which reports that the number of steps required to terminate was k. We say that u (observationally) improves t, i.e.

$$t \sqsubseteq^{\mathrm{imp}} u \ \text{ if for all contexts } C, \forall k \geq 0\,.\, C\langle t\rangle\,\Downarrow^k \Rightarrow \exists k' \leq k.\ C\langle u\rangle\,\Downarrow^{k'}\,.$$

Despite its intuitive appeal, this notion of improvement is considerably less studied than the standard preorders above. Perhaps one reason for this is the absence of a rich theory of denotational semantics capturing intensional information.

Denotational Semantics & Equational Theories. Denotational semantics gives an interpretation to program phrases by representing them as elements of a mathematical structure [ ], often a morphism in some category. This representation induces an equational theory on program phrases. In almost all cases, the interpretation of a phrase in the model is defined in a *compositional* fashion, so that the interpretation of a program is constructed from those of its constituent phrases. Moreover, the interpretation is typically *invariant under computation*: in the λ-calculus, for example, terms related by β-reductions will receive equal denotations. Thus the induced inequational theory is not sensitive to reduction steps, and does not capture notions of improvement. Nevertheless some denotational models—notably games models and relational models—do contain intensional information beyond what is captured by the contextual preorder.

Intensional Semantics & Quantitative Information. Game semantics and relational semantics therefore offer the tantalising possibility of capturing quantitative information in a model and hence in the induced theory. Relational models record the interaction of a function with its arguments via a *multiset* of input items; game models go further and record the full sequence of interactions between a program and its execution context. Despite this, the application of these models to quantitative analysis of program execution has been limited. The original impetus behind work on game semantics of programming languages was the so-called full abstraction problem for PCF, so attention was firmly on the *extensional* behaviour of programs: in the fully-abstract models of PCF [1,20], all intensional information is explicitly quotiented out, to achieve the desired full abstraction theorems. In games models of more powerful languages, such as those involving stateful computation, the intensional information in the model turns out to be exactly the same as what is captured by the standard contextual preorder. These lines of work therefore do not appear to offer a denotational account of an improvement-style ordering. Indeed, in both game and relational semantics, the intensional information revealed by the semantics of a term is invariant under reduction of that term: the exposed information concerns only the interactions that a term may have with its context, and not what happens inside the term. One may draw an analogy with communications in distributed systems and client-server protocols [20]: game and relational semantics focus their attention on external communications. Purely internal steps are not exposed, thus allowing the semantics to remain invariant under reductions. This distinction mirrors the observation that local computations within a client or a server are comparatively fast or easily optimized, whereas the communication between them is the true performance bottleneck. Nevertheless this focus appears to make the idea of an improvement ordering inapplicable.

One approach to recovering an improvement-style ordering is demonstrated by Ghica's *slot games* [21] which introduce a global counter, exposing internal reductions at the top level, and as a result achieve full abstraction with respect to the improvement ordering. A similar example of instrumenting the semantics globally, in the case of relational semantics, appears in [29], though no improvement-style characterization is given. The question of understanding the intensional information available in the models (without such global counters) remains unanswered.

Our focus in this paper is the relational semantics of the λ-calculus, which we study in this paper in the form of a system of non-idempotent intersection types. Non-idempotent intersection type systems are able to provide complexity analyses of λ-terms, yielding evaluation time bounds, as pioneered by de Carvalho [15] and later refined by Accattoli et al. [1], as well as execution space bounds [9]. This quantitative information, for the most part, only appears at the level of type derivations—the construction of the denotation of a term—and are invisible in the final denotation and hence in the inequational theory. However, this inequational theory in general does not coincide with the observational preorder. The following questions then arise naturally:

What is the inequational theory induced by these semantics?
And is it cost-aware as the type system suggests?

Counting Interactions. This paper is about reconciling the two notion of program optimizations and denotational semantics. Specifically, we characterize the preorder induced by the relational semantics as an improvement-style preorder. Our approach makes use of the quantitative analysis of programs introduced in [ ]. This novel theory aims to account for the distinction between internal reductions and external interactions, as described above. Rather than counting *all* computations steps, as Sands's improvement theory would, it only counts the *interaction* steps between a term and its context. To achieve the distinction between interaction steps and internal steps, we use the *checkers calculus* of [ ] which is a bichromatic version of the untyped λ-calculus. Checkers terms are color-annotated λ-terms, where the abstraction and the application constructors are tagged by a color, either black or white. Intuitively, programs are players uniformly painted black, and contexts are opponents uniformly painted white. As a term is executed in a context, the black and white parts meet and mix. This allows us to define notions of quantitative observational equivalence, preorder and *interaction improvement*, akin to Sands's improvement, but counting only computation steps that involve a white constructor interacting with a black one.

Characterizing The Costs at Play Behind Relational Semantics. Interaction improvement makes explicit, in particular, the quantitative aspects underlying the preorder induced by the relational model of [ ]. The interpretation in this model can be described via a type system based on multi types, also known as non-idempotent intersection types [,]. The preorder induced by this model has been characterized as the Böhm tree preorder up to η-reductions [ ]. However, the execution costs involved in this relational model have so far remained unclear.

In this paper, we introduce a colored relational semantics for the checkers calculus (Figure 2) and define an ordering on interpretations that compares their elements in a refined manner which is both qualitative and quantitative. In our main contribution (Theorem 5) we show that, when restricted to regular λ-terms, this ordering coincides with both the relational preorder—equivalently, with the Böhm tree preorder up to η-reductions—and interaction improvement.

It turns out to be quite challenging to show that the ordering we introduce on interpretations is *compositional*. Though denotational approaches are typically compositional by definition, the fact that it presents difficulties here is perhaps not surprising given the intensional information we are tracking: similar challenges arise in intensional models such as game semantics [,] and operational approaches such as applicative and normal form bisimulations [,] (or so-called operational game semantics [,]). Here we establish compositionality using a novel technique which bears comparison to game semantics.

Our relational semantics annotates the ordinary relational model in two ways: numerical annotations track the steps needed to reduce a term, while color annotations on the types track the colorings of terms and their potential contexts. An unannotated entry in the relational semantics of a term may have several valid annotations in our system. *Improvements* result not only from lower numerical annotations, but also from better matching of color annotations between a term and its context. The delicate interplay between colors and counts requires a careful analysis. Once the appropriate preorder has been defined (Definition 10), we must show that it is *compositional* (Proposition 4). To prove this property, we treat typings dynamically: we study how modifications to the context colorings of a typing affect the corresponding typings of a term—a process we call *repainting* (Lemma 6). When an improved term is placed in a context, these color changes propagate between the term and its context, in a manner reminiscent of the composition of strategies in game semantics. Once this propagation stabilizes, a new typing emerges, and the preorder is preserved.

2   The Checkers Calculus

This section is devoted to presenting the *checkers calculus* from [ ], a bichromatic variant of the λ-calculus designed to count the reductions that occur during the interaction between a λ-term (the *player*) and a context (the *opponent*).

2.1   Its Syntax and Operational Semantics

In the checkers calculus, each application $\cdot^c$ and each abstraction λ_c is assigned a color $c \in \{\circ, \bullet\}$, either white or black, depending on whether the constructor 'belongs' to the player or the opponent. We use $c^\perp$ to denote the opposite color, namely $\circ^\perp := \bullet$ and $\bullet^\perp := \circ$. We consider fixed a countable set VAR of *variables*.

Definition 1. *The set $\Lambda_{\circ\bullet}$ of* checkers terms *is inductively defined as follows:*

$$\Lambda_{\circ\bullet} \quad \ni \quad t, u, s ::= x \mid \lambda_c x.t \mid t \cdot^d u, \qquad \textit{for } x \in \mathrm{VAR} \textit{ and } c, d \in \{\circ, \bullet\}.$$

We do not color variables, as they can be substituted by arbitrary terms. As usual, we assume that application associates to the left and has a higher precedence than abstraction. When the colors are actually specified, we simply write $t \circ u$ (and $t \bullet u$) for $t \cdot^\circ u$ (and $t \cdot^\bullet u$). In case of many consecutive abstractions or applications, we shorten the notations to $\lambda_{\vec{c}}\, \vec{x}.t$ and $t \cdot^{\vec{c}} \vec{u}$, respectively. If needed, we expand the former to $\lambda_{c_1 \cdots c_k} x_1 \ldots x_k.t$, and the latter to $t \cdot^{c_1 \cdots c_k} u_1 \cdots u_k$.

The checkers calculus inherits a number of notions from the usual λ-calculus. We consider checkers terms modulo α-conversion. We denote by $\mathsf{fv}(t)$ the set of *free variables* of t, and by $t\{x := u\}$ the capture-free substitution of u for all free occurrences of x in t. The operational semantics of the checkers calculus is defined by taking appropriate context closure of basic rewriting rules. Intuitively, a checkers context is a checkers term containing one occurrence of a 'hole' $\langle \cdot \rangle$.

Definition 2.   *(i) The set $\mathcal{C}_{\circ\bullet}$ of* checkers contexts *and its subset $\mathcal{H}_{\circ\bullet}$ of* checkers head contexts *are defined by the following grammars:*

$$\mathcal{C}_{\circ\bullet} \ni C ::= \langle \cdot \rangle \mid \lambda_c x.C \mid C \cdot^c u \mid t \cdot^c C$$

$$\mathcal{H}_{\circ\bullet} \ni H ::= \lambda_{c_1 \cdots c_n} x_1 \ldots x_n. \langle \cdot \rangle \cdot^{d_1 \cdots d_k} t_1 \cdots t_k$$

(ii) Given $C \in \mathcal{C}_{\circ\bullet}$ and $t \in \Lambda_{\circ\bullet}$, we denote by $C\langle t\rangle$ the checkers term obtained by substituting t for $\langle\cdot\rangle$ in C, possibly capturing free variables in t.

(iii) The contextual closure of a binary relation $\mathcal{R}$ on $\Lambda_{\circ\bullet}$ is the least relation $\mathcal{C}_{\circ\bullet}\langle\mathcal{R}\rangle$ such that $t\,\mathcal{C}_{\circ\bullet}\langle\mathcal{R}\rangle\,u$ entails $C\langle t\rangle\,\mathcal{C}_{\circ\bullet}\langle\mathcal{R}\rangle\,C\langle u\rangle$, for all $C \in \mathcal{C}_{\circ\bullet}$. Its head-contextual closure $\mathcal{H}_{\circ\bullet}\langle\mathcal{R}\rangle$ is defined analogously, by taking $C \in \mathcal{H}_{\circ\bullet}$.

$$
\begin{aligned}
\text{S\scriptsize ILENT} \quad (\lambda_{\mathsf{c}} x.t)\cdot^{\mathsf{c}} u &\;\mapsto_{\beta_\tau}\; t\{x:=u\} \\
\text{I\scriptsize NTERACTION} \quad (\lambda_{\mathsf{c}} x.t)\cdot^{\mathsf{c}^{\perp}} u &\;\mapsto_{\beta_\bullet}\; t\{x:=u\} \\[4pt]
\text{S\scriptsize ILENT } \beta \quad \to_{\beta_\tau} &:= \mathcal{C}_{\circ\bullet}\langle\mapsto_{\beta_\tau}\rangle \\
\text{I\scriptsize NTERACTION } \beta \quad \to_{\beta_\bullet} &:= \mathcal{C}_{\circ\bullet}\langle\mapsto_{\beta_\bullet}\rangle \\
\text{C\scriptsize HECKERS } \beta \quad \to_{\beta_{\circ\bullet}} &:= \;\to_{\beta_\tau}\,\cup\,\to_{\beta_\bullet} \\
\text{S\scriptsize ILENT HEAD} \quad \to_{\mathsf{h}_\tau} &:= \mathcal{H}_{\circ\bullet}\langle\mapsto_{\beta_\tau}\rangle \\
\text{I\scriptsize NTERACTION HEAD} \quad \to_{\mathsf{h}_\bullet} &:= \mathcal{H}_{\circ\bullet}\langle\mapsto_{\beta_\bullet}\rangle \\
\text{C\scriptsize HECKERS HEAD} \quad \to_{\mathsf{h}_{\circ\bullet}} &:= \;\to_{\mathsf{h}_\tau}\,\cup\,\to_{\mathsf{h}_\bullet}
\end{aligned}
$$

Fig. 1. The operational semantics of the checkers calculus.

There are two kinds of colored β-redexes $(\lambda_{\mathsf{c}} x.t)\cdot^{\mathsf{d}} u$, the *silent* ones and the *interaction* ones. Silent redexes are β-redexes where the color c of the abstraction matches the color d of the application. Intuitively, these steps are internal to each player's world. In interaction redexes, instead, the color of the abstraction and the color of the application are different, *i.e.* $\mathsf{c} \neq \mathsf{d}$. This represents the scenario where the two players interact with each other, which, from each player's perspective, amounts to interacting with the external world. When the contracted redex occurs in head position, we say that it is a *head interaction step*.

Definition 3. *The reductions of the checkers calculus are given in Figure 1. Given a reduction $\to_R$, we say that: a checkers term t is in R-normal form (R-nf) if t does not contain any R-redex; t has a R-nf if it reduces to a u in R-nf.*

Despite our earlier intuition of associating black with the player and white with the opponent, notice that the checkers calculus is entirely symmetrical.

Example 1. Let $\mathbf{I}_\bullet := \lambda_\bullet x.x$ be the black identity and $\mathbf{D}_\circ := \lambda_\circ y.\lambda_\circ x.x \circ (y \circ x)$.

1. Their black-application gives rise to a silent step $\mathbf{I}_\bullet \bullet \mathbf{D}_\circ \to_{\beta_\tau} \mathbf{D}_\circ$, while their white-application gives rise to an interaction step, namely $\mathbf{I}_\bullet \circ \mathbf{D}_\circ \to_{\beta_\bullet} \mathbf{D}_\circ$.
2. $\mathbf{D}_\circ \bullet \mathbf{I}_\bullet \bullet \mathbf{I}_\bullet \to_{\mathsf{h}_\bullet} (\lambda_\circ x.x \circ (\mathbf{I}_\bullet \circ x)) \bullet \mathbf{I}_\bullet \to_{\beta_\bullet} (\lambda_\circ x.x \circ x) \bullet \mathbf{I}_\bullet \to_{\mathsf{h}_\bullet} \mathbf{I}_\bullet \circ \mathbf{I}_\bullet \to_{\mathsf{h}_\bullet} \mathbf{I}_\bullet$.
3. A monochromatic checkers term displays the same behavior as the underlying term: $\mathbf{D}_\circ \circ \mathbf{D}_\circ \to_{\beta_\tau} \lambda_\circ x.x \circ (\mathbf{D}_\circ \circ x) \to_{\beta_\tau} \lambda_\circ x.x \circ (\lambda_\circ z.z \circ (x \circ z))$ and $\mathbf{I}_\bullet \bullet \mathbf{I}_\bullet \to_{\beta_\tau} \mathbf{I}_\bullet$.

There are in fact two "copies" of the set Λ of λ-terms within $\Lambda_{\circ\bullet}$: one obtained by painting all λ-terms black, and the other by painting them white. Formally:

Definition 4. *Given a color* $\mathsf{c} \in \{\circ, \bullet\}$, *the* c-*painting of ordinary* λ-*terms, is the mapping* $\overline{\cdot}^{\mathsf{c}} : \Lambda \to \Lambda_{\circ\bullet}$ *defined by induction as follows:*

$$\overline{x}^{\mathsf{c}} := x, \qquad \overline{\lambda x.t}^{\mathsf{c}} := \lambda_{\mathsf{c}} x.\overline{t}^{\mathsf{c}}, \qquad \overline{tu}^{\mathsf{c}} := \overline{t}^{\mathsf{c}} \cdot^{\mathsf{c}} \overline{u}^{\mathsf{c}}.$$

Given a checkers term $t \in \Lambda_{\circ\bullet}$, *the* color washing *map* $\underline{\cdot} : \Lambda_{\circ\bullet} \to \Lambda$ *is defined by:*

$$\underline{x} := x, \qquad \underline{\lambda_{\mathsf{c}} x.t} := \lambda x.\underline{t}, \qquad \underline{t \cdot^{\mathsf{c}} u} := \underline{t} \cdot \underline{u}.$$

Note that β-reductions $\to_\beta$ (resp. head reductions $\to_{\mathsf{h}}$) on ordinary λ-terms correspond to silent (head) reductions.

Lemma 1 (Correspondence [, Prop. 3.13-14]). *Let* $t \in \Lambda_{\circ\bullet}$ *and* $u, u' \in \Lambda$.

(i) If $u \to_\beta u'$, *then* $\overline{u}^{\mathsf{c}} \to_{\beta_\tau} \overline{u'}^{\mathsf{c}}$. *Similarly, if* $u \to_{\mathsf{h}} u'$, *then* $\overline{u}^{\mathsf{c}} \to_{\mathsf{h}_\tau} \overline{u'}^{\mathsf{c}}$.
(ii) If $\underline{t} \to_{\mathsf{R}} u$ *then* $\exists t' \in \Lambda_{\circ\bullet}$ *such that* $\underline{t'} = u$ *and* $t \to_{\mathsf{R}_{\bullet\circ}} t'$ *for* $\mathsf{R} \in \{\beta, \mathsf{h}\}$.

So, *checkers head normal forms* ($\mathsf{h}_{\circ\bullet}$-nfs) have the shape $h = \lambda_{\vec{\mathsf{c}}} \, \vec{x}.y \cdot^{\mathsf{d}_1 \cdots \mathsf{d}_k} t_1 \cdots t_k$.

Theorem 1 (Confluence [ ]). *Reductions* $\to_{\beta_{\circ\bullet}}$, $\to_{\beta_\tau}$, *and* $\to_{\beta_\bullet}$ *are confluent.*

Remark 1. The checkers calculus has no analogue of the η-reduction $\lambda x.tx \to_\eta t$, with $x \notin \mathsf{fv}(t)$. Consider the candidate black η-expansion $\mathbf{1}_\bullet := \lambda_\bullet x.\lambda_\bullet y.x \bullet y$ of the black identity $\mathsf{I}_\bullet$. Then, assuming $\mathbf{1}_\bullet \to_{\bullet\eta} \mathsf{I}_\bullet$, would break confluence: $\mathbf{1}_\bullet \circ z \circ w \to_{\mathsf{h}_\bullet} (\lambda_\bullet y.z \bullet y) \circ w \to_{\mathsf{h}_\bullet} z \bullet w$, while $\mathbf{1}_\bullet \circ z \circ w \to_{\bullet\eta} \mathsf{I}_\bullet \circ z \circ w \to_{\mathsf{h}_\bullet} z \circ w$.

2.2 Interaction Improvement

Observational preorders have been introduced to express that one program is more defined than another: when placed in any context, if the former terminates, then so does the latter. By varying the notion of termination being observed, one obtains different preorders that nonetheless share the same qualitative nature. To achieve a *quantitative* notion of observation, one can compare two programs by considering the number of steps required for termination (see Sands [11,10,39]). However, this approach is often too fine-grained, since it also distinguishes terms that represent the same program at different stages of evaluation. In the checkers calculus, the introduction of colors makes it possible to compare programs by counting only the interaction steps between the program and its context, ignoring silent steps. In this paper, as in [ ], we adopt head-reduction as our operational semantics, leading to the following definitions.

Given $t \in \Lambda_{\circ\bullet}$, we write $\Downarrow_{\mathsf{h}_{\circ\bullet}}^{\bullet k}$ whenever t has a $\mathsf{h}_{\circ\bullet}$-nf h, and the reduction sequence $t \to_{\mathsf{h}_{\circ\bullet}} \cdots \to_{\mathsf{h}_{\circ\bullet}} h$ contains k interaction head steps $\to_{\mathsf{h}_\bullet}$. Since $\to_{\mathsf{h}_{\circ\bullet}}$ is a deterministic strategy, such a sequence is unique and so is the number k. Silent head steps are ignored in the calculation of k, as expected.

Definition 5 (Checkers interaction improvement and equivalence [ ]).
On checkers terms $t, u \in \Lambda_{\circ\bullet}$, we define the following relations:

(i) Interaction preorder $\sqsubseteq_{\mathcal{O}}^{\mathrm{int}}$.
$$t \sqsubseteq_{\mathcal{O}}^{\mathrm{int}} u \ \text{if} \ \forall C \in \mathcal{C}_{\circ\bullet}, k \in \mathbb{N} \,.\, [\, C\langle t\rangle \Downarrow_{\mathrm{h}_{\circ\bullet}}^{\mathcal{O}k} \ \Rightarrow \ C\langle u\rangle \Downarrow_{\mathrm{h}_{\circ\bullet}}^{\mathcal{O}k} \,].$$

(ii) Interaction improvement (preorder) $\sqsubseteq_{\mathcal{O}}^{\mathrm{imp}}$.
$$t \sqsubseteq_{\mathcal{O}}^{\mathrm{imp}} u \ \text{if} \ \forall C \in \mathcal{C}_{\circ\bullet}, k \in \mathbb{N} \,.\, [\, C\langle t\rangle \Downarrow_{\mathrm{h}_{\circ\bullet}}^{\mathcal{O}k} \Rightarrow C\langle u\rangle \Downarrow_{\mathrm{h}_{\circ\bullet}}^{\mathcal{O}k'} \ \text{for some } k' \leq k];$$

(iii) Interaction equivalence $\equiv_{\mathcal{O}}^{\mathrm{int}}$. *It is the equivalence relation induced by* $\sqsubseteq_{\mathcal{O}}^{\mathrm{int}}$:
$$t \equiv_{\mathcal{O}}^{\mathrm{int}} u \ \text{if} \ t \sqsubseteq_{\mathcal{O}}^{\mathrm{int}} u \ \text{and} \ u \sqsubseteq_{\mathcal{O}}^{\mathrm{int}} t.$$

It is easy to see that $\sqsubseteq_{\mathcal{O}}^{\mathrm{int}} \subsetneq \sqsubseteq_{\mathcal{O}}^{\mathrm{imp}}$. The strictness of the inclusion is due to the fact that $\mathrm{I}_{\bullet} \circ \mathrm{I}_{\circ} \sqsubseteq_{\mathcal{O}}^{\mathrm{imp}} \mathrm{I}_{\circ} := \lambda_{\circ} x.x$, but $\mathrm{I}_{\bullet} \circ \mathrm{I}_{\circ} \not\sqsubseteq_{\mathcal{O}}^{\mathrm{int}} \mathrm{I}_{\circ}$, as both checkers terms have $\mathrm{I}_{\circ}$ as $\mathrm{h}_{\circ\bullet}$-nf, but $\mathrm{I}_{\bullet} \circ \mathrm{I}_{\circ}$ requires an additional interaction step to reach it. Observe that the equivalence relation induced by $\sqsubseteq_{\mathcal{O}}^{\mathrm{imp}}$ coincides with $\equiv_{\mathcal{O}}^{\mathrm{int}}$.

$$\frac{}{x:[\mathsf{L}] \vdash_{\mathcal{O}}^{0} x:\mathsf{L}} \ \mathrm{ax} \qquad \frac{(\Gamma_i \vdash_{\mathcal{O}}^{k_i} t:\mathsf{L}_i)_{i \in I} \quad I \text{ finite}}{\biguplus_{i \in I} \Gamma_i \vdash_{\mathcal{O}}^{\sum_{i \in I} k_i} t:[\mathsf{L}_i]_{i \in I}} \ \text{many} \ \Bigg| \ \frac{\Gamma \vdash_{\mathcal{O}}^{k_1} t:\mathsf{M} \overset{\mathsf{c}}{\to} \mathsf{L} \quad \Delta \vdash_{\mathcal{O}}^{k_2} u:\mathsf{M}}{\Gamma \uplus \Delta \vdash_{\mathcal{O}}^{k_1+k_2} t \cdot^{\mathsf{c}} u:\mathsf{L}} \ @_{\tau}$$

$$\frac{\Gamma, x:\mathsf{M} \vdash_{\mathcal{O}}^{k} t:\mathsf{L}}{\Gamma \vdash_{\mathcal{O}}^{k} \lambda_{\mathsf{c}} x.t:\mathsf{M} \overset{\mathsf{c}}{\to} \mathsf{L}} \ \lambda \quad \frac{\Gamma \vdash_{\mathcal{O}}^{k_1} t:\mathsf{M} \overset{\mathsf{c}}{\to} \mathsf{L} \quad \Delta \vdash_{\mathcal{O}}^{k_2} u:\mathsf{M}}{\Gamma \uplus \Delta \vdash_{\mathcal{O}}^{k} t \cdot^{\mathsf{d}} u:\mathsf{L}} \ @ \ \Bigg| \ \frac{\Gamma \vdash_{\mathcal{O}}^{k_1} t:\mathsf{M} \overset{\mathsf{c}}{\to} \mathsf{L} \quad \Delta \vdash_{\mathcal{O}}^{k_2} u:\mathsf{M}}{\Gamma \uplus \Delta \vdash_{\mathcal{O}}^{k_1+k_2+1} t \cdot^{\mathsf{c}^{\perp}} u:\mathsf{L}} \ @_{\mathcal{O}}$$

where in the rule @ we have $k = k_1 + k_2$ when $\mathsf{c} = \mathsf{d}$, and $k = k_1 + k_2 + 1$ otherwise. Note that the rule @ compactly represents the rules $@_{\tau}$ and $@_{\mathcal{O}}$.

Fig. 2. Checkers multi type system $\vdash_{\mathcal{O}}$.

Example 2. Consider $\mathrm{I}_{\bullet}$ and $\mathrm{D}_{\circ}$ from Example 1, and $\mathbf{1}_{\bullet}$ from Remark 1.

- Since $\Omega := (\lambda x.xx)(\lambda y.yy)$ has no hnf, any of its colorings will be a bottom element w.r.t. $\sqsubseteq_{\mathcal{O}}^{\mathrm{int}}$ and $\sqsubseteq_{\mathcal{O}}^{\mathrm{imp}}$. E.g. $\overline{\Omega}^{\bullet} \sqsubseteq_{\mathcal{O}}^{\mathrm{int}} \mathrm{I}_{\bullet}$ and $\overline{\Omega}^{\circ} \sqsubseteq_{\mathcal{O}}^{\mathrm{int}} \mathbf{1}_{\bullet}$.
- More generally, $t \equiv_{\mathcal{O}}^{\mathrm{int}} u$ holds whenever $t, u \in \Lambda_{\circ\bullet}$ have no $\mathrm{h}_{\circ\bullet}$-nf.
- To see that $\lambda_{\circ} x.x \not\sqsubseteq_{\mathcal{O}}^{\mathrm{imp}} \lambda_{\bullet} x.x$, just take the context $C := \langle \cdot \rangle \bullet \mathrm{I}_{\bullet}$. Indeed, $C\langle \lambda_{\circ} x.x\rangle \Downarrow_{\mathrm{h}_{\circ\bullet}}^{\mathcal{O}1}$ while $C\langle \lambda_{\bullet} x.x\rangle \Downarrow_{\mathrm{h}_{\circ\bullet}}^{\mathcal{O}0}$ because the head step is silent.
- $\mathbf{1}_{\bullet} \not\sqsubseteq_{\mathcal{O}}^{\mathrm{int}} \mathrm{I}_{\bullet}$, as they are separated by $C = \langle \cdot \rangle \circ z \circ w$, but $\mathbf{1}_{\bullet} \sqsubseteq_{\mathcal{O}}^{\mathrm{imp}} \mathrm{I}_{\bullet}$ holds. This is a consequence of our main Theorem 5.
- Recall that $\mathrm{Y} \in \Lambda$ is a fixed point combinator (fpc) if $\mathrm{Y} =_{\beta} \lambda f.f(\mathrm{Y}f)$ and that all fpcs Y, Y' share the same Böhm tree. By [, Theorem 8.6], $\overline{\mathrm{Y}}^{\circ} \equiv_{\mathcal{O}}^{\mathrm{int}} \overline{\mathrm{Y}'}^{\circ}$.
- The combinator $\mathrm{J} \in \Lambda$ from [ ] satisfying $\mathrm{J}x =_{\beta} \lambda z.x(\mathrm{J}z)$ is an "infinite η-expansion" of $\lambda x.x$. To check that $\overline{\mathrm{J}}^{\circ} \not\sqsubseteq_{\mathcal{O}}^{\mathrm{int}} \lambda_{\circ} x.x$, take once again $C := \langle \cdot \rangle \bullet \mathrm{I}_{\bullet}$. The fact that $\overline{\mathrm{J}}^{\circ} \sqsubseteq_{\mathcal{O}}^{\mathrm{imp}} \lambda_{\circ} x.x$ follows from Theorem 5 (by $\circ/\bullet$-symmetry).

In this paper, we focus on the interaction improvement $\sqsubseteq_{\mathcal{O}}^{\mathrm{imp}}$, which was introduced in [ ], but not previously characterized either by relational semantics or in terms of a tree-like ordering.

3 Checkers Relational Improvement Semantics

This section is devoted to presenting a denotational semantics of the checkers calculus, which can be seen as an annotated version of the relational semantics of the λ-calculus [,]. Since the work of [ ], it has been known that—just as filter models can be presented as intersection type systems—relational models can be presented via multitype systems. Thanks to the absence of weakening and contraction, and the presence of multisets of types $[\mathsf{L}_1, \ldots, \mathsf{L}_n]$, these type systems are resource-aware: intuitively, a λ-term $\lambda x.t$ having type $[\mathsf{L}_1, \ldots, \mathsf{L}_n] \to \mathsf{L}'$ needs to consume n copies of its argument to produce a result of type L'. During evaluation, the argument is used once with type L_1, once with type L_2, etc. This makes it possible to infer intensional properties of a program, like the amount of steps needed to reach its head normal form, by examining its type derivations.

3.1   Checkers Type System

We present a multi type system in which arrows are decorated with a color $\mathsf{c} \in \{\circ, \bullet\}$, intuitively matching the color of the outer λ-abstraction (of its head normal form), as in $\lambda_{\mathsf{c}} x.t : \mathsf{M} \xrightarrow{\mathsf{c}} \mathsf{L}$.

Definition 6. *1.* Linear types L, L' *and* multi types M, N *are generated by:*

$$\text{LINEAR TYPES } \mathsf{L}, \mathsf{L}' ::= \mathsf{A} \mid \mathsf{M} \xrightarrow{\mathsf{c}} \mathsf{L} \qquad \textit{where } \mathsf{c} \in \{\circ, \bullet\} \textit{ and } \mathsf{A} \textit{ is an atom}$$
$$\text{MULTI TYPES } \mathsf{M}, \mathsf{N} ::= [\mathsf{L}_1, \ldots, \mathsf{L}_n] \qquad n \geq 0$$

  We use T, T' *as meta-variables denoting either linear or multi types.*
2. *Given multi types* M, N, *we denote by* $\mathsf{M} + \mathsf{N}$ *their multiset union, and by* $\mathbf{0}$ *the empty multi type. In other words,* $\mathbf{0}$ *denotes the neutral element of* $+$.
3. *A type environment* Γ *is a map from* VAR *to multi types having a finite support* $\operatorname{supp}(\Gamma) := \{x \in \text{VAR} \mid \Gamma(x) \neq \mathbf{0}\}$. *We let* $x_1 : \mathsf{M}_1, \ldots, x_n : \mathsf{M}_n$ *denote the environment* $\Gamma(y) = \mathsf{M}_i$ *if* $y = x_i$, *otherwise* $\Gamma(y) = \mathbf{0}$.
4. *The empty environment* $\Gamma(x) = \mathbf{0}$, *for all* $x \in$ VAR, *is denoted* $\emptyset$ *or omitted.*
5. *Typing judgements are quadruples denoted by* $\Gamma \vdash_{\mathsf{e}}^{k} t : \mathsf{T}$, *where* $k \in \mathbb{N}$, *and can be derived by applying the rules given in Figure* 2.

Remark 2. The relational semantics of the λ-calculus can be recovered from our system by erasing all color and index annotations, or embedded in our system by coloring all term-constructors and types black and setting all indices to zero.

We shall prove that a checkers term is typable $\Gamma \vdash_{\mathsf{e}}^{k} t : \mathsf{L}$ exactly when it is head normalizable, and the index k gives an upper bound on the number of head interaction steps from t to its $\mathsf{h}_{\circ\bullet}$-nf (Theorem 2(1)). This explains the rule @ where a checkers term $t : \mathsf{M} \xrightarrow{\mathsf{c}} \mathsf{L}$ can always be d-applied to a term $u : \mathsf{M}$, but the index k associated with $t \cdot^{\mathsf{d}} u : \mathsf{L}$ must be incremented whenever $\mathsf{c} \neq \mathsf{d}$.

Remark 3. The type system $\vdash_{\mathsf{e}}^{k}$ presented in Fig. 2 is strongly related to the one used in [ ]. In the latter, a second color d is annotated on the arrow type $\xrightarrow{\mathsf{cd}}$, and in the rule @ such color needs to match the color of the application $\cdot^{\mathsf{d}}$. The second color is needed to have a notion of *tight type*, but otherwise redundant.

We now present the main properties of our system. As in the relational case [16], it enjoys *quantitative* versions of subject reduction (SR) and expansion (SE).

Definition 7 (Applicative size). *Given a derivation π of $\Gamma \vdash^k_{\mathbf{\odot}} t : T$, in symbols $\pi \rhd \Gamma \vdash^k_{\mathbf{\odot}} t : T$, its* applicative size $|\pi|_{@}$ *is the number of @-rules in π.*

The resource awareness of the type system allows us to prove that the applicative size of $\pi \rhd \Gamma \vdash^k_{\mathbf{\odot}} t : T$ decreases along each head step $\to_{\mathrm{h}_{\mathrm{o}\bullet}}$. Moreover, if the step is an interaction one, then the index k decreases by exactly 1; it is stable otherwise.

Proposition 1. *Let $t, t' \in \Lambda_{\mathrm{o}\bullet}$ be such that $t \to_{\mathrm{h}_{\mathrm{o}\bullet}} t'$.*

1. Quantitative subject reduction*: if $\pi \rhd \Gamma \vdash^k_{\mathbf{\odot}} t : \mathsf{L}$ then there is a derivation $\pi' \rhd \Gamma \vdash^{k'}_{\mathbf{\odot}} t' : \mathsf{L}$ such that $|\pi'|_{@} = |\pi|_{@} - 1$. Moreover:*
   (i) if $t \to_{\mathrm{h}_{\mathbf{\odot}}} t'$ then $k' = k - 1$;
   (ii) otherwise, if $t \to_{\mathrm{h}_\tau} t'$ then $k' = k$.
2. Subject expansion*: if $\pi' \rhd \Gamma \vdash^{k'}_{\mathbf{\odot}} t' : \mathsf{L}$ then there is a derivation $\pi \rhd \Gamma \vdash^k_{\mathbf{\odot}} t : \mathsf{L}$ for some k.*

The quantitative subject reduction entails the soundness of the checkers type system, and the typability of hnfs gives its completeness via subject expansion.

Theorem 2 (Typability characterizes head normalization). *Let $t \in \Lambda_{\mathrm{o}\bullet}$.*

1. Soundness*: if $\pi \rhd \Gamma \vdash^k_{\mathbf{\odot}} t : \mathsf{L}$ then there exists $k' \leq k$ such that $t \Downarrow^{\mathbf{\odot}k'}_{\mathrm{h}_{\mathrm{o}\bullet}}$.*
2. Completeness*: if $t \Downarrow^{\mathbf{\odot}k}_{\mathrm{h}_{\mathrm{o}\bullet}}$ then $\Gamma \vdash^k_{\mathbf{\odot}} t : \mathsf{L}$ is derivable for some Γ, L.*

3.2 Checkers Type Interpretation and Preorders

As previously mentioned, multi type systems are often employed to present relational models of the λ-calculus [16,34,12] since the relational interpretation of a λ-term is isomorphic to the set of its 'typings' (Γ, L). Analogously, we can define the colored interpretation of a checkers term t as the set of its typings that in this setting are triples (Γ, L, k) such that $\pi \rhd \Gamma \vdash^k_{\mathbf{\odot}} t : \mathsf{L}$.

Definition 8. *The* colored interpretation *of a checkers term $t \in \Lambda_{\mathrm{o}\bullet}$ is given by:*

$$[\![t]\!]^{\mathbf{\odot}} := \{(\Gamma, \mathsf{L}, k) \mid \exists \pi \rhd \Gamma \vdash^k_{\mathbf{\odot}} t : \mathsf{L}\}.$$

Two terms t, u are type equivalent *if they have the same colored interpretation.*

Remark 4. By Proposition 1, the interpretation $[\![\cdot]\!]^{\mathbf{\odot}}$ is invariant along silent head reductions. However, since the index k is taken into account in the interpretation, we may have $t \to_{\mathrm{h}_{\mathbf{\odot}}} t'$ with $[\![t]\!]^{\mathbf{\odot}} \neq [\![t']\!]^{\mathbf{\odot}}$. As an example, we have $\mathrm{I}_\bullet \circ x \to_{\mathrm{h}_{\mathbf{\odot}}} x$ and $[\![\mathrm{I}_\bullet \circ x]\!]^{\mathbf{\odot}} \cap [\![x]\!]^{\mathbf{\odot}} = \emptyset$, *e.g.* $(x:[\mathsf{A}], \mathsf{A}, 0) \in [\![x]\!]^{\mathbf{\odot}} - [\![\mathrm{I}_\bullet \circ x]\!]^{\mathbf{\odot}}$.

There are several possible ways of comparing these semantic interpretations. The most natural one is set theoretical inclusion $[\![t]\!]^{\circledcirc} \subseteq [\![u]\!]^{\circledcirc}$, which leads to the checkers type preorder studied in [ ]. In this paper we want to capture the interaction improvement $\sqsubseteq_{\circledcirc}^{\mathrm{imp}}$, so we need to refine the comparison. A naive attempt would be to check that t and u have the same types in the same environments, but allowing the index k' associated with u to be smaller. Formally:

$$t \leq u \text{ if } \forall \Gamma,\ \mathsf{L},\ k,\ \Gamma \vdash_{\circledcirc}^{k} t : \mathsf{L} \implies \exists k' \leq k,\ \Gamma \vdash_{\circledcirc}^{k'} u : \mathsf{L}$$

Unfortunately, this definition does not work, since $\sqsubseteq_{\circledcirc}^{\mathrm{imp}}$ validates η-reduction on monochromatic checkers terms (cf. Example 2), while this comparison suffers from the same issue as the inclusion. Consider the black η-expansion of x:

$$\dfrac{\dfrac{}{x:[\mathbf{0} \xrightarrow{\circ} \mathsf{L}] \vdash_{\circledcirc}^{0} x:\mathbf{0} \xrightarrow{\circ} \mathsf{L}}\ \mathsf{ax} \qquad \dfrac{}{\vdash_{\circledcirc}^{0} y:\mathbf{0}}\ \mathsf{many}}{\dfrac{x:[\mathbf{0} \xrightarrow{\circ} \mathsf{L}] \vdash_{\circledcirc}^{1} x \bullet y : \mathsf{L}}{x:[\mathbf{0} \xrightarrow{\circ} \mathsf{L}] \vdash_{\circledcirc}^{1} \lambda_{\bullet}y.x \bullet y : \mathbf{0} \xrightarrow{\bullet} \mathsf{L}}\ \lambda}\ @_{\circledcirc}$$

and note that $\lambda_{\bullet}y.x \bullet y \sqsubseteq_{\circledcirc}^{\mathrm{imp}} x$ holds, whereas $x:[\mathbf{0} \xrightarrow{\circ} \mathsf{L}] \vdash_{\circledcirc}^{k'} x:\mathbf{0} \xrightarrow{\bullet} \mathsf{L}$ cannot hold, because of the mismatch between the colors. Indeed, the rule ax requires that both occurrences of $\mathbf{0} \to \mathsf{L}$ share the same color.

Whitening Types. The above example suggests that, if we wish to define a preorder $\sqsubseteq_{\circledcirc}^{\mathrm{pwc}}$ that coincides with $\sqsubseteq_{\circledcirc}^{\mathrm{imp}}$ on the interpretations of black terms, we need a comparison between typings that can reduce the cost of derivations and whiten certain arrows in the types. Let us analyze the example further:

$$(x : [\mathsf{M} \xrightarrow{\mathsf{c}} \mathsf{L}], \mathsf{M} \xrightarrow{\mathsf{d}} \mathsf{L}, k) \in [\![\lambda_{\bullet}y.x \bullet y]\!]^{\circledcirc}$$

iff $\mathsf{d} = \bullet$ (forced by the $\lambda_{\bullet}y$) and either $\mathsf{c} = \bullet$ and $k = 0$, or $\mathsf{c} = \circ$ and $k = 1$. A triple of this kind[5] belongs to the interpretation of x whenever $\mathsf{c} = \mathsf{d}$ and $k = 0$. In particular, there is a whiter and cheaper typing $(x : [\mathsf{M} \xrightarrow{\circ} \mathsf{L}], \mathsf{M} \xrightarrow{\circ} \mathsf{L}, 0) \in [\![x]\!]^{\circledcirc}$.

This can be generalized. We shall prove (Lemma 4) that whenever $\Gamma \vdash_{\circledcirc}^{k} t:\mathsf{L}$ and t is a black η-expansion of u, we can derive $\Gamma' \vdash_{\circledcirc}^{k'} u:\mathsf{L}'$ where (Γ', L') is a whiter version of (Γ, L) and $k' \leq k$, yielding a derivation that is both whiter and cheaper. Moreover, we show that whitening must be applied only to arrows of positive polarity. We begin by defining a polarized whitening relation on types.

$$\dfrac{\mathsf{p} \in \{+, -\}}{\mathsf{A} \leq_{0}^{\mathsf{p}} \mathsf{A}} \qquad \dfrac{\mathsf{M}' \leq_{k_1}^{-} \mathsf{M} \quad \mathsf{L}' \leq_{k_2}^{+} \mathsf{L}}{\mathsf{M}' \xrightarrow{\circ} \mathsf{L}' \leq_{k_1+k_2+1}^{+} \mathsf{M} \xrightarrow{\bullet} \mathsf{L}} \qquad \dfrac{\mathsf{M}' \leq_{k_1}^{\neg \mathsf{p}} \mathsf{M} \quad \mathsf{L}' \leq_{k_2}^{\mathsf{p}} \mathsf{L}}{\mathsf{M}' \xrightarrow{\mathsf{c}} \mathsf{L}' \leq_{k_1+k_2}^{\mathsf{p}} \mathsf{M} \xrightarrow{\mathsf{c}} \mathsf{L}}$$

$$\dfrac{\mathsf{L}_1' \leq_{k_1}^{\mathsf{p}} \mathsf{L}_1 \quad \cdots \quad \mathsf{L}_n' \leq_{k_n}^{\mathsf{p}} \mathsf{L}_n}{[\mathsf{L}_1', \ldots, \mathsf{L}_n'] \leq_{k_1+\cdots+k_n}^{\mathsf{p}} [\mathsf{L}_1, \ldots, \mathsf{L}_n]}$$

Fig. 3. Polarized whitening of a type.

Definition 9 (Polarized Whitening).

   (i) Given a polarity $\mathsf{p} \in \{+, -\}$ *we denote by* $\neg\mathsf{p}$ *the opposite polarity.*

   (ii) For all $k \in \mathbb{N}$, *define* $\mathsf{T}' \leq_k^+ \mathsf{T}$ *and* $\mathsf{T}' \leq_k^- \mathsf{T}$ *by mutual induction in Fig.* 3.

   (iii) When $\mathsf{T}' \leq_k^+ \mathsf{T}$ *(resp.* $\mathsf{T}' \leq_k^- \mathsf{T}$) *holds we say that* T' *is* k-whiter *than* T *on positively (resp. negatively) occurring arrows.*

---

[5] Note that $(x\!:\![\mathsf{A}], \mathsf{A}, 0) \in [\![x]\!]^{\circledcirc} - [\![\lambda_\bullet y.x \bullet y]\!]^{\circledcirc}$. This is expected as $x \not\sqsubseteq_{\circledcirc}^{\mathrm{imp}} \lambda_\bullet y.x \bullet y$.

   (iv) The relations above extend to environments Γ *and pairs* $\langle \Gamma, \mathsf{L} \rangle$ *as follows:*

$$\frac{}{\emptyset \leq_0^{\mathsf{p}} \emptyset} \qquad \frac{\Gamma' \leq_{k_1}^{\mathsf{p}} \Gamma \quad \mathsf{M}' \leq_{k_2}^{\mathsf{p}} \mathsf{M}}{\Gamma', x : \mathsf{M}' \leq_{k_1+k_2}^{\mathsf{p}} \Gamma, x : \mathsf{M}} \qquad \frac{\Gamma' \leq_{k_1}^{\neg\mathsf{p}} \Gamma \quad \mathsf{L}' \leq_{k_2}^{\mathsf{p}} \mathsf{L}}{\langle \Gamma', \mathsf{L}' \rangle \leq_{k_1+k_2}^{\mathsf{p}} \langle \Gamma, \mathsf{L} \rangle}$$

Intuitively, $\mathsf{T}' \leq_k^{\mathsf{p}} \mathsf{T}$ holds if T' is obtained from T by whitening k arrows occurring with polarity p. In particular, the underlying uncolored type must be the same. Note the absence, in Fig. 3, of a rule that allows one to infer $\mathsf{M}' \overset{\circ}{\to} \mathsf{L}' \leq_k^- \mathsf{M} \overset{\bullet}{\to} \mathsf{L}$. This omission is intentional, as it prevents the whitening of arrows that occur negatively in the type that would break the polarization of the change of color.

   Everything is now in place to introduce the preorder $\sqsubseteq_{\circledcirc}^{\mathrm{pwc}}$ on checkers terms.

Definition 10 (Polarized Whiter-Cheaper Improvement). *For all checkers terms* t, u *we define* $t \sqsubseteq_{\circledcirc}^{\mathrm{pwc}} u$ *if and only if*

$$\forall (\Gamma, \mathsf{L}, k) \in [\![t]\!]^{\circledcirc}, \exists (\Gamma', \mathsf{L}', k') \in [\![u]\!]^{\circledcirc}, \exists d, \langle \Gamma', \mathsf{L}' \rangle \leq_d^+ \langle \Gamma, \mathsf{L} \rangle \text{ such that } k \geq k' + d$$

Summing up, this means that for all derivations of $\Gamma \vdash_{\circledcirc}^k t : \mathsf{L}$, there exists one of $\Gamma' \vdash_{\circledcirc}^{k'} u : \mathsf{L}'$ which is whiter, and cheaper by at least the amount of whitenings.

   We now investigate some properties of the whitening relations.

Lemma 2. *For all environments* Γ, Γ' *and linear types* L, L', *we have:*

   (i) 0-whitening is equality: *for* $\mathsf{p} \in \{+, -\}$, *we have that:*

      — $\mathsf{L} \leq_0^{\mathsf{p}} \mathsf{L}$ *if and only if* $\mathsf{L} = \mathsf{L}'$;

      — $\Gamma \leq_0^{\mathsf{p}} \Gamma'$ *if and only if* $\Gamma = \Gamma'$;

      — *Therefore,* $\langle \Gamma, \mathsf{L} \rangle \leq_0^{\mathsf{p}} \langle \Gamma', \mathsf{L}' \rangle$ *if and only if* $\Gamma = \Gamma'$ *and* $\mathsf{L} = \mathsf{L}'$.

   (ii) Inversion: *for* $\mathsf{p} \in \{+, -\}, \mathsf{c} \in \{\circ, \bullet\}$ *and* $k \in \mathbb{N}$, *we have that:*

$$\langle \Gamma', x : \mathsf{M}', \mathsf{L}' \rangle \leq_k^{\mathsf{p}} \langle \Gamma, x : \mathsf{M}, \mathsf{L} \rangle \text{ if and only if } \langle \Gamma', \mathsf{M}' \overset{\mathsf{c}}{\to} \mathsf{L}' \rangle \leq_k^{\mathsf{p}} \langle \Gamma, \mathsf{M} \overset{\mathsf{c}}{\to} \mathsf{L} \rangle.$$

   (iii) Transitivity of polarized whitening: *for* $\mathsf{p} \in \{+, -\}$, *we have that:*

$$\text{if } \langle \Gamma, \mathsf{L} \rangle \leq_{k_1}^{\mathsf{p}} \langle \Gamma', \mathsf{L}' \rangle \leq_{k_2}^{\mathsf{p}} \langle \Gamma'', \mathsf{L}'' \rangle \text{ then } \langle \Gamma, \mathsf{L} \rangle \leq_{k_1+k_2}^{\mathsf{p}} \langle \Gamma'', \mathsf{L}'' \rangle.$$

4   Checkers Semantics for the λ-calculus

In the introduction, we argued that the checkers calculus is interesting in itself, but in this paper we mainly use it to infer properties of the standard λ-calculus.

As discussed in Section 2, ordinary λ-terms can be embedded into the checkers calculus via the coloring map $\overline{(\cdot)}^{\bullet} : \Lambda \to \Lambda_{\circ\bullet}$ from Definition 4, and then compared using the various preorders we defined on checkers terms.

Definition 11 (Preorders on λ-terms).   *For all λ-terms $t, u \in \Lambda$, define:*

(i) Interaction preorder ($\sqsubseteq^{\mathrm{int}}$). $t \sqsubseteq^{\mathrm{int}} u$ *iff* $\overline{t}^{\bullet} \sqsubseteq^{\mathrm{int}}_{\bullet} \overline{u}^{\bullet}$;

(ii) Interaction improvement ($\sqsubseteq^{\mathrm{imp}}$). $t \sqsubseteq^{\mathrm{imp}} u$ *iff* $\overline{t}^{\bullet} \sqsubseteq^{\mathrm{imp}}_{\bullet} \overline{u}^{\bullet}$;*f*

(iii) Polarized whiter-cheaper preorder ($\sqsubseteq^{\mathrm{pwc}}$). $t \sqsubseteq^{\mathrm{pwc}} u$ *iff* $\overline{t}^{\bullet} \sqsubseteq^{\mathrm{pwc}}_{\bullet} \overline{u}^{\bullet}$.

To help the reader avoid any confusion, we reserve the subscript '$\bullet$' for relations between checkers terms, while it is omitted in relations between λ-terms. However, be aware that the quantification in $\sqsubseteq^{\mathrm{int}}$ and $\sqsubseteq^{\mathrm{imp}}$ still ranges over all checkers contexts, and the interaction steps are being counted.

The preorder $\sqsubseteq^{\mathrm{int}}$ has been characterized as Böhm tree inclusion [ ], and semantically corresponds to the preorder induced by Plotkins model $\mathcal{P}\omega$ [ ]. We have seen that, although there is no analogue of η-reduction in the checkers calculus (Remark 1), the inequation $\mathbf{1}_{\bullet} \sqsubseteq^{\mathrm{imp}}_{\bullet} \mathbf{I}_{\bullet}$ holds (Example 2), which entails $\lambda y.ty \sqsubseteq^{\mathrm{imp}} t$ for λ-terms (with $y \notin \mathtt{fv}(t)$). In other words, $\sqsubseteq^{\mathrm{imp}}$ validates η-reduction in the sense that $t \to_\eta u$ implies $t \sqsubseteq^{\mathrm{imp}} u$.

We shall see that the situation is even subtler: $\sqsubseteq^{\mathrm{imp}}$ also validates possibly infinite η-reductions. This notion has been formalized by Lassen [ ] as follows. We write $t \Downarrow_{\mathrm{h}} h$ to mean that t is head normalizable and h is its hnf.

Definition 12 (Böhm tree preorder up to η-reductions). *The Böhm preorder up to η-reductions $\sqsubseteq_{\mathcal{B}\eta^\infty_{\mathrm{red}}}$ is defined coinductively on λ-terms t, u as the largest relation $t \sqsubseteq_{\mathcal{B}\eta^\infty_{\mathrm{red}}} u$ closed under the following clauses:*

(bot) $t \not\Downarrow_{\mathrm{h}}$ i.e. t has no head normal form.

$(\mathrm{H}\eta_{\mathrm{red}})$ $t \Downarrow_{\mathrm{h}} \lambda x_1 \ldots x_{n+p}.y\, t_1 \cdots t_{k+p}$ and $u \Downarrow_{\mathrm{h}} \lambda x_1 \ldots x_n.y\, u_1 \cdots u_k$, with $p \geq 0$,

y is equally bound or free in both terms, $(t_i \sqsubseteq_{\mathcal{B}\eta^\infty_{\mathrm{red}}} u_i)_{i=1}^{k}$, $(t_i \sqsubseteq_{\mathcal{B}\eta^\infty_{\mathrm{red}}} x_i)_{i=k+1}^{k+p}$ and $(x_i)_{i=k+1}^{k+p}$ are chosen by α-conversion not to occur free in $y u_1 \cdots u_k$.

The Böhm preorder up to possibly infinite η-conversion $\sqsubseteq_{\mathcal{B}\eta^\infty}$ is defined similarly, by adding a clause catching η-expansions in u, thus symmetric to $(\mathrm{H}\eta_{\mathrm{red}})$.

The definition of $t \sqsubseteq_{\mathcal{B}\eta^\infty_{\mathrm{red}}} u$ requires a certain level of complexity to take into account possibly infinite η-expansions occurring in the Böhm tree of t. Breuvart et al. [ ] characterized $\sqsubseteq_{\mathcal{B}\eta^\infty_{\mathrm{red}}}$ as the preorder induced by the relational graph model $\mathcal{E}$ [ ], Ronchi Della Rocca as the one induced by the filter model $\mathcal{D}_{\mathrm{BCD}}$ [ ].

Example 3. The $\mathsf{J} := \mathsf{Y}(\lambda zxy.x(zy))$ from Example 2 produces no η-redexes:

$$\mathsf{J} =_\beta \lambda xy_0.x(\mathsf{J}y_0) =_\beta \lambda xy_0.x(\lambda y_1.x\mathsf{J}y_1) =_\beta \lambda xy_0.x(\lambda y_1.x(\lambda y_2.y_1(\mathsf{J}y_2))) =_\beta \cdots$$

but it is an infinite η-expansion of $\mathsf{I} := \lambda x.x$, and in fact $\mathsf{J} \sqsubseteq_{\mathcal{B}\eta^\infty_{\mathrm{red}}} \mathsf{I}$. This example clarifies that the freshness condition in $(\mathsf{H}\eta_{\mathrm{red}})$ does not need to hold at any finite step, but holds at the limit. By contextuality, we get $\lambda y.x(\lambda z.y(\mathsf{J}z)) \sqsubseteq_{\mathcal{B}\eta^\infty_{\mathrm{red}}} \lambda y.xy$, and it is interesting to compare their interpretations once painted black. E.g.,

$$
\cfrac{
  \cfrac{x:[[0 \xrightarrow{\bullet} \mathsf{L}'] \xrightarrow{\mathsf{d}} \mathsf{L}] \vdash^0_{\circledcirc} x:[0 \xrightarrow{\bullet} \mathsf{L}'] \xrightarrow{\mathsf{d}} \mathsf{L}}{}\ \mathsf{ax}
  \qquad
  \cfrac{
    \cfrac{
      \cfrac{
        \cfrac{y:[0 \xrightarrow{\mathsf{c}} \mathsf{L}'] \vdash^0_{\circledcirc} y:0 \xrightarrow{\mathsf{c}} \mathsf{L}' \quad \vdash^0_{\circledcirc} \overrightarrow{\mathsf{J}z}^\bullet:0}{y:[0 \xrightarrow{\mathsf{c}} \mathsf{L}'] \vdash^{\delta^\perp_{\mathsf{c},\bullet}}_{\circledcirc} y \bullet (\overrightarrow{\mathsf{J}z}^\bullet):\mathsf{L}'}\ @
      }{y:[0 \xrightarrow{\mathsf{c}} \mathsf{L}'] \vdash^{\delta^\perp_{\mathsf{c},\bullet}}_{\circledcirc} \lambda_\bullet z.y \bullet (\overrightarrow{\mathsf{J}z}^\bullet):0 \xrightarrow{\bullet} \mathsf{L}'}\ \lambda
    }{y:[0 \xrightarrow{\mathsf{c}} \mathsf{L}'] \vdash^{\delta^\perp_{\mathsf{c},\bullet}}_{\circledcirc} \lambda_\bullet z.y \bullet z:[0 \xrightarrow{\bullet} \mathsf{L}']}\ \mathsf{m}
  }{}
}{
  \cfrac{x:[[0 \xrightarrow{\bullet} \mathsf{L}'] \xrightarrow{\mathsf{d}} \mathsf{L}], y:[0 \xrightarrow{\mathsf{c}} \mathsf{L}'] \vdash^{(\delta^\perp_{\mathsf{c},\bullet}+\delta^\perp_{\mathsf{d},\bullet})}_{\circledcirc} x \bullet \lambda_\bullet z.y \bullet (\overrightarrow{\mathsf{J}z}^\bullet):\mathsf{L}}{x:[[0 \xrightarrow{\bullet} \mathsf{L}'] \xrightarrow{\mathsf{d}} \mathsf{L}] \vdash^{(\delta^\perp_{\mathsf{c},\bullet}+\delta^\perp_{\mathsf{d},\bullet})}_{\circledcirc} \lambda_\bullet y.x \bullet (\lambda_\bullet z.y \bullet (\overrightarrow{\mathsf{J}z}^\bullet)):[0 \xrightarrow{\bullet} \mathsf{L}'] \xrightarrow{\bullet} \mathsf{L}}\ \lambda}\ @_{\circledcirc}
$$

where m is short for many, and $\delta^\perp_{\cdot,\cdot}$ is (the dual of) Kronecker's delta, i.e. $\delta^\perp_{\mathsf{c},\mathsf{d}} := 1$ if $\mathsf{c} \neq \mathsf{d}$, and $\delta^\perp_{\mathsf{c},\mathsf{d}} := 0$ otherwise. By taking $\mathsf{c} = \circ$, one obtains a typing that is not suitable for $\lambda y.xy$ since $x:[[0 \xrightarrow{\bullet} \mathsf{L}'] \xrightarrow{\mathsf{d}} \mathsf{L}] \not\vdash^k_{\circledcirc} \lambda_\bullet y.x \bullet \lambda_\bullet z.y \bullet z:[0 \xrightarrow{\circ} \mathsf{L}'] \xrightarrow{\bullet} \mathsf{L}$.

There exists however a positively 1-whiter typing:

$$
\cfrac{
  \cfrac{x:[[0 \xrightarrow{\circ} \mathsf{L}'] \xrightarrow{\mathsf{d}} \mathsf{L}] \vdash^0_{\circledcirc} x:[0 \xrightarrow{\circ} \mathsf{L}'] \xrightarrow{\mathsf{d}} \mathsf{L}}{}\ \mathsf{ax}
  \qquad
  \cfrac{
    \cfrac{y:[0 \xrightarrow{\circ} \mathsf{L}'] \vdash^0_{\circledcirc} y:0 \xrightarrow{\circ} \mathsf{L}'}{y:[0 \xrightarrow{\circ} \mathsf{L}'] \vdash^0_{\circledcirc} y:[0 \xrightarrow{\circ} \mathsf{L}']}\ \mathsf{many}
  }{}
}{
  \cfrac{x:[[0 \xrightarrow{\circ} \mathsf{L}'] \xrightarrow{\mathsf{d}} \mathsf{L}], y:[0 \xrightarrow{\circ} \mathsf{L}'] \vdash^{\delta^\perp_{\mathsf{d},\bullet}}_{\circledcirc} x \bullet y:\mathsf{L}}{x:[[0 \xrightarrow{\circ} \mathsf{L}'] \xrightarrow{\mathsf{d}} \mathsf{L}] \vdash^{\delta^\perp_{\mathsf{d},\bullet}}_{\circledcirc} \lambda_\bullet y.x \bullet y:[0 \xrightarrow{\circ} \mathsf{L}'] \xrightarrow{\bullet} \mathsf{L}}\ \lambda}\ @_{\circledcirc}
$$

that is 1-cheaper since $\delta^\perp_{\mathsf{d},\bullet} < \delta^\perp_{\mathsf{c},\bullet} + \delta^\perp_{\mathsf{d},\bullet} = \delta^\perp_{\mathsf{d},\bullet} + 1$. As a result of our main Theorem 5, we shall see that $\lambda y.x(\lambda z.y(\mathsf{J}z)) \sqsubseteq^{\mathrm{pwc}} \lambda y.xy$ holds.

4.1   Two inclusions for the Böhm preorder up to η-reductions

The remainder of the paper is devoted to showing that the preorders $\sqsubseteq^{\mathrm{imp}}$, $\sqsubseteq^{\mathrm{pwc}}$, and $\sqsubseteq_{\mathcal{B}\eta^\infty_{\mathrm{red}}}$ coincide. In this subsection, we relate the Böhm preorder up to η-reductions to the polarized whiter-cheaper improvement ($\sqsubseteq_{\mathcal{B}\eta^\infty_{\mathrm{red}}} \subseteq \sqsubseteq^{\mathrm{pwc}}$) and to interaction improvement ($\sqsubseteq^{\mathrm{imp}} \subseteq \sqsubseteq_{\mathcal{B}\eta^\infty_{\mathrm{red}}}$). To prove $\sqsubseteq_{\mathcal{B}\eta^\infty_{\mathrm{red}}} \subseteq \sqsubseteq^{\mathrm{pwc}}$, we need to study first the case of a possibly infinite η-expansion of a single variable.

Lemma 3. *Let $t \in \Lambda$ and $x \in \mathrm{VAR}$ be such that $t \sqsubseteq_{\mathcal{B}\eta^\infty_{\mathrm{red}}} x$.*

1. *If $\Gamma \vdash^k_{\circledcirc} \vec{t}^\bullet : \mathsf{L}$ then there exist $\mathsf{L}', \mathsf{L}''$ such that $\Gamma = x:[\mathsf{L}']$, $\mathsf{L}'' \leq^-_{k'} \mathsf{L}'$ and $\mathsf{L}'' \leq^+_{k''} \mathsf{L}$ with $k = k' + k''$;*

2. *If $\Gamma \vdash_{\mathbf{e}}^{k} \vec{t}^{\bullet} : \mathsf{M}$ then there exist $\mathsf{M}', \mathsf{M}''$ such that $\Gamma = x : [\mathsf{M}'], \mathsf{M}'' \leq_{k'}^{-} \mathsf{M}'$ and $\mathsf{M}'' \leq_{k''}^{+} \mathsf{M}$ with $k = k' + k''$.*

We now generalize the result to any pair of λ-terms related by $\sqsubseteq_{\mathcal{B}\eta_{\mathrm{red}}^{\infty}}$.

Lemma 4. *Assume $t \sqsubseteq_{\mathcal{B}\eta_{\mathrm{red}}^{\infty}} u$ and $\Gamma \vdash_{\mathbf{e}}^{k} \vec{t}^{\bullet} : \mathsf{L}$. Then $\Gamma' \vdash_{\mathbf{e}}^{k'} \vec{u}^{\bullet} : \mathsf{L}'$ with $\langle \Gamma', \mathsf{L}' \rangle \leq_{p}^{+} \langle \Gamma, \mathsf{L} \rangle$, for $0 \leq p = k - k'$.*

Corollary 1. *For all $t, u \in \Lambda$, $t \sqsubseteq_{\mathcal{B}\eta_{\mathrm{red}}^{\infty}} u$ entails $t \sqsubseteq^{\mathrm{pwc}} u$.*

Completeness of the Böhm preorder up to η-reductions. To prove the inclusion $\sqsubseteq^{\mathrm{imp}} \subseteq \sqsubseteq_{\mathcal{B}\eta_{\mathrm{red}}^{\infty}}$ it is sufficient to slightly adapt the proof that the interaction preorder entails the Böhm tree preorder [ ]. The proof exploits the famous Böhm out technique, used by Hyland to construct a λ-calculus context C separating t and u whenever $t \not\sqsubseteq_{\mathcal{B}\eta^{\infty}} u$, i.e., $C\langle t \rangle \Downarrow_{\mathrm{h}}$ and $C\langle u \rangle \not\Downarrow_{\mathrm{h}}$ [ ]. Since we count interaction steps, we can also separate η-convertible λ-terms like $\mathbf{1}_{\bullet}$ and $\mathbf{I}_{\bullet}$, using the white context $\langle \cdot \rangle \circ (\lambda_{\circ} x.x)$. We extend the Definition 4 of white painting $\overline{(\cdot)}^{\circ}$ to λ-calculus contexts C by adding the case $\overline{\langle \cdot \rangle}^{\circ} = \langle \cdot \rangle$.

Lemma 5 (Interaction Böhm-out).   *Let $t, u \in \Lambda$ be such that $t \sqsubseteq_{\mathcal{B}\eta^{\infty}} u$ and $t \not\sqsubseteq_{\mathcal{B}\eta_{\mathrm{red}}^{\infty}} u$. Then, there exists a λ-calculus context C such that $\overline{C}^{\circ}\langle \vec{t}^{\bullet} \rangle \Downarrow_{\mathrm{h}_{\circ}\bullet}^{\mathbf{e}i}$ and $\overline{C}^{\circ}\langle \vec{u}^{\bullet} \rangle \Downarrow_{\mathrm{h}_{\circ}\bullet}^{\mathbf{e}i'}$ with $i' > i$.*

Theorem 3 (Completeness).   *Let $t, u \in \Lambda$. If $t \sqsubseteq^{\mathrm{imp}} u$ then $t \sqsubseteq_{\mathcal{B}\eta_{\mathrm{red}}^{\infty}} u$.*

Proof. Assume $t \not\sqsubseteq_{\mathcal{B}\eta_{\mathrm{red}}^{\infty}} u$, towards a contradiction. There are two cases:

- If $t \not\sqsubseteq_{\mathcal{B}\eta^{\infty}} u$, then there is a context C such that $C\langle t \rangle \Downarrow_{\mathrm{h}}$, while $C\langle u \rangle \not\Downarrow_{\mathrm{h}}$ [ ]. Since head reductions can be simulated by $\rightarrow_{\mathrm{h}_{\circ}\bullet}$ by Lemma 1.(ii), it follows that $\overline{C}^{\circ}\langle \vec{t}^{\bullet} \rangle \Downarrow_{\mathrm{h}_{\circ}\bullet}$, while $\overline{C}^{\circ}\langle \vec{u}^{\bullet} \rangle \not\Downarrow_{\mathrm{h}_{\circ}\bullet}$. This shows $t \not\sqsubseteq^{\mathrm{imp}} u$.
- If $t \sqsubseteq_{\mathcal{B}\eta^{\infty}} u$ then $t \not\sqsubseteq^{\mathrm{imp}} u$ follows directly from Böhm out (Lemma 5).   □

5   Compositionality of Polarized Whiter-Cheaper

Our goal is to use the polarized whiter-cheaper preorder to characterize the interaction improvement ordering, which is a contextual preorder. We therefore need to show that our preorder is preserved when placing related terms into contexts, *i.e.* that it is *compositional*. This section, which is the main technical work of the paper, is devoted to establishing this fact.

Let us first describe the difficulty we encounter when proving compositionality. We are trying to show the following:

$$\text{If } t \sqsubseteq_{\mathbf{e}}^{\mathrm{pwc}} u \text{ and } s \sqsubseteq_{\mathbf{e}}^{\mathrm{pwc}} r \text{ then } t \cdot^{\mathrm{c}} s \sqsubseteq_{\mathbf{e}}^{\mathrm{pwc}} u \cdot^{\mathrm{c}} r \text{ for any } \mathrm{c}.$$

Let us start by considering an arbitrary derivation of $t \cdot^c s$:

$$\frac{\Gamma_1 \vdash^{k_1}_{\pmb{\circ}} t : \mathsf{M} \xrightarrow{\mathsf{d}} \mathsf{L} \quad \Gamma_2 \vdash^{k_2}_{\pmb{\circ}} s : \mathsf{M}}{\Gamma \uplus \Delta \vdash^{k_1+k_2+\delta^{\perp}_{c,d}}_{\pmb{\circ}} t \cdot^c s : \mathsf{L}} \; @$$

By hypothesis ($t \sqsubseteq^{\mathrm{pwc}}_{\pmb{\circ}} u$ and $s \sqsubseteq^{\mathrm{pwc}}_{\pmb{\circ}} r$), we know that there are whiter and cheaper derivations for the premises of the application rule, but we have no way of knowing if they will still be able to be combined into an application rule:

$$\frac{\Gamma'_1 \vdash^{k'_1}_{\pmb{\circ}} u : \mathsf{M}' \xrightarrow{\mathsf{d}'} \mathsf{L}' \quad \Gamma'_2 \vdash^{k'_2}_{\pmb{\circ}} r : \mathsf{M}''}{\Gamma'_1 \uplus \Gamma'_2 \vdash^{k'_1+k'_2+\delta^{\perp}_{c,d'}}_{\pmb{\circ}} u \cdot^c r : \mathsf{L}'} \; @(\text{if } \mathsf{M}' = \mathsf{M}'')$$

where $\delta^{\perp}_{c,d'}$ is (the dual of) Kronecker's delta (as in Example 3). We shall show that we can indeed build a multi type M''' on which u and r agree. We know that M, M' and M'' are related by the polarized whitening relation in the sense that:

$$\mathsf{M}' \leq^{-}_{l} \mathsf{M} \text{ and } \mathsf{M}'' \leq^{+}_{h} \mathsf{M} \text{ for some } h, l \text{ such that } k'_1 + l \leq k_1 \text{ and } k'_2 + l \leq k_2.$$

Example 4. 1. It does happen that the argument type M is unchanged, as is the case of $\lambda_{\bullet}y.x \bullet y \sqsubseteq^{\mathrm{pwc}}_{\pmb{\circ}} x$ applied to an argument $s \sqsubseteq^{\mathrm{pwc}}_{\pmb{\circ}} s$.

$$\frac{x : [\mathsf{M} \xrightarrow{\circ} \mathsf{L}] \vdash^{1}_{\pmb{\circ}} \lambda_{\bullet}y.x \bullet y : \mathsf{M} \xrightarrow{\bullet} \mathsf{L} \quad \Gamma \vdash^{k}_{\pmb{\circ}} s : \mathsf{M}}{\Gamma, x : [\mathsf{M} \xrightarrow{\circ} \mathsf{L}] \vdash^{1+k+\delta^{\perp}_{\bullet,d}}_{\pmb{\circ}} (\lambda_{\bullet}y.x \bullet y) \cdot^d s : \mathsf{L}} \; @ \qquad \frac{x : [\mathsf{M} \xrightarrow{\circ} \mathsf{L}] \vdash^{0}_{\pmb{\circ}} x : \mathsf{M} \xrightarrow{\circ} \mathsf{L} \quad \Gamma \vdash^{k}_{\pmb{\circ}} s : \mathsf{M}}{\Gamma, x : [\mathsf{M} \xrightarrow{\circ} \mathsf{L}] \vdash^{k+\delta^{\perp}_{\circ,d}}_{\pmb{\circ}} x \cdot^d s : \mathsf{L}} \; @$$

What if we were to use another term than s as argument, say r such that $s \sqsubseteq^{\mathrm{pwc}}_{\pmb{\circ}} r$? We have that there exists $\langle \Gamma', \mathsf{M}' \rangle \leq^{+}_{d} \langle \Gamma, \mathsf{M} \rangle$ such that $\Gamma' \vdash^{k'}_{\pmb{\circ}} r : \mathsf{M}'$ with $k \geq k' + d$. Then it is easy to change the typing for x so that it fits M'.

$$\frac{x : [\mathsf{M}' \xrightarrow{\circ} \mathsf{L}] \vdash^{0}_{\pmb{\circ}} x : \mathsf{M}' \xrightarrow{\circ} \mathsf{L} \quad \Gamma' \vdash^{k'}_{\pmb{\circ}} r : \mathsf{M}'}{\Gamma', x : [\mathsf{M}' \xrightarrow{\circ} \mathsf{L}] \vdash^{k'+\delta^{\perp}_{\circ,d}}_{\pmb{\circ}} x \cdot^d s : \mathsf{L}} \; @$$

Note that $\langle \Gamma', x : [\mathsf{M} \xrightarrow{\circ} \mathsf{L}], \mathsf{L} \rangle \leq^{+}_{d} \langle \Gamma, x : [\mathsf{M}' \xrightarrow{\circ} \mathsf{L}], \mathsf{L} \rangle$. The change from $x : [\mathsf{M} \xrightarrow{\circ} \mathsf{L}] \vdash^{0}_{\pmb{\circ}} x : \mathsf{M} \xrightarrow{\circ} \mathsf{L}$ to $x : [\mathsf{M}' \xrightarrow{\circ} \mathsf{L}] \vdash^{0}_{\pmb{\circ}} x : \mathsf{M}' \xrightarrow{\circ} \mathsf{L}$ possibly whitens in negative and positive positions, but the negative changes do not appear in the conclusion of the derivation as they are consumed in the application rule.

2. Consider now $\lambda_{\bullet}x.x \bullet \lambda_{\bullet}z.y \bullet z \sqsubseteq^{\mathrm{pwc}}_{\pmb{\circ}} \lambda_{\bullet}x.x \bullet y$, and let us apply a term s on both sides. We have a derivation for $(\lambda_{\bullet}x.x \bullet \lambda_{\bullet}z.y \bullet z) \cdot^d s$:

$$\frac{\dfrac{\dfrac{x : [[\mathsf{M} \xrightarrow{\bullet} \mathsf{L}] \xrightarrow{c} \mathsf{L}'] \vdash^{0}_{\pmb{\circ}} x : [\mathsf{M} \xrightarrow{\bullet} \mathsf{L}] \xrightarrow{c} \mathsf{L}' \quad y : [\mathsf{M} \xrightarrow{\circ} \mathsf{L}] \vdash^{1}_{\pmb{\circ}} \lambda_{\bullet}z.y \bullet z : [\mathsf{M} \xrightarrow{\bullet} \mathsf{L}]}{y : [\mathsf{M} \xrightarrow{\circ} \mathsf{L}], x : [[\mathsf{M} \xrightarrow{\bullet} \mathsf{L}] \xrightarrow{c} \mathsf{L}'] \vdash^{1+\delta^{\perp}_{c,\bullet}}_{\pmb{\circ}} x \bullet \lambda_{\bullet}z.y \bullet z : \mathsf{L}'} \; @}{y : [\mathsf{M} \xrightarrow{\circ} \mathsf{L}] \vdash^{1+\delta^{\perp}_{c,\bullet}}_{\pmb{\circ}} \lambda_{\bullet}x.x \bullet \lambda_{\bullet}z.y \bullet z : [[\mathsf{M} \xrightarrow{\bullet} \mathsf{L}] \xrightarrow{c} \mathsf{L}'] \xrightarrow{\bullet} \mathsf{L}'} \; \lambda \quad \Gamma \vdash^{k}_{\pmb{\circ}} s : [[\mathsf{M} \xrightarrow{\bullet} \mathsf{L}] \xrightarrow{c} \mathsf{L}']}{\Gamma, y : [\mathsf{M} \xrightarrow{\circ} \mathsf{L}] \vdash^{1+k+\delta^{\perp}_{\bullet,d}+\delta^{\perp}_{c,\bullet}}_{\pmb{\circ}} (\lambda_{\bullet}x.x \bullet \lambda_{\bullet}z.y \bullet z) \cdot^d s : \mathsf{L}'} \; @$$

388 A. Lancelot, G. Manzonetto, G. McCusker, G. Vanoni

There exists a positive whiter-cheaper derivation for $\lambda_{\bullet}x.x \bullet y$ but it is hard to use for the application to s afterwards. See the incomplete derivation:

$$
\dfrac{
\dfrac{
\dfrac{x : [[M \xrightarrow{\circ} L] \xrightarrow{c} L'] \vdash_{\oplus}^{0} x : [M \xrightarrow{\circ} L] \xrightarrow{c} L' \qquad y : [M \xrightarrow{\circ} L] \vdash_{\oplus}^{0} y : [M \xrightarrow{\circ} L]}
{y : [M \xrightarrow{\circ} L], x : [[M \xrightarrow{\circ} L] \xrightarrow{c} L'] \vdash_{\oplus}^{\delta^{\perp}_{c,\bullet}} x \bullet y : L'} \; @
}
{y : [M \xrightarrow{\circ} L] \vdash_{\oplus}^{\delta^{\perp}_{c,\bullet}} \lambda_{\bullet}x.x \bullet y : [[M \xrightarrow{\circ} L] \xrightarrow{c} L'] \xrightarrow{\bullet} L'} \; \lambda
\qquad \Gamma \vdash_{\oplus}^{k} s : [[M \xrightarrow{\bullet} L] \xrightarrow{c} L']
}{
\Gamma, y : [M \xrightarrow{\circ} L] \vdash_{\oplus}^{k+\delta^{\perp}_{\bullet,\mathsf{d}}+\delta^{\perp}_{c,\bullet}} (\lambda_{\bullet}x.x \bullet y) \cdot^{\mathsf{d}} s : L'
} \; @?
$$

We therefore need to change the black arrow in the type of s, which appears in a negative position. Such a change is possible, but may lead to more positive changes in $\langle \Gamma, [[M \xrightarrow{\bullet} L] \xrightarrow{c} L'] \rangle$. This reasoning is exactly the object of the repainting mechanism described in Lemma 6.

Repainting Negative Occurrences. Negative color occurrences in a typing judgment $\Gamma \vdash_{\oplus}^{k} t : L$ can be seen as *unspecified* colors, not directly determined by those appearing in the syntax of the term. However, they cannot be recolored arbitrarily. We formally specify how one may *repaint* a negatively occurring arrow white, by constructing a new typing derivation that either also includes one positively occurring arrow repainted white, or whose index k is shifted by 1.

Lemma 6 (Repainting). *Suppose $\Gamma \vdash_{\oplus}^{k} t : L$ and that $\langle \Gamma', L' \rangle \leq_{1}^{-} \langle \Gamma, L \rangle$. Then there exists a derivation of $\Gamma'' \vdash_{\oplus}^{k'} t : L''$ such that one of the following holds:*

(i) $\langle \Gamma'', L'' \rangle \leq_{1}^{+} \langle \Gamma', L' \rangle$ and $k = k'$; or
(ii) $\langle \Gamma'', L'' \rangle = \langle \Gamma', L' \rangle$ and $|k' - k| = 1$.

Equivalently, Point (i) and (ii) may be rephrased as: there exists $0 \leq i \leq 1$ such that $\langle \Gamma'', L'' \rangle \leq_{i}^{+} \langle \Gamma', L' \rangle$ and $|k - k'| \leq 1 - i$.

This first repainting lemma only applies to changes of a singular color in a singular arrow type, but we might need to repaint several. With the help of a commutation property for the $\leq_{1}^{-}$ and $\leq_{1}^{+}$ relations (Lemma 7 below), we show that one can apply repeatedly the repainting lemma, obtaining Proposition 2.

Lemma 7. *Let $\Gamma, \Gamma^{\oplus}, \Gamma_{\ominus}$ and $L, L^{\oplus}, L_{\ominus}$ such that*

$$\langle \Gamma_{\ominus}, L_{\ominus} \rangle \leq_{1}^{-} \langle \Gamma, L \rangle \quad \text{and} \quad \langle \Gamma^{\oplus}, L^{\oplus} \rangle \leq_{1}^{+} \langle \Gamma, L \rangle.$$

Then there are $\Gamma_{\ominus}^{\oplus}, L_{\ominus}^{\oplus}$ such that

$$\langle \Gamma_{\ominus}^{\oplus}, L_{\ominus}^{\oplus} \rangle \leq_{1}^{+} \langle \Gamma_{\ominus}, L_{\ominus} \rangle \quad \text{and} \quad \langle \Gamma_{\ominus}^{\oplus}, L_{\ominus}^{\oplus} \rangle \leq_{1}^{-} \langle \Gamma^{\oplus}, L^{\oplus} \rangle.$$

Proof. The typing $\langle \Gamma_{\ominus}, L_{\ominus} \rangle$ arises by repainting one negatively-occurring arrow in $\langle \Gamma, L \rangle$ from black to white; $\langle \Gamma^{\oplus}, L^{\oplus} \rangle$ arises by repainting one positively occurring arrow in $\langle \Gamma, L \rangle$. Construct $\langle \Gamma_{\ominus}^{\oplus}, L_{\ominus}^{\oplus} \rangle$ by making both these repaintings. $\quad \square$

Proposition 2 (Sequence of Repainting).   *Suppose $\Gamma \vdash_{\mathbf{e}}^{k} t : \mathsf{L}$ and that $\langle \Gamma', \mathsf{L}' \rangle \leq_{k_1}^{-} \langle \Gamma, \mathsf{L} \rangle$ for some $k_1 \geq 0$. Then there exists a derivation of $\Gamma'' \vdash_{\mathbf{e}}^{k'} t : \mathsf{L}''$ and k_2 with $0 \leq k_2 \leq k_1$ such that*

- *$\langle \Gamma'', \mathsf{L}'' \rangle \leq_{k_2}^{+} \langle \Gamma', \mathsf{L}' \rangle$; and*
- *$|k - k'| \leq k_1 - k_2$.*

Back to the Compositionality Proof. Now that we have introduced an appropriate notion of repainting, we return to the proof of compositionality, where the following proposition provides the main argument for the application case.

Proposition 3. *Given $\Gamma \vdash_{\mathbf{e}}^{k} t : \mathsf{M} \xrightarrow{\mathsf{c}} \mathsf{L}$ and $\Delta \vdash_{\mathbf{e}}^{l} u : \mathsf{N}$ with either $\mathsf{M} \leq_{d}^{-} \mathsf{N}$ or $\mathsf{N} \leq_{d}^{+} \mathsf{M}$, there exists a typing $\Gamma' + \Delta' \vdash_{\mathbf{e}}^{m} t \cdot^{\mathsf{d}} u : \mathsf{L}'$ and d' with $0 \leq d' \leq d$ such that $\langle \Gamma' + \Delta', \mathsf{L}' \rangle \leq_{d'}^{+} \langle \Gamma + \Delta, \mathsf{L} \rangle$ and $m \leq k + l + \delta_{\mathsf{c},\mathsf{d}}^{\perp} + d - d'$.*

Proposition 4 (Compositionality of Polarized Whiter-Cheaper Improvement). *If $t \sqsubseteq_{\mathbf{e}}^{\mathrm{pwc}} u$ then $C\langle t \rangle \sqsubseteq_{\mathbf{e}}^{\mathrm{pwc}} C\langle u \rangle$ for any checkers context C.*

Proof. Note that it is sufficient (by transitivity of $\sqsubseteq_{\mathbf{e}}^{\mathrm{pwc}}$) to prove for all t, u, s such that $t \sqsubseteq_{\mathbf{e}}^{\mathrm{pwc}} u$, we have that for all $\mathsf{c} \in \{\circ, \bullet\}$:

$$\lambda_{\mathsf{c}} x.t \sqsubseteq_{\mathbf{e}}^{\mathrm{pwc}} \lambda_{\mathsf{c}} x.u; \qquad t \cdot^{\mathsf{c}} s \sqsubseteq_{\mathbf{e}}^{\mathrm{pwc}} u \cdot^{\mathsf{c}} s; \qquad \text{and } s \cdot^{\mathsf{c}} t \sqsubseteq_{\mathbf{e}}^{\mathrm{pwc}} s \cdot^{\mathsf{c}} u.$$

We prove here the first application case; the other cases are easier.

Suppose $(\Gamma, \mathsf{L}, k) \in [\![t \cdot^{\mathsf{d}} s]\!]^{\mathbf{e}}$. Then we must have $(\Gamma_1, \mathsf{M} \xrightarrow{\mathsf{c}} \mathsf{L}, k_1) \in [\![t]\!]^{\mathbf{e}}$ and $(\Gamma_2, \mathsf{M}, k_2) \in [\![s]\!]^{\mathbf{e}}$ with $k = k_1 + k_2 + \delta_{\mathsf{c},\mathsf{d}}^{\perp}$ and $\Gamma = \Gamma_1 + \Gamma_2$. Since $t \sqsubseteq_{\mathbf{e}}^{\mathrm{pwc}} u$ we can find $(\Gamma_1', \mathsf{M}' \xrightarrow{\mathsf{c}'} \mathsf{L}', k_1') \in [\![u]\!]^{\mathbf{e}}$ with $\langle \Gamma_1', \mathsf{M}' \xrightarrow{\mathsf{c}'} \mathsf{L}' \rangle \leq_{d}^{+} \langle \Gamma_1, \mathsf{M} \xrightarrow{\mathsf{c}} \mathsf{L} \rangle$ and $k_1 \geq k_1' + d$. This implies that there are d_1, d_2 such that $d = d_1 + d_2 + \delta_{\mathsf{c},\mathsf{c}'}^{\perp}$ and $\langle \Gamma_1', \mathsf{L}' \rangle \leq_{d_1}^{+} \langle \Gamma_1, \mathsf{L} \rangle$ and $\mathsf{M}' \leq_{d_2}^{-} \mathsf{M}$. By Proposition 3 there exists $(\Gamma_1'' + \Gamma_2', \mathsf{L}'', m) \in [\![u \cdot^{\mathsf{d}} s]\!]^{\mathbf{e}}$ with $\langle \Gamma_1'' + \Gamma_2', \mathsf{L}'' \rangle \leq_{d'}^{+} \langle \Gamma_1' + \Gamma_2, \mathsf{L}' \rangle$ and $m \leq k_1' + k_2 + \delta_{\mathsf{c}',\mathsf{d}}^{\perp} + d_2 - d'$. So we have

$$\langle \Gamma_1'' + \Gamma_2', \mathsf{L}'' \rangle \leq_{d'}^{+} \langle \Gamma_1' + \Gamma_2, \mathsf{L}' \rangle \leq_{d_1}^{+} \langle \Gamma_1 + \Gamma_2, \mathsf{L} \rangle$$

and hence $\langle \Gamma_1'' + \Gamma_2', \mathsf{L}'' \rangle \leq_{d'+d_1}^{+} \langle \Gamma_1 + \Gamma_2, \mathsf{L} \rangle = \langle \Gamma, \mathsf{L} \rangle$. It only remains to show the easy inequation $k \geq m + d' + d_1$. $\qquad\square$

By compositionality, soundness and completeness of the checkers multi type system, we are able to relate $\sqsubseteq_{\mathbf{e}}^{\mathrm{pwc}}$ and $\sqsubseteq_{\mathbf{e}}^{\mathrm{imp}}$.

Theorem 4 (PWC Improvement implies Interaction Improvement). *For all $t, u \in \Lambda$, $\overline{t}^{\bullet} \sqsubseteq_{\mathbf{e}}^{\mathrm{pwc}} \overline{u}^{\bullet}$ entails $t \sqsubseteq_{\mathbf{e}}^{\mathrm{imp}} u$.*

Proof. Suppose $\overline{t}^{\bullet} \sqsubseteq_{\mathbf{e}}^{\mathrm{pwc}} \overline{u}^{\bullet}$. We have to show that for any checkers-context C, if $C\langle \overline{t}^{\bullet} \rangle \Downarrow_{\mathrm{h}\circ\bullet}^{\mathbf{e}k}$ for some k then $C\langle \overline{u}^{\bullet} \rangle \Downarrow_{\mathrm{h}\circ\bullet}^{\mathbf{e}k'}$ for some $k' \leq k$. So suppose $C\langle \overline{t}^{\bullet} \rangle \Downarrow_{\mathrm{h}\circ\bullet}^{\mathbf{e}k}$, by completeness of the checkers type system (Theorem 2.2), we have some $(\Gamma, \mathsf{L}, k) \in [\![C\langle \overline{t}^{\bullet} \rangle]\!]^{\mathbf{e}}$. Proposition 4 tells us that $C\langle \overline{t}^{\bullet} \rangle \sqsubseteq_{\mathbf{e}}^{\mathrm{pwc}} C\langle \overline{u}^{\bullet} \rangle$ so by definition of $\sqsubseteq_{\mathbf{e}}^{\mathrm{pwc}}$ there is some $(\Gamma', \mathsf{L}', k') \in [\![C\langle u^{\bullet} \rangle]\!]^{\mathbf{e}}$ with $k' \leq k$. By soundness of the type system (Theorem 2.1), there is $k'' \leq k'$ such that $C\langle \overline{u}^{\bullet} \rangle \Downarrow_{\mathrm{h}\circ\bullet}^{\mathbf{e}k''}$ as required. $\qquad\square$

Wrapping Up. We can at last state our final theorem, completing the characterization of interaction improvement. Indeed, interaction improvement also characterizes the preorder induced by relational semantics $\sqsubseteq^{\mathrm{rel}}$, i.e., the inequational theory of the model $\mathcal{E}$ in Breuvart et al. [ ]. We focus on this model because it can be presented via a multi type system very similar to the one of Fig. 2. In fact, checkers multi types are obtained by coloring the well-known multi types for head evaluation [,]. If one removes the colors from the types in the interpretation $[\![\overline{t}^{\bullet}]\!]^{\mathcal{O}}$ of a term then one recovers exactly the interpretation $[\![t]\!]^{\mathrm{rel}}$ in relational semantics (as presented by multi types). However, it is not true that $[\![t]\!]^{\mathrm{rel}} \subseteq [\![u]\!]^{\mathrm{rel}}$ implies $[\![\overline{t}^{\bullet}]\!]^{\mathcal{O}} \subseteq [\![\overline{u}^{\bullet}]\!]^{\mathcal{O}}$ (because of the various ways of coloring types). This is exactly the reason why we introduced the $\sqsubseteq_{\mathcal{O}}^{\mathrm{pwc}}$ preorder to compare checkers type interpretations instead of the vanilla set inclusion.

Theorem 5. *For $t, u \in \Lambda$, the following are equivalent:*

1. Böhm tree preorder up to η-reductions: $t \sqsubseteq_{\mathcal{B}\eta_{\mathrm{red}}^{\infty}} u$;
2. Polarized Whiter-Cheaper Type Improvement: $t \sqsubseteq^{\mathrm{pwc}} u$;
3. Interaction Improvement: $t \sqsubseteq^{\mathrm{imp}} u$;
4. Plain Type Preorder: $t \sqsubseteq^{\mathrm{rel}} u$.

Proof. $(1 \Rightarrow 2)$. By Corollary 1.
$(2 \Rightarrow 3)$. By Theorem 4.
$(3 \Rightarrow 1)$. By Theorem 3.
$(1 \Leftrightarrow 4)$. By [, Theorem 5.6]. $\qquad\qquad\square$

White head contexts are enough. Our result of completeness states that interaction improvement is included in the Böhm tree up to η-reductions preorder. Due to the particular shape of the separating context built in Lemma 5, our result has stronger consequences. The *white applicative interaction improvement* can be defined analogously to the standard interaction improvement, but restricted to head contexts consisting only of white applications and lambda abstractions. In this sense, these contexts can be viewed as uniformly "whitened" head contexts from the plain λ-calculus. Formally, we write $t \sqsubseteq^{\circ\mathrm{imp}} u$ if:

$\forall$ head contexts C such that $\overline{C}^{\circ}\langle \overline{t}^{\bullet}\rangle \Downarrow_{\mathrm{h}_{\circ\bullet}}^{\mathcal{O}i}$, we have $\overline{C}^{\circ}\langle \overline{u}^{\bullet}\rangle \Downarrow_{\mathrm{h}_{\circ\bullet}}^{\mathcal{O}i'}$ for some $i' \leq i$.

It is evident that the class of applicative white contexts is a subset of the unrestricted general checker contexts $\mathcal{C}_{\circ\bullet}$. Therefore, the inclusion $\sqsubseteq^{\mathrm{imp}} \subseteq \sqsubseteq^{\circ\mathrm{imp}}$ holds trivially. More precisely:

$$t \sqsubseteq^{\mathrm{imp}} u \xRightarrow{\mathrm{triv.}} t \sqsubseteq^{\circ\mathrm{imp}} u \xRightarrow{\mathrm{L.5}} t \sqsubseteq_{\mathcal{B}\eta_{\mathrm{red}}^{\infty}} u \tag{1}$$

Since we have already proved that $\sqsubseteq_{\mathcal{B}\eta_{\mathrm{red}}^{\infty}} \subseteq \sqsubseteq^{\mathrm{imp}}$, it follows immediately from (1) that it suffices to compare terms within white contexts to determine the interaction improvement between two programs. The applicative restriction was not emphasized in [ ], even though the separation argument there already used an applicative context. Here, the restriction is particularly relevant because other proof methods, such as Howes method [,], do not easily establish that applicative contexts alone are sufficient. Notably, prior to this work, it could not even be proven that $t \to_{\eta} u$ implies $t \sqsubseteq^{\mathrm{imp}} u$; this was only conjectured in [ ].

6   Conclusions

The main contribution of this paper is to make precise in what sense the relational semantics conveys quantitative information about programs. We establish this by characterizing the relational preorder in terms of interaction improvement—a quantitative refinement of the contextual preorder that measures the number of interactions between a term and its context. Our key technical tool is the checkers calculus of [ ], which, in the spirit of game semantics, treats programs and environments as first-class entities.

Future Work. Several research directions stem from this work:

- *Towards more applied calculi.* We have carried out our analysis in the world of the untyped call-by-name λ-calculus. We would like to extend our results on the one hand to typed calculi, such as PCF, and on the other one to calculi with sharing, such as call-by-value and call-by-need.
- *Revisiting improvement theory.* Sands and collaborators employ improvement theory to establish several quantitative results on functional program transformations (see, *e.g.*, [, ,]). We aim to explore whether our framework can also be applied in this setting. To do so, it may be necessary to relax the constraint on the difference in interaction steps in the definition of $\sqsubseteq^{\mathrm{imp}}$, for instance by allowing a linear overhead.
- *Categorical analysis; relationship with other models.* We aim to investigate how our annotated semantics can be understood at a more abstract, categorical level. While Cartesian closed categories are part of the picture, we expect that additional structure will be required to capture the colors of the constructs and their operational interpretation. A more abstract formulation of the model may allow us to clarify its relationship with game-theoretic approaches to program semantics [, , , ,], which originally inspired this work, and to explore the broader applicability of our techniques.

Acknowledgements. We thank Beniamino Accattoli for interesting discussions and the anonymous reviewers for their careful reading and useful suggestions.

References

1. Abramsky, S., Jagadeesan, R., Malacaria, P.: Full abstraction for PCF. Inf. Comput. **163**(2), 409–470 (2000). https://doi.org/10.1006/INCO.2000.2930
2. Abramsky, S., Ong, C.L.: Full abstraction in the lazy lambda calculus. Inf. Comput. **105**(2), 159–267 (1993). https://doi.org/10.1006/INCO.1993.1044
3. Accattoli, B., Dal Lago, U., Vanoni, G.: Multi types and reasonable space. Proc. ACM Program. Lang. **6**(ICFP), 799–825 (2022). https://doi.org/10.1145/3547650
4. Accattoli, B., Graham-Lengrand, S., Kesner, D.: Tight typings and split bounds, fully developed. J. Funct. Program. **30**, e14 (2020). https://doi.org/10.1017/S095679682000012X
5. Accattoli, B., Lancelot, A., Manzonetto, G., Vanoni, G.: Interaction equivalence. Proc. ACM Program. Lang. **9**(POPL) (Jan 2025). https://doi.org/10.1145/3704891, https://doi.org/10.1145/3704891
6. Alcolei, A., Clairambault, P., Laurent, O.: Resource-tracking concurrent games. In: Bojanczyk, M., Simpson, A. (eds.) Foundations of Software Science and Computation Structures - 22nd International Conference, FOSSACS 2019, Held as Part of the European Joint Conferences on Theory and Practice of Software, ETAPS 2019, Prague, Czech Republic, April 6-11, 2019, Proceedings. Lecture Notes in Computer Science, vol. 11425, pp. 27–44. Springer (2019). https://doi.org/10.1007/978-3-030-17127-8_2
7. Barendregt, H.: The Lambda Calculus – Its Syntax and Semantics, Studies in logic and the foundations of mathematics, vol. 103. North-Holland (1984)
8. Berry, G.: Some syntactic and categorical constructions of lambda-calculus models. Research Report RR-0080, INRIA (1981), https://inria.hal.science/inria-00076481
9. Biernacki, D., Lenglet, S., Polesiuk, P.: A complete normal-form bisimilarity for state. In: Bojanczyk, M., Simpson, A. (eds.) Foundations of Software Science and Computation Structures - 22nd International Conference, FOSSACS 2019, Held as Part of the European Joint Conferences on Theory and Practice of Software, ETAPS 2019, Prague, Czech Republic, April 6-11, 2019, Proceedings. Lecture Notes in Computer Science, vol. 11425, pp. 98–114. Springer (2019). https://doi.org/10.1007/978-3-030-17127-8_6, https://doi.org/10.1007/978-3-030-17127-8_6
10. Biernacki, D., Lenglet, S., Polesiuk, P.: A Complete Normal-Form Bisimilarity for Algebraic Effects and Handlers. In: Ariola, Z.M. (ed.) 5th International Conference on Formal Structures for Computation and Deduction (FSCD 2020). Leibniz International Proceedings in Informatics (LIPIcs), vol. 167, pp. 7:1–7:22. Schloss Dagstuhl–Leibniz-Zentrum für Informatik, Dagstuhl, Germany (2020). https://doi.org/10.4230/LIPIcs.FSCD.2020.7, https://drops.dagstuhl.de/opus/volltexte/2020/12329
11. Biernacki, D., Lenglet, S., Polesiuk, P.: Proving soundness of extensional normal-form bisimilarities. Electronic Notes in Theoretical Computer Science **336**, 41–56 (2018). https://doi.org/https://doi.org/10.1016/j.entcs.2018.03.015, https://www.sciencedirect.com/science/article/pii/S1571066118300185, the Thirty-third Conference on the Mathematical Foundations of Programming Semantics (MFPS XXXIII)
12. Breuvart, F., Manzonetto, G., Ruoppolo, D.: Relational graph models at work. Log. Methods Comput. Sci. **14**(3) (2018). https://doi.org/10.23638/LMCS-14(3:2)2018, https://doi.org/10.23638/LMCS-14(3:2)2018

13. Bucciarelli, A., Ehrhard, T., Manzonetto, G.: Not enough points is enough. In: Duparc, J., Henzinger, T.A. (eds.) Computer Science Logic, 21st International Workshop, CSL 2007, 16th Annual Conference of the EACSL, Lausanne, Switzerland, September 11-15, 2007, Proceedings. Lecture Notes in Computer Science, vol. 4646, pp. 298–312. Springer (2007). https://doi.org/10.1007/978-3-540-74915-8_24, https://doi.org/10.1007/978-3-540-74915-8_24

14. Bucciarelli, A., Kesner, D., Ventura, D.: Non-idempotent intersection types for the lambda-calculus. Log. J. IGPL **25**(4), 431–464 (2017). https://doi.org/10.1093/JIGPAL/JZX018

15. de Carvalho, D.: Sémantiques de la logique linéaire et temps de calcul. Thèse de doctorat, Université Aix-Marseille II (2007)

16. de Carvalho, D.: Execution time of λ-terms via denotational semantics and intersection types. Math. Struct. Comput. Sci. **28**(7), 1169–1203 (2018). https://doi.org/10.1017/S0960129516000396

17. Clairambault, P.: Causal Investigations in Interactive Semantics. Habilitation à diriger des recherches, Aix-Marseille Université, Marseille, France (2024), https://tel.archives-ouvertes.fr/tel-04523273

18. Dal Lago, U., Gavazzo, F.: Effectful Normal Form Bisimulation. In: ESOP 2019 - European Symposium on Programming. Prague, Czech Republic (Apr 2019), https://hal.inria.fr/hal-02386004

19. Ehrhard, T., Pagani, M., Tasson, C.: Full abstraction for probabilistic PCF. J. ACM **65**(4), 23:1–23:44 (2018). https://doi.org/10.1145/3164540, https://doi.org/10.1145/3164540

20. Fokkink, W.J.: Modelling Distributed Systems. Springer, Berlin, Heidelberg (2007). https://doi.org/10.1007/978-3-540-49321-0

21. Ghica, D.R.: Slot games: a quantitative model of computation. In: Palsberg, J., Abadi, M. (eds.) Proceedings of the 32nd ACM SIGPLAN-SIGACT Symposium on Principles of Programming Languages, POPL 2005, Long Beach, California, USA, January 12-14, 2005. pp. 85–97. ACM (2005). https://doi.org/10.1145/1040305.1040313

22. Howe, D.J.: Proving congruence of bisimulation in functional programming languages. Inf. Comput. **124**(2), 103–112 (1996). https://doi.org/10.1006/inco.1996.0008, https://doi.org/10.1006/inco.1996.0008

23. Hyland, M.: A syntactic characterization of the equality in some models for the lambda calculus. Journal of the London Mathematical Society **s2-12**(3), 361–370 (1976). https://doi.org/https://doi.org/10.1112/jlms/s2-12.3.361

24. Hyland, M., Nagayama, M., Power, J., Rosolini, G.: A category theoretic formulation for engeler-style models of the untyped lambda. In: Seda, A.K., Hurley, T., Schellekens, M.P., an Airchinnigh, M.M., Strong, G. (eds.) Proceedings of the Third Irish Conference on the Mathematical Foundations of Computer Science and Information Technology, MFCSIT 2004, Dublin, Ireland, July 22-23, 2004. Electronic Notes in Theoretical Computer Science, vol. 161, pp. 43–57. Elsevier (2004). https://doi.org/10.1016/J.ENTCS.2006.04.024, https://doi.org/10.1016/j.entcs.2006.04.024

25. Hyland, M., Ong, C.L.: On full abstraction for PCF: I, II, and III. Inf. Comput. **163**(2), 285–408 (2000). https://doi.org/10.1006/INCO.2000.2917

26. Ker, A.D., Nickau, H., Ong, C.L.: Innocent game models of untyped lambda-calculus. Theor. Comput. Sci. **272**(1-2), 247–292 (2002). https://doi.org/10.1016/S0304-3975(00)00353-4, https://doi.org/10.1016/S0304-3975(00)00353-4

27. Ker, A.D., Nickau, H., Ong, C.L.: Adapting innocent game models for the böhm tree λ-theory. Theor. Comput. Sci. **308**(1-3), 333–366 (2003). https://doi.org/10.1016/S0304-3975(02)00849-6

28. Koutavas, V., Lin, Y.Y., Tzevelekos, N.: Fully abstract normal form bisimulation for call-by-value pcf. In: 2023 38th Annual ACM/IEEE Symposium on Logic in Computer Science (LICS). pp. 1–13 (2023). https://doi.org/10.1109/LICS56636.2023.10175778

29. Laird, J., Manzonetto, G., McCusker, G., Pagani, M.: Weighted relational models of typed lambda-calculi. In: 28th Annual ACM/IEEE Symposium on Logic in Computer Science, LICS 2013, New Orleans, LA, USA, June 25-28, 2013. pp. 301–310. IEEE Computer Society (2013). https://doi.org/10.1109/LICS.2013.36, https://doi.org/10.1109/LICS.2013.36

30. Lancelot, A., Manzonetto, G., McCusker, G., Vanoni, G.: Interaction improvement (2026), https://arxiv.org/abs/2601.01638

31. Lassen, S.B.: Bisimulation in untyped lambda calculus: Böhm trees and bisimulation up to context. Electronic Notes in Theoretical Computer Science **20**, 346–374 (1999). https://doi.org/10.1016/S1571-0661(04)80083-5

32. Levy, P.B., Staton, S.: Transition systems over games. In: Proceedings of the Joint Meeting of the Twenty-Third EACSL Annual Conference on Computer Science Logic (CSL) and the Twenty-Ninth Annual ACM/IEEE Symposium on Logic in Computer Science (LICS). CSL-LICS '14, Association for Computing Machinery, New York, NY, USA (2014). https://doi.org/10.1145/2603088.2603150, https://doi.org/10.1145/2603088.2603150

33. Morris, J.H.: Lambda-calculus Models of Programming Languages. Ph.D. thesis, Massachusetts Institute of Technology (1968), https://books.google.is/books?id=Dk1AAQAAIAAJ

34. Paolini, L., Piccolo, M., Ronchi Della Rocca, S.: Essential and relational models. Math. Struct. Comput. Sci. **27**(5), 626–650 (2017). https://doi.org/10.1017/S0960129515000316, https://doi.org/10.1017/S0960129515000316

35. Patrignani, M., Ahmed, A., Clarke, D.: Formal approaches to secure compilation: A survey of fully abstract compilation and related work. ACM Comput. Surv. **51**(6), 125:1–125:36 (2019). https://doi.org/10.1145/3280984

36. Pitts, A.M.: Howe's method for higher-order languages. In: Sangiorgi, D., Rutten, J.J.M.M. (eds.) Advanced Topics in Bisimulation and Coinduction, Cambridge tracts in theoretical computer science, vol. 52, pp. 197–232. Cambridge University Press (2012). https://doi.org/10.1017/CBO9780511792588.006

37. Plotkin, G.D.: A powerdomain construction. SIAM J. Comput. **5**(3), 452–487 (1976). https://doi.org/10.1137/0205035, https://doi.org/10.1137/0205035

38. Ronchi Della Rocca, S.: Characterization theorems for a filter lambda model. Inf. Control. **54**(3), 201–216 (1982). https://doi.org/10.1016/S0019-9958(82)80022-3, https://doi.org/10.1016/S0019-9958(82)80022-3

39. Sands, D.: Proving the correctness of recursion-based automatic program transformations. Theor. Comput. Sci. **167**(1&2), 193–233 (1996). https://doi.org/10.1016/0304-3975(96)00074-6

40. Sands, D.: Total correctness by local improvement in the transformation of functional programs. ACM Trans. Program. Lang. Syst. **18**(2), 175234 (mar 1996). https://doi.org/10.1145/227699.227716

41. Sands, D.: Improvement theory and its applications, p. 275306. Cambridge University Press, USA (1999)

42. Scott, D., Strachey, C.: Toward a mathematical semantics for computer languages. In: MRI Symposium Proceedings, vol. 21: Proceedings of the Symposium on Computers and Automata. pp. 19–46. Fox, J. (Ed.). Polytechnic Press, Polytechnic Institute of Brooklyn, New York (1971)
43. Støvring, K., Lassen, S.B.: A complete, co-inductive syntactic theory of sequential control and state. In: Proc. 34th Annual ACM Symposium on Principles of Programming Languages. pp. 161–172. Nice, France (2007), http://doi.acm.org/10.1145/1190215.1190244
44. Wadsworth, C.P.: The Relation Between Computational and Denotational Properties for Scott's $\mathcal{D}_\infty$-Models of the Lambda-Calculus. SIAM J. Comput. 5(3), 488–521 (1976). https://doi.org/10.1137/0205036

From Trees to Tree-Like: Distribution and Synthesis for Asynchronous Automata

Mathieu Lehaut[1] *, Anca Muscholl[2] **†, and
Nir Piterman[3] ★ ★ ★†

[1] University of Warsaw, Poland
[2] LaBRI, University of Bordeaux, France
[3] University of Gothenburg and Chalmers University of Technology, Gothenburg,
Sweden

Abstract. We revisit constructions for distribution and synthesis of
Zielonka's asynchronous automata in restricted settings. We show first a
simple, quadratic, distribution construction for asynchronous automata,
where the process architecture is tree-like. An architecture is tree-like if
there is an underlying spanning tree of the architecture and communications are local on the tree. This quadratic distribution result generalizes
the known construction for tree architectures and improves on an older,
exponential construction for triangulated dependence alphabets. Lastly
we consider the problem of distributed controller synthesis and show that
it is decidable for tree-like architectures. This extends the decidability
boundary from tree architectures to tree-like keeping the same $\text{Tower}_d(n)$
complexity bound, where n is the size of the system and $d \geq 0$ the depth
of the process tree.

Keywords: distributed synthesis, Zielonka automata, language distribution

1 Introduction

We consider a canonical formalism for distributed systems with a fixed communication structure – Zielonka's *asynchronous automata*. These are a well-known
model that supports distributed synthesis under a fixed communication topology, rooted in the theory of Mazurkiewicz traces [21]. In Zielonka automata
processes are connected to a specific set of channels, each process to their own

---

* Supported by Swedish research council (VR) project (No. 2020-04963) and NCN
grant 2021/41/B/ST6/00535.

** Supported by ANR-23-CE48-0005 grant PaVeDyS.

★ ★ ★ Supported by Swedish research council (VR) project (No. 2020-04963) and the Wallenberg AI, Autonomous Systems and Software Program (WASP) funded by the
Knut and Alice Wallenberg Foundation.

† Part of this research was conducted while visiting the Simons Institute for the Theory
of Computing.

N. Bertrand and S. Milius (Eds.): FoSSaCS 2026, LNCS 16503, pp. 396–417, 2026.
https://doi.org/10.1007/978-3-032-22730-0_19

subset of the channels. Processes communicate by synchronizing on channels they are connected to. During communication, all involved processes share their local states and then, based on this mutually shared information, move to new states. In addition, this model is rendez-vous: communication occurs only if all participants (processes connected to the channel) agree to it. The model is asynchronous as communications on two channels that do not have a process in common can be performed in parallel. Highlighting the importance of this model is Zielonka's seminal result about distribution of regular languages. Given a communication architecture – which process is connected to which channels – and a regular language that respects the asynchrony of the architecture, it is always possible to distribute this language into an asynchronous automaton [27].

Zielonka's result is a prominent, and rare, example of distributed synthesis, yet computationally expensive. Starting with a regular language accepted by an automaton with n states and a topology with p processes, an equivalent asynchronous automaton of size $O(4^{p^4} n^{p^2})$ can be constructed [11]. Note that the exponential is only due to the number of processes (which is part of the input), but this is indeed needed, as shown by a lower bound in [11]. The construction is quite involved and, even after nearly 40 years, it is not widely or easily understood. It has been the source for much research on how to understand, explain, and improve it (e.g. [22,13,11,2]). Recently, an alternative construction for asynchronous automata has been proposed using a partial-order variant of Propositional Dynamic Logic [1].

The complexity of the general Zielonka result prompted researchers to look at simpler cases where distribution could be easier. One notable example is the construction of Krishna and Muscholl [16] who show that when the communication architecture is restricted to a tree – every channel is listened to by at most two processes and there are no cycles – then every process needs a quadratic number of states in that of the original sequential automaton. Muscholl and Diekert [6] showed a simpler construction, however exponential in the number of states of the sequential automaton, for triangulated dependence alphabets.

The asynchronous automata model is notable also for its usage in control, similar to Ramadge and Wonham's supervisory control of discrete event systems [25,18]. For systems communicating synchronously by shared variables, it has been established early on that distributed control is undecidable [24], and only decidable for very restricted communication architectures, moreover with very high complexity [17,8]. In contrast, for distributed control based on asynchronous communication using Zielonka automata researchers have found more complex architectures for which the problem was still decidable. Notably, control for ω-regular winning conditions was shown to be decidable when communication is restricted to a tree [23] (see also [12] for local reachability conditions). Their construction has Tower(d) complexity in the depth d of the tree, but works for an architecture that is undecidable in the synchronous shared-variable world. Similarly, for *uniformly well-connected* architectures, the problem is again decidable but this time in exponential space [9]. On a general note, distributed control based on Zielonka automata, also known as asynchronous games with causal memory, is equivalent to so-called Petri games [3]. However, in spite of earlier hopes emanating from these general architectures having a decidable control problem, the general distributed synthesis has recently been established as undecidable [14].

Recently, Hausmann et al. studied distribution in the model of reconfigurable asynchronous automata [15]. Their model generalizes asynchronous automata by allowing processes to change the set of channels they communicate on at runtime. They established that the simple distribution construction of Krishna and Muscholl [16] is applicable also to a more general architecture they call *tree-like*. That is, processes are allowed to communicate more freely as long as their communication has an underlying spanning tree and communications are local on the tree. Here we present their construction in the context of asynchronous automata removing the complex notations related to the reconfiguration. While their construction was exponential in the number of processes in order to follow the structure of the tree, when the communication architecture is fixed this is no longer required and the quadratic construction of Krishna and Muscholl emerges. Furthermore, we show that this construction also improves a previously known exponential construction by Diekert and Muscholl in the context of triangulated dependence graphs [6], as they turn out to be equivalent to tree-like architectures. Finally, we revisit the distributed synthesis result for tree architectures [23] and show that this construction as well can be generalized to the case of tree-like architectures. Omitted proofs can be found in the long version of this paper [19].

For convenience, technical terms and notations in the electronic version of this manuscript are hyper-linked to their definitions (cf. `https://ctan.org/pkg/knowledge`).

2 Preliminaries

2.1 Deterministic Finite Automata

A deterministic finite automaton (DFA) over alphabet Σ is denoted as $\mathcal{A} = (\Sigma, S, \Delta, s_0, F)$, where S is the finite set of states, $\Delta : S \times \Sigma \to S$ the partial transition function, $s_0 \in S$ the initial state, and $F \subseteq S$ the set of accepting states. Given a word $w = a_0 \cdots a_{n-1}$, an initial run of A on w is a path $s_0 \xrightarrow{a_0} s_1 \xrightarrow{a_1} \ldots \xrightarrow{a_{n-1}} s_n$ of $\mathcal{A}$, i.e. for every $0 \le i < n$ we have $s_{i+1} = \Delta(s_i, a_i)$. An initial run is accepting if $s_n \in F$ and then w is accepted by $\mathcal{A}$. The language of $\mathcal{A}$, denoted by $\mathcal{L}(\mathcal{A})$, is the set of words accepted by $\mathcal{A}$. Often we abuse notation and write $\Delta(s, u)$ for the state s' reached from s on the word u (or just $s \xrightarrow{u} s'$).

An *independence relation* is a symmetric, irreflexive relation $I \subseteq \Sigma \times \Sigma$. Two words $u, v \in \Sigma^*$ are said to be I-indistinguishable, denoted by $u \sim_I v$, if one can start from u, repeatedly switch two consecutive independent letters, and end up with v. That is, $\sim_I$ is the transitive closure of the relation $\{(uabv, ubav) \mid (a, b) \in I, u, v \in \Sigma^*\}$. We denote by $[u]_I$ the I-equivalence class of a word $u \in \Sigma^*$. This is a.k.a. a Mazurkiewicz trace [21]. Let $\mathcal{A} = (\Sigma, S, \Delta, s_0, F)$ be a deterministic automaton over Σ. We say that $\mathcal{A}$ is *I-diamond* if for all pairs of independent letters $(a, b) \in I$ and all states $s \in S$, we have $\Delta(s, ab) = \Delta(s, ba)$. If $\mathcal{A}$ has this property, then a word u is accepted by $\mathcal{A}$ if and only if all words in $[u]_I$ are accepted.

2.2 Asynchronous Automata

A *communication architecture* $\mathbb{C} : \Sigma \to 2^{\mathbb{P}}$ associates with each letter the subset of processes reading it. We put $\mathbb{C}^{-1}(p) = \{a \in \Sigma \mid p \in \mathbb{C}(a)\}$. A *distributed alphabet* is $(\Sigma, \mathbb{C})$, where $\mathbb{C}$ is a communication architecture. It induces an independence relation $I(\mathbb{C}) \subseteq \Sigma \times \Sigma$ by $(a, b) \in I(\mathbb{C})$ iff $\mathbb{C}(a) \cap \mathbb{C}(b) = \emptyset$. The complement of the independence relation is a dependence relation $D(\mathbb{C}) = \Sigma \times \Sigma \setminus I(\mathbb{C})$. It is simple to see that $(a, b) \in D(\mathbb{C})$ iff $\mathbb{C}(a) \cap \mathbb{C}(b) \neq \emptyset$. A dependence relation D induces a graph $G_D = (\Sigma, D)$, where Σ is the set of nodes and D is the set of edges. We say that $\mathbb{C}$ is binary if for each $a \in \Sigma$ we have $|\mathbb{C}(a)| \leq 2$. A binary $\mathbb{C}$ induces a tree if the graph $(\mathbb{P}, E)$, where $(p, q) \in E$ iff $\mathbb{C}(a) = \{p, q\}$ with $p \neq q$ for some $a \in \Sigma$, is a tree. For any process $p \in \mathbb{P}$, let $\Sigma_p = \{a \in \Sigma \mid \{p\} = \mathbb{C}(a)\}$ be the set of p-*local actions*, that is the set of actions involving only p.

An *asynchronous automaton* (in short: AA) [27] over a distributed alphabet $(\Sigma, \mathbb{C})$ and processes $\mathbb{P}$ is a tuple $\mathcal{B} = ((S_p)_{p \in \mathbb{P}}, (s_p^0)_{p \in \mathbb{P}}, (\delta_a)_{a \in \Sigma}, Acc)$ such that:

- S_p is the finite set of states for process p, and $s_p^0 \in S_p$ is its initial state,
- $\delta_a : \prod_{p \in \mathbb{C}(a)} S_p \to \prod_{p \in \mathbb{C}(a)} S_p$ is a partial transition function for letter a that only depends on the states of processes in $\mathbb{C}(a)$ and changes them,
- $Acc \subseteq \prod_{p \in \mathbb{P}} S_p$ is a set of accepting states.

A global state of $\mathcal{B}$ is $\mathbf{s} = (s_p)_{p \in \mathbb{P}}$, giving the state of each process. For a global state $\mathbf{s}$ and a subset $P \subseteq \mathbb{P}$, we denote by $\mathbf{s} \downarrow_P = (s_p)_{p \in P}$ the part of $\mathbf{s}$ consisting of states from processes in P.

An initial run of $\mathcal{B}$ on a word $a_1 a_2 \dots a_n$ is a path $\mathbf{s}_0 \xrightarrow{a_1} \mathbf{s}_1 \xrightarrow{a_2} \dots \xrightarrow{a_n} \mathbf{s}_n$ of the product (global) automaton from the initial state $\mathbf{s}_0 = (s_p^0)_{p \in \mathbb{P}}$: for all $0 < i \leq n$, $\mathbf{s}_i \in \prod_{p \in \mathbb{P}} S_p$, $a_i \in \Sigma$, satisfying $\mathbf{s}_i \downarrow_{\mathbb{C}(a_i)} = \delta_{a_i}(\mathbf{s}_{i-1} \downarrow_{\mathbb{C}(a_i)})$ and $\mathbf{s}_i \downarrow_{\mathbb{P} \setminus \mathbb{C}(a_i)} = \mathbf{s}_{i-1} \downarrow_{\mathbb{P} \setminus \mathbb{C}(a_i)}$. An initial run is accepting if $\mathbf{s}_n$ belongs to Acc. The word $a_1 a_2 \dots a_n$ is accepted by $\mathcal{B}$ if such an accepting run exists (note that automata are deterministic but runs on certain words may not exist). The language of $\mathcal{B}$, denoted by $\mathcal{L}(\mathcal{B})$, is the set of words accepted by $\mathcal{B}$. Let $\mathrm{Runs}(\mathcal{B})$ denote the set of runs of $\mathcal{B}$, and $\mathrm{Runs}_p(\mathcal{B})$ their projection on the p-component of the state.

We say that a language $\mathcal{L} \subseteq \Sigma^*$ over $(\Sigma, \mathbb{C})$ is *distributively recognized* if there exists an asynchronous automaton $\mathcal{B}$ such that $\mathcal{L}(\mathcal{B}) = \mathcal{L}$.

Theorem 1 ([27,11,16]). *Given an $I(\mathbb{C})$-diamond deterministic automaton $\mathcal{A}$, there exists an AA $\mathcal{B}$ that distributively recognizes the language of $\mathcal{A}$. In general, if $\mathcal{A}$ has n states then every process of $\mathcal{B}$ has $O(4^{|\mathbb{P}|^4} n^{|\mathbb{P}|^2})$ states. If $\mathbb{C}$ induces a tree, then every process of $\mathcal{B}$ has $O(n^2)$ states.*

The size of of an AA $\mathcal{B}$ (written as $|\mathcal{B}|$) is defined as $\max_{p \in \mathbb{P}} |S_p|$. This standard notion of size takes into account the maximal local memory (state space) of a process, because an AA is a *distributed* device.

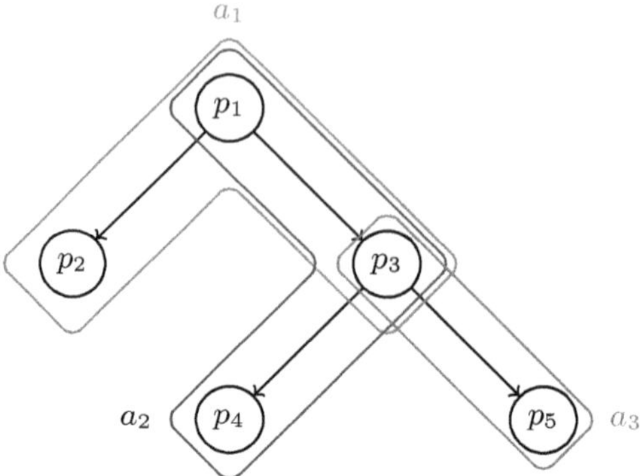

Fig. 1. A tree-like communication architecture: the tree is given by the black edges, while the communication architecture is drawn with one color group for each letter.

2.3 Tree-like communication architectures

As before, let $\mathbb{P}$ be a set of processes and Σ an alphabet, both finite. A communication architecture is tree-like if there is a tree that spans letters that synchronize more than two processes. Formally, we have the following.

Let $|\mathbb{P}| = n$. A *tree* over $\mathbb{P}$ is a $T = (\mathbb{P}, r, E)$ where $\mathbb{P}$ is the set of nodes, of which $r \in \mathbb{P}$ is the *root*, and $E \subseteq \mathbb{P}^2$ is the set of *edges* that is cycle free and connected.

Definition 1. *We say that* $\mathfrak{A} = (\mathbb{C}, T)$, *where* $\mathbb{C}$ *is a communication architecture and* T *is a tree, is a* tree-like *communication architecture (short: TCA) if*

1. *for all letters* $a \in \Sigma$, *the set* $\mathbb{C}(a)$ *is connected in* T, *i.e. if* $p, q \in \mathbb{C}(a)$, *then all processes along the path from* p *to* q *in* T *are also in* $\mathbb{C}(a)$, *and*
2. *if* $(p, q) \in E$, *then there is a letter* $a \in \Sigma$ *such that* $p, q \in \mathbb{C}(a)$.

Condition 1 ensures that a communication on a given letter is always "local" in the tree. Condition 2 ensures that the tree does not consist of disconnected parts. In case that condition 2 does not hold we say that $\mathfrak{A}$ is *forest-like*.

Example 1. Fix sets $\mathbb{P} = \{p_1, \ldots, p_5\}$ of processes and $\Sigma = \{a_1, a_2, a_3\}$ of letters. Then $\mathfrak{A} = ((\mathbb{C}(a_1), \mathbb{C}(a_2), \mathbb{C}(a_3)), T)$ as given in Figure 1 is a tree-like communication architecture rooted in p_1:

1. Every letter is connected in T.
2. All edges in T are covered by at least one letter.

Notions of parents, children, descendants, leaves, neighbors, and subtrees are defined as usual.

3 Asynchronous Automata with Tree-like Architectures

Hausmann et al. [15] showed that *reconfigurable asynchronous automata* can be distributed when they communicate over tree-like architectures. Reconfigurable asynchronous automata change their communication architectures during the operation. Thus, they are more general than asynchronous automata. Here, we present their construction for the case of a *fixed* tree-like architecture. Equivalently, we show that the construction of Krishna and Muscholl [16] for tree architectures extends, with more complicated transitions, to tree-like architectures.

We extend the definition of independence to sequences and to sets of letters. First we set $C_1, C_2 \subseteq \Sigma$ as independent (and write $(C_1, C_2) \in I(\mathbb{C})$) if for every $a_1 \in C_1$ and $a_2 \in C_2$ we have $(a_1, a_2) \in I(\mathbb{C})$. Equivalently, $C_1 \times C_2 \subseteq I(\mathbb{C})$. We set $w_1, w_2 \in \Sigma^*$ as independent (and write $(w_1, w_2) \in I(\mathbb{C})$) if there exist sets $C_1, C_2 \subseteq \Sigma$ such that $w_1 \in C_1^*$, $w_2 \in C_2^*$, and $(C_1, C_2) \in I(\mathbb{C})$.

Fix a DFA $\mathcal{A} = (\Sigma, S, \Delta, s_0, F)$ and $\mathfrak{A} = (\mathbb{C}, T)$ such that $\mathcal{A}$ is $I(\mathbb{C})$-diamond. If w_1 and w_2 are two $I(\mathbb{C})$-independent words, then we have that $\Delta(s, w_1 w_2) = \Delta(s, w_2 w_1) = \Delta(s, w)$ for any w that is an interleaving of w_1 and w_2. This can be verified by induction on the length of w.

We now study what information is needed in order to compute $\Delta(s, w_1 w_2)$ without memorizing w_1 and w_2 themselves.

Fix a state s of $\mathcal{A}$ and let w_1 and w_2 be two $I(\mathbb{C})$-independent words. Let $s_1 = \Delta(s, w_1)$, $s_2 = \Delta(s, w_2)$, and let C_1, C_2 be sets such that $w_1 \in C_1^*$, $w_2 \in C_2^*$ and $(C_1, C_2) \in I(\mathbb{C})$. The next lemma shows that the state $\Delta(s, w_1 w_2)$ can be computed knowing only s, s_1, s_2, and C_2 (but not the exact words w_1, w_2).

Lemma 1 ([5]). *Let $\mathcal{A}$ be an $I(\mathbb{C})$-diamond DFA. There exists a (partially-defined) function* Diam $: S^3 \times 2^\Sigma \to S$ *such that if* $s, s_1, w_1, s_2, w_2, C_2$ *are all defined as described above, then* $\mathrm{Diam}(s, s_1, s_2, C_2) = \Delta(s, w_1 w_2)$.

Proof. Let C_1' be the maximal set of letters such that $(C_1', C_2) \in I(\mathbb{C})$. It follows that $C_1 \subseteq C_1'$. Consider some arbitrary words w_1' and w_2' such that $w_1' \in C_1'^*$, $w_2' \in C_2^*$ and such that $s_i = \Delta(s, w_i) = \Delta(s, w_i')$ for $i \in \{1, 2\}$. Let us show that $\Delta(s, w_1' w_2') = \Delta(s, w_1 w_2)$.

First, by diamond property and hypothesis we have that $s' := \Delta(s, w_1 w_2) = \Delta(s, w_2 w_1) = \Delta(s_2, w_1)$. We also know that $\Delta(s, w_2') = s_2$, so $s' = \Delta(s, w_2' w_1)$. Then, reusing the diamond property: $s' = \Delta(s, w_2' w_1) = \Delta(s, w_1 w_2')$. We conclude reusing the hypothesis: $s' = \Delta(s, w_1 w_2') = \Delta(s, w_1' w_2')$.

Finally we can set $\mathrm{Diam}(s, s_1, s_2, C_2)$ as the state $\Delta(s, w_1' w_2')$, for some $w_1' \in C_1'^*, w_2' \in C_2^*$. $\square$

In [16], the Diam function has been used for TCA $\mathfrak{A} = (\mathbb{C}, T)$ with binary $\mathbb{C}$ and $I(\mathbb{C})$-diamond DFA $\mathcal{A}$ in the following way. On a communication between a parent and one of its children, let s be the most recent state of $\mathcal{A}$ known by both the parent and the child, s_1 and s_2 be the most recent states known by the parent and the child respectively, and C_2 the set of letters with domains in the subtree rooted in the child. Then one can combine the information known by the parent and the child after the communication by computing $\mathrm{Diam}(s, s_1, s_2, C_2)$. We do not provide more details because we show below the more general case of non-binary $\mathbb{C}$.

In our tree-like setting, communications can include more than two participants. To that end, we naturally extend the Diam function to work on subtrees. Let T be a tree where each node is labeled by a pair (s, t) of states and a set $C^{\downarrow}$ of letters, with the intuition that s is the most recent state of $\mathcal{A}$ known by both the process at this node and its parent, t is the most recent state of $\mathcal{A}$ known by this process, and $C^{\downarrow}$ is the set of letters that can be found in the subtree rooted in that node and that are independent from those shared with the parent. Let r be the root of T, labeled by $(s_r, t_r, C_r^{\downarrow})$, and assume that r has k children labeled $(s_i, t_i, C_i^{\downarrow})$ and leading to tree T_i for $i \in \{1, \ldots, k\}$. Then we define the function TDiam recursively in the following way:

$$
\text{TDiam}(T) = \begin{cases} t_r & \text{if } k = 0, \\ \text{Diam}(s_k, \text{TDiam}(T \setminus T_k), \text{TDiam}(T_k), C_k^{\downarrow}) & \text{otherwise.} \end{cases}
$$

Note that the state(p) function from [16] is equivalent to $\text{TDiam}(T_p)$ where T_p is the tree rooted in p.

Consider a letter $a \in \Sigma$, where $\mathbb{C}(a) = \{p_1, \ldots, p_k\}$. We define the *tree of a*, denoted by T_a, as the subtree of T whose set of nodes is exactly $\mathbb{C}(a)$; by definition of a TCA this is indeed a tree. Consider now a second letter $b \neq a$ and some p such that $p \in \mathbb{C}(a)$ but $p \notin \mathbb{C}(b)$. Again by definition of a TCA, $\mathbb{C}(b)$ occupies a connected area of T, and this area does not contain p. Therefore, it must either lie "above" p, meaning that the (unique) path from p to this area goes through the parent of p, or "below" p, meaning that it goes through one of p's children. We define $C_p^{\downarrow}$ to be the set of letters for which the second case holds. Now assume that each $p_i \in \mathbb{C}(a)$ is associated with a pair of states (s_i, t_i) of $\mathcal{A}$. Then we label each node p_i of T_a by $(s_i, t_i, C_{p_i}^{\downarrow})$ and define the state $s_a = \text{TDiam}(T_a)$. Intuitively, that means we combine the information of all nodes in T_a to compute the most up-to-date state s_a.

We are now ready to state our construction. We distribute $\mathcal{A} = (\Sigma, S, \Delta, s_0, F)$ by defining $\mathcal{B} = ((S_p)_{p \in \mathbb{P}}, (s_p^0)_{p \in \mathbb{P}}, (\delta_a)_{a \in \Sigma}, Acc)$. Namely, for each $p \in \mathbb{P}$ we define the set S_p and its initial state s_p^0, the transition function δ_a for each $a \in \Sigma$ and the acceptance set Acc. For each $p \in \mathbb{P}$ we set $S_p = S \times S$ and $s_p^0 = (s_0, s_0)$.

For $a \in \Sigma$ we describe now the transition $\delta_a : \prod_{p \in \mathbb{C}(a)} S_p \to \prod_{p \in \mathbb{C}(a)} S_p$. The transition computes the diamond closure over the states held by all other processes in the correct order. Based on this computation each process updates its state.

Let $\mathbb{C}(a) = \{p_1, \ldots, p_k\}$ and for all $1 \leq i \leq k$ let (s_i, t_i) be the state of p_i. Let $s_a = \text{TDiam}(T_a)$ as described earlier, and let $s' = \Delta(s_a, a)$. We set

$$
\delta_a((s_1, t_1), \ldots, (s_k, t_k)) = ((u_1, v_1), \ldots, (u_k, v_k))
$$

where the new state of process p_i is $u_i = \text{IS-ROOT}_a^{p_i}?s_i:s'$ and $v_i = s'$ with

$$
\text{IS-ROOT}_a^p?s:t = \begin{cases} s & \text{if } p \text{ is the root of } T_a, \\ t & \text{otherwise.} \end{cases}
$$

If $\Delta(s_a, a)$ is undefined, then so is δ_a.

We now define Acc. Consider a global state $\mathbf{s} = ((s_1, t_1), \ldots (s_n, t_n))$. Recall that the TCA is $\mathfrak{A} = (\mathbb{C}, T)$. For the tree T, consider the labeling of every node i in T by the pair (s_i, t_i) and the set $C_i^{\downarrow}$ of letters found "below" node i, as defined earlier. We then set $\mathbf{s} \in Acc$ if and only if $\mathrm{TDiam}(T) \in F$.

We state our main result below.

Theorem 2. *If $\mathcal{A}$ is $\mathbb{C}$-diamond and $\mathbb{C}$ is tree-like, then $\mathcal{B}$ recognizes distributively $\mathcal{L}(\mathcal{A})$. The size of $\mathcal{B}$ is in $O(n^2)$, where n is the number of states of $\mathcal{A}$.*

3.1 Proof of correctness

This section is dedicated to the proof of Theorem 2.

First, and completely independently from the construction, we need to define what is the view of a (set of) process(es), as in [16], and then show a useful property relating TDiam and those views. Let $\mathcal{A}$ be a $\mathbb{C}$-diamond DFA, $w \in \Sigma^*$ a word, $\rho = s_0 \xrightarrow{a_1} \ldots \xrightarrow{a_k} s_k$ the initial run of $\mathcal{A}$ on $w = a_1 \ldots a_k$, and $X \subseteq \mathbb{P}$ a set of processes. As processes in X are not necessarily part of every communication in w, there may be some parts of ρ that are happening outside of X's knowledge and processes in X may not know the state reached at the end of ρ just by themselves. We define recursively the *view* of X on w, which intuitively corresponds to the subword of w that processes in X can see through shared actions, and denote it by $\mathrm{view}_X(w)$:

$$\mathrm{view}_X(\varepsilon) = \varepsilon$$

$$\mathrm{view}_X(w \cdot a_k) = \begin{cases} \mathrm{view}_{X \cup \mathbb{C}(a_k)}(w) \cdot a_k & \text{if } X \cap \mathbb{C}(a_k) \neq \emptyset, \\ \mathrm{view}_X(w) & \text{otherwise.} \end{cases}$$

We compute what is also known as the causal past of processes in X: We start from the last communication in which at least one process in X participated. Then, the $\mathrm{view}_X(w)$ computes backwards the subsequence of w seen by processes in X through shared actions. Abusing notations, we identify any tree $T = (\mathbb{P}, r, E)$ with its set of processes $\mathbb{P}$ and write $\mathrm{view}_T(w)$ instead of $\mathrm{view}_{\mathbb{P}}(w)$. We also define the state of $\mathcal{A}$ that those processes can compute as $\sigma_X(w) = \Delta(s_0, \mathrm{view}_X(w))$. Furthermore, with $\mathcal{A}$ being $\mathbb{C}$-diamond, we can deduce that for any letter a, $\Delta(s_k, a)$ is defined if and only if $\Delta(s_0, \mathrm{view}_{\mathbb{C}(a)}(w)a)$ is defined. That is, whether a communication with letter a can happen cannot depend on communications made outside of the view of those listening to a.

Let $p \in \mathbb{P}$ be some process with parent q in T. The *shared parent view* of p on w, denoted by $\mathrm{view}_p^{\uparrow}(w)$, is defined as $\mathrm{view}_{\{p\}}(w')$ where w' is the minimal prefix of w containing all occurrences of all letters b such that $\{p, q\} \subseteq \mathbb{C}(b)$. In other words this is the sequence of actions, as seen from p's point of view, until the last communication involving both p and its parent. This is not necessarily the same as $\mathrm{view}_{\{p\}}(w)$, because p may have communicated with its children after the end of w'. For similar reasons, $\mathrm{view}_p^{\uparrow}(w)$ is not necessarily the same as $\mathrm{view}_{\{q\}}(w)$. If p is the root of T, we set $\mathrm{view}_p^{\uparrow}(w) = \varepsilon$ for all w. Finally, we set $\sigma_p^\uparrow(w) = \Delta(s_0, \mathrm{view}_p^\uparrow(w))$ to be the state reached in $\mathcal{A}$ after reading that sequence.

The Diam function, and by extension TDiam, interplay nicely with the views of independent parts of the system. More specifically, we show that if each node in T is correctly labeled, then $\mathrm{TDiam}(T)$ correctly computes the view according to every process in T.

Lemma 2. *For all words w and for all subtrees T' of T where each node p of T' is labeled by $(\sigma_p^\uparrow(w), \sigma_{\{p\}}(w), C_p^\downarrow)$, we have $\mathrm{TDiam}(T') = \sigma_{T'}(w)$.*

Proof. By induction on the structure of T', same as in [16].

If T' is a single node p labeled by $(s_p, t_p, C_p^\downarrow)$ then $\mathrm{TDiam}(T') = t_p = \sigma_{\{p\}}(w)$ by assumption. Otherwise, assume T' is made of a root p labeled by $(s_p, t_p, C_p^\downarrow)$ with k children $p_1, \ldots, p_k$ labeled by $(s_1, t_1, C_1^\downarrow), \ldots, (s_k, t_k, C_k^\downarrow)$ and leading to subtrees $T_1, \ldots, T_k$, respectively. Assume that the property holds on those subtrees, i.e. that for all $i \leq k$, we have $\mathrm{TDiam}(T_i) = \sigma_{T_i}(w)$. Then with T_i' denoting the subtree of T' composed of p, its first i children, and their respective descendants, we recursively show that

$$(P_i) \qquad\qquad \mathrm{TDiam}(T_i') = \sigma_{T_i'}(w)$$

For the initial step $i = 0$, $\mathrm{TDiam}(T_0') = t_p = \sigma_{\{p\}}(w)$ by assumption. Now suppose (P_{i-1}) holds for $1 \leq i \leq k$. By construction

$$\mathrm{TDiam}(T_i') = \mathrm{Diam}(s_i, \mathrm{TDiam}(T_{i-1}'), \mathrm{TDiam}(T_i), C_i^\downarrow).$$

By (P_{i-1}) we know that $\mathrm{TDiam}(T_{i-1}') = \sigma_{T_{i-1}'}(w)$, and by assumption we have that $s_i = \sigma_{p_i}^\uparrow(w)$ and $\mathrm{TDiam}(T_i) = \sigma_{T_i}(w)$.

Let $w_0 = \mathrm{view}_{p_i}^\uparrow(w)$, $w_i = \mathrm{view}_{T_i}(w)$, and $w_{i-1} = \mathrm{view}_{T_{i-1}'}(w)$. By definition, w_0 is a prefix of both w_i and w_{i-1}. Thus, let $w_i = w_0 \cdot w_i'$ and $w_{i-1} = w_0 \cdot w_{i-1}'$. As w_0 contains the last communication involving p_i and its parent p, it is easy to see that w_i' and w_{i-1}' are independent, otherwise w_0 would not be their longest common prefix. Note also that $\mathrm{view}_{T_i'}(w) = w_0 v$, where v is an interleaving of w_i' and w_{i-1}'. Also, all actions of w_i' are in $C_{p_i}^\downarrow$, otherwise they would involve p. Using the definition of TDiam, we get $\mathrm{TDiam}(T_i') = \mathrm{Diam}(s_i, \sigma_{T_{i-1}'}(w), \sigma_{T_i}(w), C_i^\downarrow)$ with $s_i = \sigma_{p_i}^\uparrow(w) = \Delta(s_0, w_0)$, $\sigma_{T_{i-1}'}(w) = \Delta(s_i, w_{i-1}')$ and $\sigma_{T_i}(w) = \Delta(s_i, w_i')$. Thus, by Lemma 1, $\mathrm{TDiam}(T_i') = \sigma_{T_i'}(w)$.

We have shown that (P_i) holds for all i, and therefore for $i = k$, and as $T' = T_k'$ we have $\mathrm{TDiam}(T') = \sigma_{T'}(w)$. $\qquad\qquad\square$

Note that this lemma implies that if T' is the full tree T and is correctly labeled, then $\mathrm{TDiam}(T) = \sigma_{\mathbb{P}}(w) = \Delta(s_0, w)$. We will use this property to prove the correctness of our construction.

Let us define invariants that should be satisfied throughout a run. Let $w \in \Sigma^*$. Our first invariant states that a run ρ' of $\mathcal{B}$ on w exists if and only if a run ρ of $\mathcal{A}$ on the same word w exists.

$$(\mathbf{I}_{\mathrm{DEF}}[w]) : \text{The run } \rho \text{ on } w \text{ is defined} \Leftrightarrow \text{The run } \rho' \text{ on } w \text{ is defined.}$$

Now if both runs are undefined, then w is trivially rejected by both $\mathcal{A}$ and $\mathcal{B}$ and there is nothing left to show. So for the remaining invariants, assume that both ρ and ρ' are defined and end in states s and $\mathbf{s} = ((s_1, t_1), \ldots, (s_n, t_n))$ respectively. Then we establish the invariants relating the actual state of ρ with the pair of states that each process keeps in its local state, which corresponds to the intuition given in the previous section.

$$(\mathtt{I}_S^\uparrow[w]): \quad \forall p \in \mathbb{P}.\, s_p = \sigma_p^\uparrow(w) \qquad (\mathtt{I}_S[w]): \quad \forall p \in \mathbb{P}.\, t_p = \sigma_{\{p\}}(w)$$

Those two invariants, used with Lemma 2, are key to prove that the languages of $\mathcal{A}$ and $\mathcal{B}$ are equivalent. Let us now prove that those invariants hold inductively.

Initialization of the invariants. All invariants hold on the empty word $w = \varepsilon$. $(\mathtt{I}_{\mathrm{DEF}}[\varepsilon])$ is trivial. $(\mathtt{I}_S^\uparrow[\varepsilon]), (\mathtt{I}_S[\varepsilon])$ are easily derived from the definition of the initial state $\mathbf{s}_0$ of $\mathcal{B}$.

Invariants after a transition. Now consider a word of the form $w' = w \cdot a$, with both runs ρ and ρ' defined up to w ending in states s and $\mathbf{s}$ respectively. Assume that all invariants hold in w. We show that they still hold in w'.

For $(\mathtt{I}_{\mathrm{DEF}}[w'])$, by using $(\mathtt{I}_S^\uparrow[w])$, $(\mathtt{I}_S[w])$, and Lemma 2, we have that $s' := \mathrm{TDiam}(T_a) = \sigma_{\mathbb{C}(a)}(w)$. Therefore, $\Delta(s, a)$ is defined if and only if $\Delta(s', a)$ is defined. By definition of $\mathcal{B}$, there is an a transition from $\mathbf{s}$ iff $\Delta(s', a)$ is defined iff $\Delta(s, a)$ is defined iff there is an a transition in $\mathcal{A}$. Thus $(\mathtt{I}_{\mathrm{DEF}}[w'])$ holds.

Now assume both runs are defined for w' and let $s \xrightarrow{a} s'$ in $\mathcal{A}$ and $\mathbf{s} \xrightarrow{a} \mathbf{s}'$ in $\mathcal{B}$. Let $p \in \mathbb{P}$, its state in $\mathbf{s}$ be (s_p, t_p), and its state in $\mathbf{s}'$ be (s_p', t_p'). If p was not part of the communication on a, then its state is unchanged and by definition $\mathrm{view}_p^\uparrow(w') = \mathrm{view}_p^\uparrow(w)$ and $\mathrm{view}_{\{p\}}(w') = \mathrm{view}_{\{p\}}(w)$, so $(\mathtt{I}_S^\uparrow[w'])$ and $(\mathtt{I}_S[w'])$ are trivially obtained from $(\mathtt{I}_S^\uparrow[w])$ and $(\mathtt{I}_S[w])$ respectively.

Assume now that p was part of this communication. Let us first show $(\mathtt{I}_S[w'])$. By definition, $\mathrm{view}_{\{p\}}(w') = \mathrm{view}_{\mathbb{C}(a)}(w) \cdot a$. Again, by combining $(\mathtt{I}_S[w])$, $(\mathtt{I}_S^\uparrow[w])$, and Lemma 2, we get that $\sigma_{\mathbb{C}(a)}(w) = \mathrm{TDiam}(T_a) = s'$. Thus $\sigma_{\{p\}}(w') = \Delta(s', a) = t_p'$ and $(\mathtt{I}_S[w'])$ holds. For $(\mathtt{I}_S^\uparrow[w'])$, there are two cases. First, assume that the parent of p in T is not part of the communication on a (either because p is the root and has no parent, or because its parent does not listen to a). In that case, $\mathrm{view}_p^\uparrow(w') = \mathrm{view}_p^\uparrow(w)$ by definition and $\sigma_p^\uparrow(w) = s_p$ by $(\mathtt{I}_S^\uparrow[w])$. Moreover, p must be the root of T_a. Therefore $s_p' = \text{IS-ROOT}_a^p? s_p : \Delta(s_a, a) = s_p$ and $(\mathtt{I}_S^\uparrow[w'])$ is satisfied. Second, assume now that p's parent, say q, is part of the communication on a. Then by definition $\mathrm{view}_p^\uparrow(w') = \mathrm{view}_{\{p\}}(w') = \mathrm{view}_{\mathbb{C}(a)}(w) \cdot a$. Thus, $\sigma_p^\uparrow(w') = \Delta(s', a)$ by the same proof as for $(\mathtt{I}_S[w'])$. Thus $s_p' = \text{IS-ROOT}_a^p? s_p : \Delta(s', a) = \Delta(s', a)$ and we have $(\mathtt{I}_S^\uparrow[w'])$.

Conclusion of the proof. We have shown that the invariants hold for any word w. We use them to show that $\mathcal{L}(\mathcal{A}) = \mathcal{L}(\mathcal{B})$. Let $w \in \Sigma^*$ be a word, then w is accepted by $\mathcal{A}$ iff there is a run of $\mathcal{A}$ on w ending in a state $s \in F$. By $(\mathtt{I}_{\mathrm{DEF}}[w])$, the run of $\mathcal{A}$ on w is defined iff the run of $\mathcal{B}$ on w is defined. If both runs are undefined, then w is neither in $\mathcal{L}(\mathcal{A})$ nor $\mathcal{L}(\mathcal{B})$. Otherwise, by $(\mathtt{I}_S^\uparrow[w])$, $(\mathtt{I}_S[w])$, and Lemma 2, we have that $\text{TDiam}(T) = \text{view}_{\mathbb{P}}(w) = s$. Then $\mathcal{B}$ accepts w iff $\text{TDiam}(T) = s \in F$ iff $\mathcal{A}$ accepts w. Therefore $\mathcal{L}(\mathcal{A}) = \mathcal{L}(\mathcal{B})$, which concludes the proof of Theorem 2.

3.2 Triangulated dependence alphabets

The complexity of Zielonka's general distribution construction stems from the need to store information regarding knowledge about other processes and a time-stamping mechanism that allows to put together these pieces of knowledge [27]. Diekert and Muscholl have shown that in the case that the dependence relation between the letters in Σ is a *triangulated graph*, this complex structure of information can be avoided [6,7].

We observe that tree-like architectures give a fresh view on Diekert and Muscholl's construction, and an AA construction that is exponentially better.

Consider a graph $G = (V, E)$. A sequence of vertices $v_1, \ldots, v_n$ is a cycle if for every i we have $(v_i, v_{i+1}) \in E$ and $(v_n, v_1) \in E$. A graph $G = (V, E)$ is *triangulated* if for every cycle $v_1, \ldots, v_n$ such that $n > 3$ there exists a pair $(v_i, v_j) \in E$ such that $1 < |j - i| < n - 1$. That is, the cycle must have some chord.

Consider an architecture $\mathbb{C}$ and let $D = D(\mathbb{C})$. It turns out that an architecture is tree-like if and only if the dependency graph $G = (\Sigma, D)$ is triangulated and connected. It follows that if a dependency graph is triangulated but not connected then its architecture is forest-like.

Lemma 3 ([10]). *An architecture $\mathbb{C}$ is tree-like iff $G_D = (\Sigma, D)$ is triangulated and connected.*

In Gavril's terms, given a tree, every graph of subtrees of the tree, where edges correspond to non-empty intersection of the subtrees, is triangulated. In the other direction, every triangulated graph can be represented as the graph of intersections of subtrees of an appropriate tree. The part about connectivity is simple to add.

Note first that the construction of AA for non-connected dependency graphs easily reduces to the construction for connected dependency graphs. To see this assume that Σ is the disjoint union of two pairwise independent alphabets Σ_1, Σ_2, so $(a, b) \notin D$ for every $a \in \Sigma_1, b \in \Sigma_2$. Consider some I-diamond DFA $\mathcal{A}$ over Σ and an input word $w \in \Sigma^*$. We can use the diamond property (Lemma 1) to infer from the states reached by $\mathcal{A}$ on the projection w_1 respectively w_2 of the input w on Σ_1 and Σ_2, resp., the state reached by $\mathcal{A}$ on $w \sim_I w_1 w_2$. More formally, assume that we constructed an AA $\mathcal{B}_1$ over alphabet Σ_1 and an AA $\mathcal{B}_2$ over alphabet Σ_2. For each of $\mathcal{B}_1, \mathcal{B}_2$ we can suppose that the global state reached on the respective input determines the state of $\mathcal{A}$ reached on that word. We obtain an AA $\mathcal{B}$ by composing $\mathcal{B}_1, \mathcal{B}_2$ in parallel (because they share no letter). The global state reached by $\mathcal{B}$ on $w \sim_I w_1 w_2$ determines the two states $s_1 = \Delta(s_0, w_1), s_2 = \Delta(s_0, w_2)$ of $\mathcal{A}$, and $\text{Diam}(s_0, s_1, s_2, \Sigma_2)$ says which global states are accepting.

The result we improve on in Theorem 2 is stated in the next theorem. The exponential size comes from the fact that the construction relies on transition monoids.

Theorem 3 ([5,6]). *Given an I-diamond DFA $\mathcal{A}$ over a triangulated dependency graph $G = (\Sigma, D)$, an AA $\mathcal{S}$ that recognizes the same language can be constructed with $|\mathcal{S}| = n!$, where $n = |\mathcal{A}|$.*

4   Distributed Synthesis

We extend now the result of decidability of the control problem for asynchronous automata from binary-tree architectures [23] to tree-like architectures. We first recall the definitions and adapt them to our notations.

4.1   Controllers and Control Problem

Let us fix some distributed alphabet $(\Sigma, \mathbb{C})$. As in the setting of [25], Σ is partitioned into System (*controllable*) actions Σ_{sys} and Environment (*uncontrollable*) actions Σ_{env}. Remember that Σ_p denotes the set of p-local actions, then let $\Sigma_p^{sys} = \Sigma_p \cap \Sigma_{sys}$ and similarly for Σ_p^{env}. As in [23] we require that all communication actions, i.e., actions shared by at least 2 processes, are uncontrollable, in other words all controllable actions are local: $\Sigma_{sys} \subseteq \cup_{p\in\mathbb{P}}\Sigma_p$. This is not a restriction, as the general setting where communication actions can be controllable reduces to this one, as shown in Proposition 6 of [23]. Furthermore, there should be at least one controllable (thus local) action from each state. This is of course not a restriction because one can always add local, controllable self-loops when needed. This assumption will make the construction a bit simpler.

For control it is common to use local acceptance conditions, instead of global ones, because more systems are controllable with local conditions. That is, we replace the *Acc* part of the definition of AA by a set $\{Acc_p\}_{p\in\mathbb{P}}$. Each Acc_p is of the form (F_p, Ω_p) with $F_p \subseteq S_p$ and $\Omega_p : S_p \to \mathbb{N}$ a local parity function. The idea is that a finite run is accepted by p if it ends in a state in F_p, and an infinite run is accepted by p if it satisfies the (max) parity condition for p. Then a run is accepted by $\mathcal{A}$ if it is accepted by all processes p.

We say that a run over w is *maximal* if there is no way to decompose w into $w = u \cdot v$ and no $a \in \Sigma$ with $\mathbb{C}(a) \cap \mathbb{C}(v) = \emptyset$ such that the run over $u \cdot a \cdot v$ is defined. In plain words, it means we cannot extend the run on w by some action involving processes that perform only a finite number of actions in w.

Now we define the notion of controllers. For the rest of this section, let us fix some AA $\mathcal{A} = ((S_p)_{p\in\mathbb{P}}, (s_p^0)_{p\in\mathbb{P}}, (\delta_a)_{a\in\Sigma}, (Acc_p)_{p\in\mathbb{P}})$. For simplicity we will not use the general definition of controllers but go directly to what is actually called covering controllers for $\mathcal{A}$ in [23], and simply call them controllers. It is shown in Lemma 10 of [23] that the Control Problem for covering controllers is equivalent to the one for general controllers, and so we focus only on the former.

The intuition for a controller is that it is a machine (described by an AA without acceptance condition) guiding the System in choosing which controllable action(s) to follow in order to get only runs accepted by $\mathcal{A}$. On the other hand, as Environment actions are uncontrollable, the controller should not be able to restrict them from happening. To do that, a controller can enrich states of $\mathcal{A}$ by adding extra information and use this to choose a subset of controllable actions that are possible from this state, and enable only this subset of actions.

Formally, a *controller* for $\mathcal{A}$ is itself an AA $\mathcal{C} = ((C_p)_{p \in \mathbb{P}}, (c_p^0)_{p \in \mathbb{P}}, (\Delta_a)_{a \in \Sigma})$ together with a projection π that maps each state $c_p \in C_p$ of $\mathcal{C}$ to a state $s_p \in S_p$ of $\mathcal{A}$ and that satisfies the following requirements:

- for all $a \in \Sigma$ with $\mathbb{C}(a) = \{p_1, \ldots, p_k\}$, and $\mathbf{c}, \mathbf{c}' \in C_{p_1} \times \cdots \times C_{p_k}$, we have that $\Delta_a(\mathbf{c}) = \mathbf{c}'$ implies $\delta_a(\pi(\mathbf{c})) = \pi(\mathbf{c}')$ (with π applied component-wise),
- for all $a \in \Sigma_{env}$, if $\delta_a(\pi(\mathbf{c}))$ is defined then so is $\Delta_a(\mathbf{c})$,
- for all $p \in \mathbb{P}$, $\pi(c_p^0) = s_p^0$.

Thanks to the projection π, a controller $\mathcal{C}$ inherits the acceptance condition of $\mathcal{A}$. A controller $\mathcal{C}$ is *winning* for $\mathcal{A}$ if every maximal run of $\mathcal{C}$ is accepting.

The *Control Problem* asks, for a given input $\mathcal{A}$, whether there exists a winning controller $\mathcal{C}$ for $\mathcal{A}$ (and if so to build it explicitly).

Example 2. Consider a server-client setup where one process, the server, communicates with several processes, the clients. The server broadcasts to all clients that it wants a task to be solved. Each client starts working independently on the task requested, and communicates back to the server when it has finished. When the server gets two answers back from its children, it then sends another broadcast telling all clients to stop working.

Formally, we take $\mathbb{P} = \{p, c_1, \ldots, c_n\}$ with p the server and $c_1, \ldots, c_n$ the clients. We have two actions t_1, t_2 representing requests for two different *tasks*, $\mathbb{C}(t_1) = \mathbb{C}(t_2) = \mathbb{P}$. Each process c_i has two local actions p_1^i, p_2^i to *progress* those tasks, $\mathbb{C}(p_1^i) = \mathbb{C}(p_2^i) = \{c_i\}$. Then there is a shared action e_i to communicate back to the server that the task has *ended*, $\mathbb{C}(e_i) = \{c_i, p\}$. Finally, there is an action r that can broadcast to all clients to *reset* their state when the task has been successfully completed, $\mathbb{C}(r) = \mathbb{P}$. The architecture is tree-like with the server p at its root, and all clients c_i are the children of p. The automata of the server and the clients are illustrated in Figure 2. For the acceptance condition, we require that the server visits its initial state infinitely often, and that each client performs action e_i infinitely often. For simplicity of presentation, finite runs are not accepted (clients do not crash).

Now for the Control Problem, we say that all local actions (i.e. actions p_k^i) are controllable, while all non-local ones are uncontrollable by definition. The naive controller that only enables the correct progressing action on every process, e.g. p_1^i after receiving a t_1 broadcast, is not winning. Indeed, since Environment controls which processes get to perform their e_i, it could choose the same two processes for every task. Then those not chosen never get their turn at performing e_i and thus will not be accepting. However we can still build a winning controller in a round-robin manner. To do this, we simply add to the state of each c_i a count of how many tasks have been requested since the beginning modulo n, so we have $C_{c_i} = S_{c_i} \times \{0, \ldots, n-1\}$ and the projection π is simply the projection on the first component. If the count is $k - 1$, then the controllers for processes c_k and c_{k+1} are set to enable only the correct progress action, while other processes enable nothing. After that, Environment has no choice but to perform the corresponding e_k and e_{k+1} to keep the run going. Then the server sends a reset broadcast, the next task is requested, k is incremented, and each process gets to perform its e_i action in turn, thus this controller is winning.

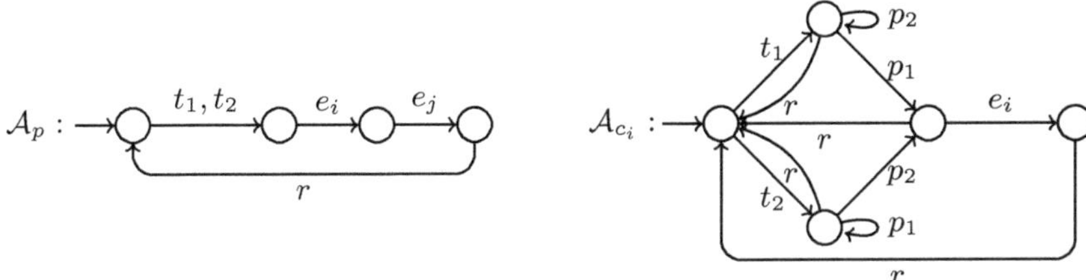

Fig. 2. A server-client protocol for distributing and executing tasks.

4.2   Leaf Process Simulated by Parent

In [23], it is shown that in an architecture that forms a tree where all channels are binary, a leaf process can be simulated by its parent for the purpose of control w.r.t. local parity conditions. This is done by abstracting away all local actions of the leaf into a finite summary, and any synchronization between the leaf and the parent can then be simulated locally by the parent. This simulation allows to deduce an upper complexity bound for the Control Problem that is a tower of exponentials in the depth of the tree.

We show here that the proof of [23] extends to tree-like architectures. The key observation is that, in TCA, by the continuity condition, any communication involving a leaf and other processes necessarily includes the parent of the leaf as well. Therefore, if some process relies on the state of the to-be-removed leaf for some transition, it will still be able to access that information after the removal because it will instead be provided by the parent that is simulating the leaf.

For the rest of this subsection we consider an AA $\mathcal{A}$ and a TCA where $\mathbb{P}$ is the set of processes, ℓ is a leaf process, and p is the parent of ℓ. The first step is to ensure that processes communicate "frequently". This will provide a finite description of what happens locally on leaf ℓ between two consecutive actions shared by ℓ with other processes.

Definition 2. *An AA $\mathcal{A}$ is ℓ-short if there is some bound $B \in \mathbb{N}$ such that in every local ℓ-run of $\mathcal{A}$, the number of local actions that ℓ can do in a row is at most B.*

Lemma 4 (Th.10, [23]). *For every AA $\mathcal{A}$ we can construct an ℓ-short automaton $\mathcal{A}^{\circledS}$ such that $\mathcal{A}$ is controllable iff $\mathcal{A}^{\circledS}$ is controllable. The state space of any $q \neq \ell$ does not change from $\mathcal{A}$ to $\mathcal{A}^{\circledS}$; the states of ℓ in $\mathcal{A}^{\circledS}_{\ell}$ are simple, local paths of $\mathcal{A}_{\ell}$.*

Although the proof of Lemma 4 is quite involved, it does not rely on the communication architecture being a tree. The main idea behind is that every parity game is equivalent to a finite cycle game.

It follows that to show that a leaf process can be simulated by its parent (for the purposes of control) it is enough to work with $\mathcal{A}^{\circledS}$, so in particular, with an ℓ-short AA. From Lemma 4 we know that the bound B can be set to the number of ℓ-states in $\mathcal{A}$, so $B = |S_{\ell}|$ with S_{ℓ} the set of states of ℓ in $\mathcal{A}$.

We are now ready to show that given $\mathcal{A}$ we can construct an AA $\mathcal{A}^{\backslash \ell}$ whose set of processes is $\mathbb{P} \setminus \{\ell\}$ and such that $\mathcal{A}$ is controllable iff $\mathcal{A}^{\backslash \ell}$ is controllable. Moreover, for every $p' \neq p$ we have $\mathcal{A}^{\backslash \ell}_{p'} = \mathcal{A}_{p'}$ and the size of $\mathcal{A}^{\backslash \ell}_{p}$ is proportional to $|\mathcal{A}_p| \cdot |\Sigma^{sys}_p| \cdot exp(|\mathcal{A}_{\ell}|)$.

Let $\mathcal{A} = ((S_p)_{p \in \mathbb{P}}, (s^0_p)_{p \in \mathbb{P}}, (\delta_a)_{a \in \Sigma}, (Acc_p)_{p \in \mathbb{P}})$, $\ell \in \mathbb{P}$ a leaf and $p \in \mathbb{P}$ its parent, and assume that $\mathcal{A}$ is ℓ-short thanks to Lemma 4. With $\mathbb{P}^{\backslash \ell} = \mathbb{P} \setminus \{\ell\}$, we build $\mathcal{A}^{\backslash \ell} = ((S^{\backslash \ell}_p)_{p \in \mathbb{P}^{\backslash \ell}}, (s^{0 \backslash \ell}_p)_{p \in \mathbb{P}^{\backslash \ell}}, (\delta^{\backslash \ell}_a)_{a \in \Sigma^{\backslash \ell}}, (Acc^{\backslash \ell}_p)_{p \in \mathbb{P}^{\backslash \ell}})$. The new alphabet $\Sigma^{\backslash \ell}$ will be defined shortly after. For all processes $q \neq p$, we have $S^{\backslash \ell}_q = S_q$, $s^{0 \backslash \ell}_q = s^0_q$, and $Acc^{\backslash \ell}_q = Acc_q$. For all actions $a \in \Sigma$ such that $\mathbb{C}(a) \cap \{\ell, p\} = \emptyset$, we have $\delta^{\backslash \ell}_a = \delta_a$. That is, all components that are not related to either p or ℓ remain unchanged.

Let us define $S^{\backslash \ell}_p$. To that end, we define ℓ-local strategies first. Remember that $\mathcal{A}$ is ℓ-short, so let B be the corresponding bound. Recall also that Σ_{ℓ} is the set of ℓ-local actions. For presentation purposes, we write $s \xrightarrow{a_1 \ldots a_n} s'$ as a stand in for $s' = \delta_{a_n}(\ldots (\delta_{a_1}(s)))$ when $a_1, \ldots, a_n \in \Sigma_{\ell}$ and s, s' are states of ℓ. We write $s \xrightarrow{a_1 \cdots a_n}$ when there exists a state s' such that $s \xrightarrow{a_1 \cdots a_n} s'$. An *$\ell$-local strategy* from a given state $s_{\ell} \in S_{\ell}$ is a partial function $f : (\Sigma_{\ell})^{\leq B} \to \Sigma^{sys}_{\ell}$ such that whenever $f(w) = a$ then $s_{\ell} \xrightarrow{w \cdot a}$ is defined in $\mathcal{A}_{\ell}$. If $s_{\ell} \xrightarrow{a} s'_{\ell}$ and f is an ℓ-local strategy from s_{ℓ}, then $f_{|a}$ is the ℓ-local strategy from s'_{ℓ} defined as $f_{|a}(w) = f(a \cdot w)$. With this defined, a state in $S^{\backslash \ell}_p$ of the parent p of ℓ is of one of 3 types:

$$(s_p, s_{\ell}), (s_p, s_{\ell}, f), (s_p, a, s_{\ell}, f)$$

where $s_p \in S_p$, $s_{\ell} \in S_{\ell}$ are the simulated states of p and ℓ respectively, f is an ℓ-local strategy from s_{ℓ}, and $a \in \Sigma^{sys}_p$ is a controllable local action of the parent p. Since $\mathcal{A}$ is ℓ-short, ℓ-local strategies can be seen simply as paths of $\mathcal{A}_{\ell}$ without repetitions, and thus there are only $exp(|\mathcal{A}_{\ell}|)$ many of them. Finally, we let $s^{0 \backslash \ell}_p = (s^0_p, s^0_{\ell})$ be the initial state of p.

We now precisely define the new alphabet $\Sigma^{\backslash \ell}$ before defining the new transitions. All actions that were previously ℓ-local are changed into uncontrollable p-local actions, regardless of whether they were controllable or not in Σ. Similarly, all actions that were p-local are now uncontrollable in $\Sigma^{\backslash \ell}$, but for each

p-local action $a \in \Sigma_p^{sys}$ that was controllable we add a new action $ch(a)$ in $\Sigma^{\backslash \ell}$ that is also p-local and controllable. We also add new actions $ch(f)$ that are p-local and controllable for each ℓ-local strategy f. Any non-local action that included ℓ in Σ remains in $\Sigma^{\backslash \ell}$ but with ℓ removed from its domain. All remaining actions in Σ, i.e. those that are local to some process not in $\{p, \ell\}$ or those that are non-local and do not intersect with ℓ, remain unchanged.

Let us now define transitions over actions involving p. First, from a state of p of the form (s_p, s_ℓ), there is only one kind of transition available

$$(s_p, s_\ell) \xrightarrow{ch(f)} (s_p, s_\ell, f) \text{ for all } \ell\text{-local strategies } f \text{ from } s_\ell$$

where System fixes a local strategy for ℓ. Then from these states there is again only one kind of transition possible

$$(s_p, s_\ell, f) \xrightarrow{ch(a)} (s_p, a, s_\ell, f) \text{ for all } a \in \Sigma_p^{sys} \text{ enabled from } s_p$$

where System chooses one p-local controllable (in Σ_p) action[4], provided there is a transition from s_p with this action. Then there are multiple choices from here. The two actions

$$(s_p, a, s_\ell, f) \xrightarrow{b_p} (s'_p, s_\ell, f) \text{ if } (s_p, s'_p) \in \delta_{b_p} \text{ and } b_p = a \text{ or } b_p \in \Sigma_p^{env}$$

$$(s_p, a, s_\ell, f) \xrightarrow{b_\ell} (s_p, a, s'_\ell, f_{|b_\ell}) \text{ if } (s_\ell, s'_\ell) \in \delta_{b_\ell} \text{ and } b_\ell \text{ is enabled by } f$$

are p-local actions (all uncontrollable) which either progress p by its pre-selected (previously) controllable action a, progress p by any local uncontrollable action b_p, or progress ℓ by any action enabled by the pre-selected strategy f, i.e. any uncontrollable ℓ-local action or the one controllable action output by f on ε. Then there are also non-local actions where p communicates with other processes.

$$((s_p, a, s_\ell, f), s_{q_1}, \ldots, s_{q_k}) \xrightarrow{b_{p+}} ((s'_p, s_\ell, f), s'_{q_1}, \ldots, s'_{q_k}) \text{ if}$$
$$p \in \mathbb{C}(b_{p+}), \ell \notin \mathbb{C}(b_{p+}), ((s_p, s_{q_1}, \ldots, s_{q_k}), (s'_p, s'_{q_1}, \ldots, s'_{q_k})) \in \delta_{b_{p+}}$$

$$((s_p, a, s_\ell, f), s_{q_1}, \ldots, s_{q_k}) \xrightarrow{b_{p\ell+}} ((s'_p, s'_\ell), s'_{q_1}, \ldots, s'_{q_k}) \text{ if}$$
$$\ell, p \in \mathbb{C}(b_{p\ell+}), ((s_p, s_\ell, s_{q_1}, \ldots, s_{q_k}), (s'_p, s'_\ell, s'_{q_1}, \ldots, s'_{q_k})) \in \delta_{b_{p\ell+}}$$

The first kind is for when ℓ is excluded from the communication, so while s_p gets updated to a new state and the choice of a is reset, s_ℓ and f remain unchanged. The second kind occurs when ℓ is part of the communication, so both s_p and s_ℓ are read and updated, and the choices of a and f are reset. Note that $k = 0$ covers the case where a communication involving only p and ℓ occurs. Also, since we assumed that the architecture is tree-like, it is not possible for a communication to involve ℓ and other processes without p being included, which is why this case is not covered. Note that this is the only major difference with the original

---

[4] The assumption that in each state there is at least one enabled controllable action is used here.

construction from [23], and it is the reason that the proofs of correctness are mostly similar.

The last part of $\mathcal{A}^{\backslash \ell}$ is the acceptance conditions $Acc_p^{\backslash \ell} = (F_p^{\backslash \ell}, \Omega_p^{\backslash \ell})$. For $F_p^{\backslash \ell}$ it is simple: any state of the form $(s_p, s_\ell) \in F_p \times F_\ell$, and any $(s_p, s_\ell, f), (s_p, a, s_\ell, f)$ where $s_p \in F_p$ and any possible maximal continuation respecting f from s_ℓ eventually reaches a state in F_ℓ, are in this set. Defining $\Omega_p^{\backslash \ell}$ is trickier. First, an infinite run of p must satisfy the parity condition of Ω_p when restricted to the s_p component of the state. Then there are two cases. If the projection on the s_ℓ component is infinite and satisfies Ω_ℓ then the run must be accepted, which amounts to a conjunction of two parity conditions. This can be translated into a single parity condition, and we will discuss the complexity in Section 4.3. Otherwise, we require that for the last f and s_ℓ reached (uniquely defined as the only components seen infinitely often in the run), any possible continuation respecting f from s_ℓ eventually reaches a state in F_ℓ. This amounts to a single parity condition, the one on p.

Note that we do not define what happens when the run on ℓ is infinite but the run on p is finite, because this is not compatible with the ℓ-short assumption.

Now that $\mathcal{A}^{\backslash \ell}$ is defined, we are ready to prove the following theorem (proof in appendix).

Theorem 4. *For every ℓ-short AA $\mathcal{A}$ with local acceptance condition: there is a winning controller for $\mathcal{A}$ iff there is a winning controller for $\mathcal{A}^{\backslash \ell}$.*

4.3 Solving the Control Problem on Tree-like Architectures

Theorem 4 generalizes the leaf-elimination procedure of [23] from binary channels to tree-like architectures. Iterating this step reduces the distributed control problem to one involving a single process, so to a usual parity game. It is argued in [23] that the size of the resulting parity game is $\mathrm{Tower}_d(n)$, where n is the size of the AA and $d \geq 0$ the depth of the process tree. A matching lower bound for the control problem is provided in [12]. For the upper bound we provide below a detailed argument based on the following lemma:

Lemma 5. *Given $n+1$ parity conditions p_0 and $p_1, \ldots, p_n$, there is a deterministic parity automaton with $O\left((p_0 + p) \cdot \frac{p! \cdot (p_0+1)^p}{(\prod_{i>0}(p_i!))}\right)$ states and $p_0 + p$ priorities, where $p = \sum_{i>1} p_i$, reading tuples of $n+1$ priorities and accepting iff all $n+1$ parity conditions are accepting.*

Proof. We use a variant of the index appearance record construction [26,20] to convert $n+1$ parity conditions to one condition. We start by using the index appearance record to convert the conditions $p_1, \ldots, p_n$ to one parity condition with p priorities. This is a straightforward application of the index appearance record, noticing that the order within each parity condition is always maintained. Thus, we keep a permutation of p indices, but where all the indices belonging to a single parity condition are always ordered in decreasing order. Thus, there are $\frac{p!}{(\prod_{i>0}(p_i!))}$ such permutations. It remains to keep the respective order of the elements in p_0 among the sequence of p indices. As the elements of p_0 themselves are ordered in decreasing order, it is sufficient to keep track of how many of them appear before every element in the sequence of p indices. Thus, we multiply the number of options by $(p_0 + 1)^p$. Thus, $\frac{p! \cdot (p_0+1)^p}{(\prod_{i>0}(p_i!))}$ possible values represent all possible sequences of $p_0 + p$ indices, where the indices of each parity condition itself are ordered in decreasing order. Every transition of the created parity automaton needs to signal a priority between $p_0 + p$ and 1. For that, we keep a record of what was the left-most index that is visited in a transition from one $p_0 + p$ sequence to its successor. This brings the total number of states to the stated in the Lemma. We note that p_0 does not appear in the exponent.   □

We note that the complexity of solving games where the winning condition is a conjunction of parity conditions is known to be co-NP-complete [4]. Thus, a much simpler conversion to parity games than the one described in the previous lemma is not possible.

We also note that the upper bound of the previous construction [23] was assuming that the acceptance condition is not part of the input.

Corollary 1. *The Control Problem for asynchronous automata over tree-like architectures is decidable in d-EXPTIME where d is the depth of the tree.*

Proof. We apply Theorem 4 iteratively. Choose a parent p with leaves $\ell_1, \ldots, \ell_n$. Apply the theorem to each one of the leaves $\ell_1, \ldots, \ell_n$. We end up with an equi-realizable control problem, where $\ell_1, \ldots, \ell_n$ are missing and p is replaced by $p^{\backslash \ell_1, \ldots, \ell_n}$. The number of states of $p^{\backslash \ell_1, \ldots, \ell_n}$ is proportional to $|S_p| \cdot |\Sigma|^n \cdot \prod_{i>0} exp(|S_{\ell_i}|)$. Furthermore, the infinitary condition of $p^{\backslash \ell_1, \ldots, \ell_n}$ is a conjunction of $n + 1$ parity conditions. By Lemma 5, if the number of parities of p and $\ell_1, \ldots, \ell_n$ are $p_0, \ldots, p_n$, respectively, then by multiplying the number of states of $p^{\backslash \ell_1, \ldots, \ell_n}$ by $O\left((p_0 + p) \cdot \frac{p! \cdot (p_0+1)^p}{(\prod_{i>0}(p_i!))} \right)$ we get a parity condition with $\sum_{i \geq 0} p_i$ priorities.

It follows that by eliminating leaves level by level, we get an asynchronous automaton whose size is a tower of exponentials of height d, where d is the depth of the eliminated tree, and whose parity index is the sum of all parity indices of all processes in the tree.

When the entire tree is reduced to a single process, we get a control problem for one process and the corollary follows.   □

The d-EXPTIME complexity bound stated in Corollary 1 is matched by a result from [12] showing that how to reduce acceptance of a d-EXPSPACE bounded Turing machine to a Control Problem on a *tree architecture* of depth $O(d)$ (although not exactly d, so some gap remains).

5 Conclusion

We revisited simpler constructions of Zielonka asynchronous automata, and the control problem, in the case of tree-like process architectures. We showed that the quadratic construction of Zielonka automata generalizes from trees to tree-like architectures, and also showed how this improves on a previous, exponential construction of asynchronous automata for triangulated dependence alphabets. Finally we showed that distributed controller synthesis generalizes smoothly to tree-like architectures, too.

While causal synchronisation like in the Zielonka model may appear far from real systems, it represents a first step to handle communication. Communication has different flavor than rendez-vous mechanisms like in the Zielonka model, e.g. it lacks the symmetry of Zielonka synchronisations since receivers are better informed than senders. At the same time, causal dependencies, like those arising in the Zielonka model, are important and can be approximated by practical forms of communication (e.g., local broadcast). In such a context, we believe that tree-like process architectures may provide more potential for future applications involving communication, than tree architectures. Indeed, tree-like architectures allow for real multi-party communication as channels are not restricted to binary connections. It adds the requirement to inform all the "region" affected by a communication, suggesting possible uses where interaction is connected to physical location. In our future work, we are working on an implementation of a communication infrastructure that supports a relaxed version of Zielonka type synchronization. Our framework ensures that actions happen in the same partial order as in the theoretical (Zielonka) model. The communication framework currently enacts asynchronous automata that are modelled by hand. However, if the communication architecture is tree-like we could use an automated distribution of centrally crafted plans based on similar ideas as in this paper. At a theoretical level, we plan to investigate which central plans can be distributed into a tree-like communication framework with bounded communication channels.

References

1. Bharat Adsul, Paul Gastin, Shantanu Kulkarni, and Pascal Weil. An expressively complete local past propositional dynamic logic over Mazurkiewicz traces and its applications. In Pawel Sobocinski, Ugo Dal Lago, and Javier Esparza, editors, *Proceedings of the 39th Annual ACM/IEEE Symposium on Logic in Computer Science, LICS 2024, Tallinn, Estonia, July 8-11, 2024*, pages 2:1–2:13. ACM, 2024. `doi:10.1145/3661814.3662110`.

2. S. Akshay, Ionut Dinca, Blaise Genest, and Alin Stefanescu. Implementing realistic asynchronous automata. In *FSTTCS'13*, LIPIcs, pages 213–224. Schloss Dagstuhl - Leibniz-Zentrum fuer Informatik, 2013.

3. Raven Beutner, Bernd Finkbeiner, and Jesko Hecking-Harbusch. Translating asynchronous games for distributed synthesis. In *International Conference on Concurrency Theory (CONCUR'19)*, volume 140 of *LIPIcs*, pages 26:1–26:16. Schloss Dagstuhl - Leibniz-Zentrum für Informatik, 2019.

4. Krishnendu Chatterjee, Thomas A. Henzinger, and Nir Piterman. Generalized parity games. In Helmut Seidl, editor, *Foundations of Software Science and Computational Structures, 10th International Conference, FOSSACS 2007, Held as Part of the Joint European Conferences on Theory and Practice of Software, ETAPS 2007, Braga, Portugal, March 24-April 1, 2007, Proceedings*, volume 4423 of *Lecture Notes in Computer Science*, pages 153–167. Springer, 2007. `doi:10.1007/978-3-540-71389-0_12`.

5. Robert Cori, Yves Métivier, and Wieslaw Zielonka. Asynchronous mappings and asynchronous cellular automata. *Inf. Comput.*, 106(2):159–202, 1993. URL: `https://doi.org/10.1006/inco.1993.1052`, `doi:10.1006/INCO.1993.1052`.

6. Volker Diekert and Anca Muscholl. A note on Métivier's construction of asynchronous automata for triangulated graphs. *Fundam. Informaticae*, 25(3):241–246, 1996. `doi:10.3233/FI-1996-253402`.

7. Volker Diekert and Grzegorz Rozenberg, editors. *The Book of Traces*. World Scientific, 1995. `doi:10.1142/2563`.

8. Bernd Finkbeiner and Sven Schewe. Uniform distributed synthesis. In *20th Annual IEEE Symposium on Logic in Computer Science (LICS' 05)*, pages 321–330, 2005. `doi:10.1109/LICS.2005.53`.

9. Paul Gastin, Nathalie Sznajder, and Marc Zeitoun. Distributed synthesis for well-connected architectures. *Formal Methods Syst. Des.*, 34(3):215–237, 2009. URL: `https://doi.org/10.1007/s10703-008-0064-7`, `doi:10.1007/S10703-008-0064-7`.

10. Fănică Gavril. The intersection graphs of subtrees in trees are exactly the chordal graphs. *Journal of Combinatorial Theory, Series B*, 16(1):47–56, 1974. URL: `https://www.sciencedirect.com/science/article/pii/009589567490094X`, `doi:https://doi.org/10.1016/0095-8956(74)90094-X`.

11. Blaise Genest, Hugo Gimbert, Anca Muscholl, and Igor Walukiewicz. Optimal Zielonka-type construction of deterministic asynchronous automata. In *Automata, Languages and Programming, ICALP 2010*, volume 6199 of *Lecture Notes in Computer Science*, pages 52–63. Springer, 2010. `doi:10.1007/978-3-642-14162-1_5`.

12. Blaise Genest, Hugo Gimbert, Anca Muscholl, and Igor Walukiewicz. Asynchronous games over tree architectures. In Fedor V. Fomin, Rusins Freivalds, Marta Z. Kwiatkowska, and David Peleg, editors, *Automata, Languages, and Programming - 40th International Colloquium, ICALP 2013, Riga, Latvia, July 8-12, 2013, Proceedings, Part II*, volume 7966 of *Lecture Notes in Computer Science*, pages 275–286. Springer, 2013. `doi:10.1007/978-3-642-39212-2_26`.

13. Blaise Genest and Anca Muscholl. Constructing exponential-size deterministic Zielonka automata. In *Automata, Languages and Programming, ICALP 2006*, pages 565–576. Springer, 2006.

14. Hugo Gimbert. Distributed asynchronous games with causal memory are undecidable. *Logical Methods in Computer Science*, Volume 18, Issue 3, 09 2022. `doi:10.46298/lmcs-18(3:30)2022`.

15. Daniel Hausmann, Mathieu Lehaut, and Nir Piterman. Distribution of reconfiguration languages maintaining tree-like communication topology. In *Automated Technology for Verification and Analysis - 22nd International Symposium, ATVA 2024, Kyoto, Japan, October 21-24, 2024, Proceedings*, Lecture Notes in Computer Science. Springer, 2024.

16. Siddharth Krishna and Anca Muscholl. A quadratic construction for Zielonka automata with acyclic communication structure. *Theor. Comput. Sci.*, 503:109–114, 2013. `doi:10.1016/j.tcs.2013.07.015`.

17. Orna Kupferman and Moshe Y. Vardi. Synthesizing distributed systems. In *Proceedings of the 16th Annual IEEE Symposium on Logic in Computer Science*, LICS '01, page 389, USA, 2001. IEEE Computer Society.

18. Stephane Lafortune, Karen Rudie, and Stavros Tripakis. Thirty years of the Ramadge-Wonham theory of supervisory control: A retrospective and future perspectives [conference reports]. *IEEE Control Systems Magazine*, 38(4):111–112, 2018. `doi:10.1109/MCS.2018.2830083`.

19. Mathieu Lehaut, Anca Muscholl, and Nir Piterman. From trees to tree-like: Distribution and synthesis for asynchronous automata, 2026. URL: `https://arxiv.org/abs/2601.14078`, `arXiv:2601.14078`.

20. Christoph Löding. Methods for the transformation of automata: Complexity and connection to second order logic. Master's thesis, Christian-Albrechts-University of Kiel, Kiel, Germany, 1999. Available at `https://www.lics.rwth-aachen.de/global/show_document.asp?id=aaaaaaaaabcqdty`.

21. Antoni Mazurkiewicz. Concurrent program schemes and their interpretations. *DAIMI Report Series*, (78), 1977.

22. Madhavan Mukund and Milind A. Sohoni. Keeping track of the latest gossip in a distributed system. *Distributed Comput.*, 10(3):137–148, 1997. `doi:10.1007/S004460050031`.

23. Anca Muscholl and Igor Walukiewicz. Distributed synthesis for acyclic architectures. *arXiv preprint arXiv:1402.3314*, 2014. Conference version FST&TCS 2014.

24. A. Pnueli and R. Rosner. Distributed reactive systems are hard to synthesize. In *Proceedings [1990] 31st Annual Symposium on Foundations of Computer Science*, pages 746–757 vol.2, 1990. `doi:10.1109/FSCS.1990.89597`.

25. Peter Ramadge and Walter Wonham. Supervisory control of a class of discrete event systems. *SIAM Journal on Control and Optimization*, 25:206–230, 01 1987. `doi:10.1137/0325013`.

26. Shmuel Safra. Exponential determinization for omega-automata with strong-fairness acceptance condition (extended abstract). In S. Rao Kosaraju, Mike Fellows, Avi Wigderson, and John A. Ellis, editors, *Proceedings of the 24th Annual ACM Symposium on Theory of Computing, May 4-6, 1992, Victoria, British Columbia, Canada*, pages 275–282. ACM, 1992. `doi:10.1145/129712.129739`.

27. Wieslaw Zielonka. Notes on finite asynchronous automata. *RAIRO Theor. Informatics Appl.*, 21(2):99–135, 1987. URL: `https://doi.org/10.1051/ita/1987210200991`.

Quantum Coherence Spaces Revisited: A von Neumann (Co)Algebraic Approach

Thea Li and Vladimir Zamdzhiev

Université Paris-Saclay, CNRS, ENS Paris-Saclay, Inria, Laboratoire Méthodes
Formelles, 91190, Gif-sur-Yvette, France
`thea.li@inria.fr, vladimir.zamdzhiev@inria.fr`

Abstract. We describe a categorical model of MALL (Multiplicative
Additive Linear Logic) inspired by the Heisenberg-Schrödinger duality of
finite-dimensional quantum theory. Proofs of formulas with positive logical polarity correspond to CPTP (completely positive trace-preserving)
maps in our model, i.e. the quantum operations in the Schrödinger
picture, whereas proofs of formulas with negative logical polarity correspond to CPU (completely positive unital) maps, i.e. the quantum
operations in the Heisenberg picture. The mathematical development
is based on noncommutative geometry and finite-dimensional von Neumann (co)algebras, which can be defined as special kinds of (co)monoid
objects internal to the category of finite-dimensional operator spaces.

Keywords: Quantum Theory, Linear Logic, Categorical Semantics

1 Introduction

Linear Logic [18] was discovered by Girard while studying the mathematical
model of *coherence spaces* [17,18]. Coherence spaces "played an essential role
in the discovery of linear logic" [10, p. 967] and clearly they have had a major
impact on the study of (semantics of) Linear Logic (LL). Later, *Probabilistic
Coherence Spaces* (PCSs) [10] were described and they give a model of LL which
is suitable for modeling discrete probability. The study of PCSs has resulted in
impressive results, such as full abstraction for probabilistic PCF [14] and also
full abstraction for a more complicated probabilistic programming language with
recursive types [15].

A natural next step in the development of these models is to consider *quantum
coherence spaces* (QCSs) with the intention of taking the semantic study of LL
to quantum theory. Girard already proposed models of QCSs [19,20,21] and
they were later studied in [2], but this approach has already been criticised
by Selinger [32] as being inappropriate for quantum theory. One of the main
problems with this proposal is that the morphisms between QCSs do not appear
to correspond to *completely positive maps* on relevant homsets (see also [2, p. 2]).
This is problematic from the point of view of quantum theory, because *quantum
operations* (also known as *quantum channels*) are modelled mathematically as
certain kinds of completely positive maps.

N. Bertrand and S. Milius (Eds.): FoSSaCS 2026, LNCS 16503, pp. 418–439, 2026.
https://doi.org/10.1007/978-3-032-22730-0_20

Inspired by the success of (probabilistic) coherence spaces, in this work we set out to paint a new picture of what a quantum coherence space ought to be. We consider finite-dimensional quantum theory and we describe a model of MALL based on the following natural ideas: (1) proofs $P \vdash R$ of formulas with *positive* logical polarities should admit an interpretation as CPTP (completely positive trace-preserving) maps, i.e. as the quantum operations in the *Schrödinger picture* of quantum theory; (2) proofs $M \vdash N$ of formulas with *negative* logical polarities should admit an interpretation as CPU (completely positive unital) maps, i.e. as the quantum operations in the *Heisenberg picture* of quantum theory; (3) the LL duality should coincide with the Heisenberg-Schrödinger duality of quantum theory on polarised formulas. These desiderata naturally lead to the theory of operator spaces [13,6,31] which has excellent categorical properties [29] and which can be used to construct a model of (full) LL whose duality is compatible with the Heisenberg-Schrödinger duality on the level of objects/formulas [30]. However, in [30], the morphisms/proofs do *not* correspond to the quantum operations in either picture. We address this problem (which is suggested for future work in [30]) for *finite-dimensional* (f.d.) quantum theory and MALL.

In §2, we provide relevant background on operator spaces. We begin our contributions in §3, where we show that **FdOS**, the category of f.d. operator spaces, is a model of MALL, and we prove other relevant categorical properties. The category has a very important monoidal structure $(\textbf{FdOS}, \mathbb{C}, \overset{\text{h}}{\otimes})$ that is outside the scope of LL – it is given by the Haagerup tensor [13, §9], [6, pp. 30 – 34], [31, §5], which is one of the highlights of operator space theory. This monoidal structure allows us to give a new, but equivalent, definition of a f.d. *von Neumann algebra* (vN-algebra) in §4 as a certain kind of internal monoid object in $(\textbf{FdOS}, \mathbb{C}, \overset{\text{h}}{\otimes})$. It is well-known that CPU maps may be defined on vN-algebras and this allows us to identify the subcategory of **FdOS** corresponding to negative logical polarity and the Heisenberg picture. The novel categorical definition of f.d. vN-algebras, and the self-duality of the Haagerup tensor, allow us to easily dualise this definition and we introduce f.d. *von Neumann coalgebras* (vN-coalgebras) in §5 as certain kinds of internal comonoid objects in $(\textbf{FdOS}, \mathbb{C}, \overset{\text{h}}{\otimes})$. We show that CPTP maps are a natural notion of morphism between vN-coalgebras and so we identify the subcategory of **FdOS** corresponding to positive logical polarity and the Schrödinger picture. In §6 we prove that there is a categorical duality between the subcategories, which gives a categorical formulation of the Heisenberg-Schrödinger duality. Finally, in §7, we apply the semantic technique of *gluing and orthogonality* of Hyland and Schalk [22] to **FdOS** which results in a model of MALL that satisfies the desiderata (1) – (3), see Figure 1. Note that the proofs omitted in this paper can be found in our companion paper [28].

2 Background on Operator Spaces and Banach Spaces

Operator spaces are widely seen as the "non-commutative" or "quantised" generalisation of Banach spaces. In this section we recall some basic background material on these topics from several books on the matter [13], [6], [31].

2.1　Normed Spaces and Banach Spaces

A *normed space* is a pair $(X, \|\cdot\|_X)$ where X is a complex vector space and $\|\cdot\|_X$ a norm on it. A *Banach space* is a normed space that is complete with respect to the induced norm topology. In this paper we work with finite-dimensional normed spaces and every such space is necessarily a Banach space. We usually simply write X to refer to a Banach space and we just write $\|\cdot\|$ for its norm, which should be clear from context. Recall that, if X and Y are two Banach spaces and $\varphi\colon X \to Y$ is a linear map between them, then its *operator norm* is defined by the assignment $\|\varphi\| \triangleq \sup\{\|\varphi(x)\| \ : \ x \in X, \|x\| \leq 1\}$. Then φ is continuous iff φ is bounded iff $\|\varphi\| < \infty$. A linear map $f\colon X \to Y$ between two Banach spaces is called an *isometry* if $\|f(x)\| = \|x\|$ for each $x \in X$ and it is called a *contraction* whenever $\|f\| \leq 1$, equivalently when $\|f(x)\| \leq \|x\|$ for every $x \in X$. We write **FdBan** for the category whose objects are finite-dimensional Banach spaces and whose morphisms are the linear contractions between them. In order to gain some intuition for the constructions that follow, it is useful to recall some of the categorical properties of **FdBan**. It has a zero object given by the zero-dimensional Banach space 0. The binary categorical product of X and Y is given by the Banach space $X \overset{\infty}{\oplus} Y \triangleq (X \oplus Y, \|\cdot\|_{\ell^\infty})$, where $\|(x, y)\|_{\ell^\infty} \triangleq \max(\|x\|, \|y\|)$ is simply the ℓ^∞-norm and $X \oplus Y$ is the usual direct sum of vector spaces. The binary categorical coproduct of X and Y is given by the Banach space $X \overset{1}{\oplus} Y \triangleq (X \oplus Y, \|\cdot\|_{\ell^1})$, where $\|(x, y)\|_{\ell^1} \triangleq \|x\| + \|y\|$ is simply the ℓ^1-norm. The category **FdBan** is symmetric monoidal closed with respect to the *projective* tensor product $\overset{\gamma}{\otimes}$ (details omitted) with internal hom given by $B(X, Y)$, the space of bounded linear maps between X and Y. In fact, **FdBan** is a model of MALL [5, §2.6]: the additive conjunction (disjunction) is modelled by the categorical (co)products; the multiplicative conjunction (disjunction) corresponds to $\overset{\gamma}{\otimes}$ ($\overset{\epsilon}{\otimes}$), where $\overset{\epsilon}{\otimes}$ stands for the *injective* tensor product; negation corresponds to the dual Banach space $X^* \triangleq B(X, \mathbb{C})$.

2.2　Operator Spaces and their Morphisms

An (abstract) operator space may be seen as a vector space X equipped with additional norms on matrices with entries in X that satisfy some conditions.

Definition 1 (Matrix Space). *Given a complex vector space X, we write $\mathbb{M}_{n,m}(X)$ for the vector space consisting of the $n \times m$ matrices with entries in X. The vector space structure is defined in the obvious way, i.e. componentwise. We also use the shorthand notation $\mathbb{M}_n(X) \triangleq \mathbb{M}_{n,n}(X)$ and $\mathbb{M}_n \triangleq \mathbb{M}_n(\mathbb{C})$.*

Definition 2 (Matrix Norm [13, p. 20]). *A matrix norm on a vector space X is an assignment of a norm $\|\cdot\|_n$ on $\mathbb{M}_n(X)$ for each $n \in \mathbb{N}$. Given such a norm, we write $M_n(X)$ for the normed space $(\mathbb{M}_n(X), \|\cdot\|_n)$.*

Example 3. We write $M_{n,m}$ for the normed space $(\mathbb{M}_{n,m}, \|\cdot\|_{op})$, where $\|\cdot\|_{op}$ is the usual operator norm on $\mathbb{M}_{n,m}$. The normed spaces $M_n \triangleq M_{n,n}$ for $n \in \mathbb{N}$ determine a matrix norm on $\mathbb{C}$ that we use throughout the paper.

Definition 4 (Operator Space [31, pp. 34–35]). *An* (abstract) *operator space is a complex vector space X equipped with a matrix norm that satisfies:*

(B) *The pair $(\mathbb{M}_1(X), \|\cdot\|_1)$ is a Banach space;*
(M1) $\|x \oplus y\|_{m+n} = \max\{\|x\|_m, \|y\|_n\}$
(M2) $\|\alpha x \beta\|_m \leq \|\alpha\| \|x\|_m \|\beta\|$

for $n, m \in \mathbb{N}$, $x \in M_m(X)$, $y \in M_n(X)$, $\alpha, \beta \in M_m$. We write $x \oplus y \triangleq \begin{bmatrix} x & 0 \\ 0 & y \end{bmatrix} \in M_{m+n}(X)$ for the indicated matrix and $\alpha x \beta \in M_m(X)$ for the matrix that results from the obvious generalisation of matrix multiplication.

Note that α, β in the above definition are scalar (i.e. complex) matrices, but x need not be one. A matrix norm that satisfies the above criteria is called an *operator space structure* (o.s.s.) on X. The obvious isomorphism $M_1(X) \cong X$ determines a norm on X for which we write $\|\cdot\|_X$ or simply $\|\cdot\|$ in the sequel. If $x \in M_n(X)$ we often simply write $\|x\|$ for $\|x\|_n$. Standard textbook results show that for each operator space X, both $M_n(X)$ and X are Banach spaces. In this paper we work with *finite-dimensional* operator spaces, i.e. operator spaces whose underlying vector space is finite-dimensional. The condition (B) is then automatically satisfied.

Example 5. The complex numbers $\mathbb{C}$ is an operator space when equipped with the matrix norm from Example 3. More generally, each vector space $\mathbb{M}_{n,m}$ can be equipped with an o.s.s by defining a norm on $\mathbb{M}_k(\mathbb{M}_{n,m})$ via the linear isomorphism $\mathbb{M}_k(\mathbb{M}_{n,m}) \cong M_{nk,mk}$ and the norm on the latter space (see Example 3). We write $M_{n,m}$ for this operator space. Note that this notation is compatible with Example 3, i.e. the norm $\|\cdot\|_{M_{n,m}}$ on $M_{n,m}$ is exactly the operator norm. The operator spaces $M_n \triangleq M_{n,n}$ are fundamental for the theory of operator spaces and also for the results that we present in this paper.

Example 6. For an operator space X, each of the spaces $M_n(X)$ has a canonical o.s.s that is determined by the linear isomorphism $\mathbb{M}_m(M_n(X)) \cong M_{mn}(X)$ and the already given norm on the latter space. More generally, each matrix space $\mathbb{M}_{m,n}(X)$ also has a canonical o.s.s that can be defined via the linear embedding $\mathbb{M}_{m,n}(X) \hookrightarrow M_p(X)$, where $p = \max(m, n)$, and the o.s.s of the latter space [13, §2.1] and we write $M_{m,n}(X)$ for this operator space.

Next, we introduce important classes of morphisms between operator spaces.

Definition 7 ([13, §2.2]). *Let X and Y be two vector spaces and $\varphi \colon X \to Y$ a linear map between them. The n-th amplification of φ is the linear function $\varphi_n : \mathbb{M}_n(X) \to \mathbb{M}_n(Y) :: [x_{ij}] \mapsto [\varphi(x_{ij})]$, i.e. the componentwise application of φ to the matrix $[x_{ij}]$.*

Definition 8 ([13, §2.2]). *Let X and Y be operator spaces and $\varphi \colon X \to Y$ a linear map. Consider the assignment $\|\varphi\|_{cb} \triangleq \sup\{\|\varphi_n\| : n \in \mathbb{N}\}$, where $\|\varphi_n\|$ is the operator norm of $\varphi_n \colon M_n(X) \to M_n(Y)$. We say that φ is*

- *a completely bounded map (c.b) if $\|\varphi\|_{cb} < \infty$;*
- *a complete contraction (c.c) if $\|\varphi\|_{cb} \leq 1$, i.e. each φ_n is a contraction;*
- *a complete isometry (c.i) if each φ_n is an isometry;*
- *a complete quotient map (c.q) if each φ_n is a quotient map, i.e. φ_n maps the open unit ball of $M_n(X)$ onto the open unit ball of $M_n(Y)$.*
- *a completely isometric isomorphism (c.i.i) if φ is a surjective c.i.*

Every linear map $\varphi\colon X \to Y$ between two finite-dimensional operator spaces is (completely) bounded [13, Corollary 2.2.4] and therefore φ and all of its amplifications φ_n are continuous (and bounded). The morphisms that are central to our development are the complete contractions. We write **(Fd)OS** for the category whose objects are the (f.d.) operator spaces with morphisms the complete contractions between them. We write **FdVect** for the category of finite-dimensional complex vector spaces with linear maps as morphisms and we write $U\colon \mathbf{FdOS} \to \mathbf{FdBan}$ and $V\colon \mathbf{FdOS} \to \mathbf{FdVect}$ for the forgetful functors.

The next proposition is useful because it simplifies some norm computations.

Proposition 9 ([13, §2.2, §3.3]). *Let $\varphi\colon X \to Y$ be a c.b. map. If $X = \mathbb{C}$ or $Y = \mathbb{C}$, then $\|\varphi\|_{cb} = \|\varphi\|$, i.e. the cb-norm and operator norm coincide.*

The vector space $\mathrm{CB}(X,Y)$, consisting of the completely bounded maps between X and Y, becomes a Banach space when equipped with the cb-norm $\|\cdot\|_{cb}$ and it becomes an operator space when equipped with the matrix norm given by the linear isomorphism $\mathbb{M}_n(\mathrm{CB}(X,Y)) \cong \mathrm{CB}(X, M_n(Y))$ and the Banach space structure of the latter space [13, §3.2]. Note that in **FdOS**, the underlying vector space of $\mathrm{CB}(X,Y)$ is exactly $L(X,Y)$, the vector space of linear maps between X and Y, i.e. $V(\mathrm{CB}(X,Y)) = L(X,Y)$.

If X is an operator space, then its (operator space) *dual* is the operator space $X^* \triangleq \mathrm{CB}(X,\mathbb{C})$. This is consistent with the notation for the Banach space dual, i.e. we have a Banach space equality $U(X^*) = (UX)^*$ [13, §3.2]. It is also consistent with the vector space dual for *finite-dimensional* spaces. That is, for $Y \in \mathbf{FdVect}$, we define $Y^* \triangleq L(Y,\mathbb{C})$ and then for $X \in \mathbf{FdOS}$ we also have $V(X^*) = (VX)^*$. In the sequel, we show that $\mathrm{CB}(X,Y)$ is the internal hom of **FdOS** (just like it is for **OS** [30]) and X^* is the dual in the sense of $*$-autonomy.

Example 10. The vector space $\mathbb{M}_n$ can be equipped with another important o.s.s. via the linear isomorphism $\mathbb{M}_n \cong M_n^*\ ::\ a \mapsto (b \mapsto \mathrm{tr}(ab))$ and the o.s.s. of the latter space. We write T_n for the resulting operator space. Here, tr stands for the usual trace functional and ab for matrix multiplication. The induced norm on T_n coincides with the *trace norm* and we have a completely isometric isomorphism $T_n \cong M_n^*$. This is known as the *trace class* o.s.s. [6, 1.4.5]. It follows that $\mathrm{tr}\colon T_n \to \mathbb{C}$ is a complete contraction, but $\mathrm{tr}\colon M_n \to \mathbb{C}$ is *not* one for $n > 1$.

Remark 11. Complete isometries and complete quotient maps are dual notions. A c.b. map $\varphi\colon X \to Y$ is a complete isometry (complete quotient map) iff $\varphi^*\colon Y^* \to X^*$ is a complete quotient map (complete isometry) [6, 1.4.3]. The complete isometries coincide with the strong monomorphisms and the complete quotient maps coincide with the strong epimorphisms in **FdOS** [29, §4.3].

Since $V(M_n) = \mathbb{M}_n = V(T_n)$, the usual definition of a completely positive map makes sense for linear maps $M_n \to M_m$ and $T_n \to T_m$. The next example shows that the quantum operations in the Heisenberg and Schrödinger pictures provide two important classes of completely contractive maps.

Example 12. Every CPTP map $\varphi\colon T_n \to T_m$ is a complete contraction and every CPU map $\psi\colon M_m \to M_n$ is also a complete contraction with $\|\varphi\|_{cb} = 1 = \|\psi\|_{cb}$.

2.3   Direct Sums and (Co)products

The category **FdOS** has finite (co)products which follows immediately from [29, §4.2]. We recall these constructions, which are actually completely analogous and compatible with the (co)products in **FdBan**. The category **FdOS** has a zero object given by the zero-dimensional operator space 0. If $X, Y \in$ **FdOS**, then the vector space $X \oplus Y$ can be equipped with a matrix norm via the linear isomorphism $\mathbb{M}_n(X \oplus Y) \cong M_n(X) \overset{\infty}{\oplus} M_n(Y)$ and the Banach space structure of the latter space (see §2.1). This gives an o.s.s. on $X \oplus Y$ and we write $X \overset{\infty}{\oplus} Y$ for the resulting operator space, which is the categorical product of X and Y. The notation is consistent with the one in **FdBan**, i.e. $U(X \overset{\infty}{\oplus} Y) = UX \overset{\infty}{\oplus} UY$ as Banach spaces. The categorical coproduct of X and Y in **FdOS**, written $X \overset{1}{\oplus} Y$, is the vector space $X \oplus Y$ together with the o.s.s inherited via the linear isomorphism $X \oplus Y \cong (X^* \overset{\infty}{\oplus} Y^*)^*$ and the o.s.s. of the latter space. This is also consistent with the notation in **FdBan**, i.e. $U(X \overset{1}{\oplus} Y) = UX \overset{1}{\oplus} UY$ as Banach spaces. Note that $V(X \overset{1}{\oplus} Y) = VX \oplus VY = V(X \overset{\infty}{\oplus} Y)$, i.e. V sends (co)products of **FdOS** to biproducts in **FdVect**. The functorial action of $\overset{1}{\oplus}$ and $\overset{\infty}{\oplus}$ coincides with that of $\oplus$ on morphisms, i.e. $f \overset{1}{\oplus} g = f \oplus g = f \overset{\infty}{\oplus} g$.

2.4   Tensor Products of Operator Spaces

We recall the construction of three operator space tensor products that we use in the sequel. The first two are the completely projective tensor $\widehat{\otimes}$ and the completely injective one $\widecheck{\otimes}$. They are analogous to their non-complete Banach space counterparts $\overset{\gamma}{\otimes}$ and $\overset{\epsilon}{\otimes}$ in that $\widehat{\otimes}$ $(\widecheck{\otimes})$ is the largest (smallest) tensor o.s.s that can be assigned on the vector space tensor product $X \otimes Y$, but we elide the details of what this means. The third tensor, called the Haagerup tensor, is rather unique to operator space theory and it allows us to define von Neumann (co)algebras as certain kinds of internal (co)monoid objects in the sequel. We define all three tensors for finite-dimensional operator spaces for simplicity.

Remark 13. The completely projective tensor is usually referred to as the "projective tensor" of operator spaces and the completely injective one as the "minimal" or "injective" tensor of operator spaces. We have chosen the alternative names in order to avoid confusion with their Banach space counterparts.

Definition 14 ([6, (1.5.1)]). *Let $X, Y \in \mathbf{FdOS}$. The* completely injective *tensor product, written $X \,\check{\otimes}\, Y$, is the vector space $X \otimes Y$ with the o.s.s inherited via the linear isomorphism $X \otimes Y \cong \mathrm{CB}(Y^*, X)$ and the o.s.s of the latter space.*

We show in the sequel that the above tensor product serves as the multiplicative disjunction and that the next one serves as the multiplicative conjunction.

Definition 15. *Let $X, Y \in \mathbf{FdOS}$. The* completely projective tensor product, *written $X \,\widehat{\otimes}\, Y$, is the vector space $X \otimes Y$ with the o.s.s inherited via the linear isomorphism $X \otimes Y \cong (X^* \,\check{\otimes}\, Y^*)^*$ and the o.s.s of the latter space.*

The above definition is not the standard one in the literature, but for *finite-dimensional* operator spaces, we prove that the two are equivalent [28, Appendix B]. Before we may recall the third tensor, we need an auxiliary definition.

Definition 16 ([13, §9.1]). *Given $x \in \mathbb{M}_{n,r}(X)$ and $y \in \mathbb{M}_{r,m}(Y)$ the* matrix inner product *of x and y is the matrix $x \odot y \in \mathbb{M}_{n,m}(X \otimes Y)$ defined as*

$$(x \odot y)_{i,j} \triangleq \sum_{k=1}^{r} x_{i,k} \otimes y_{k,j}.$$

Definition 17 ([13, §9.2]). *Let $X, Y \in \mathbf{FdOS}$. The* Haagerup tensor product, *written $X \overset{\mathrm{h}}{\otimes} Y$, is the vector space $X \otimes Y$ together with the o.s.s defined for $v \in \mathbb{M}_n(X \otimes Y)$ by:*

$$\|v\|_h \triangleq \inf\{\|x\|\|y\| : v = x \odot y, x \in M_{n,r}(X), y \in M_{r,n}(Y), r \in \mathbb{N}\}.$$

An interesting property of the Haagerup tensor is that it is self-dual in $\mathbf{FdOS}$.

Proposition 18 ([13, Cor. 9.4.8]). *Given $X, Y \in \mathbf{FdOS}$, the linear isomorphism $\theta : X^* \overset{\mathrm{h}}{\otimes} Y^* \cong (X \overset{\mathrm{h}}{\otimes} Y)^* :: f \otimes g \mapsto (x \otimes y \mapsto f(x)g(y))$ is a c.i.i.*

Another fact that distinguishes it (again) from the other two tensors is that the Haagerup tensor is *not* symmetric in $\mathbf{FdOS}$. What they do have in common is that $V(X \,\widehat{\otimes}\, Y) = V(X \overset{\mathrm{h}}{\otimes} Y) = V(X \,\check{\otimes}\, Y) = VX \otimes VY$, i.e. the underlying vector space of all three tensors coincides with the vector space tensor product. Furthermore, the functorial action of all three tensors on morphisms coincides with $\otimes$, i.e. $(f \,\widehat{\otimes}\, g) = (f \overset{\mathrm{h}}{\otimes} g) = (f \,\check{\otimes}\, g) = f \otimes g$. The identity map (on $X \otimes Y$) gives complete contractions $X \,\widehat{\otimes}\, Y \to X \overset{\mathrm{h}}{\otimes} Y \to X \,\check{\otimes}\, Y$, because the respective norms on the three spaces satisfy $\|\cdot\|_\wedge \geq \|\cdot\|_h \geq \|\cdot\|_\vee$. The next proposition (special case of [11, Theorem 6.1]) shows even more.

Proposition 19. *Given operator spaces $A, B, C, D \in \mathbf{FdOS}$, the permutation*

$$u_{ABCD} : (A \otimes B) \otimes (C \otimes D) \to (A \otimes C) \otimes (B \otimes D) :: (a \otimes b) \otimes (c \otimes d) \mapsto (a \otimes c) \otimes (b \otimes d)$$

gives the two complete contractions $v_{ABCD} = u_{ABCD} = w_{ABCD}$ with types

$$w_{ABCD} : (A \overset{\mathrm{h}}{\otimes} B) \,\widehat{\otimes}\, (C \overset{\mathrm{h}}{\otimes} D) \to (A \,\widehat{\otimes}\, C) \overset{\mathrm{h}}{\otimes} (B \,\widehat{\otimes}\, D)$$

$$v_{ABCD} : (A \,\check{\otimes}\, B) \overset{\mathrm{h}}{\otimes} (C \,\check{\otimes}\, D) \to (A \overset{\mathrm{h}}{\otimes} C) \,\check{\otimes}\, (B \overset{\mathrm{h}}{\otimes} D)$$

In operator space theory, maps of the above form are known as *shuffle maps*, whereas in category theory they are known as *interchange laws*.

2.5   Conjugate Operator Space and Opposite Operator Space

In the sequel, we need the operation $(\cdot)^*$ of taking the conjugate transpose of a matrix. This is not a linear map, nor a complete contraction $M_n \to M_n$ (when $n > 1$). We deal with this by recalling two involutive constructions.

An anti-linear map $f \colon V \to W$ between two complex vector spaces may be equivalently described as a linear map $f \colon V_c \to W$, where V_c is the *conjugate* vector space of V. Recall that V_c is a complex vector space with the same underlying set as V and same additive structure as V, but whose scalar multiplication is defined by $\lambda * v \triangleq \overline{\lambda} v$. This makes sense for operator spaces as well.

Definition 20 ([31, Sec. 2.9]). *Let X be an operator space. The conjugate vector space X_c can be equipped with an o.s.s. defined by $\|x\|_{M_n(X_c)} \triangleq \|x\|_{M_n(X)}$ for $x \in \mathbb{M}_n(X_c)$. We say that X_c is the* conjugate operator space *of X.*

Definition 21 ([31, Sec. 2.10]). *Let X be an operator space. The* opposite operator space *of X, written X_o, has the same underlying vector space but its o.s.s. is defined by $\|[x_{ji}]\|_{M_n(X_o)} \triangleq \|[x_{ij}]\|_{M_n(X)}$, for $x \in \mathbb{M}_n(X_o)$.*

For convenience, we also sometimes write X^o or X^c for the opposite/conjugate space, e.g. we write M_n^o instead of $(M_n)_o$.

Example 22. The conjugate transpose is a completely isometric isomorphism $(\cdot)^* \colon M_n^c \cong M_n^o$. This is just a special case of a result for C*-algebras [31, p. 65].

3   Categorical Properties of FdOS

We begin our contributions by proving relevant categorical properties of **FdOS**. Our first proposition follows easily from existing results (see [28, Appendix B]). Our first theorem, given after it, shows the relevance of **FdOS** to Linear Logic.

Proposition 23. *The category **FdOS** is a monoidal category when equipped with any of the tensors $\widehat{\otimes} \,/\, \widecheck{\otimes} \,/\, \overset{\mathrm{h}}{\otimes}$ as monoidal product and $\mathbb{C}$ as monoidal unit. The first two monoidal structures are symmetric and in all cases the (symmetric) monoidal natural isomorphisms coincide with those in **FdVect**, i.e. the forgetful functor $V \colon$ **FdOS** $\to$ **FdVect** is a strict (symmetric) monoidal functor.*

Theorem 24. *The category **FdOS** has finite (co)products, it is $*$-autonomous with relevant data given in Table 1, and is therefore a model of MALL.*

The Haagerup tensor is outside the scope of Linear Logic, but **FdOS** together with this tensor and the maps from Proposition 19 constitute a *BV-category (with negation)* in the sense of [7] (see [28, Appendix B]). The binary connective $\lhd$ from

MALL / BV	**FdOS**	**FdBan**	**FdVect**
$X \multimap Y$	$\mathrm{CB}(X,Y)$	$B(X,Y)$	$L(X,Y)$
$X \,\&\, Y$	$X \stackrel{\infty}{\oplus} Y$	$X \stackrel{\infty}{\oplus} Y$	$X \oplus Y$
$X \oplus Y$	$X \stackrel{1}{\oplus} Y$	$X \stackrel{1}{\oplus} Y$	$X \oplus Y$
$X \otimes Y$	$X \stackrel{\wedge}{\otimes} Y$	$X \stackrel{\gamma}{\otimes} Y$	$X \otimes Y$
$X \parr Y$	$X \stackrel{\vee}{\otimes} Y$	$X \stackrel{\epsilon}{\otimes} Y$	$X \otimes Y$
$X \lhd Y$	$X \stackrel{h}{\otimes} Y$		$X \otimes Y$

Table 1. MALL and BV categorical structure.

Table 1 stands for the "seq" tensor of BV-categories, which corresponds to $\stackrel{h}{\otimes}$ in **FdOS**. In the sequel, we show that $\stackrel{h}{\otimes}$ has even more categorical properties.

Next, we consider some of the categorical properties of the assignments $(\cdot)_c$ and $(\cdot)_o$ of taking conjugate/opposite operator spaces. The second equality from our next proposition follows from [9, Lemma 2.1] where it is shown that the identity map is a c.i.i (and therefore an equality). The first equality is easy.

Proposition 25. *Given operator spaces X and Y, we have operator space equalities* $\mathrm{CB}(X,Y)_c = \mathrm{CB}(X_c, Y_c)$ *and* $\mathrm{CB}(X,Y)_o = \mathrm{CB}(X_o, Y_o)$.

Since $\mathbb{C} = \mathbb{C}_o$ and $\mathbb{C}_c \cong \mathbb{C}$, with conjugation serving as the completely isometric isomorphism, we have as a special case that $(X_c)^* \cong (X^*)_c$ and $(X_o)^* = (X^*)_o$. Moreover, we can see the assignments $(\cdot)_c \colon$ **FdOS** $\to$ **FdOS** and $(\cdot)_o \colon$ **FdOS** $\to$ **FdOS** as (covariant) functors by simply defining $f_o \triangleq f$ and $f_c \triangleq f$. It is easy to see that $(\cdot)_c \circ (\cdot)_c = \mathrm{Id} = (\cdot)_o \circ (\cdot)_o$ and $(\cdot)_c \circ (\cdot)_o = (\cdot)_o \circ (\cdot)_c$ and therefore we have a pair of commuting functorial involutions on **FdOS** which are (set-theoretic) identities on morphisms, but not on objects. Even more, these functors behave well with other functorial constructions we have seen.

Proposition 26. *Given operator spaces $X, Y \in$ **FdOS**, we have:*

$$- (X \stackrel{\wedge}{\otimes} Y)_c = X_c \stackrel{\wedge}{\otimes} Y_c \qquad - (X \stackrel{1}{\oplus} Y)_c = X_c \stackrel{1}{\oplus} Y_c \qquad - (X \stackrel{h}{\otimes} Y)_c = X_c \stackrel{h}{\otimes} Y_c$$

$$- (X \stackrel{\vee}{\otimes} Y)_c = X_c \stackrel{\vee}{\otimes} Y_c \qquad - (X \stackrel{\infty}{\oplus} Y)_c = X_c \stackrel{\infty}{\oplus} Y_c \qquad - a \quad c.i.i., \quad called \quad \gamma,$$

$$- (X \stackrel{\wedge}{\otimes} Y)_o = X_o \stackrel{\wedge}{\otimes} Y_o \qquad - (X \stackrel{1}{\oplus} Y)_o = X_o \stackrel{1}{\oplus} Y_o \qquad (X \stackrel{h}{\otimes} Y)_o \cong Y_o \stackrel{h}{\otimes} X_o$$

$$- (X \stackrel{\vee}{\otimes} Y)_o = X_o \stackrel{\vee}{\otimes} Y_o \qquad - (X \stackrel{\infty}{\oplus} Y)_o = X_o \stackrel{\infty}{\oplus} Y_o \qquad \gamma(x \otimes y) \triangleq (y \otimes x).$$

Proof. Some of these facts are already known and the rest are in [28, Appendix B].

Therefore, even though the Haagerup tensor is *not* symmetric in **FdOS**, we can still talk about the usual symmetry γ, provided we use the involution $(\cdot)_o$. We also use γ to denote its own inverse, i.e. $\gamma : Y_o \stackrel{h}{\otimes} X_o \cong (X \stackrel{h}{\otimes} Y)_o$, since both γ and γ^{-1} are defined in the same way, namely, via swapping.

We also briefly use the category $\mathbf{FdOS}_{cb}$ which has f.d. operator spaces as objects, but whose morphisms are the c.b. maps between them. All the categorical properties from this section hold verbatim for $\mathbf{FdOS}_{cb}$ and the subcategory inclusion $\mathbf{FdOS} \hookrightarrow \mathbf{FdOS}_{cb}$ preserves this data. However, as we already explained, the linear and c.b. maps coincide for f.d. operator spaces, so it follows that the forgetful functor $V\colon \mathbf{FdOS}_{cb} \to \mathbf{FdVect}$ gives an equivalence of categories $\mathbf{FdOS}_{cb} \simeq \mathbf{FdVect}$. Because of this, $\mathbf{FdOS}_{cb}$ is a less interesting, and more degenerate (from a MALL perspective) category compared to $\mathbf{FdOS}$, e.g. the three tensors $\widehat{\otimes}, \overset{h}{\otimes}, \overset{\vee}{\otimes}$ are naturally isomorphic to each other, via the identity map, in $\mathbf{FdOS}_{cb}$ and each of them gives a compact closed structure on $\mathbf{FdOS}_{cb}$. Of course, this is not true in $\mathbf{FdOS}$, which is the primary category of interest.

4 Novel Categorical Approach to von Neumann Algebras

Operator spaces alone do not have sufficient structure for quantum computation. In the Heisenberg picture, we can use *von Neumann algebras (vN-algebras)* to define the relevant quantum operations. Every such algebra can be equipped with a canonical o.s.s., so we may think of them as operator spaces with additional structure. In this section, we consider f.d. vN-algebras and show that they may be equivalently defined as certain kinds of involutive monoid objects in $(\mathbf{FdOS}, \mathbb{C}, \overset{h}{\otimes})$. This serves two purposes: (1) it can help readers who are familiar with category theory; (2) it makes it easy to dualise the definition and formulate *von Neumann coalgebras* (§5) which we use for the Schrödinger picture.

We are only concerned with *finite-dimensional* vN-algebras. They coincide with the finite-dimensional C*-algebras and our next definition is standard [34,4].

Definition 27. *A f.d.* vN-algebra *is a f.d. complex vector space A together with:*

1. *a bilinear binary operation, called* multiplication, *and written via juxtaposition, which is associative, i.e. $a(bc) = (ab)c$;*
2. *an element $1 \in A$, called* unit, *such that $1a = a = a1$;*
3. *an anti-linear unary operation $(\cdot)^*$, called* involution, *such that $(a^*)^* = a$ and $(ab)^* = b^*a^*$;*
4. *a (necessarily complete) norm which is* submultiplicative, *i.e. $\|ab\| \leq \|a\|\|b\|$, and which satisfies the* C*-identity, *i.e. $\|a^*a\| = \|a\|^2$.*

If A is a f.d. vN-algebra, then so is $\mathbb{M}_n(A)$ – the multiplication, unit and involution are defined in the obvious way and then there is a *unique* way to define a norm on $\mathbb{M}_n(A)$ such that it becomes a vN-algebra [6, 1.2.3]. These norms give the canonical o.s.s of A [6, 1.2.3] and henceforth we view vN-algebras as operator spaces with additional structure.

Example 28. The operator space M_n is a vN-algebra with unit given by the identity matrix, multiplication by matrix multiplication, and involution given by $(\cdot)^*$. The operator space $\overset{\infty}{\oplus}_{1 \leq i \leq n} M_{k_i}$ is also a vN-algebra with unit, multiplication, and involution defined pointwise. In fact, modulo a vN-algebra isomorphism (defined below), every f.d. vN-algebra is of this form [34, p. 50].

For our categorical definition of a f.d. vN-algebra, we take inspiration from *reversing involutive monoids* [23] and *star algebras* in a bar category [3]. More importantly, the (f.d.) unital operator algebras have been characterised as the monoid objects w.r.t. the Haagerup tensor in $(\mathbf{Fd})\mathbf{OS}$ [31, Theorem 6.1], [8]. Since f.d. vN-algebras are special kinds of unital operator algebras, it is natural to consider them as special kinds of monoid objects in $(\mathbf{FdOS}, \mathbb{C}, \overset{h}{\otimes})$. We do so in our subsequent definition.

Definition 29. *A f.d.* vN-algebra *is a f.d. operator space A together with*

1. *a complete contraction $\mu\colon A \overset{h}{\otimes} A \to A$, called* multiplication*;*
2. *a complete contraction $\eta\colon \mathbb{C} \to A$, called* unit*;*
3. *a complete contraction $i\colon A_c \to A_o$, called* involution*,*

such that (A, η, μ) is a monoid object in $(\mathbf{FdOS}, \mathbb{C}, \overset{h}{\otimes})$, the following diagrams

$$
\begin{array}{ccc}
A \!=\!=\!=\!=\!=\!=\!=\! A & \qquad & A_c \overset{h}{\otimes} A_c \!=\!=\! (A \overset{h}{\otimes} A)_c \xrightarrow{\;\mu\;} A_c \\
\Big\| \qquad\qquad\qquad \Big\| & & {\scriptstyle i\otimes i}\Big\downarrow \qquad\qquad\qquad\qquad \Big\downarrow{\scriptstyle i} \\
A_{cc} \xrightarrow[i_c]{} A_{oc} = A_{co} \xrightarrow[i_o]{} A_{oo} & & A_o \overset{h}{\otimes} A_o \xrightarrow[\;\gamma\;]{} (A \overset{h}{\otimes} A)_o \xrightarrow[\;\mu\;]{} A_o
\end{array}
\tag{1}
$$

commute, and for every complete isometry (i.e. strong mono) $a\colon \mathbb{C} \to A$, the following composite in $\mathbf{FdOS}$ is also a complete isometry (i.e. strong mono)

$$
\mathbb{C} \cong \mathbb{C}_o \overset{h}{\otimes} \mathbb{C}_c \xrightarrow{a\otimes a} A_o \overset{h}{\otimes} A_c \xrightarrow{A_o \otimes i} A_o \overset{h}{\otimes} A_o \xrightarrow{\gamma} (A \overset{h}{\otimes} A)_o \xrightarrow{\mu} A_o.
\tag{2}
$$

Theorem 30. *Definitions 27 and 29 are equivalent, i.e. the data in one definition uniquely determines the data in the other.*

The proof of this theorem can be found in the companion paper [28, Appendix C] and it shows that the two (equivalent) notions may be used interchangeably. We do so in the sequel.

Definition 31. *Let* $\mathbf{vNAlg_{fd}}$ *denote the subcategory of* $\mathbf{FdOS}$ *whose objects are f.d. vN-algebras and whose morphisms $\varphi\colon A \to B$ are multiplicative, unital, and* involutive*, i.e. the following diagrams in* $\mathbf{FdOS}$ *commute*[1].

$$
\begin{array}{ccccccc}
A \overset{h}{\otimes} A \xrightarrow{\varphi\otimes\varphi} B \overset{h}{\otimes} B & \qquad & & \mathbb{C} & & \qquad & A_c \xrightarrow{\varphi_c} B_c \\
{\scriptstyle \mu_A}\Big\downarrow \qquad\qquad \Big\downarrow{\scriptstyle \mu_B} & & {\scriptstyle \eta_A}\nearrow & & \nwarrow{\scriptstyle \eta_B} & & {\scriptstyle i_A}\Big\downarrow \qquad \Big\downarrow{\scriptstyle i_B} \\
A \xrightarrow[\;\varphi\;]{} B & & A & \xrightarrow[\;\varphi\;]{} & B & & A_o \xrightarrow[\;\varphi_o\;]{} B_o
\end{array}
$$

In the literature, these morphisms are known as "*unital $*$-homomorphisms*", but we simply call them vN-algebra morphisms. It is already known that (possibly infinite-dimensional) vN-algebras form a symmetric monoidal category with small products [27], so the next propositions follow easily. However, we present them in our categorical style, as this helps for the development in §5.

---

[1] It suffices to assume φ is linear and the three diagrams imply it is also c.c.

Proposition 32. *If $A, B \in \mathbf{vNAlg_{fd}}$, then the operator space $A \,\breve{\otimes}\, B$ with*

- *unit $\eta \triangleq \left(\mathbb{C} \cong \mathbb{C} \,\breve{\otimes}\, \mathbb{C} \xrightarrow{\eta_A \otimes \eta_B} A \,\breve{\otimes}\, B \right)$*
- *involution $i \triangleq \left((A \,\breve{\otimes}\, B)_c = A_c \,\breve{\otimes}\, B_c \xrightarrow{i_A \otimes i_B} A_o \,\breve{\otimes}\, B_o = (A \,\breve{\otimes}\, B)_o \right)$*
- *mult. $\mu \triangleq \left((A \,\breve{\otimes}\, B) \stackrel{h}{\otimes} (A \,\breve{\otimes}\, B) \xrightarrow{v} (A \stackrel{h}{\otimes} A) \,\breve{\otimes}\, (B \stackrel{h}{\otimes} B) \xrightarrow{\mu_A \otimes \mu_B} A \,\breve{\otimes}\, B \right)$*

is also a vN-algebra. Also, the operator space $A \stackrel{\infty}{\oplus} B$ is a vN-algebra with

- *unit $\eta \triangleq \left(\mathbb{C} \xrightarrow{\langle id, id \rangle} \mathbb{C} \stackrel{\infty}{\oplus} \mathbb{C} \xrightarrow{\eta_A \oplus \eta_B} A \stackrel{\infty}{\oplus} B \right)$*
- *involution $i \triangleq \left((A \stackrel{\infty}{\oplus} B)_c = A_c \stackrel{\infty}{\oplus} B_c \xrightarrow{i_A \oplus i_B} A_o \stackrel{\infty}{\oplus} B_o = (A \stackrel{\infty}{\oplus} B)_o \right)$*
- *mult. $\mu \triangleq \left((A \stackrel{\infty}{\oplus} B) \stackrel{h}{\otimes} (A \stackrel{\infty}{\oplus} B) \xrightarrow{v'} (A \stackrel{h}{\otimes} A) \stackrel{\infty}{\oplus} (B \stackrel{h}{\otimes} B) \xrightarrow{\mu_A \oplus \mu_B} A \stackrel{\infty}{\oplus} B \right)$*

where $v' \triangleq \left((A \stackrel{\infty}{\oplus} B) \stackrel{h}{\otimes} (A \stackrel{\infty}{\oplus} B) \xrightarrow{\langle (\pi_A \otimes \pi_A), (\pi_B \otimes \pi_B) \rangle} (A \stackrel{h}{\otimes} A) \stackrel{\infty}{\oplus} (B \stackrel{h}{\otimes} B) \right)$

Proposition 33. *The category $\mathbf{vNAlg_{fd}}$ is symmetric monoidal and has finite products with the action of $\breve{\otimes}$ and $\stackrel{\infty}{\oplus}$ extended as in Proposition 32.*

Quantum Operations. The quantum operations in the Heisenberg picture of f.d. quantum theory are given by the completely positive unital (CPU) maps. We already defined unital maps, so now we recall completely positive maps. Given $A \in \mathbf{vNAlg_{fd}}$, we say that $p \in A$ is a *positive element* if there exists $a \in A$, such that $p = a^* a$. An equivalent way of saying this is that the diagram in $\mathbf{FdOS_{cb}}$

$$
\begin{array}{ccccccccc}
\mathbb{C} & \xrightarrow{\cong} & \mathbb{C}_o \stackrel{h}{\otimes} \mathbb{C}_c & \xrightarrow{a \otimes a} & A_o \stackrel{h}{\otimes} A_c & \xrightarrow{A_o \otimes i} & A_o \stackrel{h}{\otimes} A_o & \xrightarrow{\gamma} & (A \stackrel{h}{\otimes} A)_o \\
\Big\| & & & & & & & & \Big\downarrow \mu \\
\mathbb{C}_o & & & & \xrightarrow{\hspace{3cm} p \hspace{3cm}} & & & & A_o
\end{array}
\tag{3}
$$

commutes, where we have made the obvious identification of elements of A as morphisms in $\mathbf{FdOS_{cb}}$. A linear map $\varphi \colon A \to B$ between two f.d. vN-algebras is called *positive* if it preserves positive elements and *completely positive (c.p.)* if $M_n \,\breve{\otimes}\, \varphi \colon M_n \,\breve{\otimes}\, A \to M_n \,\breve{\otimes}\, B$ is positive for every $n \in \mathbb{N}$.

Finally, we write $\mathbf{H}$ for the category whose objects are f.d. vN-algebras and whose morphisms are the CPU maps. The next proposition, which is a special case of [13, Corollary 5.1.2], shows that $\mathbf{H}$ is a subcategory of $\mathbf{FdOS}$. The proposition after it summarises the categorical properties of $\mathbf{H}$ relevant to MALL.

Proposition 34. *Let $\varphi \colon A_1 \to A_2$ be a linear unital map between two von Neumann algebras. Then, φ is completely positive iff φ is completely contractive.*

Proposition 35. *The category $\mathbf{H}$ is symmetric monoidal (w.r.t. $\breve{\otimes}$), has finite products (given by $\stackrel{\infty}{\oplus}$), and the inclusions $\mathbf{vNAlg_{fd}} \hookrightarrow \mathbf{H} \hookrightarrow \mathbf{FdOS}$ preserve this data.*

5   von Neumann Coalgebras

In [11], a vN-coalgebra structure is *defined* as the predual of the structure of a vN-algebra. Instead of doing this, we introduce f.d. vN-coalgebras *independently* of vN-algebras and we *derive* the duality later. Thanks to the preparatory categorical work we did in Sections 3–4, and keeping in mind that the Haagerup tensor is self-dual (Proposition 18), we can now easily dualise the definition of a f.d. vN-algebra to obtain that of a f.d. vN-coalgebra.

Definition 36. *A f.d. vN-coalgebra is a f.d. operator space C together with*

1. *a complete contraction $\delta \colon C \to C \overset{h}{\otimes} C$, called* comultiplication;
2. *a complete contraction $\varepsilon \colon C \to \mathbb{C}$, called* counit;
3. *a complete contraction $j \colon C_o \to C_c$, called* involution,

such that (C, ε, δ) is a comonoid object in $(\mathbf{FdOS}, \mathbb{C}, \overset{h}{\otimes})$, the following diagrams

$$
\begin{array}{ccc}
C \;=\!=\!=\!=\!=\!=\; C & \qquad & C_o \xrightarrow{\;j\;} C_c \xrightarrow{\;\delta\;} (C \overset{h}{\otimes} C)_c \\
\| \qquad\qquad \| & & {\scriptstyle\delta}\downarrow \qquad\qquad\qquad \| \\
C_{oo} \xrightarrow{\;j_o\;} C_{co} = C_{oc} \xrightarrow{\;j_c\;} C_{cc} & & (C \overset{h}{\otimes} C)_o \xrightarrow{\;\gamma\;} C_o \overset{h}{\otimes} C_o \xrightarrow{\;j\otimes j\;} C_c \overset{h}{\otimes} C_c
\end{array}
\tag{4}
$$

commute, and for every complete quotient map (i.e. strong epi) $e \colon C \to \mathbb{C}$, the following composite in $\mathbf{FdOS}$ is a complete quotient map (i.e. strong epi)

$$
C_o \xrightarrow{\;\delta\;} (C \overset{h}{\otimes} C)_o \xrightarrow{\;\gamma\;} C_o \overset{h}{\otimes} C_o \xrightarrow{\;C_o \otimes j\;} C_o \overset{h}{\otimes} C_c \xrightarrow{\;e \otimes e\;} \mathbb{C}_o \overset{h}{\otimes} \mathbb{C}_c \cong \mathbb{C}.
\tag{5}
$$

We dualise Definition 31 in order to obtain the next one.

Definition 37. *Let $\mathbf{vNCoalg_{fd}}$ denote the subcategory of $\mathbf{FdOS}$ whose objects are f.d. vN-coalgebras and whose morphisms $\varphi \colon C \to D$ are comultiplicative, counital, and involutive, i.e. the following diagrams in $\mathbf{FdOS}$ commute.*

$$
\begin{array}{ccc}
\begin{array}{ccc}
C & \xrightarrow{\;\varphi\;} & D \\
{\scriptstyle\delta_C}\downarrow & & \downarrow{\scriptstyle\delta_D} \\
C \overset{h}{\otimes} C & \xrightarrow{\;\varphi\otimes\varphi\;} & D \overset{h}{\otimes} D
\end{array}
& \quad
\begin{array}{c}
C \xrightarrow{\;\varphi\;} D \\
{\scriptstyle\varepsilon_C}\searrow \quad \swarrow{\scriptstyle\varepsilon_D} \\
\mathbb{C}
\end{array}
& \quad
\begin{array}{ccc}
C_o & \xrightarrow{\;\varphi_o\;} & D_o \\
{\scriptstyle j_C}\downarrow & & \downarrow{\scriptstyle j_D} \\
C_c & \xrightarrow{\;\varphi_c\;} & D_c
\end{array}
\end{array}
$$

We also dualise the construction of tensors and direct sums from Proposition 32 and we now use the remaining shuffle map from Proposition 19.

Proposition 38. *If $C, D \in \mathbf{vNCoalg_{fd}}$, then the operator space $C \widehat{\otimes} D$ with*

- *counit $\varepsilon \triangleq \left(C \widehat{\otimes} D \xrightarrow{\;\varepsilon_C \otimes \varepsilon_D\;} \mathbb{C} \widehat{\otimes} \mathbb{C} \cong \mathbb{C} \right)$*
- *involution $j \triangleq \left((C \widehat{\otimes} D)_o = C_o \widehat{\otimes} D_o \xrightarrow{\;j_C \otimes j_D\;} C_c \widehat{\otimes} D_c = (C \widehat{\otimes} D)_c \right)$*
- *comult. $\delta \triangleq \left(C \widehat{\otimes} D \xrightarrow{\;\delta_C \otimes \delta_D\;} (C \overset{h}{\otimes} C) \widehat{\otimes} (D \overset{h}{\otimes} D) \xrightarrow{\;w\;} (C \widehat{\otimes} D) \overset{h}{\otimes} (C \widehat{\otimes} D) \right)$*

is also a vN-coalgebra. Also, the operator space $C \overset{1}{\oplus} D$ is a vN-coalgebra with

$$- \ \text{counit } \varepsilon \triangleq \left(C \overset{1}{\oplus} D \xrightarrow{\varepsilon_C \oplus \varepsilon_D} \mathbb{C} \overset{1}{\oplus} \mathbb{C} \xrightarrow{[id,id]} \mathbb{C} \right)$$

$$- \ \text{involution } j \triangleq \left((C \overset{1}{\oplus} D)_o = C_o \overset{1}{\oplus} D_o \xrightarrow{j_C \oplus j_D} C_c \overset{1}{\oplus} D_c = (C \overset{1}{\oplus} D)_c \right)$$

$$- \ \text{comult. } \delta \triangleq \left(C \overset{1}{\oplus} D \xrightarrow{\delta_C \oplus \delta_D} (C \overset{h}{\otimes} C) \overset{1}{\oplus} (D \overset{h}{\otimes} D) \xrightarrow{w'} (C \overset{1}{\oplus} D) \overset{h}{\otimes} (C \overset{1}{\oplus} D) \right)$$

$$\text{where } w' \triangleq \left((C \overset{h}{\otimes} C) \overset{1}{\oplus} (D \overset{h}{\otimes} D) \xrightarrow{[(\iota_C \otimes \iota_C),(\iota_D \otimes \iota_D)]} (C \overset{1}{\oplus} D) \overset{h}{\otimes} (C \overset{1}{\oplus} D) \right)$$

Proposition 39. *The category* **vNCoalg$_{\text{fd}}$** *is symmetric monoidal and has finite coproducts with the action of* $\widehat{\otimes}$ *and* $\overset{1}{\oplus}$ *extended as in Proposition 38.*

Example 40. The operator space T_n is a vN-coalgebra with: (1) counit given by the trace functional $\varepsilon \triangleq \text{tr} : T_n \to \mathbb{C} :: t \mapsto \text{tr}(t)$; (2) involution given by the conjugate transpose $j \triangleq (\cdot)^* : T_n^o \to T_n^c$; (3) comultiplication given by $\delta : T_n \to T_n \overset{h}{\otimes} T_n :: e_{ij} \mapsto \sum_k (e_{kj} \otimes e_{ik})$, where $e_{ij} \in T_n$ is the matrix that has 1 in position (i,j) and 0 elsewhere. Using Proposition 38, we see that $\overset{1}{\oplus}_{1 \leq i \leq n} T_{k_i}$ is also a vN-coalgebra and using Theorem 44 and Example 28 it follows that each f.d. vN-coalgebra is of this form, modulo vN-coalgebra isomorphism.

Quantum Operations. We use f.d. vN-coalgebras to define the quantum operations in the Schrödinger picture, i.e. the CPTP maps. This can be done by dualising the CPU maps from §4. The dual of a unital map is obviously a counital one. Unfortunately, in quantum information theory, the term "trace-preserving" is standard, so we simply say that a linear map $\varphi : C \to D$ between two vN-coalgebras is *trace-preserving* if it is counital. We explain why this is justified. Note that if $C = T_n$ and $D = T_m$, then from Example 40, we see that a counital map is precisely a trace-preserving one. More generally, the direct sum $\overset{1}{\oplus}_i T_{k_i}$ can be linearly embedded block-diagonally in a sufficiently large matrix space $\mathbb{M}_m$ and then counitality is equivalent to trace-preservation with respect to this view. Using Example 40, this covers all f.d. vN-coalgebras, modulo isomorphism.

We proceed with the (dual) definition of a completely positive map between two f.d. vN-coalgebras. We work again in the category **FdOS$_{\text{cb}}$**. Given a vN-coalgebra C, we say that a linear functional $p : C \to \mathbb{C}$ is *positive* if there exists a linear functional $a : C \to \mathbb{C}$, such that the following diagram:

$$
\begin{array}{ccccccccc}
(C \overset{h}{\otimes} C)_o & \xrightarrow{\gamma} & C_o \overset{h}{\otimes} C_o & \xrightarrow{C_o \otimes j} & C_o \overset{h}{\otimes} C_c & \xrightarrow{a \otimes a} & \mathbb{C}_o \overset{h}{\otimes} \mathbb{C}_c & \xrightarrow{\cong} & \mathbb{C} \\
{\scriptstyle \delta} \uparrow & & & & & & & & \| \\
C_o & & & & \xrightarrow{ p } & & & & \mathbb{C}_o
\end{array}
\tag{6}
$$

commutes. Note that (6) is dual to (3). Continuing in this fashion, we say that a linear map $\varphi : C \to D$ between vN-coalgebras is *positive* if it reflects positive linear functionals, i.e. $p \circ \varphi \colon C \to \mathbb{C}$ is a positive linear functional on C for every positive linear functional $p \colon D \to \mathbb{C}$. Finally, we say φ is *completely positive* if $T_n \widehat{\otimes} \varphi \colon T_n \widehat{\otimes} C \to T_n \widehat{\otimes} D$ is positive for every $n \in \mathbb{N}$. This notion of (complete) positivity is equivalent to the usual/concrete one (see [28, Appendix D]).

We write $\mathbf{S}$ for the category whose objects are f.d. vN-coalgebras and whose morphisms are the CPTP maps between them. Finally, we show that $\mathbf{S}$ is a subcategory of $\mathbf{FdOS}$ and summarise its structure relevant to MALL.

Proposition 41. *Let $\varphi \colon C_1 \to C_2$ be a linear trace-preserving map between f.d. vN-coalgebras. Then, φ is completely positive iff φ is completely contractive.*

Proposition 42. *The category $\mathbf{S}$ is symmetric monoidal (w.r.t. $\widehat{\otimes}$), has finite coproducts (given by $\overset{1}{\oplus}$), and the inclusions $\mathbf{vNCoalg_{fd}} \hookrightarrow \mathbf{S} \hookrightarrow \mathbf{FdOS}$ preserve this data.*

6   Heisenberg-Schrödinger Duality and vN-(co)algebras

In f.d. quantum theory, the quantum operations in the Heisenberg picture are precisely the CPU maps, i.e. the morphisms of $\mathbf{H}$, whereas the quantum operations in the Schrödinger picture are precisely the CPTP maps, i.e. the morphisms of $\mathbf{S}$. Both $\mathbf{H}$ and $\mathbf{S}$ are subcategories of $\mathbf{FdOS}$ and the dual functor $(\cdot)^* \colon \mathbf{FdOS} \to \mathbf{FdOS}$ can be defined also on vN-(co)algebras, as we show next.

Proposition 43. *If $A \in \mathbf{vNAlg_{fd}}$, then the o.s. dual A^* is a vN-coalgebra with counit $\varepsilon \triangleq \left(A^* \xrightarrow{\eta^*} \mathbb{C}^* \cong \mathbb{C} \right)$, involution $j \triangleq \left((A^*)_o = (A_o)^* \xrightarrow{i^*} (A_c)^* \cong (A^*)_c \right)$, and comultiplication $\delta \triangleq \left(A^* \xrightarrow{\mu^*} (A \overset{h}{\otimes} A)^* \cong A^* \overset{h}{\otimes} A^* \right)$. Conversely, if $C \in \mathbf{vNCoalg_{fd}}$, then the o.s. dual C^* is a vN-algebra with unit $\eta \triangleq \left(\mathbb{C} \cong \mathbb{C}^* \xrightarrow{\varepsilon^*} C^* \right)$, involution $i \triangleq \left((C^*)_c \cong (C_c)^* \xrightarrow{j^*} (C_o)^* = (C^*)_o \right)$, and multiplication given by $\mu \triangleq \left(C^* \overset{h}{\otimes} C^* \cong (C \overset{h}{\otimes} C)^* \xrightarrow{\delta^*} C^* \right)$.*

Theorem 44. *We have equivalences of categories $(\cdot)^* \colon \mathbf{S} \simeq \mathbf{H}^{\mathrm{op}} \colon (\cdot)^*$ and also $(\cdot)^* \colon \mathbf{vNCoalg_{fd}} \simeq \mathbf{vNAlg_{fd}}^{\mathrm{op}} \colon (\cdot)^*$, where the action on objects of $(\cdot)^*$ is defined as in Proposition 43 and on morphisms in the usual way, i.e. as in $\mathbf{FdOS}$. Moreover, these equivalences are strong monoidal (and (co)product preserving).*

Example 45. The c.i.i. from Example 10 is a vN-coalgebra isomorpism $T_n \cong M_n^*$, which determines a vN-algebra isomorphism $T_n^* \cong M_n$ by duality.

Theorem 44 gives a categorical formulation of the Heisenberg-Schrödinger duality and shows precisely how the MALL related structure of $\mathbf{vNCoalg_{fd}}$ and $\mathbf{S}$ from §5 is dual to that of $\mathbf{vNAlg_{fd}}$ and $\mathbf{H}$ from §4, and vice-versa. The duality $\mathbf{vNCoalg_{fd}} \simeq \mathbf{vNAlg_{fd}}^{\mathrm{op}}$ shows that vN-(co)algebras are dual to each other in a strong mathematical sense, whereas the duality $\mathbf{S} \simeq \mathbf{H}^{\mathrm{op}}$ is more relevant for quantum computation and we use it in the next section.

7 Revisiting Quantum Coherence Spaces

The preparatory categorical work in Sections 3 – 6 shows that the (sub)categories $\mathbf{S} \hookrightarrow \mathbf{FdOS} \hookleftarrow \mathbf{H}$ have the right categorical structure for a model of MALL whose duality is induced by the Heisenberg-Schrödinger duality. However, there is one problem: neither $\mathbf{S}$, nor $\mathbf{H}$, is a *full* subcategory of $\mathbf{FdOS}$. This means that we have morphisms between vN-algebras that are not CPU maps and morphisms between vN-coalgebras that are not CPTP maps in $\mathbf{FdOS}$. For example, the map $a \mapsto -a : M_2 \to M_2$ is not positive but it is in $\mathbf{FdOS}(M_2, M_2)$. Luckily, there is a simple solution: a semantic technique of Hyland and Schalk [22], based on *gluing and orthogonality*, allows us to carve out a category $\mathbf{Q}$ from $\mathbf{FdOS}$, such that we get fully faithful inclusions $\mathbf{S} \hookrightarrow \mathbf{Q} \hookleftarrow \mathbf{H}$, ensuring that $\mathbf{Q}$ contains precisely the quantum operations in the relevant homsets, while still preserving all the MALL structure. The main idea, in a nutshell, is to ensure trace-preservation (unitality) for morphisms between vN-coalgebras (vN-algebras) and then fullness is guaranteed by Propositions 34 and 41. This is precisely what we do by defining a suitable notion of orthogonality/polarity based on ideas from [22]. The resulting model has an obvious resemblance to (probabilistic) coherence spaces.

More specifically, we are interested in pairs (X, S), where X is an operator space and $S \subseteq \mathrm{Ball}(X)$ is a subset of the unit ball of X, which satisfies an additional condition that we proceed to explain. Our next definition has similarities with *polar sets* from functional and convex analysis, which motivates its name.

Definition 46 (Polar). *Given an operator space X and a subset $S \subseteq \mathrm{Ball}(X)$, we define the* polar *of S to be $S^\circ \triangleq \{f \in \mathrm{Ball}(X^*) \mid \forall s \in S.\ f(s) = 1\}$. We say that S is* bipolar *if $S = d^{-1}[S^{\circ\circ}]$, where $d : X \cong X^{**}$ is the canonical c.i.i.*

To avoid notational overhead, when dealing with multiple applications of the polar construction, we implicitly consider the image of such polar sets to be taken under d^{-1}. With this convention, we simply say that S is bipolar if $S = S^{\circ\circ}$. As expected, the polar introduces a Galois connection [22, §5.1] and we have: for $R \subseteq S \subseteq \mathrm{Ball}(X)$, it follows that $S^\circ \subseteq R^\circ$, $S \subseteq S^{\circ\circ}$, and $S^{\circ\circ\circ} = S^\circ$. We can now introduce the main category that is computationally relevant.

Definition 47. *Let $\mathbf{Q}$ be the category whose objects are pairs (X, S) with $X \in \mathbf{FdOS}$ and $S \subseteq \mathrm{Ball}(X)$ a bipolar set, i.e. $S = S^{\circ\circ}$, and whose morphisms $f : (X, S) \to (Y, R)$ are complete contractions $f : X \to Y$ such that $f[S] \subseteq R$.*

Our notion of polarity is instead called "orthogonality" in [22] and it enjoys useful properties, e.g. it is *tight, stable, focused, precise*, in the sense of [22]. By using results from [22], it follows that the MALL structure of $\mathbf{FdOS}$ is preserved.

Theorem 48. *The category $\mathbf{Q}$ is $*$-autonomous and has finite (co)products: duals are given by $(X, S)^* \triangleq (X^*, S^\circ)$; the initial object by $(0, \varnothing)$; the terminal object by $(0, \{0\})$; binary products by $(X, S) \overset{\infty}{\oplus} (Y, R) \triangleq (X \overset{\infty}{\oplus} Y, S \times R)$; binary coproducts by $(X, S) \overset{1}{\oplus} (Y, R) \triangleq (X \overset{1}{\oplus} Y, (S^\circ + R^\circ)^\circ)$, where*

$$S^\circ + R^\circ \triangleq \{[s', r'] : X \overset{1}{\oplus} Y \to \mathbb{C} \mid s' \in S^\circ, r' \in R^\circ\} \subseteq \mathrm{Ball}((X \overset{1}{\oplus} Y)^*);$$

Schrödinger Picture	$\mathbf{S} \overset{\text{full}}{\hookrightarrow} \mathbf{Q}$	LL_+
System description	$\mathcal{C}, \mathcal{D}$	P, R
Quantum composition	$\mathcal{C} \mathbin{\widehat{\otimes}} \mathcal{D}$	$P \otimes R$
Classical composition	$\mathcal{C} \mathbin{\overset{1}{\oplus}} \mathcal{D}$	$P \oplus R$
Quantum operation	$\mathcal{C} \xrightarrow{\text{CPTP}} \mathcal{D}$	$P \vdash R$

Heisenberg Picture	$\mathbf{H} \overset{\text{full}}{\hookrightarrow} \mathbf{Q}$	LL_-
System description	$\mathcal{A}, \mathcal{B}$	N, M
Quantum composition	$\mathcal{A} \mathbin{\widecheck{\otimes}} \mathcal{B}$	$N \,⅋\, M$
Classical composition	$\mathcal{A} \mathbin{\overset{\infty}{\oplus}} \mathcal{B}$	$N \mathbin{\&} M$
Quantum operation	$\mathcal{B} \xrightarrow{\text{CPU}} \mathcal{A}$	$M \vdash N$

Fig. 1. Schrödinger/Heisenberg picture and positive/negative logical polarity.

multiplicative conjunction is given by $(X, S) \mathbin{\widehat{\otimes}} (Y, R) \triangleq (X \mathbin{\widehat{\otimes}} Y, (S \otimes R)^{\circ\circ})$, *where* $S \otimes R \triangleq \{s \otimes r \mid s \in S, \ r \in R\}$; *multiplicative disjunction by* $(X, S) \mathbin{\widecheck{\otimes}} (Y, R) \triangleq (X \mathbin{\widecheck{\otimes}} Y, (S^\circ \otimes R^\circ)^\circ)$; *tensor unit for both tensors is given by* $(\mathbb{C}, \{1\})$.

The polarity that we introduced behaves very well w.r.t. vN-(co)algebras. In order to see this, we first introduce some notation. For a vN-coalgebra C, we say that an element $p \in C$ is *positive* if the map $1 \mapsto p \colon \mathbb{C} \to C$ is positive and we write $p \geq 0$ to indicate this. The *density operators* on C are defined to be the set $P_C \triangleq \{p \in C \mid \varepsilon(p) = 1 \text{ and } p \geq 0\}$. Note that if $C = T_n$, then P_C is precisely the set D_n of $n \times n$ density matrices, i.e. positive matrices whose trace is 1.

Theorem 49. *If* $A \in \mathbf{vNAlg_{fd}}$, *then* $(A, \{1_A\}) \in \mathrm{Ob}(\mathbf{Q})$. *Moreover, the functor* $H \colon \mathbf{H} \to \mathbf{Q}$ *defined by* $H(A) \triangleq (A, \{1_A\})$ *and* $H(f) \triangleq f$, *is fully faithful, strict monoidal w.r.t* $\widecheck{\otimes}$, *and it strictly preserves finite products.*

Theorem 50. *If* $C \in \mathbf{vNCoalg_{fd}}$, *then* $(C, P_C) \in \mathrm{Ob}(\mathbf{Q})$. *Moreover, the functor* $S \colon \mathbf{S} \to \mathbf{Q}$ *defined by* $S(C) \triangleq (C, P_C)$ *and* $S(f) \triangleq f$, *is fully faithful, strict monoidal w.r.t* $\widehat{\otimes}$, *and it strictly preserves finite coproducts.*

Example 51. For a unitary matrix $u \in \mathbb{M}_{2^n}$, the c.i.i. $a \mapsto uau^* : T_{2^n} \to T_{2^n}$ describes a unitary evolution of the system with respect to u in the Schrödinger picture. It preserves density operators and is thus a map in $\mathbf{Q}((T_{2^n}, D_{2^n}), (T_{2^n}, D_{2^n}))$. The corresponding c.i.i. in the Heisenberg picture, $a \mapsto u^*au : M_{2^n} \to M_{2^n}$, preserves the unit and is thus a map in $\mathbf{Q}((M_{2^n}, \{1_{2^n}\}), (M_{2^n}, \{1_{2^n}\}))$. Since $(T_2, D_2) \mathbin{\widehat{\otimes}} \cdots \mathbin{\widehat{\otimes}} (T_2, D_2) \cong (T_{2^n}, D_{2^n})$ and likewise $(M_2, \{1_2\}) \mathbin{\widecheck{\otimes}} \cdots \mathbin{\widecheck{\otimes}} (M_2, \{1_2\}) \cong (M_{2^n}, \{1_{2^n}\})$, these objects of $\mathbf{Q}$ can be understood as representing an array of n qubits in the Schrödinger/Heisenberg picture.

We are now justified in presenting the summary provided by Figure 1, on which we elaborate. In our model $\mathbf{Q}$, formulas in MALL admit interpretations as objects of $\mathbf{Q}$ and proofs are interpreted as morphisms of $\mathbf{Q}$. Formulas with positive (negative) logical polarities admit natural interpretations as vN-coalgebras $\mathcal{C}, \mathcal{D}$ (vN-algebras $\mathcal{A}, \mathcal{B}$) and proofs between such formulas correspond precisely to the CPTP (CPU) maps, i.e. the quantum operations in the Schrödinger (Heisenberg) picture. Moreover, this correspondence is preserved by classical composition (spacewise composition of systems where only a limited amount of classical interactions are possible) and by quantum composition (spacewise composition with the full range of quantum interactions possible, e.g. entanglement), in both pictures, whose interpretations are provided by the respective tables.

Pure Quantum Computation. We showed that $\mathbf{Q}$ captures *mixed state* quantum computation in both pictures. In fact, $\mathbf{Q}$ also has interesting properties that are relevant to *pure state* quantum computation, where unitarity is very important. If M is a f.d. vN-algebra and $u \in M$ a unitary element, i.e. $uu^* = 1_M = u^*u$, then $(M, \{u\}) \in \mathrm{Ob}(\mathbf{Q})$, i.e. the set $\{u\}$ is bipolar in M. We can reason in $\mathbf{Q}$ about interesting higher-order (pure state) maps, such as the pure state *quantum switch*. This is the linear map $\mathrm{qsw}\colon M_n \widehat{\otimes} M_n \to M_{2n}$ defined by $\mathrm{qsw}(a \otimes b) \triangleq (|0\rangle\langle 0| \otimes (ab)) + (|1\rangle\langle 1| \otimes (ba))$ and it is a complete contraction (special case of [30, §IV]). For any $n \times n$ unitary matrices u and v, we have object equalities $(M_n, \{u\}) \widehat{\otimes} (M_n, \{v\}) = (M_n \widehat{\otimes} M_n, \{u \otimes v\}^{\circ\circ}) = (M_n \widehat{\otimes} M_n, \{u \otimes v\})$ and qsw becomes a morphism of our model $\mathbf{Q}$ (see [28, Appendix F] for a proof) with type $\mathrm{qsw}\colon (M_n, \{u\}) \widehat{\otimes} (M_n, \{v\}) \to (M_{2n}, \{|0\rangle\langle 0| \otimes (uv) + |1\rangle\langle 1| \otimes (vu)\})$, where all three bipolar subsets consist of unitary matrices. Therefore qsw may be recognised as a valid (unitary-preserving) higher-order map in $\mathbf{Q}$. The "higher-order" description of qsw is justified by the fact that in pure state computation, we think of the unitary elements $u, v \in M_n$ as first-order (reversible) functions.

One of the reasons qsw is interesting, is that it does not admit any reasonable multi-linear decomposition, i.e. there are no complete contractions $\varphi_1\colon M_n \widehat{\otimes} M_n \to M_{2n,m}$ and $\varphi_2\colon M_n \widehat{\otimes} M_n \to M_{m,2n}$ such that $\mathrm{qsw}(a \otimes b) = \varphi_1(a)\varphi_2(b)$. In fact, this can be determined by replacing $\widehat{\otimes}$ with $\overset{\mathrm{h}}{\otimes}$ and checking if qsw is still a complete contraction (see [30, §IV] for more details). The main reason why this is the case is that the Haagerup tensor actually *characterises* maps that do admit similar multi-linear decompositions [13, §9.4]. Such reasoning can also be done in $\mathbf{Q}$, because we have a valid object $(M_n \overset{\mathrm{h}}{\otimes} M_n, \{u \otimes v\}) \in \mathrm{Ob}(\mathbf{Q})$, but $\mathrm{qsw}\colon (M_n \overset{\mathrm{h}}{\otimes} M_n, \{u \otimes v\}) \to (M_{2n}, \{|0\rangle\langle 0| \otimes (uv) + |1\rangle\langle 1| \otimes (vu)\})$ is *not* a complete contraction, so we can conclude following the same arguments as in [30, §IV]. This shows our model $\mathbf{Q}$ can be used to reason about complex higher-order behaviour via the Haagerup tensor, which is outside the scope of Linear Logic, and suggests a new direction for future work.

8   Discussion and Future Work

We described a new approach to quantum coherence spaces, in which the proofs of formulas with positive logical polarity correspond precisely to CPTP maps and proofs of formulas with negative logical polarity correspond precisely to CPU maps, remedying the issues identified by Selinger [32] in the approach proposed by Girard [19,20,21]. Other works related to quantum theory and gluing and orthogonality include [33] which describes a model of MALL and [35] which describes a model of LL. Our model exhibits two main differences: first, all of our polarised formulas have physical interpretations and a mathematical formulation based on mathematical physics; second, complete positivity is not assumed a priori, but it is *derived* only where necessary, i.e. for *mixed state* quantum computation. Note that [33] is a follow-up work to the MLL model of [26] which is also based on gluing and orthogonality. We showed in §7 that our model has the advantage that it can also be used to reason about *pure state* quantum computation, where complete positivity makes no sense. Our model can be used to interpret a quantum lambda calculus using standard techniques, and since it can talk about both pure state and mixed state quantum computation, one direction for future work is to use it to extract suitable design for type systems and/or logics that can combine classical and quantum control.

Another direction for future work is related to general recursion – our morphisms should be seen as "total" rather than "partial" from the point of view of domain theory [1]. To get more interesting domain-theoretic properties, one should replace trace-preserving maps with *trace-non-increasing* (TNI) maps in the Schrödinger picture, replace unital maps with *subunital* (SU) ones in the Heisenberg picture. Indeed, in finite dimensions, the Loewner order gives both the CPTNI maps and the CPSU maps the structure of a continuous domain [32,25], which is a very nice kind of dcpo from a domain-theoretic point of view. To do this, we would need to modify the polarity/orthogonality in §7 accordingly. We hope this might be achieved through the noncommutative generalisation of convexity, called *matrix convexity* [13, §5.5][12]. This would hopefully strengthen the links to probabilistic coherence spaces, which can be used to model recursion.

We note that a categorical approach to (infinite dimensional) *pre-C^*-algebras* as monoid objects in a monoidal category with an involution is presented in [16], however, their definition requires them to work with unbounded linear maps. By comparison, our approach based on using the Haagerup tensor does not have the same issue, even if we were to use infinite-dimensional spaces, because this tensor may be used to *characterise* unital operator algebras [31,8]. Note that we still get some structure relevant to categorical probability theory as subcategories of (co)commutative vN-(co)algebras in $\mathbf{vNAlg_{fd}}^{\mathrm{op}}(\simeq \mathbf{vNCoalg_{fd}})$ and $\mathbf{H}^{\mathrm{op}}(\simeq \mathbf{S})$ form Markov categories w.r.t. the completely injective (projective) tensors.

Another direction for future work would be to try to generalise the categorical definition of vN-algebras and vN-coalgebras to infinite dimensions, where the dual and predual do not coincide. Note that this would also require us to adapt the choice of some of the monoidal structures used, as in [11], because the Haagerup tensor is not self-dual in a De Morgan sense for infinite-dimensional operator spaces (in general). Developing the theory of infinite-dimensional vN-coalgebras can hopefully be used to further strengthen the observations in [30] regarding the connections between polarised LL and the Heisenberg-Schrödinger duality. Finally, one could investigate if and how our linear logic approach to the Heisenberg-Schrödinger duality relates to the categorical logic approach of state-and-effect triangles [24], given that $\mathbf{H}^{\mathrm{op}} \simeq \mathbf{S}$ is indeed an effectus.

Acknowledgements. We thank the anonymous reviewers for their feedback which led to multiple improvements of the paper. We also thank Benoît Valiron, Bert Lindenhovius, Titouan Carette, and James Hefford for discussions and/or useful feedback. This work has been partially funded by the French National Research Agency (ANR) within the framework of "Plan France 2030", under the research projects EPIQ ANR-22-PETQ-0007, HQI-Acquisition ANR-22-PNCQ-0001, and HQI-R&D ANR-22-PNCQ-0002.

References

1. Abramsky, S., Jung, A.: Domain theory (1994)
2. Baratella, S.: Quantum coherent spaces and linear logic. RAIRO-Theoretical Informatics and Applications **44**(4), 419–441 (2010). https://doi.org/10.1051/ita/2010021
3. Beggs, E.J., Majid, S.: Bar categories and star operations (2009). https://doi.org/10.1007/s10468-009-9141-x, https://arxiv.org/abs/math/0701008
4. Blackadar, B.: Operator algebras: theory of C*-algebras and von Neumann algebras, vol. 122. Springer Berlin, Heidelberg (2006). https://doi.org/10.1007/3-540-28517-2
5. Blanco, N.: Bifibrations of polycategories and classical multiplicative linear logic. CoRR **abs/2305.15139** (2023). https://doi.org/10.48550/ARXIV.2305.15139
6. Blecher, D.P., Le Merdy, C.: Operator Algebras and Their Modules: An operator space approach. Oxford University Press (10 2004). https://doi.org/10.1093/acprof:oso/9780198526599.001.0001
7. Blute, R., Panangaden, P., Slavnov, S.: Deep inference and probabilistic coherence spaces. Applied Categorical Structures **20**, 209 – 228 (2010). https://doi.org/10.1007/s10485-010-9241-0
8. Chirvasitu, A., Thompson, I.: Local presentability and monadicity of forgetful functors for operator algebraic categories (2025), https://arxiv.org/abs/2507.23152
9. Choi, Y.: A twisted inclusion between tensor products of operator spaces (2020), https://arxiv.org/abs/1606.06287
10. Danos, V., Ehrhard, T.: Probabilistic coherence spaces as a model of higher-order probabilistic computation. Inf. Comput. **209**(6), 966–991 (2011). https://doi.org/10.1016/J.IC.2011.02.001
11. Effros, E.G., Ruan, Z.J.: Operator space tensor products and Hopf convolution algebras. Journal of Operator Theory **50**(1), 131–156 (2003), http://www.jstor.org/stable/24718935
12. Effros, E.G., Winkler, S.: Matrix convexity: Operator analogues of the bipolar and Hahn-Banach theorems. Journal of Functional Analysis **144**(1), 117–152 (1997). https://doi.org/https://doi.org/10.1006/jfan.1996.2958
13. Effros, E., Ruan, Z.: Operator Spaces. London Mathematical Society monographs, Clarendon Press (2000), https://books.google.fr/books?id=v7mj8Dy84k8C
14. Ehrhard, T., Pagani, M., Tasson, C.: Full abstraction for probabilistic PCF. J. ACM **65**(4), 23:1–23:44 (2018). https://doi.org/10.1145/3164540
15. Ehrhard, T., Tasson, C.: Probabilistic call by push value. Log. Methods Comput. Sci. **15**(1) (2019). https://doi.org/10.23638/LMCS-15(1:3)2019
16. Fritz, T., Lorenzin, A.: Involutive Markov categories and the quantum de Finetti theorem (2025), https://arxiv.org/abs/2312.09666
17. Girard, J.: The system F of variable types, fifteen years later. Theor. Comput. Sci. **45**(2), 159–192 (1986). https://doi.org/10.1016/0304-3975(86)90044-7
18. Girard, J.: Linear logic. Theor. Comput. Sci. **50**, 1–101 (1987). https://doi.org/10.1016/0304-3975(87)90045-4
19. Girard, J.Y.: Between Logic and Quantic: a Tract, p. 346381. London Mathematical Society Lecture Note Series, Cambridge University Press (2004)
20. Girard, J.Y.: Le point aveugle II: Cours de logique, Vers l'imperfection. Visions des sciences (2007)

21. Girard, J.Y.: Truth, modality and intersubjectivity. Mathematical Structures in Computer Science **17**(6), 11531167 (2007). https://doi.org/10.1017/S0960129507006342
22. Hyland, M., Schalk, A.: Glueing and orthogonality for models of linear logic. Theoretical Computer Science **294**(1), 183–231 (2003). https://doi.org/https://doi.org/10.1016/S0304-3975(01)00241-9
23. Jacobs, B.: Involutive categories and monoids, with a gns-correspondence. Foundations of Physics **42**(7), 874895 (Sep 2011). https://doi.org/10.1007/s10701-011-9595-7
24. Jacobs, B.: A recipe for state-and-effect triangles. Logical Methods in Computer Science **Volume 13, Issue 2** (May 2017). https://doi.org/10.23638/lmcs-13(2:6)2017, `http://dx.doi.org/10.23638/LMCS-13(2:6)2017`
25. Jia, X., Kornell, A., Lindenhovius, B., Mislove, M.W., Zamdzhiev, V.: Semantics for variational quantum programming. Proc. ACM Program. Lang. **6**(POPL), 1–31 (2022). https://doi.org/10.1145/3498687, `https://doi.org/10.1145/3498687`
26. Kissinger, A., Uijlen, S.: A categorical semantics for causal structure. Log. Methods Comput. Sci. **15**(3) (2019). https://doi.org/10.23638/LMCS-15(3:15)2019, `https://doi.org/10.23638/LMCS-15(3:15)2019`
27. Kornell, A.: Quantum collections. International Journal of Mathematics **28**(12), 1750085 (2017). https://doi.org/10.1142/S0129167X17500859
28. Li, T., Zamdzhiev, V.: Quantum coherence spaces revisited: A von Neumann (co)algebraic approach (2026), `https://arxiv.org/abs/2601.15832`
29. Lindenhovius, B., Zamdzhiev, V.: The category of operator spaces and complete contractions. CoRR **abs/2412.20999** (2024). https://doi.org/10.48550/ARXIV.2412.20999
30. Lindenhovius, B., Zamdzhiev, V.: Operator spaces, linear logic and the Heisenberg-Schrödinger duality of quantum theory. In: 2025 40th Annual ACM/IEEE Symposium on Logic in Computer Science (LICS). pp. 870–883 (2025). https://doi.org/10.1109/LICS65433.2025.00071
31. Pisier, G.: Introduction to Operator Space Theory. London Mathematical Society Lecture Note Series, Cambridge University Press (2003)
32. Selinger, P.: Towards a semantics for higher-order quantum computation. In: Proceedings of the 2nd International Workshop on Quantum Programming Languages, TUCS General Publication. vol. 33, pp. 127–143 (2004)
33. Simmons, W., Kissinger, A.: Higher-order causal theories are models of BV-logic. In: Szeider, S., Ganian, R., Silva, A. (eds.) 47th International Symposium on Mathematical Foundations of Computer Science, MFCS 2022, August 22-26, 2022, Vienna, Austria. LIPIcs, vol. 241, pp. 80:1–80:14. Schloss Dagstuhl - Leibniz-Zentrum für Informatik (2022). https://doi.org/10.4230/LIPICS.MFCS.2022.80
34. Takesaki, M.: Theory of Operator Algebras I. Springer New York, NY (1979). https://doi.org/10.1007/978-1-4612-6188-9
35. Tsukada, T., Asada, K.: Enriched presheaf model of quantum FPC. Proc. ACM Program. Lang. **8**(POPL), 362–392 (2024). https://doi.org/10.1145/3632855

Well-quasi-orderings on word languages

Nathan Lhote[1], Aliaume Lopez[3], and
Lia Schütze[2]

1 Aix-Marseille University
2 Max Planck Institute for Software Systems
3 INP Bordeaux, LaBRI, CNRS

Abstract. The set of finite words over a well-quasi-ordered set is itself
well-quasi-ordered. This seminal result by Higman is a cornerstone of
the theory of well-quasi-orderings and has found numerous applications
in computer science. However, this result is based on a specific choice
of ordering on words, the (scattered) subword ordering. In this paper,
we describe to what extent other natural orderings (prefix, suffix, and
infix) on words can be used to derive Higman-like theorems. More specifically, we are interested in characterizing *languages* of words that are
well-quasi-ordered under these orderings, and explore their properties
and connections with other language theoretic notions. We furthermore
give decision procedures when the languages are given by various computational models such as automata, context-free grammars, and automatic
structures.

⊡ This document uses knowledge: a notion points to its *definition*.

1   Introduction

A *well-quasi-ordered* set is a set X equipped with a quasi-order $\preceq$ such that
every infinite sequence $(x_n)_{n\in\mathbb{N}}$ of elements taken in X contains an increasing
pair $x_i \preceq x_j$ with $i < j$. Well-quasi-orderings serve as a core combinatorial
tool powering many termination arguments, and were successfully applied to
the verification of infinite state transition systems [2,1]. One of the appealing
properties of well-quasi-orderings is that they are closed under many operations,
such as taking products, finite unions, and finite powerset constructions [13].
Perhaps more surprisingly, the class of well-quasi-ordered sets is also stable under
the operation of taking finite words and finite trees labeled by elements of a
well-quasi-ordered set [20,23].

Note that in the case of finite words and finite trees, the precise choice of
ordering is crucial to ensure that the resulting structure is well-quasi-ordered.
The celebrated result of Higman states that the set of finite words over a well-
quasi-ordered alphabet $(X, \preceq)$ is well-quasi-ordered by the so-called subword
embedding relation [20]. Let us recall that the subword relation for words over
$(X, \preceq)$ is defined as follows: a word u is a *subword* of a word v, written $u \leq^* v$,

© The Author(s) 2026
N. Bertrand and S. Milius (Eds.): FoSSaCS 2026, LNCS 16503, pp. 440–461, 2026.
https://doi.org/10.1007/978-3-032-22730-0_21

if there exists an increasing function $f\colon \{1,\ldots,|u|\} \to \{1,\ldots,|v|\}$ such that $u_i \preceq v_{f(i)}$ for all $i \in \{1,\ldots,|u|\}$.

However, there are many other natural orderings on words that could be considered in the context of well-quasi-orderings, even in the simplified setting of a finite alphabet Σ equipped with the equality relation. In this setting, the three alternatives we consider are the *prefix relation* ($u \sqsubseteq_{\mathsf{pref}} v$ if there exists w with $uw = v$), the *suffix relation* ($u \sqsubseteq_{\mathsf{suff}} v$ if there exists w such that $wu = v$), and the *infix relation* ($u \sqsubseteq_{\mathsf{infix}} v$ if there exist w_1, w_2 such that $w_1 u w_2 = v$). Note that these three relations straightforwardly generalize to infinite quasi-ordered alphabets. Unfortunately, it is easy to see that none of these relations yield well-quasi-ordered sets as soon as the alphabet contains two distinct letters: for instance, the infinite sequence of words $(ab^n a)_{n \in \mathbb{N}}$ is well-quasi-ordered by the subword relation but by neither the prefix relation, nor the suffix relation, nor the infix relation.

While this dooms well-quasi-orderedness of these relations in the general case, there may be *subsets* of Σ^* which are well-quasi-ordered by these relations. As a simple example, take the case of finite sets of (finite) words which are all well-quasi-ordered regardless of the ordering considered. This raises the question of characterizing exactly which subsets $L \subseteq \Sigma^*$ are well-quasi-ordered with respect to the prefix relation (respectively, the suffix relation or the infix relation), and designing suitable decision procedures.

Let us argue that these decision procedures fit a larger picture in the research area of well-quasi-orderings. Indeed, there have been recent breakthroughs in deciding whether a given order is a well-quasi-order, for instance in the context of the verification of infinite state transition systems [19] or in the context of logic [7]. In the graph theory community, recent works have studied classes of graphs that are well-quasi-ordered by the induced subgraph relation using similar language theoretic techniques [12,27,6]. Furthermore, a previous work by Kuske shows that any *reasonable*[4] partially ordered set $(X, \leq)$ can be embedded into $\{a,b\}^*$ with the infix relation [25, Lemma 5.1]. Phrased differently, one can encode a large class of partially ordered sets as subsets of $\{a,b\}^*$. As a consequence, the following decision problem provides a reasonable abstract framework for deciding whether a given partially ordered set is well-quasi-ordered: given a language $L \subseteq \Sigma^*$, decide whether L is well-quasi-ordered by the infix relation.

The runtime of an algorithm based on well-quasi-orderings is deeply related to the "complexity" of the underlying quasi-order $(X, \leq)$ [31]. One way to measure this complexity is to consider its so-called ordinal invariants: for instance, the maximal order type (or m.o.t., $\mathfrak{o}(X)$), originally defined by De Jongh and Parikh [21], is the order type of the maximal linearization of a well-quasi-ordered set. In the case of a finite set, the m.o.t. is precisely the size of the set. Better runtime bounds were obtained by considering two other parameters [32]: the ordinal height $(\mathfrak{h}(X))$[30], and the ordinal width $(\mathfrak{w}(X))$ [26]. Therefore, when characterizing well-quasi-ordered languages, we will also be interested in deriving upper bounds on their ordinal invariants. This analysis also allows us to better compare the

---

[4] This will be made precise in Lemma 7.

well-quasi-orderings. We refer to Section 2 for a more detailed introduction to these parameters and ordinal computations in general.

Contributions We focus on languages over a finite alphabet Σ. In this setting, we first characterize languages that are well-quasi-ordered by the prefix relation (and symmetrically, by the suffix relation), and derive tight bounds on their ordinal invariants. These generic results are then used to devise a decision procedure for checking whether a language is well-quasi-ordered by the prefix relation, provided the language is given as input as a finite automaton (Corollary 4). A summary of these results can be found in Figure 1.

L	Characterisation	$\mathfrak{w}(L)$	$\mathfrak{o}(L)$
arbitrary	Theorem 5: finite unions of chains	$< \omega$	$< \omega^2$
regular	Corollary 4: finite unions of regular chains	$< \omega$	$< \omega^2$

Fig. 1: Summary of results for the prefix relation (and symmetrically, for the suffix relation).

We then turn our attention to the infix relation. In this case, we notice that Lemma 5.1 from [25] implies that there are well-quasi-ordered languages for the infix relation that have arbitrarily large ordinal invariants (except for the ordinal height, which is always at most ω). Therefore, we focus on two natural semantic restrictions on languages: on the one hand, we consider bounded languages, that is, languages included in some $w_1^* \cdots w_k^*$ for some finite choice of words $w_1, \ldots, w_k$; on the other hand, we consider downward closed languages, that is, languages closed under taking infixes. In both cases, we provide a very precise characterization of well-quasi-ordered languages by the infix relation, and derive tight bounds on their ordinal invariants. These results are summarized in Figure 2. We furthermore notice that for downward closed languages that are well-quasi-ordered by the infix relation, being bounded is the same as being regular (Lemma 33), and that a bounded language is well-quasi-ordered by the infix relation if and only if its downwards closure is well-quasi-ordered by the infix relation (Corollary 15). This shows that, for bounded languages, being well-quasi-ordered implies that their downwards closure is a regular language, which is a weakening of the usual result that the downwards closure of *any language* for the scattered subword relation is always a regular language.

Turning our attention to decision procedures, we consider two computational models respectively tailored to downward closed languages and to bounded languages. For downward closed languages, we consider a model based on representations of infinite words (Section 5.2), for which we provide a decision procedure (Theorem 27). The model used to represent these infinite words is based on automatic sequences and morphic sequences [11], which are well-studied in the context of symbolic dynamics. For bounded languages, we consider the

L	Characterisation	$\mathfrak{w}(L)$	$\mathfrak{o}(L)$
arbitrary	Lemma 7: countable well-quasi orders with finite initial segments	$< \omega_1$	$< \omega_1$
bounded	Theorem 8: finite union of products of chains for the prefix and suffix relations	$< \omega^2$	$< \omega^3$
downward closed	Theorem 20: finite union of infixes of ultimately uniformly recurrent words	$< \omega^2$	$< \omega^3$

Fig. 2: Summary of results for the infix relation, the bounds on $\mathfrak{w}(L)$ and $\mathfrak{o}(L)$ are tight, and respectively proven in Corollary 14 and Corollary 26.

model of amalgamation systems [5], which is an abstract computational model that encompasses many classical ones, such as finite automata, context-free grammars, and Petri nets [5]. We show that if a language recognized by an amalgamation system is well-quasi-ordered by the infix relation, then it is a bounded language (Theorem 29), and is therefore regular. Furthermore, we show that we can decide whether a given language recognized by an amalgamation system is well-quasi-ordered by the infix relation (Theorem 30). We defer the introduction of amalgamation systems to Section 6.1.

Related work The study of alternative well-quasi-ordered relations over finite words is far from new. For instance, orders obtained by so-called *derivation relations* were already analysed by Bucher, Ehrenfeucht, and Haussler [9], and were later extended by D'Alessandro and Varricchio [16,17]. However, in all those cases the orderings are *multiplicative*, that is, if $u_1 \preceq v_1$ and $u_2 \preceq v_2$ then $u_1 u_2 \preceq v_1 v_2$. This assumption does not hold for the prefix, suffix, and infix relations.

A similar question was studied by Atminas, Lozin, and Moshkov [6], in the hope of finding characterizations of classes of *finite graphs* that are well-quasi-ordered by the *induced subgraph relation* [6, Section 7]. In this setting, it is common to refer to classes of graphs via a list of *forbidden patterns*, which are finite graphs that cannot be found as induced subgraphs in the class. Applying this reasoning to finite words with the infix relation, they provide an efficient decision procedure for checking whether a language $L \subseteq \Sigma^*$ is well-quasi-ordered by the infix relation whenever said language is given as input via a list of *forbidden factors* [6, Theorem 1, Theorem 2]. The key construction of their paper is to study languages L that are *regular* (recognized by some finite deterministic automata), for which they can decide whether L is well-quasi-ordered by the infix relation [6, Theorem 1]. Because it is easy to transform a list of forbidden factors into a regular language [6, Theorem 1], this yields the desired decision procedure. Our work extends this result in several ways: first, we also consider the prefix relation and the suffix relation, then we consider non-regular languages, and finally, we provide very precise descriptions of the well-quasi-ordered languages, as well as tight bounds on their ordinal invariants.

Outline We introduce in Section 2 the necessary background on well-quasi-orders and ordinal invariants. In Section 3, which is relatively self-contained, we study the prefix relation and prove in Theorem 5 the characterization of well-quasi-ordered languages by the prefix relation. In Section 4, we obtain the infix analogue of Theorem 5 specifically for bounded languages (Theorem 8). In Section 5, we study the downward closed languages, characterize them using a notion of ultimately uniformly recurrent words borrowed from symbolic dynamics (Theorem 20), and compute bounds on their ordinal invariants in Corollary 26. Finally, we generalize these results to all amalgamation systems in Section 6 in (Theorem 29), and provide a decision procedure for checking whether a language is well-quasi-ordered by the infix relation (resp. prefix and suffix) in this context (Theorem 30).

Acknowledgements We would like to thank participants of the 2024 edition of Autobóz for their helpful comments and discussions. We would also like to thank Vincent Jugé for his pointers on word combinatorics.

2   Preliminaries

Finite words. In this paper, we use upper Greek letters Σ, Γ to denote finite alphabets, Σ^* to denote the set of finite words over Σ, and ε for the empty word in Σ^*. In order to give some intuition on the decision problems, we will sometimes use the notion of *finite automata, regular languages,* and Monadic Second Order logic (MSO) over finite words, and assume the reader to be familiar with them. We refer to the textbook of [33] for a detailed introduction. However, we will require no prior knowledge on word combinatorics.

Orderings and Well-Quasi-Orderings. A *quasi-order* is a reflexive and transitive binary relation, it is a *partial order* if it is furthermore antisymmetric. A *total order* is a partial order where any two elements are comparable. Let us now introduce some notations for well-quasi-orders. A sequence $(x_i)_{n \in \mathbb{N}}$ in a set X is *good* if there exist $i < j$ such that $x_i \le x_j$. It is *bad* otherwise. Therefore, a well-quasi-ordered set is a set where every infinite sequence is good. A *decreasing sequence* is a sequence $(x_i)_{n \in \mathbb{N}}$ such that $x_{i+1} < x_i$ for all i, a *chain* is a sequence such that $x_i \le x_{i+1}$ for all i, and an *antichain* is a set of pairwise incomparable elements. An equivalent definition of a well-quasi-ordered set is that it contains no infinite decreasing sequences, nor infinite antichains. We refer to [13] for a detailed survey on well-quasi-orders.

The prefix relation (resp. the suffix relation and the infix relation) on Σ^* are always *well-founded*, i.e., there are no infinite decreasing sequences for this ordering. In particular, for a language $L \subseteq \Sigma^*$ to be well-quasi-ordered with respect to one of these orderings, it suffices to prove that it contains no infinite antichain.

A useful operation on quasi-ordered sets is to compute the *upwards closure* of a set S for a relation $\preceq$, which is defined as $\uparrow_\preceq S \triangleq \{y \in \Sigma^* \mid \exists x \in S. x \preceq y\}$. In this paper, we will also use the symmetric notion of *downwards closure*:

$\downarrow_{\preceq} S \triangleq \{y \in \Sigma^* \mid \exists x \in S.y \preceq x\}$. Abusing notations, we will write $\uparrow w$ and $\downarrow w$ for the upwards and downwards closure of a single element w, omitting the ordering relation when it is clear from the context. A set S is called *downward closed* if $\downarrow S = S$.

Ordinal Invariants. An *ordinal* is a well-founded totally ordered set. We use α, β, γ to denote ordinals, and use ω to denote the first infinite ordinal, i.e., the set of natural numbers with the usual ordering. We also use ω_1 to denote the first *uncountable* ordinal. We only assume superficial familiarity with ordinal arithmetic, and refer to the books of Kunen [24] and Krivine [22, Chapter II] for a detailed introduction to this domain. Given a tree T whose branches are all finite we can define an ordinal α_T inductively as follows: if T is a leaf then $\alpha_T = 0$, if T has children $(T_i)_{i \in I}$ then $\alpha_T = \sup\{\alpha_{T_i} + 1 \mid i \in I\}$. We say that α_T is the *rank* of T.

Let $(X, \leq)$ be a well-quasi-ordered set. One can define three well-founded trees from X: the tree of bad sequences, the tree of decreasing sequences, and the tree of antichains. The nodes of these trees are respectively the bad sequences, the decreasing sequences, and the antichains of X, and the ancestor relation is the prefix relation on sequences (or subset relation on antichains). The rank of these trees are called respectively the *maximal order type* of X written $\mathfrak{o}(X)$ [21], the *ordinal height* of X written $\mathfrak{h}(X)$ [30], and the *ordinal width* of X written $\mathfrak{w}(X)$ [26]. These three parameters are called the *ordinal invariants* of a well-quasi-ordered set X. As an example, for $(\mathbb{N}, \leq)$, all bad sequences are descending and antichains have size at most 1. In fact, $(\mathbb{N}, \leq)$ is itself an ordinal, namely ω. Hence it is its own maximal order type and ordinal height, and its ordinal width is 1. We refer to the survey of [15] for a detailed discussion on these concepts and their computation on specific classes of well-quasi-ordered sets.

We will use the following inequality between ordinal invariants, due to [26], and that was recalled in [15, Theorem 3.8]: $\mathfrak{o}(X) \leq \mathfrak{h}(X) \otimes \mathfrak{w}(X)$, where $\otimes$ is the *commutative ordinal product*, also known as the *Hessenberg product*. We will not recall the definition of this product here, and refer to [15, Section 3.5] for a detailed introduction to this concept. The only equalities we will use are $\omega \otimes \omega = \omega^2$ and $\omega^2 \otimes \omega = \omega^3$.

3   Prefixes and Suffixes

In this section, we study the well-quasi-ordering of languages under the prefix relation. Let us immediately remark that the map $u \mapsto u^R$ that reverses a word is an order-bijection between $(X^*, \sqsubseteq_{\mathsf{pref}})$ and $(X^*, \sqsubseteq_{\mathsf{suff}})$, that is, $u \sqsubseteq_{\mathsf{pref}} v$ if and only if $u^R \sqsubseteq_{\mathsf{suff}} v^R$. Therefore, we will focus on the prefix relation in the rest of this section, as $(L, \sqsubseteq_{\mathsf{pref}})$ is well-quasi-ordered if and only if $(L^R, \sqsubseteq_{\mathsf{suff}})$ is.

The next remark we make is that Σ^* is not well-quasi-ordered by the prefix relation as soon as Σ contains two distinct letters a and b. As an example of infinite antichain, we can consider the set of words $a^n b$ for $n \in \mathbb{N}$. As mentioned in the introduction, there are however some languages that are well-quasi-ordered by the prefix relation. A simple example being the (regular) language $a^* \subseteq \{a, b\}^*$, which is order-isomorphic to natural numbers with their usual orderings $(\mathbb{N}, \leq)$.

In order to characterize the existence of infinite antichains for the prefix relation, we will introduce the following tree.

Definition 1. *The* tree of prefixes *over a finite alphabet Σ is the infinite tree T whose nodes are the words of Σ^*, and such that the children of a word w are the words wa for all $a \in \Sigma$.*

We will use this tree of prefixes to find simple witnesses of the existence of infinite antichains in the prefix relation for a given language L, namely by introducing antichain branches.

Definition 2. *An* antichain branch *for a language L is an infinite branch B of the tree of prefixes such that from every point of the branch, one can reach a word in $L \setminus B$. Formally: $\forall u \in B, \exists v \in \Sigma^*, uv \in L \setminus B$.*

Let us illustrate the notion of antichain branch over the alphabet $\Sigma = \{a, b\}$, and the language $L = a^*b$. In this case, the set a^* (which is a branch of the tree of prefixes) is an antichain branch for L. This holds because for any a^k, the word $a^k b$ belongs to $L \setminus a^*$. In general, the existence of an antichain branch for a language L implies that L contains an infinite antichain, and because the alphabet Σ is assumed to be finite, one can leverage the fact that the tree of prefixes is finitely branching to prove that the converse holds as well.

Lemma 3. *Let $L \subseteq \Sigma^*$ be a language. Then, L contains an infinite antichain if and only if there exists an antichain branch for L.*

One immediate application of Lemma 3 is that antichain branches can be described inside the tree of prefixes by a monadic second order formula (MSO-formula), allowing us to leverage the decidability of MSO over infinite binary trees [29, Theorem 1.1]. This result will follow from our general decidability result (Theorem 30) but is worth stating on its own for its simplicity.

Corollary 4. *If L is regular, then the existence of an infinite antichain is decidable.*

Let us now go further and fully characterize languages L such that the prefix relation is well-quasi-ordered, without any restriction on the decidability of L itself.

Theorem 5. *A language $L \subseteq \Sigma^*$ is well-quasi-ordered by the prefix relation if and only if L is a finite union of chains.*

As an immediate consequence, we have a very fine-grained understanding of the ordinal invariants of such well-quasi-ordered languages, which can be leveraged in bounding the complexity of algorithms working on such languages.

Corollary 6. *Let $L \subseteq \Sigma^*$ be a language that is well-quasi-ordered by the prefix relation. Then, the maximal order type of L is strictly smaller than ω^2, the ordinal height of L is at most ω, and its ordinal width is finite. Furthermore, these bounds are tight.*

Proof. The upper bounds follow from the fact that L is a finite union of chains. The tightness can be obtained by considering the languages $L_k \triangleq \bigcup_{i=0}^{k-1} a^i b^*$ for $k \in \mathbb{N}$, which are well-quasi-ordered by the prefix relation (as they are finite unions of chains), and satisfy that $\mathfrak{w}(L_k) = k$, $\mathfrak{h}(L_k) = \omega$, and therefore $\mathfrak{o}(L_k) = k \cdot \omega$. $\qquad\square$

4   Infixes and Bounded Languages

In this section, we study languages equipped with the infix relation. As opposed to the prefix and suffix relations, the infix relation can lead to very complicated well-quasi-ordered languages. Formally, the upcoming Lemma 7 due to Kuske shows that *any* countable partial-ordering with finite initial segments can be embedded into the infix relation of a language. To make the former statement precise, let us recall that an *order embedding* from a quasi-ordered set $(X, \preceq)$ into a quasi-ordered set $(Y, \preceq')$ is a function $f \colon X \to Y$ such that for all $x, y \in X$, $x \preceq y$ if and only if $f(x) \preceq' f(y)$. When such an embedding exists, we say that X *embeds into* Y. Recall that a quasi-ordered set $(X, \preceq)$ is a partial ordering whenever the relation $\preceq$ is antisymmetric, that is $x \preceq y$ and $y \preceq x$ implies $x = y$.

Lemma 7. *[25, Lemma 5.1] Let $(X, \preceq)$ be a partially ordered set, and Σ be an alphabet with at least two letters. Then the following are equivalent:*

1. *X embeds into $(\Sigma^*, \sqsubseteq_{\mathsf{infix}})$,*
2. *X is countable, and for every $x \in X$, its downwards closure $\downarrow_{\preceq} x$ is finite (that is, $(X, \preceq)$ has finite initial segments).*

As a consequence of Lemma 7, we cannot replay proofs of Section 3, and will actually need to leverage some regularity of the languages to obtain a characterization of well-quasi-ordered languages under the infix relation. This regularity will be imposed through the notion of *bounded languages*, i.e., languages $L \subseteq \Sigma^*$ such that there exists words $w_1, \ldots, w_n$ satisfying $L \subseteq w_1^* \cdots w_n^*$. Let us now state the main theorem of this section.

Theorem 8. *Let L be a bounded language of Σ^*. Then, L is a well-quasi-order when endowed with the infix relation if and only if it is included in a finite union of products $S_i \cdot P_i$ where S_i is a chain for the suffix relation, and P_i is a chain for the prefix relation, for all $1 \leq i \leq n$.*

Let us first remark that if S is a chain for the suffix relation and P is a chain for the prefix relation, then SP is well-quasi-ordered for the infix relation. This proves the (easy) right-to-left implication of Theorem 8.

In order to prove the (difficult) left-to-right implication of Theorem 8, we will rely heavily on the combinatorics of periodic words. Let us use a slightly non-standard notation by saying that a non-empty word $w \in \Sigma^+$ is *periodic* with period $x \in \Sigma^*$ if there exists a $p \in \mathbb{N}$ such that $w \sqsubseteq_{\mathsf{infix}} x^p$. The *periodic length* of a word u is the minimal length of a period x of u.

The reason why periodic words built using a given period $x \in \Sigma^+$ are interesting for the infix relation is that they naturally create chains for the prefix and suffix relations. Indeed, if $x \in \Sigma^+$ is a finite word, then $\{x^p \mid p \in \mathbb{N}\}$ is a chain for the infix relation. Note that in general, the downwards closure of a chain is *not* a chain (see Remark 9). However, for the chains generated using periodic words, the downwards closure $\downarrow_{\sqsubseteq_{\mathsf{infix}}} \{x^p \mid p \in \mathbb{N}\}$ is a *finite union* of chains. Because this set will appear in bigger equations, we introduce the shorter notation $\mathsf{P}{\downarrow}(x)$ for the set of infixes of words of the form x^p, where $p \in \mathbb{N}$.

Remark 9. Let $(X, \preceq)$ be a quasi-ordered set, and $L \subseteq X$ be such that $(L, \preceq)$ is well-quasi-ordered. It is not true in general that $(\downarrow L, \preceq)$ is well-quasi-ordered. In the case of $(\Sigma^*, \sqsubseteq_{\mathsf{infix}})$ a typical example is to start from an infinite antichain A, together with an enumeration $(w_i)_{i \in \mathbb{N}}$ of A, and build the language $L \triangleq \{\prod_{i=0}^{n} w_i \mid n \in \mathbb{N}\}$. By definition, L is a chain for the infix ordering, hence well-quasi-ordered. However, $\downarrow_{\sqsubseteq_{\mathsf{infix}}} L$ contains A, and is therefore not well-quasi-ordered.

Lemma 10. *Let $x \in \Sigma^+$ be a word. Then $\mathsf{P}{\downarrow}(x)$ is a finite union of chains for the infix, prefix and suffix relations.*

The following combinatorial Lemma 12 connects the property of being well-quasi-ordered to a property of the periodic lengths of words in a language, based on the assumption that some factors can be iterated. It is the core result that powers the analysis done in Theorems 8 and 29. It is fundamentally based on a classical result of combinatorics on words (Lemma 11) that we recall here for the sake of completeness.

Lemma 11 ([18, Theorem 1]). *Let $u, v \in \Sigma^+$ be two words and $n = \gcd(|u|, |v|)$. If there exists $p, q \in \mathbb{N}$ such that u^p and v^q have a common prefix of length at least $|uv| - n$, then there exists $z \in \Sigma^+$ such that u and v are powers of z, and in particular z has length at most $\min\{|u|, |v|\}$.*

Lemma 12. *Let $L \subseteq \Sigma^*$ be a language that is well-quasi-ordered by the infix relation. Let $k \in \mathbb{N}$, $u_1, \cdots, u_{k+1} \in \Sigma^*$, and $v_1, \cdots, v_k \in \Sigma^+$ be such that $w[n] \triangleq (\prod_{i=1}^{k} u_i v_i^{n_i}) u_{k+1}$ belongs to L for vectors $\boldsymbol{n} \in \mathbb{N}^k$ with all coordinates arbitrarily large. Then, there exist $x, y \in \Sigma^+$ of size at most $\max\{|v_i| \mid 1 \leq i \leq k\}$ such that for all $\boldsymbol{n} \in \mathbb{N}^k$ one of the following holds: $w[n] \in \mathsf{P}{\downarrow}(x) u_i \, \mathsf{P}{\downarrow}(y)$ for some $1 \leq i \leq k+1$.*

Lemma 13. *Let $L \subseteq \Sigma^*$ be a bounded language that is well-quasi-ordered by the infix relation. Then, there exists a finite subset $E \subseteq (\Sigma^*)^3$, such that:*

$$L \subseteq \bigcup_{(x,u,y) \in E} \mathsf{P}{\downarrow}(x) u \, \mathsf{P}{\downarrow}(y) \quad .$$

Proof (Proof of Theorem 8 as stated on page 8). We apply Lemma 13, and conclude because $P{\downarrow}(x)$ is a finite union of chains for the prefix, suffix and infix relations (Lemma 10). $\qquad\square$

Corollary 14. *Let L be a bounded language of Σ^* that is well-quasi-ordered by the infix relation. Then, the ordinal width of L is less than ω^2, its ordinal height is at most ω, and its maximal order type is less than ω^3. Furthermore, those three bounds are tight.*

Proof. Upper bounds are a direct consequence of Theorem 8. For the tightness, remark that the ordinal width, ordinal height and maximal order type of the language $a * b*$ are respectively ω, ω and ω^2. Thus, by using finite unions of languages of the form $a_i * b_i*$, one can reach the desired bounds $\omega \cdot k$, ω and $\omega^2 \cdot k$.[5] $\qquad\square$

5   Infixes and Downwards Closed Languages

Let us now discuss another classical restriction that can be imposed on languages when studying well-quasi-orders, that of being downward closed. Indeed, Lemma 7 crucially relies on constructing languages that are *not* downward closed, and we have shown in Remark 9 that the downwards closure of a well-quasi-ordered language is not necessarily well-quasi-ordered.

5.1   Characterization of Well-Quasi-Ordered Downward Closed Languages

An immediate consequence of Theorem 8 is that if L is a bounded language, then considering L or its downwards closure $\downarrow_{\sqsubseteq_{\mathsf{infix}}} L$ is equivalent with respect to being well-quasi-ordered by the infix relation, as opposed to the general case illustrated in Remark 9.

Corollary 15. *Let L be a bounded language of Σ^*. Then, L is a well-quasi-order when endowed with the infix relation if and only if $\downarrow_{\sqsubseteq_{\mathsf{infix}}} L$ is.*

Corollary 15 is reminiscent of a similar result for the subword embedding, stipulating that for any language $L \subseteq \Sigma^*$, the downwards closure $\downarrow_{\leq_*} L$ is described using finitely many excluded subwords, hence is regular. However, this is not the case for the infix relation, even with bounded languages, as we will now illustrate with the following example.

Example 16. Let $L \triangleq a^* b^* \cup b^* a^*$. This language is bounded, is downward closed for the infix relation, is well-quasi-ordered for the infix relation, but is characterized by an *infinite* number of excluded infixes, respectively of the form $ab^k a$ and $ba^k b$ where $k \geq 1$.

---

[5] Note that one can encode these languages using a binary alphabet.

To strengthen Example 16, we will leverage the *Thue-Morse sequence* $\mathbf{t} \in \{0,1\}^{\mathbb{N}}$, which we will use as a black-box for its two main characteristics: it is cube-free and uniformly recurrent. Being *cube-free* means that no (finite) word of the form uuu is an infix of $\mathbf{t}$, and being *uniformly recurrent* means that for every word u that is an infix of $\mathbf{t}$, there exists $k \geq 1$ such that u occurs as an infix of every k-sized infix $v \sqsubseteq_{\mathsf{infix}} \mathbf{t}$. We refer the reader to a nice survey of Allouche and Shallit for more information on this sequence and its properties [4]. We are using the notations $\Sigma^{\mathbb{N}}$ for the set of infinite words over the alphabet Σ, and will be using $\Sigma^{\mathbb{Z}}$ for the set of bi-infinite words over Σ.

Theorem 17. *Let $w \in \Sigma^{\mathbb{N}}$ be a uniformly recurrent word. Then, the set of finite infixes of w is well-quasi-ordered for the infix relation.*

Proof. Let L be the set of finite infixes of w. Consider a sequence $(u_i)_{i \in \mathbb{N}}$ of words in L. Without loss of generality, we may consider a subsequence such that $|u_i| < |u_{i+1}|$ for all $i \in \mathbb{N}$. Because $\mathbf{t}$ is uniformly recurrent, there exists $k \geq 1$ such that u_1 is an infix of every word v of size at least k. In particular, u_1 is an infix of u_k, hence the sequence $(u_i)_{i \in \mathbb{N}}$ is good. $\quad\square$

Lemma 18. *The language $I_{\mathbf{t}}$ of infixes of the Thue-Morse sequence is downward closed for the infix relation, well-quasi-ordered for the infix relation, but is not bounded.*

Proof. By construction $I_{\mathbf{t}}$ is downward closed for the infix relation, and by Theorem 17, it is well-quasi-ordered.

Assume by contradiction that $I_{\mathbf{t}}$ is bounded. In this case, there exist words $w_1, \ldots, w_k \in \Sigma^*$ such that $I_{\mathbf{t}} \subseteq w_1^* \cdots w_k^*$. Since $I_{\mathbf{t}}$ is infinite and downward closed, there exists a word $u \in I_{\mathbf{t}}$ such that $u = w_i^3$ for some $1 \leq i \leq k$. This is a contradiction, because $u \sqsubseteq_{\mathsf{infix}} \mathbf{t}$, which is cube-free. $\quad\square$

One may refine our analysis of the Thue-Morse sequence to obtain precise bounds on the ordinal invariants of its language of infixes.

Lemma 19. *Under $\sqsubseteq_{\mathsf{infix}}$, the maximal order type of $I_{\mathbf{t}}$ is ω, the ordinal height of $I_{\mathbf{t}}$ is ω, the ordinal width of $I_{\mathbf{t}}$ is ω.*

Proof. We first show that ω is an upper bound for each of these measures, before showing that the bounds are tight.

Let us prove that these are upper bounds for the ordinal invariants of $I_{\mathbf{t}}$. The bound of the ordinal height holds for any language L, as the length of a decreasing sequence of words is bounded by the length of its first element. For the maximal order type, we remark that the uniform recurrence of $\mathbf{t}$ means that the maximal length of a bad sequence is determined by its first element, hence that it is at most ω. Finally, because the ordinal width is at most the maximal order type (as per Section 2, using for instance the results of [26] or [15, Theorem 3.8] stating $\mathfrak{o}(X) \leq \mathfrak{h}(X) \otimes \mathfrak{w}(X)$): we conclude that the ordinal width is also at most ω.

Now, let us prove that these bounds are tight. It is clear that $\mathfrak{h}(I_\mathbf{t}) = \omega$: given any number $n \in \mathbb{N}$, one can construct a decreasing sequence of words in $I_\mathbf{t}$ of length n, for instance by considering the first n prefixes of the Thue-Morse sequence by decreasing size. Let us now prove that $\mathfrak{w}(I_\mathbf{t}) = \omega$. To that end, we can leverage the fact that the number of infixes of size n in $I_\mathbf{t}$ is bounded below by a non-constant affine function in n [34], and that two words of length n are comparable for the infix relation if and only if they are equal. Hence, there cannot be a finite bound on the size of an antichain in $I_\mathbf{t}$, and we conclude that $\mathfrak{w}(I_\mathbf{t}) = \omega$. Finally, because the ordinal width is at most the maximal order type, we conclude that the maximal order type of $I_\mathbf{t}$ is also ω. $\square$

We prove in the upcoming Theorem 20 that the status of the Thue-Morse sequence is actually representative of downward closed languages for the infix relation. To that end, let us introduce the notation $\mathsf{Infixes}(w)$ for the set of finite infixes of a (possibly infinite or bi-infinite) word $w \in \Sigma^* \cup \Sigma^\mathbb{N} \cup \Sigma^\mathbb{Z}$. We say that an infinite word $w \in \Sigma^\mathbb{N}$ is *ultimately uniformly recurrent* if there exists a bound $N_0 \in \mathbb{N}$ such that $w_{\geq N_0}$ is uniformly recurrent. We extend this notion to finite words by considering that they all are ultimately uniformly recurrent, and to bi-infinite words by considering that they are ultimately uniformly recurrent if and only if both their left-infinite and right-infinite parts are.

Theorem 20. *Let L be a well-quasi-ordered language for the infix relation that is downward closed. Then, there exist finitely many ultimately uniformly recurrent words $w_1, \ldots, w_n \in \Sigma^* \cup \Sigma^\mathbb{N} \cup \Sigma^\mathbb{Z}$ such that $L = \bigcup_{i=1}^n \mathsf{Infixes}(w_i)$.*

To connect infixes of a (bi)-infinite word to downward closed languages, a useful notion is that of directed sets. A subset $I \subseteq X$ is *directed* if, for every $x, y \in I$, there exists $z \in I$ such that $x \leq z$ and $y \leq z$. Given a well-quasi-order $(X, \leq)$, one can always decompose X into a finite union of *order ideals*, that is, non-empty sets $I \subseteq X$ that are downward closed and directed for the relation $\leq$. In our case, a well-quasi-ordered order ideal for the infix relation is the set of finite infixes of a finite, infinite, or bi-infinite word $w \in \Sigma^* \cup \Sigma^\mathbb{N} \cup \Sigma^\mathbb{Z}$ (Lemma 21).

Lemma 21. *Let $L \subseteq \Sigma^*$ be an order ideal for the infix relation. Then L is the set of finite infixes of a finite, infinite or bi-infinite word w.*

Lemma 22. *Let $w \in \Sigma^\mathbb{N}$ be an infinite word. Then, the set of finite infixes of w is well-quasi-ordered for the infix relation if and only if w is ultimately uniformly recurrent.*

Lemma 23. *Let $w \in \Sigma^\mathbb{Z}$ be a bi-infinite word. Then, the set of finite infixes of w is well-quasi-ordered for the infix relation if and only if w is ultimately uniformly recurrent as a bi-infinite word.*

We are now ready to conclude the proof of Theorem 20.

Proof (Proof of Theorem 20 as stated on page 12). It is clear that the set of finite infixes of a finite, infinite or bi-infinite ultimately uniformly recurrent word is well-quasi-ordered for the infix relation thanks to Lemma 22.

Conversely, let us consider a well-quasi-ordered language L that is downward closed for the infix relation. Because it is a well-quasi-ordered set, it can be written as a finite union of order ideals $L = \bigcup_{i=1}^{n} L_i$.

For every such ideal L_i, we can apply Lemma 21, and conclude that L_i is the set of finite infixes of a finite, infinite or bi-infinite word w_i. Because the languages L_i are well-quasi-ordered, we can apply Lemma 22, and conclude that w_i is ultimately uniformly recurrent. $\qquad\square$

Finally, we comment on the ordinal invariants of the set of finite infixes of an ultimately uniformly recurrent infinite word, from which the bounds of Corollary 26 naturally follow.

Lemma 24. *Let $w \in \Sigma^{\mathbb{N}}$ be an ultimately uniformly recurrent word. Then, the set of finite infixes of w has ordinal width less than $\omega \cdot 2$. Furthermore, this bound is tight.*

Lemma 25. *Let $w \in \Sigma^{\mathbb{Z}}$ be an ultimately uniformly recurrent bi-infinite word. Then, the ordinal width of the set of finite infixes of w is less than $\omega \cdot 3$, and this bound is tight.*

Thanks to Theorem 20, and by analysing the ordinal invariants of infixes of an ultimately uniformly recurrent infinite word w (Lemma 22), we conclude that the ordinal invariants of a well-quasi-ordered downward closed language are relatively small.

Corollary 26. *Let L be a well-quasi-ordered downward closed language for the infix relation. Then, the maximal order type of L is strictly less than ω^3, its ordinal height is at most ω, and its ordinal width is at most ω^2.*

Furthermore, those bounds are tight.

5.2   Decision Procedures

As we have demonstrated, infinite (or bi-infinite words) can be used to represent languages that are well-quasi-ordered for the infix relation by considering their set of finite infixes. Let us formalise the representation of languages by sets of bi-infinite words that we will use in this section, following the characterization of Lemma 21. A *sequence representation* of a language $L \subseteq \Sigma^*$ is a finite set of triples $(w_i^-, a_i, w_i^+)_{1 \leq i \leq n}$ where $w_i^-, w_i^+ \in \Sigma^{\mathbb{N}} \cup \Sigma^*$ are two potentially infinite words, and $a_i \in \Sigma$ is a letter, such that

$$L = \bigcup_{i=1}^{n} \mathsf{Infixes}(\mathsf{reversed}(w_i^-) a_i w_i^+) \quad .$$

Given an effective representation of sequences, one obtains an effective representation of languages via sequence representations. In this section, we will rely on definitions originating from the area of symbolic dynamics, that precisely study infinite words whose generation follows from a finitely described process.

However, we will not assume that the reader is familiar with this domain, and we will use as black-boxes key results from this area.

A first model that one can use to represent infinite words is the model of *automatic sequences*. In this case, the infinite word w is described by a finite state automaton, that can compute the i-th letter of the word w given as input the number i written in some base $b \in \mathbb{N}$. An example of such a sequence is the Thue-Morse sequence that can be described by a finite automaton using a binary representation of the indices. The good algorithmic properties of automatic sequences come from the fact that a Presburger definable property that uses letters of the sequence can be (trivially) translated into a finite automaton that reads the base b representation of the free variables (that are indices of the sequence). In particular, it follows that one can decide if an automatic sequence is ultimately uniformly recurrent. Based on this, we now prove:

Theorem 27. *Given a sequence representation of a language $L \subseteq \Sigma^*$ where all infinite words are automatic sequences, one can decide whether L is well-quasi-ordered for the infix relation.*

In fact, automatic sequences are part of a larger family of sequences studied in symbolic dynamics, called morphic sequences. Let us first recall that a *morphism* is a function $f \colon \Sigma^* \to \Gamma^*$ such that for every $u, v \in \Sigma^*$, $f(uv) = f(u)f(v)$. A *morphic sequence* w is an infinite word obtained by iterating a morphism $f \colon \Sigma^* \to \Sigma^*$ on a letter $a \in \Sigma$ such that $f(a)$ starts with a, and then applying a homomorphism $h \colon \Sigma^* \to \Gamma^*$. The infinite word $f^\omega(a)$ is the limit of the sequence $(f^n(a))_{n \in \mathbb{N}}$, which is well-defined because $f(a)$ starts with a, and the morphic sequence is $w \triangleq h(f^\omega(a))$.

Every automatic sequence is a morphic sequence, but not the other way around. We refer the reader to a short survey of [3] for more details on the possible variations on the definition of morphic sequences and their relationships. It was relatively recently proven that one can decide whether a morphic sequence is uniformly recurrent [14, Theorem 1]. We were not able to find in the literature whether one can decide ultimate uniform recurrence, but conjecture that it is the case, which would allow us to decide whether a language represented by morphic sequences is well-quasi-ordered for the infix relation.

Conjecture 28. Given a morphic sequence $w \in \Sigma^\mathbb{N}$, one can decide whether it is ultimately uniformly recurrent.

6   Infixes and Amalgamation Systems

In the previous section, we have represented languages that are downward closed by the infix relation as infixes of infinite words. However, there are many other natural ways to represent languages, such as finite automata or context-free grammars. In this section, we are going to show that our results on bounded languages can be applied to a large class of systems, called amalgamation systems, that includes as particular examples finite automata and context-free grammars.

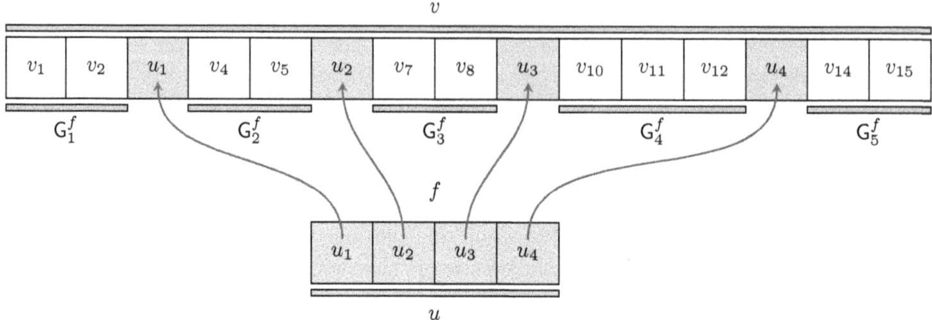

Fig. 3: The gap words resulting from a subword embedding between two finite words.

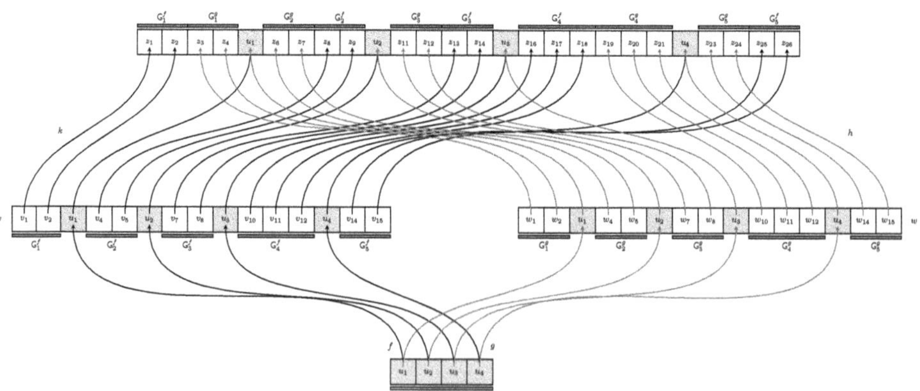

Fig. 4: We illustrate how embeddings f and g between runs of an amalgamation system can be glued together, seen on their canonical decomposition.

Our first result, of theoretical nature, is that amalgamation systems cannot define well-quasi-ordered languages that are not bounded. This implies that all the results of Section 4, and in particular Theorem 8, can safely be applied to amalgamation systems.

Theorem 29. *Let $L \subseteq \Sigma^*$ be a language recognized by an amalgamation system. If L is well-quasi-ordered by the infix relation then L is bounded.*

Our second focus is of practical nature: we want to give a decision procedure for being well-quasi-ordered. This will require us to introduce *effectiveness assumptions* on the amalgamation systems. While most of them will be innocuous, an important consequence is that we have to consider *classes of languages* rather than individual ones, for instance: the class of all regular languages, or the class of all context-free languages. Such classes will be called effective amalgamative classes (Section 6.1). In the following theorem, we prove that under such assumptions, testing well-quasi-ordering is inter-reducible to testing whether a language of the class is empty, which is usually the simplest problem for a computational model.

Theorem 30. *Let $\mathcal{C}$ be an effective amalgamative class of languages. Then the following are equivalent:*

1. *Well-quasi-orderedness of the infix relation is decidable for languages in $\mathcal{C}$.*
2. *Well-quasi-orderedness of the prefix relation is decidable for languages in $\mathcal{C}$.*
3. *Emptiness is decidable for languages in $\mathcal{C}$.*

6.1 Amalgamation Systems

Let us now formally introduce the notion of amalgamation systems, and recall some results from [5] that will be useful for the proof of Theorem 29. The notion of amalgamation system is tailored to produce *pumping arguments*, which is exactly what our Lemma 12 talks about. At the core of a pumping argument, there is a notion of a *run*, which could for instance be a sequence of transitions taken in a finite state automaton. Continuing on the analogy with finite automata, there is a natural ordering between runs, i.e., a run is smaller than another one if one can "delete" loops of the larger run to obtain the other. Typical pumping arguments then rely on the fact that *minimal* runs are of finite size, and that all other runs are obtained by "gluing" loops to minimal runs. Generalizing this notion yields the notion of amalgamation systems.

Let us recall that over an alphabet $(\Sigma, =)$ a subword embedding between two words $u \in \Sigma^*$ and $v \in \Sigma^*$ is an order-preserving function $\rho \colon [1, |u|] \to [1, |v|]$ such that $u_i = v_{\rho(i)}$ for all $i \in [1, |u|]$. We write $\mathsf{Hom}^*(u, v)$ the set of all subword embeddings between u and v. It may be useful to notice that the set of finite words over Σ forms a category when we consider subword embeddings as morphisms, which is a fancy way to state that $\mathrm{id} \in \mathsf{Hom}^*(u, u)$ and that $f \circ g \in \mathsf{Hom}^*(u, w)$ whenever $g \in \mathsf{Hom}^*(u, v)$ and $f \in \mathsf{Hom}^*(v, w)$, for any choice of words $u, v, w \in \Sigma^*$.

Given a subword embedding $f\colon u \to v$ between two words $u = u_1 \cdots u_k$ and v, there exists a unique decomposition $v = \mathsf{G}_0^f\, u_1\, \mathsf{G}_1^f \cdots \mathsf{G}_{k-1}^f\, u_k\, \mathsf{G}_k^f$ where $\mathsf{G}_i^f = v_{f(i)+1} \cdots v_{f(i+1)-1}$ for all $1 \le i \le k-1$, $\mathsf{G}_k^f = v_{f(k)+1} \cdots v_{|v|}$, and $\mathsf{G}_0^f = v_1 \cdots v_{f(1)-1}$. We say that G_i^f is the i-th *gap word* of f. We encourage the reader to look at Figure 3 to see an example of the gap words resulting from a subword embedding between two words. These gap words will be useful to describe how and where runs of a system (described by words) can be combined.

Definition 31. *An* amalgamation system *is a tuple* $(\Sigma, R, \mathsf{can}, E)$ *where Σ is a finite alphabet, R is a set of so-called* runs*, $\mathsf{can}\colon R \to (\Sigma \uplus \{\#\})^*$ *is a function computing a* canonical decomposition *of a run, and E describes the so-called* admissible embeddings *between runs: If ρ and σ are runs from R, then $E(\rho, \sigma)$ is a subset of the subword embeddings between $\mathsf{can}(\rho)$ and $\mathsf{can}(\sigma)$. We write $\rho \trianglelefteq \sigma$ if $E(\rho, \sigma)$ is non-empty. If we want to refer to a specific embedding $f \in E(\rho, \sigma)$, we also write $\rho \trianglelefteq_f \sigma$. Given a run $r \in R$, and $i \in [0, |\mathsf{can}(r)|]$, the* gap language *of r at position i is $\mathsf{L}_i^r \triangleq \{\mathsf{G}_i^f \mid \exists s \in R.\exists f \in E(r, s)\}$. An amalgamation system furthermore satisfies the following properties:*

1. *(R, E)* Forms a Category. *For all $\rho, \sigma, \tau \in R$, $\mathrm{id} \in E(\rho, \rho)$, and whenever $f \in E(\rho, \sigma)$ and $g \in E(\sigma, \tau)$, then $g \circ f \in E(\rho, \tau)$.*
2. Well-Quasi-Ordered System. *$(R, \trianglelefteq)$ is a well-quasi-ordered set.*
3. Concatenative Amalgamation. *Let ρ_0, ρ_1, ρ_2 be runs with $\rho_0 \trianglelefteq_f \rho_1$ and $\rho_0 \trianglelefteq_g \rho_2$. Then for all $0 \le i \le |\mathsf{can}(\rho_0)|$, there exists a run $\rho_3 \in R$ and embeddings $\rho_1 \trianglelefteq_{g'} \rho_3$ and $\rho_2 \trianglelefteq_{f'} \rho_3$ satisfying two conditions: (a) $g' \circ f = f' \circ g$ (we write h for this composition) and (b) for every $0 \le j \le |\rho_0|$, the gap word G_j^h is either $\mathsf{G}_j^f\, \mathsf{G}_j^g$ or $\mathsf{G}_j^h = \mathsf{G}_j^g\, \mathsf{G}_j^f$. Specifically, for i we may fix $\mathsf{G}_i^h = \mathsf{G}_i^f\, \mathsf{G}_i^g$. We refer to Figure 4 for an illustration of this property.*

The yield *of a run is obtained by projecting away the separator symbol $\#$ from the canonical decomposition, i.e. $\mathsf{yield}(\rho) = \pi_\Sigma(\rho)$. The language recognized by an amalgamation system is $\mathsf{yield}(R)$.*

We say a language L is an amalgamation language *if there exists an amalgamation system recognizing it.*

Intuitively, the definition of an amalgamation system allows the comparison of runs, and the proper "gluing" of runs together to obtain new runs. A number of well-known language classes can be seen to be recognized by amalgamation systems, e.g., regular languages [5, Theorem 5.3], reachability and coverability languages of VASS [5, Theorem 5.5], and context-free languages [5, Theorem 5.10].

We can now show a simple lemma that illuminates much of the structure of amalgamation systems whose language is well-quasi-ordered by $\sqsubseteq_{\mathsf{infix}}$. Note that Lemma 32 uses Lemma 12 in its proof, and our Theorem 29 follows from it.

Lemma 32. *Let L by an amalgamation language recognized by $(\Sigma, R, E, \mathsf{can})$ that is well-quasi-ordered by $\sqsubseteq_{\mathsf{infix}}$. Let ρ be a run with $\rho = a_1 \cdots a_n$, and let σ, τ be runs with $\rho \trianglelefteq_f \sigma$ and $\rho \trianglelefteq_g \tau$.*

For any $0 \le \ell \le n$, we have $\mathsf{G}_\ell^f \sqsubseteq_{\mathsf{infix}} \mathsf{G}_\ell^g$ or vice versa.

If we additionally assume that such a language is closed under taking infixes, we obtain an even stronger structure: All such languages are regular!

Lemma 33. *Let $L \subseteq \Sigma^*$ be a downward closed language for the infix relation that is well-quasi-ordered. Then, the following are equivalent:*

(i) L is a regular language,
(ii) L is recognized by some amalgamation system,
(iii) L is a bounded language,
(iv) There exists a finite set $E \subseteq (\Sigma^*)^3$ such that $L = \bigcup_{(x,u,y)\in E} \mathsf{P}{\downarrow}(x)u\,\mathsf{P}{\downarrow}(y)$.

Combining Lemmas 18 and 33, we can conclude that the collection of infixes of the Thue-Morse sequence cannot be recognized by *any* amalgamation system.

To construct a decision procedure for well-quasi-orderedness under $\sqsubseteq_{\mathsf{infix}}$, we need our amalgamation systems to satisfy certain *effectiveness assumptions*. We require that for an amalgamation system $(\Sigma, R, E, \mathsf{can})$, R is recursively enumerable, the function $\mathsf{can}(\cdot)$ is computable, and for any two runs $\rho, \sigma \in R$, the set $E(\rho, \sigma)$ is computable. Additionally, we require the class to be effectively closed under rational transductions [8, Chapter 5, page 64].

Under these assumptions, one can improve on Theorem 8 into an effective procedure, using pumping arguments from [5, Section 4.2], which, in turn, allows us to prove Theorem 30. Since the class $\mathcal{C}_{\mathsf{aut}}$ of regular languages and the class $\mathcal{C}_{\mathsf{cfg}}$ of context-free languages are examples of effective amalgamative classes, the following corollary is immediate.

Corollary 34. *Let $\mathcal{C} \in \{\mathcal{C}_{aut}, \mathcal{C}_{cfg}\}$. It is decidable whether a language in $\mathcal{C}$ is well-quasi-ordered by the infix relation. Furthermore, whenever it is well-quasi-ordered by the infix relation, it is a bounded language.*

7 Conclusion

We have described the landscapes of well-quasi-ordered languages for the natural orderings on finite words: prefix, suffix, and infix relations. While the prefix and suffix relation exhibit very simple behaviours, the infix relation can encode many complex quasi-orders (and even simulate the subword ordering). In the case of languages that are described by simple computational models, or languages that are "structurally simple" (bounded languages, downward closed languages), we showed that only very simple well-quasi-orders can be obtained: they are essentially isomorphic to disjoint unions of copies of finite sets, $(\mathbb{N}, \leq)$, and $(\mathbb{N}^2, \leq)$. Finally, under effectiveness assumptions on the language (such as being recognized by an amalgamation system, or being the set of infixes of an automatic sequence), we proved the decidability of being well-quasi-ordered for the infix relation. We believe that these very encouraging results pave the way for further research on deciding which sets are well-quasi-ordered for other orderings. Let us now discuss some possible research directions and remarks.

Towards infinite alphabets In this paper, we restricted our attention to *finite* alphabets, having in mind the application to regular languages. However, the conclusions of Theorem 8, Corollary 26, and Theorem 5 could be conjectured to hold in the case of infinite alphabets (themselves equipped with a well-quasi-ordering). This would require new techniques, as the finiteness of the alphabet is crucial to all of our positive results.

Lexicographic orderings There is another natural ordering on words, the *lexicographic ordering*, which does not fit well in our current framework because it is always of ordinal width 1. However, the order-type of the lexicographic ordering over regular languages has already been investigated in the context of infinite words [10], and it would be interesting to see if one can extend these results to decide whether such an ordering is well-founded for languages recognized by amalgamation systems.

Factor Complexity Let us conclude this section with a few remarks on the notion of factor complexity of languages. Recall that the *factor complexity* of a language $L \subseteq \Sigma^*$ is the function $f_L : \mathbb{N} \to \mathbb{N}$ such that $f_L(n)$ is the number of distinct words of size n in L. We extend the notion of factor complexity to finite, infinite, and bi-infinite words as the factor complexity of their set of finite infixes. For the prefix relation and the suffix relation, all well-quasi-ordered languages have a bounded factor complexity, since they are finite unions of chains.

While there clearly are languages with low factor complexity that are not well-quasi-ordered for the infix relation, such as the language $L \triangleq\ \downarrow ab^*a$; one would expect that languages that are well-quasi-ordered for the infix relation would have a low factor complexity.

In some sense, our results confirm this intuition in the case of languages described by a simple computational model. For languages recognized by amalgamation systems, being well-quasi-ordered implies being a bounded language, and therefore being included in some finite union of languages of the form $w_1^* w_2 w_3^*$. Hence, these languages have at most a linear factor complexity. This is also the case for languages described as the infixes of a finite set of pairs of morphic sequences. Indeed, the factor complexity of a morphic sequence that is uniformly recurrent is linear [28, Theorem 24], therefore the factor complexity of a language given by sequence representation using morphic sequences is at most linear.

However, there are downward closed languages that are well-quasi-ordered for the infix relation but have an exponential factor complexity: the $(5,3)$-Toeplitz word is uniformly recurrent [11, p. 499], and has exponential factor complexity [11, Theorem 5]. This shows that our computational models somehow fail to capture vast classes of well-quasi-ordered languages with a high factor complexity. It would be interesting to understand which new proof techniques would be required to obtain decidability for these languages.

To conclude on a positive note for the infix relation, our results show that for downward closed and well-quasi-ordered languages, there is a strong connection between the factor complexity and the ordinal width: it is the same to have bounded factor complexity and finite ordinal width.

References

1. Abdulla, P.A., Jonsson, B.: Verifying networks of timed processes. In: Proceedings of TACAS'98. vol. 1384, pp. 298–312. Springer (1998). https://doi.org/10.1007/BFb0054179
2. Abdulla, P.A., Čerāns, K., Tsay, B.J., Yih-Kuen: General decidability theorems for infinite-state systems. In: Proceedings of LICS'96. pp. 313–321. IEEE (1996). https://doi.org/10.1109/LICS.1996.561359
3. Allouche, J.P., Cassaigne, J., Shallit, J., Zamboni, L.Q.: A taxonomy of morphic sequences (2017), https://arxiv.org/abs/1711.10807
4. Allouche, J.P., Shallit, J.: The ubiquitous prouhet-thue-morse sequence. Discrete Mathematics and Theoretical Computer Science p. 1–16 (1999). https://doi.org/10.1007/978-1-4471-0551-0_1, http://dx.doi.org/10.1007/978-1-4471-0551-0_1
5. Anand, A., Schmitz, S., Schütze, L., Zetzsche, G.: Verifying unboundedness via amalgamation. In: Proceedings of the 39th Annual ACM/IEEE Symposium on Logic in Computer Science. LICS '24, Association for Computing Machinery, New York, NY, USA (2024). https://doi.org/10.1145/3661814.3662133, https://doi.org/10.1145/3661814.3662133
6. Atminas, A., Lozin, V., Moshkov, M.: Wqo is decidable for factorial languages. Information and Computation **256**, 321–333 (Oct 2017). https://doi.org/10.1016/j.ic.2017.08.001, http://dx.doi.org/10.1016/j.ic.2017.08.001
7. Bergsträßer, P., Ganardi, M., Lin, A.W., Zetzsche, G.: Ramsey quantifiers in linear arithmetics. Proc. ACM Program. Lang. **8**(POPL), 1–32 (2024). https://doi.org/10.1145/3632843, https://doi.org/10.1145/3632843
8. Berstel, J.: Transductions and Context-Free Languages. Vieweg+Teubner Verlag (1979). https://doi.org/10.1007/978-3-663-09367-1, http://dx.doi.org/10.1007/978-3-663-09367-1
9. Bucher, W., Ehrenfeucht, A., Haussler, D.: On total regulators generated by derivation relations. In: International Colloquium on Automata, Languages, and Programming. vol. 40, pp. 71–79 (1985). https://doi.org/10.1016/0304-3975(85)90162-8, https://www.sciencedirect.com/science/article/pii/0304397585901628, eleventh International Colloquium on Automata, Languages and Programming
10. Carton, O., Colcombet, T., Puppis, G.: An algebraic approach to mso-definability on countable linear orderings. The Journal of Symbolic Logic **83**(3), 1147–1189 (2018), https://www.jstor.org/stable/26600366
11. Cassaigne, J., Karhumäki, J.: Toeplitz words, generalized periodicity and periodically iterated morphisms. European Journal of Combinatorics **18**(5), 497–510 (Jul 1997). https://doi.org/10.1006/eujc.1996.0110, http://dx.doi.org/10.1006/eujc.1996.0110
12. Daligault, J., Rao, M., Thomassé, S.: Well-quasi-order of relabel functions. Order **27**(3), 301–315 (September 2010). https://doi.org/10.1007/s11083-010-9174-0, http://dx.doi.org/10.1007/s11083-010-9174-0
13. Demeri, S., Finkel, A., Goubault-Larrecq, J., Schmitz, S., Schnoebelen, P.: Algorithmic aspects of wqo theory (mpri course) (2012), https://cel.archives-ouvertes.fr/cel-00727025, course notes
14. Durand, F.: Decidability of uniform recurrence of morphic sequences. International Journal of Foundations of Computer Science **24**(01), 123–146 (2013). https://doi.org/10.1142/S0129054113500032

15. Džamonja, M., Schmitz, S., Schnoebelen, P.: On Ordinal Invariants in Well Quasi Orders and Finite Antichain Orders, pp. 29–54. Springer International Publishing, Cham (2020). https://doi.org/10.1007/978-3-030-30229-0_2, https://doi.org/10.1007/978-3-030-30229-0_2
16. D'Alessandro, F., Varricchio, S.: On well quasi-orders on languages, pp. 230–241. Springer Berlin Heidelberg (2003). https://doi.org/10.1007/3-540-45007-6_18
17. D'Alessandro, F., Varricchio, S.: Well quasi-orders, unavoidable sets, and derivation systems. RAIRO - Theoretical Informatics and Applications 40(3), 407–426 (Jul 2006). https://doi.org/10.1051/ita:2006019, http://dx.doi.org/10.1051/ita:2006019
18. Fine, N.J., Wilf, H.S.: Uniqueness theorems for periodic functions. Proceedings of the American Mathematical Society 16(1), 109–114 (Feb 1965). https://doi.org/10.1090/s0002-9939-1965-0174934-9, http://dx.doi.org/10.1090/S0002-9939-1965-0174934-9
19. Finkel, A., Gupta, E.: The well structured problem for presburger counter machines. In: Chattopadhyay, A., Gastin, P. (eds.) 39th IARCS Annual Conference on Foundations of Software Technology and Theoretical Computer Science, FSTTCS 2019, December 11-13, 2019, Bombay, India. LIPIcs, vol. 150, pp. 41:1–41:15. Schloss Dagstuhl - Leibniz-Zentrum für Informatik (2019). https://doi.org/10.4230/LIPICS.FSTTCS.2019.41, https://doi.org/10.4230/LIPIcs.FSTTCS.2019.41
20. Higman, G.: Ordering by divisibility in abstract algebras. Proceedings of the London Mathematical Society 3, 326–336 (1952). https://doi.org/10.1112/plms/s3-2.1.326
21. de Jongh, D., Parikh, R.: Well-partial orderings and hierarchies. Indagationes Mathematicae (Proceedings) 80(3), 195–207 (1977). https://doi.org/10.1016/1385-7258(77)90067-1, http://dx.doi.org/10.1016/1385-7258(77)90067-1
22. Krivine, J.L.: Introduction to Axiomatic Set Theory. Springer Netherlands (1971). https://doi.org/10.1007/978-94-010-3144-8
23. Kruskal, J.B.: The theory of well-quasi-ordering: A frequently discovered concept. Journal of Combinatorial Theory, Series A 13(3), 297–305 (Nov 1972). https://doi.org/10.1016/0097-3165(72)90063-5, http://dx.doi.org/10.1016/0097-3165(72)90063-5
24. Kunen, K.: Set Theory. Elsevier (1980). https://doi.org/10.1016/s0049-237x(08)x7037-5, http://dx.doi.org/10.1016/s0049-237x(08)x7037-5
25. Kuske, D.: Theories of orders on the set of words. RAIRO Theor. Informatics Appl. 40(1), 53–74 (2006). https://doi.org/10.1051/ITA:2005039, https://doi.org/10.1051/ita:2005039
26. Kříž, I., Thomas, R.: Ordinal Types in Ramsey Theory and Well-Partial-Ordering Theory, p. 57–95. Springer Berlin Heidelberg (1990). https://doi.org/10.1007/978-3-642-72905-8_7, http://dx.doi.org/10.1007/978-3-642-72905-8_7
27. Lopez, A.: Labelled Well Quasi Ordered Classes of Bounded Linear Clique-Width. In: Gawrychowski, P., Mazowiecki, F., Skrzypczak, M. (eds.) 50th International Symposium on Mathematical Foundations of Computer Science (MFCS 2025). Leibniz International Proceedings in Informatics (LIPIcs), vol. 345, pp. 70:1–70:17. Schloss Dagstuhl – Leibniz-Zentrum für Informatik, Dagstuhl, Germany (2025). https://doi.org/10.4230/LIPIcs.MFCS.2025.70, https://drops.dagstuhl.de/entities/document/10.4230/LIPIcs.MFCS.2025.70
28. Nicolas, F., Pritykin, Y.: On uniformly recurrent morphic sequences. International Journal of Foundations of Computer Science 20(05), 919–940 (Oct 2009). https://doi.org/10.1142/s0129054109006966, http://dx.doi.org/10.1142/S0129054109006966

29. Rabin, M.O.: Decidability of second-order theories and automata on infinite trees. Transactions of the American Mathematical Society **141**, 1–35 (1969), `http://www.jstor.org/stable/1995086`

30. Schmidt, D.: The relation between the height of a well-founded partial ordering and the order types of its chains and antichains. Journal of Combinatorial Theory, Series B **31**(2), 183–189 (Oct 1981). `https://doi.org/10.1016/s0095-8956(81)80023-8`, `http://dx.doi.org/10.1016/S0095-8956(81)80023-8`

31. Schmitz, S.: Algorithmic Complexity of Well-Quasi-Orders. Habilitation à diriger des recherches, École normale supérieure Paris-Saclay (Nov 2017), `https://theses.hal.science/tel-01663266`

32. Schmitz, S.: The parametric complexity of lossy counter machines. In: Baier, C., Chatzigiannakis, I., Flocchini, P., Leonardi, S. (eds.) 46th International Colloquium on Automata, Languages, and Programming (ICALP 2019). Leibniz International Proceedings in Informatics (LIPIcs), vol. 132, pp. 129:1–129:15. Schloss Dagstuhl – Leibniz-Zentrum für Informatik, Dagstuhl, Germany (2019). `https://doi.org/10.4230/LIPIcs.ICALP.2019.129`, `https://drops.dagstuhl.de/entities/document/10.4230/LIPIcs.ICALP.2019.129`

33. Thomas, W.: Languages, automata, and logic. In: Rozenberg, G., Salomaa, A. (eds.) Handbook of formal languages, pp. 389–455. Springer (1997). `https://doi.org/10.1007/978-3-642-59136-5`

34. Tromp, J., Shallit, J.: Subword complexity of a generalized thue-morse word. Information Processing Letters **54**(6), 313–316 (Jun 1995). `https://doi.org/10.1016/0020-0190(95)00074-m`, `http://dx.doi.org/10.1016/0020-0190(95)00074-M`

Composition Theorems for f-Differential Privacy

Natasha Fernandes[1], Annabelle McIver[1], and Parastoo Sadeghi[2]

[1] Macquarie University, Sydney
natasha.fernandes@mq.edu.au, annabelle.mciver@mq.edu.au
[2] UNSW, Canberra
p.sadeghi@unsw.edu.au

Abstract. f-differential privacy (f-DP) is a recent definition for privacy which can offer improved predictions of "privacy loss". It has been used to analyse specific privacy mechanisms, such as the popular Gaussian mechanism.

In this paper we show how f-DP's foundation in statistical hypothesis testing implies equivalence to the channel model of Quantitative Information Flow (QIF). We demonstrate this equivalence as a Galois connection between two partially-ordered sets, namely f-DP's trade-off functions, and a class of information channels. This equivalence enables novel general composition theorems for f-DP, supporting improved analysis for complex privacy designs. We apply our results to the popular privacy amplification mechanisms of sub-sampling and purification, to produce novel f-DP profiles for these general privacy-enhancing algorithms.

Keywords: Quantitative information flow · semantics for probabilistic programs · compositional analyses for privacy.

1 Introduction

Since the first definitions for privacy were introduced [29, 14, 24] the principles underlying privacy protections have become steadily more refined. It is now generally recognised that a privacy definition should be based on a measurement of "information leakage", however defined, and that the measurement should satisfy some version of the "data processing inequality", namely that post-processing after a data release should only improve privacy. A second important property of a privacy definition is that it should admit good composition laws, because most implementations of privacy algorithms usually comprise a composition of several different privacy-enhancing schemes, as illustrated in Alg. 1 (below) [19].

The purpose of Alg. 1 is to "boost" the privacy properties of an input mechanism M applied to a secret value x. This requires a fine-grained analysis of the steps in the computation so that the information leaks due to M on its own can be compared to the overall information leaks when used in conjunction with Alg. 1. The challenge here is that the only information about M is the privacy definition it satisfies. The traditional approach to verifying such implementations is to develop techniques based on refinement and abstraction, so that

N. Bertrand and S. Milius (Eds.): FoSSaCS 2026, LNCS 16503, pp. 462–483, 2026.
https://doi.org/10.1007/978-3-032-22730-0_22

Algorithm 1 Privacy purification

Require: Mechanism $M : \mathcal{X} \to \mathcal{Y}$ satisfying (ϵ, δ)-differential privacy, Private input x,
    Parameters: $r \in [0, 1]$, $c' > 0$
Ensure: Satisfies a "pure differential privacy constraint". (See comment after (1).)
    $v \leftarrow U[0, 1]$; $\triangleright$ Choose a value uniformly from $[0, 1]$
    if $(v < r)$ **then** $\triangleright$ Hidden probabilistic choice with r bias
       $y \leftarrow M(x)$
    else
       $y \leftarrow U[\mathcal{Y}']$ $\triangleright$ Choose a value uniformly from $\mathcal{Y}'$
    end if
    $z \leftarrow y + G_{\epsilon'}(0)$; $\triangleright$ Add Geometric perturbation using parameter ϵ'
    Output z; $\triangleright$ Output sanitised result

---

abstractions of program components can be analysed efficiently, with the resulting analysis also applicable to more detailed implementations via refinement.

In this paper we investigate how to do that for privacy properties using a recent notion called f-*differential privacy* [13] (f-DP). Similar to the more established (ϵ, δ)-differential privacy (or (ϵ, δ)-DP for short), which is based on "indistinguishability" between related scenarios, f-DP is based on statistical hypothesis testing, which turns out to support more nuanced evaluations of privacy risks than does traditional DP. Although f-DP has the potential to provide more accurate privacy assessments, a significant drawback is that it does not appear to be associated with simple composition laws, making it difficult to use in practice, except for specific mechanisms.

Our goal is to show how to enable accurate, general analysis of algorithms wrt. f-DP by establishing an equivalence between f-DP and information channels, using the theory of Quantitative Information Flow (QIF). Here, QIF provides an extensive theory for fine-grained analysis of information flow in programs [1], and is therefore suited to modelling combinations of general privacy-enhancing schemes. Moreover it supports analysis at different levels of abstraction through its information refinement order.

In this paper we provide a detailed foundational analysis of f-DP in terms of its information leakage properties via QIF. With that understanding, we show how f-DP admits a number of universal composition laws, supporting detailed analysis of privacy-preserving algorithms. **Our contributions** are:

1. We establish a Galois connection (Thm 2) between two partially ordered sets: the set of f-DP's trade-off functions under pointwise less-than $(\mathbb{F}, \leq)$, and the set of QIF's two-row information channels ordered by refinement, $(\mathbb{C}_2, \sqsubseteq)$. The Galois connection is defined by two order-preserving mappings $\mathcal{T}$ and $\mathcal{C}$ in Fig. 1. We discover (Thm 1) a novel relationship between the leakage measurements in QIF and so-called "hockey-stick divergence", as indicated by $(\mathbb{C}_2, \leq_h)$ in Fig. 1.
2. We study (§5) the behaviour of f-DP wrt. a range of probabilistic program constructors via their interpretation in $\mathbb{C}_2$, including hidden and visible prob-

abilistic choices, probabilistic perturbation and pre-processing, producing a number of new universal composition laws (§6).

3. We demonstrate (§7) our techniques on some popular algorithms, including purification (Alg. 1) and sub-sampling (Alg. 3).

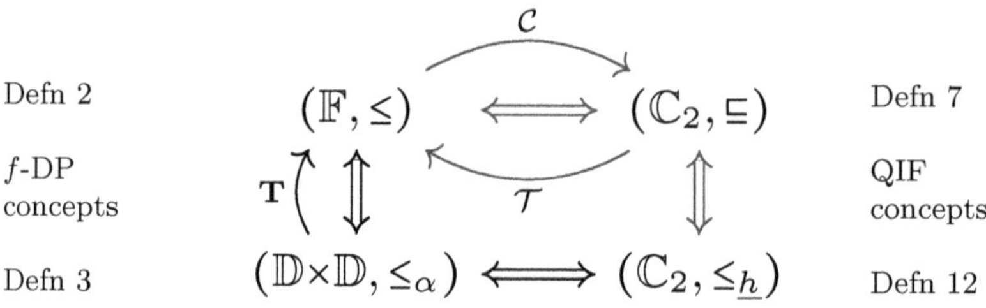

Fig. 1. A summary of the relationships between hypothesis testing and quantitative information flow. Our contributions in this paper are highlighted in blue.

2   Preliminaries: privacy and information flow

2.1   Standard Differential Privacy

A differential privacy mechanism is one that sanitises results of queries to datasets by adding a random perturbation before publication. We model such mechanisms $\mathcal{M}$ as a function $\mathcal{D} \to \mathbb{D}\mathcal{Y}$, where $\mathcal{D}$ denotes the set of datasets, and $\mathbb{D}\mathcal{Y}$ is the set of probability distributions over set $\mathcal{Y}$. (For simplicity we work under the assumption that $\mathcal{Y}$ is discrete, showing how to remove it in the appendix.) A dataset itself consists of a "set of records"; standard *differential privacy* enforces a constraint on $\mathcal{M}$ wrt. non-negative parameters (ϵ, δ), and pairs of "adjacent datasets". We say that $D, D' \in \mathcal{D}$ are *adjacent* "if they differ by a single record", which we define here as: $D \subseteq D'$ and $|D' - D| = 1$, or vice versa. Mechanism $\mathcal{M}$ is said to satisfy (ϵ, δ)-differential privacy if, for any $Y \subseteq \mathcal{Y}$, and any pair of *adjacent* datasets:

$$\mathcal{M}(D)(Y) \;\leq\; e^\epsilon \mathcal{M}(D')(Y) + \delta \;. \tag{1}$$

We say that $\mathcal{M}$ satisfies a **"pure" differential privacy** definition if $\delta = 0$.

We typically visualise this definition as follows. Let $\mathcal{Y} = \{y_0, \ldots, y_n\}$, and D_0, D_1 be a pair of adjacent datasets. We write $p_i = \mathcal{M}(D_0)(\{y_i\})$ (and $q_i = \mathcal{M}(D_1)(\{y_i\})$) for the probability that the output y_i is observed when the input to $\mathcal{M}$ is D_0 (or D_1). This yields the following "channel", M (see §2.3 below):

$$M \;=\; \begin{bmatrix} p_0 \; p_1 \; \cdots \; p_{n-1} \; p_n \\ q_0 \; q_1 \; \cdots \; q_{n-1} \; q_n \end{bmatrix}, \quad where \quad \sum_i p_i = \sum_i q_i = 1 \;. \tag{2}$$

Wlog, assume M_{D_0} is the first row and M_{D_1} is the second row of this channel. Then, M satisfies (ϵ, δ)-DP if for all $S \subset \{0, \cdots, n\}$, $p_S \leq e^\epsilon q_S + \delta$, and $q_S \leq e^\epsilon p_S + \delta$, where $p_S = \sum_{i \in S} p_i$, $q_S = \sum_{i \in S} q_i$. Not all perturbation methods satisfy (ϵ, δ)-DP for given parameters. When ϵ and δ are small, then the probability pairs (p_i, q_i) are more similar to each other, and so it makes it harder to distinguish between the (D_0, D_1) inputs for any observed output. The idea is that making it hard to distinguish between adjacent datasets, means making it hard to determine whether any particular record has been exposed. Notice that (ϵ, δ)-DP takes a "worst-case" approach, in that the level of "indistinguishability" between (D_0, D_1) as defined in Eqn (1) must hold whatever the output y, however unlikely its occurrence. In common scenarios where a mechanism is repeatedly applied to the same dataset, this worst-case measurement quickly becomes severe. For example, the standard (ϵ, δ)-DP composition theorem says that $\mathcal{M} \circ \mathcal{M}$ satisfies $(2\epsilon, 2\delta)$-privacy, if $\mathcal{M}$ satisfies (ϵ, δ) privacy. It is now recognised that more nuanced definitions of privacy can yield more realistic predictions of privacy risks, with f-DP being a recent proposal [28].

2.2 f-differential privacy

f-DP uses hypothesis testing as the means to distinguish between inputs. Let $\mathcal{M}(D_0) = p$ $\mathcal{M}(D_1) = q$; further let H_0, H_1 be hypotheses:

$$H_0 : \text{the input dataset is } D_0 \qquad H_1 : \text{the input dataset is } D_1 \ .$$

Definition 1. *A test is a mapping $\phi : \mathcal{Y} \to [0, 1]$ where $\phi(y) = 0$ means H_0 is accepted, $\phi(y) = 1$ means H_1 is accepted, and $\phi(y) = c$ means H_0 is accepted with probability c. For p and q probability distributions, as above,* the significance level *of test ϕ is $\alpha_\phi := \mathbb{E}_p[\phi]$. The* power *of test ϕ is $1 - \beta_\phi := \mathbb{E}_q[\phi]$.*

The quantities α_ϕ, β_ϕ are known respectively as Type I and Type II errors, or false negative and false positive rates. It turns out that the most effective test for distinguishing between distributions, in terms of minimising the false positive and negative rates, is the simple *likelihood ratio test*. The celebrated Neyman-Pearson lemma sets out the details.

Lemma 1 (Neyman-Pearson [18]). *Let p, q be distributions as in Defn 1. A test $\phi : \mathcal{Y} \to [0, 1]$ is the most powerful at significance level α, i.e. $\mathbb{E}_p[\phi] = \alpha$, if there are two constants $h \in [0, \infty]$ and $c \in [0, 1]$ such that the test has the form:*

$$\phi(y) = \begin{cases} 1, & \text{if } q_y > h p_y \\ c & \text{if } q_y = h p_y \\ 0, & \text{if } q_y < h p_y \end{cases} \ . \tag{3}$$

A trade-off function details the relation between Type I/II errors wrt. most powerful tests.

Definition 2 (Trade-off function). *Let p, q be as above. The trade-off function $\mathbf{T}(p, q) : [0, 1] \to [0, 1]$ is defined:*

$$\mathbf{T}(p, q)(\alpha) = \inf_\phi \{ \beta_\phi : \alpha_\phi \le \alpha \} \ . \tag{4}$$

We say that $(p, q) \le_\alpha (p', q')$ if $\mathbf{T}(p, q)(\alpha) \le \mathbf{T}(p', q')(\alpha)$ for $0 \le \alpha \le 1$.

It turns out that, for fixed distributions, $\mathbf{T}(p, q)$ is convex and satisfies $\mathbf{T}(p, q)(\alpha) \leq 1-\alpha$. The definition of f-DP uses *abstract trade-off functions* to describe indistinguishability via hypothesis testing.

Definition 3 (Abstract trade-off functions). *The set of abstract trade-off functions $(\mathbb{F}, \leq)$ defines $f \in \mathbb{F}$ if it is a convex function $[0, 1] \to [0, 1]$, and satisfies $f(\alpha) \leq (1-\alpha)$. Abstract trade-off functions are ordered pointwise, i.e $f \leq f'$ if and only if $f(\alpha) \leq f'(\alpha)$ for all $0 \leq \alpha \leq 1$.*

The pointwise maximum of two abstract trade-off functions is $f \sqcup f'$ (and is a trade-off function). We define the trade-off minimum $f \sqcap f'$ to be $\sqcup\{g \mid g \leq f \text{ and } g \leq f'\}$. The minimum trade-off function is the constant zero function, and the maximum trade-off function takes α to $1-\alpha$.

Definition 4 (f-DP). *For $f \in \mathbb{F}$, we say that mechanism $\mathcal{M}$ satisfies f-DP if $f \leq \mathbf{T}(\mathcal{M}(D), \mathcal{M}(D'))$, for any pair of adjacent datasets D, D'.*

It was shown in [13], that a mechanism $\mathcal{M}$ satisfies (ϵ, δ)-DP if and only if it satisfies $f_{\epsilon, \delta}$-DP, where $f_{\epsilon, \delta}$ is depicted in Fig. 2, and defined:

$$f_{\epsilon, \delta}(\alpha) = \max\{0, \ -e^{\epsilon}\alpha + 1 - \delta, \ -e^{-\epsilon}\alpha + e^{-\epsilon}(1-\delta)\} . \tag{5}$$

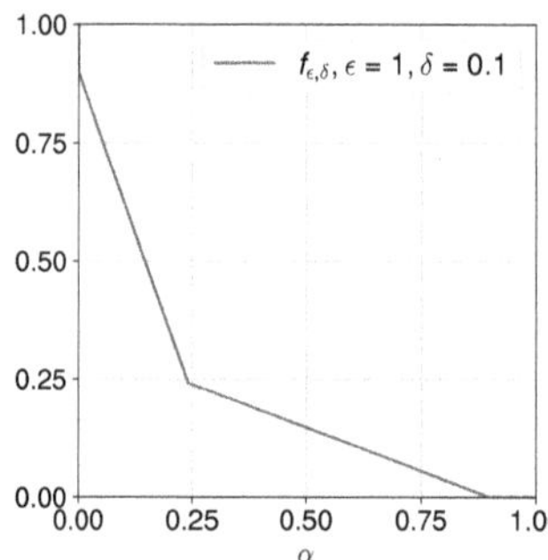

Notice that the parameters for standard differential privacy can be read-off from the plot via gradients. If the gradient $f_{\epsilon, \delta}$ at significance level α is ϵ', then it means that there is a test that can distinguish between M_{D_0} and M_{D_1} consistent with (pure) ϵ'-DP. If there is some α for which the gradient is 0 or ∞ then it means that there is some non-zero probability for which there is a test which exactly distinguishes between M_{D_0} and M_{D_1}. For the plot at left this occurs for $1-\delta \leq \alpha \leq 1$, where the gradient of $f_{\epsilon, \delta}$ is 0.

Fig. 2. Trade-off function $f_{\epsilon, \delta}$.

Unlike standard differential privacy, the details of the trade-off function f make it challenging to determine composition theorems that reflect f-DP accurately, not only for the basic composition $\mathcal{M} \circ \mathcal{M}$ mentioned above, but for other compositions that typically arise in privacy-enhancing algorithms. We set out a systematic approach for defining a range of composition theorems for f-DP, by demonstrating an equivalence between $(\mathbb{F}, \leq)$ and a class of information channels, for which compositions can be straightforwardly defined.

2.3   Information channels and QIF

QIF [1] is a framework for quantifying information leaks in programs. It features information channels as the basic model, together with the "g-leakage" semantics for assessing associated security risks, depending on the scenario. We set out components of QIF, and summarise its mathematical properties which are needed here.

An information channel C maps inputs (secrets) $x \in \mathcal{X}$ to observations $y \in \mathcal{Y}$ according to a distribution in $\mathbb{D}\mathcal{Y}$. In the discrete case, such channels are $\mathcal{X} \times \mathcal{Y}$ matrices C whose row-x, column-y element $C_{x,y}$ is the probability that input x produces observation y. The x-th row $C_{x,-}$ is thus a discrete distribution in $\mathbb{D}\mathcal{Y}$. For example the channel M at Eqn (2) is displayed as a channel where $\mathcal{X}$ consists of adjacent datasets $\{D_0, D_1\}$.

We can use Bayes rule to model an adversary who uses their observations from a channel to (optimally) update their knowledge about the secrets $\mathcal{X}$. Given a prior distribution $\pi : \mathbb{D}\mathcal{X}$ (representing an adversary's prior knowledge) and channel C, we can compute a joint distribution $J : \mathbb{D}(\mathcal{X} \times \mathcal{Y})$ where $J_{x,y} = \pi_x C_{x,y}$. Marginalising down columns yields the y-marginals $\Pr(y) = \sum_x \pi_x C_{x,y}$ each having a posterior over $\mathcal{X}$ corresponding to the posterior probabilities $P_{X|y}(x)$, computed as $J_{x,y}/\Pr(y)$ (when $\Pr(y)$ is non-zero). We denote by δ^y the posterior distribution $P_{X|y}(X|y)$ corresponding to the observation y. The set of posterior distributions and the corresponding marginals can be used to compute the adversary's posterior knowledge after making an observation from the channel.

Definition 5 (Refinement of channels). *Let $C \in \mathcal{X} \to \mathbb{D}\mathcal{Y}$ and $C' \in \mathcal{X} \to \mathbb{D}\mathcal{Z}$ be channels; we say C is refined by C' or $C \sqsubseteq C'$ if there is a channel $W \in \mathcal{Y} \to \mathbb{D}\mathcal{Z}$ such that $C \cdot W = C'$. We call W the witness to the refinement.*

We model information leakage using *gain functions.*

Definition 6 (Leakage semantics). *A gain function is a mapping $\mathcal{A} \times \mathcal{X} \to \mathbb{R}$, where $\mathcal{A}$ is a set of actions. Given a gain function g, we can define a vulnerability $V_g : \mathbb{D}\mathcal{X} \to \mathbb{R}$, defined $V_g[\pi] := \max_{a \in \mathcal{A}} \sum_{x \in \mathcal{X}} \pi_x \times g(a, x)$. The conditional vulnerability wrt. channel C and prior π is given by $V_g[\pi \triangleright C] := \sum_{y \in \mathcal{Y}} \Pr(y) V_g[\delta^y]$.*

We focus on the following class of channels.

Definition 7 (Two-row channels). *Let $(\mathbb{C}_2, \sqsubseteq)$ be the set of 2-row channels, ordered by refinement.*

Summary of QIF properties established elsewhere [1]. Observe that the leakage semantics is based on a generalisation of the notion of "entropy" and we can use it to determine how much information is leaked by comparing the prior vulnerability $V_g[\pi]$ to the posterior vulnerability $V_g[\pi \triangleright C]$, with the greater the difference corresponding to a greater amount of leaked information relative to the gain g. We summarise the leakage properties we need here; more details can be found elsewhere [1].

(I) $V_g[\pi \triangleright C]$ is independent of the column labels of C; this means that we can re-order the columns of C without changing its leakage semantics.

(II) $C \sqsubseteq C'$ *iff* $V_g[u \triangleright C] \geq V_g[u \triangleright C']$, for all gain functions g, and u the uniform prior on $\mathcal{X}$.

(III) We can render a channel $C \in \mathbb{C}_2$ as the corresponding "hyper-distribution" $[u \triangleright C]$ as a convex sum $\sum_{y \in \mathcal{Y}} \Pr(y)\delta^y$, where we are considering δ^y as a 1-summing vector in $[0,1] \times [0,1]$. This means that we can depict the posteriors using a Barycentric representation, as illustrated in Fig. 3.

(IV) It turns out that if $C \in \mathbb{C}_2$, having exactly two posteriors (i.e., over two outputs), then $C \sqsubseteq C'$ if and only if all of the posteriors of $[u \triangleright C']$ lie in the convex hull of the two posteriors of $[u \triangleright C]$.

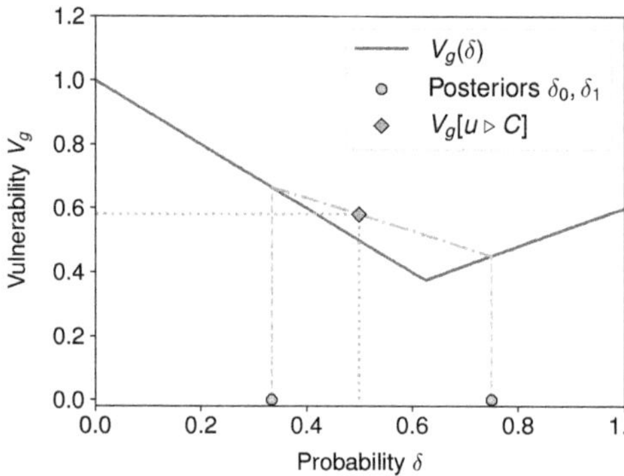

Fig. 3. Barycentric representation of $[u \triangleright C]$ for $C = \begin{bmatrix} 2/5 & 3/5 \\ 4/5 & 1/5 \end{bmatrix}$, showing posteriors $(1/3, 2/3)$ and $(3/4, 1/4)$, rendered as orange points on the horizontal axis to indicate the probability of the first component. Observe that $V_g[u \triangleright C]$ corresponds to the intersection of the vertical at the mid-point, and the line connecting $V_g(\delta_0)$ and $V_g(\delta_1)$, as C has only two posteriors.

3   Modelling f-DP in QIF

Let $M \in \mathbb{C}_2$ be a channel; recall that its two rows denoted by M_{D_0} M_{D_1} define two distributions over output $\mathcal{Y}$. We can therefore map two row channels to trade-off functions using Defn 2.

Definition 8 (Distinguishability profile). *Let $M \in \mathbb{C}_2$; define the distinguishability profile as a mapping $\mathcal{T} : \mathbb{C}_2 \to \mathbb{F}$:*

$$\mathcal{T}(M) := \mathbf{T}(M_{D_0}, M_{D_1}) \ .$$

Our aim in this section is to show that indistinguishability properties defined by trade-off functions can be modelled exactly in terms of information leakage properties of $\mathbb{C}_2$. We demonstrate that hypothesis testing at significance α corresponds to a class of 2×2 channels, and that the Neyman-Pearson Lemma(1) suggests a distinguishing class of gain functions called "hockey-stick" functions.

$\mathcal{T}(M)(\alpha)$ **defines a refinement:** Recall the channel M in Eqn (2); by QIF property (I) above we may assume that the columns are ordered with increasing ratio q_i/p_i:

$$M \ = \ \begin{bmatrix} p_0 \cdots p_{k-1} & p_k \cdots p_n \\ q_0 \cdots q_{k-1} & q_k \cdots q_n \end{bmatrix} \tag{6}$$

The Neyman-Pearson Lemma says that the most effective test to achieve a given significance level α is defined by some $h{\geq}0$; just above we have indicated a line separating the $k{-}1$'th column from the k'th column corresponding to such a test, where to the right of the line we have $q_r/p_r{\geq}h$, and to the left we have $q_l/p_l{<}h$. But now, referring to Defn 1 the corresponding significance level $\alpha = \sum_{r\geq k} p_r$, and power $\mathcal{T}(M)(\alpha) = \sum_{l<k} q_l$. We can summarise these observations using the refinement $M^\alpha = M \cdot R^h$, where the post-processing channel is given by:

$$R^h_{i,0} = \begin{cases} 0, & \text{if} \quad q_i \geq hp_i \\ 1, & \text{else.} \end{cases} \qquad\qquad R^h_{i,1} = \begin{cases} 1, & \text{if} \quad q_i \geq hp_i \\ 0, & \text{else .} \end{cases}$$

More generally, we have the following direct definition.

Definition 9 (Trade-off channel). *Given a channel $M{\in}\mathbb{C}_2$, define the trade-off channel at significance level α to be:*

$$M^\alpha \quad : \quad = \quad \begin{bmatrix} 1-\alpha & \alpha \\ \mathcal{T}(M)(\alpha) & 1-\mathcal{T}(M)(\alpha) \end{bmatrix}$$

For a given h, we can compute the error probability for the test it defines.

Definition 10 (Error function). *The error probability α of M at level h is:*

$$\alpha = err_M(h) \quad = \sum_{q_i-hp_i \geq 0} p_i \quad and \quad \mathcal{T}(M)(err_M(h)) \quad = \quad 1- \sum_{q_i-hp_i \geq 0} q_i \ .$$

Test at level h defines a gain function: Next, for $h{\geq}0$ we can define a class of "hockey-stick" gain functions, so-called because they give rise to vulnerabilities that resemble a hockey stick, as illustrated in Fig. 4, below.

Definition 11 (Hockey-stick gain). *Given $h{\geq}0$, we define the hockey-stick gain function $\underline{h}$:*

$$\underline{h}(a_1,d) \quad := \quad 1 \ \ if \ \ d = D_1 \ \ else \ \ -h \ , \qquad \underline{h}(a_0,d) \quad := \quad 0 \ \ if \ \ d{\in}\{D_0,D_1\} \ .$$

The associated vulnerability $V_{\underline{h}}$ is called a "hockey-stick" vulnerability.

Finally, we can use hockey-stick functions to define a partial order on channels which, we will see below in Thm 1, enables us to prove an equivalence between trade-off functions and channels.

Definition 12 (Hockey-stick order). *We define the hockey-stick order on channels: we say $C \leq_h M$ whenever $V_{\underline{h}}[u \rhd C] \leq V_{\underline{h}}[u \rhd M]$, for all $h{\geq}0$.*

3.1 Trade-off functions, hockey sticks and refinement

We illustrate, briefly, the concepts we have so far defined. In particular we demonstrate how hypothesis testing can be represented in QIF terms by encoding most powerful tests as hockey-stick functions. This establishes the connection between the partial order of $\mathbb{F}$ and the refinement relation of $\mathbb{C}_2$.

Consider channels $C, M \in \mathbb{C}_2$, and recall that each defines a trade-off function $f_C = \mathcal{T}(C), f_M = \mathcal{T}(M) \in \mathbb{F}$. For a given α, we construct the trade-off channels C^α, M^α, as per Defn 9. An example is shown below for $\alpha = 0.1$.

$$M^\alpha = \begin{bmatrix} 9/10 & 1/10 \\ 4/5 & 1/5 \end{bmatrix} \qquad C^\alpha = \begin{bmatrix} 9/10 & 1/10 \\ 1/2 & 1/2 \end{bmatrix}$$

In our example, $f_C(\alpha) < f_M(\alpha)$. This implies that $C^\alpha \sqsubseteq M^\alpha$, as depicted in Fig. 4 (below). As noted in Fig. 3, computing $V_g[u \triangleright M]$, where M has only two columns corresponds to a simple construction on the Barycentric representation. Applied here to hockey-stick functions and M^α, C^α, we can see clearly that for $C^\alpha \sqsubseteq M^\alpha$, the construction shows that $V_{\underline{h}}[u \triangleright C^\alpha] \geq V_{\underline{h}}[u \triangleright M^\alpha]$, indicated in Fig. 4 by the grey point on the orange diagonal line (corresponding to $V_{\underline{h}}[u \triangleright C^\alpha]$) lying above the grey point on the blue diagonal line ($V_{\underline{h}}[u \triangleright M^\alpha]$).

Whilst Fig. 4 illustrates the idea that hockey-stick gain functions characterise refinement of 2×2 channels, Fig. 5 shows how these observations can be transferred to refinement more generally in $\mathbb{C}_2$. The plots show two equivalent methods for computing $V_{\underline{h}}[u \triangleright C]$: on the left each posterior is evaluated and then averaged by their marginal, on the right the averaging happens first; by linearity the final values are the same. The trick here is to note that provided that the h corresponds to the most powerful Neyman-Pearson test for the given α, the averaging on the right corresponds to the refinement to C^α. Therefore we can deduce that $V_{\underline{h}}[u \triangleright C] = \frac{1}{2}\left(\sum_{i=k}^{n} q_i - h p_i\right)$ which is also equal to $\frac{1}{2}(1 - f_C(\alpha) - h\alpha)$, where $\alpha = err_C(h)$, which in turn is also equal to $V_{\underline{h}}[u \triangleright C^\alpha]$.

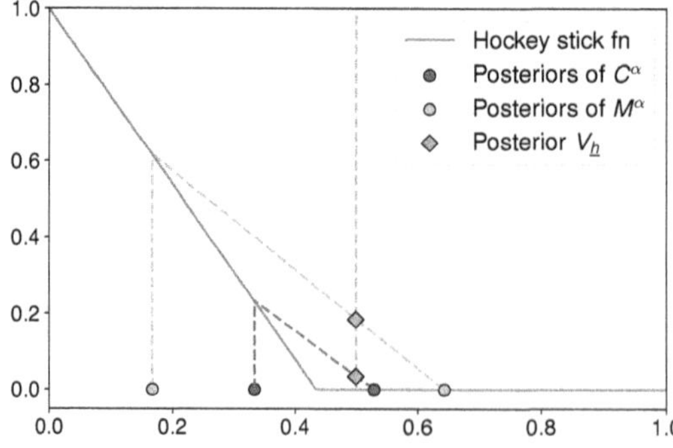

Fig. 4. Illustration of refinement: The posteriors (orange points) of C^α lie outside the posteriors (blue points) of M^α indicating by (IV) that $C^\alpha \sqsubseteq M^\alpha$. For every hockey stick function (green line), the orange diagonal line will lie above (or on) the blue diagonal line, indicating that $V_{\underline{h}}[u \triangleright C^\alpha] \geq V_{\underline{h}}[u \triangleright M^\alpha]$ for any h. The grey diamonds correspond to the particular $V_{\underline{h}}$ values for C^α and M^α in this example.

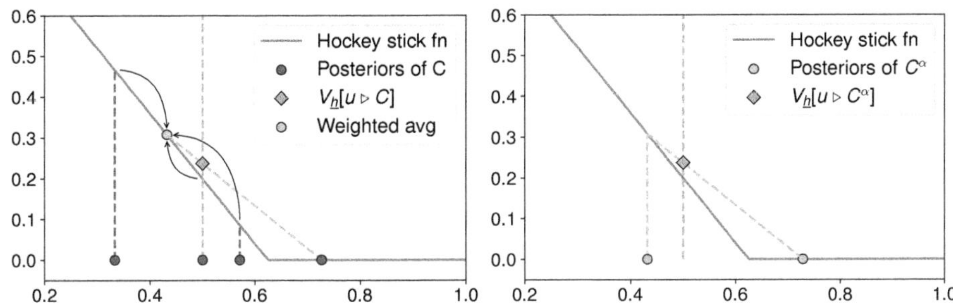

Fig. 5. Two equivalent methods for computing $V_{\underline{h}}[u \triangleright C]$: the left plot first computes $V_{\underline{h}}[\delta]$ for each (blue) posterior of C, and then averages, the right plot first averages the (blue) posteriors (equivalent to taking a refinement) and then computes $V_{\underline{h}}[\delta']$ on the two (orange) results.

The next theorem (proved in [16]) shows that these ideas illustrated in Fig. 4 and Fig. 5 hold in general, in particular that hockey stick vulnerabilities are sufficient to characterise refinement in $\mathbb{C}_2$.

Theorem 1 (Hypothesis testing in QIF). *For $M, C \in \mathbb{C}_2$, the following holds:*

1. *If $\mathcal{T}(C) \leq \mathcal{T}(M)$, then $C^\alpha \sqsubseteq M^\alpha$.*
2. *If $M^\alpha \sqsubset C^\alpha$, then there exists an h such that we have $V_{\underline{h}}[u \triangleright C] < V_{\underline{h}}[u \triangleright M]$,*
3. *$C \leq_h M$ iff $M \sqsubseteq C$.*

As a corollary, we have that $\mathcal{T}$ is monotone.

Corollary 1. *If $C \sqsubseteq M$, then $\mathcal{T}(C) \leq \mathcal{T}(M)$.*

Proof. Suppose, for contradiction that there is some α with $\mathcal{T}(C)(\alpha) > \mathcal{T}(M)(\alpha)$; this means that $M^\alpha \sqsubset C^\alpha$. Thus, by Thm 1 part 2, we can find h such that $V_{\underline{h}}[u \triangleright C] < V_{\underline{h}}[u \triangleright M]$, contradicting the refinement assumption.

4   Equivalence of two-row channels and trade-off functions

We show that $(\mathbb{C}_2, \sqsubseteq)$ and $(\mathbb{F}, \leq)$ are equivalent. We begin by defining a converse to Defn 8.

Definition 13 (Least f-private channel). *Define a mapping $\mathcal{C} : \mathbb{F} \to \mathbb{C}_2$ as:*

$$\mathcal{C}(f) \;:=\; \min_{\alpha \in [0,1]} \begin{bmatrix} 1-\alpha & \alpha \\ f(\alpha) & 1-f(\alpha) \end{bmatrix} ,$$

where $\min$ *is the greatest lower bound operator in* $(\mathbb{C}_2, \sqsubseteq)$.

Lemma 2 ($\mathcal{C}$ is well-defined). *Given any trade-off function $f \in \mathbb{F}$, $\mathcal{C}(f)$ in Defn 13 is a well-defined channel in $\mathbb{C}_2$. (Refer to [16] for the proof.)*

As an example, if $\alpha_0 < \alpha_1$ then the greatest lower bound is given by:

$$\begin{bmatrix} 1-\alpha_0 & \alpha_0 \\ f(\alpha_0) & 1-f(\alpha_0) \end{bmatrix} \min \begin{bmatrix} 1-\alpha_1 & \alpha_1 \\ f(\alpha_1) & 1-f(\alpha_1) \end{bmatrix} = \begin{bmatrix} 1-\alpha_1 & \alpha_1-\alpha_0 & \alpha_0 \\ f(\alpha_1) & f(\alpha_0)-f(\alpha_1) & 1-f(\alpha_0) \end{bmatrix},$$

where note that the glb of two channels with 2 columns results in a channel with 3 columns. More generally, the glb of n 2×2 channels results in a channel with multiple columns, sometimes as many as $n+1$.

In Alg. 2 (and more generally in [16]) we show how the min above extends more generally in $\mathbb{C}_2$. For now, using the glb property, we have immediately that $\mathcal{C}$ is monotone.

Lemma 3 (Monotonicity of $\mathcal{C}$). *Let $f \leq g$ be trade-off functions in $\mathbb{F}$. Then $\mathcal{C}(f) \sqsubseteq \mathcal{C}(g)$.*

Proof. For any $\alpha \in [0,1]$, then

$$\mathcal{C}(f) \sqsubseteq \begin{bmatrix} 1-\alpha & \alpha \\ f(\alpha) & 1-f(\alpha) \end{bmatrix} \sqsubseteq \begin{bmatrix} 1-\alpha & \alpha \\ g(\alpha) & 1-g(\alpha) \end{bmatrix},$$

where the first refinement follows from Defn 13, and the second since $f \leq g$. The result now follows from the greatest lower bound property for $\mathcal{C}(g)$.

Next, we have the important result that the space of trade-off functions corresponds to the set of two-row channels via the following Galois connection.

Theorem 2 (Galois connection). *Let $f \in \mathbb{F}$ and $M \in \mathbb{C}_2$, then:*

$$f \leq \mathcal{T}(M) \quad \Leftrightarrow \quad \mathcal{C}(f) \sqsubseteq M . \tag{7}$$

Proof. (Sketch) Assume that $f \leq \mathcal{T}(M)$, then by Thm 1 part (1),

$$f^\alpha = \begin{bmatrix} 1-\alpha & \alpha \\ f(\alpha) & 1-f(\alpha) \end{bmatrix} \sqsubseteq \begin{bmatrix} 1-\alpha & \alpha \\ \mathcal{T}(M)(\alpha) & 1-\mathcal{T}(M)(\alpha) \end{bmatrix} = M^\alpha , \tag{8}$$

hence $\mathcal{C}(f) \sqsubseteq \mathcal{C}(\mathcal{T}(M)) = M$. See [16] §?? for the last equality.

Now, suppose that $\mathcal{C}(f) \sqsubseteq M$ (so that for all hockey-stick functions we must have $V_{\underline{h}}[u \triangleright M] \leq V_{\underline{h}}[u \triangleright \mathcal{C}(f)]$). Assume by contradiction that there is some α such that $f(\alpha) > \mathcal{T}(M)(\alpha)$. This means that $M^\alpha \sqsubset f^\alpha$, with the refinement being strict. Then using Thm 1 part (2), there must be an h such that

$$V_{\underline{h}}[u \triangleright \mathcal{C}(f)] < V_{\underline{h}}[u \triangleright M] .$$

which is a contradiction of the assumption $\mathcal{C}(f) \sqsubseteq M$.

As a corollary, we have that the greatest lower bound (glb) operator in $\mathbb{C}_2$ corresponds to the lattice minimum of $\mathbb{F}$.

Corollary 2. *Let $C, C' \in \mathbb{C}_2$. Then $\mathcal{T}(C \min C') = \mathcal{T}(C) \sqcap \mathcal{T}(C')$.*

Proof. Since $C \min C' \sqsubseteq C, C'$, it follows by Cor. 1 that $\mathcal{T}(C \min C') \leq \mathcal{T}(C)$ and $\mathcal{T}(C \min C') \leq \mathcal{T}(C')$. Hence, $\mathcal{T}(C \min C') \leq \mathcal{T}(C) \sqcap \mathcal{T}(C')$.

Next, since $\mathcal{T}(C) \sqcap \mathcal{T}(C') \leq \mathcal{T}(C), \mathcal{T}(C')$, by Thm 2, we must have $\mathcal{C}(\mathcal{T}(C) \sqcap \mathcal{T}(C')) \sqsubseteq C, C'$, and therefore by the glb property, $\mathcal{C}(\mathcal{T}(C) \sqcap \mathcal{T}(C')) \sqsubseteq C \min C'$ also. Finally, appealing to Thm 2 again, it follows that $\mathcal{T}(C) \sqcap \mathcal{T}(C') \leq \mathcal{T}(C \min C')$.

4.1   Finite channels and piecewise linear trade-off functions.

It turns out that when M has a finite number of columns, $\mathcal{T}(M)$ is *piecewise linear*, which means that the domain $[0,1]$ can be split into finitely many disjoint sub-intervals, such that $\mathcal{T}(M)$ is linear on each sub-interval. We denote the subset of piecewise linear trade-off functions by $\mathbb{F}^{PL}$; for each such f, there are correspondingly a finite set of increasing α_i, $(0 \leq i \leq n)$ such that f is linear on the sub-intervals $[\alpha_i, \alpha_{i+1}]$. We call these α_i "facet" points of f. Next, we can show directly that when $\alpha_i \leq \alpha \leq \alpha_{i+1}$ that $(f^{\alpha_i} \min f^{\alpha^{i+1}}) \sqsubseteq f^{\alpha}$ (where f^{α} is as in Eqn (8)), and therefore we deduce that $\mathcal{C}(f)$ is determined by the facet points for f. This then supports Alg. 2 for computing $\mathcal{C}(f)$ by forming the greatest lower bound of increasing α_i, i.e.: $\mathcal{C}(f) = (f^{\alpha_0} \min f^{\alpha_1}) \min f^{\alpha_2}) \cdots \min f^{\alpha_n}))$.

Algorithm 2 Computing the channel $\mathcal{C}(f)$ for a given $f \in \mathbb{F}^{PL}$

Require: $f \in \mathbb{F}$; $0 = \alpha_0 < \ldots < \alpha_n = 1$ correspond to the facet points of f
Ensure: Channel $\begin{bmatrix} C_{-,0} \ldots C_{-,n} \end{bmatrix} \in \mathbb{C}_2$ such that C is equal to $\mathcal{C}(f)$

$\quad C_{-,n} \leftarrow \begin{bmatrix} \alpha_0 \\ 1-f(\alpha_0) \end{bmatrix}$ $\triangleright$ Compute the last column of of $\mathcal{C}(f)$

$\quad i \leftarrow 0;$
$\quad$**while** $i < n$ **do**

$\quad\quad C_{-,n-i-1} \leftarrow \begin{bmatrix} \alpha_{i+1}-\alpha_i \\ f(\alpha_i)-f(\alpha_{i+1}) \end{bmatrix}$ $\triangleright$ Compute the $n{-}i{-}1$'th column

$\quad\quad i \leftarrow i+1;$
$\quad$**end while**

To see Alg. 2 in action, recall the trade-off function $f_{\epsilon,\delta}$ depicted in Fig. 2, which we observe is in $\mathbb{F}^{PL}$ with facet points $\alpha_0=0$, $\alpha_1=\frac{1-\delta}{e^\epsilon+1}$, $\alpha_2=1-\delta$, $\alpha_3 = 1$. Alg. 2 computes the columns of $\mathcal{C}(f_{\epsilon,\delta})$ in order of increasing α_i as follows:

$$\begin{bmatrix} 1-(1-\delta) & (1-\delta)-(\frac{1-\delta}{e^\epsilon+1}) & \frac{1-\delta}{e^\epsilon+1}-0 & 0 \\ f_{\epsilon,\delta}(1-\delta)-f_{\epsilon,\delta}(1) & f_{\epsilon,\delta}(\frac{1-\delta}{e^\epsilon+1})-f_{\epsilon,\delta}(1-\delta) & f_{\epsilon,\delta}(0)-f_{\epsilon,\delta}(\frac{1-\delta}{e^\epsilon+1}) & 1-f_{\epsilon,\delta}(0) \end{bmatrix},$$

yielding,

$$\mathcal{C}(f_{\epsilon,\delta}) = C_{\epsilon,\delta} \;\; = \;\; \begin{bmatrix} \delta & (1-\delta)e^\epsilon/1+e^\epsilon & (1-\delta)/1+e^\epsilon & 0 \\ 0 & (1-\delta)/1+e^\epsilon & (1-\delta)e^\epsilon/1+e^\epsilon & \delta \end{bmatrix} . \tag{9}$$

By Thm 2 this is the greatest lower bound in the $\sqsubseteq$ order for channels in $\mathbb{C}_2$ that satisfy $f_{\epsilon,\delta}$-DP. Thus we have the following corollary:

Corollary 3 (Canonical (ϵ,δ) channel). *Let $M \in \mathbb{C}_2$. Then M satisfies (ϵ,δ)-DP if and only if $C_{\epsilon,\delta} \sqsubseteq M$.*

5   Compositions of channels define compositions for f-DP

With Thm 2, we can now obtain composition rules for f-DP, by using compositions defined on channels [1]. We do this for typical compositions used to implement or analyse privacy mechanisms.

5.1 Parallel composition

A typical scenario for analysis is repeated application of a mechanism $\mathcal{M}$ to the same dataset. This assumes that the output of $\mathcal{M} \circ \mathcal{M}(D)$ is a pair (y_0, y_1), one for each (independent) application of $\mathcal{M}$. In $\mathbb{C}_2$, this corresponds to parallel composition: if $C : \mathcal{X} \to \mathcal{Y}$ and $M : \mathcal{X} \to \mathcal{Z}$, then the parallel composition $C \parallel M$ outputs a pair from $\mathcal{Y} \times \mathcal{Z}$ as follows [1]:

$$(C \parallel M)_{x,(y,z)} = C_{x,y} \times M_{x,z} . \tag{10}$$

We can obtain an exact privacy profile for parallel composition by using Thm 2 to express channels as the min of their trade-off channels (Defn 9), and then Cor. 2 which says that min in $\mathbb{C}_2$ corresponds to $\sqcap$ in $\mathbb{F}$. Let $\mathcal{T}(C) \geq f$, and $\mathcal{T}(M) \geq f'$, then:

$$\mathcal{T}(C \parallel M) \;\geq\; \prod_{\alpha,\alpha'} \mathcal{T}\left(\begin{bmatrix} 1-\alpha & \alpha \\ f(\alpha) & 1-f(\alpha) \end{bmatrix} \parallel \begin{bmatrix} 1-\alpha' & \alpha' \\ f'(\alpha') & 1-f'(\alpha') \end{bmatrix} \right) . \tag{11}$$

As an example, Fig. 6, illustrates how the f-DP rule gives a better measurement for privacy loss than does the standard (ϵ, δ)-DP composition rule.

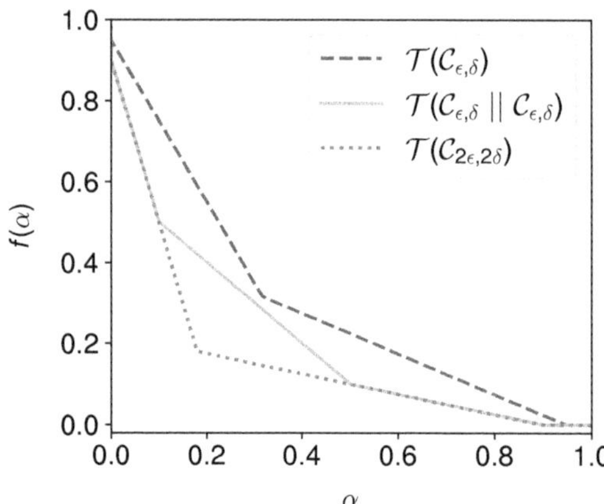

Fig. 6. Detailed privacy profiles for composition. Notice how the orange plot lies above the green plot, and has a gradient of −1 for α between 0.1 and 0.4 (approx.) indicating no privacy leakage for these tests, unlike the green plot which corresponds to the standard DP-composition, giving an over approximation for privacy loss.

5.2 Visible probabilistic choice

Visible probabilistic choice models the situation where the choice between applying M or C is made randomly, but which one was applied can be deduced from the output. This occurs when the outputs of the two channels are drawn from non-intersecting sets. Given channels $C : \mathcal{X} \to \mathcal{Y}, M : \mathcal{X} \to \mathcal{Z}$, where $\mathcal{Y} \cap \mathcal{Z} = \phi$ we define the *visible* probabilistic choice $C \,_r\!\oplus M$ as follows:

$$\begin{bmatrix} p_0 \cdots p_n \\ q_0 \cdots q_n \end{bmatrix} \,_r\!\oplus \begin{bmatrix} p'_0 \cdots p'_m \\ q'_0 \cdots q'_m \end{bmatrix} \;=\; \begin{bmatrix} rp_0 \cdots rp_n \; (1-r)p'_0 \cdots (1-r)p'_m \\ rq_0 \cdots rq_n \; (1-r)q'_0 \cdots (1-r)q'_m \end{bmatrix} .$$

In the (abstract) channel at right, all columns are derived from either C or M, but scaled by r or $1-r$ depending on whether the column originated from C or M. It turns out that the visible probabilistic choice composition rule for f-DP is determined by the error function Defn 10. As before, let $\mathcal{T}(C) \geq f$, and $\mathcal{T}(M) \geq f'$, and let $\alpha = err_{C(f)}(h)$, and $\alpha' = err_{C(f')}(h)$, then:

$$\mathcal{T}(C_{\,r}\oplus M)(r{\times}\alpha + (1{-}r){\times}\alpha') \;\geq\; r{\times}f(\alpha) + (1{-}r){\times}f'(\alpha') \,. \tag{12}$$

The effect on the corresponding trade-off functions is illustrated in Fig. 7 below.

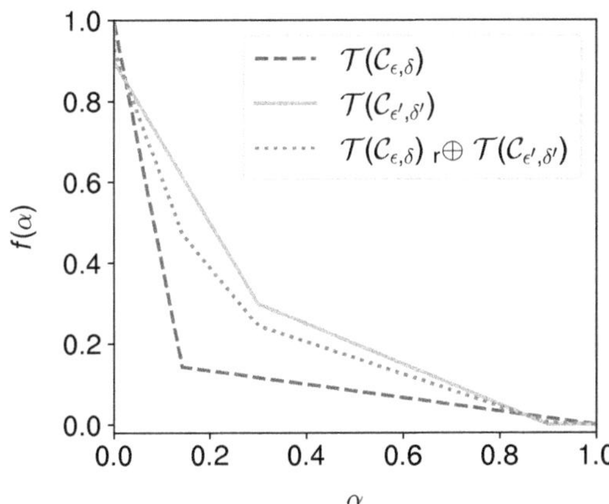

Fig. 7. Taking $C = C_{\epsilon,\delta}$ and $M = C_{\epsilon',\delta'}$ in Eqn (12), we can see the effect of visible choice. The green dotted line is the r-weighted average of the trade-off functions for $C_{\epsilon,\delta}$ and $C_{\epsilon',\delta'}$.

5.3   Hidden probabilistic choice

Hidden probabilistic choice is similar to visible probabilistic choice in that the choice between applying M or C is made randomly, but *unlike* visible choice it *cannot* be determined by looking at the outputs which one was applied. This situation occurs in implementations such as Alg. 1, as described in §7.

For example, assume that C, M both have $n{+}1$ columns with the columns labelled with outputs $\mathcal{Y} = \{y_0, \ldots y_n\}$. The hidden probabilistic choice with bias r produced by combining C and M in this way is:

$$\begin{bmatrix} p_0 \cdots p_n \\ q_0 \cdots q_n \end{bmatrix} {}_r\boxplus \begin{bmatrix} p'_0 \cdots p'_n \\ q'_0 \cdots q'_n \end{bmatrix} \;=\; \begin{bmatrix} rp_0{+}(1{-}r)p'_0 \cdots rp_n{+}(1{-}r)p'_n \\ rq_0{+}(1{-}r)q'_0 \cdots rq_n{+}(1{-}r)q'_n \end{bmatrix} .$$

Interestingly, the f-DP composition rule for hidden probabilistic choice does not have a direct definition because hidden probabilistic choice is sensitive to the precise outputs of the two mechanisms, and this information is not recorded in trade-off functions. However it is still the case that we can compute the indistinguishability profile for algorithms that use hidden probabilistic choice, by applying $\mathcal{T}$ to the channel composition, as in right-hand-side channel above.

5.4 Pre-processing as a composition

Our final composition is pre-processing, which arises when the inputs to a privacy mechanism are processed in some way before being presented to the mechanism. This is the case for the common example of sub-sampling in machine learning applications [6]. Given a mechanism M, a pre-processing process P is applied before applying the mechanism M. As a channel, this is modelled as a pre-matrix multiplication $P \cdot M$, thus its privacy profile becomes $\mathcal{T}(P \cdot M)$.

6 Universal Properties of Compositions

Finally we study relationships between privacy profiles of the different operators.

Theorem 3 (Composition theorems). *The following inequalities hold:*

1. *(Parallel composition)* $\mathcal{T}(C \parallel D) \le \mathcal{T}(C) \sqcap \mathcal{T}(D)$
2. *(Visible choice)* $\mathcal{T}(C) \sqcap \mathcal{T}(D) \le \mathcal{T}(C \,_p{\oplus} D) \le \mathcal{T}(C) \sqcup \mathcal{T}(D)$
3. *(Pre-processing)* $\mathcal{T}(C) \le \mathcal{T}(C \cdot Q)$
4. *(Visible and hidden choice)* $\mathcal{T}(C \,_p{\oplus} D) \le \mathcal{T}(C \,_p{\boxplus} D)$
5. *Let* $\mathcal{Y}_1, \mathcal{Y}_2$ *be two disjoint sets. Let* $D, E : \mathcal{X} \to \mathcal{Y}_1$ *and* $C : \mathcal{X} \to \mathcal{Y}_2$. *Then*

$$\mathcal{T}((C \,_p{\oplus} D) \,_r{\boxplus} E) = \mathcal{T}(C \,_{rp}{\oplus} (D \,_{\frac{r(1-p)}{r(1-p)+(1-r)}} {\boxplus} E))$$

Proof. All results follow from well-known channel refinements and Thm 2 and Cor. 1. For example, (1) follows since $C \parallel D \sqsubseteq C, D$. *For (2) we have,*

$$(C \min D) = (C \min D) \,_p{\oplus} (C \min D) \sqsubseteq C \,_p{\oplus} D \ ,$$

with the last inequality following from the observation that $C \min D \sqsubseteq C, D$, *and* $_p{\oplus}$ *is a monotone operator. The remaining inequalities follow similarly.*

We also have the following monotonicity results, which again follow from Thm 2, Cor. 1 and refinement properties of channels [1].

Theorem 4 (Monotonicity results). *Let* $C \sqsubseteq C'$ *and* $Q \sqsubseteq Q'$, *then the following refinements hold:*

1. *(Parallel Composition)* $\mathcal{T}(C \parallel Q) \le \mathcal{T}(C' \parallel Q)$
2. *(Visible choice)* $\mathcal{T}(C \,_p{\oplus} Q) \le \mathcal{T}(C' \,_p{\oplus} Q)$
3. *(Pre-processing)* $\mathcal{T}(C \cdot Q) \le \mathcal{T}(C \cdot Q')$

We end this section by demonstrating a canonical representation for a common form of trade-off function, namely symmetric, piecewise linear functions.

Definition 14 (Symmetric, piecewise linear trade-off functions). *A trade-off function is symmetric if* $f(\alpha) = f \circ f(\alpha)$, *for all* $0 \le \alpha \le 1$. *It is finite, piecewise linear if it is linear almost everywhere, except for a finite number of facet points.*

Recall $f_{\epsilon,\delta}$ which is piecewise linear and symmetric; we noted above that there is a canonical representation of $\mathcal{C}(f_{\epsilon,\delta})$ as a visible probabilistic choice over channels $C_{\epsilon,0}$ and $C_{\infty,0}$. It turns out that this is true generally for all symmetric, piecewise linear trade-off functions.

Lemma 4 (Canonical representation for symmetric trade-off functions).
Let f be a finite, symmetric trade-off function, with $N+1$ facet points. Then there are $\epsilon_0, \ldots \epsilon_{\lfloor N/2 \rfloor}$ reals such that $\mathcal{C}(f)$ is a visible probabilistic choice over $C_{\epsilon_i,0}$.

Proof. Direct consequence of Alg. 2, using the symmetric condition of f. In particular, let M be the result of applying Alg. 2. This means that M has $N+1$ columns and two rows, such that $M_{x,n} = M_{1-x,N-n}$. We define $\epsilon_n = M_{0,n}/M_{1,n}$ for $0 \leq n \leq N/2$, and observe that the two columns at $M_{-n}, M_{-(N-n)}$ correspond to $C_{\epsilon_n,0}$, scaled by (visible) probability $M_{0,n}+M_{0,(N-n)}$.

We note finally that many of our analyses can be usefully carried out using the canonical mechanisms satisfying f-DP, since the monotonicity results of Thm 4 imply a (tight) lower bound for the class of mechanisms satisfying the given f-DP constraint. As usual, refinement ensures that a property holds generally.

7   Implementations of privacy-enhancing mechanisms

Privacy purification We can now use the operators in §5 to describe the privacy semantics of Alg. 1 in $\mathbb{C}_2$, and then use Thm 2 to compute a tight privacy profile. Recall that the algorithm takes as input a mechanism M that we assume satisfies some $f_{\epsilon,\delta}$-differential privacy specification, and the objective is to ensure that the profile of the output corresponds to a pure differential privacy constraint. In this case it means that $\mathcal{T}(Alg.\ 1)$ has gradient always bounded away from 0 and ∞. This is equivalent to $f_{\epsilon',0} \leq \mathcal{T}(Alg.\ 1)$, for some $\epsilon' \geq 0$. We first show how to model Alg. 1 using channel compositions.

We observe that the first six lines of Alg. 1 correspond to a probabilistic choice with bias r i.e. M (is applied to x) with probability r, or, with probability $1-r$, x is ignored and a random value is reported. This choice is a combination of hidden and visible choice depending on overlap of $\mathcal{Y}$ and $\mathcal{Y}'$. [3] When they overlap entirely, this is a hidden choice; when they overlap partially then it is a combination of visible and hidden choice. For example, if we assume that $\mathcal{Y}' \subset \mathcal{Y}$, then we can rewrite M as $M_1{}_p\oplus M_2$, where the output of M_1 is $\mathcal{Y}'$; we can then use Thm 3(5). to compute the trade-off function of $M{}_r\boxplus U[\mathcal{Y}']$.

The final line of the algorithm, before the output is effectively a post-processing by a Geometric perturbation, which we denote by $G_{\epsilon'}$. Overall we can model the effect of the purification algorithm as $(M{}_r\boxplus U[\mathcal{Y}']) \cdot G_{\epsilon'}$.

In Fig. 8 we consider the two scenarios when $\mathcal{Y}$ and $\mathcal{Y}'$ do, or when they don't, coincide. We assume that $M = C_{\epsilon,\delta}$, and that each of the four columns of M are selected uniformly by $U[\mathcal{Y}]$. The left-hand plot shows the profiles $\mathcal{T}(M)$, $\mathcal{T}(M{}_r\boxplus U[\mathcal{Y}])$, and $\mathcal{T}((M{}_r\boxplus U[\mathcal{Y}]) \cdot G_{\epsilon'})$. We see that both steps of Alg. 1

---

[3] We observe that for well-definedness $\mathcal{Y}'$ must be bounded.

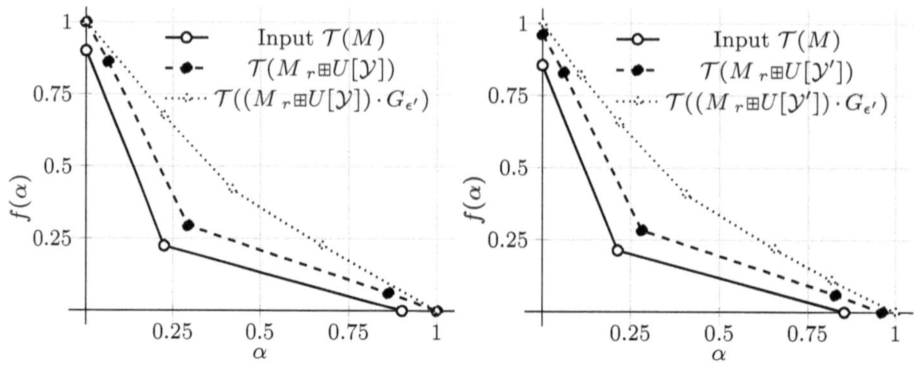

Fig. 8. Results at left assume $\mathcal{Y}' = \mathcal{Y}$, and at right that $\mathcal{Y}' \subset \mathcal{Y}$

improve the privacy of M; but when $\mathcal{Y} = \mathcal{Y}'$, the final post-processing is not necessary for purification since $\mathcal{T}((M\,_r\boxplus U[\mathcal{Y}])$ has finite, non-zero gradient.

This contrasts with the right-hand plot where we assume that $\mathcal{Y}' \subset \mathcal{Y}$. In this case the plot $\mathcal{T}(M\,_r\boxplus U[\mathcal{Y}'])$ crosses the horizontal axis at a point strictly less than 1 so that there is a (non-trivial) gradient of 0. This means that the final post-processing step is crucial to ensure that the the overall sanitisation satisfies the desired pure differential privacy property.

Interestingly, if we know that the input mechanism M commutes with the final line of Alg. 1 using our composition rule, we can deduce the following:

$$G_{\epsilon'} \sqsubseteq M \cdot G_{\epsilon'} \sqsubseteq ((M\,_r\boxplus U[\mathcal{Y}]) \cdot G_{\epsilon'} \ ,$$

and therefore $f_{\epsilon',0} = \mathcal{T}(G_{\epsilon'}) \leq \mathcal{T}(((M\,_r\boxplus U[\mathcal{Y}]) \cdot G_{\epsilon'})$, leading to clause of Alg. 1 being tightened to "Satisfies $f_{\epsilon',0}$-DP". This happens when M is a simple perturbation, i.e. $x \leftarrow x + \mu(0)$, where μ is some probability distribution.

Sub-sampling is a technique to implement private training in machine learning [27], and Alg. 3 is an example. A private dataset D is to be used for training using a privacy-preserving process M. Rather than using the entire dataset D, only a sample $d \subseteq D$ is used as input to M. The while-loop in Alg. 3 implements probabilistic sampling known as "Poisson sampling", where each record in D is included in the sample d with probability γ. Fig. 9 illustrates the effect.

8   Related Work

f-DP was introduced by Dong et al. [13] as a way to obtain more accurate analyses of privacy, showcasing the benefits largely for the Gaussian perturbation mechanism. It is also finding success for auditing privacy in complex mechanisms [25], and many of the definitions given in §2.2 are drawn from there. Dong

---

Algorithm 3 Privacy-preservation with sub-sampling

Require: Mechanism $M : \mathcal{D} \to \mathcal{Y}$, Private dataset D; Parameters: $\gamma \in [0,1]$
Ensure: z is a sanitised output for D, with better privacy than M without sub-sampling.
  $i, d \leftarrow |D|, \phi$;
  while **do** $i > 0$
    $v \leftarrow U[0,1]$;   ▷ Choose a value uniformly from $[0,1]$
    if $(v < \gamma)$ **then**   ▷ Probabilistic choice with γ bias, used as a pre-processor
      $d \leftarrow d \cup D_i$
    end if
    $i \leftarrow i-1$;
  end while
  $z \leftarrow M(d)$;   ▷ Apply the mechanism to the sample d
  Output z;   ▷ Output sanitised result

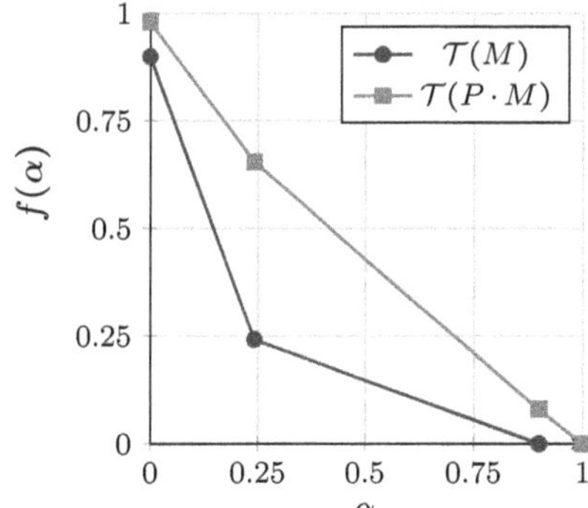

Thus sub-sampling becomes a pre-processing of M; for Poisson sampling it can be shown that the pre-processing channel $P = \begin{bmatrix} 1 & 0 \\ 1-\gamma & \gamma \end{bmatrix}$, so that the privacy profile for Alg. 3 is $\mathcal{T}(P \cdot M)$. Moreover, by Thm 4(3), any mechanism $C_{\epsilon,\delta} \sqsubseteq M'$ satisfies $\mathcal{T}(P \cdot C_{\epsilon,\delta})$-differential privacy, defined by the red line in the plot. Note that other methods of sub-sampling, such as sampling with replacement, can be modelled in a similar way, although the details of the pre-processing is different.

Fig. 9. Privacy profile for Alg. 3

et al. showed that for every trade-off function f there exists a distribution μ such that $f = \mathbf{T}(U, \mu)$, where U is the uniform distribution over the (bounded) domain of μ. Our Thm 2, and its extension to general trade-off functions shows that Dong's construction is actually unique up to equivalence under refinement. Su [28] also provides more insight about the connection to Blackwell's distinguishability by hypothesis testing [10]. What is interesting here, is that Blackwell [10] suggests that it does not seem possible to extend hypothesis testing using hockey-stick gain functions to more complex privacy scenarios that require the analysis of many secrets [4,3]. This means that channels $\mathbb{C}_2$ corresponding to trade-off functions represents a worst-case scenario of maximal separation between eg. datasets. Moreover, Blackwell's observation is consistent with the fact that (abstract) channels over multiple secrets do not form a lattice [1].

Awan et al. [5] have looked at sampling mechanisms for implementing exact f-DP constraints, especially for symmetric trade-off functions. Their proposal for efficient sampling for general $f_{\epsilon,\delta}$, for example could potentially be extended via our canonical channel representation for symmetric f-DP mechanisms (Thm 4).

Privacy purification was proposed by Lin et al. [19]; however our Alg. 1 uses Geometric rather than Laplace noise as the final perturbation. Nevertheless our lower bound still applies to Lin's (by Thm 4), since the Laplace refines the Geometric perturber [15]. Sub-sampling has also been investigated by Balle et al.[6] wrt. an (ϵ, δ)-DP definition and Wang et al. [30] for Renyi-DP. Our analysis provides a full privacy profile, similar to the analysis in Dong et al. [13].

Quantitative Information Flow for analysing security leaks in programs was first proposed by Clark et al. [12, 11] Malacaria [20], and Smith [26]; the g-leakage framework was introduced by Alvim et al. [2] for general quantitative information flow analysis, and McIver et al. [21] for programs. Many of these ideas have been extended and developed by Alvim et al. [1]. As mentioned above, the lattice structure of $\mathbb{C}_2$ does not extend to channels with more than 2 rows, as was shown in [1], Chap 12.

A quantitative information flow semantics for programs was given by Gibbons et al. [17]. Other verification techniques for privacy have been given in [8], [22],[9] and [7]; a light-weight language for the verification of traditional differential privacy has been developed by Zhang and Kifer [31]. These methods use the "worst-case" features of differential privacy, and the simplifications in the analyses that it brings.

9   Discussion and Future work

We have demonstrated the equivalence between distinguishability via hypothesis testing and the QIF operational model for an information-aware program semantics. This enables novel composition rules for f-DP that can be used in the analysis for implementations for privacy-enhancing mechanisms. We have also provided new analyses for privacy purification and sub-sampling.

One important application of this work is to connect f-DP constraints to implementations of privacy-enhancing algorithms. For example the programming language *Kuifje* [17] has an information-flow aware programming semantics based on the channel model of QIF [1] and HMM's [23]. This suggests the possibility of using a verification technique based on f-DP constraints as specifications. For example, if we would like to verify that a program P satisfies a privacy constraint defined by f, then we can convert it into a refinement problem:

$$\mathcal{C}(f) \sqsubseteq [\![P]\!] \, , \tag{13}$$

where $[\![P]\!]$ refers to the interpretation of program P in the information-flow semantics of *Kuifje* [17, 23]. If the refinement holds, then this means that the information leaks of P are within those prescribed by the constraint f.

In future work we would like to make such an approach more accessible, for example by combining *Kuifje*'s automated refinement analysis with our Alg. 2 which creates the lhs of the refinement inequality in Eqn (13).

References

1. M. Alvim, K. Chatzikokolakis, A.K. McIver, C.C. Morgan, G.S. Smith, and C. Palamidessi. *The Science of Quantitative Information Flow*. Information Security and Cryptography. 2020.
2. M. S. Alvim, K. Chatzikokolakis, C. Palamidessi, and G.S. Smith. Measuring information leakage using generalized gain functions. In *Proc. 25th IEEE Computer Security Foundations Symposium (CSF 2012)*, pages 265–279, June 2012.
3. Mário S. Alvim, Miguel E. Andrés, Konstantinos Chatzikokolakis, Pierpaolo Degano, and Catuscia Palamidessi. On the information leakage of differentially-private mechanisms. *Journal of Computer Security*, 23(4):427–469, 2015.
4. Mário S. Alvim, Miguel E. Andrés, Konstantinos Chatzikokolakis, and Catuscia Palamidessi. On the relation between differential privacy and quantitative information flow. In *Automata, Languages and Programming - 38th International Colloquium, ICALP 2011, Zurich, Switzerland, July 4-8, 2011, Proceedings, Part II*, pages 60–76, 2011.
5. Jordan Awan and Salil Vadhan. Canonical noise distributions and private hypothesis tests, 2023.
6. Borja Balle, Gilles Barthe, and Marco Gaboardi. Privacy profiles and amplification by subsampling. 10, Jan. 2020.
7. Gilles Barthe, Marco Gaboardi, Emilio Jesus Gallego Arias, Justin Hsu, Cesar Kunz, and Pierre-Yves Strub. Proving Differential Privacy in Hoare Logic . In *2014 IEEE 27th Computer Security Foundations Symposium (CSF)*, pages 411–424. IEEE Computer Society, 2014.
8. Gilles Barthe, Marco Gaboardi, Benjamin Grégoire, Justin Hsu, and Pierre-Yves Strub. Proving differential privacy via probabilistic couplings. LICS '16, pages 749–758. Association for Computing Machinery, 2016.
9. Gilles Barthe, Boris Köpf, Federico Olmedo, and Santiago Zanella-Béguelin. Probabilistic relational reasoning for differential privacy. *ACM Trans. Program. Lang. Syst.*, 35(3), November 2013.
10. David Blackwell. Equivalent comparisons of experiments. *The Annals of Mathematical Statistics*, 24(2):265–272, 1953.
11. David Clark, Sebastian Hunt, and Pasquale Malacaria. Quantitative analysis of the leakage of confidential data. *Electr. Notes Theor. Comput. Sci.*, 59(3):238–251, 2001.
12. David Clark, Sebastian Hunt, and Pasquale Malacaria. A static analysis for quantifying information flow in a simple imperative language. *J. Comput. Secur.*, 15(3):321–371, 2007.
13. Jinshuo Dong, Aaron Roth, and Weijie J. Su. Gaussian differential privacy. *Journal of the Royal Statistical Society Series B: Statistical Methodology*, 84(1):3–37, 02 2022.
14. Cynthia Dwork. Differential privacy. In *Proc. 33rd International Colloquium on Automata, Languages, and Programming (ICALP 2006)*, pages 1–12, 2006.
15. Natasha Fernandes, Annabelle McIver, and Carroll Morgan. The laplace mechanism has optimal utility for differential privacy over continuous queries. In *Proceedings of the 36th Annual ACM/IEEE Symposium on Logic in Computer Science*, LICS '21. IEEE Press, 2021.
16. Natasha Fernandes, Annabelle McIver, and Parastoo Sadeghi. Composition theorems for f-differential privacy, 2025. arXiv:2512.21358 [cs.CR].

17. Jeremy Gibbons, Annabelle McIver, Carroll Morgan, and Tom Schrijvers. *Quantitative Information Flow with Monads in Haskell*, pages 391–448. Cambridge University Press, 2020.

18. Neyman Jerzy and Pearson Egon Sharpe. On the problem of the most efficient tests of statistical hypotheses. *Philosophical Transactions of the Royal Society of London*, 1933.

19. Yingyu Lin, Erchi Wang, Yi-An Ma, and Yu-Xiang Wang. Purifying approximate differential privacy with randomized post-processing, 2025.

20. P. Malacaria. Risk assessment of security threats for looping constructs. *Journal of Computer Security*, 18(2):191–228, 2010.

21. Annabelle McIver, Larissa Meinicke, and Carroll Morgan. Compositional closure for Bayes Risk in probabilistic noninterference. In *Proceedings of the 37th international colloquium conference on Automata, languages and programming: Part II*, volume 6199 of *ICALP'10*, pages 223–235, Berlin, Heidelberg, 2010.

22. Annabelle McIver and Carroll Morgan. Proving that programs are differentially private. In Anthony Widjaja Lin, editor, *Programming Languages and Systems*, pages 3–18, Cham, 2019. Springer International Publishing.

23. Annabelle McIver, Carroll Morgan, and Tahiry Rabehaja. Abstract Hidden Markov Models: a monadic account of quantitative information flow. In *Proc. LiCS 2015*, 2015.

24. Ilya Mironov. Rényi Differential Privacy . In *2017 IEEE 30th Computer Security Foundations Symposium (CSF)*, pages 263–275. IEEE Computer Society, 2017.

25. Milad Nasr, Jamie Hayes, Thomas Steinke, Borja Balle, Florian Tramèr, Matthew Jagielski, Nicholas Carlini, and Andreas Terzis. Tight auditing of differentially private machine learning. In *Proceedings of the 32nd USENIX Conference on Security Symposium*, SEC '23, USA, 2023. USENIX Association.

26. Geoffrey Smith. On the foundations of quantitative information flow. In Luca de Alfaro, editor, *Proc. 12th International Conference on Foundations of Software Science and Computational Structures (FoSSaCS '09)*, volume 5504 of *Lecture Notes in Computer Science*, pages 288–302, 2009.

27. Thomas Steinke. Composition of differential privacy & privacy amplification by subsampling, 2022.

28. Weijie J. Su. A statistical viewpoint on differential privacy: Hypothesis testing, representation, and blackwell's theorem. *Annual Review of Statistics and Its Application*, 12(Volume 12, 2025):157–175, 2025.

29. Latanya Sweeney. k-anonymity: a model for protecting privacy. *Int. J. Uncertain. Fuzziness Knowl.-Based Syst.*, 10(5):557–570, October 2002.

30. Yu-Xiang Wang, Borja Balle, and Shiva Prasad Kasiviswanathan. Subsampled renyi differential privacy and analytical moments accountant. In Kamalika Chaudhuri and Masashi Sugiyama, editors, *Proceedings of the Twenty-Second International Conference on Artificial Intelligence and Statistics*, volume 89 of *Proceedings of Machine Learning Research*, pages 1226–1235. PMLR, 16–18 Apr 2019.

31. Danfeng Zhang and Daniel Kifer. Lightdp: towards automating differential privacy proofs. *SIGPLAN Not.*, 52(1):888–901, January 2017.

On Reversibility in Petri Nets [*]

Hernán Melgratti[1], Claudio Antares Mezzina[2],
and G. Michele Pinna[3]

[1] ICC - Universidad de Buenos Aires, Argentina
[2] Dipartimento di Scienze Pure e Applicate, Università di Urbino, Urbino, Italy
[3] Dipartimento di Matematica e Informatica, Università di Cagliari, Italy

Abstract. Reversible semantics for Petri nets have received increasing
attention, yet a fully satisfactory account for general nets is still missing. Existing approaches apply either to restricted subclasses, such as
occurrence nets, or to general nets studied via unfoldings, which often
produce infinite models even for finite nets. Other approaches adopt a
weaker notion of reversibility, based solely on the ability to restore an
initial marking, without undoing individual computation steps.

In this paper, we propose a reversible semantics for general Petri nets
that supports stepwise reversal and enforces causal consistency. Our approach combines explicit identifiers to track executed transitions with
an enriched token model. This combination ensures that only executed
transitions can be reversed and that reversibility respects causal dependencies.

Keywords: Reversibility · Petri Nets · Concurrency

1 Introduction

Reversible computation has been studied since the 1960s, when Landauer showed
that logically reversibility is fundamental to low-energy computation [5]. Today,
reversible computing has numerous applications, including biochemical reaction
modeling [9,17], parallel discrete-event simulation [25], robotics [13], control theory [8], fault-tolerant systems [21], and concurrent program debugging [6,14].
Reversible systems support both forward and backward computation, allowing
previous steps to be undone. However, the meaning of reversibility varies across
domains.

In this paper, we study reversibility in a concurrent setting, focusing specifically on Petri nets. Petri nets play a central role in concurrency theory and

* This work has been supported by the Italian MUR PRIN 2022 project *DeKLA*
(F53D23004840006), PRIN 2022 DeLiCE (F53D23009130001), the INdAM-GNCS
project CUP_E53C24001950001 *MARQ*, and the European Union - NextGenerationEU program Research and Innovation Program PE00000014 *SEcurity and RIghts
in the CyberSpace* (SERICS), projects STRIDE and SWOPS and the MSCA SE
project QCOMICAL (Grant Agreement ID: 101182520). The second author also ac-
knowledges partial support of the Japan Society for the Promotion of Science (JSPS)
through the Invitational Fellowship Program, Grant No. S25016.

N. Bertrand and S. Milius (Eds.): FoSSaCS 2026, LNCS 16503, pp. 484–504, 2026.
https://doi.org/10.1007/978-3-032-22730-0_23

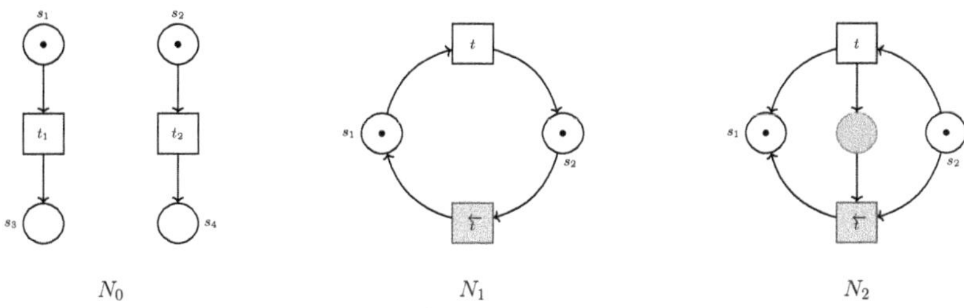

Fig. 1: Local reversibility

are often used as a unifying and comparative formalism for other models. Likewise, we envision their reversible counterparts to play an analogous role in the study of reversible concurrency. Despite their importance, there is currently no fully satisfactory semantics for reversible Petri nets. Our work is concerned with *local reversibility*, which addresses the step-by-step undoing of transition firings [19,22,16,20,4,23]. This notion contrasts with *global reversibility*, where the emphasis is on the ability to return to the initial marking from any reachable marking [2,7].

Local reversibility varies in granularity, from undoing individual transition firings [19,22,16,20,23] to reversing entire steps, i.e., multisets of concurrently executed transitions [4]. We focus on the most fine-grained notion, where local reversibility is realised at the transition level by associating with each transition t a backward transition $\overleftarrow{t}$ that undoes the effects of t by consuming tokens from its postset and producing tokens in its preset. This is illustrated by the net N_1 at the centre of Figure 1, where the transition $\overleftarrow{t}$ (in blue) is designated as the *backward* of t. When introduced naively, this modelling does not necessarily enforce reversibility and exhibits several drawbacks. First, the backward transition $\overleftarrow{t}$ may be executable even if the forward transition t has not occurred. This issue can be resolved by introducing an explicit dependency between t and $\overleftarrow{t}$, represented by a shared initially unmarked place, ensuring that the backward transition can fire only after the corresponding forward transition has occurred, as illustrated by the net N_2 on the right of Figure 1. Still, this is insufficient to enforce that reverse transitions are executed in a suitable order. Addressing this requires carefully choosing the semantics that determine when a reverse transition is enabled, as noted in reversible process calculi [24,3,10]

One natural interpretation, inspired by the sequential setting, treats reverse execution as backtracking, where the most recently fired transition must be reversed first. In a concurrent setting, however, reversibility is generally expected to be less restrictive, allowing independent threads of execution to be undone independently. As an example, consider the Petri net N_0 shown on the left of Figure 1, which consists of two independent (concurrent) transitions, t_1 and t_2. A forward execution may fire t_1 followed by t_2, yielding the sequence $t_1; t_2$. Under rollback semantics, this execution must be reversed in the opposite order, namely by undoing t_2 first and then t_1. Such semantics can be implemented using a global memory (i.e., a stack) where the execution of each forward transition pushes it onto the stack, and a backward transition is permitted only if it matches the transition at the top of the stack, which is then popped [23].

Remarkably, backtracking is well suited to the interleaving semantics of Petri nets. However, from a true-concurrency perspective, the particular forward interleaving in the sequence $t_1; t_2$ is incidental. Consequently, any reversal that preserves the underlying partial order of causality should be permitted, meaning that any Mazurkiewicz-equivalent sequence [15] should be admissible. In our example, the sequence $t_1; t_2$ should therefore admit reversing executions that undo either t_1 or t_2 first. This notion is known as *causal-consistent reversibility*.

Causal-consistent reversibility is usually implemented by using distributed memories [3]. Any reversible semantics that uses a linear memory discipline (stack, log, sequence) will inevitably serialize concurrency and therefore break causal consistency. To support causal reversibility, the memory must be capable of representing partial orders and allowing unordered removal of incomparable events, e.g. graph-shaped histories, or cause-tracking structures. In our approach, memories are represented by key places and the causal dependencies induced by the structure of the net.

Existing approaches to causally-consistent, local reversibility are mostly restricted to occurrence nets [20,18], where causal relations are explicitly encoded and backward steps are unambiguous. As a result, reversible semantics for general nets are defined only indirectly, via their unfoldings. While these methods successfully introduce backward transitions, the unfolding often produces an infinite model, even when the original Petri net is finite. Unfoldings can be avoided by using colored tokens to encode the potentially long causal history, though the required pattern matching for firing backward transitions may degrade performance [20]. Certain proposals [23] achieve reversibility by adding elements like memories and bonds, which arguably modify the computational model of Petri nets.

The **contribution** of this work is a direct and general model for local reversibility in Petri nets, based on the principle that reversibility should explicitly account for how each individual forward step is undone. Our approach builds on two established ideas: (i) using *key places*, analogous to the key identifiers in reversible process calculi, to track executed transitions, and (ii) leveraging *structured tokens* that carry computational history, drawn from the concurrent semantics of Petri nets. We formalize our model using *Dynamic Petri Nets* (DPNs) [1], which extend classical Petri nets with mechanisms for dynamic creation of places and transitions. This dynamic capability allows the net to evolve during execution, enabling the modeling of systems with mobility, dynamic topologies, or processes that generate new components. In our approach, we exploit this feature to generate at runtime the machinery needed to reverse a fired transition. We note that we use DPNs primarily for convenience in defining our model; however, thanks to the results in [1], which show that DPNs can be encoded in standard Petri nets, our approach remains fully within the Petri net formalism.

We investigate two reversible semantics differing on the definition of backward transitions:

- **Local reversibility using key places:** enables step-by-step undoing but does not guarantee causal consistency.
- **History-aware structure:** integrates key places with key-aware tokens, guaranteeing reversibility that is both local and causally consistent.

In both cases, backward transitions can fire only after their corresponding forward transitions, preventing indefinite backward computations. The novelty of our approach is that backward transitions explicitly depend on their forward counterparts. To establish causal consistency, we adopt the framework of [12] and show that our system satisfies its required basic properties.

Organisation of the paper. In the next section, we recall some basic notions. In Section 3, we present our first approach to reversibility in Petri nets, based on the use of key places. This is achieved by defining reversible nets in terms of expansible nets, a subclass of dynamic nets. The resulting model ensures local reversibility, but does not enforce causal reversibility. In Sections 4 and 5, we introduce a refined approach where tokens carry additional information to enforce causal reversibility while avoiding the need to record their full causal history. We formally establish that the approach satisfies causal consistency using the framework of [12]. We conclude with a discussion of our findings.

2   Preliminaries and notations

The symbol $\mathbb{N}$ indicates the set of natural numbers. A *coloured multiset* over a set A of elements and a set C of colours is a mapping $f : A \to C \to \mathbb{N}$. Multisets are assumed to be equipped with the standard operations of union ($\oplus$) and difference ($\ominus$): $(f \oplus g)(a)(c) = f(a)(c) + g(a)(c)$, and similarly $(f \ominus g)(a)(c) = f(a)(c) - g(a)(c)$ provided that $f(a)(c) \geq g(a)(c)$. We write $f \subseteq f'$ if $f(a)(c) \leq f'(a)(c)$ for all $a \in A$ and $c \in C$. The multiset $\lfloor\!\lfloor f \rfloor\!\rfloor$ is defined such that $\lfloor\!\lfloor f \rfloor\!\rfloor(a)(c) = 1$ if $f(a)(c) > 0$ and $\lfloor\!\lfloor f \rfloor\!\rfloor(a)(c) = 0$ otherwise. We may confuse a multiset f with the set $\{(a,c) \mid a \in A, c \in C . f(a)(c) \neq 0\}$ when $f = \lfloor\!\lfloor f \rfloor\!\rfloor$. In such cases $(a,c) \in f$ is used for $f(a)(c) \neq 0$. The set of all multisets over A and C is denoted as $\mu(A, C)$; the symbol 0 stands for the unique multiset defined such that $\lfloor\!\lfloor 0 \rfloor\!\rfloor = \emptyset$. We sometimes denote a multiset f as $\oplus_{(a,c)\in\lfloor\!\lfloor f \rfloor\!\rfloor} f(a)(c) \cdot a(c)$, thus $a(c)\oplus 2b(d)$ would be the multiset f such that $f(a)(c) = 1$ and $f(b)(d) = 2$. When C is a singleton we identify multisets f on A and C as mappings $f : A \to \mathbb{N}$ and we avoid mentioning the unique element of C. The operations on these multisets are the same and, with abuse of notation, we will use μA to denote the set of these multisets and $\lfloor\!\lfloor f \rfloor\!\rfloor$ will be the multiset $\lfloor\!\lfloor f \rfloor\!\rfloor(a) = 1$ whenever $f(a) > 0$ and $\lfloor\!\lfloor f \rfloor\!\rfloor(a) = 0$ otherwise. When $f \in \mu(A, C)$, $A \subseteq A', C \subseteq C'$ we consider f a multiset in $\mu(A', C')$ which is such that $f(a)(c) = 0$ for all $a \in A' \setminus A$ and $c \in C' \setminus C$.

Given a partial function $f : A \to B$, with $dom(f)$ we denote the subset of elements of A where $f(a)$ is defined.

$$\frac{t = f \;\triangleright\; g \in T \qquad m'' \in \mu S}{\langle S, T, f \oplus m'' \rangle \xrightarrow{t} \langle S, T, g \oplus m'' \rangle}$$

Fig. 2: Operational semantics of nets.

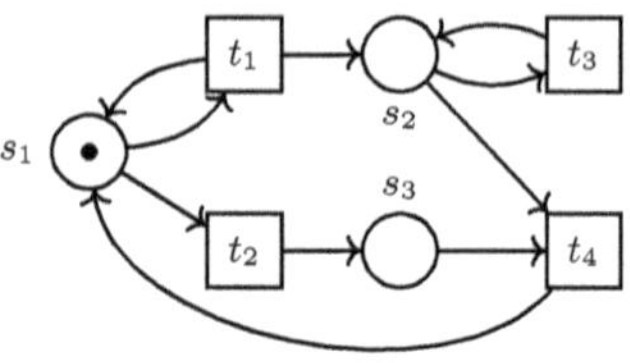

Fig. 3: A net N

3 Nets

We recall the basic notions about nets. Usually nets are presented as tuple (S, T, F, m) where S and T are two disjoint sets of *places* and of *transitions*, $F : (S \times T) \cup (T \times S) \to \mathbb{N}$ is a mapping associating to each arc connecting places and transitions a weight and $m : S \to \mathbb{N}$ is a *marking*. If we consider a transition $t \in T$ with ${}^\bullet t$ we denote the multiset $F(t, -)$ and with $t^\bullet$ the multiset $F(-, t)$. Here we focus on transitions, that are considered to be *pairs* of multisets on places, i.e. elements of μS. A transition t is then the pair (f, g) with ${}^\bullet t = f$ and $t^\bullet = g$ and we will denote it with $f \triangleright g$.

Definition 1. *A net is the triple* $N = \langle S, T, m_0 \rangle$*, where* S *is a set of* places, $T \subseteq \mu S \times \mu S$ *is a set of* transitions, *and* $m_0 : S \to \mathbb{N}$ *is the* initial *marking. The set of nets is denoted by* $\mathcal{N}$.

Given a net $N = \langle S, T, m_0 \rangle$ and a place s, with abuse of notation, we write ${}^\bullet s = \{t \in T \mid t^\bullet(s) > 0\}$ and $s^\bullet = \{t \in T \mid {}^\bullet t(s) > 0\}$. Places in a net represent resources, while transitions specify how these resources are manipulated— namely, consumed and produced. This interaction is typically described by the so-called *token game*, where resource instances, represented as tokens, reside within places. The state of a net is captured by the distribution of tokens across its places, a concept formally defined as a *marking*. A marking thus represents the nets state, while the token game describes the evolution of these states over time. A marking is a total function from places to natural numbers, explicitly indicating the number of tokens present in each place.

Given a net $\langle S, T, m_0 \rangle$, we say that a transition t is *enabled* at a marking m if ${}^\bullet t \subseteq m$. This can be rewritten by saying that $m = {}^\bullet t \oplus m''$ for $m'' \in \mu S$. If a transition t is enabled at a marking $m = {}^\bullet t \oplus m''$ then it may *fire* producing a marking m', which is $m'' \oplus t^\bullet$. The firing of a transition is formalized in rule in Figure 2, where, to be consistent with the rest of the paper, we keep the whole net rather than considering just the markings. Thus, fixed a net $N = \langle S, T, m_0 \rangle$, $\to \;\subseteq\; \mathcal{N} \times T \times \mathcal{N}$.

Example 1. Consider the net shown in Figure 3. Places are represented by circles, transitions by boxes, and arrows connecting places and transitions represent pre- and postsets: there is an arc from each place s such that ${}^\bullet t(s) \neq 0$ and the transition t, possibly annotated with a *weight* in case ${}^\bullet t(s) > 1$, and similarly there is an arc from the transition t to each place s such that $t^\bullet(s) \neq 0$. Tokens are depicted as bullets inside places. The transition t_1 has s_1 in its preset and s_1, s_2 in its postset, hence $t_1 = (s_1, s_1 \oplus s_2)$. Transition $t_4 = (s_2 \oplus s_3, s_1)$. Transition $t_3 = (s_2, s_2)$ and $t_2 = (s_1, s_3)$. The initial marking is defined by $m_0(s_1) = 1$ and $m_0(s_i) = 0$ for $i \in \{2, 3\}$. Transition t_1 is enabled at m_0 since $^\bullet t_1 = s_1 \subseteq m_0$. When fired, it produces a marking m where $m(s_1) = m(s_2) = 1$ and $m(s_3) = 0$. At marking m, both transitions t_1 and t_2 are enabled as $^\bullet t_1 = {}^\bullet t_2 = s_1$. The firing of t_2 produces the marking m' such that $m'(s_1) = 0$ and $m'(s_2) = m'(s_3) = 1$.

3.1 Firing sequences

The computations of a net are described in terms of *sequences* of transitions, that are called *firing sequences*.

Definition 2 (Firing sequence). *A firing sequence (shortened to fs) of a net* $N = \langle S, T, m_0 \rangle$ *is a sequence of transitions* $\vec{t} = t_0; t_1; t_2; \dots$ *such that there exists a sequence of nets* $N_i = \langle S, T, m_i \rangle$ *in* $\mathcal{N}$ *such that* $N_i \xrightarrow{t_i} N_{i+1}$ *for all* i *and* $N = N_0$. *When the sequence* $\vec{t} = t_0; t_1; t_2; \cdots; t_n$ *is finite, we write* $N \xRightarrow{\vec{t}} N'$ *where* $N' = N_{n+1}$.

Given a fs $\vec{t} = t_0; t_1; t_2; \dots$, we say $t \in \vec{t}$ if there exists an index j and $t = t_j$. We observe that the notion of firing sequence we adopt is essentially the same as the standard one, besides the fact that we keep the structure of the net.

Example 2. Consider the net in Figure 3. Two possible finite firing sequences are $\vec{t} = t_1; t_2$ and $\vec{t'} = t_1; t_2; t_3; t_4$. Among several possibilities, we can also construct an infinite sequence: $\vec{t''} = t_1; t_2; t_3; t_4; t_1; t_2; t_3; t_4; t_1; t_2; t_4; \cdots$ where the transitions t_1, t_2 and t_4 are fired in sequence forever.

4 Enriching nets

We focus on a restricted class of the dynamic nets of Asperti and Busi [1], where the firing of a transition in a net N may extend the net with new places and transitions. We exploit this mechanism of net growth to ensure that a forward transition can produce a subnet that will later enable the corresponding backward transition.

We fix $\mathcal{P}$ as an infinite set of *place* names. We begin with the following definition.

Definition 3. *The set* EN *is the* least set *of triple* $\langle S, T, m \rangle$ *satisfying the recursive equation:*

$$\text{EN} = \{\langle S, T, m \rangle \mid S \subseteq \mathcal{P} \wedge T \subseteq \mu\mathcal{P} \times \text{EN} \wedge m \in \mu\mathcal{P}\}.$$

Note that transitions in T are pairs consisting of a multiset of places and an element of the set EN. That is, a transition consumes a multiset of tokens but produces a net. However, the simplest elements of EN are markings of the form $\langle \emptyset, \emptyset, m \rangle$. Hence, a standard transition is simply one that does not generate any new places or transitions.

Note that the initial marking may assign tokens to places that are not part of the net itself; that is, the initial marking is not necessarily a multiset over the places of the net. Note that this is essential to let transitions to mark places that are not created by the transition itself

Definition 4 (Defined and Free names). *The set of defined names in a marking m is* $\mathsf{dn}(m) = \lfloor\!\lfloor m \rfloor\!\rfloor$, *i.e., names appearing in place position. Given $N = \langle S, T, m \rangle \in$ EN, the set of defined (dn) and free (fn) names of transitions, sets of transitions, and nets are defined as follow:*

$$\mathsf{dn}(m \triangleright N) = \mathsf{dn}(m) \qquad\qquad \mathbf{fn}(m \triangleright N) = \mathsf{dn}(m) \ \cup \ \mathbf{fn}(N)$$
$$\mathsf{dn}(T) = \textstyle\bigcup_{t \in T} \mathsf{dn}(t) \qquad\qquad \mathbf{fn}(T) = \textstyle\bigcup_{t \in T} \mathbf{fn}(t) \ \setminus \ \mathsf{dn}(T)$$
$$\mathsf{dn}(N) = S \qquad\qquad \mathbf{fn}(\langle S, T, m \rangle) = (\mathbf{fn}(T) \cup \mathsf{dn}(m)) \setminus S$$

Although the postset of a transition may be a net that reference places defined in the original net (i.e., contain free names), we will only consider completely defined nets, i.e., nets that are closed.

Definition 5 (Expansible Net). $N \in$ EN *is an* expansible net *if $\mathbf{fn}(N) = \emptyset$.*

Nets are considered up-to α-conversion on S. As an example $\langle \{s\}, \emptyset, s \rangle$ and $\langle \{s'\}, \emptyset, s' \rangle$ are α-equivalent, whereas $\langle \emptyset, \emptyset, s \rangle$ and $\langle \emptyset, \emptyset, s' \rangle$ are not.

To define the firing of a transition, we first introduce the notion of a *net extension*, which is obtained by adding the net produced by the fired transition, provided that the added places are *new*.

Definition 6 (Extension of a net). *Let $N = \langle S, T, m \rangle$ and $N' = \langle S', T', m' \rangle$ be nets in EN. The extension of N with N' is defined as*

$$N + N' \ = \ \langle S \cup S', \ T \cup T', \ m \oplus m' \rangle,$$

provided that $S \cap S' = \emptyset$ and $\mathbf{fn}(N) \cap S' = \emptyset$.

Note that the places S' of the added net N' are required to be distinct from those in S. This requirement can always be satisfied because nets are considered up to α-conversion.

Lemma 1. *Let $N = \langle S, T, m \rangle$ be an expansible net and $N' = \langle S', T', m' \rangle$ an element of EN such that $\mathbf{fn}(N') \subseteq S$. Then, the extension of N with N', i.e., $N + N'$, is an expansible net.*

We now formalize the revised token game in expansible nets, where transitions can produce nets, as illustrated in Figure 4.

$$\text{(EN-FIRING)} \quad \frac{t = f \, \rhd \, N \in T \qquad m \in \mu S}{\langle S, T, f \oplus m \rangle \xrightarrow{t} \langle S, T, m \rangle + N}$$

Fig. 4: Operational semantics of expansible nets

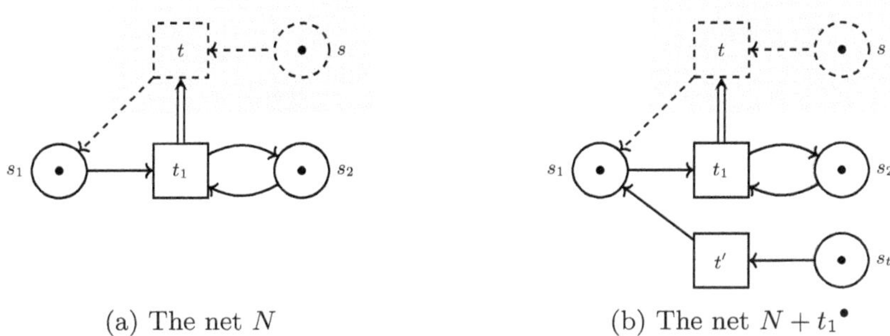

(a) The net N (b) The net $N + t_1^{\bullet}$

Fig. 5: Expansible nets

Example 3. Figure 5(a) depicts the expansible net

$$N = \langle \{s_1, s_2\}, \{s_1 \, \rhd \, \langle \{k\}, \{k \, \rhd \, \langle \emptyset, \emptyset, s_1 \rangle\}, k \rangle\}, s_1 \oplus s_2 \rangle$$

The dashed place s and dashed transition t belong to the net $t_1^{\bullet}$. Dashed elements indicate components that can be *freshly* generated by firing a transition: once t_1 fires, a new instance of $t_1^{\bullet}$ is generated and added to the current net. Consequently, after firing t_1, we obtain the net shown in Figure 5(b), i.e.,

$$N + t_1^{\bullet} = \langle \{s_1, s_2, k_1\}, \{s_1 \, \rhd \, \langle \{k\}, \{s \, \rhd \, \langle \emptyset, \emptyset, s_1 \rangle, k_1 \, \rhd \, \langle \emptyset, \emptyset, s_1 \rangle\}, k \rangle\}, k_1 \oplus s_2 \rangle.$$

The notion of firing sequences is analogous to that for nets and is denoted in the same way. Thus, executing the firing sequence $\vec{t}$ on an expansible net N produces the net N', denoted as $N \stackrel{\vec{t}}{\Longrightarrow} N'$. When the specific firing sequence $\vec{t}$ is not relevant, we omit it and simply write $N \Longrightarrow N'$.

A net $\langle S, T, m_0 \rangle \in \mathcal{N}$ can be viewed as an element of EN. Consider $\langle S, T, m_0 \rangle$, where $T = \{f \, \rhd \, g \mid f, g \in \mu S\}$ and $m_0 \in \mu S$. It can equivalently be written as $\langle S, T', m_0 \rangle$, where $T' = \{f \, \rhd \, \langle \emptyset, \emptyset, g \rangle \mid f \, \rhd \, g \in T\}$. In this form, it is clear that firing a transition simply modifies the marking. Observe that $\langle S, T', m_0 \rangle \in$ EN and, furthermore, it is an expansible one.

Definition 7 (Standard Net). *Let* $N = \langle S, T, m_0 \rangle \in$ EN *be an expansible net. We say that it is* standard *when* $T \subseteq \{f \, \rhd \, \langle \emptyset, \emptyset, g \rangle\}$.

4.1 Adding reversibility

We start from a standard net $N = \langle S, T, m_0 \rangle$, which only has *forward* transitions, i.e., transitions of the form $f \, \rhd \, \langle \emptyset, \emptyset, g \rangle$. To add reversibility, we view this net as

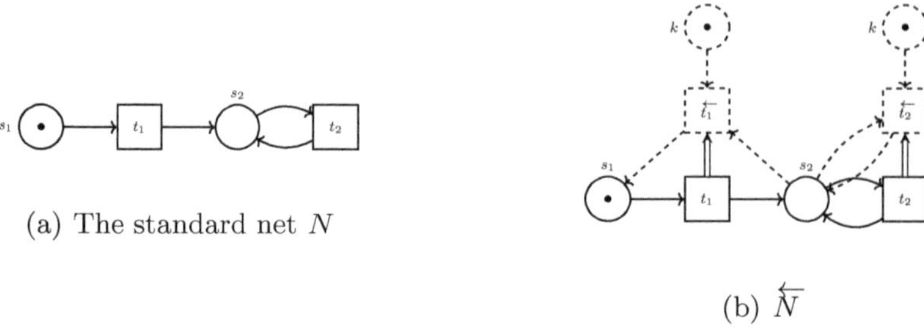

(a) The standard net N

(b) $\overleftarrow{N}$

Fig. 6: Reversible net.

one in which each forward transition, besides consuming and producing tokens in the proper places, *creates* a backward transition along with a key place.

For instance, a forward transition $t = f \rhd \langle \emptyset, \emptyset, g \rangle$ is transformed into

$$(f, \langle \{k\}, \{g \oplus k \rhd \langle \emptyset, \emptyset, f \rangle\}, k \oplus g \rangle),$$

so that each time this transition is executed, it creates a copy of the net with a new place k_1 instantiated with a fresh name. This instantiated name serves as the *key* place of the transition. In addition, a backward transition $g \oplus k_1 \rhd \langle \emptyset, \emptyset, f \rangle$ is added to the net.

Definition 8 (Reversing of a net). *Let $N = \langle S, T, m \rangle$ be a standard net. Then the corresponding reversing expansible net $\overleftarrow{N}$ is the net $\langle S, [\![T]\!], m \rangle$ where $[\![T]\!] = \{[\![t]\!] \mid t \in T\}$ and the encoding of the transition $t = f \rhd \langle \emptyset, \emptyset, g \rangle$ is is given by $[\![t]\!] = f \rhd \langle \{k_1\}, \{g \oplus k_1 \rhd \langle \emptyset, \emptyset, f \rangle\}, k_1 \rangle, k_1 \oplus g \rangle$.*

Let $\overleftarrow{N} = \langle S, [\![T]\!], m \rangle$ be a reversible net, and consider the net N' such that $\overleftarrow{N} \overset{\vec{t}}{\Longrightarrow} N' = \langle S', T', m' \rangle$. The places in $S' \cap S$ are *standard* places, and the places in $S' \setminus S$ are the *key* places. Furthermore it is easy to see that if s is a key place, then $^{\bullet}s = \emptyset$ and there exists a unique transition $t = s \oplus f \rhd N'$ consuming the token in s.

The following lemma guarantees that a reversible net is an expansible one, and its proof is trivial as for each backward transition $[\![t]\!] = {}^{\bullet}t \rhd N_t$ we have that $\mathtt{fn}(N_t) \subseteq S$.

Lemma 2. *Let $N = \langle S, T, m \rangle$ be a standard net. Then $\overleftarrow{N} = \langle S, [\![T]\!], m \rangle$ is an expansible net.*

Example 4. In the Figure 6 we depict a standard net N (Figure 6a) and the associated reversible net $\overleftarrow{N}$ (Figure 6b). The name of the dashed places k are actually *binders* for the fresh names created when the transitions t_1 or t_2 are executed. The net $\overleftarrow{N}$ has just two transitions, and are both forward.

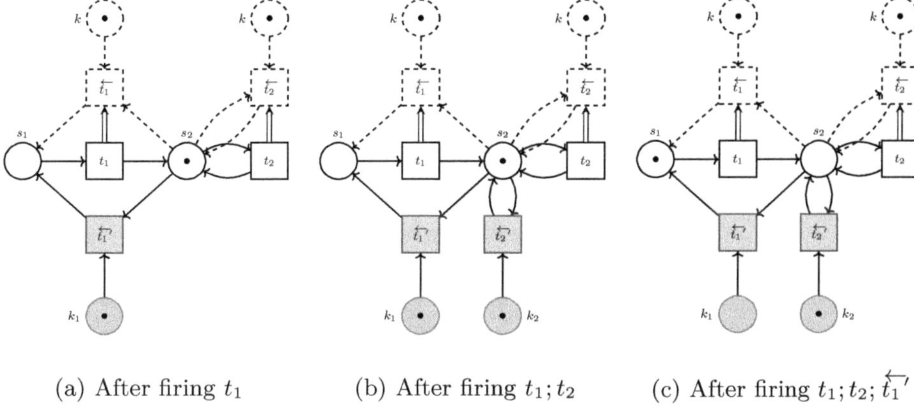

(a) After firing t_1 (b) After firing $t_1; t_2$ (c) After firing $t_1; t_2; \overleftarrow{t_1}'$

Fig. 7: The reversible net $\overleftarrow{N}$ of Figure 6 and the firing of t_1, t_2 and $\overleftarrow{t_1}'$

The following lemma states that a backward transition belong to the net only when as the reversing part of the transition is actually created when the forward transition is executed.

Lemma 3. *Let $N = \langle S, T, m_0 \rangle$ be a standard net and $\overleftarrow{N} = \langle S, [\![T]\!], m_0 \rangle$ be the associated reversible net. Let $\overleftarrow{N} \xRightarrow{\vec{t}} N'$ for some fs $\vec{t}$.*

1. *A transition t is a forward transition of N' iff it is a forward transition of N.*
2. *If t' is a backward transition of N', then there exists a forward transition t in N' such that $[\![t]\!] = f \,\triangleright\, \langle \{k_1\}, \{k_1 \oplus g \,\triangleright\, \langle \emptyset, \emptyset, f \rangle\}, g \oplus k_1 \rangle$ such that $t' = s' \oplus g \,\triangleright\, \langle \emptyset, \emptyset, f \rangle$ for some key place s'.*

A consequence of this lemma is that if $\overleftarrow{N} \xRightarrow{\vec{t}} N' \xrightarrow{t'} N''$ with t' backward transition then there exists a forward transition $t \in \vec{t}$ such that t' is a backward transition corresponding to one execution of t.

Example 5. Take the reversible net $\overleftarrow{N}$ of Figure 6b. In $\overleftarrow{N}$ there is only one enabled transition, that is t_1. Executing such transition $\overleftarrow{N} \xrightarrow{t_1} N_1$ will lead to the net depicted in Figure 7a, which is the $\overleftarrow{N} + t_1{}^\bullet$. Let us note that the firing of t_1 produces two tokens, along with the transition $\overleftarrow{t_1}'$. The first token is used to mark the key place of the backward transition. Suppose that from N_1 we execute the firing sequence: $t_2; \overleftarrow{t_1}'$, we respectively obtain nets if Figure 7b, Figure 7c .

One may wonder whether the machinery introduced so far is sufficient to guarantee causal-consistent reversibility. One possible interpretation of causal-consistent reversibility is that any computation–including those obtained by interleaving forward and backward transitions–must always lead to causally consistent states, that is, states that could also be reached by purely forward computations. Consider the example of Figure 7 and focus on the marking/state reached in the net in Figure 7c. We can see that this state cannot be reached by any combination of forward computations. The reason why this happens is because formally t_1 causes t_2, but then t_1 is undone before t_2, leaving the net in a *spurious* state. Aside from this interpretation, one could see that in Figure 7b two backward transitions are enabled: $\overleftarrow{t_1}'$ and $\overleftarrow{t_2}'$. According to [12], one property that has to hold, in order to prove causal-consistency, is that *backward transitions are independent* (or BTI). BTI can be seen as a backward diamond property: if from a state two backward transitions are enable then one can close the diamond. Clearly we see that both $\overleftarrow{t_1}'$ and $\overleftarrow{t_2}'$ are enabled, but one disables the other, meaning that they are not independent. Hence, BTI does not hold with expansible net. This happens because our tokens are anonymous, and they do not bring enough causal information in order to force the order of execution of backward computations. We then will resort to more complex tokens which carry enough causal information to enable causal consistent reversibility. To this end we will rely on colours.

5   Dynamic nets

We use dynamic nets, first introduced by Asperti and Busi [1]. Dynamic nets generalise expansible nets by allowing coloured tokens that can refer to places within the net, thereby enabling reconfigurability. We define the set $\mathcal{C} = \mathcal{P} + \mathcal{V}$ of colours.

Definition 9 (Dynamic nets).   *The set* DN *is the* least set *satisfying the recursive equation:*

$$\mathrm{DN} = \{\langle S, T, m\rangle \mid S \subseteq \mathcal{P} \ \wedge \ T \subseteq \mu(\mathcal{P}, \mathcal{C}^*) \times \mathrm{DN} \ \wedge m \in \mu(\mathcal{C}, \mathcal{C}^*) \ \}$$

Observe that, when we restrict $\mathcal{C}$ to $\mathcal{P}$ and $\mathcal{C}^*$ to the set containing only the empty sequence, we obtain precisely EN.

Definition 10 (Received names of a transition).   *The* set of colours of a multiset $m \in \mu(S, \mathcal{C}^*)$ *is defined as* $col(m) = \cup_{a(c_1,\ldots,c_j) \in m}\{c_1, \ldots, c_j\}$. *The* set of places *is* $col_{\mathcal{P}}(m) = col(m) \cap \mathcal{P}$, *and the* set of variables *(or received names) of a multiset is* $\mathsf{rn}(m) = col_{\mathcal{V}}(m) = col(m) \cap \mathcal{V}$. *Given a transition* $t = m \triangleright N$, *the* received names *of t are defined as* $\mathsf{rn}(t) = col_{\mathcal{V}}(m)$.

Note that $col(m)$ for $m \triangleright N$ denotes the formal parameters of a transition. The subset $col_{\mathcal{V}}(m)$ represents the set of variables, which are instantiated when the transition fires. In contrast, $col_{\mathcal{P}}(m)$ consists of constant (non-variable) colours and provides a basic mechanism for pattern matching, since a transition is enabled only when the consumed tokens carry the same constant colours as those specified in the preset of the transition.

Definition 11 (Defined and Free names). *The set of* defined names in $m \in \mu(S, \mathcal{C}^*)$ *is defined as* $\mathsf{dn}(m) = \cup_{a(c_1,\ldots,c_j) \in m}\{a\}$, *i.e., the names appearing in* *place positions. Given $N = \langle S, T, m \rangle \in$ DN, the sets of defined (dn) and free (fn) names of transitions, sets of transitions, and nets are defined as follows:*

$$\mathsf{dn}(m \triangleright N) = \mathsf{dn}(m) \qquad \mathbf{fn}(m \triangleright N) = \mathsf{dn}(m) \cup col_{\mathcal{P}}(m) \cup (\mathbf{fn}(N) \setminus \mathsf{rn}(m))$$

$$\mathsf{dn}(T) = \bigcup_{t \in T} \mathsf{dn}(t) \qquad \mathbf{fn}(T) = \bigcup_{t \in T} \mathbf{fn}(t) \setminus \mathsf{dn}(T)$$

$$\mathsf{dn}(N) = S \qquad \mathbf{fn}(\langle S, T, m \rangle) = (\mathbf{fn}(T) \cup \mathsf{dn}(m) \cup col(m)) \setminus S$$

Definition 12 (Dynamic Net). $N \in$ DN *is a* dynamic net *if* $\mathbf{fn}(N) = \emptyset$.

The above definition states that a dynamic net is closed, i.e., it does not refer to places outside itself. The condition $\mathbf{fn}(N) = \emptyset$ ensures that tokens are always generated within the net, as markings are bound to its own places, distinct from those of other nets.

Definition 13 (Substitution on colours). *Let $\sigma : \mathcal{V} \rightharpoonup \mathcal{C}$ be a partial function from variables to colours. The application of σ to a colour expression and coloured multisets is defined as follows:*

$$x\sigma = \begin{cases} \sigma(x) & \textit{if } x \in \mathrm{dom}(\sigma), \\ x & \textit{otherwise}, \end{cases} \qquad \textit{for } x \in \mathcal{V} \cup \mathcal{C},$$

$$(c_1, \ldots, c_n)\sigma = (c_1\sigma, \ldots, c_n\sigma), \qquad \textit{for } (c_1, \ldots, c_n) \in (\mathcal{V} \cup \mathcal{C})^*,$$

$$(m\sigma)(s)(c) = \sum_{t\sigma = s, d\sigma = c} m(t)(d), \qquad \textit{for } m \in \mu(\mathcal{V} \cup \mathcal{C}, (\mathcal{V} \cup \mathcal{C})^*).$$

Definition 14 (Instantiation of a net). *Let $\sigma : \mathcal{V} \rightarrow \mathcal{C}$ be a substitution. The instantiation of a transition $t = m \triangleright N$ with σ, provided that $\mathsf{rn}(t) \cap \mathrm{dom}(\sigma) = \emptyset$, is defined as $t\sigma = m \triangleright N\sigma$. Given a dynamic net $N = \langle S, T, m \rangle$, the instantiation of N with σ, provided that $\mathrm{dom}(\sigma) \cap S = \emptyset$, is defined as $N\sigma = \langle S, \{t\sigma \mid t \in T\}, m\sigma \rangle$.*

The side conditions on the substitution ensure that no free names occurring in σ are captured. If these conditions are not satisfied, one may apply an α-conversion on the places of the net (or on the received names of the transition) beforehand.

The operational semantics is presented in Figure 8. Rule DYN-FIRING describes the firing of a transition t when the marking contains an instance of the preset of t (for a suitable substitution on colours σ). The resulting net consists of the original net, with the consumed tokens removed, together with a new instance of N (i.e., the postset of t). The extension of dynamic nets, is the same as for expansible nets, and is denoted in the same way. We remark that the definition of $+$ guarantees that the names of the added components are fresh, as we add the proper α-equivalent net.

Along the same lines of the proof of Lemma 1, we have the following lemma.

Lemma 4. *If N is a dynamic net and $N \xrightarrow{t} N'$ then N' is a dynamic net.*

(DYN-FIRING)

$$\frac{t = m \,\triangleright\, N \in T \qquad m'' \in \mu(S, \mathcal{C}^*)}{\langle S, T, m\sigma \oplus m''\rangle \xrightarrow{t} \langle S, T, m''\rangle + N\sigma} \qquad \begin{array}{l} dom(\sigma) = \mathsf{rn}(t), \text{ and} \\ \sigma(v) \in S \text{ for } v \in dom(\sigma) \end{array}$$

Fig. 8: Operational semantics of dynamic nets.

5.1 Adding reversibility

Definition 15 (Reversing of a net). *Let $N = \langle S, T, m\rangle$ be a standard net. The corresponding reversing dynamic net is defined as $\overleftarrow{N} = \langle S, \llbracket T \rrbracket, \llbracket m \rrbracket \rangle$ where $\llbracket T \rrbracket = \{\llbracket t \rrbracket \mid t \in T\}$ and each transition $t = a_0 \oplus \ldots \oplus a_m \,\triangleright\, b_0 \oplus \ldots \oplus b_n \in T$ is encoded as*

$$\llbracket t \rrbracket \; = \; a_0(x_0) \oplus \ldots \oplus a_m(x_m) \,\triangleright\, \langle \{k\}, \{\overleftarrow{t}\}, \; k(k, x_0, \ldots, x_m) \oplus b_0(k) \oplus \ldots \oplus b_n(k)\rangle$$

with $\overleftarrow{t} = k(y, x_0, \ldots, x_m) \oplus b_0(y) \oplus \ldots \oplus b_n(y) \,\triangleright\, a_0(x_0) \oplus \ldots \oplus a_m(x_m)$. The initial marking is encoded as

$$\llbracket m \rrbracket(s)(c) \; = \; \begin{cases} m(s), & \text{if } c = \bullet, \\ 0, & \text{if } c \neq \bullet. \end{cases}$$

Intuitively, each transition is encoded to both perform the token game of the original net and generate an associated backward transition $\overleftarrow{t}$ with its key. The encoded transition $\llbracket t \rrbracket$ uses colours to control the enabling of $\overleftarrow{t}$: it places a token in the created place k carrying the key k and the colours of consumed tokens, while all tokens in the postset are assigned the new key, i.e., $b_i(k)$. The backward transition $\overleftarrow{t}$ consumes all these tokens, with the variable y in its preset ensuring they share the same key. The token in the key place stores information needed to restore the original colours during backward execution.

Example 6. In Figure 9 we depict a standard net N (Figure 9a) and the associated reversible dynamic net $\overleftarrow{N}$ (Figure 9b). Observe that the dashed places have now more information with respect to the ones in Figure 6b.

We start by stating some properties about reversing nets that will be useful for showing that reversing nets are actual causally reversible.

Lemma 5. *Let N be a standard net, then $\overleftarrow{N}$ is a dynamic net.*

Given a standard net $N = \langle S, T, m\rangle$ and a dynamic net $N' = \langle S', T', m'\rangle$ such that $\overleftarrow{N} \implies N'$, we introduce the following notions. For $t' \in T'$, t' is *forward* if there exists $t \in T$ such that $t' = \llbracket t \rrbracket$, and *backward* otherwise. A place $s \in S'$ is a *standard place* if $s \in S$, and a *key* otherwise, i.e., if $s \notin S$. We denote by $m'_{|K}$ the restriction of m' to the set of keys in N'.

We state some properties concerning the computation of our reversible nets. We begin by characterising the form of the forward and backward transitions of

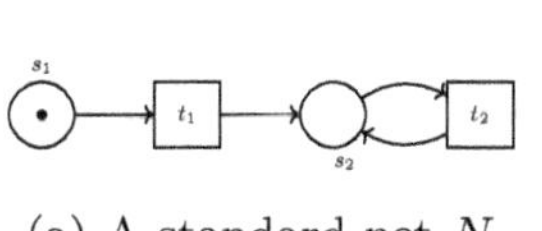

(a) A standard net N.

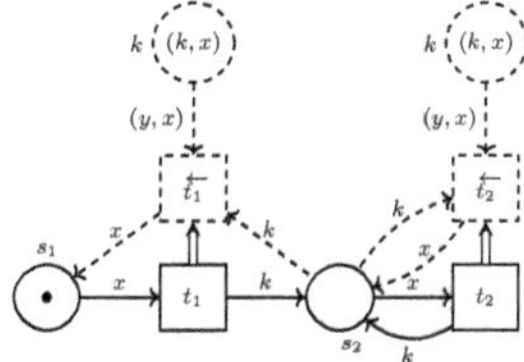

(b) The reversible dynamic net $\overleftarrow{N}$.

Fig. 9: Dynamic encoding example with successive transitions.

nets that can be reached from a reversible net. This characterisation is instrumental for proving the following auxiliary properties.

Lemma 6. *Let N be a standard net, and suppose that $\overleftarrow{N} \Longrightarrow N'$.*

1. *If t' is a backward transition of N', then there exists a forward transition t of N',*

$$[\![t]\!] = a_0(x_0) \oplus \ldots \oplus a_m(x_m) \triangleright \langle \{k\}, \{\overleftarrow{t}\}, k(k, x_0, \ldots, x_m) \oplus b_0(k) \oplus \ldots \oplus b_n(k)\rangle$$

such that $t' = k(y, x_0, \ldots, x_m) \oplus b_0(y) \oplus \ldots \oplus b_n(y) \triangleright a_0(x_0) \oplus \ldots \oplus a_m(x_m) = \overleftarrow{t}$.
2. *A transition t is a forward transition of N if and only if it is a forward transition of N'.*
3. *for all key place s of N', ${}^{\bullet}s = \emptyset$ and $|s^{\bullet}| = 1$.*

The following lemma characterises the colour of the tokens in a reversible net.

Lemma 7. *Let N be a standard net, and suppose that $\overleftarrow{N} \Longrightarrow N' = \langle S', T', m'\rangle$, and let $s \in S'$.*

1. *If s is a standard place and $m'(s)(c) > 0$, then $c = \bullet$ or c is a key place of N'.*
2. *If s is a key place and $m'(s)(c) > 0$, then $c = (s, c_1, \ldots, c_n)$, and for all i, $c_i = \bullet$ or c_i is a key place of N'.*

The following lemma states that during a computation, there are only finitely many tokens in key places (i.e., finitely many transitions that can be undone), key places are 1-safe (that is, they contain at most 1 token); and moreover, that this number decreases whenever a transition is reversed.

Lemma 8. *Let N be a standard net, and suppose that $\overleftarrow{N} \Longrightarrow N' = \langle S', T', m'\rangle$.*

1. *The restriction of the marking to key places is finite, i.e., $|m'_{|K}| < \infty$.*

2. *If $N' \xrightarrow{t} \langle S'', T'', m''\rangle$ with t a backward transition, then $|m''_{|K}| = |m'_{|K}| - 1$.*
3. *Key places are 1-safe, i.e., for all key place s of N', $|m'_{|s}| \leq 1$.*

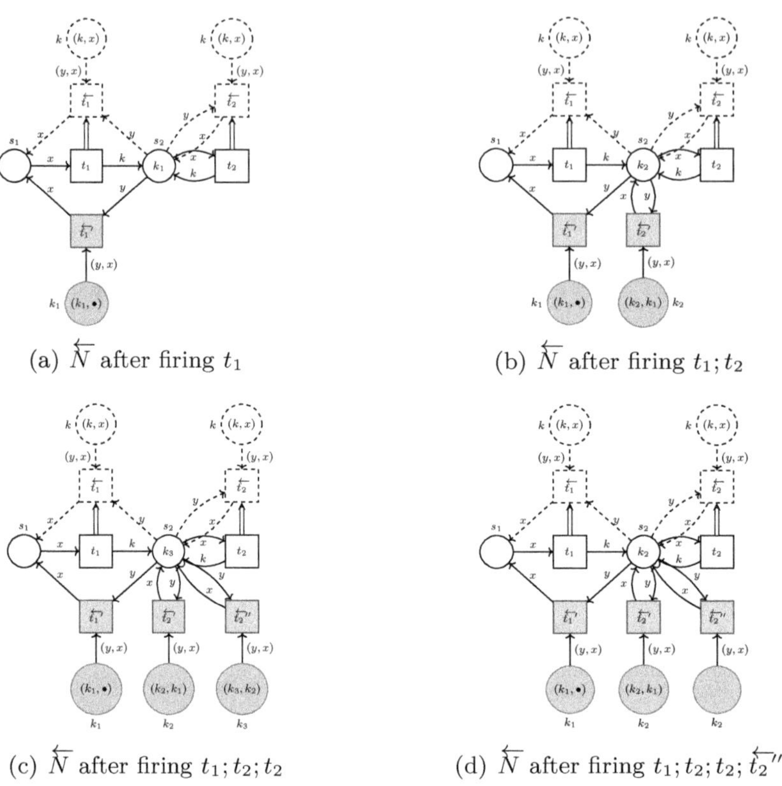

(a) $\overleftarrow{N}$ after firing t_1

(b) $\overleftarrow{N}$ after firing $t_1; t_2$

(c) $\overleftarrow{N}$ after firing $t_1; t_2; t_2$

(d) $\overleftarrow{N}$ after firing $t_1; t_2; t_2; \overleftarrow{t_2}''$

Fig. 10: Firing forward and backward transition starting from $\overleftarrow{N}$.

Example 7. Take the reversible net $\overleftarrow{N}$ of Figure 9b. In $\overleftarrow{N}$ there is only one enabled transition, that is t_1. Executing such transition $\overleftarrow{N} \xrightarrow{t_1} N_1$ will lead to the net depicted in Figure 10a. Let us note that the firing of t_1 produces two tokens along with the transition $\overleftarrow{t_1'}$. The first token is used to mark the key place of the backward transition and contains the information about the continuation of the computation (key k_1) and the past of the computation (the original token $\bullet$). The second token contains just the information about the continuation of the computation, that is the key k_1. To fire transition $\overleftarrow{t_1}$ the presence of the token k_1 is required in the place s_2. This is possible thanks to a pattern-matching like semantics of dynamic net. Suppose that from N_1 we execute the following firing sequence: $t_2; t_2; \overleftarrow{t_2''}$, we respectively obtain nets if Figure 10b, Figure 10c and Figure 10d.

If we compare the net Figure 7b, in which BTI fails, and the net in Figure 10b we can see that colours add the needed information such that $\overleftarrow{t_1}$ cannot happen before $\overleftarrow{t_2}$.

5.2 Properties of a reversing net

One way to prove that a reversible semantics satisfies causal consistency is to follow the framework introduced in [12]. Rather than reproving causal consistency as a whole as done in the original proof of [3], this approach decomposes the proof into three fundamental axioms: the *Square Property* (SP), the *Backward Transition Independence* (BTI), and the *Well-Foundedness* (WF). Additionally, [12] assumes the system is modeled as a *Labelled Transition System with an Independence relation* (LTSI), where the Loop Lemma holds (e.g., each transition can be undone).

Before proving the Loop Lemma, we first start by establishing the underlying LTS on which our model will be analyzed. In particular, we consider the LTS induced by the firing semantics. Formally, we study nets up to the equivalence induced by lemma 6(3). Note that, since a key place constitutes a sort of minimal element: an unmarked key place cannot become marked again in any subsequent computation once it has been unmarked. Consequently, the unique transition in its postset can no longer be fired. Hence, we may consider nets up-to the removal of useless key places and related backward transition. Given a net $N \Longrightarrow N' = \langle S, T', m \rangle$ and an unmarked key place k, we define

$$N' \setminus k = \langle S \setminus \{k\},\, T' \setminus \{k^\bullet\},\, m \rangle.$$

As established by the following lemma, such nets are sequentially equivalent and we denote this equivalence with $\equiv$.

Lemma 9. *Let N be a standard net and suppose that $\overleftarrow{N} \Longrightarrow N'$, and k is a key place of N' such that $m'(k)(c) = 0$ for all c. Then, $N' \xrightarrow{t} N''$ iff $N' \setminus \{k\} \xrightarrow{t} N'' \setminus \{k\}$.*

We shall write $N \xrightarrow{t}_\equiv N'$ for there exists N'' such that $N \xrightarrow{t} N''$ and $N'' \equiv N'$. Notation $N \Longrightarrow_\equiv N'$ and $N \xRightarrow{\vec{t}}_\equiv N'$ is defined analogously.

Lemma 10 (Loop Lemma). *Let N be a standard net and suppose that $\overleftarrow{N} \Longrightarrow N'$. Then:*

- *If $\overleftarrow{t}$ is a backward transition of N', then there exists a forward transition t of N' such that $N' \xrightarrow{\overleftarrow{t}} N''$ and $N'' \xrightarrow{t}_\equiv N'$.*
- *If t is a forward transition of N' and $N' \xrightarrow{t} N''$, then there exists a backward transition $\overleftarrow{t}$ of N'' such that $N'' \xrightarrow{\overleftarrow{t}}_\equiv N'$.*

It is worth noting that the loop lemma does not ensure that the resulting state coincides exactly with the initial one, but only that it is equivalent to it–typically after the removal of garbage associated to dead transitions related to unmarked keys. This formulation is reminiscent of analogous results in related frameworks, where, for instance, distinct generated keys can still yield configurations that are regarded as equivalent. This is the case of CCSK [24] or the rho-π [11], were configurations that differ only in the identities of freshly generated keys are treated as equivalent under structural congruence. Moreover, the second item is somewhat unconventional: the backward transition corresponding to an executed forward transition cannot be determined *a priori*, but only *a posteriori*, once the firing has occurred, since it is at this time when the corresponding backward transition is generated. Nevertheless, this peculiarity does not compromise the applicability of the framework [12], as for example it is one of the main feature or reversible calculi.

We now introduce a notion of independence for dynamic nets. In standard Petri nets, the independence relation among transitions is defined in terms of disjointness of their presets, that is, two transitions are independent if the firing of one does not disable the firing of the other. In the context of dynamic nets, where tokens are annotated with names, the notion of independence must also account for the fact that there should be no overlap for any possible instantiation of their presets.

Definition 16. *Two transitions t_1 and t_2 of a dynamic net $N = \langle S, T, m \rangle$ are* independent *if* $\mathsf{rn}(t_1) \cap \mathsf{rn}(t_2) = \emptyset$ *and, for all substitutions σ such that* $\mathsf{rn}(t_1)\sigma \subseteq m$ *and* $\mathsf{rn}(t_2)\sigma \subseteq m$, *it holds that* $\lfloor\!\lfloor\, {}^\bullet t_1\sigma \rfloor\!\rfloor \cap \lfloor\!\lfloor\, {}^\bullet t_2\sigma \rfloor\!\rfloor = \emptyset$.

The requirement $\mathsf{rn}(t_1) \cap \mathsf{rn}(t_2) = \emptyset$ can always be satisfied since we consider transitions up to α-renaming of received names. Independence means that any instantiation of the presets of the two transitions, if found in the marking, are disjoint. Intuitively, this means that there is no marking in which tokens used to fire one of the transitions can also be used to fire the other.

We now address the three properties which are required to achieve causal consistency. We shall start form Square property, stating that two independent transitions can commute.

Property 1 (Square property). Let N be a standard net and suppose that $\overleftarrow{N} \Longrightarrow N'$. If there exists two independent transitions t_1 and t_2 such that $N' \xrightarrow{t_1} N'_1$ and $N' \xrightarrow{t_2} N'_2$, then $N'_2 \xrightarrow{t_1} N''$ and $N'_1 \xrightarrow{t_2} N''$.

Proof. Since the two transitions are independent, there is no overlap in the consumed tokens. Then we can write $N = \langle S, T, m_1\sigma_1 \oplus m_2\sigma_2 \oplus m'' \rangle$ with $t_i = m_i \triangleright N_i \in T$ for $i \in \{1, 2\}$. We have then that

$$N \xrightarrow{t_1} \langle S, T, m_2\sigma_2 \oplus m'' \rangle + N_1\sigma_1 \xrightarrow{t_2} \langle S, T, m_2 \rangle + N_1\sigma_1 + N_2\sigma_2$$

$$N \xrightarrow{t_2} \langle S, T, m_1\sigma_1 \oplus m'' \rangle + N_2\sigma_2 \xrightarrow{t_1} \langle S, T, m_2 \rangle + N_2\sigma_2 + N_1\sigma_1$$

and we are done.

Property 2 (BTI). Let N be a standard net and suppose that $\overleftarrow{N} \Longrightarrow N'$. If there exists two different backward transitions $\overleftarrow{t_1} \neq \overleftarrow{t_2}$ such that $N' \xrightarrow{\overleftarrow{t_1}} N'_1$ and $N' \xrightarrow{\overleftarrow{t_1}} N'_2$, then $N'_2 \xrightarrow{\overleftarrow{t_1}} N''$ and $N'_1 \xrightarrow{\overleftarrow{t_2}} N''$.

The following property states that reachable nets have backward computations.

Property 3 (WF). Let N be a standard net and suppose that $\overleftarrow{N} \Longrightarrow N'$. There exists no infinite sequence $N'_i \overset{\overleftarrow{t_i}}{\to} N'_{i+1}$ for all the $i = 0, 1, 2, \cdots$.

Proof. By observing that number of tokens in key-places is finite (Lemma 8(1)), and that backward transitions always consume a key-place token so the total number of token in key-places is decreased (Lemma 8(3)).

Before stating causal-consistency, we need to define a causal equivalence on firing sequence. Two firing sequences $\vec{t_1}$ and $\vec{t_2}$ are said to be coinitial if they start from the same net (up to $\equiv$), and cofinal if they finish in the same net (up to $\equiv$). In a firing sequence $\vec{t_1} = t_0; \cdots ; t_n$ such that $N_0 \overset{t_0}{\to} \cdots \overset{t_n}{\to} N_n$ we refer to N_0 as the source of the firing sequence and to N_n as the target of the sequence. By abusing of the notation we indicate with $\vec{t_1}; \vec{t_2}$ the firing sequence obtained by concatenating $\vec{t_1}$ and $\vec{t_2}$ indicating that the target of $\vec{t_1}$ is the source of $\vec{t_2}$. Also, if a firing sequence $\vec{t}$ is made of forward transitions we indicate with $\overleftarrow{t}$ its inverse, which are the corresponding backward transitions followed in the reverse order, i.e. $\overleftarrow{t} = \overleftarrow{t_n}; \cdots ; \overleftarrow{t_1}$.

Definition 17 (Causal equivalence). *Let $\asymp$ be the smallest equivalence on firing sequences closed under composition and satisfying:*

cancellation *two consecutive opposite transitions can be removed;*
swap *two consecutive independent transitions can be swapped.*

Intuitively, firing sequences are causal equivalent if they differ only for swapping independent transitions and for adding do-undo or undo-redo pairs of transitions. The following allows to see any firing sequence as a composition of two firing sequence: a backward one and a forward one. For example, if the take the net in Figure 10d, which is achieved from the net in Figure 9b with the firing sequence $t_1; t_2; t_2; t''_2$, we have that $t_1; t_2; t_2; \overleftarrow{t''_2} \asymp t_1; t_2$, and indeed the net in Figure 10d is equivalent (in terms of $\equiv$) to the net in Figure 10b.

Lemma 11 (Parabolic Lemma). *Let N a dynamic net. For any firing sequence $\vec{t}$ there exists two firing sequences $\vec{t_1}$ and $\vec{t_2}$ made just by forward transitions such that $\vec{t} \asymp \overleftarrow{t_1}; \vec{t_2}$ and $|\vec{t_1}| + |\vec{t_2}| \leq |\vec{t}|$.*

Proof. It follows from the square property and BTI thanks to [12].

Theorem 1 (Causal Consistency). *Let N be a dynamic net. For any two firing sequences $\vec{t_1}$ and $\vec{t_2}$, they are coinitial and cofinal iff $\vec{t_1} \asymp \vec{t_2}$.*

Proof. It follows from WF and the parabolic lemma thanks to [12].

Theorem 1 shows that causal equivalence characterizes a space for admissible rollbacks that are (i) correct as they do not lead to states not reachable by some forward computation and (ii) flexible enough to allow undo operations to be rearranged with respect to the order in which the undone concurrent transitions were originally performed. This implies that the states reached by any backward computation could be reached by performing forward computations only. Therefore, we can conclude that the semantics of dynamic nets meets causal reversibility.

6   Conclusions

In this work, we have proposed a systematic and semantically grounded approach to incorporating reversibility into Petri nets, aiming to ensure causal consistency, a critical property for many applications in fault-tolerant and reversible computing. Our methodology begins with a straightforward augmentation of standard Petri nets through the introduction of key places, allowing transitions to be reversed only if they have been previously executed. However, we demonstrated that this construction alone is insufficient to guarantee causally consistent reversibility. To overcome this limitation, we refined the model by enriching tokens with computational history, a technique inspired by reversible process calculi. This enrichment allows each token to encode not only its presence but also its provenance, thereby enabling precise control over when a transition and its effects can be undone. The resulting semantics is powerful enough to support reversibility that respects the causal dependencies among transitions.

In the last years there have been several attempts to provide a causally consistent reversible semantics to Petri nets either by using unfolding into Occurrence net [20], or by using inhibitor arcs [18] or using a complex semantics [23]. The approach is simpler and more general as it requires just few modifications to Petri nets semantics.

There are multiple promising directions for future work. First, we could explore tokens with reduced computational history to improve efficiency. Moreover, our approach naturally aligns with the non-sequential semantics of prime event structures. Hence, a further comparison among our work and Reversible Prime Event Structure [26] is left as future work. Also, reversible dynamic nets could be used to give a truly concurrent semantics to the reversible π-calculus [11] or to endow join-calculus with reversible semantics. Finally, there are several interesting properties of [12] that we could prove in our setting.

References

1. Andrea Asperti and Nadia Busi. Mobile Petri nets. *Math. Struct. Comput. Sci.*, 19(6):1265–1278, 2009. doi:10.1017/S0960129509990193.
2. Kamila Barylska, Maciej Koutny, Lukasz Mikulski, and Marcin Piatkowski. Reversible computation vs. reversibility in Petri nets. *Sci. Comput. Program.*, 151:48–60, 2018. doi:10.1016/J.SCICO.2017.10.008.

3. Vincent Danos and Jean Krivine. Reversible communicating systems. In Philippa Gardner and Nobuko Yoshida, editors, *CONCUR 2004 - 15th International Conference*, volume 3170 of *Lecture Notes in Computer Science*, pages 292–307. Springer, 2004. `doi:10.1007/978-3-540-28644-8_19`.

4. David de Frutos-Escrig, Maciej Koutny, and Lukasz Mikulski. Investigating reversibility of steps in Petri nets. *Fundam. Informaticae*, 183(1-2):67–96, 2021. `doi:10.3233/FI-2021-2082`.

5. Iulia Georgescu. 60 years of landauer's principle. *Nature Reviews Physics*, 3(12):770–770, 2021. `doi:10.1038/s42254-021-00400-8`.

6. Elena Giachino, Ivan Lanese, and Claudio Antares Mezzina. Causal-consistent reversible debugging. In *Fundamental Approaches to Software Engineering - 17th International Conference, FASE 2014,*, volume 8411 of *Lecture Notes in Computer Science*, pages 370–384. Springer, 2014. `doi:10.1007/978-3-642-54804-8_26`.

7. Thomas Hujsa and Raymond Devillers. On deadlockability, liveness and reversibility in subclasses of weighted Petri nets. *Fundam. Informaticae*, 161(4):383–421, 2018. `doi:10.3233/FI-2018-1708`.

8. Dimitrios Kouzapas, Constantinos Skitsas, Taqwa Saeed, Vassos Soteriou, Marios Lestas, Anna Philippou, Sergi Abadal, Christos Liaskos, Loukas Petrou, Julius Georgiou, and Andreas Pitsillides. Towards fault adaptive routing in metasurface controller networks. *J. Syst. Archit.*, 106:101703, 2020. `doi:10.1016/J.SYSARC.2019.101703`.

9. Stefan Kuhn and Irek Ulidowski. Modelling of DNA mismatch repair with a reversible process calculus. *Theor. Comput. Sci.*, 925:68–86, 2022. `doi:10.1016/J.TCS.2022.06.009`.

10. Ivan Lanese, Claudio Antares Mezzina, and Jean-Bernard Stefani. Reversing higher-order pi. In *CONCUR 2010 - Concurrency Theory*, volume 6269 of *Lecture Notes in Computer Science*, pages 478–493. Springer, 2010. `doi:10.1007/978-3-642-15375-4_33`.

11. Ivan Lanese, Claudio Antares Mezzina, and Jean-Bernard Stefani. Reversibility in the higher-order π-calculus. *Theor. Comput. Sci.*, 625:25–84, 2016. `doi:10.1016/J.TCS.2016.02.019`.

12. Ivan Lanese, Iain C. C. Phillips, and Irek Ulidowski. An axiomatic theory for reversible computation. *ACM Trans. Comput. Log.*, 25(2):11:1–11:40, 2024. `doi:10.1145/3648474`.

13. Ivan Lanese, Ulrik Pagh Schultz, and Irek Ulidowski. Reversible execution for robustness in embodied AI and industrial robots. *IT Prof.*, 23(3):12–17, 2021. `doi:10.1109/MITP.2021.3073757`.

14. Ivan Lanese, Ulrik Pagh Schultz, and Irek Ulidowski. Reversible computing in debugging of Erlang programs. *IT Prof.*, 24(1):74–80, 2022. `doi:10.1109/MITP.2021.3117920`.

15. Antoni Mazurkiewicz. Concurrent program schemes and their interpretations. *DAIMI Report Series*, 6(78), Jul. 1977. URL: `https://tidsskrift.dk/daimipb/article/view/7691`, `doi:10.7146/dpb.v6i78.7691`.

16. Hernán C. Melgratti, Claudio Antares Mezzina, Iain Phillips, G. Michele Pinna, and Irek Ulidowski. Reversible occurrence nets and causal reversible prime event structures. In Ivan Lanese and Mariusz Rawski, editors, *Reversible Computation - 12th International Conference, RC 2020*, volume 12227 of *Lecture Notes in Computer Science*, pages 35–53. Springer, 2020. `doi:10.1007/978-3-030-52482-1_2`.

17. Hernán C. Melgratti, Claudio Antares Mezzina, and G. Michele Pinna. A Petri net view of covalent bonds. *Theor. Comput. Sci.*, 908:89–119, 2022. `doi:10.1016/J.TCS.2022.01.013`.

18. Hernán C. Melgratti, Claudio Antares Mezzina, and G. Michele Pinna. A reversible perspective on Petri nets and event structures. *ACM Trans. Comput. Log.*, 25(4):1–38, 2024. `doi:10.1145/3686154`.

19. Hernán C. Melgratti, Claudio Antares Mezzina, and Irek Ulidowski. Reversing P/T nets. In Hanne Riis Nielson and Emilio Tuosto, editors, *Coordination Models and Languages - 21st IFIP WG 6.1 International Conference, COORDINATION*, volume 11533 of *Lecture Notes in Computer Science*, pages 19–36. Springer, 2019. `doi:10.1007/978-3-030-22397-7_2`.

20. Hernán C. Melgratti, Claudio Antares Mezzina, and Irek Ulidowski. Reversing place transition nets. *Log. Methods Comput. Sci.*, 16(4), 2020. URL: `https://lmcs.episciences.org/6843`.

21. Claudio Antares Mezzina, Francesco Tiezzi, and Nobuko Yoshida. Checkpoint-based rollback recovery in session programming. *Log. Methods Comput. Sci.*, 21(1):2, 2025. `doi:10.46298/LMCS-21(1:2)2025`.

22. Lukasz Mikulski and Ivan Lanese. Reversing unbounded Petri nets. In Susanna Donatelli and Stefan Haar, editors, *Application and Theory of Petri Nets and Concurrency - 40th International Conference, Petri NETS 2019*, volume 11522 of *Lecture Notes in Computer Science*, pages 213–233. Springer, 2019. `doi:10.1007/978-3-030-21571-2_13`.

23. Anna Philippou and Kyriaki Psara. Reversible computation in nets with bonds. *J. Log. Algebraic Methods Program.*, 124:100718, 2022. `doi:10.1016/J.JLAMP.2021.100718`.

24. Iain C. C. Phillips and Irek Ulidowski. Reversing algebraic process calculi. *J. Log. Algebraic Methods Program.*, 73(1-2):70–96, 2007. `doi:10.1016/J.JLAP.2006.11.002`.

25. Sudip K. Seal and Kalyan S. Perumalla. Reversible parallel discrete event formulation of a tlm-based radio signal propagation model. *ACM Trans. Model. Comput. Simul.*, 22(1):4:1–4:23, 2011. `doi:10.1145/2043635.2043639`.

26. Irek Ulidowski, Iain Phillips, and Shoji Yuen. Reversing event structures. *New Gener. Comput.*, 36(3):281–306, 2018. `doi:10.1007/S00354-018-0040-8`.

A Complete Propositional Dynamic Logic for Regular Expressions with Lookahead

Yoshiki Nakamura[1,2]

[1] Chiba University, Japan
[2] Institute of Science Tokyo, Japan
nakamura.yoshiki.ny@gmail.com

Abstract. We consider (logical) reasoning for *regular expressions with lookahead* (REwLA). In this paper, we give an axiomatic characterization for both the (match-)language equivalence and the largest substitution-closed equivalence that is sound for the (match-)language equivalence. To achieve this, we introduce a variant of propositional dynamic logic (PDL) on finite linear orders, extended with two operators: the restriction to the identity relation and the restriction to its complement. Our main contribution is a sound and complete Hilbert-style finite axiomatization for the logic, which captures the equivalences of REwLA. Using the extended operators, the completeness is established via a reduction into an *identity-free variant of PDL* on finite strict linear orders. Moreover, the extended PDL has the same computational complexity as REwLA.

Keywords: Regular expressions with lookahead · PDL · Completeness · Löb's axiom · Kleene algebra with antidomain.

1 Introduction

While classical *regular expressions* (RE) [26] are built from constants and the operators: concatenation (;), union (+), and Kleene star ($\cdot^*$), various extensions are implemented in real-world *regexes* (see, *e.g.*, [18, 21]). To optimize regexes (*e.g.*, *w.r.t.*, its length or the complexity of matching algorithms), we are interested in transforming a given expression into an equivalent one. In classical RE, there is a finite (and quasi-equational) algebraic axiomatization [27], known as Kleene algebra, which is sound and complete for language equivalence. Another framework is *propositional dynamic logic* (PDL) of RE with rich tests [20, 23]. PDL also enjoys a sound and complete axiomatization [22, 44] and embeds the language equivalence. A natural (naive) question is whether such sound and complete axiomatic systems can be extended with the operators employed in regexes.

As a first step, we consider *regular expressions with lookahead* (REwLA) [33, 34]. *Lookahead* allows us to assert that a certain pattern is satisfied in the future of the current position. For instance, the expression `((?!ab)(a|b))*` expresses the set $\{b^n a^m \mid n, m \geq 0\}$, where the *negative lookahead* `(?!ab)` asserts that the next two characters are not `ab` and the symbol "`|`" expresses the union (+).

N. Bertrand and S. Milius (Eds.): FoSSaCS 2026, LNCS 16503, pp. 505–526, 2026.
https://doi.org/10.1007/978-3-032-22730-0_24

Unlike RE, the language equivalence of REwLA is *not* closed under substitutions, so there is no sound and complete set of (substitution-closed) axiom schemas. For instance, although `((?!ab)(a|b))*` and `b*a*` have the same language as above, substituting `b` with `a` yields `((?!aa)(a|a))*` and `a*a*`, which define the sets $\{\varepsilon, a\}$ and $\{a^n \mid n \geq 0\}$, breaking the language equivalence.[3] For that reason, we also study the *largest substitution-closed equivalence* that is sound for the language equivalence, which has a sound and complete set of axiom schemas (Thm. 3.3) via a slight encoding (Prop. 2.3). Such equivalences are also useful from the perspective of reusability. This equivalence is alternatively characterized by the relational semantics on finite linear orders, based on the semantics of slices of strings [32], where the valuations are filled (Prop. 2.2).

Contributions The main contribution of this paper is to present an axiomatic characterization for REwLA *w.r.t.* both (i) the substitution-closed equivalence (Thm. 3.3), and (ii) the standard language equivalence (Thm. 7.2). While several algebraic equational properties have been investigated (*e.g.*, [33, Definition 2.2 "Kleene algebra with lookahead"] [32, Lemma 11]), no complete axiomatizations have yet been presented, to our knowledge (*cf.*, *e.g.*, [32, p. 92:10]).

In this paper, we present a finite axiomatization for an extended PDL on finite linear orders, which embeds the equivalences above for REwLA (§ 3). More precisely, we introduce $\mathrm{PDL}_{\mathrm{REwLA}+}$: PDL with the restriction to the identity relation ($\cdot^{\cap_1}$) and the restriction to the complement of the identity relation ($\cdot^{\cap_{\overline{I}}}$). Using these operators, we can decompose the relational semantics into the identity-part and the identity-free-part (§ 5). We then can give a reduction from the completeness theorem of $\mathrm{PDL}_{\mathrm{REwLA}+}$ on finite linear orders to the completeness theorem of an *identity-free* variant of PDL (denoted by PDL^-) on finite *strict* linear orders (§ 6). While axiomatizations for fragments/variants of PDL on finite trees [1, 30, 31] (or equivalently, on finite strict linear orders via bisimulation) were presented [5, 6, 43], no axiomatization for full PDL have yet been presented, to our knowledge. A key of their axiomatizations is to employ *Löb's axiom*. Also for PDL^-, we can employ Löb's axiom, thanks to the absence of the identity-part. Our approach eliminating the identity-part is inspired by Brunet's reduction [8], which shows the completeness of the equational theory of (reversible) Kleene lattices interpreted as algebras of languages from that of identity-free Kleene lattices [19]. Here, in our reduction, the operator $\cdot^{\cap_{\overline{I}}}$ is introduced for employing Löb's axiom forcibly (Fig. 2).

Moreover, the extension above does not increase the complexity, in that the theory of $\mathrm{PDL}_{\mathrm{REwLA}+}$ on finite linear orders ($\mathsf{GREL}_{\leq_{\mathrm{fin\text{-}lin}}}$) and the embedded substitution-closed equivalence problems of REwLA are ExpTime-complete (Thm. 8.1), and the theory of $\mathrm{PDL}_{\mathrm{REwLA}+}$ on a subclass of finite linear orders ($\mathsf{GREL}_{\leq^{\mathrm{st}}_{\mathrm{fin\text{-}lin}}}$) and the embedded standard (match-)language equivalence problems of REwLA are PSpace-complete (Thm. 8.2), respectively (§ 8).

---

[3] Another instance is `(?=a)b` $= \emptyset$, where the *positive lookahead* `?=` asserts that the next character is `a` and the symbol $\emptyset$ expresses the empty language, *cf.*, `(?=a)a` $\neq \emptyset$.

Organization In § 2, we give basic definitions of REwLA and PDL. In § 3, we introduce $\mathrm{PDL}_{\mathrm{REwLA}+}$ and its Hilbert-style axiomatization $\mathcal{H}^{\mathrm{PDL}_{\mathrm{REwLA}+}}_{\leq_{\mathrm{fin\text{-}lin}}}$ on finite linear orders. In § 4 to 6, we prove the completeness theorem of $\mathcal{H}^{\mathrm{PDL}_{\mathrm{REwLA}+}}_{\leq_{\mathrm{fin\text{-}lin}}}$. After introducing PDL^- and its Hilbert-style axiomatization $\mathcal{H}^{\mathrm{PDL}^-}_{<_{\mathrm{fin\text{-}lin}}}$ on finite strict linear orders in § 4, we provide the reduction from the completeness theorem of $\mathcal{H}^{\mathrm{PDL}^-}_{<_{\mathrm{fin\text{-}lin}}}$ to that of $\mathcal{H}^{\mathrm{PDL}_{\mathrm{REwLA}+}}_{\leq_{\mathrm{fin\text{-}lin}}}$ in § 5. In § 6, we prove the completeness theorem of $\mathcal{H}^{\mathrm{PDL}^-}_{<_{\mathrm{fin\text{-}lin}}}$. In § 7, we also give an axiomatic characterization for the standard (match-)language equivalence. In § 8, we consider the computational complexity. In § 9, we conclude this paper with future work. A long version is available at [42].

2 Preliminaries

We write $\mathbb{N}$ for the set of non-negative integers. For $l, r \in \mathbb{N}$, we write $[l..r]$ for the set $\{i \in \mathbb{N} \mid l \leq i \leq r\}$. For a set X, we write $\wp(X)$ for the power set of X and write $\#(X)$ for the cardinality of X. We often use $\sqcup$ to denote that the set union $\cup$ is disjoint.

For a set X, we write X^* for the set of all strings over X. We write ε for the empty string. For a string $w = a_1 \ldots a_n$, we write $\|w\|$ for the *length* n of w.

Given two disjoint sets $\mathbb{A}$ (for term variables) and $\mathbb{P}$ (for formula variables), we use $a, b, \ldots \in \mathbb{A}$ to denote *term variables* and use $p, q, \ldots \in \mathbb{P}$ to denote *formula variables*. We will use $t, s, u, \ldots$ to denote *terms*, use $\varphi, \psi, \rho, \ldots$ to denote *formulas*, and use $E, F, G, \ldots$ to denote *expressions*, *i.e.*, terms or formulas. An *equation* $t = s$ is a pair of terms. We denote by $t \leq s$ the equation $t + s = s$. For an expression E, we write $\mathbb{A}(E)$ for the set of term variables occurring in E and $\mathbb{P}(E)$ for the set of formula variables occurring in E, respectively.

A *frame* $\mathfrak{F}$ is a tuple $\langle |\mathfrak{F}|, U^{\mathfrak{F}} \rangle$, where its *universe* $|\mathfrak{F}|$ is a non-empty set and its *universal relation* $U^{\mathfrak{F}} \subseteq |\mathfrak{F}|^2$ is a binary relation. Given two disjoint sets $\mathbb{A}$ and $\mathbb{P}$, a *generalized structure* $\mathfrak{A}$ on a frame $\mathfrak{F}$ is a tuple $\langle |\mathfrak{A}|, U^{\mathfrak{A}}, \{a^{\mathfrak{A}}\}_{a \in \mathbb{A}}, \{p^{\mathfrak{A}}\}_{p \in \mathbb{P}} \rangle$, where $\langle |\mathfrak{A}|, U^{\mathfrak{A}} \rangle = \mathfrak{F}$, $a^{\mathfrak{A}} \subseteq U^{\mathfrak{A}}$ is a binary relation for each $a \in \mathbb{A}$, and $p^{\mathfrak{A}} \subseteq |\mathfrak{A}|$ is a unary relation for each $p \in \mathbb{P}$. We say that $\mathfrak{A}$ is a *structure* if $U^{\mathfrak{A}} = |\mathfrak{A}|^2$. We write GREL (*resp.*, REL) for the class of all generalized structures (*resp.*, structures). We also write $\mathrm{GREL}_{\underset{\sim}{\leq}}$ (*resp.*, $\mathrm{GREL}_{\leq_{\mathrm{fin\text{-}lin}}}$, $\mathrm{GREL}_{<_{\mathrm{fin\text{-}lin}}}$) for all $\mathfrak{A} \in$ GREL *s.t.* $U^{\mathfrak{A}}$ is a preorder (*resp.*, finite linear order, finite strict linear order).

Given an $\mathfrak{A} \in$ GREL, the *semantics* $[\![\cdot]\!]^{\mathfrak{A}} : \Pi^{\mathbb{A}, \mathbb{P}} \to \wp(U^{\mathfrak{A}}) \sqcup \Phi^{\mathbb{A}, \mathbb{P}} \to \wp(|\mathfrak{A}|)$ is partially[4] defined as the unique *homomorphism* (as the two sort algebra) extending the *valuation* $(\lambda a.a^{\mathfrak{A}}) \sqcup (\lambda p.p^{\mathfrak{A}}) : (\mathbb{A} \to \wp(U^{\mathfrak{A}})) \sqcup (\mathbb{P} \to \wp(|\mathfrak{A}|))$. Here, $\Pi^{\mathbb{A}, \mathbb{P}}$ (*resp.*, $\Phi^{\mathbb{A}, \mathbb{P}}$) denotes the class of all terms (*resp.*, formulas).

Let $\mathcal{C} \subseteq$ GREL be a class *s.t.* $[\![\cdot]\!]^{\mathfrak{A}}$ is well-defined for all $\mathfrak{A} \in \mathcal{C}$. We say that an equation $t = s$ (*resp.*, formula φ) is *valid* on $\mathcal{C}$ if $[\![t]\!]^{\mathfrak{A}} = [\![s]\!]^{\mathfrak{A}}$ (*resp.*, $[\![\varphi]\!]^{\mathfrak{A}} = |\mathfrak{A}|$) for all $\mathfrak{A} \in \mathcal{C}$; we denote them by $\mathcal{C} \models t = s$ (*resp.*, $\mathcal{C} \models \varphi$). The *equational theory* (*resp.*, *theory*) on $\mathcal{C}$ is the class of all equations (*resp.*, formulas) valid on $\mathcal{C}$.

---

[4] Note that $[\![\cdot]\!]^{\mathfrak{A}}$ may be ill-defined, when $[\![t]\!]^{\mathfrak{A}} \not\subseteq U^{\mathfrak{A}}$ for some t.

2.1   REwLA: Regular Expressions with LookAhead

Regular expressions with lookahead (REwLA) [33, 34] are generated by the following grammar:[5]

$$t, s, u \in \Pi^{\mathbb{A}}_{\text{REwLA}} ::= \qquad a \qquad [\textit{character}, \text{term variable } a \in \mathbb{A}]$$
$$\mid 1 \quad [\textit{empty string}] \qquad \mid 0 \quad [\textit{empty set}]$$
$$\mid s\,;\,u \quad [\textit{concatenation}] \qquad \mid s + u \quad [\textit{union}]$$
$$\mid s^+ \quad [\textit{Kleene plus}] \qquad \mid s^{\mathtt{a}} \quad [\textit{negative lookahead "?!s"}]$$

We usually abbreviate $t\,;\,s$ to ts. We use parentheses in ambiguous situations. We write $\sum_{i=1}^{n} t_i$ for the term $0 + t_1 + \cdots + t_n$, and write $\overset{\bullet}{,}{}_{i=1}^{n} t_i$ for the term $1\,;\,t_1\,;\,\cdots\,;\,t_n$. We use the following abbreviations:

$$t^* := 1 + t^+ \quad [\textit{Kleene star}] \qquad\qquad t^n := \overset{n}{\underset{i=1}{\overset{\bullet}{,}}}\, t \quad [n\text{-th }\textit{iteration} \ (n \in \mathbb{N})]$$

$$t^{\mathtt{d}} := (t^{\mathtt{a}})^{\mathtt{a}} \quad [\textit{positive lookahead "?=t"}]$$

2.2   Relational Semantics, Match-Languages, and Languages

We recall the algebraic semantics from *slices of strings* [32].[6] Below, we slightly reformulate in terms of algebras of binary relations. Given a set X, we consider the following operators on binary relations on X:

$$1 := \triangle_X := \{\langle c, c\rangle \mid c \in X\} \qquad\qquad\qquad [\textit{identity relation}]$$
$$0 := \emptyset \qquad\qquad\qquad\qquad\qquad\qquad\qquad\qquad [\textit{empty relation}]$$
$$R\,;\,S := \{\langle c, e\rangle \mid \exists d \in X, \langle c, d\rangle \in R \text{ and } \langle d, e\rangle \in S\} \quad [\textit{relational composition}]$$
$$R + S := R \cup S \qquad\qquad\qquad\qquad\qquad\qquad\qquad [\textit{union}]$$
$$R^+ := \{\langle c_0, c_n\rangle \mid \exists n \geq 1, \exists c_1, \ldots, \exists c_{n-1}, \forall i < n, \langle c_i, c_{i+1}\rangle \in R\} \qquad [\textit{TC}]$$
$$R^{\mathtt{a}} := \{\langle c, c\rangle \in \triangle_X \mid \forall d \in X, \langle c, d\rangle \notin R\} \qquad\qquad [\textit{antidomain}]$$

Additionally, we define the following operators:

$$R^* := 1 + R^+ \quad [\textit{RTC}] \qquad R^n := \overset{n}{\underset{i=1}{\overset{\bullet}{,}}}\, R \quad [n\text{-th }\textit{iteration}] \qquad R^{\mathtt{d}} := (R^{\mathtt{a}})^{\mathtt{a}} \quad [\textit{domain}]$$

Given an $\mathfrak{A} \in \mathsf{GREL}_{\lesssim}$ (on preorder), the *semantics* $[\![\cdot]\!]^{\mathfrak{A}}$ of REwLA is well-defined, where each operator on $\wp(U^{\mathfrak{A}})$ [7] is interpreted as above.

---

[5] For later convenience, we use Kleene plus $(\cdot^+)$ instead of Kleene star $(\cdot^*)$ as a primitive operator. The notations for negative lookahead $(\mathtt{a})$ and positive lookahead $(\mathtt{d})$ are based on antidomain and domain (in the context of Kleene algebra with (anti)domain [15–17]), where we use the superscript notation for short.

[6] *Cf. matching relation* [10], where assignments for backward references are forgotten.

[7] Each REwLA operator on $\wp(U^{\mathfrak{A}})$ is well-defined, because $U^{\mathfrak{A}}$ is a preorder. Note that 1 requires reflexivity of $U^{\mathfrak{A}}$ and $;$ requires transitivity of $U^{\mathfrak{A}}$.

The *string structure* $\mathfrak{A}^w$ of a string $w = a_1 \dots a_n \in \mathbb{A}^*$ is the generalized structure defined by $U^{\mathfrak{A}^w} := \{\langle i, j \rangle \in [0..n]^2 \mid i \leq j\}$, $b^{\mathfrak{A}^w} := \{\langle i, i+1 \rangle \mid i \in [0..n-1] \text{ and } a_{i+1} = b\}$ for $b \in \mathbb{A}$, and $p^{\mathfrak{A}^w} = \emptyset$ for $p \in \mathbb{P}$.

The *match-language* $\mathcal{M}(t)$ and the *language* $\mathcal{L}(t)$ are defined as follows [32]:

$$\mathcal{M}(t) := \{\langle w, i, j \rangle \mid w \in \mathbb{A}^*, \langle i, j \rangle \in [\![t]\!]^{\mathfrak{A}^w}\}, \quad \mathcal{L}(t) := \{w \in \mathbb{A}^* \mid \langle 0, \|w\| \rangle \in [\![t]\!]^{\mathfrak{A}^w}\}.$$

Let $\mathsf{GREL}_{\leq^{\mathrm{st}}_{\mathrm{fin\text{-}lin}}}$ denote the class of all $\mathfrak{B} \in \mathsf{GREL}_{\leq_{\mathrm{fin\text{-}lin}}}$ (with finite linear order) that are isomorphic to $\mathfrak{A}^w$ for some w after replacing $p^{\mathfrak{B}}$ with $\emptyset$ for each $p \in \mathbb{P}$. Since REwLA terms do not contain any formula variables, we have:

$$\mathcal{M}(t) = \mathcal{M}(s) \quad \Longleftrightarrow \quad \{\mathfrak{A}^w \mid w \in \mathbb{A}^*\} \models t = s \quad \Longleftrightarrow \quad \mathsf{GREL}_{\leq^{\mathrm{st}}_{\mathrm{fin\text{-}lin}}} \models t = s.$$

We can embed the language equivalence into the match-language equivalence via a slight encoding, where the term $u_\$$ is defined to express the end of the string.

Proposition 2.1. $\mathcal{L}(t) = \mathcal{L}(s)$ *iff* $\mathsf{GREL}_{\leq^{\mathrm{st}}_{\mathrm{fin\text{-}lin}}} \models tu_\$ = su_\$$, *where* $u_\$:=$ $\left(\sum_{a \in \mathbb{A}(t\,s)} a \right)^{\mathrm{a}}$.

We thus consider $\mathsf{GREL}_{\leq^{\mathrm{st}}_{\mathrm{fin\text{-}lin}}}$ for the standard (match)-language equivalence.

2.3   Substitution-Closed Equivalences

For a term t and a substitution Θ mapping each term variable to a term, we write $t[\Theta]$ for the term obtained by substituting each term variable x with $\Theta(x)$.

A binary relation R on terms is *substitution-closed* if, for all terms t, s and all substitutions Θ, if $\langle t, s \rangle \in R$, then $\langle t[\Theta], s[\Theta] \rangle \in R$. In this paper, we consider the *largest* substitution-closed equivalence relation contained in match-language equivalence (*resp.*, language equivalence); henceforth, just *the substitution-closed equivalence sound for (match-)language equivalence* or the *substitution-closed (match-)language equivalence*. By definition, they are characterized by the equivalence of the *match-languages with substitutions* $\mathcal{M}_{\mathrm{s}}(t)$ (*resp.*, the *languages with substitutions* $\mathcal{L}_{\mathrm{s}}(t)$), defined as follows:

$$\mathcal{M}_{\mathrm{s}}(t) := \bigcup_{\Theta \text{ a substitution}} \{\Theta\} \times \mathcal{M}(t[\Theta]), \qquad \mathcal{L}_{\mathrm{s}}(t) := \bigcup_{\Theta \text{ a substitution}} \{\Theta\} \times \mathcal{L}(t[\Theta]).$$

For every string w and relation $R \subseteq U^{\mathfrak{A}^w}$, one can see that $R = [\![u]\!]^{\mathfrak{A}^w}$ holds by some REwLA u. Thus, every $\mathfrak{A} \in \mathsf{GREL}_{\leq_{\mathrm{fin\text{-}lin}}}$ can be "represented" by some $\mathfrak{A}^w$ with some substitution. From this observation, we see the following:

Proposition 2.2. $\mathcal{M}_{\mathrm{s}}(t) = \mathcal{M}_{\mathrm{s}}(s)$ *iff* $\mathsf{GREL}_{\leq_{\mathrm{fin\text{-}lin}}} \models t = s$.

Additionally, by a similar encoding as in Prop. 2.1 where we use the operator $\cap_{\overline{1}}$ (see REwLA+ in § 3), the substitution-closed language equivalence also can be characterized, as follows:

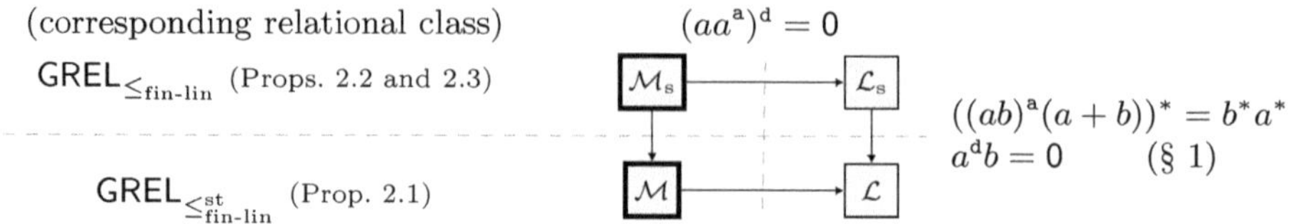

Fig. 1. Inclusions among the four language equivalences. Each arrow from X to Y means that the equivalence induced by X is a subset of that induced by Y.

Proposition 2.3. $\mathcal{L}_{\mathrm{s}}(t) = \mathcal{L}_{\mathrm{s}}(s)$ *iff* $\mathsf{GREL}_{\le_{\text{fin-lin}}} \models tu_\$ = su_\$$, *where* $u_\$:=$ $\left(\sum_{a \in \mathbb{A}(t\,s)} a^{\cap_{\overline{\mathbb{I}}}} \right)^{\mathsf{a}}$.

We thus consider $\mathsf{GREL}_{\le_{\text{fin-lin}}}$ for the substitution-closed (match-)language equivalence. Such equivalences are useful from the perspective of reusability, as well as when extending our system with additional operators (*cf.*, the axiom (PDL) in Fig. 2 and the axiom (Prop) in Fig. 3).

Fig. 1 summarizes the inclusions among the equivalences; they are strict by the given equations. For $(aa^{\mathsf{a}})^{\mathsf{d}} = 0$, observe that $\langle d_M, d_M \rangle \notin [\![(aa^{\mathsf{a}})^{\mathsf{d}}]\!]^{\mathfrak{A}}$ where d_M is maximum on $\mathfrak{A}$. Hence, $\mathcal{L}_{\mathrm{s}}((aa^{\mathsf{a}})^{\mathsf{d}}) = \emptyset$, although clearly $\mathcal{M}((aa^{\mathsf{a}})^{\mathsf{d}}) \ne \emptyset$.

2.4   PDL

We recall *propositional dynamic logic* (PDL) of regular programs with rich tests [20,23]. The set of *formulas* and *terms* are defined by the following grammar:

$$\varphi, \psi, \rho \in \Phi_{\mathrm{PDL}}^{\mathbb{A},\mathbb{P}} ::= p \mid \varphi \to \psi \mid \mathbf{F} \mid [t]\varphi \qquad [\text{modal logic, } p \in \mathbb{P}] \qquad (\textit{formulas})$$

$$t, s, u \in \Pi_{\mathrm{PDL}}^{\mathbb{A},\mathbb{P}} ::= a \mid s\,;u \mid s + u \mid s^{+} \qquad [\text{regular programs, } a \in \mathbb{A}]$$

$$\mid \varphi? \qquad\qquad\qquad [\textit{test}] \qquad\qquad\qquad (\textit{terms})$$

We use the following abbreviations as usual: $\neg\varphi := \varphi \to \mathbf{F}$, $\varphi \lor \psi := \neg\varphi \to \psi$, $\varphi \land \psi := \neg((\neg\varphi)\lor(\neg\psi))$, $(\varphi \leftrightarrow \psi) := (\varphi \to \psi)\land(\psi \to \varphi)$, $\mathbf{T} := \neg\mathbf{F}$, $\langle t\rangle\varphi := \neg[t]\neg\varphi$, $1 := \mathbf{T}?$, and $0 := \mathbf{F}?$. We call $[t]$ the *box operator* and $\langle t\rangle$ the *diamond operator*.

Given a set X, we consider the following operators on unary relations on X:

$$A \to B := (X \setminus A) \cup B, \qquad\qquad\qquad \mathbf{F} := \emptyset,$$

$$[R]A := \{c \in X \mid \forall d, \langle c, d\rangle \in R \text{ implies } d \in A\}, \quad A? := \{\langle c, c\rangle \mid c \in A\},$$

Given an $\mathfrak{A} \in \mathsf{GREL}_{\le}^{\mathbb{A},\mathbb{P}}$, the *semantics* $[\![\cdot]\!]^{\mathfrak{A}}$ of PDL is well-defined, where the operators are interpreted as above and as in REwLA.

PDL has a sound and complete Hilbert-style finite axiomatization on REL [22,44]:[8] $\mathsf{REL} \models \varphi$ *iff* $\vdash_{\mathcal{H}\mathrm{PDL}} \varphi$, where $\vdash_{\mathcal{H}\mathrm{PDL}} \varphi$ if φ is derivable in the system. As $U^{\mathfrak{A}}$ is not used in the evaluation, we also have: $\mathsf{REL} \models \varphi$ *iff* $\mathsf{GREL}_{\le} \models \varphi$.

---

[8] The axioms were introduced by Segerberg [47] and the following completeness was shown independently by Gabbay [22] and Parikh [44]; see [23]. Here, precisely, our syntax is a minor variant using Kleene plus ($\cdot^{+}$).

3   PDL for REwLA+

In this section, we introduce PDL for a slight extension of REwLA. *Extended regular expressions with lookahead* (REwLA+) are generated by the following grammar:

$$t, s, u \in \Pi^{\mathbb{A}}_{\text{REwLA+}} ::= a \mid s\,;u \mid s+u \mid s^{+} \mid s^{\mathsf{a}} \qquad\qquad \text{[syntax of REwLA]}$$
$$\mid s^{\cap_1} \qquad\qquad\qquad\qquad \text{[restriction to the identity relation]}$$
$$\mid s^{\cap_{\bar{1}}} \qquad \text{[restriction to the complement of the identity relation]}$$

Given a set X, we define the following two operators on $\wp(X^2)$ as follows:[9]

$$R^{\cap_1} := R \cap \triangle_X, \qquad\qquad R^{\cap_{\bar{1}}} := R \setminus \triangle_X.$$

For the PDL of REwLA+, denoted by $\text{PDL}_{\text{REwLA+}}$, the set of *formulas* and *terms* are mutually defined by the following grammar:

$$\varphi, \psi, \rho \in \Phi^{\mathbb{A},\mathbb{P}}_{\text{PDL}_{\text{REwLA+}}} ::= p \mid \varphi \to \psi \mid \mathbf{F} \mid [t]\varphi \qquad\qquad (formulas)$$
$$t, s, u \in \Pi^{\mathbb{A},\mathbb{P}}_{\text{PDL}_{\text{REwLA+}}} ::= a \mid s\,;u \mid s+u \mid s^{+} \mid s^{\mathsf{a}} \mid s^{\cap_1} \mid s^{\cap_{\bar{1}}} \mid \varphi? \quad (terms)$$

Let $\mathcal{E}^{\mathbb{A},\mathbb{P}}_{\text{PDL}_{\text{REwLA+}}} := \Phi^{\mathbb{A},\mathbb{P}}_{\text{PDL}_{\text{REwLA+}}} \sqcup \Pi^{\mathbb{A},\mathbb{P}}_{\text{PDL}_{\text{REwLA+}}}$ denote the set of *expressions*. Given an $\mathfrak{A} \in \text{GREL}^{\mathbb{A},\mathbb{P}}_{\underset{\sim}{\leq}}$, the *semantics* $\llbracket \cdot \rrbracket^{\mathfrak{A}}$ of $\text{PDL}_{\text{REwLA+}}$ is well-defined, where the operators are interpreted as in REwLA+ and PDL. We use the notations as PDL (in § 2.4).

3.1   Embedding equivalences

The theories of $\text{PDL}_{\text{REwLA+}}$ embed (substitution-closed | standard) (match-language | language) equivalences, because we can embed equations into formulas as an analog of the embedding in PDL [20, §5], as follows.

Proposition 3.1. *Let* $\mathcal{C} \in \{\text{GREL}_{\leq_{\text{fin-lin}}}, \text{GREL}_{\leq^{\text{st}}_{\text{fin-lin}}}\}$. *For all REwLA+,* t *and* s, *we have:*

$$\mathcal{C} \models t = s \quad\Longleftrightarrow\quad \mathcal{C} \models [t]p \leftrightarrow [s]p,$$

where p *is a fresh formula variable.*

Unlike the complexity gap between PDL (ExpTime-complete [20]) and RE (PSpace-complete [49]), $\text{PDL}_{\text{REwLA+}}$ does not increase the complexity from REwLA+, because the antidomain (negative lookahead) can express the box and (unary) negation (see also, *e.g.*, [15] [45, Def. 6]).

Proposition 3.2. *There exists some polynomial-time reduction from the theory of* $\text{PDL}_{\text{REwLA+}}$ *on* $\text{GREL}_{\leq_{\text{fin-lin}}}$ *(resp.* $\text{GREL}_{\leq^{\text{st}}_{\text{fin-lin}}}$*) to the equational theory of* REwLA+ *on* $\text{GREL}_{\leq_{\text{fin-lin}}}$ *(resp.* $\text{GREL}_{\leq^{\text{st}}_{\text{fin-lin}}}$*).*

---

[9] The operators $\cdot^{\cap_1}$ and $\cdot^{\cap_{\bar{1}}}$ can be found, *e.g.*, in [12, strong loop predicate] [38, graph loop] and in [37], respectively. The notation is based on [37].

Rules:

$$\frac{\varphi \qquad \varphi \to \psi}{\psi} \ (\text{MP}) \qquad\qquad \frac{\varphi}{[t]\varphi} \ (\text{Nec})$$

Axioms:

PDL axioms   All substitution-instances of valid PDL formulas (on REL)   (PDL)

Axiom for a $\qquad\qquad [t^{\mathbf{a}}]\varphi \leftrightarrow [[t]\mathbf{F}?]\varphi \qquad\qquad$ (a)

Axioms for $\cap_1$ and $\cap_{\bar 1}$

$[t^{\cap_1}]\varphi \leftrightarrow [\langle t^{\cap_1}\rangle \mathbf{T}?]\varphi$	$(\cap_1\text{-T})$	$[t]\varphi \leftrightarrow [t^{\cap_1} + t^{\cap_{\bar 1}}]\varphi$	$(\cap_1\text{-}+\text{-}\cap_{\bar 1})$
$[(ts)^{\cap_1}]\varphi \leftrightarrow [t^{\cap_1} s^{\cap_1}]\varphi$	$(\cap_1\text{-};)$	$[(ts)^{\cap_{\bar 1}}]\varphi \leftrightarrow [t^{\cap_{\bar 1}} s^{\cap_1} + t^{\cap_1} s^{\cap_{\bar 1}} + t^{\cap_{\bar 1}} s^{\cap_{\bar 1}}]\varphi$	$(\cap_{\bar 1}\text{-};)$
$[(t+s)^{\cap_1}]\varphi \leftrightarrow [t^{\cap_1} + s^{\cap_1}]\varphi$	$(\cap_1\text{-}+)$	$[(t+s)^{\cap_{\bar 1}}]\varphi \leftrightarrow [t^{\cap_{\bar 1}} + s^{\cap_{\bar 1}}]\varphi$	$(\cap_{\bar 1}\text{-}+)$
$[(t^+)^{\cap_1}]\varphi \leftrightarrow [t^{\cap_1}]\varphi$	$(\cap_1\text{-}\cdot^+)$	$[(t^+)^{\cap_{\bar 1}}]\varphi \leftrightarrow [(t^{\cap_{\bar 1}})^+]\varphi$	$(\cap_{\bar 1}\text{-}\cdot^+)$
$[(t^{\mathbf{a}})^{\cap_1}]\varphi \leftrightarrow [t^{\mathbf{a}}]\varphi$	$(\cap_1\text{-}\cdot^{\mathbf{a}})$	$[(t^{\mathbf{a}})^{\cap_{\bar 1}}]\varphi \leftrightarrow \mathbf{T}$	$(\cap_{\bar 1}\text{-}\cdot^{\mathbf{a}})$
$[(t^{\cap_1})^{\cap_1}]\varphi \leftrightarrow [t^{\cap_1}]\varphi$	$(\cap_1\text{-}\cdot^{\cap_1})$	$[(t^{\cap_1})^{\cap_{\bar 1}}]\varphi \leftrightarrow \mathbf{T}$	$(\cap_{\bar 1}\text{-}\cdot^{\cap_1})$
$[(t^{\cap_{\bar 1}})^{\cap_1}]\varphi \leftrightarrow \mathbf{T}$	$(\cap_1\text{-}\cdot^{\cap_{\bar 1}})$	$[(t^{\cap_{\bar 1}})^{\cap_{\bar 1}}]\varphi \leftrightarrow [t^{\cap_{\bar 1}}]\varphi$	$(\cap_{\bar 1}\text{-}\cdot^{\cap_{\bar 1}})$
$[(\psi?)^{\cap_1}]\varphi \leftrightarrow [\psi?]\varphi$	$(\cap_1\text{-}?)$	$[(\psi?)^{\cap_{\bar 1}}]\varphi \leftrightarrow \mathbf{T}$	$(\cap_{\bar 1}\text{-}?)$

(Restricted) Löb's axiom $\qquad [(t^{\cap_{\bar 1}})^+]([(t^{\cap_{\bar 1}})^+]\varphi \to \varphi) \to [(t^{\cap_{\bar 1}})^+]\varphi \qquad (\text{Löb-}\cdot^{\cap_{\bar 1}+})$

Fig. 2. $\mathcal{H}^{\mathrm{PDL_{REwLA+}}}_{\leq_{\text{fin-lin}}}$: Rules and axioms for $\mathrm{PDL_{REwLA+}}$ on finite linear orders.

3.2 Soundness and completeness

We write $\vdash_{\mathcal{H}^{\mathrm{PDL_{REwLA+}}}_{\leq_{\text{fin-lin}}}} \varphi$ if the formula φ is derivable in the proof system of Fig. 2. Intuitively, the axioms are given based on as follows.

- The axiom for a is for replacing the antidomain $t^{\mathbf{a}}$ with the box test $[t]\mathbf{F}?$;
- The axioms for $\cap_1$ and $\cap_{\bar 1}$ are for transforming the formula so that both $\cdot^{\cap_1}$ and $\cdot^{\cap_{\bar 1}}$ only apply to term variables;
- A variant of the *Löb's axiom* $(\text{Löb-}\cdot^{\cap_{\bar 1}+})$ is employed for the well-foundedness of $\mathrm{GREL}_{\leq_{\text{fin-lin}}}$.

In this paper, we prove the following completeness result.

Theorem 3.3 (Main theorem). *For every* $\mathrm{PDL_{REwLA+}}$ *formula* φ, *we have:*

$$\mathrm{GREL}_{\leq_{\text{fin-lin}}} \models \varphi \quad\Longleftrightarrow\quad \vdash_{\mathcal{H}^{\mathrm{PDL_{REwLA+}}}_{\leq_{\text{fin-lin}}}} \varphi.$$

The soundness ($\Longleftarrow$) is straightforward. Note that some axioms only hold on $\mathrm{GREL}_{\leq_{\text{fin-lin}}}$. For instance, the axiom $(\cap_1\text{-};)$ fails on REL; consider the following case: $\overset{t}{\underset{s}{\rightleftarrows}}$ (which does not appear in $\mathrm{GREL}_{\leq_{\text{fin-lin}}}$ by antisymmetricity). This axiom $(\cap_1\text{-};)$ is based on [2, (3.1)] for language Kleene lattices. Also, the axiom $(\text{Löb-}\cdot^{\cap_{\bar 1}+})$ fails on non-well-founded relations. For the completeness ($\Longrightarrow$), after an identity-free variant of PDL on finite strict linear orders (denoted by $\mathrm{GREL}_{<_{\text{fin-lin}}}$) is introduced in § 4, it is shown by the following two steps:

1. We give a reduction from the completeness of the identity-free PDL. (§ 5)
2. We show the completeness of the identity-free PDL. (§ 6)

> **Rules:** (MP) and (Nec)
>
> -
>
> **Axioms:**
> Prop. axioms   All substitution-instances of valid propositional formulas   (Prop)
> Normal modal logic axiom + variants of Segerberg's axioms + Löb's axiom
>
> $$[ts]\varphi \leftrightarrow [t][s]\varphi \qquad (;) \qquad\qquad [t+s]\varphi \leftrightarrow [t]\varphi \wedge [s]\psi \qquad\qquad (+)$$
>
> $$[t^+]\varphi \leftrightarrow ([t]\varphi \wedge [t][t^+]\varphi) \quad (\cdot^+) \qquad ([t]\varphi \wedge [t^+](\varphi \to [t]\varphi)) \to [t^+]\varphi \quad (\cdot^+\text{-Ind})$$
>
> $$[\psi?\,;t]\varphi \leftrightarrow (\psi \to [t]\varphi) \qquad (?\text{-L}) \qquad [t\,;\psi?]\varphi \leftrightarrow [t](\psi \to \varphi) \qquad\qquad (?\text{-R})$$
>
> $$[t](\varphi \to \psi) \to ([t]\varphi \to [t]\psi) \quad (\mathrm{K}) \qquad [t^+]([t^+]\varphi \to \varphi) \to [t^+]\varphi \qquad (\text{Löb-}\cdot^+)$$

Fig. 3. $\mathcal{H}^{\mathrm{PDL}^-}_{<_{\mathrm{fin\text{-}lin}}}$: Rules and axioms for PDL^- on finite strict linear orders.

4   Identity-Free PDL on Finite Strict Linear Orders

In this section, we define *identity-free* PDL[10] (PDL^-, for short), as a syntax fragment of PDL. The *formulas* and *terms* are mutually generated by the following grammar:

$$\varphi, \psi, \rho \in \Phi^{\mathbb{A},\mathbb{P}}_{\mathrm{PDL}^-} \;::=\; p \mid \varphi \to \psi \mid \mathrm{F} \mid [t]\varphi \qquad\qquad (\textit{formulas})$$

$$t, s, u \in \Pi^{\mathbb{A},\mathbb{P}}_{\mathrm{PDL}^-} \;::=\; a \mid s\,;u \mid s+u \mid s^+ \mid \varphi?\,;u \mid s\,;\varphi? \qquad (\textit{terms})$$

Let $\mathcal{E}^{\mathbb{A},\mathbb{P}}_{\mathrm{PDL}^-} := \Phi^{\mathbb{A},\mathbb{P}}_{\mathrm{PDL}^-} \sqcup \Pi^{\mathbb{A},\mathbb{P}}_{\mathrm{PDL}^-}$ denote the set of *expressions*. Given a generalized structure $\mathfrak{A}$ on a *transitive* frame, the *semantics* $[\![\cdot]\!]^{\mathfrak{A}}$ of PDL^- is well-defined, where the operators are interpreted based on PDL. We use the same notations as with PDL (in § 2.4).

Our syntax restriction is given so that if each term variable a is "identity-free" (*i.e.*, $[\![a]\!]^{\mathfrak{A}} \cap \triangle_{|\mathfrak{A}|} = \emptyset$), then each PDL^- term t is also identity-free. Thus, for the well-definedness of the semantics $[\![\cdot]\!]^{\mathfrak{A}}$ of PDL^-, we do not have to require the *reflexivity* for $U^{\mathfrak{A}}$ in PDL^-, contrary to PDL.[11]

We write $\vdash_{\mathcal{H}^{\mathrm{PDL}^-}_{<_{\mathrm{fin\text{-}lin}}}} \varphi$ if φ is derivable in the system of Fig. 3. The rules $(?\text{-L})$ and $(?\text{-R})$ are variants of the axiom for tests in PDL. The rule $(\text{Löb-}\cdot^+)$ is a variant of the Löb's axiom. Since $\mathcal{H}^{\mathrm{PDL}^-}_{<_{\mathrm{fin\text{-}lin}}}$ has $(\mathrm{Nec})(\mathrm{K})$, from the completeness result of the modal logic $\mathbf{K}$, we have:

For each valid φ on REL with modalities restricted to $[a]$ for a fixed $a \in \mathbb{A}$,

all substitution-instances of φ are derivable in $\mathcal{H}^{\mathrm{PDL}^-}_{<_{\mathrm{fin\text{-}lin}}}$. $\qquad\qquad$ (**K**)

This system is sound and complete *w.r.t.* $\mathsf{GREL}_{<_{\mathrm{fin\text{-}lin}}}$, which will be shown (§ 6).

---

[10] A similar fragment is found in [13], which, in our setting, can be regarded as "test-free" identity-free PDL. See also [28, §2.2] for identity-free Kleene algebra.

[11] Yet, PDL^- formulas have the same expressive power as PDL formulas (on $\mathsf{GREL}_{\lesssim}$).

Theorem 4.1 (§ 6). *For every* PDL^- *formula* φ*, we have:*

$$\mathsf{GREL}_{<\text{fin-lin}} \models \varphi \quad\Longleftrightarrow\quad \vdash_{\mathcal{H}^{\mathrm{PDL}^-}_{<\text{fin-lin}}} \varphi.$$

5 Reduction from the Completeness of Identity-Free PDL

In this section, assuming the completeness theorem for PDL^- on $\mathsf{GREL}_{<\text{fin-lin}}$ (Thm. 4.1), we prove the completeness theorem for $\mathrm{PDL}_{\mathrm{REwLA}+}$ on $\mathsf{GREL}_{\leq\text{fin-lin}}$ (Thm. 3.3). To this end, we transform $\mathrm{PDL}_{\mathrm{REwLA}+}$ into a normal form such that

- the antidomain does not occur (by using the rule (**a**)),
- each term variable a occurs either in the form $a^{\cap_1}$ or the form $a^{\cap_{\overline{1}}}$ (by using the rules for $\cap_1$ and $\cap_{\overline{1}}$).

First, the following congruence rules (Cong) are derivable in $\vdash_{\mathcal{H}^{\mathrm{PDL}_{\mathrm{REwLA}+}}_{\leq\text{fin-lin}}}$ inheriting those in PDL, where ψ in $\{\cdot\}_\psi$ ranges over any formulas:

$$\begin{array}{c}
\dfrac{\varphi_1 \leftrightarrow \psi_1 \quad \varphi_2 \leftrightarrow \psi_2}{\varphi_1 \to \varphi_2 \leftrightarrow \psi_1 \to \psi_2} \qquad \dfrac{\varphi \leftrightarrow \psi}{[t]\varphi \leftrightarrow [t]\psi} \qquad \dfrac{\psi \leftrightarrow \rho}{[\psi?]\varphi \leftrightarrow [\rho?]\varphi} \\[2ex]
\dfrac{\{[t_1]\psi \leftrightarrow [s_1]\psi\}_\psi \quad \{[t_2]\psi \leftrightarrow [s_2]\psi\}_\psi}{[t_1 \star t_2]\varphi \leftrightarrow [s_1 \star s_2]\varphi} \ \star \in \{;,+\} \qquad \dfrac{\{[t]\psi \leftrightarrow [s]\psi\}_\psi}{[t^+]\varphi \leftrightarrow [s^+]\varphi}
\end{array} \quad \text{(Cong)}$$

We define the following translation. Intuitively, $t^{\heartsuit_1}$ and $t^{\heartsuit_2}$ express the identity-part and the identity-free-part of a term t, respectively (Prop. 5.2).

Definition 5.1. *The function* $\cdot^\heartsuit$ *of type*

$$(\Phi_{\mathrm{PDL}_{\mathrm{REwLA}+}} \to \Phi_{\mathrm{PDL}_{\mathrm{REwLA}+}}) \sqcup (\Pi_{\mathrm{PDL}_{\mathrm{REwLA}+}} \to (\Phi_{\mathrm{PDL}_{\mathrm{REwLA}+}} \times \Pi_{\mathrm{PDL}_{\mathrm{REwLA}+}})),$$

is defined as follows, where we write $\langle t^{\heartsuit_1}, t^{\heartsuit_2}\rangle := t^\heartsuit$ *for* $t \in \Pi_{\mathrm{PDL}_{\mathrm{REwLA}+}}$*:*

$$p^\heartsuit := p, \ \ (\psi \to \rho)^\heartsuit := \psi^\heartsuit \to \rho^\heartsuit, \ \ \mathrm{F}^\heartsuit := \mathrm{F}, \ \ ([t]\psi)^\heartsuit := (t^{\heartsuit_1} \to \psi^\heartsuit) \wedge [t^{\heartsuit_2}]\psi^\heartsuit,$$

$$a^{\heartsuit_1} := \langle a^{\cap_1}\rangle\mathrm{T}, \ (s\,;u)^{\heartsuit_1} := s^{\heartsuit_1} \wedge u^{\heartsuit_1}, \ (s+u)^{\heartsuit_1} := s^{\heartsuit_1} \vee u^{\heartsuit_1}, \ (s^+)^{\heartsuit_1} := s^{\heartsuit_1},$$

$$(s^\mathbf{a})^{\heartsuit_1} := \neg s^{\heartsuit_1} \wedge [s^{\heartsuit_2}]\mathrm{F}, \ \ (s^{\cap_1})^{\heartsuit_1} := s^{\heartsuit_1}, \ \ (s^{\cap_{\overline{1}}})^{\heartsuit_1} := \mathrm{F}, \ \ (\varphi?)^{\heartsuit_1} := \varphi^\heartsuit,$$

$$a^{\heartsuit_2} := a^{\cap_{\overline{1}}}, \ (s\,;u)^{\heartsuit_2} := s^{\heartsuit_2}u^{\heartsuit_1}? + s^{\heartsuit_1}?u^{\heartsuit_2} + s^{\heartsuit_2}u^{\heartsuit_2}, \ (s+u)^{\heartsuit_2} := s^{\heartsuit_2} + u^{\heartsuit_2},$$

$$(s^+)^{\heartsuit_2} := (s^{\heartsuit_2})^+, \ (s^\mathbf{a})^{\heartsuit_2} := 0, \ (s^{\cap_1})^{\heartsuit_2} := 0, \ (s^{\cap_{\overline{1}}})^{\heartsuit_2} := s^{\heartsuit_2}, \ (\varphi?)^{\heartsuit_2} := 0.$$

Proposition 5.2. *For every expression* $E \in \mathcal{E}_{\mathrm{PDL}_{\mathrm{REwLA}+}}$*, we have the following.*

1. *If* $E = \varphi$ *is a formula,* $\vdash_{\mathcal{H}^{\mathrm{PDL}_{\mathrm{REwLA}+}}_{\leq\text{fin-lin}}} \varphi^\heartsuit \leftrightarrow \varphi$*.*

2. *If* $E = t$ *is a term,* $\vdash_{\mathcal{H}^{\mathrm{PDL}_{\mathrm{REwLA}+}}_{\leq\text{fin-lin}}} [t^{\heartsuit_1}?]\rho \leftrightarrow [t^{\cap_1}]\rho$*, where* ρ *is any formula.*

3. *If* $E = t$ *is a term,* $\vdash_{\mathcal{H}^{\mathrm{PDL}_{\mathrm{REwLA}+}}_{\leq\text{fin-lin}}} [t^{\heartsuit_2}]\rho \leftrightarrow [t^{\cap_{\overline{1}}}]\rho$*, where* ρ *is any formula.*

Proof. By easy induction on E, using the axioms for $\cap_1$ and $\cap_{\overline{1}}$ with (Cong). $\square$

Let $\mathbb{P}'$ and $\mathbb{A}'$ be sets disjoint from $\mathbb{P}$ and $\mathbb{A}$ (and having the same cardinality as $\mathbb{A}$), let π_1 be a bijection from $\mathbb{P}'$ to $\mathbb{A}$, and let π_2 be a bijection from $\mathbb{A}'$ to $\mathbb{A}$. Let Θ_0 be the substitution mapping each $p \in \mathbb{P}$ to itself, each $p \in \mathbb{P}'$ to $\langle \pi_1(p)^{\cap_1} \rangle \mathrm{T}$, and each $a \in \mathbb{A}'$ to $\pi_2(a)^{\cap_{\overline{1}}}$. For each PDL^- expression $F \in \mathcal{E}_{\mathrm{PDL}^-}^{\mathbb{A}',\mathbb{P}\sqcup\mathbb{P}'}$, we write $F[\Theta_0] \in \mathcal{E}_{\mathrm{PDL_{REwLA+}}}^{\mathbb{A},\mathbb{P}}$ for the $\mathrm{PDL_{REwLA+}}$ expression obtained from F by applying Θ_0. We observe that for every $\varphi \in \Phi_{\mathrm{PDL_{REwLA+}}}^{\mathbb{A},\mathbb{P}}$, there is some $\psi \in \Phi_{\mathrm{PDL}^-}^{\mathbb{A}',\mathbb{P}\sqcup\mathbb{P}'}$ such that $\varphi^{\heartsuit} = \psi[\Theta_0]$. By construction, we have the following lemma.[12]

Lemma 5.3. *For every formula* $\varphi \in \Phi_{\mathrm{PDL}^-}^{\mathbb{A}',\mathbb{P}\sqcup\mathbb{P}'}$*, we have:*

$$\mathsf{GREL}_{\leq \mathrm{fin\text{-}lin}}^{\mathbb{A},\mathbb{P}} \models \varphi[\Theta_0] \quad \Longrightarrow \quad \mathsf{GREL}_{<\mathrm{fin\text{-}lin}}^{\mathbb{A}',\mathbb{P}\sqcup\mathbb{P}'} \models \varphi.$$

Proof. Let $\mathfrak{A} \in \mathsf{GREL}_{<\mathrm{fin\text{-}lin}}^{\mathbb{A}',\mathbb{P}\sqcup\mathbb{P}'}$. We define the $\mathfrak{B} \in \mathsf{GREL}_{\leq\mathrm{fin\text{-}lin}}^{\mathbb{A},\mathbb{P}}$ as follows:

$$|\mathfrak{B}| := |\mathfrak{A}|, \qquad U^{\mathfrak{B}} := U^{\mathfrak{A}} \cup \triangle_{|\mathfrak{A}|},$$

$$p^{\mathfrak{B}} := p^{\mathfrak{A}} \text{ for } p \in \mathbb{P}, \qquad a^{\mathfrak{B}} := [\![\pi_1^{-1}(a)^{\cap_1}]\!]^{\mathfrak{A}} \cup [\![\pi_2^{-1}(a)^{\cap_{\overline{1}}}]\!]^{\mathfrak{A}} \text{ for } a \in \mathbb{A}.$$

Then, $p^{\mathfrak{A}} = [\![\langle \pi_1(p)^{\cap_1} \rangle \mathrm{T}]\!]^{\mathfrak{B}}$ for $p \in \mathbb{P}'$ and $a^{\mathfrak{A}} = [\![\pi_2(a)^{\cap_{\overline{1}}}]\!]^{\mathfrak{B}}$ for $a \in \mathbb{A}'$. By easy induction on ψ, we have $[\![\psi]\!]^{\mathfrak{A}} = [\![\psi[\Theta_0]]\!]^{\mathfrak{B}}$ for all $\psi \in \Phi_{\mathrm{PDL}^-}^{\mathbb{A}',\mathbb{P}\sqcup\mathbb{P}'}$. By $[\![\varphi[\Theta_0]]\!]^{\mathfrak{B}} = |\mathfrak{B}|$, we have $[\![\varphi]\!]^{\mathfrak{A}} = |\mathfrak{A}|$. Hence, this completes the proof. $\square$

We also have the following lemma.

Lemma 5.4. *For every formula* $\varphi \in \Phi_{\mathrm{PDL}^-}^{\mathbb{A}',\mathbb{P}\sqcup\mathbb{P}'}$*, we have:*

$$\vdash_{\mathcal{H}_{<\mathrm{fin\text{-}lin}}^{\mathrm{PDL}^-}} \varphi \quad \Longrightarrow \quad \vdash_{\mathcal{H}_{\leq\mathrm{fin\text{-}lin}}^{\mathrm{PDL_{REwLA+}}}} \varphi[\Theta_0].$$

Proof. By induction on the derivation tree of $\vdash_{\mathcal{H}_{<\mathrm{fin\text{-}lin}}^{\mathrm{PDL}^-}} \varphi$. Since $\varphi[\Theta_0]$ is a substitution-instance of φ and each rule is closed under substitutions (*w.r.t.* both formula variables and term variables), the crucial cases are for (?-L), (?-R), and (Löb-$\cdot^+$), which are not equipped in $\mathrm{PDL_{REwLA+}}$. For (?-L) and (?-R), we can show these cases easily by (PDL). The remaining case is for (Löb-$\cdot^+$). We prepare the following claims.

Claim. For every $t \in \Pi_{\mathrm{PDL}^-}^{\mathbb{A}',\mathbb{P}\sqcup\mathbb{P}'}$, $\vdash_{\mathcal{H}_{\leq\mathrm{fin\text{-}lin}}^{\mathrm{PDL_{REwLA+}}}} t[\Theta_0]^{\heartsuit_1} \leftrightarrow \mathrm{F}$.

Proof. By induction on t. Below, we show some selected cases. **Case** $t = a$ By $(a[\Theta_0])^{\heartsuit_1} = (\pi_2(a)^{\cap_{\overline{1}}})^{\heartsuit_1} = \mathrm{F}$. **Case** $t = \varphi?; u$ By $((\varphi?;u)[\Theta_0])^{\heartsuit_1} = (\varphi?[\Theta_0])^{\heartsuit_1} \wedge (u[\Theta_0])^{\heartsuit_1} \leftrightarrow_{\mathrm{IH}} (\varphi?[\Theta_0])^{\heartsuit_1} \wedge \mathrm{F} \leftrightarrow_{(\mathrm{PDL})} \mathrm{F}$. $\square$

---

[12] We may explicitly write, *e.g.*, $\mathsf{GREL}_{\leq\mathrm{fin\text{-}lin}}^{\mathbb{A},\mathbb{P}}$ instead of $\mathsf{GREL}_{\leq\mathrm{fin\text{-}lin}}$, for clarity.

516   Y. Nakamura

Claim. For every $t \in \Pi_{\mathrm{PDL}^-}^{\mathbb{A}', \mathbb{P} \sqcup \mathbb{P}'}$, $\vdash_{\mathcal{H}_{\leq \mathrm{fin}\text{-}\mathrm{lin}}^{\mathrm{PDL_{REwLA+}}}} ([t[\Theta_0]^+]\rho)^{\heartsuit} \leftrightarrow [(t[\Theta_0]^{\cap_{\overline{1}}})^+]\rho^{\heartsuit}$, where ρ is any formula.

Proof. We have:

$$([t[\Theta_0]^+]\rho)^{\heartsuit} \leftrightarrow_{\substack{\mathrm{claim} \\ \mathrm{above}}} ((\mathrm{F} \to \rho^{\heartsuit}) \wedge [(t[\Theta_0]^+)^{\heartsuit_2}]\rho^{\heartsuit}) \leftrightarrow_{(\mathrm{PDL})} [(t[\Theta_0]^+)^{\heartsuit_2}]\rho^{\heartsuit}$$

$$\leftrightarrow_{\mathrm{Prop.\ 5.2}} [(t[\Theta_0]^+)^{\cap_{\overline{1}}}]\rho^{\heartsuit} \leftrightarrow_{(\cap_{\overline{1}}.^+)} [(t[\Theta_0]^{\cap_{\overline{1}}})^+]\rho^{\heartsuit}. \qquad \square$$

Now, let $\varphi = [t^+]([t^+]\psi \to \psi) \to [t^+]\psi$. To prove $\vdash_{\mathcal{H}_{\leq \mathrm{fin}\text{-}\mathrm{lin}}^{\mathrm{PDL_{REwLA+}}}} \varphi[\Theta_0]$, by Prop. 5.2, it suffices to prove $\vdash_{\mathcal{H}_{\leq \mathrm{fin}\text{-}\mathrm{lin}}^{\mathrm{PDL_{REwLA+}}}} (\varphi[\Theta_0])^{\heartsuit}$. By the claim above, the formula $(\varphi[\Theta_0])^{\heartsuit}$ is equivalent to:

$$[(t[\Theta_0]^{\cap_{\overline{1}}})^+]([(t[\Theta_0]^{\cap_{\overline{1}}})^+](\varphi[\Theta_0])^{\heartsuit} \to (\varphi[\Theta_0])^{\heartsuit}) \to [(t[\Theta_0]^{\cap_{\overline{1}}})^+](\varphi[\Theta_0])^{\heartsuit}.$$

As this formula is exactly (Löb-$\cdot^{\cap_{\overline{1}}+}$), this completes the proof. $\qquad \square$

Proof of Thm. 3.3 assuming Thm. 4.1. Summarizing the lemmas above, by letting ψ be such that $\psi[\Theta_0] = \varphi^{\heartsuit}$, we have:

$$\mathrm{GREL}_{\leq \mathrm{fin}\text{-}\mathrm{lin}}^{\mathbb{A}, \mathbb{P}} \models \varphi \implies_{\substack{\mathrm{Prop.\ 5.2} \\ \mathrm{with\ soundness}}} \mathrm{GREL}_{\leq \mathrm{fin}\text{-}\mathrm{lin}}^{\mathbb{A}, \mathbb{P}} \models \psi[\Theta_0] \implies_{\mathrm{Lem.\ 5.3}}$$

$$\mathrm{GREL}_{< \mathrm{fin}\text{-}\mathrm{lin}}^{\mathbb{A}', \mathbb{P} \sqcup \mathbb{P}'} \models \psi \implies_{\substack{\mathrm{Thm.\ 4.1} \\ (\mathrm{later})}} \vdash_{\mathcal{H}_{< \mathrm{fin}\text{-}\mathrm{lin}}^{\mathrm{PDL}^-}} \psi \implies_{\mathrm{Lem.\ 5.4}} \vdash_{\mathcal{H}_{\leq \mathrm{fin}\text{-}\mathrm{lin}}^{\mathrm{PDL_{REwLA+}}}} \psi[\Theta_0]$$

$$\implies_{\mathrm{Prop.\ 5.2}} \vdash_{\mathcal{H}_{\leq \mathrm{fin}\text{-}\mathrm{lin}}^{\mathrm{PDL_{REwLA+}}}} \varphi \implies_{\mathrm{soundness}} \mathrm{GREL}_{\leq \mathrm{fin}\text{-}\mathrm{lin}}^{\mathbb{A}, \mathbb{P}} \models \varphi. \qquad \square$$

Example: showing substitution-closed equivalence in $\mathcal{H}_{\leq \mathrm{fin}\text{-}\mathrm{lin}}^{\mathrm{PDL_{REwLA+}}}$

Finally, we give examples of our derivation system to show the substitution-closed equivalence in REwLA. First, for equations without $\cap_1$ and $\cap_{\overline{1}}$ and not requiring (Löb-$\cdot^{\cap_{\overline{1}}+}$), we can easily prove them by (PDL) after removing **a**.

Example 5.5. We prove $\mathrm{GREL}_{\leq \mathrm{fin}\text{-}\mathrm{lin}} \models (t + s)^{\mathsf{a}} = t^{\mathsf{a}} \,; s^{\mathsf{a}}$ [16, §9 (8)] [33, Def. 2.2 (4)], which is equivalent to show $[(t+s)^{\mathsf{a}}]p \leftrightarrow [t^{\mathsf{a}}; s^{\mathsf{a}}]p$ (Prop. 3.1). By transforming it into $[[t + s]\mathrm{F}?]p \leftrightarrow [[t]\mathrm{F}? \,; [s]\mathrm{F}?]p$ via (**a**)(Cong), this is a consequence of (PDL).

By the same argument, we can also show, for example, the following:[13]

$$t^{\mathsf{d}}s^{\mathsf{d}} = s^{\mathsf{d}}t^{\mathsf{d}}, \quad t^{\mathsf{d}}t^{\mathsf{d}} = t^{\mathsf{d}}, \quad (t^{\mathsf{d}})^+ = t^{\mathsf{d}}, \quad (t + s)^{\mathsf{d}} = t^{\mathsf{d}} + s^{\mathsf{d}}, \quad (t^{\mathsf{d}}s)^{\mathsf{d}} = t^{\mathsf{d}}s^{\mathsf{d}},$$

$$(ts^{\mathsf{d}})^{\mathsf{d}} = (ts)^{\mathsf{d}}, \quad t^{\mathsf{d}} + t^{\mathsf{a}} = 1, \quad t^{\mathsf{d}}t^{\mathsf{a}} = 0, \quad (t_1^{\mathsf{d}}s_1)^{\mathsf{d}}t_2^{\mathsf{d}}s_2 = (t_1^{\mathsf{d}}t_2^{\mathsf{d}})s_1^{\mathsf{d}}s_2.$$

Particularly, we can also show the axioms in boolean domain semiring [16, 17]:

$$t^{\mathsf{a}}t = 0, \quad (ts)^{\mathsf{a}} \leq (ts^{\mathsf{d}})^{\mathsf{a}}, \quad t^{\mathsf{d}} + t^{\mathsf{a}} = 1.$$

For equations requiring (Löb-$\cdot^{\cap_{\overline{1}}+}$), to apply this axiom, we transform formulas using the translation $\cdot^{\heartsuit}$.

---

[13] They are the variants of [32, Lemma 11] where ?> has been replaced with ?=.

Example 5.6. We prove $\mathsf{GREL}_{\leq_{\text{fin-lin}}} \models u \leq uu^{\mathsf{a}}$ where $u := (ab^{\mathsf{a}})^{+}ab^{\mathsf{d}}$,[14] in our system. Similar to Example 5.5, we show $[u ; [u]\mathbf{F}?]p \to [u]p$. By (PDL), it suffices to show $[u]([u]p \to p) \to [u]p$. Note that

$$u^{\heartsuit_1} \;\leftrightarrow\; a^{\heartsuit_1} \wedge (b^{\mathsf{a}})^{\heartsuit_1} \wedge a^{\heartsuit_1} \wedge (\neg(b^{\mathsf{a}})^{\heartsuit_1} \wedge [(b^{\mathsf{a}})^{\heartsuit_1}]\mathbf{F}) \;\leftrightarrow\; \mathbf{F} \qquad \text{(By (PDL))}$$

$$[u^{\heartsuit_2}]\psi \;\leftrightarrow\; [((ab^{\mathsf{a}}a)^{+})^{\heartsuit_2}][(b^{\mathsf{d}})^{\heartsuit_1}?]\psi \qquad\qquad\qquad \text{(By } (b^{\mathsf{d}})^{\heartsuit_2} \leftrightarrow \mathbf{F})$$

$$\leftrightarrow\; [((ab^{\mathsf{a}}a)^{\cap_{\overline{\mathbf{I}}}})^{+}][(b^{\mathsf{d}})^{\heartsuit_1}?]\psi. \qquad\qquad \text{(By Prop. 5.2 and } (\cap_{\overline{\mathbf{I}}}\text{-}\cdot^{+}))$$

Thus by (PDL)(Cong), the formula above is replaced with:

$$[((ab^{\mathsf{a}}a)^{\cap_{\overline{\mathbf{I}}}})^{+}]([((ab^{\mathsf{a}}a)^{\cap_{\overline{\mathbf{I}}}})^{+}][(b^{\mathsf{d}})^{\heartsuit_1}?]p \to [(b^{\mathsf{d}})^{\heartsuit_1}?]p) \to [((ab^{\mathsf{a}}a)^{\cap_{\overline{\mathbf{I}}}})^{+}][(b^{\mathsf{d}})^{\heartsuit_1}?]p.$$

Hence, by (Löb-$\cdot^{\cap_{\overline{\mathbf{I}}}+}$), we have shown the equation above.

6 Completeness of Identity-Free PDL

In this section, we finally prove the completeness theorem for PDL^{-} on finite strict linear orders. Our proof is based on [29]. Particularly, Lem. 6.6 and the direction ($\Longrightarrow$) of Lem. 6.7.3 are based on [5, Section 4] [6, Section 4.4]. We first defined the closure for PDL^{-} based on that for standard PDL (*cf.* [20, 29]).

Definition 6.1. *The* FL-closure $\mathrm{cl}(\varphi)$ *of a* PDL^{-} *formula* φ *is the smallest set of* PDL^{-} *formulas closed under the following rules:*

$$\varphi \in \mathrm{cl}(\varphi), \qquad\qquad\qquad \psi \to \rho \in \mathrm{cl}(\varphi) \Longrightarrow \psi, \rho \in \mathrm{cl}(\varphi),$$

$$[t]\psi \in \mathrm{cl}(\varphi) \Longrightarrow \psi \in \mathrm{cl}(\varphi), \qquad [t+s]\psi \in \mathrm{cl}(\varphi) \Longrightarrow [t]\psi, [s]\psi \in \mathrm{cl}(\varphi),$$

$$[t^{+}]\psi \in \mathrm{cl}(\varphi) \Longrightarrow [t][t^{+}]\psi \in \mathrm{cl}(\varphi), \qquad [t ; s]\psi \in \mathrm{cl}(\varphi) \Longrightarrow [t][s]\psi \in \mathrm{cl}(\varphi),$$

$$[\rho? ; t]\psi \in \mathrm{cl}(\varphi) \Longrightarrow \rho \to [t]\psi \in \mathrm{cl}(\varphi), \quad [t ; \rho?]\psi \in \mathrm{cl}(\varphi) \Longrightarrow [t](\rho \to \psi) \in \mathrm{cl}(\varphi).$$

Additionally, we let $\mathrm{tcl}(\varphi) = \{t \mid [t]\psi \in \mathrm{cl}(\varphi)\}$. ⌟

We observe that $\#\,\mathrm{cl}(\varphi)$ is finite and polynomial in the size of φ.

For a finite formula set $\Gamma = \{\varphi_1, \ldots, \varphi_n\}$, we write $\widehat{\Gamma}$ for the formula $\bigwedge_{i=1}^{n} \varphi_i$. For a finite set $\mathscr{A}$ of finite formula sets, we write $\bigvee \mathscr{A}$ for the formula $\bigvee_{\Gamma \in \mathscr{A}} \widehat{\Gamma}$.

A formula φ is *consistent* if $\nvdash_{\mathcal{H}^{\mathrm{PDL}^{-}}_{\leq_{\text{fin-lin}}}} \neg\varphi$. A finite formula set Γ is *consistent* if the formula $\widehat{\Gamma}$ is consistent. An *atom* of a finite formula set $\Gamma = \{\varphi_1, \ldots, \varphi_n\}$ is a consistent set $\{\psi_1, \ldots, \psi_n\}$, where each ψ_i is either φ_i or $\neg\varphi_i$. We use $\alpha, \beta, \gamma, \ldots$ to denote atoms. For a formula φ, we write $\mathrm{at}(\varphi)$ for the set of all atoms of $\mathrm{cl}(\varphi)$. Note that $\vdash_{\mathcal{H}^{\mathrm{PDL}^{-}}_{\leq_{\text{fin-lin}}}} \bigvee \mathrm{at}(\varphi)$.

Below, we list some basic properties of atoms. The following proposition gives saturation arguments to obtain an atom from a consistent set Γ, which are easily shown by (MP)(**K**).

---

[14] E.g., for $a = (\mathsf{a} + \mathsf{b})^{*}$ and $b = \mathsf{b}$, on $\mathsf{GREL}_{\leq^{\text{st}}_{\text{fin-lin}}}$, the term u matches when a is read at least once and the next character is b. By taking the farthest position that u matches (such position always exists on $\mathsf{GREL}_{\leq_{\text{fin-lin}}}$ with finite linear order), we can not match uu^{a} from the position, and thus $u \leq uu^{\mathsf{a}}$ holds on $\mathsf{GREL}_{\leq_{\text{fin-lin}}}$.

Proposition 6.2. *For every formula φ, ψ, ρ and term t, the following hold.*

1. *If φ is consistent, either $\varphi \wedge \psi$ or $\varphi \wedge \neg\psi$ is consistent.*
2. *If $\varphi \wedge \langle t \rangle \psi$ is consistent, either $\varphi \wedge \langle t \rangle (\psi \wedge \rho)$ or $\varphi \wedge \langle t \rangle (\psi \wedge \neg\rho)$ is consistent.*

The following proposition gives properties of atoms, which are shown in the same manner as [29, Lemma 2].

Proposition 6.3. *For each formula φ_0 and atom $\alpha \in \mathrm{at}(\varphi_0)$, the following hold.*

1. *For every $\psi \to \rho \in \mathrm{cl}(\varphi_0)$, $\psi \to \rho \in \alpha$ iff $\psi \notin \alpha$ or $\rho \in \alpha$.*
2. $\mathrm{F} \notin \alpha$.
3. *For every $[s + u]\psi \in \mathrm{cl}(\varphi_0)$, $[s + u]\psi \in \alpha$ iff $[s]\psi \in \alpha$ and $[u]\psi \in \alpha$.*
4. *For every $[s \, ; u]\psi \in \mathrm{cl}(\varphi_0)$, $[s \, ; u]\psi \in \alpha$ iff $[s][u]\psi \in \alpha$.*
5. *For every $[s^+]\psi \in \mathrm{cl}(\varphi_0)$, $[s^+]\psi \in \alpha$ iff $[s]\psi \in \alpha$ and $[s][s^+]\psi \in \alpha$.*
6. *For every $[\rho? \, ; s]\psi \in \mathrm{cl}(\varphi_0)$, $[\rho? \, ; s]\psi \in \alpha$ iff $\rho \to [s]\psi \in \alpha$.*
7. *For every $[s \, ; \rho?]\psi \in \mathrm{cl}(\varphi_0)$, $[s \, ; \rho?]\psi \in \alpha$ iff $[s](\rho \to \psi) \in \alpha$.*

The following proposition gives the properties of the consistency, which are shown in the same manner as [29, Lemma 1].

Proposition 6.4. *Let φ_0 be a formula. Let $\alpha, \alpha' \in \mathrm{at}(\varphi_0)$ be atoms. Let s, u be terms and $\rho \in \mathrm{cl}(\varphi_0)$ be a formula. Then the following hold.*

1. *If $\widehat{\alpha} \wedge \langle s + u \rangle \widehat{\alpha}'$ is consistent, $\widehat{\alpha} \wedge \langle s \rangle \widehat{\alpha}'$ is consistent or $\widehat{\alpha} \wedge \langle u \rangle \widehat{\alpha}'$ is consistent.*
2. *If $\widehat{\alpha} \wedge \langle s \, ; u \rangle \widehat{\alpha}'$ is consistent, there is some atom α'' such that $\widehat{\alpha} \wedge \langle s \rangle \widehat{\alpha}''$ and $\widehat{\alpha}'' \wedge \langle u \rangle \widehat{\alpha}'$ are consistent.*
3. *If $\widehat{\alpha} \wedge \langle s^+ \rangle \widehat{\alpha}'$ is consistent, there are $n \geq 1$ and atoms $\alpha_0'', \dots, \alpha_n''$ with $\alpha_0'' = \alpha$ and $\alpha_n'' = \alpha'$ such that $\widehat{\alpha}_i'' \wedge \langle s \rangle \widehat{\alpha}_{i+1}''$ is consistent for every $i < n$.*
4. *If $\widehat{\alpha} \wedge \langle \rho? \, ; s \rangle \widehat{\alpha}'$ is consistent, $\rho \in \alpha$ and $\alpha \wedge \langle s \rangle \alpha'$ is consistent.*
5. *If $\widehat{\alpha} \wedge \langle s \, ; \rho? \rangle \widehat{\alpha}'$ is consistent, $\alpha \wedge \langle s \rangle \alpha'$ is consistent and $\rho \in \alpha'$.*

6.1 Ordering Atoms

In this subsection, we give a finite linear order on the atoms of φ_0, which will be used to construct a canonical model in the class $\mathsf{GREL}_{<_{\text{fin-lin}}}$.

Using (Löb-$\cdot^{\cap\overline{\mathrm{I}}+}$), we have the following lemma.

Lemma 6.5. *Let φ_0 be a formula. If $\mathscr{A} \subseteq \mathrm{at}(\varphi_0)$ is a non-empty set, then there is some $\alpha \in \mathscr{A}$ such that $\widehat{\alpha} \wedge [t^+] \bigvee (\mathrm{at}(\varphi_0) \setminus \mathscr{A})$ is consistent.*

Proof. Towards a contradiction, assume that $\vdash_{\mathcal{H}^{\mathrm{PDL}-}_{<_{\text{fin-lin}}}} [t^+] \bigvee (\mathrm{at}(\varphi_0) \setminus \mathscr{A}) \to \neg\widehat{\alpha}$ for all $\alpha \in \mathscr{A}$. Combining them and $\vdash_{\mathcal{H}^{\mathrm{PDL}-}_{<_{\text{fin-lin}}}} [t^+]((\bigvee \mathrm{at}(\varphi_0) \setminus \mathscr{A}) \leftrightarrow \neg \bigvee \mathscr{A})$ (by (Prop)(Nec)) with (MP)(**K**) yields $\vdash_{\mathcal{H}^{\mathrm{PDL}-}_{<_{\text{fin-lin}}}} [t^+](\neg \bigvee \mathscr{A}) \to \neg \bigvee \mathscr{A}$. By applying the *Löb's rule* $\dfrac{[t^+]\varphi \to \varphi}{\varphi}$, which is derivable from (Löb-$\cdot^+$), we have $\vdash_{\mathcal{H}^{\mathrm{PDL}-}_{<_{\text{fin-lin}}}} \neg \bigvee \mathscr{A}$. Hence, each $\widehat{\alpha}$ is not consistent for every $\alpha \in \mathscr{A}$. This reaches a contradiction, since each atom is consistent and $\mathscr{A}$ is non-empty. $\qquad\square$

Using this lemma, we obtain an appropriate finite linear order on atoms.

Lemma 6.6. *Let φ_0 be a formula. There is a sequence $\alpha_1 \ldots \alpha_n$ of pairwise distinct atoms with $\mathrm{at}(\varphi_0) = \{\alpha_1, \ldots, \alpha_n\}$ such that, for all $[t]\psi \in \mathrm{cl}(\varphi_0)$, if $[t]\psi \notin \alpha_i$, then there is a $j > i$ such that $\widehat{\alpha_i} \wedge \langle t \rangle \widehat{\alpha_j}$ is consistent and $\psi \notin \alpha_j$.*

Proof. By applying Lem. 6.5 (with $t = \sum \mathbb{A}(\varphi_0)$), iteratively, there is a sequence $\alpha_1 \ldots \alpha_n$ of pairwise distinct atoms with $\mathrm{at}(\varphi_0) = \{\alpha_1, \ldots, \alpha_n\}$ such that $\widehat{\alpha_i} \wedge [(\sum \mathbb{A}(\varphi_0))^+] \bigvee_{j>i} \widehat{\alpha_j}$ is consistent for each $i \in [n]$. By $\vdash_{\mathcal{H}^{\mathrm{PDL}-}_{<\mathrm{fin\text{-}lin}}} [(\sum \mathbb{A}(\varphi_0))^+]\rho \to [t]\rho$, for every t with $\mathbb{A}(t) \subseteq \mathbb{A}(\varphi_0)$, the formula $\widehat{\alpha_i} \wedge [t] \bigvee_{j>i} \widehat{\alpha_j}$ is also consistent for each $i \in [n]$. Suppose that there is no $j > i$ such that $\widehat{\alpha_i} \wedge \langle t \rangle \widehat{\alpha_j}$ is consistent and $\psi \notin \alpha_j$; namely, $\vdash_{\mathcal{H}^{\mathrm{PDL}-}_{<\mathrm{fin\text{-}lin}}} \widehat{\alpha_i} \to \neg \langle t \rangle (\neg \psi \wedge \widehat{\alpha_j})$ for all $j > i$. Then $\vdash_{\mathcal{H}^{\mathrm{PDL}-}_{<\mathrm{fin\text{-}lin}}} \widehat{\alpha_i} \to [t](\psi \vee \neg \bigvee_{j>i} \widehat{\alpha_j})$ by (MP)(**K**). As $\widehat{\alpha_i} \wedge [t] \bigvee_{j>i} \widehat{\alpha_j}$ is consistent, $\widehat{\alpha_i} \wedge [t]\psi$ is consistent by (MP)(**K**). Hence $[t]\psi \in \alpha_i$, which contradicts the assumption. $\square$

6.2   Canonical Model

Let φ_0 be a formula. Let $\alpha_1, \ldots, \alpha_n \in \mathrm{at}(\varphi_0)$ be the linearly ordered atoms obtained from Lem. 6.6. The *canonical model* $\mathfrak{A}^{\varphi_0}$ is the generalized structure defined as follows:

$$|\mathfrak{A}^{\varphi_0}| := \{\alpha_1, \ldots, \alpha_n\}, \qquad U^{\mathfrak{A}^{\varphi_0}} := \{\langle \alpha_i, \alpha_j \rangle \mid 1 \le i < j \le n\},$$

$$a^{\mathfrak{A}^{\varphi_0}} := \{\langle \alpha, \alpha' \rangle \in U^{\mathfrak{A}^{\varphi_0}} \mid \widehat{\alpha} \wedge \langle a \rangle \widehat{\alpha'} \text{ is consistent}\}, \quad p^{\mathfrak{A}^{\varphi_0}} := \{\alpha \in |\mathfrak{A}^{\varphi_0}| \mid p \in \alpha\}.$$

By definition, we have $\mathfrak{A}^{\varphi_0} \in \mathsf{GREL}_{<\mathrm{fin\text{-}lin}}$. We have the following truth lemma.

Lemma 6.7 (Truth lemma). *Let φ_0 be a formula. For every expression E, we have the following:*

1. *If $E = \varphi \in \mathrm{cl}(\varphi_0)$ is a formula, then for all $\beta \in \mathrm{at}(\varphi_0)$, $\varphi \in \beta$ iff $\beta \in [\![\varphi]\!]^{\mathfrak{A}^{\varphi_0}}$.*
2. *If $E = t \in \mathrm{tcl}(\varphi_0)$ is a term, then for all $\langle \beta, \beta' \rangle \in U^{\mathfrak{A}^{\varphi_0}}$, if $\widehat{\beta} \wedge \langle t \rangle \widehat{\beta'}$ is consistent, then $\langle \beta, \beta' \rangle \in [\![t]\!]^{\mathfrak{A}^{\varphi_0}}$.*
3. *If $E = t$ is a term and ψ is a formula s.t. $[t]\psi \in \mathrm{cl}(\varphi_0)$, then for all $\beta \in \mathrm{at}(\varphi_0)$, $[t]\psi \notin \beta$ iff there is some β' such that $\langle \beta, \beta' \rangle \in [\![t]\!]^{\mathfrak{A}^{\varphi_0}}$ and $\psi \notin \beta'$.*

Proof. By induction on E.

For 1 We distinguish the following cases. **Case** $\varphi = p$ By the definition of $p^{\mathfrak{A}^{\varphi_0}}$. **Case** $\varphi = \psi \to \rho$, **Case** $\varphi = \mathsf{F}$ By Prop. 6.3, with IH. **Case** $\varphi = [t]\psi$ We have: $[t]\psi \notin \beta$ iff there is some $\beta' \in \mathrm{at}(\varphi_0)$ s.t. $\langle \beta, \beta' \rangle \in [\![t]\!]^{\mathfrak{A}^{\varphi_0}}$ and $\psi \notin \beta'$ (by IH w.r.t. t) iff $\beta \notin [\![[t]\psi]\!]^{\mathfrak{A}^{\varphi_0}}$ (by IH w.r.t. ψ).

For 2 We distinguish the following cases. **Case** $t = a$ By the definition of $a^{\mathfrak{A}^{\varphi_0}}$, **Case** $t = s + u$, **Case** $t = s \,;\, u$, **Case** $t = s^+$, **Case** $t = \rho? \,;\, s$, **Case** $t = s \,;\, \rho?$ By Prop. 6.4 with IH for each case.

For $\Longrightarrow$ of 3 Let $\beta = \alpha_i$. By Lem. 6.6, there is some $j > i$ such that $\widehat{\alpha_i} \wedge \langle t \rangle \widehat{\alpha_j}$ is consistent and $\psi \notin \alpha_j$. By IH of 2, $\langle \alpha_i, \alpha_j \rangle \in [\![t]\!]^{\mathfrak{A}^{\varphi_0}}$.

For $\Longleftarrow$ of 3   We distinguish the following cases. Below are shown in the same manner as [29, Lemma 2]. In each case, Def. 6.1 is crucial for using IH.

Case $t = a$ By the definition of $a^{\mathfrak{A}^{\varphi_0}}$, the formula $\widehat{\beta} \wedge \neg[a]\neg\widehat{\beta}'$ is consistent. By $\neg\psi \in \beta'$ with $(\mathbf{K})$, $\beta \wedge \neg[a]\psi$ is consistent. We thus have $[a]\psi \notin \beta$.

Case $t = s + u$ By $\langle \beta, \beta' \rangle \in [\![s + u]\!]^{\mathfrak{A}^{\varphi_0}}$, we have $\langle \beta, \beta' \rangle \in [\![s]\!]^{\mathfrak{A}^{\varphi_0}}$ or $\langle \beta, \beta' \rangle \in [\![u]\!]^{\mathfrak{A}^{\varphi_0}}$. By IH, we have $[s]\psi \notin \beta$ or $[u]\psi \notin \beta$. By Prop. 6.3, in either case, we have $[s + u]\psi \notin \beta$.

Case $t = s \,; u$ By $\langle \beta, \beta' \rangle \in [\![s \,; u]\!]^{\mathfrak{A}^{\varphi_0}}$, there is some β'' such that $\langle \beta, \beta'' \rangle \in [\![s]\!]^{\mathfrak{A}^{\varphi_0}}$ and $\langle \beta'', \beta' \rangle \in [\![u]\!]^{\mathfrak{A}^{\varphi_0}}$. By IH, $[u]\psi \notin \beta''$. By IH, $[s][u]\psi \notin \beta$. By Prop. 6.3, we have $[s \,; u]\psi \notin \beta$.

Case $t = s^+$ By $\langle \beta, \beta' \rangle \in [\![s^+]\!]^{\mathfrak{A}^{\varphi_0}}$, there are some $n \geq 1$ and $\beta_0, \ldots, \beta_n$ with $\beta_0 = \beta$ and $\beta_n = \beta'$ such that $\langle \beta_i, \beta_{i+1} \rangle \in [\![s]\!]^{\mathfrak{A}^{\varphi_0}}$ for all $i < n$. By induction on i from $n - 1$ to 0, we show $[s^+]\psi \notin \beta_i$.

Case $i = n - 1$ By IH *w.r.t.* s, $[s]\psi \notin \beta_{n-1}$. By Prop. 6.3, $[s^+]\psi \notin \beta_{n-1}$.

Case $i < n - 1$ By IH *w.r.t.* i, $[s][s^+]\psi \notin \beta_i$. By Prop. 6.3, $[s^+]\psi \notin \beta_i$.

Hence, we have $[s^+]\psi \notin \beta$.

Case $t = \rho? \,; s$ By $\langle \beta, \beta' \rangle \in [\![\rho? \,; s]\!]^{\mathfrak{A}^{\varphi_0}}$, we have $\beta \in [\![\rho]\!]^{\mathfrak{A}^{\varphi_0}}$ and $\langle \beta, \beta' \rangle \in [\![s]\!]^{\mathfrak{A}^{\varphi_0}}$. By IH *w.r.t.* s, we have $[s]\psi \notin \beta$. By IH *w.r.t.* ρ, we have $\rho \in \beta$. Thus by Prop. 6.3, we have $[\rho? \,; s]\psi \notin \beta$.

Case $t = s \,; \rho?$ By $\langle \beta, \beta' \rangle \in [\![s \,; \rho?]\!]^{\mathfrak{A}^{\varphi_0}}$, we have $\langle \beta, \beta' \rangle \in [\![s]\!]^{\mathfrak{A}^{\varphi_0}}$ and $\beta' \in [\![\rho]\!]^{\mathfrak{A}^{\varphi_0}}$. By IH *w.r.t.* ρ, we have $\rho \in \beta'$. We thus have $\rho \to \psi \notin \beta'$. By IH *w.r.t.* s, we have $[s](\rho \to \psi) \notin \beta$. Thus by Prop. 6.3, we have $[s \,; \rho?]\psi \notin \beta$. $\square$

By Lem. 6.7, we are now ready to prove the completeness theorem.

Proof of Thm. 4.1. **Soundness** $(\Longleftarrow)$ Easy. **Completeness** $(\Longrightarrow)$ We prove the contrapositive. Suppose $\nvdash_{\mathcal{H}^{\mathrm{PDL}-}_{<\mathrm{fin\text{-}lin}}} \varphi$, namely $\neg\varphi$ is consistent, by (Prop). By Prop. 6.2, there is an atom $\alpha \in \mathrm{at}(\varphi)$ such that $\neg\varphi \in \alpha$. By Lem. 6.7, $\alpha \in [\![\neg\varphi]\!]^{\mathfrak{A}^{\varphi}}$. Hence, $\mathsf{GREL}_{<\mathrm{fin\text{-}lin}} \nvDash \varphi$. $\square$

6.3   Remark on identity-free PDL on REL

We write $\vdash_{\mathcal{H}^{\mathrm{PDL}-}} \varphi$ if φ is derivable in the system of Fig. 3 without the axiom (Löb-$\cdot^+$). As a corollary, we also have the following completeness theorem:

Corollary 6.8 (*Cf.* Thm. 4.1).   *For all* PDL^- *formulas* φ,

$$\mathsf{REL} \models \varphi \quad \Longleftrightarrow \quad \vdash_{\mathcal{H}^{\mathrm{PDL}-}} \varphi.$$

Proof. **Soundness** $(\Longleftarrow)$ Easy. **Completeness** $(\Longrightarrow)$ Similar to § 6.2 where we forget linear ordering (*i.e.*, we set the full relation $U^{\mathfrak{A}^{\varphi}} := \{\langle \alpha_i, \alpha_j \rangle \mid i, j \in [n]\}$), we can construct a canonical model (without (Löb-$\cdot^+$)). $\square$

$$\boxed{\quad \langle a^{\cap_1}\rangle \mathbf{T} \leftrightarrow \mathbf{F} \;\; (\cap_1 a) \qquad\qquad \langle a\rangle\varphi \to [a]\varphi \;\; (\text{Det-1}) \qquad\qquad \langle a\rangle\varphi \to [b]\psi \;\; \text{for } a \neq b \;\; (\text{Det-2}) \quad}$$

Fig. 4. Additional axioms for $\mathcal{H}^{\mathrm{PDL_{REwLA+}}}_{\leq^{\mathrm{st}}_{\mathrm{fin\text{-}lin}}}$. Here, a, b are term variables.

7   Completeness for the (Match-)Language Equivalence

In this section, we moreover show the completeness for the standard match-language equivalence and language equivalence.

We write $\vdash_{\mathcal{H}^{\mathrm{PDL_{REwLA+}}}_{\leq^{\mathrm{st}}_{\mathrm{fin\text{-}lin}}}} \varphi$ if the formula φ is derivable in the system of Fig. 2 with the additional axioms in Fig. 4. The axiom $(\cap_1 a)$ states that the identity-part of a is empty. The axiom (Det-1) states that the binary relation denoted by a is *deterministic* (*i.e.*, the number of outgoing edges labelled by a is at most one), *cf. deterministic PDL* [3, §6.1 (8)]. The axiom (Det-2) states that the number of outgoing labels is at most one. Note that these additional axioms are *not substitution-closed* (*w.r.t.* term variables).

We also write $\vdash_{\mathcal{H}^{\mathrm{PDL^-}}_{<^{\mathrm{st}}_{\mathrm{fin\text{-}lin}}}} \varphi$ if φ is derivable in the system of Fig. 3 with (Det-1) and (Det-2). Let $\mathsf{GREL}_{<^{\mathrm{st}}_{\mathrm{fin\text{-}lin}}}$ denote the class of all $\mathfrak{A} \in \mathsf{GREL}_{\leq^{\mathrm{st}}_{\mathrm{fin\text{-}lin}}}$ in which $U^{\mathfrak{A}}$ has been replaced with $U^{\mathfrak{A}} \setminus \triangle_{|\mathfrak{A}|}$. We then have:

Theorem 7.1 (*Cf.* Thm. 4.1). *For every* PDL^- *formula* φ,

$$\mathsf{GREL}_{<^{\mathrm{st}}_{\mathrm{fin\text{-}lin}}} \models \varphi \quad\Longleftrightarrow\quad \vdash_{\mathcal{H}^{\mathrm{PDL^-}}_{<^{\mathrm{st}}_{\mathrm{fin\text{-}lin}}}} \varphi.$$

Proof Sketch. **Soundness** ($\Longleftarrow$)   Easy. **Completeness** ($\Longrightarrow$)   By the construction as in § 6.2 with pruning unnecessary edges and vertices using (Det-1) and (Det-2), we can construct a canonical model isomorphic to $\mathfrak{A}^w$ for some w. $\square$

From this, we have the following completeness theorem.

Theorem 7.2 (*Cf.* Thm. 3.3). *For every* $\mathrm{PDL_{REwLA+}}$ *formula* φ, *we have:*

$$\mathsf{GREL}_{\leq^{\mathrm{st}}_{\mathrm{fin\text{-}lin}}} \models \varphi \quad\Longleftrightarrow\quad \vdash_{\mathcal{H}^{\mathrm{PDL_{REwLA+}}}_{\leq^{\mathrm{st}}_{\mathrm{fin\text{-}lin}}}} \varphi.$$

Proof. **Soundness** ($\Longleftarrow$)   Easy. **Completeness** ($\Longrightarrow$)   We can give a reduction from Thm. 7.1 by the same argument as in § 5, where the translation $\cdot^{\blacktriangledown}$ (Def. 5.1) is redefined by $a^{\blacktriangledown_1} := \mathbf{F}$, using $(\cap_1 a)$. $\square$

8   On the Complexity

In this section, we consider the complexity of the (match-)language equivalences. The proof of Thm. 3.3 only gives the 2NExpTime upper bound; observe that the formula transformation in § 5 makes an exponential blowup *w.r.t.* the size of the formula. Below, we show that extending REwLA with the two operators $\cdot^{\cap_1}$ and $\cdot^{\cap_{\bar{1}}}$ (thus reaching REwLA+) does not increase the complexity (Cors. 8.3 and 8.4). The following theorems are shown by an analog of the standard automata construction for modal logics.

Theorem 8.1. *For* $\mathrm{PDL}_{\mathrm{REwLA+}}$ *(resp.,* PDL*), the theory is* ExpTime*-complete on* $\mathsf{GREL}_{\leq_{\mathrm{fin\text{-}lin}}}$.

Proof Sketch. **Lower bound** PDL on $\mathsf{GREL}_{\leq_{\mathrm{fin\text{-}lin}}}$ is ExpTime-hard, by the same reduction as in [20, Section 4], which encodes the non-membership problem of alternating polynomial space Turing machines (see also [48, p. 18] [1, Theorem 2]). **Upper bound** By a variant of the "tree unwinding" argument, we can give a polynomial-time reduction to the theory of $\mathrm{PDL}_{\mathrm{REwLA+}}$ on finite trees. Thus by an analog of *e.g.*, [9, 14, 51], we can give a polynomial-time reduction to the emptiness problem of alternating finite tree automata, which is in ExpTime [11]. $\square$

Theorem 8.2. *For* $\mathrm{PDL}_{\mathrm{REwLA+}}$ *(resp.,* PDL*), the theory is* PSpace*-complete on* $\mathsf{GREL}_{\leq^{\mathrm{st}}_{\mathrm{fin\text{-}lin}}}$.

Proof Sketch. **Lower bound** The language equivalence is PSpace-hard for RE [49]. By the reduction of Prop. 3.1, the theory of PDL on $\mathsf{GREL}_{\leq^{\mathrm{st}}_{\mathrm{fin\text{-}lin}}}$ is PSpace-hard. **Upper bound** We can give a reduction to the emptiness problem of alternating finite *string* automata, similar to Thm. 8.1, which is in PSpace [25]. $\square$

We thus have the following complexity results.

Corollary 8.3. *The substitution-closed (match-language | language) equivalences for* (REwLA+ | REwLA) *are all* ExpTime*-complete.*

Proof. **Lower bound** Similar to Thm. 8.1, as a corollary of the reduction of [20, Section 4], they are also shown to be ExpTime-hard. **Upper bound** By Thm. 8.1 with Props. 2.2, 2.3 and 3.1. $\square$

Corollary 8.4. *The (match-language | language) equivalences for* (REwLA+ | REwLA) *are all* PSpace*-complete.*[15]

Proof. **Lower bound** The language equivalence is PSpace-hard already for RE [49]. By Prop. 2.1, the match-language equivalence is also PSpace-hard. **Upper bound** By Thm. 8.2 with Props. 2.1 and 3.1. $\square$

Remark 8.5. For the decidability results in this section, it is crucial that we consider $\mathsf{GREL}_{\leq_{\mathrm{fin\text{-}lin}}}$ (*resp.,* $\mathsf{GREL}_{\leq^{\mathrm{st}}_{\mathrm{fin\text{-}lin}}}$), not REL. On REL, the equational theory of REwLA+ (and also RE with $\cdot^{\cap_{\bar{1}}}$) is Π^0_1-hard [38, 41] (more precisely, it is shown from the reduction of [41, Corollary 7.2], by replacing each hypothesis of the form $t \leq 1$ with $t^{\cap_{\bar{1}}} = 0$), *cf.*, decidable for RE with $\cap$ [35, 39] and for RE with $\bar{1}$ [36].

---

[15] For the language equivalence of REwLA, the PSpace upper bound can also be derived form the polynomial-time reduction to the language equivalence of alternating string automata (which is in PSpace [25]) presented by Morihata [34].

9   Conclusion and Future Work

We have introduced $\mathrm{PDL_{REwLA+}}$ and presented a sound and complete Hilbert-style finite axiomatization (Thm. 3.3), which characterizes the substitution-closed equivalence for both match-language and language. Moreover, we have presented a sound and complete axiomatization (Thm. 7.2), characterizing the match-language equivalence and language equivalence. Additionally, $\mathrm{PDL_{REwLA+}}$ is ExpTime-complete for the substitution-closed equivalences (Thm. 8.1) and is PSpace-complete for the standard equivalences (Thm. 8.2).

A near future work is to present a direct (quasi-)equational axiomatization for REwLA+, in the style of test algebra [24] or Kleene algebra with (anti)domain [15, 17, 45]. This paper's work in the paper would possibly be a first step towards this direction, as the boolean sort of test algebras is heavily based on PDL [24] and the completeness of Kleene algebras with domain shown in [45] is based on the completeness of test algebras. Extending the syntax with *lookbehind* [32] (and with the *leftmost anchor* and *rightmost anchor*) would also be interesting. For REwLA with *backreferences*, we have no recursive axiomatization, because the standard language emptiness is undecidable [10, Theorem 4] [50, Theorem 4] (precisely, Π_1^0-hard). An open question is whether we can extend the completeness and decidability for REwLA+ with full intersection $\cap$ on $\mathrm{GREL}_{\leq_{\mathrm{fin\text{-}lin}}}$, *cf.* decidable and ExpSpace-complete for the equational theory of RE with $\cap$ (and mirror image) [7] on algebras of languages, which is equivalent to the equational theory on "RSUB" a variant of $\mathrm{GREL}_{\leq_{\mathrm{fin\text{-}lin}}}$ [40] (if mirror image is omitted). Additionally, it would also be interesting to investigate the guarded-Kleene-algebra-like fragments of our deterministic PDLs on $\mathrm{GREL}_{\leq_{\mathrm{fin\text{-}lin}}^{\mathrm{st}}}$, *cf.* [4].

In this paper, using the two operators $\cdot^{\cap_1}$ and $\cdot^{\cap_{\overline{1}}}$, we have shown the completeness for $\mathrm{PDL_{REwLA+}}$, via the completeness of identity-free PDL. Another open question is whether we can give a direct axiomatization for pure PDL on $\mathrm{GREL}_{\leq_{\mathrm{fin\text{-}lin}}}$ (without $\cdot^{\cap_1}$ or $\cdot^{\cap_{\overline{1}}}$). For instance, the following formula derived from modal Grzegorczyk logic [46, p. 96], is valid on $\mathrm{GREL}_{\leq_{\mathrm{fin\text{-}lin}}}$.

$$[t^*]([t^*](\varphi \to [t^*]\varphi) \to \varphi) \to \varphi \qquad \text{(Grz-}\cdot^*\text{)}$$

Acknowledgments. We are grateful to Paul Brunet for introducing his technique [8] for eliminating the identity in algebras of languages. We also thank anonymous reviewers for their helpful and insightful comments. This work was supported by JSPS KAKENHI Grant Number JP25K14985.

References

1. Afanasiev, L., Blackburn, P., Dimitriou, I., Gaiffe, B., Goris, E., Marx, M., de Rijke, M.: PDL for ordered trees. Journal of Applied Non-Classical Logics **15**(2), 115–135 (2005). https://doi.org/10.3166/jancl.15.115-135
2. Andréka, H., Mikulás, S., Németi, I.: The equational theory of Kleene lattices. Theoretical Computer Science **412**(52), 7099–7108 (2011). https://doi.org/10.1016/J.TCS.2011.09.024

3. Ben-Ari, M., Halpern, J., Pnueli, A.: Deterministic propositional dynamic logic: Finite models, complexity, and completeness. Journal of Computer and System Sciences **25**(3), 402–417 (1982). https://doi.org/10.1016/0022-0000(82)90018-6

4. Benevides, M., Gomes, L., Lopes, B.: Towards determinism in PDL: relations and proof theory. Journal of Logic and Computation **35**(5), exae022 (2025). https://doi.org/10.1093/logcom/exae022

5. Blackburn, P., Meyer-Viol, W.: Linguistics, logic and finite trees. Bulletin of the IGPL **2**(1), 3–29 (1994). https://doi.org/10.1093/jigpal/2.1.3

6. Blackburn, P., Meyer-Viol, W., de Rijke, M.: A proof system for finite trees. In: CSL. LNCS, vol. 1092, pp. 86–105. Springer (1996). https://doi.org/10.1007/3-540-61377-3_33

7. Brunet, P.: Reversible Kleene lattices. In: MFCS. LIPIcs, vol. 83, pp. 66:1–66:14. Schloss Dagstuhl (2017). https://doi.org/10.4230/LIPICS.MFCS.2017.66

8. Brunet, P.: A complete axiomatisation of a fragment of language algebra. In: CSL. LIPIcs, vol. 152, p. 11:1–11:15. Schloss Dagstuhl (2020). https://doi.org/10.4230/LIPIcs.CSL.2020.11

9. Calvanese, D., De Giacomo, G., Lenzerini, M., Vardi, M.: An automata-theoretic approach to regular XPath. In: DBPL. LNISA, vol. 5708, pp. 18–35. Springer (2009). https://doi.org/10.1007/978-3-642-03793-1_2

10. Chida, N., Terauchi, T.: On lookaheads in regular expressions with backreferences. IEICE Transactions on Information **E106-D**(5), 959–975 (2023). https://doi.org/10.1587/transinf.2022EDP7098

11. Comon, H., Dauchet, M., Gilleron, R., Jacquemard, F., Lugiez, D., Löding, C., Tison, S., Tommasi, M.: Tree Automata Techniques and Applications (2007), https://inria.hal.science/hal-03367725

12. Danecki, R.: Propositional dynamic logic with strong loop predicate. In: MFCS. LNCS, vol. 176, pp. 573–581. Springer (1984). https://doi.org/10.1007/BFb0030342

13. Das, A., Girlando, M.: Cyclic proofs, hypersequents, and transitive closure logic. In: IJCAR. LNAI, vol. 13385, pp. 509–528. Springer (2022). https://doi.org/10.1007/978-3-031-10769-6_30

14. De Giacomo, G., Vardi, M.: Linear temporal logic and linear dynamic logic on finite traces. In: IJCAI. pp. 854–860. AAAI Press (2013), https://www.ijcai.org/Proceedings/13/Papers/132.pdf

15. Desharnais, J., Möller, B., Struth, G.: Kleene algebra with domain. ACM Transactions on Computational Logic **7**(4), 798–833 (2006). https://doi.org/10.1145/1183278.1183285

16. Desharnais, J., Struth, G.: Modal semirings revisited. In: MPC. LNTCS, vol. 5133, pp. 360–387. Springer (2008). https://doi.org/10.1007/978-3-540-70594-9_19

17. Desharnais, J., Struth, G.: Internal axioms for domain semirings. Science of Computer Programming **76**(3), 181–203 (2011). https://doi.org/10.1016/j.scico.2010.05.007

18. Developers, T.P.: Perl-compatible regular expressions (revised api: PCRE2), https://www.pcre.org/current/doc/html/pcre2syntax.html, last updated: 14 October 2025

19. Doumane, A., Pous, D.: Completeness for identity-free Kleene lattices. In: CONCUR. LIPIcs, vol. 118, pp. 18:1–18:17. Schloss Dagstuhl (2018). https://doi.org/10.4230/LIPICS.CONCUR.2018.18

20. Fischer, M., Ladner, R.: Propositional dynamic logic of regular programs. Journal of Computer and System Sciences **18**(2), 194–211 (1979). https://doi.org/10.1016/0022-0000(79)90046-1

21. Friedl, J.: Mastering Regular Expressions. O'Reilly Media, Inc., 3 edn. (2006), https://www.oreilly.com/library/view/mastering-regular-expressions/0596528124/

22. Gabbay, D.: Axiomatizations of logics of programs (1977), note: unpublished manuscript

23. Harel, D., Kozen, D., Tiuryn, J.: Dynamic Logic. The MIT Press (2000). https://doi.org/10.7551/mitpress/2516.001.0001

24. Hollenberg, M.: Equational axioms of test algebra. In: CSL. LNCS, vol. 1414, pp. 295–310. Springer (1998). https://doi.org/10.1007/BFb0028021

25. Jiang, T., Ravikumar, B.: A note on the space complexity of some decision problems for finite automata. Information Processing Letters **40**(1), 25–31 (1991). https://doi.org/10.1016/S0020-0190(05)80006-7

26. Kleene, S.: Representation of events in nerve nets and finite automata. Tech. rep., RAND Corporation (1951), https://www.rand.org/pubs/research_memoranda/RM704.html

27. Kozen, D.: A completeness theorem for Kleene algebras and the algebra of regular events. In: LICS. pp. 214–225. IEEE (1991). https://doi.org/10.1109/LICS.1991.151646

28. Kozen, D.: Typed Kleene algebra. Tech. rep., Cornell University (1998), https://hdl.handle.net/1813/7323

29. Kozen, D., Parikh, R.: An elementary proof of the completeness of PDL. Theoretical Computer Science **14**(1), 113–118 (1981). https://doi.org/10.1016/0304-3975(81)90019-0

30. Kracht, M.: Syntactic codes and grammar refinement. Journal of Logic, Language and Information **4**(1), 41–60 (1995). https://doi.org/10.1007/BF01048404

31. Kracht, M.: Inessential features. In: Logical Aspects of Computational Linguistics. pp. 43–62. Springer (1997). https://doi.org/10.1007/BFb0052150

32. Mamouras, K., Chattopadhyay, A.: Efficient matching of regular expressions with lookaround assertions. Proc. ACM Program. Lang. **8**(POPL), 92:2761–92:2791 (2024). https://doi.org/10.1145/3632934

33. Miyazaki, T., Minamide, Y.: Derivatives of regular expressions with lookahead. Journal of Information Processing **27**, 422–430 (2019). https://doi.org/10.2197/ipsjjip.27.422

34. Morihata, A.: Translation of regular expression with lookahead into finite state automaton (written in Japanese). Computer Software **29**(1), 1_147–1_158 (2012). https://doi.org/10.11309/jssst.29.1_147

35. Nakamura, Y.: Partial derivatives on graphs for Kleene allegories. In: LICS. pp. 1–12. IEEE (2017). https://doi.org/10.1109/LICS.2017.8005132

36. Nakamura, Y.: Existential calculi of relations with transitive closure: Complexity and edge saturations. In: LICS. pp. 1–13. IEEE (2023). https://doi.org/10.1109/LICS56636.2023.10175811

37. Nakamura, Y.: On the finite variable-occurrence fragment of the calculus of relations with bounded dot-dagger alternation. In: MFCS. LIPIcs, vol. 272, p. 69:1–69:15. Schloss Dagstuhl (2023). https://doi.org/10.4230/LIPIcs.MFCS.2023.69

38. Nakamura, Y.: Undecidability of the positive calculus of relations with transitive closure and difference: Hypothesis elimination using graph loops. In: RAMICS. LNTCS, vol. 14787, pp. 207–224. Springer (2024). https://doi.org/10.1007/978-3-031-68279-7_13

39. Nakamura, Y.: Derivatives on graphs for the positive calculus of relations with transitive closure. Logical Methods in Computer Science **Volume 21, Issue 4** (2025), https://doi.org/10.46298/lmcs-21(4:27)2025
40. Nakamura, Y.: Finite relational semantics for language Kleene algebra with complement. In: CSL. LIPIcs, vol. 326, pp. 37:1–37:23. Schloss Dagstuhl (2025). https://doi.org/10.4230/LIPIcs.CSL.2025.37
41. Nakamura, Y.: Undecidability of the emptiness problem of deterministic propositional while programs with graph loop: Hypothesis elimination using loops (2025), http://arxiv.org/abs/2504.20415v1
42. Nakamura, Y.: A complete propositional dynamic logic for regular expressions with lookahead (2026). https://doi.org/10.48550/arXiv.2601.15214
43. Palm, A.: Propositional tense logic for finite trees. In: MOL 6, 6th Biennial Conference on Mathematics of Language (1999), https://citeseerx.ist.psu.edu/document?doi=934688af7e9150b78a0e1035ccef8c4b95a05e94
44. Parikh, R.: The completeness of propositional dynamic logic. In: MFCS. LNCS, vol. 64, pp. 403–415. Springer (1978). https://doi.org/10.1007/3-540-08921-7_88
45. Sedlár, I.: On the complexity of Kleene algebra with domain. In: RAMICS. LNCS, vol. 13896, pp. 208–223. Springer (2023). https://doi.org/10.1007/978-3-031-28083-2_13
46. Segerberg, K.: An Essay in Classical Modal Logic. Ph.D. thesis, Uppsala universitet (1971)
47. Segerberg, K.: A completeness theorem in the modal logic of programs. Banach Center Publications **9**, 31–46 (1982). https://doi.org/10.4064/-9-1-31-46
48. Spaan, E.: Complexity of Modal Logics. Ph.D. thesis, University of Amsterdam (1993), https://eprints.illc.uva.nl/id/eprint/1846/
49. Stockmeyer, L., Meyer, A.: Word problems requiring exponential time (preliminary report). In: STOC. pp. 1–9. ACM (1973). https://doi.org/10.1145/800125.804029
50. Uezato, Y.: Regular expressions with backreferences and lookaheads capture NLOG. In: ICALP. LIPIcs, vol. 297, pp. 155:1–155:20. Schloss Dagstuhl (2024). https://doi.org/10.4230/LIPIcs.ICALP.2024.155
51. Vardi, M., Wolper, P.: Automata-theoretic techniques for modal logics of programs. Journal of Computer and System Sciences **32**(2), 183–221 (1986). https://doi.org/10.1016/0022-0000(86)90026-7

The Modal Logic of Abstraction Refinement

Jakob Piribauer ⬡, Vinzent Zschuppe

Technische Universität Dresden, Germany
`jakob.piribauer@tu-dresden.de, vinzent.zschuppe@mailbox.tu-dresden.de`

Abstract. Iterative abstraction refinement techniques are one of the most prominent paradigms for the analysis and verification of systems with large or infinite state spaces. This paper investigates the changes of truth values of system properties expressible in computation tree logic (CTL) when abstractions of transition systems are refined. To this end, the paper utilizes modal logic by defining *alethic modalities* expressing possibility and necessity on top of CTL: The modal operator $\Diamond$ is interpreted as "there is a refinement, in which ..." and $\Box$ is interpreted as "in all refinements, ...". Upper and lower bounds for the resulting *modal logics of abstraction refinement* are provided for three scenarios: 1) when considering all finite abstractions of a transition system, 2) when considering all abstractions of a transition system, and 3) when considering the class of all transition systems. Furthermore, to prove these results, generic techniques to obtain upper bounds of modal logics using novel types of so-called control statements are developed.

Acknowledgments. This work was partly funded by the DFG Grant 389792660 as part of TRR 248 (Foundations of Perspicuous Software Systems) and by the BMBF (Federal Ministry of Education and Research) in DAAD project 57616814 (SECAI, School of Embedded and Composite AI) as part of the program Konrad Zuse Schools of Excellence in Artificial Intelligence.

Extended version. An extended version with full proofs is available [32].

1   Introduction

Verification techniques like model checking take a mathematical model of a system as well as a formal specification expressing the intended behavior and automatically verify whether the specification is met by the system model. A key challenge in practical applications is the potentially huge size of the state space of system models (see, e.g., [13]). One of the most important approaches to tackle this problem is the use of abstractions.

Abstractions. In a nutshell, the idea of an abstraction is to group *concrete* states of a system together into *abstract* states. For every transition between concrete states, the abstraction contains a transition between the corresponding abstract states (this is called an *existential abstraction* in [15]). The more states are grouped together the more details of the original system are lost. A refinement of an abstraction is obtained by breaking up some of the abstract states

N. Bertrand and S. Milius (Eds.): FoSSaCS 2026, LNCS 16503, pp. 527–548, 2026.
https://doi.org/10.1007/978-3-032-22730-0_25

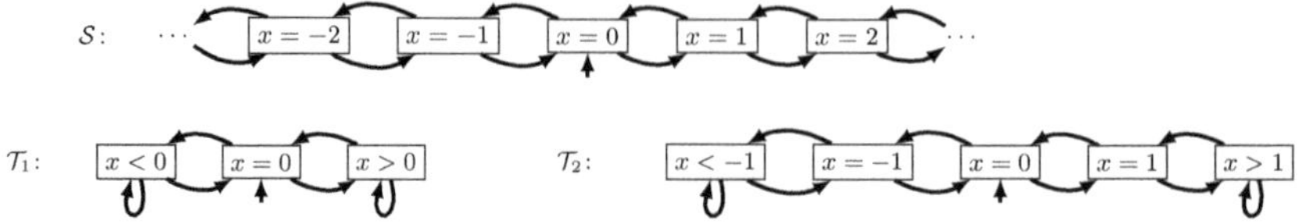

Fig. 1: A transition system $\mathcal{S}$ and abstractions $\mathcal{T}_1$ and $\mathcal{T}_2$ where $\mathcal{T}_2$ refines $\mathcal{T}_1$.

into finer abstract states. For an illustration, consider Figure 1. The transition system $\mathcal{S}$ increases or decreases x by 1 in each step. A possible abstraction $\mathcal{T}_1$ and a refinement $\mathcal{T}_2$ of $\mathcal{T}_1$ grouping together states in a finer manner are depicted.

The set of executions in an abstraction forms an over-approximation of the set of executions in the concrete system. So, if all executions in the abstraction satisfy a linear-time property φ, then all executions in the underlying system satisfy φ; satisfaction of linear-time properties is *preserved* under abstraction refinement. If there are executions in an abstraction that do not satisfy φ, however, no immediate conclusions about the underlying system are possible.

In contrast, the situation for branching-time logics such as *computation tree logic* (CTL), is different: CTL uses existential and universal quantification over executions in a nested manner. Due to the existential quantification, there is no general preservation result for the satisfaction of CTL formulas Φ under refinement. Consider for example the CTL formula $\Phi = \exists X \exists X \exists X (x = 0)$ stating that there is a sequence of three transitions such that afterwards the value of x is 0. This formula is true at the initial state of abstraction $\mathcal{T}_1$ in Figure 1. By refining the abstraction to $\mathcal{T}_2$, we see that the satisfaction of Φ is not preserved. Only for the universally quantified fragment ACTL of CTL, satisfaction in an abstraction implies satisfaction in the underlying system in general [10,11].

Extending CTL with alethic modalities. To address whether an abstraction permits conclusions about the underlying system, we propose extending CTL with *alethic modalities* expressing possibility and necessity. For a CTL formula Φ, we let the modal operator $\Box\Phi$ indicate that Φ is necessarily true, meaning it holds in all refinements of the current abstraction. Conversely, $\Diamond\Phi$ expresses that Φ is possibly true, signifying that it holds in at least one refinement. More formally, we use transition systems as system models and consider the relation $\rightsquigarrow$, where $\mathcal{T}_1 \rightsquigarrow \mathcal{T}_2$ means $\mathcal{T}_2$ is a refinement of $\mathcal{T}_1$, on the class of transition systems. In this way, we obtain a directed graph, called a *Kripke frame*, with transition systems as *worlds*. The Kripke frame can be extended to a Kripke model by a valuation V assigning to atomic propositions p a set of worlds at which p holds. Modal formulas are then evaluated at a world: $\Box\varphi$ holds at a world if φ holds at all successors of the current node; $\Diamond\varphi$ holds if φ holds at some successor.

As we aim to investigate how the satisfaction of branching-time properties evolves under abstraction refinement, we restrict the valuations to range over sets of transition systems sharing a CTL-expressible property. We choose CTL as arguably the simplest prominent branching-time logic. With this restriction, we obtain a *general* Kripke frame. The modal formulas valid on the general Kripke frame are those that hold at any world under any admissible valuation. These are

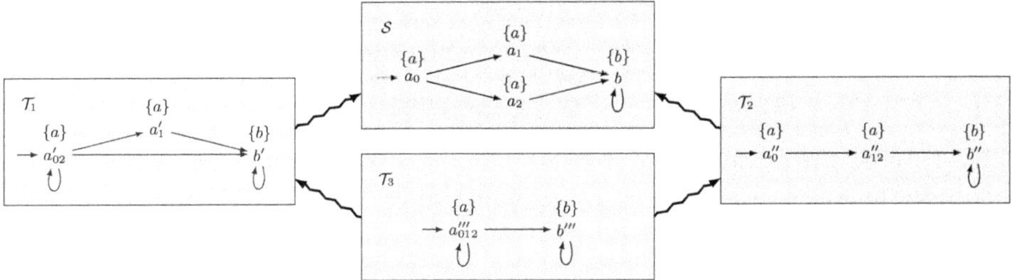

Fig. 2: Example of the general frame $\mathcal{A}_{\mathcal{S}}$ $(= \mathcal{F}_{\mathcal{S}})$ resulting from transition system $\mathcal{S}$ and—up to isomorphism—all its (finite) abstractions. The accessibility relation is the transitive reflexive closure of the indicated refinement arrows $\rightsquigarrow$ between the worlds. The transition systems $\mathcal{S}$ and $\mathcal{T}_2$ are not distinguishable by CTL as indicated by the filling of the rectangles.

the general principles according to which truth values of CTL-properties change under abstraction refinement. Informally, propositional variables can now be understood as "placeholders" for CTL formulas.

Example 1. Consider the transition system $\mathcal{S}$ depicted in Figure 2. The sets next to the states indicate the labels. We require abstractions to only group together states with the same label. An abstraction of $\mathcal{S}$ hence might collapse some of the states a_0, a_1, a_2 as they are labelled with a. If a_0 and a_1, or a_0 and a_2, are collapsed to one state, we obtain (up to isomorphism) the abstraction $\mathcal{T}_1$. Collapsing a_1 and a_2 leads to the abstraction $\mathcal{T}_2$. Finally, collapsing all three states results in $\mathcal{T}_3$. So, the worlds of the general Kripke frame $\mathcal{A}_{\mathcal{S}}$ $(= \mathcal{F}_{\mathcal{S}})$ consisting of all (finite) abstractions of $\mathcal{S}$ are $\mathcal{S}, \mathcal{T}_1, \mathcal{T}_2, \mathcal{T}_3$.

Further, we observe that $\mathcal{S}$ and $\mathcal{T}_2$ are not distinguishable by CTL-formulas. Hence, any valuation V admissible on this general Kripke frame satisfies the following: For each atomic proposition p, the set $V(p)$ contains either both $\mathcal{S}$ and $\mathcal{T}_2$ or neither $\mathcal{S}$ nor $\mathcal{T}_2$. For example, a valuation might set $V(p)$ to the set of all transition systems that satisfy the CTL formula $\Phi = \exists \mathsf{X} b$ (i.e., whose initial states satisfy Φ). This results in $V(p) = \{\mathcal{T}_1, \mathcal{T}_3\}$ in the example.

Now, e.g., the modal formula $\Box \Diamond p \rightarrow \Diamond \Box p$ is valid on the general frame in Figure 2: No matter which CTL-state formula Φ corresponds to the valuation of p, either Φ is not true in (the initial state of) $\mathcal{S}$ and so p does not hold at the world $\mathcal{S}$ making $\Box \Diamond p$ false everywhere, or $\mathcal{S}$ satisfies Φ and so p holds at the world $\mathcal{S}$ and so $\Diamond \Box p$ is true at every world of the frame.

In this paper, we focus on determining the modal tautologies that are always true no matter which CTL-expressible properties are plugged in for the atomic propositions. We call the resulting modal logics *modal logics of abstraction refinement (MLARs)*—which depend on the precise class of transition systems under consideration. We consider three cases:

1. $\mathrm{MLAR}_{\mathcal{T}}^{fin}$ on the class $\mathcal{F}_{\mathcal{T}}$ of all finite abstractions of a transition system $\mathcal{T}$,
2. $\mathrm{MLAR}_{\mathcal{T}}^{all}$ on the class $\mathcal{A}_{\mathcal{T}}$ of all abstractions of a transition system $\mathcal{T}$, and
3. MLAR on the class of all transition systems.

(T)	$= p \to \Diamond p$	(4)	$= \Diamond\Diamond p \to \Diamond p$
(.2)	$= \Diamond\Box p \to \Box\Diamond p$	(.1)	$= \Box\Diamond p \to \Diamond\Box p$
S4	$= \mathsf{K} + (\mathsf{T}) + (4)$	S4.2	$= \mathsf{K} + (\mathsf{T}) + (4) + (.2)$
S4.1	$= \mathsf{K} + (\mathsf{T}) + (4) + (.1)$	S4.2.1	$= \mathsf{K} + (\mathsf{T}) + (4) + (.2) + (.1)$

Table 1: Important axioms and normal modal logics. For formulas $\alpha_1, \ldots, \alpha_n$, we denote the smallest normal modal logic containing $\alpha_1, \ldots, \alpha_n$ by $\mathsf{K} + \alpha_1 + \cdots + \alpha_n$.

Example 2. The formula $(\mathsf{T}) = p \to \Diamond p$ belongs to all three MLARs. The validity translates to "For any transition system $\mathcal{T}$ and any CTL-state formula Φ: if $\mathcal{T} \vDash \Phi$, then $\mathcal{T}$ has a refinement $\mathcal{S}$ with $\mathcal{S} \vDash \Phi$." This is true because $\mathcal{T} \rightsquigarrow \mathcal{T}$, i.e., $\mathcal{T}$ is a refinement of itself, for any $\mathcal{T}$. The formula $p \to \Box p$ does not belong to any of the MLARs: In Figure 1, let the valuation of p be the set of transition systems satisfying $\Phi = \exists\mathsf{X}\exists\mathsf{X}\exists\mathsf{X}(x = 0)$. Then, p holds at $\mathcal{T}_1$, but $\Box p$ does not hold at $\mathcal{T}_1$ as $\mathcal{T}_2$ is a refinement of $\mathcal{T}_1$ not satisfying Φ.

Contributions. We provide lower and upper bounds for the MLARs in all three cases. The modal logics important for our main result are the well known logics S4.1, S4.2 and S4.2.1 as well as the logic S4FPF denoting the modal logic of *finite partial function posets*, which will be introduced in Section 4.2. The axioms of S4.1, S4.2, and S4.2.1 are presented in Table 1. Our main results, also summarized in Table 2, are as follows:

1. For any $\mathcal{T}$, we have $\mathsf{S4.2} \subseteq \mathsf{MLAR}_{\mathcal{T}}^{fin}$. There is a $\mathcal{T}$ with $\mathsf{S4.2} = \mathsf{MLAR}_{\mathcal{T}}^{fin}$.
2. For any $\mathcal{T}$, we have $\mathsf{S4.2.1} \subseteq \mathsf{MLAR}_{\mathcal{T}}^{all}$. There is a $\mathcal{T}$ with $\mathsf{S4.2.1} = \mathsf{MLAR}_{\mathcal{T}}^{all}$.
3. We have $\mathsf{S4.1} \subseteq \mathsf{MLAR} \subseteq \mathsf{S4.2.1} \cap \mathsf{S4FPF}$ where S4FPF denotes the modal logic of *finite partial function posets* introduced in Section 4.2.

To prove the upper bounds, we make use of so-called *control statements*, introduced in [20]. Control statements in our context are CTL-formulas that follow certain truth value patterns. E.g., a *pure weak button*[1] is a CTL-formula that stays true in all refinements once it is true and a *switch* can always be made true and be made false by further refinement. While we rely on the results of [20] to prove the upper bound S4.2 using the existence of pure buttons and switches, we introduce new types of control statements, namely:

1. *restricted switches* to prove the upper bound S4.2.1 (Theorem 3),
2. *decisions* to prove the upper bound S4FPF (Theorem 6).

Restricted switches can be made true and false only as long as some pure button restricting the switches is not true yet. Decisions on the other hand are pairs of mutually exclusive pure weak buttons. So, by choosing refinements, one can *decide* which of the pure weak buttons in the decision to make true forcing the other pure weak button to be false also in all subsequent refinements. We show general results on how to prove these upper bounds using these new types of control statements. These results are applicable also for other classes of structures, relations between these structures, and languages describing properties of these structures and hence are of independent interest.

---

[1] Weakness refers to the fact that the button might stay false forever, while pureness expresses that it stays true as soon as it is true for the first time.

class of transition systems	logic	lower bound	upper bound
$\mathcal{F}_{\mathcal{T}}$: all finite abstractions of transition system $\mathcal{T}$	$\mathsf{MLAR}_{\mathcal{T}}^{fin}$	S4.2 (Prop. 1)	$\exists\mathcal{T}$ with $\mathsf{MLAR}_{\mathcal{T}}^{fin}$ = S4.2 (Thm. 2)
$\mathcal{A}_{\mathcal{T}}$: all abstractions of transition system $\mathcal{T}$	$\mathsf{MLAR}_{\mathcal{T}}^{all}$	S4.2.1 (Prop. 2)	$\exists\mathcal{T}$ with $\mathsf{MLAR}_{\mathcal{T}}^{all}$ = S4.2.1 (Thm. 4)
all transition systems	MLAR	S4.1 (Prop. 3)	$\subseteq$ S4.2.1 $\cap$ S4FPF (Thm. 5, 7)

Table 2: Overview of the bounds on the modal logics of abstraction refinement.

Related work. Techniques on how to refine abstractions in case the current abstraction is too coarse have been developed for decades (see [15] for an overview). In particular, seminal work on counter-example guided abstraction refinement (CEGAR) [10,11], where abstractions are refined along executions π violating a desirable property in order to potentially remove these counter-example executions from the abstraction, has inspired an ever-growing line of research (see, e.g., [9,12,30,27]). Typically these approaches deal with linear-time properties of systems expressible in the universally quantified fragment ACTL of CTL.

In work by Bozzelli et al. [5,6], the idea to add a modality expressing truth in some/all refinements to a logic describing properties of Kripke structures is also employed. In this work, multi-agent Kripke structures modeling the epistemic state of agents are used. The notion of refinement, however, differs from our notion. Bozzelli et al. use the "reverse" of simulation as the notion of refinement meaning that a system has to simulate the systems refining it. This means that there is no restriction forbidding to remove transitions in the refinement. Further, there is—in contrast to our work—no restriction stating that states have to have successors and so there always is a "greatest refinement" without any transitions. This also implies that their refinement relation is always directed, a.k.a, confluent. In contrast, our refinement relation is not directed on the class of all transition systems. Consequently, the axiom (.2) is not part of MLAR in our case as shown by the fact that (.2) is not in S4FPF. Only restricted to the class of abstractions of a fixed transition system, our refinement relation is directed.

Furthermore, Bozelli et al. investigate the full logic obtained from adding the refinement modalities to a multi-modal logic, while we use CTL-definability to restrict the admissible valuations and investigate the resulting uni-modal logic only using the refinement modality. This leads to the fact that the logic Bozelli et al. is not a normal modal logic (beginning of section 5 in [5]), while we show that on the classes of finite or arbitrary abstractions of transition systems our modal logic is the normal modal logic S4.2 or S4.2.1, respectively.

Further, there are several approaches aiming at preservation results for branching-time logics, which are orthogonal to our work. Circumstances under which the preservation of satisfaction of non-universal formulas can be guaranteed are investigated in [14]. Furthermore, three-valued logics evaluating formulas in an abstraction with truth values *true*, *false*, and *don't know*, indicating what can be concluded about the underlying system, have been studied (see, e.g., [19]). In the same spirit, *Kripke modal transition systems* (KMTS) as abstractions simultaneously maintain an over- and an under-approximation of the set of possible executions of a system by using *may*- and *must*-transitions (see, e.g., [29,28,25]).

Conceptually, abstraction refinement shares similarities with the *generalized model-checking problem* asking whether there is a concretization of a transition system, in which not all atomic propositions are specified, that satisfies a property [7,17,16,18]. Here, a concretization adds more information on the atomic propositions. There is, however, no direct relation to abstraction refinement.

The idea to use modal logic to describe the principles according to which the truth values of properties of structures can change under a model construction was introduced in the context of set-theoretic forcing, a construction to extend models of set theory, by Hamkins and Löwe [20]. Subsequently, a series of work investigated the modal logic of forcing and of further relations between set-theoretic models [21,34,23,31,26,22]. These ideas have been transferred to other mathematical areas such as graph theory [24] and group theory [2].

2 Preliminaries

In the sequel, we introduce our notation. For details on transition systems, abstractions, and temporal logics, see [1]. For details on modal logic, see [4].

Transition systems. A *transition system* is a tuple $\mathcal{T} = (S, \rightarrow, I, \mathsf{AP}, L)$ where S is a non-empty set of states, $\rightarrow\; \subseteq S \times S$ is a binary transition relation, $I \subseteq S$ is a set of initial states, AP is a set of atomic propositions, and $L\colon S \rightarrow 2^{\mathsf{AP}}$ is a labeling function. We require that there are no terminal states, i.e., for each $s \in S$, there is a $t \in S$ with $s \rightarrow t$. A *path* in a transition system is a sequence $s_0\, s_1\, s_2 \cdots \in S^\omega$ with $s_i \rightarrow s_{i+1}$ for all $i \in \mathbb{N}$. We denote the set of paths starting in a state s by $Paths_\mathcal{T}(s)$ or $Paths(s)$ if $\mathcal{T}$ is clear from context. The *trace* of a path $\pi = s_0\, s_1\, s_2 \ldots$ is the sequence $L(\pi) = L(s_0)\, L(s_1)\, L(s_2) \ldots$.

Computation tree logic. *Computation tree logic (CTL)* is a branching-time logic whose syntax contains state and path formulas. *CTL state formulas* over the set of atomic propositions AP are formed according to the grammar $\Phi ::= \top \mid a \mid \Phi \wedge \Phi \mid \neg\Phi \mid \exists\varphi \mid \forall\varphi$ where $a \in \mathsf{AP}$ is an atomic proposition and φ is a *CTL path formula* formed according to the grammar $\varphi ::= \mathsf{X}\Phi \mid \Phi\, \mathsf{U}\, \Phi$ where Φ represents CTL state formulas and X and U are the next-step and the until operator, respectively. In a transition systems $\mathcal{T} = (S, \rightarrow, I, \mathsf{AP}, L)$, the semantics of CTL state formulas over AP is given by the following recursive definition of the satisfaction relation for states $s \in S$:

$$
\begin{array}{llll}
s \vDash \top & \text{always,} & s \vDash \Phi \wedge \Psi & \text{iff } s \vDash \Phi \text{ and } s \vDash \Psi, \\
s \vDash a & \text{iff } a \in L(s), & s \vDash \exists\varphi & \text{iff } \pi \vDash \varphi \text{ for some } \pi \in Paths(s), \\
s \vDash \neg\Phi & \text{iff not } s \vDash \Phi, & s \vDash \forall\varphi & \text{iff } \pi \vDash \varphi \text{ for all } \pi \in Paths(s),
\end{array}
$$

with the satisfaction of path formulas φ on paths $\pi = s_0\, s_1\, s_2 \ldots$ given by

$$
\begin{array}{ll}
\pi \vDash \mathsf{X}\Phi & \text{iff } s_1 \vDash \Phi, \\
\pi \vDash \Phi\, \mathsf{U}\, \Psi & \text{iff there is a } j \geq 0 \text{ with } s_j \vDash \Psi \text{ and } s_i \vDash \Phi \text{ for all } i < j.
\end{array}
$$

We use the usual Boolean abbreviations as well as the temporal abbreviations *eventually* F and *always* G defined as $\mathsf{F}\Phi = \top\, \mathsf{U}\, \Phi$ as well as $\exists\mathsf{G}\Phi = \neg\forall\mathsf{F}\neg\Phi$ and $\forall\mathsf{G}\Phi = \neg\exists\mathsf{F}\neg\Phi$. Further, we write $\forall\mathsf{X}^i\Phi$ as an abbreviation for $\forall\mathsf{X}\ldots\forall\mathsf{X}\Phi$ where $\forall\mathsf{X}$ is repeated i times. We say that a transition system $\mathcal{T}$ satisfies a CTL-state formula Φ if all initial states of $\mathcal{T}$ satisfy Φ.

Abstractions and refinements. Let $\mathcal{T}_i = (S_i, \rightarrow_i, I_i, \mathsf{AP}, L_i)$, $i = 1, 2$ be transition systems. We call $\mathcal{T}_1$ an *abstraction* of $\mathcal{T}_2$ and $\mathcal{T}_2$ a *refinement* of $\mathcal{T}_1$ if there is a surjective so-called *abstraction function* $f : S_2 \rightarrow S_1$ s.t. the following conditions hold:

1. For all $s \in S_2$, $L_1(f(s)) = L_2(s)$.
2. $\rightarrow_1$ is the smallest relation satisfying that if $s \rightarrow_2 s'$ then $f(s) \rightarrow_1 f(s')$ for all $s, s' \in S_2$.
3. $I_1 = \{f(s) \mid s \in I_2\}$.

We write $\mathcal{T}_1 \rightsquigarrow \mathcal{T}_2$ to denote that $\mathcal{T}_2$ is a refinement of $\mathcal{T}_1$. In this case, the set of traces in $\mathcal{T}_1$ is a superset of the set of traces in $\mathcal{T}_2$. The function f induces an equivalence relation on S_2 relating states mapped to the same state in S_1.

If $\mathcal{T}_1$ and $\mathcal{T}_2$ are abstractions of some transition system $\mathcal{T}_0$ with abstraction functions f_{01} and f_{02}, we call $(\mathcal{T}_1, f_{01})$ an *equivalence class preserving refinement* of $(\mathcal{T}_2, f_{02})$ if there is an abstraction function f_{12} with $f_{02} = f_{12} \circ f_{01}$. We write $(\mathcal{T}_2, f_{02}) \rightsquigarrow^{\equiv} (\mathcal{T}_1, f_{01})$ in this case. Equivalence-class preserving means that the refinement and the corresponding abstraction function can be obtained by further splitting up the equivalence classes in $\mathcal{T}_2$ induced by f_{02}.

Modal logic. Formulas of *modal logic* over propositional variables in a countably infinite set Π are given by the grammar $\varphi ::= \top \mid p \mid \varphi \wedge \varphi \mid \neg\varphi \mid \Diamond\varphi$ where $p \in \Pi$ is a propositional variable. We use the usual Boolean abbreviations as well as the abbreviation $\Box\varphi = \neg\Diamond\neg\varphi$.

A *normal modal logic* is a set Λ of modal formulas that contains all propositional tautologies as well as the K-axiom $\Box(p \rightarrow q) \rightarrow (\Box p \rightarrow \Box q)$ and that is closed under modus ponens (if $\varphi \in \Lambda$ and $\varphi \rightarrow \psi \in \Lambda$, then $\psi \in \Lambda$), uniform substitution (if $\varphi \in \Lambda$, then also $\varphi[p/\chi] \in \Lambda$ where $\varphi[p/\chi]$ is obtained from φ by replacing every occurrence of $p \in \Pi$ in φ with the modal formula χ), and generalization (if $\varphi \in \Lambda$, then $\Box\varphi \in \Lambda$).

A *Kripke frame* is a pair $F = (W, R)$ where W is a non-empty set and $R \subseteq W \times W$ is a binary relation on W. We call W the *domain* or *universe* of the frame F. The elements of W are called *states* or *worlds*. The relation R is called the *accessibility relation*. If we have $(w, v) \in R$ for $w, v \in W$, we also write Rwv and say that w *can access* or *sees* v, and that v is a successor of w. We call a world r that can see every world of F a *root* of F. A *Kripke model* is a tuple $M = (W, R, V)$ where (W, R) is a frame and $V \colon \Pi \rightarrow 2^W$ is a valuation. We recursively define what it means for a modal formula φ to be *true* or *satisfied* at a world $w \in W$, which we denote by $M, w \Vdash \varphi$:

$M, w \Vdash \top$ always, $\qquad M, w \Vdash \psi \wedge \theta$ iff $M, w \Vdash \psi$ and $M, w \Vdash \theta$,

$M, w \Vdash p$ iff $w \in V(p)$, $\qquad M, w \Vdash \Diamond\psi$ iff there is v with Rwv and $M, v \Vdash \psi$,

$M, w \Vdash \neg\psi$ iff $M, w \nVdash \psi$.

A modal formula φ is *valid at a state w* of a frame $F = (W, R)$, written $F, w \Vdash \varphi$, if we have $(F, V), w \Vdash \varphi$ for all valuations $V \colon \mathsf{AP} \rightarrow 2^W$ on F. The formula φ is *valid* on F, written $F \Vdash \varphi$ if it is valid at all states $w \in W$. A set of formulas Γ is valid on a frame F, written $F \Vdash \Gamma$, if all $\varphi \in \Gamma$ are valid on F. A (set of) formula(s) φ is valid on a class of frames $\mathcal{C}$ if φ is valid on all frames $F \in \mathcal{C}$.

Finally, we call the set of formulas that is valid on a class of frames $\mathcal{C}$ the *logic of $\mathcal{C}$*, denoted by $\Lambda_{\mathcal{C}}$, which always is a normal modal logic.

General Kripke frames. A *general Kripke frame* is a tuple $G = (W, R, \mathcal{A})$ where (W, R) is a Kripke frame and $\mathcal{A} \subseteq 2^W$ is a collection of *admissible valuations* that is closed under finite unions and complements. A modal formula φ is valid on G at a world $w \in W$, written $G, w \Vdash \varphi$, if $M, w \Vdash \varphi$ for all Kripke models $M = (W, R, V)$ with $V(p) \in \mathcal{A}$ for all $p \in \Pi$. It is valid on G, written $G \Vdash \varphi$ if it is valid at all $w \in W$. We denote the set of formulas valid at w in G by $\Lambda_G(w)$ and the set of formulas that are valid on G by Λ_G. Note that $\Lambda_G \supseteq \Lambda_{(W,R)}$ where $\Lambda_{(W,R)}$ is the set of formulas valid on the frame (W, R) because validity on a general frame is weaker than on a Kripke frame.

3 Modal logic of abstraction refinement (MLAR)

The idea behind MLARs is to interpret $\Diamond$ and $\Box$ evaluated at a transition system $\mathcal{T}$ as "there is a refinement of $\mathcal{T}$, in which ..." and "in all refinements of $\mathcal{T}$, ...", respectively. For atomic propositions, we allow valuations that correspond to CTL-expressible properties. For a class of transition systems $\mathcal{C}$, we obtain a general frame in this way: The states are the elements of $\mathcal{C}$, the accessibility relation is the refinement relation $\rightsquigarrow$ and the admissible valuations are defined by CTL-formulas. We call the set of modal formulas valid on such a general frame a MLAR. In the sequel, we define three types of MLARs on different classes of transition systems and afterwards provide lower bounds on these logics.

3.1 Defining the modal logic

First, we will describe a general set-up for the modal logic of a relation $\rightsquigarrow$ on a class of structures $\mathcal{C}$ with respect to a *language $\mathcal{L}$* expressing properties of these structures. By language, we mean a formalism containing formulas ϕ that express properties $\mathcal{P} = \{c \in \mathcal{C} \mid c \vDash \phi\} \subseteq \mathcal{C}$. We require that the properties expressible in $\mathcal{L}$ are closed under complement and finite union, i.e., that $\mathcal{L}$ can express negations and disjunctions.

Definition 1. *Let $\mathcal{C}$ be a class of structures, $\rightsquigarrow$ be a reflexive transitive binary relation on $\mathcal{C}$, $\mathcal{L}$ be a language and $G_{\mathcal{C}, \rightsquigarrow, \mathcal{L}} = (\mathcal{C}, \rightsquigarrow, \mathrm{Admiss}(\mathcal{L}))$ be the general Kripke frame where*

$$\mathrm{Admiss}(\mathcal{L}) = \{\mathcal{P} \subseteq \mathcal{C} \mid \text{there is a form. } \phi \text{ in } \mathcal{L} \text{ s.t. for all } c \in \mathcal{C} : c \vDash \phi \text{ iff } c \in \mathcal{P}\}.$$

The modal logic $\Lambda_{\mathcal{C}, \rightsquigarrow, \mathcal{L}}$ of the relation $\rightsquigarrow$ w.r.t. language $\mathcal{L}$ on the class $\mathcal{C}$ is the modal logic of the general frame $G_{\mathcal{C}, \rightsquigarrow, \mathcal{L}}$, i.e., $\Lambda_{\mathcal{C}, \rightsquigarrow, \mathcal{L}} = \{\phi \mid G_{\mathcal{C}, \rightsquigarrow, \mathcal{L}} \Vdash \phi\}$. For any $c \in \mathcal{C}$, $\Lambda_{\mathcal{C}, \rightsquigarrow, \mathcal{L}}(c)$ is the set of all valid formulas on $G_{\mathcal{C}, \rightsquigarrow, \mathcal{L}}$ at c, i.e., $\Lambda_{\mathcal{C}, \rightsquigarrow, \mathcal{L}}(c) = \{\phi \mid G_{\mathcal{C}, \rightsquigarrow, \mathcal{L}}, c \Vdash \phi\}$.

Example 3. For an example instantiation of this definition consider the modal logic of Abelian groups studied in [2]: The class $\mathcal{C}$ is the class of all Abelian groups. The relation $A \hookrightarrow B$ expresses that A is a subgroup of B. The language $\mathcal{L}$ is first-order logic. A formula in the resulting modal logic is a formula that holds for any Abelian group when $\Diamond$ is interpreted as "there is a super-group in which" and atomic propositions are replaced by any first-order sentences in the language of groups. In [2], it is shown that the resulting logic is S4.2.

Now, we can formally define our *modal logics of abstraction refinement*:

1. For a transition system $\mathcal{T}$, the logic $\mathsf{MLAR}_{\mathcal{T}}^{fin}$ is the modal logic of the refinement relation $\rightsquigarrow$ w.r.t. the language CTL on the class $\mathcal{F}_{\mathcal{T}}$ of all finite abstractions of $\mathcal{T}$, i.e., $\mathsf{MLAR}_{\mathcal{T}}^{fin} = \Lambda_{\mathcal{F}_{\mathcal{T}}, \rightsquigarrow, \mathrm{CTL}}$.
2. For a transition system $\mathcal{T}$, $\mathsf{MLAR}_{\mathcal{T}}^{all}$ is the modal logic of $\rightsquigarrow$ w.r.t. CTL on the class $\mathcal{A}_{\mathcal{T}}$ of all abstractions of $\mathcal{T}$, i.e., $\mathsf{MLAR}_{\mathcal{T}}^{all} = \Lambda_{\mathcal{A}_{\mathcal{T}}, \rightsquigarrow, \mathrm{CTL}}$
3. The logic MLAR is the modal logic of $\rightsquigarrow$ w.r.t. CTL on the class $\mathfrak{A}$ of all transition systems, i.e., $\mathsf{MLAR} = \Lambda_{\mathfrak{A}, \rightsquigarrow, \mathrm{CTL}}$

Note that the choice of the language determines the admissible valuations and hence influences the MLARs. A more expressive logic leads to a smaller MLAR. Recall that we choose CTL as arguably the simplest prominent branching-time logic. For an illustration of this definition, we refer back to Example 1.

Remark 1. In the first two cases in the list above, it is also reasonable to consider equivalence-class preserving refinements instead. To do that for $\mathsf{MLAR}_{\mathcal{T}}^{fin}$ for example, we would consider the class $\mathcal{F}'_{\mathcal{T}}$ of pairs $(\mathcal{S}, f)$ of finite abstractions of $\mathcal{T}$ together with a corresponding abstraction function f. Further, we would consider the relation $\rightsquigarrow^{\equiv}$ on this class. In fact, all arguments in the sequel also work for this view. While we will not spell this out in detail, we will comment on this at the critical places in the proofs. Consequently, the bounds we provide also apply to $\Lambda_{\mathcal{F}'_{\mathcal{T}}, \rightsquigarrow^{\equiv}, \mathrm{CTL}}$ and for the analogously defined $\Lambda_{\mathcal{A}'_{\mathcal{T}}, \rightsquigarrow^{\equiv}, \mathrm{CTL}}$.

3.2 Lower bounds

The frame $(\mathcal{C}, \rightsquigarrow)$ is always reflexive and transitive as transition systems are refinements of themselves and abstraction functions can be composed to show transitivity. The modal logic S4 is valid on all reflexive and transitive frames (see [4]) and hence also on all general reflexive and transitive frames. For any class of transition systems $\mathcal{C}$, we hence have $\mathsf{S4} \subseteq \Lambda_{\mathcal{C}, \rightsquigarrow, \mathrm{CTL}}$. For other axioms, the validity depends on the class of transition systems we consider. More precisely, the structure of the relation $\rightsquigarrow$ on $\mathcal{C}$ allows us to conclude stronger lower bounds.

For our first lower bound of S4.2 for $\mathsf{MLAR}_{\mathcal{T}}^{fin}$ for any $\mathcal{T}$, we use that S4.2 is valid on reflexive, transitive, directed frames (see [4]). A frame (W, R) is directed if for all $w, v, u \in W$ with Rwv and Rwu, there is a $z \in W$ with Rvz and Ruz. Showing directedness of $\rightsquigarrow$ on $\mathcal{F}_{\mathcal{T}}$ is straightforward (full proofs can be found in the extended version [32]).

Proposition 1. *For any transition system $\mathcal{T}$, we have* $\mathsf{S4.2} \subseteq \mathsf{MLAR}_{\mathcal{T}}^{fin}$.

For $\mathsf{MLAR}_{\mathcal{T}}^{all}$, the lower bound S4.2.1 follows as $\mathcal{A}_{\mathcal{T}}$ contains the element $\mathcal{T}$ that is a refinement of all transition systems in $\mathcal{A}_{\mathcal{T}}$. So, directedness and the validity of (.2) is immediate. The validity of $(.1) = \Box\Diamond p \to \Diamond\Box p$ on directed frames is equivalent to the existence of a greatest element, which is $\mathcal{T}$ in this case.

Proposition 2. *For any transition system $\mathcal{T}$, we have* $\mathsf{S4.2.1} \subseteq \mathsf{MLAR}_{\mathcal{T}}^{all}$.

The axiom (.1) is in MLAR as we can always go to a refinement for which further refinements do not affect the truth of CTL-formulas anymore by making everything except for one path starting from each initial state unreachable. If p corresponds to a CTL-property Φ s.t. $\Box\Diamond p$ holds at some transition system $\mathcal{T}$, then it has to hold at these "maximal" refinements and hence $\Diamond\Box p$ is true, too.

Proposition 3. *We have* $\mathsf{S4.1} \subseteq \mathsf{MLAR}$.

4   Upper bounds

In this section, we prove upper bounds for the MLARs. An overview of the results can be found in Table 2. The key technical vehicle for the proofs are so-called *control statements*. Given a general frame $G_{\mathcal{C},\hookrightarrow,\mathcal{L}} = (\mathcal{C}, \hookrightarrow, \mathrm{Admiss}(\mathcal{L}))$ as in Def. 1, control statements are formulas in the language $\mathcal{L}$ that exhibit certain patterns of truth values under the relation $\hookrightarrow$. We summarize the types of control statements introduced in [20,21] in Section 4.1. These control statements are sufficient to prove an upper bound of S4.2 in Section 4.2. For the upper bounds S4.2.1 and S4FPF, we develop new control statements in Section 4.2.

4.1   Control statements

The definitions and results given in this subsection are adapted from [20] and [21]. We define three different types of control statements, namely *pure buttons*, a more general version of them called *pure weak buttons*, and *switches*. Intuitively, a pure button can always be made true by moving to another structure via the relation $\hookrightarrow$ and stays true once it is true. In comparison, a pure weak button might eventually have to stay false forever, but also has to stay true once it is true. Lastly, a switch can always be made true and be made false via $\hookrightarrow$.

Pure buttons and switches. Formally, let G be the general Kripke frame wrt. some class of structures $\mathcal{C}$, relation $\hookrightarrow$, and language $\mathcal{L}$ as defined in Def. 1 and let $c \in \mathcal{C}$. A sentence $\beta \in \mathcal{L}$ is a *pure button* in G at c if, for any valuation V with $V(b) = \{d \in \mathcal{C} \mid d \vDash \beta\}$, we have $(G, V), c \Vdash \Box(b \to \Box b)$ and $(G, V), c \Vdash \Box\Diamond b$. For an illustration, consider Figure 3a and view the root as the world c: From every world reachable from c, a world at which the pure button is true is reachable and worlds where the pure button is true can only reach such worlds. A pure button β is *pushed* at $e \in \mathcal{C}$ with $c \hookrightarrow e$ if $(G, V), e \Vdash b$ and *unpushed* otherwise.

A sentence $\lambda \in \mathcal{L}$ is a *pure weak button* in G at c if, for any valuation V with $V(b) = \{d \in \mathcal{C} \mid d \vDash \lambda\}$, we have $(G, V), c \Vdash \Box(b \to \Box b)$ and $(G, V), c \Vdash \Diamond b$. So, it is not necessary that the pure weak button can always be made true. The

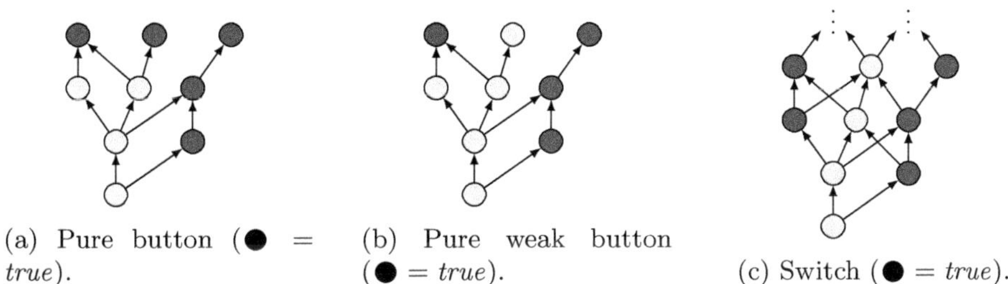

(a) Pure button ($\bullet$ = *true*).

(b) Pure weak button ($\bullet$ = *true*).

(c) Switch ($\bullet$ = *true*).

Fig. 3: Illustration of the truth value patterns for different control statements. In all cases, the relation between worlds is meant to be reflexive and transitive.

truth value pattern is illustrated in Figure 3b. A pure weak button λ is *pushed* at $e \in \mathcal{C}$ with $c \hookrightarrow e$ if $(G, V), e \Vdash b$ and *unpushed* otherwise. A pure weak button λ is *intact* at $e \in \mathcal{C}$ with $c \hookrightarrow e$ if $(G, V), e \Vdash \Diamond b$ and *broken* otherwise.

A sentence $\sigma \in \mathcal{L}$ is a *switch* in G at c if, for any valuation V with $V(s) = \{d \in \mathcal{C} \mid d \vDash \sigma\}$, we have $(G, V), c \Vdash \Box(\Diamond s \wedge \Diamond \neg s)$. A switch σ is *on* at $e \in \mathcal{C}$ with $c \hookrightarrow e$ if $(G, V), e \Vdash s$ and *off* otherwise. Figure 3c shows the truth value pattern: Every world can access worlds at which the switch is on and off, respectively.

Independence. Different control statements might interfere with one another. For instance, to push a pure button it might be necessary to also push another pure button; however, to prove upper bounds, it is often required to be able to operate control statements *independently*. Formally, we define this as follows:

Let $n, m \in \mathbb{N}$. Let $\Phi = \{\beta_i \mid 0 \le i \le n - 1\}$ be a set of unpushed pure buttons in G at c and let $\Psi = \{\sigma_j \mid 0 \le j \le m - 1\}$ be a set of switches in G at c. We call $\Phi \cup \Psi$ *independent* if for every $I_0 \subseteq I_1 \subseteq \{0, 1, \ldots, n - 1\}$ and $J_0, J_1 \subseteq \{0, 1, \ldots, m - 1\}$, and any V with $V(b_i) = \{d \in \mathcal{C} \mid d \vDash \beta_i\}$ and $V(s_j) = \{d \in \mathcal{C} \mid d \vDash \sigma_j\}$,

$$(G, V), c \Vdash \Box\big[\big(\textstyle\bigwedge_{i \in I_0} b_i \wedge \bigwedge_{i \notin I_0} \neg b_i \wedge \bigwedge_{j \in J_0} s_j \wedge \bigwedge_{j \notin J_0} \neg s_j\big)$$
$$\to \Diamond\big(\textstyle\bigwedge_{i \in I_1} b_i \wedge \bigwedge_{i \notin I_1} \neg b_i \wedge \bigwedge_{j \in J_1} s_j \wedge \bigwedge_{j \notin J_1} \neg s_j\big)\big].$$

This formula says that no matter which set of pure buttons I_0 and switches J_0 is currently true, there is an accessible world at which an arbitrary other combination of pure buttons $I_1 \supseteq I_0$ and switches J_1 is true. So, it is possible to change the truth values of any control statement without influencing the truth values of the others (besides the fact that pushed pure buttons have to remain true). Independent families of control statements are a powerful tool to prove upper bounds on the modal logics of model constructions as shown in [20]:

Theorem 1 ([20, Section 2]). *Let G be the general Kripke frame w.r.t. some $\mathcal{C}, \hookrightarrow$ and $\mathcal{L}$ as in Def. 1 and let $c \in \mathcal{C}$. If for any $n, m \in \mathbb{N}$, there is a set of unpushed pure buttons $\Phi = \{\beta_i \mid 0 \le i \le n - 1\}$ in G at c and a set of switches $\Psi = \{\sigma_j \mid 0 \le j \le m - 1\}$ in G at c s.t. $\Phi \cup \Psi$ is independent, then $\Lambda_G(c) \subseteq$ S4.2.*

To prove such theorems the notion of *F-labeling* is developed in [20,21]. Let $F = (W, \preccurlyeq)$ be a finite Kripke frame with a root $w_0 \in W$. An F-labeling for c in G assigns a sentence ϕ_w in the language $\mathcal{L}$ to each node $w \in W$ such that

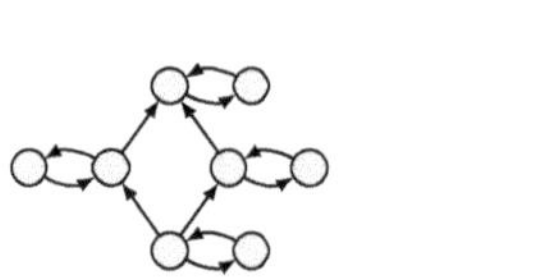

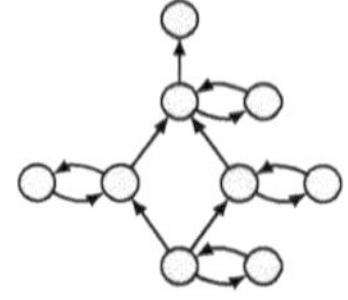

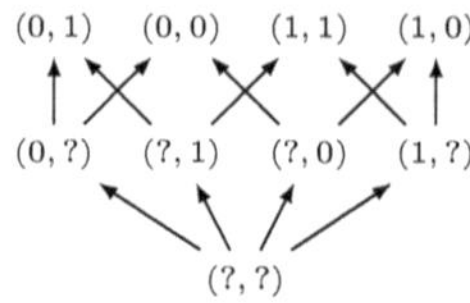

(a) Pre-Boolean algebra.   (b) Inverted lollipop.   (c) Partial function poset.

Fig. 4: Illustration of different types of transitive and reflexive frames.

logic	sound and complete with respect to	control statements for upper bounds
S4.2	finite pre-Boolean algebras [20]	pure buttons, switches [20]
S4.2.1	inverted lollipops [26]	pure buttons, restr. switches (Thm. 3)
S4FPF	finite partial function posets (by def.)	decisions (Theorem 6)

Table 3: Soundness and completeness results and control statements.

1. For all $d \in \mathcal{C}$ with $c \hookrightarrow d$, there is exactly one $w \in W$ such that $d \vDash \phi_w$
2. For all $d \in \mathcal{C}$ with $c \hookrightarrow d$, if $d \vDash \phi_w$, then there is an $e \in \mathcal{C}$ with $d \hookrightarrow e$ and $e \vDash \phi_v$ if and only if $w \preccurlyeq v$.
3. $c \vDash \phi_{w_0}$.

In other words, an F-labeling assigns sentences ϕ_w to the worlds w of F such that grouping together the structures in $\mathcal{C}$ that make the respective formulas ϕ_w true results exactly in the structure of the frame F.

Lemma 1 ([21, Lem. 9]). *Let $F = (W, \preccurlyeq)$ be a finite Kripke frame with a root $w_0 \in W$ and $w \mapsto \phi_w$ be an F-labeling for c in G. For any formula ϕ, if $F, w_0 \nVdash \phi$ then $\phi \notin \Lambda_G(c)$.*

Proof. Let V be a valuation on F such that $(F, V), w_0 \nVdash \phi$. Consider the valuation V' on G with $V'(p) = \{d \in \mathcal{C} : d \vDash \bigvee_{w \in V(p)} \phi_w\}$. Then, $(G, V'), c \nVdash \phi$. □

Theorem 1 can be shown using F-labelings by showing that sufficiently many independent pure buttons and switches over a structure c allow to provide an F-labeling for any finite pre-Boolean algebra F. A finite Boolean algebra is a partial order isomorphic to a powerset 2^X ordered by inclusion $\subseteq$ for some finite set X. A finite pre-Boolean algebra, in turn, is a partial pre-order, which is obtained by replacing the nodes of a finite Boolean algebra by clusters of states that can see each other as illustrated in Figure 4a. Together with the soundness and completeness result stating that the set of modal formulas valid on all finite pre-Boolean algebras is precisely S4.2, this shows Theorem 1 (see [20,21]).

4.2   Upper bounds

To prove the upper bounds for the modal logics of abstraction refinements, we will start by utilizing Thm. 1 to prove $\mathsf{MLAR}^{fin}_{\mathcal{T}} \subseteq \mathsf{S4.2}$ for some $\mathcal{T}$. Afterwards, we prove new results to obtain a technique to prove upper bounds of S4.2.1 and S4FPF using new control statements. The soundness and completeness results these proofs rely on and the respective control statements are shown in Table 3.

Upper bound S4.2. Using Theorem 1, we prove:

Theorem 2. *There is a transition system $\mathcal{T}$ with* $\mathsf{MLAR}_{\mathcal{T}}^{fin} \subseteq \mathsf{S4.2}$.

Proof (Proof sketch). We provide a transition system $\mathcal{T}$ and an infinite independent set of pure buttons and switches in $G_{\mathcal{F}_{\mathcal{T}},\rightsquigarrow,\mathrm{CTL}}$ at any $\mathcal{S}$. By Thm. 1, this implies $\mathsf{MLAR}_{\mathcal{T}}^{fin}(\mathcal{S}) \subseteq \mathsf{S4.2}$ for all $\mathcal{S} \in \mathcal{F}_{\mathcal{T}}$. A sketch of $\mathcal{T}$ is depicted in Fig. 5 where we, notably, have different parts $\mathcal{T}^{\beta}$, $\mathcal{T}^{\sigma}$ for the different types of control statements. So, an abstraction of $\mathcal{T}$ can be viewed as independently abstracting $\mathcal{T}^{\beta}$ and $\mathcal{T}^{\sigma}$, which guarantees that we do not have a dependence between a pure button and a switch. We will give full details on $\mathcal{T}^{\beta}$ and the pure buttons, but only sketch the more involved switches (for the full proof, see [32]).

Pure buttons. We define $\mathcal{T}^{\beta} = (S, \rightarrow, \{s\}, \{s, f, a\}, L)$ sketched in Fig. 5a:

- The state space is $S = \{s, f\} \cup \{(i, h, k) \in \mathbb{N}^3 \mid i \geq 2, i \geq h \geq 1\}$.
- The relation $\rightarrow$ is given by
    - $s \rightarrow (i, 1, k)$ for all $i \geq 2$ and all k,
    - $(i, h, k) \rightarrow (i, h + 1, k)$ for all $i \geq 2$, all $h < i$, and all k,
    - $(i, i, k) \rightarrow f$ for all $i \geq 2$ and all k, and $f \rightarrow f$.
- The labeling over atomic propositions $\{s, f, a\}$ is given by $L(s) = \{s\}$, $L(f) = \{f\}$, and $L((i, h, k)) = \{a\}$ for all i, h, k.

Given a finite abstraction $\mathcal{S}$ of $\mathcal{T}^{\beta}$ with abstraction function f, we denote the set of states mapped to the same state as (i, h, k) by $[(i, h, k)]$. Overloading the notation, we identify the state $f((i, h, k))$ with this equivalence class $[(i, h, k)]$.

The idea behind the ith pure button is roughly to say that some path along i states labelled with a from s to f has been "isolated" and not merged with paths of other length. For any $i \geq 2$, we define the CTL-sentence

$$\beta_i = \exists\mathsf{X}(\forall\mathsf{X}^{i-1}(a \wedge \forall\mathsf{X}f)).$$

In a finite abstraction $\mathcal{S}$ of $\mathcal{T}^{\beta}$, β_i holds iff there exists $k \geq 1$ such that $a^i f^{\omega}$ is the only trace of paths starting in $[(i, 1, k)]$. Once the only trace of paths starting in $[(i, 1, k)]$ is $a^i f^{\omega}$, this cannot be changed in a finer abstraction of $\mathcal{T}^{\beta}$.

Further, whenever β_i is unpushed at an abstract transition system $\mathcal{S}_0$ then there is an accessible abstraction $\mathcal{S}_1$ of $\mathcal{T}^{\beta}$ where β_i is pushed. In a finite abstraction there is always a path $[(i, 1, k)], \ldots, [(i, i, k)]$ for some $k \geq 1$ that only visits infinite equivalence classes. Otherwise we would have infinitely many finite equivalence classes which leads to a contradiction. Thus, it is possible to safely (without pushing other pure buttons) split off the path $(i, 1, k), \ldots, (i, i, k)$ by going to a refinement $\mathcal{S}_1$ that is even equivalence class preserving. Note that at any finite transition system $\mathcal{S}$ only finitely many β_i can be pushed. So, for any $n \in \mathbb{N}$, there is a $I \subset \mathbb{N}$ with $|I| = n$ such that for any $i \in I$, β_i is unpushed.

Switches. In $\mathcal{T}^{\sigma}$ we add an additional "dimension": we not only have infinitely many copies of paths of equal length, but put them into infinitely many "groups" of infinite size. Precisely such a group is depicted in Figure 5b for length 3. Paths within a group are connected via special states allowing them to "see" all other paths in that group but themselves. Then, a switch σ_j is on whenever there is a group containing paths of length j where exactly one path is isolated. Again,

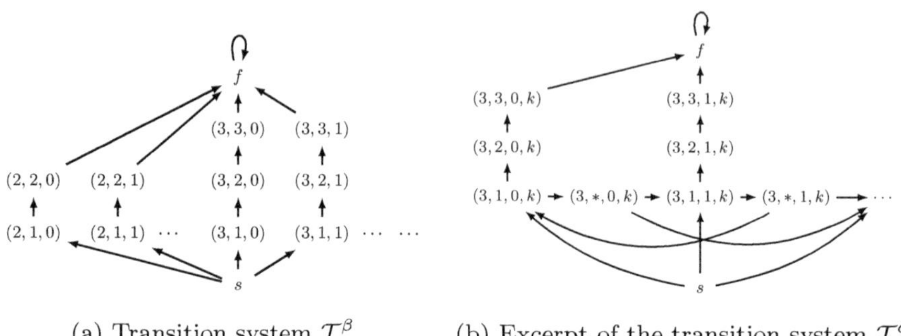

(a) Transition system $\mathcal{T}^\beta$.   (b) Excerpt of the transition system $\mathcal{T}^\sigma$.

Fig. 5: The components of the transition system $\mathcal{T}$ in the proof of Theorem 2 sharing the initial state s.

because of the infinitely many copies, we can ensure that a switch can always be turned on and that we also have independence.   □

Example 4. By providing an upper bound for $\mathsf{MLAR}_{\mathcal{T}}^{fin}$, we show that all formulas φ not in S4.2 are falsifiable in the sense that there is a transition system $\mathcal{T}$, a finite abstraction $\mathcal{S}$, and CTL formulas P for each atomic proposition p of φ such that φ does not hold at $\mathcal{S}$ in $\mathcal{F}_{\mathcal{T}}$ when making each p true at all transition systems in $\mathcal{F}_{\mathcal{T}}$ satisfying the corresponding P. The proof provides the necessary transition system and abstractions: Consider, e.g., the formula $(.1) = \Box\Diamond p \to \Diamond\Box p \notin$ S4.2. It can be falsified on a pre-Boolean algebra F with just one cluster of two states. A single switch is then sufficient to obtain an F-labelling. Now, letting $\mathcal{S}$ be any finite abstraction of the transition system $\mathcal{T}$ constructed in the proof of Theorem 2 and making p true at all transition systems that satisfy one of the switches σ, we have that $\Box\Diamond p$ holds at $\mathcal{S}$, but $\Diamond\Box p$ does not hold by the definition of a switch.

Upper bound S4.2.1. The modal logic S4.2.1 is the set of formulas valid on all *inverted lollipops* as shown in [26]. An inverted lollipop is a pre-Boolean algebra equipped with an additional single top element that is larger than any element in the pre-Boolean algebra as illustrated in Figure 4b. We introduce new control statements, called **B***-restricted switches* for a pure button **B**, to make use of this characterization result to prove the upper bound S4.2.1 for $\mathsf{MLAR}_{\mathcal{T}}^{all}$ for some $\mathcal{T}$ and for MLAR. Intuitively, **B**-restricted switches differ from regular switches by allowing them to break similarly to how pure weak buttons break by pushing **B**.

Formally, let G be the general Kripke frame with respect to some $\mathcal{C}$, $\hookrightarrow$ and $\mathcal{L}$ as defined in Definition 1 and let $c \in \mathcal{C}$. Let **B** be an unpushed pure button in G at c. A sentence $\sigma^{\mathbf{B}} \in \mathcal{L}$ is a **B***-restricted switch* in G at c if, for any valuation V with $V(s) = \{d \in \mathcal{C} \mid d \vDash \sigma^{\mathbf{B}}\}$ and $V(B) = \{d \in \mathcal{C} \mid d \vDash \mathbf{B}\}$, we have $(G, V), c \Vdash \Box(\neg B \to (\Diamond(s \land \neg B) \land \Diamond(\neg s \land \neg B)))$. A **B**-restricted switch $\sigma^{\mathbf{B}}$ is *on* at $e \in \mathcal{C}$ with $c \hookrightarrow e$ if $(G, V), e \Vdash s$ and *off* otherwise. Intuitively, it must be possible to turn a **B**-restricted switch on and off as long as **B** is not true yet. As soon as **B** is true, the truth value of the restricted switch does not

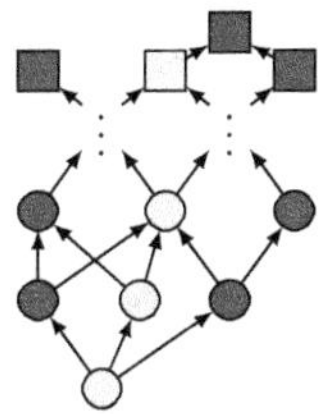 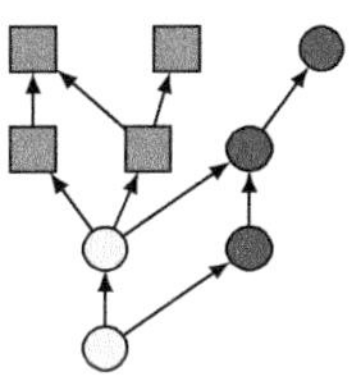

(a) Restricted switch: true at worlds depicted as ●,■ and restricted by a pure button true at worlds depicted as ■, ▨.

(b) Decision consisting of two pure weak buttons true at worlds depicted as ● and ▨, respectively.

Fig. 6: Illustration of the truth value patterns for the new control statements. In both cases, the relation between worlds is the reflexive transitive closure of the relation indicated by the arrows.

matter. This pattern is illustrated in Figure 6a where the restricted switch is true at worlds filled in black and squares represent worlds where the restricting pure button is true. Further, a $\mathbf{B}$-restricted switch $\sigma^{\mathbf{B}}$ is *intact* at $e \in \mathcal{C}$ with $c \hookrightarrow e$ if $(G, V), e \Vdash \neg B$ and *broken* otherwise.

We again need a notion of independence between different $\mathbf{B}$-restricted switches; also in combination with pure buttons. As $\mathbf{B}$-restricted switches are only meaningful as long as they are intact we define independence only until $\mathbf{B}$ is pushed.

Let $n, m \in \mathbb{N}$. Let $\Phi = \{\beta_i \mid 0 \leq i \leq n - 1\} \cup \{\mathbf{B}\}$ be a set of unpushed pure buttons in G at c and let $\Psi = \{\sigma^{\mathbf{B}}{}_j \mid 0 \leq j \leq m - 1\}$ be a set of $\mathbf{B}$-restricted switches in G at c. We call $\Phi \cup \Psi$ *independent until* $\mathbf{B}$ if for any $I_0 \subseteq I_1 \subseteq \{0, 1, \ldots, n-1\}$ and $J_0, J_1 \subseteq \{0, 1, \ldots, m-1\}$, and any V with $V(b_i) = \{d \in \mathcal{C} \mid d \vDash \beta_i\}$, $V(s_j) = \{d \in \mathcal{C} \mid d \vDash \sigma^{\mathbf{B}}{}_j\}$ and $V(B) = \{d \in \mathcal{C} \mid d \vDash \mathbf{B}\}$,

$$(G, V), c \Vdash \square\left[\left(\bigwedge_{i \in I_0} b_i \wedge \bigwedge_{i \notin I_0} \neg b_i \wedge \bigwedge_{j \in J_0} s_j \wedge \bigwedge_{j \notin J_0} \neg s_j \wedge \neg B\right)\right.$$
$$\left. \rightarrow \Diamond\left(\bigwedge_{i \in I_1} b_i \wedge \bigwedge_{i \notin I_1} \neg b_i \wedge \bigwedge_{j \in J_1} s_j \wedge \bigwedge_{j \notin J_1} \neg s_j \wedge \neg B\right)\right].$$

So, we require that it is possible to change truth values of switches and push the pure buttons independently (without pushing $\mathbf{B}$) only as long as $\mathbf{B}$ is not true. Once $\mathbf{B}$ is true, the truth values of the control statements are irrelevant.

Theorem 3. *Let G be the general frame w.r.t. some $\mathcal{C}$, $\hookrightarrow$ and $\mathcal{L}$ as in Def. 1 and let $c \in \mathcal{C}$. If for any $n, m \in \mathbb{N}$, there are sets of unpushed pure buttons $\Phi = \{\beta_i \mid 0 \leq i \leq n-1\} \cup \{\mathbf{B}\}$ in G at c and of $\mathbf{B}$-restricted switches $\Psi = \{\sigma^{\mathbf{B}}{}_j \mid 0 \leq j \leq m-1\}$ in G at c s.t. $\Phi \cup \Psi$ is independent until $\mathbf{B}$, then $\Lambda_G(c) \subseteq$ S4.2.1.*

Proof. Without loss of generality, we can assume all $\mathbf{B}$-restricted switches are off at c. Since S4.2.1 is characterized by the class of finite inverted lollipops [26], if a formula $\phi \notin$ S4.2.1, then there is an inverted lollipop $L = (W, \preccurlyeq)$, a valuation V on L and a initial state $w_0 \in W$ such that $(L, V), w_0 \nVdash \phi$. L has an underlying pre-Boolean algebra containing n atom clusters for some n. Furthermore, by copying states, we can assume that each cluster consists of 2^m states for some m. Thus, we can write $W = \mathcal{P}(n) \times \mathcal{P}(m) \cup \{\star\}$ where $\star$ is greater than any other element, i.e., for all $(I, J) \in \mathcal{P}(n) \times \mathcal{P}(m)$, $(I, J) \preccurlyeq \star$, and for all $(I, J), (I', J') \in \mathcal{P}(n) \times \mathcal{P}(m)$, $(I, J) \preccurlyeq (I', J')$ if and only if $I \subseteq I'$. We define formulas $\phi_w \in \mathcal{L}$ for every $I \subseteq \{0, 1, \ldots, n-1\}$ and $J \subseteq \{0, 1, \ldots, m-1\}$

$$\phi_{(I,J)} = \left(\bigwedge_{i \in I} \beta_i\right) \wedge \left(\bigwedge_{j \in J} \sigma_j\right) \wedge \neg \mathbf{B} \quad \text{and} \quad \phi_\star = \mathbf{B}.$$

We check that $w \mapsto \phi_w$ satisfies the conditions of an L-labeling for c in G:

1. Clearly, every $d \in \mathcal{C}$ satisfies exactly one of these formulas.
2. Since the pure buttons and $\mathbf{B}$-restricted switches are independent at c, for every $I, I' \subseteq \{0, 1, \ldots, n-1\}$, $J, J' \subseteq \{0, 1, \ldots, m-1\}$ and $d \in \mathcal{C}$, if $d \vDash \phi_{(I,J)}$ then there is an $e \in \mathcal{C}$ with $d \hookrightarrow e$ and $e \vDash \phi_{(I',J')}$ iff $I \subseteq I'$. Further, by pushing $\mathbf{B}$ there is always an extension where $\phi_\star$ holds and as $\mathbf{B}$ is a pure button every further extension also satisfies $\phi_\star$.
3. As all pure buttons are unpushed and the $\mathbf{B}$-restricted switches are all off at c., $c \vDash \phi_{(\emptyset, \emptyset)}$

By Lemma 1, since $(L, V), w_0 \nVDash \phi$, we have $\phi \notin \Lambda_G(c)$. So, for any formula $\phi \notin$ S4.2.1, we have $\phi \notin \Lambda_G(c)$ and thus, $\Lambda_G(c) \subseteq$ S4.2.1. $\square$

Corollary 1. *Let G be the general frame w.r.t. some $\mathcal{C}, \hookrightarrow$ and $\mathcal{L}$ as in Def. 1. If for any $n, m \in \mathbb{N}$, there are a state $c \in \mathcal{C}$ and sets of unpushed pure buttons $\Phi = \{\beta_i \mid 0 \leq i \leq n-1\} \cup \{\mathbf{B}\}$ and of $\mathbf{B}$-restricted switches $\Psi = \{\sigma^{\mathbf{B}}_j \mid 0 \leq j \leq m-1\}$ in G at c s.t. $\Phi \cup \Psi$ is independent until $\mathbf{B}$, then $\Lambda_G \subseteq$ S4.2.1.*

Proof. By the same argument, for any $n, m \in \mathbb{N}$, there exists a state $c \in \mathcal{C}$ such that $w \mapsto \phi_w$ is a L-labeling for c in G. By Lemma 1, since $(L, V), w_0 \nVDash \phi$, we have $\phi \notin \Lambda_G(c)$ and thus, $\phi \notin \Lambda_G$. So, for any formula $\phi \notin$ S4.2.1, we have $\phi \notin \Lambda_G$ and thus, $\Lambda_G \subseteq$ S4.2.1. $\square$

Using Thm. 3, we now prove the upper bound S4.2.1 for $\mathsf{MLAR}^{all}_{\mathcal{T}}$ for some $\mathcal{T}$.

Theorem 4. *There is a transition system $\mathcal{T}$ with $\mathsf{MLAR}^{all}_{\mathcal{T}} \subseteq$ S4.2.1.*

Proof (Proof sketch). We provide transition systems $\mathcal{T}$ and $\mathcal{S} \in \mathcal{A}_{\mathcal{T}}$ with arbitrarily many independent unpushed pure buttons and $\mathbf{B}$-restricted switches in $G_{\mathcal{A}_{\mathcal{T}}, \rightsquigarrow, \mathrm{CTL}}$ at $\mathcal{S}$. For each pure button, we have a pair of states in $\mathcal{T}$ that is collapsed in $\mathcal{S}$. By splitting it apart, the respective pure button gets pushed. Similarly, we use infinitely many collapsed triples of states for each $\mathbf{B}$-restricted switch. If one of the triples is refined to two states, the switch is on; otherwise, it is off. So, a switch can be turned on by splitting a fully collapsed triple into two parts and off by refining all triples split into two to three states. As long as infinitely many triples are still collapsed to a single state, the switch can hence be turned on and off at will. The restricting pure button $\mathbf{B}$ using an infinite chain of states from which the triples are reachable now expresses that all but finitely many triples are at least partially refined. As long as $\mathbf{B}$ is not true, the switch is intact. The full proof can be found in the extended version [32]. $\square$

To also prove $\mathsf{MLAR} \subseteq \mathsf{S4.2.1}$, we are not able to provide a single transition system $\mathcal{S}$ with $\mathsf{MLAR}(\mathcal{S}) \subseteq \mathsf{S4.2.1}$. Instead we rely on Cor. 1 and provide a family $\{\mathcal{S}_{n,m}\}_{n,m\in\mathbb{N}}$ of transition systems with $n+1$ unpushed pure buttons (including some $\mathbf{B}$) and m $\mathbf{B}$-restricted switches. The proof is provided in the extended version [32].

Theorem 5. $\mathsf{MLAR} \subseteq \mathsf{S4.2.1}$.

Upper bound S4FPF. To further improve the upper bound for MLAR, we define the logic $\mathsf{S4FPF}$ semantically as the logic of finite partial function posets.

Definition 2. *For any $n \in \mathbb{N}$, let F be the set of partial functions $f \colon \{0, 1, \ldots, n-1\} \to \{0, 1\}$ and $\preccurlyeq$ the partial order on F with $g \preccurlyeq h$ iff $g = h\!\restriction_{\mathrm{dom}(g)}$ where $\mathrm{dom}(g)$ is the domain of g for all f, g. We call $(F, \preccurlyeq)$ a finite partial function (FPF) poset on n elements. The modal theory $\mathsf{S4FPF}$ is the set of all formulas valid on the class of all FPF posets.*

A FPF poset with $n = 2$ is illustrated in Figure 4c. There, a partial function $f \colon \{0, 1\} \to \{0, 1\}$ is depicted as pair $(f(0), f(1))$ with ? indicating that the value is not defined. Before we proceed, we relate the logic $\mathsf{S4FPF}$ to well-known logics. Grzegorczyk's logic $\mathsf{Grz} = \mathsf{K} + \square(\square(p \to \square p) \to p) \to p$ is valid on all reflexive, transitive, weakly conversely well-founded frames (also called Noetherian partial orders), i.e., frames in which all infinite paths take only one self-loop from some point on (see, e.g., [8, Section 3.8]). As FPF posets have this property, $\mathsf{S4FPF}$ is an extension of $\mathsf{Grz} \supseteq \mathsf{S4.1}$. As $\mathsf{Grz} \not\subseteq \mathsf{S4.2.1}$, we conclude $\mathsf{S4FPF} \not\subseteq \mathsf{S4.2.1}$ (see, e.g., [3] for the relation of Grz to other extensions of $\mathsf{S4}$). Further, as FPF posets are not directed, the axiom $(.2) = \Diamond\square p \to \square\Diamond p$ is not included in $\mathsf{S4FPF}$. So:

Proposition 4. *We have $\mathsf{Grz} \subseteq \mathsf{S4FPF}$, $\mathsf{S4FPF} \not\subseteq \mathsf{S4.2.1}$, and $\mathsf{S4.2.1} \not\subseteq \mathsf{S4FPF}$.*

To prove an upper bound of $\mathsf{S4FPF}$, we introduce *decisions* as new control statements. A decision consists of two mutually exclusive pure weak buttons. So, there is a dependence: by pushing one of them, we break the other one. Formally, let G be the general frame with respect to some $\mathcal{C}$, $\hookrightarrow$ and $\mathcal{L}$ as defined in Def. 1 and let $c \in \mathcal{C}$. A pair of unpushed pure weak buttons (λ, δ) is a *decision* in G at c if $\lambda \vee \delta$ is an unpushed pure button in G at c and for any V with $V(l) = \{d \in \mathcal{C} : d \vDash \lambda\}$ and $V(r) = \{d \in \mathcal{C} : d \vDash \delta\}$, $(G, V), c \Vdash \square(\neg l \vee \neg r)$ and $(G, V), c \Vdash \square((\Diamond l \wedge \Diamond r) \vee l \vee r)$. A decision (λ, δ) is *pushed* or *decided* at $e \in \mathcal{C}$ with $c \hookrightarrow e$ if $(G, V), e \Vdash l \vee r$ and *unpushed* or *undecided* otherwise. An illustration can be found in Figure 6b.

While decisions are internally dependent, we again require independence among different decisions. Let $n \in \mathbb{N}$. Let $\Phi = \{(\lambda_i, \delta_i) \mid 0 \leq i \leq n - 1\}$ be a set of unpushed decisions in G at c. We call Φ independent if for every $I_0 \subseteq I_1 \subseteq \{0, 1, \ldots, n - 1\}$ and $J_0 \subseteq J_1 \subseteq \{0, 1, \ldots, n - 1\} \setminus I_1$, and any V with $V(l_i) = \{d \in \mathcal{C} \mid d \vDash \lambda_i\}$ and $V(r_i) = \{d \in \mathcal{C} \mid d \vDash \delta_i\}$,

$$(G, V), c \Vdash \square\Big[\Big(\bigwedge_{i \in I_0} l_i \wedge \bigwedge_{i \notin I_0} \neg l_i \wedge \bigwedge_{i \in J_0} r_i \wedge \bigwedge_{i \notin J_0} \neg r_i\Big)$$
$$\to \Diamond\Big(\bigwedge_{i \in I_1} l_i \wedge \bigwedge_{i \notin I_1} \neg l_i \wedge \bigwedge_{i \in J_1} r_i \wedge \bigwedge_{i \notin J_1} \neg r_i\Big)\Big].$$

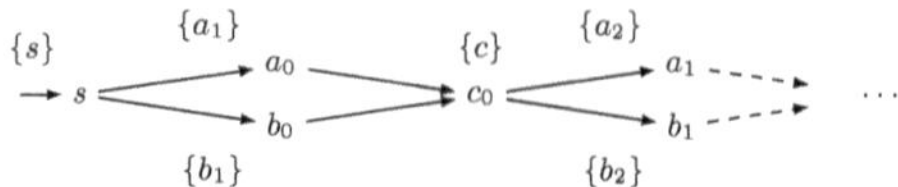

Fig. 7: The transition system $\mathcal{S}$ in the proof of Theorem 7.

Theorem 6. *Let G be the general Kripke frame w.r.t. some $\mathcal{C}$, $\hookrightarrow$ and $\mathcal{L}$ as in Def. 1 and let $c \in \mathcal{C}$. If for any $n \in \mathbb{N}$, there is a set of unpushed decisions $\Phi = \{(\lambda_i, \delta_i) \mid 0 \leq i \leq n - 1\}$ in G at c s.t. Φ is independent, then $\Lambda_G(c) \subseteq$ S4FPF.*

The proof works analogously to the proof of Thm. 3 and is given in [32].

Theorem 7. MLAR $\subseteq$ S4FPF.

Proof. We will show that there is a transition system $\mathcal{S}$ such that MLAR($\mathcal{S}$) $\subseteq$ S4FPF. By Theorem 6, it is sufficient provide a independent set of unpushed decisions $\Phi = \{(\lambda_i, \delta_i) \mid 0 \leq i \leq n - 1\}$ in $G_{\mathfrak{A}, \rightsquigarrow, \text{CTL}}$ at $\mathcal{S}$ for any $n \in \mathbb{N}$.

We define $\mathcal{S} = (S, \rightarrow, \{s\}, \mathsf{AP}, L)$ depicted also in Figure 7 where

- The state space is $S = \{s\} \cup \{a_i \mid i \in \mathbb{N}\} \cup \{b_i \mid i \in \mathbb{N}\} \cup \{c_i \mid i \in \mathbb{N}\}$.
- $s \rightarrow a_1$, $s \rightarrow b_1$, $a_i \rightarrow c_i$, $b_i \rightarrow c_i$, $c_i \rightarrow a_{i+1}$, $c_i \rightarrow b_{i+1}$ for all i.
- The set of atomic propositions is $\mathsf{AP} = \{s, c\} \cup \{a_i \mid i \in \mathbb{N}\} \cup \{b_i \mid i \in \mathbb{N}\}$.
- The labeling over AP is given by $L(s) = \{s\}$, $L(a_i) = \{a_i\}$ for all i, $L(b_i) = \{b_i\}$ for all i and $L(c_i) = \{c\}$ for all i.

Decisions. For any i, we define the CTL-sentences $\lambda_i = \forall G \neg a_i$ and $\delta_i = \forall G \neg b_i$ expressing that a_i or b_i is not reachable anymore. They are pure weak buttons at $\mathcal{S}_{n,m}$ since it is possible to simply remove a_i or b_i, respectively, by going to a refinement where the respective state is not reachable from the initial state. Further, if λ_i gets pushed, δ_i breaks and vice-versa, meaning that (λ_i, δ_i) is a decision. Independence is clear as we use different labels. $\qquad\square$

In conclusion, we showed that S4.1 $\subseteq$ MLAR $\subseteq$ S4.2.1 $\cap$ S4FPF. As shown in Prop. 4, S4.2.1 $\cap$ S4FPF is a stronger bound than S4FPF or S4.2.1 on their own.

5   Conclusion

We defined the modal logics of abstraction refinement in three settings: MLAR$_{\mathcal{T}}^{fin}$ and MLAR$_{\mathcal{T}}^{all}$ are defined on the frame of all finite and all abstractions of a transition system $\mathcal{T}$, respectively, while MLAR is defined on the class of all transition systems. In the first two cases, we proved matching upper and lower bounds of S4.2 and S4.2.1, respectively. These results also unveil the computational complexity of the respective MLARs as checking validity and satisfiability for both S4.2 and S4.2.1 are PSPACE-complete [33]. Determining the precise logic MLAR for which we know S4.1 $\subseteq$ MLAR $\subseteq$ S4.2.1 $\cap$ S4FPF remains as future work. The fact that S4FPF is defined semantically raises the question whether this logic can be finitely axiomatized.

The MLARs of course depend on the used notion of abstraction and the use of CTL to describe properties. Some observations are immediate: For example, for any logic at least as expressive as CTL such as CTL*, the upper bounds shown here hold as well. For the proof of the upper bounds, we only need that the respective control statements are expressible in the logic. The lower bounds, on the other hand, only depend on the structure of the refinement relation and hence apply no matter which logic is used instead of CTL. Regarding the notion of abstraction, we considered the well-established notion of *existential* abstraction [15] here. The investigation of different abstraction and refinement relations such as predicate abstraction might be an interesting direction for future research.

Discussion of the results. Seminal abstraction refinement paradigms such as CEGAR are limited to branching-time properties expressed in syntactic fragments of logics like the universally quantified fragment of CTL and CTL* (subsuming implicitly universally quantified LTL-properties). One of the goals of investigating the MLAR is to determine to which extent such abstraction refinement techniques can be extended to more general branching-time properties.

To some degree, the results presented in this paper are negative: There are transition systems for which $\mathsf{MLAR}_{\mathcal{T}}^{fin}$ and $\mathsf{MLAR}_{\mathcal{T}}^{all}$ coincide with the lower bounds obtained from the structure of the refinement relation. So, the evolution of CTL-expressible properties is governed by the same general principles as arbitrary properties. The only principles that can be exploited in general are the validity of $(.2) = \Diamond\Box p \to \Box\Diamond p$ and $(.1) = \Box\Diamond p \to \Diamond\Box p$ in the case of $\mathsf{MLAR}_{\mathcal{T}}^{all}$. Nevertheless, some potentially useful, although simple, observations are possible: From $(.2)$, the formula $\Diamond\Box p \wedge \Diamond\Box q \to \Diamond\Box(p \wedge q)$ follows. So, given an abstraction $\mathcal{S}$ of a system $\mathcal{T}$, to show that a conjunction $\Phi \wedge \Psi$ of properties holds in $\mathcal{T}$, it is sufficient to find separate refinements $\mathcal{S}_1$ and $\mathcal{S}_2$ of $\mathcal{S}$ such that $\Box\Phi$ holds in $\mathcal{S}_1$ and $\Box\Psi$ holds in $\mathcal{S}_2$. Formula $(.1)$ means that it is sufficient to show a property can always be made true by refinement in order to show that it holds in $\mathcal{T}$.

The results, however, do not mean that, for any transition system $\mathcal{T}$ and any abstraction $\mathcal{S}$, the logics $\mathsf{MLAR}_{\mathcal{T}}^{fin}(\mathcal{S})$ and $\mathsf{MLAR}_{\mathcal{T}}^{all}(\mathcal{S})$ coincide with the lower bounds. This opens up a path for future investigations on conditions on systems $\mathcal{T}$ and abstractions $\mathcal{S}$ under which more modal principles are valid, which in turn could be exploited in abstraction refinement-based verification methods.

Further, the key concept of control statements is worth exploring further. On the one hand, identifying which kind of CTL-formulas can act as which kind of control statements could be useful to tailor abstraction refinement to certain types of specifications. In particular, investigating when a CTL-formula (e.g., in a certain syntactic fragment) acts as a pure button is an interesting direction for future research as pure buttons have the desirable property that it is sufficient to find a refinement in which they are true in order to conclude that they are true in the underlying system model. On the other hand, the new generic results on how to use restricted switches and decisions to prove upper bounds of S4.2.1 and S4FPF have the potential to be applied in a wide variety of settings.

References

1. Christel Baier and Joost-Pieter Katoen. *Principles of Model Checking*. MIT Press, 2008.
2. Sören Berger, Alexander Christensen Block, and Benedikt Löwe. The modal logic of abelian groups. *Algebra universalis*, 84(3):25, 2023. `doi:10.1007/s00012-023-00821-9`.
3. Guram Bezhanishvili, Nick Bezhanishvili, Joel Lucero-Bryan, and Jan van Mill. Tree-like constructions in topology and modal logic. *Arch. Math. Log.*, 60(3-4):265–299, 2021. URL: `https://doi.org/10.1007/s00153-020-00743-6`, `doi:10.1007/S00153-020-00743-6`.
4. Patrick Blackburn, Maarten de Rijke, and Yde Venema. *Modal Logic*, volume 53 of *Cambridge Tracts in Theoretical Computer Science*. Cambridge University Press, 2001. `doi:10.1017/CBO9781107050884`.
5. Laura Bozzelli, Hans van Ditmarsch, Tim French, James Hales, and Sophie Pinchinat. Refinement modal logic. *Inf. Comput.*, 239:303–339, 2014. URL: `https://doi.org/10.1016/j.ic.2014.07.013`, `doi:10.1016/J.IC.2014.07.013`.
6. Laura Bozzelli, Hans van Ditmarsch, and Sophie Pinchinat. The complexity of one-agent refinement modal logic. *Theor. Comput. Sci.*, 603:58–83, 2015. URL: `https://doi.org/10.1016/j.tcs.2015.07.015`, `doi:10.1016/J.TCS.2015.07.015`.
7. Glenn Bruns and Patrice Godefroid. Generalized model checking: Reasoning about partial state spaces. In Catuscia Palamidessi, editor, *CONCUR 2000 — Concurrency Theory*, pages 168–182, Berlin, Heidelberg, 2000. Springer Berlin Heidelberg. `doi:10.1007/3-540-44618-4_14`.
8. Alexander Chagrov and Michael Zakharyaschev. *Modal logic*. Oxford University Press, 1997.
9. Pankaj Chauhan, Edmund M. Clarke, James H. Kukula, Samir Sapra, Helmut Veith, and Dong Wang. Automated abstraction refinement for model checking large state spaces using SAT based conflict analysis. In Mark D. Aagaard and John W. O'Leary, editors, *Formal Methods in Computer-Aided Design, 4th International Conference, FMCAD 2002, Portland, OR, USA, November 6-8, 2002, Proceedings*, volume 2517 of *Lecture Notes in Computer Science*, pages 33–51. Springer, 2002. `doi:10.1007/3-540-36126-X_3`.
10. Edmund M. Clarke, Orna Grumberg, Somesh Jha, Yuan Lu, and Helmut Veith. Counterexample-guided abstraction refinement. In E. Allen Emerson and A. Prasad Sistla, editors, *Computer Aided Verification, 12th International Conference, CAV 2000, Chicago, IL, USA, July 15-19, 2000, Proceedings*, volume 1855 of *Lecture Notes in Computer Science*, pages 154–169. Springer, 2000. `doi:10.1007/10722167_15`.
11. Edmund M. Clarke, Orna Grumberg, Somesh Jha, Yuan Lu, and Helmut Veith. Counterexample-guided abstraction refinement for symbolic model checking. *J. ACM*, 50(5):752–794, 2003. `doi:10.1145/876638.876643`.
12. Edmund M. Clarke, Anubhav Gupta, James H. Kukula, and Ofer Strichman. SAT based abstraction-refinement using ILP and machine learning techniques. In Ed Brinksma and Kim Guldstrand Larsen, editors, *Computer Aided Verification, 14th International Conference, CAV 2002, Copenhagen, Denmark, July 27-31, 2002, Proceedings*, volume 2404 of *Lecture Notes in Computer Science*, pages 265–279. Springer, 2002. `doi:10.1007/3-540-45657-0_20`.
13. Edmund M. Clarke, William Klieber, Milos Novácek, and Paolo Zuliani. Model checking and the state explosion problem. In Bertrand Meyer and Martin Nor-

dio, editors, *Tools for Practical Software Verification, LASER, International Summer School 2011, Elba Island, Italy, Revised Tutorial Lectures*, volume 7682 of *Lecture Notes in Computer Science*, pages 1–30. Springer, 2011. `doi:10.1007/978-3-642-35746-6_1`.

14. Dennis Dams, Rob Gerth, and Orna Grumberg. Abstract interpretation of reactive systems. *ACM Trans. Program. Lang. Syst.*, 19(2):253–291, 1997. `doi:10.1145/244795.244800`.

15. Dennis Dams and Orna Grumberg. Abstraction and abstraction refinement. In Edmund M. Clarke, Thomas A. Henzinger, Helmut Veith, and Roderick Bloem, editors, *Handbook of Model Checking*, pages 385–419. Springer, 2018. URL: `https://doi.org/10.1007/978-3-319-10575-8_13`, `doi:10.1007/978-3-319-10575-8_13`.

16. P. Godefroid and M. Huth. Model checking vs. generalized model checking: semantic minimizations for temporal logics. In *20th Annual IEEE Symposium on Logic in Computer Science (LICS' 05)*, pages 158–167, 2005. `doi:10.1109/LICS.2005.28`.

17. Patrice Godefroid and Radha Jagadeesan. Automatic abstraction using generalized model checking. In Ed Brinksma and Kim Guldstrand Larsen, editors, *Computer Aided Verification, 14th International Conference, CAV 2002, Copenhagen, Denmark, July 27-31, 2002, Proceedings*, volume 2404 of *Lecture Notes in Computer Science*, pages 137–150. Springer, 2002. `doi:10.1007/3-540-45657-0_11`.

18. Patrice Godefroid and Nir Piterman. LTL generalized model checking revisited. *Int. J. Softw. Tools Technol. Transf.*, 13(6):571–584, 2011. URL: `https://doi.org/10.1007/s10009-010-0169-3`, `doi:10.1007/S10009-010-0169-3`.

19. Orna Grumberg, Martin Lange, Martin Leucker, and Sharon Shoham. *Don't Know* in the μ-calculus. In Radhia Cousot, editor, *Verification, Model Checking, and Abstract Interpretation, 6th International Conference, VMCAI 2005, Paris, France, January 17-19, 2005, Proceedings*, volume 3385 of *Lecture Notes in Computer Science*, pages 233–249. Springer, 2005. `doi:10.1007/978-3-540-30579-8_16`.

20. Joel Hamkins and Benedikt Löwe. The modal logic of forcing. *Transactions of the American Mathematical Society*, 360(4):1793–1817, 2008. `doi:10.1090/S0002-9947-07-04297-3`.

21. Joel David Hamkins, George Leibman, and Benedikt Löwe. Structural connections between a forcing class and its modal logic. *Israel Journal of Mathematics*, 207(2):617–651, 2015. `doi:10.1007/s11856-015-1185-5`.

22. Joel David Hamkins and Øystein Linnebo. The modal logic of set-theoretic potentialism and the potentialist maximality principles. *The Review of Symbolic Logic*, 15(1):1–35, 2022. `doi:10.1017/s1755020318000242`.

23. Joel David Hamkins and Benedikt Löwe. Moving up and down in the generic multiverse. In *Logic and Its Applications: 5th Indian Conference, ICLA 2013, Chennai, India, January 10-12, 2013. Proceedings 5*, pages 139–147. Springer, 2013. `doi:10.1007/978-3-642-36039-8_13`.

24. Joel David Hamkins and Wojciech Aleksander Wołoszyn. Modal model theory. *Notre Dame Journal of Formal Logic*, 65(1):1–37, 2024. `doi:10.1215/00294527-2024-0001`.

25. Michael Huth, Radha Jagadeesan, and David A. Schmidt. Modal transition systems: A foundation for three-valued program analysis. In David Sands, editor, *Programming Languages and Systems, 10th European Symposium on Programming, ESOP 2001 Held as Part of the Joint European Conferences on Theory and Practice of Software, ETAPS 2001 Genova, Italy, April 2-6, 2001, Proceedings*, volume 2028 of *Lecture Notes in Computer Science*, pages 155–169. Springer, 2001. `doi:10.1007/3-540-45309-1_11`.

26. Tanmay Inamdar and Benedikt Löwe. The modal logic of inner models. *The Journal of Symbolic Logic*, 81(1):225–236, 2016. doi:10.1017/jsl.2015.67.
27. Himanshu Jain, Franjo Ivancic, Aarti Gupta, Ilya Shlyakhter, and Chao Wang. Using statically computed invariants inside the predicate abstraction and refinement loop. In Thomas Ball and Robert B. Jones, editors, *Computer Aided Verification, 18th International Conference, CAV 2006, Seattle, WA, USA, August 17-20, 2006, Proceedings*, volume 4144 of *Lecture Notes in Computer Science*, pages 137–151. Springer, 2006. doi:10.1007/11817963_15.
28. Kim Guldstrand Larsen. Modal specifications. In Joseph Sifakis, editor, *Automatic Verification Methods for Finite State Systems, International Workshop, Grenoble, France, June 12-14, 1989, Proceedings*, volume 407 of *Lecture Notes in Computer Science*, pages 232–246. Springer, 1989. doi:10.1007/3-540-52148-8_19.
29. Kim Guldstrand Larsen and Bent Thomsen. A modal process logic. In *Proceedings of the Third Annual Symposium on Logic in Computer Science (LICS '88), Edinburgh, Scotland, UK, July 5-8, 1988*, pages 203–210. IEEE Computer Society, 1988. doi:10.1109/LICS.1988.5119.
30. Kenneth L. McMillan and Nina Amla. Automatic abstraction without counterexamples. In Hubert Garavel and John Hatcliff, editors, *Tools and Algorithms for the Construction and Analysis of Systems, 9th International Conference, TACAS 2003, Held as Part of the Joint European Conferences on Theory and Practice of Software, ETAPS 2003, Warsaw, Poland, April 7-11, 2003, Proceedings*, volume 2619 of *Lecture Notes in Computer Science*, pages 2–17. Springer, 2003. doi:10.1007/3-540-36577-X_2.
31. Jakob Piribauer. The modal logic of generic multiverses, 2017. M. Sc. thesis, Universiteit van Amsterdam.
32. Jakob Piribauer and Vinzent Zschuppe. The modal logic of abstraction refinement, 2026. URL: https://arxiv.org/abs/2601.05897, arXiv:2601.05897.
33. Ilya Shapirovsky. On PSPACE-decidability in transitive modal logic. *Advances in Modal Logic*, 5:269–287, 2004.
34. Ur Ya'ar. The modal logic of -centered forcing and related forcing classes. *J. Symb. Log.*, 86(1):1–24, 2021. URL: https://doi.org/10.1017/jsl.2019.41, doi:10.1017/JSL.2019.41.

A Coalgebraic Approach to Infinite Games

Benjamin Plummer and Corina Cîrstea

School of Electronics and Computer Science, University of Southampton
{bjp1g19@,cc2@ecs.}soton.ac.uk

Abstract. We study infinite outcomes of two-player games, and strategies therein, finding both to be captured by a categorical limit. Outcomes of strategies generalise traces in labelled transition systems, allowing us to link our work to, and even unify, various coalgebraic approaches to infinite trace semantics. We obtain largest homomorphism characterisations of outcomes in both strategies and games. For practical applications, we show how infinite outcomes can be approximated by a least fixed point, which may be viewed as maximally permissive controller synthesis.

Keywords: Two-player game · Coalgebra · Trace semantics

1 Introduction

Two-player games are ubiquitous throughout theoretical computer science – they are deeply connected to logic, algorithms, automata, complexity and verification. Our interest in two-player games stems from their use in automated verification, where *controller synthesis* can be reduced to solving an infinite game between a controller and its environment. Typical objectives for the controller include parity objectives, in which case synthesising winning controller strategies involves some form of (nested) fixed point computation. Our ultimate goal is to give a general account of infinite two-player games, that both covers several types of games of interest in verification (qualitative *and* quantitative), and lends itself well to the study of such fixed point computations.

To achieve this level of generality, one can model two-player games as coalgebras. Previous work [19] established a link between strategies in two-player games and coalgebraic trace semantics: it was shown in loc. cit. that an existing coalgebraic formalism for finite trace semantics [12] can be applied to two-player games, and the semantics recovers the outcomes of *finitely completing* controller strategies. The resulting least fixpoint characterisation of the (completed) outcome of a game can be exploited to compute completing strategies inductively.

The caveat with the work [19] is that it only deals with *finite* behaviour, whereas for the purpose of verification and synthesis system behaviour is viewed as infinite: many verification tasks deal with properties that must hold *forever*, e.g. safety properties, and these can be more naturally expressed when system behaviour is viewed as infinite. The present work extends loc. cit. to also include *infinite* plays and strategies.

© The Author(s) 2026
N. Bertrand and S. Milius (Eds.): FoSSaCS 2026, LNCS 16503, pp. 549–570, 2026.
https://doi.org/10.1007/978-3-032-22730-0_26

Tackling this problem using coalgebraic machinery means that our results have the potential to be generalised to other types of games (like the work [19], which is already applicable to more than one type of game). It also allows us to make explicit the similarities with single-player models, such as labelled transition systems. Conversely, it is valuable to explore whether coalgebraic techniques can be used to capture infinite traces in this richer setting. One of the contributions of our work is to link existing coalgebraic approaches to infinite trace semantics [13,5,15,21], and to show that any of them can be applied to games.

To the best of our knowledge, infinite outcomes of games have received little attention even outside the coalgebraic literature, with the closest results being work on (alternating-time) temporal logic [8], wherein *path effectivity models* and the notion of *limit-closure* are used to describe game outcomes.

Related Work The work of Hasuo, Jacobs and Sokolova [12] gives a method to obtain finite trace semantics for systems with branching behaviour, by modelling such systems as coalgebras. The typical model is a labelled transition system, in which the trace semantics of a state is given by the set of *traces* (finite sequences of observations) that can be observed from that state. To capture such linear-time semantics with coalgebra, one has to move to a category whose morphisms only preserve the linear-time semantics. In the classical case of labelled transition systems, this means moving from the category of sets and functions (wherein transitions systems can be naturally modelled as coalgebras) to the category **Rel** of sets and relations. This works smoothly for *finite* traces since, under suitable assumptions, the domain of finite traces forms both an initial algebra and a terminal coalgebra. On the other hand, the domain of *infinite* traces is much less well-behaved (as witnessed by several existing approaches to infinite trace semantics [13,5,21], none of which matches the simplicity of [12]).

Previous work [19] showed that the above method can be applied to the setting of two-player games. Games have a more complex branching structure than transition systems, given the alternation between controller and environment moves. As a consequence, the resulting trace map has a different type: it assigns to each state a *set of subsets* of traces. The main result of loc. cit. states that the trace semantics assigned to each state is the set of precisely those subsets of finite traces which a controller strategy can force. In this setting, the trace map comes "for free": it is the unique coalgebra homomorphism.

Our Setting We focus purely on two non-deterministic players whose turns alternate (unlike [19], where probabilistic opponents are also considered). This is to simplify matters, as even in the classical case of labelled transition systems there are several, and previously unrelated, coalgebraic frameworks which can capture infinite traces [13,5,21,18].

Our main focus is on exploring the interplay between different definitions of *strategy outcome* and *game outcome*. The outcome of a controller strategy from a controller state is the set of plays that can arise when the controller follows said strategy from that state. A *play* records the controller states visited and the observable outcomes of basic interactions; the latter come in two forms: continuing or terminating. The outcome of a game at a controller state is the set of outcomes of controller strategies starting in that state. We give an example to demonstrate.

We denote controller states with squares and environment states with thick black dots. In the small example, we have a single controller state, and two environment states. There is a set $A = \{a\}$ of continuing observations and a set $B = \{b_1, b_2\}$ of terminating observations. Our coalgebraic model of games abstracts the environment, leaving just a set X of controller states (in this example, a singleton $\{x\}$). The game corresponds to a set $\{\{b_1\}, \{b_2, (a, x)\}\}$. This represents that the controller can either choose to terminate with a b_1, or let the environment choose to either terminate with a b_2, or to output a and remain in state x. There are an infinite number of controller strategies, each with their own outcome; they correspond to when the controller chooses to go left. For example, the strategy which chooses to go right first, and then left if the environment chooses to go back to x, has an outcome $\{xb_2, xaxb_1\}$. The outcome of the game is the set of all outcomes of strategies: $\{\{xb_1\}, \{xb_2, xaxb_1\}, \{xb_2, xaxb_2, xaxaxb_1\}, \dots, \{xaxa \dots\}\}$.

We make a few assumptions for technical reasons, that also make practical sense. Firstly, we assume that the environment cannot deadlock, so any strategy has a non-empty outcome (note that there *can* be no strategies, i.e. the outcome of a game at a state can be empty). Secondly, our games are *convex*: the controller has strategies where choices are left undetermined. In technical terms, this means that all sets of subsets are closed under arbitrary union. In the first example, this means that we have an element $\{b_1, b_2, (a, x)\}$ which corresponds to the controller leaving the decision to go left or right undetermined. Our reason for assuming convexity is threefold. Firstly, it is key to obtaining a monad structure, allowing us to associatively iterate games. Secondly, it is a natural assumption for the induced trace relation, for example it gives a subset inclusion between the semantics of the top two states in the games above. Intuitively, the right game is better for the controller than the left, as the environment choice in the left game becomes a controller choice in the right game. Without convexity, the traces at the top states in these games are incomparable. Finally, convexity ensures that a *maximally permissive strategy* exists from every state, which allows strategy refinement to be performed in an inductive fashion (see work such as [20,2,3]). Other categorical formulations of infinite games (e.g. [10]) make the stronger assumptions that sets of subsets are *upwards closed*.

Road Map Section 2.1 introduces game-theoretic notions, and how they are modelled using coalgebras. Section 2.2 discusses *linear functors*, which model the kind of observations made in the games we consider. Section 2.3 introduces the monads we use. In Section 2.4, we recall some facts about **Rel**, and adjunctions therein. Section 2.5 introduces some terminology related to cones. We finish the preliminaries in Section 2.6, by discussing related work on infinite traces.

We begin in Section 3.1 by showing how to define the outcome of a strategy; we find that the most natural definition is as a limit in **Set**. This idea is novel, but a similar definition was suggested in [18, Remark 4] for labelled transition systems. Furthermore, we can see the outcome of a strategy as a colimit in a corresponding subcategory of relations, by duality. It is natural to consider whether outcomes form colimits in the full category **Rel** (we will see that a strategy is naturally a chain in **Rel**). In Section 3.2, inspired by [16,17], we find that a strategy outcome is not a colimit in **Rel**, but that a weaker universal property, stating that there are largest mediating maps out the cocone, does exist. The following subsections, 3.3 and 3.4, discuss using other coalgebraic approaches, [13,21] and [5] respectively, to capture outcomes of strategies. Section 3.5 presents our main result in Section 3, Theorem 1, giving a correspondence between homomorphisms into the terminal coalgebra (lifted to **Rel**) and cones in **Set**. To establish this correspondence, we impose that cones are jointly monic and componentwise epi, to make them resemble sets of infinite paths. We profit from this result in Section 3.6, by showing that all our definitions of strategy outcome are equivalent. Furthermore we offer a final characterisation, which employs the notion of *limit closure*, linking to work in temporal logic [6]. Roughly, a subset of paths is limit closed when it is determined by its finite length approximations.

Outcomes of games are studied in Section 4. Section 4.1 provides a definition of game outcomes as limits in **Set**. This involves obtaining a cochain in **Set**, built from unfolding the coalgebra structure in the Kleisli category of the game monad. Proposition 15 explains how cones over this cochain in **Set** are collections of strategies in the game. Section 4.2 shows that we can take the largest homomorphism in a Kleisli homset, similarly to [13,21], and can build cones from them. In Section 4.3 we discuss how the homomorphism approach can *not* be used to give the outcome of a game in Example 3. We rectify this in Proposition 20 by restricting the homomorphisms to ones whose image contains limit closed subsets only, and then axiomatise cones to establish order-preserving isomorphism between such cones and limit-closed homomorphisms (Theorem 2). Section 4.4 shows how to approximate infinite game outcomes with a least fixed point. This resembles permissive controller strategy refinement, where it is assumed that a maximally permissive controller strategy exists, which can then be refined in a backwards fashion (see approaches such as [20,2,3]).

Section 5 provides a comparison between the monad we use to model games, and the monotone neighbourhood monad, used in other coalgebraic approaches to games. We conclude the paper with a discussion of future work in Section 6.

Contributions

1. Give definitions of outcomes of both strategies and games as limits.
2. Show that several existing approaches to infinite traces [13,5,18,21] coincide in the case of labelled transition systems.

3. Show that the largest mediating map approach to infinite traces in [5] only provides the expected result when intermediate states are recorded.
4. Show that the largest homomorphism approach of [13,21] is not suitable for defining outcomes of games, giving evidence for the superiority of the limit-based definition.
5. Obtain a greatest fixed-point characterisation of infinite outcomes in games by suitably restricting the homomorphisms considered. The new homomorphisms can be viewed as computing sets of strategies.
6. Theoretically ground permissive controller strategy synthesis as a least fixed point computation, and show how it approximates infinite strategies.

2 Preliminaries

2.1 Games, Plays, and Strategies

Given a set X, $P(X)$ denotes the powerset of X, and $Q(X)$ denotes the non-empty powerset of X. We let $\mathcal{G}(X)$ denote the set of sets of non-empty subsets which are closed under union, $\mathcal{G}(X) := \{\mathcal{U} \subseteq Q(X) \mid \forall \mathcal{V} \subseteq \mathcal{U} : \bigcup \mathcal{V} \in \mathcal{U}\}$.

A *controller-versus-environment game*, or simply a *game*, is a function $\delta : X \to \mathcal{G}(A \times X + B)$, where $+$ denotes the disjoint union of sets. The game is from the controllers perspective; each set $U \in c(x)$ is a controller move. The set $U \subseteq A \times X + B$ collects all valid environment moves following the controller move. These are either pairs of a *continuing observation* from A and a controller state, or *terminating observations* in B.

For technical reasons, we work with a coalgebra $\gamma : X \to \mathcal{G}(X \times (A \times X + B))$ built from δ, which maps $x \xmapsto{\gamma} \{\{(x, u) \mid u \in U\} \mid U \in \delta(x)\}$. This records state information, which is a useful assumption that is exploited throughout our work. The use of $\mathcal{G}(X)$, rather than $PQ(X)$, make our games *convex*: for any non-empty collection of controller moves $U_i \in \gamma(x)$, there is a move which consists of the union $\bigcup_{i \in I} U_i$. As discussed in the introduction, this is a common assumption which allows for maximally permissive controllers to be synthesised. Furthermore, it comes "for free" out of the categorical theory of weak distributive laws [7,9], which give that $\mathcal{G}$ is a *composite monad*, giving us nice mathematical properties which, for instance, allow us to compose games associatively.

An *n-step partial play* is an element of $(XA)^n X$ such that for each prefix $x_0 a_0 \cdots x_i a_{i+1} x_{i+1}$ we have that $a_{i+1} x_{i+1} \in U \in c(x_i)$. A *maximal play* on (X, γ) is an element of $z \in (XA)^\omega + (XA)^* XB$ such that each prefix is a partial play, and if $z = x_0 \ldots x_n b$, then $b \in U \in \gamma(x_n)$.

A *strategy* is a partial function $\sigma : (XA)^* X \to Q(A \times X + B)$ taking partial plays to successors: $\sigma(x_0 a_0 \ldots x_n) \in c(x_n)$. We say a partial play $x_0 a_0 \ldots x_n$ *conforms* to a strategy σ when $x_{i+1} \in \sigma(x_0 a_0 \ldots x_i)$ for all $i < n$. Analogously, we say a maximal play conforms to σ, when every partial prefix of the play conforms to σ, and if the maximal play has shape τb, then $b \in \sigma(\tau)$. We impose the condition that a strategy is defined over *exactly* the partial plays which conform to it (the standard approach is to define it on *at least* the partials plays which conform to it). The *n-step partial outcome of a strategy* is the set of all partial n-step plays and maximal plays of length less than n which a strategy can force. The *outcome of a strategy* is the set of maximal plays that a strategy can force. Finally, the *outcome of a game* at a state is the set of all outcomes of strategies which start in x (are defined only over the singleton play $\{x\}$).

Games generalise labelled transition systems (LTSs) in two ways, corresponding to the two players (essentially coming from the two monad morphisms discussed in Section 2.3). Either, we take each move $\{u\} \in \gamma(x)$ from a transition $x \to u$ in an LTS (and close under convexity), or we take a single move $U = \{u \mid x \to u\} \in \gamma(x)$ at each state $x \in X$. The first method shows how games generalise LTSs: subsets of executions in the LTS give rise to strategies in the game. Using the second method, we see that there is a single memoryless strategy on a game constructed in this way, thus it makes sense to consider strategies as a generalisation of LTSs (in our work, ones that do not deadlock).

2.2   Linear Functors

The collection of *linear* functors captures the observable behaviour of plays in the games we consider. We use $[-]$ to denote a constant functor, and 1 to denote the terminal object in **Set** (a singleton). The collection can be specified as the least collection such that $[1]$ is linear, if F_i is linear then so is $\coprod_{i \in I} F_i$, and if F is linear then so is $F \times \mathrm{id}_{\mathbf{Set}}$. The general form we shall use here is $[A] \times (-) + [B]$, which shall be referred to as F throughout the paper.

To capture plays in games, we modify F to also capture states. This is achieved by defining a new linear functor $F_X := [X] \times F$, allowing plays in a game, with state space X, to also record intermediate states.

The terminal coalgebra for F is given by iterating the terminal sequence in **Set**, and taking the limit, denoted by Z, of the chain. We use (Z_X, ζ_X) to denote the terminal coalgebra for F_X. Using the standard characterisation of limits in **Set**, we obtain:

$$1 \xleftarrow{!} F_X(1) \xleftarrow{F_X(!)} F_X^2(1) \xleftarrow{F_X^2(!)} \cdots$$

with projections p_0, p_1, p_2 into Z_X.

$$Z_X \cong \{(z_0, z_1, z_2, \dots) \mid z_n \in F_X^n(1),\ z_n = F_X^n(!)(z_{n+1})\} \cong (XA)^\omega + (XA)^* XB$$

Stream Operations We will require fine-grained control over elements in Z_X. We use the projection $\pi_1 : F_X(X) \to X$. The map $F_X^n(\pi_1) : F_X^{n+1}(X) \to F_X^n(X)$ provides us a function which forgets the final observable behaviour. For succinctness, we define an operation $\star : F(Y) \times Z \to F(Z)$ which maps $(ay, z) \mapsto az$ and $(b, z) \mapsto b$. We also use a map $\mathrm{snip}_n^m : F^m(X) \to F^n(X)$ for $n \leq m$, which snips off the first $m - n$ observations: $a_1 \dots a_m x \mapsto a_{m-n+1} \dots a_m x$, when we have a terminating observation $b \in B$ in the first $m - n$ indexes, we return that b.

2.3   Powerset Monads

Recall that $T : \mathbf{Set} \to \mathbf{Set}$ is a *monad* when it comes equipped with natural transformations $\mu^T : TT \to T$ and $\eta^T : 1 \to T$ which allow functions of the shape $X \to T(Y)$ to be composed associatively (and for $\eta_X : X \to T(X)$ to be the unit of composition). The category where this composition takes place is called the *Kleisli category* of the monad, and will be denoted $\mathbf{Kl}(T)$.

Two monads that we use frequently are the powerset monad (P, μ^P, η^P) and the non-empty powerset monad (Q, μ^Q, η^Q). It is well-known that $\mathbf{Kl}(P) \cong \mathbf{Rel}$, the category of sets with relations between them. We also have that $\mathbf{Kl}(Q) \cong \mathbf{Rel}_{\mathrm{lt}}$, the category of sets and *left-total relations*. We also have *Eilenberg-Moore categories* $\mathbf{EM}(P) \cong \mathbf{CJSL}$ (*complete join semi-lattices*), and $\mathbf{EM}(Q) \cong \mathbf{AJSL}$ (*affine join semi-lattices*), where only non-empty joins are required to exist.

We use $\rightarrowtail$ to denote relations, aka morphisms in $\mathbf{Rel}$ and $\mathbf{Rel}_{\mathrm{lt}}$. We will use $\odot$ to denote relation composition, i.e. composition in both $\mathbf{Rel}$ and $\mathbf{Rel}_{\mathrm{lt}}$. An element in a category is a morphism from 1. Thus, given an element $U : 1 \rightarrowtail X$, we can write $f \odot U : 1 \rightarrowtail Y$ for some $f : X \rightarrowtail Y$. Defining $f \odot U$ for all such U is equivalent to defining the *Kleisli extension* of a monad, which is equivalent data to the multiplication and unit.

We use $\overline{(-)} : \mathbf{Set} \to \mathbf{Rel}$ for the left adjoint of the Kleisli adjunction, it lets us view a function as a relation by mapping $f \mapsto \eta^P \circ f$. We also use, for some $\mathbf{Set}$ endofunctor G, $\overline{G} : \mathbf{Rel} \to \mathbf{Rel}$ to denote a *lifting* of G to $\mathbf{Rel}$, meaning that it commutes with $\overline{(-)}$, i.e. $\overline{G}(\overline{f}) = \overline{G(f)}$.

The Game Monad Recall from Section 2.1, that $\mathcal{G}(X)$ is the set of sets of non-empty subsets of X which are closed under arbitrary union. It is easy to see that $\mathcal{G}$ is a functor $\mathbf{Set} \to \mathbf{Set}$ with $\mathcal{G}(f)(\mathcal{U}) = \{\{f(x) \mid x \in U\} \mid U \in \mathcal{U}\}$. It turns out it is a monad formed by composing P with Q with a *weak distributive law* $\delta : QP \to PQ$. We do not go into details here, but present a novel description of the Kleisli extension in the following proposition. We denote morphisms in $\mathbf{Kl}(\mathcal{G})$ with $\rightarrowtail$, and composition with $\odot\!\!\!\!\bigcirc$.

Proposition 1. *Let $f : X \rightarrowtail Y$, we have $f \odot\!\!\!\!\bigcirc \mathcal{U} = \{\bigcup_{x \in U} V_x \mid U \in \mathcal{U}, \forall x \in U : V_x \in f(x)\}$. The unit of $\mathcal{G}$ maps $x \mapsto \{\{x\}\}$.*

Denote the left adjoint of the Kleisli adjunction by $\overline{\overline{(-)}} : \mathbf{Set} \to \mathbf{Kl}(\mathcal{G})$, and a lifting of a $\mathbf{Set}$ endofunctor G, with $\overline{\overline{G}} : \mathbf{Kl}(\mathcal{G}) \to \mathbf{Kl}(\mathcal{G})$. There is a natural transformation $\mathsf{cl} : PQ \to \mathcal{G}$ that closes a set of subsets under arbitrary non-empty unions. We have monad morphisms $\eta^P : Q \to \mathcal{G}$ and $\mathsf{cl} \circ P(\eta^Q) : P \to \mathcal{G}$.

2.4   Relations

Recall that $\mathbf{Rel}$ is a dagger category, it is equipped with an involution $(-)^\dagger : \mathbf{Rel}^{op} \to \mathbf{Rel}$. This maps a morphism $r : C \rightarrowtail D$ to a morphism $r^\dagger : D \rightarrowtail C$, defined as $r^\dagger(d) := \{c \in C \mid d \in r(x)\}$.

Definition 1. *Let $r, r' : C \rightarrowtail D$ and $l, l' : D \rightarrowtail C$. We say that*

- *l and r form an adjunction $l \dashv r$ iff $l \odot r \sqsubseteq \mathrm{id}_C$ and $\mathrm{id}_D \sqsubseteq r \odot l$*
- *$l \dashv r$ is a reflection iff $l \dashv r$ and $\mathrm{id}_C \sqsubseteq l \odot r$*

- $l \dashv r$ *is a coreflection iff* $l \dashv r$ *and* $r \odot l \sqsubseteq \mathsf{id}_D$
- l *is left total iff* $\forall c \in C, \exists d \in D : c \in l(d)$
- l *is deterministic iff* $\forall d \in D, \forall c, c' \in C : c \in l(d)$ *and* $c' \in l(d) \implies c = c'$
- l *is functional iff* l *is left total and deterministic*
- r *is right total iff* $\forall d \in D, \exists c \in C : d \in r(c)$
- r *is separating iff* $\forall c, c' \in C, \forall d \in D : d \in r(c)$ *and* $d \in r(c') \implies c = c'$
- r *is cofunctional iff* r *is right total and separating*

Proposition 2. *Let* $l : D \twoheadrightarrow C$ *and* $r : C \twoheadrightarrow D$. *We have* $l \dashv r$ *iff* l *is functional iff* r *is cofunctional. We have that* $l \dashv r$ *is a coreflection iff* l *is functional and right total iff* l *is a surjective function iff* r *is cofunctional and left total.*

Thus, $(-)^\dagger$ *restricts to isomorphisms* $\mathbf{Rel}_{\mathrm{co}}^{\mathrm{op}} \cong \mathbf{Set}$ *and* $\mathbf{Rel}_{\mathrm{co+lt}}^{\mathrm{op}} \cong \mathbf{Set}_{\mathrm{surj}}$.

2.5 Cones

Let $D : \mathbb{I} \to \mathbf{Set}$. A cone $c : [C] \to D$ is *monic* when the collection $c_i : C \to D(i)$ is *jointly monic*. This means $\forall x, x' \in C : (\forall i \in \mathbb{I} : c_i(x) = c_i(x')) \implies x = x'$.

Similarly, if $D : \mathbb{I} \to \mathbf{Rel}$, a cocone $c : D \to [C]$ is *deterministic* when the collection $c_i : D(i) \to C$ is *jointly deterministic*: for any indexed collection $y_i \in D(i)$, and $x, x' \in C$, if $x \in c_i(y_i)$ and $x' \in c_i(y_i)$ for all $i \in \mathbb{I}$, then $x = x'$.

A *cone morphism* $f : (C, c) \to (C', c')$ is a morphism $f : C \to C'$ such that $c' \circ [f] = c$. We say that (C', c') is *larger than* (C, c) when there is a mono $C' \rightarrowtail C$ which is a cone morphism. We use this order to talk about order-preserving isomorphisms to and from collections of cones.

2.6 Infinite Traces

The work in [13] is the first to give a coalgebraic account of infinite traces in labelled transition systems. Recall that the central challenge in infinite traces is that we do not have a unique coalgebra morphism $(X, \gamma) \twoheadrightarrow (Z, \bar\zeta)$ in $\mathbf{Rel}$. This is exemplified in the transition system on the right, where there are non-canonical morphisms such as $x_0 \mapsto \{w \in \{a_1, a_2\}^\omega \mid a_1 \text{ occurs in } w \text{ infinitely often}\}$.

To remedy this, the solution in [13] is to take the *largest homomorphism*. This is equivalent to taking a join in the complete lattice $\mathbf{Rel}(X, Z)$.

This viewpoint is elaborated in [21], where they work with the operator $\Phi : (X \twoheadrightarrow Z) \to (X \twoheadrightarrow Z)$ mapping $f \mapsto \overline{\zeta^{-1}} \odot f \odot \gamma$. Given an order structure $\sqsubseteq$ on morphisms, a lax coalgebra morphism $f : (X, \gamma) \to (Y, \delta)$ is one s.t. $F(f) \circ \gamma \sqsubseteq \delta \circ f$, and an oplax coalgebra morphism is one s.t. $F(f) \circ \gamma \sqsupseteq \delta \circ f$. There is a correspondence between fixed points/pre-fixed points/post-fixed points of Φ and homomorphisms/lax homomorphisms/oplax homomorphisms $(X, \gamma) \twoheadrightarrow (Z, \bar\zeta)$. By Φ being a monotone operator (this follows from the enrichment of $\mathbf{Rel}$, and F being locally monotone) we can apply the Knaster-Tarski theorem to obtain a complete lattice of homomorphisms, where the smallest homomorphism is the least lax homomorphism, and the largest homomorphism is the largest oplax homomorphism. This perspective shift will be present throughout our work.

Largest Mediating Map Approach We briefly recall the approach in [5], which uses that the limiting cone $(Z, p_n : Z \to F^n(1))$ lifts into **Rel** and forms a *weak limit* over $1 \xleftarrow{\overline{!}} F(1) \xleftarrow{\overline{F(!)}} \cdots$. We can build a cone over this final sequence by unfolding a coalgebra structure $X \to F(X) \colon X \xrightarrow{\gamma_n} F^n(X) \xrightarrow{\overline{F^n(!)}} F^n(1)$. Furthermore, we can take *largest maps* into Z which commute with the cone projections, again because **Rel** is enriched in **CJSL**. This serves as an alternative definition of infinite traces in [5]. We will discuss in Section 3.4 why this approach works better for executions, where states are recorded too, rather than traces.

3   Strategies

3.1   Definition

To begin, recall the definition of a strategy in [19], as a chain of maps in $\mathbf{Rel}_{\mathrm{lt}}$.

Definition 2 ([19]).   *Let* $\gamma : X \to \mathcal{G}F_X(X)$. *A strategy* σ *in* (X, γ) *consists of a family of morphisms* $\{\sigma_n\}_{n \in \omega}$ *in* $\mathbf{Rel}_{\mathrm{lt}}$, *with* $\sigma_0 : 1 \to X$ *and* $\sigma_{n+1} :$ $\mathsf{Im}(\sigma_n) \to F_X^{n+1}(X)$, *satisfying the following conditions:*

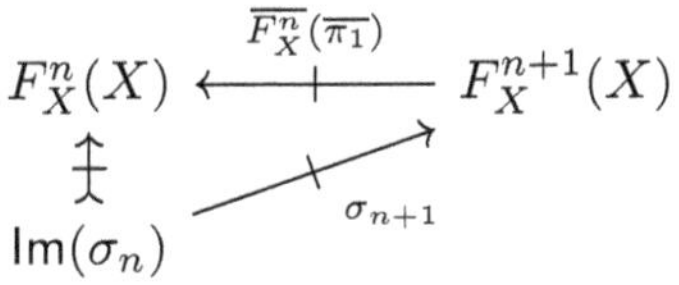

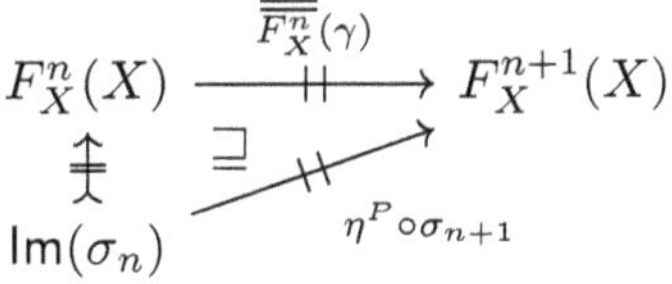

$$
\begin{array}{ccc}
F_X^n(X) \xleftarrow{\overline{F_X^n(\pi_1)}} F_X^{n+1}(X) & & F_X^n(X) \xrightarrow{\overline{\overline{F_X^n(\gamma)}}} F_X^{n+1}(X) \\
\updownarrow \qquad \nearrow \sigma_{n+1} & & \updownarrow \quad \sqsupseteq \nearrow \eta^P \circ \sigma_{n+1} \\
\mathsf{Im}(\sigma_n) & & \mathsf{Im}(\sigma_n)
\end{array}
$$

for all $n \in \omega$. *We say that a strategy* σ *starts in* x *when* $\sigma_0(*) = \{x\}$, *and use* $\Sigma_\gamma(x)$ *to denote the set of such strategies.*

This definition coincides with the definition of strategy presented in Section 2.1. The left diagram ensures strategies *extend* partial plays, and the right diagram ensures that moves chosen are from the game. Note in this definition the strategy is defined over exactly the partial plays it forces.

The Outcome of a Strategy In Section 2.1, we have seen a definition of outcome of a strategy σ: all the plays which conform to σ. We can make this definition categorical with a limit in **Set**. We prove that the chain $1 \xrightarrow{\sigma_0} \mathsf{Im}(\sigma_0) \xrightarrow{\sigma_1} \cdots$ in $\mathbf{Rel}_{\mathrm{lt}}$, lives in the subcategory $\mathbf{Rel}_{\mathrm{co+lt}}$. So by the isomorphism $\mathbf{Rel}_{\mathrm{co+lt}}^{\mathrm{op}} \cong \mathbf{Set}_{\mathrm{surj}}$ from Proposition 2, we have a surjective cochain a **Set**.

In fact, we have a nice characterisation of morphisms in this cochain in **Set**. We summarise this in the proposition below, note we abuse notation and use σ_{n+1} to refer to the relation after restricting its codomain to $\mathsf{Im}(\sigma_{n+1})$.

Proposition 3. *Each* $\sigma_{n+1} : \mathsf{Im}(\sigma_n) \to \mathsf{Im}(\sigma_{n+1})$ *is right total, separating, and left total. Furthermore, the corresponding surjective function* $s_{n+1} :$ $\mathsf{Im}(\sigma_{n+1}) \twoheadrightarrow \mathsf{Im}(\sigma_n)$ *is a restriction of* $F_X^n(\pi_1) :$ $F_X^{n+1}(X) \to F_X^n(X)$ *to* $\mathsf{Im}(\sigma_{n+1}) \to \mathsf{Im}(\sigma_n)$.

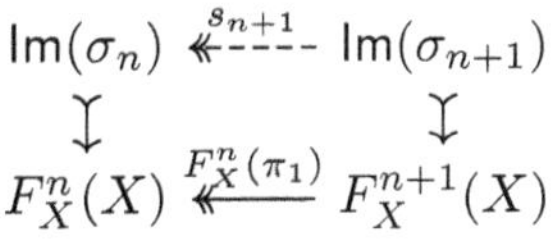

$$
\begin{array}{ccc}
\mathsf{Im}(\sigma_n) & \xleftarrow{s_{n+1}} & \mathsf{Im}(\sigma_{n+1}) \\
\downarrow & & \downarrow \\
F_X^n(X) & \xleftarrow{F_X^n(\pi_1)} & F_X^{n+1}(X)
\end{array}
$$

Definition 3. *Denote the functor which picks out the chain* $\mathsf{Im}(\sigma_0) \xrightarrow{\sigma_1} \dots$ *with* $D : \omega \to \mathbf{Rel}$. *The outcome of a strategy is the limit* $(\mathsf{Out}(\sigma), \pi)$ *of* $D^\dagger = (-)^\dagger \circ D : \omega \to \mathbf{Set}$, *the cochain* $\mathsf{Im}(\sigma_0) \leftarrow \mathsf{Im}(\sigma_1) \leftarrow \cdots$.

$\mathsf{Out}(\sigma)$ is isomorphic to a set with elements $(u_0, u_1, \dots)$ with $u_n \in \mathsf{Im}(\sigma_n)$ and $u_{n+1} \in \sigma_n(u_n)$ for all $n \in \omega$, thus precisely the plays which conform to the strategy are captured.

Notice that a limit in $\mathbf{Set}$ is a colimit in $\mathbf{Rel}_{\mathrm{co}}$, so $\mathsf{Out}(\sigma)$ is a colimit in this subcategory of relations. This leads to a question of whether $\mathsf{Out}(\sigma)$ is a colimit in $\mathbf{Rel}$, which we tackle now.

3.2   Colimits in Rel

This section is more technical than the rest, and is not directly relied upon in the rest of paper, so some readers may wish to skip to Section 3.3.

We start with an observation: $\mathsf{Out}(\sigma)$ is not a colimit in $\mathbf{Rel}$. The counterexample we present is inspired by a counterexample in [16,17], where it is shown that in general colimits of ω-chains do not exist in $\mathbf{Rel}$.

The flavour of this counterexample can be summarised by the slogan: "measuring" finite prefixes of plays does not determine measurements on infinite plays. Note that $\pi^\dagger$ refers to the cocone legs of $\mathsf{Out}(\sigma)$ in $\mathbf{Rel}$. These have components $\pi_n^\dagger : \mathsf{Im}(\sigma_n) \twoheadrightarrow \mathsf{Out}(\sigma)$, and extend an n-step prefix to an infinite play.

Counterexample 1 *The cocone* $(\mathsf{Out}(\sigma), \pi^\dagger)$ *is not a colimit in* $\mathbf{Rel}$.

Proof. Consider the (memoryless) strategy (in a game with a single observation): $\overset{\curvearrowright}{\underset{x}{\boxdot}} \,\, \cdots \overset{\curvearrowright}{\underset{y}{\boxdot}}$. We compute that $\mathsf{Out}(\sigma) = x^\omega + x^* y^\omega$, with $\pi_n^\dagger$ mapping: $x^n \mapsto$

$\{ x^k y^\omega \mid k \geq n \} \cup \{ x^\omega \}$ and $x^k y^{n-k} \mapsto \{ x^k y^\omega \}$ for $0 \leq k < n$. Define a cocone $C := \{0, 1\}$ with $c_n : \mathsf{Im}(\sigma_n) \twoheadrightarrow C$ mapping $x^n \mapsto \{0, 1\}$ and $x^k y^{n-k} \mapsto \{1\}$ for $0 \leq k < n$. We claim there are two mediating maps $\mathsf{Out}(\sigma) \twoheadrightarrow C$: the first maps $x^\omega \mapsto \{0\}$ and $x^\omega \mapsto \{0, 1\}$, the second maps $x^* y^\omega \mapsto \{1\}$ and $x^* y^\omega \mapsto \{1\}$.

Despite this, one can show that $\mathsf{Out}(\sigma)$ has a 2-categorical universal property. This is again inspired by [16,17], however we work in a less general setting which allows to give us a short proof. We refer the reader to loc. cit. for more discussion, where this flavour of colimit is termed a *lax coop adjoint colimit*. Our presentation is in terms of adjunctions to keep it self-contained.

An *oplax cocone* $c : D \to [C]$ has legs $c_n : \mathsf{Im}(\sigma_n) \twoheadrightarrow C$, which commute oplaxly: $c_{n+1} \sqsubseteq \sigma_{n+1} \odot c_n$, the category of these objects is denoted $\mathbf{Oplax}(D, [C])$.

Proposition 4. *We have an adjunction:* $\mathbf{Rel}(\mathsf{Out}(\sigma), C) \underset{!}{\overset{[-]\circ\pi^\dagger}{\rightleftarrows}} \mathbf{Oplax}(D, [C])$

for all sets C. *We can restrict the right side to strict cocones (as* $\pi^\dagger$ *is strict),* *and obtain a reflection:* $\mathbf{Rel}(\mathsf{Out}(\sigma), C) \underset{!}{\overset{[-]\circ\pi^\dagger}{\rightleftarrows}} \mathbf{Nat}(D, [C])$ *with* $[!(c)] \circ \pi^\dagger = c$.

Proof. $[-] \circ \pi^\dagger$ is a strict cocone by calculation. As **Rel** is enriched in **CJSL** we have $[-] \circ \pi^\dagger$ preserves joins (and is hence a functor), thus it has a right adjoint by the Adjoint Functor Theorem: for some oplax cocone $c : D \to [C]$: $!(c) = \bigsqcup \{h : \mathsf{Out}(\sigma) \nrightarrow C \mid [h] \circ \pi^\dagger \sqsubseteq c\}$. We prove $[!(c)] \circ \pi^\dagger = c$ in Proposition 6.

By the above proof, for some (possibly oplax) cocone (C, c), there exists a canonical mediating map h out of $(\mathsf{Out}(\sigma), \pi^\dagger)$, which is the largest map which "laxly commutes" ($[h] \circ \pi^\dagger \sqsubseteq c$) with the cone (C, c). Next, we give a second description of $!$ as a countable intersection. It allows us to prove the bottom adjunction is a reflection (although the characterisation holds in both adjunctions).

Proposition 5. *The right adjoint $!$ be defined as* $!(c)((u_n)_{n \in \omega}) = \bigcap_{n \in \omega} c_n(u_n)$.

With this characterisation we can prove the counit is an equality, meaning that any strict cocone $c : D \to [C]$ factors through $!(c)$ strictly (not just laxly).

Proposition 6. *Let $c : D \to [C]$ be a cocone, we have $[!(c)] \circ \pi^\dagger = c$.*

3.3 Homomorphism Approach

As described in Section 2.6, there is a nice view of infinite traces in a labelled transition system as a largest homomorphism (a greatest fixed point). Given strategies generalise transition systems, this view can be applied to the outcomes of strategies too, with some work. Our key insight is to use the *unravelling* of a strategy: we equip the set of finite prefixes which conform to the strategy with a coalgebra structure (this technique is familiar to temporal logicians).

We collect prefixes occurring along $\mathsf{Im}(\sigma_0) \nrightarrow \mathsf{Im}(\sigma_1) \nrightarrow \cdots$ in $\mathsf{Pref}(\sigma) := \coprod_{n \in \omega} \mathsf{Im}(\sigma_n)$, a coproduct in **Set** (and **Rel**). Recall that the strategy chain resides in $\mathbf{Rel}_{\mathrm{lt}}$, so the following coalgebra structure is a left-total relation.

Definition 4. *Define* $\mathsf{unravel}(\sigma) : \mathsf{Pref}(\sigma) \nrightarrow F_X(\mathsf{Pref}(\sigma))$ *by each component*

$$\mathsf{unravel}(\sigma)_n := \mathsf{Im}(\sigma_n) \xrightarrow{\sigma_{n+1}} \mathsf{Im}(\sigma_{n+1}) \xrightarrow{\overline{\mathsf{peek}_{n+1}}} F_X(\mathsf{Im}(\sigma_{n+1})) \xrightarrow{\overline{F_X(\mathsf{in}_{n+1})}} F_X(\mathsf{Pref}(\sigma))$$

where $\mathsf{peek}_n := (\mathsf{Im}(\sigma_n) \xrightarrow{\langle \mathsf{snip}_1^n, \mathsf{id}\rangle} F_X(X) \times \mathsf{Im}(\sigma_n) \xrightarrow{\star} F_X(\mathsf{Im}(\sigma_n)))$.

That is, the states of the coalgebra are partial plays, and the transition map relates an n-step partial play to all the $n + 1$-partial plays which extend it.

We can now use the approach in Section 2.6 to obtain an infinite trace semantics for the $\overline{F_X}$-coalgebra $(\mathsf{Pref}(\sigma), \mathsf{unravel}(\sigma))$.

Proposition 7. *Let $\Phi : (\mathsf{Pref}(\sigma) \nrightarrow Z_X) \to (\mathsf{Pref}(\sigma) \nrightarrow Z_X)$ be defined on some $h : \mathsf{Pref}(\sigma) \nrightarrow Z_X$ as mapping $h \mapsto \zeta^{-1} \odot \overline{F_X}(h) \odot \mathsf{unravel}(\sigma)$. This operator is a monotone operator on a complete lattice.*

Let $u \in \mathsf{Im}(\sigma_n)$, we have $\Phi(h)(u) = \bigcup_{u' \in \sigma_{n+1}(u)} \mathsf{snip}_1^{n+1}(u') \star h(u')$.

So we have another potential definition of the outcome of a strategy: as $h(*)$, where h the largest homomorphism $h : (\mathsf{Pref}(\sigma), \mathsf{unravel}(\sigma)) \nrightarrow (Z_X, \overline{\zeta_X})$. In Section 3.6, we prove that this definition is equivalent to Definition 3.

3.4 Largest Mediating Map

The weak limit approach in [5] gives us another possible definition of the infinite outcome of a strategy. Let $\sigma_{0n} : 1 \twoheadrightarrow \mathsf{Im}(\sigma_n)$ denote $\sigma_n \circ \cdots \circ \sigma_0$. We can lift the final sequence of F_X into **Rel**, and take a cone with projections

$$1 \xrightarrow{\sigma_{0n}} \mathsf{Im}(\sigma_n) \rightarrowtail F_X^n(X) \xrightarrow{\overline{F_X}(!)} F_X^n(1) \ .$$

Proposition 8. $(1, \overline{F_X}(!) \odot \sigma_{0n})$ *forms a cone over* $1 \xleftarrow{\ \bar{!}\ } F_X(1) \xleftarrow{\ \overline{F_X}(!)\ } \cdots$ *in* **Rel**.

The greatest cone morphism $1 \twoheadrightarrow Z_X$ gives a subset of Z_X, which we can also take as the outcome of a strategy. We show in Section 3.6 that this is equivalent to Definition 3. This is somewhat surprising, when you consider that this approach gives too many traces when you do not record intermediate states.

Example 1. Consider the two memoryless strategies depicted on the right.

The largest mediating morphism $1 \twoheadrightarrow Z$, takes the greatest subset which agrees with the finite approximations provided by the cone in Proposition 8. In both strategies we see that $a^*b + a^\omega$ is the largest mediating map $1 \twoheadrightarrow Z$. The right strategy should have as an outcome a^*b: it is impossible for a^ω to be observed.

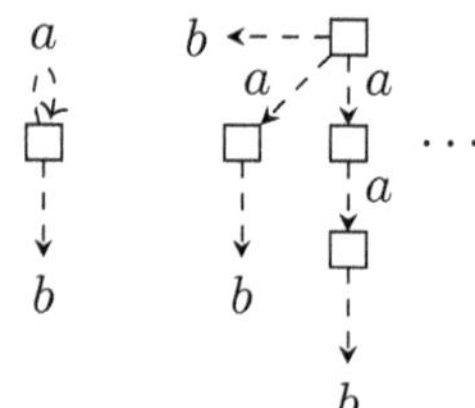

3.5 Correspondence Result

We present a correspondence result which identifies conditions under which cones in **Set**, cocones in **Rel**, coalgebra morphisms $\mathsf{Pref}(\sigma) \twoheadrightarrow Z_X$, and mediating maps into Z_X viewed as a weak limit in **Rel**, coincide.

To begin, recall that $\mathbf{Rel}_{\mathrm{co+lt}}$, the category of sets with left-total, right-total, and separating relations, is dually equivalent to $\mathbf{Set}_{\mathrm{surj}}$, the category of sets and surjective functions. We can restrict D to have type $\omega \to \mathbf{Rel}_{\mathrm{co+lt}}$, by Proposition 3. We see a cone $[C] \to D^\dagger$ in **Set** as representing a collection of infinite plays, via sequences of partial plays they induce. We do not want duplicate copies of infinite plays, so we impose that our cones are monic (i.e. the cone legs are jointly monic). The dual condition to this in **Rel** is that cocone legs are *jointly deterministic*, this is defined in Section 2.5.

Proposition 9. *Let* $c^\dagger : D \to [C]$ *be a cocone in* $\mathbf{Rel}_{\mathrm{co}}$. *We have that* $c^\dagger$ *is deterministic iff* c *is monic.*

We also want the cone to be "full", in the sense that every finite prefix should be accounted for, this is captured by cone legs being surjective functions (the dual condition in **Rel** is legs being left total).

It turns out that the most laborious direction is verifying that cocones give us homomorphisms. This is because a cocone leg $c_n^\dagger : \mathsf{Im}(\sigma_n) \twoheadrightarrow C$ (morally) extends a prefix to an infinite play, whereas a homomorphism $h : \mathsf{Pref}(\sigma) \twoheadrightarrow Z_X$, gives the plays which conform to the strategy which complete the prefix. To help with this bookkeeping issue, we introduce a map which interrogates a cone C to obtain n-step prefixes with corresponding suffixes.

Definition 5. *Let $(C, c : [C] \to D^\dagger)$ be a cone in **Set**. $(Z_X, p_n : Z_X \to F_X^n(1))$ is a limiting cone for the final sequence of F_X and $F_X : **Set** \to **Set**$ preserves limits; so $F_X^n(Z_X)$ with projections $F_X^n(p_m) : F_X^n(Z_X) \to F_X^{n+m}(1)$ is a limiting cone for all $n \in \omega$. We define a map $\mathsf{obs}_n : C \to F_X^n(Z_X)$ as the unique cone morphism from C with legs $\ C \xrightarrow{c_{n+m}} \mathsf{Im}(\sigma_{n+m}) \rightarrowtail F_X^{n+m}(X) \xrightarrow{F_X^{n+m}(!)} F_X^{n+m}(1)$ into $F_X^n(Z_X)$.*

Theorem 1. *Let $\sigma \in \Sigma_\gamma(x)$ be a strategy, the following are equivalent:*

1. *$\overline{F_X}$-coalgebra morphisms $(\mathsf{Pref}(\sigma), \mathsf{unravel}(\sigma)) \twoheadrightarrow (Z_X, \overline{\zeta_X})$*
2. *Fixed points of $\Phi(\sigma) : (\mathsf{Pref}(\sigma) \twoheadrightarrow Z_X) \to (\mathsf{Pref}(\sigma) \twoheadrightarrow Z_X)$*
3. *Mediating maps $(1, \overline{F_X^n}(!) \odot \sigma_{0n})_{n\in\omega} \twoheadrightarrow (Z_X, \overline{p_n})_{n\in\omega}$ in **Rel***
4. *Deterministic cocones $D \to [C]$ in $**Rel**_{co+lt}$*
5. *Monic cones $[C] \to D^\dagger$ in $**Set**_{surj}$.*

Furthermore, these translations are order preserving.

Proof (Sketch). **(1)** $\Longleftrightarrow$ **(2)** See Section 2.6. **(4)** $\Longleftrightarrow$ **(5)** Proposition 9.

(1) $\Longrightarrow$ **(3)** A coalgebra morphism $h : \mathsf{Pref}(\sigma) \twoheadrightarrow Z_X$ has a component $h_0 : 1 \cong \mathsf{Im}(\sigma_0) \twoheadrightarrow Z_X$, which defines a mediating map into $(Z_X, \overline{p_n})_{n\in\omega}$ in **Rel**.

(3) $\Longrightarrow$ **(5)** Take a mediating map $f : 1 \twoheadrightarrow Z_X$. The set $\mathsf{Im}(f)$ with cone legs $Z_X \xrightarrow{p_{n+1}} F_X^{n+1}(1) \xrightarrow{F_X^n(\pi_1)} F_X^n(X)$ (restricted to a map $\mathsf{Im}(f) \to \mathsf{Im}(\sigma_n)$) form a deterministic cone in $**Set**_{surj}$.

(4) $\Longrightarrow$ **(1)** Given a deterministic cocone $(C, c_n^\dagger : \mathsf{Im}(\sigma_n) \twoheadrightarrow C)$, the morphism $[\overline{\mathsf{snip}_0^n} \odot \overline{\mathsf{obs}}_n \odot c_n^\dagger]_{n\in\omega}$ is a $\overline{F_X}$-coalgebra morphism.

3.6 Equivalent Characterisations of Strategy Outcomes

Theorem 1 says that homomorphisms $\mathsf{Pref}(\sigma) \twoheadrightarrow Z_X$ give rise to cones over the strategy chain. We have not established that defining a strategy outcome as a limit is equivalent to defining it as the greatest homomorphism. We do know that the greatest homomorphism will be the largest monic cone, as the translation is order preserving. Thus, all we must establish is that the largest monic cone is the limiting cone. This is provided by the following proposition, which we have not found a proof of in the literature, but imagine is known.

Proposition 10. *Take a diagram $D : \mathbb{I} \to **Set**$, a largest monic cone is a limiting cone.*

This is an important characterisation of the largest monic cone, there is also one for the greatest homomorphism, which we discuss now. It involves the concept of *limit closure*, which arose when studying the semantics of temporal logics [1,6]. We present a definition here for executions, i.e. elements of Z_X.

Definition 6. *Let $U \subseteq Z_X$, we say that U is limit closed precisely when for any $z \in Z_X$, if $(\forall n \in \omega, \exists \rho \in Z_X : z_n \star \rho \in U)$ then $z \in U$.*

A subset of Z_X which forms a cone in $**Set**_{surj}$ over the strategy chain, is maximal iff it is limit-closed. We can exploit this to prove the following result.

Proposition 11. *Let* $h : \mathsf{Pref}(\sigma) \twoheadrightarrow Z_X$ *be a homomorphism for some* $\sigma \in \Sigma_c(x)$, h is the greatest homomorphism iff $h(x)$ is limit closed.

Corollary 1. *We identify (under the transformations in Theorem 1): the* $\overline{F_X}$-*coalgebra homomorphism with a limit-closed image; the greatest* $\overline{F_X}$-*coalgebra homomorphism; the greatest fixed point of* Φ; *the greatest mediating map into* $(Z_X, \overline{p_n})_{n\in\omega}$; *the largest deterministic cocone for* D; *the largest monic cone for* $D^\dagger$; *and the limiting cone for* $D^\dagger$.

Thus, any of these can serve as a definition for the outcome of a strategy.

Remark 1. This result connects various coalgebraic approaches to infinite trace semantics in the literature, in the case of the powerset monad with a simple behaviour functor. The greatest homomorphism, like in [13,21], is equivalent to taking the largest mediating map, like in [5]. But also, applying the approach to trace semantics in [15,18] to a memoryless strategy, we will obtain the sequence $(\mathsf{Im}(\sigma_n))_{n\in\omega}$ which are called *pretraces* in loc. cit. (we include the states in the behaviour). In [15, Remark 4], it is suggested to take a limit over a sequence of spans which are the *immediate prefix relations*. This is similar to the idea we use here, as we take a limit over $\mathsf{Im}(\sigma_0) \xleftarrow{s_0} \mathsf{Im}(\sigma_1) \xleftarrow{s_1} \cdots$. We keep more data, as n-step executions are recorded rather than just traces, which allows a more economic presentation using functions rather than relations (this can be seen in Proposition 3, with the key property being that strategies are separating). We will see shortly that this idea can also be applied to outcomes in games.

4 Games

4.1 Definition of Outcome

Recall that a game is a coalgebra $\gamma : X \twoheadrightarrow F_X(X)$, i.e. a function $X \to \mathcal{G}F_X(X)$.

Our first task is to define the *outcome of a game*. In [19], we defined the (finite) outcome of a game via a map $X \to \mathcal{G}(Z_X)$ (for a slightly different monad $\mathcal{G}$ subject to certain finiteness restrictions), assigning to each state those subsets of $(XA)^*XB$ (i.e. sets of finite completed plays) which can be forced by some controller strategy. It was shown in loc. cit. that this map arises as an instance of the coalgebraic framework for finite trace semantics in [12]. Here the situation is more subtle, since we are dealing with infinite trace semantics, for which no general account via initiality or finality exists.

A functor describing how we can drop morphisms from $\mathbf{Kl}(\mathcal{G})$ into $\mathbf{Rel}$ will be required. Recall that $\mathcal{G}$ is built from a weak distributive law, and while normally we automatically get a functor $\mathbf{Kl}(\mathcal{G}) \to \mathbf{Rel}$, here we do not[1].

Proposition 12. *There is a functor* $K : \mathbf{Kl}(\mathcal{G}) \to \mathbf{Rel}$ *mapping* $f : X \twoheadrightarrow Y$ *to a relation* $QX \twoheadrightarrow QY$, *it is defined as* $K(f)(U) := \{ \bigcup_{x \in U} V_x \mid \forall x \in U : V_x \in f(x)\}$. *For some* $g : X \to Y$, *we have* $K(\overline{\overline{g}}) = \overline{Q}(\overline{g}) = \overline{Q(g)}$.

---

[1] For standard distributive laws, $\overline{Q}$ is a monad on $\mathbf{Kl}(P)$ and $\mathbf{Kl}(\overline{Q}) \cong \mathbf{Kl}(PQ)$; however, here $\overline{Q}$ is only a semi-monad.

The only definition for the outcome of a game that we could use out-of-the-box is the greatest homomorphism approach of [13], however as we will show in Section 4.2, this does not give the expected outcome. For now, we show how to apply the limit approach of Section 3.1 to games.

Our starting point is the unfolding of γ in $\mathbf{Kl}(\mathcal{G})$: $X \overset{\gamma}{\longrightarrow\!\!\!\!\!\rightarrow} F_X(X) \overset{\overline{\overline{F_X}}(\gamma)}{\longrightarrow\!\!\!\!\!\rightarrow} \cdots$

We first drop this chain into $\mathbf{Rel}$: $X \overset{\gamma}{\longrightarrow\!\!\!\!\!\rightarrow} QF_X(X) \overset{K(\overline{\overline{F_X}}(\gamma))}{\longrightarrow\!\!\!\!\!\rightarrow} QF_X^2(X) \overset{K(\overline{\overline{F_X^2}}(\gamma))}{\longrightarrow\!\!\!\!\!\rightarrow} \cdots$ (γ is readily seen as a morphism in $\mathbf{Rel}$, for the rest we use K). Define $\gamma_n : X \longrightarrow\!\!\!\!\!\rightarrow QF_X^n X$ by $\gamma_1 = \gamma$ and $\gamma_{n+1} = K(\overline{\overline{F_X^n}}(\gamma)) \odot \gamma_n$. We then take the image of these maps (in $\mathbf{Rel}$): we let $\mathsf{Im}(\gamma_0) = X$, and $e_n^\dagger : \mathsf{Im}(\gamma_n) \longrightarrow\!\!\!\!\!\rightarrow \mathsf{Im}(\gamma_{n+1})$ be $K(\overline{\overline{F_X^n}}(\gamma))$ suitably restricted. Hence, we have a chain $\mathsf{Im}(\gamma_0) \overset{e_0^\dagger}{\longrightarrow\!\!\!\!\!\rightarrow} \mathsf{Im}(\gamma_1) \overset{e_1^\dagger}{\longrightarrow\!\!\!\!\!\rightarrow} \cdots$ The notation is justified by the following proposition.

Proposition 13. $\mathsf{Im}(\gamma_n) = \bigcup_{x \in X} \gamma_n(x)$.

Notice also, that each $e_n^\dagger$ is right total and separating. Like in Proposition 3, separation follows because we are recording state information. Note that, unlike before, these maps are not left-total (due to possible controller deadlocking).

Proposition 14. $e_n^\dagger : \mathsf{Im}(\gamma_n) \longrightarrow\!\!\!\!\!\rightarrow \mathsf{Im}(\gamma_{n+1})$ *is separating. The left adjoint* $e_n : \mathsf{Im}(\gamma_{n+1}) \to \mathsf{Im}(\gamma_n)$ *restricts* $QF_X^n(\pi_1) : QF_X^{n+1}(X) \to QF_X^n(X)$.

We are again left with a cochain in $\mathbf{Set}$: $\mathsf{Im}(\gamma_0) \overset{e_0}{\longleftarrow} \mathsf{Im}(\gamma_1) \overset{e_1}{\longleftarrow} \cdots$.

Definition 7. *The* outcome of the game γ, *denoted* $\mathsf{Out}(\gamma)$, *is the limit of the above cochain in* $\mathbf{Set}$.

Strategy outcomes can now be recovered as special cones over this cochain.

Proposition 15. *Cones from* 1 *over the cochain* $\mathsf{Im}(\gamma_0) \leftarrow \mathsf{Im}(\gamma_1) \leftarrow$ *are in bijection with strategies. We have* $\mathsf{Out}(\gamma) \cong \{\mathsf{Out}(\sigma) \mid x \in X, \sigma \in \Sigma_\gamma(x)\}$.

Recall that the cones over $\mathsf{Im}(\gamma_0) \leftarrow \mathsf{Im}(\gamma_1) \leftarrow \cdots$ are equivalently cocones under $\mathsf{Im}(\gamma_0) \longrightarrow\!\!\!\!\!\rightarrow \mathsf{Im}(\gamma_1) \longrightarrow\!\!\!\!\!\rightarrow \cdots$ with right-total and separating legs (but not necessarily left total, again because of deadlocking).

4.2 Homomorphism Approach

We discuss the connection with the largest homomorphism approach to infinite traces in [13,21]. The first thing we must establish is that the operator $\Psi : (X \longrightarrow\!\!\!\!\!\rightarrow Z_X) \to (X \longrightarrow\!\!\!\!\!\rightarrow Z_X)$ built from $f \mapsto \overline{\overline{\zeta_X^{-1}}} \circledcirc \overline{\overline{F_X}}(f) \circledcirc \gamma$ is monotone. This follows from $\mathbf{Kl}(\mathcal{G})$ being $\mathbf{Pos}$-enriched and $\overline{F_X}$ being locally monotone. Note that the conditions of [21, Proposition 4.1] are met, meaning that we can give a constructive proof of the existence of a greatest fixed point, using the transfinite induction proof in [4]. We present the version using the non-constructive Knaster-Tarksi Theorem here. Note it was shown in [19] that $\mathbf{Kl}(\mathcal{G})$ is not $\mathbf{CJSL}$-enriched.

Proposition 16. $\mathbf{Kl}(\mathcal{G})$ *is* $\mathbf{Pos}$*-enriched,* $\overline{\overline{F_X}}$ *is locally monotone (w.r.t. this enrichment), and* $\mathbf{Kl}(\mathcal{G})(X,Y)$ *is a complete lattice (inducing the same order as the enrichment).*

So we are in a position where we have least and greatest homomorphisms $(X,\gamma) \twoheadrightarrow (Z_X, \overline{\overline{\zeta_X}})$. It turns out that the greatest homomorphism gives too many sets of subsets, as the next example demonstrates.

Example 2. There is a homomorphism: $x \mapsto \{x{\cdot}\mathsf{GF}x, x{\cdot}\mathsf{GF}y, Z_X\}$ and $y \mapsto \{y{\cdot}\mathsf{GF}x, y{\cdot}\mathsf{GF}y, Z_X\}$ in the right game, where $A = \{*\}$. Here, $\mathsf{GF}x \subseteq Z_X$ is the set of infinite streams over $\{x,y\}$ that contain infinitely many xs.

Notice that in our example, the problematic subsets are not limit-closed. Restricting homomorphisms to only contain limit-closed subsets of plays, lets us recover $\mathsf{Out}(\gamma)$ as the largest limit-closed homomorphism. This is proven in the next subsection. For now, we state that all homomorphisms give rise to cones.

Lemma 1. *For any oplax coalgebra morphism* $f : (X,\gamma) \twoheadrightarrow (Z_X, \overline{\overline{\zeta_X}})$*, we have:* $\overline{\overline{F_X^n}}(\overline{\overline{\pi_1}}) \odot \overline{\overline{p_{n+1}}} \odot f \sqsubseteq \gamma_n$ *for all* $n \in \omega$.

Proposition 17. *Let* $f : X \twoheadrightarrow Z_X$ *be a coalgebra morphism. The set* $\mathsf{Im}(f)$ *with legs* $Q(Z_X) \xrightarrow{Q(p_{n+1})} Q(F_X^{n+1}(1)) \xrightarrow{QF_X^n(\pi_1)} QF_X^n(X)$ *restricted to maps* $\mathsf{Im}(f) \to \mathsf{Im}(\gamma_n)$ *form a cone over* $\mathsf{Im}(\gamma_0) \leftarrow \mathsf{Im}(\gamma_1) \leftarrow \cdots$.

4.3   Cones to Homomorphisms

We now answer the question: when does a cone over $\mathsf{Im}(\gamma_0) \leftarrow \mathsf{Im}(\gamma_1) \leftarrow \cdots$, give rise to a homomorphism $(X,\gamma) \twoheadrightarrow (Z_X, \overline{\overline{\zeta_X}})$? In Section 3, the cones were "full" in the sense they had epic components, however here we cannot assume that the legs are surjective, because the controller can deadlock.

Instead, we axiomatise cones directly. We introduce the following terminology for a cone (C, c): say $y \in C$ is *anchored* over a state $x \in X$, when $c_0(y) = x$. The set of elements of C which are anchored over x is denoted $C(x)$. We extend this terminology to some $u \in F_X(X)$, and notate the subset of C whose elements are anchored at u with $C(u) = \{y \in C \mid \mathsf{snip}_0^1(u) = c_0(y)\}$. Note that it only makes sense to talk about $C(u)$ when u is a partial play (ends in a state).

Definition 8. *Let* (C, c) *be a cone over* $\mathsf{Im}(\gamma_0) \leftarrow \mathsf{Im}(\gamma_1) \leftarrow \cdots$*. We say that* (C, c) *is closed under decomposition if when* $y \in C$

$$\exists \{y_u \in C(u)\}_{u \in c_1(y)}, \forall n \in \omega : \bigcup_{u \in c_1(y)} c_n(y_u) = Q(\mathsf{snip}_n^{n+1}) \circ c_{n+1}(y)$$

Say (C, c) *is closed under composition, if when* $U \in \gamma(x)$ *and* $\{y_u \in C(u)\}_{u \in U}$:

$$\exists y \in C(x), \forall n \in \omega : c_{n+1}(y) = \bigcup_{u \in U} u \star c_n(y)$$

Say a cone is convex closed if for any indexing set I and state $x \in X$:

$$\{y_i \in C(x)\}_{i \in I} \implies \exists y \in C, \forall n \in \omega : c_n(y) = \bigcup_{i \in I} c_n(y_i)$$

We say a cone is homomorphic precisely when it is closed under decomposition, composition, and convexity.

Proposition 18. *The construction in Proposition 17, from homomorphisms to cones, yield monic homomorphic cones.*

Proposition 19. *Given a cone (C, c) closed under convexity, define a morphism $f : X \twoheadrightarrow Z_X$ as $f(x) := \{\mathsf{lim}(\{x\} \leftarrow c_1(y) \leftarrow \cdots) \mid y \in C(x)\}$.*
  If (C, c) is closed under decomposition, then f is an oplax coalgebra morphism. If (C, c) is closed under composition, then f is a lax coalgebra morphism. Thus, if (C, c) is homomorphic, then f is a coalgebra morphism.

We now have constructions from monic homomorphic cones to coalgebra morphisms, and back. These constructions are not mutual inverses, as we show in the example below. Note that mapping a monic homomorphic cone to a coalgebra morphism and back, does yield the original cone.

Example 3. Recall Example 2. There is a homomorphism $x \mapsto \{x \cdot \mathsf{GF}x\}$ and $y \mapsto \{y \cdot \mathsf{GF}y\}$. This maps into a cone with two elements:

$$(x, \{xx, xy\}, \{xxx, xyx, xxy, xyy\}, \dots), (y, \{yx, yy\}, \{yxx, yyx, yxy, yyy\}, \dots)$$

Mapping this back to a homomorphism yields $x \mapsto \{x \cdot Z_X\}$ and $y \mapsto \{y \cdot Z_X\}$.

To obtain a bijection, we restrict the homomorphisms to *limit-closed homomorphisms*, i.e. those whose underlying $f : X \to \mathcal{G}(Z_X)$ is such that every $V \in f(x)$ is limit closed. Let $\mathcal{L}(Z_X)$ denote the union-closed sets of limit-closed subsets of executions: $\mathcal{L}(Z_X) := \{\mathcal{V} \in \mathcal{G}(Z_X) \mid \forall V \in \mathcal{V} : V \text{ is limit closed}\}$.

Proposition 20. *The operator $\Psi : (X \twoheadrightarrow Z_X) \to (X \twoheadrightarrow Z_X)$ restricts to $(X \to \mathcal{L}(Z_X)) \to (X \to \mathcal{L}(Z_X))$. The set $\mathcal{L}(Z_X)$ is a complete lattice. Thus the collection of limit-closed homomorphisms form a complete lattice.*

Theorem 2. *There is an order-preserving isomorphism between monic homomorphic cones and limit-closed homomorphisms.*

Theorem 2 gives us that the largest monic homomorphic cone is equivalently the largest limit-closed homomorphism $X \twoheadrightarrow Z_X$. To show this is the limit of the cochain $\mathsf{Im}(\gamma_0) \leftarrow \mathsf{Im}(\sigma_1) \cdots$, we would like to employ Proposition 10, but for this we need a final proposition which states that the largest monic homomorphic cone is the largest monic cone.

Proposition 21. *Any monic cone can be closed under decomposition, composition and convexity, and turned into a homomorphic monic cone.*

Thus, the largest homomorphic monic cone is the limiting cone, meaning we have a secondary definition of the infinite outcome of a game: as the largest limit-closed homomorphism.

Remark 2. The homomorphism definition of infinite outcome generalises the definition of infinite traces in transition systems given in [21]. Recall that a one-player model $X \to PF_X(X)$ embeds into a game via $\mathsf{cl} \circ P(\eta^Q)$. It is possible to show that the induced embedding $\mathbf{Rel}(X, Z_X)$ into $\mathbf{Kl}(\mathcal{G})(X, Z_X)$ is monotone and preserves fixed points (this is almost an instance of [14, Proposition 4.11]), and maps into limit-closed morphisms. Thus the largest limit-closed homomorphism comes from the greatest homomorphism $X \nrightarrow Z_X$ in $\mathbf{Rel}$.

4.4   Computing Strategies from Below

Finally, we tackle approximating strategies from below in the lattice $X \to \mathcal{G}(Z_X)$. Define $\mathsf{prune}_\gamma : P(X) \to P(X)$ as mapping U to $\{x \in X \mid \exists V \in \gamma(x) : \mathsf{snip}_0^1(V) \subseteq U\}$. Let Y be the set of states in the greatest fixed point of prune_γ. The $\mathcal{G}F_X$-coalgebra (X, γ) restricts to a coalgebra (Y, δ), where $\delta : Y \to \mathcal{G}F_X(Y)$ maps $x \mapsto \{V \in \gamma(x) \mid \mathsf{snip}_0^1(V) \subseteq Y\}$. By construction, this coalgebra is serial: $\delta(y) \neq \emptyset$ for any $y \in Y$. We now define $\underline{\delta} : Y \nrightarrow F_X(Y)$ as mapping $x \mapsto \bigcup \delta(x)$. We take $Z_{\gamma,x}$ to be all the valid traces in $\underline{\delta}$ from x. The following proposition states that $Z_{\gamma,x}$ is the outcome of the most permissive strategy in γ at x.

Proposition 22. *If $x \in Y$, there exists a strategy $\tau \in \Sigma_\gamma(x)$ with $\mathsf{Out}(\tau)(x) = Z_{\gamma,x}$. Moreover, for all $\sigma \in \Sigma_\gamma(x)$, $\mathsf{Out}(\sigma) \subseteq \mathsf{Out}(\tau)$.*

A simple corollary is that $x \in Y$ precisely if there is a strategy from x in γ.

Proposition 23. *Let $\mathcal{X}_{\gamma,x} := \{\mathcal{U} \in \mathcal{G}(X) \mid Z_{\gamma,x} \in \mathcal{U}\}$. The set of dependent functions $(x \in X) \to \mathcal{X}_{\gamma,x}$ is a complete lattice, with bottom $\bot_\gamma(x) = \{Z_{\gamma,x}\}$. Furthermore, Ψ restricts to $\Xi_\gamma : ((x \in X) \to \mathcal{X}_{\gamma,x}) \to ((x \in X) \to \mathcal{X}_{\gamma,x})$.*

Lemma 2. *Take a prefix point f of Ξ_γ, we have that for all $x \in X$, for any $\sigma \in \Sigma_c(x)$, and $n \in \omega$, that $\bigcup_{u \in \mathsf{Im}(\sigma_n)} u \star Z_{\gamma, \mathsf{snip}_0^n(u)} \in f(x)$.*

Proposition 24. *Let $l : (x \in X) \to \mathcal{X}_{\gamma,x}$ be the least fixed point of Ξ. Denote closing $(l(x), \subseteq)$ under meets (intersection) of descending ω-chains of sets with $\mathsf{close}(l(x))$. We have that $\mathsf{close}(l(x)) = \mathsf{Out}(\gamma)(x)$.*

Note in the least fixed point of Ξ_γ we do not get all strategies, i.e. we do really need to close.

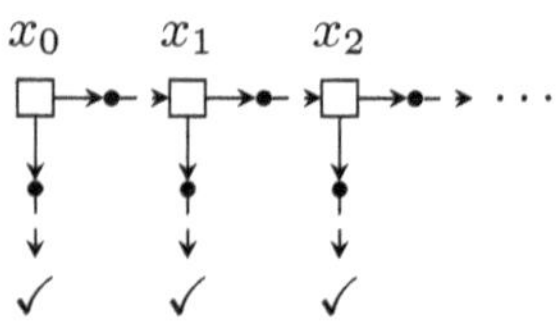

Example 4. Consider the game on the right. The least fixed point does not contain the set $\{x_0 x_1 x_2 \cdots\}$ at x_0. We can show this formally by verifying that $x_n \mapsto \mathsf{Out}(\gamma)(x_n) \setminus \{x_n x_{n+1} \ldots\}$ is a prefix point of Ξ_γ. We have $x_0 \ldots x_n \cdot \{x_{n+1}\checkmark, x_{n+1}x_{n+2}\checkmark, \ldots\}$ in the least fixed point, so close adds the meet:

$$\bigcap_{n \in \omega} x_0 \ldots x_n \cdot \{x_{n+1}\checkmark, x_{n+1}x_{n+2}\checkmark, \ldots\} = \{x_0 x_1 \ldots\}$$

5   The Monotone Neighbourhood Monad

As we mentioned in the introduction, previous work in coalgebra [10,11,14] models the branching structure in a game with the *monotone neighbourhood monad* $\mathcal{M}$: **Set** $\to$ **Set**. The analogous variant to our $\mathcal{G}$ maps a set X to the set of upwards closed subsets of X, i.e. fixed points of $\uparrow_X : PQ(X) \to PQ(X)$. Its unit $\eta_X^{\mathcal{M}}$ maps $x \mapsto \uparrow \{\{x\}\}$. Many aspects of our approach (especially in Section 3) will apply immediately to $\mathcal{M}F_X$-coalgebras, however some parts do not. In particular, the clear analogue of the functor $K : \mathbf{Kl}(\mathcal{G}) \to \mathbf{Rel}$ in Proposition 12 will not preserve units, and hence only be a *semi-functor*.

What is clear, is that the two monads are closely related. The composition in $\mathbf{Kl}(\mathcal{M})$ can in fact be written with an identical expression as Proposition 1, although can be simplified a little from upwards closure. Let $f : X \to \mathcal{M}(Y)$ and $\mathcal{U} \subseteq Q(X)$, we have $f \odot^{\mathcal{M}} \mathcal{U} = \{V \subseteq X \mid \exists U \in \mathcal{U}, \forall x \in U : V \in f(x)\}$.

If we restrict $\uparrow_X$ to $\mathcal{G}(X) \to \mathcal{M}(X)$, we obtain a monad morphism (notice that this does not have injective components). Conversely, the inclusion $\mathcal{M}(X) \rightarrowtail \mathcal{G}(X)$ is a *semi-monad* morphism: it doesn't commute with the monads units. This semi-monad morphism allows us to map from $\mathcal{M}F_X$-coalgebras to $\mathcal{G}F_X$-coalgebras without destroying any information.

Proposition 25. *We have the following monad morphisms, where $\rightarrowtail$ denotes an injective monad morphism. The arrow labelled* semi *is only a semi-monad morphism. The two triangles in the diagram commute. Furthermore, we have that $\uparrow_X \dashv \text{semi}_X$ is a coreflection.*

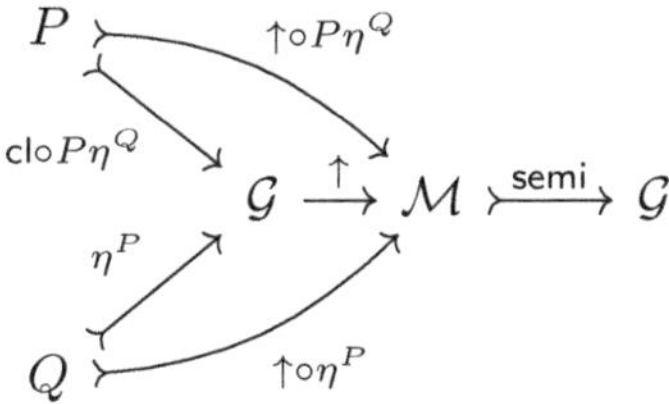

6   Future Work

We have presented a coalgebraic account of infinite outcomes in strategies and games. We showed that the outcomes of both are naturally captured as limits in **Set**, and can equivalently be characterised as largest homomorphisms. We also showed how infinite outcomes can be approximated using least fixed points.

Several avenues for future work remain. First, we aim to extend our results to games with probabilistic and weighted opponents by varying the underlying monad. It would also be interesting to investigate whether a unified framework can capture outcomes of games modelled both by our game monad and by the monotone neighbourhood monad. Finally, a natural next step is to generalise the coalgebraic approach to parity automata developed in [22] to the setting of games.

Acknowledgement. The first author thanks Jessica Newman for drawing their attention to the notion of limit-closure.

References

1. Abrahamson, K.R.: Decidability and expressiveness of logics of processes. Ph.D. thesis, USA (1980), aAI8109709
2. Bernet, J., Janin, D., Walukiewicz, I.: Permissive strategies : from parity games to safety games. RAIRO - Theoretical Informatics and Applications - Informatique Théorique et Applications **36**(3), 261–275 (2002). https://doi.org/10.1051/ita:2002013, https://www.numdam.org/articles/10.1051/ita:2002013/
3. Bouyer, P., Markey, N., Olschewski, J., Ummels, M.: Measuring permissiveness in parity games: Mean-payoff parity games revisited. In: Bultan, T., Hsiung, P.A. (eds.) Automated Technology for Verification and Analysis. pp. 135–149. Springer Berlin Heidelberg, Berlin, Heidelberg (2011)
4. Cousot, P., Cousot, R.: Constructive versions of Tarski's fixed point theorems. Pacific Journal of Mathematics **81**(1), 43–57 (1979)
5. Cîrstea, C.: Maximal traces and path-based coalgebraic temporal logics. Theoretical Computer Science **412**(38), 5025–5042 (2011). https://doi.org/https://doi.org/10.1016/j.tcs.2011.04.025, https://www.sciencedirect.com/science/article/pii/S0304397511003239, cMCS Tenth Anniversary Meeting
6. Emerson, E.: Alternative semantics for temporal logics. Theoretical Computer Science **26**(1), 121–130 (1983). https://doi.org/https://doi.org/10.1016/0304-3975(83)90082-8, https://www.sciencedirect.com/science/article/pii/0304397583900828
7. Garner, R.: The vietoris monad and weak distributive laws. Applied Categorical Structures **28**(2), 339–354 (2020)
8. Goranko, V., Jamroga, W.: State and path coalition effectivity models of concurrent multi-player games. Autonomous Agents and Multi-Agent Systems **30**(3), 446–485 (2016)
9. Goy, A., Petrişan, D., Aiguier, M.: Powerset-Like Monads Weakly Distribute over Themselves in Toposes and Compact Hausdorff Spaces. In: Bansal, N., Merelli, E., Worrell, J. (eds.) 48th International Colloquium on Automata, Languages, and Programming (ICALP 2021). Leibniz International Proceedings in Informatics (LIPIcs), vol. 198, pp. 132:1–132:14. Schloss Dagstuhl – Leibniz-Zentrum für Informatik, Dagstuhl, Germany (2021). https://doi.org/10.4230/LIPIcs.ICALP.2021.132, https://drops.dagstuhl.de/entities/document/10.4230/LIPIcs.ICALP.2021.132
10. Hansen, H.H., Kupke, C., Marti, J., Venema, Y.: Parity games and automata for game logic. In: Madeira, A., Benevides, M. (eds.) Dynamic Logic. New Trends and Applications. pp. 115–132. Springer International Publishing, Cham (2018)
11. Hasuo, I.: Generic weakest precondition semantics from monads enriched with order. Theoretical Computer Science **604**, 2–29 (2015). https://doi.org/https://doi.org/10.1016/j.tcs.2015.03.047, https://www.sciencedirect.com/science/article/pii/S0304397515002947, coalgebraic Methods in Computer Science
12. Hasuo, I., Jacobs, B., Sokolova, A.: Generic trace semantics via coinduction. Logical Methods in Computer Science **Volume 3, Issue 4**, 11 (Nov 2007). https://doi.org/10.2168/LMCS-3(4:11)2007, https://lmcs.episciences.org/864

13. Jacobs, B.: Trace semantics for coalgebras. Electronic Notes in Theoretical Computer Science **106**, 167–184 (2004). https://doi.org/https://doi.org/10.1016/j.entcs.2004.02.031, `https://www.sciencedirect.com/science/article/pii/S1571066104051746`, proceedings of the Workshop on Coalgebraic Methods in Computer Science (CMCS)
14. Kojima, R., Cirstea, C.: Continuation semantics for fixpoint modal logic and computation tree logics. Electronic Notes in Theoretical Informatics and Computer Science **Volume 5 - Proceedings of MFPS XLI**, 13 (Dec 2025). https://doi.org/10.46298/entics.16653, `https://entics.episciences.org/16653`
15. Kurz, A., Milius, S., Pattinson, D., Schröder, L.: Simplified Coalgebraic Trace Equivalence, pp. 75–90. Springer International Publishing, Cham (2015). https://doi.org/10.1007/978-3-319-15545-6₈, `https://doi.org/10.1007/978-3-319-15545-6_8`
16. Milius, S.: Relations in categories. York University Toronto, Ontario (2000)
17. Milius, S.: On colimits in categories of relations. Applied Categorical Structures **11**(3), 287–312 (2003)
18. Milius, S., Pattinson, D., Schröder, L.: Generic Trace Semantics and Graded Monads. In: Moss, L.S., Sobocinski, P. (eds.) 6th Conference on Algebra and Coalgebra in Computer Science (CALCO 2015). Leibniz International Proceedings in Informatics (LIPIcs), vol. 35, pp. 253–269. Schloss Dagstuhl – Leibniz-Zentrum für Informatik, Dagstuhl, Germany (2015). https://doi.org/10.4230/LIPIcs.CALCO.2015.253, `https://drops.dagstuhl.de/entities/document/10.4230/LIPIcs.CALCO.2015.253`
19. Plummer, B., Cirstea, C.: Traces via strategies in two-player games. Electronic Notes in Theoretical Informatics and Computer Science **Volume 5 - Proceedings of MFPS XLI**, 18 (Dec 2025). https://doi.org/10.46298/entics.16816, `https://entics.episciences.org/16816`
20. Ramadge, P.J., Wonham, W.M.: Supervisory control of a class of discrete event processes. In: Bensoussan, A., Lions, J.L. (eds.) Analysis and Optimization of Systems. pp. 475–498. Springer Berlin Heidelberg, Berlin, Heidelberg (1984)
21. Urabe, N., Hasuo, I.: Coalgebraic Infinite Traces and Kleisli Simulations. In: Moss, L.S., Sobocinski, P. (eds.) 6th Conference on Algebra and Coalgebra in Computer Science (CALCO 2015). Leibniz International Proceedings in Informatics (LIPIcs), vol. 35, pp. 320–335. Schloss Dagstuhl – Leibniz-Zentrum für Informatik, Dagstuhl, Germany (2015). https://doi.org/10.4230/LIPIcs.CALCO.2015.320, `https://drops.dagstuhl.de/entities/document/10.4230/LIPIcs.CALCO.2015.320`
22. Urabe, N., Shimizu, S., Hasuo, I.: Coalgebraic Trace Semantics for Buechi and Parity Automata. In: Desharnais, J., Jagadeesan, R. (eds.) 27th International Conference on Concurrency Theory (CONCUR 2016). Leibniz International Proceedings in Informatics (LIPIcs), vol. 59, pp. 24:1–24:15. Schloss Dagstuhl – Leibniz-Zentrum für Informatik, Dagstuhl, Germany (2016). https://doi.org/10.4230/LIPIcs.CONCUR.2016.24, `https://drops.dagstuhl.de/entities/document/10.4230/LIPIcs.CONCUR.2016.24`

Active Learning Techniques for Pomset Recognizers

Adrien Pommellet[1], Amazigh Amrane[1], Edgar Delaporte[1], Geoffroy Du Prey[1], and Oscar Peyron[12]

[1] LRE, EPITA, Le Kremlin-Bicêtre, France
[2] École Polytechnique, France

Abstract. Series-parallel pomsets are a promising mathematical formalism for concurrent programs, as they can be recognized by simple algebraic structures known as pomset recognizers. Active learning consists in inferring a formal model of a system by interactively probing its behavior through queries to a Minimally Adequate Teacher (MAT). We improve existing learning algorithms for pomset recognizers by 1. designing a new counterexample analysis procedure that is in the best case scenario exponentially more efficient than existing techniques, 2. introducing and implementing a new algorithm PL^λ that extends the state-of-the-art L^λ algorithm to pomset recognizers, minimizing the impact of exceedingly verbose counterexamples and removing redundant queries, and 3. designing a suitable finite test suite that ensures equivalence between two pomset recognizers by adapting the well-known W-method.

Keywords: Active learning · Concurrency · Pomsets

1   Introduction

Finite state automata are a straightforward model for terminating sequential systems. Runs are described by a *total* order relation: an execution is an ordered, linear sequence of events. But concurrent programs require richer structures.

Indeed, two threads may be acting in parallel, neither of them preceding nor following the other. In this case, runs may be modelled using a *partial* order: concurrent events cannot be relatively ordered, but sequential ones can. *Series-parallel partially ordered multisets* [26] (SP pomsets) offer a convenient linear description of such executions based on sequential and parallel composition of letters.

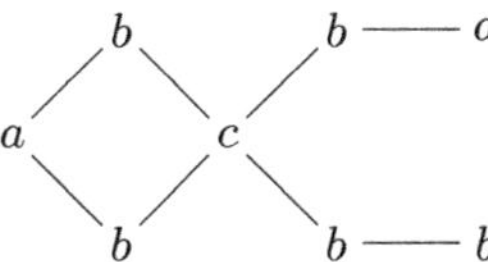

Fig. 1: The series-parallel pomset $a(b \parallel b)c(ba \parallel bb)$.

As an example, Figure 1 displays the Hasse diagram of the SP pomset $a(b \parallel b)c(ba \parallel bb)$. This model is in some cases exponentially more succinct than words: a single pomset $a_1 \parallel \ldots \parallel a_n$ describes the interleaving semantics of n parallel threads and subsumes the $n!$ possible linearized traces of the interwoven threads.

N. Bertrand and S. Milius (Eds.): FoSSaCS 2026, LNCS 16503, pp. 571–592, 2026.
https://doi.org/10.1007/978-3-032-22730-0_27

Languages of pomsets can be efficiently recognized by a special class of deterministic bottom-up tree automata known as *pomset recognizers*, which are derived from an algebraic framework studied by Lodaya and Weil [21] called *bimonoids* [5]: sets equipped with two internal operations, one associative and the other associative and commutative, with a neutral element for each operation.

Active learning (AL) consists in inferring a formal model of a black-box system that can be dynamically queried. Under *the Minimally Adequate Teacher* (MAT) framework, interactions with the black-box system are twofold: *membership queries* (MQs) to ask whether a given trace can be generated by the system, and *equivalence queries* (EQs) to determine whether a given formal model (known as the *hypothesis*) accurately represents all executions of the system, returning a counterexample that refines the hypothesis if the answer is negative.

One of the earliest AL algorithms is Angluin's L^* [3] for rational languages. In a FoSSaCS 2021 article, van Heerdt et al. [31] applied L^* to the class of recognizable pomset languages (we denote this extension PL^*). Over the years, various improvements have been brought to the original L^*:

- New algorithms such as TTT [15], $L^\#$ [29], or L^λ [14] have been shown to significantly reduce the number of MQs performed, e.g. through the use of redundancy-free discrimination trees [17].
- Since the counterexample's length m returned by the MAT can be arbitrarily large, it may dominate the AL process. Rivest and Schapire [27] proposed an algorithm that refines the hypothesis using only $\mathcal{O}(\log(m))$ MQs.
- Using EQs makes little practical sense as it assumes that the MAT knows the very formal model M of the system being inferred; Chow [6] and Vasilevskii [32] independently showed that, given a bound on $|M| - |H|$, a test suite of polynomial size w.r.t. $|H|$ can subsume equivalence between M and H.

However, very few AL algorithms for intrinsically concurrent models have been designed so far. In order to advance automated reasoning about concurrent behaviours, we adapt and extend state-of-the-art techniques to recognizable pomset languages. Our new contributions are the following:

A new counterexample analysis algorithm. Assume that a counterexample has a tree representation of depth d that features m nodes. Existing techniques infer a refinement in $\mathcal{O}(m)$ MQs. We introduce a method that only requires $\mathcal{O}(d)$ queries instead (hence $\mathcal{O}(\log(m))$ for balanced trees).

Extending L^λ [14] to recognizable pomset languages. Algorithm L^λ is a state-of-the-art active learning algorithm for rational languages which has shown better empirical results than L^*. We design an expansion PL^λ to recognizable pomset languages that shares the same empirical behavior.

Adapting the W-method [6,32]. We design a finite test suite that can conditionally replace EQs.

A C++ tool. We implemented and benchmarked our algorithms as well as van Heerdt et al.'s [31] PL^* (that, until now, had yet to be implemented).

Finally, it is worth noting that these algorithms can be applied as is to binary tree automata, associativity and commutativity constraints notwithstanding.

1.1   Related works

Pomset languages. Automata over series-parallel pomsets, known as *branching automata*, were first introduced by Lodaya and Weil [20,21], as well as a generalization of regular expressions [19,22] to pomset languages. *Pomset automata* were then introduced in [16]; it was later shown in [4] that branching automata and pomset automata are effectively equivalent. From an algebraic perspective, Lodaya and Weil investigated the recognizability of languages of SP pomsets using *bimonoids*. They proved that recognizable languages adhere to a Myhill-Nerode-like theorem (hence, amenable to learning under a MAT) and are also recognized by branching automata, but that the converse does not hold.

Alur et al. [2] recently introduced synchronized series-parallel graphs, a variant of series-parallel pomsets that allows ordered and unordered parallel compositions, and a corresponding class of automata and a logical characterization. They posit that these graphs could be used to model distributed data streams.

Active learning algorithms for finite state machines. The original L^* AL algorithm [3] maintains a table structure called an *observation table* that it fills by calling MQs. TTT [15] instead features a *discrimination tree* that has been experimentally shown to reduce the number of MQs needed. $L^\#$ [29] operates directly on a prefix tree that stores MQs and tries to establish apartness, a constructive form of non-equivalence. L^λ [14] relies on partition refinement; its peculiarity is not adding substrings of counterexamples to its data structures.

The last three algorithms have all been shown to be competitive and a net improvement over L^*. We chose to adapt L^λ due to its generic, unifying framework that is not intrinsically tied to finite words and rational languages.

Active learning algorithms for parallel models. Pomset recognizers can be seen as a special case of deterministic bottom-up tree automata. Drewes and Högberg [9] proposed an AL algorithm for tree automata, which later inspired the approach of van Heerdt et al. [31] for pomset recognizers. Whereas the latter directly infers new distinguishers from counterexamples, the former computes them inductively in a fashion similar to PL^λ, although they failed to account for sharpness issues in the partition it maintains.

Moreover, both Drewes and Högberg's algorithm as well as PL^* rely on a counterexample analysis procedure of linear complexity w.r.t. the input tree's number of nodes, whereas FindEBP's complexity instead depends instead on its depth, hence, is logarithmic if the input tree is balanced.

In [1,13,18,23,25,28], *compositional* learning algorithms for concurrent programs are introduced: the system under learning is decomposable as a parallel product of components individually modelled by Mealy or Moore machines. Rather than learning the entire system as a single composite product, each component is instead learnt in isolation. Depending on the algorithm considered, the alphabets of the various components may or may not be known beforehand.

Compositional approaches are theoretically more frugal but less expressive than our pomset-based approach, as the resulting model learnt remains finite state. Indeed, given w_1, w_2 in the set $\mathrm{SP}(\Sigma)$ of SP pomsets, we can inductively define a *linearization* operator $\pi : \mathrm{SP}(\Sigma) \to 2^{\Sigma^*}$ such that $\pi(a) = a$ for any letter $a \in \Sigma$, $\pi(w_1 \cdot w_2) = \pi(w_1) \cdot \pi(w_2)$ and $\pi(w_1 \parallel w_2) = \pi(w_1) \, ш \, \pi(w_2)$, where $ш$ is the *shuffle* product on finite words. Given a partial order on a labelled set (i.e. a pomset), π lists the total orders (i.e. finite words) on the same labelled set compatible with the aforementioned partial order. It allows us to directly compare the expressiveness of recognizable pomset languages and regular languages over finite words.

Thanks to the pumping lemma, one can prove that there exists a recognizable pomset language $L \subseteq \mathrm{SP}(\Sigma)$ such that $\pi(L) \subseteq \Sigma^*$ is not regular. Indeed, Example 1 defines one such language. Intuitively, its inductive definition allows for an arbitrary number of nested parallel branches; it models dynamic nested thread creation.

2   Preliminary Definitions

2.1   Series-parallel pomsets

Definition 1 (Series-parallel terms). *It is the set* $\mathsf{ST}(\Sigma)$ *over a finite alphabet* Σ *defined by the following grammar:*

$$t ::= \varepsilon \;\mid\; a \in \Sigma \;\mid\; t \cdot t \;\mid\; t \parallel t$$

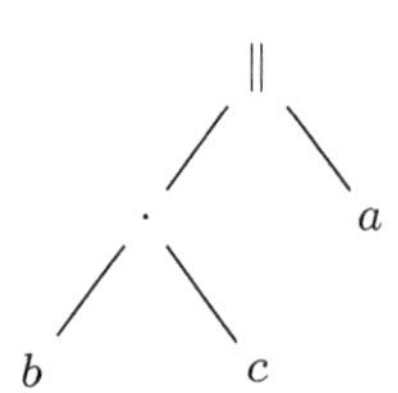

Series-parallel terms (from now on, *terms*) are a particular class of full binary trees. The operators $\cdot$ and $\parallel$ are respectively called *sequential composition* (or concatenation) and *parallel composition*. Term ε is called the *empty term*. A prefix traversal of a term yields a string known as its *linear description*: syntactically, $\cdot$ is made implicit, $\cdot$ has priority over $\parallel$, and $\cdot$ and $\parallel$ are left associative. A term is shown in Figure 2.

Fig. 2: The term $bc \parallel a$.

Definition 2 (Series-parallel pomsets). *The free algebra of series-parallel pomsets [5] is the quotient space* $\mathrm{SP}(\Sigma) = \mathsf{ST}(\Sigma) \, / \equiv_{\mathsf{ST}(\Sigma)}$ *where* $\equiv_{\mathsf{ST}(\Sigma)}$ *is the congruence relation such that* ε *is a neutral element for* $\cdot$ *and* $\parallel$, $\cdot$ *and* $\parallel$ *are associative, and* $\parallel$ *is commutative w.r.t.* $\equiv_{\mathsf{ST}(\Sigma)}$.

Intuitively, $a \cdot b$ stands for b occurring after a, whereas $a \parallel b$ means that a and b are happening concurrently (hence the latter operator's commutativity). A series-parallel pomset (from now on, *pomset*) models an execution trace of a concurrent program; a term is a description of this trace.

We conventionally describe a pomset with the linear description of some representative: $a \parallel b = b \parallel a = a \cdot \varepsilon \parallel b$ all refer to the same pomset. Given $w \in \mathrm{SP}(\Sigma)$, $\mathsf{ST}(w) \subseteq \mathsf{ST}(\Sigma)$ stands for the set of terms representing the same pomset w (that is, the pomset itself, were we to view it as an congruence class w.r.t. $\equiv_{\mathsf{ST}(\Sigma)}$). Terms in $\mathsf{ST}(w)$ all share the same non-ε leaves.

We define the set $\mathrm{SP}^+(\Sigma) = \mathrm{SP}(\Sigma) \setminus \{\varepsilon\}$ of *non-empty pomsets*.

A term of w is said to be *minimal* if it is ε-free (or consists of ε alone), of minimal depth w.r.t. $\mathsf{ST}(w)$, and its subterms (that is, its subtrees) are minimal too. The *depth $\delta(w)$* (resp. *size $|w|$*) of w is the minimum of the tree depth (resp. number of nodes) function on $\mathsf{ST}(w)$. Given $w_1, w_2 \in \mathrm{SP}(\Sigma)$, w_1 is a *subpomset* of w_2 if there exist $t_1 \in \mathsf{ST}(w_1)$ and $t_2 \in \mathsf{ST}(w_2)$ such that t_2 is a subterm of t_1.

Historically, the word pomset [26] stands for *partially ordered multiset*: an isomorphism class of labelled, partially ordered sets. The *parallel composition* $w_1 \parallel w_2$ (resp. *sequential composition $w_1 \cdot w_2$*) of two pomsets w_1 and w_2 consists in assuming their elements are pairwise incomparable (resp. every element of w_2 is greater than every element of w_1). The set $\mathrm{SP}(\Sigma)$ is the smallest set containing $\{\varepsilon\}$ and Σ, closed under sequential and parallel composition. It coincides with *N-free* pomsets [30]: there are no x_1, x_2, x_3, x_4 such that $x_1 < x_2, x_3 < x_2, x_3 < x_4$ but x_1 and x_4 cannot be compared.

2.2   Pomset recognizers

Definition 3 (Bimonoids [5]). *It is a tuple $(M, \odot, \oplus, e)$ such that M is a set equipped with two internal associative operations $\odot$ and $\oplus$, $\oplus$ being commutative as well, and a neutral element e common to $\odot$ and $\oplus$.*

Definition 4 (Pomset recognizers). *The tuple $\mathcal{R} = (R, \odot, \oplus, e, i, F)$ is said to be a* pomset recognizer *(PR) over Σ if $(R, \odot, \oplus, e)$ is a finite bimonoid, $i \colon \Sigma \to R$, and $F \subseteq R$. The set R (resp. F) is also called the set of* states *(resp.* accepting *or* final *states) of $\mathcal{R}$.*

We define the *evaluation function $i^\sharp : \mathsf{ST}(\Sigma) \to R$* of $\mathcal{R}$ inductively: $i^\sharp(\varepsilon) = e$, if $a \in \Sigma$, $i^\sharp(a) = i(a)$, $i^\sharp(t_1 \cdot t_2) = i^\sharp(t_1) \odot i^\sharp(t_2)$ and $i^\sharp(t_1 \parallel t_2) = i^\sharp(t_1) \oplus i^\sharp(t_2)$. If $t_1 \equiv_{\mathsf{ST}(\Sigma)} t_2$, then $i^\sharp(t_1) = i^\sharp(t_2)$: this property follows from the *freeness* of $\mathrm{SP}(\Sigma)$ [5] and from $\mathcal{R}$ being a bimonoid. Thus, we can define $i^\sharp : \mathrm{SP}(\Sigma) \to R$ where, given $w \in \mathrm{SP}(\Sigma)$, $i^\sharp(w) = i^\sharp(t)$ for any $t \in \mathsf{ST}(w)$. The *language* of $\mathcal{R}$ is the set $\mathcal{L}(\mathcal{R}) = \{w \in \mathrm{SP}(\Sigma) \mid i^\sharp(w) \in F\}$. If $w \in \mathcal{L}(\mathcal{R})$, then $\mathcal{R}$ *accepts* w; we then write $\mathcal{R}(w) = 1$. If $w \notin \mathcal{L}(\mathcal{R})$ $\mathcal{R}(w) = 0$.

Two PRs $\mathcal{R}_1$ and $\mathcal{R}_2$ over a common alphabet Σ are *equivalent* if $\mathcal{L}(\mathcal{R}_1) = \mathcal{L}(\mathcal{R}_2)$. A set (or *language*) $L \subseteq \mathrm{SP}(\Sigma)$ is said to be *recognizable* if there exists a PR $\mathcal{R}$ such that $L = \mathcal{L}(\mathcal{R})$. Intuitively, PRs act as bottom-up deterministic finite tree automata on terms with desirable algebraic properties: two terms representing the same pomset share the same acceptance.

Example 1. Let L be the language containing singleton c and every pomset $(a \parallel bu)$ where $u \in L$, i.e. $L = \{c, a \parallel (bc), a \parallel (b(a \parallel (bc))), \dots\}$.

This language is accepted by the PR $\mathcal{R} = (R, \odot, \oplus, e, i, F)$ where $R = \{r_a, r_b, r_c, r_{bc}, r_0, e\}$, $i(x) = r_x$ for $x \in \{a, b, c\}$, $F = \{r_c\}$, and $\oplus$ and $\odot$ are such that $r_b \odot r_c = r_{bc}$, $r_a \oplus r_{bc} = r_c$, e is the neutral element for both operations, and all the other possible products return r_0.

2.3 Contexts

We define pomset patterns potentially featuring placeholder symbols denoted $\square_j$ in lieu of letters that can be replaced by pomsets.

Definition 5 (Multi-contexts). *For $m \in \mathbb{N} \setminus \{0\}$, let $\Xi = \{\square_1, \ldots, \square_m\}$ be a set of m distinct letters such that $\Xi \cap \Sigma = \emptyset$. The set of m-contexts $\mathrm{C}_m(\Sigma)$ is the subset of $\mathrm{SP}(\Sigma \cup \Xi)$ of pomsets containing exactly one element labelled by $\square_j$ for all $j \in \{1, \ldots, m\}$.*

Given $c \in \mathrm{C}_m(\Sigma)$ and $w_1, \ldots, w_m \in \mathrm{SP}(\Sigma)$, $c[w_1, \ldots, w_m]$ denotes the pomset where $\square_j$ has been replaced by w_j, that is, a term $t_j \in \mathrm{ST}(w_j)$ has been inserted in a term of $t_c \in \mathrm{ST}(c)$ in place of $\square_j$. The latter operation trivially yields the same pomset regardless of the representatives t_j and t_c chosen.

We write $\mathrm{SP}(\Sigma) = \mathrm{C}_0(\Sigma)$. We simply call 1-contexts *contexts*, and always denote their placeholder symbol $\square$. Given $c_1, c_2 \in \mathrm{C}_1(\Sigma)$, $c_1[c_2] \in \mathrm{C}_1(\Sigma)$ stands for the context obtained by replacing $\square$ with c_2 in c_1. Context c_2 is then said to be a *subcontext* of $c_1[c_2]$.

For $w \in \mathrm{SP}(\Sigma)$, a *split* of w is a pair $(c, t) \in \mathrm{C}_1(\Sigma) \times \mathrm{ST}(z)$ for some $z \in \mathrm{SP}(\Sigma)$ such that $w = c[z]$. Note that z is a subpomset of w. Given $C \subseteq \mathrm{C}_1(\Sigma)$ and $A \subseteq \mathrm{SP}(\Sigma) \cup \mathrm{C}_1(\Sigma)$, we define the set $C[A] = \{c[z] \mid c \in C, z \in A\}$.

2.4 A Myhill-Nerode theorem

Given $L \subseteq \mathrm{SP}(\Sigma)$, we consider the *congruence relation* $\sim_L$ such that for $u, v \in \mathrm{SP}(\Sigma)$, $u \sim_L v$ if and only if for all $c \in \mathrm{C}_1(\Sigma)$, $c[u] \in L \iff c[v] \in L$. It is an equivalence relation on $\mathrm{SP}(\Sigma)$ compatible with $\cdot$ and $\|$. Set $[w]_{\sim_L}$ denotes the equivalence class of w in the quotient space $\mathrm{SP}(\Sigma) / \sim_L$. There exists a Myhill-Nerode characterization of recognizable languages of $\mathrm{SP}(\Sigma)$:

Theorem 1 ([21]). *A language $L \subseteq \mathrm{SP}(\Sigma)$ is recognizable if and only if $\sim_L$ has finite index, i.e. $\mathrm{SP}(\Sigma) / \sim_L$ is finite.*

Given a pomset language L and $w_1, w_2 \in \mathrm{SP}(\Sigma)$, we say that $c \in \mathrm{C}_1(\Sigma)$ is a *distinguishing context* in L for w_1 and w_2 if $c[w_1] \in L \iff c[w_2] \notin L$, that is, one of $c[w_1]$ and $c[w_2]$ is in L while the other is not. If L is recognized by a PR $\mathcal{R}$, this necessarily implies that $m_1 = i^\sharp(c[w_1]) \neq m_2 = i^\sharp(c[w_2])$: one state must be in F while the other is not. We then say that c *distinguishes* states m_1 and m_2 (resp. pomsets w_1 and w_2). If there is no such c, m_1 (resp. w_1) and m_2 (resp. w_2) are said to be *indistinguishable*. We write $\sim_{\mathcal{R}} = \sim_{\mathcal{L}(\mathcal{R})}$.

Example 2. Consider Example 1 again. Context $\square \parallel c$ distinguishes a from ε, hence states r_a and e, as $a \parallel c \notin L$ but $\varepsilon \parallel c = c \in L$.

Definition 6 (Reachable, distinguished, minimal pomset recognizers). *A pomset recognizer $\mathcal{R} = (R, \odot, \oplus, 1, i, F)$ is said to be* reachable *if, for all $m \in R$, there exists $w \in \mathrm{SP}(\Sigma)$ such that $i^\sharp(w) = m$; w is said to be an* access pomset *of m. $\mathcal{R}$ is* distinguished *if for all $w_1, w_2 \in \mathrm{SP}(\Sigma)$ such that $i^\sharp(w_1) \neq i^\sharp(w_2)$, there exists $c \in \mathrm{C}_1(\Sigma)$ such that $\mathcal{R}(c[w_1]) \neq \mathcal{R}(c[w_2])$. $\mathcal{R}$ is* minimal *if it is both reachable and distinguished.*

If L is recognizable, $\sim_L$ induces an obvious, unique minimal pomset recognizer $\mathcal{R}_L = (\mathrm{SP}(\Sigma)\,/\sim_L, \cdot, \|, [\varepsilon]_{\sim_L}, i_L, F_L)$ such that $\forall a \in \Sigma$, $i_L(a) = [a]_{\sim_L}$, $F_L = \{[w]_{\sim_L} \in \mathrm{SP}(\Sigma)\,/\sim_L \mid w \in L\}$ and $[w_1]_{\sim_L} \circ [w_2]_{\sim_L} = [w_1 \circ w_2]_{\sim_L}$ for $\circ \in \{\cdot, \|\}$. The resulting evaluation function $i^{\sharp}$ is the syntactic homomorphism $\mathrm{SP}(\Sigma) \to \mathrm{SP}(\Sigma)\,/\sim_L$ that matches pomsets to their congruence class. As a consequence, knowledge of a minimal PR that recognizes L and knowledge of L's congruence relation and classes are one and the same.

3   An Active Learning Algorithm for Pomset Recognizers

Consider a recognizable pomset language L on an alphabet Σ. Let the *model* $\mathcal{M}$ be the unique minimal PR such that $\mathcal{L}(\mathcal{M}) = L$. *Active learning* (AL) is a game between a learner and a *minimally adequate teacher* (MAT) that consists in the former guessing $\mathcal{M}$ by asking two types of queries to the latter:

Membership queries. Given $w \in \mathrm{SP}(\Sigma)$, does $w \in L$, i.e. what is $\mathcal{M}(w)$?

Equivalence queries Given a pomset recognizer $\mathcal{H}$ (called the hypothesis) on Σ, does $\mathcal{L}(\mathcal{H}) = L = \mathcal{L}(\mathcal{M})$? If it does not, return a *counterexample* $w \in \mathrm{SP}(\Sigma)$ such that $\mathcal{H}(w) \neq \mathcal{M}(w)$.

We introduce in this section PL^{λ}, a new AL algorithm for PRs.

3.1   Data structures

The ability to infer $\mathcal{M}$ from queries stems from Theorem 1. AL algorithms compute an *under-approximation* $\sim_{\mathcal{H}}$ of $\sim_L = \sim_{\mathcal{M}}$ such that $w_1 \not\sim_{\mathcal{H}} w_2 \implies w_1 \not\sim_L w_2$. Obviously, $w_1 \sim_{\mathcal{H}} w_2 \implies w_1 \sim_L w_2$ may not hold if the hypothesis is too coarse, thus, $\mathcal{H}$ may have to be refined several times. Nevertheless, each refinement increases the number of congruence classes of $\sim_L$ distinguished by $\sim_{\mathcal{H}}$, until the classes of $\sim_{\mathcal{H}}$ are exactly the classes of $\sim_L$, at which point $\sim_{\mathcal{H}} = \sim_{\mathcal{M}} = \sim_L$ and $\mathcal{L}(\mathcal{H}) = \mathcal{L}(\mathcal{M}) = L$; PL^{λ} then terminates.

Covering the congruence classes. PL^{λ} maintains a finite set S of pomsets called the set of *access pomsets*, meant to store representatives of $\sim_L$'s congruence classes. By design, S will be closed by the subpomset relation and contain the empty pomset ε. The classes of $\sim_L$ are the states of the model $\mathcal{M}$.

However, knowledge of its evaluation function $i_{\mathcal{M}}$ and its internal operations $\cdot_{\mathcal{M}}$ and $\|_{\mathcal{M}}$ is also required to fully infer $\mathcal{M}$. Thus, we also introduce a *frontier* set $S^+ = (\Sigma \cup \{u \circ v \mid \circ \in \{\cdot, \|\}, u, v \in S\}) \setminus S$ that contains single letters and combinations of elements of S so that we can infer their congruence classes.

A *pack of components* $\mathcal{B} = \{B_1, \ldots, B_m\}$ partitions $S \cup S^+$ in such a manner that each component contains at least one $s \in S$. For $s \in S \cup S^+$, $\mathcal{B}_s$ stands for the only component of $\mathcal{B}$ s belongs to. Given $B \in \mathcal{B}$, $\alpha_{\mathcal{B}}(B) = S \cap B$ is called the set of *access pomsets* of B. For $s \in S \cup S^+$, we define $\alpha_{\mathcal{B}}(s) = \alpha_{\mathcal{B}}(\mathcal{B}_s)$. $\mathcal{B}$'s purpose is to under-approximate the classes of $\sim_{\mathcal{M}}$ and classify elements of $S \cup S^+$.

Distinguishing the congruence classes. To compute $\mathcal{B}$ and posit which class of $\sim_L$ an element of $S \cup S^+$ belongs to, we introduce a *discrimination tree* (DT) $\mathcal{D}$. It is a full binary tree (see Figures 3 and 5 for examples): its inner nodes are labelled by contexts in $C_1(\Sigma)$; in particular, its root is labelled by $\square$; its leaves are either unlabelled or labelled by a component of $\mathcal{B}$, in such a fashion $\mathcal{D}$'s set of labelled leaves is in bijection with $\mathcal{B}$.

The labels of $\mathcal{D}$'s inner nodes form a set of contexts $\mathcal{C}$. Given $B \in \mathcal{B}$, $\mathcal{C}_B$ is defined as the set of contexts that appear along the branch that runs from the root of $\mathcal{D}$ to the leaf labelled by B. Note that for all $B \in \mathcal{B}$, $\square \in \mathcal{C}_B$.

For any pomset $w \in \mathrm{SP}(\Sigma)$, we define the *sifting* operation of w through $\mathcal{D}$: starting from the root of $\mathcal{D}$, at every inner node of $\mathcal{D}$ labelled by a context c, we perform a MQ on $c[w]$ then branch towards the right (resp. left) child if $c[w] \in L$ (resp. $c[w] \notin L$). We iterate this procedure until a leaf is reached: the matching component $B \in \mathcal{B}$ is the result of the sifting operation. We define $\mathcal{D}(w) = B$. $\mathcal{D}(w)$ may be undefined if w is sifted into an unlabelled leaf, at which point a new component with access pomset w is created (see Algorithm 1). Thus, $\mathcal{D}$ can be viewed as a partial function $\mathrm{SP}(\Sigma) \to \mathcal{B}$.

Sifting requires a number of MQs bounded by the height of $\mathcal{D}$. Intuitively, DTs are used to classify pomsets: pomsets that behave similarly w.r.t. the finite set of distinguishing contexts $\mathcal{C}_B$ are sifted into the same component $B \in \mathcal{B}$; thus, $\mathcal{B}_w = \mathcal{D}(w)$. Conversely, if two elements of $S \cup S^+$ belong to two different components B_1 and B_2, then there are distinguished by the deepest common ancestor of leaves B_1 and B_2 in $\mathcal{D}$.

Example 3. Let us sift b through the DT shown in Figure 5. First, we query $\square[b] = b$. It does not belong to L; we branch left then query $(\square \cdot c)[b] = bc$. The answer is negative and we branch left again. Finally, we reach a leaf labelled by $B_a = \mathcal{D}(b)$, having performed two MQs.

Properties of the partition. $\mathcal{B}$ is said to be *consistent* if for any $B_1, B_2 \in \mathcal{B}$, $u_1, v_1 \in \alpha_\mathcal{B}(B_1)$, $u_2, v_2 \in \alpha_\mathcal{B}(B_2)$, and $\circ \in \{\cdot, \|\}$, $u_1 \circ u_2$ and $v_1 \circ v_2$ belong to the same component of $\mathcal{B}$: no matter the elements of $\alpha_\mathcal{B}(B_1)$ and $\alpha_\mathcal{B}(B_2)$ we consider, their composition must belong to the same component. Moreover, $\mathcal{B}$ is $\circ$-*associative* for $\circ \in \{\cdot, \|\}$ if for any $s_1, s_2, s_3 \in S$ and $s_l \in \alpha_\mathcal{B}(s_1 \circ s_2)$, $s_r \in \alpha_\mathcal{B}(s_2 \circ s_3)$, $\mathcal{B}_{s_l \circ s_3} = \mathcal{B}_{s_1 \circ s_r}$. The *sharpness index* of $\mathcal{B}$ is the smallest integer $\rho(\mathcal{B})$ such that for any $B \in \mathcal{B}$, $|S \cap B| \le \rho(\mathcal{B})$. $\mathcal{B}$ is said to be *sharp* if $\rho(\mathcal{B}) = 1$.

If $\mathcal{B}$ is consistent and associative, we can extend the operators $\cdot$ and $\|$ to components of $\mathcal{B}$, and the resulting laws will be internal and associative. Finally, for any $B \in \mathcal{B}$, since $\square \in \mathcal{C}_B$ and for any $u \in B$, $\square[u] = u$, $\mathcal{M}$ is constant on B: this shared value is written $\mathcal{M}(B)$.

3.2   Building the hypothesis

Defining the hypothesis. If $\mathcal{B}$ is *consistent*, $\cdot$-*associative*, and $\|$-*associative*, then we design a minimal hypothesis $\mathcal{H} = (H, \cdot_\mathcal{H}, \|_\mathcal{H}, e_\mathcal{H}, i_\mathcal{H}, F_\mathcal{H})$ as follows:

  – $H = \mathcal{B}$. $\mathcal{H}$'s states are the postulated congruence classes of $\sim_L$.

- Given $u, v \in S$, since $\mathcal{B}$ is *consistent*, we can define $\mathcal{B}_u \cdot_{\mathcal{H}} \mathcal{B}_v = \mathcal{B}_{u \cdot v}$ (resp. $\mathcal{B}_u \parallel_{\mathcal{H}} \mathcal{B}_v = \mathcal{B}_{u \parallel v}$). We use S^+ and $\mathcal{B}$ to build $\mathcal{H}$'s internal operations.
- $e_{\mathcal{H}} = \mathcal{B}_\varepsilon$. The neutral element is the class of the empty pomset.
- Given $a \in \Sigma$, $i_{\mathcal{H}}(a) = \mathcal{B}_a$. We rely on $\Sigma \subseteq S \cup S^+$ to build a pomset homomorphism.
- $F_{\mathcal{H}} = \{B \in \mathcal{B} \mid \mathcal{M}(B) = 1\}$. A component is accepting if its members are accepted by $\mathcal{M}$. $F_{\mathcal{H}}$ contains the labelled leaves of $\mathcal{D}$ belonging to its right subtree.

For any $w \in \mathrm{SP}(\Sigma)$, the state $i_{\mathcal{H}}^{\sharp}(w)$ pomset w evaluates to is a component of $\mathcal{B}$ we denote $\mathcal{B}_w$; its set of access pomsets is $\alpha_{\mathcal{H}}(w) = \alpha_{\mathcal{B}}(\mathcal{B}_w)$. This notation is compatible with the earlier definition of $\mathcal{B}_w$ for $w \in S \cup S^+$ due to $\mathcal{H}$ being reachable. The hypothesis handles pomsets and their access pomsets similarly:

Property 1 (Substitution by access pomsets). $\forall c \in \mathrm{C}_1(\Sigma)$, $\forall w \in \mathrm{SP}(\Sigma)$, $\forall p \in \alpha_{\mathcal{H}}(w)$, $\mathcal{H}(c[w]) = \mathcal{H}(c[p])$ and $i_{\mathcal{H}}^{\sharp}(c[w]) = i_{\mathcal{H}}^{\sharp}(c[p])$.

Example 4. Let us learn the pomset recognizer of Example 1. As discussed later in Section 3.3, initially, $S = \{\varepsilon\}$ and the DT $\mathcal{D}$ features a single inner node labelled by $\square$. As $\varepsilon \notin L$, $\mathcal{D}$'s left child is labelled by $\mathcal{B}_\varepsilon$. Sifting the elements of S^+ through $\mathcal{D}$ results in a new component $\mathcal{B}_c$ being created and S, S^+ being updated accordingly. The resulting $\mathcal{B}$ is consistent and associative, yielding a first hypothesis $\mathcal{H}_1$ (see Figure 3).

$$S = \{\varepsilon, c\}$$
$$S^+ \setminus S = \{a, b, cc, c \parallel c\}$$
$$\mathcal{B}_\varepsilon = \{\boxed{\varepsilon}, a, b, cc, c \parallel c\}$$
$$\mathcal{B}_c = \{\boxed{c}\}$$

$$H = \{\mathcal{B}_\varepsilon, \mathcal{B}_c\}, \; F_{\mathcal{H}} = \{\mathcal{B}_c\},$$
$$i_{\mathcal{H}}(a) = i_{\mathcal{H}}(b) = \mathcal{B}_\varepsilon, \; i_{\mathcal{H}}(c) = \mathcal{B}_c,$$
$$\mathcal{B}_\varepsilon \circ \mathcal{B}_c = \mathcal{B}_c \circ \mathcal{B}_\varepsilon = \mathcal{B}_c \text{ and}$$
$$\mathcal{B}_\varepsilon \circ \mathcal{B}_\varepsilon = \mathcal{B}_c \circ \mathcal{B}_c = \mathcal{B}_\varepsilon \text{ for } \circ \in \{\cdot_{\mathcal{H}}, \parallel_{\mathcal{H}}\}.$$

Fig. 3: First pack of components, a discrimination tree, and the resulting hypothesis.

Compatibility of the hypothesis. Given a set $X \subseteq \mathrm{SP}(\Sigma)$ of pomsets, hypothesis $\mathcal{H}$ is *X-compatible* if for any $w \in X$, $\mathcal{H}(w) = \mathcal{M}(w)$. $\mathcal{H}$ is said to be *compatible* with $\mathcal{B}$ if it is compatible with $\bigcup_{B \in \mathcal{B}} \{c[s] \mid s \in B, c \in \mathcal{C}_B\}$.

AL algorithms such as TTT [15], $L^{\#}$ [29], or van Heerdt et al.'s adaptation of L^* [31] to PRs may not always immediately result in a compatible hypothesis. However, incompatibilities provide a 'free' counterexample $c[s]$ such that $\mathcal{H}(c[s]) \neq \mathcal{M}(c[s])$ without requiring an extra MQ or EQ due to $\mathcal{M}(c[s])$ having already been queried when s was sifted into $\mathcal{B}_s$. Note that compatibility of $\mathcal{H}$ implies its minimality.

3.3 Adapting the L^λ Algorithm

Inspired by L^λ [14], we outline in this section algorithm PL^λ and detail in Figure 4 the various functions that compose it.

Updating data structures. S, S^+, $\mathcal{B}$, and $\mathcal{D}$ are updated through calls to two functions: **Expand** and **Refine**, respectively Algorithms 1 and 2 of Figure 4.

Algorithm 1 inserts a new pomset w belonging to S^+ or equal to ε into the set S of access pomsets then updates S^+ and $\mathcal{B}$, generating a new component if a pomset is sifted into an unlabelled leaf of $\mathcal{D}$ which is then updated accordingly.

Algorithm 2 refines a component B into two new components B_0 and B_1 (thus increasing $|\mathcal{B}|$), assuming a context c distinguishes two access pomsets of B. S, $\mathcal{B}$ and $\mathcal{D}$ are updated accordingly. In particular, $\mathcal{D}$'s leaf labelled by B is replaced by an inner node labelled by c with children B_0 and B_1.

Intuitively, Algorithm 1 is used to expand the set of access pomsets (hence, the frontier as well) while Algorithm 2 creates a new congruence classes by using a context to refine an existing class.

Finding conflicts. Consistency and associativity issues result in a new distinguishing context being inductively built from an existing element of $\mathcal{C}$ by functions **MakeConsistent** (Algorithm 3) or **MakeAssoc** (Algorithm 4).

Algorithm 3 refines partition $\mathcal{B}$ whenever it encounters a consistency issue, e.g. class B contains two representatives p_1 and p_2 such that $p_1 \circ p$ and $p_2 \circ p$ in $S \cup S^+$ for some $p \in S$ and $\circ \in \{\cdot, \|\}$ do not belong to the same component. This inconsistency yields a context $c[\square \circ p]$ that distinguishes p_1 and p_2, where $c \in \mathcal{C}$ is the label of the deepest common ancestor in $\mathcal{D}$ of $p_1 \circ p$ and $p_2 \circ p$. This algorithm returns Boolean $\top$ if and only if $\mathcal{B}$ was already consistent in the first place. For brevity, we omit a similar symmetric case involving $p \circ p_1$ and $p \circ p_2$.

Similarly, Algorithm 4 refines partition $\mathcal{B}$ whenever it encounters an associativity issue and returns Boolean $\top$ if and only if $\mathcal{B}$ is already associative.

Lazy refinement. If no consistency or associativity issue is found, PL^λ builds hypothesis $\mathcal{H}$ defined in Section 3.2 then submits an EQ to the teacher. The answer is either positive and the algorithm ends with $\mathcal{H} = \mathcal{M}$, or negative and a counterexample w is returned; a new algorithm **FindEBP** detailed in Section 4 infers from w a $c \in C_1(\Sigma)$ and $p \in S^+$ such that c distinguishes p from a $p' \in \alpha_\mathcal{B}(p)$, hence proving that class B_p should be refined.

However, rather than immediately using c to refine B_p, PL^λ performs a signature *delayed refinement*: p is added to S, S^+ and $\mathcal{B}$ are updated accordingly, but c, being a context of arbitrary size w.r.t. $\mathcal{M}$, is not inserted in $\mathcal{D}$. This optimization leads to significant improvements of the algorithm's symbol complexity that we discuss in Sections 5 and 7.

Instead, function **AnalyzeCE** (Algorithm 5) waits for an associativity or consistency defect to necessarily arise in order to inductively update $\mathcal{C}$ and $\mathcal{D}$, rather than directly use c. Moreover, $c[p]$ and $c[p']$ are added to a counterexample pool $\mathcal{E}$ (Line 4) for any representative $p' \in \alpha_\mathcal{H}(p)$; indeed, as long as p and p' are not distinguished, either $c[p]$ or $c[p']$ will remain a counterexample: if $p, p' \in S$ belong to the same component B then $\mathcal{H}(c[p]) = \mathcal{H}(c[p'])$; but $\mathcal{M}(c[p]) \neq \mathcal{M}(c[p'])$. This loop eventually depletes $\mathcal{E}$ and ends with $\mathcal{B}$ being sharp.

Fig. 4: The various components of the PL^λ algorithm.

Algorithm 1 $\mathtt{Expand}(w)$ where $w \in S^+$ or $w = \varepsilon$ if $\mathcal{B} = \emptyset$

1: $S \leftarrow S \cup \{w\}$
2: **for** $p \in \{w\} \cup \{p' \circ w, w \circ p' \mid \circ \in \{\cdot, \|\}, p' \in S\} \cup \Sigma$ **do**
3:     **if** p does not belong to any class of $\mathcal{B}$ **then**
4:         $B \leftarrow \mathcal{D}(p)$
5:         **if** B is defined **then**
6:             $B \leftarrow B \cup \{p\}$
7:         **else**
8:             $B_p \leftarrow \{p\}$
9:             $\mathcal{B} \leftarrow \mathcal{B} \cup \{B_p\}$
10:            $\mathtt{UpdateTreeLeaf}(\mathcal{D}, p, B_p)$
11:            $\mathtt{Expand}(p)$

Algorithm 2 $\mathtt{Refine}(B, c)$ where $B \in \mathcal{B}, c \in C_1(\Sigma)$, and $\exists z_1, z_2 \in B, \mathcal{M}(c[z_1]) \neq \mathcal{M}(c[z_2])$

1: $B_0 \leftarrow \{w \in B \mid \mathcal{M}(c[w]) = 0\}$
2: $B_1 \leftarrow \{w \in B \mid \mathcal{M}(c[w]) = 1\}$
3: $\mathcal{B} \leftarrow (\mathcal{B} \setminus \{B\}) \cup \{B_0, B_1\}$
4: $\mathtt{RefineTree}(\mathcal{D}, B, c, B_0, B_1)$
5: **if** $S \cap B_0 = \emptyset$ **then** $\mathtt{Expand}(p_0)$ for some $p_0 \in B_0$
6: **if** $S \cap B_1 = \emptyset$ **then** $\mathtt{Expand}(p_1)$ for some $p_1 \in B_1$

Algorithm 3 $\mathtt{MakeConsistent}()$

1: $\mathtt{already\_consistent} \leftarrow \top$
2: **while** $\exists \circ \in \{\cdot, \|\}, \exists B \in \mathcal{B}, \exists p_1, p_2 \in \alpha_B(B), \exists p \in S, \mathcal{D}(p_1 \circ p) \neq \mathcal{D}(p_2 \circ p)$ **do**
3:     Let $c \in \mathcal{C}$ be such that $\mathcal{M}(c[p_1 \circ p]) \neq \mathcal{M}(c[p_2 \circ p])$.
4:     $\mathtt{Refine}(B, c[\square \circ p])$
5:     $\mathtt{already\_consistent} \leftarrow \bot$
6: **return** $\mathtt{already\_consistent}$

Algorithm 4 $\mathtt{MakeAssoc}()$

1: $\mathtt{already\_assoc} \leftarrow \top$
2: **while** $\exists \circ \in \{\cdot, \|\}, \exists s_1, s_2, s_3 \in S, \exists s_l \in \alpha_\mathcal{B}(s_1 \circ s_2), \exists s_r \in \alpha_\mathcal{B}(s_2 \circ s_3), \mathcal{D}(s_1 \circ s_r) \neq \mathcal{D}(s_l \circ s_3)$ **do**
3:     Let $c \in \mathcal{C}$ be such that $\mathcal{M}(c[s_1 \circ s_r]) \neq \mathcal{M}(c[s_l \circ s_3])$.
4:     $\mathtt{query} \leftarrow \mathcal{M}(c[s_1 \circ s_2 \circ s_3])$
5:     **if** $\mathcal{M}(c[s_l \circ s_3]) \neq \mathtt{query}$ **then**
6:         $\mathtt{Refine}(\mathcal{B}_{s_1 \circ s_2}, c[\square \circ s_3])$
7:     **else**
8:         $\mathtt{Refine}(\mathcal{B}_{s_2 \circ s_3}, c[s_1 \circ \square])$
9:     $\mathtt{already\_assoc} \leftarrow \bot$
10: **return** $\mathtt{already\_assoc}$

Algorithm 5 $\mathtt{AnalyzeCE}(w)$ where $w \in \mathrm{SP}^+(\Sigma)$ is such that $\mathcal{H}(w) \neq \mathcal{M}(w)$

1: $\mathcal{E} \leftarrow \{w\}$
2: **while** $\exists u \in \mathcal{E}, \mathcal{M}(u) \neq \mathcal{H}(u)$ **do**
3:     $(c, p) \leftarrow \mathtt{FindEBP}(\square, t)$ where t is a minimal term of u
4:     $\mathcal{E} \leftarrow \mathcal{E} \cup \{c[p]\} \cup \{c[p'] \mid p' \in \alpha_\mathcal{H}(p)\}$
5:     $\mathtt{Expand}(p)$
6:     **repeat**
7:     **until** $\mathtt{MakeConsistent}() \wedge \mathtt{MakeAssoc}()$
8:     $\mathcal{H} \leftarrow \mathtt{BuildHypothesis}(S, \mathcal{B})$

Algorithm 6 $\mathtt{Learn}()$

1: $S, \mathcal{B}, \mathcal{D} \leftarrow \emptyset, \emptyset, \mathtt{Tree}(\square)$
2: $\mathtt{Expand}(\varepsilon)$
3: $\mathcal{H} \leftarrow \mathtt{BuildHypothesis}(S, \mathcal{B})$
4: **while** $\exists w \in \mathrm{SP}(\Sigma), \mathcal{H}(w) \neq \mathcal{M}(w)$ **do**
5:     $\mathtt{AnalyzeCE}(w)$
6:     **while** $\exists B \in \mathcal{B}, \exists s \in B, \exists c \in \mathcal{C}_B, \mathcal{H}(c[s]) \neq \mathcal{M}(c[s])$ **do**
7:         $\mathtt{AnalyzeCE}(c[s])$
8: **return** $\mathcal{H}$

The main loop. Function **Learn** (Algorithm 6) first initializes $S = \{\varepsilon\}$ and $\mathcal{C} = \{\Box\}$, then build the pack of components $\mathcal{B}$ and a first hypothesis $\mathcal{H}$ by expanding the empty pomset ε. It then submits $\mathcal{H}$ to the teacher.

If the EQ returns a counterexample w, a call to **AnalyzeCE** identifies new components and $\mathcal{H}$ is refined accordingly. Otherwise, a model $\mathcal{H}$ equivalent to $\mathcal{M}$ has been learnt and the algorithm returns $\mathcal{H}$. Lines 6 and 7 guarantee that $\mathcal{H}$ is compatible before submitting an EQ.

Example 5. Consider counterexample $w = (cc)(a \parallel c)$ to Example 4 in the context of Example 1's learning. A call to **FindEBP** (later described in Example 6) returns context $\Box \parallel c$ and pomset $a \in S^+$. A subsequent call to **Expand** results in a being added to S and S^+ being consequently extended.

The resulting (second) pack of components (shown in Figure 5) is neither sharp nor consistent: $\varepsilon, a \in B_\varepsilon \cap S$, $\varepsilon \cdot c = c \in B_c$ yet $ac \in B_\varepsilon$. B_ε is thus refined by $\Box[\Box \cdot c] = \Box \cdot c$, leading to the creation of a new component B_a.

The resulting (third) pack of component $\mathcal{B}$ is then consistent and associative.

$$S = \{\varepsilon, c, a\}$$
$$S^+ \setminus S = \{b, cc, aa, ca, ac,$$
$$a \parallel a, a \parallel c, c \parallel c\}$$
$$B_\varepsilon = \{\boxed{\varepsilon}, \boxed{a}, b, cc, aa, ca, ac,$$
$$a \parallel c, c \parallel c, a \parallel a\}$$
$$B_c = \{\boxed{c}\}$$

$$S = \{\varepsilon, c, a\}$$
$$S^+ \setminus S = \{b, cc, aa, ca, ac,$$
$$a \parallel c, a \parallel a, c \parallel c\}$$
$$B_\varepsilon = \{\boxed{\varepsilon}\}$$
$$B_a = \{\boxed{a}, b, cc, aa, a \parallel a, ca,$$
$$ac, c \parallel c, a \parallel c\}$$
$$B_c = \{\boxed{c}\}$$

Fig. 5: Second and third packs of components, second discrimination tree.

4   A New Counterexample Analysis Algorithm

AL algorithms infer from a counterexample w to hypothesis $\mathcal{H}$ a *refining pair* $(c, p) \in \mathrm{C}_1(\Sigma) \times S^+$ such that $\exists p' \in \alpha_\mathcal{H}(p)$, $\mathcal{M}(c[p']) \neq \mathcal{M}(c[p])$. It proves that p and p' are distinguished by c in $\mathcal{M}$, yet belong to the same component $\mathcal{B}_p$, thus highlighting a necessary refinement of $\mathcal{B}_p$. Inspired by Drewes et al. [9], van Heerdt et al. [31] designed **HandleCounterExample** (from now on, **HCE**), an algorithm that infers a refining pair from w in $\mathcal{O}(\rho(\mathcal{B}) \cdot |w|)$ MQs.

We introduce in this section one of our main contributions, **FindEBP**: a new counterexample analysis algorithm that requires only $\mathcal{O}(\rho(\mathcal{B}) \cdot \delta(w))$ MQs.

4.1   Breaking points

Inspired by the work of Rivest and Schapire [27], we are trying to infer *breaking points* (BPs) from counterexamples: witnesses to the model and hypothesis no longer being in agreement over a computation. Given a consistent, associative $\mathcal{B}$ from which $\mathcal{H}$ is inferred, we extend these definitions to pomsets and terms.

Definition 7 (Agreement). *Given $c \in C_1(\Sigma)$ and $z \in SP(\Sigma)$, we define the agreement predicate $\mathcal{A}(c,z) = \text{``}\forall p \in \alpha_{\mathcal{H}}(z), \mathcal{H}(c[p]) = \mathcal{M}(c[p])\text{''}$.*

Definition 8 (Breaking point). *Given $w \in SP^+(\Sigma)$, a split $(c,t) \in C_1(\Sigma) \times ST(z)$ of w is an Effective Breaking Point (EBP) if $\mathcal{A}(c,z) = 1$ and either:*

1. *$z \in S^+$ and $\mathcal{H}(w) \neq \mathcal{M}(w)$.*
2. *$t = t_1 \circ t_2$ for some $\circ \in \{\cdot, \|\}$, $t_i \in ST(z_i)$ for $i \in \{1,2\}$, and:*
   (a) $\mathcal{A}(c[\square \circ z_2], z_1) = 0$; this equality implies that $\exists p_1 \in \alpha_{\mathcal{H}}(z_1), \mathcal{H}(c[p_1 \circ z_2]) \neq \mathcal{M}(c[p_1 \circ z_2])$;
   (b) $\mathcal{A}(c[p_1 \circ \square], z_2) = 0$.
   If only (a) holds, the BP is said to be tentative (TBP).

Note that computing $\mathcal{A}(c,z)$ requires at most $\rho(\mathcal{B})$ MQs. We seek EBPs to induce a refinement of $\mathcal{B}$:

Lemma 1. *If (c,t) is an EBP of $\mathcal{H}$ w.r.t. $\mathcal{M}$, then $\mathcal{B}$ admits a refining pair.*

Proof. Indeed, $\mathcal{A}(c,z) = 1$ implies $\mathcal{H}(c[p']) = \mathcal{M}(c[p'])$ for every $p' \in \alpha_{\mathcal{H}}(z)$. Case 1. (resp. 2.) of Definition 8 implies $\mathcal{H}(c[p]) \neq \mathcal{M}(c[p])$ for $p = z$ (resp. $p = p_1 \circ p_2 \in S^+$ where $\mathcal{A}(c[p_1 \circ \square], z_2) = 0$ yields a $p_2 \in \alpha_{\mathcal{H}}(z_2)$). By Property 1, $\mathcal{H}(c[p]) = \mathcal{H}(c[p'])$. Thus, $\mathcal{M}(c[p]) \neq \mathcal{M}(c[p'])$ and c distinguishes p from p'. $\qquad\square$

4.2   The algorithm `FindEBP`

Instead of performing a prefix traversal of a term t like `HCE`, Algorithm 7 recursively descends along a branch of t until it finds an EBP, using TBPs to orient itself and prune irrelevant branches.

Indeed, if $z \in S^+$, by Definition 8, (c,t) is an EBP. In particular, note that $\Sigma \subseteq S \cup S^+$. Thus, the algorithm terminates if it reaches a leaf. However, if $t = t_1 \circ t_2$ for some $\circ \in \{\cdot, \|\}$, we first try to determine whether (c,t) is a TBP:

- If $\mathcal{A}(c[\square \circ z_2], t_1) = 1$, we can recursively call $\texttt{FindEBP}(c[\square \circ z_2], t_1)$ as its preconditions still hold: t's leftmost branch is explored until a TBP is found.
- However, if $\mathcal{A}(c[\square \circ z_2], t_1) = 0$, by Definition 7, $\exists p_1 \in \alpha_{\mathcal{H}}(z_1), \mathcal{H}(c[p_1 \circ z_2]) \neq \mathcal{M}(c[p_1 \circ z_2])$ **(i)**. We then determine whether (c,t) is effective or not:
  - If $\mathcal{A}(c[p_1 \circ \square], t_2) = 0$, (c,t) is effective and the algorithm ends.
  - Otherwise $\mathcal{A}(c[p_1 \circ \square], t_2) = 1$ and by **i**, $c[p_1 \circ z_2]$ is a counterexample. We can recursively call $\texttt{FindEBP}(c[p_1 \circ \square], t_2)$ as its preconditions hold. Practically speaking, if the BP turns out not to be effective, we prune t's left subterm t_1, replace it with an access pomset p_1, then switch branch and recursively explore t's right subterm t_2 instead.

Example 6. Let us resume Example 5. Consider a call to $\texttt{FindEBP}(\square, t)$ where term $t \in ST(cc(a \| c))$ is displayed in Figure 6. Initially, we try replacing the left subterm cc rooted in node **1** with its access pomset ε. Since $\mathcal{A}(\square \cdot (a \| c), cc) = 0$, we found a TBP. We now try to replace subterm $a \| c$ rooted in node **2** with its access pomset c. But $\mathcal{A}(\varepsilon \cdot \square, a \| c) = 1$ and the BP is not effective.

Algorithm 7 FindEBP(c, t) where $c \in C_1(\Sigma)$, t is a minimal term of $z \in \mathrm{SP}^+(\Sigma)$, $\mathcal{H}(c[z]) \neq \mathcal{M}(c[z])$ and $\mathcal{A}(c, z) = 1$

1: **if** $z \in S^+$ **then return** (c, z)
2: **else if** $t = t_1 \circ t_2$, t_i being a minimal term of $z_i \in \mathrm{SP}^+(\Sigma)$ for $i \in \{1, 2\}$ **then**
3:     **if** $\mathcal{A}(c[\square \circ z_2], t_1)$ **then**
4:         **return** FindEBP$(c[\square \circ z_2], t_1)$
5:     **else**
6:         Let $p_1 \in \alpha_{\mathcal{H}}(z_1)$ be such that $\mathcal{H}(c[p_1 \circ z_2]) \neq \mathcal{M}(c[p_1 \circ z_2])$.
7:         **if** $\mathcal{A}(c[p_1 \circ \square], t_2)$ **then**
8:             **return** FindEBP$(c[p_1 \circ \square], t_2)$
9:         **else**
10:             Let $p_2 \in \alpha_{\mathcal{H}}(z_2)$ be such that $\mathcal{H}(c[p_1 \circ p_2]) \neq \mathcal{M}(c[p_1 \circ p_2])$.
11:             **return** $(c, p_1 \circ p_2)$

We therefore replace cc with ε and we switch to a new branch, exploring node **2** by calling FindEBP$(\varepsilon \cdot \square, a \parallel c)$. We try replacing subterm a rooted in node **3** with its access pomset ε. As $\mathcal{A}(\varepsilon \cdot (\square \parallel c), a) = 1$, we recursively call FindEBP$(\varepsilon \cdot (\square \parallel c), a)$ and explore node 3.

We have reached a base case: $a \in S^+$ is distinguished from its access pomset $\varepsilon \in S$ by context $\varepsilon \cdot (\square \parallel c) = \square \parallel c$.

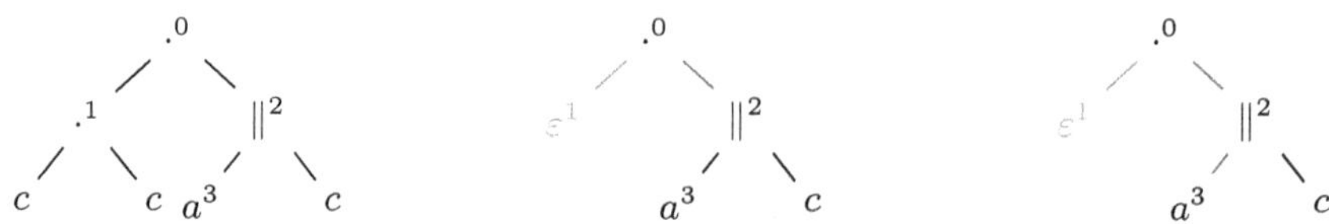

Fig. 6: Running FindEBP on counterexample $(cc)(a \parallel c)$ yields $(\square \parallel c, a)$.

5   Termination, Correction and Complexity

5.1   Algorithm FindEBP

A proof of Algorithm 7 can be performed by induction on t's depth: the base case holds thanks to Line 1, while Lines 4 and 8 handle the inductive case, as both calls preserve FindEBP's preconditions.

If t is of depth d, FindEBP$(\square, t)$ performs at most d recursive calls, and the agreement predicate is called twice per call (Lines 3 and 7). Given a minimal term t of a counterexample w, FindEBP$(\square, t)$ thus performs $\mathcal{O}(\rho(\mathcal{B}) \cdot \delta(w))$ MQs. Moreover, if $\mathcal{B}$ is sharp, e.g. in the context of van Heerdt et al.'s PL^*, then $\mathcal{O}(\delta(w))$ queries are required. Finally, if we assume t is a full binary tree, only $\mathcal{O}(\log |w|)$ queries are needed. By comparison, HCE performs $\mathcal{O}(\rho(\mathcal{B}) \cdot |w|)$ MQs.

5.2   The full algorithm PL^λ

Correction. As mentioned in Section 3.3, correction of PL^λ stems from $\sim_\mathcal{H}$ being an under-approximation of $\sim_\mathcal{M}$ on $S \cup S^+$: the discrimination tree $\mathcal{D}$ explicitly distinguishes the components of $\mathcal{B}$. Once both congruence relations have the same index over $S \cup S^+$, we can fully infer $\mathcal{M}$ from $\mathcal{B}$.

Termination. Algorithms 3 and 4 refine $\mathcal{B}$ each time a consistency or associativity defect is found. By design, $\mathcal{B}$ features at most $|\mathrm{SP}(\Sigma) / \sim_\mathcal{M}|$ classes: thus, these algorithms terminate with a consistent, associative pack of components.

In a similar fashion, each call to $\mathtt{FindEBP}(\square, t)$ in Algorithm 5 returns a new refining pair. Thus, only a finite number of iterations of the **while** loop can be performed, and Algorithm 5 terminates. Moreover, it ends with $\mathcal{B}$ being sharp: if $p, p' \in S$ belong to the same component B, $\mathcal{H}(c[p]) = \mathcal{H}(c[p'])$; but $\mathcal{M}(c[p]) \neq \mathcal{M}(c[p'])$, hence either $c[p]$ or $c[p']$ remains a counterexample.

Finally, as each counterexample returned by the teacher results in at least one new class of $\sim_L$ being identified, Algorithm 6 ends.

Query complexity. Let $|\mathcal{M}| = n$, $\sigma = |\Sigma|$, m and d the maximal size and depth of counterexamples returned by the MAT. Let us analyze PL^λ's complexity.

PL^λ performs at most $\mathcal{O}(n)$ EQs. Thus, there are at most n calls to $\mathtt{FindEBP}$, and each call needs $\mathcal{O}(d \cdot n)$ MQs: due to $\mathcal{B}$ possibly not being sharp, $1 \leq \rho(\mathcal{B}) \leq n$. Computing $\mathcal{B}$ requires $\mathcal{O}(n \cdot (n^2 + \sigma))$ MQs: since $|\mathcal{C}| \leq n$, sifting a pomset requires at most n queries, and $|S| \leq n$, $|S^+| \leq n^2 + \sigma$.

Finally, PL^λ with $\mathtt{FindEBP}$ performs $\mathcal{O}(n \cdot (n^2 + \sigma) + d \cdot n^2)$ MQs, whereas PL^* [31] with $\mathtt{HCE}$ requires $\mathcal{O}(n \cdot (n^2 + \sigma) + m \cdot n)$ queries, or $\mathcal{O}(n \cdot (n^2 + \sigma) + d \cdot n)$ were we to replace $\mathtt{HCE}$ with $\mathtt{FindEBP}$. Similarly to L^λ [14], we show that this unfavourable comparison does not empirically hold in Section 7.

Symbol complexity. A mere estimate of the number of queries overlooks the fact that the actual execution time of a MQ depends on the size of its input. Thus, we consider the *symbol complexity* of PL^λ, i.e. the sum of the sizes of the pomsets queried. Since S is built inductively from elements of S^+, for any $s \in S \cup S^+$, $|s| = \mathcal{O}(2^n)$. Similarly, $\mathcal{C}$ is built inductively by combining some existing element of $\mathcal{C}$ with context $\square \circ s$ for some $s \in S$ and $\circ \in \{\cdot, \|\}$; thus, for any $c \in \mathcal{C}$, $|c| = \mathcal{O}(2^n \cdot n)$. A query $\mathcal{M}(c[s])$ is therefore of size $\mathcal{O}(2^n \cdot n)$. Computing $\mathcal{B}$ thus requires $\mathcal{O}(2^n \cdot n^2 \cdot (n^2 + \sigma))$ symbols.

The pomset inductively manipulated by $\mathtt{FindEBP}$ is of size $\mathcal{O}(m + d \cdot 2^n)$, and so are queries performed by $\mathtt{FindEBP}$: out of the original term of size m, up to one subterm per level may be replaced by an access pomset of size at most 2^n. Assuming it relies on $\mathtt{FindEBP}$, PL^λ's symbol complexity is therefore $\mathcal{O}(2^n \cdot n^2 \cdot (n^2 + \sigma) + d \cdot n^2 \cdot (m + d \cdot 2^n))$.

By comparison, PL^* infers its contexts directly from counterexamples: they are therefore of size $\mathcal{O}(m)$, and a query $\mathcal{M}(c[s])$, of size $\mathcal{O}(m + 2^n)$. PL^*'s symbol complexity is $\mathcal{O}((m + 2^n) \cdot n \cdot (n^2 + \sigma) + d \cdot n \cdot (m + d \cdot 2^n))$ if we assume it also relies on $\mathtt{FindEBP}$. Thus, the symbol complexity of PL^λ's computation of $\mathcal{B}$ does not depend on m, although the counterexample analysis procedure obviously does. This property results in sizeable complexity gains, as shown in Section 7.

6   Generating Test Suites for Equivalence Queries

We remedy the use of EQs by designing a suitable finite test suite ensuring general equivalence between two pomsets recognizers, assuming that the size of the system is bounded w.r.t. the hypothesis we submit. This test suite extends the W-method [6,32,24] to recognizable languages of pomsets.

Definition 9 (Equivalence on a test suite). *Let $Z \subseteq \mathrm{SP}(\Sigma)$. Two PRs $\mathcal{R}_1$ and $\mathcal{R}_2$ are said to be Z-equivalent, written $\mathcal{H} \equiv_Z \mathcal{M}$, if for any $z \in Z$, $\mathcal{H}(z) = \mathcal{M}(z)$.*

Given a hypothesis $\mathcal{H} = (H, \cdot_{\mathcal{H}}, \|_{\mathcal{H}}, e_{\mathcal{H}}, i^{\sharp}_{\mathcal{H}}, F_{\mathcal{H}})$ of size $|H| = n$, and a model $\mathcal{M} = (M, \cdot_{\mathcal{M}}, \|_{\mathcal{M}}, e_{\mathcal{M}}, i^{\sharp}_{\mathcal{M}}, F_{\mathcal{M}})$ sharing the same alphabet such that $\mathcal{H}$ and $\mathcal{M}$ are minimal, assume that a bound k such that $0 \le |M| - |H| \le k$ is known.

6.1   Computing a state cover

We seek a test suite that covers and distinguishes the states of $\mathcal{M}$.

Definition 10 (State cover). *A set $P \subseteq \mathrm{SP}(\Sigma)$ is a state cover of a PR $\mathcal{R} = (R, \odot, \oslash, e, i, F)$ if $\varepsilon \in P$ and every $r \in R$ admits an access pomset $p \in P$.*

Definition 11 (Characterization set). *A set of contexts $W \subseteq \mathrm{C}_1(\Sigma)$ is a characterization set of a PR $\mathcal{R} = (R, \odot, \oslash, e, i, F)$ if $\square \in W$ and for any $r_1, r_2 \in R$, if r_1 and r_2 are distinguishable, then $\exists c \in W, \mathcal{R}(c[r_1]) \ne \mathcal{R}(c[r_2])$.*

Based on bound k and our knowledge that states distinguished in $\mathcal{H}$ are still distinguished in $\mathcal{M}$, we use P to design a state cover of the unknown PR $\mathcal{M}$.

Definition 12 (Extended state cover). *Given a state cover P of $\mathcal{H}$, it is the set $L_i^P = \{c[p] \mid m \in \mathbb{N}, c \in \mathrm{C}_m(\Sigma), \delta(c) \le i, p \in P^m\}$.*

Intuitively, L_i^P consists of all pomsets obtained by inserting access pomsets of P in a multi-context of height equal to or smaller than i. Then:

Theorem 2. *Let P be a state cover of $\mathcal{H}$ and W a characterization set of $\mathcal{H}$ such that $\mathcal{H} \equiv_{W[P]} \mathcal{M}$. Then L_k^P is a state cover of $\mathcal{M}$.*

Intuitively, branches of an access term to a state of $\mathcal{M}$ can be shortened in such fashion subterms below depth k are replaced by access terms to $\mathcal{H}$.

6.2   Exhaustivity of the test suite

Using P and W, our goal is to design a *complete* test suite Z, i.e. such that Z-equivalence must imply full equivalence of $\mathcal{H}$ and $\mathcal{M}$. To do so, we extend the notion of bisimulation to PRs:

Definition 13 (Bisimulation relation). *A* bisimulation relation $\sim$ *between two PRs* $\mathcal{R}_1 = (R_1, \odot_1, \oplus_1, e_1, i_1, F_1)$ *and* $\mathcal{R}_2 = (R_2, \odot_2, \oplus_2, e_2, i_2, F_2)$ *is a binary relation* $R_1 \times R_2$ *such that:*

1. $r_1 \sim r_2$ *implies that* $r_1 \in F_1 \iff r_2 \in F_2$.
2. $r_1 \sim r_2$ *and* $r_1' \sim r_2'$ *implies that* $r_1 \circ_1 r_2 \sim r_1' \circ_2 r_2'$ *for* $\circ \in \{\odot, \oplus\}$.

Lemma 2. *Given a bisimulation relation* $\sim$ *between* $\mathcal{R}_1$ *and* $\mathcal{R}_2$*, if* $\mathcal{R}_1$ *and* $\mathcal{R}_2$ *share the same alphabet* Σ*, the same neutral element* e*, and for any* $x \in \Sigma \cup \{e\}$*,* $i_{\mathcal{R}_1}^{\sharp}(x) \sim i_{\mathcal{R}_2}^{\sharp}(x)$*, then* $\mathcal{R}_1$ *and* $\mathcal{R}_2$ *are equivalent.*

We design a set Z such that that Z-equivalence results in bisimilarity:

Theorem 3. *Let* P *be a minimal (w.r.t.* $\subseteq$*) state cover of* $\mathcal{H}$*,* W *be a minimal characterization set of* $\mathcal{H}$ *and* $Z = W[L_{k+1}^P]$*. We introduce the binary relation* $\sim$ *on* $H \times M$*:*

$$\sim\; = \{(i_{\mathcal{H}}^{\sharp}(l), i_{\mathcal{M}}^{\sharp}(l)) \mid l \in L_k^P\}$$

If $\mathcal{H}$ *and* $\mathcal{M}$ *are* Z*-equivalent, then* $\sim$ *is a bisimulation relation.*

Corollary 1. $\mathcal{H}$ *and* $\mathcal{M}$ *are* Z*-equivalent if and only if they are equivalent.*

We can infer a state cover P and a characterization set W from PL^λ's S and $\mathcal{C}$; thus, as long as parameter k is known, we can substitute EQs with MQs over Z. Counting terms in L_{k+1}^P results in the following bound:

Theorem 4. $|Z| \leq n \cdot (|\Sigma| + n)^{2^{k+1}}$.

7   Experimental Results

We implemented PL^λ and `FindEBP` as well as van Heerdt et al.'s PL^* [31] and HCE in a C++ prototype[3]. Since `FindEBP` and HCE can be used by PL^λ and PL^* alike, we ran different experiments in order to independently assess how the various learning algorithms and counterexample analysis procedures impact query and symbol complexity. To evaluate the impact of arbitrary long counterexamples, minimal counterexamples returned by the MAT were artificially lengthened by pumping loops in the product PR $\mathcal{H} \oplus \mathcal{M}$ (built in a similar fashion to a product tree automaton [7]).

We benchmarked the four algorithm combinations against a sample of 236 randomly generated PRs, as shown in Figure 8. Points above the diagonal line

---

[3] `https://gitlab.lre.epita.fr/adrien/treelearn`

correspond to experiments where PL^λ and FindEBP outperformed another combination of options. Table 7 displays a summary of these benchmarks.

	Mem.	Eq.	Symb.
PL^λ + FindEBP	296.17	2.93	5763.8
PL^λ + HCE	325.49	2.92	9594.1
PL^* + FindEBP	486.38	2.73	30911
PL^* + HCE	515.05	2.70	40943

Fig. 7: Summary of the arithmetic mean of the **Mem**bership, **Symb**ol, and **E**quivalence complexities.

Algorithm PL^λ (resp. FindEBP) on average outperforms PL^* (resp. HCE) in terms of membership and symbol complexity. In particular, the combination of PL^λ and FindEBP outperforms PL^* and HCE in every single instance save two, and the latter's membership (resp. symbol) complexity is on average 1.74 (resp. 7.10) times higher. However, PL^λ and FindEBP's equivalence complexity is on average 1.08 times higher, a meagre difference explained by compatibility tests.

The introduction of PL^λ yields the greatest benefits. Nevertheless, the use of FindEBP instead of HCE also results in a significant symbol complexity improvement—likely due to entire subterms of the counterexample being pruned by FindEBP —as well as more modest membership complexity gains.

Assuming the learning algorithm used is PL^λ, it's worth noting that HCE yields marginal membership complexity gains in 28% of all cases, a possible explanation being that HCE performs bottom-up substitutions by access pomsets while FindEBP prunes branches in a top-down fashion; thus, if a refinement can be directly inferred from the leftmost leaf, HCE will outperform FindEBP. Nevertheless, FindEBP's theoretical complexity guarantees a significantly better worst case scenario.

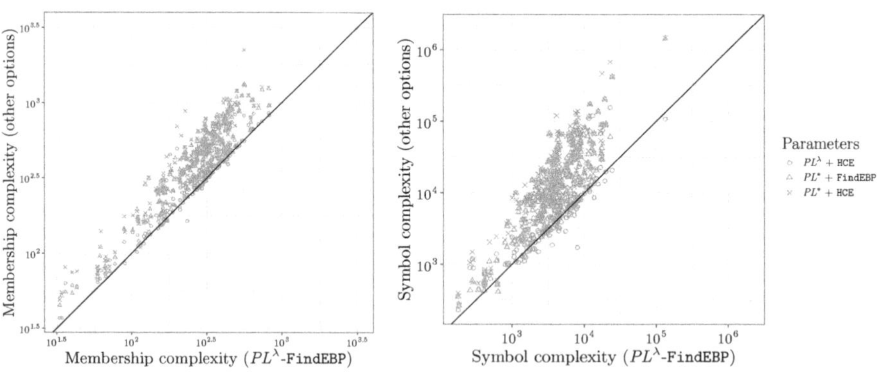

Fig. 8: Comparing PL^λ to PL^* and FindEBP to HCE.

8   Conclusion and Further Developments

Our active learning algorithm PL^λ and our new counterexample analysis algorithm FindEBP have been empirically shown to yield significant membership and symbol complexity gains. But the following practical issues we successfully tackled or are currently dealing with are also worth mentioning:

Representing series-parallel pomsets. We settled on a *canonical* representation of pomsets as binary trees that guarantees minimal depth and prevents duplicate queries.

Generating a benchmark. We used a SAT-based approach to create a test sample of minimal, reachable pomset recognizers.

Benchmarking the W-method. Our W-method has yet to be implemented. It remains to be seen whether its high cost dominates the learning process. We plan on adapting various improvements to the W-method, such as the H-method [8], to PRs. Random walks are also a lead worth exploring.

Optimizing counterexample handling. `FindEBP` benefits from the use of canonical terms of minimal depth. However, its complexity remains linear w.r.t. the input's size if the canonical term is a linear tree. We may therefore try to develop another counterexample handling procedure optimized for linear trees, and even seek one that is truly logarithmic.

Another lead consists in determining whether `FindEBP` and PL^λ can enhance AL for general (non-binary) tree languages [9]. We also plan on exploring the *passive learning* problem for pomset samples: given two non-empty sets $Z^+ \subseteq \mathrm{SP}(\Sigma)$ and $Z^- \subseteq \mathrm{SP}(\Sigma)$ such that $Z^+ \cap Z^- = \emptyset$, find a PR that accepts all the elements of Z^+ and rejects all the elements of Z^-.

Moreover, some use cases such as producer-consumer systems require modelling pomsets that feature N patterns. These scenarios can be effectively formalized using *interval pomsets* [11] (that is, pomsets such that $x_1 < x_2$ and $x_3 < x_4$ implies $x_3 < x_2$ or $x_1 < x_4$) recognized by *higher-dimensional automata* (HDA) [10]. The extension of Myhill-Nerode's theorem to languages of HDA [12] opens up the possibility of an AL algorithm for HDA. The class of HDA supplements rather than subsumes PRs: indeed, N-shaped pomsets are interval but not series-parallel, while pomset $ab \parallel cd$ is series-parallel but not interval.

Finally, designing a practical oracle capable of processing pomset-shaped queries is a significant technical challenge: submitting linearized (as mentioned in Section 1.1) pomset queries to an ordinary sequential oracle is not a suitable solution as it may cause the algorithm's membership complexity to blow up.

Data availability statement. The data that support the findings of this article are openly available at: `https://doi.org/10.5281/zenodo.18175343`

References

1. Omar al Duhaiby and Jan Friso Groote. Active learning of decomposable systems. In *Proceedings of the 8th International Conference on Formal Methods in Software Engineering*, FormaliSE '20, page 1–10, New York, NY, USA, 2020. Association for Computing Machinery.
2. Rajeev Alur, Caleb Stanford, and Christopher Watson. A robust theory of series parallel graphs. *Proc. ACM Program. Lang.*, 7(POPL):1058–1088, 2023.

3. Dana Angluin. Learning regular sets from queries and counterexamples. *Inf. Comput.*, 75(2):87–106, 1987.

4. Nicolas Bedon. Branching automata and pomset automata. In Mikolaj Bojanczyk and Chandra Chekuri, editors, *41st IARCS Annual Conference on Foundations of Software Technology and Theoretical Computer Science, FSTTCS 2021, December 15-17, 2021, Virtual Conference*, volume 213 of *LIPIcs*, pages 37:1–37:13. Schloss Dagstuhl - Leibniz-Zentrum für Informatik, 2021.

5. S.L. Bloom and Z. Ésik. Free shuffle algebras in language varieties. *Theoretical Computer Science*, 163(1-2):55–98, 1996.

6. Tsun S. Chow. Testing software design modeled by finite-state machines. *IEEE Trans. Software Eng.*, 4(3):178–187, 1978.

7. Hubert Comon, Max Dauchet, Rémi Gilleron, Florent Jacquemard, Denis Lugiez, Christof Löding, Sophie Tison, and Marc Tommasi. Tree automata techniques and applications, 2008.

8. Rita Dorofeeva, Khaled El-Fakih, and Nina Yevtushenko. An improved conformance testing method. In Farn Wang, editor, *Formal Techniques for Networked and Distributed Systems - FORTE 2005*, pages 204–218, Berlin, Heidelberg, 2005. Springer Berlin Heidelberg.

9. Frank Drewes and Johanna Högberg. Learning a regular tree language from a teacher. In Zoltán Ésik and Zoltán Fülöp, editors, *Developments in Language Theory, 7th International Conference, DLT 2003, Szeged, Hungary, July 7-11, 2003, Proceedings*, volume 2710 of *Lecture Notes in Computer Science*, pages 279–291. Springer, 2003.

10. Uli Fahrenberg, Christian Johansen, Georg Struth, and Krzysztof Ziemianski. A kleene theorem for higher-dimensional automata. In Bartek Klin, Slawomir Lasota, and Anca Muscholl, editors, *33rd International Conference on Concurrency Theory, CONCUR 2022, September 12-16, 2022, Warsaw, Poland*, volume 243 of *LIPIcs*, pages 29:1–29:18. Schloss Dagstuhl - Leibniz-Zentrum für Informatik, 2022.

11. Uli Fahrenberg, Christian Johansen, Georg Struth, and Krzysztof Ziemianski. Posets with interfaces as a model for concurrency. *Inf. Comput.*, 285(Part):104914, 2022.

12. Uli Fahrenberg and Krzysztof Ziemianski. Myhill-nerode theorem for higher-dimensional automata. *Fundam. Informaticae*, 192(3-4):219–259, 2024.

13. Léo Henry, Mohammad Reza Mousavi, Thomas Neele, and Matteo Sammartino. Compositional active learning of synchronizing systems through automated alphabet refinement. In Patricia Bouyer and Jaco van de Pol, editors, *36th International Conference on Concurrency Theory, CONCUR 2025, Aarhus, Denmark, August 26-29, 2025*, volume 348 of *LIPIcs*, pages 20:1–20:22. Schloss Dagstuhl - Leibniz-Zentrum für Informatik, 2025.

14. Falk Howar and Bernhard Steffen. Active automata learning as black-box search and lazy partition refinement. In Nils Jansen, Mariëlle Stoelinga, and Petra van den Bos, editors, *A Journey from Process Algebra via Timed Automata to Model Learning - Essays Dedicated to Frits Vaandrager on the Occasion of His 60th Birthday*, volume 13560 of *Lecture Notes in Computer Science*, pages 321–338. Springer, 2022.

15. Malte Isberner, Falk Howar, and Bernhard Steffen. The TTT algorithm: A redundancy-free approach to active automata learning. In Borzoo Bonakdarpour and Scott A. Smolka, editors, *Runtime Verification*, pages 307–322, Cham, 2014. Springer International Publishing.

16. Tobias Kappé, Paul Brunet, Bas Luttik, Alexandra Silva, and Fabio Zanasi. On series-parallel pomset languages: Rationality, context-freeness and automata. *J. Log. Algebraic Methods Program.*, 103:130–153, 2019.

17. M. J. Kearns and U. V. Vazirani. *An Introduction to Computational Learning Theory*. MIT Press, Cambridge, MA, USA, 1994.

18. Faezeh Labbaf, Jan Friso Groote, Hossein Hojjat, and Mohammad Reza Mousavi. Compositional learning for interleaving parallel automata. In Orna Kupferman and Pawel Sobocinski, editors, *Foundations of Software Science and Computation Structures*, Lecture Notes in Computer Science (LNCS), pages 413–435, Germany, April 2023. Springer. 26th International Conference on Foundations of Software Science and Computation Structures, FoSSaCS 2023.

19. K. Lodaya and P. Weil. A Kleene iteration for parallelism. In *Foundations of Software Technology and Theoretical Computer Science*, pages 355–366, 1998.

20. Kamal Lodaya and Pascal Weil. Series-parallel posets: algebra, automata and languages. In *Annual Symposium on Theoretical Aspects of Computer Science*, pages 555–565. Springer, 1998.

21. Kamal Lodaya and Pascal Weil. Series–parallel languages and the bounded-width property. *Theoretical Computer Science*, 237(1-2):347–380, 2000.

22. Kamal Lodaya and Pascal Weil. Rationality in algebras with a series operation. *Information and Computation*, 171(2):269–293, 2001.

23. Joshua Moerman. Learning product automata. In Olgierd Unold, Witold Dyrka, and Wojciech Wieczorek, editors, *Proceedings of The 14th International Conference on Grammatical Inference 2018*, volume 93 of *Proceedings of Machine Learning Research*, pages 54–66. PMLR, feb 2019.

24. Joshua Moerman. *Nominal Techniques and Black Box Testing for Automata Learning*. PhD thesis, Radboud University, 2019.

25. Thomas Neele and Matteo Sammartino. Compositional automata learning of synchronous systems. In Leen Lambers and Sebastián Uchitel, editors, *Fundamental Approaches to Software Engineering*, pages 47–66, Cham, 2023. Springer Nature Switzerland.

26. Vaughan R. Pratt. Modeling concurrency with partial orders. *Journal of Parallel Programming*, 15(1):33–71, 1986.

27. Ronald L. Rivest and Robert E. Schapire. Inference of finite automata using homing sequences. *Inf. Comput.*, 103(2):299–347, 1993.

28. Mahboubeh Samadi, Aryan Bastany, and Hossein Hojjat. Compositional learning for synchronous parallel automata. In Artur Boronat and Gordon Fraser, editors, *Fundamental Approaches to Software Engineering*, pages 101–121, Cham, 2025. Springer Nature Switzerland.

29. Frits Vaandrager, Bharat Garhewal, Jurriaan Rot, and Thorsten Wißmann. A new approach for active automata learning based on apartness. In Dana Fisman and Grigore Rosu, editors, *Tools and Algorithms for the Construction and Analysis of Systems*, pages 223–243, Cham, 2022. Springer International Publishing.

30. Jacobo Valdes, Robert Endre Tarjan, and Eugene L. Lawler. The recognition of series parallel digraphs. *SIAM J. Comput.*, 11(2):298–313, 1982.

31. Gerco van Heerdt, Tobias Kappé, Jurriaan Rot, and Alexandra Silva. Learning pomset automata. In Stefan Kiefer and Christine Tasson, editors, *Foundations of Software Science and Computation Structures - 24th International Conference, FoSSaCS 2021, Held as Part of the European Joint Conferences on Theory and Practice of Software, ETAPS 2021, Luxembourg City, Luxembourg, March 27 - April 1, 2021, Proceedings*, volume 12650 of *Lecture Notes in Computer Science*, pages 510–530. Springer, 2021.

32. M. P. Vasilevskii. Failure diagnosis of automata. *Cybernetics*, 9(4):653–665, 1973.

A 2-categorical approach to the semantics of dependent type theory with computation axioms

Matteo Spadetto [iD]

Nantes Université
`matteo.spadetto.42@gmail.com`

Abstract. Axiomatic type theory is a dependent type theory without computation rules. The term equality judgements that usually characterise these rules are replaced by computation *axioms*, i.e., additional term judgements that are typed by identity types. This paper is devoted to providing an effective description of its semantics, from a higher categorical perspective: given the challenge of encoding intensional type formers into 1-dimensional categorical terms and properties, a challenge that persists even for axiomatic type formers, we adopt Richard Garner's approach in the 2-dimensional study of dependent types. We prove that the type formers of axiomatic theories can be encoded into natural 2-dimensional category theoretic data, obtaining a presentation of the semantics of axiomatic type theory via 2-categorical models called *display map 2-categories*. In the axiomatic case, the 2-categorical requirements identified by Garner for interpreting intensional type formers are relaxed. Therefore, we obtain a presentation of the semantics of the axiomatic theory that generalises Garner's one for the intensional case. Our main result states that the interpretation of axiomatic theories within display map 2-categories is well-defined and enjoys the soundness property. We use this fact to provide a semantic proof that the computation rule of intensional identity types is not admissible in axiomatic type theory. This is achieved via a revisitation of Hofmann and Streicher's *groupoid model* that believes axiomatic identity types but does not believe intensional ones.

Keywords: Identity types · Computation axioms · 2-categorical models.

1 Introduction and motivation

In recent decades, dependent type theory has become a central framework for the foundations of mathematics, offering a highly expressive language for formulating and proving mathematical statements. Beyond its theoretical importance, it has found practical use in areas such as software verification through proof assistants and the theory of programming languages. From a semantic perspective, the study of dependent type theory has been deeply shaped by category-theoretic and homotopy-theoretic approaches, which have proven crucial in providing models for its interpretation and for understanding its essence [16,44,21,31,30,8,23,4,34]. This holds true especially when one aims to capture

© The Author(s) 2026
N. Bertrand and S. Milius (Eds.): FoSSaCS 2026, LNCS 16503, pp. 593–615, 2026.
https://doi.org/10.1007/978-3-032-22730-0_28

the notion of a model of dependent type theory through simple categorical properties, instead of mirroring the syntax directly. However, as discussed in Subsection 1.1, the difficulty of this problem depends on the level of extensionality of the dependent type theory under consideration: while the category theory of the model theory of extensional type theory—where type formers contain both computation and expansion rules—is clear and naturally one-dimensional [26,20,37,17], this is no longer the case for intensional and axiomatic type theories—whose type formers do not have the expansion rule. This specific aspect is the focus of the present paper, which lies within the analysis of the semantics of dependent types from a categorical perspective: our aim is to contextualise and discuss it within the study of *axiomatic type theory*, namely, intensional type theory with propositional computation rules, presented in detail in Subsection 1.2.

1.1   The problem of the semantics of intensional type formers

In formulating the semantics of theories of dependent types, we typically encounter two distinct, but related, approaches: *syntactic* and *categorical*. The syntactic approach directly mirrors the structure of the syntax of the theory, defining how judgements in the conclusions of inference rules are interpreted based on how those in the premises are: each inference rule is translated by explicitly assigning a choice function that maps interpretations of its premises to an interpretation of its conclusion. In contrast, the categorical approach equips the model with structure that automatically yields such choice functions.

For example, given a display map category $(\mathbf{C}, \mathcal{D})$—a category $\mathbf{C}$ equipped with a distinguished class $\mathcal{D}$ of arrows, called *display maps*, which is stable under pullback, as stated formally in Definition 1—consider the case of extensional =-types. To model a theory featuring this type former, we may just replicate its syntax within the "language" of $(\mathbf{C}, \mathcal{D})$, where a *display map* interprets a type judgement $A : \text{TYPE}$ and its *sections* interpret the corresponding term judgements $t : A$. Suppose that any display map interpreting a given type $A : \text{TYPE}$ is equipped with the choice of another display map interpreting the identity type $\lfloor x, x' : A \rfloor\ x = x' : \text{TYPE}$. This assignment will act as a choice function validating the *formation rule* for extensional =-types. Analogously, other choice functions will validate the remaining rules of the extensional =-type former. However, as explained in [32], one may simply require that the diagonal arrow $\Gamma.A \to \Gamma.A.A[P_A]$ be isomorphic to some display map over $\Gamma.A.A[P_A]$—where $\Gamma.A.A[P_A]$ represents the variable context $\lfloor \gamma, x, x' : A(\gamma) \rfloor$ with γ the possibly empty context of variables represented by Γ—whenever P_A is a display map $\Gamma.A \to \Gamma$ in $\mathcal{D}$: this categorical condition on $(\mathbf{C}, \mathcal{D})$ is equivalent to equipping it with the necessary choice functions to validate the extensional =-type former. To provide the reader with another example, a categorical condition characterising the semantics of extensional Σ-types requires that display maps be closed under composition up to isomorphism. This categorical characterisation of Σ-types is the basis of the analysis contained in [44] and [26].

These are examples of how extensional type formers admit a concise categorical characterisation of their models: it consists of *closure properties on the class of their display maps*. However, in case we drop extensionality and still ask ourselves how to handle $=$- and Σ-types such a transparent characterisation of their semantics is harder to find. To simplify the problem and recover a categorical description of the semantics of intensional type formers, Garner proposes a 2-categorical formulation in [24]. In this scenario, a synthetic and conceptually simple characterisation of identities, dependent sums, and other type constructors *exists*. This stems from the wider range of potential closure properties—whether strong or weak—to require on the class $\mathcal{D}$, that we can access: properties that enable us to distinguish between varying strengths of type formers. Notably, if dimension 2 were omitted, these properties would collapse into the ones we have just discussed, recovering the extensional version of our type constructors. In this paper, we adopt Garner's perspective to study the semantics of *axiomatic type theory* from a categorical point of view, specifically through a 2-categorical one.

1.2   Axiomatic dependent type theory

In recent years, there has been a growing interest in *weakened* forms of dependent type theories, particularly those relaxing the computation rules of type formers. A type former encodes logical structure—such as existential quantification for Σ-types—through its formation, introduction, and elimination rules, together with computation rules that describe how introduction and elimination interact. In both extensional and intensional type theory (ETT and ITT), computation rules are expressed as *judgemental equalities* $t \equiv t'$, where t and t' are two terms of the same type. Alternatively, if the theory includes some flavour of the identity type former, one may judge a specific term judgement $p : t = t'$, typed by the *identity type* $t = t'$: TYPE, to state *propositionally* that "there exists a proof p that the two terms t and t' are equal". Judgemental equality is at least as strong as propositional equality due to the $=$-introduction rule. Type formers in ETT also feature expansion (η) rules, together with computation (β) rules, again in judgemental form—while for intensional type formers they can typically be derived only as propositional equalities [51]—and this is equivalent to saying that in ETT judgemental equality and propositional equality coincide. In the intensional setting, propositional equality is weaker—as shown e.g. by the groupoid model [29,30,49], which lacks uniqueness of identity proofs, believed by every extensional model, and thus does not entail $t \equiv t'$ whenever $t = t'$ is inhabited. This weakening of the $=$-type former yields the computational advantages of ITT, namely the decidability of judgemental equality and of type checking. In this paper, we are interested in a setting in which propositional equality is even weaker, that is, when $=$-types are merely *axiomatic*.

   Axiomatic $=$-types are "very intensional" $=$-types. They retain the usual rules of intensional identity types, but replace the judgemental equality in the computation rule with a *propositional* one. Whenever we are given judgements $\lfloor x, x' : A; \ p : x = x' \rfloor \ C(x, x', p)$: TYPE and $\lfloor x : A \rfloor \ c(x) : C(x, x, r(x))$,

$$\text{Form rl} \frac{A : \textsc{Type}}{\lfloor x, x' : A \rfloor \; x = x' : \textsc{Type}} \qquad \text{Elim rl} \frac{\begin{array}{c} A : \textsc{Type} \\ \lfloor x, x'; \; p : x = x' \rfloor \; C(x, x', p) : \textsc{Type} \\ \lfloor x \rfloor \; c(x) : C(x, x, \mathsf{r}(x)) \end{array}}{\lfloor x, x'; \; p \rfloor \; \mathsf{J}(c, x, x', p) : C(x, x', p)}$$

$$\text{Intro rl} \frac{A : \textsc{Type}}{\lfloor x : A \rfloor \; \mathsf{r}(x) : x = x} \qquad \text{Comp ax} \frac{\begin{array}{c} A : \textsc{Type} \\ \lfloor x, x'; \; p : x = x' \rfloor \; C(x, x', p) : \textsc{Type} \\ \lfloor x \rfloor \; c(x) : C(x, x, \mathsf{r}(x)) \end{array}}{\lfloor x \rfloor \; \mathsf{H}(c, x) : \mathsf{J}(c, x, x, \mathsf{r}(x)) = c(x)}$$

Fig. 1. Axiomatic =-types

instead of requiring $\mathsf{J}(c, x, x, \mathsf{r}(x)) \equiv c(x)$, one only asks for an additional term judgement $\mathsf{H}(c, x) : \mathsf{J}(c, x, x, \mathsf{r}(x)) = c(x)$, i.e. a *computation axiom* stating that the predicate $\mathsf{J}(c, x, x, \mathsf{r}(x)) = c(x)$ holds. This is depicted in detail in Figure 1. In general, when a dependent type theory has a type former whose computation rules are replaced by computation axioms, one says that the type former is in *axiomatic* form. This paper is about semantics of *axiomatic type theory* (ATT), a weakening of ITT with computation axioms in place of computation rules. We will be focusing on the following type formers: =-types, Σ-types, Π-types, function extensionality, 0-types, 1-types, 2-types, and N-types. We refer the reader to work by Otten and by the author [41,46], as well as to the extended version of the present paper [45], for a more detailed treatment of ATT. There are several motivations and advantages to work on ATT:

≻ Axiomatic type theory (ATT) is *objective* [42,10], meaning that operations like addition or identity proof composition are defined without ambiguity. Unlike ITT, where such operations may satisfy some equalities judgementally and others propositionally, in ATT all these equalities only hold propositionally. Although it has a single notion of equality like ETT, ATT does not imply uniqueness of identity proofs, remaining compatible with homotopy type theory and extensions of ITT.

≻ Axiomatic type formers are *homotopy invariant*, as explained in [6,7]. Whenever a type is homotopy equivalent to one derived from the formation rule of a given type former, it adheres to all the rules associated with that type former.

≻ As shown in [10], by replacing computation rules with computation axioms, the *type checking*, i.e. the decidability of the derivability of a term judgement $t : A$, holds and can be done in quadratic time. This improves the non-elementary one of ordinary ITT.

≻ Compared to ITT, axiomatic type theory (ATT) admits a *broader concept of semantics*: every ITT model is an ATT model, but not vice versa, since— as we show in Section 5—some ATT models fail to satisfy computation rules, as in the case of the cubical set model [18,12]. This relaxation of judgemental equalities makes constructing concrete models—e.g. for independence results—

simpler. Despite this, ATT remains as expressive as ITT, encompassing full intuitionistic predicate logic.

➤ Accordingly, several *conservativity results* relate intensional and extensional type theories to the axiomatic one. In particular, ETT is conservative over ATT with function extensionality and uniqueness of identity proofs [53,46]. Building on Hofmann's work [27], conservativity properties have been explored by Bocquet [13], Boulier and Winterhalter [15,53], and the author [46], showing that ATT does not lose much deductive power with respect to ITT and ETT. Following this perspective, our interest in ATT lies primarily in its role as a *semantic framework* for ITT. Although the conservativity of ITT over ATT remains conjectural, it is widely expected to hold—what is known can be regarded as a truncated version of that conjecture. If confirmed, this would imply that the semantics of ATT provides an equivalent, but technically simpler, semantics for ITT, as fewer equalities need to be verified.

Related work. Hints of the emergence of an interest in propositional computation rules, or computation axioms, are scattered everywhere in the homotopy type theoretic literature, starting from the work of Awodey, Gambino, Sojakova [6,7], that presented an initial study of an axiomatic type former, namely axiomatic well-founded tree types, or axiomatic W-types. Bezem, Cohen, Coquand, Danielsson, Huber, and Mörtberg [19,18,12] conducted initial investigations into axiomatic =-types, in connection with the development of cubical type theory. Following this, the type constructor has been thoroughly examined by van den Berg and Moerdijk [9,11], who introduced and explored a semantic concept for dependent type theories with propositional identity types, using the notion of a *path category* and showing its link with axiomatic =-types. Otten and the author [41] explain how this notion of semantics is actually an instance of the usual one via display map categories (formulated as full comprehension categories). That work constitutes the homotopy-theoretic counterpart of the present one: while the latter captures ATT models via higher-categorical data, the former describes the semantics of ATT from a homotopy-theoretic perspective—namely, by formulating models as specific categories equipped with a primitive notion of weak equivalence [43]—and, in doing so, generalises the corresponding result of Clairambault and Dybjer [17]. Among other works on axiomatic type formers, we note the previously mentioned studies on conservativity over such type theories by Bocquet [13], by Boulier and Winterhalter [15], and by the author [46], Bocquet's work on the coherence property of the class of models of axiomatic type formers [14], and Vidmar's work on extending natural models [5] to type theories with propositional expansion rules, or expansion axioms [52].

Contributions. In this paper, we adopt Garner's perspective [24] to study the semantics of ATT from a categorical point of view, specifically from a 2-categorical one. In detail, we prove that a *display map 2-category* with some categorical structure—obtained reformulating the syntax of the axiomatic type formers in 2-categorical terms—is sufficient to recover the semantic counterparts of these type constructors as in the *syntactic approach*, particularly inducing an ordinary (split) display map category—i.e. an actual model of the structural rules of dependent type theory—that models the theory of dependent types in axiomatic form—see Section 4 and Theorem 2. This provides an answer to the problem of formulating models of ATT in categorical terms, i.e. within the *categorical approach*, in a way that includes and generalises the class of intensional models identified by Garner: display maps are not necessarily normal isofibrations, as in the intensional case, but merely cloven isofibrations and the axiomatic form of Σ-types does not require that display maps be closed under composition up to injective equivalence, as for intensional models, but merely up to homotopy equivalence. In broad terms, this work falls within the field of categorical semantics of dependent type theories, and aims to contribute to the central objective of developing a general and feasible notion of semantics for variants and generalisations of dependent type theory.

Outline. In Section 2 we recall the notion of a (split) display map category and briefly explain how such a structure provides a model of the structural rules of dependent type theory. In Section 3 we specialise this notion to model ATT, using the syntactic approach. In Section 4, we formulate the notion of a display map 2-category as a structure that encodes the structural and logical aspects of ATT as 2-categorical data on display maps. Accordingly, we show that fulfilling such data allows a display map 2-category to induce a model of ATT as described in Section 3. In Section 5, we use this result to identify a model of ATT—based on the groupoid model—that is not a model of ITT. In Section 6, we comment on the completeness of this and related notions of semantics, and discuss the position of this work within contemporary research on the categorical semantics of generalised dependent type theory.

2   Basics of display map categories

In this section, we recall the notion of display map categories [50,32,39] one of several equivalent categorical structures that provide a sound and complete semantics for dependent type theories. We specialise it to provide such a notion of semantics for ATT with =-types. We leave the semantics of the other axiomatic type formers for the appendix (as well as for an extended version).

Definition 1. *A (split) display map category* $(\mathbf{C}, \mathcal{D})$ *is a category* $\mathbf{C}$ *with a chosen terminal object* ϵ*, together with a class* $\mathcal{D}$ *of arrows of* $\mathbf{C}$*, called* display maps *and denoted as* $\Gamma.A \to \Gamma$ *and labeled as* P_A*, such that:*
≻ for every display map $\Gamma.A \to \Gamma$ *and every arrow* $f : \Delta \to \Gamma$*, there is a choice of a display map* $\Delta.A[f] \to \Delta$ *and of a square:*

$$\begin{array}{ccc} \Delta.A[f] & \to & \Gamma.A \\ \downarrow & \lrcorner & \downarrow \\ \Delta & \xrightarrow{\ f\ } & \Gamma \end{array}$$

which is a pullback; the arrow $\Delta.A[f] \to \Gamma.A$ *will be denoted as* $f.A$*, or as* $f^{\bullet}$ *in absence of ambiguity; we will indicate* $A[P_A]$ *as* $A^{\mathbf{v}}$*;*

$\succ$ $P_{A[1_A]} = P_A$, $(1_\Gamma).A = 1_{\Gamma.A}$, $P_{A[fg]} = P_{A[f][g]}$, and $(f.A)(g.A[f]) = (fg).A$ for every choice of composable arrows f and g and every display map $\Gamma.A \to \Gamma$, where Γ is the target of f.

In this paper, the term "display map category" specifically refers to a "*split* display map category". That is, we always assume that the display map categories we discuss satisfy the second of the two conditions in Definition 1. Under this assumption, the notion of semantics provided by display map categories is entirely equivalent to those based on other structures commonly used as models of dependent type theories, such as categories with attributes [16,38,33], categories with families [17,21,25,28], and full split comprehension categories [31,36]. Such a display map category $(\mathbf{C}, \mathcal{D})$ in fact constitutes a sound model of the structural part of dependent type theory: via the language—mentioned in Section 1—according to which a context $\lfloor \gamma : \Gamma \rfloor$ is encoded as an object Γ of $\mathbf{C}$, a type $A(\gamma)$ as a display map P_A of codomain Γ, and a term $a(\gamma) : A(\gamma)$ as a section $a : \Gamma \to \Gamma.A$ of P_A, the conclusions of the fundamental structural rules:

$$\frac{}{\lfloor\_\rfloor} \qquad \frac{\lfloor \gamma : \Gamma \rfloor \; A(\gamma) : \text{TYPE}}{\lfloor \gamma : \Gamma, x : A(\gamma) \rfloor} \qquad \frac{\lfloor \gamma : \Gamma \rfloor \; A(\gamma) : \text{TYPE}}{\lfloor \gamma : \Gamma, x : A(\gamma) \rfloor \; x : A(\gamma)}$$

are validated by the terminal object ϵ, by the domain $\Gamma.A$ of P_A, and by the unique section $\delta_A : \Gamma.A \to \Gamma.A.A^{\blacktriangledown}$ of $P_{A^{\blacktriangledown}}$ such that $P_A^{\bullet}\delta_A = 1_{\Gamma.A}$, respectively.

Additionally, substitution—and hence weakening—is interpreted by using the cleavage of $(\mathbf{C}, \mathcal{D})$: if the arrow $f : \Delta \to \Gamma$ is the interpretation of a substitution $\lfloor \delta : \Delta \rfloor \; f(\delta) : \Gamma$, then the type $A(f(\delta)) : \text{TYPE}$ and the term $a(f(\delta)) : A(f(\delta))$ will be interpreted by the display map $P_{A[f]}$ and by its unique section $a[f] : \Delta \to \Delta.A[f]$ such that $f^{\bullet}a[f] = af$, respectively.

Finally, if b is another section of $P_A : \Gamma.A \to \Gamma$, interpreting a term $b(\gamma) : A(\gamma)$, then the substitution $\lfloor \gamma : \Gamma \rfloor \; \gamma, a(\gamma), b(\gamma) : \Gamma.A.A^{\blacktriangledown}$ is interpreted by the unique arrow $a;b : \Gamma \to \Gamma.A.A^{\blacktriangledown}$ whose postcompositions via $P_{A^{\blacktriangledown}}$ and $P_A^{\bullet}$ are a and b, respectively.

In the next section we show how to use display map categories to model axiomatic type theory via the syntactic approach.

3 Syntactic approach to the semantics of axiomatic type theory

In this section we specialise the notion of display map categories to make them into models of axiomatic type theory. This consists in endowing a display map category with additional structure constituting the semantic counterpart of the type former. As we mentioned in Section 1, this additional structure may be formulated in alignment with the syntax, by means of choice functions, each associated to a given rule of the theory, assigning to the interpretation of the premises of that rule an interpretation of its consequence. These choice functions constitute an encoding of type formers into the given display map categories.

For example, referring to the rules of Figure 1 for axiomatic =-types, we give the following:

Definition 2 (Semantics of axiomatic =-types—syntactic formulation).
Let $(\mathbf{C}, \mathcal{D})$ be a display map category. Let us assume that, for every object Γ and every display map P_A of codomain Γ, there is a choice of:
- *(Form Rule) a display map $\Gamma.A.A^{\blacktriangledown}.\mathsf{Id}_A \to \Gamma.A.A^{\blacktriangledown}$;*
- *(Intro Rule) a section refl_A of the display map $\Gamma.A.\mathsf{Id}_A[\delta_A] \to \Gamma.A$;*

and that, for every object Γ, every display map P_A of codomain Γ, every display map P_C of codomain $\Gamma.A.A^{\blacktriangledown}.\mathsf{Id}_A$, and every section $c : \Gamma.A \to \Gamma.A.C[\mathsf{r}_A]$ of $P_{C[\mathsf{r}_A]}$—where r_A is $\Gamma.A - \mathsf{refl}_A \to \Gamma.A.\mathsf{Id}_A[\delta_A] - \delta_A^{\bullet} \to \Gamma.A.A^{\blacktriangledown}.\mathsf{Id}_A$—there is a choice of:
- *(Elim Rule) a section J_c of the display map $\Gamma.A.A^{\blacktriangledown}.\mathsf{Id}_A.C \to \Gamma.A.A^{\blacktriangledown}.\mathsf{Id}_A$;*
- *(Comp Axiom) a section H_c of the display map $\Gamma.A.\mathsf{Id}_{C[\mathsf{r}_A]}[\mathsf{J}_c[\mathsf{r}_A]; c] \to \Gamma.A$,*

where $\mathsf{J}_c[\mathsf{r}_A]; c$ is built as in Section 2.
Let us assume that, for every arrow $f : \Delta \to \Gamma$, the following stability conditions:

$$\mathsf{Id}_A[f^{\bullet\bullet}] = \mathsf{Id}_{A[f]} \qquad \mathsf{J}_c[f^{\bullet\bullet\bullet}] = \mathsf{J}_{c[f^\bullet]} \qquad \mathsf{refl}_A[f^\bullet] = \mathsf{refl}_{A[f]} \qquad \mathsf{H}_c[f^\bullet] = \mathsf{H}_{c[f^\bullet]}$$

hold[1]*. Then we say that $(\mathbf{C}, \mathcal{D})$ is* endowed with axiomatic =-types.

Summarising Definition 2, to identify a model of axiomatic =-types from a display map category $(\mathbf{C}, \mathcal{D})$ is to provide choice functions:

Form	Intro	Elim	Comp	
Id	r	J	H	axiomatic =-types

interpreting the inference rules of Figure 1 within the language of $(\mathbf{C}, \mathcal{D})$. Analogously, in a display map category endowed with axiomatic =-types, we may define appropriate choice functions:

Form	Intro	Elim	Comp	
Σ	pair	split	σ	axiomatic Σ-types
Π	abst	ev	β	axiomatic Π-types
	funext		β^{Π} , η^{Π}	axiomatic fun ext
0		ind^0		axiomatic 0-types
1	$\star$	ind^1	β^1	axiomatic 1-types
2	$\perp$, $\top$	ind^2	$\beta^{2,\perp}$, $\beta^{2,\top}$	axiomatic 2-types
$\mathbb{N}$	0 , succ	$\mathsf{ind}^{\mathbb{N}}$	$\beta^{\mathbb{N},0}$, $\beta^{\mathbb{N},\mathsf{s}}$	axiomatic $\mathbb{N}$-types

corresponding to the inference rules for axiomatic Σ- and Π-types, axiomatic function extensionality, axiomatic 0-, 1-, 2-, and $\mathbb{N}$-types, respectively—see the full version [45, Section 3].

If, in this way, a display map category $(\mathbf{C}, \mathcal{D})$ is endowed with such a function for every logical rule in the theory, i.e. it is *endowed with axiomatic =-types, axiomatic Σ-types, axiomatic Π-types & axiomatic function extensionality, and axiomatic 0-, 1-, 2-, $\mathbb{N}$-types*—see Definition 2 and the corresponding notions in the extended version [45, Section 3]—then the interpretation in $(\mathbf{C}, \mathcal{D})$ of every type judgement and every term judgement of the theory is defined and

---

[1] We refer the reader to the extended version [45, Appendix A] for additional details on the type-checking of the stability conditions.

respects all the type equality judgements and all the term equality judgements of the theory. In other words, a notion of interepretation of ATT—with $=$-, Σ-, Π-, 0-, 1-, 2-, $\mathbb{N}$-types, and function extensionality—in $(\mathbf{C}, \mathcal{D})$ is defined in the sense of Section 2, meaning that the interpetations of judgements $\lfloor \gamma : \Gamma \rfloor$, $\lfloor \gamma : \Gamma \rfloor\ A(\gamma) : \textsc{Type}$, $\lfloor \gamma : \Gamma \rfloor\ a(\gamma) : A(\gamma)$ consist of an object Γ, a display map P_A of codomain Γ, and a section a of P_A, respectively. Additionally, this notion of interpretation is *sound*:

Theorem 1 (Soundness property). *The interpretation of* ATT *in* $(\mathbf{C}, \mathcal{D})$
is sound, i.e. if ATT *infers:*

the judgement $\lfloor \gamma : \Gamma \rfloor$,   *then the object* Γ *is defined;*
the judgement $\lfloor \gamma : \Gamma \rfloor \equiv \lfloor \gamma' : \Gamma' \rfloor$,   *then* Γ *and* Γ' *coincide;*
the judgement $\lfloor \gamma : \Gamma \rfloor\ A(\gamma) : \textsc{Type}$,   *then the display map* P_A *is defined;*
the judgement $\lfloor \gamma : \Gamma \rfloor\ A(\gamma) \equiv A'(\gamma)$, *then* P_A *and* $P_{A'}$ *coincide;*
the judgement $\lfloor \gamma : \Gamma \rfloor\ a(\gamma) : A(\gamma)$,   *then the section* $a : \Gamma \to \Gamma.A$ *is defined;*
the judgement $\lfloor \gamma : \Gamma \rfloor\ a(\gamma) \equiv a'(\gamma)$, *then* a *and* a' *coincide.*

The proof of this fact is standard and adapts the argument provided by Streicher [48] and Hofmann [28]. We also refer the reader to [16,47,25] for further details. The argument consists in defining an a priori partial interpretation function, whose domain consists of the *pre-judgements* of the raw syntax of ATT, and which can be shown by induction on derivations to be well-defined on the actual judgements of ATT. We refer the reader to the extended version [45, Subsection 3.1] for a full proof as well as for a list of the inductive clauses satisfied by the interpretation function when restricted to the judgements of ATT.

In light of this result, we may call such a display map category $(\mathbf{C}, \mathcal{D})$ equipped with axiomatic $=$-types, axiomatic Σ-types, axiomatic Π-types & axiomatic function extensionality, and axiomatic 0-, 1-, 2-, $\mathbb{N}$-types a *model of* ATT. As anticipated, in the next section we show how to provide a categorical presentation of axiomatic type constructors.

4   Categorical formulation of the semantics of axiomatic type theory

As mentioned in the introduction, the issue with the formulation presented in Section 3 regarding the semantics of axiomatic type formers is that, in this form, it may be regarded as impractical, depending on the context—for example, when seeking concrete models of ATT. As we pointed out, a more practical formulation should rely less on equipping the base structure with choice functions and more on universal properties, which allow these choice functions to be derived automatically. Universal properties are generally easier to identify in mathematical practice. However, the challenge with axiomatic type formers is that one-dimensional universal properties fail to uniquely characterise them. These properties collapse intensional and axiomatic type formers into extensional ones, making it impossible to distinguish between them. Consequently, they effectively characterise only the semantic transcription of the rules of ETT.

This is why the notion we introduce below is 2-dimensional: while 1-cells, as usual, represent substitutions, 2-cells represent (equivalence classes of) *context propositional equalities*, or propositional equalities between substitutions: lists, indicated as $\lfloor \gamma : \Gamma \rfloor\, p(\gamma) : f(\gamma) = g(\gamma)$, of term judgements of the form:

$$\lfloor \gamma : \Gamma \rfloor\, p_1(\gamma) : f_1(\gamma) = g_1(\gamma)$$
$$\lfloor \gamma : \Gamma \rfloor\, p_2(\gamma) : f_2(\gamma) = p_1(\gamma)^* g_2(\gamma)$$
$$\lfloor \gamma : \Gamma \rfloor\, p_3(\gamma) : f_3(\gamma) = (p_1(\gamma), p_2(\gamma))^* g_3(\gamma)$$
$$\cdots$$
$$\lfloor \gamma : \Gamma \rfloor\, p_n(\gamma) : f_n(\gamma) = (p_1(\gamma), ..., p_{n-1}(\gamma))^* g_n(\gamma)$$

where $\lfloor \gamma : \Gamma \rfloor\, f(\gamma) : \Delta$ and $\lfloor \gamma : \Gamma \rfloor\, g(\gamma) : \Delta$ are parallel substitutions of ATT and transport operations along multiple identity proofs are defined by the elimination rule for axiomatic identity types.

Following the approach devised by Garner [24], the presence of an additional dimension enables us to impose more fine-grained universal properties on such a structure. These properties allow us to identify axiomatic type formers without conflating them with extensional ones.

Definition 3. *A (split) display map 2-category $(\mathcal{C}, \mathcal{D})$ is a (2,1)-category $\mathcal{C}$ with a chosen 2-terminal object ϵ, together with a class $\mathcal{D}$ of 1-cells of $\mathcal{C}$, called* display maps *and denoted as $\Gamma.A \to \Gamma$ and labeled as P_A, such that:*

$\succ$ *for every display map $\Gamma.A \to \Gamma$ and every arrow $f : \Delta \to \Gamma$, there is a choice of a display map $\Delta.A[f] \to \Delta$ and of a square:*

$$\Delta.A[f] \to \Gamma.A$$
$$\downarrow \quad \lrcorner \quad \downarrow$$
$$\Delta - f \to \Gamma$$

which is a pullback and a 2-pullback; the arrow $\Delta.A[f] \to \Gamma.A$ will be denoted as $f.A$, or as $f^\bullet$ in absence of ambiguity; we will indicate $A[P_A]$ as $A^\bullet$;

$\succ$ *every display map $\Gamma.A \to \Gamma$ is a cloven isofibration i.e. for every 2-cell p as in the diagram below, there is a choice of a 1-cell t_g^p and of a 2-cell τ_g^p as in the diagram below, in such a way that the equality:*

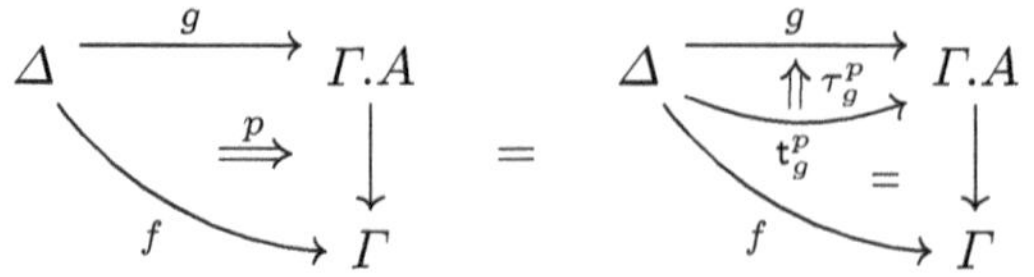

$$\Delta \xrightarrow{\;g\;} \Gamma.A \qquad \qquad \Delta \xrightarrow[\Uparrow \tau_g^p]{\;g\;} \Gamma.A$$

between the 2-cells p and $P_A * \tau_g^p$ holds, and $\mathsf{t}_{gh}^{p*h} = \mathsf{t}_g^p h$ and $\tau_{gh}^{p*h} = \tau_g^p * h$ for every 1-cell $h : \Omega \to \Delta$;

$\succ$ *the equalities:*

$$P_{A[1_A]} = P_A \quad (1_\Gamma).A = 1_{\Gamma.A} \quad P_{A[fg]} = P_{A[f][g]} \quad (f.A)(g.A[f]) = (fg).A$$

hold for every choice of composable arrows f and g and every display map $\Gamma.A \to \Gamma$, where Γ is the target of f;

$\succ$ *if* $\Gamma.A_1 \to \Gamma$, $\Gamma.A_1.A_2 \to \Gamma.A_1$, ..., $\Gamma.A_1.\ldots.A_{n-1}.A_n \to \Gamma.A_1.\ldots.A_{n-1}$ *are composable display maps and* $f : \Delta \to \Gamma$ *is a 1-cell, and if we are given a display map* $\Gamma.A.C \to \Gamma.A$ [2] *and commutative squares:*

$$
\begin{array}{ccc}
\Omega' \xrightarrow{\ f'\ } \Omega & \qquad & \Omega' \xrightarrow{\ f'\ } \Omega \\
g' \downarrow \qquad \downarrow g & & h' \downarrow \qquad \downarrow h \\
\Delta.A[f].C[f.A] \xrightarrow{\ f.A.C\ } \Gamma.A.C & & \Delta.A[f] \xrightarrow{\ f.A\ } \Gamma.A
\end{array}
$$

such that $P_A P_C g = P_A h$ *and* $P_{A[f]} P_C g' = P_{A[f]} h'$ *and a 2-cell* $p : h \Rightarrow P_C g$ *such that* $P_A * p = 1_{P_A h}$, *then the following diagram of 2-cells:*

$$
\begin{array}{ccc}
& \Omega' \xrightarrow{\quad f' \quad} \Omega & \\
\mathsf{t}^{p[f]}_{g'} \Big(\xRightarrow{\ \tau^{p[f]}_{g'}\ } \Big) g' & & \mathsf{t}^{p}_{g} \Big(\xRightarrow{\ \tau^{p}_{g}\ } \Big) g \\
& \Delta.A[f].C[f.A] \xrightarrow{\ f.A.C\ } \Gamma.A.C &
\end{array}
$$

commutes, i.e. the equalities $\mathsf{t}^{p}_{g}[f.A] = \mathsf{t}^{p[f]}_{g'}$ *and* $\tau^{p}_{g}[f.A] = \tau^{p[f]}_{g'}$ *hold.*[3]

The notion of a display map 2-category captures the structural aspects of dependent type theory, combined with fragments of the specific logical aspects of ATT. Specifically, the cloven isofibration structure with which the display maps are equipped provides a 2-dimensional interpretation of a consequence of the elimination rule and the computation *axiom* for =-types: transport—along one or more identity proofs. In [24], display maps in a model of ITT are endowed with a *normal* isofibration structure, which in our setting corresponds to the additional *normality* requirements that $\mathsf{t}^{1_{P_A g}}_{g} = g$ and $\tau^{1_{P_A g}}_{g} = 1_g$. However, dropping these requirements and reducing display maps to mere cloven isofibrations is fundamental to prevent models of ATT from being models of ITT. Let us briefly analyse how the cloven isofibration structure is induced in the context of the syntax of ATT itself.

If we are given a type judgement $\lfloor \gamma : \Gamma \rfloor\ A(\gamma) : \textsc{Type}$ and substitutions $\lfloor \delta : \Delta \rfloor\ f(\delta) : \Gamma$ and $\lfloor \delta : \Delta \rfloor\ g(\delta) : \Gamma.A$—i.e. $g(\delta)$ is given by $g_1(\delta) : \Gamma$, $g_2(\delta) : A(g_1(\delta))$—with a context identity proof $\lfloor \delta : \Delta \rfloor\ p(\delta) : f(\delta) = g_1(\delta)$, then t^{p}_{g} is the substitution $\lfloor \delta : \Delta \rfloor\ f(\delta) : \Gamma, p(\delta)^* g_2(\delta) : A(f(\delta))$ and τ^{p}_{g} is the context identity proof $\lfloor \delta : \Delta \rfloor\ f(\delta), p(\delta)^* g_2(\delta) = g_1(\delta), g_2(\delta)$ provided by the list:

$$
\begin{array}{lll}
\lfloor \delta : \Delta \rfloor & p(\delta) : & f(\delta) = g_1(\delta) \\
\lfloor \delta : \Delta \rfloor\ \mathsf{r}(p(\delta)^* g_2(\delta)) : & p(\delta)^* g_2(\delta) = p(\delta)^* g_2(\delta)
\end{array}
$$

---

[2] Here $A_1.\ldots.A_n$ and $A_1[f].A_2[f^\bullet].A_3[f^{\bullet\bullet}].\ldots.A_n[f^{\bullet\bullet\cdots\bullet}]$ are indicated as A and $A[f]$ respectively, and $P_{A_n} P_{A_{n-1}} \ldots P_{A_1}$ is indicated as P_A. Re-indexings on 1- and 2-cells are defined using the 1- and 2-pullback structure, analogously to display map categories.

[3] This requirement states that, by re-indexing along $\lfloor \delta \rfloor\ f(\delta) : \Gamma$ the isofibration structure that $\lfloor \gamma, x : A(\gamma), z : C(\gamma, x) \rfloor\ \gamma, x$ is endowed with, the isofibration structure that we obtain on $\lfloor \delta, x' : A(f(\delta)), z' : C(f(\delta), x') \rfloor\ \delta, x'$ is the one that the latter is already endowed with.

of identity proofs. Now, if $f(\delta) \equiv g_1(\delta)$ and $p(\delta) \equiv \mathsf{r}(g_1(\delta))$, then t_g^p is $\lfloor \delta : \Delta \rfloor\, g_1(\delta) : \Gamma$, $\mathsf{r}(g_1(\delta))^*g_2(\delta) : A(g_1(\delta))$ and τ_g^p is the list:

$$\lfloor \delta : \Delta \rfloor \qquad \mathsf{r}(g_1(\delta)) : \qquad g_1(\delta) = g_1(\delta)$$
$$\lfloor \delta : \Delta \rfloor\; \mathsf{r}(\mathsf{r}(g_1(\delta))^*g_2(\delta)) :\; \mathsf{r}(g_1(\delta))^*g_2(\delta) = \mathsf{r}(g_1(\delta))^*g_2(\delta)$$

hence in general in this case t_g^p is *not* $g(\delta)$ and τ_g^p is *not* its identity 2-cell: in ATT we can infer that $\mathsf{r}(g_1(\delta))^*g_2(\delta) = g_2(\delta)$—it is a fragment of the computation axiom for =-types—but not that $\mathsf{r}(g_1(\delta))^*g_2(\delta) \equiv g_2(\delta)$. In other words, the display map associated to the type A is a cloven isofibration but not necessarily a normal isofibration.

This fact explains why, in a general display map 2-category, display maps are cloven isofibrations but not normal isofibrations: this is what happens when we consider the syntax of ATT. As we will see—Subsection 4.2—the cloven isofibration structure on display maps enables the interpretation of computation axioms within a display map 2-category. However, the fact that they are generally *not* normal isofibrations is crucial to ensuring that these computation axioms are validated without necessarily validating the corresponding computation rules: this weakening prevents the interpretation of terms H_c and σ_c from collapsing into reflexivities, hence the respective propositional equalities from being judgemental.

We specialise the notion of a display map 2-category to obtain a version that incorporates a 2-dimensional semantic interpretation of axiomatic =-types. We require that each display map P_A be equipped with a *homotopy* arrow object: whereas for an ordinary arrow object there is an isomorphism between 2-cells into P_A and 1-cells into its arrow object itself, in the homotopy case this is weakened to an equivalence of categories, preventing models from validating higher-dimensional extensionality (discreteness) principles—for additional details see the extended version [45, Subsection 4.1 and Section 6].

Definition 4 (Semantics of axiomatic =-types—categorical formulation). *Let $(\mathcal{C}, \mathcal{D})$ be a display map 2-category. Let us assume that, for every object Γ and every display map P_A of codomain Γ, there is a choice of a display map $\Gamma.A.A^{\blacktriangledown}.\mathsf{Id}_A \to \Gamma.A.A^{\blacktriangledown}$ and of a homotopy arrow object for P_A, i.e. a 2-cell:*

$$\Gamma.A.A^{\blacktriangledown}.\mathsf{Id}_A \xrightarrow[P_A^{\bullet}P_{\mathsf{Id}_A}]{\overset{P_A^{\blacktriangledown}P_{\mathsf{Id}_A}}{\Downarrow \alpha_A}} \Gamma.A \xrightarrow{P_A} \Gamma$$

of $\mathcal{C}/\Gamma$ together with a functor $\tilde{\ } : (\mathcal{C}/\Gamma)(h, P_A)^{\to} \to (\mathcal{C}/\Gamma)(h, P_A P_A^{\blacktriangledown}P_{\mathsf{Id}_A})$ for all 1-cells $h : \Delta \to \Gamma$, such that the postcomposition via α_A is a surjective equivalence $(\mathcal{C}/\Gamma)(h, P_A P_A^{\blacktriangledown}P_{\mathsf{Id}_A}) \to (\mathcal{C}/\Gamma)(h, P_A)^{\to}$ and $\tilde{\ }$ is a section of it and the stability conditions $\mathsf{Id}_A[f^{\bullet\bullet}] = \mathsf{Id}_{A[f]}$ and $\alpha_A[f] = \alpha_{A[f]}$ and $\tilde{\ }[f] = \tilde{\ }[f]$ hold for every 1-cell $f : \Delta \to \Gamma$. Then we say that $(\mathcal{C}, \mathcal{D})$ is endowed with axiomatic =-types.

Notation and properties. Let $P_A : \Gamma.A \to \Gamma$ be a display map, let $h : \Delta \to \Gamma$ be a substitution, and let a, b be 1-cells $h \to P_A$ i.e. arrows $\Delta \to \Gamma.A$ such that $P_A a = P_A b = h$. Let p be an arrow $a \Rightarrow b$ in the category $(\mathcal{C}/\Gamma)(h, P_A)$. Then, being α_A a homotopy arrow object, there is a choice of an arrow $\tilde{p} : \Delta \to \Gamma.A.A^{\blacktriangledown}.\mathsf{Id}_A$ in $(\mathcal{C}/\Gamma)(h, P_A P_{A^{\blacktriangledown}} P_{\mathsf{Id}_A})$ such that $\alpha_A * \tilde{p} = p$. In particular $P_{A^{\blacktriangledown}} P_{\mathsf{Id}_A} \tilde{p} = a$ and $P_A^{\bullet} P_{\mathsf{Id}_A} \tilde{p} = b$ which means that $P_{\mathsf{Id}_A} \tilde{p} = a; b$. We denote as $\{p\}$ the unique section of $P_{\mathsf{Id}_A[a;b]}$ such that:

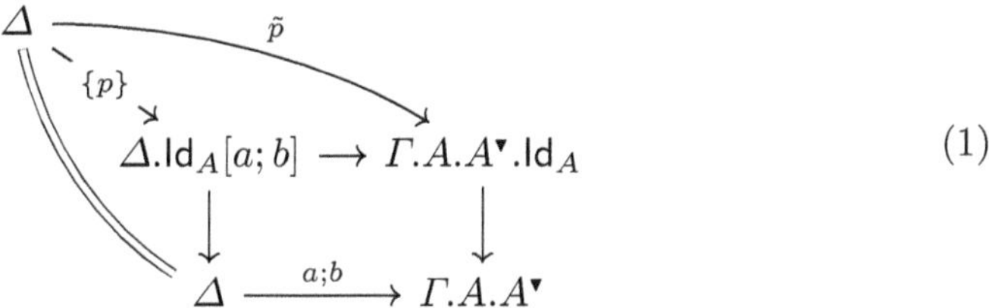

$$(1)$$

commutes. In the case where h is 1_Γ and hence a, b are sections of P_A—a situation that can be achieved by re-indexing $p : a \Rightarrow b$ via h itself—and f is any substitution $\Omega \to \Gamma$, the stability of $\tilde{}$ under re-indexing propagates to the stability of $\{\ \}$ itself under re-indexing: the equality $\{p\}[f] = \{p[f]\}$ holds for every arrow p in the category of the section of P_A. For additional details, we refer to the extended version [45, Subsection 4.1], where this condition is proven in the special case in which homotopy arrow objects are ordinary arrow objects: in that case, the functor $\tilde{}$ is uniquely determined and its stability under re-indexing is automatic. In the general case, the argument is exactly the same, except that the stability of $\tilde{}$ under re-indexing is given by definition.

Let $(\mathcal{C}, \mathcal{D})$ be a display map 2-category. Clearly $(\mathcal{C}, \mathcal{D})$ is in particular a display map category $(\mathbf{C}, \mathcal{D})$, where the category $\mathbf{C}$ is the underlying category of $\mathcal{C}$: the class $\mathcal{D}$ of display maps over $\mathcal{C}$ according to Definition 3 is in particular a class of display maps over $\mathbf{C}$ according to Definition 1.

Now, let us assume that the display map 2-category $(\mathcal{C}, \mathcal{D})$ is endowed with axiomatic $=$-types—Definition 4—and let P_A be a display map of codomain Γ. The next paragraphs show that the display map category $(\mathbf{C}, \mathcal{D})$ induced by $(\mathcal{C}, \mathcal{D})$ is endowed with appropriate choice functions validating axiomatic $=$-types—as in Definition 2.

Form Rule for $=$-types. A choice of a display map $\Gamma.A.A^{\blacktriangledown}.\mathsf{Id}_A \to \Gamma.A.A^{\blacktriangledown}$ is given by definition.

Intro Rule for $=$-types. The identity 2-cell $1_{1_{\Gamma.A}}$ of $1_{\Gamma.A}$ factors through the homotopy arrow object α_A as a 1-cell $\mathsf{r}_A := 1_{1_{\Gamma.A}} : \Gamma.A \to \Gamma.A.A^{\blacktriangledown}.\mathsf{Id}_A$. We define refl_A as the corresponding section $\{1_{1_{\Gamma.A}}\} : \Gamma.A \to \Gamma.A.\mathsf{Id}_A[1_{\Gamma.A}; 1_{\Gamma.A}] = \Gamma.A.\mathsf{Id}_A[\delta_A]$ of $P_{\mathsf{Id}_A[\delta_A]}$ such that $\delta_A^{\bullet}\{1_{1_{\Gamma.A}}\} = \mathsf{r}_A$.

Elim Rule for $=$-types. Let $\Gamma.A.A^{\blacktriangledown}.\mathsf{Id}_A.C \to \Gamma.A.A^{\blacktriangledown}.\mathsf{Id}_A$ be a display map and let c be a section $\Gamma.A \to \Gamma.A.C[\mathsf{r}_A]$ of the display map $P_{C[\mathsf{r}_A]}$. The pair $(\alpha_A, 1_{P_A^{\bullet} P_{\mathsf{Id}_A}})$ constitutes an arrow from α_A to $\alpha_A * \mathsf{r}_A(P_A^{\bullet} P_{\mathsf{Id}_A})$ in the category

$(\mathcal{C}/\Gamma)(P_A P_{A^{\blacktriangledown}} P_{\mathsf{Id}_A}, P_A)^{\rightarrow}$. Hence, it is induced, by post-composition via α_A, by a unique arrow $\varphi_A : 1_{\Gamma.A.A^{\blacktriangledown}.\mathsf{Id}_A} \implies \mathsf{r}_A(P_A^{\bullet} P_{\mathsf{Id}_A}) : P_A P_{A^{\blacktriangledown}} P_{\mathsf{Id}_A} \to P_A P_{A^{\blacktriangledown}} P_{\mathsf{Id}_A}$. Therefore $P_{A^{\blacktriangledown}} P_{\mathsf{Id}_A} * \varphi_A * \mathsf{r}_A = \alpha_A * \mathsf{r}_A = 1_{1_{\Gamma.A}} = P_{A^{\blacktriangledown}} P_{\mathsf{Id}_A} * 1_{\mathsf{r}_A}$ and $P_A^{\bullet} P_{\mathsf{Id}_A} * \varphi_A * \mathsf{r}_A = 1_{P_A^{\bullet} P_{\mathsf{Id}_A}} * \mathsf{r}_A = 1_{1_{\Gamma.A}} = P_A^{\bullet} P_{\mathsf{Id}_A} * 1_{\mathsf{r}_A}$ hence the arrows $\varphi_A * \mathsf{r}_A$ and 1_{r_A} induce, by post-composition via α_A, the same arrow $1_{1_{\Gamma.A}} \to 1_{1_{\Gamma.A}}$ in $(\mathcal{C}/\Gamma)(P_A, P_A)^{\rightarrow}$—namely the pair $(1_{1_{\Gamma.A}}, 1_{1_{\Gamma.A}})$. Being post-composition via α_A faithful, we conclude that $\varphi_A * \mathsf{r}_A = 1_{\mathsf{r}_A}$. Now, let $\tilde{\mathsf{J}}_c := (\mathsf{r}_A.C)cP_A^{\bullet} P_{\mathsf{Id}_A}$ and observe that $\varphi_A : 1_{\Gamma.A.A^{\blacktriangledown}.\mathsf{Id}_A} \implies \mathsf{r}_A P_A^{\bullet} P_{\mathsf{Id}_A} = \mathsf{r}_A P_{C[\mathsf{r}_A]} c P_A^{\bullet} P_{\mathsf{Id}_A} = P_C \tilde{\mathsf{J}}_c$ and therefore, using the cloven isofibration structure on P_C, we obtain a section $\mathsf{t}_{\tilde{\mathsf{J}}_c}^{\varphi_A} : \Gamma.A.A^{\blacktriangledown}.\mathsf{Id}_A \to \Gamma.A.A^{\blacktriangledown}.\mathsf{Id}_A.C$ of P_C, as well as a 2-cell $\tau_{\tilde{\mathsf{J}}_c}^{\varphi_A} : \mathsf{t}_{\tilde{\mathsf{J}}_c}^{\varphi_A} \implies \tilde{\mathsf{J}}_c$ such that $P_C * \tau_{\tilde{\mathsf{J}}_c}^{\varphi_A} = \varphi_A$. We define $\mathsf{J}_c := \mathsf{t}_{\tilde{\mathsf{J}}_c}^{\varphi_A}$.

Comp Axiom for =-types. Referring to the diagram:

$$
\begin{array}{ccc}
\Gamma.A & \xrightarrow{\;\;\mathsf{r}_A\;\;} & \Gamma.A.A^{\blacktriangledown}.\mathsf{Id}_A \\
{\scriptstyle \mathsf{J}_c[\mathsf{r}_A]}\downarrow \;\; \downarrow{\scriptstyle c} & & \mathsf{J}_c\downarrow \Rightarrow \downarrow\tilde{\mathsf{J}}_c \\
\Gamma.A.C[\mathsf{r}_A] & -\;\mathsf{r}_A^{\bullet}\;\to & \Gamma.A.A^{\blacktriangledown}.\mathsf{Id}_A.C \\
{\scriptstyle P_{C[\mathsf{r}_A]}}\downarrow & \lrcorner & \downarrow{\scriptstyle P_C} \\
\Gamma.A & \xrightarrow{\;\;\mathsf{r}_A\;\;} & \Gamma.A.A^{\blacktriangledown}.\mathsf{Id}_A
\end{array}
\tag{2}
$$

and observing that $c = \tilde{\mathsf{J}}_c[\mathsf{r}_A]$ [4] and that $P_C * \tau_{\tilde{\mathsf{J}}_c}^{\varphi_A} * \mathsf{r}_A = P_C * \tau_{\tilde{\mathsf{J}}_c \mathsf{r}_A}^{\varphi_A * \mathsf{r}_A} = \varphi_A * \mathsf{r}_A = 1_{\mathsf{r}_A} = \mathsf{r}_A * 1_{1_{\Gamma.A}}$ we conclude, by the 2-universal property of $\Gamma.A.C[\mathsf{r}_A]$, that there is unique a 2-cell $h_c : \mathsf{J}_C[\mathsf{r}_A] \Rightarrow c$ such that $P_{C[\mathsf{r}_A]} * h_c = 1_{1_{\Gamma.A}}$ and $\mathsf{r}_A^{\bullet} * h_c = \tau_{\tilde{\mathsf{J}}_c}^{\varphi_A} * \mathsf{r}_A$. Then we define H_C as the unique section $\{h_c\}$ of $\Gamma.A.\mathsf{Id}_{C[\mathsf{r}_A]}[\mathsf{J}_c[\mathsf{r}_A]; c] \to \Gamma.A$ making the Diagram (1) commute—where h is the identity and p is h_c.

Proposition 1. *Let $(\mathcal{C}, \mathcal{D})$ be a display map 2-category endowed with axiomatic =-types. Referring to the data* $\mathsf{Id}, \mathsf{refl}, \mathsf{J}, \mathsf{H}$ *defined in paragraphs* **Form Rule, Intro Rule, Elim Rule, and Comp Axiom, for =-types,** *the* stability conditions *of Definition 2 hold for every arrow* $f : \Delta \to \Gamma$*. Hence the associated display map category* $(\mathbf{C}, \mathcal{D})$ *is endowed with axiomatic =-types.*

Proof. See the extended version [45, Appendix B].

4.1   Some glances at axiomatic Σ-, Π-, 0-, 1-, 2-, and $\mathbb{N}$-types

We can specialise the notion of a display map 2-category with axiomatic =-types to obtain a version that incorporates a 2-dimensional semantic interpretation of axiomatic Σ-types [45, Subsection 4.2] and a version that incorporates 2-dimensional semantic interpretation of axiomatic Π-types & axiomatic function extensionality [45, Subsection 4.3].

The condition we require, for Σ-types, to be satisfied is a generalisation of the condition characterising the semantics of extensional Σ-types: in place of

---

[4] Since $P_C \tilde{\mathsf{J}}_c \mathsf{r}_A = P_C \mathsf{r}_A^{\bullet} c P_A^{\bullet} P_{\mathsf{Id}_A} \mathsf{r}_A = P_C \mathsf{r}_A^{\bullet} c = \mathsf{r}_A P_{C[\mathsf{r}_A]} c = \mathsf{r}_A 1_{\Gamma.A}$ and $P_{C[\mathsf{r}_A]} c = 1_{\Gamma.A}$ and $\mathsf{r}_A^{\bullet} c = \mathsf{r}_A^{\bullet} c P_A^{\bullet} P_{\mathsf{Id}_A} \mathsf{r}_A = \mathsf{J}_c \mathsf{r}_A$.

asking that display maps are closed under composition up to isomorphism [32], we require that they are closed by composition up to *equivalence*. The 2-cell components of such an equivalence induce the interpretation of the corresponding Comp Axiom. We observe a significant difference between the intensional case [24] and the axiomatic one: *such an equivalence is a mere equivalence, i.e. it need not consist of a* deformation retract. Similarly, the condition we require, for Π-types, to be satisfied is a generalisation of the condition characterising the semantics of extensional Π-types: right adjoints [32] to weakening functors are demoted to *right biadjoints*. In particular, we observe the same difference between the intensional case [24] and the axiomatic one, that we observe for Σ-types: *such a right biadjoint need not be a* retract *right biadjoint*.

Finally, we can specialise the notion of a display map 2-category with axiomatic =-types to obtain a version that incorporates a 2-dimensional semantic interpretation of axiomatic 0-, 1-, 2-, and $\mathbb{N}$-types [45, Subsection 4.4]. In this case, the conditions that we require to be satisfied consist of the existence of *bireflections to specific forgetful functors*. Again, we observe the usual weakening from the intensional case [24] to the very intensional, axiomatic one: *retract bireflections are demoted to mere bireflections*.

Full details on these notions are developed in the extended version [45, Section 4], where we prove that, if a display map 2-category $(\mathcal{C}, \mathcal{D})$ is endowed with axiomatic Σ- (resp. Π-, 0-, etc.) types, then the display map category $(\mathbf{C}, \mathcal{D})$ induced by $(\mathcal{C}, \mathcal{D})$ is endowed with appropriate choice functions validating axiomatic Σ- (resp. Π-, 0-, etc.) types as explained in Section 3 and in the extended version [45, Section 3]. Analogues of Proposition 1 are provided. By these results, we infer the following:

Theorem 2. *Every* display map 2-category $(\mathcal{C}, \mathcal{D})$ *endowed with axiomatic =-types, axiomatic Σ-types, axiomatic Π-types & axiomatic function extensionality, and axiomatic 0-, 1-, 2-, $\mathbb{N}$-types[5] induces a* display map category $(\mathbf{C}, \mathcal{D})$ *endowed with axiomatic =-types, axiomatic Σ-types, axiomatic Π-types & axiomatic function extensionality, and axiomatic 0-, 1-, 2-, $\mathbb{N}$-types[5] i.e. a model of* ATT.

and, by combining Theorem 2 with Theorem 1, we deduce that an interpretation of ATT in any display map 2-category—endowed with axiomatic =-types, axiomatic Σ-types, axiomatic Π-types & axiomatic function extensionality, and axiomatic 0-, 1-, 2-, $\mathbb{N}$-types—is well-defined and sound.

4.2   The role of the cloven isofibration structure

Let $(\mathcal{C}, \mathcal{D})$ be a display map 2-category equipped with axiomatic =-types. As proven in the paragraph **Elim Rule for =-types**, the structure of axiomatic =-types associated with the display maps in the class $\mathcal{D}$ enables the construction of an *elimination pseudo-term* $\tilde{\mathsf{J}}_c$ *for =-types*—for every section c of the display

---

[5] See Definition 4 (resp. 2) and the corresponding notions in the extended version [45, Section 4 (resp. 3)].

map $P_{C[r_A]}$. This is not a genuine section of P_C, as the composition $P_C\tilde{J}_c$ equals the identity of $\Gamma.A.A^\mathbf{v}.\mathsf{Id}_A$ merely up to the 2-cell φ_A, making it the interpretation of a pseudo-term of type C. The structure of cloven isofibrations on display maps plays a crucial role at this point, as it enables us to "strictify" the pseudo-term $\tilde{J}_c$—by transporting them back along the context identity proof φ_A—into the genuine elimination term J_c. In other words, it becomes an actual section of P_C at the cost of introducing an additional 2-cell $J_c \Rightarrow \tilde{J}_c$. The paragraph **Comp Axiom for =-types** shows how, essentially, this 2-cell represents the associated computation axiom: it leads, through re-indexing and further exploiting the homotopy arrow object structure—the one of the display map $P_{C[r_A]}$—to the identification of a section H_c and of the display map associated to the type $\mathsf{Id}_{C[r_A]}[J_c[r_A]; c]$. This section, therefore, provides an interpretation for the computation axiom for =-types.

As mentioned above, since the isofibration structure on display maps is cloven but not necessarily normal, this ensures that the computation axioms are satisfied, but not necessarily that the computation rules are. In other words, $J_c[r_A]$ does not, in general, coincide with the term c. To clarify this point, let us start by briefly analysing what happens if display maps in $(\mathcal{C}, \mathcal{D})$ are normal, as isofibrations. Referring to Diagram (2), if P_C is normal, then:

$$J_c r_A = t^{\varphi_A * r_A}_{\tilde{J}_c r_A} = t^{1_{r_A}}_{\tilde{J}_c r_A} = \tilde{J}_c r_A$$

$$(J_c \Rightarrow \tilde{J}_c) * r_A = \tau^{\varphi_A * r_A}_{\tilde{J}_c r_A} = \tau^{1_{r_A}}_{\tilde{J}_c r_A} = 1_{\tilde{J}_c r_A}$$

implying that $J_c[r_A]$ is in fact c and that h_c is the identity 1-cell of c, showing, in turn, that H_c is $\mathsf{refl}_{C[r_A]}[c]$. Similar considerations apply to the other type formers of ATT. For example, if $(\mathcal{C}, \mathcal{D})$ is endowed with axiomatic Σ- or 1-types with the assumption that the 2-cells β^B_A and β^1 are identities, and if the respective display map P_C is normal, then the same argument referred to Diagrams (4) and (9) in the extended version [45, Section 4] shows that $\mathsf{split}_c[\mathsf{p}^B_A]$ and $\mathsf{ind}^1_c[\star]$ coincide with the respective term c and the corresponding computation axioms are refl terms. Therefore $(\mathcal{C}, \mathcal{D})$ is actually a model of the intensional version of these type formers. In general, we have the following:

Theorem 3. *Let $(\mathcal{C}, \mathcal{D})$ be a display map 2-category endowed with axiomatic 0-, 1-, 2-, $\mathbb{N}$-, =-, Σ-, Π-types, and function extensionality—Definition 4 and [45, Section 4]. Let us assume that the (families of) 2-cells β^B_A, $\beta^{A,B}$, β^1, β^2, $\beta^{\mathbb{N}}$ are identity 2-cells and the display maps in $\mathcal{D}$ are normal, as isofibrations. Then $(\mathcal{C}, \mathcal{D})$ (induces a display map category $(\mathbf{C}, \mathcal{D})$ which) is a model of ITT.*

However, in the general case where display maps are *not* required to be normal as isofibrations, but only cloven, this situation is not necessary: this is what we show in the next section.

5   Revisiting the groupoid model

The groupoid model, devised by Hofmann and Streicher [29,30,49] as a model of ITT in which uniqueness of identity proofs fails, offers a semantic proof of its non-admissibility in ITT. Seen as a prelude towards homotopy type theory, it also exemplifies Garner's notion of semantics [24], as other models according to this notion can be viewed as generalisations of it. Notably, formulated as a display map 2-category, its display maps are normal isofibrations. Building on this foundational model, subsequent work—such as North [40] and Altenkirch and Neumann [3]—has demoted the groupoids to mere categories, yielding models for directed type theory. Following a similar but orthogonal trajectory, we propose a weakening of the groupoid model as well: demoting display maps to mere cloven isofibrations—chosen so as to maintain split re-indexing and thus preserve the integrity of a genuine model—in order to naturally reduce it to a model of ATT.

Let $\mathbf{Grpd}$ be the $(2,1)$-category made of groupoids, functors, and natural transformations—i.e. natural isomorphisms—with a chosen groupoid 1 made of one object and one morphism, as a specified 2-terminal object of $\mathbf{Grpd}$. Let $\mathcal{D}$ be the class of 1-cells whose elements are the functors of the form $P_A : \Gamma.A \to \Gamma$ obtained by applying the Grothendieck construction to the *pseudofunctors* $(A, \phi^A, \psi^A) : \Gamma \to \mathbf{Grpd}$, where Γ is a groupoid and ϕ^A and ψ^A are the *coherent* families of natural isomorphisms making A into a pseudofunctor. Hence $\Gamma.A$ is the category of the elements of (A, ϕ^A, ψ^A)—composition, identity arrows, and inverses are defined using ϕ^A and ψ^A and their coherence laws[6]—and the functor P_A is the projection on the first component.

Cloven isofibration structure on display maps. If we are given a 2-cell $\pi : P_A g \Rightarrow f$, where $g = (g_1, g_2)$ is a functor $\Delta \to \Gamma.A$ then the mappings:

$$\delta \mapsto (f\delta \, , \, A_{\pi_\delta^{-1}} g_2 \delta) \quad \text{and} \quad (\delta \xrightarrow{p} \delta') \mapsto (fp \, , \, A_{fp} A_{\pi_\delta^{-1}} g_2 \delta \xrightarrow{\phi^A} A_{(fp)\pi_\delta^{-1}} g_2 \delta =$$

$$A_{\pi_{\delta'}^{-1}(g_1 p)} g_2 \delta \xrightarrow{(\phi^A)^{-1}} A_{\pi_{\delta'}^{-1}} A_{(g_1 p)} g_2 \delta \xrightarrow{A_{\pi_{\delta'}^{-1}} g_2 p} A_{\pi_{\delta'}^{-1}} g_2 \delta')$$

define a functor $\mathsf{t}_g^\pi : \Delta \to \Gamma.A$ whose post-composition via P_A is f. Moreover, the arrow $(\pi_\delta \, , \, A_{\pi_\delta} A_{\pi_\delta^{-1}} g_2 \delta \xrightarrow{\phi^A} A_{1_{g_1\delta}} g_2 \delta \xrightarrow{\psi^A} g_2 \delta)$ from $(f\delta, A_{\pi_\delta^{-1}} g_2 \delta)$ to $(g_1\delta, g_2\delta)$ is the δ-component of a natural isomorphism $\tau_g^\pi : \mathsf{t}_g^\pi \Rightarrow g$ whose post-composition via P_A is π. These choices of t_g^π and τ_g^π define a cloven isofibration structure on P_A, which in general is *not* normal: if π is the identity and hence f coincides with g_1, then the mapping $\delta \mapsto (f\delta \, , \, A_{\pi_\delta^{-1}} g_2 \delta)$ coincides with $\delta \mapsto (g_1\delta \, , \, A_{1_{g_1\delta}} g_2 \delta)$ which in general is different from $(g_1\delta, g_2\delta)$, since the pseudofunctor A can be non-strict, hence in general t_g^π is different form g. The takeaway is that allowing all pseudofunctors into $\mathbf{Grpd}$ among the semantic types, rather than only strict ones as in the original groupoid model, broadens the class $\mathcal{D}$ of display maps to include cloven isofibrations that are not necessarily normal.

Re-indexing structure and arrow object structure on display maps. The re-indexing of a display map $P_A : \Gamma.A \to \Gamma$ along a 1-cell $f : \Delta \to \Gamma$ of $\mathbf{Grpd}$ is obtained by applying the Grothendieck construction to the pseudofunctor $\Delta - f \to \Gamma - A \to \mathbf{Grpd}$, where A is the pseudofunctor inducing P_A—in this

---

[6] See the extended version [45, Section 5 and Appendix C] for additional details.

way the requirements in Definition 3 are satisfied.[6] Additionally, if P_{Id_A} is the display map induced by the (pseudo)functor $\mathsf{Id}_A : \Gamma.A.A^{\mathbf{v}} \to \mathbf{Grpd}$ mapping $(\gamma, x, y) \mapsto A_\gamma(x, y)$—where $A_\gamma(x, y)$ is a discrete groupoid—and the (γ, x, y, p)-component of the 2-cell α_A is the arrow $(1_\gamma, A_{1_\gamma} x \xrightarrow{\psi^A} x \xrightarrow{p} y) : (\gamma, x) \to (\gamma, y)$, we identify an arrow object for P_A.[6] Following the construction at paragraphs **Elim Rule** and **Comp Axiom for** $=$**-types**, one can now reconstruct the choice functions r, φ, and J and observe that J_c acts *on objects* as the mapping $(\gamma, x, y, p) \mapsto (\gamma, x, y, p, C_{\varphi_A^{-1} c(\gamma, y)})$ whenever $(\gamma, x) \mapsto (\gamma, x, c(\gamma, x))$ is the action on object of a section c of $P_{C[\mathsf{r}_A]}$, for some display map P_C over $\Gamma.A.A^{\mathbf{v}}.\mathsf{Id}_A$. This allows to conclude that $\mathsf{J}_c[\mathsf{r}_A]$ acts on objects as the mapping $(\gamma, x) \mapsto (\gamma, x, C_{1_{(\gamma, x, x, 1_x)}} c(\gamma, x))$ and therefore, unless C was chosen to be a strict functor $\Gamma.A.A^{\mathbf{v}}.\mathsf{Id}_A \to \mathbf{Grpd}$, in general $\mathsf{J}_c[\mathsf{r}_A]$ and c *will not coincide*. We conclude that:

Theorem 4. *The pair* $(\mathbf{Grpd}, \mathcal{D})$ *is a display map 2-category endowed with axiomatic* $=$*-type, which, as a display map category—see Theorem 2—is a model of axiomatic* $=$*-types that does not validate Comp Rule for* $=$*-types. In particular Comp Rule for* $=$*-types is not admissible in* ATT.

6 Conclusions, remarks on completeness, and future work

In this paper we provided a 2-categorical procedure to construct models of ATT. We applied this procedure to identify a weakening of the groupoid model that models ATT without believing Comp Rule for $=$-types. We briefly comment on the extent to which the semantics provided by this class of models is complete with respect to ATT. In Garner's notion of semantics for ITT [24] every model validates the discreteness rule—namely, every model believes that every type is a 1-type. This follows from the *strict* arrow object structure imposed on display maps. The same phenomenon arises in both the original and our weakened versions of the groupoid model, as well as in the versions of the category model proposed in [40] and [3]—using directed $=$-types, or *hom-types*, in place of $=$-types. Conversely, adding to ATT a 2-dimensional discreteness rule asserting the equality of any two proofs of the identity of two parallel identity proofs yields completeness of the interpretation of this theory in display map 2-categories with strict arrow objects—see [45, Section 6]. However, a general model of axiomatic $=$-types in our setting does not necessarily feature strict arrow objects, but rather *homotopy* arrow objects. We believe that in general this prevents the automatic validation of the discreteness rule in a model, suggesting that one may find models without any higher-dimensional truncation. Even in this case, however, homotopy arrow objects validate choice principles for identity types that are not normally derivable in ATT. Hence, the interpretation of ATT cannot be complete. Nevertheless, it is natural to view our work as a step toward developing higher-dimensional semantics for ATT, in a broader sense involving $(\infty, 1)$-categorical structures equipped with a class of display maps—for example, locally cartesian closed $(\infty, 1)$-categories. The goal is to clarify how models constructed according to this semantics differ from the corresponding models of ITT, and how ATT can serve as an alternative language for higher groupoid theory. In a display map 2-category with strict arrow objects, two parallel 2-cells are either distinct or coincide. However, in a *display map higher category*, we may talk about homotopies between two 2-cells, or propositional equalities between identity proofs, without necessarily collapsing provably identical identity proofs into the same 2-cell. Moreover, every identity proof between identity proofs is itself in particular an identity proof: this leads to the idea that the third—or higher—dimension should not be explicitly used to define the interpretation, which, like for display map 2-categories, will only rely on 2-cells—to be later converted into sections using the arrow object. Yet, as mentioned, we will not have the issue of having to collapse every two parallel 2-cells that are not distinct into the same 2-cell. This is why we believe that the interpretation of ATT in appropriate display map higher categories might be complete.

This work belongs to the field of research in the categorical semantics of dependent type theories, along with their variants and generalisations—such as ATT itself—and aims to advance one of the main objectives of the area: to establish a concept of a general semantics that is both broad enough to encompass highly general notions of dependent type theories and functional enough to be effectively applied in their study. We argue that this work can be contextualised within the bicategorical approach to the semantics of dependent type theory proposed by Ahrens, North, and van der Weide [1,2], which is also related to the approach of Fiore and Saville [22] in the non-dependent case. This approach aims to provide a general framework for describing the semantics of the structural part of dependent type theory, as well as its generalisation towards directed type theory. Specifically, we believe that display map 2-categories constitute a particular instance of the notion of a *display map bicategory* introduced in that work, under the condition that reductions are symmetric. In fact, in a display map 2-category, the 2-cells are independent of the notion of a =-type former and can, a priori, be regarded as *reduction judgements*—although not directed: it is the semantic =-type former that enables us to interpret and convert them as identity proofs.

We are also interested in investigating whether Garner's approach of encoding =-types via arrow objects—here applied to axiomatic =-types—could be generalised to encode the semantics of directed identity types as proposed by North [40], by Altenkirch and Neumann [3], or by Laretto, Loregian, and Veltri [35]. This would aim to address a problem posed by Ahrens, North, and van der Weide [2] themselves: extending the notion of comprehension bicategory to accommodate the interpretation of the hom-type former à la North. More generally, we believe there is still much to explore regarding the semantics of various extensions and variations of dependent type theory. This is especially true in relation to the weakenings of the identity type constructor—both with respect to its elimination rule and its computation rule—that are currently prominent, and their relationships. We believe the notion of semantics introduced in this article provides a contribution to this important line of research in type theory.

Acknowledgments. This research was supported by a School of Mathematics full-time EPSRC Doctoral Training Partnership Studentship 2019/2020, by the Italian MUR PRIN 2022 "STENDHAL", and by eOTP RECIPROG.

The author is grateful to Nicola Gambino, Federico Olimpieri, Benedikt Ahrens, Daniël Otten, Marino Miculan, and Niels van der Weide for useful discussions on the subject.

Disclosure of Interests. The author has no competing interests to declare that are relevant to the content of this article.

References

1. Ahrens, B., North, P.R., van der Weide, N.: Semantics for two-dimensional type theory. In: Proceedings of the 37th Annual ACM/IEEE Symposium on Logic in Computer Science. pp. [Article 12], 14. ACM, New York ([2022] ©2022). https://doi.org/10.1145/3531130.3533334, https://doi.org/10.1145/3531130.3533334

2. Ahrens, B., North, P.R., van der Weide, N.: Bicategorical type theory: semantics and syntax. Math. Structures Comput. Sci. **33**(10), 868–912 (2023). https://doi.org/10.1017/s0960129523000312, https://doi.org/10.1017/s0960129523000312

3. Altenkirch, T., Neumann, J.: Synthetic 1-categories in directed type theory. arXiv:2410.19520 (2024)

4. Arndt, P., Kapulkin, K.: Homotopy-theoretic models of type theory. In: Typed lambda calculi and applications. 10th international conference, TLCA 2011, Novi Sad, Serbia, June 1–3, 2011. Proceedings, pp. 45–60. Berlin: Springer (2011). https://doi.org/10.1007/978-3-642-21691-6_7

5. Awodey, S.: Natural models of homotopy type theory. Math. Structures Comput. Sci. **28**(2), 241–286 (2018). https://doi.org/10.1017/S0960129516000268, https://doi.org/10.1017/S0960129516000268

6. Awodey, S., Gambino, N., Sojakova, K.: Inductive types in homotopy type theory. In: Proceedings of the 2012 27th Annual ACM/IEEE Symposium on Logic in Computer Science. pp. 95–104. IEEE Computer Soc., Los Alamitos, CA (2012). https://doi.org/10.1109/LICS.2012.21, https://doi.org/10.1109/LICS.2012.21

7. Awodey, S., Gambino, N., Sojakova, K.: Homotopy-initial algebras in type theory. J. ACM **63**(6), Art. 51, 45 (2017). https://doi.org/10.1145/3006383, https://doi.org/10.1145/3006383

8. Awodey, S., Warren, M.A.: Homotopy theoretic models of identity types. Math. Proc. Cambridge Philos. Soc. **146**(1), 45–55 (2009). https://doi.org/10.1017/S0305004108001783, https://doi.org/10.1017/S0305004108001783

9. van den Berg, B.: Path categories and propositional identity types. ACM Trans. Comput. Log. **19**(2), Art. 15, 32 (2018). https://doi.org/10.1145/3204492, https://doi.org/10.1145/3204492

10. van den Berg, B., den Besten, M.: Quadratic type checking for objective type theory. arXiv:2102.00905 (2021)

11. van den Berg, B., Moerdijk, I.: Exact completion of path categories and algebraic set theory. Part I: Exact completion of path categories. J. Pure Appl. Algebra **222**(10), 3137–3181 (2018). https://doi.org/10.1016/j.jpaa.2017.11.017, https://doi.org/10.1016/j.jpaa.2017.11.017

12. Bezem, M., Coquand, T., Huber, S.: A model of type theory in cubical sets. In: 19th International Conference on Types for Proofs and Programs, LIPIcs. Leibniz Int. Proc. Inform., vol. 26, pp. 107–128. Schloss Dagstuhl. Leibniz-Zent. Inform., Wadern (2014)
13. Bocquet, R.: Coherence of strict equalities in dependent type theories. arXiv:2010.14166 (2020)
14. Bocquet, R.: Strictification of weakly stable type-theoretic structures using generic contexts. In: 27th International Conference on Types for Proofs and Programs, LIPIcs. Leibniz Int. Proc. Inform., vol. 239, pp. Art. No. 3, 23. Schloss Dagstuhl. Leibniz-Zent. Inform., Wadern (2022). https://doi.org/10.4230/lipics.types.2021.3, https://doi.org/10.4230/lipics.types.2021.3
15. Boulier, S., Winterhalter, T.: Weak type theory is rather strong. 30th International Conference on Types for Proofs and Programs (2019), https://www.ii.uib.no/~bezem/abstracts/TYPES_2019_paper_18
16. Cartmell, J.: Generalised Algebraic Theories and Contextual Categories. Ph.D. thesis, University of Oxford (1978)
17. Clairambault, P., Dybjer, P.: The biequivalence of locally cartesian closed categories and Martin-Löf type theories. Math. Structures Comput. Sci. **24**(6), e240606, 54 (2014). https://doi.org/10.1017/S0960129513000881, https://doi.org/10.1017/S0960129513000881
18. Cohen, C., Coquand, T., Huber, S., Mörtberg, A.: Cubical type theory: a constructive interpretation of the univalence axiom. In: 21st International Conference on Types for Proofs and Programs, LIPIcs, vol. 69 (2018)
19. Coquand, T., Danielsson, N.A.: Isomorphism is equality. Indag. Math. (N.S.) **24**(4), 1105–1120 (2013). https://doi.org/10.1016/j.indag.2013.09.002, https://doi.org/10.1016/j.indag.2013.09.002
20. Curien, P.L., Garner, R., Hofmann, M.: Revisiting the categorical interpretation of dependent type theory. Theoret. Comput. Sci. **546**, 99–119 (2014). https://doi.org/10.1016/j.tcs.2014.03.003, https://doi.org/10.1016/j.tcs.2014.03.003
21. Dybjer, P.: Internal type theory. In: Types for Proofs and Programs: International Workshop, TYPES'95, Torino, Italy, June 5-8, 1995 Selected Papers. vol. 1158, p. 120. Springer Science & Business Media (1996)
22. Fiore, M., Saville, P.: A type theory for Cartesian closed bicategories. In: Proceedings of the 2019 34th annual ACM/IEEE symposium on logic in computer science, LICS 2019, Vancouver, Canada, June 24–27, 2019, p. 13. Piscataway, NJ: IEEE Press (2019). https://doi.org/10.1109/LICS.2019.8785708, dl.acm.org/doi/10.5555/3470152.3470190, id/No 38
23. Gambino, N., Garner, R.: The identity type weak factorisation system. Theoretical Computer Science **409**(1), 94–109 (2008). https://doi.org/10.1016/j.tcs.2008.08.030, http://dx.doi.org/10.1016/j.tcs.2008.08.030
24. Garner, R.: Two-dimensional models of type theory. Math. Structures Comput. Sci. **19**(4), 687–736 (2009). https://doi.org/10.1017/S0960129509007646, https://doi.org/10.1017/S0960129509007646
25. Hofmann, M.: Extensional concepts in intensional type theory. Ph.D. thesis, University of Edinburgh (1995)
26. Hofmann, M.: On the interpretation of type theory in locally cartesian closed categories. In: Pacholski, L., Tiuryn, J. (eds.) Computer Science Logic. pp. 427–441. Springer Berlin Heidelberg, Berlin, Heidelberg (1995)
27. Hofmann, M.: Conservativity of equality reflection over intensional type theory. In: Berardi, S., Coppo, M. (eds.) Types for Proofs and Programs. pp. 153–164. Springer Berlin Heidelberg, Berlin, Heidelberg (1996)

28. Hofmann, M.: Syntax and semantics of dependent types, pp. 13–54. Springer, London (1997). https://doi.org/10.1007/978-1-4471-0963-1_2, https://doi.org/10.1007/978-1-4471-0963-1_2

29. Hofmann, M., Streicher, T.: The groupoid model refutes uniqueness of identity proofs. In: Proceedings of the Ninth Annual IEEE Symposium on Logic in Computer Science (LICS 1994). pp. 208–212. IEEE Computer Society Press (July 1994)

30. Hofmann, M., Streicher, T.: The groupoid interpretation of type theory. In: Twenty-five years of constructive type theory (Venice, 1995), Oxford Logic Guides, vol. 36, pp. 83–111. Oxford Univ. Press, New York (1998)

31. Jacobs, B.: Comprehension categories and the semantics of type dependency. Theoret. Comput. Sci. **107**(2), 169–207 (1993). https://doi.org/10.1016/0304-3975(93)90169-T, https://doi.org/10.1016/0304-3975(93)90169-T

32. Jacobs, B.: Categorical logic and type theory, Studies in Logic and the Foundations of Mathematics, vol. 141. North-Holland Publishing Co., Amsterdam (1999)

33. Kapulkin, K., Lumsdaine, P.L.: Homotopical inverse diagrams in categories with attributes. Journal of Pure and Applied Algebra **225**(4), 106563 (2021). https://doi.org/https://doi.org/10.1016/j.jpaa.2020.106563, https://www.sciencedirect.com/science/article/pii/S0022404920302644

34. Kapulkin, K., Lumsdaine, P.L.: The homotopy theory of type theories. Adv. Math. **337**, 1–38 (2018). https://doi.org/10.1016/j.aim.2018.08.003

35. Laretto, A., Loregian, F., Veltri, N.: Di- is for directed: First-order directed type theory via dinaturality. Proc. ACM Program. Lang. **10**(POPL) (Jan 2026). https://doi.org/10.1145/3776703, https://doi.org/10.1145/3776703

36. Lumsdaine, P.L., Warren, M.A.: The local universes model: an overlooked coherence construction for dependent type theories. ACM Trans. Comput. Log. **16**(3), Art. 23, 31 (2015). https://doi.org/10.1145/2754931, https://doi.org/10.1145/2754931

37. Maietti, M.E.: Modular correspondence between dependent type theories and categories including pretopoi and topoi. Math. Structures Comput. Sci. **15**(6), 1089–1149 (2005). https://doi.org/10.1017/S0960129505004962, https://doi.org/10.1017/S0960129505004962

38. Moggi, E.: A category-theoretic account of program modules. Math. Structures Comput. Sci. **1**(1), 103–139 (1991). https://doi.org/10.1017/S0960129500000074, https://doi.org/10.1017/S0960129500000074

39. Moss, S.K., von Glehn, T.: Dialectica models of type theory. In: 33rd Annual ACM/IEEE Symposium on Logic in Computer Science. p. 739–748. Association for Computing Machinery, New York, NY, USA (2018)

40. North, P.R.: Towards a directed homotopy type theory. In: Proceedings of the Thirty-Fifth Conference on the Mathematical Foundations of Programming Semantics. Electron. Notes Theor. Comput. Sci., vol. 347, pp. 223–239. Elsevier Sci. B. V., Amsterdam (2019). https://doi.org/10.1016/j.entcs.2019.09.012, https://doi.org/10.1016/j.entcs.2019.09.012

41. Otten, D., Spadetto, M.: The biequivalence of path categories and axiomatic Martin-Löf type theories. In: 34th EACSL Annual Conference on Computer Science Logic (CSL 2026). Leibniz International Proceedings in Informatics (LIPIcs) (2026). https://drops.dagstuhl.de/entities/document/10.4230/LIPIcs.CSL.2026.38

42. Paulson, L.C.: Formalising mathematics in simple type theory. In: Reflections on the foundations of mathematics—univalent foundations, set theory and general thoughts, Synth. Libr., vol. 407, pp. 437–453. Springer, Cham ([2019] ©2019). https://doi.org/10.1007/978-3-030-15655-8_20, https://doi.org/10.1007/978-3-030-15655-8_20

43. Riehl, E.: Categorical homotopy theory, New Math. Monogr., vol. 24. Cambridge: Cambridge University Press (2014). https://doi.org/10.1017/CBO9781107261457
44. Seely, R.A.G.: Locally Cartesian closed categories and type theory. Math. Proc. Cambridge Philos. Soc. **95**(1), 33–48 (1984). https://doi.org/10.1017/S0305004100061284, https://doi.org/10.1017/S0305004100061284
45. Spadetto, M.: A 2-categorical approach to the semantics of dependent type theory with computation axioms (extended version). arXiv:2507.07208 (2025)
46. Spadetto, M.: Relating homotopy equivalences to conservativity in dependent type theories with computation axioms. Logical Methods in Computer Science **Volume 21, Issue 3**, 32 (Sep 2025). https://doi.org/10.46298/lmcs-21(3:32)2025, https://lmcs.episciences.org/11565
47. Streicher, T.: Correctness of the Interpretation of the Calculus of Constructions in Doctrines of Constructions, pp. 156–220. Birkhäuser Boston, Boston, MA (1991). https://doi.org/10.1007/978-1-4612-0433-6_4, https://doi.org/10.1007/978-1-4612-0433-6_4
48. Streicher, T.: Semantics of type theory — Correctness, completeness and independence results. Progress in Theoretical Computer Science, Birkhäuser Boston, Inc., Boston, MA (1991). https://doi.org/10.1007/978-1-4612-0433-6, https://doi.org/10.1007/978-1-4612-0433-6
49. Streicher, T.: The genesis of the groupoid model. Math. Structures Comput. Sci. **31**(9), 1003–1005 (2021). https://doi.org/10.1017/S0960129520000286, https://doi.org/10.1017/S0960129520000286
50. Taylor, P.: Practical foundations of mathematics, Cambridge Studies in Advanced Mathematics, vol. 59. Cambridge University Press (1999)
51. Univalent Foundations Program, T.: Homotopy Type Theory: Univalent Foundations of Mathematics. https://homotopytypetheory.org/book, Institute for Advanced Study (2013)
52. Vidmar, J.: Polynomial functors and W-types for groupoids. Ph.D. thesis, University of Leeds (2018), https://etheses.whiterose.ac.uk/22517/
53. Winterhalter, T.: Formalisation and meta-theory of type theory. Ph.D. thesis, Université de Nantes (2020)

The Value Problem for Weighted Timed Games with Two Clocks is Undecidable

Quentin Guilmant [iD], Joël Ouaknine*[iD],
and Isa Vialard [iD]

Max Planck Institute for Software Systems, Saarland Informatics Campus,
Saarbrücken, Germany

Abstract. We prove that the Value Problem for weighted timed games (WTGs) with two clocks and non-negative integer weights is undecidable – even under a time bound. The Value Problem for weighted timed games (WTGs) consists in determining, given a two-player weighted timed game with a reachability objective and a rational threshold, whether or not the value of the game exceeds the threshold. This problem was shown to be undecidable some ten years ago for WTGs making use of at least three clocks, and is known to be decidable for single-clock WTGs. Our reduction encodes a deterministic two-counter machine using two clocks and uses punishment gadgets that let the opponent detect and penalize any incorrect simulation. This closes one of the last remaining major gaps in our algorithmic understanding of WTGs.

1   Introduction

Real-time systems are not only ubiquitous in modern technological society, they are in fact increasingly pervasive in critical applications – from embedded controllers in automotive and avionics platforms to resource-constrained communication protocols. In such systems, exacting timing constraints and quantitative objectives must often be met simultaneously. Weighted Timed Games (WTGs), introduced over two decades ago [15,2,13,1,4], provide a powerful modelling framework for the automatic synthesis of controllers in such settings: they combine the expressiveness of Alur and Dill's clock-based timed automata with **Min-Max** gameplay and non-negative integer weights on both locations and transitions, enabling one to reason about quantitative aspects such as energy consumption, response times, or resource utilisation under adversarial conditions.

A central algorithmic task for WTGs is the *Value Problem*: given a two-player, turn-based WTG with a designated start configuration and a rational threshold c, determine whether Player **Min** can guarantee reaching a goal location with cumulative cost at most c, despite best adversarial play by Player

---

* Joël Ouaknine is also affiliated with Keble College, Oxford as emmy.network Fellow, and is supported by ERC grant DynAMiCs (101167561) and DFG grant 389792660 as part of TRR 248.

N. Bertrand and S. Milius (Eds.): FoSSaCS 2026, LNCS 16503, pp. 616–635, 2026.
https://doi.org/10.1007/978-3-032-22730-0_29

Max. This problem lies at the heart of quantitative controller synthesis and performance analysis for real-time systems.

Unfortunately, fundamental algorithmic barriers are well known. In particular, the Value Problem is known to be undecidable for WTGs making use of three or more clocks [5]. In fact, even approximating the value of three-clock WTGs with arbitrary (positive and negative) weights is known to be computationally unsolvable [12]. On the positive side, the Value Problem for WTGs making use of a single clock is decidable, regardless of whether weights range over $\mathbb{N}$ or $\mathbb{Z}$ [6,17]. There is a voluminous literature in this general area; for a comprehensive overview and discussion of the state of the art, we refer the reader to [11]. See also Fig. 1 in which we summarise some of the key existing results.

Clocks	Weights in	Value Problem	Existence Problem
1	$\mathbb{N}$	decidable [6]	decidable [7]
	$\mathbb{Z}$	decidable [17]	
2	$\mathbb{N}$	**undecidable**	**undecidable**
	$\mathbb{Z}$	**undecidable**	undecidable [10]
3+	$\mathbb{N}$	undecidable [5]	undecidable [3]
	$\mathbb{Z}$	inapproximable [12]	undecidable

Fig. 1. State of the art on the Value Problem for weighted timed games. Approximability for 2-clock WTGs, and 3-clock WTGs with weights in $\mathbb{N}$, remain open. This paper's main contribution (undecidability for WTGs with two clocks and weights in $\mathbb{N}$) is highlighted in boldface blue.

The case of WTGs with exactly two clocks (and non-negative weights) has remained stubbornly open. Resolving this question is essential, since two clocks suffice to encode most practical timing constraints (e.g., deadline plus cooldown), and efficient single-clock algorithms cannot in general be lifted to richer timing scenarios. The main contribution of this paper is to close this gap by establishing undecidability:

Theorem 1. *The Value Problem for two-player, turn-based, time-bounded, two-clock, weighted timed games with non-negative integer weights is undecidable. The same holds for weighted timed games over unbounded time otherwise satisfying the same hypotheses.*

Our reduction is from the Halting Problem for deterministic two-counter machines and proceeds via a careful encoding of counter values in clock valuations, combined with "punishment" gadgets that enforce faithful simulation or allow the adversary to drive the accumulated cost upwards. Key technical novelties include:

- Counter-Evolution Control (CEC) modules, which enforce precise proportional delays for encoding the incrementation and decrementation of counters.
- Multiplication-Control (MC) gadgets, which enable the adversary to verify whether simulated counter updates correspond to exact multiplication factors.

Together, these constructions fit within the two-clock timing structure and utilize only non-negative integer weights, thereby demonstrating that even the two-clock fragment previously the only remaining decidability candidate admits no algorithmic solution to the Value Problem. As a complementary result, we also show that the related Existence Problem (i.e., does **Min** have a strategy to achieve a cost at most c?) is undecidable under the same hypotheses.

Finally, we note that our reduction is implemented via WTGs with bounded duration by construction. This is particularly noteworthy given that many algorithmic problems for real-time and hybrid systems, which are known to be undecidable over unbounded time, become decidable in a time-bounded setting. See, for example, [18,19,14,8,9].

A full version of this paper, with detailed proofs, can be found at `https://arxiv.org/abs/2507.10550`.

2   Weighted Timed Games

Let $\mathcal{X}$ be a finite set of **clocks**. **Clock constraints** over $\mathcal{X}$ are expressions of the form $x \sim n$ or $x - y \sim n$, where $x, y \in \mathcal{X}$ are clocks, $\sim \in \{<, \leq, =, \geq, >\}$ is a comparison symbol, and $n \in \mathbb{N}$ is a natural number. We write $\mathcal{C}$ to denote the set of all clock constraints over $\mathcal{X}$. A **valuation** on $\mathcal{X}$ is a function $\nu : \mathcal{X} \to \mathbb{R}_{\geq 0}$. For $d \in \mathbb{R}_{\geq 0}$ we denote by $\nu + d$ the valuation such that, for all clocks $x \in \mathcal{X}$, $(\nu + d)(x) = \nu(x) + d$. Let $X \subseteq \mathcal{X}$ be a subset of all clocks. We write $\nu[X := 0]$ for the valuation such that, for all clocks $x \in X$, $\nu[X := 0](x) = 0$, and $\nu[X := 0](y) = \nu(y)$ for all other clocks $y \notin X$. For a set $C \subseteq \mathcal{C}$ of clock constraints over $\mathcal{X}$, we say that the valuation ν **satisfies** C, denoted $\nu \models C$, if and only if all the comparisons in C hold when replacing each clock x by its corresponding value $\nu(x)$.

Definition 1. *A **(turn-based) weighted timed game** is given by a tuple $\mathcal{G} = (L_{\mathsf{Min}}, L_{\mathsf{Max}}, G, \mathcal{X}, T, w)$, where:*

- *L_{Min} and L_{Max} are the (disjoint) sets of **locations** belonging to Players* Min *and* Max *respectively; we let $L = L_{\mathsf{Min}} \cup L_{\mathsf{Max}}$ denote the set of all locations. (In drawings, locations belonging to* Min *are depicted by blue circles, and those belonging to* Max *are depicted by red squares.)*
- *$G \subseteq L_{\mathsf{Min}}$ are the **goal locations**.*
- *$\mathcal{X}$ is a set of clocks.*
- *$T \subseteq (L \setminus G) \times 2^{\mathcal{C}} \times 2^{\mathcal{X}} \times L$ is a set of **(discrete) transitions**. A transition $\ell \xrightarrow{C,X} \ell'$ enables moving from location ℓ to location ℓ', provided all clock*

constraints in C are satisfied, and afterwards resetting all clocks in X to zero.

$-\ w : (L \setminus G) \cup T \to \mathbb{Z}$ *is a **weight function**.*

In the above, we assume that all data (set of locations, set of clocks, set of transitions, set of clock constraints) are finite.

Let $\mathcal{G} = (L_{\mathsf{Min}}, L_{\mathsf{Max}}, G, \mathcal{X}, T, w)$ be a weighted timed game. A **configuration** over $\mathcal{G}$ is a pair (ℓ, ν), where $\ell \in L$ and ν is a valuation on $\mathcal{X}$. Let $d \in \mathbb{R}_{\geq 0}$ be a **delay** and $t = \ell \xrightarrow{C,X} \ell' \in T$ be a discrete transition. One then has a valid **delayed transition** (or simply a **transition** if the context is clear) $(\ell, \nu) \xrightarrow{d,t} (\ell', \nu')$ provided that $\nu + d \models C$ and $\nu' = (\nu + d)[X := 0]$. Intuitively, control remains in location ℓ for d time units, after which it transitions to location ℓ', resetting all the clocks in X to zero in the process. The **weight** of such a delayed transition is $d \cdot w(\ell) + w(t)$, taking account both of the time spent in ℓ as well as the weight of the discrete transition t.

As noted in [11], without loss of generality one can assume that no configuration (other than those associated with goal locations) is deadlocked; in other words, for any location $\ell \in L \setminus G$ and valuation $\nu \in \mathbb{R}_{\geq 0}^{\mathcal{X}}$, there exists $d \in \mathbb{R}_{\geq 0}$ and $t \in T$ such that $(\ell, \nu) \xrightarrow{d,t} (\ell', \nu')$.[1]

Let $k \in \mathbb{N}$. A **run** ρ of length k over $\mathcal{G}$ from a given configuration (ℓ_0, ν_0) is a sequence of matching delayed transitions, as follows:

$$\rho = (\ell_0, \nu_0) \xrightarrow{d_0, t_0} (\ell_1, \nu_1) \xrightarrow{d_1, t_1} \cdots \xrightarrow{d_{k-1}, t_{k-1}} (\ell_k, \nu_k) \,.$$

The **weight** of ρ is the cumulative weight of the underlying delayed transitions:

$$\mathsf{weight}(\rho) = \sum_{i=0}^{k-1} (d_i \cdot w(\ell_i) + w(t_i)) \,.$$

An infinite run ρ is defined in the obvious way; however, since no goal location is ever reached, its weight is defined to be infinite: $\mathsf{weight}(\rho) = +\infty$.

A run is **maximal** if it is either infinite or cannot be extended further. Thanks to our deadlock-freedom assumption, finite maximal runs must end in a goal location. We refer to maximal runs as **plays**.

We now define the notion of **strategy**. Recall that locations of $\mathcal{G}$ are partitioned into sets L_{Min} and L_{Max}, belonging respectively to Players Min and Max. Let Player $\mathsf{P} \in \{\mathsf{Min}, \mathsf{Max}\}$, and write $\mathcal{FR}_{\mathcal{G}}^{\mathsf{P}}$ to denote the collection of all non-maximal finite runs of $\mathcal{G}$ ending in a location belonging to Player P. A **strategy**

---

[1] This can be achieved by adding unguarded transitions to a sink location for all locations controlled by Min and unguarded transitions to a goal location for the ones controlled by Max (noting that in all our constructions, Max-controlled locations always have weight 0). Nevertheless, in the interest of clarity we omit such extraneous transitions and locations in our representation of WTGs; we merely assume instead that neither player allows him- or herself to end up in a deadlocked situation, unless a goal location has been reached.

for Player P is a mapping $\sigma_\mathsf{P} : \mathcal{FR}_\mathcal{G}^\mathsf{P} \to \mathbb{R}_{\geq 0} \times T$ such that for all finite runs $\rho \in \mathcal{FR}_\mathcal{G}^\mathsf{P}$ ending in configuration (ℓ, ν) with $\ell \in L_\mathsf{P}$, the delayed transition $(\ell, \nu) \xrightarrow{d,t} (\ell', \nu')$ is valid, where $\sigma_\mathsf{P}(\rho) = (d, t)$ and (ℓ', ν') is some configuration (uniquely determined by $\sigma_\mathsf{P}(\rho)$ and ν).

Let us fix a starting configuration (ℓ_0, ν_0), and let σ_{Min} and σ_{Max} be strategies for Players Min and Max respectively (one speaks of a *strategy profile*). We write $\mathsf{play}_\mathcal{G}((\ell_0, \nu_0), \sigma_{\mathsf{Min}}, \sigma_{\mathsf{Max}})$ to denote the unique maximal run starting from configuration (ℓ_0, ν_0) and unfolding according to the strategy profile $(\sigma_{\mathsf{Min}}, \sigma_{\mathsf{Max}})$: in other words, for every strict finite prefix ρ of $\mathsf{play}_\mathcal{G}((\ell_0, \nu_0), \sigma_{\mathsf{Min}}, \sigma_{\mathsf{Max}})$ in $\mathcal{FR}_\mathcal{G}^\mathsf{P}$, the delayed transition immediately following ρ in $\mathsf{play}_\mathcal{G}((\ell_0, \nu_0), \sigma_{\mathsf{Min}}, \sigma_{\mathsf{Max}})$ is labelled with $\sigma_\mathsf{P}(\rho)$.

Recall that the objective of Player Min is to reach a goal location through a play whose weight is as small possible. Player Max has an opposite objective, trying to avoid goal locations, and, if not possible, to maximise the cumulative weight of any attendant play. This gives rise to the following two symmetrical definitions:

$$\overline{\mathsf{Val}}_\mathcal{G}(\ell_0, \nu_0) = \inf_{\sigma_{\mathsf{Min}}} \left\{ \sup_{\sigma_{\mathsf{Max}}} \left\{ \mathsf{weight}(\mathsf{play}_\mathcal{G}((\ell_0, \nu_0), \sigma_{\mathsf{Min}}, \sigma_{\mathsf{Max}})) \right\} \right\} \text{ and}$$

$$\underline{\mathsf{Val}}_\mathcal{G}(\ell_0, \nu_0) = \sup_{\sigma_{\mathsf{Max}}} \left\{ \inf_{\sigma_{\mathsf{Min}}} \left\{ \mathsf{weight}(\mathsf{play}_\mathcal{G}((\ell_0, \nu_0), \sigma_{\mathsf{Min}}, \sigma_{\mathsf{Max}})) \right\} \right\} .$$

$\overline{\mathsf{Val}}_\mathcal{G}(\ell_0, \nu_0)$ represents the smallest possible weight that Player Min can possibly achieve, starting from configuration (ℓ_0, ν_0), against best play from Player Max, and conversely for $\underline{\mathsf{Val}}_\mathcal{G}(\ell_0, \nu_0)$: the latter represents the largest possible weight that Player Max can enforce, against best play from Player Min.[2] As noted in [11], turn-based weighted timed games are *determined*, and therefore $\overline{\mathsf{Val}}_\mathcal{G}(\ell_0, \nu_0) = \underline{\mathsf{Val}}_\mathcal{G}(\ell_0, \nu_0)$ for any starting configuration (ℓ_0, ν_0); we denote this common value by $\mathsf{Val}_\mathcal{G}(\ell_0, \nu_0)$.

We can now state:

Definition 2 (Value Problem). *Given a WTG $\mathcal{G}$ with starting location ℓ_0 and a threshold $c \in \mathbb{Q}$, the **Value Problem** asks whether $\mathsf{Val}_\mathcal{G}(\ell_0, \mathbf{0}) \leq c$.*

The Value Problem differs subtly but importantly from the *Existence Problem*:

Definition 3 (Existence Problem). *Given a WTG $\mathcal{G}$ with starting location ℓ_0 and a threshold $c \in \mathbb{Q}$, the **Existence Problem** asks whether **Min** has a strategy σ_{Min} such that*

$$\sup_{\sigma_{\mathbf{Max}}} \left\{ \mathsf{weight}(\mathsf{play}_\mathcal{G}((\ell_0, \mathbf{0}), \sigma_{\mathbf{Min}}, \sigma_{\mathbf{Max}})) \right\} \leq c .$$

Remark 1. The Existence problem is undecidable for two-clock WTGs when arbitrary integer (positive and negative) weights are allowed [10].

---

[2] Technically speaking, these values may not be literally achievable; however given any $\varepsilon > 0$, both players are guaranteed to have strategies that can take them to within ε of the optimal value.

3   Undecidability

To establish undecidability, we reduce the Halting Problem for two-counter machines to the Value Problem. A two-counter machine is a tuple $\mathcal{M} = (Q, q_i, q_h, T)$ where Q is a finite set of states, $q_i, q_h \in Q$ are the initial and final state and $T \subseteq (Q \times \{c, d\} \times Q) \cup (Q \times \{c, d\} \times Q \times Q)$ is a set of transitions. A two-counter machine is deterministic if for any state q there is at most one transition $t \in T$ which has q as first component. As its name suggests, a two-counter machine comes equipped with two counters, c and d, which are variables with values in $\mathbb{N}$. The semantics is as follows: a transition (q, e, q') increases the value of counter $e \in \{c, d\}$ by 1 and moves to state q'. A transition (q, e, q', q'') moves to q' if $e = 0$ and to q'' otherwise. In the latter case, it also decreases the value of e by 1. The Halting Problem for (deterministic) two-counter machines is known to be undecidable (see [16, Thm. 14-1]).

3.1   Overview of the reduction

Let $\mathcal{M}$ be a two-counter machine with counters c and d. We construct a WTG $\mathcal{G}_\mathcal{M}$ using two clocks, x and y. Player **Min** is responsible for simulating the behavior of $\mathcal{M}$, while **Max** is given the opportunity to punish any incorrect simulation. Each punishment by **Max** leads directly to the goal state, thereby terminating the game.

Our construction satisfies the following proposition:

Proposition 1. *Let $\mathcal{M} = (Q, q_i, q_h, T)$ be a deterministic two-counter machine.*

- *If $\mathcal{M}$ does not halt, then the WTG $\mathcal{G}_\mathcal{M}$, starting from the configuration $(q_i, \mathbf{0})$, has value at most 64.*
- *If $\mathcal{M}$ halts in at most N steps, then $\mathcal{G}_\mathcal{M}$, starting from the configuration $(q_i, \mathbf{0})$, has value at least $64 + \frac{11}{12 \times 30^{5N}}$.*

Intuitively, if $\mathcal{M}$ does not halt, then **Min** can faithfully simulate its infinite execution for an arbitrary number of steps. The longer she plays, the closer the accumulated weight is to 64 when she eventually exits to a goal location.

On the other hand, if $\mathcal{M}$ halts in N steps, **Min** has only two options: either simulate the execution faithfully for at most N steps, and exit, yielding a value strictly greater than 64, or attempt to cheat in order to make the game longer. The key point is that we show that in order to push the computation further, she will eventually cheat by some quantity bounded from below depending on N. Detecting this, **Max** is given the opportunity to enforce a punishment reaching a weight of at least $64 + \frac{11}{12 \times 30^{5N}}$.

In the encoding of $\mathcal{M}$ into $\mathcal{G}_\mathcal{M}$, control states of $\mathcal{M}$ become locations of weight 30 in $L_{\mathbf{Min}}$. When entering one of these locations, a faithful encoding of the counters c and d is represented by a clock valuation

$$ x = 1 - \frac{1}{2^c 3^d 5^n}, \quad y = 0, $$

where n denotes the number of simulated steps of $\mathcal{M}$ so far.

Note that when $x = 1 - \frac{1}{2^c 3^d 5^n}$, reaching a configuration where $x = 1 - \frac{1}{2^{c+1} 3^d 5^{n+1}}$ (e.g., when simulating an increment of counter c, which also increments n) requires waiting for $\frac{9}{10}(1 - x)$ time units. Similarly, to simulate a decrement of counter c, or an increment/decrement of counter d, or a simple increment of n, one must wait $\alpha \cdot (1 - x)$ time units, where α is one of $\frac{3}{5}, \frac{14}{15}, \frac{2}{5}$ or $\frac{4}{5}$, respectively.

Hence, each increasing transition $(q, e, q') \in T$ is simulated using a module of the following structure: **Min** selects a delay to update x, and then **Max** is given an opportunity to punish her if the delay is incorrect (see Fig. 2 [3]).

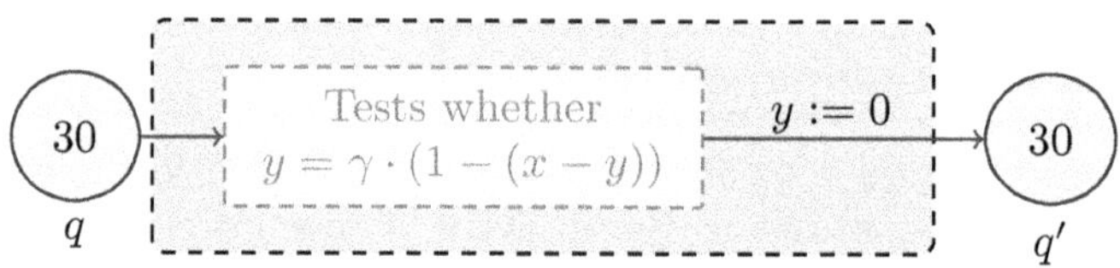

Fig. 2. Transition module for increments. Here $x - y$ is the previous value of x, and y contains the delay **Min** waited in q.

A branching transition $(q, e, q_z, q_{nz}) \in T$, where $q, q_z, q_{nz} \in Q$ and $e \in \{c, d\}$, is simulated in two steps: first, **Min** decides whether to move toward q_z or toward q_{nz}. In both cases, **Max** is given the opportunity to punish if the encoding in x is not valid with respect to **Min**'s choice. Then, **Min** updates x, and **Max** is again given the opportunity to punish her for an invalid update.

---

[3] In this paper, we follow the conventions below to represent WTGs: Blue circles represent locations controlled by **Min**. Red squares represent locations controlled by **Max**. Green circles are goal locations. Grey rectangles represent modules (i.e., subgames): a transition entering a module transfers control to its starting location. Modules may have outgoing edges. Orange rectangles provide the specification of modules. Numbers attached to locations denote their respective weights. Numbers in grey boxes attached to arrows denote the weight of the corresponding transition. Transitions without such boxes have weight 0. Green boxes attached to transitions are comments (or assertions) on the values of the clocks, which hold upon taking the transition. These may be complemented by orange boxes, which represent assertions on the corresponding cost incurred. Some locations are decorated with a numbered flag for ease of reference in proofs.

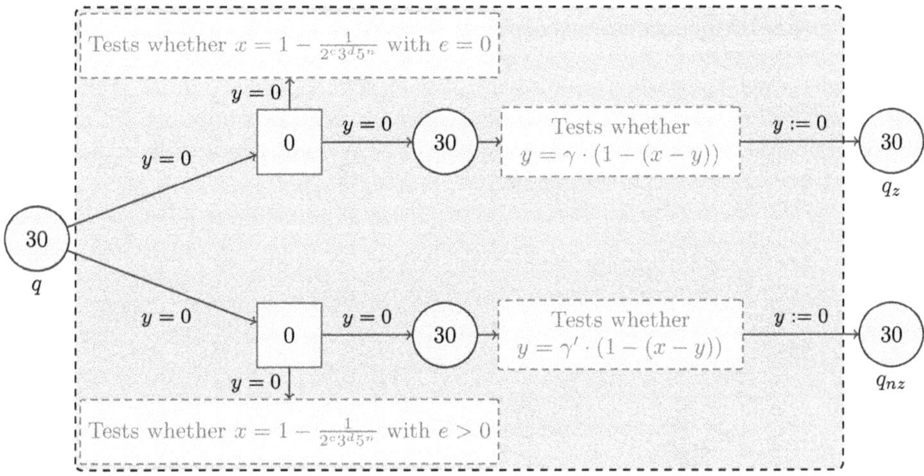

Fig. 3. Transition module for branching decrement of counter $e \in \{c, d\}$.

Finally, for every $q \in Q$ (including q_h), we add the following exit module for **Min**:

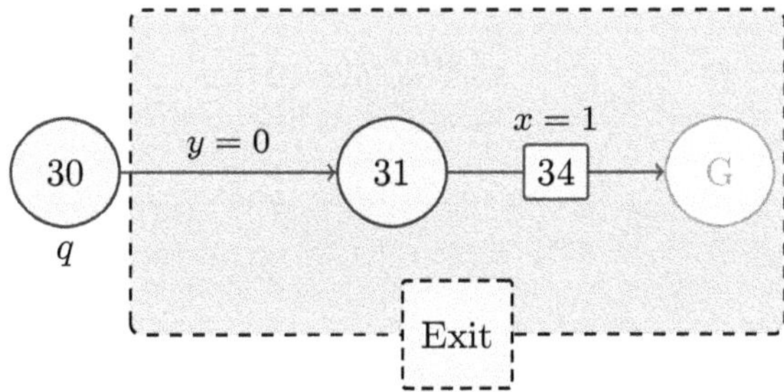

Fig. 4. **Min**'s exit module. Yields a final cost of at most $64 + 1 - x$.

Note that, while faithfully simulating an infinite execution of $\mathcal{M}$, x will approach 1 arbitrarily closely, since the encoding $\frac{1}{2^c 3^d 5^n}$ tends to 0 as the number of steps n increases. Therefore, when **Min** decides to exit, the accumulated cost will be $30x + 31(1 - x) + 34 = 64 + 1 - x$.

In Section 3.2, we present the punishment module that enforces $y = \alpha(1 - x)$. Section 3.3 then introduces punishment modules for cases where **Min** cheats on branching decisions. Finally, in Section 3.4, we combine these modules to construct $\mathcal{G}_\mathcal{M}$ and prove Proposition 1.

3.2   Controlling counter evolution

Let us introduce the gadget behind the punishment module "Punish if $y \neq \alpha(1 - x)$": the *CEC* (Counter Evolution Control) module. The module $CEC^M_{\alpha,\beta}(x,y)$, depicted in Fig. 5, requires that $0 \leq y \leq x < 1$. We will denote the initial values of x and y as $a + b$ and b, respectively.

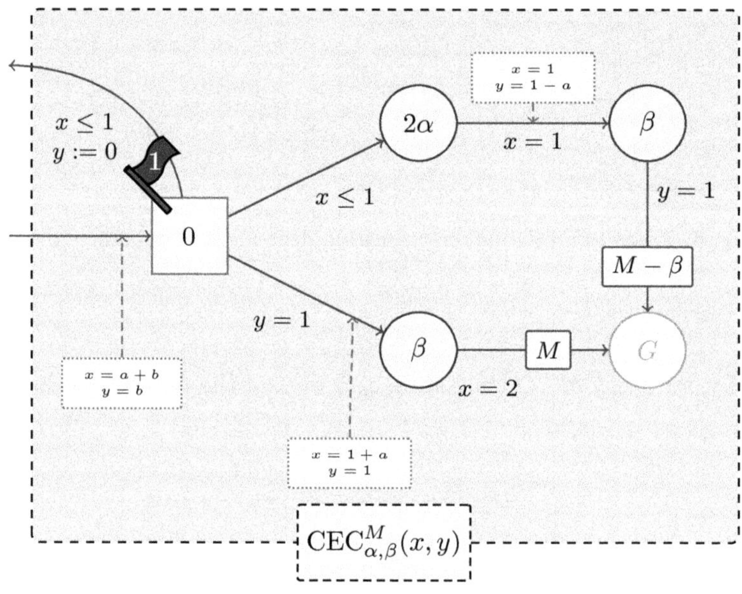

Fig. 5. The CEC (Counter Evolution Control) module. Enforces $b = \gamma(1 - a)$ with $\gamma = 1 - \frac{\beta}{\alpha}$.

Proposition 2. *Let $0 \leq \beta < \alpha$. Let t be the time **Max** spends in state ⚑ of $CEC^M_{\alpha,\beta}(x,y)$. Provided that, upon entering the state ⚑ in $CEC^M_{\alpha,\beta}(x,y)$, $(x,y) = (a + b, b)$ with $0 \leq b \leq a + b \leq 1$, and assuming that the overall cost accumulated so far is $\alpha(a + b) + E$, then **Max** can either end the game with a final cost of*

$$\alpha \left(1 + \left| b - \left(1 - \frac{\beta}{\alpha} \right) (1 - a) \right| \right) + E + M$$

or exit the module with $x = a + b + t$, $y = 0$ and accumulated cost $\alpha(a + b + t) + (E - \alpha t)$.

Proof. **Max** can either:

1. exit the module after waiting t time units.
2. take the upper path, which ends the game with cost

$$(\alpha - \beta)(1 - a) - \alpha b - 2\alpha t + \alpha + \beta + M - \beta + E$$
$$= (\beta - \alpha)a - \alpha b - 2\alpha t + 2\alpha + M - \beta + E .$$

3. take the lower path, which ends the game with cost

$$\alpha b - (\alpha - \beta)(1 - a) + \alpha + M + E = \alpha b - (\beta - \alpha)a + \beta + M + E . \quad \square$$

Thus, if we take $\alpha = 30$ and $\beta \in \{18, 12, 6, 3, 2\}$, the value of b minimising the cost of **Max** reaching the goal location is indeed

$$b = \gamma(1 - a) \text{ for } \gamma \in \left\{ \frac{2}{5}, \frac{3}{5}, \frac{4}{5}, \frac{9}{10}, \frac{14}{15} \right\} .$$

For $\alpha = 30$, $E \leq 0$ and $M = 34$, note that when $b = \gamma(1 - a)$ the cost of **Max**'s punishment yields cost at most 64, thus **Max** has no incentive to punish **Min** when she has not cheated.

3.3 Controlling whether a counter is zero

We now address the case in which **Min** has moved to the wrong state when simulating a zero-test. To handle this, depending on the situation, **Max** has access to either a control-if-zero (ZC) or control-if-not-zero (NZC) module, which checks whether a counter is indeed zero or non-zero by forcing a series of multiplications that can reach 1 only if **Min** chose the right branch. We begin by presenting the module that controls the multiplication (MC), depicted in Fig. 6

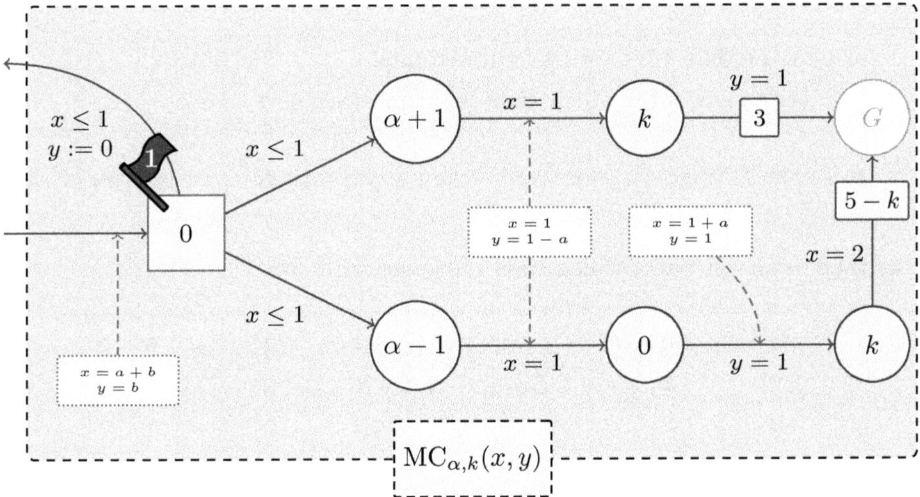

Fig. 6. The MC (Multiplication Control) module. Enforces $x = ka$ for $k \in \{2, 3, 5\}$.

$MC_{\alpha,k}(x,y)$ operates similarly to the CEC module:

- It requires that $0 \leq y \leq x < 1$. We will denote the initial values of x and y as $a + b$ and b, respectively.
- Whereas the CEC module enforces that $b = \gamma(1-a)$ for some γ, $MC_{\alpha,k}(x,y)$ ensures that $a + b = ka$. This is because the MC module is designed to multiply encodings of the form $\frac{1}{2^c 3^d 5^n}$ by $k \in 2, 3, 5$.

Proposition 3. *Let $0 \leq \beta \leq \alpha$. Let t be the time* **Max** *spends in state* ✦ *of $MC_{\alpha,k}(x,y)$. Provided that, upon entering the state* ✦ *in $MC_{\alpha,k}(x,y)$, $(x,y) = (a+b,b)$ with $0 \leq b \leq a+b \leq 1$, and assuming that the overall cost accumulated so far is $\alpha(a+b) + E$,* **Max** *can either end the game with final cost*

$$\alpha + 4 + E + k + |b - (k-1)a|$$

or exit the module with $x = a + b + t$, $y = 0$ and an accumulated cost of $\alpha(a + b + t) + (E - \alpha t)$.

Proof. **Max** can either:

1. exit the module after waiting t time units.
2. take the upper path, which ends the game with cost

$$\alpha(a+b)+E+(\alpha+1)(1-a-b-t)+ka+3 = \alpha+4+E-(\alpha+1)t-b+(k-1)a\,.$$

3. take the lower path, which ends the game with cost

$$\alpha(a + b) + E + (\alpha - 1)(1 - a - b - t) + k(1 - a) + 5 - k$$
$$= \alpha + 4 + E - (\alpha - 1)t + b + k - (k - 1)a\,. \qquad \square$$

Here we see clearly that the MC module enforces multiplication: the cost of **Max** ending the game is minimised for $b = (k - 1)a$, hence $x = a + b = ka$.

We now introduce modules to control whether a counter is indeed 0 or not.

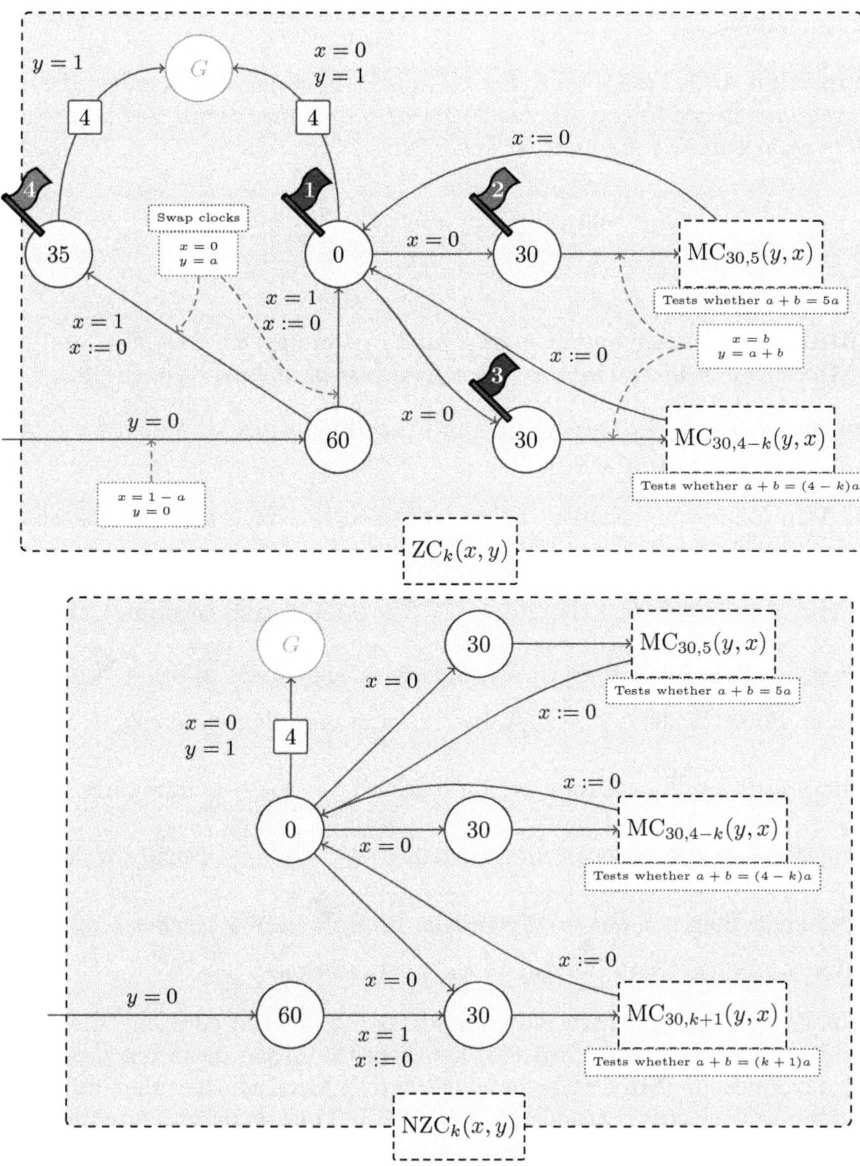

Fig. 7. The Zero-Control and Non-Zero-Control modules, for $k \in \{1, 2\}$.

In these two modules, note that MC is always invoked as $MC(y, x)$, i.e., by swapping the roles of the two clocks. This is because it is easier to translate the encoding $(1-a, 0)$ into $(0, a)$ rather than $(a, 0)$. This translation is the first step in both modules.

Proposition 4. *Let $k \in \{1, 2\}$. Let $1 - a \in [0, 1)$ be the initial value of x upon entering the module $ZC_k(x, y)$. Let $30(1 - a) + E$ be the overall cost accumulated to date upon entering the module. Let*

$$\mu = \min \left\{ \left| \frac{1}{(4-k)^d 5^n} - a \right| : d, n \in \mathbb{N} \right\} .$$

Then

- **Min** *has a strategy that ensures a final cost of at most $64 + E + 5\mu$.*
- **Max** *has a strategy that ensures a final cost of at least $64 + E + \mu$.*

Proof sketch. In the ZC module, a valid encoding is any valuation for y of the form $\dfrac{1}{(4-k)^d 5^n}$ with $d, n \in \mathbb{N}$.

   If **Min** follows the strategy below, she ensures a final cost of at most $64 + E + 5\mu$:

- First, if a is closer to 1 than to $\dfrac{1}{4-k}$ (the largest valid encoding), then **Min** takes the transition to State ⚑(0); otherwise, she moves to State ⚑(1).
- From State ⚑(1), let $\dfrac{1}{(4-k)^d 5^n}$ be the valid encoding closest to a. If $n \geq 1$, move to State ⚑(2) and wait until y reaches $\dfrac{1}{(4-k)^d 5^{n-1}}$. Similarly, if $d \geq 1$, move to State ⚑(3) and wait until y reaches $\dfrac{1}{(4-k)^{d-1} 5^n}$. Finally, if the closest valid encoding is 1 but $y < 1$, wait in State ⚑(3) until y reaches 1.
- When $y = 1$ in State ⚑(0), exit to the goal location.

   Under this strategy, **Min** always updates y to a valid encoding. Hence, any "error" in the MC module arises either from the initial deviation μ or from a delay introduced by **Max** in the previous step. If **Max** punishes the initial deviation, the resulting cost is at most $64 + E + 5\mu$. If **Max** punishes a later deviation, the corresponding punishment cost is offset by the "lost" cost incurred by **Max** while waiting in a zero-weight location in the previous step. Indeed, when only **Min** delays, the accumulated weight is $30 + E + 30x$ for the current value of clock x; any delay introduced by **Max** subtracts from this $30x$. Therefore, by punishing his own mistakes, **Max** can only obtain a total cost of at most $64 + E$.

   Conversely, if **Max** follows the strategy below, he ensures a final cost of at least $64 + E + \mu$:

   Upon entering a MC module with clock valuation (x, y), let

$$\eta = |x - (k' - 1)(y - x)| ,$$

where $k' = 5$ or $4 - k$ depending on whether the module is entered from State ⚑
or State ⚑. Then, if $\eta \geq \mu$, then punish along the path in the MC module that
maximises the cost. Otherwise, return to State ⚑ without delay.

We claim that, in order to reach valuation $(0, 1)$, **Min** must eventually incur
an error of magnitude $\eta \geq \mu$, at which point **Max** can punish her, resulting in
a final cost of at least $64 + E + \mu$. □

Proposition 5. *Let $k \in \{1, 2\}$. Let $1 - a \in [0, 1)$ be the initial value of x upon
entering module $NZC_k(x, y)$. Let $30(1 - a) + E$ be the overall cost accumulated
thus far when entering the module. Let*

$$\mu = \min \left\{ \left| \frac{1}{(k + 1)^c (4 - k)^d 5^n} - a \right| : c, d, n \in \mathbb{N} \quad c > 0 \right\}.$$

Then

- **Min** *has a strategy that ensures a final cost of at most $64 + E + 5\mu$.*
- **Max** *has a strategy that ensures a final cost of at least $64 + E + \mu$.*

Proof. The proof is almost identical to that of Prop. 4. The main difference between modules ZC_k and NZC_k is that NZC_k forces **Min** to do a multiplication
by $k + 1$ before multiplying by $k + 1$, $4 - k$, and 5 arbitrarily many times. □

3.4   Combining modules

We can now implement explicitly the modules given in Figures 2 and 3 with the
CEC, ZC and NZC modules:

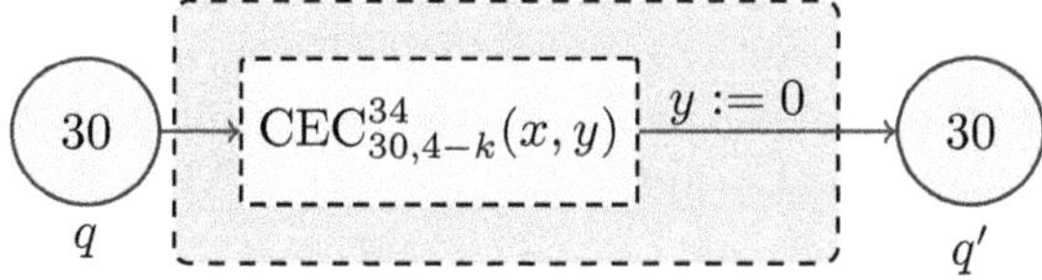

Fig. 8. Transition module simulating an increment transition $(q, e, q') \in T$, with $q, q' \in
Q$. Here $k = 1$ or 2 when $e = c$ or d, respectively.

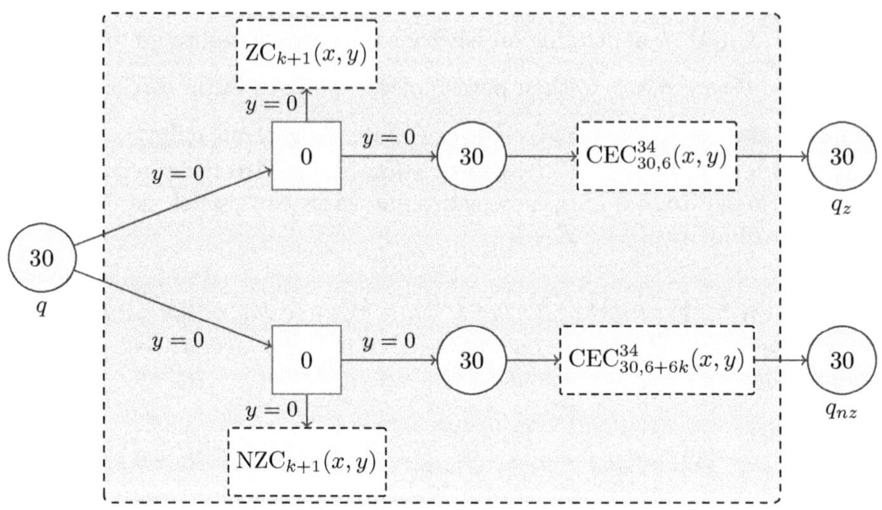

Fig. 9. Transition module simulating a branching transition $(q, e, q_z, q_{nz}) \in T$, with $q, q_z, q_{nz} \in Q$. Here $k = 1$ or 2 when $e = c$ or d, respectively.

Proposition 1. *Let $\mathcal{M} = (Q, q_i, q_h, T)$ be a deterministic two-counter machine.*

- *If $\mathcal{M}$ does not halt, then the WTG $\mathcal{G}_{\mathcal{M}}$, starting from the configuration $(q_i, \mathbf{0})$, has value at most 64.*
- *If $\mathcal{M}$ halts in at most N steps, then $\mathcal{G}_{\mathcal{M}}$, starting from the configuration $(q_i, \mathbf{0})$, has value at least $64 + \frac{11}{12 \times 30^{5N}}$.*

Proof sketch. If $\mathcal{M}$ does not halt, then **Min** can faithfully simulate the execution of $\mathcal{M}$ for an arbitrary number of steps and then exit via the exit module with x arbitrarily close to 1. The resulting accumulated cost is at most $64 + 1 - x$, which tends to 64 as $x \to 1$. If **Max** decides to punish **Min** during this run, the punishment module yields a cost of at most 64.

Note that **Max** can perturb the simulation by waiting in the transition modules; however, this is disadvantageous for **Max**. Let a_p and a_{p+1} be the encodings of the counters of $\mathcal{M}$ at steps p and $p+1$. Assume that **Min** has just updated x to a_p, with accumulated cost at most $30a_p$, and that **Max** waits for some $\varepsilon_p > 0$. Then:

- If $a_p + \varepsilon_p \leq a_{p+1}$, **Min** can update x to a_{p+1} with accumulated cost at most $30(a_{p+1} - \varepsilon_p) < 30a_{p+1}$. In effect, **Max** has waited ε_p in a location of weight 0 instead of allowing **Min** to realize the same delay in her location of weight 30, strictly decreasing the total cost. If **Max** then punishes her, the extra cost of the CEC module will be offset by the time **Max** wasted waiting ε_p in a weight-0 location.

– If $a_p + \varepsilon_p > a_{p+1}$, **Min** can no longer update x to the next encoding. In that case she can take the exit module, yielding an accumulated cost

$$64 - 30\varepsilon_p + 1 - (a_p + \varepsilon_p) \leq 64.$$

The extra cost of leaving early is offset by the time **Max** wasted waiting ε_p in a weight-0 location.

Conversely, if $\mathcal{M}$ halts in at most N steps, consider the following strategy for **Max**: never wait in the transition modules (as we have just seen, waiting is not to his advantage), and punish **Min** whenever she makes an error of at least $\dfrac{1}{30^{5N+1}}$ while updating the clocks, or when she takes the wrong transition in a zero-test. Then:

– If **Min** faithfully simulates the execution of $\mathcal{M}$, she is forced to exit in at most N simulation steps, with accumulated cost $\geq 64 + \dfrac{11}{12 \times 30^{5N}}$.
– To avoid this outcome, she may attempt to cheat to obtain an arbitrarily long run. However, she has only N simulation steps in which to distribute her cheating. Hence we claim that either she makes an error of at least $\dfrac{1}{30^{5N+1}}$ while updating the clocks, or she cheats on a zero-test when the value of clock x is within $\dfrac{1}{12 \times 30^{5N}}$ of a faithful encoding of the simulation. In either case, **Max** can punish her, yielding a final accumulated cost $\geq 64 + \dfrac{11}{12 \times 30^{5N}}$. $\square$

Theorem 1. *The Value Problem for two-player, turn-based, time-bounded, two-clock, weighted timed games with non-negative integer weights is undecidable. The same holds for weighted timed games over unbounded time otherwise satisfying the same hypotheses.*

Proof. Let $\mathcal{M}$ be a deterministic two-counter machine. Note that in $\mathcal{G}_{\mathcal{M}}$, outside of control modules, clock x is never reset and always upper-bounded by 1. Therefore any play has duration at most 1 time unit plus the total time spent in the control modules, which is at most 2 time units. In other words, by construction the WTG $\mathcal{G}_{\mathcal{M}}$ requires at most 3 time units for any execution. Prop. 1 asserts that the halting problem for a deterministic two-counter machine reduces to the Value problem for 2-clock weighted timed games, which concludes the proof. $\square$

Theorem 2. *The Existence Problem for two-player, turn-based, time-bounded, two-clock, weighted timed games with non-negative integer weights is undecidable. The same holds for weighted timed games over unbounded time otherwise satisfying the same hypotheses.*

Proof sketch. Let $\mathcal{M}$ be a deterministic two-counter machine. We define $\mathcal{G}_{\mathcal{M}}^{E}$ similarly to $\mathcal{G}_{\mathcal{M}}$, except that we add a "soft-exit" module, reachable from the halting state of $\mathcal{M}$. This soft-exit module is simply the exit module where the location cost of 31 has been replaced by 30. Then **Min** has a strategy to enforce a cost of at most 64 if and only if $\mathcal{M}$ halts.

Indeed, if $\mathcal{M}$ halts, then **Min** can reach a halting state without cheating (i.e., she faithfully simulates $\mathcal{M}$), and exits at cost 64 through a soft-exit module. If **Max** decides to exit through a CEC or MC module, this also yields a cost of at most 64. As seen in Prop. 1, every delay taken by **Max** in a CEC or MC module is a net negative for **Max**. On the other hand, if $\mathcal{M}$ does not halt, the only way for **Min** to exit the game through a soft-exit module is to cheat in order to reach a halting state. The normal exit module for **Min** yields cost strictly greater than 64. Consider the strategy where **Max** punishes any cheating by exiting the game. Then either **Min** never cheats, and the game never ends, thereby incurring cost $+\infty$, or **Min** cheats and is punished, in which case the game ends with cost strictly above 64. $\qquad\square$

4   Commentary

How does this reduction compare to the reduction to 3-clock WTGs with positive weights in [5]?

- First, in [5], the encoding is of the form $\frac{1}{2^c 3^d}$. Clocks x and y are used to store the previous and new counter encodings at each step (they alternate). A third clock, t, acts as a ticking clock, ensuring that exactly n time units are spent in a module, for some $n \in \mathbb{N}$. This allows, for instance, **Min** to assign a new value to y while keeping x unchanged. Our encoding $1 - \frac{1}{2^c 3^d 5^n}$ has the advantage that it only increases, allowing us to update it without losing the previous value, even without a ticking clock.
- A key property of the 3-clock construction is that it yields an *almost non-Zeno* WTG $\mathcal{G}_\mathcal{M}$:

Definition 4. *A WTG $\mathcal{G}$ with non-negative weights is said to be* non-Zeno *if all its cycles* [4] *have weight at least 1. It is said to be* almost non-Zeno *if all its cycles have weight either exactly* 0 *or at least* 1.

Indeed, all cycles of $\mathcal{G}_\mathcal{M}$ lie in the part of the game simulating the execution of $\mathcal{M}$; once we enter a punishment module, no further cycles are possible. In the three-clock reduction, the only positive weights occur in the punishment modules.

However, in our two-clock construction, $\mathcal{G}_\mathcal{M}$ is not almost non-Zeno. In fact, it was recently shown in [20] that the Value problem is decidable for almost non-Zeno two-clock WTGs with positive weights.

Let us now give a simple intuition for why the two-clock reduction requires weight > 0 in each location of $Q \subseteq L_\mathbf{Min}$: When entering a CEC module with clock configuration $(x, y) = (a + b, b)$, we want to offer **Max** a punishment module whose weight function is of the form $|b - \gamma(1 - a)|$ (modulo some multiplicative and additive constants), for some $\gamma \in \mathbb{Q}$.

---

[4] Here, we refer to cycles in the *region game* of $\mathcal{G}$; see [5] for the definition.

It is straightforward to construct a punishment module with weight functions of the form $k \cdot (1 - a - b)$ or $k \cdot a$ (for $k \in \mathbb{N}$), and to combine them using max and $+$ operations. However, we cannot obtain a function of the form $k \cdot b$ directly. Therefore, b must be reflected in the accumulated weight *before* entering the punishment module. This can only happen by taking in weight when **Min** waits b time units.

- Our construction is timed-bounded, while the three-clock construction is not (every simulation step takes an integer amount of time). This is due to our $1 - \frac{1}{2^c 3^d 5^n}$ encoding, but also to the constraint above: since **Min** accumulates weight while updating the clocks, we must ensure that the time spent on clock updates remains bounded.

References

1. Alur, R., Bernadsky, M., Madhusudan, P.: Optimal reachability for weighted timed games. In: Proceedings of the 31st International Colloquium on Automata, Languages and Programming (ICALP04). pp. 122–133. Springer Berlin Heidelberg, Berlin, Heidelberg (2004)
2. Alur, R., Henzinger, T.A.: Modularity for timed and hybrid systems. In: CONCUR. Lecture Notes in Computer Science, vol. 1243, pp. 74–88. Springer (1997)
3. Bouyer, P., Brihaye, T., Markey, N.: Improved undecidability results on weighted timed automata. Information Processing Letters **98**(5), 188–194 (2006). `https://doi.org/https://doi.org/10.1016/j.ipl.2006.01.012`, `https://www.sciencedirect.com/science/article/pii/S0020019006000652`
4. Bouyer, P., Cassez, F., Fleury, E., Larsen, K.G.: Optimal strategies in priced timed game automata. In: FSTTCS 2004: Foundations of Software Technology and Theoretical Computer Science. pp. 148–160. Springer Berlin Heidelberg, Berlin, Heidelberg (2005)
5. Bouyer, P., Jaziri, S., Markey, N.: On the Value Problem in Weighted Timed Games. In: Aceto, L., de Frutos Escrig, D. (eds.) Proceeding of the 26th International Conference on Concurrency Theory (CONCUR 2015). Leibniz International Proceedings in Informatics (LIPIcs), vol. 42, pp. 311–324. Schloss Dagstuhl – Leibniz-Zentrum für Informatik, Dagstuhl, Germany (2015). `https://doi.org/10.4230/LIPIcs.CONCUR.2015.311`, `https://drops.dagstuhl.de/entities/document/10.4230/LIPIcs.CONCUR.2015.311`
6. Bouyer, P., Larsen, K.G., Markey, N., Rasmussen, J.I.: Almost optimal strategies in one clock priced timed games. In: Arun-Kumar, S., Garg, N. (eds.) FSTTCS 2006: Foundations of Software Technology and Theoretical Computer Science, 26th International Conference, Kolkata, India, December 13-15, 2006, Proceedings. Lecture Notes in Computer Science, vol. 4337, pp. 345–356. Springer (2006). `https://doi.org/10.1007/11944836_32`, `https://doi.org/10.1007/11944836_32`
7. Brihaye, T., Bruyère, V., Raskin, J.F.: On optimal timed strategies. In: Formal Modeling and Analysis of Timed Systems. pp. 49–64. Springer Berlin Heidelberg, Berlin, Heidelberg (2005)
8. Brihaye, T., Doyen, L., Geeraerts, G., Ouaknine, J., Raskin, J., Worrell, J.: On reachability for hybrid automata over bounded time. In: Aceto, L., Henzinger, M., Sgall, J. (eds.) Automata, Languages and Programming - 38th International Colloquium, ICALP 2011, Zurich, Switzerland, July 4-8, 2011, Proceedings, Part

II. Lecture Notes in Computer Science, vol. 6756, pp. 416–427. Springer (2011). https://doi.org/10.1007/978-3-642-22012-8_33, https://doi.org/10.1007/978-3-642-22012-8_33

9. Brihaye, T., Doyen, L., Geeraerts, G., Ouaknine, J., Raskin, J., Worrell, J.: Time-bounded reachability for monotonic hybrid automata: Complexity and fixed points. In: Hung, D.V., Ogawa, M. (eds.) Automated Technology for Verification and Analysis - 11th International Symposium, ATVA 2013, Hanoi, Vietnam, October 15-18, 2013. Proceedings. Lecture Notes in Computer Science, vol. 8172, pp. 55–70. Springer (2013). https://doi.org/10.1007/978-3-319-02444-8_6, https://doi.org/10.1007/978-3-319-02444-8_6

10. Brihaye, T., Geeraerts, G., Narayanan Krishna, S., Manasa, L., Monmege, B., Trivedi, A.: Adding negative prices to priced timed games. In: Baldan, P., Gorla, D. (eds.) CONCUR 2014 – Concurrency Theory. pp. 560–575. Springer Berlin Heidelberg, Berlin, Heidelberg (2014)

11. Busatto-Gaston, D., Monmege, B., Reynier, P.: Optimal controller synthesis for timed systems. Log. Methods Comput. Sci. **19**(1) (2023)

12. Guilmant, Q., Ouaknine, J.: Inaproximability in Weighted Timed Games. In: Majumdar, R., Silva, A. (eds.) 35th International Conference on Concurrency Theory (CONCUR 2024). Leibniz International Proceedings in Informatics (LIPIcs), vol. 311, pp. 27:1–27:15. Schloss Dagstuhl – Leibniz-Zentrum für Informatik, Dagstuhl, Germany (2024). https://doi.org/10.4230/LIPIcs.CONCUR.2024.27, https://drops.dagstuhl.de/entities/document/10.4230/LIPIcs.CONCUR.2024.27

13. Henzinger, T.A., Horowitz, B., Majumdar, R.: Rectangular hybrid games. In: CONCUR. Lecture Notes in Computer Science, vol. 1664, pp. 320–335. Springer (1999)

14. Jenkins, M., Ouaknine, J., Rabinovich, A., Worrell, J.: Alternating timed automata over bounded time. In: Proceedings of the 25th Annual IEEE Symposium on Logic in Computer Science, LICS 2010, 11-14 July 2010, Edinburgh, United Kingdom. pp. 60–69. IEEE Computer Society (2010). https://doi.org/10.1109/LICS.2010.45, https://doi.org/10.1109/LICS.2010.45

15. Maler, O., Pnueli, A., Sifakis, J.: On the synthesis of discrete controllers for timed systems (an extended abstract). In: STACS. Lecture Notes in Computer Science, vol. 900, pp. 229–242. Springer (1995)

16. Minsky, M.L.: Computation: Finite and Infinite Machines. Prentice-Hall Series in Automatic Computation, Prentice-Hall (1967)

17. Monmege, B., Parreaux, J., Reynier, P.A.: Decidability of One-Clock Weighted Timed Games with Arbitrary Weights. In: Klin, B., Lasota, S., Muscholl, A. (eds.) 33rd International Conference on Concurrency Theory (CONCUR 2022). Leibniz International Proceedings in Informatics (LIPIcs), vol. 243, pp. 15:1–15:22. Schloss Dagstuhl – Leibniz-Zentrum für Informatik, Dagstuhl, Germany (2022). https://doi.org/10.4230/LIPIcs.CONCUR.2022.15, https://drops.dagstuhl.de/entities/document/10.4230/LIPIcs.CONCUR.2022.15

18. Ouaknine, J., Rabinovich, A., Worrell, J.: Time-bounded verification. In: Bravetti, M., Zavattaro, G. (eds.) CONCUR 2009 - Concurrency Theory, 20th International Conference, CONCUR 2009, Bologna, Italy, September 1-4, 2009. Proceedings. Lecture Notes in Computer Science, vol. 5710, pp. 496–510. Springer (2009). https://doi.org/10.1007/978-3-642-04081-8_33, https://doi.org/10.1007/978-3-642-04081-8_33

19. Ouaknine, J., Worrell, J.: Towards a theory of time-bounded verification. In: Abramsky, S., Gavoille, C., Kirchner, C., auf der Heide, F.M., Spirakis, P.G. (eds.)

Automata, Languages and Programming, 37th International Colloquium, ICALP 2010, Bordeaux, France, July 6-10, 2010, Proceedings, Part II. Lecture Notes in Computer Science, vol. 6199, pp. 22–37. Springer (2010). https://doi.org/10.1007/978-3-642-14162-1_3, https://doi.org/10.1007/978-3-642-14162-1_3
20. Vialard, I.: Deciding the value of two-clock almost non-zeno weighted timed games (2025), https://arxiv.org/abs/2508.00014

Generalized Kantorovich-Rubinstein Duality beyond Hausdorff and Kantorovich [*]

Paul Wild[1] , Lutz Schröder[1] , Karla Messing[2] , Barbara König[2] , and Jonas Forster[1]

[1] Friedrich-Alexander-Universität Erlangen-Nürnberg, Erlangen, Germany
`{paul.wild,lutz.schroeder,jonas.forster}@fau.de`
[2] Universität Duisburg-Essen, Duisburg, Germany
`{karla.messing,barbara_koenig}@uni-due.de`

Abstract. The classical Kantorovich-Rubinstein duality guarantees coincidence between metrics on the space of probability distributions defined on the one hand via transport plans (*couplings*) and on the other hand via price functions. Both constructions have been lifted to the level of generality of set functors, with the construction based on couplings referred to as the *Wasserstein* or simply the *coupling-based* lifting, and the price-function-based construction as the *Kantorovich* or *codensity* lifting, both based on a choice of quantitative modalities for the given functor. It is known that every coupling-based lifting can be expressed as a price-function-based lifting; however, the latter in general needs to use additional modalities. We give an example showing that this cannot be avoided in general. We refer to cases in which the same modalities can be used as satisfying the *generalized Kantorovich-Rubinstein duality*. We establish the generalized Kantorovich-Rubinstein duality in this sense for two important cases: The Lévy-Prokhorov distance on distributions, which finds wide-spread applications in machine learning due to its favourable stability properties, and the standard metric on convex sets of distributions that arises by combining the Hausdorff and Kantorovich-Rubinstein distances.

1 Introduction

Measuring behavioural distances between probabilistic systems requires notions of distance between probability distributions (e.g. [35]). One well-established metric on the set of distributions over a metric space is variously termed the *Kantorovich-Rubinstein*, *Wasserstein*, or *Hutchinson* metric. It can be calculated either by minimizing over the expected value of *transport plans* between, or *couplings* of, the given distributions, or by maximizing over the difference of expectations taken over all nonexpansive *price functions*. The coincidence of these

[*] The authors acknowledge support by the Deutsche Forschungsgemeinschaft (DFG, German Research Foundation). The second author has been supported by project number 531706730 (CoRSA), while the remaining authors have been supported by project number 434050016 (SpeQt).

N. Bertrand and S. Milius (Eds.): FoSSaCS 2026, LNCS 16503, pp. 636–657, 2026.
https://doi.org/10.1007/978-3-032-22730-0_30

two values is the classical *Kantorovich-Rubinstein duality* [36, Theorem 5.10]. Intuitively speaking, a transport plan or coupling is a way to transform one distribution into another by shifting around weight, and its cost, to be minimized, is determined by how much weight is shifted over which distances. On the other hand, a price function determines a price for some commodity at given points; nonexpansiveness of the price function means that no profit can be made from full-cost transport. The difference between the expected values of a price function under the given distributions is the profit to be made by having the commodity transported, and hence the amount one can offer to a logistics provider when outsourcing the transport.

Both the coupling-based definition and the price-function-based definition have been generalized categorically to construct liftings of set functors to the category of (pseudo-)metric spaces [4] and quantitative lax extensions of set functors [38], where the latter are distinguished by applying to unrestricted quantitative relations instead of only to (pseudo-)metrics. Metric functor liftings and lax extensions in particular serve to give a general treatment of *behavioural distances* on quantitative systems such as probabilistic, weighted, or metric [2] transition systems in the framework of *universal coalgebra* [31]. In this framework, functors serve as parameters determining a type of systems as their coalgebras; for instance, coalgebras for the distribution functor are Markov chains. In the generalized setting, the coupling-based construction is often referred to as the *Wasserstein* lifting or extension, and the price-function-based one as the *Kantorovich* lifting or extension, but other names are found in the literature, such as *coupling-based* [22] and *codensity* [24] lifting, respectively. Both constructions are parametrized over a choice of quantitative modalities; the classical case involves, on both sides, only one modality, the expectation modality.

The interest in having a price-function-based presentation of a given functor lifting or lax extension lies inter alia in the fact that one obtains a quantitative Hennessy-Milner property for the quantitative modal logic generated by the respective modalities [26,38]. This property states coincidence of the behavioural distance induced by the given lifting or extension with the logical distance induced by the respective quantitative modal logic; thus, high distance between two states can always be *certified* by means of a modal formula, a prominent principle in the study of behavioural distances (e.g. [35,10,38,26,13] and more recently [30,34]). It has been shown that every metric functor lifting that preserves isometries [14] and every quantitative lax extension [38] *is Kantorovich*, i.e. can be presented via the generalized price-function-based construction, using however a rather large (in particular typically infinite) set of modalities called the *Moss modalities*.

An important open point that remains is thus the question of what we term *generalized Kantorovich-Rubinstein duality*: In which cases does a coupling-based distance given by a choice of modalities for a functor coincide with the price-function-based distance *for the same modalities*? Known positive examples include, as mentioned, the classical Kantorovich-Rubinstein distance on distributions, but also the Hausdorff distance on the powerset. We begin our analysis by giving an example of a natural coupling-based metric for which generalized Kantorovich-Rubinstein duality in this sense fails, namely the standard *p-Wasserstein* metric for $p > 1$ (which minimizes p-th roots of the expectation of p-th powers of couplings). As our main contribution, we then provide two new positive examples, namely the *Lévy-Prokhorov* distance on probability distributions [29] and the standard distance on the *convex powerset*, whose elements are convex sets of probability distributions. The Lévy-Prokhorov distance has seen a recent rise in popularity due to its favourable robustness properties that make it suitable for tasks in machine learning (such as conformal prediction [3] and corruption resistance [5]); moreover, it has been shown in recent work [11] to induce precisely the behavioural distance defined by ε-bisimilarity [10]. In this case, the relevant modality is precisely the *generally* modality used in work on fuzzy description logics as an alternative formal correspondent of the natural-language term 'probably' [33]. The convex powerset plays a central role in the distribution semantics of Markov decision processes (or probabilistic automata) [7]. Its standard metric is just the composition of the Hausdorff and Kantorovich-Rubinstein metrics, and as such is again given via the coupling-based construction for a modality composed from the expectation modality and the standard fuzzy diamond modality [38]. We show that the generalized Kantorovich-Rubinstein duality holds w.r.t. this modality. Beyond the mentioned benefits regarding characteristic quantitative modal logics, we demonstrate in this case that the new price-function-based description also allows for more efficient computation of distances.

Proofs are sometimes omitted or only sketched; a full version with all proofs is available [39].

Related Work Categorical distance constructions based on couplings were first considered in work on monoidal topology [20,21]. The categorical price-function-based construction is predated by several constructions for specific functors in work on stochastic games [17]. Our price-function-based presentation of convex powerset complements earlier results on a coupling-based characterization [38] and a quantitative-algebraic presentation [27]. The modality underlying the distance on the convex powerset functor relates to Goubault-Larrecq's *previsions* [16]; distances between such previsions are also expressed through a combination of (topological versions of) the Hausdorff and Kantorovich-Rubinstein-distances [17,18,19]. There has been recent work on what are termed *correspondences* between the generalized coupling-based and price-function-based constructions where a single modality is assumed for the former, while an associated set of modalities is considered for the latter [22]. Correspondences in this sense thus lie between dualities as considered here, where we insist on the same modalities being used on both sides, and general theorems on Kantorovich presentations of lax extensions [38] and classes of functors [14] that use very large sets of liftings. Results are obtained for functors constructed from the main known instances (distributions with the standard Kantorovich-Rubinstein metric, powerset) by applying coproduct and product. The price-function-based construction is sometimes referred to as the *codensity* construction [24,25,22], and as such has been used for logical characterizations of behavioural distances as mentioned above [25] but also for game characterizations [24]. The problem of generalized Kantorovich-Rubinstein duality has been stated already in work introducing coalgebraic behavioural distances [4], and a simple counterexample has been given; the counterexample we give here is distinguished by involving a quantitative modality that satisfies an analogue of two-valued *separation* [28].

2   Preliminaries

We discuss preliminaries on (pseudo-)metric spaces and coalgebras. Generally, we assume basic familiarity with category theory [1].

Metric spaces We write $\oplus, \ominus\colon [0,1] \times [0,1] \to [0,1]$ for truncated addition and subtraction, i.e. $a \oplus b = \min\{1, a+b\}$ and $a \ominus b = \max\{0, a-b\}$. A *(1-bounded) pseudometric space* is a pair (X, d_X), where X is a set and $d_X\colon X \times X \to [0,1]$ is a function, which for all $x, y, z \in X$ is subject to the conditions of *reflexivity* $d_X(x,x) = 0$, *triangle inequality* $d_X(x,z) \le d_X(x,y) + d_X(y,z)$ and *symmetry*, that is $d_X(x,y) = d_X(y,x)$. A *metric space* is then a pseudometric space which is separated: If $d_X(x,y) = 0$ then $x = y$. The *Euclidean distance* $d_e(a,b) = |b-a|$ makes $[0,1]$ into a metric space.

A function between the underlying sets $f\colon X \to Y$ of two pseudometric spaces (X, d_X), (Y, d_Y) is *nonexpansive* if distances are not increased by f, explicitly if for all $x, y \in X$ it holds that $d_Y(f(x), f(y)) \le d_X(x,y)$. Pseudometric spaces and nonexpansive functions between them form a category **PMet**. The full subcategory of **PMet** spanned by metric spaces is denoted by **Met**.

Coalgebra Our main results derive some of their interest from their relevance to behavioural distances in coalgebras. Generally, the framework of *universal coalgebra* [31] is based on abstracting state-based systems as F-*coalgebras* for a functor $F\colon \mathbf{C} \to \mathbf{C}$ on a category $\mathbf{C}$, with F determining the *type* of the system. Specifically, an F-coalgebra is a pair (C, γ) consisting of a $\mathbf{C}$-object C, thought of as an object of *states*, and a morphism $\gamma\colon C \to FC$ determining *transitions* from states to structured collections of successor states, with the structure determined by F. A *homomorphism* between F-coalgebras (C, γ) and (D, δ) is a $\mathbf{C}$-morphism $h\colon C \to D$ such that $\delta \circ h = Fh \circ \gamma$.

We list some common functors that will be useful in the further technical development and their associated coalgebras.

Example 1.   1. The (covariant) *powerset functor* $\mathcal{P}\colon \mathbf{Set} \to \mathbf{Set}$ sends each set to its powerset. On functions, $\mathcal{P}$ acts by taking images: For $A \in \mathcal{P}X$ and $f\colon X \to Y$ we have $\mathcal{P}f(A) = f[A]$. Its coalgebras are precisely sets equipped with a binary relation, i.e. *transition systems* or *Kripke frames*.

2. The *finitely supported probability distribution functor* $\mathcal{D}\colon \mathbf{Set} \to \mathbf{Set}$ sends a set X to the set

$$\mathcal{D}X = \{\mu\colon X \to [0,1] \mid \mu(x) > 0 \text{ for finitely many } x \in X \text{ and } \sum_{x \in X} \mu(x) = 1\}.$$

On a function $f\colon X \to Y$ the functor $\mathcal{D}$ measures probabilities of preimages:

$$\mathcal{D}f(\mu)(y) = \sum_{x \in f^{-1}(y)} \mu(x).$$

The coalgebras of $\mathcal{D}$ are precisely (discrete-time) Markov chains.

3   Dual Characterizations of Metrics

A central question in the study of state-based systems at large is whether two states exhibit the same behaviour. In universal coalgebra, answers for this type of question are provided by such concepts as Aczel-Mendler bisimulation or behavioural equivalence. When the behaviour of states has quantitative aspects, however, such as probabilistic transitions or outputs in a metric space, small deviations in these quantities immediately render two states behaviourally distinct under such two-valued notions. When one prefers to retain the information that these states differ only slightly, an established approach, discussed next, is to switch from behavioural equivalence relations to the more robust concept of *behavioural metrics*, equipping the state space with a pseudometric structure to describe how dissimilar individual states are in their behaviour.

A central role in the general coalgebraic treatment of behavioural distances is played by the concept of a *functor lifting*.

Definition 2. Let $F\colon \mathbf{Set} \to \mathbf{Set}$ and $U\colon \mathbf{C} \to \mathbf{Set}$ be functors. A *lifting* of F along U is a functor $\overline{F}\colon \mathbf{C} \to \mathbf{C}$ such that the following diagram commutes.

$$
\begin{array}{ccc}
\mathbf{C} & \xrightarrow{\ \overline{F}\ } & \mathbf{C} \\
\downarrow{\scriptstyle U} & & \downarrow{\scriptstyle U} \\
\mathbf{Set} & \xrightarrow{\ F\ } & \mathbf{Set}
\end{array}
$$

When $\mathbf{C}$ is the category of pseudometric spaces and U is the forgetful functor, i.e. the functor that maps pseudometric spaces to their underlying sets, the fibres above any set X (the collection of pseudometric spaces carried by X) form a complete lattice under the pointwise order; we denote this lattice by $\mathbf{C}_X$. Given an F-coalgebra (X, γ) and a functor lifting $\overline{F}$ we can construct a monotone function $\Phi_\gamma\colon \mathbf{C}_X \to \mathbf{C}_X$ on this complete lattice, sending a pseudometric $d_X\colon X \times X \to [0,1]$ to $\Phi_\gamma(d_X)$ given by

$$\Phi_\gamma(d_X)(x,y) = d_{\overline{F}(X,d_X)}(\gamma(x),\gamma(y)).$$

The *behavioural distance* [4] on (X, γ) is then defined as the least fixpoint $\mu\Phi_\gamma$ of the function Φ_γ, which exists by the Knaster-Tarski fixpoint theorem.

Example 3.   1. The *Hausdorff lifting* $\overline{\mathcal{P}}\colon \mathbf{PMet} \to \mathbf{PMet}$ equips the powerset $\mathcal{P}X$ of the carrier of a metric space (X, d_X) with the Hausdorff metric $\delta^{\mathsf{H}}(d_X)$. The distance of two subsets $A, B \in \mathcal{P}X$ is then given by

$$\delta^{\mathsf{H}}(d_X)(A, B) := \max(\sup_{x \in A} \inf_{y \in B} d_X(x, y), \sup_{y \in B} \inf_{x \in A} d_X(x, y))$$

2. The *Kantorovich-Rubinstein lifting* $\overline{\mathcal{D}}\colon \mathbf{PMet} \to \mathbf{PMet}$ of $\mathcal{D}$ equips the set $\mathcal{D}X$ with the *Kantorovich-Rubinstein distance* $\delta^{\mathsf{KR}}(d_X)$, which is defined as

$$\delta^{\mathsf{KR}}(d_X)(\mu, \nu) := \sup\{\mathbb{E}_\nu(f) - \mathbb{E}_\mu(f) \mid f\colon (X, d_X) \to ([0, 1], d_e) \text{ nonexpansive}\},$$

where $\mathbb{E}_\mu(f) = \sum_{x \in X} \mu(x) \cdot f(x)$ denotes the expected value of f under μ.

The examples above can be seen as instances of more general constructions introduced below, which are parametric in a **Set**-endofunctor and a set of $[0, 1]$-valued predicate liftings.

Definition 4.   1. A $[0, 1]$-*valued predicate lifting* for a functor $F\colon \mathbf{Set} \to \mathbf{Set}$ is a natural transformation of type $\lambda\colon [0, 1]^- \Rightarrow [0, 1]^{F-}$. A predicate lifting λ is *well-behaved* if the following conditions hold:
- *Monotonicity*: If $f \le g$, then $\lambda_X(f) \le \lambda_X(g)$, where the ordering on functions is computed pointwise.
- *Subadditivity*: for $f, g \in [0, 1]^X$, we have $\lambda_X(f \oplus g) \le \lambda_X(f) \oplus \lambda_X(g)$, where the sum of two functions is calculated pointwise.
- *Zero preservation*: $\lambda_X(0_X) = 0_{FX}$, where 0_X, 0_{FX} are the constant zero functions on the respective sets.

2. Let λ be a predicate lifting for F. The *price-function-based lifting* of F to the category of pseudometric spaces sends a metric d_X to $K_\lambda(d_X)\colon FX \times FX \to [0, 1]$ defined by

$$K_\lambda(d_X)(s, t) := \sup\{|\lambda_X(f)(t) - \lambda_X(f)(s)| \mid f\colon (X, d_X) \to ([0, 1], d_e) \text{ nonexp.}\}$$

If Λ is a set of predicate liftings, we put $K_\Lambda = \sup_{\lambda \in \Lambda} K_\lambda$ (pointwise).
   3. Let $s \in FX$ and $t \in FY$. The set of *couplings* $\Gamma(s, t)$ is defined as

$$\Gamma(s, t) := \{c \in F(X \times Y) \mid F\pi_1(c) = s \text{ and } F\pi_2(c) = t\}.$$

4. Let λ be a well-behaved predicate lifting for F and assume that F preserves weak pullbacks. The *coupling-based lifting* of F to the category of pseudometric spaces is defined as the lifting that equips FX with $W_\lambda(d_X)$ where

$$W_\lambda(d_X)(s, t) = \inf\{\lambda_{X \times X}(d_X)(c) \mid c \in \Gamma(s, t)\}.$$

While the price-function-based lifting assumes no conditions on supplied structures, the coupling-based lifting is significantly more particular, requiring both pullback preservation of the underlying functor and that the predicate lifting be well-behaved, to ensure that $W_\lambda(d_X)$ is a pseudometric whenever d_X is [4,6].

Remark 5. It is well known that predicate liftings correspond to simple morphisms, sometimes dubbed *evaluation functions*, by the Yoneda lemma [32]. In the concrete instance of $[0,1]$-valued predicate liftings, we have that natural transformations of the form $\lambda\colon [0,1]^- \Rightarrow [0,1]^{F-}$ are in bijection with morphisms of type $\mathsf{ev}_\lambda\colon F[0,1] \to [0,1]$. Then the condition of a predicate lifting being well-behaved translates roughly to the corresponding evaluation function being well-behaved [4,38].

Example 6.   1. Let $F = \mathcal{P}$ and $\lambda\colon [0,1]^X \to [0,1]^{\mathcal{P}-}$ be the natural transformation whose components calculate suprema of images: For $A \subseteq X$ and $f \in [0,1]^X$ we define $\lambda_X(f)(A) = \sup f[A]$. Then $\delta^{\mathsf{H}} = K_\lambda = W_\lambda$.
  2. Let $F = \mathcal{D}$ and let λ be the predicate lifting calculating expected values: For $\mu \in \mathcal{D}X$ and $f \in [0,1]^X$ we have $\lambda_X(f)(\mu) = \mathbb{E}_\mu(f)$. Then $\delta^{\mathsf{KR}} = K_\lambda = W_\lambda$.

If, like in the two examples above, the categorical price-function-based and coupling-based constructions K_λ and W_λ coincide, we say that *generalized Kantorovich-Rubinstein duality* holds. This name is motivated by the particular case of Example 6.2, the classical *Kantorovich-Rubinstein duality* that dates back to the beginnings of transportation theory [23]. It is important to note that generalized duality in this sense may fail, as demonstrated by the case of p-Wasserstein distance that we discuss later. It is a general fact [4, Theorem 5.27] that $K_\lambda \leq W_\lambda$, so duality hinges on the inequality $W_\lambda \leq K_\lambda$.

Transportation Theory The names of the two constructions are motivated by the probabilistic case as well. The coupling-based presentation is closely related to optimal transportation theory: In the case of the Kantorovich-Rubinstein lifting, one may view probability distributions μ, ν as producers and consumers of a resource respectively (with the amount of the resource produced/consumed being fixed to 1). Now one wants to transport the produced resources to the consumers in the economically most efficient possible way, minimizing the average distance each unit of resource needs to travel. Couplings c of μ and ν can then be seen as *transport plans*, with $\mathbb{E}_c(d_X)$ giving the total cost of carrying out the plan. Then the Kantorovich-Rubinstein distance gives us by definition the minimal possible cost.

In the distribution case (Example 6.2), the price-function-based presentation can be explained via the analogy of "outsourcing the transport" by defining a function f that assigns a price to each $x \in X$. This function must satisfy the requirement that it is nonexpansive, i.e., that the difference of prices assigned to x, y is always at most $d_X(x, y)$ (meaning that no extra profit can be made from such a transport). The overall profit under such a price function is then the income obtained from the consumers ($\mathbb{E}_\nu(f)$) minus the cost paid to the producers ($\mathbb{E}_\mu(f)$). Taking the maximum over all such f gives us the value of the Kantorovich-Rubinstein lifting.

Expressive Logics The price-function-based lifting is closely related to characteristic multi-valued modal logics: These are logics in which formulae φ receive semantics in coalgebras (X, γ), inducing an interpretation function

$$[\![\varphi]\!]_\gamma \colon X \to [0, 1].$$

The logical distance of two states is then the supremum of all distances witnessed by such formulas.

The semantics of a modal operator L in these types of logics is usually given as a predicate lifting $\lambda \colon [0, 1]^- \Rightarrow [0, 1]^{F^-}$, with $[\![L\varphi]\!]_\gamma$ being inductively defined as $\lambda_X([\![\varphi]\!]_\gamma) \circ \gamma$. Then *expressivity* of the logic (the fact that the behavioural distance can be witnessed by formulae of the logic arbitrarily closely) can be shown by exploiting the price-function-based presentation of the lifting. In fact, the interpretation $\lambda_X([\![\varphi]\!]_\gamma) \circ \gamma$ can be viewed as emulating one step of the functional Φ_γ, under the condition that the interpretations of formulae $[\![\varphi]\!]_\gamma$ are able to approximate any nonexpansive function $f \colon X \to [0, 1]$ arbitrarily closely [13].

Hence, the duality can be used very fruitfully: One can use the coupling-based view to compute (an under-approximation of) the behavioural distance and switch to price functions to determine the distinguishing formula witnessing this distance.

The p-Wasserstein Distance There also exists a parametrized version of the Kantorovich-Rubinstein distance, called the *p-Wasserstein distance* for some real parameter $p \geq 1$. In categorical terms, it is given by the predicate lifting $\lambda_p(f)(\mu) = (\mathbb{E}_\mu(f^p))^{\frac{1}{p}}$ (where f^p takes the p-th power pointwise), that is:

$$W_{\lambda_p}(d)(\mu, \nu) = \inf\{(\mathbb{E}_\rho(d^p))^{\frac{1}{p}} \mid \rho \text{ is a coupling of } \mu \text{ and } \nu\}.$$

For $p = 1$ this is just the usual Kantorovich-Rubinstein distance and duality holds. If $p > 1$, however, then duality may fail, and the corresponding price-function-based construction K_{λ_p} may be strictly below W_{λ_p}. This gives witness to the general idea from transportation theory that $p = 1$ constitutes a special case among the family of Wasserstein distances:

Example 7. Let $p = 2$. Let (X, d) be a two-element discrete metric space, that is $X = \{0, 1\}$ and $d(0, 1) = d(1, 0) = 1$, and let $\mu = \frac{2}{3} \cdot 0 + \frac{1}{3} \cdot 1$ and $\nu = \frac{1}{3} \cdot 0 + \frac{2}{3} \cdot 1$. Then we have $K_{\lambda_p}(d)(\mu, \nu) \leq \frac{1}{3} < \frac{1}{\sqrt{3}} = W_{\lambda_p}(d)(\mu, \nu)$.

Proof. We begin by showing the last equality. We note that $d^2 = d$, so that

$$W_{\lambda_2}(d)(\mu, \nu) = (\delta^{\mathsf{KR}}(d)(\mu, \nu))^{\frac{1}{2}} = (\tfrac{1}{3})^{\frac{1}{2}} = \frac{1}{\sqrt{3}},$$

where in the first step we used that $(-)^{\frac{1}{2}}$ is monotone and continuous.

For the first inequality, let $f \colon X \to [0, 1]$ be nonexpansive and put $a = f(0)$ and $b = f(1)$. We show that $\lambda_2(f)(\nu) - \lambda_2(f)(\mu) \leq \frac{1}{3}$; the proof that $\lambda_2(f)(\mu) - \lambda_2(f)(\nu) \leq \frac{1}{3}$ is analogous. We may assume wlog. that $a < b$, as otherwise the left hand side of our target inequality is nonpositive. Define $g \colon [0, 1] \to \mathbb{R}$ via $g(t) =$

$\sqrt{(1-t)a^2 + tb^2}$. Then we have $g(0) = a$, $g(\frac{1}{3}) = \lambda_2(f)(\mu)$, $g(\frac{2}{3}) = \lambda_2(f)(\nu)$ and $g(1) = b$. As the function g is concave, we also have

$$g(0) + g(1) = (\tfrac{1}{3} \cdot g(0) + \tfrac{2}{3} \cdot g(1)) + (\tfrac{2}{3} \cdot g(0) + \tfrac{1}{3} \cdot g(1))$$
$$\leq g(\tfrac{1}{3} \cdot 0 + \tfrac{2}{3} \cdot 1) + g(\tfrac{2}{3} \cdot 0 + \tfrac{1}{3} \cdot 1) = g(\tfrac{1}{3}) + g(\tfrac{2}{3}). \quad (1)$$

Additionally, nonexpansiveness of f implies that

$$b^2 - a^2 = (b-a)(a+b) = (g(1) - g(0))(a+b) \leq (1-0)(a+b) = a + b. \quad (2)$$

Therefore we have:

$$\begin{aligned}
(g(\tfrac{2}{3}) - g(\tfrac{1}{3})) \cdot (g(\tfrac{2}{3}) + g(\tfrac{1}{3})) &= g(\tfrac{2}{3})^2 - g(\tfrac{1}{3})^2 \\
&= (\tfrac{1}{3}a^2 + \tfrac{2}{3}b^2) - (\tfrac{2}{3}a^2 + \tfrac{1}{3}b^2) \\
&= \tfrac{1}{3}(b^2 - a^2) \\
&\leq \tfrac{1}{3}(g(0) + g(1)) && (2) \\
&\leq \tfrac{1}{3}(g(\tfrac{2}{3}) + g(\tfrac{1}{3})). && (1)
\end{aligned}$$

Our earlier assumption that $a < b$ implies that $g(\tfrac{2}{3}) + g(\tfrac{1}{3})$ is positive, so we can divide by it on both sides, which results in the claimed inequality. $\qquad\square$

4   Lévy-Prokhorov Distance

The Lévy-Prokhorov distance provides an alternative to the Kantorovich-Rubinstein distance when it comes to measuring the distance between probability distributions. If (X, d) is a pseudometric space, and $\mu, \nu \in \mathcal{D}X$ are (discrete) probability distributions, then we define

$$\delta^{\mathsf{LP}}(d)(\mu, \nu) = \inf\{\varepsilon \mid \forall A \subseteq X.\ \mu(A) \leq \nu(A_\varepsilon^d) + \varepsilon\},$$

where $A_\varepsilon^d = \{y \in X \mid \inf_{x \in A} d(x, y) \leq \varepsilon\}$. The definition of the Lévy-Prokhorov distance sometimes includes the mirrored condition $\forall B \subseteq X.\ \nu(B) \leq \mu(B_\varepsilon^d) + \varepsilon$, but this second clause is redundant and does not actually change the induced pseudometric.

The Lévy-Prokhorov distance has recently been investigated by Desharnais and Sokolova [11], who prove that it is a functor lifting, but not a monad lifting, and that it characterizes the notion of ε-bisimulation [10].

The Lévy-Prokhorov distance admits a representation based on couplings (cf. Section 3), or equivalently in terms of pairs of (not necessarily independent) random variables that are distributed according to the given distributions, known as the *Ky Fan metric* [12]. The predicate lifting λ underlying this representation is given by

$$\lambda_X(f)(\mu) = \inf\{\varepsilon \geq 0 \mid \mu(\{x \in X \mid f(x) > \varepsilon\}) \leq \varepsilon\}, \quad (3)$$

and using this predicate lifting we have $\delta^{\mathsf{LP}}(d) = W_\lambda(d)$ for every pseudometric d, explicitly:

$$\delta^{\mathsf{LP}}(d)(\mu,\nu) = \inf\{\inf\{\varepsilon \geq 0 \mid \rho(\{(x,y) \in X \times Y \mid d(x,y) > \varepsilon\}) \leq \varepsilon\}$$
$$\mid \rho \in \Gamma(\mu,\nu)\}.$$

The predicate lifting from (3) has independently been used under the name *'generally'* in work on fuzzy description logics [33], and it admits a number of equivalent representations. Intuitively, all of these representations amount to the statement that the value $\lambda_X(f)(\mu)$ is high if the value of f is high with high probability when sampling according to the distribution μ.

Lemma 8. *Let X be a set, let $f\colon X \to [0,1]$, and let $\mu \in \mathcal{D}X$. Then we have:*

1. *$\lambda_X(f)(\mu) = \inf_{\varepsilon \geq 0} \max(\mu(\{x \in X \mid f(x) > \varepsilon\}), \varepsilon)$*
2. *$\lambda_X(f)(\mu) = \sup_{\varepsilon \geq 0} \min(\mu(\{x \in X \mid f(x) > \varepsilon\}), \varepsilon)$*
3. *$\lambda_X(f)(\mu) = \sup\{\varepsilon \geq 0 \mid \mu(\{x \in X \mid f(x) > \varepsilon\}) \geq \varepsilon\}$*

All these identities, and also (3), remain true if $f(x) > \varepsilon$ is replaced by $f(x) \geq \varepsilon$.

It follows from Lemma 8 that λ is *self-dual* (equal to its own dual):

Lemma 9. *For every X, f and μ we have $\lambda_X(f)(\mu) = 1 - \lambda_X(1 - f)(\mu)$.*

The coupling-based representation above is justified by the fact that the predicate lifting is well-behaved:

Lemma 10. *The predicate lifting λ as per (3) is well-behaved.*

4.1 Duality

Next, we show that the Lévy-Prokhorov distance admits a dual representation using the same predicate lifting λ, that is, we have $K_\lambda = W_\lambda$. We prove this duality in the more general setting where the two constructions apply to fuzzy relations that need not be pseudometrics. Recall that a *fuzzy relation $r\colon X \nrightarrow Y$* between sets X and Y is a function $r\colon X \times Y \to [0,1]$. The coupling-based construction applies to fuzzy relations the same way it does to pseudometrics, while the price-function-based construction is defined in terms of pairs of functions that satisfy a nonexpansiveness condition with respect to the given fuzzy relation. They are therefore both examples of *(fuzzy) relational liftings* or *relators* (e.g. [15] and references therein), as they lift fuzzy relations of type $X \nrightarrow Y$ to relations of type $FX \nrightarrow FY$:

Definition 11. Let λ be a monotone predicate lifting for a set functor F, and let $r\colon X \nrightarrow Y$.

1. The *relational coupling-based lifting $W_\lambda(r)\colon FX \nrightarrow FY$* is defined as

$$W_\lambda(r)(s,t) = \inf\{\lambda_{X \times Y}(r)(c) \mid c \in \Gamma(s,t)\}$$

   for every $s \in FX$ and $t \in FY$.

2. An *r-nonexpansive pair* is a pair of functions (f, g) where $f \colon X \to [0,1]$, $g \colon Y \to [0,1]$ and $g(y) - f(x) \le r(x, y)$ for all $x \in X$ and $y \in Y$.
3. The *relational price-function-based lifting* $K_\lambda^{\mathsf{rel}} \colon FX \nrightarrow FY$ is defined as

$$K_\lambda^{\mathsf{rel}}(r)(s, t) = \sup\{\lambda_Y(g)(t) \ominus \lambda_X(f)(s) \mid (f, g) \ r\text{-nonexpansive}\}$$

for every $s \in FX$ and $t \in FY$. Additionally, put $K_\Lambda^{\mathsf{rel}} = \sup_{\lambda \in \Lambda} K_\lambda^{\mathsf{rel}}$ if Λ is a set of predicate liftings.

Both of these constructions satisfy certain laws (that we will not restate here) making them *lax extensions*. Wild and Schröder [38] give results that relate K^{rel} to its pseudometric counterpart. The most relevant consequence of these results for our purposes is the following:

Lemma 12. *If λ is a self-dual predicate lifting, then $K_\lambda^{\mathsf{rel}}(d) = K_\lambda(d)$ for every pseudometric d.*

Out of the two representations of the Lévy-Prokhorov distance discussed earlier, the second, being based on couplings, readily generalizes to fuzzy relations. Therefore, we define the *relational Lévy-Prokhorov lifting* δ^{LP} to be the assignment that maps each fuzzy relation $r \colon X \nrightarrow Y$ to $\delta^{\mathsf{LP}}(r) = W_\lambda(r) \colon \mathcal{D}X \nrightarrow \mathcal{D}Y$.

We discussed in Section 3 that the inequality '$\le$' follows from the general theory of coupling-based and price-function-based liftings. The same is true for the respective lax extensions [37, Lemma 5.22], so that it suffices to prove the converse inequality '$\ge$'. In the proof of the classical Kantorovich-Rubinstein duality (e.g [36, Theorem 5.10]), this direction amounts to, given an optimal transport plan in the shape of a coupling between the distributions at hand, constructing two price functions that correspond to the optimal cost, in the sense that they form a nonexpansive pair that witnesses the supremum in the definition of the relational price-function-based lifting $K_\mathbb{E}$. In our proof of Lévy-Prokhorov duality we use a similar approach, which means that we should first understand how to phrase computation of the Lévy-Prokhorov distance in terms of a transport problem.

Let $r \colon X \nrightarrow Y$ and let $\mu \in \mathcal{D}X$ and $\nu \in \mathcal{D}Y$. The coupling-based representation $\delta^{\mathsf{LP}}(r)(\mu, \nu) = W_\lambda(r)(\mu, \nu)$ can be rewritten for this purpose. For $\varepsilon \ge 0$, define $r^\varepsilon(x, y) = 0$ if $r(x, y) < \varepsilon$ and $r^\varepsilon(x, y) = 1$ otherwise. Then $\rho(\{(x, y) \mid r(x, y) \ge \varepsilon\}) = \mathbb{E}_\rho(r^\varepsilon)$ for every $\rho \in \mathcal{D}(X \times Y)$. We may now use the representation of λ from Lemma 8.1, replacing the strict inequality with a non-strict one, and then swap the infimum over couplings inside to obtain

$$W_\lambda(r)(\mu, \nu) = \inf\{\inf_{\varepsilon \ge 0} \max(\varepsilon, \mathbb{E}_\rho(r^\varepsilon)) \mid \rho \in \Gamma(\mu, \nu)\} = \inf_{\varepsilon \ge 0} \max(\varepsilon, W_\mathbb{E}(r^\varepsilon)(\mu, \nu)).$$

This means that Lévy-Prokhorov distance is determined by the solutions to the transport problems for the r^ε. As each such r^ε is a crisp relation (i.e. only has 0 and 1 entries), the optimal price functions can be made crisp as well:

Lemma 13. *Let $r \colon X \times Y \to \{0, 1\}$ be a crisp relation and let $\mu \in \mathcal{D}X$, $\nu \in \mathcal{D}Y$. Then there exist functions $f \colon X \to \{0, 1\}$ and $g \colon Y \to \{0, 1\}$ such that (f, g) is an r-nonexpansive pair and $\mathbb{E}_\nu(g) - \mathbb{E}_\mu(f) \ge W_\mathbb{E}(r)(\mu, \nu)$.*

This allows us to establish duality:

Theorem 14. *For every* $r\colon X \twoheadrightarrow Y$ *and every* $\mu \in \mathcal{D}X$, $\nu \in \mathcal{D}Y$,

$$\delta^{\mathsf{LP}}(r)(\mu,\nu) = K_\lambda^{\mathsf{rel}}(r)(\mu,\nu) = W_\lambda(r)(\mu,\nu).$$

Proof (sketch). As mentioned before, we only need to prove $W_\lambda(r) \leq K_\lambda^{\mathsf{rel}}(r)$. Assume $\varepsilon < W_\lambda(r)(\mu,\nu)$. Then there is an r^ε-nonexpansive pair (p,q) of crisp price functions witnessing the transport cost wrt. r^ε. One then replaces the function values 0 and 1 by $\mathbb{E}_\mu(p)$ and $\mathbb{E}_\mu(p) + \varepsilon$ to arrive at an r-nonexpansive pair (f,g). Using this pair we show

$$K_\lambda^{\mathsf{rel}}(r)(\mu,\nu) \geq \lambda_Y(g)(\nu) - \lambda_X(f)(\mu) \geq (\mathbb{E}_\mu(p) + \varepsilon) - \mathbb{E}_\mu(p) = \varepsilon. \qquad \square$$

Theorem 15. *For every pseudometric* d *we have* $\delta^{\mathsf{LP}}(d) = K_\lambda(d) = W_\lambda(d)$.

Proof. We only need to show that $K_\lambda^{\mathsf{rel}}(d) = K_\lambda(d)$, which follows by Lemmas 9 and 12. $\qquad \square$

5   Convex Powerset Functor

We will next tackle duality for the case of the convex powerset functor, a functor that has been studied in-depth for modelling systems combining probability and non-determinism (e.g., [7]). A non-empty set $D \subseteq \mathcal{D}X$ of probability distributions is *convex* if for all $\mu_1, \mu_2 \in D$ it also holds that $\mu_1 +_p \mu_2 := p \cdot \mu_1 + (1-p) \cdot \mu_2 \in D$ (where $p \in [0,1]$). For a set X, we define

$$\mathcal{C}X = \{\emptyset \neq D \subseteq \mathcal{D}X \mid D \text{ is convex}\}.$$

Of course, we have $\mathcal{C}X \subseteq \mathcal{P}\mathcal{D}X$ for every set X. In fact, it is easily verified that every map $\mathcal{P}\mathcal{D}f$ preserves convex sets, so that we obtain a subfunctor $\mathcal{C}$ of the composite functor $\mathcal{P}\mathcal{D}$.

A straightforward – but futile – approach to prove duality for $\mathcal{C}$ would be to observe that it holds for the powerset and distribution functor, and then apply a compositionality result. However, the studied liftings (price-function-based and coupling-based) are quite fragile when it comes to compositionality, i.e. it does not hold in general that the composition of liftings of functors F, G based on the predicate liftings λ^F, λ^G is the lifting of the composite FG (based on the obvious combined modality $\lambda_X^{FG} = \lambda_{GX}^F \circ \lambda_X^G$) [4]. While it is known that the coupling-based lifting of the convex powerset functor arises by combining the coupling-based liftings of the component functors [38], this is incorrect for the price-function-based lifting [9]. In fact, the given counterexample uses a set that is *not* convex, thus suggesting that the problem might disappear if we restrict to convex sets.

We use the sup modality for the powerset functor and the expectation ($\mathbb{E}$) modality for the distribution functor. Our aim is to study the convex powerset functor $\mathcal{C}$ and establish the rather non-trivial result that the combined modality $\lambda_X(f)(A) = \sup\{\mathbb{E}_\mu(f) \mid \mu \in A\}$ is indeed expressive on its own.

As before, $K_\lambda \leq W_\lambda$ holds in categorical generality, so the main task is to prove the converse inequality $W_\lambda \leq K_\lambda$. As discussed earlier, it is known that the coupling-based representation in terms of λ decomposes into the coupling-based representations in terms of $\sup$ and $\mathbb{E}$, i.e. the Hausdorff and Kantorovich-Rubinstein liftings, respectively:

$$W_\lambda(d) = W_{\sup}(W_\mathbb{E}(d)) = \delta^\mathsf{H}(\delta^\mathsf{KR}(d)) =: \delta^\mathsf{HK}(d)$$

To achieve the duality result, it will be convenient to pass from pseudometric spaces to metric spaces. Recall that the *metric quotient* of a pseudometric space (X, d) is the metric space $(X_\sim, d_\sim)$ where $X_\sim$ is the set of equivalence classes of the equivalence relation $x \sim y \iff d(x, y) = 0$, and $d_\sim([x], [y]) = d(x, y)$ for any two equivalence classes $[x], [y] \in X_\sim$.

Lemma 16. *Let (X, d) be a pseudometric space, let $(X_\sim, d_\sim)$ be its metric quotient, and let $\pi\colon X \to X_\sim, x \mapsto [x]$. We then have, for every $A, B \in \mathcal{C}X$,*

$$K_\lambda(d)(A, B) = K_\lambda(d_\sim)(A_\sim, B_\sim) \quad and \quad W_\lambda(d)(A, B) = W_\lambda(d_\sim)(A_\sim, B_\sim),$$

where $A_\sim = \mathcal{C}\pi(A)$ and $B_\sim = \mathcal{C}\pi(B)$.

Using the above lemma, we may therefore from now on assume that we are working over a metric space (X, d). We may also assume that $X \neq \emptyset$, as otherwise $\mathcal{C}X = \emptyset$ and both coupling-based and price-function-based distance are the empty metric, hence equal.

The main intuition behind the proof is best understood in the case where $X = \{x_1, \ldots, x_n\}$ is finite, even though the proof will work for arbitrary X. In this case we may view probability distributions and fuzzy predicates on X as vectors in $\mathbb{R}^n$, and the expectation modality simply computes the dot product between two such vectors: $\mathbb{E}_\mu(f) = \mu(x_1) \cdot f(x_1) + \cdots + \mu(x_n) \cdot f(x_n)$. If $A, B \in \mathcal{C}X$ satisfy $\delta^\mathsf{HK}(d)(A, B) > \varepsilon$, then this means, by the definition of the Hausdorff distance, that there must be some $\mu \in A$ such that $\delta^\mathsf{KR}(\mu, \nu) > \varepsilon$ for all $\nu \in B$ (or we are in the symmetric situation with A and B swapped). This implies that B and the ε-ball around μ are disjoint convex sets, so we can apply the hyperplane separation theorem to find a hyperplane H such that the two sets lie on opposite sides of that plane (more precisely, B may intersect the hyperplane, but the ε-ball may not). A price function witnessing distance at least ε under the combined modality λ can now be constructed from the normal vector of H.

Our proof mostly follows the outline above, but because we may now be working with infinite-dimensional vector spaces, some functional analysis will be required. We leverage this added complexity in Section 5.2, where we show that the duality result remains true when passing from discrete probability measures to Borel probability measures.

We fix a point $x_0 \in X$ and consider the vector space $\mathsf{Lip}_0(X)$ consisting of the real-valued Lipschitz functions on X vanishing at x_0:

$$\mathsf{Lip}_0(X) = \{f\colon X \to \mathbb{R} \mid f(x_0) = 0, \sup_{x \neq y} \tfrac{f(y) - f(x)}{d(x, y)} < \infty\}.$$

This is a Banach space with norm given by $\|f\|_{\mathsf{Lip}} = \sup_{x \neq y} \frac{f(y)-f(x)}{d(x,y)}$. We will construct our price function in this space, which is made possible by the fact that the set of probability distributions can be mapped into its dual:

Lemma 17. *The set $\mathcal{D}X$ embeds into the continuous dual space $\mathsf{Lip}_0(X)^*$, that is, every discrete probability measure μ gives rise to a continuous linear functional $L_\mu \colon \mathsf{Lip}_0(X) \to \mathbb{R}$, which may explicitly be given by $L_\mu(f) = \mathbb{E}_\mu(f)$.*

In what follows, we often do not distinguish between μ and L_μ and treat $\mathcal{D}X$ as a subset of $\mathsf{Lip}_0(X)^*$. We equip $\mathsf{Lip}_0(X)^*$ with the *weak-* topology*, which is the weakest topology on $\mathsf{Lip}_0(X)^*$ making all the maps $\psi \mapsto \psi(f)$ for $f \in \mathsf{Lip}_0(X)$ continuous. Equivalently, this is the *initial topology* wrt. the maps $\psi \mapsto \psi(f)$. Crucially, this topology coincides with the one given by the Kantorovich-Rubinstein distance:

Lemma 18. *Let $(\mu_n)_{n \in \mathbb{N}}$ be a sequence in $\mathcal{D}X$ and let $\mu \in \mathcal{D}X$. Then $\mu_n \to \mu$ in the topology given by $\delta^{\mathsf{KR}}(d)$ iff $L_{\mu_n} \to L_\mu$ in the weak-* topology.*

The space $\mathsf{Lip}_0(X)^*$ is normed via the *operator norm* $\|\psi\|_{\mathsf{op}} = \sup_{\|f\|_{\mathsf{Lip}} \leq 1} \psi(f)$. This norm relates to the Kantorovich-Rubinstein metric as follows:

Lemma 19. *For any $\mu, \nu \in \mathcal{D}X$ we have $\delta^{\mathsf{KR}}(d)(\mu, \nu) = \|\nu - \mu\|_{\mathsf{op}}$.*

We are now in a position to state and prove the duality result:

Theorem 20. *For any two convex sets $A, B \in \mathcal{C}X$,*

$$\delta^{\mathsf{HK}}(d)(A, B) = K_\lambda(d)(A, B).$$

Proof (sketch). The inequality $K_\lambda \leq \delta^{\mathsf{HK}}$ follows from previous results [4], hence it is sufficient to show $\delta^{\mathsf{HK}}(d)(A, B) \leq K_\lambda(d)(A, B)$.

Let $0 < \varepsilon < \delta^{\mathsf{HK}}(d)(A, B)$. As outlined above, we may assume wlog. that there exists some $\mu \in A$ such that $\delta^{\mathsf{KR}}(d)(\mu, \nu) > \varepsilon$ for every $\nu \in B$. Let C be the closed ε-ball around μ, shifted by $-\mu$, and let D be the closure of B, also shifted by $-\mu$. These two sets are closed and convex, so by the Hahn-Banach separation theorem there exists a continuous linear functional $g \colon \mathsf{Lip}_0(X)^* \to \mathbb{R}$ and $c \in \mathbb{R}$ such that

$$\sup_{\nu \in C} g(\nu) < c \leq \inf_{\nu \in D} g(\nu),$$

and because we are in the weak-* topology, the functional g can be represented in the form $\mu \mapsto \mathbb{E}_\mu(f)$ for some $f \in \mathsf{Lip}_0(X)$. We replace f by $f_1 = -f/\|f\|_{\mathsf{Lip}}$, which results in a nonexpansive function for which

$$\left| \sup_{\nu \in B} \mathbb{E}_\nu(f_1) - \sup_{\nu \in A} \mathbb{E}_\nu(f_1) \right| \geq \varepsilon.$$

The range of f_1 is not necessarily contained in $[0, 1]$, but it must be contained in some subinterval of $\mathbb{R}$ of length at most 1 by nonexpansiveness and because d is 1-bounded. As expectation is linear, we may simply shift f_1 by a suitable amount to extract the desired price function. $\square$

Remark 21. A natural question to ask is whether one can, like in the previous section, also obtain a fuzzy-relational version of the duality result. We expect the answer to be 'yes', but that it will be necessary to additionally consider the dual predicate lifting $\kappa_X(f)(A) = \inf\{\mathbb{E}_\mu(f) \mid \mu \in A\}$, resulting in the duality result $K^{\mathsf{rel}}_{\{\lambda,\kappa\}} = W_\lambda$. This would be reflective of the situation that arises in the case of the Hausdorff extension, where $K^{\mathsf{rel}}_{\{\sup,\inf\}} = W_{\sup} = \delta^{\mathsf{H}}$ holds [38]. We leave this question open for now.

Remark 22 (Compositionality). It has been shown in work on *correspondences* between price-function-based and coupling-based representations of metric liftings [22] (cf. Section 1 under *related work*) that such correspondences can be combined along sums and product of functors, so that one arrives at general correspondence results for classes of functors obtained by closing given basic building blocks (originally constant functors, identity, powerset, and distributions) under sum and product. One thus obtains correspondences for composite system types such as labelled Markov chains [22, Example 41]. The correspondences produced in this way are not generalized Kantorovich-Rubinstein dualities in the strict sense we use here, as the transition from the coupling-based presentation to the price-function-based presentation in general involves the introduction of additional modalities. In particular, this happens for products, where one needs to introduce separate modalities for the factors (indeed, this is what is morally behind the fact that generalized Kantorovich-Rubinstein duality fails for the squaring functor [4]). Nevertheless, our results on generalized Kantorovich-Rubinstein duality for Lévy-Prokhorov distance and convex powerset imply that these functors can now be used as additional basic building blocks in this framework.

5.1   Algorithmic considerations

A nice aspect of the duality result for the convex powerset functor is that it can be used as the basis of an algorithm to compute values of the distance $\delta^{\mathsf{HK}}(d)$. Explicitly, if (X, d) is a *finite* pseudometric space, and $A_0, B_0 \subseteq \mathcal{D}X$ are *finite* sets of probability measures, the problem is to compute the distance $\delta^{\mathsf{HK}}(d)(A, B)$, where $A = \mathsf{conv}(A_0)$ and $B = \mathsf{conv}(B_0)$. The distance expands as follows:

$$\delta^{\mathsf{HK}}(d)(A, B) = \max(\sup_{\mu \in A} \inf_{\nu \in B} \delta^{\mathsf{KR}}(d)(\mu, \nu), \sup_{\nu \in B} \inf_{\mu \in A} \delta^{\mathsf{KR}}(d)(\mu, \nu)).$$

As the map $\mu \mapsto \inf_{\nu \in B} \delta^{\mathsf{KR}}(d)(\mu, \nu)$ is convex, the left supremum can instead be taken over A_0 without changing the value, and similarly we may take the right supremum over B_0. It is however not in general true that the infima may be taken over B_0 and A_0, respectively.

Example 23. Let $X = \{x, y, z\}$, and assume $d(x, y) = d(x, z) = d(y, z) = 1$. Let $A_0 = \{\mu_0, \mu_1\}$, $B_0 = \{\mu_2, \mu_3\}$, $\mu_0 = \frac{1}{3} \cdot x + \frac{1}{3} \cdot y + \frac{1}{3} \cdot z$, $\mu_1 = \frac{2}{3} \cdot x + \frac{1}{3} \cdot y$, $\mu_2 = \frac{2}{3} \cdot y + \frac{1}{3} \cdot z$, $\mu_3 = \frac{2}{3} \cdot z + \frac{1}{3} \cdot x$. Then the minimal distance from μ_1 to B

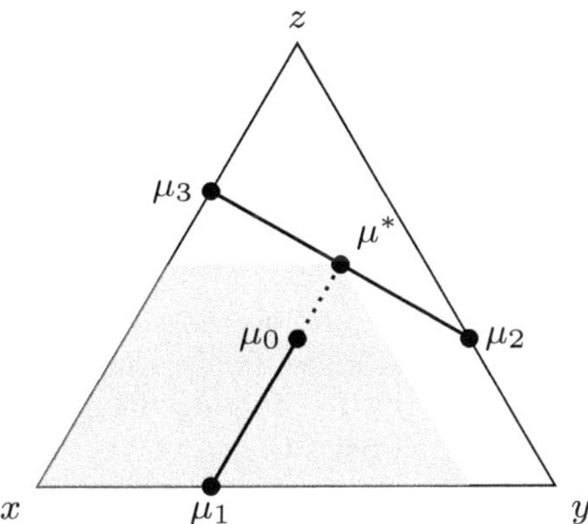

Fig. 1. Illustration of Example 23. The thick line segments correspond to the sets $A = \mathsf{conv}(A_0)$ and $B = \mathsf{conv}(B_0)$, while the shaded region shows the radius-$\frac{1}{2}$-ball around μ_1, which is part of a regular hexagon centered at μ_1.

is witnessed by $\mu^* = \frac{1}{2} \cdot \mu_2 + \frac{1}{2} \cdot \mu_3 = \frac{1}{6} \cdot x + \frac{1}{3} \cdot y + \frac{1}{2} \cdot z$. See Figure 1 for an illustration.

To compute the inner infimum, therefore, a more sophisticated approach is required. Çelik et al. [8] consider the problem of solving an optimal transport problem between probability distributions μ and ν, where μ is fixed and ν ranges over some algebraic variety. Their methods can be adapted to our setting, where ν instead ranges over a convex set B:

Consider the typical flow network that one constructs to solve the optimal transport problem, i.e. a complete bipartite graph whose partitions are two copies of the set X, which we think of as representing the two probability distributions μ and ν. A transport plan consists of assigning weights to the edges of this graph in such a way that the sum of weights of incident edges for each vertex matches its probability. The key idea is that there always exists an optimal transport plan that is acyclic in the sense that the edges with positive weight do not contain any cycle. Moreover, if the tree of edges used by the transport plan is known, then the weights of the plan are uniquely determined by the probabilities in μ and ν and can be computed by a depth-first (or breadth-first) traversal of the tree. This means that the distance between μ and ν can be computed by enumerating all spanning trees of the complete bipartite graph, computing the weights, and taking the least value over all trees where the weights are all non-negative. If $\mu \in A_0$ is fixed and ν ranges over B, then the weights in the coupling corresponding to a given spanning tree are linear combinations of the probabilities in ν. The conditions that these weights are non-negative define a linear program whose variables are the coefficients in the convex combination of the elements of B_0 and whose constraints state that the weights in the tree are all non-negative, and the distance can be found by solving all these linear programs.

The algorithm we just described requires solving exponentially many linear programs in the size of X, as there is one such linear program for each spanning tree. If we expand the dual representation of the distance between A and B instead, we obtain

$$K_\lambda(d)(A,B) = \sup\{|\sup_{\nu \in B} \mathbb{E}_\nu(f) - \sup_{\mu \in A} \mathbb{E}_\mu(f)| \mid f\colon (X,d) \to (\mathbb{R}, d_e) \text{ nonexp.}\}.$$

This quantity is much easier to compute, as there are no nested suprema/infima. Similar to before, as expectation is linear, the two suprema above may instead be taken over A_0 and B_0, respectively. To compute the distance, we can employ the following algorithm. Loop over all pairs $(\mu_0, \nu_0) \in A_0 \times B_0$. Given μ_0 and ν_0, the subproblem is then to find the supremum above for all the nonexpansive f such that $\mathbb{E}_{\mu_0}(f) = \sup_{\mu \in A} \mathbb{E}_\mu(f)$ and $\mathbb{E}_{\nu_0}(f) = \sup_{\nu \in B} \mathbb{E}_\nu(f)$. This subproblem can be rephrased as a linear program over the variables $(f_x)_{x \in X}$ corresponding to the function values of f. Nonexpansiveness of f, the constraints $\mathbb{E}_{\mu_0}(f) \geq \mathbb{E}_\mu(f)$ for $\mu \in A_0$ and the similar constraints for the $\nu \in B_0$ are all easily expressed as linear inequalities between the f_x. The objective function is $|\mathbb{E}_{\nu_0}(f) - \mathbb{E}_{\mu_0}(f)|$, which is the maximum of two linear expressions over the f_x, so we can simply solve the linear program twice, once for each of the two expressions. This algorithm has a runtime complexity that is polynomial in $|X|$, $|A_0|$ and $|B_0|$, a clear improvement over the exponential complexity for the previous approach.

5.2 Borel Measures

As the categorical coupling-based and price-function-based constructions are typically considered in the shape of liftings or lax extensions of set endofunctors, their probabilistic instances are restricted to dealing with discrete probability distributions by necessity. The (probabilistic) Kantorovich-Rubinstein duality, meanwhile, is known to hold for much larger classes of probability distributions, such as Radon measures on metric spaces [36]. In this section we show that this is also true for the convex powerset duality.

For a pseudometric space (X,d) we denote by $\mathsf{Bor}(X,d)$ the set of Borel probability measures, i.e. the probability measures defined on the σ-algebra generated by the open balls $B_\varepsilon^d(x)$. Every convex combination of Borel measures is itself a Borel measure. We can therefore also define a functor $\mathcal{C}^{\mathsf{Bor}}\colon \mathbf{PMet} \to \mathbf{PMet}$ where $\mathcal{C}^{\mathsf{Bor}}(X,d)$ is the set of non-empty convex subsets of $\mathsf{Bor}(X,d)$, equipped with Hausdorff-Kantorovich distance $\delta^{\mathsf{HK}}(d) = \delta^{\mathsf{H}}(\delta^{\mathsf{KR}}(d))$, where both δ^{H} and δ^{KR} are defined as before. Note that every nonexpansive map on (X,d) is Borel-measurable, so that no issues arise when taking expected values.

$$\delta^{\mathsf{KR}}(d)(\mu,\nu) = \sup\{|\textstyle\int_X f \, d\nu - \int_X f \, d\mu| \mid f\colon (X,d) \to ([0,1], d_e) \text{ nonexp.}\}$$

$$\delta^{\mathsf{H}}(d)(A,B) = \max(\sup_{x \in A} \inf_{y \in B} \, d(x,y), \sup_{y \in B} \inf_{x \in A} \, d(x,y))$$

We emphasize that $\mathcal{C}^{\mathsf{Bor}}$ is not the lifting of a set functor, as the definition of the underlying set depends on the pseudometric on the base space.

Due to a general result by Goncharov et al. [14], we know that $\mathcal{C}^{\mathsf{Bor}}$ admits a Kantorovich representation for some suitable class of predicate liftings. By [14, Theorem 5.3] it suffices to show that it preserves *initial morphisms*. In the real-valued setting these correspond to *isometries*, i.e. nonexpansive maps $f\colon (X, d_1) \to (Y, d_2)$ such that $d_2(f(x), f(x')) = d_1(x, x')$ for all $x, x' \in X$.

Lemma 24. $\mathcal{C}^{\mathsf{Bor}}$ *preserves isometries.*

The class of predicate liftings one obtains is quite large; we show that one can in fact make do with just a single predicate lifting, which strengthens the corresponding instance of the coalgebraic quantitative Hennessy-Milner theorem [14, Corollary 5.10] by providing a compact explicit syntax for the expressive logic:

Theorem 25. *Let* (X, d) *be a pseudometric space and* $A, B \in \mathcal{C}^{\mathsf{Bor}}(X, d)$*. Then*

$$\delta^{\mathsf{HK}}(d)(A, B) = \sup\{|\sup_{\nu \in B} \int_X f \, d\nu - \sup_{\mu \in A} \int_X f \, d\mu \,| \,|$$
$$f\colon (X, d) \to ([0, 1], d_e) \ nonexp.\}.$$

In the proof, we follow the same steps as before, first passing from pseudometrics to metrics and then leveraging linear algebra to obtain coincidence of the two distances.

Remark 26. Theorem 25 establishes that the composition of the Hausdorff and Kantorovich-Rubinstein distances can be expressed in price-function form with respect to the single modality $A \mapsto \sup_{\mu \in A} \int_X f \, d\mu$. To extend this into a full-blown duality result for continuous measures, stating that δ^{HK} also coincides with the corresponding coupling-based construction, two additional steps are needed: First, we need to have duality for the two individual distances. In the case of the Hausdorff distance, no additional requirements are necessary, but in the case of Kantorovich-Rubinstein distance one needs to restrict from Borel measures over general pseudometric spaces to one of the settings in which duality is known to hold, such as Polish spaces or Radon measures [36]. Second, one needs to establish a compositionality result for the respective coupling-based constructions, generalizing the one from the discrete case [38, Example 6.8.3]. We leave the details of such a duality result for future work.

6   Conclusions and Future Work

We have proved *generalized Kantorovich-Rubinstein duality*, i.e. coincidence of coupling-based (or Wasserstein) and price-function-based (or Kantorovich or codensity) presentations of functor liftings induced by a given choice of modalities, for two important and non-trivial cases: the Lévy-Prokhorov distance on distributions, and the standard distance on convex sets of distributions that arises from composing the Hausdorff and Kantorovich-Rubinstein metrics. In both cases, we obtain a characterization of the respective distance by means of quantitative modal logics defined by the given modalities; for the case of the

Lévy-Prokhorov distances, this logic is (up to restriction of the propositional base) the logic of *generally* previously studied in context of fuzzy description logics [33], and in the second case the involved modality is just the composite of the usual fuzzy diamond and the expectation modality [38]. In the case of convex powerset, we demonstrate additionally that the price-function-based presentation plays out algorithmic advantages in the actual computation of distances.

We leave several key open problems, among them on the one hand the extension of the duality result for the Lévy-Prokhorov metric from discrete to Borel probability distributions, and on the other hand the extension of the duality result for convex powerset to unrestricted fuzzy relations in place of pseudometrics (already established in our result on the Lévy-Prokhorov metric). The latter generalization will amount to a duality result for the known coupling-based lax extension of convex powerset [38]. Also, we aim to capitalize on the present result in the design of algorithms that actually compute distinguishing formulae as witnesses of lower bounds on behavioural distance, complementing recent results on behavioural distance under the Kantorovich-Rubinstein distance of distributions [30,34].

References

1. Adámek, J., Herrlich, H., Strecker, G.E.: Abstract and concrete categories: The joy of cats. John Wiley & Sons Inc. (1990), `http://tac.mta.ca/tac/reprints/articles/17/tr17abs.html`, republished in: Reprints in Theory and Applications of Categories, No. 17 (2006) pp. 1–507
2. de Alfaro, L., Faella, M., Stoelinga, M.: Linear and branching system metrics. IEEE Trans. Software Eng. **35**(2), 258–273 (2009). https://doi.org/10.1109/TSE.2008.106
3. Aolaritei, L., Wang, O., Zhu, J., Jordan, M., Marzouk, Y.: Conformal prediction under Lévy-Prokhorov distribution shifts: Robustness to local and global perturbations. In: Neural Information Processing Systems, NeurIPS 2025 (2025), to appear. Preprint available at `https://arxiv.org/abs/2502.14105`
4. Baldan, P., Bonchi, F., Kerstan, H., König, B.: Coalgebraic behavioral metrics. Log. Methods Comput. Sci. **14**(3) (2018). https://doi.org/10.23638/LMCS-14(3:20)2018
5. Bennouna, M.A., Lucas, R., Parys, B.P.G.V.: Certified robust neural networks: Generalization and corruption resistance. In: Krause, A., Brunskill, E., Cho, K., Engelhardt, B., Sabato, S., Scarlett, J. (eds.) International Conference on Machine Learning, ICML 2023. Proc. Machine Learning Res., vol. 202, pp. 2092–2112. PMLR (2023), `https://proceedings.mlr.press/v202/bennouna23a.html`
6. Bonchi, F., König, B., Petrişan, D.: Up-to techniques for behavioural metrics via fibrations. Mathematical Structures in Computer Science **33**(4–5), 182–221 (2023). https://doi.org/10.1017/S0960129523000166
7. Bonchi, F., Silva, A., Sokolova, A.: The Power of Convex Algebras. In: Meyer, R., Nestmann, U. (eds.) Concurrency Theory, CONCUR 2017. Leibniz International Proceedings in Informatics (LIPIcs), vol. 85, pp. 23:1–23:18. Schloss Dagstuhl–Leibniz-Zentrum fuer Informatik, Dagstuhl, Germany (2017). https://doi.org/10.4230/LIPIcs.CONCUR.2017.23

8. Çelik, T.Ö., Jamneshan, A., Montúfar, G., Sturmfels, B., Venturello, L.: Optimal transport to a variety. In: Slamanig, D., Tsigaridas, E.P., Zafeirakopoulos, Z. (eds.) Mathematical Aspects of Computer and Information Sciences, MACIS 2019. LNCS, vol. 11989, pp. 364–381. Springer (2019). https://doi.org/10.1007/978-3-030-43120-4_29

9. D'Angelo, K., Gurke, S., Kirss, J.M., König, B., Najafi, M., Różowski, W., Wild, P.: Behavioural metrics: Compositionality of the Kantorovich lifting and an application to up-to techniques. In: Concurrency Theory, CONCUR 2024. LIPIcs, vol. 311, pp. 20:1–20:19. Schloss Dagstuhl – Leibniz Center for Informatics (2024), `https://doi.org/10.4230/LIPIcs.CONCUR.2024.20`

10. Desharnais, J., Laviolette, F., Tracol, M.: Approximate analysis of probabilistic processes: Logic, simulation and games. In: Quantitative Evaluation of Systems, QEST 2008. pp. 264–273. IEEE Computer Society (2008). https://doi.org/10.1109/QEST.2008.42

11. Desharnais, J., Sokolova, A.: ε-distance via Lévy-Prokhorov lifting. In: Computer Science Logic, CSL 2026 (2026), to appear. Preprint available at `https://arxiv.org/abs/2507.10732`

12. Dudley, R.M.: Real Analysis and Probability. Cambridge Studies in Advanced Mathematics, Cambridge University Press, 2 edn. (2002)

13. Forster, J., Goncharov, S., Hofmann, D., Nora, P., Schröder, L., Wild, P.: Quantitative Hennessy-Milner theorems via notions of density. In: Klin, B., Pimentel, E. (eds.) Computer Science Logic, CSL 2023. LIPIcs, vol. 252, pp. 22:1–22:20. Schloss Dagstuhl – Leibniz-Zentrum für Informatik (2023). https://doi.org/10.4230/LIPIcs.CSL.2023.22

14. Goncharov, S., Hofmann, D., Nora, P., Schröder, L., Wild, P.: Kantorovich functors and characteristic logics for behavioural distances. In: Kupferman, O., Sobocinski, P. (eds.) Foundations of Software Science and Computation Structures, FoSSaCS 2023. LNCS, vol. 13992, pp. 46–67. Springer (2023). https://doi.org/10.1007/978-3-031-30829-1_3

15. Goncharov, S., Hofmann, D., Nora, P., Schröder, L., Wild, P.: Relators and notions of simulation revisited. In: Logic in Computer Science, LICS 2025. pp. 776–789. IEEE (2025). https://doi.org/10.1109/LICS65433.2025.00064

16. Goubault-Larrecq, J.: Continuous previsions. In: Duparc, J., Henzinger, T.A. (eds.) Computer Science Logic, 21st International Workshop, CSL 2007, 16th Annual Conference of the EACSL, Lausanne, Switzerland, September 11-15, 2007, Proceedings. Lecture Notes in Computer Science, vol. 4646, pp. 542–557. Springer (2007). https://doi.org/10.1007/978-3-540-74915-8_40

17. Goubault-Larrecq, J.: Simulation hemi-metrics between infinite-state stochastic games. In: Amadio, R.M. (ed.) Foundations of Software Science and Computational Structures, 11th International Conference, FoSSaCS 2008. Lecture Notes in Computer Science, vol. 4962, pp. 50–65. Springer (2008). https://doi.org/10.1007/978-3-540-78499-9_5

18. Goubault-Larrecq, J.: Isomorphism theorems between models of mixed choice. Math. Struct. Comput. Sci. **27**(6), 1032–1067 (2017). https://doi.org/10.1017/S0960129515000547

19. Goubault-Larrecq, J.: Kantorovich-Rubinstein quasi-metrics i: Spaces of measures and of continuous valuations. Topology and its Applications **295**, 107673 (2021). https://doi.org/10.1016/j.topol.2021.107673

20. Hofmann, D.: Topological theories and closed objects. Adv. Math. **215**(2), 789 – 824 (2007). https://doi.org/https://doi.org/10.1016/j.aim.2007.04.013

21. Hofmann, D., Seal, G., Tholen, W.: Monoidal Topology: A Categorical Approach to Order, Metric, and Topology, vol. 153. Cambridge University Press (2014). https://doi.org/10.1017/CBO9781107517288, http://dx.doi.org/10.1017/CBO9781107517288

22. Humeau, S., Petrisan, D., Rot, J.: Correspondences between codensity and coupling-based liftings, a practical approach. In: Endrullis, J., Schmitz, S. (eds.) Computer Science Logic, CSL 2025. LIPIcs, vol. 326, pp. 29:1–29:18. Schloss Dagstuhl – Leibniz-Zentrum für Informatik (2025). https://doi.org/10.4230/LIPICS.CSL.2025.29

23. Kantorovich, L.V.: The mathematical method of production planning and organization. Management Science 6(4), 363–422 (1939)

24. Komorida, Y., Katsumata, S., Hu, N., Klin, B., Hasuo, I.: Codensity games for bisimilarity. In: Logic in Computer Science, LICS 2019. pp. 1–13. IEEE (2019). https://doi.org/10.1109/LICS.2019.8785691

25. Komorida, Y., Katsumata, S., Kupke, C., Rot, J., Hasuo, I.: Expressivity of quantitative modal logics: Categorical foundations via codensity and approximation. In: Logic in Computer Science, LICS 2021. pp. 1–14. IEEE (2021). https://doi.org/10.1109/LICS52264.2021.9470656

26. König, B., Mika-Michalski, C.: (Metric) bisimulation games and real-valued modal logics for coalgebras. In: Concurrency Theory, CONCUR 2018. LIPIcs, vol. 118, pp. 37:1–37:17. Schloss Dagstuhl – Leibniz Center for Informatics (2018). https://doi.org/10.4230/LIPICS.CONCUR.2018.37

27. Mio, M., Vignudelli, V.: Monads and quantitative equational theories for nondeterminism and probability. In: Konnov, I., Kovács, L. (eds.) Concurrency Theory, CONCUR 2020. LIPIcs, vol. 171, pp. 28:1–28:18. Schloss Dagstuhl – Leibniz-Zentrum für Informatik (2020). https://doi.org/10.4230/LIPICS.CONCUR.2020.28

28. Pattinson, D.: Expressive logics for coalgebras via terminal sequence induction. Notre Dame J. Formal Log. 45(1), 19–33 (2004). https://doi.org/10.1305/NDJFL/1094155277

29. Prokhorov, Y.V.: Convergence of random processes and limit theorems in probability theory. Theory of Probability & Its Applications 1(2), 157–214 (1956). https://doi.org/10.1137/1101016

30. Rady, A., van Breugel, F.: Explainability of probabilistic bisimilarity distances for labelled Markov chains. In: Kupferman, O., Sobocinski, P. (eds.) Foundations of Software Science and Computation Structures, FoSSaCS 2023. LNCS, vol. 13992, pp. 285–307. Springer (2023). https://doi.org/10.1007/978-3-031-30829-1_14

31. Rutten, J.J.M.M.: Universal coalgebra: a theory of systems. Theor. Comput. Sci. 249(1), 3–80 (2000). https://doi.org/10.1016/S0304-3975(00)00056-6

32. Schröder, L.: Expressivity of coalgebraic modal logic: The limits and beyond. Theor. Comput. Sci. 390(2), 230–247 (2008). https://doi.org/10.1016/j.tcs.2007.09.023

33. Schröder, L., Pattinson, D.: Description logics and fuzzy probability. In: Walsh, T. (ed.) International Joint Conference on Artificial Intelligence, IJCAI 2011. pp. 1075–1081. IJCAI/AAAI (2011). https://doi.org/10.5591/978-1-57735-516-8/IJCAI11-184

34. Turkenburg, R., Beohar, H., van Breugel, F., Kupke, C., Rot, J.: Constructing witnesses for lower bounds on behavioural distances. In: Computer Science Logic, CSL 2026 (2026), to appear. Preprint available at https://arxiv.org/abs/2504.08639

35. van Breugel, F., Worrell, J.: A behavioural pseudometric for probabilistic transition systems. Theor. Comput. Sci. **331**(1), 115–142 (2005). https://doi.org/10.1016/J.TCS.2004.09.035
36. Villani, C.: Optimal Transport – Old and New. Springer (2009). https://doi.org/10.1007/978-3-540-71050-9
37. Wild, P.: The Model Theory of Quantitative Coalgebraic Modal Logics. Ph.D. thesis, Friedrich-Alexander-Universität Erlangen-Nürnberg (2024). https://doi.org/10.25593/open-fau-480
38. Wild, P., Schröder, L.: Characteristic logics for behavioural hemimetrics via fuzzy lax extensions. Log. Methods Comput. Sci. **18**(2) (2022). https://doi.org/10.46298/lmcs-18(2:19)2022
39. Wild, P., Schröder, L., Messing, K., König, B., Forster, J.: Generalized Kantorovich-Rubinstein duality beyond Hausdorff and Kantorovich (2025). https://doi.org/10.48550/arXiv.2510.23552

Author Index

N. Bertrand and S. Milius (Eds.): FoSSaCS 2026, LNCS 16503, pp. 659–660, 2026.
https://doi.org/10.1007/978-3-032-22730-0

GPSR Compliance
The European Union's (EU) General Product Safety Regulation (GPSR) is a set
of rules that requires consumer products to be safe and our obligations to
ensure this.

If you have any concerns about our products, you can contact us on

ProductSafety@springernature.com

In case Publisher is established outside the EU, the EU authorized
representative is:

Springer Nature Customer Service Center GmbH
Europaplatz 3
69115 Heidelberg, Germany

www.ingramcontent.com/pod-product-compliance
Ingram Content Group UK Ltd.
Pitfield, Milton Keynes, MK11 3LW, UK
UKHW020813080726
473059UK00007B/2225